BRAIN EDEMA 1993

Brain Edema IX

Proceedings of the Ninth International Symposium
Tokyo, May 16–19, 1993

Edited by

U. Ito, A. Baethmann, K.-A. Hossmann, T. Kuroiwa,
A. Marmarou, H.-J. Reulen, K. Takakura

Acta Neurochirurgica
Supplementum 60

Springer-Verlag Wien New York

Umeo Ito, M.D.
Director and Neurosurgeon in Chief, Department of Neurosurgery, Musashino Red-Cross Hospital, Tokyo, Japan

Alexander Baethmann, M.D.
Professor, Institut für Chirurgische Forschung, Ludwig-Maximilians-Universität, München,
Federal Republic of Germany

Konstantin-A. Hossmann, M.D.
Professor, Max-Planck Institut für Neurologische Forschung, Köln, Federal Republic of Germany

Toshihiko Kuroiwa, M.D.
Associate Professor, Department of Neuropathology, Tokyo Medical and Dental University, Tokyo, Japan

Anthony Marmarou, Ph.D.
Professor, Division of Neurosurgery, Medical College of Virginia, Richmond, Virginia, U.S.A.

Hans-J. Reulen, M.D.
Professor, Neurochirurgische Klinik, Ludwig-Maximilians-Universität, München, Federal Republic of Germany

Kintomo Takakura, M.D.
Professor, Department of Neurosurgery, Tokyo Women's Medical College, Tokyo, Japan

Proceedings of the Ninth International Symposium on Brain Edema

Organization
Honorary Member: I. Klatzo
Organizers: U. Ito (Secretary General), K. Kogure, K. Takakura, A. Tamura, M. Tomita
Advisory Board: A. Baethmann, J. T. Hoff, K.-A. Hossmann, H. E. James, T. Kirino, T. Kuroiwa,
A. Marmarou, H.-J. Reulen, A. R. Shakhnovich
Local Organizer
Senior Members: T. Hayakawa, K. Hirakawa, H. Nagai, R. Okeda, K. Sato, Y. Shinohara, A. Terashi, T. Tsubokawa
Junior Members: N. Hayashi, J. Ikeda, Y. Ikeda, M. Ishikawa, Y. Katayama, K. Kawai, K. Nishimoto, K. Ohno,
N. Saito, T. Sasaki, T. Shigeno, R. Suzuki, K. Tanaka, H. Tomita, O. Tone, T. Yamaguchi

With 281 partly coloured Figures

ISSN 0065-1419 (Acta Neurochirurgica/Suppl.)

ISBN-13:978-3-7091-9336-5   e-ISBN-13:978-3-7091-9334-1
DOI: 10.1007/978-3-7091-9334-1

Preface

The first international symposium on brain edema was held in Vienna/Austria in 1965 followed by altogether eight meetings since. The most recent was organized in Yokohama by the Department of Neurosurgery of the Musashino Red Cross Hospital, Tokyo. The continuing interest of both, clinicians and experimental scientists alike may be attributable to the fact that brain edema is a common denominator of many cerebral disorders, which under acute conditions threatens life and well-being of afflicted patients. Although progress in understanding as well as treatment can be recognized since 1965 many problems remain, particularly concerning the control of brain edema under acute conditions, as in trauma or ischemia. A quantum leap was the distinction of the cytotoxic and vasogenic brain edema prototypes as advanced by Igor Klatzo, providing for transition from a morphological to functional understanding now. The recent brain edema meetings were certainly benefiting from developments of both, molecular neurobiology on the one hand side and functional brain imaging at an ever-increasing resolution on the other, such as magnetic resonance imaging or positron emission tomography. The international symposium in San Diego 1996 may witness further breakthroughs, hopefully also of effective treatment modalities.

The symposium in Yokohama was dedicated to the "Legacy of 28 Years of Brain Edema Research" as a reminder of accomplishments as well as remaining challenges. Major guidelines were: (1) blood-parenchymal cell border injury, (2) neuron-glial interactions and injury, (3) formation, propagation and resolution of edema, and (4) treatment. The sessions organized around these topics were introduced by key-note lectures of internationally high-ranking experts and frequently completed by panel sessions involving the general audience. An increasing understanding of the complex pathophysiology, such as blood-brain barrier disruption or cell swelling and damage, could be recognized in almost all areas which were covered. Recent emphasis is given to endothelial factors, such as vascular/endothelial growth factor (VEGF), which is highly expressed in the fetal brain during angiogenesis, but low in the adult brain after this process has been terminated. VEGF is considered as a permeability factor, since embryonic brain capillaries are leaky. It may, therefore, play a role in extravasation of brain edema around tumors, where formation of the molecule is upregulated. A continuing matter of concern is the pathological role of mediator compounds associated with the many brain edema phenomena, such as barrier dysfunction, cell swelling, or disturbances of the microcirculation among others. Newcomers are the vasoactive endothelins taking part in a complex mediator network, for example by virtue to release arachidonic acid from endothelium which in turn raises barrier permeability. Clearly, the mediator concept is useful from a clinical point as long as results have an impact on treatment. If, however, release or activation of an ever-growing host of mediator substances in the damaged brain cannot be therapeutically controlled due to the sheer complexity, different approaches might be required. Major attention was also paid to the significance of neuron-glial interactions in cell injury and edema in vivo and in vitro. Different behavior and susceptibility, respectively of neurons and glial cells might be relevant with regard to the edematogenic role of glutamate, acidosis, hypertonicity, or as to the vulnerability from ischemia.

Further, most advanced technologies reaching from e.g. in situ hybridization to NMR diffusion imaging can currently be recognized to pervade almost all scientific approaches related to the general subject. These developments notwithstanding, basic aspects of formation, propagation, and resolution of brain edema remained a focus of the symposium. Edema conditions studied with respect to this subject range from brain tumor to head injury. Computerized tomography provides an important methodological basis for analysis of the kinetics of brain edema spread and resolution in patients. Mathematical modeling for exploration of edema dynamics indicates that both, diffusion and convection characteristics play a role as validated by MRI studies. In this context, "isotropic" and "anisotropic" tissue properties can be distinguished with regard to the orientation of the nerve bundles in white matter, which in fact is reviving former conclusions on spread of edema in relation to the fibre direction elucidated by morphology. New data obtained on early but transient barrier opening question a major role of the vasogenic edema type in diffuse traumatic brain injury. To the contrary, occurrence of a reduced tissue blood volume simultaneously with an increased water content indicates significant involvement of cell swelling. The importance of astrocytic swelling was thus extensively evaluated and discussed to reflect either uptake of extracellular material as a result of glial clearance functions or, more likely, representing a manifestation of cytotoxic cell damage. Treatment of brain edema continues to acquire a central position, as shown not only by the impressive number of contributions, but also by the diversity of conditions, which were studied. The more provocative and apparently progressive approaches of therapy were selected for a round table discussion, providing participants with up-to-date information by experts on particularly promising modalities.

The rising attendance of scientists from all over the world and growth of oral and poster contributions demonstrate the significance as well as well-being of this internationally prominent symposium. An "International Society for Brain Edema Research" was consequently established in Yokohama in order to better focus and structure purpose and interests for this interdisciplinary area of research. The importance to advance understanding of brain edema up to the molecular level and to accomplish breakthroughs for treatment, particularly for life-threatening edema manifestations were recognized as major challenges of the future. The Society sincerely welcomes members, particularly of the younger generation, who are expected to carry on the legacy of brain edema research into the next century. The present Proceedings are therefore dedicated to our Society, simultaneously inviting for Brain Edema 1996 in San Diego. Finally, it is a great pleasure to gratefully acknowledge the competent collaboration by R. Petri-Wieder of Springer, Vienna, in concluding our preparations of this volume.

January 1994   Umeo Ito
                                                                                                and Coeditors

Supported by
The Commemorative Association for the Japan World Exposition (1970).

Contents

Cell Swelling and Cell Damage in vivo and in vitro

Mediator Mechanisms in Vasogenic and Cytotoxic Brain Edema

Blood-Brain Barrier Function

Formation, Propagation and Resolution of Brain Edema

Ischemic Brain Edema: Pathophysiology and Nuclear Magnetic Resonance

Ischemic Brain Edema: Pathobiochemistry and Brain Edema

Ischemic Brain Edema: Treatment

Contents

Tumor and Brain Edema: Pathophysiology, Laboratory Studies

Tumor and Brain Edema: Pathophysiology, Clinical Studies

Tumor and Brain Edema: Treatment

Traumatic Brain Edema: Pathophysiology, Laboratory Studies

Traumatic Brain Edema: Pathophysiology, Clinical Studies

Traumatic Brain Edema: Treatment

Intracerebral Hematoma and Various Edema Pathophysiology

Listed in Current Contents

Cell Swelling and Cell Damage in vivo and in vitro

Acta Neurochir (1994) [Suppl] 60: 3–6

Evolution of Brain Edema Concepts

I. Klatzo

Stroke Branch, National Institute of Neurological Disorders and Stroke, National Institutes of Health, Bethesda, MD, U.S.A.

Summary

The evolution of brain edema research is outlined from early experimental studies to the development of present concepts and interpretations, as well as the establishment of criteria for the two main types of edema, cytotoxic and vasogenic.

Keywords: Brain edema; cytotoxic; vasogenic; blood-brain barrier.

Brain edema (BE) has been recognized from the beginnings of modern neurology and neurosurgery as a common complication that may aggravate the severity of a primary pathological process. Despite recent great advances in the neurological sciences, BE remains a serious clinical problem demanding the design of new effective methods for its control. According to definition, *BE is an abnormal accumulation of fluid within the brain parenchyma producing a volumetric enlargement of the brain tissue.* This excludes volumetric enlargement based on an increase in blood volume alone, as for example due to an acute engorgement of the vascular bed without an increase in brain tissue water. Also, a mere shift of water between extra- and intracellular compartments does not constitute BE.

A first major attempt to interpret the pathophysiological basis for BE was undertaken by Reichardt[21], who classified BE into two types, i.e. *Hirnödem* (brain edema) and *Hirnschwellung* (brain swelling). This classification was based primarily on the appearance of cut brain surfaces at autopsy, with wet and soft surfaces indicating *Hirnödem* and tough and dry tissue indicating *Hirnschwellung*. These differential features, as well as some of Reichardt's other physico-chemical criteria pertaining mainly to relative amounts of dry brain weight and water, proved to be ambiguous and misleading. Reichardt's classification has subsequently failed to gain wide acceptance, although Zülch[27] provided a description of the histopathological features for each type of BE and Ambo[1] claimed success in experimentally reproducing the two types of BE. Nevertheless, the most pertinent question, concerning the location of the edema fluid and derivation, remains to be clearly answered. Expectations that the advent of electron microscopy (EM) would clarify these questions were initially disappointed because the early inadequate EM fixation methods frequently led to erroneous evaluation of EM images. This prompted some electron microscopists[3,4,14,17] to advance the monastic concept of BE as an entirely intracellular phenomenon, with primarily intraglial localization of edema fluid. The existence of functional extracellular spaces was virtually denied, assuming that even small intercellular spaces were occupied by a homogeneous electron-dense substance, presumably of a mucopolysaccharide nature[3]. However, with progress in a proper understanding of the extracellular compartment, based on the use of the elegant freeze-substitution EM technic developed by Van Harreveld *et al.*[25,26] as well as on basic electrophysiological assays and in vivo space determinations for some extracellular substances, it became evident that edema fluid may accumulate either in injured brain cellular elements or extracellularly in association with an increased vascular permeability.

This realization that edema is associated with either intra- or extracellular accumulation of abnormal fluid led to the classification of BE into two main types, i.e. *cytotoxic*, related to the abnormal uptake of water by injured brain cells, and *vasogenic*, associated with dysfunction of the blood-brain barrier (BBB) allowing abnormal passage of proteins and water into the extracellular compartment[6]. It should be emphasized, however, that in a particular case of BE, both cytotoxic and vasogenic mechanisms are usually operative, their relative contributions sometimes undergoing wide, temporary variations. A good example of variability in the character of edematous changes can be observed in cerebral ischemia. Although interruption of energy supply rapidly leads to the cessation of ionic pumps and the intracellular flow of sodium and water, this change per se is easily reversible and severe cellular injury is required for persistent cytotoxic swelling of the cells. The vasogenic component of ischemic BE may have a biphasic character, related to the biphasic openings of the BBB[13]. The first opening, hemodynamic in nature, follows a recirculation of previously ischemic regions and appears to be related to a loss of autoregulation in previously ischemic regions. Although this first opening may be of limited duration and may not contribute much to the development of edema, increased permeability of the BBB for even a brief duration may allow the entry of various exogenous substances that may in turn influence postischemic pathophysiology. The second BBB opening following the release of ischemic occlusion appears to be directly related to the breakdown of the BBB as it is observed at the margins of cold lesions or ischemic infarcts and may be associated with a progressive increase in the size of ischemic infarcts. This occurs in a "vicious circle" fashion, where an increase in osmolarity within a necrotic infarct due to the rupture of dying cell membranes combined with an exudation of water and proteins from the affected vessels, results in a marked increase in local tissue pressure leading to a depression of regional blood flow below the critical thresholds of viability in penumbral regions and to further extension of the territory that undergoes irreversible tissue damage[8,9].

Vasogenic BE, a common complication of many brain disorders such as cerebrovascular insult, trauma, tumors and inflammatory conditions, has received a great deal of attention and its dynamic features have started to be explored since the introduction of the experimental cold injury model[14]. Concerning the dynamics of vasogenic edema, the Starling equation, referring to the fluid flow from the blood into the tissue; Jv (flow) $= Lp \, [(P^{plasma} - P^{tissue}) - \sigma \, (\Pi^{plasma} - \Pi^{tissue})]$, seems to be helpful in interpreting the various phases of the edematous process. Thus, at the site of increased vascular permeability, Lp (the hydraulic conductivity of the endothelial membrane) reflecting the status of the BBB and P^{plasma} (the intravascular hydrostatic pressure related to systemic blood pressure) appear to be the predominant factors in the spread of edema. In this respect, the effects of systemic blood pressure on the rate of edema spread were clearly established in the cyrogenic edema model[10]. On the other hand, in the regions of edematous white matter, where the BBB to proteins remains intact, the flow of the fluid (Jv) across the vascular wall is influenced by the difference in osmotic pressures between plasma (Π^{plasma}) and edematous tissue (Π^{tissue}). Since the intact BBB in such areas prevents the reentry of extravasated serum proteins into the vascular lumen, the proteins become trapped in extracellular spaces, exerting their colloidal osmotic pressure and reducing the movement of water into the vessels.

The relationship between the presence of extravasated serum proteins and the dynamics of vasogenic edema has been demonstrated in a number of investigations. In studies in which an increased BBB permeability was produced without any noticeable injury to brain parenchyma, it was demonstrated that during the period when the BBB opened to proteins, the progressive increase in the amount of extravasated ^{125}I-albumin was paralleled by an increase in water content in such areas[12]. Furthermore, Marmarou[15,16] demonstrated that an experimental infusion edema produced with mock CSF (cerebrospinal fluid) completely resolved after 72 h, whereas there was only a negligible clearance of water in areas infused with serum. The interrelationship between water content and the extravasated proteins was also clearly established using specific gravity measurements and immunostaining for proteins[11]. In these studies, the disappearance of serum proteins from extracellular spaces coincided with the return of water content to normal values, while astrocytic cells showed vigorous intracytoplasmic uptake of extravasated proteins. The latter observation suggests that uptake of proteins by the glial cells is one of the factors involved in the resolution of vasogenic brain edema.

Using the cold injury model, Reulen *et al.*[22,23] made a significant advance in the understanding of the dynamics of vasogenic edema by elucidating the main features of its spread. They showed that overpowering the initial resistance of cellular structures permits the progression of edema and is associated with a rise in interstitial tissue pressure (IFP) and the development of pressure gradients. Edema spreads by bulk flow, preferentially through the white matter, which structurally offers less tissue resistance. Reulen *et al.* demonstrated that the clearance of edema fluid into the CSF occurs mainly because of the pressure gradients between the edematous subependymal regions and the CSF in the ventricles. This was further supported by the observation that reduction of CSF pressure accelerates the clearance of edema fluid into the ventricle[22].

The origin of *cytotoxic* BE is primarily related to disturbances of cellular osmoregulation, resulting in the intracellular uptake of water. The most commonly encountered cytotoxic edema is observed in cerebral ischemia, where an interruption of energy supply leads to ionic pump failure and osmoregulation, resulting in an intracellular increase in sodium and water. Among the main elements of postischemic injury, the *neurotoxic effects of excitatory amino acids* and particularly of glutamate draw considerable attention at present[20]. Glutamate neurotoxicity has been associated with the intracellular influx of Ca^{++} through membrane channels, which are predominantly gated by receptors of the N-methyl-D-aspartate (NMDA) subtype[2]. Morphologically, it is characterized by early astrocytic and, especially, dendritic swelling[24]. In cerebral ischemia associated with cardiac arrest, the cells most sensitive to swelling appear to be GABA-ergic neurons of the nucleus reticularis thalami, where extreme cytoplasmic vacuolation leads to severe loss of these neurons[5]. In addition to a direct excitotoxic effect on the neurons, an excessive release of excitatory acids affects the endothelial permeability of the BBB, allowing the escape of proteins into extracellular spaces and their uptake into presynaptic vesicles and postsynaptic dendrites[18,19].

The current rapid and spectacular progress in various branches of the neurosciences provides an opportunity to get a deep insight into mechanisms operative in neurotoxicity and increased permeability of the BBB – the basic events in both types of edema. With regard to cytotoxicity, there is a rapidly accumulating wealth of knowledge on molecular changes involving ionic channels, second messengers, altered genetic expression, etc, which may have a direct bearing on osmoregulation and the development of delayed or chronic changes.

Concerning the changes in BBB permeability, there is now information on interactions between blood vessel walls and the surrounding parenchyma. Elucidation of the biochemical reactivity of the cerebral endothelium and the role of various hematogenous cellular elements immensely increases our comprehension of how the function of the BBB can be disturbed, resulting in an abnormal accumulation of edema fluid.

It can be hoped that rational measures to successfully control and treat brain edema will evolve from this immense store of new knowledge.

References

1. Ambo H (1961) Kritische Bemerkungen über die Hirnschwellung. Folia Psychiatr Neurolog Japan 15: 42–52
2. Choi DW (1988) Calcium-mediated neurotoxicity: relationship to specific channel types and role of ischemic damage. Trends Neurosci 11: 465–469
3. DeRoberties EDP, Gerschenfeld HM (1961) Submicroscopic morphology and function of glial cells. In: Pfeiffer CC, Smythies JR (eds) Intern Rev Neurobiol 3: 1–13
4. Gerschenfeld HM, Wald F, Zadunaisky JA, DeRoberties EPD (1950) Function of astroglia in the water-ion metabolism of the central nervous system. An electron microscope study. Neurology 9: 412–427
5. Kawai K, Nitecka L, Ruetzler CA, Nagashima G, Joo F, Mies G, Nowak TS Jr, Saito N, Lohr JM, Klatzo I (1992) Global cerebral ischemia associated with cardiac arrest in the rat: I. dynamics of early neuronal changes. J Cereb Blood Flow Metab 12: 238–249
6. Klatzo I (1967) Presidential address. Neuropathological aspects of brain edema. J Neuropathol Exp Neurol 26: 1–14
7. Klatzo I, Piraux A, Laskowski EJ (1958) The relationship between edema, blood-brain barrier and tissue elements in a local brain injury. J Neuropathol Exp Neurol 17: 548–564
8. Klatzo I (1985) Concepts of ischemic injury associated with brain edema. In: Inaba Y, Klatzo I, Spatz M (eds) Brain edema. Springer, Berlin Heidelberg New York Tokyo, pp 1–5
9. Klatzo I (1985) Brain oedema following brain ischaemia and the influence of therapy. Br J Anaesth 57: 18–22
10. Klatzo I, Wisniewski H, Steinwall O, Streicher E (1967) Dynamics of cold injury edema. In: Klatzo I, Seitelberger F (eds) Brain edema. Springer, Berlin Heidelberg New York, pp 554–563
11. Klatzo I, Chui E, Fujiwara K, Spatz M (1980) Resolution of vasogenic brain edema. Adv Neurol 28: 359–374
12. Kuriowa T, Cahn R, Juhler M, Goping G, Campbell G, Klatzo I (1985) Role of extracellular proteins in the dynamics of vasogenic brain edema. Acta Neuropath 66: 3–11
13. Kuroiwa T, Ting P, Martinez H, Klatzo I (1985) The biphasic opening of the blood-brain barrier to proteins following temporary middle cerebral artery occlusion. Acta Neuropath 68: 122–129
14. Luse SA, Harris B (1960) Electron microscopy of the brain in experimental edema. J Neurosurg 17: 439–447
15. Marmarou A, Nakamura T, Tanaka T, Hochwald GM (1984) The time course and distribution of water in the resolution phase of infusion edema. In: Go KG, Baethmann A (eds) Recent progress in the study and therapy of brain edema. Plenum, New York, pp 37–44
16. Marmarou A, Tanaka K, Schulman K (1982) The brain response to infusion edema: dynamics of fluid resolution. In: Hartmann A, Brock M (eds) Treatment of cerebral edema. Springer, Berlin Heidelberg New York, pp 11–18
17. Niessing K, Vogell W (1960) Elektronenmikroskopische Untersuchungen über Strukturveränderungen in der Hirnrinde beim Oedem und ihre Bedeutung für die Grundsubstanz. Z Zellforsch 52: 136–151
18. Nitsch C, Goping G, Laursen H, Klatzo I (1986) The blood-brain barrier to horseradish peroxidase at the onset of bicuculline-induced seizures in hypothalamus, pallidum, hippocampus and other selected regions of the rabbit. Acta Neuropath 69: 1–16
19. Nitsch C, Goping G, Klatzo I (1986) Pathophysiological aspects of the blood-brain barrier permeability in epileptic seizures. In: Schwarcz R, Ben-Ar Y (eds) Excitatory amino acids and epilepsy. Plenum, New York, pp 175–189
20. Olney JW (1978) Neurotoxicity of excitatory amino-acids. In: McGeer EG, Olney JW, McGeer PL (eds) Kainic acid as tool in neurobiology. Raven, New York, pp 95–121

21. Reichardt M (1905) Zur Entstehung des Hirndrucks. Dtsch Z Nervenheilkunde 28: 306–355
22. Reulen HJ, Graham R, Fenske A, Tsuyumu M, Klatzo I (1976) The role of tissue pressure and bulk flow in the formation and resolution of cold-induced edema. In: Pappius HM, Feindel W (eds) Dynamics of brain edema. Springer, Berlin Heidelberg New York, pp 103–112
23. Reulen HJ, Graham R, Spatz M, Klatzo I (1977) Role of pressure gradients and bulk flow in dynamics of vasogenic brain edema. J Neurosurg 46: 24–35
24. Sloviter RS, Dempster DW (1985) "Epileptic" brain damage is replicated qualitatively in the rat hippocampus by central injection of glutamate or aspartate but not GABA or acetylcholine. Brain Res Bull 15: 39–60
25. Van Harreveld A, Crowell J, Malhotra SK (1965) A study in extracellular space in the central nervous tissue by freeze-substitution. J Cell Biol 25: 117–137
26. Van Harreveld A, Collewijn H, Malhotra SK (1966) Electrolytes and extracellular space in hydrated and dehydrated brains. Am J Physiol 120: 251–269
27. Zülch KJ (1959) Störungen des intrakraniellen Druckes. In: Handbuch der Neurochirurgie, vol I, part I. Springer, Berlin Göttingen Heidelberg, p 208–303

Correspondence: Igor Klatzo, M.D., Bldg. 36, Room 4D04, Stroke Branch, NINDS, NIH, 9000 Rockville Pike, Bethesda, MD 20892, U.S.A.

Acta Neurochir (1994) [Suppl] 60: 7–11
© Springer-Verlag 1994

Neuron-Glial Interaction During Injury and Edema of the CNS

O. S. Kempski and **C. Volk**

Institute for Neurosurgical Pathophysiology, Johannes Gutenberg-Universität Mainz, Mainz, Federal Republic of Germany

Summary

During injury and ischemia of the CNS mediator compounds are released or activated which cause secondary swelling and damage of nerve cells. Such mediators are glutamate, acidosis, free fatty acids, or high extracellular potassium. Glial homeostatic mechanisms are activated to prevent the secondary injury from these mediators. The glial clearance mechanisms have been studied in detail using in vitro systems allowing for a close control of the glial environment. Current evidence suggests glial swelling to occur together with glutamate uptake or in response to extracellular acidosis. Glial swelling, therefore, is rather the result of homeostatic mechanisms than an indication of glial demise.

Keywords: Glia-neuron interaction; cell volume; glutamate; lactacidosis.

Introduction

In *neuro*science, neurons have always received the primary attention, interest and funding, whereas glia has only gradually been recognized as an essential, functional counterpart of neuronal elements. Virchow's 150 year old idea of glia as a connective tissue, a cement or putty glueing neurons in place ("Nervenkitt") has finally been overcome. A homeostatic function of glia has now been accepted, delegating specific 'housekeeping functions' to astrocytes such as potassium uptake, transmitter inactivation or nutrient supply to neurons. More recently it became evident that astrocytes are even involved in information processing by fine-tuning of receptor sensitivity, or by release of modulators of nerve cell function. We are just in the process of understanding the manyfold functions of an ever growing diversity of glial cell subtypes in the normal, healthy brain. Therefore the malfunction of these poorly understood physiological mechanisms *during injury and edema* is even less clear. Moreover, the acute interactions of glial and nerve cells during injury are overlapped by more gradual processes responsible for regeneration, or its prevention, by widely unknown mechanisms allowing for glial and neuronal plasticity. This short summary therefore can only highlight some aspects which must be subjective and incomplete. For further readings two recent volumes dealing exclusively with glial-neuronal interaction are recommended (Glial-Neuronal Interaction (Abbott NJ, ed) Ann NY Acad Sci Vol 633, 1991; Neuronal-Astrocytic Interactions (Yu ACH, Hertz L, Norenberg MD, Sykova E, Waxman S, eds), Progress in Brain Res, Vol 94, Elsevier, Amsterdam, 1992).

Glial-Neuronal Interactions in Excitotoxicity

Glutamate is an excitatory transmitter in the CNS[9], which under pathophysiologic conditions assumes neurotoxic properties[23]. Glutamate is ubiquitous in the central nervous system but strictly compartmentalized in the intracellular space with intracellular concentrations up to 15 mM, more than any other free amino acid in the brain. The neurotoxic potential of glutamic acid has been proven for cerebral ischemia and brain injury and is commonly addressed as 'excitotoxicity'[23]. It is now widely accepted that glutamate may destroy neurons by excessive activation of excitatory receptors. On the other hand glutamate is the most important excitatory transmitter with the widest distribution in the central nervous system. Hence, under physiological conditions glutamate is highly compartmentalized, and there are efficient uptake systems both in the glial and in the neuronal membrane to clear the extracellular space from glutamate[4–6,11,12,28]. This process is coupled to a Na^+-downhill influx into the cell and thus is energy dependent. "High-affinity" sodium-dependent glutamate uptake is known since more than 20 years, and recently the transporter molecule has been cloned[25].

Glial glutamate uptake is accompanied by an inward membrane current which is abolished if extracellular sodium or potassium are removed. Respective studies conclude that glutamate is taken up by an electrogenic uptake carrier which cotransports 1 molecule of glutamate together with 2–3 sodium ions into the cell and countertransports 1 potassium ion out of the cell[4,5]. The ATP-dependent transformation of glutamate to glutamine via the glial enzyme glutamine synthetase may further accelerate energy expenditure in energy-deprived conditions.

A still operative glial glutamate uptake in the vicinity of an ischemic focus, i.e. in the ischemic penumbra, may compromise central nervous function by causing glial swelling. Swelling is a direct consequence of glutamate accumulation together with Na[+]-ions. This is corroborated by own data which proved glial swelling (astrocytes as well as C6 glioma cells) during glutamate exposure[27] in an invitro model of cytotoxic edema[14,15,19]. The model permits a close control of the extracellular compartment together with high-precision determination of cell volume changes of suspended cells. A significant increase of glial cell volume was indeed shown, first in a preliminary study using a high glutamate level of 15 mM[14] and recently in a more detailed investigation by Schneider et al.[27] studying glutamate in concentrations from 50 µM to 10 mM. Taken together the data underline that glial swelling by glutamate occurs together with glutamate uptake. Glutamate uptake as well as swelling are only seen if the transmembrane sodium gradient of the glial cells is maintained. It should be mentioned, that under pathophysiologic conditions glutamate may reach concentrations in the extracellular space which are well comparable to the highest levels studied. With tissue concentrations of 10–12 mM, and a 20% interstitial space, an intracellular concentration of 14–15 mM has to be assumed. In synaptic vesicles even concentrations up to 100 mM have been postulated[8]. Therefore, in the vicinity of a traumatic or necrotic focus, where cellular constituents gain access to the interstitial space, extracellular glutamate can be expected to temporarily rise to 5–10 mM. 1 mM of glutamate has actually been measured in edema fluid collected from cat brain with a freezing injury and additional ischemia[3].

Glial Swelling and Cytotoxic Edema

Glial swelling in the course of an uptake process (described above for glutamate), is hence the *result of* *glial homeostasis* to protect neurons and their function. This statement, although trivial on the first view, deserves some attention, since glial swelling has long been thought to be an indicator of glial dysfunction, even a first step towards glial decay, and cell death. The swelling of cellular elements in the brain is referred to as 'cytotoxic brain edema'. So far, it has been the general understanding that with impaired energy supply, cell swelling results from the failure of Na[+]/K[+]-ATPase according to the pump-leak model of cell volume regulation[22]. Glial swelling in cerebral ischemia occurs fast, within 5 min after interruption of the energy supply. In vitro the onset of swelling depends on the volume of the incubation medium, as elegantly demonstrated by Ames and Nesbett[1] using the isolated retina. Their observations are best explained by assuming that toxic compounds accumulating in the extracellular environment mediate cell swelling, and also nerve cell death. Ames's findings are nicely supported by our own data using the in vitro model of 'cytotoxic' glial swelling (see above). The model was used to evaluate whether anoxia or inhibition of the cellular energy supply without any further changes of the extracellular environment would be accompanied by swelling[16]. Anoxia was induced by discontinuation of the oxygen supply to the membrane oxygenator in an incubation chamber containing the suspended test cells (C6 glioma), and subsequent ventilation with nitrogen. Anoxia was maintained for 2 hours – and cell swelling was *never* observed. The increase of lactate indicated that anaerobic glycolysis was sufficient to maintain ion gradients and cell volume control. Therefore, iodoacetate was used to inhibit glycolysis in addition to anoxia. This treatment successfully abolished the intracellular/extracellular Na[+]-gradient within 120 min, but, again, cell volume was unaffected[16].

We conclude that in the narrow, restricted extracellular space of the brain in vivo, mediators of secondary brain damage released or activated by pathophysiologic events have to accumulate to cause glial swelling and, eventually, nerve and glial cell death. In another study this conclusion found further support. Here, the sodium-potassium-pump was blocked by the cardiac glycoside ouabain. The Na[+]-gradient was lost by this treatment within 90 min in C6 cells[18]. Simultaneously a steep drop in the i.c. potassium content from 190 to 30×10^{-15} mol/cell was observed. Just as with complete anoxia, the inhibition of the Na[+]/K[+]-pump *did not cause cell swelling*[18].

Again, the data support the contention that failure of

the energy supply per se, or of sodium transport by Na$^+$/K$^+$-ATPase do not suffice to induce glial swelling. The results are another indication that glial swelling, and, therefore, a major part of cytotoxic edema, is the result of secondary events caused by mediators of secondary brain damage[2] such as glutamate, high K$^+$, free fatty acids, or lactacidosis. As discussed above for excitotoxins, evidence is accumulating, that these mediators act in a dual fashion: on the one hand they compromise neuronal function or even survival, and on the other hand they cause glial swelling as a consequence of glial homeostatic activity.

Glial Swelling in Acidosis

Lactacidosis from anaerobic metabolism is regularly found during cerebral ischemia, seizures and head injury[29]. A marked decrease of brain tissue pH has been demonstrated in cerebral ischemia particularly in hyperglycemic subjects, where lactic acid may accumulate to 20–30 mM[26,13]. Tissue pH may drop down to pH 5.5[7] or allegedly even as low as 3.9 in selected cells[21]. Acidosis has long been suspected as a mediator of brain damage[29]. Using the in vitro system we were able to confirm the swelling inducing capacity of extracellular acidosis induced either by an inorganic acid[17] or by lactic acid[30]. A pH range between pH 7.6 and 4.2 was studied under strict maintenance of isotonicity. Glial volume was found to increase if the e.c. pH was titrated to 6.8 or below. From this level downward, the extent of swelling depended on the degree of acidosis and the duration of exposure. Lactacidosis of pH 6.2 for 60 min led to a 24.5% increase in cell volume, while pH 5.0 or 4.2 increased cell size for 51.1% or 90.9%, respectively. Cell viability was hardly affected down to pH 6.2. At pH 5.6 cell viability remained normal (89%) for 30 min but then decreased to 73% after 60 min. At pH 4.2 only 21% of the cells survived 1 h of acidosis[30]. Cell swelling could be inhibited by replacement of Na$^+$ and bicarbonate in the medium by use of choline chloride and HEPES as buffer.

From these results acidosis-induced swelling was explained as a consequence of mechanisms to maintain a normal intracellular pH: Addition of acid to bicarbonate buffered media leads to the formation of carbonic acid which immediately dissociates to CO_2 and water. In turn, intracellular acidosis develops, since CO_2 freely passes the cell membrane, and, catalyzed by carbonic anhydrase forms carbonic acid. To prevent

acidosis, the resulting H$^+$-ions are exchanged against sodium ions (amiloride-inhibitable). As acidosis-induced glial swelling could be partially blocked by stilbenes, involvement of Cl$^-$/HCO$_3$$^-$-exchange or additional Na$^+$/HCO$_3$$^-$-cotransport can be concluded. The resulting influx of Na$^+$ and Cl$^-$-ions accompanied by water would explain swelling.

This concept implies pH-regulation even under conditions of severe extracellular acidosis. Opposed to this are observations of Kraig and Chesler[21], who found the glial intracellular pH to decrease tremendously – and far below total tissue pH – during hyperglycemic ischemia in vivo. The authors explain their observation by a combination of three mechanisms: (a) the continued acid production in glia, (b) slow acid efflux as a result of maintained membrane integrity, and (c) acid inhibition of antiport mechanisms. Obviously the two proposed mechanisms do not fit together: The continued, amiloride-inhibitable swelling at pH levels as low as 4.6 speaks against an inhibition of the antiports by low pH. Second, a low extracellular pH is not necessarily lethal for glia, which in vitro may survive pH 5.6 for more than 30 min. Therefore we currently attempt to combine all available data in order to find a 'unifying' hypothesis.

To do so we measured intracellular pH in extracellular acidosis by the fluorochrome BCECF. The fluorescence of BCECF changes in response to pH alterations. pH is obtained from the ratio of the fluorescence signals emitted by stimulation at 442 and 495 nm. In brief, after loading of adherent glial cells with the fluorochrome, the cells are exposed to media buffered by either HEPES or bicarbonate. After a control period at pH 7.35, acidosis (pH 6.4) is induced and i.c. pH changes are monitored.

The results (Volk et al. in preparation) prove that, in contrast to our original concept, extracellular acidosis is in fact rapidly followed by intracellular acidification, which occurs even faster in the presence of bicarbonate buffer. The latter fits the concept that CO_2 may enter the cell freely to form carbonic acid, and then to dissociate to protons and bicarbonate. The intriguing observation, however, is a gradual acidification for 30 min nearly down to the extracellular pH of 6.4 in HEPES buffer. Although Na$^+$/H$^+$ exchange is activated by intracellular acidification – demonstrated by acid loading experiments – this does not suffice to maintain a normal pH in etracellular acidosis. Inhibitors such as amiloride or stilbenes do not affect acidification. However, when using Zn^{2+} in high potassium media (pH 6.4), the acidification process could be slowed down (Volk *et al.* in preparation). Zn^{2+} is known to inhibit voltage dependent proton currents as demonstrated for neurons[10] but not studied in glia so far. With the postulated proton channels in the glial membrane, the resulting leak of H^+-ions (following the electrochemical gradient) into glia during acidosis would explain intracellular acidification together with swelling caused by the futile pH regulatory attempts of respective transport systems.

The results enable us to expand the model of the glial cell as a control organ for the cerebral extracellular space. The model is schematically outlined in Fig. 1. It is based on a proposed *functional polarization of astrocytes* in vivo, with ion channels and transport systems unevenly distributed over the cell membrane (or alternatively: activated unevenly) enabling 'compartmental buffering' of the extracellular pH. The hypothetic model is similar to mechanisms proposed by Newman[24] and Kraig *et al.*[20], and could explain many of the observed phenomena: with enhanced neuronal activity glial cells would depolarize, and reduce the normally outward directed Na^+/HCO_3^--cotransport. This could explain the often observed glial alkalization with neuronal activity, and would also cause a pH-induced increase of lCBF to remove acid equivalents, and to increase the supply with nutrients. In case of acid production by neurons, glial proton channels would permit a rapid 'enlargement' of the virtual

buffer space since the interior of the astrocyte would be accessible for H^+-ions. These would predominantly be extruded towards the capillary bed by respective transporters as long as there is blood flow to remove acid equivalents. For the Na^+/HCO_3^--cotransporter such a unilateral localization has been proposed at least for the retina[24]. No matter, whether acid removal occurs via Na^+/H^+-exchange or Na^+/HCO_3^--cotransport, an elevation of intracellular sodium is the consequence. The resulting osmotic load could explain glial swelling as observed in vitro. An enlarged glial volume, on the other hand, would provide a larger buffer space.

The model is in line with the neuropathological observation that cytotoxic glial swelling is first observed in glial endfeet around small blood vessels. So far it was not clear why cells which would receive nutrients and oxygen most easily should suffer first. With the proposed model glial endfeet should be rich in Na^+/H^+-exchange or Na^+/HCO_3^--cotransport systems: glial endfeet would accumulate sodium, and hence osmoles to explain early swelling. Likewise the model would explain why amiloride analogues interfere with cytotoxic swelling from acidosis.

In vitro, however, the anatomically uneven distribution of transporters is abolished, and only net effects of all phenomena involved can be measured as change of cell volume, membrane potential, or intracellular pH. By careful use of specific inhibitors it should nevertheless be possible to study glial pH homeostasis in vitro.

Conclusion

Some of the postulated interactions are still hypothetical. They can only function efficiently if all the cited transporters and ion channels cooperate under close control. Such regulatory principles are far from being understood in detail. Although it may be too early to draw far reaching conclusions, it is quite evident that a close interaction of glial and nerve cells provides a basis not only for normal brain function but also for its preservation and restauration during conditions of injury. Most important, the sequence of events in injury and edema of the CNS has become more clear: as depicted in Fig. 2, we now know that glial swelling is the result of homeostatic functions, which may secondarily be impeded by the sequelae of ischemia. Dendritic swelling and nerve cell death, on the other hand, clearly are secondary manifestations of cerebral injury.

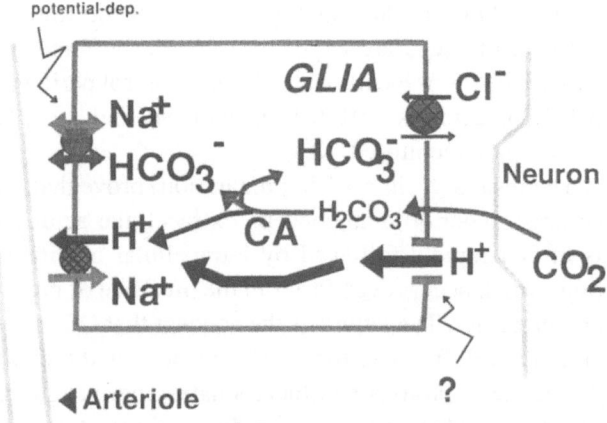

Fig. 1. Model of neuronal-glial interaction during neuronal excitation and ischemia as explained by a postulated polar orientation of glial cells between neurons and blood vessels. Nerve cell excitation followed by glial depolarization would alkalinize glia via influx of sodium and bicarbonate from the vessel side. There a local pH drop would mediate an increase of lCBF. In ischemia, glia would provide a sink for protons. In the absence of a functional circulation the pile-up of acid equivalents stimulates pH-regulatory mechanisms leading to an influx of sodium and chloride ions, and osmotically obliged water

CEREBRAL INJURY

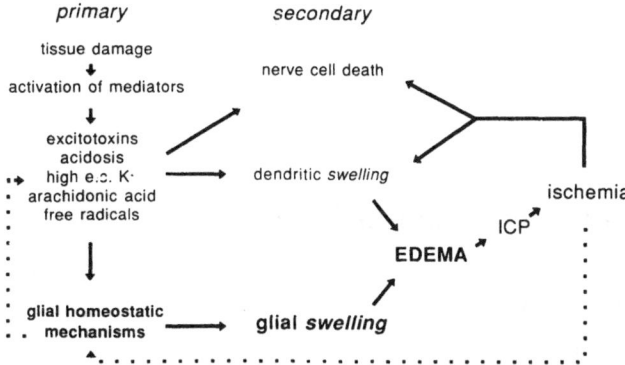

Fig 2. Scheme of the sequence of events in cerebral injury. In a primary focus the release or activation of mediators leads to (i) secondary injury with dendritic swelling and nerve cell death, and (ii) to the activation of glial homeostatic mechanisms. These are accompanied by glial swelling. Glial and dendritic swelling contribute to cerebral edema with its consequences, increases of ICP and secondary ischemia. Ischemia may finally impede glial homeostasis (dotted line)

References

 1. Ames A, Nesbett FB (1983) Pathophysiology of ischemic cell death I–III. Stroke 14: 219–240
 2. Baethmann A, Oettinger W, Rothenfusser W, Kempski O, Unterberg A, Geiger R (1980) Brain edema factors: current state with particular reference to glutamate. Adv Neurol 28: 171–195
 3. Baethmann A, Maier-Hauff K, Schürer L, Lange M, Guggenbichler C, Vogt W, Jakob K, Kempski O (1989) Release of glutamate and free fatty acids in vasogenic brain edema. J Neurosurg 70: 578–591
 4. Barbour B, Brew H, Attwell D (1988) Electrogenic glutamate uptake in glial cells is activated by intracellular potassium. Nature 335: 433–435
 5. Barbour B, Brew H, Attwell D (1991) Electrogenic uptake of glutamate and aspartate into glial cells isolated from the salamander (ambystoma) retina. J Physiol 436:169–193
 6. Benjamin AM, Quastel JH (1976) Cerebral uptakes and exchange diffusion in vitro of L- and D-glutamates. J Neurochem 26: 431–441
 7. Chopp M, Welch KMA, Tidwell CD, Helpern JA (1988) Global cerebral ischemia and intracellular pH during hyperglycemia and hypoglycemia in cats. Stroke 19: 1383–1387
 8. Danbolt NC, Ottersen OP, Storm-Mathisen J (1993) Anatomy of glutamate: an amino acid studied by immunocytochemistry. In: Kempski O (ed) Glutamate-transmitter and toxin. Zuckschwerdt, Munich
 9. Fonnum F (1984) Glutamate: a neurotransmitter in mammalian brain – short review. J Neurochem 42: 1–11
10. Frey G, Schlue W-R (1993) pH recovery from intracellular alkalinization in Retzius neurones of the leech central nervous system. J Physiol 462: 627–643
11. Henn FA, Goldstein MN, Hamberger A (1974) Uptake of the neurotransmitter candidate glutamate by glia. Nature 249: 663–664
12. Hertz L, Schousboe A, Boechler N, Mukerji S, Fedoroff S (1978) Kinetic characteristics of the glutamate uptake into normal astrocytes in cultures. Neurochem Res 3: 1–14
13. Katsura K, Ekholm A, Asplund F, Siesjö BK (1991) Extracellular pH in the brain during ischemia: relationship to the severity of lactic acidosis. J Cereb Blood Flow Metab 11: 597–599
14. Kempski O, Gross U, Baethmann A (1982) An in vitro model of cytotoxic brain edema: cell volume and metabolism of cultivated glial and nerve cells. In: Driesen W, Brock M, Klinger M (eds) Advances in neurosurgery. Springer, Berlin Heidelberg New York, pp 254–258
15. Kempski O, Chaussy L, Gross U, Zimmer M, Baethmann A (1983) Volume regulation and metabolism of suspended C6 glioma cells. Brain Res 279: 217–228
16. Kempski O, Zimmer M, Neu A, v Rosen F, Baethmann A (1987) Control of glial cell volume in anoxia. Stroke 18: 623–628
17. Kempski O, Staub F, Jansen M, Schödel M, Baethmann A (1988a) Glial swelling during extracellular acidosis in-vitro. Stroke 19: 385–392
18. Kempski O, Staub F, v Rosen F, Zimmer M, Neu A, Baethmann A (1988b) Molecular mechanisms of glial swelling in vitro. Neurochem Pathol 9: 109–125
19. Kempski O, v Rosen F, Weigt H, Staub F, Peters J, Baethmann A (1991) Glial ion transport and volume control. Ann NY Acad Sci 633: 306–317
20. Kraig RP, Pulsinelli WA, Plum F (1985) Heterogenous distribution of hydrogen and bicarbonate ions during complete brain ischemia. Prog Brain Res 163: 155–165
21. Kraig RP, Chesler M (1988) Glial acid-base homeostasis in brain ischemia. In: Norenberg M, Hertz L, Schousboe (eds) Biochemical pathology of astrocytes. New York, pp 365–376
22. Macknight ADC, Leaf A (1977) Regulation of cellular volume. Physiol Rev 57: 510–573
23. Mayer ML, Westbrook GL (1987) Cellular mechanisms underlying excitotoxicity. TINS 10: 59–61
24. Newman E (1991) Sodium-bicarbonate cotransport in retinal Muller (glial) cells of the salamander. J Neurosci 11: 3972–3983
25. Pines G, Danbolt NC, Bjoras M, Zhang Y, Bendahan A, Eide L, Koepsell H, Storm-Mathisen J, Seeberg E, Kanner BI (1992) Cloning and expression of a rat brain L-glutamate transporter. Nature 360: 464–467
26. Rehncrona S, Siesjö BK, Smith DS (1980) Reversible ischemia of the brain: Biochemical factors influencing restitution. Acta Physiol Scand [Suppl] 492: 135–140
27. Schneider G-H, Baethmann A, Kempski O (1993) Mechanisms of glial swelling induced by glutamate. Can J Physiol Pharmacol 70: S334–S343
28. Schousboe A, Svenneby G, Hertz L (1977) Uptake and metabolism in astrocytes cultured from dissociated mouse brain hemispheres. J Neurochem 29: 999–1005
29. Siesjö BK (1981) Cell damage in the brain: a speculative synthesis. J Cereb Blood Flow Metab 1: 155–185
30. Staub F, Baethmann A, Peters J, Weigt H, Kempski O (1990) Effects of lactacidosis on glial cell volume and viability. J Cereb Blood Flow Metab 10: 866–876

Correspondence: Oliver S. Kempski, M.D., Institute for Neurosurgical Pathophysiology, Johannes Gutenberg-Universität Mainz, Langenbeckstrasse 1, D-55101 Mainz, Federal Republic of Germany.

Acta Neurochir (1994) [Suppl] 60: 12–14

Mechanisms of Glutamate Induced Swelling in Astroglial Cells

E. Hansson[1], B. B. Johansson[3], I. Westergren[3], and L. Rönnbäck[1,2]

[1] Institute of Neurobiology, [2] Department of Neurology, University of Göteborg, and [3] Department of Neurology, University of Lund, Sweden

Summary

Relative changes in volume were registered in single cells by using a microspectrofluorometric equipment and the fluorescent probe fura-2/AM, excited at its isosbestic point. At this wavelength the probe is ion-insensitive and the fluorescent signals emitted is dependent on variations in the concentration of the dye. Variations in cell volume thus lead to changes in fluorescence intensity as the probe concentration is changed in the lightened delimited zone selected for each cell. When changing the excitation wavelength Ca^{2+} transients can be recorded.

Glutamate (Glu) induced swelling of type I astroglial cells in primary culture and a parallel intracellular Ca^{2+} increase was obtained. A Glu induced swelling was obtained even after blockade of the Glu ionotropic receptors with NBQX, suggesting that activation of ionotropic receptors might not be necessary for swelling to occur. On the other hand, blockade of the Glu carrier, or of pertussis toxin sensitive G-proteins reduced the Glu induced swelling. Blockade of Ba^{2+} or TEA sensitive K^+ channels completely blocked the Glu induced swelling as did also blockade with furosemide of the Na^+/K^+/Cl^- co-transporter. Glu induced swelling occurred in parallel with intracellular Ca^{2+} transients but extracellular Ca^{2+} did not seem necessary for swelling to occur.

Keywords: Astroglia; glutamate; cell volume; calcium.

Introduction

Astrocytes constitute a prominent part of the brain cell volume. The cells possess ion and amino acid transport systems[6,8,16] which participate in the regulation of the extracellular milieu, including the water content in the extracellular space. There is ultrastructural evidence that astroglial swelling is an early and important event in the development of brain edema in various pathologic states such as hypoglycemia, status epilepticus, ischemia, head trauma and hepatic encephalopathy[1,9]. An excessive release of glutamate (Glu) to the extracellular space is observed in some of these conditions and Glu is one important substance which causes swelling of astrocytes[2,4,10].

Calcium is believed to play a fundamental role in cellular volume regulation[11], although the molecular details of its involvement are poorly understood. Recently we described how to register variations in relative volume changes in single cells by microspectrofluorimetry after loading the cells with the highly fluorescent intracellular probe fura-2/AM[4]. At its isosbestic point, the probe was ion-insensitive and the fluorescent signals emitted related only to the intracellular dye concentration. By varying the excitation wavelengths, changes in intracellular Ca^{2+} transients could be recorded simultaneously with the relative volume variations of the individual cells. As astroglial properties are difficult to study in the intact nervous system, we used astrocytes in primary culture to evaluate molecular mechanisms of Glu induced swelling.

Materials and Methods

Cell culturing: The primary cultures were prepared from newborn rat cerebral cortex, as previously described[7] with some modifications[5].

Microspectrofluorimetry: The cells were incubated at 37°C with 10 μM fura-2/AM (dissolved in dimethyl sulfoxide; DMSO) for 45 min in HEPES-buffered Hank's balanced salt solution (HHBSS), pH 7.4 according to Hansson and co-workers[5]. The coverslip was mounted on a thermostated (37°C) perfusion chamber. After HHBSS rinsing, microscopic control of the probe loading of the cells was performed and individual type 1 astrocytes were located. A SPEX CM-X microspectrofluorimetric equipment interfaced with a Nikon diaphot inverted microscope was used. All cells selected were shown to emit fluorescence along a stable baseline. To measure volume variations and Ca^{2+} transients simultaneously, the isosbestic point, 358 nm, was used for volume measurements and 340 nm was used for Ca^{2+} transients. Calibration experiments were performed to correlate the variations in fluorescence intensity with the concentration of the pentapotassium salt of fura-2[4].

Calibration of relative changes in cell volume were made from series of data where cells were exposed in different hypo/hyperosmotic solutions. Changes in fluorescence intensity were proportional to changes in osmotic pressure and thereby to changes in cell volume.

Results and Discussion

The cells tested were morphologically identical to, and characterized as, type 1 astrocytes[12,13]. They were polygonal in shape and immunopositive for glial fibrillary acidic protein (GFAP+; Fig. 1a) and immunonegative for A2B5.

The cells were incubated with fura-2 and 0.1 mM Glu was added to induce swelling. In the top curve of Fig. 1b, the fluorescence intensity decreased by 12%, measured at the isosbestic point of fura-2, indicating a

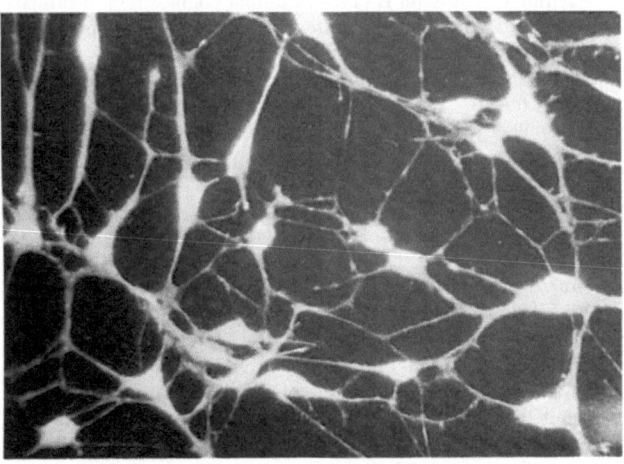

Fig. 1a. Astroglial enriched culture from rat cerebral cortex stained with antibodies against GFAP. Most cells are positive indicating their astroglial origin

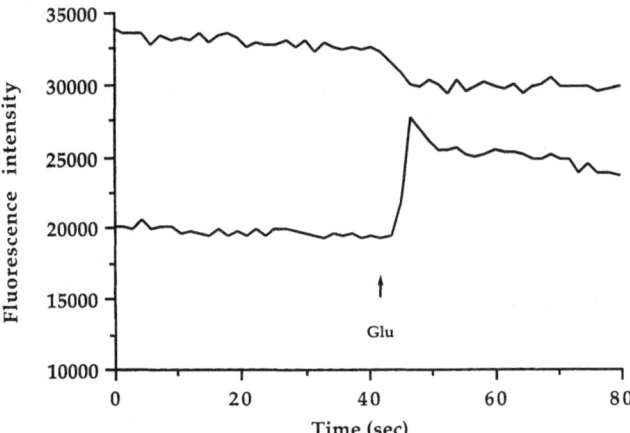

Fig. 1b. A single type I astroglial cell incubated with 0.1 mM Glu. Variations in cell volume and Ca²⁺ transients were measured simultaneously. Top curve shows registration at the isosbestic point, 358 nm (relative volume measurements; a decrease means an increased volume) and in the bottom curve Ca²⁺ transients were registered at 340 nm

volume increase of approximately 12%. The bottom curve represents the Ca^{2+} response, which increased simultaneously as the astrocyte swelled.

It should be noted that all cells selected were shown to emit fluorescence along a stable baseline for at least 200 s before they were stimulated to check that there was no leakage of the probe, nor any significant bleaching. At its isosbestic point, 358 nm for fura-2, the probe is Ca^{2+}-insensitive and the fluorescent signal emitted by the trapped probe is taken to depend solely on the variations of concentration induced by the cellular volume changes. Variations in cell volume will lead to changes in fluorescence intensity as the probe concentration is changed in the lightened delimited zone selected for each cell. Rather small amounts of the probe is trapped within membranes[14], and the probe was therefore assumed to have an isotropic distribution within the cell. Calibration experiments were performed to correlate the variations in the fluorescent signal with the osmotic pressure of the incubation medium[4]. Variations in cell volume down to 3% could be detected.

Mechanisms of Glu induced swelling were studied. Blockade of ionotropic Glu receptors with NBQX still gave an increase in astroglial cell volume suggesting that the ionotropic Glu receptors are not directly involved in swelling mechanisms (Table 1). This is interesting as NBQX has been shown to reduce Glu-induced brain edema in vivo[17]. On the other hand, we found a partial blockade of the Glu induced astroglial swelling by the Glu uptake inhibitor dihydroaspartate (Table 1). This finding indicates that the Glu carrier, whereby 1 Glu and 2 Na^+ and 1 H^+ or 3 Na^+ is transported into the cell and 1K^+ is transported out of the cell, is involved in swelling. It is assumed that water is transported together with Na^+. Glu induced swelling was also blocked by furosemide, a blocker of the $Na^+/K^+/Cl^-$ co-transporter, and by $BaCl_2$ or TEA, indicating that opening of inward and outward K^+ channels[5,15], respectively, are necessary for swelling to occur (Table 1).

Furthermore, the Glu induced astroglial swelling was partially blocked by pertussis toxin indicating that a G-protein[3] is involved in the Glu induced volume increase (Table 1).

In our experiments intracellular Ca^{2+} transients were seen in parallel with cell swelling (Table 1). However, swelling occurred even in the absence of extracellular Ca^{2+}. This is interesting in view of earlier suggestions that Ca^{2+} must enter the cell to initiate cellular volume regulation as cellular, rather than extracellular, Ca^{2+}

Table 1. *Relative Changes in Cell Volume (Registered at 358 nm) and Ca^{2+} Concentration (at 340 nm) Induced by 0.1 mM Glu Alone or in Combination with Other Substances*

Substance	Volume increase	Ca^{2+} transient
Glu	+	+
+ NBQX	+	+
+ dihydroaspartate	(+)	+
+ BaCl$_2$	–	+
+ TEA	–	+
+ furosemide	–	+
+ pertussis toxin	(+)	+
– Ca^{2+}$_{(e)}$	+	+

+   Volume increase and/or Ca^{2+} transient can be elicited.
(+) Partial blockade of volume increase.
–   No volume increase.

activates the ion fluxes that underlie volume regulation[11].

Our data thus suggest that Glu induced swelling is a complex process at the molecular level including the activation of pertussis toxin sensitive G-proteins, different K$^+$ channels, at least one co-transporter, the Glu carrier and probably requires an increase in intracellular Ca^{2+} but not necessarily presence of extracellular Ca^{2+}. It is well known that this swelling is of pathophysiological significance. However, it could be asked whether the Glu induced cell swelling has any physiological role. Astroglial swelling leads to a decrease of the extracellular volume which in turn might change the concentration of substances in the extracellular space. Thus, it might be speculated that astroglial volume changes with secondary changes of the extracellular space might be one way to modulate neuronal excitability.

Acknowledgements

The skillful work of Maria Wagberg is greatly appreciated. The work was supported by grants from the Swedish Medical Research Council (grant no 12X-06812), Magnus Bergvall's Foundation, and Syskonen Svensson's Foundation.

References

1. Beal MF (1992) Mechanisms of excitotoxicity in neurologic diseases. FASEB J 6: 3338–3344
2. Chan PH, Chu L (1989) Ketamine protects cultured astrocytes from glutamate-induced swelling. Brain Res 487: 380–383
3. Cockcroft S (1992) G-protein-regulated phospholipases C, D and A$_2$-mediated signalling in neutrophils. Biochim Biophys Acta 1113: 135–160
4. Eriksson PS, Nilsson M, Wagberg M, Rönnbäck L, Hansson E (1992) Volume regulation of single astroglial cells in primary culture. Neurosci Lett 143:195–199
5. Hansson E, Johansson B, Westergren I, Rönnbäck L (1993) Molecular mechanisms underlying glutamate induced astroglial swelling. Submitted
6. Hansson E, Rönnbäck L (1991) Receptor regulation of the glutamate, GABA and taurine high-affinity uptake into astrocytes in primary culture. Brain Res 548: 215–221
7. Hansson E, Rönnbäck L, Persson LI, Lowenthal A, Noppe M, Alling C, Karlsson B (1984) Cellular composition of primary cultures from cerebral cortex, striatum, hippocampus, brain stem and cerebellum. Brain Res 300: 9–18
8. Kimelberg HK, Norenberg MD (1989) Astrocytes. Sci Am 260: 66–76
9. Kimelberg HK, Ransom BR (1986) Physiological and pathological aspects of astrocytic swelling. In: Fedoroff S, Vernadakis A (eds) Astrocytes, vol 3. Academic Press, Orlando, FL, pp 129–166
10. Koyama Y, Sugimoto T, Shigenaga Y, Baba A, Iwata H (1991) A morphological study on glutamate-induced swelling of cultured astrocytes: involvement of calcium and chloride ion mechanisms. Neurosci Lett 124: 235–238
11. McCarty NA, O'Neil RG (1992) Calcium signaling in cell volume regulation. Physiol Rev 72:1037–1061
12. McCarthy KD, Salm A, Lerea LS (1988) Astroglial receptors and their regulation of intermediate filament protein phosphorylation. In: Kimelberg HK (ed) Glial cell receptors. Raven, New York, pp 1–22
13. Raff MC, Miller RH (1984) Glial cell development in the rat optic nerve. Trends Neurosci 7: 469–472
14. Roe MW, Lemasters JJ, Herman B (1990) Assessment of fura-2 for measurements of cytosolic free calcium. Cell Calcium 11: 63–73
15. Tse FW, Fraser DD, Duffy S, MacVicar BA (1992) Voltage activated K$^+$ currents in acutely isolated hippocampal astrocytes. J Neurosci 12: 1781–1788
16. Walz W (1989) Role of glial cells in the regulation of the brain ion microenvironment. Prog Neurobiol 33: 309–333
17. Westergren I, Johansson BB (1992) NBQX, an AMPA antagonist, reduces glutamate-mediated brain edema. Brain Res 573: 324–326

Correspondence: Elisabeth Hansson, M.D., Institute of Neurobiology, University of Göteborg, Medicinaregatan 5, S-413 90 Göteborg, Sweden.

Acta Neurochir (1994) [Suppl] 60: 15–19

Neuroprotective NMDA Antagonists: the Controversy over Their Potential for Adverse Effects on Cortical Neuronal Morphology

R. J. Hargreaves, R. G. Hill, and **L. L. Iversen**

Merck, Sharp and Dohme Research Laboratories, Neuroscience Research Centre, Terlings Park, Eastwick Road, Harlow, Essex, U. K.

Summary

It has been reported that several uncompetitive NMDA receptor ion channel blocking agents (phencyclidine, ketamine, dizocilpine, dextrorphan) cause transient reversible vacuolation in neurons in the posterior cingulate cortex of rats. Similar effects have also been observed with competitive glutamate antagonists such as CPP, CGS 19755 and CGP 37849. This transient morphological change has been noted to be coincident anatomically with brain regions showing hypermetabolism after administration of uncompetitive NMDA receptor ion channel blockers and competitive glutamate antagonists. These results therefore indicate that the functional consequences of NMDA receptor blockade with competitive glutamate and uncompetitive channel antagonists are ultimately the same. These changes do not appear to be a prelude to irreversible damage except after relatively high doses of the receptor ion channel antagonists but they have given rise to concern over the safety in use of NMDA antagonists as neuroprotective agents. In contrast, vacuolation has not yet been demonstrated with agents acting at the glycine (L-687,414) or polyamine (eliprodil) modulatory sites of the NMDA receptor complex suggesting that agents acting at these sites may have a greater potential therapeutic window.

Keywords: NMDA antagonists; vacuolation; metabolism.

Introduction

Release of excitatory amino acids during cerebral ischaemia is thought to result in an overactivation of N-methyl-D-asparate (NMDA) receptors and cause an excessive calcium influx into neurons that leads to cell death. Antagonists acting at the NMDA receptor complex have been shown to be protective in various experimental models of focal ischaemia and to protect against excitotoxin-induced neuronal loss. NMDA antagonists act at several distinct sites within the NMDA receptor complex (a) within the ion channel of the receptor, producing a dose-dependent uncompetitive blockade, (b) competitively at the glutamate recogni-tion site, or (c) competitively at allosteric modulatory sites where glycine and polyamines act to enhance NMDA receptor function.

NMDA Receptor Ion Channel Antagonists

Olney and colleagues first reported that acute dosing of several uncompetitive NMDA receptor ion channel antagonists (phencyclidine, tiletamine and dizocilpine) induced dose-dependent reversible swelling and vacuolation in medium and large sized neurons in the rat posterior cingulate cortex[1]. It has recently been suggested that there are age-related changes in the susceptibility of rats to this phenomenon with responsivity being first observed at 30 days increasing to 90 days of age[2]. In our studies with dizocilpine[3,4], we replicated the findings of reversible neuronal vacuolation and showed additionally that chronic administration of dizocilpine (1 mg/kg/day for 14 days) to mature Sprague-Dawley rats was not associated with neuronal cell loss. Only after high (5 mg/kg s.c.) acute doses of dizocilpine were a few neurons (< 5%) found to be irreversibly damaged when tissues were examined 48 h after dosing. More detailed studies from another group[5] on the time sequence of effects with dizocilpine have also shown that at 1.3 mg/kg s.c. in rats vacuolation is present at 4 h but absent at 24 h after dosing. However, in these experiments a small number of dark neurons were detected 24 h after dizocilpine that in sections examined at 3 and 7 days after dosing were associated with phagocytes suggesting that the neurons were necrotic. At 30 days after dosing increased numbers of microglia were observed in the target brain regions. No neuronal morphometry to quantify cell loss was however reported. Electron microscopy during these studies showed that the vacuoles were characterized by dilated rough endoplasmic reticulum and perinuclear cisternae but (unlike the observations of Olney) the mitochondria in the cytoplasm of affected cells appeared swollen but not lysed. It is not yet known if neurons vacuolated at 4 h reverse to normal morphology or go on to become necrotic.

In our recent studies with dizocilpine given intravenously at a steady plasma level for 4 h using a dose-regime optimised for neuroprotection (0.12 mg/kg bolus + 1.8 µg/kg/min) we have shown that there is little margin between doses giving wanted neuroprotective effects in the rat MCAO model of focal ischaemia and unwanted vacuolation in cortical neurons[6]. In this series of studies we also investigated the effects of dizocilpine on cerebral glucose metabolism (CMRglc). The results showed that there was a pronounced increase in CMRglc throughout limbic pathways whilst metabolism was decreased in the auditory system. As the hippocampus contains a high density of NMDA receptors it has been hypothesised that blockade in this region had evoked increased levels of firing in cells with an input into this area in an attempt to overcome the block. Since antagonism with NMDA receptor ion channel blockers is both dose-dependent and uncompetitive this increased firing then leads to an intensification of the blockade and an increased CMRglc back around the limbic Papez circuit. Very high rates of CMRglc were observed in the posterior cingulate cortex, the area of the brain susceptible to the neuronal vacuolar changes. The significance of this anatomical coincidence of hypermetabolic and morphological changes with NMDA antagonists is considered further below (see Mechanisms and antagonism of vacuolation).

Recent time-course experiments with dextrorphan[7], another NMDA receptor ion channel blocker, have shown that doses of 400 mg/kg s.c. produce reversible vacuolation typical of this class of agent. However at later times up to 30 days post-dosing there was no evidence for neuronal cell loss as judged by morphology or neuronal counting. Whilst the doses producing vacuolation with dizocilpine and dextrorphan seem highly disparate for agents acting at the same site, it should be remembered that these doses are influenced both by their relative affinity for binding in the receptor ion channel and also pharmacodynamic considerations such as duration and degree of brain penetration. These factors are reflected in vivo in a 100 fold difference in the mouse MCAO model of focal ischaemia[8].

Competitive NMDA Glutamate Antagonists

Competitive NMDA antagonists acting at the glutamate recognition site (CPP, CGS 19755 and CGP 37849) have been shown by Olney[9] and ourselves[10] to cause neuronal vacuolation but unlike the situation with dizocilpine, this appears to occur only at higher doses (CPP: 50 mg/kg i.v.; CGS 19755 100 mg/kg i.p.; CGP 37849 100 mg/kg i.p.) relative to their neuroprotective dose levels. Thus, this class of agents may have a greater potential therapeutic margin than the ion channel blocking agents. We have shown[10] that the competitive glutamate antagonist (CGP 37849) at doses causing neuronal vacuolation appears to increase CMRglc selectively within the limbic system in a pattern similar to that observed with dizocilpine. This suggests that the level of blockade that can be obtained with competitive glutamate antagonist can approach that attained by uncompetitive ion channel blocking agents when the doses given are high enough. It is noteworthy that by comparison with the high brain extraction of dizocilpine the brain entry of the competitive agents is slow[11,12]. Whilst this may be a disadvantage for speed of onset it is likely that their egress from the brain will also be lower than that of dizocilpine and thus their cerebral concentration profile and duration of action may be relatively prolonged . Interestingly at lower doses of GGP37849 (30 mg/kg) a fall in CMRglc was observed selectively in the auditory pathways indicating that transmission in this system is very sensitive to NMDA receptor blockade.

Glycine and Polyamine Modulatory Site Antagonists

We have shown recently that the NMDA glycine site antagonist L-687,414 ((+)-cis-4-methyl-HA-966), when given by constant intravenous infusion to maintain steady plasma levels that were neuroprotective in the rat MCAO model, did not alter cortical neuronal morphology nor increase CMRglc in the limbic system[6]. When infused at a higher rate producing some 2–3 fold higher plasma levels, there was still no evidence for vacuolation although CMRglc in the auditory pathways was reduced. Scatton and colleagues have shown also that eliprodil (SL 82.0715), an antagonist at the polyamine modulatory site does not evoke cortical vacuolar changes nor increases in CMRglc in the rat limbic system at doses in excess of those that are neuroprotective doses in the mouse MCAO model[13]. There is thus a clear window between neuroprotection and neuromorphological effects with the NMDA modulatory site antagonists in contrast to the NMDA receptor ion channel blocker dizocilpine.

It is noteworthy that the CMRglc effects at the high dose of L-687,414 were similar to those observed with a low dose of the competitive glutamate antagonist CGP 37849, whilst CMRglc after higher doses of CGP 37849 resembled the pattern of effects after giving the ion channel blocker dizocilpine. This series of observations comparing effects at neuroprotective doses suggests that there is a continuum across the various classes of NMDA antagonists for their effects on CMRglc in the auditory and limbic systems, and indeed this also appears to be the case for effects on cortical neuronal morphology. The rank order indicates that NMDA antagonists acting through the modulatory sites are less potent in producing these side-effects than competitive glutamate antagonists acting at the glutamate recognition site with NMDA receptor ion channel blockers producing the greatest response. This suggests that NMDA antagonists acting through the glycine and polyamine modulatory sites will have a greater therapeutic window than other classes of NMDA antagonists since neither vacuolation nor activation of CMRglc and central sympathomimetic activity have been reported but significant neuroprotection is still observed.

Interpretation of the actions of the glycine site antagonist L-687,414 are complicated by its partial agonist nature. The demonstration of weak agonist properties of L-687,414 at high concentrations (see [6]) suggests that its side-effect liability may be limited pharmacologically since a complete blockade equivalent to a receptor ion channel blocker will not be achievable. Further examination of the possibility of site-related differences in the potential therapeutic window with NMDA antagonists awaits the testing of a full glycine site antagonist at dose-levels at and above those giving significant neuroprotection for its morphological and metabolic effects in the CNS.

Mechanisms and Antagonism of Cortical Vacuolation

The mechanism of vacuolation in posterior cingulate cortical neurons is not yet known. The coincidence anatomically of vacuolation and increase in CMRglc has implicated hypermetabolism in the response. Several lines of evidence can be given in support of this view. Kurumaji et al.[14] showed that after repeated dosing the hypermetabolic response to dizocilpine in the posterior cingulate cortex was markedly reduced. In line with this lower sensitivity to dizocilpine-induced changes in CMRglc, the vacuolation response to acute doses of dizocilpine was absent after chronic dizocilpine dosing[4]. McCulloch and Iversen[15] have drawn parallels to the morphological changes in neurons after hypermetabolic seizure episodes again suggesting that the two events may be in some way linked. Importantly also the studies cited above with the glycine and polyamine modulatory site antagonists L-687,414 and eliprodil show that NMDA antagonists that do not induce increases in CMRglc in the posterior cingulate cortex do not cause vacuolation. It has been argued that the posterior cingulate cortex has a relatively low density of NMDA receptors (compared for example to hippocampus) and that effects in this area are thus unlikely to arise as a result of a direct action on cells in this area but rather from innervation of the vacuolated cells by processes from neurons that are rich in NMDA receptors. It would appear from antagonism of the vaculation response (see below) that the cells projecting to the vacuolated cells may be both cholinergic and subject to GABAergic inhibition. This illustrates an important point in the interpretation of functional autoradiographic maps obtained with the 2-deoxyglucose technique in that it should always be remembered that high activity may not reflect actions directly within a particular brain region but effects on neurons that send projections into ⁺his area whilst having cell bodies that are located elsewhere within a neuroanatomical circuit.

The cortical morphological effects of NMDA receptor ion channel blockers have been shown to be antagonised by muscarinic antagonists and drugs acting at GABAergic receptors and an elegant scheme to illustrate the possible mechanisms involved has been proposed by Olney[9]. This suggestion of an involvement of muscarinic pathways has been supported recently by the observation that vacuolation following administration of dextrorphan, an NMDA receptor ion channel blocker, is blocked by the brain penetrant muscarinic antagonist scopolamine[16]. McCulloch[15] has noted that the pattern of changes of regional cerebral blood flow and CMRglc after dizocilpine resemble those following stimulation of the fastigial nucleus that is thought to activate ascending cholinergic pathways and this again suggests a link between these neurotransmitter systems in the functional response to NMDA antagonists. It is noteworthy that dizocilpine and ketamine cause profound increases in cerebral blood flow in the posterior cingulate cortex that is far in excess of that expected for the increase in CMRglc in this region[17].

The precise mechanism of the reversible cortical vacuolation is unknown. It may reflect hydropic changes that result from water entry triggered by disturbances in the intracellular ionic environment because of a "relative hypoxia" that occurs during the intensive activation of these cells. The hypermetabolic response (in particular increased glycolysis) may thus be an attempt to maintain ionic homeostasis, which when restored results in the "disappearance" of the vacuoles, and the hyperemia a "waste disposal" mechanism for accumulated lactate. Further studies to test these hypotheses are now needed.

Sharp *et al.* have shown[18] that neurons vacuolated by dizocilpine and ketamine also later express heat shock protein (HSP 72). They have suggested that drug induced vacuolar changes and altered intracellular proteins act as the stimulus for the production of heat shock protein. It is important to note that heat shock protein expression occurs primarily in neurons that are destined to survive. The transient and reversible vacuolation can thus perhaps be viewed, like increases in CMRglc, as an early marker of cellular metabolic stress. Production of heat shock protein has been shown[9] to be prevented by agents that prevent vacuolation supporting the idea that their pathogenesis is linked to a common mechanism. Recent reports[19] on the inhibition of heat shock protein expression by high doses of haloperidol (> 5 mg/kg) and rimcazole (> 60 mg/kg) have been suggested to support the involvement of other receptor systems in the cortical morphological response to dizocilpine. Such doses are however likely to have caused generalised CNS depression. If this factor is considered alongside the report that pretreatment with gaseous anaesthetics prevents the increase in CMRglc[20] and vacuolation (Hargreaves and Rigby unpublished observations) associated with dizocilpine administration then it would appear that nonspecific depression of CNS activity may also antagonise the cortical morphological effects of the NMDA receptor ion channel blockers. It should be noted however that in other studies[9] when halothane was administered after dizocilpine then protection from vacuolation was not observed.

Conclusions

It remains to be determined whether the apparent differences in potential therapeutic window for NMDA antagonists acting at different sites within the

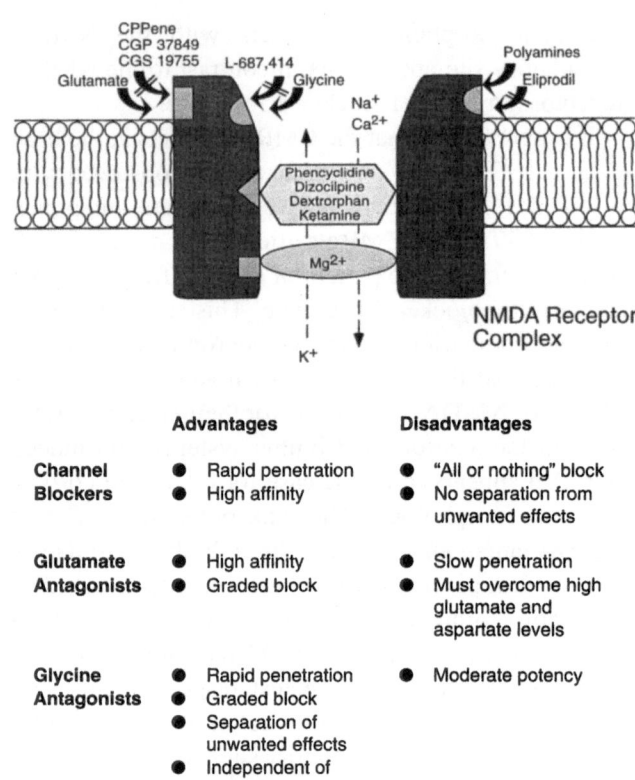

Fig. 1. Summary of the advantages and disadvantages of the various classes of neuroprotectants that act through antagonism of the NMDA receptor complex

receptor complex are due to true differences in their side-effect pharmacology or reflect a greater ability to 'fine tune' the level of blockade so that the neuroprotective and neuromorphological effects can be separated. Figure 1 summarizes with examples some of the advantages and disadvantages of the various classes of NMDA antagonists. By analogy with agents acting at the GABAergic receptors it is clear that compounds acting at different sites within the same receptor complex can have diverse spectra of pharmacological effects. Clearly however the actions of competitive antagonists acting at the NMDA glutamate, glycine and polyamine sites will be influenced critically by the prevailing concentrations of the endogenous ligands unlike the uncompetitive ion channel blocking agents. In ischaemic brain, extracellular glutamate and aspartate levels are known to be increased whilst glycine levels are relatively unchanged (see [8]). Thus, in ischaemia relatively high concentrations of competitive glutamate antagonists will be required to achieve the same level of NMDA receptor blockade that can be attained with an ion channel blocking agent. Similarly, since under normal conditions endogenous glycine concentrations are near saturation for the NMDA glycine site, it is likely to be very difficult to achieve a level of antagonism through this modulatory site that is equivalent to a complete channel blockade. The effects of competition with endogenous substrates is reflected in the moderate apparent potency of the competitive antagonists, by comparison with ion channel blocking agents such as dizocilpine, in models of ischaemia-induced neurodegeneration in vivo. In considering the therapeutic margin it should also be remembered that for all classes of NMDA antagonists it is very much greater when anticonvulsant rather than neuroprotective doses are used for comparison to the doses producing morphological effects. The eventual clinical utility of NMDA antagonists as neuroprotective agents may not be limited by their efficacy but by their profile of CNS side-effects. The studies so far suggest that the potential therapeutic windows are greater with competitive antagonists rather than with receptor ion channel blocking agents and that neuroprotection mediated by NMDA receptor blockade could potentially be achieved without cortical morphological effects by antagonism at the glycine or polyamine modulatory sites.

References

1. Olney JW, Labruyere J, Price MT (1989) Pathological changes induced in cerebrocortical neurons by phencyclidine and related drugs. Science 244: 1360–1362
2. Farber NB, Price MT, Labruyere J, Fuller TA, Olney JW (1992) Age-dependency of NMDA antagonist neurotoxicity. Soc Neurosci Abs 18: 1148
3. Allen HL, Iversen LL (1990) Phencyclidine, dizocilpine and cerebrocortical neurons. Science 247: 221
4. Allen HL, Smith DW, Hargreaves RJ, Iversen LL (1991) The effects of acute and chronic treatment with dizocilpine on neuronal morphology in the rat CNS. In: Hunter AJ, Clark M (eds) Neurodegeneration. Academic Press, London
5. Anderson TD, Arceo R, Ruben Z, Feldman D (1993) Detection and confirmation of neuronal necrosis in toxicity studies. Vet Pathol 29: 465
6. Hargreaves RJ, Rigby M, Smith D, Hill RG (1993) Lack of effect of L-687,414 ((+)-cis-4-methyl-HA-966), an NMDA receptor antagonist acting at the glycine site, on cerebral glucose metabolism and neuronal morphology. Br J Pharmacol, in press
7. Arceo RJ, Feldman D, Ruben Z, Anderson TD (1993) A time sequence study of neuronal vacuoles induced by dextrorphan hydrochloride (DH) in the rat. Toxicol Pathol, in press
8. Edvinsson L, Mackenzie ET, McCulloch J (1993) Cerebral blood flow and metabolism. Raven New York, pp 483–493
9. Olney JW, Labruyere J, Wang G, Wozniak DF, Price MT, Sesma MA (1991) NMDA antagonist neurotoxicity: mechanism and prevention. Science 254: 1515–1518
10. Hargreaves RJ, Rigby, M, Smith D, Hill RG, Iversen LL (1993) Competitive as well as uncompetitive N-methyl-D-aspartate receptor antagonists affect cortical neuronal morphology and cerebral glucose metabolism. Neurochem Res 18: 1263–1269
11. Chen M, Bullock R, Graham DI, Frey P, Lowe D, McCulloch J (1991) Evaluation of a competitive NMDA antagonist (D-CPPene) in feline focal cerebral ischaemia. Ann Neurol 30: 62–70
12. Hogan MJ, Gjedde A, Hakim AM (1992) In vivo distribution of CGS 19755 within brain in a model of focal ischaemia. J Neurochem 58: 186–191
13. Duval D, Roome N, Gauffeny C, Nowicki JP, Scatton B (1992) SL 82.0715 an NMDA antagonist acting at the polyamine site does not induce neurotoxic effects on rat cortical neurons. Neurosci Lett 137: 193–197
14. Kurumaji A, Ikeda M, Dewar D, MacCormack AG, McCulloch J (1991) Effects of chronic administration of MK-801 upon local cerebral glucose utilisation and ligand binding to the NMDA receptor complex. Brain Res 563: 57–65
15. McCulloch J, Iversen LL (1991) Autoradiographic assessment of the effects of N-methyl-D-aspartate (NMDA) receptor antagonists in vivo. Neurochem Res 16: 951–961
16. Ruben Z, Anderson TD, Arceo RJ, Kaufmann LS (1993) Scopolamine (SC) prevents neuronal cytoplasmic vacuoles induced by dextrorphan (DX) in retrosplenial/posterior cingulate cortex of rat brain. Toxicologist 13: 213
17. Nehls DG, Park CK, MacCormack AG, McCulloch J (1990) The effects of N-methyl-D-aspartate receptor blockade with MK-801 upon the relationship between cerebral blood flow and glucose utilisation. Brain Res 511: 271–279
18. Sharp FR, Jasper P, Hall J, Noble L, Sagar SM (1991) MK-801 and ketamine induce heat shock protein HSP72 in injured neurons in posterior cingulate and retrosplenial cortex. Ann Neurol 30: 801–809
19. Sharp FR, Wang S, Butman J, Koistanaho J, Graham S, Noble L, Sagar SM (1992) Haloperidol and rimcazole prevent induction of HSP70 heat shock gene in neurons injured by phencyclidine and MK-801. Soc Neurosci Abs 18: 1145
20. Kurumaji A, McCulloch J (1989) Effects of MK-801 upon local cerebral glucose utilisation in conscious rats and rats anaesthetised with halothane. J Cereb Blood Flow Metab 9: 786–794

Correspondence: R. J. Hargreaves, M.D., Department of Pharmacology, Merck, Sharp and Dohme Research Laboratories, Neuroscience Research Centre, Terlings Park, Eastwick Raod, Harlow, Essex, CM20 2QR, U. K.

Acta Neurochir (1994) [Suppl] 60: 20–23

Mechanisms of Glial Swelling by Arachidonic Acid

F. Staub[1], **A. Winkler**[1], **J. Peters**[1], **O. Kempski**[2], and **A. Baethmann**[1]

[1] Institute for Surgical Research, Ludwig-Maximilians-University, München and [2] Institute for Neurosurgical Pathophysiology, Johannes Gutenberg-University, Mainz, Federal Republic of Germany

Summary

The effect of arachidonic acid (AA, 20:4) was analyzed in vitro by employment of C6 glioma cells and astrocytes from primary culture. The cells were suspended in an incubation chamber under continuous control of pH, pO_2, and temperature. Cell swelling was quantified by flow cytometry. After a control period, the suspension was added with AA at concentrations of 0.01 to 1.0 mM. Administration of AA induced an immediate, dose dependent swelling in C6 glioma cells or astrocytes. AA-concentrations of 0.01 mM led to an increase of the glial cell volume to $103.0 \pm 1.0\%$ of control, 0.1 mM to $110.0 \pm 1.5\%$, and 1.0 mM to $118.8 \pm 1.5\%$ within 10 min. The swelling response to linoleic acid (18 : 2) was only about half of what was found when AA was administered at a concentration of 0.1 mM, whereas stearic acid (18 : 0) did not induce any cell volume changes. Inhibition of the cyclo- and lipoxygenase pathway by BW 755C did not prevent glial swelling from AA, whereas it was reduced by SOD, or almost completely abolished by the aminosteroid U-74389F, an antagonist of lipid peroxidation. Replacement of Na^+- and Cl^--ions in the suspension medium by choline chloride was also associated with complete abolishment of cell swelling from AA. The results demonstrate an impressive efficacy of arachidonic acid to induce glial swelling which might be attributable to activation of lipid peroxidation by the fatty acid, leading to an increased Na^+-permeability and subsequent influx of water into the cells.

Keywords: Cytotoxic cell swelling; arachidonic acid; C6 glioma cells; astrocytes.

Introduction

Arachidonic acid (AA, 20 : 4) is a major constituent of membrane phospholipids in brain tissue. Normally, the free fatty acid is present only in a small amount, but it accumulates under adverse conditions, such as ischemia or brain injury[1,9]. The release of free fatty acids involves activation of phospholipases and breakdown of membrane phospholipids. AA in particular is considered to mediate pathological processes. The polyunsaturated compound is precursor of prostaglandins, leukotrienes, and oxygen-derived free radicals[11].

In cerebral ischemia concentrations of free AA of up to 0.5 mM/kg were found in brain tissue[9].

Noxious properties of AA have been observed in many investigations. For instance, it was demonstrated that lipid peroxidation or disturbances of respiration in mitochondria are induced by the fatty acid[2,6]. Moreover, AA has a significant role in brain edema. The fatty acid was found to induce swelling of cerebral tissue slices in vitro and to increase permeability of the blood-brain barrier in vivo[3,10]. Since AA may also be involved in cytotoxic brain edema formation, we have currently investigated the glial cell volume response during administration of the fatty acid. Additional experiments were performed to analyze, whether respective effects can be inhibited by blocking of the cyclo- and lipoxygenase pathway, by inhibition of free radical formation or of lipid peroxidation. To assess specificity of AA (20:4) comparative experiments were made using linoleic acid (18 : 2) or stearic acid (18 : 0). The study was carried out in vitro under control of relevant parameters to examine the significance of a single pathophysiological factor in isolation[8]. For that purpose, suspensions of C6 glioma cells or of astrocytes from primary culture were incubated under continuous recording of pH, pO_2, and temperature with assessment of the cell volume by flow cytometry.

Materials and Methods

C6 glioma cells were cultivated as monolayers in Petri dishes using Dulbecco's modified minimal essential medium (DMEM) with addition of 25 mM bicarbonate. The medium was supplemented with 10% fetal calf serum (FCS) and 100 IU/ml penicillin G and 50 µg/ml streptomycin. The cells were grown in a humidified atmosphere of 5% CO_2 and room air at 37°C. Glial cells from primary culture were prepared from 3-day-old rats according to a modified method of Frangakis and Kimelberg[5]. The culture conditions were identical with those described above. The cells were harvested for the experiment with 0.05% trypsin/0.02% EDTA in phosphate-buffered saline and washed twice thereafter. After resuspension in serum-free medium the glial cells were transferred to a plexiglas incubation chamber which was supplied with electrodes for control of pH, temperature, and pO_2. Details of the method have been published elsewhere[8].

The volume of the glial cells was determined by flow cytometry using an advanced coulter system with hydrodynamic focusing[7]. The experiments with administration of AA were performed after a 45 minute control period used for measurements of normal cell volume and medium osmolality. Subsequently, the suspension was added with AA (20 : 4) in a dose range of 0.01 mM to 1.0 mM (final concentration). Cell volume and viability were measured for 90 min during incubation with AA. Comparative studies using linoleic- (18 : 2) or stearic acid (18 : 0) were made at concentrations of 0.1 mM. Further experiments with AA at this concentration were conducted with inhibition of the cyclo- and lipoxygenase pathway by BW755C (0.2 mM), with scavenging of superoxide radicals by superoxide dismutase (SOD, 300 U/ml), or during blocking of lipid peroxidation by the aminosteroid U-74389F (0.1 mM). Inhibitors were administered to the suspension 15 min prior to addition of AA. In further experiments cell swelling by 0.1 mM AA was studied in Na^+-free medium, where Na^+-ions were replaced for that purpose by choline, and bicarbonate by 10 mM HEPES.

Results and Discussion

Administration of AA to the suspension caused an immediate dose-dependent swelling of C6 glioma cells. AA-concentrations as low as 0.01 mM led to an increase of the cell volume to $103.0 \pm 1.0\%$ (mean $\pm$ SEM) of control within 10 min ($p < 0.01$). A volume increase to $110.0 \pm 1.5\%$ (Table 1) or $118.8 \pm 1.5\%$ of control was obtained, when the cells were administered with AA at 0.1 or 1.0 mM, respectively ($p < 0.001$). After initial and rapid swelling the increased cell volume was largely constant for the remaining observation period (Table 1). Swelling of C6 glioma cells by AA was confirmed in experiments using astrocytes from primary culture. Addition of AA at a dose level of 0.05 or 0.1 mM led to swelling of astrocytes ($p < 0.01$), which was comparable to that of C6 glioma cells. In order to assess specificity of the AA-induced glial swelling, cell swelling inducing properties of 0.1 mM linoleic (18 : 2) or of 0.1 mM stearic acid (18 : 0) were tested (Table 1). While stearic acid was not found to

Table 1. *Volume Response of C6 Glioma Cells to Administration of AA (0.1 mM) to the Suspension Medium.* Cell swelling was significant ($p < 0.01$) during the entire observation period as compared to a control group without addition of AA (not shown). The cell volume response to AA is further shown during inhibition of the cyclo- and lipoxygenase pathway by BW755C (0.2 mM), in the presence of SOD (300 U/ml), U-74389F (0.1 mM), or in Na^+-free incubation medium. Cell volume is given in percent of the cell size obtained during the last 15 min of the control period. Mean $\pm$ SEM of 4–7 experiments per group are shown

Time	Control			Incubation with free fatty acids					
	−15	−10	−5	1	5	10	30	60	90
AA 0.1 mM	100.40	99.93	99.67	104.45	110.04	112.17	111.00	109.58	106.85
	0.26	0.17	0.25	1.16	1.49	1.90	2.23	2.20	1.72
LA 0.1 mM	100.36	99.63	100.01	103.05	106.53	105.88 *	105.66	108.22	105.53
	0.61	0.35	0.75	0.71	0.82	0.76	0.82	1.38	1.41
SA 0.1 mM	100.63	99.07	100.30	100.07 **	100.45 **	99.42 **	99.36 **	99.46 **	100.25 **
	0.37	0.64	0.28	0.54	0.96	0.26	1.44	2.13	2.23
AA 0.1 mM + BW 755 C	99.88	99.90	100.23	103.84	109.60	111.18	112.36	107.57	106.91
	0.29	0.20	0.33	1.77	1.58	2.04	2.32	2.58	3.86
AA 0.1 mM + SOD	99.84	100.25	99.91	104.59	108.28	109.42	106.79	103.19 **	103.05
	0.09	0.44	0.43	2.03	1.98	1.17	0.82	1.03	0.80
AA 0.1 mM + U-74389F	100.27	99.40	100.33	99.41 **	100.18 **	98.93 **	99.47 **	100.97 **	101.39 **
	0.33	0.49	0.35	0.70	1.29	1.35	1.37	1.11	1.53
AA 0.1 mM Na^+-free M.	98.63	101.25	100.12	100.87 **	101.65 **	103.77 **	100.01 **	102.00 **	103.46
	0.75	0.40	0.44	0.46	0.41	0.86	1.82	1.99	2.51

AA arachidonic acid, *LA* linoleic acid, *SA* stearic acid.
* $p < 0.05$ vs. AA 0.1 mM.
** $p < 0.01$ vs. AA 0.1 mM.

induce volume changes of the glioma cells, cell volume was significantly increased by linoleic acid (p < 0.01). Nevertheless, the volume increase to linoleic acid was only about 50% of what was found when AA was administered at the same concentration (p < 0.05; Table 1). Additional experiments were performed to analyze mechanisms of the AA-induced glial swelling. Inhibition of the metabolization of AA by cyclo- and lipoxygenase using the dual pathway inhibitor BW755C did not affect glial swelling from AA (Table 1). Administration of SOD to scavenge superoxide radicals did not influence the initial volume response but reduced glial swelling by AA significantly at 60 min after addition of the fatty acid (p < 0.01). Preincubation with the aminosteroid U-74389F, however, prevented cell swelling from AA practically completely (p < 0.01). Similar results were obtained when the experiments were conducted with choline for replacement of Na+-ions in the suspension medium and HEPES as buffer compound instead of bicarbonate (Table 1).

The present results demonstrate a powerful potential of AA to induce glial swelling. Exposure of glial cells to the fatty acid led to a dose-dependent cell volume increase at concentrations, which have been observed in brain tissue in vivo under pathophysiological conditions, such as focal injury or ischemia[1,9].

Not only AA but also its metabolites, such as prostaglandins or leukotrienes, can be considered to have mediator functions in the brain under respective circumstances[11]. The present results on inhibition of the cyclo- and lipoxygenase pathway by BW755C suggest, however, that AA itself was the swelling inducing agent but not these metabolites. On the other hand, since administration of the aminosteroid U-74389F was almost completely inhibiting cell swelling from AA, lipid peroxidation must be taken into consideration as major factor of AA-induced cell swelling. Lipid peroxidation commences, if oxygen-derived free radicals accumulate in the presence of free fatty acids[2,4]. It is conceivable that superoxide radicals generated by the conversion of AA were interacting with AA as substrate, thereby initiating the formation of lipid- and lipidperoxide radicals. The marginal success of SOD to reduce glial swelling from AA suggests that superoxide radicals formed in the intracellular compartment were not available for the enzyme, since SOD as a large, hydrophilic molecule is unlikely to penetrate the plasma membrane. Accumulation of lipid- and lipidperoxide radicals, however, may result in a chain reaction, mainly acting on fatty acids of cell membranes,

consequently leading to damage of the double-lipid layer[2]. The results indicate further that the swelling-inducing properties of free fatty acids are related with their number of non-saturated chemical bonds. Accordingly, the production of superoxide- and lipid radicals from free fatty acids by glial cells appears to be directly correlated with the number of non-saturated bonds. This is supported albeit indirectly by findings that formation of reactive radical species is minimal from saturated fatty acids as precursors[4]. Finally, glial swelling from AA was more or less completely prevented, if the experiments were conducted in a Na+-free suspension medium making obvious the significance of the cellular uptake of Na+, and thereby water as ultimate mechanism of the AA-induced cell volume increase.

Taken together, AA causes swelling of C6 glioma cells and of astrocytes from primary culture, probably by activation of lipid peroxidation among others. The lipid peroxides by damaging lipid bilayers of cell membranes are likely to increase Na+-permeability and, thus influx of water into the cells. The resulting cellular accumulation of Na+ together with Cl− and water must be viewed as the final step of glial cell swelling.

Acknowledgements

The excellent technical and secretarial assistance of Ingrid Kölbl, Helga Kleylein, and Monika Stucky is gratefully acknowledged. We appreciate the generous gifts of BW755C by Dr. Cooke, Wellcome Research Laboratories, Beckenham, Kent, U.K., and of U-74389F by Dr. McCall, Upjohn Laboratories, Kalamazoo, MI, U.S.A.

References

1. Baethmann A, Maier-Hauff K, Schürer L, Lange M, Guggenbichler C, Vogt W, Jacob K, Kempski O (1989) Release of glutamate and of free fatty acids in vasogenic brain edema. J Neurosurg 70: 578–591
2. Braughler JM, Hall ED (1989) Central nervous system trauma and stroke. I. Biochemical considerations for oxygen radical formation and lipid peroxidation. Free Rad Biol Med 6: 289–301
3. Chan PH, Fishman RA (1978) Brain edema: induction in cortical slices by polyunsaturated fatty acids. Science 201: 358-360
4. Chan PH, Chen SF, Yu ACH (1988) Induction of intracellular superoxide radical formation by arachidonic acid and by polyunsaturated fatty acids in primary astrocytic culture. J Neurochem 50: 1185–1193
5. Frangakis MV, Kimelberg HK (1984) Dissociation of neonatal rat brain by dispase for preparation of primary astrocyte cultures. Neurochem Res 9: 1689–1698
6. Hillered L, Chan PH (1988) Role of arachidonic acid and other free fatty acids in mitochondrial dysfunction in brain ischemia. J Neurosci Res 20: 451–456

7. Kachel V, Glossner E, Kordwig G, Ruhenstroth-Bauer G (1977) Fluvo-Metricell, a combined cell volume and cell fluorescence analyzer. J Histochem Cytochem 25: 804–812

8. Kempski O, Chaussy L, Groß U, Zimmer M, Baethmann A (1983) Volume regulation and metabolism of suspended C6 glioma cells: an in vitro model to study cytotoxic brain edema. Brain Res 279: 217–228

9. Kinouchi H, Imaizumi 5, Yoshimoto T, Motomiya M (1990) Phenytoin affects metabolism of free fatty acids and nucleotides in rat cerebral ischemia. Stroke 21: 1326–1332

10. Unterberg A, Wahl M, Hammersen F, Baethmann A (1987) Permeability and vasomotor response of cerebral vessels during exposure to arachidonic acid. Acta Neuropathol 73: 209–219

11. Wolfe LS (1982) Eicosanoids: postaglandins, thromboxanes, leukotrienes, and other derivates of carbon-20 unsaturated fatty acids. J Neurochem 38:1–14

Correspondence: Frank Staub, M. D., Institut für Chirurgische Forschung, Klinikum Großhadern, Marchioninistrasse 15, D-81366 München, Federal Republic of Germany.

Acta Neurochir (1994) [Suppl] 60: 24–27
© Springer-Verlag 1994

Astrocyte Swelling in Liver Failure: Role of Glutamine and Benzodiazepines

M. D. Norenberg and **A. S. Bender**

Laboratory of Neuropathology, Departments of Pathology, Biochemistry and Molecular Biology, University of Miami School of Medicine and Veterans Affairs Medical Center, Miami, Florida, U.S.A.

Summary

We examined the effect of methionine sulfoximine (MSO) and peripheral benzodiazepine (BZD) ligands on ammonia-induced swelling of primary astrocyte cultures. Swelling was completely abolished by co-treatment with MSO, an inhibitor of glutamine synthetase. We also established that many of the effects caused by ammonia, including the reduction in K^+ uptake, increase in Cl^- uptake and reduction in *myo*-inositol uptake were diminished by co-treatment with MSO. Agonists of the peripheral-type BZD receptor aggravated ammonia-induced swelling, whereas a peripheral BZD receptor blocker, PK 11195, diminished the extent of swelling. Our findings implicate glutamine and the peripheral-type benzodiazepine receptor in the pathogenesis of the edema associated with fulminant hepatic failure.

Keywords: Astrocyte; ammonia; glutamine; fulminant hepatic failure.

Introduction

Brain edema represents a major complication of acute hepatic encephalopathy (HE) or fulminant hepatic failure (FHF). It has an extremely poor prognosis with a mortality in the range of 80–90%[4]. The only effective treatment currently available is liver transplantation. Ammonia is believed to play an important role in the pathogenesis of liver failure and we have recently shown that ammonia can directly cause astrocyte swelling in primary cultures[12]. The mechanism for this effect is not known. Brusilow and colleagues have proposed that the accumulation of glutamine derived from ammonia "detoxification" may serve as a major organic osmolyte in hyperammonemic states leading to cell swelling[3]. Another factor implicated in the pathogenesis of HE are endogenous benzodiazepine (BZD) ligands[1].

To test the role of glutamine as an accumulating osmolyte in ammonia-induced swelling, we examined the effect of methionine sulfoximine (MSO), a glutamine synthetase (GS) inhibitor, as this agent brings about a marked reduction in astrocyte glutamine content. We also determined whether MSO influences organic and inorganic osmolyte transport that might explain any protective effect of this agent. Additionally, we examined whether agonists and antagonists of the peripheral BZD receptor affected ammonia-induced swelling.

Materials and Methods

Cultures were prepared from cerebral cortices of 1–2 day old rats. The cortex was dissociated and plated in 35 mm dishes at a density of 0.5×10^6. After two weeks, the cells were maintained with dibutyryl cyclic AMP. Based on immunohistochemistry of GFAP and glutamine synthetase, at least 95% of the cell population were astrocytes. Cells 3–5 weeks old were used for experiments.

The intracellular water space was measured utilizing 3-O-methyl-[^{3}H]-glucose (3-OMG) (specific activity 86.7 Ci/mmol) as described by Kletzien *et al.*[9] Cultured cells were incubated in DMEM media containing 1 mM 3-OMG and 0.5 µCi/ml of ^{3}H-OMG for various time intervals up to 72 hrs in the absence (control) or presence of 5 mM NH_4Cl.

K^+ uptake was determined by utilizing 86-RbCl, a K^+ analog. Cells were incubated in bicarbonate buffered media containing 2 µCi/ml of 86-Rb^+ for 1 h in the absence or presence of 5 mM NH_4Cl. Cl^- uptake was studied using 36-Cl^-. Cells were incubated in bicarbonate buffered media with 0.5 µCi/ml 36-Cl^- for 1 h in the absence (control) or presence of 5 mM NH_4Cl.

To study *myo*-inositol uptake, the cells were incubated with DMEM media containing 20 µM *myo*-inositol and 0.2 µCi [^{3}H]-inositol per dish in the absence (control) or presence of 5 or 10 mM NH_4Cl for various time periods up to 72 h.

Results

We have previously shown that ammonia treatment results in glial swelling[12]. As shown in Fig. 1, cell swelling was totally prevented by concurrent treatment with MSO. We have extended these observations to examine the effects of MSO on the flux of organic and

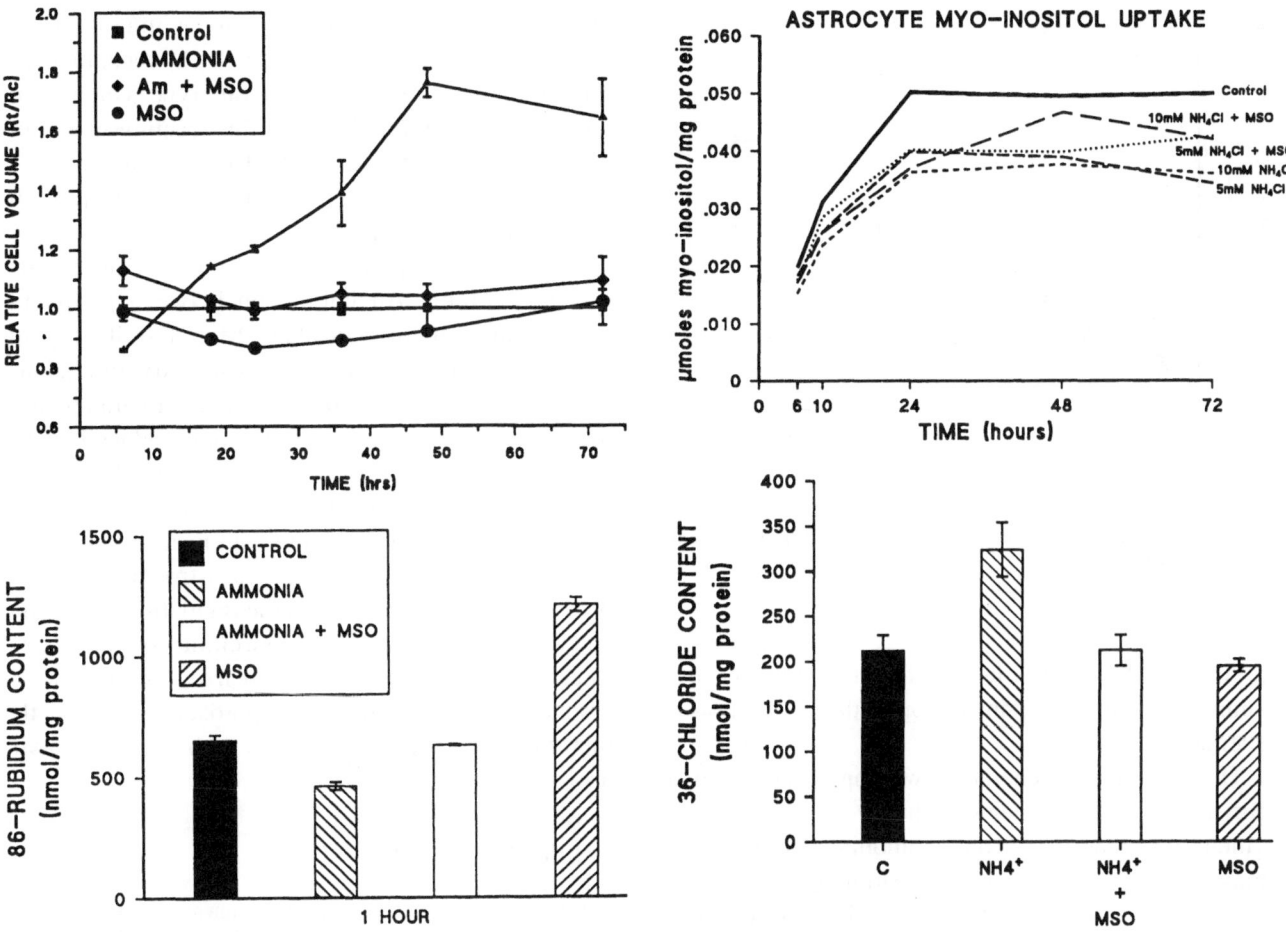

Fig. 1. Upper left: Time course of the effect of MSO (3 mM) on ammonium chloride (5 mM) induced astrocyte swelling, expressed as relative to control values (mean ± SEM of 6 separate determinations). Upper right: Time course of *myo*-inositol uptake by ammonia treatment and the effect of MSO (mean ± SEM of 5 separate determinations). Lower left: Effect of ammonia and MSO on 86-rubidium uptake (mean ± SEM of 6 separate determinations). Lower right: Effect of ammonia and MSO on 36-chloride uptake (mean ± SEM of 6 separate determinations)

inorganic osmolytes. We investigated the possible role of ammonia on *myo*-inositol transport, as the latter has been shown to act as a regulatory osmolyte[13]. We found that a 3-day treatment of astrocytes with ammonium chloride (5 mM) decreased *myo*-inositol uptake by approximately 50% (Fig. 1). The effect of ammonia on myo-inositol uptake was partially reversed (by 45%) in the presence of 3 mM MSO (Fig. 1). One hour treatment of astrocytes with 5 mM NH_4Cl resulted in a 30% decrease in potassium accumulation (measured as 86-rubidium) (Fig. 1) which was blocked by MSO. Similarly, chloride influx was accelerated by ammonia treatment and prevented by MSO (Fig. 1).

Treatment of ammonia-treated astrocytes with the peripheral BZD receptor agonist, Ro5-4864 (10 μM) exacerbated the swelling by 20%, whereas treatment with a receptor antagonist, PK 11195 (10 μM) diminished the swelling by 23% (Fig. 2).

Discussion

The mechanisms involved in the production of FHF-associated edema are controversial as both vasogenic and cytotoxic mechanisms have been proposed[2]. However, most animal and human studies have emphasized astrocyte swelling (i.e., cytotoxic edema) as a major component of the edema associated with HE. See Blei[2] for review.

The factors in FHF responsible for the generation of glial swelling have not been fully elucidated. In vivo studies have shown a good correlation between the extent of swelling and levels of ammonia[14]. Ganz *et al.*[6] demonstrated in tissue slices that the swelling was mostly confined to astrocytes. Cell culture studies of ammonia-treated astrocytes also show cell swelling[12]. Taken together, there is strong evidence to indicate that the main mechanism for brain edema in FHF is cyto-

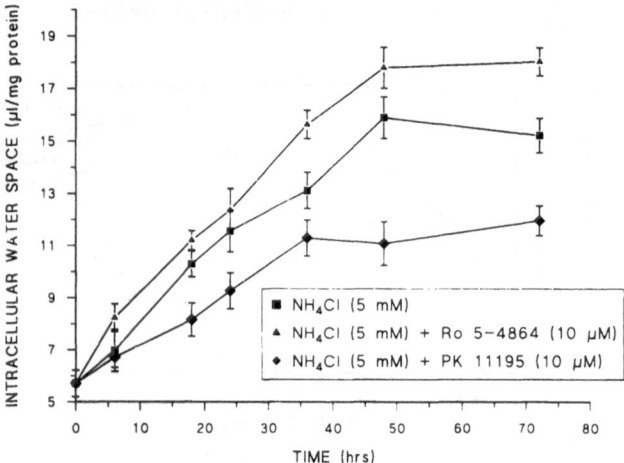

Fig. 2. Time course of the effect of peripheral benzodiazepine ligands on ammonia-induced swelling (mean ± SEM of 4 separate determinations)

toxic, that the fluid is largely in astrocytes, and that ammonia is an important factor in the production of edema.

How ammonia generates swelling is not known. Brusilow and colleagues have suggested that glutamine, generated from ammonia detoxification, may serve as a major organic osmolyte in hyperammonemic states leading to cell swelling[3]. Supporting this concept is that ammonia-induced swelling was prevented when the animals were pretreated with MSO[15]. Additionally, Swain et al. showed a correlation between edema and brain glutamine levels[14]. Our findings are in agreement with this view in that MSO totally abolished ammonia-induced swelling of cultured astrocytes. Moreover, the ionic fluxes associated with ammonia treatment were also significantly diminished with MSO treatment. Nevertheless it remains to be established whether this protective effect of MSO can be explained by inhibition of GS alone or by some other unknown effect of MSO.

myo-inositol appears to be a major osmolyte involved in the maintenance of intracellular osmolarity[13]. We found that ammonia decreased myo-inositol uptake by 50%. This finding is in accordance with the report of a 50% reduction in myo-inositol content in patients with HE as determined by proton magnetic resonance spectroscopy[10]. Such change may be reflective of an adaptive response to cell swelling. The effect of ammonia on myo-inositol uptake was partially reversed (by 45%) in the presence of MSO.

Our studies also disclosed that benzodiazepines can affect ammonia-induced swelling. Benzodiazepines

are currently under intense scrutiny as important factors in HE. Astrocytes possess BZD receptors, although they are of the "peripheral" type. Ducis et al. have shown an upregulation of this receptor in ammonia-treated astrocytes[5], while Butterworth and colleagues identified increased numbers of receptors in portocaval-shunted rats[7].

The benzodiazepine system may contribute to HE-related swelling. An interaction between peripheral BZD ligands and diuretics have been described[11], and we have recently shown that hypotonicity upregulates the peripheral BZD receptor[8]. Treatment with the agonist Ro 5-4864 exacerbated ammonia-induced swelling whereas use of a peripheral antagonist, PK 11195, diminished the extent of swelling. How these BZD-induced changes are brought about is not known.

In summary, ammonia causes astrocyte swelling, probably by abnormalities in the regulation of organic and inorganic osmolytes. The swelling could be ameliorated by MSO and peripheral benzodiazepine blockers suggesting potential new approaches towards the therapy of the edema associated with acute liver failure.

Acknowledgements

This work was supported by the Department of Veterans Affairs (Merit Review and GRECC) and NIH grants DK38153 and NS30291.

References

1. Basile AS, Jones EA, Skolnick P (1991) The pathogenesis and treatment of hepatic encephalopathy: evidence for the involvement of benzodiazepine receptor ligands. Pharmacol Rev 43: 27–71
2. Blei AT (1991) Cerebral edema and intracranial hypertension in acute liver failure: distinct aspects of the same problem. Hepatology 13: 376–379
3. Brusilow SW, Traystman RJ (1986) Letter to editor. N Engl J Med 314: 786
4. Capocaccia L, Angelico M (1991) Fulminant hepatic failure: clinical features, etiology, epidemiology, and current management. Dig Dis Sci 36: 775–779
5. Ducis I, Norenberg LOB, Norenberg MD (1989) Effect of ammonium chloride on the astrocyte benzodiazepine receptor. Brain Res 499: 362–365
6. Ganz R, Swain M, Traber P, DalCanto M, Butterworth RF, Blei AT (1989) Ammonia-induced swelling of rat cerebral cortical slices: implications for the pathogenesis of brain edema in acute hepatic failure. Metab Brain Dis 4: 213–223
7. Giguere JF, Hamel E, Butterworth RF (1992) Increased densities of binding sites for the 'peripheral- type' benzodiazepine receptor ligand [3H] PK 11195 in rat brain following portacaval anastomosis. Brain Res 585: 295–298
8. Itzhak Y, Norenberg MD (1993) Hypotonicity regulates the peripheral-type benzodiazepine receptor in cultured astrocytes. Soc Neurosci (abstract), in press

9. Kletzien RF, Pariza MW, Becker JE, Potter VR (1975) A method using 3-O-methyl-D-glucose and phloretin for the determination of intracellular water space of cells in monolayer culture. Anal Biochem 68: 537–544

10. Kreis R, Ross BD, Farrow NA, Ackerman Z (1992) Metabolic disorders of the brain in chronic hepatic encephalopathy detected with 1-H MR spectroscopy. Radiology 182: 19–27

11. Lukeman DS, Fanestil DD (1987) Interactions of diuretics with the peripheral-type benzodiazepine receptor in rat kidney. J Pharmacol Exp Ther 241: 950–955

12. Norenberg MD, Baker L, Norenberg L-OB, Blicharska J, Bruce-Gregorios JH, Neary JT (1991) Ammonia-induced astrocyte swelling in primary culture. Neurochem Res 16: 833–836

13. Sherman WR (1989) Inositol homeostasis, lithium and diabetes. In: Michell RH, Drummond AH, Downes CP (eds) Inositol lipids in cell signalling. Academic Press, London, pp 39–79

14. Swain M, Butterworth RF, Blei AT (1992) Ammonia and related amino acids in the pathogenesis of brain edema in acute ischemic liver failure in rats. Hepatology 15: 449–453

15. Takahashi H, Koehler RC, Brusilow SW, Traystman RJ (1991) Inhibition of brain glutamine accumulation prevents cerebral edema in hyperammonemic rats. Am J Physiol Heart Circ Physiol 261: H825–H829

Correspondence: Michael D. Norenberg, M.D., Department of Pathology (D-33), University of Miami School of Medicine, P.O. Box 016960, Miami, Fl 33101, U.S.A.

Acta Neurochir (1994) [Suppl] 60: 28–30
© Springer-Verlag 1994

The Role of K+ Influx on Glutamate Induced Astrocyte Swelling: Effect of Temperature

A. S. Bender and **M. D. Norenberg**

Laboratory of Neuropathology, Departments of Pathology, Biochemistry and Molecular Biology, University of Miami School of Medicine and Veterans Affairs Medical Center, Miami, Florida, U.S.A.

Summary

We studied factors involved in glutamate-stimulated astrocyte swelling and the role of temperature in this process. Glutamate-induced astrocyte swelling was reduced by 74% when temperature was lowered from 37°C to 24°C. Reduction in temperature to 24°C also resulted in inhibition of glutamate-stimulated K^+ uptake. When both extracellular and intracellular Ca^{2+} were removed with BAP-TA/AM (50 μM) and 0.1 mM EGTA and in the absence of extracellular Ca^{2+}, glutamate was not able to stimulate the K^+ uptake or swelling. Dantrolene (20 μM), MK-801 (0.5 mM) and ouabain (0.5 mM) inhibited glutamate-induced K^+ uptake as well as the glutamate effects on swelling. These findings show the importance of Ca^{2+} in the mechanism of glutamate swelling and suggests that glutamate induces swelling through stimulation of K^+ influx and that the diminished effect of glutamate swelling with decrease in temperature may be due to a decrease in K^+ uptake

Keywords: Astrocytes; glutamate; hypothermia; swelling.

Introduction

It is well established that hypothermia (30–31°C) protects the brain from ischemic and traumatic brain injury[5] and reduces the degree of postischemic edema[4]. The mechanism of this protective effect is largely unknown. In various models of brain injury there is a massive increase in extracellular levels of the excitatory amino acid glutamate[6], which could be responsible for astrocyte swelling and brain edema.

While glutamate is known to induce astrocytic swelling, the mechanism of swelling is not well understood[1,8,11,12,14]. We wished to determine factors involved in glutamate-induced swelling, especially as they relate to temperature. While glutamate-induced swelling is a Na^+ dependent process[1,8,14], glutamate induced Na^+ influx into astrocytes is not temperature dependent[9]. Another Na^+ dependent effect of glutama-

te is K^+ uptake[8]. We therefore decided to examine the role of K^+ uptake in glutamate-induced swelling and to determine if this could be affected by changes in temperature. Since glutamate swelling is also Ca^{2+} dependent[11,12], we also examined the role of Ca^{2+} on K^+ uptake.

Materials and Methods

Primary cultures of rat astrocytes were prepared as described by Gregorios *et al.*[7] Briefly, cultures were prepared from cerebral cortices of 1–2 day old rats. The tissue was dissociated and plated in 35 mm dishes at a density of 0.5×10^6. After two weeks, the cells were maintained with dibutyryl cyclic AMP. Based on immuno-histochemistry of GFAP and glutamine synthetase, at least 95% of the cell population were astrocytes. Cells 3–5 weeks old were used for the experiments. The intracellular water space was measured utilizing 3-O-methyl-[3H]-glucose (New England Nuclear Products, Boston, MA; specific activity 86.7 Ci/mmol) as described by Kletzien *et al.*[10] and Chan *et al.*[2] K^+ uptake was studied utilizing 2 μCi/ml 86RbCl, a K^+ analog (New England Nuclear Products), as described by Kimelberg[8].

Results

Cultured astrocytes were incubated in bicarbonate-buffered media in the presence or absence of 1 mM L-glutamate, for 1 h at 37°C and 24°C. The results on cell volume are shown in Fig. 1A. Glutamate increased astrocytes cell volume to 197% from 6.6 ± 0.3 to 12.9 ± 0.3 μl/mg protein (n = 9; p < 0.0001) during the 1 hr incubation period at 37°C. At 24°C glutamate increased the volume to only 123% from 7.3 ± 0.4 to 9.1 ± 0.3 μl/mg protein (n = 9; p = 0.0039). Therefore, hypothermia reduced glutamate-induced swelling by 74%.

Figure 1B shows the effects of glutamate-stimulated K^+ uptake at 37°C and 24°C. In the presence of 1 mM

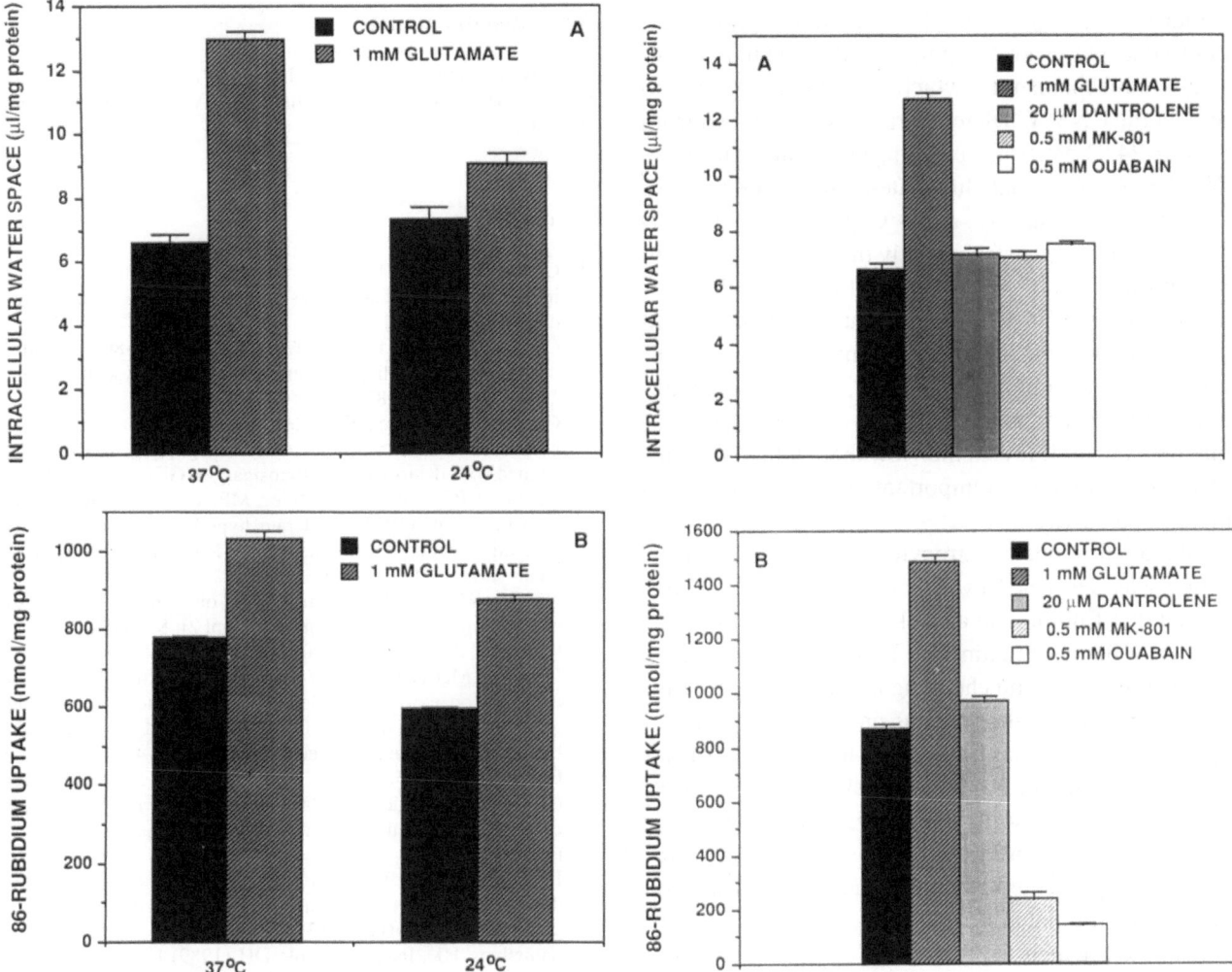

Fig. 1. Effect of temperature on L-glutamate induced swelling (A) and on L-glutamate stimulated K⁺ uptake (B) in cultured astrocytes. Results represent means ± S.E.M. of nine separate determinations

Fig. 2. Effect of dantrolene, MK-801 and ouabain on L-glutamate induced swelling (A) and L-glutamate stimulated K⁺ uptake (B) at 37°C in cultured astrocytes. Results represent means ± S.E.M. of six separate determinations

glutamate, the K⁺ equilibrated content, as measured with ^{86}Rb (a K⁺ analog) at 1 h at 37°C was increased by 30% from 776 ± 5.8 to 1028.7 ± 22.1 nmol/mg protein. At 24°C, the K⁺ equilibrated content decreased to 595.4 nmol/mg protein while 1 mM L-glutamate increased it to 876.6 ± 10.5 nmol/mg protein. Thus glutamate stimulated K⁺ content at 24°C is equivalent to that of control at 37°C.

Glutamate induced swelling as well as K⁺ influx are Ca^{2+}-dependent processes. Removal of extracellular Ca^{2+} partially inhibited both glutamate-induced swelling and K⁺ uptake by 30 and 53%, respectively. When EGTA (0.1 mM) was added to the medium a decrease by 61 and 86% in glutamate-induced swelling and K⁺ uptake, respectively, was observed . When extracellular Ca^{2+} was absent from the media and in the presence of 0.1 mM EGTA and 50 µM BAPTA/AM, glutamate

was not able to induce swelling or to stimulate K⁺ uptake (results not shown).

Figure 2A shows that 20 µM dantrolene, 0.5 mM MK-801 and 0.5 mM ouabain completely prevented glutamate-induced swelling. Figure 2B shows that these agents, at the same concentration, are able to prevent glutamate-stimulated K⁺ uptake.

Discussion

The protective effect of reducing body temperature against CNS injury due to anoxia/ischemia or trauma is of great potential interest as it may offer a therapeutic approach towards diminishing the extent of CNS injury from variety of insults. It appears that even small reductions in body temperature (~ 2°C) can protect the brain from anoxic/ischemic injury[4,5].

Our results suggest that the reduction in glutamate-induced astrocyte swelling by hypothermia may represent one mechanism by which hypothermia is able to protect the CNS. When temperature was reduced from 37°C to 24°C, astrocytic swelling was reduced by 74%. We also observed that glutamate-stimulated K$^+$ uptake is temperature sensitive. At 24°C, glutamate-stimulated K$^+$ uptake was approximately the same as observed in controls at 37°C, indicating that lowering the temperature from 37°C to 24°C prevented glutamate from stimulating K$^+$ uptake. Since glutamate-stimulated Na$^+$ influx is temperature insensitive[9], while glutamate-induced K$^+$ influx and astrocyte swelling are temperature dependent, it favors the view that glutamate-stimulated K$^+$ uptake is an important component of glutamate-induced swelling.

Glutamate-induced astrocyte swelling Ca^{2+} dependency was described by Koyama et al.[11,12] We confirmed that finding and also showed that the effect of Ca^{2+} may be due to stimulate K$^+$ uptake. Removal of extracellular Ca^{2+} and chelating intracellular Ca^{2+} with BAPTA prevented the swelling and K$^+$ influx induced by glutamate. Ca^{2+} has been shown to play a significant role in modulating K$^+$ homeostasis in astrocytes as described by Latzkovits et al.[13] Dantrolene, which is known to inhibit intracellular Ca^{2+} release in astrocytes[3], also prevented glutamate-induced swelling and glutamate-stimulated K$^+$ uptake.

MK-801, which is known to inhibit glutamate-induced astrocyte swelling[1,2], also potently inhibited K$^+$ uptake in the presence of glutamate. This is a first report showing that MK-801 is able to affect K$^+$ homeostasis in astrocytes. Perhaps the mechanism of MK-801 prevention of glutamate-induced swelling may be related to its inhibition of K$^+$ uptake into astrocytes. Ouabain is a known inhibitor of Na$^+$/K$^+$-ATPase and thus inhibits K$^+$ uptake into astrocytes. Ouabain prevented glutamate swelling, as previously reported by Schneider et al.[14]

In summary, we have shown that glutamate-induced astrocyte swelling and glutamate-stimulated K$^+$ uptake are temperature sensitive. These two processes are also Ca^{2+} dependent. Pharmacological agents such as dantrolene, MK-801 and ouabain, which inhibited K$^+$ influx, also blocked astrocyte swelling. Our results suggests hypothermia may exert its beneficial effects in CNS injury through a reduction of astrocyte swelling by reducing K$^+$ influx into astrocytes.

Acknowledgements

This research was supported by USPHS grants NS-30291 and DK-38153, and the Department of Veterans Affairs and GRECC.

References

1. Chan PH, Chu L (1990) Mechanisms underlying glutamate-induced swelling of astrocytes in primary culture. Acta Neurochir (Wien) [Suppl] 51: 7–10
2. Chan PH, Chu L, Chen S (1990) Effects of MK-801 on glutamate-induced swelling of astrocytes in primary cell culture. J Neurosci Res 25: 87–93
3. Charles AC, Dirksen ER, Merrill JE, Sanderson MJ (1993) Mechanisms of intracellular calcium signaling in glial cells studied with dantrolene and thapsigargin. Glia 7: 134–145
4. Dempsey RJ, Combs DJ, Maley ME, Cowe DE, Roy MW, Donaldson DL (1987) Moderate hypothermia reduces postischemic edema development and leukotriene production. Neurosurgery 21: 177–181
5. Dietrich WD (1992) The importance of brain temperature in cerebral injury. J Neurotrauma 9 [Suppl 2]: S474–S485
6. Globus MYT, Busto R, Martinez E, Valdes I, Dietrich WD, Ginsberg MD (1991) Comparative effect of transient global ischemia on extracellular levels of glutamate, glycine, and gamma-aminobutyric acid in vulnerable and nonvulnerable brain regions in the rat. J Neurochem 57: 470–478
7. Gregorios JB, Mozes LW, Norenberg LOB, Norenberg MD (1985) Morphological effects of ammonia on primary astrocyte cultures. I. Light microscopic studies. J Neuropathol Exp Neurol 44: 397–403
8. Kimelberg HK (1987) Anisotonic media and glutamate-induced ion transport and volume responses in primary astrocyte cultures. J Physiol (Paris) 82: 294–303
9. Kimelberg HK, Pang S, Treble DH (1989) Excitatory amino acid-stimulated uptake of ^{22}Na$^+$ in primary astrocyte cultures. J Neurosci 9: 1141–1149
10. Kletzien RF, Pariza MW, Becker JE, Potter VR (1975) A method using 3-O-methyl-D-glucose and phloretin for the determination of intracellular water space of cells in monolayer culture. Anal Biochem 68: 537–544
11. Koyama Y, Baba A, Iwata H (1991) L-Glutamate-induced swelling of cultured astrocytes is dependent on extracellular Ca^{2+}. Neurosci Lett 122: 210–212
12. Koyama Y, Sugimoto T, Shigenaga Y, Baba A, Iwata H (1991) A morphological study on glutamate-induced swelling of cultured astrocytes: involvement of calcium and chloride ion mechanisms. Neurosci Lett 124: 235–238
13. Latzkovits L, Rimanoczy A, Juhasz A, Torday C, Sensenbrenner M (1982) Control of cation transport in cultured glial cells by external Ca^{++}: a possible signal in glial-neuronal interaction. Dev Neurosci 5: 92–100
14. Schneider GH, Baethmann A, Kempski O (1992) Mechanisms of glial swelling induced by glutamate. Can J Physiol Pharmacol 70: S334–S343

Correspondence: Alex S. Bender, Ph. D., Department of Pathology (D-33), University of Miami School of Medicine, P.O.Box 016960, Miami, Fl 33101, U.S.A.

Acta Neurochir (1994) [Suppl] 60: 31–33
© Springer-Verlag 1994

Differential Behavior of Glial and Neuronal Cells Exposed to Hypotonic Solution

M. Tomita, Y. Fukuuchi and **S. Terakawa**[1]

Department of Neurology, School of Medicine, Keio University, Tokyo, and [1] Department of Cell Physiology, National Institute for Physiological Sciences, Okazaki, Japan

Summary

The comparative modes of swelling of glial and neuronal cells were examined by video-enhanced differential interference contrast (VEC) microscopy. When exposed to hypotonic solution, C6 cells swelled slowly to 5 times their normal volume and burst, whereas N18 cells swelled more rapidly, forming blebs, and then burst partially. The time to burst was 410.6 ± 45.7 s (mean $\pm$ SD, n = 5) for C6, and 69.3 ± 10.4 s (n = 5, p < 0.01) for N18, respectively. The present findings suggest a great difference in physical strength of the cell membrane between glial and neuronal cells: the former cell membrane appears to be elastic and tolerant to high tension, while the latter cell membrane is relatively weak.

Keywords: Glial swelling; neuronal swelling; VEC microscopy.

Introduction

The most pronounced histopathological changes occurring at the early stage of cerebral ischemia are swelling of astrocytes in the perikarya and perivascular processes[2,4]. In contrast, the morphological changes in the neurons are not regular, involving swelling, no change or even shrinking, and usually occur later[4,5]. Advances in technology based on video microscopy and digital image processing have made it possible to visualize cultured glial and neuronal cells with a spatial resolution which is greatly increased from that of conventional light microscopy. The present study compares the modes of swelling of glial and neuronal cells, as revealed by such video-enhanced differential interference contrast (VEC) microscopy[1,3,6].

Materials and Methods

In a previous communication[7], we demonstrated that the modes of swelling of cells under hypotonic conditions (osmosis) and swelling due to full opening of all ionic channels at the cell membrane (osmosis with no osmotic gradient) were quite similar in terms of their speed and the magnitude of the fluid shift. This suggests that investigations of cell swelling by hypotonic shock

could provide a clue for gaining a better insight into the dynamics of specific astroglial and neuronal alterations occurring in the cerebral cortex during ischemia.

Cultured glial cells (C6; astroglioma) and cultured neuronal cells (N18; neuroblastoma) were inoculated into a flask containing 10% Dulbecco's modified Eagle's medium plus 10% Gibco newborn calf serum (DMEM), and then incubated at 37°C in 5% CO_2 plus oxygen. Two days before the experiment, cells were harvested from the flask, reseeded and cultured further on a collagen-coated thin coverglass placed in a small dish containing the DMEM medium. On the day of the experiment, the cells were observed and video-taped in a microscopic system devised by one of the authors (S.T.): the coverslip on which the cells grew was taken out, and secured watertightly with grease on a slide glass having a square window so that a microscopic observation chamber with the cells at the bottom was created. The cells were maintained at 37°C in the microscopic observation chamber by superfusion with either DMEM or distilled water. The morphological changes of the cells were observed during exposure to hypotonic solution under an inverted Nomarski microscope equipped with a ×100 DIC objective lens and a ×2.5

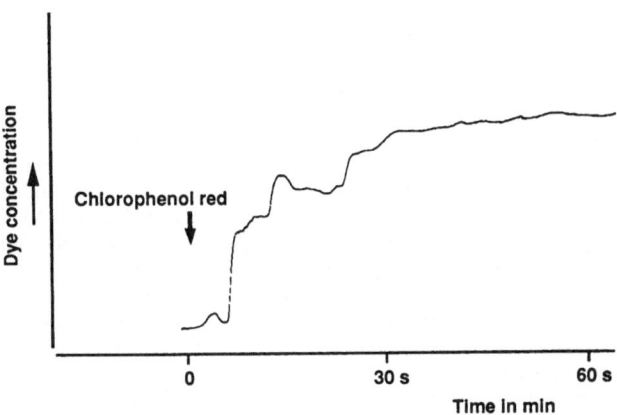

Fig. 1. Dye concentration curve. Optical density changes were monitored using a small silicon photodiode placed on the monitor screen at the center of the cell. At the downward arrow, a concentrated dye solution of chlorophenol red was introduced into the tip of the feeding tube for the superfusion fluid. Approximately 20 s was required for the dye to reach the cell in the observation chamber. This time lag was subtracted from the total time taken for the cell to burst

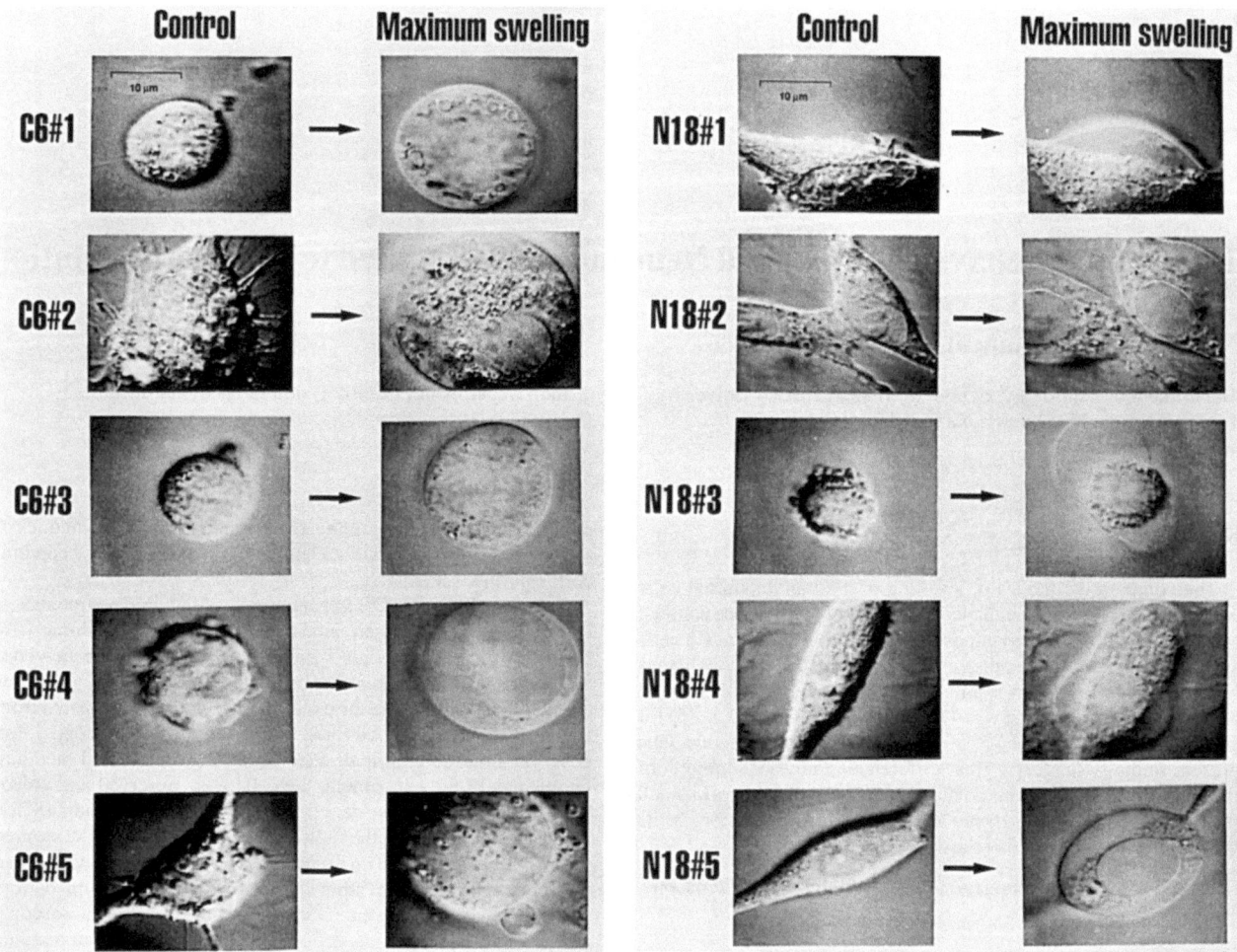

Fig. 2. Morphological changes of C6 cells at the control stage (left column) and at just about the time to burst (right column). Note the full moon appearance of the C6 cells in which all intracellular structures appear to be stretched with cell membrane enlargement

Fig. 3. Morphological changes of N18 cells at the control stage (left column) and at just about the time to burst (right column). Note the hydropic changes of the cell membrane which contains only lucid fluid

insertion lens. Optical images were obtained using a CCD (charge-coupled device) camera (TI-23P, NEC, Tokyo). The video signals from the small detection area (5×7 µm) of this camera were monitored by a 14-inch monitor screen at a final magnification of ×12,000. The video image was contrast-enhanced with a digital image processor (PIP-4000, ADS, Osaka) and recorded on S-VHS tape. Individual responses of 5 glial cells and 5 neuronal cells with respect to the temporal profiles of the cell shape as well as intracellular organelles under hypotonic conditions, with application of distilled water, were observed and the time to osmotic bursting of the cells was measured. The time delay of the hypotonic solution through the feeding tube was estimated using chlorophenol red as a dye, and was subtracted to obtain the burst time (Fig. 1).

Results and Discussion

Both the C6 and N18 cells contained a large nucleus, mitochondria, lysosomes, and numerous fine granules which were clearly visible in the cells as dark or bright spots moving slowly or rapidly. Frequently, the gran-

ules exhibited gliding or travelling movements over fairly long distances at a various but roughly individually constant velocity.

When the hypotonic solution was introduced into the superfusing fluid, the C6 cells (n = 5) swelled slowly. The nucleus as well as the intracellular organelles, such as mitochondria and granular particles, also swelled together with the cell body. Gradual development of lucid fluid areas was observed in the cytoplasm which contained rapidly moving particles (Brownian movement). The cells swelled to 5 times their normal volume, and became like a full moon. The shape changes of each of the C6 cells from the control (left column) to maximum swelling (right column) are illustrated in Fig. 2. The time from onset of hypotonic stress to bursting was 470 s for #1, 324 s for #2, 527 s for #3, 257 s for #4 and 475 s for #5, respectively. The mean ±

SD was 410.6 ± 45.7 s. The membrane retained its balloon-like contours even after the cell had burst.

The N18 cells (n = 5) swelled more rapidly, forming blebs which tended to join together. The cell membrane became detached from the intracellular structures showing hydropic changes (Fig. 3) and then burst partially. The time from onset to bursting was 59.8 ± 13.3 s (55, 45, 65, 80, and 54 s). The time difference between C6 and N18 was statistically significant (p < 0.01).

The present findings suggest that a great difference in physical strength of the cell membrane exists between glial and neuronal cells: the former cell membrane appears to be elastic and tolerant to a high tension, while the latter cell membrane is relatively weak.

References

1. Allen RD, Travis JL, Allen NS, Yilmaz H (1981) Video-enhanced contrast polarization (AVEC-POL) microscopy: a new method applied to the detection of birefringence in the motile reticulopodial network of Allogromia laticollaris. Cell Motil 1: 275–279
2. Hirano A (1980) Fine structure of edematous encephalopathy. Adv Neurol 28: 83–97
3. Inoue S (1981) Video image processing greatly enhances contrast, quality, and speed in polarization-based microscopy. J Cell Biol 89: 346–356
4. Jenkins LW, Becker DP, Coburn TH (1984) A quantitative analysis of glial swelling and ischemic neuronal injury following complete cerebral ischemia. In: Go KG, Baethmann A (eds) Recent progress in the study and therapy of brain edema. Plenum, New York, pp 523–537
5. Mchedlishvili G, Tsitsishvili A, Sikharulidze N (1989) Dynamics of changes in astrocytes and neurons of rabbit cerebral cortex induced by temporary blood supply deficiency. Neuropat Pol 27: 25–37
6. Terakawa S, Fan J, Kumakura K, Ohara-Imaizumi M (1991) Quantitative analysis of exocytosis directly visualized in living chromaffin cells. Neurosci Lett 123: 82–86
7. Tomita M, Gotoh F (1992) Cascade of cell swelling-thermodynamic potential discharge of brain cells after membrane injury. Am J Physiol 262: H603–H610

Correspondence: Minoru Tomita, M.D., Department of Neurology, School of Medicine, Keio University, 35 Shinanomachi, Shinjuku-ku, Tokyo 160, Japan.

Acta Neurochir (1994) [Suppl] 60: 34–37
© Springer-Verlag 1994

The Regulation of Intracellular pH is Strongly Dependent on Extracellular pH in Cultured Rat Astrocytes and Neurons

P. Mellergård[1,2], **Y.-B. Ouyang**[1], and **B. K. Siesjö**[1]

[1] Laboratory for Experimental Brain Research and [2] Department of Neurosurgery, Lund University Hospital, Sweden

Summary

We studied the mechanisms regulating intracellular pH (pH_i) in cultured rat astrocytes and neurons, with particular reference to the influence of extracellular pH (pH_e) on these mechanisms, using microspectrofluorometric monitoring from single cells, loaded with the pH-sensitive fluorophore BCECF. The pH regulatory mechanisms differ between neurons and astrocytes. The experimental data suggest the presence of a Na^+/H^+ and a Na^+-*in*dependent HCO_3^-/Cl^- exchanger in both types of cells, while astrocytes, in addition, utilise a Na^+-dependent HCO_3^-/Cl^- exchanger for regulating acid transients. In both cell types the pH regulatory mechanisms are strongly dependent on pH_e. Thus, at pH_e 6.85 or below, there was no recovery of pH_i. Steady state pH_i was also strongly dependent on pH_e, in both astrocytes and neurons. The pH_i recovery following normalisation of pH_e was very rapid, (indicating that a prolonged exposure to a low pH stimulates pH regulating mechanisms), and was inhibited by 4,4'-diisothiocyanatostilbene-2,2'-disulphonic acid (DIDS) and amiloride, or in the absence of Na^+. The results challenge the concept of a H^+-regulatory site solely at the internal side of the exchanger regulating pH_i to a constant value.

Keywords: Astrocytes; neurons; intracellular pH; pH regulation.

Introduction

A number of pathological conditions in the central nervous system, including e.g. cerebral ischemia and status epilepticus, are accompanied by severe disturbances in the acid-base balance[8]. These disturbances in pH homeostasis probably influence cellular volume regulation, and are likely to have a decisive effect on the final outcome. The successful treatment of such pathological conditions requires a better understanding of the mechanisms responsible for pH regulation. We, therefore, studied the pH regulation of neurons and astrocytes in vitro, evaluating the recovery of intracellular pH (pH_i) following artificial inductions of acid or alkaline loads.

It is generally assumed that the activity of the mechanisms responsible for regulation of intracellular pH are completely dependent upon, and modulated by, the value of pH_i itself. However, recent observations have shown that the pH-regulation of brain cells may be strongly dependent on the extracellular pH. Such observations are interesting from a pure physiological point of view, since they challenge the concept of a H^+ regulatory site solely on the intracellular side of H^+ transporting mechanisms, regulating pH_i to a constant value. However, the findings are also interesting from a pathophysiological viewpoint. Thus, a number of pathological conditions are accompanied by large reductions of pH_e. If pH_e is a decisive parameter for regulation of pH_i, this would suggest that regulation of pH_e is a necessary prerequisite for the recovery of pH_i, which in turn may have implications for the understanding of pathophysiological events during e.g. ischemia.

Methods

The methods have previously been described[3,5]. Briefly, primary cultures of rat astrocytes or rat cortical neurons were grown on coverslips, under standard cultivating conditions. To certify the character of cultures, they were regularly stained for glial fibrillary acidic protein (GFAP) and neuron-specific enolase, respectively.

The coverslips were mounted in a closed, temperature controlled (37°C) perfusion chamber, and placed on an inverted microscope (Nikon Diaphot TMD), connected to a SPEX CM1T11E microspectrofluorometer. The cells were loaded with 2 μM of the acetoxymethyl ester of the pH-sensitive dye 2',7'-bis(carboxyethyl-)-5,6-carboxyfluorescein (BCECF-AM, Molecular Probes, USA). Following a resting period, the fluorescence at 520 nm was monitored from single cells, excited with alternating dual monochromator excitation at 490 nm and 440 nm. Calibrations were made with the nigericin-method at the end of each experiment. The basic bicarbonate buffer used had the following composition (in mM):

140 Na^+, 123 Cl^-, 5 K^+, 1 Ca^{2+}, 1 Mg^{2+}, 5 glucose and 20 HCO_3^-, equilibrated with $\approx$ 5% CO_2 (38 ± 2 mm Hg) and 95% air. The buffer used to acidify the cells had the same composition, except that it contained 93 mM Cl^-, 50 mM HCO_3^- and was equilibrated with $\approx$ 15% CO_2 (96 ± 2 mm Hg). When nothing else is stated the pH of the solutions was 7.35 ± 0.02. Whenever required the pH of these two basic buffers was varied by varying the HCO_3^- content. For every mM HCO_3^- added an equal amount of Cl^- was removed, and vice versa. Sodium-free solutions were made by replacing Na^+ with choline. In the chloride free solutions all Cl^- salts were replaced by gluconate salts. In some experiments a nominally bicarbonate-free, HEPES-buffered solution was used, similar in composition to the 5% CO_2-buffer described above, except that it contained 20 mM HEPES instead of HCO_3^-. The net H^+ fluxes (J_H+) were calculated as the product of the total buffering capacity (β_t) and the rate of pH_i change (see [7]). The intrinsic buffer capacities of cultured astrocytes and neurons were investigated in a separate study[2]. This study showed β_i to be virtually constant in the pH_i interval 6.0–7.0.

Results

pH Regulation at Normal pH_e

As shown in Table 1, the steady state pH_i was different in neurons and astrocytes, and was dependent on the presence of CO_2/HCO_3^-. (The differences were significant, when compared with ANOVA). The maintenance of steady state pH_i required the presence of extracellular Na^+. In bicarbonate-free solution the addition of amiloride (an inhibitor of the Na^+/H^+ exchanger) was followed by a large reduction of steady state pH_i in both neurons and astrocytes (for more details, see [4,6].

Cells were exposed to acid (and alkaline) loads by the NH_4-prepulse technique, or by rapid alterations in PCO_2[7]. Both neurons and astrocytes efficiently recovered from these loads. In nominally bicarbonate-free solution, the recovery following an acid load was similar in both types of cells, J_H+ being 2.6 ± 0.46 and 2.6 ± 0.35 mM · min^{-1}, respectively. This recovery was inhibited by amiloride (1 mM), and required the presence of extracellular Na^+. These observations suggest the presence of a Na^+/H^+ exchanger in both neurons and astrocytes[3,4,6].

In neurons the recovery from an acid load was completely independent of bicarbonate, while astrocytes recovered much more effiently from an acid load in

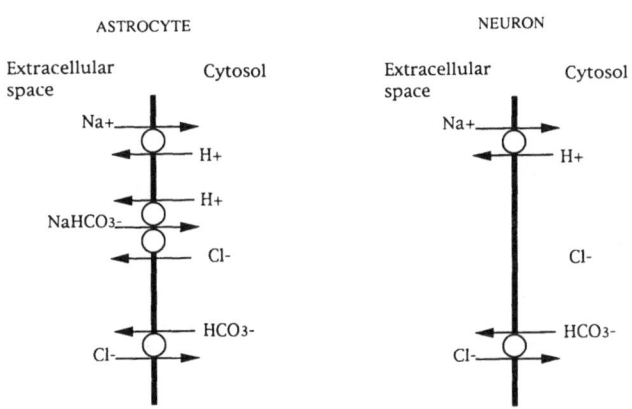

Fig. 1. The data indicate the presence of Na^+/H^+ antiporter and Na^+-independent HCO_3^-/Cl^- antiporter in both cell types. In addition astrocytes probably express a Na^+-dependent HCO_3^-/Cl^- exchanger

the presence of CO_2/HCO_3^-, J_H+ then increasing to 6.3 ± 0.36 mM · min^{-1}. This bicarbonate-dependent pH_i recovery in astrocytes was completely dependent on the presence of extracellular Na^+, was influenced by intracellular chloride, and was partially inhibited by 4,4'-diisothiocyanatostilbene-2,2'-disulphonic acid (DIDS, a classical inhibitor of HCO_3^-/Cl^- exchange). These observations suggest the presence of a Na^+ dependent HCO_3^-/Cl^- exchanger in astrocytes as pH-regulatory mechanism which is not found in neurons.

In both neurons and astrocytes the recovery from an alkaline load was influenced by the presence of chloride, and could be partially inhibited by DIDS. The recovery was not inhibited by amiloride, or by the removal of extracellular Na^+. These observations suggest that, in both neurons and astrocytes, the recovery from an alkaline load is dependent on a Na^+-independent HCO_3^-/Cl^- exchanger.

We, thus, would like to propose that pH_i regulation is achieved by the following mechanisms, in astrocytes and neurons, respectively (see Fig. 1).

The Influence of Extracellular pH on Intracellular pH

As described above, both neurons and astrocytes rapidly regulated pH_i back to normal following an acid transient. However, this was observed only when pH_e was normal, i.e. when it was maintained around 7.40[3,5]. If pH_e was lowered simultaneously with the acid transient, regulation of pH_i was impaired. At pH_e 6.8, and below, pH_i recovery was completely blocked. Similar observations were made in both neurons and astroyctes, and in both nominally bicarbonate-free and CO_2/HCO_3^- containing solution.

Table 1

	HEPES	CO_2/HCO_3^-
Neurons	7.00 ± 0.03	7.09 ± 0.02
Astrocytes	6.90 ± 0.02	7.00 ± 0.02

(± SEM).

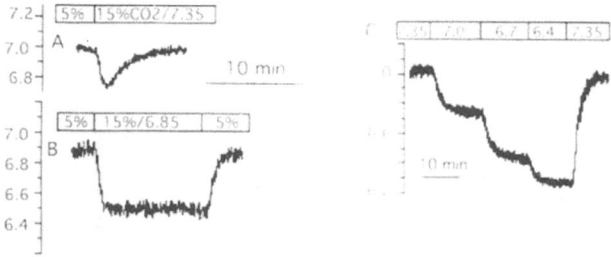

Fig. 2. Original recordings from astrocytes that extracellular pH influences both regulation of intracellular pH following an acid transient (A, B), and steady state pH_i (C)

The steady state pH_i was also dependent on pH_e. Thus, when pH_e was lowered in steps of 0.3 or 0.8 units, there was a parallel fall in pH_i. Similarly, when pH_e was increased, there was a parallel increase in pH_i (see Fig. 2).

When pH_e was normalized, pH_i always recovered to the normal, steady state level. This pH_i recovery was not the result of passive H^+ fluxes through an extremely leaky membrane, since it was inhibited by classical blockers of acid extrusion. Thus, when pH_e was increased from 6.45 to 7.35 in the presence of DIDS and amiloride, or in the absence of Na^+, there was a pronounced inhibition of pH_i recovery. The introduction of albumin was followed by a slight increase in steady state pH_i, while the presence of cadmium partly inhibited the pH_i recovery, following pH_e dependent acidifi-

cation. However, neither albumin nor cadmium influenced the pH_e dependent acidification. This was neither critically influenced by DIDS, amiloride, barium, TEA (tetra-ethyl-ammonium), TTX (tetrodotoxin), 2,3-dihydroxy-6-nitro-7-sulfamoyl-benzo(F)-quinoxaline (NBQX), kynurenic acid, 5-OH-saclofen, bicucullin, i.e. agents that would block the most obvious routes of possible H^+ entrance.

The slope of the curve describing the relationship between pH_e and pH_i was rather steep, and similar in neurons and astrocytes. Thus, over the interval examined, pH_i changed with approximately 0.8 units, for a 1.0 unit change in extracellular pH (see Fig. 3).

Thus, it seems clear that cultivated rat neurons and astrocytes are both very sensitive to changes in extracellular pH, both for maintaining a normal steady state pH_i, and for recovery of pH_i following an acid transient. The data are probably best interpreted in terms of an altered pump-leak relationship for H^+/HCO_3^-, associated with changes in pH_e. Three major possibilities are envisaged: First, a reduction in pH_e could increase the permeability for H^+ influx (or HCO_3^- efflux), to the extent that the passive leak outstrips the capacity of the pumps. Second, the reduction of pH_e may lower the rate of H^+ extrusion, by a direct influence on the H^+ pumps. The third, and most exciting possibility, is that rather than regulating pH_i to constancy, the H^+ extrusion mechanisms "sense" changes in pH_e, and regulate pH_i accordingly, perhaps to a constant pH_e/pH_i difference.

It is becoming increasingly clear, that not only pathological states, but also normal cerebral activity, is accompanied by significant changes in the pH_e of the brain. These changes may serve a genuinly modulatory function[1]. If so, pH_i regulatory mechanisms that "sense" extracellular pH may be functional.

References

1. Chesler M, Kaila K (1992) Modulation of pH by neuronal activity. TINS 15: 396–402
2. Katsura K, Mellergård P, Theander S, Ou-yang Y, Siesjö B (1993) Buffer capacity of rat cortical tissue, cultured neurons and astrocytes as evaluated by different methods. Brain Res, in press
3. Mellergård P, Ou-Yang Y, Siesjö B (1992) Regulation of intracellular pH in cultured astrocytes and neuroblastoma cells: dependence on extracellular pH. Can J Physiol 70 [Suppl]: S293–S300
4. Mellergård P, Ou-Yang Y, Siesjö B (1993) Intracellular pH regulation in cultured rat astrocytes in CO_2/HCO_3 containing media. Exp Brain Res, submitted
5. Mellergård P, Siesjö BK (1991) Astrocyte fail to regulate intracellular pH at moderately reduced extracellular pH. Neuroreport 2: 695–698

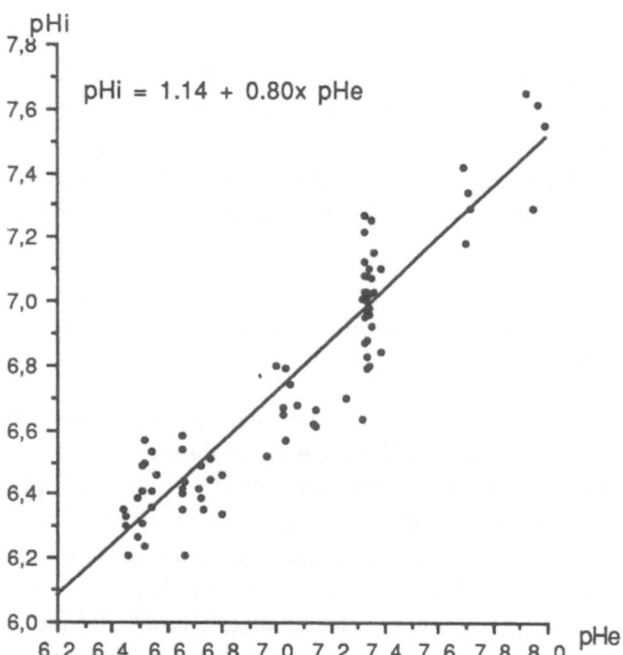

$pHi = 1.14 + 0.80x pHe$

Fig. 3. Relationship between extra and intracellular pH in astrocytes (a similar relationship was observed in neurons)

6. Ou-Yang Y, Mellergård P, Siesjö B (1993) Regulation of intracellular pH in single rat cortical neurons in vitro: a microspectrofluorometric study. J Cereb Blood Flow Metab, in press

7. Roos A, Boron WF (1981) Intracellular pH. Phys Rev 61: 296–434

8. Siesjö BK, Katsura K, Mellergård P, Ekholm A, Lundgren J, Smith ML (1993) Acidosis-related brain damage. Prog Brain Res 96: 23–48

Correspondence: Pekka Mellergård, M. D., Department of Neurosurgery, Lund University Hospital, S-221 85 Lund, Sweden.

Acta Neurochir (1994) [Suppl] 60: 38–40
© Springer-Verlag 1994

Morphological Changes of Cultured Neuronal and Endothelial Cells by Human Albumin

K. Akiyama, M. Yamamoto, M. Haida, H. Ohsuga, N. Shinohara, and **Y. Shinohara**

Department of Neurology, Tokai University School of Medicine, Isehara, Kanagawa, Japan

Summary

The role of plasma proteins in the mechanisms of brain tissue damage in ischemic events remains to be clarified. The purpose of this study was to investigate whether the presence of albumin in the extracellular fluid could induce damage to endothelial and neuronal cells. Neuronal cells from rat fetal brain (15 days) were cultured by using RPMI-1640 containing 10% Serum-Plus and endothelial cells from umbilical vein were also cultured in 96-well plates. The studies were made by using neuronal cells after 4 days of culture (N group) and endothelial cells after 4 days of culture (E4 group) and at confluence (EC group). After discarding the culture fluid, 200 μl of 5–25% human albumin was added to each well. Microscopic observations were made up to 20 minutes, and then immunostaining was done with anti-human albumin antibody. The swelling of neurons was observed immediately after application of albumin solution, and the cells became circular after 10 minutes. In the E4 group, similar morphological changes were observed, but no such changes were seen in the EC group.

Immunostaining revealed the presence of albumin in the intracellular space in both the N and E4 groups, but not the EC group. Our results suggest that albumin in extravasated fluid can induce damage to neuronal cells and endothelial cells in the non-confluent state.

Keywords: Albumin; cell swelling; neuronal cell; endothelial cell.

Introduction

The mechanism of cell swelling in ischemia is believed to involve electrolytes such as Na$^+$ ions, and changes of intracellular chemical potential by inflow of Ca^{2+} ion into cells, but the details are unknown. We have performed various studies which indicate that the effect of protein contained in the extracellular fluid is one of the mechanisms of the cell swelling. In a rat cold injury model, the protein contained in the extracellular fluid was incorporated into cells[1], and in neuronal cell culture under condition of acidosis induced by extracellular lactic acid, protein from the extracellular fluid was stained in the cells[2].

In the present study, the cytotoxic action of albumin, a plasma protein present in the extracellular fluid, was investigated using a neuronal cell culture and an endothelial cell culture.

Materials and Methods

I) Preparation of Neuronal and Endothelial Cell Cultures

Cultured neuronal cells were isolated from cerebral cortex of 15-day embryonal Wistar rats by means of the enzymatic cell-dispersion method (Cartwright et al.[3]). Cultured endothelial cells were isolated from human umbilical cord. The culture medium was RPMI-1640 (Whittaker) containing 10% Serum-Plus (Hazleton). Final cell density was 3.0×10^5/ml for both neuronal and endothelial cells. Cells were plated in 96 well micro tissue plates (96 well tissue culture cluster, flat bottom, Costar) coated with poly-DL-ornithine hydrobromide, and incubated in a CO$_2$ incubator (37°C, 5% CO$_2$ and room air). Three culture groups were prepared as follows:

(1) Four days after the seeding of neuronal cells (N group);
(2) four days after the seeding of endothelial cells (E4 group);
(3) confluent state of endothelial cells (EC group).

II) The Change of Cell Form

After removal of the medium, 200 μl of 5 to 25% human albumin (Sigma) was added. The change of cell form was observed visually for 20 minutes.

III) Immunohistochemical Staining

After 20 minutes observation, each culture was fixed with 5% ethanol – 95% acetic acid, and immunohistochemical staining with anti-human albumin antibody (Sigma) was carried out.

Results

I) The Change of Cell Form

(1) In the N group, cell swelling started immediately after the addition of 25% albumin. Within 10 minutes, the cells became round and processes became

indistinct. Some of the cells were completely destroyed 20 minutes after the addition of albumin. Similar phenomena were observed with 5% to 10% albumin, though there was a slight difference of the degree of injury.

(2) In the E4 group, cells became somewhat deformed and an intracellular granular structure appeared. By 20 minutes after addition of albumin, most of the cells had become round.

(3) In the EC group, almost no change of cell form was observed within 20 minutes after addition of 25% human albumin.

II) Immunohistochemical Staining

(1) In the N group, albumin was stained in the cytosol of all cells;

(2) in the E4 group, albumin was stained in most of the cells (Fig. 1);

(3) in the EC group, albumin staining was almost absent and almost no change of the cell form was observed (Fig. 2).

Discussion

I) Cytotoxic Activity of Albumin and Pathogenesis of Cell Swelling

Although albumin is normally present in plasma, its direct influence on neuronal or endothelial cells remains to be clarified. Our present study demonstrated that cultured neuronal cells exposed to 25% human albumin solution swelled and finally burst. A similar phenomenon was observed in 5% and 10% albumin solutions.

The immunohistochemical staining of swollen neuronal cells using anti-human albumin antibody demonstrated the existence of albumin inside the cell. Since the osmotic pressure of 25% albumin solution was demonstrated to be above 300 mmHg and the colloid osmotic pressure of brain tissue has been reported to be 200–220 mmHg, neuronal cells in 25% albumin solution should shrink. However, the opposite was the case.

Possible causes of swelling and bursting of neurons by albumin include:

(1) the contact of cell membrane and albumin may enhance membrane permeability or induce active influx of albumin;

(2) albumin imported into the intracellular space

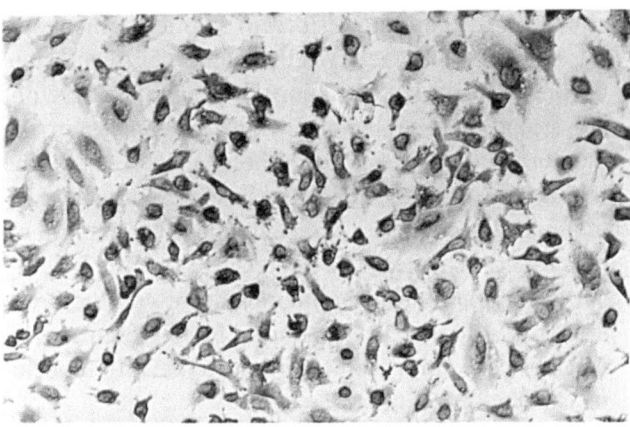

Fig. 1. Immunohistochemical staining: albumin is stained in non-confluent endothelial cells

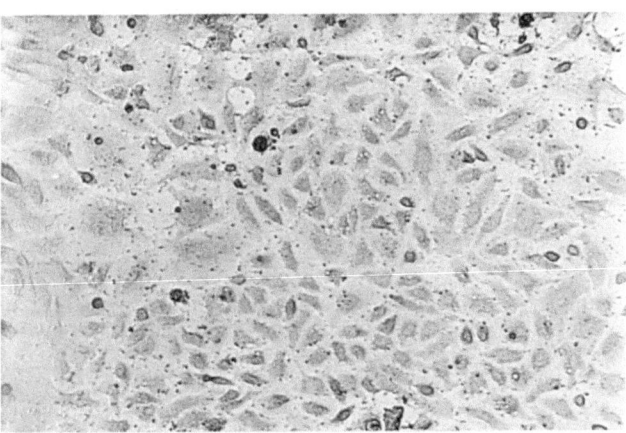

Fig 2. Immunohistochemical staining: albumin is almost absent in confluent state of endothelial cells

may induce a high neuron colloid osmotic pressure compared to that in the native state. However, the real mechanisms of neuronal cell swelling by exposure to albumin remain unknown.

Our present study raises the possibility that under conditions which cause breakdown of the blood-brain barrier, albumin extravasated into brain tissue can damage neuronal cells.

II) Resistance of Endothelial Cells to Injury

The response of non-confluent endothelial cells to albumin solution was similar to that of neuronal cells, except that bursting was not observed. The different response to albumin between neuronal cells and endothelial cells might be dependent on the extent of influx of albumin. In the confluent state of endothelial cells, little change of the cell form was observed after addition of 25% human albumin solution. This result implies that endothelial cells in the confluent state become resistant to cell injury by albumin. These observations suggest that there are differences in the membrane properties of endothelial cells between the non-confluent state and confluent state, and that endothelial cells acquire resistance to injury by albumin on reaching confluence. The existence of specific binding sites for albumin on the membrane of confluent endothelial cells has been reported by Ghitescu *et al.*[4], and these sites may well contribute to the difference of resistance to albumin.

Conclusions

(1) 5 to 25% albumin present in the extracellular fluid appears to have a cytotoxic activity;

(2) albumin is incorporated from the extracellular fluid into cultured neuronal and endothelial cells;

(3) endothelial cells in the confluent state are resistant to the cytotoxic action induced by albumin.

References

1. Shinohara Y, Takagi S, Yoshii F, *et al* (1983) Movement of extravasated protein and changes of CBF and cerebral glucose utilization BBB dysfunction in rats. J Cereb Blood Flow Metab 3 [Suppl 1]: S437
2. Ohsuga H, Shinohara Y, Akiyama K, *et al* (1991) Changes of cultured neuronal cells induced by extracellular lactic acidosis. J Cereb Blood Flow Metab 11 [Suppl 2]: S514
3. Cartwright CA, Simantov R, Kaplan PL, *et al* (1987) Alterations in pp60c-src accompany differentiation of neurons from rat embryo striatum. Mol Cell Biol 7: 1830–1840
4. Ghitescu L, Fixman A, Simionescu M, *et al* (1986) Specific binding sites for albumin restricted to plasmalemmal vesicles of continuous capillary endothelium: receptor-mediated transcytosis. J Cell Biol 102: 1304–1311

Correspondence: Yukito Shinohara, Department of Neurology, Tokai University School of Medicine, Isehara, Kanagawa, 259-11, Japan.

Acta Neurochir (1994) [Suppl] 60: 41–44
© Springer-Verlag 1994

Metabolic Alterations Accompany Ionic Disturbances and Cellular Swelling During a Hypoxic Insult to the Retina: an in vitro Study

C. Doberstein, I. Fineman, D. A. Hovda, N. A. Martin, L. Keenly, and **D. P. Becker**

Division of Neurosurgery, UCLA School of Medicine, Los Angeles, California, U.S.A.

Summary

To study the ionic, metabolic, and morphologic derangements that occur following brain injury we utilized a retina in vitro model of hypoxia. Retinas were dissected into oxygenated (95% O_2, 5% CO_2) Ames medium, a physiologic solution resembling cerebrospinal fluid, and randomly assigned to either experimental hypoxic conditions (95% N_2, 5% CO_2) or control conditions. All retinas were incubated and maintained at 37°C. Changes in extracellular K^+ and lactate concentration, intracellular incorporation of ^{45}Ca and ^{14}C-leucine, uptake of glucose using [^{14}C]-2-deoxy-D-glucose (2DG), and cell size were determined at 10, 20, 30, and 60 minute time intervals. The results show that compared to control retinas hypoxia produced: (1) an early increase in extracellular concentration of K^+ and lactate, (2) a delayed increase in the intracellular incorporation of ^{45}Ca, (3) an early onset of cellular swelling, and (4) a decrease in the intracellular incorporation of ^{14}C-leucine, and (5) increased glucose utilization. All of the results were statistically significant ($p < 0.05$) and exhibited a dose response relationship with the exception of intracellular incorporation of ^{45}Ca which did not become significantly different until 30 minutes post-hypoxia. A 16% increase in cell size was noted after 10 minutes of hypoxia. Increased hypoxic cell size persisted for 30 minutes but after 60 minutes the control retinas appeared enlarged as well. Our results suggest that ionic, metabolic, and morphologic derangements can be demonstrated utilizing an in vitro model of hypoxia which are similar to those seen following in vivo traumatic brain injury. With use of this model the mechanisms behind these ionic-metabolic relationships can be addressed at the molecular level.

Keywords: Brain injury; hypoxia; ionic flux; metabolism.

Introduction

Ischemic damage to central nervous system (CNS) cells plays an important role in the pathophysiology of traumatic brain injury. The direct effects of the primary injury as well as secondary events such as edema, raised intracranial pressure, loss of local autoregulation, and vasospasm may contribute to ischemic damage. Following injury, cells are exposed to an external environment which may compromise their function and survival. Previous in vivo studies have provided some insight into the ionic and metabolic alterations caused by brain injury and ischemia[5,8]. However, to further our understanding of these neurochemical and neurometabolic changes a more molecular approach is needed. Further characterization of the injury-induced extracellular and intracellular changes will ultimately aid the development of therapeutic methods aimed at optimizing cell recovery and survival.

Since the retina shares many distinctive characteristics of CNS tissue and can be isolated without significant cellular perturbation, precise in vitro analysis of biochemical alterations can be performed. To further clarify the dynamic biochemical processes occurring during and immediately following brain injury, we incubated retinas in hypoxic medium for progressive periods of time (10 to 60 minutes) and examined the ionic flux of K^+ and Ca^{++}, lactate production, the intracellular incorporation of ^{14}C-leucine and [^{14}C]-2-deoxy-D-glucose, and cellular morphologic changes.

Methods

Adult Sprague-Dawley rats (n = 104, wt. 250–300 g) were anesthetized using enflurane administered by mask and then lethally injected with pentobarbital (150 mg/kg, intracardiac). Both eyes were immediately enucleated and placed in pre-oxygenated (95% O_2, 5% CO_2) Ames solution (pH = 7.4). This solution was designed to simulate the normal physiologic solution which bathes the retina in vivo and is similar to CSF[1]. Each eye was hemisected at the limbus and the anterior portion of the globe, along with the vitreous humor and cornea, were discarded. Under microscopic magnification, dissection with blunt glass pipettes was used to remove the retinas. Paired retinas from each animal were then randomly assigned to either experimental or control conditions. Control retinas were gently bubbled with a 95% O_2/5% CO_2 mixture for 30 seconds and then placed in a water bath maintained at 37°C and exposed to room air. Experimental retinas were made hypoxic by placing them within an environmental chamber (Labconco) maintained at less than 2% atmospheric O_2 and gently bubbling in a gas mixture composed of 95% N_2/5% CO_2 for 30 seconds prior to placing them in a 37°C water bath. Control and hypoxic retinas were incubated for 10, 20, 30, and 60 minute intervals. In addition, a 5 minute incubation period was used in experiments to determine intracellular ^{45}Ca uptake.

Flame spectrometry (Beckman, Kline Flame) and light spectrometry (Sigma 826-B) were used to determine the potassium and lactate concentrations respectively in the extracellular fluid prior to and after incubation. Comparisons were then made between the pre- and post-incubation samples to determine K^+ efflux and lactate production.

To determine intracellular calcium accumulation, ^{45}Ca was added to form a final concentration of 5 µCi/ml prior to incubation. Following incubation, autoradiography was performed by exposing the retinas to Kodak NMC 100 film for 3 days. Intracellular ^{45}Ca accumulation was then quantified using a computer-supported image analyzer (Jandel Scientific).

Intracellular ^{14}C-leucine incorporation was used as an index of protein synthesis. After the addition of ^{14}C-leucine to form a final concentration of 1 µCi/ml, retinas were exposed to either control or experimental conditions. Retinas were rinsed with formalin and distilled water to remove any unbound isotope. Autoradiography was subsequently performed by exposing the retinas to Kodak NMC 100 film for 24 hours and the intracellular incorporation of ^{14}C-leucine was quantified using computer-supported densitometry (Jandel Scientific).

To determine 2DG uptake, 2DG was added to Ames medium for a final activity of 1 µCi/ml. Initial activity of the incubation medium was determined by a scintillation counter (Beckman). Post-incubation scintillation counts were also performed and compared to pre-incubation samples. 2DG-loaded whole retinas were exposed to Kodak NMC film for 6 hours and autoradiographs were quantified using computer-supported densitometry (Jandel Scientific).

Cell swelling was determined by examining 30 control and hypoxic retinas (100 cells per retina) using a camera lucida drawing tube. Comparisons in cell morphology were made between the control and hypoxic retinas.

Results

Ionic flux: Extracellular potassium increased at a significantly higher rate in the hypoxic preparation

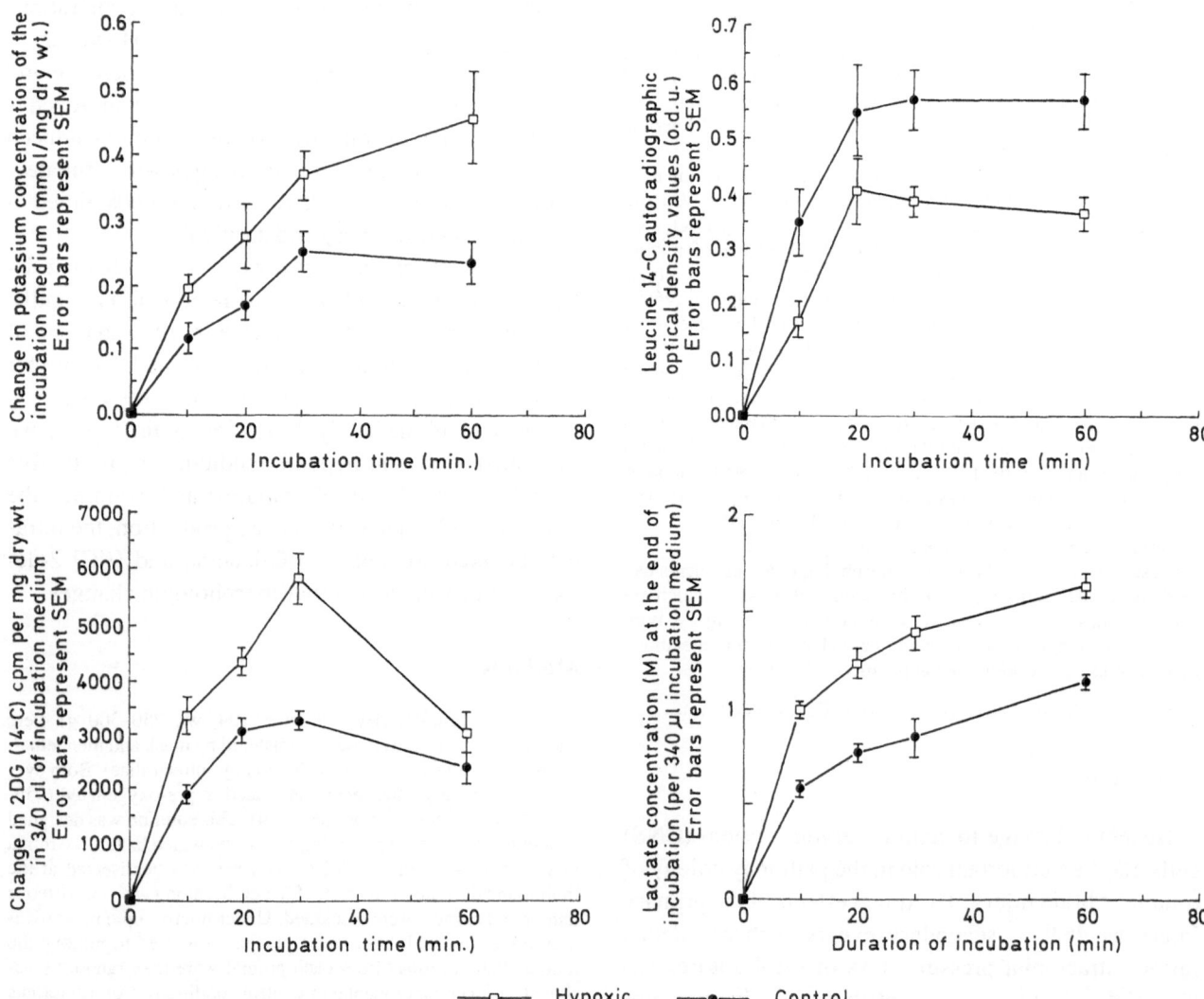

━□━ Hypoxic   ━●━ Control

Fig. 1. Control retinas represented by squares and hypoxic retinas by circles. Upper left: time course of K^+ efflux; upper right: ^{14}C-leucine incorporation; lower left: glucose utilization; lower right: lactate production

than in the control at all time intervals throughout the first hour of incubation ($p < 0.05$). The time course of potassium efflux from retinas exposed to control and environmental conditions is shown in the upper left corner of Fig. 1. The time course of disparities in ^{45}Ca accumulation between control and hypoxic retinas is shown in Fig. 2. Hypoxic retinas did not accumulate significantly more calcium until 20 to 30 minutes post-hypoxia ($p < 0.05$), suggesting that hypoxic retinas begin to lose their ability to maintain the steep trans-membrane calcium gradient at this time.

Glucose metabolism and lactate production: Hypoxic retinas demonstrated significantly increased glucose utilization and lactic acid production (lower portion, Fig. 1) starting early after hypoxia and persisting throughout the 1 hour incubation ($p < 0.05$).

Protein synthesis: Protein synthesis, as measured by ^{14}C-leucine uptake (upper right, Fig. 1), was reduced during the first 20 minutes of hypoxia compared to controls. Between 20 and 30 minutes both the control and experimental retinas ceased further uptake of ^{14}C-leucine.

Cellular edema: A 16% increase in cell size was noted after 10 minutes of hypoxia which persisted throughout the first hour of incubation at which time the control retinas also appeared swollen. Hypoxia-induced cellular edema likely results from early intracellular accumulation of ions such as Na$^+$, Ca^{++}, and Cl$^-$ which are accompanied by osmotically obliged water.

Discussion

Using a retina in vitro model of hypoxia, we examined several chemical, ionic, and metabolic changes and compared them to previous in vivo studies of brain injury. We found that extracellular K$^+$ accumulates early after hypoxia and is associated with changes in cellular metabolism as indicated by increased glucose utilization and lactate production. In contrast, Ca^{++} accumulation was delayed and occurred 20 to 30 minutes after the onset of hypoxia in conjunction with the absence of further uptake of ^{14}C-leucine. This delayed rise in intracellular Ca^{++} coupled with impaired uptake of ^{14}C-leucine likely represents irreversible cell damage.

Our findings are similar to previous in vivo studies which have shown a massive increase in extracellular K$^+$ following traumatic brain injury[4,5] which is thought to be due to a trauma-induced release of excitatory amino acids, especially glutamate. It has been pro-

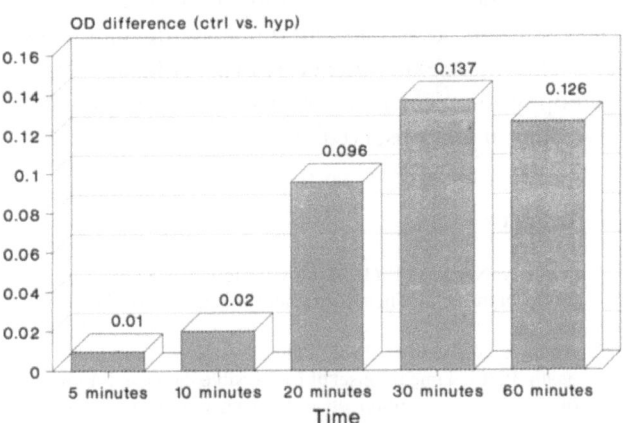

Fig. 2. Time course of the difference in ^{45}Ca accumulation between control and hypoxic retinas (o.d.^{hyp} – o.d.^{ctrl})

posed that ionic fluxes result in changes in cellular metabolism[8] and recently early increased glucose metabolism has been observed following concussive brain injury in response to a massive ionic flux[5]. The current results complement these findings and suggest that impaired K$^+$ distribution mediates increased glucose uptake early after a hypoxic insult.

In addition to K$^+$, Ca^{++} accumulation has been studied following brain injury[3]. It appears that although the efflux of K$^+$ ions occurs early after injury, Ca^{++} accumulation occurs in a more delayed fashion. Our in vitro results support this finding and are significant given that Ca^{++} accumulation has been postulated to play an important role in secondary cell death[7]. Elevated intracellular Ca^{++} may result from loss of normal Ca^{++} pump function (energy dependent), increased membrane permeability, or through the activation of ligand gated excitatory amino acid receptors. A disturbed pump/leak relationship for Ca^{++} is believed to cause cell damage through the overactivation of proteases, lipases, and endonucleases. We observed a significant increase in intracellular Ca^{++} between 20 and 30 minutes post-hypoxia. This delayed accumulation of Ca^{++} complements the findings of Ames and colleagues[2] that retina cells can recover after 20 minutes of ischemia but after 30 minutes they show extensive, irreversible damage as demonstrated in the current study by deficits in protein synthesis.

The in vitro retina model appears well-suited to study the effects of hypoxia on CNS tissue. The advantages of this model include prolonged survivability of the tissue in vitro, minimal cell trauma during dissection, and accessibility of the extracellular milieu for real-time monitoring and manipulation. Furthermore, the metabolic, ionic, and morphologic deviations described in these experiments are in accordance with the pathophysiological features of ischemia and concussive brain injury observed in vivo.

References

1. Ames A, Nesbett FB (1981) In vitro retina as an experimental model of the central nervous system. J Neurochem 37(4): 867–877
2. Ames A, Nesbett FB (1983) Pathophysiology of ischemic cell death. Part I. Time of onset of irreversible damage; importance of the different components of the ischemic insult. Stroke 14: 219–226
3. Fineman I, Hovda DA, Smith M, Yoshino A, Becker DP (1993) Concussive brain injury is associated with a prolonged accumulation of calcium: A ^{45}calcium autoradiographic study. Brain Res, in press
4. Hubschmann OR, Kornhauser D (1983) Effects of intraparenchymal hemorrhage on extracellular cortical potassium in experimental head trauma. J Neurosurg 59: 289–293
5. Katayama Y, Becker DP, Tamura T, Hovda DA (1990) Massive increases in extracellular potassium and the indiscriminate release of glutamate following brain injury. J Neurosurg 73: 889–900
6. Nedergaard M, Diemer NH (1988) Experimental cerebral ischemia: barbiturate resistant increase in regional glucose utilization. J Cereb Blood Flow Metab 8: 763–766
7. Schanne FA, Kane AB, Young EE, Farber JL (1979) Calcium dependence of toxic cell death: a final common pathway. Science 206: 700–702
8. Siejsö BK (1992) Pathophysiology and treatment of focal cerebral ischemia. Part I: pathophysiology. J Neurosurg 77: 169–184

Correspondence: David A. Hovda, Ph. D., Division of Neurosurgery, UCLA School of Medicine, 74-140 CHS, Los Angeles, CA 90024, U.S.A.

Acta Neurochir (1994) [Suppl] 60: 45–47
© Springer-Verlag 1994

Diffuse Astrocytic Swelling and Increased Second Messenger Activity Following Acute *Haemophilus influenzae* Meningitis – Evidence from a Rat Model

W. L. Maxwell[1], R. Bullock[2], A. Scott[2], Y. Kuroda[2], D. I. Graham[3], and G. Gallagher[4]

Departments of [1] Anatomy, [2] Neurosurgery, [3] Neuropathology, and [4] Bacteriology, University of Glasgow, Scotland, U.K.

Summary

Despite numerous epidemiological analyses of bacterial meningitis there is very little pathological data concerning the acute glial and neuronal responses to the disease. We have developed a safe, easily used rat model for *Haemophilus influenzae* type b meningitis. We measured cerebral blood flow, glucose utilisation and second messenger activity in this model, and carried out parallel light and ultrastructural analysis of glial and neuronal responses. Only protein kinase C activity was changed from control values. We obtained evidence for massive astrocytic swelling and neuronal degeneration. We posit that cytotoxic mechanisms may contribute to the pathology of meningitis.

Keywords: Meningitis; astrocyte; rat model.

Introduction

The mortality from bacterial meningitis remains high (25–30%), despite the use of modern antibiotic therapy, in recently presented epidemiological analyses from India[8], Australia[6], Israel[2,9] and North America[3]. In these series the commonest infective organism was *Haemophilus influenzae*. Even when patients do not die, a number of post-infective complications may occur. These commonly include sensorineural hearing loss (incidence of 8.5%), learning difficulties (12.7%), motor problems (7%), speech problems (7%), hyperactivity (4.2%), blindness (2.8%) and obstructive hydrocephalus or recurrent seizures (2.8% each) according to an Australian survey[6]. These complications are most likely due to a neuronal loss and may result in a necessity for long term care requiring major contributions from familial and health care providers. There is, however, only minimal pathological data available concerning the acute glial or neuronal responses in this disease which could lead to therapeutic strategies which

would serve to reduce brain damage. In order to provide an analysis of the acute phase of *H. influenzae* infection we have developed a safe, easily applied rat model. We have used the model to provide ultrastructural analysis of cellular responses to *H. influenzae* meningitis and to

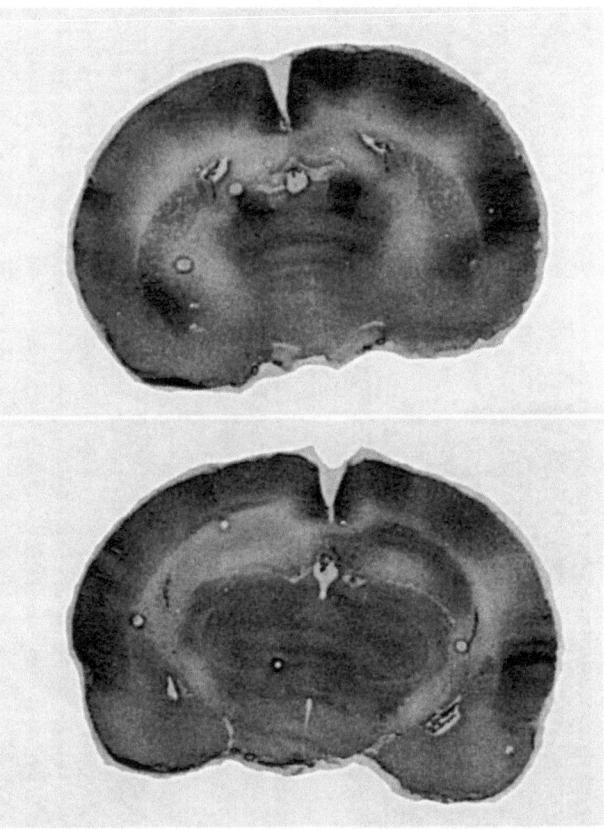

Fig. 1. [14]C iodoantipyrine autoradiograms, made 48 hours after infection with Hemophilus meningitis. Note areas of reduced CBF (32–51 ml min^{-1})

study regional blood flow and glucose use. This has provided information concerning the time course of cellular responses to infection, and thus indicate an optimal time for therapeutic intervention.

Methods and Material

Fifty adult Sprague-Dawley rats were infected, under nitrous oxide and halothane anaesthesia, by basal cisternal puncture with 25 μl of 106 colony forming units of *Haemophilus influenzae* type b organisms suspended in Ringers lactate solution at 37°C. Although none of the animals died, four required euthanasia due to dehydration, inertia and distress. Rectal temperatures ranged from 36.5 to 39.2°C (mean 38.7°). Six animals showed the "rough coat sign" and bloody nasal discharge. Cerebral blood flow (n = 6) and glucose

utilisation (n = 6) were measured autoradiographically using the iodoantipyrine (IAP) and 2-deoxyglucose (2DG)[10] methods. For measurement of second messengers, we examined adenylate cyclase activity using H3 forskolin binding (n = 4), and protein kinase C activity using H3 phorbol ester binding (n = 4)[11]. For light and electron microscopical examinations animals were perfusion fixed either with FAM (formalin acetone and methanol) (n = 10) or 3% paraformaldehyde and 2.5% glutaraldehyde in phosphate buffer (550 mOs) (n = 10). The animals were decapitated and the heads immersion fixed in the same medium for 24 hours at 4°C prior to routine processing for either paraffin or resin embedding.

Results

Leucocyte counts ranged from 900 to 20,000 per μl (mean 3,372 ± 110 polymorphs/μl) at 48 hours after

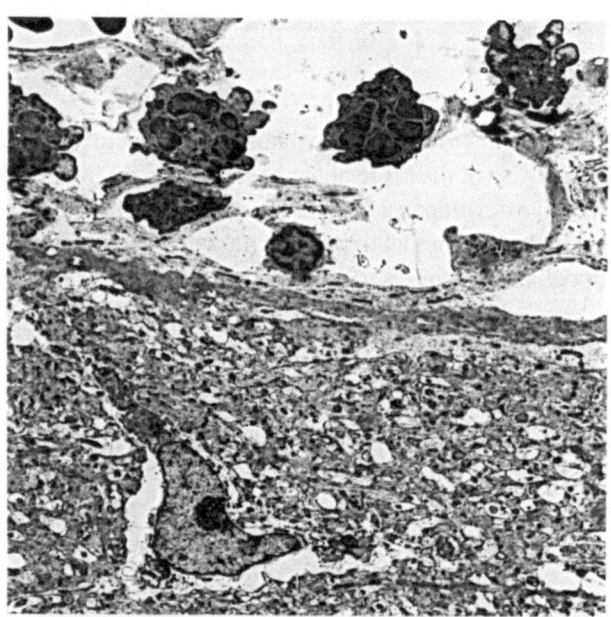

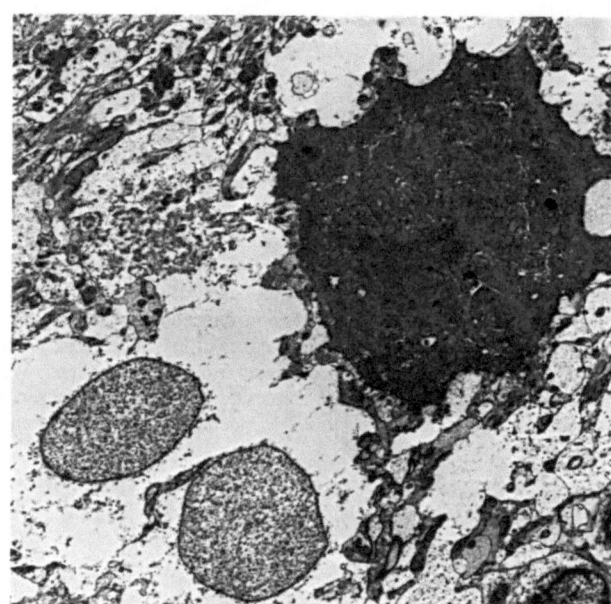

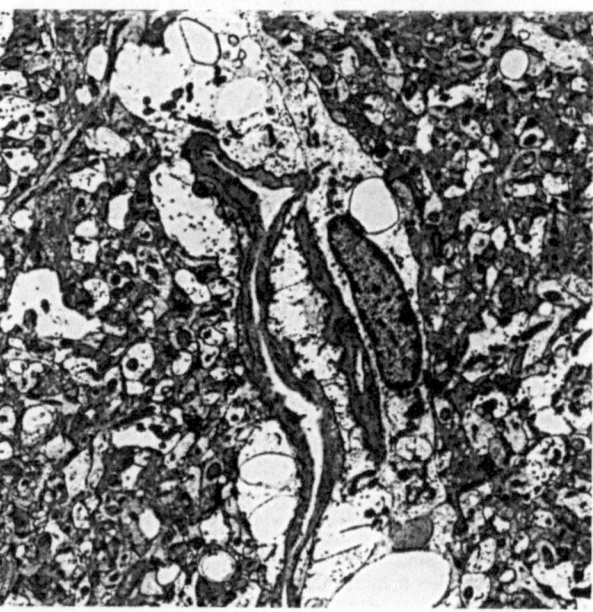

Fig. 2. Ultrastructural features in rats with Hemophilus meningitis, at 48 hours (transmission EM). (a) Polymorphonuclear leucocytes adherent to meninges (top left). (b) Neuronal pyknosis and astrocytic cytoplasm swelling (top right). (c) Perivascular astrocyte foot process swelling, causing compromise of the vessel lumen (bottom left)

infection. Colony counts in CSF ranged from 200 to 1500 per ml.

Regional CBF did not differ from control values for any of the brain regions studied in three of the four animals. In one animal, however, zones of focally reduced CBF occurred bilaterally in the striatum and hippocampus (Fig. 1). CBF in these low flow areas was 32–51 ml/100 g/min. Corresponding sections stained with haematoxylin and eosin did not show evidence for infarction.

Glucose use patterns and measured values did not differ from control values. However, there was a thin band of increased glucose use along the basal meningeal surface of the brain (99 ± 7 μmol/100 g/min. There was no alteration in glucose use over the convexity of the meninges.

Forskolin patterns and binding densities did not differ from controls. However, *phorbol ester* binding was increased in the outer layers of the frontal and parietal cortices.

Using low power electron micrographs, large numbers of leucocytes and/or macrophages in the subarachnoid space and in the perivascular space of cortical penetrator vessels (Fig. 2a) appeared. On the pial aspect of the brain enlarged, lucent astrocytes were numerous (Fig. 2c). Within the cerebral and cerebellar cortices the most obvious change from control material was the occurrence of numerous enlarged, electron lucent astrocyte cell bodies and perivascular astrocyte foot processes (Fig. 2c) on the abluminal aspect of the endothelium of the vessels. The swollen perivascular astrocyte cuffs resulted in a marked reduction of the luminal diameter of the vessels suggestive of compression or occlusion.

Where astrocyte lucency or "swelling" was most pronounced, examples of degenerating neurones were observed. These neurones were electron dense, lacked a clearly distinguishable nucleus and possessed a characteristic "scalloped" profile (Fig. 2b).

Discussion

The major findings of this study of the acute response to *Haemophilus influenzae* infection are acute "swelling" of astrocytes and degeneration of some neurones in the cerebral and cerebellar cortices. The demonstration of astrocyte swelling in this model accords with the high incidence of raised intracranial pressure, seen both at lumbar puncture, on CT, and at post mortem, in patients with severe bacterial meningitis. The cause for this astrocytic swelling is not appar-

ent from our studies. However, recent work from cell culture studies has indicated that excessive release of excitatory amino acid transmitters, eg endogenous glutamate, may be inductive of astrocyte swelling[5]. Our ultrastructural results, and those obtained from our study of the slow delivery of glutamate via a microdialysis probe[1], provided in vivo evidence to support and extend these findings.

We speculate that bacterial mediated cytokine release, or free-radical production in the meninges by neutrophils, or the release of still undefined neurotoxic factors by microglia[4] may contribute to both astrocyte swelling, and activation of second messengers such as protein kinase C during the acute phase of bacterial meningitis.

References

1. Bullock R, Landolt H, Maxwell WL, Fujisawa H (1993) Massive astrocytic swelling in response to extracellular glutamate – a possible mechanism for posttraumatic brain swelling? Acta Neurochir, in press
2. Dagan R (1992) A two-year prospective, nationwide study to determine the epidemiology and impact of invasive childhood Haemophilus influenzae type b infection in Israel. Clin Infect Dis 15: 720–725
3. Durand ML, Calderwood SB, Weber DJ, Miller SI, Southwick FS, Caviness VS, Swartz MN (1993) Acute bacterial meningitis in adults. A review of 493 episodes. N Engl J Med 328: 21–28
4. Giulian D, Vaca K, Corpuz M (1993) Brain glial release factors with opposing actions upon neuronal survival. J Neurosci 13: 29–37
5. Noble LJ, Hall JJ, Chen S, Chan PH (1992) Morphologic changes in cultured astrocytes after exposure to glutamate. J Neurotrauma 9: 255–267
6. Thomas DG (1992) Outcome of paediatric bacterial meningitis 1979–1989. Med J Aust 19: 519–20
7. Sakurada O, Kennedy C, Jehle J, et al (1978) Measurement of local cerebral blood flow with iodo [14C] antipyrine. Am J Physiol 234: H59–H66
8. Singh R, Thomas S, Kirubakaran C, Lalitha MK, Raghupathy P (1992) Occurrence of multiple antimicrobial resistance among Haemophilus influenzae type b causing meningitis. Indian J Med Res 95: 230–233
9. Slater PE, Roitman M, Costin C (1991) Epidemiology of haemophilus influenzae type b meningitis in Israel, 1981–90. Public Health Rev 18: 307–317
10. Sokoloff L, Reivich M, Kennedy C, et al (1977) The [14C] deoxyglucose method for the measurement of local glucose utilization: theory, procedure, and normal values in the conscious and anaesthetized albino rat. J Neurochem 28: 897–916
11. Worley PF, Baraban, De Souza EB, Snyder SH (1986) Mapping second messenger systems in the brain: differential localizations of adenylate cyclase and protein kinase C. Proc Natl Acad Sci USA 83: 4053–4057

Correspondence: Ross Bullock, M.D., Division of Neurological Surgery, Medical College of Virginia, Virginia Commonwealth University, MCV Station, Box 631, Richmond, Virginia 23198-0631, U.S.A.

Mediator Mechanisms in Vasogenic and Cytotoxic Brain Edema

Acta Neurochir (1994) [Suppl] 60: 51–54

Leukocyte/Endothelial Interactions and Blood-Brain Barrier Permeability in Rats During Cerebral Superfusion with LTB₄

L. Schürer[1], **S. Corvin**, **F. Röhrich**[1], **C. Abels**, and **A. Baethmann**

Department of Neurosurgery[1] and Institute for Surgical Research, Klinikum Großhadern, Ludwig-Maximilians-University, München, Federal Republic of Germany

Summary

The experimental study analyses the vasomotor response (change of diameter of pial arterioles and venules), and blood-brain barrier function of the pia-arachnoidea at the rat brain surface before, during and after cerebral superfusion with 1.5 or 15.0 nM LTB₄ in mock CSF. Leukocyte dynamics were studied by assessment of their centerline velocity, of rolling along ("roller") and attachment to ("sticker") the venular wall of white blood cells intravitally stained by Rhodamin 6G. Superfusion of the brain with LTB₄ at both dose levels led to dilation of arterioles to 130% ($p < 0.001$), while of venules to 117% ($p < 0.001$) of control. The centerline velocity of leukocytes increased from 0.7 to 0.9 mm/s, however, only after superfusion with LTB₄ at the high dose level. LTB₄ induced a dose-dependent rolling ($p < 0.01$) and sticking of leukocytes ($p < 0.001$). Yet, a delay of about 60 min between cerebral administration of LTB₄ and the maximal response of leukocyte rolling and sticking was observed. Whereas the blood-brain barrier was not opened by cerebral superfusion with 1.5 or 15.0 nM LTB₄, for i.v. Na⁺-fluorescein, barrier leakage was promptly induced by 30.0 nM. The present findings demonstrate that cerebral administration of LTB₄ by superfusion of the exposed brain surface is eliciting a pronounced vasomotor response, whereas the induction of leukocyte/endothelial interactions is less impressive.

Keywords: Leukotriene B₄; pial microcirculation; leukocyte/endothelial interactions.

Introduction

Activated polymorphonuclear leukocytes have a powerful tissue-damaging potential[29]. Evidence is accumulating that these blood cells may play a role in the pathogenesis and enhancement of adverse sequelae in cerebrovascular disorders, such as ischemic brain injury, infarction, subarachnoid hemorrhage, vasospasm, and bacterial meningitis, among others[25,18,21,19,17,26]. Activation of white blood cells under these conditions might be brought about by the release of leukotactic mediator compounds, such as leukotriene B₄ (LTB₄)[16,20].

A major requirement for identification of leukotrienes as mediator in secondary brain damage is fulfilled, namely formation and release, respectively of these agents in brain tissue under pathological conditions, as trauma, subarachnoid hemorrhage, or ischemia[8,14,30]. Further, development of postischemic brain edema was found to be associated with the formation of LTB₄ in affected brain tissue areas, and that ischemic brain swelling is attenuated by inhibition of LTB₄ synthesis[7,9]. The hypothesis may thus be advanced that the leukotactic agent is activating white blood cells under respective circumstances, resulting in endothelial damage and, thereby, enhancement of ischemic brain swelling[11]. It is generally accepted now that induction of tissue damage by leukocytes requires their interaction with the vascular endothelium. LTB₄ has been shown to stimulate respective interactions in the microvasculature of the hamster cheek pouch[3]. The findings obtained in the peripheral organ notwithstanding, respective evidence on leukocyte/endothelial interactions in cerebral blood vessels as induced by LTB₄ is not available so far.

Purpose of the present study was to examine this aspect more closely by utilization of a closed cranial window preparation, as developed by this laboratory for analysis of the pial microcirculation at the brain surface[13].

Materials and Methods

Male Sprague-Dawley rats (250–300 g body weight) were anesthetized i.v. with α-chloralose, immobilized with pancuronium bromide, and artificially ventilated. After insertion of arterial and venous catheters a closed cranial skull window was implanted above the parietal cerebral cortex for video-microscopical observations[13]. The cerebral surface was superfused with artificial CSF at an intracranial pressure in the window chamber of 3–7 mmHg adjusted by respective placement of the outflow catheter. Endexpiratory pCO_2, arterial blood pressure, intracranial pressure in the skull window chamber, and airway pressure were continuously monitored. Arterial blood gases, hematocrit, and peripheral differential blood cell counts were studied in intervals. Rhodamin 6G was i.v. administered as fluorescence marker of white blood cells and of the vascular bed of the preparation. The following microcirculatory parameters were assessed: vessel diameters, leukocyte centerline velocity, shear rate, attachment of leukocytes at the vascular endothelium ("sticker"), and frequency of leukocytes slowly rolling along the luminal vessel surface ("roller"). Following a control period the brain surface was superfused either with 1.5 or 15.0 nM LTB_4 for 20 min. Superfusion with 30.0 nM was studied in few experiments in addition. Video scenes of the brain surface were taken in intervals during 120 min following superfusion with LTB_4. The integrity of the blood-brain barrier was tested by i.v. Na^+-fluorescein at termination of the experiment.

Results

Vasomotor Response

Administration of the brain with LTB_4 (1.5 or 15.0 nM) by cerebral superfusion led to dilation of arteriolar vessels, which was more pronounced at the higher dose. Maximal dilation reached $+30 \pm 14\%$ of the control diameter at 80 min after superfusion with LTB_4. The vasomotor response of venular blood vessels was less intensive. Maximum of venous dilation of $+17 \pm 4\%$ was obtained at 100 min after superfusion of the brain with LTB_4 at 15.0 nM.

Microcirculatory Behavior of Leukocytes

As shown in Table 1, the centerline velocity of white blood cells was found to increase from 0.7 to 0.9 mm/s, however, only after superfusion of the brain with LTB_4 at 15.0 nM. Further, a dose-dependent increase of the number of rolling ($p < 0.01$) and of

sticking leukocytes ($p < 0.001$) was observed (Table 1). Intravenous injection of Na^+-fluorescein at termination of the experiment did not indicate opening of the blood-brain barrier by LTB_4 as concluded from absence of extravasation of the fluorescence marker into the cerebral parenchyma. The blood-brain barrier remained intact during administration of the lower concentrations of LTB_4 in the superfusate (i.e. 1.5 or 15.0 nM), whereas barrier opening was induced, if the preparation was superfused with LTB_4 at 30.0 nM.

Discussion

The present experiments demonstrate that superfusion of the brain with LTB_4 causes dilation of pial arterioles and venules. The vasomotor response elicited by the procedure was comparable to that observed in peripheral organs during administration of LTB_4[6]. Kamitani and coworkers[12] have reported marginal constriction (5%) of pial arterioles, although the brain surface was administered with LTB_4 at a dose as high as 3.0 μM. Other experiments with employment of a window preparation in mice and administration of LTB_4 by superfusion in a dose-range of 30 to 300 nM led to constriction of pial arterioles by 13%[22]. Obviously, the data are in contrast with the present findings, and their interpretation is difficult on account of the excessive LTB_4 concentrations which were studied. Further, since inflammatory processes of the brain are mostly associated with an increased flow in the affected microvasculature, and – most likely – enhanced LTB_4-levels, it might be concluded that dilation rather than constriction of arterioles follows exposure to the vasoactive mediator compound. It is noteworthy that the degree of leukocyte sticking in the pial venules following LTB_4 superfusion by and large is in agreement with respective data obtained in the hamster

Table 1. *Video-Fluorescence Microscopy.* Behavior of leukocytes before, during and after superfusion of 15 nM LTB_4

Time (min)	Control phase	During LTB_4	Min. after termination of LTB_4 superfusion					
			1	20	40	60	80	100
Centerline velocity ($mm \times sec^{-1}$)	0.79 ± 0.10	0.91 ± 0.08	0.79 ± 0.09	0.76 ± 0.06	0.98 ± 0.07	1.00 ± 0.10	0.90 ± 0.09	0.92 ± 0.07
Roller/venule ($min^{-1} \times 100\ \mu m^{-1}$)	2.28 ± 0.72	n.d.	2.67 ± 0.85	5.87 ± 1.86	9.62 ± 3.04	8.77 ± 2.77	10.98 ± 3.47	13.2 ± 4.12
Sticker/venule ($min^{-1} \times 100\ \mu m^{-1}$)	0.83 ± 0.25	n.d.	5.07 ± 1.53	6.06 ± 1.83	8.58 ± 2.59	6.62 ± 2.00	4.71 ± 1.42	4.97 ± 1.5

Mean ± SEM; *n.d.* not determined.

cheek pouch preparation[3,6]. In contrast to the current findings, however, the frequency of leukocyte rolling was decreasing in those studies after administration of LTB$_4$. The frequency, however, of white blood cell rolling along the vessel wall in the cheek pouch preparation was quite high already under control conditions. Recent observations of Fiebig et al.[10] indicate that surgical trauma inflicted during preparation might be involved in activation of leukocyte rolling, rendering the parameter a sensitive indicator of the quality of the surgical procedure.

Evidence could not be obtained in the present experiments that LTB$_4$ exposure of the cerebral surface causes diapedesis of white blood cells into the cerebral parenchyma, which is a characteristic response in peripheral organs, such as striated muscle or the hamster cheek pouch[3,15]. Yet, emigration of white blood cells cannot be excluded in the current studies. This is attributable to the limited resolution of the videomicroscopical technology, making it difficult to distinguish adhering white blood cells in the vessel lumen from those in the extravascular compartment. Besides diapedesis in progress was never observed. The comparatively moderate neutrophil response in the present experiments to a leukotactic agent as powerful as LTB$_4$ may relate with specific features of the brain, such as efficient defense against the mediator compound and/ or specific properties of the blood-brain barrier[5]. Nevertheless, irrespectively of whether or not the blood-brain barrier had been opened, evidence is available that diapedesis of white blood cells through the vessel wall might be independent of an increase of vascular permeability[23]. Moreover, LTB$_4$ concentrations utilized in these experiments which in peripheral organs suffice to enhance vascular permeability, might have been too low to open the blood-brain barrier. Somewhat higher LTB$_4$ concentrations, i.e. 20.0 nM, have been reported to disrupt the integrity of the barrier, if the agent is administered into cerebral parenchyma[4]. The concentration threshold of LTB$_4$ required to induce blood-brain barrier opening might be close to 20–30 nM, as confirmed by the limited number of experiments in this study (cf. Results) utilizing the latter dose level. Whether observations that injection of LTB$_4$ into brain parenchyma at a dose as high as 15.0 μM does not induce an increased water content is in conflict with this interpretation is a matter of debate[27,28]. The current results on leukocyte/endothelium interactions evolving at lower LTB$_4$ concentrations, however, indicate clearly that the phenomenon is not coupled to breakdown of the vascular barrier. This conclusion is supported by observations of others on the blood-aqueous barrier of the eye, demonstrating stimulation of leukocyte chemotaxis by LTB$_4$ at pathophysiologically relevant concentrations without opening the vascular barrier[2].

As to underlying cell biological mechanisms associated with leukocyte activation by LTB$_4$, data of Rossi et al.[24] suggest that the mediator compound causes ruffling of the cell membrane together with cell depolarization. Further, expression of CD11/CD18 adhesion receptors by LTB$_4$ has been observed, which is considered as a central process mediating adhesion of white blood cells at the vascular endothelium[16,1].

In conclusion, the present experiments demonstrate that cerebral superfusion with the leukotactic mediator compound LTB$_4$ causes leukocyte/endothelial interactions in a dose-dependent manner, although the tissue response appears not to be impressive. The limited activation of leukocytes together with preservation of the blood-brain barrier integrity to exposure to LTB$_4$ may be explained by availability of defense mechanisms of the brain parenchyma against the pathophysiologically powerful compound. Such a defense might prevent an overwhelming activation of leukocyte/endothelial interactions with its fatal sequelae under respective pathological conditions. If, however, cerebral tissue is suffering from a primary insult, such as trauma, ischemia or bacterial invasion the defense may break down, resulting in a variety of adverse sequelae by activation of leukocytes, enhancing the primary tissue damage.

Acknowledgements

The excellent secretarial and technical assistance by Ulrike Goerke, Helga Kleylein, Hildegard Lainer and Monika Stucky is gratefully appreciated. Supported by Deutsche Forschungsgemeinschaft: Ba 452/6-7.

References

1. von Andrian UH, Hansell P, Chambers D, Filho IT, Butcher EC, Arfors K-E (1992) L-Selectin is a prerequisite for beta$_2$ integrin-mediated neutrophil adhesion to inflamed venular endothelium in vivo. Am J Physiol 263: H1034–H1044
2. Bhattacherjee P (1989) The role of arachidonate metabolites in ocular inflammation. Prog Clin Biol Res 312: 211–227
3. Björk J, Arfors KE, Hedqvist P, Dahlen SE, Lindgren JA (1982) Leukotriene B$_4$ causes leukocyte emigration from postcapillary venules in the hamster cheek pouch. Microcirculation 2: 271–281
4. Black KL, Hoff JT (1985) Leuktrienes increase blood-brain barrier permeability following intraparenchymal injections in rats. Ann Neurol 18: 349–351

5. Casale TB, Abbas MK (1990) Comparison of leukotriene B_4-induced neutrophil migration through different cellular barriers. Am J Physiol 258: C639–C647

6. Dahlen SE, Björk J, Hedqvist P, Arfors KE, Hammarström S, Lindgren JA, Samuelsson B (1981) Leukotrienes promote plasma leakage and leukocyte adhesion in postcapillary venule: in vivo effects with relevance to the acute inflammatory response. Proc Natl Acad Sci USA 78: 3887–3891

7. Dempsey RJ, Roy MW, Cowen DE, Maley ME (1986a) Lipoxygenase metabolites of arachidonic acid and the development of ischemic cerebral oedema. Neurol Res 8: 53–58

8. Dempsey RJ, Roy MW, Meyer K, Cowen DE, Tai H-H (1986b) Development of cyclooxygenase and lipoxygenase metabolites of arachidonic acid after transient cerebral ischemia. J Neurosurg 64: 118–124

9. Dempsey RJ, Combs DJ, Maley ME, Cowen DE, Roy MW, Donaldson DL (1987) Moderate hypothermia reduces post-ischemic edema development and leukotriene production. Neurosurgery 21: 177–181

10. Fiebig E, Ley K, Arfors K-E (1991) Rapid leukocyte accumulation by "spontaneous" rolling and adhesion in the exteriorized rabbit mesentery. Int J Microcirc Clin Exp 10: 127–144

11. Harlan JM (1987) Neutrophil-mediated vascular injury. Acta Med Scand 715 [Suppl]: 123–129

12. Kamitani T, Little MH, Ellis EF (1985) Effect of leukotrienes, 12-HETE, histamine, bradykinin, and 5-hydroxytryptamine on in vivo rabbit cerebral arteriolar diameter. J Cereb Blood Flow Metab 5: 554–559

13. Kawamura S, Schürer L, Goetz A, Kempski O, Schmucker B, Baethmann A (1990) An improved closed cranial window technique for investigation of blood-brain barrier function and cerebral vasomotor control in the rat. Int J Microcirc Clin Exp 9: 369–383

14. Kiwak KJ, Moskowitz MA, Levine L (1985) Leukotriene production in gerbil brain after ischemic insult, subarachnoid hemorrhage, and concussive injury. J Neurosurg 62: 865–869

15. Lindbom L, Hedqvist P, Dahlen SE, Lindgren JA, Arfors KE (1982) Leukotriene B_4 induces extravasation and migration of polymorphonuclear leukocytes in vivo. Acta Physiol Scand 116: 105–108

16. Lindström P, Lerner R, Palmblad J, Patarroyo M (1990) Rapid adhesive responses of endothelial cells and of neutrophils induced by leukotriene B_4 are mediated by leukocytic adhesion protein CD18. Scand J Immunol 31: 737–744

17. Lopez JAG, Armstrong ML, Harrison DG, Piegors DJ, Heistad DD (1989) Vascular responses to leukocyte products in atherosclerotic primates. Circ Res 65: 1078–1086

18. Mercuri M, Ciuffetti G, Robinson M, Toole J (1989) Blood cell rheology in acute cerebral infarction. Stroke 20: 959–962

19. Nazar GB, Kassell NF, Povlishock JT, Lee J, Hudson S (1988) Subarachnoid hemorrhage causes adherence of white blood cells to the cerebral arterial luminal surface. In: Wilkins RH (ed) Cerebral vasospasm. Raven, New York, pp 343–356

20. Palmblad J, Lindström P, Lerner R (1990) Leukotriene B_4 induced hyperadhesiveness of endothelial cells for neutrophils. Biochem Biophys Res Comm 166: 848–851

21. Prentice RL, Szatrowski TP, Kato H, Mason MW (1982) Leukocyte counts and cerebrovascular disease. J Chron Dis 35: 703–714

22. Rosenblum WI (1985) Constricting effect of leukotrienes on cerebral arterioles of mice. Stroke 16: 262–263

23. Rosengren S, Ley K, Arfors K-E (1989) Dextran sulfate prevents LTB_4-induced permeability increase, but not neutrophil emigration, in the hamster cheek pouch. Microvasc Res 38: 243–254

24. Rossi AG, MacIntyre DE, Jones CJP, McMillan RM (1993) Stimulation of human polymorphonuclear leukocytes by leukotriene B_4 and platelet activating factor: an ultrastructural and pharmacological study. J Leukoc Biol 53: 117–125

25. Schürer L, Corvin S, Röhrich F, Abels C, Uhl E, Baethmann A (1993) Current evidence for and against an involvement of leukocytes in cerebral ischemia. Acta Neurochir (Wien) 20: 197–199

26. Tuomanen EI, Saukkonen K, Sande S, Cioffe C, Wright SD (1989) Reduction of inflammation, tissue damage, and mortality in bacterial meningitis in rabbits treated with monoclonal antibodies against adhesion-promoting receptors of leukocytes. J Exp Med 170: 959–968

27. Unterberg A, Polk T, Ellis E, Marmarou A (1990) Enhancement of infusion-induced brain edema by mediator compounds. Adv Neurol 52: 355–358

28. Unterberg A, Schmidt W, Wahl M, Ellis E, Marmarou A, Baethmann A (1991) Evidence against leukotrienes as mediators of brain edema. J Neurosurg 74: 773–780

29. Weiss SJ (1989) Tissue destruction by neutrophils. N Engl J Med 320: 365–376

30. Xu J, Hsu CY, Liu TH, Hogan EL, Perot PL, Tai H-H (1990) Leukotriene B_4 release and polymorphonuclear cell infiltration in spinal cord injury. J Neurochem 55: 907–912

Correspondence: Ludwig Schürer, M. D., Department of Neurosurgery, Klinikum Großhadern, Ludwig-Maximilians-Universität, Marchioninistrasse 15, D-81377 München, Federal Republic of Germany.

Acta Neurochir (1994) [Suppl] 60: 55–57
© Springer-Verlag 1994

The Effect of Leukotriene C4 on the Permeability of Brain Capillary Endothelial Cell Monolayer

T. Nagashima, W. Shigin, A. Mizoguchi[1], M. Arakawa[1], M. Yamaguchi[2], and N. Tamaki

Departments of Neurosurgery[1] and Anatomy, Kobe University School of Medicine, [2] School of Allied Medical Sciences, Kobe University, Japan

Summary

The role of leukotrienes as mediator of brain edema is still controversial. Recently, the ability of γ-GTP to act as enzymatic barrier and to inactivate leukotrienes in normal brain capillaries was pointed out.

A hypothesis tested in our experiments was that Leukotriene C4 (LTC4) increases permeability of a cerebral capillary endothelial monolayer which lacks γ-GTP activity.

Brain capillary endothelial cells were obtained of 10 rats from cerebral cortex by an enzymatic isolation procedure. The cells have an intact function, however, lack γ-GTP activity. The endothelial cells were cultured on an optically clear collagen membrane mounted on a plastic frame. Effects of bradykinin (1×10^{-5} M) and LTC4 (1×10^{-7} M, 1×10^{-6} M, 5×10^{-6} M, 1×10^{-5} M) were tested on permeability of the endothelial cell monolayer by measuring leakage of ^{14}C-sucrose. The effect of LTC4 and bradykinin on intracellular calcium was studied by laser scanning confocal microscopy.

LTC4 did not increase permeability of the brain capillary endothelial cell monolayer which lacked gamma-GTP activity. LTC4 did neither increase the concentration of intracellular calcium. Differences of LTC4 receptor function in normal brain capillaries and tumor capillaries remain to be studied.

Keywords: Leukotriene; endothelial cells; BBB; brain capillary.

Introduction

Leukotrienes are metabolites of arachidonic acid generated by the 5-lipoxygenase pathway. Levels of leukotrienes in brain tissue are increased during post-ischemic reperfusion, in brain tumors, subarachnoid hemorrhage, and concussive head injury. Therefore, it has been proposed that leukotrienes have a mediator function in brain edema[2,3,4]. However, the role of leukotrienes as mediator of brain edema is still discussed controversially[5,6].

LTC4 is metabolized to LTD4 by γ-GTPase, LTD4 is then converted to LTE4 by LTD4-dipeptidase. LTE4 is further metabolized to LTF4 by γ-GTPase. Recently, the ability of γ-GTP was pointed out to act as an enzymatic barrier and to inactivate leukotrienes in normal brain capillaries, because brain capillaries are rich in γ-GTP unlike capillaries in other organs. This enzymatic barrier concept seems to be quite intriguing.

A hypothesis tested by ourselves is that LTC4 increases permeability of cerebral endothelial cells lacking gamma-GTP activity.

Materials and Methods

1. Isolation of Brain Microvessel Endothelial Cells

Brain capillary endothelial cells were obtained from cerebral cortex of 10 rats by a purely enzymatic isolation procedure reported by Audus and Borchard[1]. The cortex tissue was digested by a two-step enzymatic treatment, namely dispase and collagenase/dispase. Finally, the endothelial cells were purified by centrifugation on a pre-established 50% Percoll gradient. The cells were seeded then onto collagen coated plastic plates.

The brain capillary endothelial cells in culture showed a typical cobblestone monolayer morphology. The cells were cultured by Dulbecco's MEM with 13% FCS. The endothelial cell identity was confirmed by Factor VIII immunoperoxydase and uptake of acetylated low density lipoprotein. The cells had no evidence of GFAP or smooth-muscle actin. The studies were restricted to low-passage cultures displaying morphological features of vascular endothelium in phase-contrast microscopy.

Release of prostacyclin in the supernatant was measured by assay of its stable degradation product, 6-keto-PGF1α at passage no. 2, 3, 4, and 10. γ-GTP activity was determined by a modification of a procedure based on the cell protein content.

2. Transport Studies

The endothelial cells were cultured on an optically clear collagen membrane mounted in a plastic frame. Transport characteristics of this membrane is close to that of polycarbonate filters. 10 to 14 days after seeding of the cells on the membrane, confluency is obtained.

¹⁴C-sucrose was added to the upper chamber. Leakage into the lower chamber was measured by liquid scintillation spectroscopy. Permeability was expressed as total radioactivity in the bottom chamber observed at a particular time divided by the total radio-activity added to the system.

In addition, the effect of bradykinin (1×10^{-5} M) or of LTC4 (1×10^{-7} M, 1×10^{-6} M, 5×10^{-6} M, 1×10^{-5} M) was studied on permeability of the endothelial cell monolayer. Estimation of concentrations occurring under pathological conditions are in the range of 10^{-8} to 10^{-9} M. Therefore the current concentrations were above those seen in pathological conditions. After 15 minutes incubation with bradykinin or LTC4, permeability of the endothelial cell monolayer was measured by ¹⁴C-sucrose.

3. Laser Scanning Confocal Microscopy

Recent developments of calcium fluoroprobes offers an opportunity to follow transient changes in intracellular calcium. Confocal laser scanning microscopy was employed to study effects of LTC4 on the intracellular calcium concentration. The system consists of four parts: a laser source, a laser scanning microscope, an image processing system, and a perfusion system to keep the cells in physiological conditions.

Confluent cultures were loaded with the calcium probe fluo-3 A/M, for assessment of intracellular calcium. This system has a sufficient spatial and temporal resolution for detection of intracellular calcium changes in cultured endothelial cells. After hyperosmolar mannitol of 1.8 M or LTC4 was applied to the endothelial cells, the changes of calcium were monitored every 5 seconds.

Results

1. Characterization of Endothelial Cells

Brain capillary endothelial cell produce prostacyclin in large quantities, which was unchanged by cultivation. Arachidonic acid increased the production of prostacyclin.

Freshly isolated cells contain a high level of γ-GTP activity, which decreases, however, with time in culture to a constant low level. In fact, γ-GTP activity decreased drastically during the first day after plating

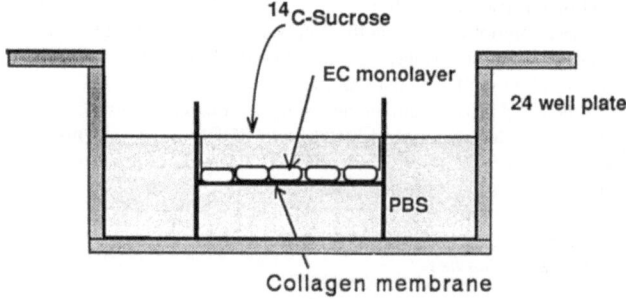

Fig. 1. Outline of transendothelial transport study design. Brain microvessel endothelial cells were seeded and grown to confluency on a collagen membrane

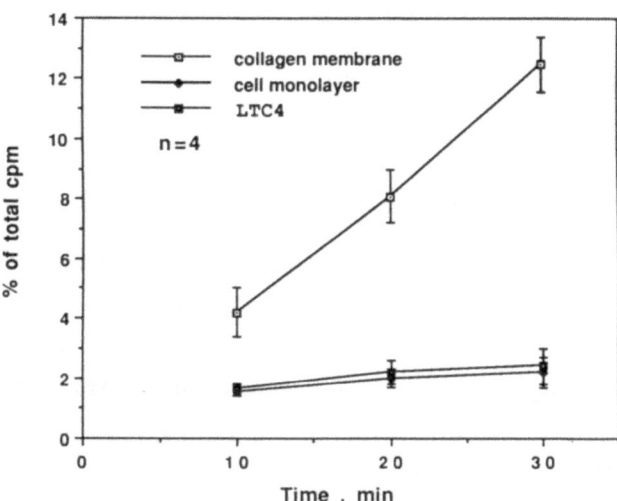

Fig. 2. Effect of LTC4 on brain microvessel endothelial cell monolayers. Concentrations of LTC4 were 1×10^{-7} M, 1×10^{-6} M, 5×10^{-6} M, 1×10^{-5} M, respectively

until a constant level of 3–6 units/mg protein was reached. This value is slightly lower than the value reported by others. The result indicates that cultured brain capillary endothelial cells lack γ-GTP activity.

2. Transport Studies

The extent of sucrose flux through endothelial cell monolayers is dependent on the age of the cell culture. Significant amount of sucrose leaking through a 12-days-old monolayer were not observed for up to 30 minutes. A greater flux of sucrose was observed in 7 days old monolayers.

Bradykinin is a mediator of brain edema. It is reported to induce an increase in cytosolic calcium in endothelial cells and to increase leakage of BBB markers. After 15 minutes incubation with 1×10^{-6} M bradykinin, permeability of endothelial cell monolayers was measured by ¹⁴C-sucrose. Bradykinin was found to induce an increase in permeability of endothelial cell monolayers.

On the other side, incubation with LTC4 for 15 min did not increase permeability of endothelial cell monolayers.

3. Confocal Laser-Scanning Microscopy

Hyperosmotic mannitol induced a remarkable increase of intracellular calcium within 5 seconds, whereas LTC4 did not increase intracellular calcium of cultured endothelial cells.

Discussion

The role of various autacoids as mediators of BBB opening and development of vasogenic bain edema has been extensively studied. However, individual contributions of the different agents is still controversially discussed. Many experimental and clinical studies have confirmed increased concentrations of leukotrienes in the brain and CSF. Recently, Black *et al.* demonstrated that intracarotid infusion of LTC4 selectively increases capillary permeability in tumors and ischemic brain but not in normal brain. A concept of selective and reversible opening of the blood-tumor barrier would be clinically important. However, our results indicate that γ-GTP is not acting as an enzymatic barrier for LTC4 in brain endothelium in vitro.

Many studies have suggested existence and heterogeneity of leukotriene receptors in tissues, however, existence of a leukotriene C4 receptor has not been conclusively demonstrated in cerebrovascular endothelium. Affinity binding studies of LTC4 in isolated brain capillaries have been made, but it is not clear whether the binding sites observed are functional receptors for LTC4 or not. The majority of LTC4 binding sites which were indentified are reported to be identical with glutathione-s-transferase.

Receptors and signal transduction processes which are well understood are concerned with LTD4. When activated by leukotriene D4 the LTD4 receptors interact with G-proteins. Via interaction with G-protein, the LTD4 receptors mobilize intracellular calcium. Although LTC4 binding sites have been identified in a variety of cells and tissues, their physiological role is far less clear as compared to that of LTD4. Despite of the heterogeneity of leukotriene receptors, the post-receptor signal transduction process is supposed to have a common pathway. Our results did not demonstrate an increase of intracellular calcium by LTC4 receptor mediated signal transduction processes. Further studies are necessary to clarify the presence of LTC4 receptors in brain capillaries.

In conclusion, γ-GTP does not act as an enzymatic barrier for LTC4. Different effects of LTC4 on pathological capillaries versus normal capillaries may be attributed to a different permeability of the vessels to LTC4, or to different LTC4 receptors on the capillaries.

References

1. Audus KL, Borchardt RT (1986) Characterization of an in vitro blood-brain barrier model system for drug transport and metabolism. Pharma Res 3: 81–87
2. Baba T, Black KL, Ikezaki K, Chen K, Becker DP (1991) Intracarotid infusion of leukotriene C4 selectively increases blood-brain permeability after focal ischemia in rats. J Cereb Blood Flow Metab 11: 638–643
3. Black KL, King WA, Ikezaki K (1990) Selective opening of the blood-tumor barrier by intracarotid infusion of leukotriene C4. J Neurosurg 72: 912–916
4. Chio CC, Baba T, Black KL (1992) Selective blood-tumor barrier disruption by leukotrienes. J Neurosurg 77: 407–410
5. Mayhan WG, Sahagun G, Spector R, Heistad D (1986) Effect of leukotriene C4 on the permeability of the cerebral microvasculature. Am J Physiol 251: H471–474
6. Unterberg A, Schmidt W, Wahl M, Ellis EF, Marmarou A, Baethmann A (1991) Evidence against leukotrienes as mediators of brain edema. J Neurosurg 74: 773–780

Correspondence: Tatsuya Nagashima, M.D., 7-5-1, Kusunoki-cho, Chuo-ku, Kobe, 650, Japan.

Acta Neurochir (1994) [Suppl] 60: 58–61
© Springer-Verlag 1994

Modulation of Edema by Dizocilpine, Kynurenate, and NBQX in Respiring Brain Slices After Exposure to Glutamate

M. T. Espanol, Y. Xu, L. Litt, L.-H. Chang, T. L. James, P. R. Weinstein, and **P. H. Chan**

Anesthesia Department, University of California, San Francisco, U.S.A.

Summary

Brain edema caused by glutamate excitotoxicity was studied in well oxygenated neonatal cerebrocortical brain slices (350 μ thick). Slices exposed to 60 minutes of 2 mM glutamate, with or without glutamate antagonists (dizocilpine, kynurenate, or NBQX), were allowed to recover for 60 minutes. The protocol was identical to that in noninvasive multinuclear NMR spectroscopy studies (^{31}P/^{1}H/^{19}F) of live slices. Percent water and swelling were determined invasively in isolated slices by wet and dry weight measurements before and after glutamate exposure. Edema was detectable within minutes in all experiments with glutamate exposures, but not in untreated control slices. Dizocilpine, kynurenate, and NBQX differently affected swelling, which correlated with PCr and ATP loss in separate NMR studies. Synaptic glutamate receptor activation appears to initiate events causing both edema and energy failure. Multiple glutamate receptor types seem to be involved. No glutamate antagonist provided greater protection against both edema and energy loss than dizocilpine. Dizocilpine might also block voltage-dependent Na$^+$ channels, and provide protection via mechanisms other than NMDA-receptor dependent channel antagonism.

Keywords: Brain edema; NMR; glutamate.

Introduction

Excessive stimulation by glutamate at central nervous system excitatory synapses appears to produce neuronal death and injury in several neuropathological states, including cerebral hypoxia/ischemia. Detailed mechanisms of such injury, termed glutamate excitotoxicity, seem to vary according to the neuropathological state. Additionally, a conceptual distinction also exists between excitotoxic injury (with or without hypoxia/ischemia) and additional injury that occurs via deleterious mechanisms subsequent to hypoxia/ischemia, during reperfusion and reoxygenation[9]. Different models of glutamate excitotoxicity have been proposed, but all are characterized by increases in extracellular glutamate, and all involve intracellular ionic derangements, including increases in free cytosolic calcium.

We recently used our tissue model, which employs perfused respiring cerebrocortical slices[6] to study energy failure noninvasively during glutamate excitotoxicity with ^{31}P/^{1}H/^{19}F NMR spectroscopy[7]. We found that energy phosphates (such as ATP and PCr) decline in the very first NMR data run after exposure to toxic extracellular glutamate levels. It has long been known that cytotoxic brain edema can be observed in vitro as cell swelling within seconds after hypoxia/ischemia[8]. We therefore decided to pursue parallel non-NMR brain edema studies of respiring cerebrocortical slices, using the same excitotoxic protocols as in the NMR studies. In these parallel studies slices were removed from the perfusate at different time points, and then examined for edema via (dry-weight) / (wet-weight) water determination. The two aims were to determine, if cerebrocortical edema occurred early after glutamate exposure, and if edema correlated with noninvasive ^{31}P NMR spectroscopy determinations of energy failure.

Methods

Our protocol for obtaining brain slices was approved by the UCSF Committee on Animal Research. Details of our method for obtaining minimally injured respiring brain slices and studying these with interleaved ^{31}P/^{1}H NMR spectroscopy have been described elsewhere[6].

The experimental design for this edema study in brain slices was based on earlier, extensive NMR spectroscopy study of the dose dependence of glutamate toxicity[7]. In the NMR studies we were specifically interested in data with an extracellular glutamate concentration of 2 millimoles (mM). 2 mM glutamate is high enough to produce toxicity, but low enough to permit rescue by pharmacologic intervention. In an earlier NMR study, protection against high energy phophorus metabolite degradation was investigated by using three different glutamate antagonists: dizocilpine (150 μM), kynurenate (1 mM), and NBQX (6 μM). In the current brain edema studies, as in the previous NMR studies, slices were exposed to glutamate together with an antagonist for 60 minutes, and then perfused for 60 minutes with glutamate-free, antagonist-free, artificial cerebrospinal fluid (ACSF). Glutamate receptor antagonists were used in both studies to target synaptic molecules, and to test thereby whether synaptic glutamate receptors are responsible for energy impairment and edema. Kynurenic Acid was obtained from Sigma Chemical Company (Missouri, USA); dizocilpine (MK-801) was graciously provided by Merck-Sharpe and Dohme (New Jersey, USA), while NBQX by Novo Nordisk, Denmark.

For each of the previous NMR studies, twenty neonatal Sprague Dawley rats (wt 10–12 gms, age 7 ± 2 days) were sacrificed by rapid decapitation. Within 30 seconds the brain was removed from the cranial cavity, and four cortical slices ($\approx 350\,\mu$ thick) were prepared (39.8 ± 1.2 mg/slice), each having only one injured side of $\approx 50\,\mu^5$. The slices were obtained from the lateral surfaces of the left and right hemispheres by sliding the brain past a blade fixed $350\,\mu$ above a flat, lubricated surface[3,12]. Recovery from post-decapitation ischemia was accomplished by immediately rinsing the slices twice with freshly oxygenated ACSF, which consisted of a modified Krebs balanced salt solution containing 124 mM NaCl, 5 mM KCl, 1.2 mM KH_2PO_4, 1.2 mM $MgSO_4$, 1.2 mM $CaCl_2$, 26 mM $NaHCO_3$ and 10 mM glucose. The ACSF solution, warmed to 37°C, was in equilibrium with a 95% O_2 / 5% CO_2 gas mixture with pH of ≈ 7.4. When Mg^{2+}-free ACSF was used, the ionic composition was adjusted to an osmolarity of ≈ 300 mOsm. After rinsing the slices in fresh oxygenated ACSF, they were immediately transferred to a 20 mm Wilmad™ NMR tube having high cylindrical symmetry and minimum magnetic inhomogeneity. Once within an NMR tube, an ensemble of slices was perfused with fresh oxy-ACSF at a flow rate of 15–20 ml/min. ^{31}P NMR spectroscopy was used to monitor and assess the intracellular energy state. For this purpose the following parameters were measured – before, during, and after glutamate exposure: intracellular pH (pH_i) and the relative concentrations of adenosine triphosphate (ATP), phosphocreatine (PCr), inorganic phosphate (P_i), phosphoethanolamine (PE), lactate and N-acetyl-aspartate (NAA). Because metabolic recovery following decapitation ischemia was always seen in NMR spectroscopy to be complete

within 120 minutes, this time period was chosen as recovery period for the non-NMR brain edema experiments.

In brain edema studies presented in this paper – conducted as separate experiments without NMR measurements – 27 slices were obtained exactly as for the NMR studies, and then exposed to glutamate and antagonists according to identical protocols. Each brain slice, however, was isolated and studied in its own separate test tube, so that the inital wet weight could be correlated with the final endpoints. At each of the following time points: t = 0, 15, 30, 45, 60, 75, 90, 105 and 120 min (during glutamate and antagonist exposure and reperfusion recovery) 3 slices were removed from perfusion, blotted lightly with filter paper to quickly eliminate surface water, and transferred to pre-weighed, clean, dry culture tubes. After measuring wet weights, slices were dried at 110°C for 16 hours, and then dry weight determinations were made. Calculations were done for each slice of the *increase in wet weight,* and the final *percent (%) water content*. Data obtained of slices removed at t = 0 (no glutamate) were then used to calculate the *percent (%) tissue swelling*. Definitions of the preceding terms and equations for calculations, described previously by one of the authors[4], are as follows:

% Increase in wet weight =
$$\frac{(\textit{final wet weight} - \textit{initial wet weight}) \times 100}{\textit{initial wet weight}}$$

% Initial/final water content =
$$\frac{(\textit{initial/final wet weight} - \textit{dry weight}) \times 100}{\textit{initial/final wet weight}}$$

% Tissue swelling =
$$\frac{(\textit{\% final water content} - \textit{\% initial water content}) \times 100}{\textit{\% initial water content}}$$

Results

Brain swelling occurred immediately after glutamate exposure, as shown in Fig. 1. Except in the control group, brain swelling was always visible in the very first data acquisition, taken 15 minutes after exposure to glutamate.

At the 120 minute time point (60 minutes after discontinuation of exposure to glutamate and/or glutamate antagonists) average values of *% increase in wet weight* and *% final water content* were: 20.2 and 92.4, glutamate only; 7.6 and 92.1 for glutamate and dizocilpine; 43.6 and 91.3 for glutamate and kynurenate; 28.9 and 94.2 for glutamate and NBQX.

In Fig. 2 swelling of brain slices of this study is compared with intracellular phosphate data from separate NMR studies, taken at corresponding time points. The purpose was to study whether 60 minutes after termination of glutamate exposure, the severity of edema correlates with intracellular energy failure.

Figure 2 suggests a relationship between the intracellular high-energy phosphates, and the severity of brain swelling. For prevention of brain edema, as for

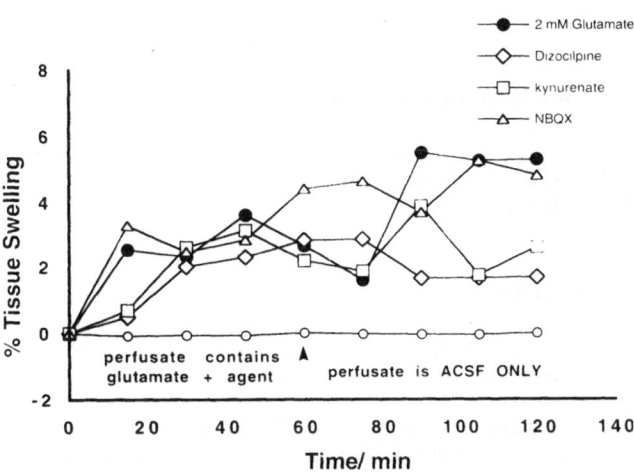

Fig. 1. Time course of *% Tissue swelling* in cerebrocortical slices. Control data refer to slices perfused with normal ACSF only. The ACSF in all other groups contained 2 mM glutamate during the first 60 minutes. In three groups, glutamate antagonists were present in the perfusate during the first 60 minutes. For all data points standard deviation corresponds to $\approx 10\%$ of the value represented on the y-axis

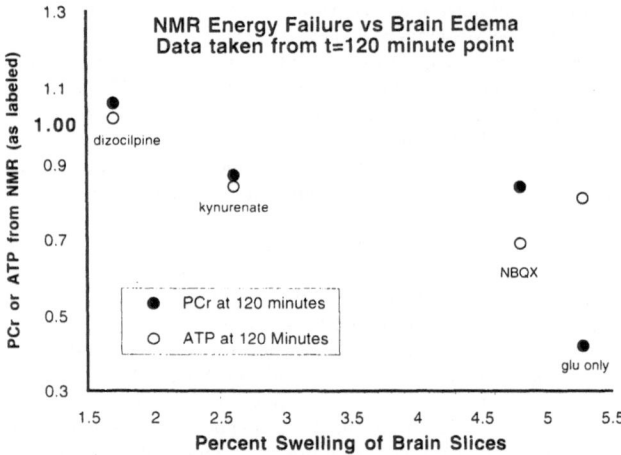

Fig. 2. Correlation of brain swelling (x-axis) and decrease of high energy phosphates studied in parallel noninvasive NMR experiments (see [7]). Standard deviations of each variable are ≈ 10% of the numerical quantity represented on corresponding axes

the prevention of PCr/ATP loss, no agent was more effective than dizocilpine. Statistically significant differences were not obtained, however, between dizocilpine and kynurenate, as to the energy loss or severity of edema. However, edema was significantly more severe in slices with NBQX, or in slices not administered with a glutamate antagonist (ie, "glutamate only" slices).

Discussion

The development of tissue swelling and water accumulation of cortical brain slices within minutes after excitotoxic exposures is in agreement with previous studies that demonstrated glutamate-induced edema in brain slices[4]. The fact that no edema occurred in control slices in parallel experiments is consistent with previous findings, indicating the integrity of the experimental setup for the NMR studies. For the NMR studies, the test tube with slices was perfused with ACSF, fed in by one hydraulic pump, and evacuated by another.

The influence of edema by the glutamate antagonists, and its apparent correlation with NMR determinations of intracellular energy failure suggests that synaptic glutamate receptor activation initiates the train of events. It has long been known that cerebral tissue swelling can be detected within seconds after toxic insults, using either light microscopy or electrical impedance measurements[8]. Although proton NMR relaxation times (T1) were seen in previous studies to increase within minutes after onset of brain edema[14],

conventional spin-echo NMR (MRI) images do not show brain edema until approximately 4 hours[1,2,10]. New "diffusion-weighted" MRI imaging methods, however, can detect cerebral edema within minutes of its onset[11,13]. In our NMR studies we tried different RF pulse repetition rates to assure that decreases in signal intensity were not attributable to edema-associated changes in ^{31}P NMR relaxation times. As an adjunct to NMR measurements, it is clear that wet-weight/dry-weight water determinations in brain slices provide as rapid an outcome measure as any other that is known – including enzymatic assays and in situ hybridization of the message of heat shock proteins or other neuronal or glial proteins.

Nevertheless, future histological procedures aided by molecularbiological and immunoassay methods are likely to be useful, as they can distinguish among glia, neurons, and endothelial cells – cell specificity of edema, as well as differential vulnerability. In the context of pharmacologic interventions, it should be possible to explore differential protection and toxicity associated with different glutamate receptor antagonists. It is conceivable that dizocilpine might reduce edema in both neurons and glia not only because of NMDA antagonism, but also because of adjunctive blocking of Na$^+$ channels[15]. We again emphasize that noninvasive NMR detection of intracellular energy failure can be made at the same rapid temporal resolution as brain edema formation. Further studies are needed to examine the extent to which cerebral edema and intracellular energy failure are causally related to each other.

Acknowledgements

NIH support was from the following grants: GM NS 34767 (LL), NS22022 (PRW), NS14543 (PHC), NS25372 (PHC), and RR03841 (TLJ).

References

1. Boisvert DPJ, Handa Y, Allen PS (1990) Proton relaxation in acute and subacute ischemic brain edema. Adv Neurol 52: 407–413
2. Brandt-Zawadski M, Pereira B, Weinstein PR, Moore S, Kucharczyk W, Berry I, McNamara M, Derugin N (1986) MR imaging of acute experimental ischemia in cats. Am J Neuroradiol 7: 7–11
3. Chan PH, Fishman RA (1978) Brain edema: induction in cortical slices by polyunsaturated fatty acids. Science 201: 358–360
4. Chan PH, Fishman RA, Lee JL, Candelise L (1979) Effects of excitatory neurotransmitter amino acids on swelling of rat brain cortical slices. J Neurochem 33: 1309–1315
5. Dingledine R (1984) Brain slices. Plenum, New York

6. Espanol MT, Litt L, Yang G-Y, Chang L-H, Chan PK, James TL, Weinstein PR (1992) Tolerance of low intracellular pH during hypercapnia by rat cortical brain slices: a ^{31}P/^{1}H NMR study. J Neurochem 59: 1820–1828

7. Espanol MT, Xu Y, Litt L, Yang G-Y, Chang L-H, James TL, Weinstein PR, Chan PK (1993) Modulation of glutamate-induced intracellular energy failure in cerebral cortical slices by kynurenic acid, dizocilpine, and NBQX. J Cereb Blood Flow Metab 13 [Suppl]: 748

8. Fishman RA (1975) Brain edema. N Engl J Med 293: 706–711

9. Ikeda Y, Long DM (1990) The molecular basis of brain injury and brain edema: the role of oxygen free radicals. Neurosurgery 27: 1–11

10. Kato H, Kogure K, Ohtomo H, Izumiyama M, Tobita M, Matsui S, Yamamoto E, Kohno H, Ikebe Y, Watanabe T (1986) Characterization of experimental ischemic brain edema utilizing proton magnetic resonance imaging. J Cereb Blood Flow Metab 6: 212–221

11. Le Bihan D, Turner R, Douek, Patronas N (1992) Diffusion MR imaging: clinical applications. Am J Roentgenol 159: 591–599

12. McIlwain H, Buddle HL (1953) Techniques in tissue metabolism. 1. A mechanical chopper. Biochem J 53: 412–420

13. Moseley ME, Kucharczyk J, Mintorovitch J, Cohen Y, Kurhanewicz J, Derugin N, Asgari H, Norman D (1990) Diffusion-weighted MR imaging of acute stroke: correlation with T2-weighted and magnetic susceptibility-enhanced MR imaging in cats. Am J Neuroradiol 11: 423–429

14. Naruse S, Horikawa Y, Tanaka C, Hirakawa K, Nishikawa H, Yosizaki K (1982) Proton magnetic resonance studies on brain edema. J Neurosurg 56: 747–752

15. Nowak LM, Wright JM (1992) Is there a role for slow voltage-dependent changes in NMDA channel open state probability? In: Simon R (ed) Excitatory amino acids. Fidia research foundation Symposium Series, Vol 9. Thieme, New York, pp 113–115

Correspondence: Lawrence Litt, Ph. D., M. D., Anesthesia Department, University of California, 521 Parnassus Avenue, San Francisco, CA 94143, U.S.A.

Acta Neurochir (1994) [Suppl] 60: 62–64

An Assessment of Progression of Brain Edema with Amino Acid Levels in Cerebrospinal Fluid and Changes in Electroencephalogram in an Adult Cat Model of Cold Brain Injury

H. Kuchiwaki, S. Inao, M. Yamamoto, K. Yoshida, and **K. Sugita**

Department of Neurosurgery, Nagoya University School of Medicine, Nagoya, Japan

Summary

We investigated the relationship between the changes of the electroencephalogram (EEG) and concentration of amino acids (AAs) in cerebrospinal fluid (CSF) using a model of cold brain injury. A cold injury was made over the motor area of anaesthetized adult cats (n = 45). The AAs in CSF from cisterna magna and in the blood were assayed by liquid chromatography. Frequency components and spike discharges/100 s in EEG were evaluated. Data were obtained before production of the lesion and every hour for 8 hours after the lesion was made. The AA-levels and EEG after the lesion was made were compared with those obtained in the controls and the sham operation group: S-group (n = 10) which were not significantly different.

Glutamate and aspartate were not detected but methionine and serine were detected in the control CSF and S-group. These AAs increased during the first 4 hours (p < 0.05) and decreased thereafter. Significant increases in spike discharge and disappearance of fast wave (p < 0.02), and increase in AAs were concurrently detected. The AAs originated from necrosis in the lesion. During the next 4 hours, increase of phenylalanine, tyrosine, and valine continued (p < 0.05). Slow wave components (p < 0.02) and precursor AAs of neurotransmitters in CSF increased in association with expansion of edema fluid.

In conclusion, our findings showed that changes in the concentration of AAs in CSF are useful indices of progression of edema associated with brain contusion.

Keywords: Cold injury; amino acid; electroencephalogram.

Introduction

The mechanism of impairment of brain function after brain contusion was assessed in the early stage of edema progression by analysing amino acid levels in cerebrospinal fluid (CSF) in relation with changes in the electroencephalogram (EEG). We hypothesized that changes of amino acid concentrations in CSF indicate changes in brain function. Experiments were made to test this concept according to the following protocol.

Materials and Methods

A cranial window was made over the motor area of adult cats (n = 55) which were anaesthetized and immobilized. Animals were separated into a cold injury group (n = 45) and a sham operated group (n = 10). A cooled (−80°C) metal bar 8 mm in diameter was placed over the dura. Amino acids in CSF were assayed by liquid chromatography, CSF was obtained from the cisterna magna using a 27G needle. Blood was also studied which was drawn from an arterial catheter. The electroencephalogram (EEG) was obtained of each hemisphere (G2: nasion, G1: parietal region). The frequency of the following wave components of the EEG was analyzed: delta: 2–3.75 Hz, theta: 4–7.75, alpha-1: 8–9.75, alpha-2: 10–12.75, beta-1: 13–19.75, beta-2: 20–25. The number of spike discharges per 100 s was recorded in addition. Data were collected before production of cold lesion, and every hour for 8 hours after the lesion. Amino acid levels and changes of the EEG after the lesion were determined and compared with the control level and findings of the sham operated group (p < 0.02–0.05). Eight of 32 amino acids which were studied in the CSF were changed in the injury group, but not significantly in the sham operation group. On the other side, EEG components were not different as compared to the control values, or to the sham operated group. The physiological parameters of these animals were within the normal limits.

Results

Amino Acids

Several amino acids were increased in CSF during the first 4 hours after production of cold injury (type I), while the remaining were increased or decreased throughout the observation period (type II).

1) Type I Amino Acid

Glutamic acid (Glu) or aspartate (Asp) were not measurable in CSF in the control period. The compounds increased rapidly, however, after lesion during one to four hours (Glu: 6.3 ± 2.6 μmol/l, Asp: 1.1 ± 0.7 μmol/l) but decreased during the next four hours thereafter. Methionine (Met: 2.8 ± 0.8 μmol/l), Serine (Ser: 31.5 ± 5.5 μmol/l) and alanine (Ala: 47.6 ± 6.8 μmol/l), however, were detected in CSF in the control phase. They increased also significantly within 4 hours. Ser was transiently decreased at 6 hours.

2) Type II Amino Acid

Phenylalanine (Phe: 11.2 ± 3.4 μmol/l) and tyrosine (Tyr: 6.1 ± 1.3 μmol/l) were present in CSF in the control phase, and increased significantly at 2 h or later periods after production of the lesion. Valine (Val: 10.5 ± 2.3 μmol/l) was also detected in the control phase, and to increase significantly at 6 hours in CSF.

EEG: Wave Components and Spike Discharge

Spike discharges increased to $30.9 \pm 12.8/100$ s (2 hours after lesion) from the control level of only $1.7 \pm 0.5/100$ s. The delta wave component (23.5–24.1% before lesion) increased significantly at 4–8 hours after injury, while the theta wave component did not change significantly throughout the experiment. The alpha (alpha-1: 9.4–8.7%, alpha-2: 7.5–7.2% before lesion) wave components were decreased at 5 hours after injury, as the beta (beta-1: 6.6–6.4%, beta-2: 1.2–0.8% before lesion) wave components, which decreased significantly at 3 or more hours after injury.

Discussion

Amino Acid Metabolism in Cold Injury Model

1) Type I Amino Acids in CSF

There are several possibilities concerning the source of origin of type I amino acids, such as Glu and Asp. One is release of the amino acids from injured tissue, another uptake through the damaged blood-brain barrier from blood into the central nervous system, and/or, loss from nerve cells by cerebral ischemia during edema formation. We believe that the excitotoxic amino acids[6] were released largely from the damaged tissue suffering from cold injury. The level of amino acids in the extracellular compartment increased by formation of the tissue necrosis which was induced by cold injury. Ala und Ser are small neutral amino acids which pass promptly through extracellular channels in the brain[1,3]. The transient increase in Ala is considered to

be a result of an excess production of pyruvate during neuronal excitation. The similar increase in Ser suggests that its metabolization to glycine was disturbed.

2) Type II Amino Acids

All type II amino acids consist of large neutral molecules. In the normal state, the level of these amino acids is adjusted by the restriction of their uptake due to transport competition through the blood-brain barrier[1,3]. Increase of type II amino acids is thought to be associated with influx and spread of edema fluid into brain parenchyma containing cytotoxic materials[2]. Injury by mediator compounds present in the edema fluid, such as free radicals may also cause impairment of neurotransmitter synthesis.

EEG and Amino Acid Metabolism

The most characteristic finding was an early increase in spike discharges. The increase in excitotoxic amino acid concentrations, as Glu and Asp occurred concomitantly with the increase in spike frequency during this period. The significant increases of inhibitory neurotransmitter-related amino acids, as Ser or Met in CSF may be suggestive of metabolic disorders of neuronal inhibitory systems.

Subsequent changes at the injured side were concerned with an increase of the delta components and a decrease of the beta-2 components. Appearance of slow waves has been reported to be induced by lesions in the white matter[4]. Slow wave activity in the present study was particularly pronounced at 4–5 hours after production of cold injury. In this period edema fluid is spreading through the white matter[5], which may impair function of the deeper structures, such as thalamus and brainstem.

Taken together, we conclude that the changes of amino acid levels in CSF are useful indices of the progression of edema resulting from brain contusion.

References

1. Betz LA, Goldstein GW, Katzman R (1989) Blood-brain-cerebrospinal fluid barrier. In: Siegel GJ, Agranoff BW, Molinoff PB (eds) Basic neurochemistry, 4th Ed. Raven, New York, pp 591–606
2. Chan PH, Longar S, Fishman RA (1987) Protective effects of liposome-entrapped superoxide dismutase on posttraumatic brain edema. Ann Neurol 21: 540–547
3. Christensen HN (1979) Developments in amino acid transport, illustrated for the blood-brain barrier. Biochem Pharmacol 28: 1989–1992

4. Gloor P, Ball G, Schaul N (1977) Brain lesion that produce delta waves in the EEG. Neurology 27: 326–333
5. Kuchiwaki H, Inao S, Nagasaka M, Andoh K, Hirano T, Sugita K (1990) The effects of cerebral hemodynamics on the progression of cold-induced oedema. Acta Neurochir (Wien) [Suppl] 51: 77–78
6. Olney JW (1969) Brain lesion, obesity, and other disturbance in mice treated with monosodium glutamate. Science 164: 719–721

Correspondence: Hiroji Kuchiwaki, M.D., Department of Neurosurgery, Nagoya University School of Medicine, 65 Tsurumai-Cho, Showa-Ku, Nagoya 466, Japan.

Acta Neurochir (1994) [Suppl] 60: 65–70
© Springer-Verlag 1994

Role of Serotonin and Prostaglandins in Brain Edema Induced by Heat Stress. An Experimental Study in the Young Rat

H. S. Sharma[1–4], **J. Westman**[2], **F. Nyberg**[3], **J. Cervos-Navarro**[1], and **P. K. Dey**[4]

[1] Institute of Neuropathology, Free University, Berlin, Federal Republic of Germany, Departments of [2] Pharmacology and [3] Human Anatomy, Biomedical Centre, Uppsala University, Uppsala, Sweden, and [4] Neurophysiology Research Unit, Department of Physiology, Institute of Medical Sciences, Banaras Hindu University, Varanasi, India

Summary

The possibility that serotonin and prostaglandins participate in edema formation following heat stress (HS) was examined in young rats. Exposure of conscious young animals (8–9 weeks old) to heat at 38°C in a biological oxygen demand (BOD) incubator (relative humidity 50–55%; wind velocity 20–25 cm/s) for 4 h resulted in marked increase in the whole brain water content (about 3%) as compared to animals kept at room temperature (21°C). A marked extravasation of Evans blue and ^{131}I-sodium occurred in the brain of heat exposed animals as compared to normal animals. Morphological examination using electron microscopy of selected brain regions of heat stressed animals showed profound cell changes. Thus perivascular edema, swollen neuronal and glial cells, membrane damage, vesiculation of myelin, axonal swelling and synaptic damage was frequent in this group of untreated animals. Pretreatment with ketanserin (a selective serotonin$_2$ receptor antagonist) or indomethacin (an inhibitor of prostaglandin synthesis) markedly reduced edema formation after 4 h HS in young animals. These heat stressed animals had considerably less extravasation of protein tracers as compared to the untreated group. Cell changes and edema at the ultrastructural level were mainly absent. Our results suggest that serotonin and prostaglandins are involved in heat stress induced breakdown of the BBB permeability, edema formation, and cell damage.

Keywords: Heat stress; edema; cell changes; blood-brain barrier.

Introduction

Heat stress (HS) is quite frequent in a majority of the world population living in temperate zones of the globe. In summer during day time, ambient temperature in this zone may reach to 35–40°C in shade and exceed even 40–45°C under the sun. Heat exposure in this environment for several hours is a serious threat especially to children whose thermoregulatory capacity is not as developed as compared to the adult[6]. There are reports that children and elderly are more susceptible to heat stress and that they may develop clinical symptoms, if suddenly exposed to this environmental condition for few hours[1].

Symptoms of HS include hyperpyrexia, delirium, mental confusion, vomiting, unconsciousness and coma[1,10]. If the body temperature exceeds 42–43°C, death may occur in more than 50% of the acute heat exposure cases[10]. Post-mortem reports generally show cerebral haemorrhage, hyperaemia, and edema in various parts of brain. Microscopical examinations revealed swollen neurones and general sponginess in many regions of the nervous system. These pathological changes appear to be more frequent in brain stem reticular formation, thalamus, and in cerebellum[10]. The mechanisms of these neuropathological alterations and of death are still obscure.

In order to evaluate the basic processes underlying cell damage and edema following HS, we developed an animal model of HS which mimics the environmental conditions of Varanasi, India during the month of June in shade[11]. This model essentially consists of the exposure of young animals to heat at 38°C for 4 h in a biological oxygen demand (BOD) incubator (relative humidity 50–55%, wind velocity 20–25 cm/s). In this model, we found a marked increase in the permeability of the blood-brain barrier (BBB), brain edema, and cell damage[11,12]. The pathophysiological changes in HS were markedly reduced by pretreatment with p-CPA (an inhibitor of serotonin synthesis) indicating that serotonin somehow plays an important role[11].

However, no single chemical compound alone can

be held responsible for any physiological or pathological reaction in vivo. Thus it seems quite likely that a spectrum of neurochemical factors is involved in the pathophysiology of brain edema and cell changes in HS[14]. There are reports that prostaglandins play a role in the pathophysiology of brain and spinal cord injury[4]. Thus, inhibition of prostaglandin synthesis before injury by indomethacin reduced edema and cell damage in the brain and spinal cord[13]. Moreover, several lines of evidence indicate a synergism between serotonin and prostaglandins in various physiological and pathological conditions, such as thermoregulation, sleep, trauma, and affective disorders[3–5,8,14,15]. Thus a possibility exists that prostaglandins and serotonin in synergy participate in the pathological processes of HS as well.

The present study was undertaken to examine whether inhibition of prostaglandin synthesis before heat exposure influences the pathophysiology of HS. Secondly, if serotonin is participating in the pathological mechanisms of HS, whether this effect is mediated via specific serotonergic receptors. To answer these questions, we pretreated separate groups of animals with indomethacin (a potent prostaglandin synthesis inhibitor) and ketanserin (a specific serotonin$_2$ receptor antagonist) and exposed them to HS. BBB permeability, brain edema and cell damage were examined in the drug treated animals and the results were compared with untreated heat exposed controls.

Materials and Methods

Animals

Experiments were carried out in 40 male Wistar rats (body weight 90–100 g, age 8–9 weeks) housed at controlled ambient temperature ($21 \pm 1°C$) with 12 h light and 12 h dark schedule. The rat food and tap water were supplied ad libitum before the experiments.

Exposure to Heat Stress

Animals were exposed to heat in a BOD incubator maintained at 38°C for 4 h. The relative humidity (50–55%) and wind velocity (20–25 cm/s) were kept constant. This experimental condition is very similar to the environmental conditions of Varanasi, India during the month of June in shade[11,12]. All stress experiments were commenced between 8:00 and 9:00 AM in order to reduce the effect of circadian variation on the stress response.

Experimental Groups

A. *Control group.* Normal animals (n = 10) maintained at room temperature ($21 \pm 1°C$) served as controls.
B. *Untreated heat exposed group.* Untreated animals (n = 10) were exposed to similar HS for 4 h.
C. *Drug treated and heat exposed groups.* In a separate group of animals ketanserin (10 mg/kg, i.p.) (n = 10) or indomethacin (10 mg/kg, i.p.) (n = 10) was administered 30 min prior to the onset of heat stress experiments. These animals were exposed to HS for 4 h.

Parameters Measured

The following parameters were measured in both heat stressed or in normal animals.

Table 1. *Stress Symptoms and Physiological Variables in Heat Stressed Animals and Their Modification by Ketanserin and Indomethacin.* Animals were exposed to heat in a BOD incubator at 38°C for 4 h. In separate groups of animals ketanserin or indomethacin were injected 30 min before heat exposure. Normal animals kept at room temperature were used as controls

Parameters measured	Control	4 h heat stressed (38°C) animals		
	n = 5	Untreated n = 5	Ketanserin treated n = 5	Indomethacin treated n = 5
Stress symptoms				
1. Rectal Temperature °C	36.86 ± 0.23	41.86 ± 0.21*	40.38 ± 0.54**	40.24 ± 0.23**
2. Salivation	nil	+++	+++	+++
3. Prostration	nil	+++	++	++
Physiological variables				
1. MABP torr	104 ± 6	76 ± 5*	78 ± 4*	80 ± 2*
2. Arterial pH	7.38 ± 0.04	7.37 ± 0.06	7.37 ± 0.08	7.37 ± 0.05
3. PaO$_2$ torr	80.34 ± 0.36	82.58 ± 0.79	81.74 + 0.82	82.34 ± 1.23
4. PaCO$_2$ torr	34.67 ± 0.33	33.34 ± 0.37	33.56 ± 0.54	33.64 ± 0.66

Values are mean ± SD, ++ = moderate, +++ = severe, * = $p < 0.05$, Significantly different from control group, ** $p < 0.05$, Significantly different from untreated heat stressed group, Student's unpaired t-test.

A. Stress symptoms. The changes in core body temperature were recorded in each animal before and after exposure to HS using a thermistor probe (inserted about 4 cm into the rectum) connected to a 12 channel telethermometer (Yellow Springfield, USA). In addition, the behavioural symptoms, such as occurrence of salivation and prostration were assessed qualitatively every 1 h during the 4 h exposure period[12].

B. Physiological variables. The mean arterial blood pressure (MABP) was recorded from an indwelling polyethylene cannula (PE 10) implanted 2 days before the experiment under ether anaesthesia and aseptical conditions. At the time of recording, the arterial catheter was connected to a pressure transducer (Statham P 23, USA) and the output was fed to a chart recorder (Electromed, UK). Immediately before sacrifice, a sample of arterial blood was withdrawn and the blood gases and pH were measured using a Radiometer apparatus (Copenhagen)[11,12].

C. Blood-brain barrier permeability. The BBB permeability was measured using Evans blue (0.3 ml/100 g of a 2% solution in physiological saline, pH 7.4) and ^{131}I-sodium (10 µCi/100 g) (Amersham, UK). Immediately after termination of HS, the animals were anaesthetised with urethane (0.8 g/kg, i.p.) and both tracers were injected into the right femoral vein through a needle puncture. The tracers were allowed to circulate for 5 min. This time and concentration of tracers is sufficient to bind to one third of serum albumin in vivo. Thus leakage of these tracers into the brain represents extravasation of a tracer-protein complex. The intravascular tracers were washed out with 0.9% saline perfusion (at 90 torr) through the heart for 30 s followed by perfusion of 4% formalin. Immediately before sacrifice, a sample of whole blood was withdrawn from the left ventricle after cardiac puncture. After perfusion, the animals were decapitated and the whole brain was taken out and examined for blue coloration. The radioactivity in the brain and the whole blood was measured according to standard protocol (energy window 500–800 KeV) in a Beckman gamma counter[11,12]. The extravasation of the tracer into the brain was expressed as percentage of the tracer concentration in the blood (CPM/mg brain over CPM/mg blood × 100).

D. Brain edema. The brain samples were dried in an oven at 90°C for 72 h. By this time the dry weight of the samples became constant. The water content calculated from the difference in the wet and dry weight was used to measure edema[11,12].

E. Morphological examination. Tissue pieces from the desired portion of brain were post fixed in osmium and embedded in epon. About one µm thick sections were cut for high resolution light microscopy and the cell changes were examined[11,12]. Ultrathin sections from the selected areas were cut using an LKB ultramicrotome (Sweden). The sections were counter-stained with uranyl acetate and lead citrate and examined under a Philips transmission electron microscope.

Statistical Analysis

The quantitative data obtained was fed to a personal computer (Apple McIntosh LC II, USA) using commercial software (Stat plus II). Unpaired Student's t-test was applied to evaluate the significance of the data obtained. A p-value less than 0.05 was considered to be significant.

Results

A. Stress Symptoms and Physiological Variables

Untreated animals subjected to 4 h HS had marked hyperthermia (41.86 ± 0.21°C) and behavioural changes (Table 1). In this group, MABP showed a mild decline (about 28 ± 2 torr), PaO_2 was increased, whereas $PaCO_2$ and arterial pH were not significantly altered from normal (Table 1). Pretreatment with ketanserin or indomethacin attenuated the hyperthermic response (about 40.31 ± 0.23°C, p < 0.01) as compared to the untreated group. However, the physiological variables were not affected by the drug treatments (Table 1).

B. BBB Permeability and Brain Edema

BBB permeability to Evans blue and radioactive iodine was increased significantly in untreated stressed animals as compared to the normal animals (Fig. 1). This increase in permeability was significantly reduced by both ketanserin and indomethacin treatment.

The water content showed a significant increase in the brain of untreated stressed animals after HS (Fig. 1). This was also markedly reduced in heat stressed animals pretreated with ketanserin or indomethacin (Fig. 1).

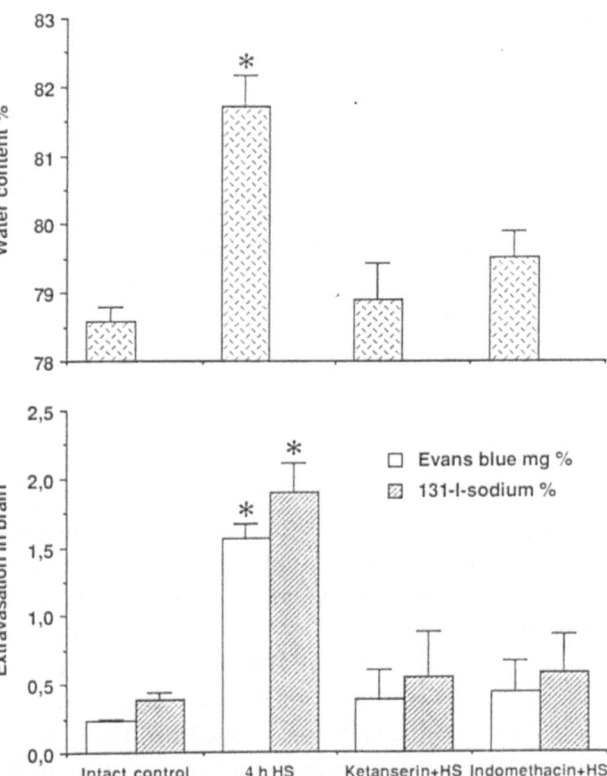

Fig. 1. Brain edema given as percent tissue water content in ml/100 g f.w. (above) and blood-brain barrier (BBB) permeability (below) in heat stressed animals and their modification by indomethacin and ketanserin pretreatment. The animals were subjected to heat stress for 4 h at 38°C in a BOD incubator. The drugs were injected in separate groups of animals 30 min prior to heat exposure. Animals kept at normal room temperature were used as controls (*p < 0.001, Student's unpaired t-test)

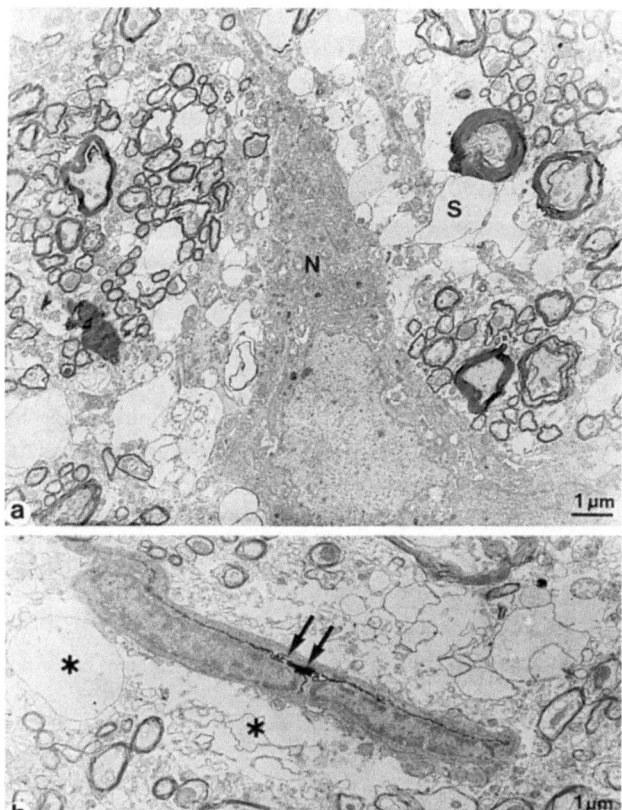

Fig. 2. Ultrastructural cell changes in the brain stem pontine region of a heat stressed animal. Dark neuron (*N*) with vacuolated cytoplasm, damage of synapse (*S*) and disruption of myelin are apparent (a). (b) A collapsed vessel with lanthanum (black particles) confined to the lumen (arrows) and perivascular edema (*) is clearly visible

C. Cell Changes

Ultrastructural examinations revealed marked cell damage in various parts of the brain in untreated stressed animals (Figs. 2 and 3). Thus, distorted neurons with vacuolated cytoplasm, damaged synapses, swollen astrocytes, and vacuolization of myelin occurred frequently (Fig. 2a). Blood vessels were mainly collapsed, perivascular edema was quite frequent (Fig. 2b). Lanthanum ions added to the fixative showed a diffuse penetration into the basement membrane, into endothelial cells, and occasionally into the neuropil (Fig. 3a).

Pretreatment with ketanserin or indomethacin markedly reduced these cell changes (Fig. 3b, c). The blood vessels appeared normal and lanthanum was mainly confined to the vessel lumen. Cell changes and edema were less evident (Fig. 3b, c).

Discussion

The present studies show that inhibition of prostaglandin synthesis by indomethacin prior to heat exposure results in a marked reduction of BBB permeability, brain edema, and cell changes. This indicates that prostaglandins are involved in the pathophysiology of HS. Furthermore, pretreatment with ketanserin (a specific serotonin₂ receptor antagonist) provided similar neuroprotection against HS. This observation suggests that the HS induced pathological changes are influenced by serotonin, probably via serotonin₂ receptors. Since both indomethacin and ketanserin treatment reduced the hyperthermic response following HS, it appears that hyperthermia directly played a role in the pathophysiology of HS. However, it is not certain whether hyperthermia per se induces cell damage, or only by other indirect mechanisms.

Pretreatment with drugs reduced the BBB permeability and edema following HS. Thus, a direct effect of hyperthermia is less likely. The BBB is a physiological dynamic barrier, which maintains a constant composition of the microenvironment of the brain[14]. Although the functional significance of BBB breakdown is still uncertain, it appears that a defective barrier will allow serum constituents to enter the cerebral compartments, which are normally protected by a tight barrier. A breakdown of the BBB may allow entrance of serum proteins into the cerebral parenchyma. This may result in an increased transport of water into the brain from blood due to changes in the osmotic gradient[11,14]. Tissue damage could then occur due to an increase in tissue pressure, local ischemia, edema and/or direct effects of various neurochemicals entering the brain because of the leaky barrier[14]. That breakdown of the BBB properties to proteins may induce vasogenic edema and cell damage is apparent from our results obtained with the drug treatments. Thus in drug treated animals with heat stress, accumulation of water, edema and cell changes were mainly absent.

The mechanisms behind the increase of BBB permeability in HS and the pathophysiological contribution of serotonin and prostaglandins are not clear. There are reports that serotonin stimulates synthesis and release of prostaglandins and vice versa[8]. Prostaglandins have the capacity to activate cAMP synthesis in cerebral vessels[2]. An increased synthesis of cAMP in the cerebral blood vessels may result in a breakdown of the BBB, probably by enhanced vesicular transport across the endothelium[9]. Cerebral vessels are capable to synthesise both prostaglandins and serotonin, and they

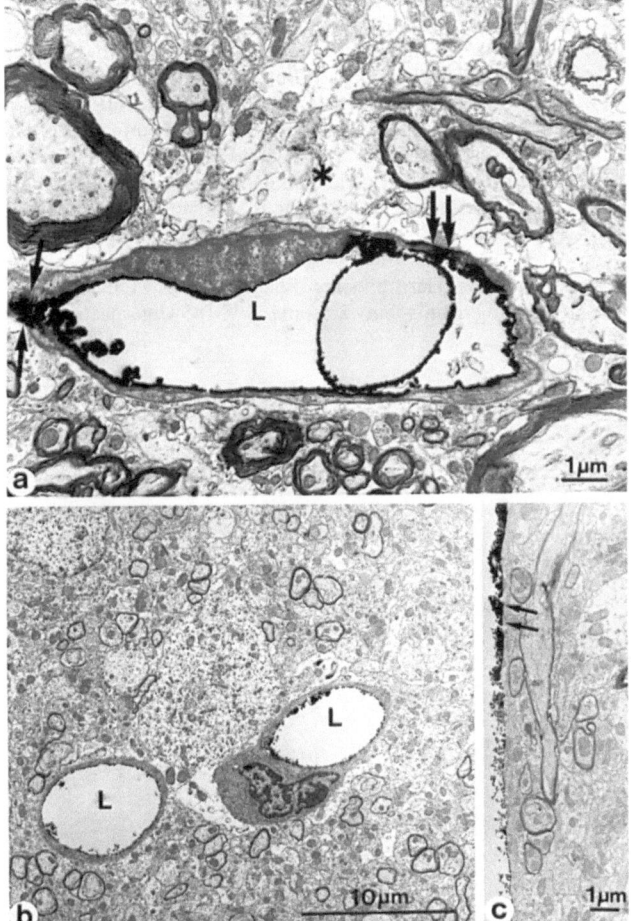

Fig. 3. Vascular permeability and cell damage in an untreated (a) and an indomethacin (b) or ketanserin (c) treated heat stressed animal. In the untreated heat stressed animal, a partially collapsed vessel shows diffuse uptake of lanthanum (black particles) into endothelial cells (arrows left side) and the basement membrane (arrows right side) (a). Perivascular edema (*) and disrupted myelin sheaths are present (a). In indomethacin (b) or ketanserin (c) treated animals, extravasation of lanthanum across cerebral endothelial cells was absent. Lanthanum is mainly confined to the lumen (L) of the vessels in the drug treated animals, and cell changes or perivascular edema is less evident (b, c)

contain receptors for serotonin and other neurotransmitters[7]. Our results obtained with ketanserin indicate that serotonin$_2$-receptors are important in mediating the BBB permeability changes in HS. Nevertheless, the results of indomethacin treatment suggest that prostaglandins are also involved in the breakdown of the BBB function. It is conceivable that serotonin through binding to serotonin$_2$ receptors induces increased synthesis of prostaglandins. The prostaglandins then stimulate cAMP synthesis in the cerebral blood vessels, causing an increased transport of material across the cerebral vessels wall.

It appears that binding of serotonin to serotonin$_2$

receptors is important for the serotonin-induced prostaglandin release and, consequently, the resulting cAMP response. This is evident by the results obtained with ketanserin, which prevented breakdown of the BBB function. However, in the case of inhibition of prostaglandin synthesis by indomethacin alone, it seems quite likely that serotonin by itself does not stimulate cAMP synthesis in cerebral vessels. The detailed mechanisms of increased transport across cerebral blood vessels in HS are not known. However, the ultrastructural findings of cerebral vessels using lanthanum as electron dense tracer in the present study suggest that transcellular transport rather than opening of junctions played an important role.

In conclusion, the current findings indicate that serotonin and prostaglandins are playing key roles in brain edema and cell damage in HS. Breakdown of the BBB in HS appears to be primarily responsible for the development of edema and the subsequent cell changes. The increased BBB permeability may result from a synergistic action of serotonin and prostaglandins at the capillary level.

Acknowledgements

This investigation is supported by grants from Alexander von Humboldt Foundation, Bonn, Germany; University Grants Commission, New Delhi, India, and Swedish Medical Research council project nos. 02710 and 9459. The skilful technical assistance of Elisabeth Scherer, Hana Plükchan, Katja Deparade, Franziska Drum, Flink Kärstin, Ingmarie Olsson and the secretarial assistance of Katherin Kern, Angela Jan and Aruna Misra are appreciated with thanks.

References

1. Amore M, Cerisoli M (1992) Heatstroke and hyperthermias. Ital J Neurol Sci 13: 337–341
2. Baca GM, Palmer GC (1978) Presence of hormonally-sensitive adenylate cyclase in capillary enriched fractions from rat cerebral cortex. Blood Vessels 15: 286–296
3. Dey PK, Feldberg W, Gupta KP, Milton AS, Wendlandt S (1974) Further studies on the role of prostaglandin in fever. J Physiol (Lond) 241: 629–646
4. Dey PK, Sharma H S (1983) Ambient temperature and development of brain edema in anaesthetised animals. Indian J Med Res 77: 554–563
5. Edvinsson E, McKenzie E T (1977) Amine mechanisms in the cerebral circulation. Pharmac Rev 28: 275-348
6. Falk B, Bar-Or O, MacDougall JD (1992) Thermoregulatory responses of pre-, mid-, and late-pubertal boys to exercise in dry heat. Med Sci Sports Exer 24: 688–694
7. Goehlert UG, Ng Ying Kin NMK, Wolfe LS (1981) Biosynthesis of prostaglandins in rat cerebral microvessels and choroid plexus. J Neurochem 36: 1192–1201
8. Haubrich DR, Perez-Cruet J, Reid WD (1973) Prostaglandin E$_1$ causes sedation and increases 5-HT turnover in rat brain. Br J Pharmacol 48: 80–87

9. Joó F, Rakonczay Z, Woolemann M (1975) cAMP mediated regulation of the permeability in the brain capillaries. Experientia 31: 582–584

10. Malamud N, Haymaker W, Custer RP (1946) Heatstroke: a clinicopathologic study of 125 fatal cases. Military Surg 99: 397–449

11. Sharma HS, Cervós-Navarro J (1990) Brain edema and cellular changes induced by acute heat stress in young rats. Acta Neurochir (Wien) [Suppl] 51: 383–386

12. Sharma HS, Kretzschmar R, Cervós-Navarro J, Ermisch A, Rühle H-J, Dey PK (1992) Age-related pathophysiology of the blood-brain barrier in heat stress. Prog Brain Res 91: 189–196

13. Sharma HS, Olsson Y, Cervós-Navarro J (1993) Early perifocal cell changes and edema in traumatic injury of the spinal cord are reduced by indomethacin, an inhibitor of prostaglandin synthesis. Acta Neuropathol (Berl) 85: 145–153

14. Wahl M, Unterberg A, Baethmann A, Schilling L (1988) Mediators of blood-brain barrier dysfunction and formation of vasogenic brain edema. J Cereb Blood Flow Metab 8: 621–634

15. Wolfe LS (1972) Prostaglandins and synaptic transmission. In: Siegel GJ, Albers RW, Katzman R, Agranoff BW (eds) Basic Neurochemistry, 2nd ed. Little, Brown, Boston, pp 263–274

Correspondence: Hari Shanker Sharma, M.D., Laboratory of Neuropathology, University Hospital, S-75185 Uppsala, Sweden.

Acta Neurochir (1994) [Suppl] 60: 71–75
© Springer-Verlag 1994

Arachidonic Acid Release and Permeability Changes Induced by Endothelins in Human Cerebromicrovascular Endothelium

D. B. Stanimirovic, N. Bertrand, R. McCarron, S. Uematsu, and M. Spatz

Stroke Branch, National Institute of Neurological Disorders and Stroke, National Institutes of Health, Bethesda, MD, U.S.A.

Summary

The vasoactive peptide endothelin-1 (ET-1) dose-dependently increased release of ^{51}Cr from human cerebromicrovascular endothelial cells (HBEC), without affecting cell viability as assessed by lactate dehydrogenase release. ET-1 also induced transient accumulation of inositol triphosphate (IP$_3$) and release of [^{3}H] arachidonic acid (AA) from HBEC. The ET-1-induced ^{51}Cr release, formation of IP$_3$, and AA release from HBEC were competitively inhibited by selective ET$_A$ subtype receptor antagonist BQ-123. ET-1-stimulated ^{51}Cr- and AA release from HBEC were potentiated by proteinkinase C (PKC) activator phorbol-myristate ester, and abolished by H7, an inhibitor of PKC. Dexamethasone, indomethacin, acetylsalicylic acid, imidazole, as well as the inhibitor of protein kinase A, H8, had no effect on ^{51}Cr release. The results suggest that ET$_A$-receptor mediated activation of PKC and increase in the HBEC 'permeability' for low molecular weight molecules in response to excessive release of endothelins from either HBEC or surrounding tissues during pathologic conditions may contribute to the formation of cerebral edema.

Keywords: Cerebromicrovascular endothelium; arachidonic acid; endothelins; permeability.

Introduction

Microvascular endothelial cells in the brain possess unique morphological and biochemical characteristics which enable them to restrict and control passage of various substances between blood and brain[16]. The blood-brain barrier (BBB) permeability for these substances can be influenced by a variety of mediators derived from either blood or the perivascular compartment (glia, neurons), or endothelium itself. It has been demonstrated that both endothelium derived relaxing factors (EDRFs, prostacyclin), and endothelium-derived contracting factors (thromboxane A$_2$, endothelins, and others) are ultimately involved in either endothelium-dependent or endothelium-independent cerebro-vasomotor responses, and possibly BBB permeability changes[8].

Endothelin-1 (ET-1), a member of a 21-amino acid peptides family of endothelins, can be derived from various cells of the cerebromicrovascular compartment (endothelium, smooth muscle)[1,17], astrocytes[12], and some neurons[6]. ET-1 is a potent and prolonged acting cerebrovasoconstrictor[10]. Increased levels of ET-1 in plasma or cerebrospinal fluid of patients with hypertension[22], stroke, head trauma[4], and subarachnoidal hemorrhage (SAH)[19] have implicated ET-1 as a mediator of cerebrovascular responses in these disorders.

In this study we demonstrate that ET-1 increases the 'permeability' of cultured human cerebromicrovascular endothelial cells for ^{51}Cr through the ET$_A$ receptor subtype-mediated activation of protein kinase C (PKC). These findings may implicate ET-1 in pathogenesis of brain edema accompanying cerebrovascular disorders.

Materials and Methods

Cell Culture

Human brain endothelial cells (HBEC) used in this study were isolated from capillaries and microvessels derived from small tissue samples of human temporal lobe removed surgically for the treatment of idiopathic epilepsy by modified method of Gerhard *et al.* (1988)[5]. These cells were identified as HBEC by positive immunocytochemical staining for Factor VIII-related antigen and incorporation of acetylated low density lipoprotein, and an absence of staining for glial cell (glial fibrillary acidic protein, galactocerebroside, and ED-2), muscle cell, and pericyte (actin and tropomysin) markers. Previous studies[1] have shown that HBEC retain their morphological and biochemical characteristics in propagated cultures up to the 15th passage.

^{51}Cr and Lactate Dehydrogenase Release

HBEC grown in 96-flat bottom well microtiter plates were prelabeled with ^{51}Cr (0.3 µCi/ml; 16 hours) in medium (Gibco M199) containing 1% of human serum, subsequently washed three times and stabilized in the same medium without radioisotope (37°C, 0.5% CO_2) for 1 hour. Cells were treated with various concentrations of endothelins in the absence or presence of other tested drugs, added to the cell medium 30 min (37°C) prior to endothelins. Both supernatants and cells (disrupted by 4 hours treatment with 0.1% Triton X-100) were counted in a gamma counter. ^{51}Cr release was expressed as the percent of ^{51}Cr released in the supernatants of total (supernatant + cells) incorporated radioactivity.

Lactate dehydrogenase (LDH) activity was determined in parallel with ^{51}Cr release in both supernatants and cell extracts by the enzymatic method described by Lowry and Passonneau[11].

[^{3}H]Arachidonic Acid Release

Release of labeled arachidonic acid (AA) in response to ET-1 was determined in [^{3}H]arachidonic acid (5 µCi/ml; 2 hours) labeled cells, thoroughly washed and replaced with physiological buffered saline (PBS). Radioactivity released into the medium after various treatments was measured by scintillation spectroscopy, consisting of both labeled AA and labeled metabolic products of arachidonic acid.

Inositol Triphosphate (IP$_3$) Formation

IP$_3$ formation was measured in [^{3}H]myoinositol prelabeled cells (2.5 µCi/ml; 16 hours) treated with endothelins in the presence of 20 mM LiCl, after separation by anion exchange chromatography (Dowex AG1X8 columns) as previously described[2].

Results and Discussion

Mechanical or metabolic (hypoxia, toxins) injury to capillary and microvascular endothelial cells comprising the blood-brain barrier may alter the normal exchange of substances and water between blood and brain. Various markers have been used to assess changes in the permeability of endothelial cells derived from peripheral vascular beds. Chopra *et al.*[3] have shown that release of ^{51}Cr from prelabeled endothelial cell cultures is a sensitive marker for reversible membrane changes accompanied with increased permeability. In contrast, larger molecular weight markers such as LDH (134 kD), commonly used for assessment of endothelial cell injury, may be retained in the cells despite significant membrane disorganization[3]. Once released, high molecular weight markers such as LDH may indicate an irreversible or lytic cell damage. Various exogenously added noxious agents, such as calcium ionophore A23187, AA (Table 1), or free radical generators xanthine/xanthine oxidase and H_2O_2[9] lead to both ^{51}Cr release, and LDH release from HBEC, indicating a substantial disruption of membrane continui-

Table 1. *Effects of Various Substances on ^{51}Cr Release and Lactate Dehydrogenase (LDH) Release from HBEC*

	^{51}Cr (%)	LDH (%)
M199	8.65 ± 0.49	5.96 ± 0.68
A23187		
1 µM	11.55 ± 0.40*	12.58 ± 0.79*
10 µM	28.02 ± 1.06**	21.15 ± 1.15**
AA		
1 µM	32.40 ± 0.51**	40.07 ± 3.43**
10 µM	65.05 ± 1.42**	68.12 ± 5.44**
PMA		
1 µM	10.65 ± 0.53*	5.20 ± 0.10
10 µM	12.23 ± 0.73**	6.01 ± 0.42
ET-1		
100 nM	12.74 ± 0.52**	5.97 ± 0.12
500 nM	13.41 ± 0.49**	5.67 ± 0.34

Values are given as means ± SEM for six replicates of a representative experiment out of three with similar results.
 * Significant difference (p < 0.01; Student's t-test) from M 199.
 ** Significant difference (p < 0.001; Student's t-test) from M 199.

ty. However, agents that primarily act to increase intracellular messengers, such as PKC stimulator, phorbol myristate ester (PMA), or receptor agonists (ET-1), induced only moderate release of ^{51}Cr, and had no effect on LDH release (Table 1). These findings indicate that ^{51}Cr release from endothelial monolayers can be a valuable test for assessment of endothelial permeability changes due to activation of various signal-transduction processes.

Using this test, we have demonstrated that ET-1, a potent cerebrovascular constrictor peptide, dose-dependently (EC$_{50}$ = 7 ± 2 nM) increased ^{51}Cr release from HBEC (Fig. 1A). The response was competitively inhibited by the selective ET$_A$ receptor antagonist BQ-123 (Fig. 1A). We have also shown that ET-1 induces a rapid increase in intracellular levels of IP$_3$ (EC$_{50}$ = 0.79 ± 0.10 nM), as well as release of labeled AA (EC$_{50}$ = 59 ± 7 nM) from HBEC (Fig. 1B, C). Both ET-1-induced IP$_3$ formation and AA release were also inhibited by BQ-123 (Fig. 1B, C).

The data indicate that HBEC express ET$_A$ subtype of receptor coupled to phospholipase C (PLC) and phospholipase A$_2$ (PLA$_2$) activation. It has previously been reported that rat cerebromicrovascular endothelium expresses both high affinity ET$_A$[21] and low affinity ET$_B$[20] receptors for ET-1. Activation of PLC upon ET-1 binding to ET$_A$ receptor in HBEC is followed by rapid hydrolysis of membrane phosphoinositides and subsequent transient formation of IP_3, which mobilizes calcium from intracellular stores, and diacylglycerol (DAG) that activates PKC. Translocated PKC is a putative activator of PLA_2. However, AA released from HBEC may have also been produced during metabolism of DAG by lipases and without a direct activation of PLA_2. Other products of phospholipid hydrolysis during this metabolic cycle may serve as classical messengers. Among them, phosphatidic acid has been shown to both mobilize calcium from intracellular stores and act as a calcium ionophore[13].

All the described ET-1-induced intracellular messengers may influence peripheral endothelial cell permeability. However, considerable differences have been found with regard to the role of different second-messengers in endothelial permeability among endothelial cells derived from various peripheral vascular beds[3,14,15]. A role for elevated intracellular calcium has been suggested by studies demonstrating effects of Ca^{2+} on endothelial cytoskeletal network[14]. The involvement of protein kinases, in particular PKC has been indicated by the ability of PKC activator phorbol esters to cause an increase in permeability for albumin and the development of intracellular gaps associated with the reorganization of F-actin filaments in endothelial cell monolayers[15]. In contrast, an increase of intracellular cyclic adenosine monophosphate (cAMP) levels by agents such as forskolin appears to be followed by a resistance to induction of permeabilization with various agonists[14]. Metabolites of AA cascade, in particular lipoxygenase products, increase endothelial cell permeability in in vitro models of BBB[9], as well as BBB permeability in brain ischemia/reperfusion[7].

It is not presently clear which of the ET-1-activated messenger systems in HBEC is primarily responsible for the observed increased permeability for ^{51}Cr. The role of PKC translocation and subsequent activation of PLA_2 in this process is strongly supported by the finding that PKC activator PMA alone increased both ^{51}Cr and AA release from HBEC and potentiated the effects of ET-1 (Fig. 2). Furthermore, PKC inhibitor H7, but not PKA inhibitor H8, prevented ET-1-induced ^{51}Cr release from HBEC and attenuated ET-1-induced AA release (Fig. 2). The effector cascade "PKC translocation $\rightarrow$ PLA_2 activation $\rightarrow$ AA release $\rightarrow$ increased ^{51}Cr permeability" is probably not the only pathway involved, since dexamethasone inhibited ET-1-induced AA release, whereas it had no effect on ^{51}Cr release from HBEC (Fig. 2). In addition, cyclooxygenase inhibitors indomethacin and acetylsalicylic acid, as well as the thromboxane synthetase inhibitor imidazole, which partially diminished ET-1-induced AA-release, also had no effect on ^{51}Cr release (Fig. 2). Although cyclooxygenase products (TxB_2, $PGF_{2\alpha}$, PGE_2, 6-keto $PGF_{1\alpha}$, and PGD_2) secreted by HBEC[18] may affect the ET-1-induced vasomotor responses of cerebromicrovascular beds, it appears that they are not directly involved in the observed 'permeability' changes of HBEC.

A role for both intracellular calcium mobilization and extracellular calcium in ET-1-induced ^{51}Cr and AA release has been suggested by the effects of ryanidine, intracellular antagonist of IP_3 receptors, and verapamil, voltage-dependent calcium channel blocker (Fig. 2). Both components of ET-1-induced intracellular Ca^{2+} "build up" can activate membrane incorporated phospholipases, and contribute to the observed AA

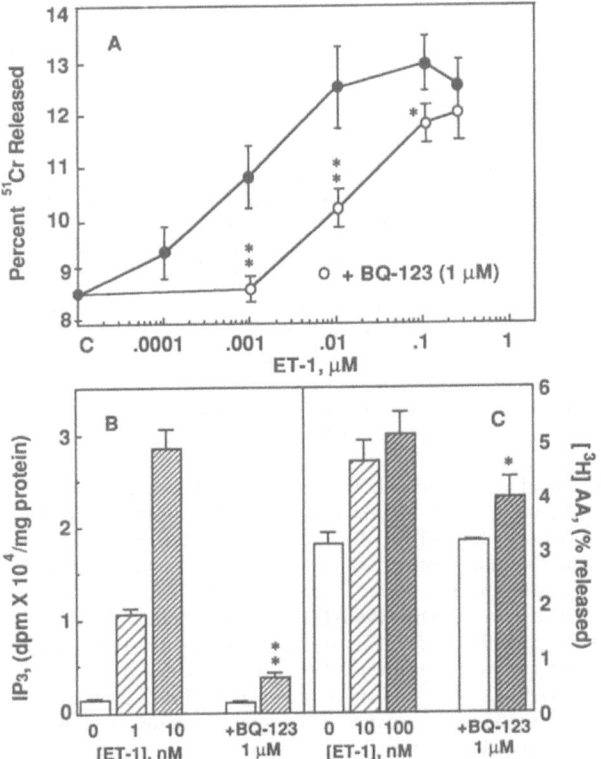

Fig. 1. (A) Concentration-dependent release of ^{51}Cr from HBEC by ET-1, and its inhibition by selective ET_A-receptor antagonist BQ-123 (1 mM). Each curve shows the data of a representative of three experiments (each point is mean ± SEM of six replicates). (B) Stimulation of IP_3 formation in HBEC by ET-1, and its inhibition by BQ-123. IP_3 was determined 10 min after ET-1 addition. (C) Stimulation of AA release from HBEC by ET-1 and its inhibition by BQ-123. AA release was measured 1 hour after ET-1 addition. Each bar represents mean ± SEM of three replicates of a representative of three experiments with similar results. * Significant difference (p < 0.01, ANOVA) from ET-1 alone

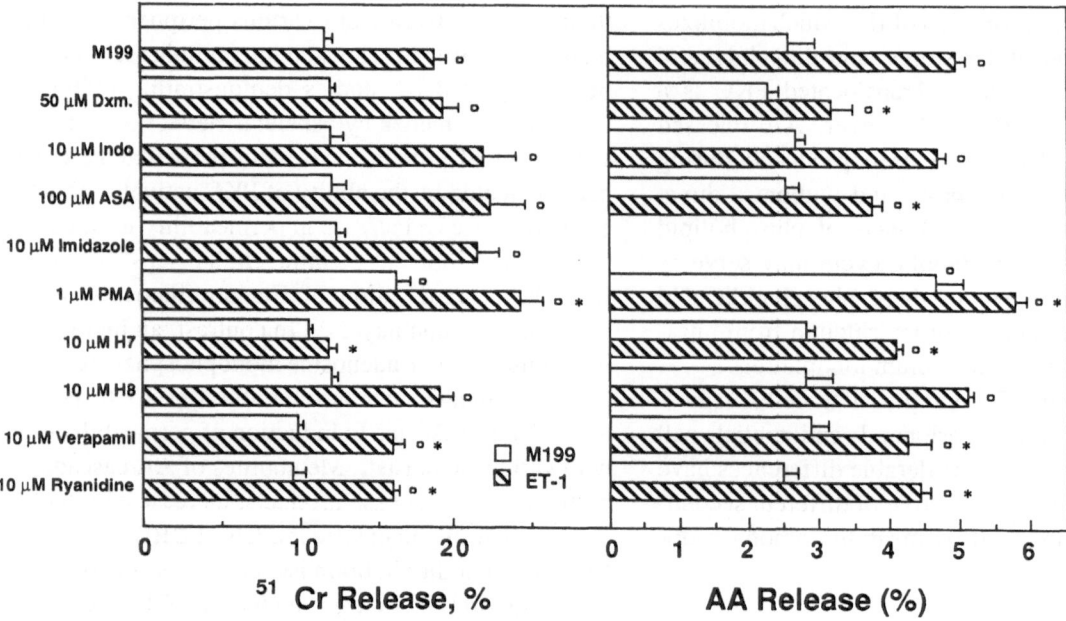

Fig. 2. Effect of various drugs (Dxm-dexamethasone, Indo-indomethacin, ASA acetylsalicylic acid, PMA-phorbol-myristate ester) on ET-1-induced ^{51}Cr release and AA release from HBEC. Drugs were added to the cells 30 min prior to ET-1 (100 nM), and ^{51}Cr and AA release were measured 1 hour after the exposure to ET-1. Each bar represents mean ± SEM of six replicates of a representative of three experiments with similar results. ° Significant difference (p < 0.01, ANOVA) from M199. * Significant difference (p < 0.01; ANOVA) from ET-1 alone

and ^{51}Cr release. It is, therefore, highly likely that multiple second messenger pathways affect the regulation of cerebromicrovascular endothelial integrity, contributing to a distinct profile of action of various agonists (mediators).

In many diseases that affect the brain, cerebral endothelium plays an active part in the disease process. An altered BBB has been reported in brain ischemia, hypertension, trauma, and neoplasia. Increased levels of ET-1 in CSF and/or plasma have been described in the same group of brain disorders[4,19,22]. The demonstrated ET-1-induced ^{51}Cr release from HBEC strongly suggests that ET-1 may contribute to the observed dysfunctions of BBB, in addition to its vasoactive effects in these diseases. The ET-1 role in altering the BBB permeability would, in part, be related to its activation of PKC in cerebromicrovascular endothelial cells through ET$_A$ type of receptor.

References

1. Bacic F, Uematsu S, McCarron RM, and Spatz M (1992) Secretion of immuno-reactive endothelin-1 by capillary and microvascular endothelium of human brain. Neurochem Res 17: 699–702
2. Berridge JM, Downes P, Hanley RM (1982) Lithium amplifies agonist-dependent phosphatidylinositol responses in brain and salivary glands. Biochem J 206: 587–595
3. Chopra J, Joist JH, Webster RO (1987) Loss of ^{51}chromium, lactate dehydrogenase, and ^{111}indium as indicators of endothelial cell injury. Lab Invest 57: 578–584
4. Ehrenreich H, Lange M, Near KA, Anneser F, Schoeller LAC, Schmid R, Winkler PA, Kehrl JH (1992) Long term monitoring of immunoreactive endothelin-1 and endothelin-3 in ventricular cerebrospinal fluid, plasma and 24-h urine of patients with subarachnoid hemorrhage. Res Exp Med 192: 257–268
5. Gerhard ZD, Broderius AM, Drewes RL (1988) Cultured human and canine endothelial cells from brain microvessels. Brain Res Bull 21: 785–793
6. Giaid A, Gibson SJ, Herrero MT, Gentleman S, Legon S, Yanagisawa M, Masaki T, Ibrahim NBN, Roberts GW, Rossi ML, Polak JM (1991) Topographical localisation of endothelin mRNA and peptides immunoreactivity in neurones of the human brain. Histochemistry 95: 303–314
7. Ikeda Y, Long DM (1990) Effects of the arachidonate lipoxigenase inhibitor BW755C on traumatic and peritumoral brain edema. Acta Neurochir (Wien) [Suppl] 51: 68–70
8. Joó F, Lengyel I, Kovacs J, Penke B (1992) Regulation of transendothelial transport in the cerebral microvessels: the role of second-messenger generating systems. Prog Brain Res 91: 177–187
9. Kempski O, Villacara A, Spatz M, Dodson RF, Corn C, Merkel N, Bembry J (1987) Cerebromicrovascular endothelial permeability. Acta Neuropath (Berl) 74: 329–334
10. Kobayashi H, Hayashi M, Kobayashi S, Kabuto M, Handa Y, Kawano H, Ide H (1991) Cerebral vasospasm and vasoconstriction caused by endothelin. Neurosurgery 28: 673–679
11. Lowry OH, Passonneau JV (1974) A flexible system of enzymatic analysis. Academic Press, New York
12. MacCumber MW, Ross CA, Snyder HS (1990) Endothelin in the brain: Receptors, mitogenesis, and biosynthesis in glial cells. Proc Natl Acad Sci USA 87: 2359–2363

13. Moolenaar WH, Kruijer W, Tilly BC, Verlaan I, Bierman AJ, DelLaat SW (1986) Growth factor-like action of phosphatidic acid. Nature 323:171–173

14. Oliver JA (1990) Adenylate cyclase and protein kinase C mediate opposite action on endothelial junction. J Cell Physiol 145: 536–542

15. Patton WF, Alexander S, Dodge AB, Patton RJ, Hecktman HB, Shepro D (1991) Mercury-arc photolysis: a method for examining second messenger regulation of endothelial cell monolayer integrity. Anal Biochem 196: 31–38

16. Rapoport SI (1976) Blood-brain barrier. In: Physiology and medicine. Raven, New York

17. Resink TJ, Hahn AWA, Scott-Burden T, Powell J, Weber E, Buhler F (1990) Inducible endothelin mRNA expression and peptides secretion in cultured human vascular smooth muscle cells. Biochem Biophys Res Comm 168: 1303–1310

18. Spatz M, Stanimirovic DB, McCarron RM (1993) Interplay between cerebro-microvascular endothelium and blood and brain cell elements. J Neurosurg, in press

19. Suzuki H, Sato S, Takekoshi K, Ishihara N, Shimoda S (1990) Increased endothelin concentration in CSF from patients with subarachnoidal hemorrhage. Acta Neurol Scand 81: 553–540

20. Vigne P, Ladoux A, Frelin C (1991) Endothelins activate Na^+/H^+ exchange in brain capillary endothelial cells via a high affinity endothelin-3 receptor that is not coupled to phospholipase C. J Biol Chem 266: 5925–5928

21. Vigne P, Marsault R, Breittmayer PJ, Frelin C (1990) Endothelin stimulates phosphatidylinositol hydrolysis and DNA synthesis in brain capillary endothelial cells. Biochem J 266: 415–420

22. Widimsky J, Horky K, Dvorakova A (1991) Plasma endothelin 1,2 levels in mild and severe hypertension. J Hypertension [Suppl] 9: S194–S195

Correspondence: Dr. Maria Spatz, Bldg. 36, Room 4D04, Stroke Branch, NINDS, NIH, 9000 Rockville Pike, Bethesda, MD 20892, U.S.A.

Acta Neurochir (1994) [Suppl] 60: 76–78

The Role of Histamine in Brain Oedema Formation

F. Joó[1], J. Kovács[2], P. Szerdahelyi[3], P. Temesvári[4], and Á. Tósaki[1]

[1] Laboratory of Molecular Neurobiology, Institute of Biophysics, Biological Research Center, Szeged, [2] Department of Pediatrics of Albert Szent-Györgyi Medical University, Szeged, [3] Central Research Laboratory, Albert Szent-Györgyi Medical University, Szeged, and [4] Department of Pediatrics, Teaching Hospital, Makó, Hungary

Summary

The effects of histamine on the cerebral endothelial cells were studied. To determine if the extent of brain oedema formation could be reduced with histamine receptor antagonists, mepyramine (H$_1$-receptor blocker), metiamide, cimetidine and ranitidine (H$_2$-receptor antagonists) were administered at a dose of 5 mg/kg body weight 4, 2 and 0 h before the onset of experimental pneumothorax induced in newborn piglets. Mepyramine and ranitidine given 2 h before the induction of EBP prevented the accumulation of water, sodium and albumin in samples taken from the parietal cortex. In other experiments, carried out on Sprague-Dawley rats of CFY strain after permanent bilateral common carotid ligation (BCCL), the accumulation of water and sodium in the ischemic brain tissue could also be prevented in a dose dependent manner by intraperitoneal injections of ranitidine given 30 min before the surgery. Taken together, these results provide pharmacological evidence for the involvement of histamine receptors in the pathogenesis of brain oedema. Consequently, the use of histamine receptor blockers both in the prevention and in the treatment of brain oedema can be recommended.

Keywords: Histamine; brain oedema; histamine receptors; antagonists.

Introduction

It is known that histamine, among other endogenous vasoactive substances, increases the permeability of peripheral capillaries, and so takes an important part in the development of oedema in the peripheral tissues. Although pharmacological data have undoubtedly suggested the sensitivity of cerebral arteries to histamine[5], there are controversial opinions in the literature as regards the hypothetized effect of histamine on brain capillaries[1,2,6,7,17,18].

The aim of our studies was to determine if histamine receptor blockers could prevent brain oedema formation.

Materials and Methods

Two animal models were applied in this study to determine the possible effects of histamine receptor blockers on ischemic brain oedema.

1. Induction of Brain Oedema in Neonatal Piglets by Cardiovascular Collapse

Thirty-six newborn piglets (either sex – birth weight: 1090–1460 g) aged between 3 and 8 h were used in the experiments. Details of the ischemic model were published earlier[4,15].

2. Induction of Ischemic Brain Oedema in Sprague-Dawley Rats of CFY Strain After Bilateral Common Carotid Artery Ligation (BCCAL)

A closed colony of randomly bred Sprague-Dawley female rats of CFY strain weighing 200–250 g was used, fed commercial food pellets and tap water. Detailed characterization of this model was published earlier[16].

Testing drug effects: To test the possible effects, different doses (1, 5, 10 and 20 mg/kg) of ranitidine were given intraperitoneally 30 minutes before BCCAL.

Measurement of tissue water and ion contents: Brain water and electrolyte contents were determined as previously described[15,16].

Measurement of Evans blue contents in the brain tissue: For the photometric determination of exuded albumin, the method of Rössner and Tempel[12] was applied.

Results

1. The Effects of Histamine Receptor Antagonists on Brain Oedema in Neonatal Piglets Induced by Cardiovascular Collapse

The effects of histamine receptor blockers on brain water, sodium and potassium content are demonstrated in Table 1. Mepyramine prevented the accumulation of

Table 1. *The Effect of Histamine Receptor Antagonists on the Brain Water, Sodium and Potassium Content 4 h After the Critical Phase*

	Water (%)	Na$^+$ (mmol/kg dry weight)	K$^+$
Control (n = 4)	83.84 ± 0.88	271.4 ± 22.4	467.2 ± 27.1
Pneumothorax only (n = 4)	86.64 ± 0.77	329.6 ± 24.0	482.8 ± 32.0
Mepyramine (n = 4)	84.40 ± 1.04*	323.0 ± 16.2	472.1 ± 35.6
Ranitidine (n = 4)	85.31 ± 0.51*	316.1 ± 12.2	496.7 ± 39.3
Metiamide (n = 4)	87.49 ± 1.92	307.5 ± 38.2	508.9 ± 41.3
Cimetidine (n = 4)	86.32 ± 1.09	321.0 ± 40.1	456.8 ± 39.9

Values are mean ± SD.
* $p < 0.01$ compared to the group "pneumothorax only".

Table 2. *The Effect of Mepyramine and Ranitidine on the Extravasation of Serum Albumin into the Brain, Measured 4 h After the Critical Phase*

		Evans Blue concentration μg/g wet weight
Control	(n = 3)	0.10 ± 0.10
Pneumothorax only	(n = 3)	0.48 ± 0.11
Mepyramine	(n = 3)	0.14 ± 0.08*
Ranitidine	(n = 3)	0.15 ± 0.03*

Values are mean ± SD.
* $p < 0.01$ compared to the group "pneumothorax only".

Table 3. *Protective Effect of Ranitidine in a Stroke Model Evoked in Sprague-Dawley CFY Rats*

	Water (%)	Na$^+$ (mmol/kg dry weight)	K$^+$
Control	81.7 ± 0.91	292.3 ± 22.4	325.1 ± 21.4
5 mg/kg ranitidine	80.4 ± 0.81**	278.2 ± 19.4*	339.4 ± 18.7
10 mg/kg ranitidine	79.4 ± 0.62***	229.4 ± 31.2***	375.7 ± 28.7***
20 mg/kg ranitidine	79.1 ± 0.53***	221.1 ± 18.9***	386.5 ± 19.8***

Values are mean ± SD.
* $p < 0.05$, ** $p < 0.01$, *** $p < 0.001$ compared to the untreated control.

water in the brain tissue but exerted no effect on the brain sodium content. Of the H$_2$-receptor antagonists used, ranitidine prevented oedematous swelling. Similarly, according to the results obtained by determining water and electrolytes, mepyramine and ranitidine also prevented leakage of serum albumin from the blood into the brain (Table 2).

2. The Effects of Histamine Receptor Antagonists on Brain Oedema in Sprague-Dawley Rats of CFY Strain Induced by BCCAL

Preischemic treatments with ranitidine protected the brain from the development of ischemic brain oedema in a dose dependent manner. Water, sodium and calcium contents of the brain could be attenuated significantly by higher doses of ranitidine (Table 3).

Discussion

The results of our studies indicate that ranitidine effectively prevents brain oedema formation. Similar effects were observed earlier[13,14] with metiamide, another histamine H$_2$-receptor antagonist, using other models.

On the basis of these findings we suppose that during tissue injury an excessive release of histamine takes place from internal (mast cells and neuronal) sources, which can activate the histamine receptors in the brain microvessels and result in the induction of brain oedema[8,9,13].

The validity of our results has been confirmed by other studies. First, the involvement of central hista-

minergic systems in the modulation of blood-brain barrier permeability[11] and the role of histamine in traumatic brain oedema have been evidenced[10]. Recently, when the transendothelial electrical resistance in brain-surface microvessels was measured in situ, a 75% decrease was observed[3] by the addition of 10^{-4} M histamine. Intraperitoneal injections of cimetidine, a potent histamine H$_2$-receptor antagonist, blocked completely the effect of histamine. It is important to note that cimetidine prevents the effect of histamine even if it is administered to the CSF-facing side of the pial vessels, providing direct evidence for the existence of abluminal H$_2$-receptors.

Prevention and treatment of brain oedema is important in many neurological conditions. The main rationale in treatment of a cerebral insult is based on the assumption that ischemically injured neurons possess an integral capacity for recovery. Therefore, further attempts should be made at revealing the molecular mechanisms underlying brain oedema formation.

Acknowledgements

The valuable work of all co-workers and collaborators, who have helped the accomplishment of this research is greatly acknowledged. The research was supported by Grants from the Hungarian Research Fund (OTKA T-915, T-5208) and the Ministry of Public Welfare (ETT- 130).

References

1. Ashton N, Cunha-Vaz JG (1965) Effect of histamine on the permeability of ocular vessels. Arch Ophtamol 73: 211–223
2. Broman T, Lindberg-Broman AM (1945) An experimental study of disorders in the permeability of the cerebral vessels ("the blood-brain barrier") produced by chemical agents. Acta Physiol Scand 10: 102–125
3. Butt AM, Jones HC (1992) Effect of histamine and antagonists on electrical resistance across the blood-brain barrier in rat brain-surface microvessels. Brain Res 569: 100–105
4. Dux E, Temesvári P, Szerdahelyi P, Nagy Á, Kovács J, Joó F (1987) Protectivce effect of antihistamines on cerebral oedema induced by experimental pneumothorax in newborn piglets. Neuroscience 22: 317–321
5. Edvinsson L, Owman Ch (1975) Pharmacologic comparison of histamine receptors in isolated extracranial and intracranial arteries in vitro. Neurology 25: 271–276
6. Földes I, Kelentei B (1954) Studies on the hematoencephalic barrier. II. The effects of histamine with special reference to the passage of antibiotics. Acta Physiol Hung 5: 149–162
7. Gross PM, Teasdale GM, Angerson WJ, Harper AM (1981) H$_2$-receptors mediate increases in permeability of the blood-brain barrier during arterial histamine infusion. Brain Res 210:396–400
8. Joó F (1986) New aspects to the function of cerebral endothelium. Nature 321: 197–198
9. Joó F, Klatzo I (1989) Role of cerebral endothelium in brain oedema. Neurol Res 11: 67–75
10. Mohanty S, Dey PK, Sharma HS, Singh S, Chansouria JPN, Olsson Y (1989) Role of histamine in traumatic brain edema. An experimental study in the rat. J Neurol Sci 90: 87–97
11. Oishi R, Baba M, Nishibori M, Itoh Y, Saeki K (1989) Involvement of central histaminergic and cholinergic systems in the morphine-induced increase in blood-brain barrier permeability to sodium fluorescein in mice. Naunyn Schmiedebergs Arch Pharmacol 339: 159–165
12. Rössner W, Tempel K (1966) Quantitative Bestimmung der Permeabilität der sogenannten Blut-Hirnschranke für Evansblau (T 1824). Med Pharmac Exp 14: 169–182
13. Sztriha L, Joó F, Szerdahelyi P (1987) Histamine H$_2$-receptors participate the formation of brain oedema induced by kainic acid in the thalamus of rat. Neurosci Lett 75: 334–338
14. Sztriha L, Joó F, Szerdahelyi P, Lelkes Z, Ádám G (1985) Kainic acid neurotoxicity: characterization of the blood-brain barrier damage. Neurosci Lett 55: 233–237
15. Temesvári P, Hencz P, Joó F, Eck E, Szerdahelyi P, Boda D (1984) Modulation of the blood-brain barrier permeability in neonatal cytotoxic brain edema: laboratory and morphological findings obtained on newborn piglets with experimental pneumothorax. Biol Neon 46: 198–208
16. Tósaki Á, Koltai M, Joó F, Ádám G, Szerdahelyi P, Leprán I, Takáts I, Szekeres L (1985) Actinomycin-D suppresses the protective effect of dexamethasone in rats affected by global cerebral ischemia. Stroke 16: 501–505
17. Westergaard E, Brightman MW (1973) Transport of proteins across cerebral arteries. J Comp Neurol 151: 17–44
18. Wolff JR, Schieweck Ch, Emmenegger H, Meier-Ruge W (1975) Cerebro-vascular ultrastructural alterations after intra-arterial infusions of ouabain, scilla-glycosides, heparine and histamine. Acta Neuropathol 31: 45–58

Correspondence: Ferenc Joó, M.D., Laboratory of Molecular Neurobiology, Institute of Biophysics, Biological Research Center, 6701-Szeged, Temesvári krt 62, P.O. Box 521, Hungary.

Acta Neurochir (1994) [Suppl] 60: 79–82

Effects of Antihistaminics on Experimental Brain Edema

L. Schilling and **M. Wahl**

Department of Physiology, Ludwig-Maximilians University, München, Federal Republic of Germany

Summary

Histamine has potent effects on cerebral blood vessels which include increased permeability and dilatation. Since its concentrations are found to be increased in brain tissue in different experimental models of brain injury, histamine may act as a mediator of secondary brain damage. Using the cold-lesion model of vasogenic brain edema the effects of application of antihistaminics were studied in rats. Neither mepyramine, an H_1 receptor blocker nor zolantidine, an H_2 blocker provided any decrease in brain swelling or water content. Experiments with application of dexamethasone yielded a small non-significant decrease of edema while the aminosteroid U74389F did not reduce swelling. The results indicate that histamine is obviously not involved in mediating cold lesion-induced brain edema. Furthermore, generation of lipid peroxides after activation of phospholipase A_2 also appears not to have a significant influence on edema in the present study.

Keywords: Cold lesion; antihistaminics; dexamethasone; aminosteroid.

Introduction

Brain blood vessels exert a rigid control of solute and water exchange between the intraluminal and the interstitial space due to the presence of the blood-brain barrier (BBB). An increase of BBB permeability is the mechanism underlying development of vasogenic brain edema, as pointed out by Klatzo[15]. Vasogenic brain edema is a threatening complication in many clinical situations such as brain tumor, incomplete ischemia, reperfusion following ischemia, injury, or inflammation. The extent of brain damage is partly determined by the direct effects of the primary insult on brain parenchyma and vasculature. In addition, it may also trigger the release and/or activation of mediator compounds which then may give rise to BBB dysfunction and development of secondary brain damage[2].

Several compounds have been suggested as mediator candidates such as bradykinin, arachidonic acid, or histamine[29]. Histamine is present in the brain in a neuronal as well as in a non-neuronal compartment which is essentially the perivascularly localized mast cells (for review see Schilling and Wahl[25]). Histamine applied from the extra- or from the intravascular side induces a nonselective increase of BBB-permeability mainly by activating H_2 receptors[8,12,24,25]. It is also a potent dilator of most cerebral arteries, including human arteries, in vitro[10,26] as well as during topical application in vivo[10,11,28]. The dilating effect of histamine is mainly due to activation of H_2 receptors on smooth muscle cells but endothelial H_1 and even H_3 receptors may also be involved[9,25,26]. Arterial dilatation may enhance the intraluminal pressure and thus secondarily enhance tracer extravasation[29]. An increase of brain histamine content or interstitial concentration has been found in a variety of brain lesion models including cold lesion[22], stab wound[19], and ischemia[1]. These results prompted us to investigate the effects of histaminergic antagonists on development of vasogenic brain edema using the cold lesion model described by Klatzo *et al.*[16].

Materials and Methods

Induction of cold lesion: Male Sprague-Dawley rats (body weight 215–350 g) were anesthetized with chloralhydrate (3.6%, 1 ml/100g ip). The subcutaneous tissue was retracted from the skull bone after a midline scalp incision. A circular craniotomy opening (6 mm diameter) was created over the left parietal cortex using a low speed dental drill. The bone flap was lifted to expose the underlying dura. Special care was taken not to damage the dura. A cold lesion was created by quickly lowering a precooled copper cylinder (5 mm diameter) 2 mm below the dural surface. The cylinder was filled with a mixture of acetone/dry ice to achieve a temperature of approximately –75°C. The contact time was 60 s. After creation of the cold lesion a local anesthetic was applied to the wound which was then closed with sutures.

The animals were brought back to their cages and allowed to awake with free access to water and food. Twentyfour hours later the rats were reanesthetized and sacrificed by heart puncture. The brains were rapidly removed and cut through the midline, and both hemispheres put into preweighed vials.

Determination of water and electrolyte content: Hemispheric and plasma water content were determined from the difference between fresh and dry weight after the samples were dried at 100°C for at least 48h. From the increase in the water content, the percentage brain swelling was calculated. Correction for unsymmetric distribution of hemispheres was performed according to a method described by von Andrian-Werburg[27].

For determination of Na+- and K+-content the dried hemispheres were carefully homogenized and a sample dissolved in 96% HNO_3. The acid was evaporated and the residue dissolved in water. The concentrations of Na+ and K+ were then measured by flame photometry and calculated in mmol/kg dry weight (dw).

Application of drugs: The animals were divided in 7 experimental groups. Animals of group I (n = 8) were sham lesioned, i.e. rats received a craniotomy and a "lesion" with the cylinder kept at room temperature. The animals of the following groups received different treatment regimens: group II (n = 9): physiological saline solution; group III (n = 8): zolantidine (0.5 mg/100 g bw), a brain permeable H_2 receptor blocker[3]; group IV (n = 9): mepyramine (0.5 mg/100 g bw), an H_1 receptor antagonist; group V (n = 9): zolantidine plus mepyramine (0.5 mg/100 g bw each); group VI (n = 7): dexamethasone (0.714 mg/100 g bw); group VII (n = 8): U74389F (0.6 mg/100 g bw). Test solutions were applied intraperitoneally every 6 hours starting 24 hours before creation of cold lesion in groups II–VI and starting 2 hours before cold lesion in group VII, respectively.

Statistics: Mean values of hemispheric water content, brain swelling and electrolyte content in the different experimental groups were compared by one-way analysis of variance followed by Duncans multiple range test. A p value < 0.05 was taken as significant. All values are given as mean ± SD.

Results

In group I (sham lesion) water content was 79.12 ± 0.26% in the left ("lesioned") and 79.13 ± 0.20% in the right (control) hemisphere. Sodium and potassium concentrations were 238.4 ± 5.2 and 438.0 ± 18.5 mmol/kg dw in the left vs 242.6 ± 5.2 and 442.8 ± 17 mmol/kg dw in the right hemisphere. Cold lesions induced a pronounced increase of water content in the left hemisphere and a small increase in the opposite side indicative of some transit of edema fluid into the contralateral hemisphere. The hemispheric water and electrolyte content obtained in groups II–VII are listed in Table 1.

Since there was no reduction of water content in the traumatized hemisphere in any of the antihistaminic treatment groups we tried to reduce edema by application of a glucocorticoid (dexamethasone) and of an aminosteroid (U74389F). The values for the hemispheric water and electrolyte content are given in Table 1 and the calculated values for brain swelling in experimental groups II–VII are depicted in Fig. 1. There was a statistically significantly lower K+ content in the left hemisphere of group VII than in controls (group II), the reason for which is unclear.

Discussion

The rationale for this study was based on the potent permeability-enhancing and dilatory effects of histamine in cerebral blood vessels and its increased concentration in cerebral tissue under different experimental conditions of brain damage (see Introduction). Furthermore, indirect evidence for an involvement of histamine in the pathogenesis of vasogenic brain edema came from results obtained by Raymond et al.[23]. These authors found after cold lesion a decrease of horseradish peroxidase extravasation produced by imidazole, a compound with H_2 receptor blocker properties. In the present study, however, no edema-reducing effect could be found by treatment with either mepyramine or zolantidine, H_1 and H_2 receptor blocking agents. In order to optimize the experimental conditions for detection of any therapeutic effect, the animals were pretreated with high concentrations of the drugs (although they were still in the range recommended for therapeutic use in humans[7]). Insufficient blockade of histamine receptors therefore appears not to be the reason for the inefficacy of antihistaminic treatment for reduction of experimental brain edema. Our negative results appear to be in contrast with results obtained by Csanda[4] who found that cimetidine (another H_2 receptor blocker) decreased brain edema

Table 1. *Mean Hemispheric Water and Electrolyte Content in Experimental Groups II–VII.* Given are mean ± SD

		Group II	Group III	Group IV	Group V	Group VI	Group VII
Water content (%)	left	81.49 ± 0.43	81.87 ± 0.36	81.75 ± 0.52	81.81 ± 0.32	81.16 ± 0.19	81.92 ± 0.34
	right	79.63 ± 0.53	79.57 ± 0.37	79.64 ± 0.33	79.66 ± 0.33	79.24 ± 0.12	79.84 ± 0.50
Na+ content mmol/kg dw	left	386.7 ± 24 5	376.5 ± 25.7	376.8 ± 34.8	359.6 ± 22.7	361.7 ± 28.8	393.2 ± 12.2
	right	272.2 ± 22.9	244.9 ± 18.5	256.1 ± 16.0	246.7 ± 14.5	252.7 ± 17.4	273.8 ± 19.0
K+ content mmol/kg dw	left	411.7 ± 37.5	385.8 ± 12.3	392.4 ± 27.0	387.1 ± 9.9	396.5 ± 17.4	366.7 ± 18.4
	right	445.0 ± 17.8	431.4 ± 11.3	444.2 ± 15.6	432.4 ± 9.8	435.1 ± 13.9	395.2 ± 24.8

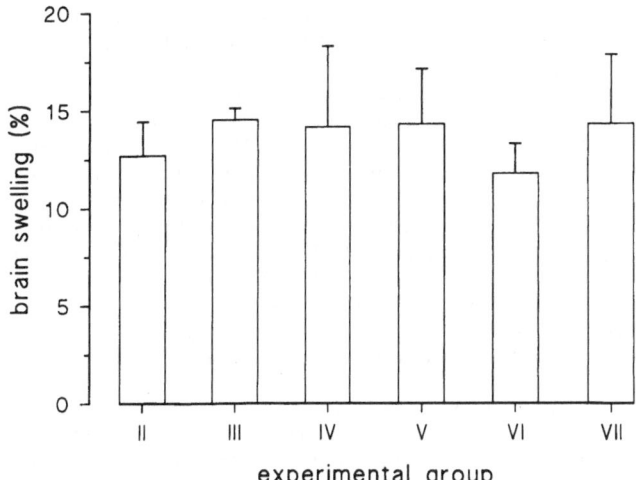

Fig. 1. Cold lesion-induced brain swelling in experimental groups II–VII. Although there was a tendency towards a decrease of brain swelling in animals treated with dexamethasone (group VI) this effect was too small to reach significance

associated with radiation. Similarly, Mohanty et al.[19] found an edema-reducing effect by cimetidine in a stab wound model of brain injury. The reason for the discrepancies in these studies is not clear but may at least in part be related to pathogenetic differences between the different models.

Since antihistaminics did not reduce cold lesion-induced edema in the present study, experiments were performed with application of dexamethasone. This glucocorticoid compound has been found to exert a positive effect (i.e. reduction of edema) in a number of studies using cold-lesion injuries[17,18,21,27] although studies reporting no therapeutic effects have also been published[6,20]. The effects of dexamethasone are mediated by inducing de novo synthesis of a protein called lipocortin[5]. In order to allow for sufficient time for protein synthesis a 24-h pretreatment period was allowed in the animals receiving dexamethasone. There was no clear-cut difference in water increase of the traumatized hemisphere although there was a tendency towards reduction of brain swelling. This may indicate that in the present study activation of phospholipase A_2 which is inhibited by dexamethasone[5] did not play a pivotal role in edema development. However, dexamethasone may exert additional protecting effects which are unrelated to phospholipase inhibition[30].

A new approach for the treatment of different brain disorders may be application of non-glucocorticoid aminosteroids. These substances act by lipid peroxide scavenging and have been found to exert protective effects in a number of experimental models of brain injury[13,14]. In the present study the compound U74389F was used which has similar properties to those of U74006F, the first compound of this class described. Pretreatment of animals with U74389F was only 2 h in contrast to dexamethasone and antihistaminics which should nevertheless be long enough to provide full therapeutic efficacy. In the present study U74389F had no effect on water accumulation or brain swelling. This is in accordance with the small effect seen with dexamethasone and may thus indicate that generation of lipid peroxides which occurs in the course of arachidonic acid metabolism is of minor importance under the conditions of the present study.

Acknowledgement

The authors thank Mrs. H. M. Hummelberger-Beck and Miss K. Lenke for expert technical assistance and Mrs. E. Held for secretarial help with the preparation of the manuscript. We wish to thank the following companies for providing drugs: Smith Kline Beecham (Frythe, England) for zolantidine, E. Merck (Darmstadt, Germany) for dexamethasone and the Upjohn Company (Kalamazoo, MI, USA) for U74389F. This study was supported by a grant from the Wilhelm-Sander Foundation.

References

1. Adachi N, Itoh Y, Oishi R, Saeki K (1992) Direct evidence for increased continuous histamine release in the striatum of conscious freely moving rats produced by middle cerebral artery occlusion. J Cereb Blood Flow Metab 12: 477–483
2. Baethmann A (1987) Mechanisms of secondary brain damage. In: Cohadon F, et al (eds) Traumatic brain edema. Liviana, Padova, pp 81–85
3. Calcutt CR, Ganellin CR, Griffiths R, Leigh BK, Maguire JP, Mitchell RC, Mylek ME, Parsons ME, Smith IR, Young RC (1988) Zolantidine (SK & F 95282) is a potent selective brain-penetrating histamine H_2-receptor antagonist. Br J Pharmacol 93: 69–78
4. Csanda E (1980) Radiation brain edema. In: Cervós-Navarro J et al (eds) Brain edema. Adv Neurol 28: 125–146
5. Davidson FF, Dennis EA (1989) Biological relevance of lipocortins and related proteins as inhibitors of phospholipase A_2. Biochem Pharmacol 38: 3645–3651
6. Dick AR, McCallum ME, Maxwell JA, Nelson SR (1976) Effect of dexamethasone on experimental brain edema in cats. J Neurosurg 45: 141–147
7. Douglas WH (1985) Histamine and 5-hydroxytryptamine (serotonin) and their antagonists. In: Goodman AG et al (eds) The pharmacological basis of therapeutics. MacMillan, New York, pp 605–638
8. Dux E, Joó F (1982) Effects of histamine on brain capillaries. Fine structural and immunohistochemical studies after intracarotid infusion. Exp Brain Res 47: 252–258
9. Ea Kim L, Javellaud J, Oudart N (1992) Endothelium-dependent relaxation of rabbit middle cerebral artery to a histamine H_3-agonist is reduced by inhibitors of nitric oxide and prostacyclin synthesis. Br J Pharmacol 105: 103–106
10. Edvinsson L, Gross PM, Mohamed A (1983) Characterization of histamine receptors in cat cerebral arteries in vitro and in situ. J Pharmacol Exp Ther 225: 168–175

11. Gross PM, Harper AM, Teasdale GM (1981) Cerebral circulation and histamine: 2. Responses of pial veins and arterioles to receptor agonists. J Cereb Blood Flow Metab 1: 219–225

12. Gross PM, Teasdale GM, Graham Dl, Angerson WJ, Harper AM (1982) Intra-arterial histamine increases blood-brain transport in rats. Am J Physiol 243: H307–H317

13. Hall ED, Braughler JM, Yonkers PA, Smith SL, Linseman KL, Means ED, Scherch HM, von Voigtlander PF, Lahti RA, Jacobsen EJ (1991) U-78517F: a potent inhibitor of lipid peroxidation with activity in experimental brain injury and ischemia. J Pharmacol Exp Ther 258: 688–694

14. Hall ED, Travis MA (1988) Inhibition of arachidonic acid-induced vasogenic brain edema by the non-glucocorticoid 21-aminosteroid U74006F. Brain Res 451: 350–352

15. Klatzo I (1967) Presidential address. Neuropathological aspects of brain edema. J Neuropathol Exp Neurol 26: 1–14

16. Klatzo I, Piraux A, Laskowski EJ (1958) The relationship between edema, blood-brain barrier and tissue elements in a local brain injury. J Neuropathol Exp Neurol 17: 548–564

17. Meinig G, Deisenroth K (1991) Dose- and time-dependent effects of dexamethasone on rat brain following cold-injury oedema. Acta Neurochir (Wien) [Suppl] 51: 100–103

18. Meinig G, Schilling L, Khalifa A (1984) Wirkung von Di-Natrium-Dexamethason-Phosphat und Di-Kalium-Triamcinolon-Azetonid-Phosphat auf das Hirnödem nach Kälteläsion. In: Grumme T (ed) Das Hirnödem. Intensivmedizinische Probleme in der Neurochirurgie. De Gruyter, Berlin, pp 33–38

19. Mohanty S, Dey PK, Sharma HS, Singh S, Chansouria JPN, Olsson Y (1989) Role of histamine in traumatic brain edema. An experimental study in the rat. J Neurol Sci 90: 87–98

20. Nelson SR (1974) Effect of drugs on experimental brain edema in mice. J Neurosurg 41: 193–199

21. Neuenfeldt D, Herrmann H-D, Loew F (1974) The influence of dexamethasone on the blood-brain-barrier and water content in experimental brain oedema. Acta Neurochir (Wien) 30: 51–57

22. Orr EL (1988) Cryogenic lesions induce a mast cell-dependent increase in cerebral cortical histamine levels in the mouse. Neurochem Pathol 8: 43–51

23. Raymond JJ, Robertson DM, Dinsdale HB, Nag S (1984) Pharmacological modification of blood-brain barrier permeability following a cold lesion. Can J Neurol Sci 11: 447–451

24. Schilling L, Ksoll E, Wahl M (1987) Effects of histamine on vasomotor response and permeability of extraparenchymal cerebral vessels. J Cereb Blood Flow Metab 7 [Suppl 1]: S506

25. Schilling L, Wahl M (1993) Histaminergic effects on cerebral hemodynamics. In: Phillis JW (ed) The regulation of cerebral blood flow. CRC, Boca Raton, pp 113–128

26. Toda N (1990) Mechanism underlying responses to histamine of isolated monkey and human cerebral arteries. Am J Physiol 258: H311–H317

27. von Andrian-Werburg U (1992) Das experimentelle vasogene Hirnödem der Ratte. Behandlungseffekte von Dexamethason, Eglin c, Glutamat und Phosphoserin. Dissertation, Universität München

28. Wahl M, Kuschinsky W (1979) The dilating effect of histamine on pial arteries of cats and its mediation by H_2 receptors. Circ Res 44: 161–165

29. Wahl M, Unterberg A, Baethmann A, Schilling L (1988) Mediators of blood-brain barrier dysfunction and formation of vasogenic brain edema. J Cereb Blood Flow Metab 8: 621–634

30. Wolfe LS, Pappius HM (1983) Involvement of arachidonic acid metabolites in functional disturbances following brain injury. In: Samuelsson B et al (eds) Advances in prostaglandin, thromboxane, and leukotriene research, vol 12. Raven, New York, pp 345–349

Correspondence: L. Schilling, M. D., Department of Physiology, University of Munich, Pettenkoferstrasse 12, D-80336 München, Federal Republic of Germany.

Acta Neurochir (1994) [Suppl] 60: 83–85
© Springer-Verlag 1994

Causative Role of Lysosomal Enzymes in the Pathogenesis of Cerebral Lesions Due to Brain Edema Under Chronic Hypertension

E. Yamada, C. H. Chue, N. Yukioka, and **F. Hazama**

Department of Pathology, Shiga University of Medical Science, Seta Tsukinowa-cho, Otsu, Japan

Summary

In order to clarify the role of lysosomal enzymes in the developmental mechanisms of cerebral lesions under chronic hypertensive conditions, we histochemically and biochemically investigated acid phosphatase, N-acetyl-β-glucosaminidase, and cathepsin B in the cerebral cortex and subcortical white matter in stroke-prone spontaneously hypertensive rats (SHRSP). Histochemical investigation showed that SHRSP had an increased number of cells with positive reaction to these enzymes in the edematous cortex and degenerated subcortical white matter. The cells with positive reaction were made up of reactive astrocytes and microglias. The activities of all enzymes in the aged SHRSP were higher than those in normotensive rats, the differences being significant at 24 weeks of age. The present study suggests that chronic hypertension or chronic edema causes increased activities of lysosomal enzymes in the cerebral cortex and subcortical white matter, and that the activated lysosomal enzymes take part in the developmental mechanisms of cystic formation as well as the diffuse degeneration of the white matter.

Keywords: Hypertension; brain edema; lysosomal enzyme.

Introduction

One of the most prominent cerebral lesions in the stroke-prone spontaneously hypertensive rats (SHRSP) is necrosis with cyst formation in the cortex and subcortical white matter[5,6,8,9]. This type of lesion resembles the lacunas observed in the brain in chronic hypertension[8]. And associated with such lacuna-like cyst formation, we often encounter the diffuse degeneration of subcortical white matter, changes which resemble the findings of Binswanger disease. In the case of cerebral infarction, macrophages play the most important role in cyst formation. However, macrophages are not so abundantly found in most of the cystic necrotic lesions in SHRSP. It is generally recognized that the major mechanism for tissue digestion, by either autolysis or heterolysis, is attributable to lysosomal enzyme activity[3,4,11,13]. The purpose of the present study was to obtain information about the changes of lysosomal enzymes in cerebral parenchyma in chronic hypertension in order to clarify their role in the development of the lacuna-like cyst formation and Binswanger-like diffuse degeneration of white matter.

Therefore we studied the activities of three lysosomal enzymes, i.e. acid phosphatase (AcPase), N-acetyl-β-glucosaminidase (NAGase) and cathepsin B (CathB) in the cerebral cortex and subcortical white matter in the SHRSP brains, enzyme-histochemically the activities of AcPase and NAGase, and immunohistochemically the location of CathB.

Materials and Methods

Sixty male SHRSP aged 8–24 weeks and age-matched normotensive Wistar rats (WKY) were used. AcPase, NAGase and CathB activities and DNA concentration were assayed with a modification of the method of Leaback and Walker[10]. All enzyme activities were expressed as unit/mg DNA. A comparison of each group was made using Student's t-test. AcPase and NAGase were enzyme-histochemically investigated with the methods of Barka-Anderson[1] and Hayashi[7], respectively. CathB was immunohistochemically studied. To identify cell species, GFAP and lectin *Ricinus communis agglutinin-1* (RCA-1) were employed as markers of astrocytes and microglia, respectively.

Results and Discussion

Intense AcPase and NAGase activities were histochemically observed in the cytoplasm of perineuronal glial cells in spongy regions of cerebral cortex of SHRSP (Fig. 1). CathB was also demonstrated in perineuronal glial cells, which proved to be microglial, because most of them were RCA-1 positive. Another change in cerebral cortex of SHRSP is massive accumulation of glial cells around fibrin deposition which often contains necrotized microvessels[2]. Such lesions

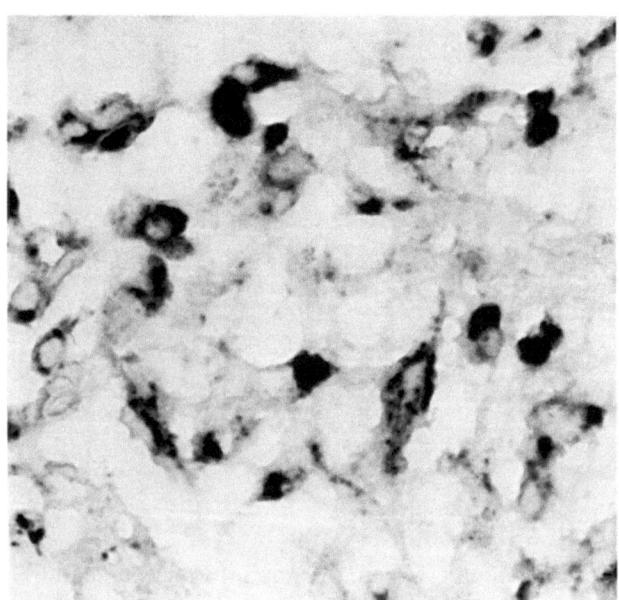

Fig. 1. Intense AcPase activity as histochemically observed in the cytoplasm of perineuronal glial cells in spongy regions of cerebral cortex of SHRSP. × 400

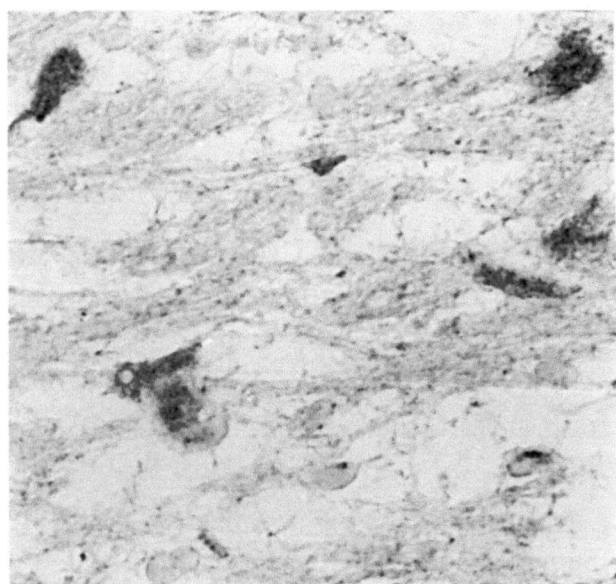

Fig. 2. Immunohistochemical micrograph of CathB in the subcortical white matter of a SHRSP. Note lytic changes of the parenchyma around the increased astrocytes with cathepsin B immunoreaction. × 400

consisted of a large number of reactive astrocytes and microglia, all of which showed intense enzyme activities of AcPase, NAGase and immunoreactivity of Cath B. These lysosomal enzyme activities in the cerebral cortex in WKY and SHRSP tended to increase with advancing age. The activities of these enzymes in SHRSP had a tendency to be higher than those in WKY, the difference being significant at 24 weeks (n = 10, p < 0.05).

In the edematous or rarefied subcortical white matter of SHRSP, an increased number of glial cells with intense AcPase and NAGase activities and CathB immunoreactivity (Fig. 2) was observed. The cells were mainly activated and swollen astrocytes which were GFAP positive. The lysosomal enzyme activities examined in the white matter, both in WKY and SHRSP were lower than those in cerebral cortex. Activities of AcPase and NAGase in SHRSP increased progressively with advancing age. The activities were significantly higher than in WKY at 24 weeks (n = 10, p < 0.01). CathB activities in SHRSP were higher than in the controls at all ages with the difference being significant at 24 weeks (n = 10, p < 0.02). This enzyme has been shown to degrade myelin constituents[12]. Thus, CathB seems to contribute to the diffuse degeneration of myelin in the subcortical white matter in SHRSP.

The findings suggest that chronic edema due to hypertension causes an increased activity of lysosomal enzymes attributable to accelerated endocytosis of edema fluid[3] into reactive glial cells in cerebral cortex and subcortical white matter, and that the activated enzymes take part in the mechanisms of formation of cystic lesions as well as the diffuse degeneration of white matter in SHRSP.

References

1. Barka T, Anderson PJ (1962) Histochemical methods for acid phosphatase using hexazonium pararosanilin as coupler. J Histochem Cytochem 10: 741–753
2. Chue C-H, Yukioka N, Yamada E, Hazama F (1993) The possible role of lysosomal enzymes in the pathogenesis of hypertensive cerebral lesion in spontaneously hypertensive rats. Acta Neuropathol (Berl) 85: 383–389
3. Dannenberg AN, Wabier PC, Kapral FA (1963) A histochemical study of phagocytic and enzymatic functions of rabbit mononuclear and polymorphonuclear exudate cells and alveolar macrophages. J Immunol 90: 448–465
4. DeDuve C (1959) Lysosomes, a new group of cytoplasmic particle. In: Hayashi T (ed) Subcellular particles. Ronald, New York, pp 128–159
5. Fredriksson K, Kalimo H, Nordborg C, Johansson BB, Olsson Y (1988) Nerve cell injury in the brain of stroke-prone spontaneously hypertensive rats. Acta Neuropathol (Berl) 76: 227–237
6. Fredriksson K, Kalimo H, Nordborg C, Olsson Y, Johansson BB (1988) Cyst formation and glial response in the brain lesions of spontaneously hypertensive rats. Acta Neuropathol (Berl) 76: 441–450
7. Hayashi M (1965) Histochemical demonstration of N-acetyl β glucosaminidase employing naphthol AS-BI N-acetyl β glucosaminide as substrate. J Histochem Cytochem 13: 355–360

8. Hazama F, Ooshima A, Tanaka T, Tomomoto K, Okamoto K (1975) Vascular lesions in the various substrains of spontaneously hypertensive rats and effects of chronic salt ingestion. Jpn Circ J 39: 7–22

9. Hazama F, Amano S, Haebara H, Yamori Y, Okamoto K (1976) Pathology and pathogenesis of cerebrovascular lesions in spontaneously hypertensive rats. In: Cervós-Navarro J, Betz E, Matakas F, Wuellenweber R (eds) The cerebral vessel wall. Raven, New York, pp 245–252

10. Leaback DH, Walker PG (1961) Studies of glucosaminidase. 4. The fluorimetric assay of N-acetyl-β-glucosaminidase. Biochem J 78: 151–156

11. Yamada E, Hazama F, Amano S, Hanakita J (1980) Cytochemical investigation of acid phosphatase activity in cerebral arteries in spontaneously hypertensive rats. Jpn Circ J 44: 467–475

12. Yanagisawa K, Sato S, Miyatake T, Kominami E, Katsunuma N (1984) Degradation of myelin proteins by cathepsin B and inhibition by E-64 analogue. Neurochem Res 9: 691–694

13. Yukioka N, Yamada E, Sasahara M, Kawai J, Hayase Y, Amano S, Hazama F (1988) Lysosomal enzyme activities in the cerebral microvessels in spontaneously and renal hypertensive rats. Exp Mol Pathol 49: 111–117

Correspondence: Eiji Yamada, M.D., Department of Pathology, Shiga University of Medical Science, Seta-Tsukinowa-cho, Otsu 520-21, Japan.

Acta Neurochir (1994) [Suppl] 60: 86–88
© Springer-Verlag 1994

A Novel Aspect of Thrombin in the Tissue Reaction Following Central Nervous System Injury

A. Nishino[1], M. Suzuki[1], T. Yoshimoto[1], H. Otani[2], and H. Nagura[2]

[1] Division of Neurosurgery, Institute of Brain Diseases, Tohoku University School of Medicine and [2] Second Department of Pathology, Tohoku University School of Medicine, Sendai, Japan

Summary

Thrombin and two different types of control solutions (buffer and albumin) were continuously infused into the rat caudoputamen by an osmotic minipump. Routine histological studies with immunohistochemistry using antibodies for BrdU, GFAP, vimentin and laminin were carried out to assess infiltration of inflammatory cells, formation of edema, cell proliferation, and reactivity of astrocytes and mesenchymal cells. The number of inflammatory cells, number of BrdU positive cells, area and number of vimentin positive astrocytes, and the area of GFAP-positive reactive astrocytes were quantitatively analyzed. In the thrombin group, pale tissue foci due to spongiosis were observed together with infiltration of inflammatory cells, proliferation of mesenchymal cells, and increase of vimentin positive astrocytes which was significantly different from the control groups. The results suggest that thrombin plays an important role in inflammation, brain edema and reactive gliosis following CNS injury.

Keywords: Brain edema; inflammation; reactive gliosis; thrombin.

Introduction

Thrombin is a multifunctional molecule. Besides acting as a coagulation factor, other functions are i) induction of mitosis in many cell types[9]; ii) chemotaxis of macrophages, monocytes, and neutrophils[1] and iii) increase of vascular permeability[4]. In central nervous system (CNS), thrombin i) facilitates mitosis in rat astroblasts[9] and human retina glial cells[10], ii) causes morphological changes of cultured rat astrocytes[8], and iii) causes neurite retraction[5].

Our preliminary studies in cryogenic CNS injury suggested that thrombin is extravasating into the brain after blood-brain barrier (BBB) breakdown (in preparation). These observations led us propose a working hypothesis that thrombin permeating after BBB breakdown provokes inflammation, brain edema and reactive gliosis in the brain parenchyma. To test this hypothesis, we developed an in vivo model with continuous intracerebral infusion of thrombin using an osmotic minipump followed by histological examinations.

Materials and Methods

1. Operative Technique

Male Sprague-Dawley rats weighing 250–300 grams were anesthetized with Nembutal (50 mg/kg, i.p.). Human thrombin, solution buffer or human serum albumin were continuously infused into the center of the right caudoputamen by an osmotic minipump (Alzet 2002), which was buried subcutaneously. Solutions were infused at 0.5 µl/h for 7 days until sacrifice.

2. Drug Infusion

Human thrombin (Sigma T6759) was administered at doses of 125 U/ml or 1000 U/ml, respectively in 10 or 11 rats. In control groups (10 animals) thrombin solvent (0.15 M sodium chloride and 0.05 M sodium citrate, pH 6.5) was administered, while 6 animals received 300 µg/ml of human albumin (Sigma A3782) having the same protein concentration as 1000 U/ml thrombin.

3. Histological Examination of Brain Tissue

Bromodeoxyuridine (BrdU; Sigma B5002) 100 µg/kg was injected into the femoral vein at the 7th postoperative day. One hour later the brain was subjected to perfusion fixation by 4% paraformaldehyde solution followed by its removal from the skull. The brain tissue was embedded then in paraffin blocks. In addition to HE staining, antibodies were used for immunohistochemical studies against: GFAP (Dako polyclonal, 1 : 25000), vimentin (Dako monoclonal, 1 : 20), laminin (Chemicon polyclonal, 1 : 500), BrdU (Amersham, 1 : 4).

4. Quantification and Statistical Analysis of Histological Changes

The number of vimentin positive astrocytes and of BrdU positive cells was counted in 12 regions (50 μm × 50 μm) in the center of the caudoputamen, where the tip of the infusion cannula was positioned. The number of neutrophils was counted likewise. Areas with GFAP or densely populated vimentin-positive astrocytes were measured by an image analyzer (Nicon V16 and Hewlett Packard 9816). To maintain constant conditions the sections used for the histological study were always within 200 μm from the center of the needle track. The area within 50 μm of the needle track was excluded to eliminate effects of tissue injury. All data were statistically analysed by employment of the Kruskal-Wallis test and the two-sample Wilcoxon test.

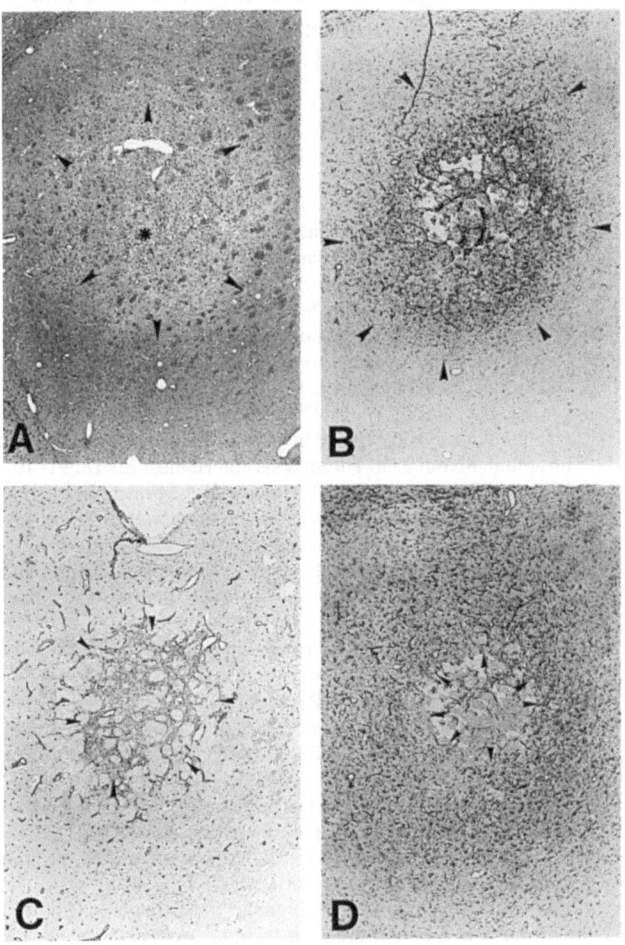

Fig. 1. (A–D) Low power photomicrograph of brain sections from animals with cerebral thrombin injection. (A) HE staining. Increased number of cells around site of injection. Paleness of HE staining of the focus (asterisk) due to spongiosis is noticeable. (B) Vimentin immunohistochemistry. Vimentin positive mesenchymal cells are present in the center, vimentin positive astrocytes are visible in the outer portion. Arrow heads show the boundary between Zone B and C. (C) Laminin immunohistochemistry. Laminin positive blood vessels and mesenchymal cells are arranged in a mesh-work pattern. Arrow heads indicate the demarcation of Zone A. (D) GFAP immunohistochemistry. Presence of GFAP positive astrocytes is revealed throughout the entire caudate nucleus outside of Zone A (arrow heads)

Results

1. HE and Immunohistochemistry

The buffer and albumin groups had no appreciable increase in cell numbers in the HE preparations. The immunohistochemial examinations disclosed an increase in GFAP positive cells only. In contrast, the thrombin group demonstrated round pale foci with reduced HE staining due to spongiosis around the injection site (Fig. 1 A). By immunohistochemistry three layers were recognized around the thrombin-injection site, each of which was characterized by different tissue reactions. The layers were termed zone A, B, C (Fig. 1 A–D). Zone A was the region closest to the injection with infiltration of inflammatory cells and an increased number of mesenchymal cells. Laminin staining (Fig. 1 C) revealed a mesh-work pattern along newly formed blood vessels. Vimentin staining (Fig. 1 B) revealed aggregation of inflammatory cells and of bipolar mesenchymal cells. In Zone B, GFAP- or vimentin-positive stellate-shaped astrocytes were present (Fig. 1 B, D). The proliferation of blood vessels was continuous with the laminin-positive structures in Zone A (Fig. 1 C). Zone C was also characterized by an increase in GFAP positive astrocytes which occupied almost the entire caudoputamen (Fig. 1 D). This change was seen in the buffer group as well, however.

A distinct area characteristic of Zone A was found in all animals of the thrombin-1000 group and in 4 of 8 in thrombin-125 group, but not in any of the buffer or albumin groups. Zone B like changes were observed in all animals of the thrombin-1000 group, in 5 of 8 in the thrombin-125 group, in 3 of 10 in the buffer group, and 1 of 6 in the albumin group. Characteristics of Zone C were present in the animals of all experimental groups.

2. Quantitative Analysis of Histological Changes

The area of vimentin-positive cells was significantly larger in the thrombin-1000 group than in the buffer and albumin groups. The thrombin groups had brain tissue areas which were significantly more densely populated by vimentin-positive astrocytes, as compared to the buffer or albumin groups (Fig. 2). The area of GFAP-positive astrocytes in the buffer group was not different from that in both thrombin groups, while it was significantly smaller in the albumin group. The number of BrdU-positive cells was significantly higher in both thrombin groups, whereas infiltration by poly-

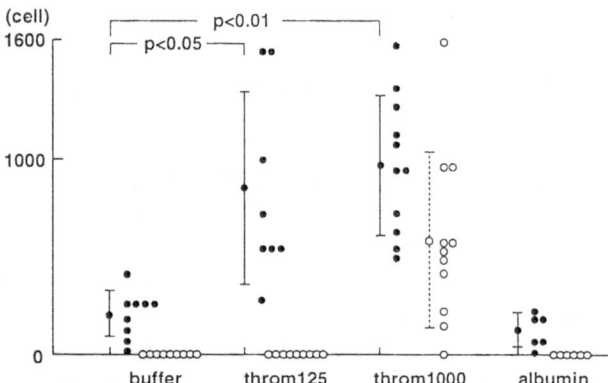

Fig. 2. Number of vimentin (+) cells (●) and neutrophils (○). Note significant differences between thrombin groups and control groups. Neutrophils are significantly increased in the thrombin-1000 group. Throm 125: Thrombin 125 U/ml, Throm 1000: Thrombin 1000 U/ml

morphonuclear leukocytes was significantly higher in the thrombin-1000 as compared to the other groups (Fig. 2).

Discussion

In this first in vivo study on effects of thrombin in the CNS the following results were obtained: i) infiltration of inflammatory cells including polymorphonuclear leukocytes, proliferation of mesenchymal cells, and angiogenesis close to the site of thrombin injection (Zone A); ii) an increase in vimentin-positive reactive astrocytes in the adjacent region (Zone B); and iii) an increase of cell proliferation. These findings suggest that thrombin is strongly involved in the inflammatory response following brain edema and reactive gliosis in the CNS.

We must distinguish the specific effects of thrombin from an unspecific infection due to the experimental procedure. The possibility of infection can be excluded because infiltration of polymorphonuclear leukocytes was significantly higher in the thrombin-1000 group although the experimental conditions were equal in all groups. Our finding of thrombin induced infiltration of inflammatory cells of CNS tissue is supported by various studies.

Thrombin activation in blood clotting is related to an inflammatory response[6]. Presence of thrombin receptors on the surface of inflammatory cells is considered as further evidence[1,3].

Adhesion molecules for inflammatory cells are up-

regulated in endothelial cells by thrombin[2]. In addition, thrombin degrades basement membranes of blood vessels by protease activity increasing vascular permeability[4]. Thrombin also induces secretion of interleukin-1 (IL-1) from inflammatory cells[6] and release of the B chain of platelet-derived growth factor (PDGF)[7]. Therefore, it is conceivable that the recruitment of inflammatory cells and release of cytokines enhances formation of brain edema and other secondary responses in addition to the direct effects of thrombin.

Thrombin may provoke various tissue reactions in vivo. Consequently a rationale may be justified for anti-thrombin agents in the treatment of CNS injury, which may suppress inflammation and, hence, brain edema and reactive gliosis, which prevents neuronal regeneration.

References

1. Bar-Shavit R, Kahn A, Mudd MS, Wilner GD, Mann KG, Fenton LW (1984) Localization of a chemotactic domain in human thrombin. Biochemistry 23: 397–400
2. Bevilacqua MP, Stengelin S, Gimbrone MA, Seed B (1989) Endothelial leukocyte adhesion molecule 1: an inducible receptor for neutrophils related to complement regulatory proteins and lectins. J Cell Physiol 128: 1160–1165
3. Carney DH, Redin W, McCroskeyn L (1992) Role of high-affinity thrombin receptors in postclotting cellular effects of thrombin. Seminars in Thrombosis and Hemostasis 18: 91–103
4. De Michele MAA, Moon DG, Fenton JW, Minnear FL (1990) Thrombin's enzymatic activity increases permeability of endothelial cell monolayers. J Appl Physiol 69: 1599–1606
5. Gurwitz D, Cunningham DD (1988) Thrombin modulates and reverses neuroblastoma neurite outgrowth. Proc Natl Acad Sci USA 85: 3440–3444
6. Jones A, Geczy CL (1990) Thrombin and factor Xa enhances the production of interleukin-1. Immunology 71: 236–241
7. Kavanaugh WM, Harsh GR, Starksen NF, Rocco CM, Williams LT (1988) Transcriptional regulation of the A and B chain genes of platelet-derived growth factor in microvascular endothelial cells. J Biol Chem 263: 8470–8472
8. Loret C, Sensenbrenner M, Labourdette G (1989) Differential phenotypic expression induced in cultured rat astroblasts by acidic fibroblast growth factor, epidermal growth factor, and thrombin. J Biol Chem 264: 8319–8327
9. Perraud F, Besnard F, Sensenbrenner M, Labourdette G (1987) Thrombin is a potent mitogen for rat astroblasts, but not for oligodendroblasts and neuroblasts in primary culture. Int J Devel Neuroscience 5: 181–188
10. Puro DG, Mano T, Chan CC, Fukuda M, Schmada H (1990) Thrombin stimulates the proliferation of human retinal glia cells. Grafes Arch Clin Exp Ophthalmol 228: 169–173

Correspondence: Akiko Nishino, Michiyasu Suzuki, M.D., Division of Neurosurgery, Institute of Brain Diseases, Tohoku University School of Medicine, 1-1 Seiryo-machi, Aoba-ku, Sendai, Japan.

Acta Neurochir (1994) [Suppl] 60: 89–93
© Springer-Verlag 1994

Proposed Toxic Oxidant Inhibitors Fail to Reduce Brain Edema

K. D. Judy, G. B. Bulkley, B. E. Hedlund, and **D. M. Long**

Department of Neurosurgery, The Johns Hopkins School of Medicine, Baltimore, Maryland and Biomedical Frontiers, Minneapolis, Minnesota, U.S.A.

Summary

Toxic oxidants (oxygen free radicals) have been implicated in the formation of brain edema from ischemia-reperfusion injury or tumor growth. We investigated the ability of an iron chelator, a calcium channel blocker, and a xanthine oxidase inhibitor to reduce formation of brain edema following a cold lesion in cats. The agents were given independently of each other in an attempt to inhibit the Haber-Weiss reaction, prevent Ca++ modulated uncoupling of oxidative phosphorylation, and inhibit the generation of toxic oxidants via xanthine oxidase, respectively. Pentastarch-deferoxamine conjugate at a dose of 50 mg/kg was given 15 minutes before and 60 minutes after the cold lesion. Nimodipine was given at a dose of 1 mg/kg 1 hour before and 2 hours after the cold lesion. Allopurinol was given at a dose of 50 mg/kg 24 hours before, at the time of the lesion and, 24 and 48 hours after the lesion. Gravimetric measurements of multiple brain areas were performed at 24 hours post-lesion in the pentastarch-deferoxamine and nimodipine groups and at 72 hours post-lesion in the allopurinol group. None of these agents led to significant reduction in brain edema formation as measured with a gravimetric column of kerosene and bromobenzene. Pentastarch-deferoxamine conjugate was utilized to avoid the confounding effects of arterial hypotension which is seen with intravenous deferoxamine. There was even a suggestion of increased edema in the periventricular white matter in animals treated with nimodipine. Taken together, independent inhibition of the Haber-Weiss reaction, of calcium channels, or of xanthine oxidase does not reduce formation of brain edema in the cold lesion model.

Keywords: Brain edema; toxic oxidants.

Introduction

Brain edema develops in association with brain tumors, cerebral infarction, and brain trauma. Controlling brain edema to reduce its mass effect has proven to be difficult. Toxic oxidants (oxygen radicals) have been implicated in the production of cold-induced brain edema[1,9]. Our laboratory has been investigating potential toxic oxidant scavenging agents in an effort to identify drugs which would reduce brain edema from the cold lesion model. We evaluated three drugs, allopurinol, pentastarch-conjugated deferoxamine, and nimodipine, which might have the potential to reduce brain edema via a toxic oxidant scavenging mechanism.

McCord[17] has stated that xanthine oxidase is the primary source of toxic oxidants in ischemia-reperfusion injury and Betz[2] has shown that significant xanthine oxidase is present in brain capillaries. Allopurinol is effective in reducing brain edema from ischemia by inhibition of xanthine oxidase[16,19].

The Haber-Weiss reaction is catalyzed by iron to form the hydroxyl radical, the most potent toxic oxidant, from the superoxide radical and hydrogen peroxide. Reduction of the catalytic agent by iron chelation with deferoxamine (DFO) reduced formation of brain edema in the cold lesion model[9]. Since DFO may cause hypotension when given intravenously, Hedlund conjugated deferoxamine with pentastarch (MPS-DFO) to reduce the hypotensive side effect[6].

In ischemia-reperfusion injury there is a massive influx of calcium, which causes uncoupling of oxidative phosphorylation in mitochondria. The rapid uptake of calcium by reoxygenated mitochondria leads to release of intracellular enzymes, which in turn promote tissue injury by the production of toxic oxidants[18]. Jacewicz reported reduction in infarct size and of edema in animals treated with nimodipine in a focal cerebral ischemia model[10].

Materials and Methods

Operative Procedure

Allopurinol

Fifteen mongrel cats weighing 3.6 to 6.0 kg were divided into two groups. Group 1 consisting of seven cats fed canned tuna 24 hours before and 24 and 48 hours after the lesion, were given an IV injection (1 ml/kg) of allopurinol diluent 1 hour prior to the lesion. Group 2 of eight cats fed canned tuna containing allopurinol (Sigma, St. Louis, Missouri) at a dose of 50 mg/kg 24 hours before and 24 and 48 hours after the lesion received an IV injection of allopurinol 50 mg/kg 1 hour prior to the lesion. Allopurinol was dissolved in a minimal volume of 1 N NaOH, the alkalotic pH of which was decreased by titration with 1 N HCl. Saline was added to yield a final concentration of 50 mg/ml.

MPS-DFO

Eight mongrel cats weighing 3.0 to 4.6 kg were divided into two groups. Group 1 consisted of three cats given an IV injection of pentastarch colloid (MPS 5 ml/kg) 15 minutes before and 60 minutes after the lesion. Group 2 consisted of five cats given an IV injection of MPS-DFO (Biomedical Frontiers, Minneapolis, Minnesota) containing 50 mg/kg of deferoxamine (10 mg/ml) 15 minutes before and 60 minutes after the lesion.

Nimodipine

Ten mongrel cats weighing 3.6 to 5.9 kg were divided into two groups. Group 1 consisted of five cats which served as controls receiving no pharmacologic treatment before or after the lesion. Group 2 consisted of five cats given nimodipine 1 mg/kg via an orogastric tube one hour before and two hours after the lesion. The liquid contents of one 30 mg capsule of nimodipine (Nimotop, Miles, West Haven, Connecticut) was diluted with propylene glycol to a final concentration of 3 mg/ml.

Anesthesia was induced in all cats with Ketamine 25 mg/kg and Acepromazine 2 mg given subcutaneously. An IV line was placed into the cephalic vein of either foreleg and 0.9% saline was given at a rate of 5 ml/h, anesthesia was maintained with Thiamylal IV. The animals were intubated to maintain an open airway but were capable of spontaneous ventilation. Right parietal 7 mm craniotomies were performed through a midline scalp incision. A cold lesion generator consisting of an insulated cup connected to a central aluminium bar was filled with dry ice and acetone which results in a temperature of –50°C at the tip of the probe[12]. A 5 mm cold lesion was made on the dura overlying the right ectomarginal gyrus by treatment for 60 seconds and the probe was released from the frozen dura with a flush of saline. The bone from the craniotomies was discarded and the skin was closed with a 2–0 silk suture.

For the MPS-DFO study we monitored blood pressure via a femoral artery catheter at baseline, during the first dose, 15 minutes after the first dose, and 30 minutes after the second dose.

The animals were then allowed to recover from the anesthesia, returned to their cages, and given free access to food and water. The animals were observed closely for any untoward effects of anesthesia.

The animals were sacrificed at 24 hours after the lesion for the MPS-DFO and nimodipine studies and at 72 hours after the lesion for the allopurinol study. The brains for the allopurinol and nimodipine studies were placed in kerosene to prevent dehydration. For the MPS-DFO study the brains were placed in a moisture chamber to prevent dehydration.

Gravimetric Measurements

A coronal slice of 3–4 mm thickness was taken through the lesion and underlying edema. Biopsies were taken with a 2 mm dermatology punch of gray and white matter of the lesioned ectomarginal gyrus, the adjacent marginal and ectosylvian gyri, distal gray matter of the caudal sylvian gyrus, and the deep periventricular white matter (Fig. 1). The biopsy samples were placed in a gravimetric

column, previously calibrated with sodium chloride solutions of known specific gravity, constructed according to Marmarou to determine specific gravities of tissue[15]. The brain specimens were allowed to stabilize their descent over 5 minutes followed by reading from the volume level of the graduate containing the gravimetric column. The brain specimen readings were converted to specific gravity data by use of the calibration curve generated by the standards.

Results

Blood Pressure

Blood pressure monitoring of the MPS-DFO study group showed no significant drop in animals given the MPS-DFO conjugate. Mean blood pressure recordings of the cats at baseline, 15 minutes after the first dose, or 30 minutes after the second dose did not reveal a significant reduction in this group by the treatment.

Specific Gravity

The cold lesion was consistently reproducible and of uniform dimensions. We ensured that results obtained with the gravimetric columns were replicable with R^2 values at 0.994–0.999.

In all cats a marked reduction in specific gravity occurred, i.e. formation of edema, in biopsy sites 5, 6, and 8 which is in agreement with brain edema spreading along white matter fiber tracts (Fig. 1). Brain edema was maximal in all cats at biopsy site 6, which was

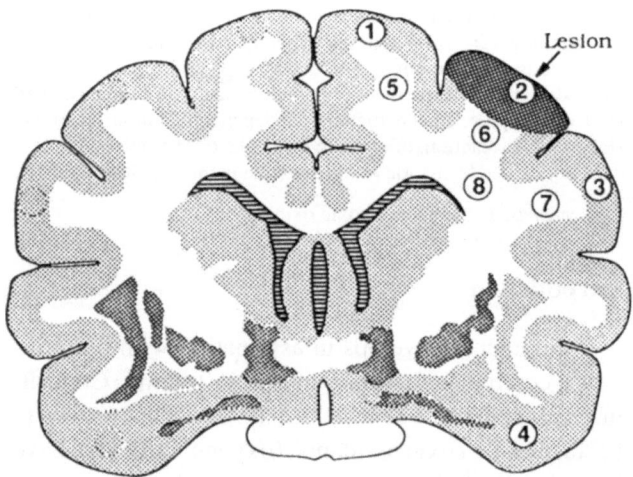

Fig. 1. Drawing of a coronal slice of a cat brain through a cold lesion. Circled numbers represent areas biopsied for gravimetric measurements. (Adapted from Ikeda *et al* (1990) Advances in neurology, Vol 52. Reprinted by permission from Raven Press, New York, USA)

ALLOPURINOL TREATED COLD LESION

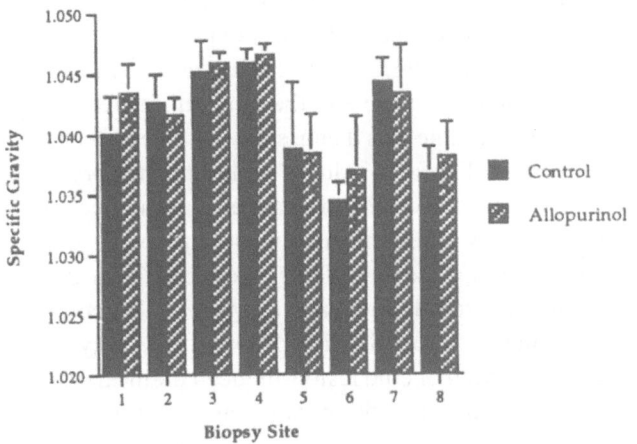

MPS-DFO TREATED COLD LESION

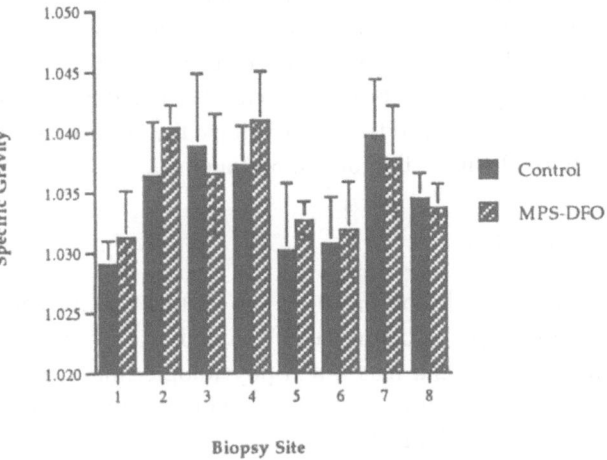

NIMODIPINE TREATED COLD LESION

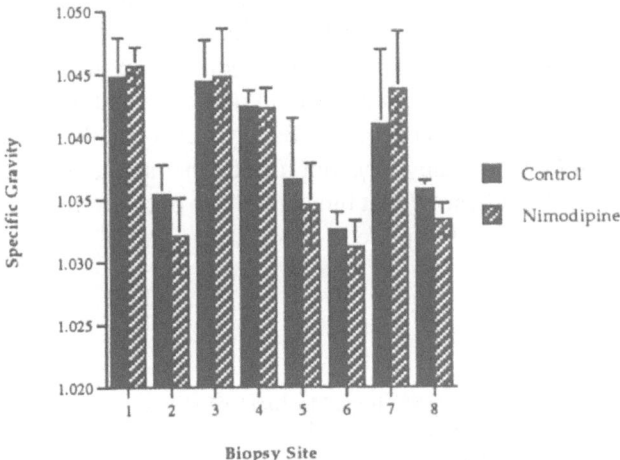

Fig. 2 . Graph of specific gravities of the brain biopsy sites of the control cats and the cats treated with the experimental agents, allopurinol, MPS-DFO, and nimodipine

white matter underlying the cold lesion. Biopsy site 7 did not exhibit the marked reduction in specific gravity found in the other white matter regions. This may be due to the path of the U-fibers of the ectosylvian gyrus which may serve as barrier to the progression of edema fluid.

The allopurinol treated cats had specific gravities which were uniformly reproducible in the gray matter regions, biopsy sites 1–4 (Fig. 2). Biopsy site 4 which is gray matter distant from the lesion had no reduction in specific gravity. The specific gravities in the allopurinol treated cats were higher than in controls in biopsy sites 6 and 8, directly underlying the cold lesion, suggesting some reduction in brain edema. However, the increase in specific gravity was statistically not significant. White matter of the adjacent gyri at biopsy sites 5 and 7 had a somewhat reduced specific gravity, implying increased edema.

The specific gravities of the MPS-DFO group, both in control and MPS-DFO cats, were lower than in the allopurinol and nimodipine studies. The lower specific gravities were due to use of the moisture chamber instead of placing the brain slices in kerosene, as was done in the other two studies. The gray matter biopsy sites 1–4 in the MPS-DFO group showed a reduction in specific gravity associated with a larger standard deviation (Fig. 2). MPS-DFO treated cats had a higher specific gravity in biopsy sites 5 and 6 with reduced specific gravity in sites 7 and 8. None of these changes associated with MPS-DFO treatment were statistically significant, however.

In the nimodipine study group there was no reduction in specific gravity in biopsy sites 1, 3, and 4 of either the control or nimodipine group (Fig. 2). Biopsy site 2, the gray matter involved by the cold lesion, exhibited a reduction in specific gravity which was not seen in the allopurinol or MPS-DFO studies. In the white matter biopsy sites 5, 6, and 8 the nimodipine treated group had a reduction in specific gravity which reached statistical significance (p = 0.007) in site 8. The biopsy site 7 of the nimodipine treated group had an increased specific gravity above the control group, although site 7 did not exhibit much edema.

Discussion

The cold lesion has been shown to be a consistent and reliable model to study brain edema[12,13]. Its main criticism is that it is a non-physiologic injury model that is most similar to a traumatic contusion when viewed histologically[4]. Due to its ease of application and reproducibility we used the cold lesion to study the brain edema response to proposed toxic oxidant reducing agents.

All animals were pretreated with one of three drugs, allopurinol, MPS-DFO (deferoxamine), or nimodipine, prior to inducing cold lesion in an attempt to maximize any edema reducing effect. Once reduction in edema was identified we had planned to investigate a therapeutical window, in which treatment is effective in successively closer time intervals to the lesion or following the lesion. We were unable, however, to identify a significant reduction in brain edema with any of the drugs.

Allopurinol is effective in reducing infarct size and brain edema in ischemia-reperfusion injuries[16,19]. Elaborating the role of xanthine oxidase in the generation of infarct size and brain edema has been difficult as some have found that xanthine oxidase is not a significant source of oxygen radicals in ischemia-reperfusion and toxic oxidant injury[3,11]. Lindsay *et al.* found that allopurinol reduced cerebral infarct by a mechanism independent of xanthine oxidase inhibition[14]. Xanthine oxidase inhibition did not afford reduction in brain edema in the cold lesion model in the present study.

Gaab[5] and Harris[7] found an increase in brain edema in animals treated with nimodipine in cold lesion and ischemia models, respectively. Nimodipine may impair autoregulation and BBB integrity and result in increased blood flow[5,7,8]. Although an increased blood flow would be beneficial in ischemic injury to minimize tissue damage, it is clear that by providing an increased blood flow to an area with focal injury, such as a cold lesion, the increased hydrostatic pressure would propagate formation of vasogenic edema. It has been shown by Klatzo that increasing blood pressure of

cats subjected to a cold lesion exacerbates vasogenic edema, whereas decreasing blood pressure reduces formation of edema[13]. This may explain why there was a significant decrease in specific gravity (increase in edema) in biopsy site 8 in the nimodipine treated cats.

Deferoxamine has been shown to be effective in reducing cold lesion-induced brain edema in pre-lesion treatment and post-lesion treatment[9]. Iron injected into brain cortex causes formation of superoxide radicals and of brain edema[21]. DFO itself can be a directly acting superoxide radical scavenger[20]. The conjugate of pentastarch (MPS) with DFO did presently not afford reduction of cold lesion-induced brain edema. The same dose of DFO, 50 mg/kg, was used for this study as in Ikeda's study in which he identified significant reduction in cold injury brain edema in pre-lesion treated cats at 6 and 24 hours and in post-lesion treated cats at 6 hours, but not at 24 hours[9]. It is possible that a transient mild drop in blood pressure at 15 minutes following the lesion seen in Ikeda's study may have been sufficient to reduce brain perfusion and thus formation of brain edema. Hallaway states that the MPS-DFO conjugate does not affect the availability of DFO[6], indicating that bioavailability of the conjugated DFO was not impaired.

References

1. Ando Y, Inoue M, Hirota M, *et al* (1989) Effect of a superoxide dismutase derivative on cold-induced brain edema. Brain Res 477: 286–291
2. Betz AL (1985) Identification of hypoxanthine transport and xanthine oxidase activity in brain capillaries. J Neurochem 44: 574–579
3. Betz AL, Randall J, Manz D (1991) Xanthine oxidase is not a major source of free radicals in focal cerebral ischemia. Am J Physiol 29: H563–H568
4. Clasen R, Cooke P, Pandolfi S, *et al* (1962) Experimental cerebral edema produced by focal freezing. J Neuropath Exp Neurol 21: 579–596
5. Gaab MR, Höllerhage HG, Walter GF, *et al* (1990) Brain edema, autoregulation, and calcium antagonism – an experimental study with nimodipine. Adv Neurol 52: 391–400
6. Hallaway PE, Eaton JW, Panter SS, Hedlund BE (1989) Modulation of deferoxamine toxicity and clearance by covalent attachment to biocompatible polymers. Proc Natl Acad Sci USA 86: 10108–10112
7. Harris RJ, Branston NM, Symon L, *et al* (1982) The effects of a calcium antagonist, nimodipine, upon physiological responses of the cerebral vasculature and its possible influence upon focal cerebral ischaemia. Stroke 13: 759–766
8. Höllerhage HG, Gaab MR, Zumkeller M, Walter GF (1988) The influence of nimodipine on cerebral blood flow autoregulation and blood-brain barrier. J Neurosurg 69: 919–922
9. Ikeda Y, Ikeda K, Long D (1989) Protective effect of the iron chelator deferoxamine on cold-induced brain edema. Neurosurgery 71: 233–238

10. Jacewicz M, Brint S, Tanabe J, Pulsinelli WA (1990) Continuous nimodipine treatment attenuates cortical infarction in rats subjected to 24 hours of focal cerebral ischemia. J Cereb Blood Flow Metab 10: 89–96

11. Joannidis M, Gstraunthaler G, Pfaller W (1990) Xanthine oxidase: evidence against a causative role in renal reperfusion injury. Am J Physiol 258: F232–F236

12. Klatzo I, Piraux A, Laskowski EJ (1958) The relationship between edema, blood-brain barrier and tissue elements in a local brain injury. J Neuropath Exp Neurol 17: 548–564

13. Klatzo I, Wisniewski H, Steinwall O, et al (1967) Dynamics of cold injury edema. In: Klatzo I, Seitelberger F (eds) Brain edema. Springer, Wien New York, pp 554–563

14. Lindsay S, Liu T, Xu J, et al (1991) Role of xanthine dehydrogenase and oxidase in focal cerebral ischemic injury to rat. Am J Physiol 261: H2051–H2057

15. Marmarou A, Pöll W, Shulman K, Bhagavan H (1978) A simple gravimetric technique for measurement of cerebral edema. J Neurosurg 49: 530–537

16. Martz D, Rayos G, Schielke G, Betz AL (1989) Allopurinol and dimethylthiourea reduce brain infarction following middle cerebral artery occlusion in rats. Stroke 20: 488–494

17. McCord J (1985) Oxygen-derived free radicals in postischemic tissue injury. N Engl J Med 312: 159–163

18. Opie LH (1989) Reperfusion injury and its pharmacologic modification. Circulation 80: 1049–1062

19. Patt A, Harken A, Burton L, et al (1988) Xanthine oxidase-derived hydrogen peroxide contributes to ischemia reperfusion-induced edema in gerbil brains. J Clin Invest 81: 1556–1562

20. Sinaceur J, Ribiere C, Nordmann J, et al (1984) Desferrioxamine: a scavenger of superoxide radicals? Biochem J 33: 1693–1694

21. Willmore LJ, Rubin J (1984) Effects of antiperoxidants on FeCl-induced lipid peroxidation and focal edema in rat brain. Exp Neurol 83: 62–70

Correspondence: Donlin M. Long, M.D., Ph.D., Department of Neurosurgery, The Johns Hopkins Hospital, Meyer 7-109, 600 N. Wolfe Street, Baltimore, Maryland 21287-7709, U.S.A.

Acta Neurochir (1994) [Suppl] 60: 94–97

Involvement of Nitric Oxide and Free Radical (O_2^-) in Neuronal Injury Induced by Deprivation of Oxygen and Glucose in vitro

J. Ikeda, L. Ma, I. Morita, and **S. Murota**

Department of Physiological Chemistry, Graduate School, Tokyo Medical and Dental University, Yushima, Bunkyo-ku, Tokyo, Japan

Summary

Nitric oxide (NO) is a free radical that has been recently proposed as a messenger molecule in the central nervous system. Since its involvement in glutamate neurotoxicity in vitro has been recently reported, using rat cortical cultures, we tested the hypothesis that NO also plays a role in neuronal injury induced by deprivation of oxygen and glucose. About 80–90% of neurons were killed in less than 12 h after a 4–6 h period of oxygen and glucose deprivation. N-nitro-L-arginine (L-NNA), an inhibitor of nitric oxidase synthase (NOS), significantly ameliorated this neuronal injury in a dose dependent manner.

Since it has been suggested that NO is inactivated in a short time period by interaction with superoxide anions (O_2^-), which are generated during ischemia-reperfusion in vivo, we further evaluated the effect of superoxide dismutase (SOD) on neuronal injury in this test system. SOD failed, however, to protect against neuronal death. Furthermore, concomitant addition of SOD and L-NNA rather reduced the beneficial effects of L-NNA. Our results suggest therefore that NO, at least in part, mediates neuronal injury secondary to deprivation of oxygen and glucose in vitro and that superoxide anions may have a protective role by inactivating NO.

Keywords: Nitric oxide; neuronal injury; superoxide anion; L-NNA.

Introduction

There is increasing evidence that the increase of intracellular calcium [Ca^{2+}], probably mediated by excessive activation of postsynaptic glutamate receptors, is involved in ischemic neuronal injury[12]. A variety of events following the Ca^{2+} increase have been hypothesized, but precise mechanisms are still unknown by which calcium overload kills neuronal cells[5].

Recent success in purification and molecular cloning of NOS revealed that this enzyme, which catalyses biosynthesis of nitric oxide from L-arginine, exists in two principal isoforms that differ with respect to their calcium dependence[4].

In central nervous system, the constitutive type of the enzyme which is rapidly activated by an increase in [Ca^{2+}], has been shown to occur in discrete neuronal populations[3]. Physiologically, NO acts as neuronal messenger by activating guanylate cyclase, resulting in the elevation of guanosine 3',5'-monophosphate (cGMP) in the target cell[8]. Since NO, however, has an unpaired electron, it is a radical compound which may have neurotoxic properties under pathological conditions.

Although beneficial effects of inhibition of NOS on glutamate neurotoxicity in vitro have been reported, results are controversial in focal ischemia in vivo[10,15]. These conflicting results may be due to simultanously beneficial and detrimental effects of the drug on cerebral blood flow and/or platelet aggregation. We decided therefore to examine inhibition of NOS on neuronal injury induced by deprivation of glucose and oxygen in vitro.

Reactions of NO with different components of biological materials are still unclear. Superoxide anion is generated during brain ischemia[9] and the interaction between NO and superoxide anion is believed to result in the generation of peroxynitrite ($ONOO^-$)[2]. Although peroxynitrite has been indicated to be a more potent radical than NO, Oury *et al.* showed recently that transgenic mice overexpressing extracellular SOD displayed enhanced O_2 toxicity. This was mitigated by a NOS inhibitor, suggesting that superoxide alleviates neuronal injury by scavenging NO[11]. To address the question on relationships between superoxide anion and NO, we studied neuronal viability following neuronal injury in vitro induced by deprivation of glucose and oxygen with or without SOD in the medium.

Materials and Methods

Cell Culture

Cortical cell cultures were prepared based on the methods described by Banker and Cowan with slight modifications[1]. After removal of the meninges, the cortices were dissected and incubated with trypsin (0.25%) and DNase (0.02%) at 37°C for 15 minutes. The cells were then washed in Dulbecco's modified Eagle's medium (DMEM) supplemented with 10% fetal bovine serum and seeded onto eight-to-ten polyethylenimine-coated coverslips that had been placed (5×10^6/ml) on a cortical glial monolayer. After 4 days in vitro, nonneuronal cell division was halted by 3-days' exposure to 5×10^{-6} M cytosine arabinoside. At 12 days, more than 80% of the cells stained positively with antibodies against neurofilament.

Deprivation of Glucose and Oxygen

Cortical cells grown on coverslips for 12 days were transferred to the wells of a 24-well dish, each well containing 300 μl of PBS supplemented with 1.8 mM calcium. Anoxia was induced by placing the dish in an airtight chamber, flushing it with 95% N_2 and 5% CO_2 for 4–6 h. The dish containing PBS had been placed in anaerobic condition overnight before transfer of the cells. Various doses of L-NNA for inhibition of NOS and SOD (500 U/ml) were added to PBS in some experiments.

Assessment of Cellular Injury

Neuronal cell injury was assessed 12 h after the end of anoxic exposure. The dish was placed on an ACAS 570 interactive laser cytometer equipped with a 5 W argon laser, an inverted phase-contrast microscope, and a 16-bit microcomputer. Surviving neurons were stained with fluorescein diacetate (FDA, 2 μg/ml) which crosses cell membranes and subsequently is hydrolysed by intracellular esterases, yielding a green fluorescence. Propidium iodide (PI, 2 μg/ml), another fluorescence dye, does not enter intact cells, but penetrates sufficiently injured cell membranes, producing red fluorescence upon interacting with nucleic acids[7]. As surviving glial cells load very little dye at the FDA concentration employed, we used digital thresholding to create a mask that allowed us to visualize only cell bodies of neuronal cells. Finally, the ratio between surviving neurons and dead ones was calculated by using a microcomputer of the ACAS.

Results and Discussion

The morphological features of neurological degeneration at 12 h after a 4–6 h period of deprivation of glucose and oxygen were loss of neuritic processes and occurrence of shrunken and irregularly shaped cells (Fig. 1). With FDA only 10–20% of the neurons were stained positively. When L-NNA (100 μM to 1 mM) was added to the medium, there was marked attenuation of neuronal damage (Fig. 2). The dose-response curve shown in Fig. 4 demonstrates that even lower concentrations of L-NNA (10 μM) significantly protect against neuronal damage induced by deprivation of glucose and oxygen.

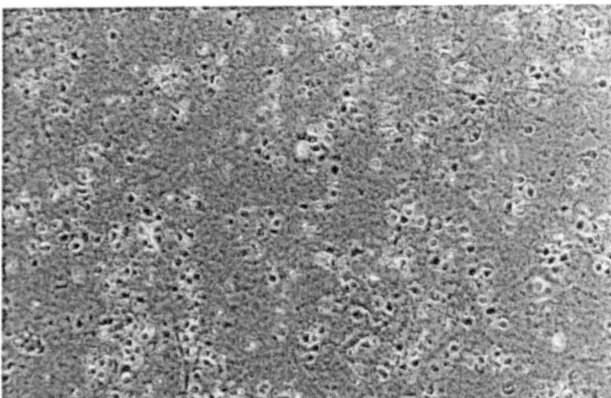

Fig. 1. Phase-contrast microphotograph showing loss of neurites and shrunken, irregularly shaped neurons after deprivation of oxygen and glucose

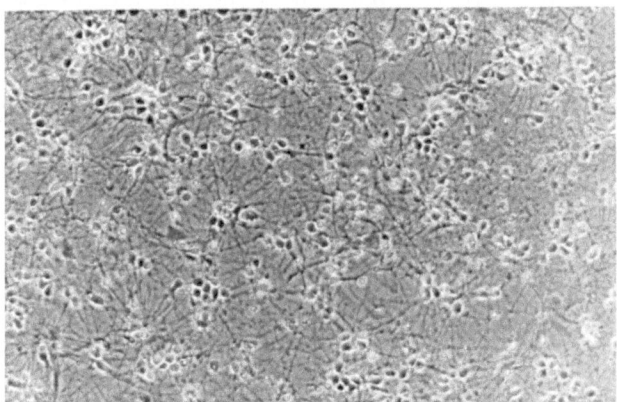

Fig. 2. When neurons were treated with L-NNA (10^{-4} M), an inhibitor of nitric oxide synthase, more than 80% of the cells were spared

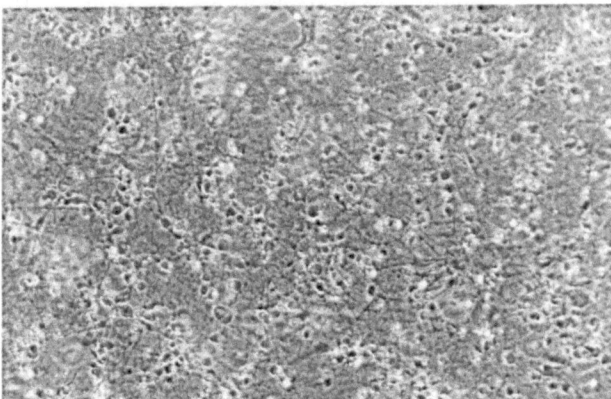

Fig. 3. Addition of SOD (500 U/ml) reduced neuronal protection by L-NNA (10^{-4} M)

96

J. Ikeda *et al.*

Unexpectedly, addition of SOD (500 U/ml) in combination with L-NNA interfered with the beneficial effect of L-NNA, especially at low concentration of the latter. Neuronal survival was decreased to approximately 44% in the cultures which were concomitantly added with L-NNA (10^{-4}M) and SOD (500 U/ml) as compared with a cell viability of 88% in the cultures with L-NNA only (Fig. 3).

NO acts as a messenger in the brain by inducing cGMP formation. On the other hand, the relatively large quantities of NO generated by activated macrophages are thought to be used for cytotoxic functions against certain target microorganisms[13]. This raises the possibility that NO synthesized by the constitutive type of the enzyme in neuronal cells can also be cytotoxic under pathological conditions. In 1991, Dawson *et al.* reported as first authors that inhibition of NOS rescues neurons from glutamate neurotoxicity. They suggested that excessive Ca^{2+} influx by stimulation of NMDA receptors activates NOS, leading to neuronal death[6].

In our experiments, L-NNA had considerable protective effects against neuronal injury. We have observed that MK-801 (10^{-5}M), a noncompetitive inhibitor of the NMDA receptor also provided considerable protection against neuronal injury in our experimental paradigm (data not shown). Therefore, Ca^{2+} influx by activation of NMDA receptors might have been responsible for the production of NO that in turn may have mediated neuronal injury in our studies. This scenario is consistent with the hypothesis raised by Dawson *et al.* on glutamate neurotoxicity[6].

Although our results indicate that NO is mediating neuronal injury, exact mechanisms are unclear. Besides its role as a free radical, NO has been suggested to cause nitrosylation of nucleic acids, resulting in breakage of DNA strands[14]. It was also hypothesized that NO binds to, and inactivates, iron- and heme-containing enzymes involved in cell respiration[13]. Which of these mechanisms finally are causing neuronal injury remains to be determined.

Surprisingly, SOD which was shown to attenuate degeneration in an ischemia-reperfusion model[9] in vivo rather reduced protection by the NOS inhibitor. Since superoxide anions are known to inactivate NO, it is possible that SOD prevented NO degradation by scavenging superoxide anions. In summary, our results show that NO mediates neuronal injury by deprivation of oxygen and glucose. Superoxide anions may have beneficial effects against this injury by inactivating NO.

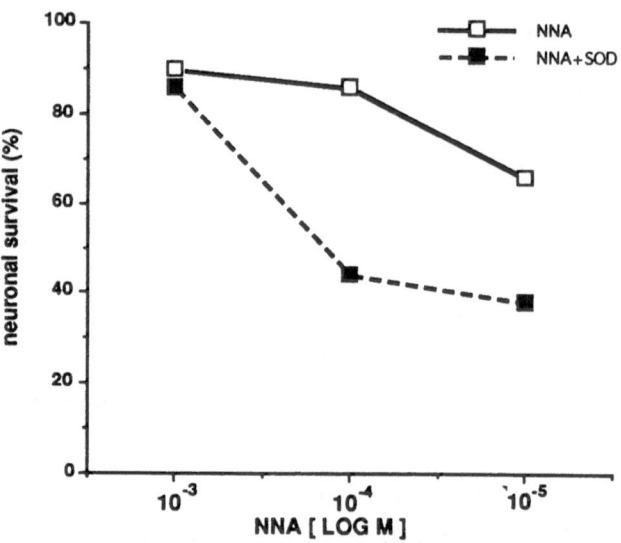

Fig. 4. Dose response of neuronal protection against O_2- and glucose deprivation by L-NNA, SOD (500 U/ml) is attenuating protection by L-NNA, especially at low concentration

References

1. Banker GA, Cowan WM (1977) Rat hippocampal neurons in dispersed cell culture. Brain Res 126: 397–425
2. Beckman JS, Beckman TW, Chen J, Marshall PA (1990) Apparent hydroxyl radical production by peroxynitrite: Implications for endothelial injury from nitric oxide and superoxide. Proc Natl Acad Sci USA 87: 1620–1624
3. Bredt DS, Glatt CE, Hwang PM, Fotuhi M (1991) Nitric oxide synthase protein and mRNA are discretely localized in neuronal populations of the mammalian CNS together with NADPH diaphorase. Neuron 7: 615–624
4. Bredt DS, Hwang PM, Glatt CE, Lowenstein C (1991) Cloned and expressed nitric oxide synthase structurally resembles cytochrome P-450 reductase. Nature 351: 714–718
5. Choi DW (1988) Glutamate neurotoxicity and diseases of the nervous system. Neuron 1: 623–634
6. Dawson VL, Dawson TM, London ED, Bredt DS (1991) Nitric oxide mediates glutamate neurotoxicity in primary cortical cultures. Proc Natl Acad Sci USA 88: 6368–6371
7. Felipo V, Minana M-D, Grisolia S (1993) Inhibitors of protein kinase C prevent the toxicity of glutamate in primary neuronal cultures. Brain Res 604: 192–196
8. Knowles RG, Palacios M, Palmer RMJ, Moncada S (1989) Formation of nitric oxide from L-arginine in the central nervous system: a transduction mechanism for stimulation of the soluble guanylate cyclase. Proc Natl Acad Sci USA 86: 5159–5162
9. Nelson CW, Wei EP, Povlishock JT, Kontos HA (1992) Oxygen radicals in cerebral ischemia. Am J Physiol 263: H1356–H1362
10. Nowicki JP, Duval D, Poignet DH, Scatton B (1991) Nitric oxide mediates neuronal death after cerebral ischemia in the mouse. Eur J Pharmacol 204: 339–340
11. Oury TD, Ho Y-S, Piantadosi CA, Crapo JD (1992) Extracellular superoxide dismutase, nitric oxide, and central nervous system O_2 toxicity. Proc Natl Sci USA 89: 9715–9719

12. Randall RD, Thayer SA (1992) Glutamate-induced calcium transient triggers delayed calcium overload and neurotoxicity in rat hippocampal neurons. J Neurosci 12: 1882–1895
13. Stuehr DJ, Nathan CF (1989) Nitric oxide, a macrophage product responsible for cytostasis and respiratory inhibition in tumor target cells. J Exp Med 169: 1543–1555
14. Wink DA, Kasprzak KS, Maragos CM, Elespuru RK (1991) DNA deaminating ability and genotoxicity of nitric oxide and its progenitors. Science 254: 1001–1003
15. Yamamoto S, Golanov EV, Berger SB, Reis DJ (1992) Inhibition of nitric oxide synthesis increases focal ischemic infarction in rat. J Cereb Blood Flow Metab 12: 717–726

Correspondence: Junichi Ikeda, M.D., Department of Physiological Chemistry, Graduate School, Faculty of Dentistry, Tokyo Medical and Dental University, 1-5-45 Yushima, Bunkyo-ku, Tokyo 113, Japan.

Acta Neurochir (1994) [Suppl] 60: 98–100
© Springer-Verlag 1994

The Endogenous Ouabain-Like Sodium Pump Inhibitor in Cold Injury-Induced Brain Edema

Z. M. Rap, W. Schoner, Z. Czernicki, G. Hildebrandt, H. W. Mueller, and **O. Hoffmann**

Institute of Biochemistry and Endocrinology, Justus-Liebig-University of Giessen, Giessen, Federal Republic of Germany

Summary

An endogenous ouabain-like factor (EOLF) was measured in brain tissue of cats 12 and 24 hrs after cold injury-induced edema. EOLF was assayed via its inhibition of $^{86}Rb^+$ uptake in human red blood cells in a fraction which was obtained from brain tissue by methanol extraction, chloroform treatment and purification of the water phase by C-18 HPLC. As compared to the contralateral hemispheres with an EOLF concentration of 605 ± 71 pmol ouabain equivalents per g wet weight, the edematous hemisphere had significantly higher concentrations: 12 hours after cold injury it was 2600 ± 1762 pmol ($p < 0.03$) and fell to 857 ± 160 pmol ouabain equivalents/g wet weight after 24 hrs. Similar kinetics were evident for the EOLF concentrations in cerebrospinal fluid. It is suggested that the increase of EOLF in the edematous brain hemisphere may participate as a mediator in the development of vasogenic brain edema in the disturbance of the sodium metabolism.

Keywords: Ouabain-like factor; brain edema.

Introduction

The findings of endogenous inhibitor(s) of the sodium pump in brain tissue and cerebrospinal fluid[1,5,7,9,11] as well as its recent identification as ouabain-like compound in plasma[8] raises the question on its physiological and pathophysiological function. The specific inhibition of the sodium pump by an endogenous ouabain-like factor (EOLF) may not only affect membrane potential, intracellular pH and calcium concentration but also water metabolism of brain cells. Brain edema is exerted by mediators which also affect the sodium pump[2,12]. Consequently, following a cold injury we studied the EOLF concentrations as a function of the developing edema. We report that EOLF increases in the edematous tissue as well as in cerebrospinal fluid.

Materials and Methods

Experimental Conditions

Spontaneously respiring cats weighing 3.5–4 kg under Nembutal anesthesia (35 mg/kg) were continuously hydrated with Ringer solution at a rate of 0.09 ml/kg min and the heart rate and blood pressure were continuously recorded. Such cats received standardized cold lesions through the intact dura mater[10]. Samples of cerebrospinal fluid (CSF), blood, and urine were taken for gasometric and water-electrolyte studies. From 10 cats, 5 were sacrificed after 12 h, 4 after 24 h, 1 died. Brains were quickly removed, frozen in dry ice and stored at $-60°C$.

Biochemical Procedures and Analysis of EOLF

Brain hemispheres were homogenized at $0°C$ for 10 min with 3 ml methanol/g tissue. The homogenate was centrifuged for 20 min at $47580 \times g$, the sediment was extracted twice with the same volume of methanol, centrifuged and the supernatants combined. After evaporation the residues were dissolved in 4.6 ml distilled water, 5 ml methanol was added and lipidous material was extracted into 6 ml chloroform. The residues of the dried water phase were dissolved in 1 ml distilled water and adsorbed to Waters C-18 cartridges. The sodium pump inhibitor was eluted therefrom with 80% methanol and dried. Residues from each brain hemisphere were taken up in 1 ml water, microfiltered (0.45 and 0.22 µm filters) and aliquots of 250 µl were applied to octadecylsilane (C-18) columns (250×4.6 mm). The columns were eluted with 30% methanol at a flow rate of 0.3 ml/min. Fractions with the identical retention time of ouabain were collected, evaporated and stored at $-18°C$ until the EOLF was measured.

Quantitation of EOLF

This was done by measuring the inhibitory effect of the brain isolate or crude CSF on the $^{86}Rb^+$ uptake by the sodium pump of human red blood cells assuming that the factor has the same affinity as ouabain (half-maximal inhibition at 4×10^{-8} M). Erythrocytes were washed twice in Ringer solution and resuspended (hematocrit: 19%). 50 µl of the erythrocytes was given into precooled Eppendorf cups which already contained 20 µl Ringer solution, 20 µl of the inhibitor fraction and 10 µl $^{86}RbCl$. The cells were allowed to pump for 60 minutes at $37°C$. Thereafter 90 µl of the cell suspension was given on top of a gradient containing 200 μl 3 M KOH and 500 μl silicon oil (from Wacker Chemie, Munich AR 200 Oil/ AR 20 oil 3 : 1). Erythrocytes were sedimented for 30 s in an Eppendorf centrifuge, the supernatant and the silicon layer were sucked off and the activity of the KOH-fraction was counted by Cerenkov radiation. All experiments were done in triplicate and the data for the edematous and control hemisphere were calculated in ouabain equivalents per g wet weight. Mean values ± standard deviation were calculated and statistical analysis was performed using the matched-pair-signal-rank-test.

Results and Discussion

When after cold injury the development of brain edema took place, CSF showed an increase in EOLF with its maximum after 6–12 h, but even after 24 h it was still higher than the initial value (Fig. 1). EOLF rose after 6 h to 230% of the initial value, went to 260% after 12 h and came to 120% after 24 h. In contrast to CSF, EOLF in serum showed only a very delayed rise. It is conceivable that EOLF in CSF originates from the edematous brain tissue. Therefore, EOLF concentrations were determined in the edematous as well as in the contralateral hemispheres. Figure 2 shows in fact that the edematous hemisphere contained more EOLF than the unaffected control hemisphere. Again, EOLF concentrations were higher at 12 h after cold injury (425% of the control hemisphere) than after 24 h (143% of the control hemisphere). Although the percentage rise in EOLF (of 425%) after 12 h was higher in the edematous hemisphere than in the crude CSF (260%), the absolute concentrations of EOLF after 12 h were in CSF 134 ± 13 nM, e.g. about 50 times higher than in the edematous tissue (2.6 ± 1.76 nmol/g wet weight). Independent from the percentage differences as well as the absolute

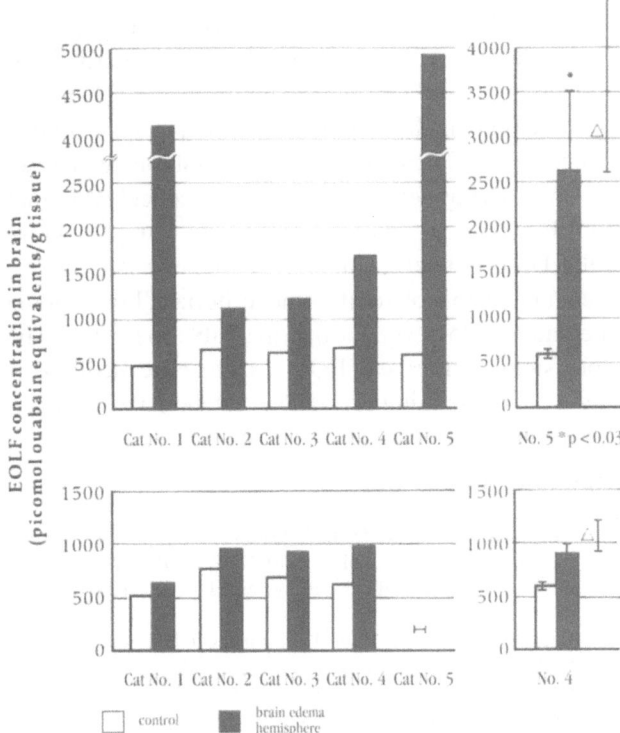

Fig. 2. EOLF concentrations in HPLC fractions from edematous and contralateral hemispheres 12 (upper) and 24 hours (lower) after cold injury. The delta values refer to the difference of the edematous and control hemispheres

concentrations, EOLF changes in CSF and edematous tissue suggest a relationship with the dynamics of vasogenic brain edema after cold injury.

The EOLF concentrations in the control hemispheres of 612 ± 86 pmol (12 h after cold injury) and 597 ± 55 pmol ouabain equivalents/g wt. wght. (24 h after cold injury) are somewhat higher than the estimates of 100 pmol ouabain equivalents/g wt. wght. by Lichtstein and Samuelov[11]. A concentration of 2600 ± 1762 pmol ouabain equivalents/g wt. wght. observed 12 h after cold injury may indeed inhibit the highly ouabain-sensitive α_3 isoform of Na^+/K^+-ATPase in brain[13]. Then, 24 h after cold injury, the slowdown of the spread of brain edema was accompanied by a fall of the EOLF concentration to 857 ± 160 pmol ouabain equivalents/g wt. wght. It is well known that mmolar concentrations of ouabain evoke cytotoxic brain damage[3] and increase the inhibition of the sodium pump at the earlier phase of brain edema[6]. Since chronic volume expansion stimulates the release of EOLF[4], one may wonder if Ringer infusion to cats may have an effect on EOLF in brain tissue and CSF. This can clearly be excluded because EOLF concentrations in control hemispheres did not change (Fig. 2).

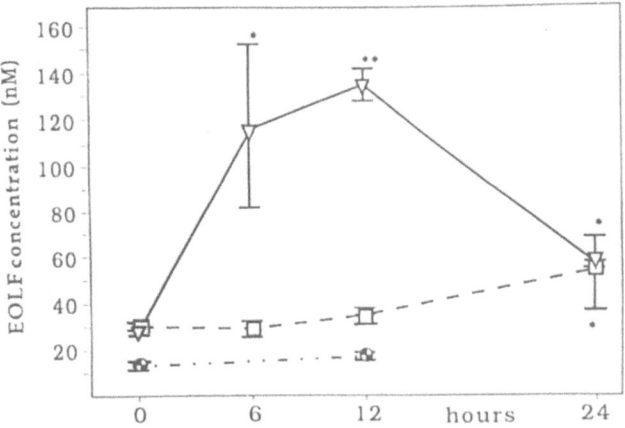

Fig. 1. Effect of cerebral cold injury in 5 Nembutal anaesthetized cats on EOLF concentrations in CSF and serum. EOLF in CSF (▽), EOLF in serum (□); EOLF in serum of 3 control animals infused with Ringer solution (◓). Statistical significances are * p < 0.05; ** p < 0.01. EOLF was measured without preceding purification

Should EOLF be considered as one of the specific brain edema mediators? Consistent with the criteria of Baethmann *et al.*[2] for such mediators we find (1) EOLF is increased in the edematous tissue; (2) the EOLF concentrations are in the range which inhibit the sodium pump in living cells. Therefore it is conceivable that EOLF affects the membrane sodium pump of brain cells. Unfortunately at this stage of our investigation we were unable to look for the other criteria of brain edema mediators[2]. However, the data available so far would be consistent with a role of EOLF as a specific mediator of brain edema relating to sodium metabolism.

References

1. Akagawa K, Hara N, Tsuka Y (1984) Partial purification and the characters of the inhibitors of Na/K-ATPase and ouabain binding in the central nervous system. J Neurochem 42: 775–780

2. Baethmann A, Maier-Hauff K, Kempski O, Unterberg A, Wahl M, Schürer L (1988) Mediators of brain edema and secondary brain damage. Crit Care Med 16: 972–978

3. Bignami A, Palladini G (1966) Experimentally produced cerebral status spongiosus and continuous pseudorhythmic electroencephalographic discharges with a membrane ATPase inhibitor in the rat. Nature 209: 413–414

4. DeWardener HE, Clarkson EM (1985) Concept of natriuretic hormone. Physiol Rev 65: 658–759

5. Fishman MC (1979) Endogenous digitalis-like activity in mammalian brain. Proc Natl Acad Sci USA 76: 4661–4663

6. Gazendam J, Go KG, Van der Meer JJ, Zuiderveen F (1979) Changes of electrical impedance in edematous cat brain during hypoxemia and after intra-cerebral ouabain injection. Exp Neurol 66: 78–87

7. Halperin JA, Riordan JF, Tosteson DC (1988) Characteristics of an inhibitor of the Na$^+$/K$^+$-ATPase pump in human cerebrospinal fluid. J Biol Chem 263: 646–651

8. Hamlyn JM, Blaustein MP, Bova S, DuCharme DW, Harris DW, Mandel F, Mathews WF, Ludens JH (1991) Identification and characterization of a ouabain-like compound from human plasma. Proc Natl Acad Sci USA 88: 6259–6263

9. Haupert GT, Carilli CT, Cantley LC (1984) Hypothalamic sodium transport inhibitor is a high-affinity reversible inhibitor of Na$^+$/K$^+$-ATPase. Am J Physiol 247: F919–F924

10. Klatzo I, Piraux A, Laskowski EJ (1958) The relationship between edema, blood-brain barrier and tissue elements in a local brain injury. J Neuropathol Exp Neurol 17: 548–564

11. Lichtstein D, Samuelov S (1982) Membrane potential changes induced by the ouabain like compound extracted from mammalian brain. Proc Natl Acad Sci USA 79: 1435–1456

12. Rigoulet M, Gerin B, Cohadon E, Vandendreissche M (1979) Unilateral brain injury in the rabbit; reversible and irreversible damage of membranal ATPase. J Neurochem 32: 535–541

13. Urayama O, Sweadner KJ (1988) Ouabain sensitivity of the alpha 3 isozyme of rat Na, K-ATPase. Biochem Biophys Res Commun 156: 796–800

Correspondence: Z. M. Rap, M. D., Institute of Biochemistry and Endocrinology, Justus-Liebig-University of Giessen, Frankfurter Strasse 100, D-35392 Giessen, Federal Republic of Germany.

Acta Neurochir (1994) [Suppl] 60: 101–103
© Springer-Verlag 1994

Effect of Steroids on Brain Lipocortin Immunoreactivity

K. G. Go[1], **F. Zuiderveen**[1], **L. De Ley**[2], **J. G. Ter Haar**[2], **L. Parente**[4], **E. Solito**[4], and **W. M. Molenaar**[3]

Departments of [1] Neurosurgery, [2] Clinical Immunology, and [3] Pathology, University of Groningen, The Netherlands, and [4] Immunobiological Research Institute, Siena, Italy

Summary

LCT- 1, LCT-2 and LCT-5 were assessed in uninjured rats and rats subjected to a cortical freezing injury or middle cerebral artery (MCA) occlusion. Apart from animals receiving no treatment, other uninjured or injured animals received methylprednisolone (2 or 30 mg/kg) or the 21-aminosteroid U-74389F (10 mg/kg) one day and 2 hours before killing. The animals were killed by decapitation 1 hour after the freezing injury or the MCA occlusion and the area containing the lesion was removed and frozen in Freon. Frozen sections were treated with rabbit polyclonal anti-LCT antibody; binding of antibody was visualized by horseradish peroxidase-conjugated swine antirabbit antibody. Without steroid pretreatment, in the uninjured brain LCT immunoreactivity was absent in the greater part of the brain, except in sporadic microglia. In steroid-pretreated animals and in the freezing lesion of both pretreated and untreated animals there was extensive immunostaining; in the freezing lesion it may be due to passage of systemic LCT across the impaired blood-brain barrier in the lesion. The cellular elements showing immunostaining were meningeal cells, neurons, ependyma, choroid plexus, oligodendroglia and capillary endothelium. It implies that also in the brain the steroid effect is consistent with LCT formation.

Keywords: Lipocortins; glucocorticosteroids; 21-aminosteroids; brain edema.

Inroduction

For many years glucocorticosteroids have been used clinically in the prophylaxis and treatment of vasogenic brain edema. While they proved to be especially beneficial (even in low dose) in brain edema associated with brain tumors, they are less effective in brain trauma and stroke[7]. Recently, however, high dose methylprednisolone has been reported capable of improving the functional result of patients with spinal cord injury[2,3]. Furthermore, a group of 21-aminosteroid compounds has been introduced that is supposed to exert a similar protective effect on traumatized neural tissue while lacking the other glucocorticoid effects[8].

It has been shown, mostly in various organs other than the brain that the effect of glucocorticosteroids upon inflammation is based upon the formation of lipocortins (LCT), proteins that inhibit the enzyme phospholipase A_2[1,10,12]. This enzyme is supposed to play a role in inflammatory processes by releasing arachidonic acid from membrane phospholipids; it is known that arachidonic acid may serve as a precursor for eicosanoids, such as prostaglandins, leukotrienes and lipoxins, substances considered as mediators of inflammation. In the brain, arachidonic acid occurs in increased amounts after various insults. Arachidonic acid itself is capable of inducing brain edema, both in vitro and in vivo following its intracerebral administration[4,5].

Lipocortins (also called annexins or calpactins) constitute a family of calcium binding proteins with a molecular weight of 32–67 kDa. Analysis of their primary sequences has identified 6 members (LCT 1-6). LCT 1–5 have in common a core with 4 repeats of 70 amino acids, whereas the core of LCT-6 contains 8 repeats. The N-terminus is variable, being longest in LCT-1 and of decreasing length in the other LCTs, in the order of LCT-2, LCT-6, LCT-3, LCT-5 and LCT-4[11]. In the brain the presence of LCTs has been demonstrated immunohistochemically in choroid plexus and ventricular ependyma only, immunoreactivity was absent in the major part of the brain[9,13]. This study aims to assess their presence in the brain, under pathological conditions, as well as after steroid administration.

Materials and Methods

Adult Wistar rats of 180 g mean body weight, anesthetized with fentanyl (0.1 mg/kg), atropine (1 mg/kg) and diazepoxide (2.5 mg/

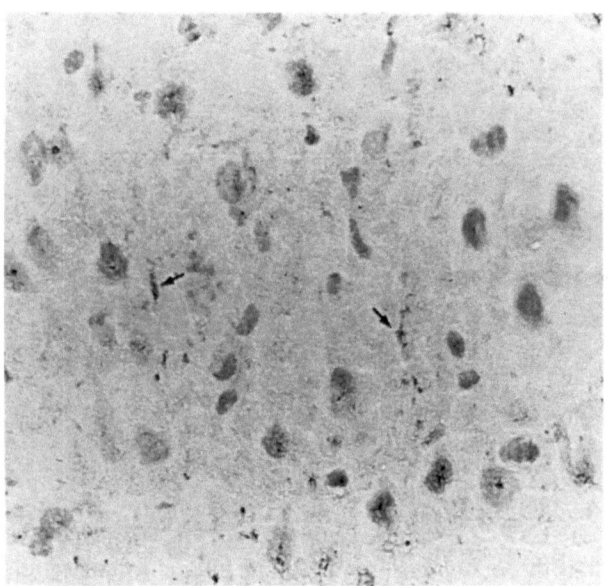

Fig. 1. Absence of LCT-1 immunostaining in brain parenchyma, except in sporadic microglia (arrows) in an untreated uninjured animal

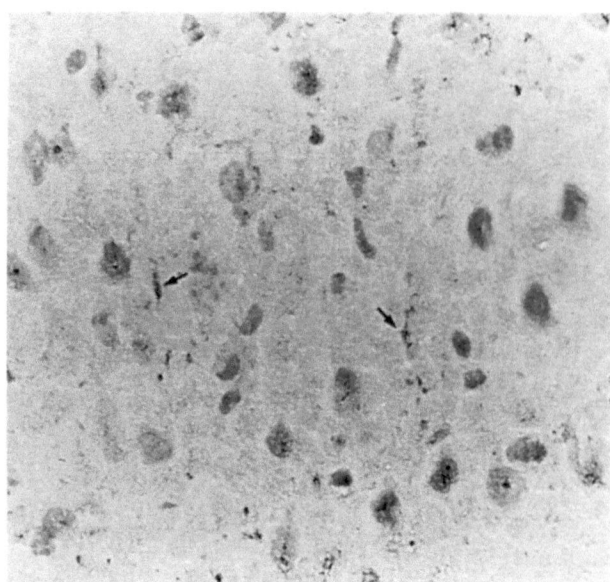

Fig. 2. LCT-1 immunostaining of neurons in cerebral cortex, many of which contain nucleoli in an uninjured rat treated with 21-aminosteroid

kg), were subjected to: (A) a freezing injury by application of a dry ice rod to the exposed cerebral cortex for 1 min; or (B) ischemia by occlusion of a middle cerebral artery according to the technique of Tamura *et al.*[14]; apart from (C) control animals that were not injured. Besides animals not receiving steroids, rats were treated with low dose (2 mg/kg) methylprednisolone, high dose (30 mg/kg) methylprednisolone, or 10 mg/kg of the 21-aminosteroid (U-74389F, Upjohn Co.), the day before and 2 hours before killing. The animals were killed by decapitation, the area containing the freezing lesion or infarction was taken from the ipsilateral cerebral hemisphere and quickly frozen in Freon. Frozen sections of 5 μm thickness were cut, and treated with antilipocortin antibody in 1 : 20 dilution. Binding of antibody was visualized by horseradish perox-idase-conjugated swine antirabbit antibody (Dako, 1 : 40) developed with 2-aminoethyl-carbazole + H_2O_2. The sections were weakly counterstained with Hematoxylin[6]. The antilipocortin anti-bodies were polyclonal antibodies, which had been raised in rabbits against recombinant human LCT-2, and against the N-terminus peptides of LCT-1 and LCT-5 (amino acids 15-31 for LCT-1, and amino acids 1–11 for LCT-5). For immunization the peptides (1 mg/ml) were cross-linked to 400 μg keyhole limpet hemocyanin by incubation with 0,5 ml of 20 mM glutaraldehyde for 30 min at room temperature. In Western blotting experiments the antibodies recognized the different proteins with no cross-reactivity. Tris buffered saline (TBS) was used as negative control.

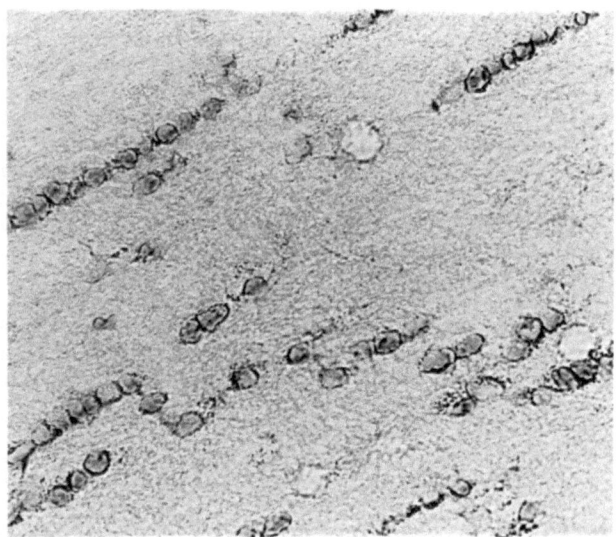

Fig. 3. LCT-1 immunoreactivity of rows of oligodendroglial cells in subcortical white matter of an uninjured rat treated with 21-aminosteroid

Results and Discussion

There was strong enhancement of LCT immuno-reactivity in the nervous parenchyma of glucocortico-steroid and 21-aminosteroid-pretreated animals in con-trast to the absence of spontaneous immunoreactivity in the majority of the brains of untreated animals. In untreated animals immunoreactivity was only local-ized in sporadic microglial cells (Fig. 1). In steroid-

pretreated animals enhanced immunoreactivity resided in neurons (Fig. 2) as well, and in oligodendrocytes (Fig. 3) in the 21-aminosteroid-pretreated animals. The choroid plexus consistently showed immunoreactivity in animals treated with steroid, as reported previously[9]. As the choroid plexus is one of the circumventricular organs possessing capillaries devoid of barrier proper-ties, the immunostaining of its stroma may reflect LCT of systemic origin, while immunostaining of choroid epithelial cells per se seems to indicate local LCT formation. Immunostaining of ventricular ependymal cells as reported previously[9,13] was seen after 21-aminosteroid pretreatment. Immunoreactivity to LCT-2 in capillary endothelium was only seen after 21-aminosteroid pretreatment.

The results after steroid pretreatment clearly indicate the ability of the steroids to cross the blood-brain barrier, which was to be expected in view of their lipophilicity. From the order of appearance of immunostaining, it seems that the following order of inducibility of LCT formation may be derived: microglia, choroid plexus epithelium, meningeal cells, neurons, ventricular ependyma, oligodendroglia, capillary endothelium. The preferential effect of the steroids on microglia is in accordance with previous findings of a preferential effect on macrophages[12]. The immunoreactivity was strongest for LCT-1, and less pronounced for LCT-2 and LCT-5.

In the present study steroids were administered 1 day and 2 hours before the experiment. When dexamethasone was administered to the animals 2 hours before killing, the pattern of LCT immunoreactivity was the same as that in untreated animals, in that it occurred in certain regions only, and was absent in the larger part of the brain[13].

In the freezing lesion as well as the infarction the diffusely scattered immunostaining of the injured parenchyma may also originate from systemic LCT, as it was evident even in untreated animals. Its increased presence within the injured cells may well reflect an unspecific tendency of injured cells to ingest proteins, as shown for other proteins such as albumin that has leaked into the parenchyma following blood-brain barrier disruption.

Although the 21-aminosteroids are reported to have no influence upon glucose metabolism, in the present study they had a strong effect in inducing LCT formation.

References

1. Blackwell GJ, Carnuccio R, Di Rosa M, Flower RJ, Parente L, Persico P (1980) Macrocortin: a polypeptide causing the antiphospholipase effect of glucocorticoids. Nature 287: 147–149

2. Bracken MB, Sheparo MJ, Collins WF, Holford TR, Young W, Baskin DS, Eisenberg HM, Flamm E, Leo-Summers L, Marcoon J, Marshall LF, Perot PL, Piepmeier J, Sonntag VKH, Wagner FC, Wilberger JE, Winn HR (1990) A randomized, controlled trial of methylprednisolone or naloxone in the treatment of acute spinal cord injury. N Engl J Med 322: 1405–1411

3. Braughler JM, Hall ED, Means ED, Waters TR, Anderson DK (1987) Evaluation of an intensive methylprednisolone sodium succinate dosing regimen in experimental spinal cord injury. J Neurosurg 61: 124–130

4. Chan PH, Fishman RA (1978) Brain edema: induction in cortical slices by polyunsaturated fatty acids. Science 201: 358–360

5. Chan PH, Fishman RA, Caronna J, Schmidley JW, Prioleau G, Lee J (1983) Induction of brain edema following intracerebral injection of arachidonic acid. Ann Neurol 13: 625–632

6. De Ley L, Poppema S, Klein Nulend J, Ter Haar JG, Schwander E, The TH (1984) Immunoperoxidase staining on frozen tissue sections as a first screening assay in the preparation of monoclonal antibodies against small cell carcinoma of the lung. Eur J Clin Oncol 20: 123–128

7. Go KG (1992) Cerebral Pathophysiology. An integral approach with some emphasis upon clinical implications. Elsevier, Amsterdam, pp 1–432

8. Hall ED, McCall JM, Yonkers PA, Chase RL, Braughler JM (1987) A nonglucocorticoid analog of methylprednisolone duplicates its high dose pharmacology in models of CNS trauma and neuronal membrane damage. J Pharmacol Exp Ther 242: 137–142

9. Johnson MD, Kamso-Pratt JM, Whetsell WO, Pepinsky RB (1989) Lipocortin-1 immunoreactivity in the normal human central nervous system and lesions with astrocytes. Am J Clin Pathol 92: 424–429

10. Parente L, Becherucci C, Perretti M, Solito E, Mugridge KG, Galeotti CL, Raugei G, Melli M, Sanso M (1990) Are the lipocortins the second messengers of the anti-inflammatory action of glucocorticoids? Cytokines and lipocortins in inflammation and differentiation. Wiley-Liss, New York, pp 55–68

11. Pepinsky RB, Tizard R, Mattaliano RJ, Sinclair LK, Miller GT, Browning JF, Chow EP, Burne C, Huang KS, Pratt D, Wachter L, Hession C, Frey AZ, Wallner BP (1988) Five distinct calcium and phospholipid binding proteins share homology with lipocortin I. J Biol Chem 263: 10799

12. Solito E, Raugei G, Melli M, Parente L (1991) Dexamethasone induces the expression of the mRNA of lipocortin 1 and 2 and the release of lipocortin 1 and 5 in differentiated, but not undifferentiated U-937 cells, FEBS Lett 291: 238–244

13. Strijbos PJLM, Tilders FJH, Carey F, Forder R, Rothwell NJ (1991) Localization of immunoreactive lipocortin-1 in the brain and pituitary gland of the rat. Effects of adrenalectomy, dexamethasone and colchicine treatment. Brain Res 553: 249–260

14. Tamura A, Graham DI, McCulloch J, Teasdale GM (1981) Focal cerebral ischaemia in the rat: 1. Description of technique and early neuropathological consequences following middle cerebral artery occlusion. J Cereb Blood Flow Metab 1: 53–60

Correspondence: Prof. K. G. Go, M.D., Afdeling Neurochirurgie, Academisch Ziekenhuis Groningen, Oostersingel 59, Postbus 30.001, NL-9700 RB, Groningen, The Netherlands.

Acta Neurochir (1994) [Suppl] 60: 104–106
© Springer-Verlag 1994

The Effects of Atrial Natriuretic Peptide on Brain Edema, Intracranial Pressure and Cerebral Energy Metabolism in Rat Congenital Hydrocephalus

J. Minamikawa, H. Kikuchi, M. Ishikawa, K. Yamamura, and **M. Kanashiro**[1]

Department of Neurosurgery, Kyoto University and [1] Nuclear Magnetic Resonance Laboratory, National Cardiovascular Center, Osaka, Japan

Summary

Atrial natriuretic peptide (ANP) regulates fluid and electrolyte homeostasis in the central nervous system. In this study, we evaluated the effects of ANP on brain edema, intracranial pressure (ICP) and cerebral energy metabolism in congenital hydrocephalus in rats. Brain edema, indicated by the longitudinal relaxation time (T1), was evaluated by [1]H-magnetic resonance imaging (MRI). The ICP was monitored with a miniature pressure-transducer with telemetric system. Cerebral energy metabolism, indicated by PCr/Pi ratio, was measured by ^{31}P-magnetic resonance spectroscopy (MRS). The rats were given 10 μl of ANP in the left cerebral ventricle. Three different concentrations of ANP were given; 0.2 (group I), 2.0 (group II) and 20.0 (group III) μg/10 μl, respectively. 10 μl of saline was injected into the ventricle of the control group rats. There were no significant changes of ICP, T1 value and PCr/Pi ratio among the control group, group I and group II. In group III, in contrast, ICP decreased significantly at 20 minutes after ANP administration and stayed at this ICP level for 60 minutes. The T1 value decreased and PCr/Pi ratio increased 30 minutes after ANP administration. This study revealed that intraventricularly administered ANP could decrease ICP, reduce brain edema and improve the cerebral energy metabolism in rats with congenital hydrocephalus.

Keywords: Atrial natriuretic peptide; brain edema; cerebral energy metabolism; congenital hydrocephalus.

Introduction

Atrial natriuretic peptide (ANP) is a unique peptide hormone that regulates blood pressure and fluid homeostasis[8]. Recent studies have proven that a group of hypothalamic neurons contained ANP in their cell bodies and nerve terminals[10], and ANP receptors were found in circumventricular structures[6] and choroid plexus[11]. ANP has also been shown to have beneficial effects on brain edema[7] and cerebrospinal fluid (CSF) production[11].

In the present study, effects of ANP administration on brain edema, intracranial pressure (ICP) and cerebral energy metabolism were evaluated in rat congenital hydrocephalus.

Materials and Methods

In this experimental study, sixty rats of HTX strain[5] aged from 19 days to 27 days after birth were used. HTX is known as a rat strain of congenital hydrocephalus with multifactorial hereditary mode. Hydrocephalus was noted with enlarged head among 30 to 50% of newborn rats. It developed progressively and animals die at about fourth or fifth week. Under light anesthesia with urethane, the left cerebral ventricle was punctured with a fine needle and 10 μl of alpha-human atrial natriuretic peptide (ANP) (Santory Corp, Japan) was injected. Three different concentrations of ANP were given; 0.2 (group I), 2.0 (group II) and 20.0 (group III) μg/10 μl, respectively. 10 μl of saline was injected into the ventricle of the control group rats. Five rats were studied in each group.

For monitoring ICP, a miniature pressure-transducer with a small cannula in the right cerebral ventricle was indwelled below the skin and the ICP was recorded with a telemetric system (PhysioTel, Data Science Inc., U.S.A.). ICP was monitored continuously for 60 minutes after ANP administration.

Experimental proton magnetic resonance imaging ([1]H-MRI) equipment (JEOL JNM-IM600, Nihon Denshi Co., Japan) was used for evaluating the brain edema. In the coronal sections of MRI, a region of interest (ROI) for calculated T1 (longitudinal relaxation time) value was placed in the parietal brain including cortex and subcortex. Data acquisition time for calculated T1 value was 21 minutes, and changes of T1 value were evaluated before and at 30, 60 minutes after ANP administration.

For evaluating cerebral energy metabolism, phosphorus-31 magnetic resonance spectroscopy (^{31}P-MRS) was done. In vivo ^{31}P-MRS spectra were obtained from a JEOL-SMR 270 spectrometer (Nihon Denshi Co., Japan; 6.34 Tesla). The ROI was placed at frontal and parietal areas. As an indicator of cellular bioenergetic status, the phosphocreatine (PCr) /inorganic phosphate (Pi) ratio was evaluated before and at 15, 30, 45 and 60 min after ANP administration.

Statistical analysis was performed by a one-way analysis of variance (ANOVA). Bonferroni's correction was used for the post-hoc test. A probability (p) of less than 0.05 was considered to be significant.

Results

1) Changes of Intracranial Pressure (ICP)

In the control group, group I and group II, changes of ICP were not statistically significant. In Group III, the ICP decreased significantly 20 minutes after ANP administration, and continued at this ICP level for 60 minutes, at the completion of observation ($p < 0.01$).

2) Changes of Brain Edema

Calculated changes in the T1 value were evaluated for brain edema for all four groups. In the control group, group I and group II, calculated changes in the T1 value were not statistically significant. In group III, it was significantly decreased at 30 minutes after ANP administration ($p < 0.01$) and lasted for 60 minutes (Fig. 1).

3) Changes of Cerebral Energy Metabolism

Serial changes of PCr/Pi were evaluated for cerebral energy metabolism. There were no statistically significant changes except in one of the groups. In group III, as in the above two studies, statistically significant changes were noted; the ratio increased at 30 minutes and stayed at an enlarged level for 60 minutes ($p < 0.01$) (Fig. 2).

Discussion

ANP has been found in a group of hypothalamic neurons in the brain[10], and binding sites for ANP have been demonstrated in the circumventricular structures[6] and the choroid plexus[11]. It might play an important role in the regulation of fluid homeostasis and cerebrospinal fluid production. In the present study, a high dose of intraventricularly-administered ANP yielded a decrease of ICP, a shortening of calculated T1 value and an increase of cerebral energy metabolism.

Steardo[11] reported ANP-induced elevation of cGMP in endothelial cells of choroid plexus in vitro, and a decrease of CSF production with intraventricularly administered ANP in rabbits. This inhibitory effect of ANP on CSF production might be a major factor for a decrease of ICP.

Bakker et al.[1] showed a good correlation between tissue water content and longitudinal relaxation time (T1) in nuclear magnetic resonance studies. As brain edema develops, water content increases and T1-value in [1]H-MRI becomes prolonged. The high dose ANP yielded a shortening of T1 in hydrocephalic rats, which suggests a beneficial effect of ANP on brain edema in hydrocephalus.

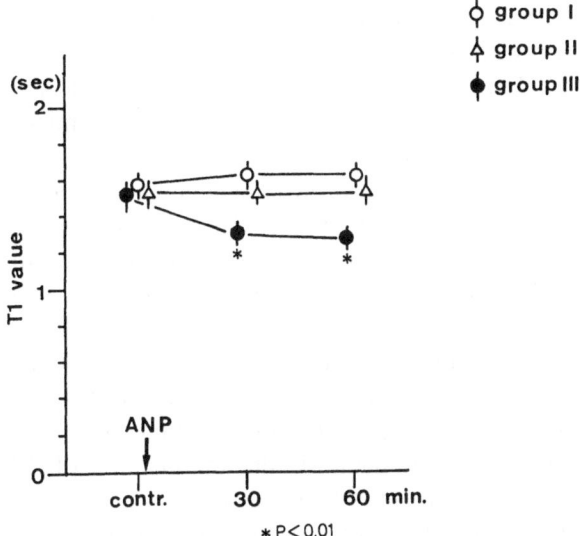

Fig. 1. Serial changes of T1 in rats of ANP administered groups

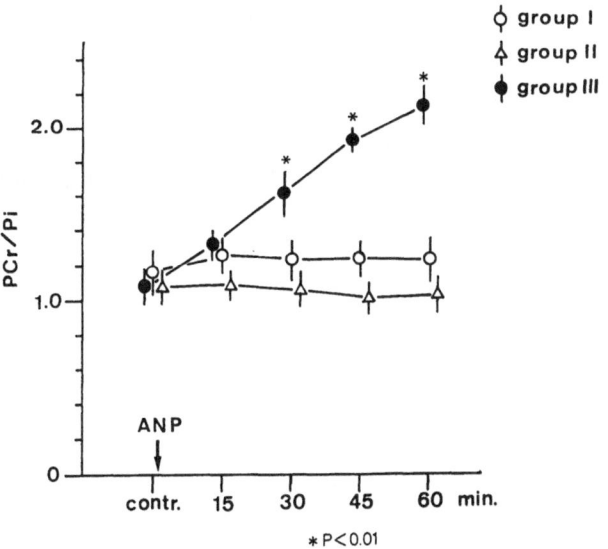

Fig. 2. Serial changes of cerebral energy metabolism in rats of ANP administered groups

Recent studies revealed binding sites for ANP in astroglia-rich brain cells[2] and brain microvessel endothelial cells[11]. ANP is reported to induce elevation of guanosine 3',5'-cyclic monophosphate (cGMP) in astroglia-rich brain cells[2]. In astrocyte, elevation of cGMP can prevent sodium influx and reduce water accumulation, which suggests that ANP may have a beneficial effect on cytotoxic edema. Naruse et al.[7] revealed beneficial effects of ANP on ischemic brain edema in proton magnetic resonance studies. In brain microvessel endothelium, cGMP has been shown to increase macromolecular transport[4], which might affect vasogenic brain edema. Although the mechanism of effectiveness on hydrocephalic edema remains to be clarified, intraventricularly administered ANP might cause a direct action on astrocytes and/or microvascular endothelium.

The PCr/Pi ratio is regarded as a good indicator of cerebral energy metabolism[3,9]. The present study revealed that intraventricularly-administered ANP improved cerebral energy metabolism in the hydrocephalic rats. In hydrocephalic rats, the brain tissue is under the condition of chronic mild cerebral ischemia due to an increase of ICP and progression of hydrocephalic edema. Improvement of cerebral energy metabolism in this study may be due to a decrease of ICP and an improvement of brain edema.

Thus, a high dose of intraventricularly administered ANP yielded a decrease of ICP, reduction of brain edema and improvement of cerebral energy metabolism. This suggests that ANP might be a new therapeutic alternative for treatment of hydrocephalus.

References

1. Bakker CJG, Vriend J (1983) Proton spin-lattice relaxation studies of tissue response to radiotherapy in mice. Phys Med Biol 28: 331–340
2. Friedl A, Harmening C, Schmalz F, Schuricht B, Schiller M, Hamprecht B (1989) Elevation by atrial natriuretic factor of cyclic GMP level in astroglia-rich cultures from murine brain. J Neurochem 52: 589–597
3. Gyulai L, Roth Z, Leigh JS, Chance B (1985) Bioenergetic studies of mitochondrial oxidative phosphorylation using 31Phosphorus NMR. J Biol Chem 260: 3947–3954
4. Joó F, Temesvari P, Dux E (1982) Regulation of the macromolecular transport in the brain microvessels: the role of cyclic GMP. Brain Res 278: 165–174
5. Kohn DF, Chinookswong N, Chou SM (1981) A new model of congenital hydrocephalus in rats. Acta Neuropathol (Berl) 54: 211–218
6. Mantyh CR, Kruger L, Brecha NC, Mantyh PW (1987) Localization of specific binding sites for atrial natriuretic factor in the central nervous system of rat, guinea pig, cat and human. Brain Res 412: 329–342
7. Naruse S, Aoki Y, Takei R, Horikawa Y, Ueda S (1991) Effects of atrial natriuretic peptide on ischemic brain edema in rats evaluated by proton magnetic resonance method. Stroke 22: 61–65
8. Needleman P, Greenwald JE (1986) A cardiac hormone intimately involved in fluid, electrolyte, and blood-pressure homeostasis. N Engl J Med 314: 828–834
9. Nilsson B, Norberg K, Siesjö BK (1975) Biochemical events in general ischemia. Br J Anesth 47: 751–760
10. Samson WK (1987) Atrial natriuretic factor and the central nervous system. Endocrinol Clin North Am 16: 145–161
11. Steardo L, Nathanson JA (1987) Brain barrier tissue: end organs for atriopeptide. Science 235: 470–473

Correspondence: Jun Minamikawa, M.D., Department of Neurosurgery, Kyoto University, 2-1-7-303, Kokumachi, Chuo-ku, Osaka 540, Japan.

Blood-Brain Barrier Function

Acta Neurochir (1994) [Suppl] 60: 109–112

Molecular Biology of Blood-Brain Barrier Ontogenesis and Function

W. Risau

Max-Planck-Institut für Psychiatrie, Martinsried, Federal Republic of Germany

Summary

The vascular system of the central nervous system is derived from capillary endothelial cells, which have invaded the early embryonic neuroectoderm. This process is called angiogenesis and is probably regulated by brain-derived factors. Vascular endothelial cell growth factor (VEGF) is an angiogenic growth factor whose expression correlates with embryonic brain angiogenesis, i.e. expression is high in the embryonic brain when angiogenesis occurs and low in the adult brain when angiogenesis is shut off under normal physiological conditions. VEGF is also a vascular permeability factor (VPF) and, therefore, its expression is also consistent with the formation of the blood-brain barrier by brain endothelial cells, i.e. capillaries are leaky in the embryonic brain but are tight in the postnatal and adult brain. Thus, VEGF/VPF may be a key factor regulating endothelial cell growth and permeability. This notion is further supported by the observation that VEGF expression is induced and strongly upregulated in human malignant glioblastoma. This tumor is characterized by vascular proliferations, vascular leakage and edema.

The differentiation of blood-brain barrier endothelial cells is probably regulated by astrocytes which form foot processes apposed to the abluminal vascular basement membrane. Blood-brain barrier endothelial cells express a set of cell surface proteins that are absent from permeable capillaries. We have characterized one such novel transmembrane glycoprotein which is a new member of the immunoglobulin superfamily. This protein and the analysis of the in vitro characteristics of brain endothelial cells may help to define the molecular mechanisms that are involved in blood-brain barrier induction and permeability.

Keywords: Blood-brain barrier; development; angiogenesis; permeability.

Introduction

The vascular system is a complex network of vessels connecting tissues and organs in the body. Endothelial cells form the inner lining of all blood vessels. These cells have common properties, for example, a non-thrombogenic luminal surface, abluminal basement membrane and von-Willebrand-Factor production. On the other hand, endothelial cells are functionally and morphologically heterogeneous in different tissues which is important for tissue and organ function. On the basis of morphology they have been subdivided into three groups: continuous, fenestrated and discontinuous capillaries.

The endothelial cells of brain capillaries that form the blood-brain barrier belong to the group of continuous capillaries. This was established by Reese and Karnovsky[25] and Brightman and Reese[8] who showed that in rat brain capillaries the permeability barrier was located at the tight junctions between brain endothelial cells (for review see [17,30]).

There are three major functions which are implicated in the term blood-brain barrier: 1) protection of the brain from the blood milieu is achieved by complex tight junctions between endothelial cells, which prevent the unspecific passage of molecules; 2) selective transport by specialized transport systems; and 3) metabolism or modification of blood- or brain-borne substances.

In certain areas of the brain, however, endothelial cells do not form tight junctions but are fenestrated and allow a free exchange of molecules between the blood and adjacent neurons. Most of these areas are situated close to the ventricle and are therefore called circumventricular organs. They include the median eminence (hypothalamus), pituitary, choroid plexus, pineal gland, subfornical organ, organum vasculosum lamina terminalis and area postrema (for review see [5,22]). In order to understand the cellular and molecular mechanisms involved in the establishment of a functional blood-brain barrier it is necessary to investigate the origin and differentiation of the cells concerned.

Development of the Brain Vascular System

A primary vascular plexus surrounds the neural tube very early in embryonic neural development (rat: day

10; chick: day 3). This vascular plexus is derived from migratory angioblasts – the precursors of embryonic blood vessels – that have invaded the head region[2]. From the perineural vascular plexus capillary sprouts invade the neuroectoderm. The endothelial cells first locally degrade the perineural basement membrane. They then migrate deep into the neuroectoderm and branch in the subependymal layer. The steps involved in embryonic brain angiogenesis are similar to tumor angiogenesis in that endothelial cells of an existing vessel degrade a basement membrane and invade into a surrounding tissue. Endothelial cell proliferation is a hallmark of vasculogenesis and angiogenesis and is tightly regulated during development. In the mouse brain, for example, endothelial turnover and sprouting is maximal at postnatal days 6–8[32]. It then ceases and the turnover is very low in the adult brain[15]. However, endothelial cells in the adult are not postmitotic, terminally differentiated cells. Upon a stimulus, e.g. during wound healing or tumor growth, they can rapidly resume cell proliferation giving rise to new capillaries.

Differentiation of Brain Endothelial Cells

There are two possible ways in which endothelial cells could differentiate into a blood-brain barrier: (1) Predetermined neuroectoderm-derived endothelial precursor cells (angioblasts) could differentiate into the blood-brain barrier endothelium, or (2) endothelial cells from outside the CNS could invade the neuroectoderm and differentiate in response to the neural environment. Morphological studies have convincingly shown that endothelial cells of brain capillaries are not derived from the neuroectoderm. Furthermore, embryonic brain tissue transplanted into ectopic sites of host embryos of a different species has been shown to induce angiogenesis from the host. The host origin of endothelial cells in the grafted tissue was demonstrated by the difference in nuclear morphology between chick and quail cells[36] and by using species specific monoclonal antibodies[29]. Moreover, the grafted neural tissue induced blood-brain barrier characteristics in the host-derived endothelial cells.

Development of the Blood-Brain Barrier

Permeability of the blood-brain barrier is usually determined using inert tracers (i.e. those not taken up by brain endothelial cell carriers or receptors) like horseradish peroxidase. By analyzing the permeability of CNS capillaries for this molecule during embryonic

chick development, it was shown that the capillaries gradually became impermeable commencing at day 13[13,38]. Similar data have been presented for mouse embryo[28]. There has been some controversy about the permeability of fetal barriers[31,33] but irrespective of some variations in the exact timing of blood-brain barrier formation it is clear that all the capillaries that are present in the brain before day 13 in the chick and day 15 in the rat lack a mature barrier. Thus, the differentiation of the blood-brain barrier seems to be independent of endothelial cell invasion and proliferation because rat brain endothelial cells in vivo commence barrier formation around embryonic day 16, while they still proliferate.

Differentiated Properties of Brain Endothelial Cells in vivo

The transition from a permeable capillary to a capillary with features of a blood-brain barrier is characterized by several changes in endothelial cell morphology, biochemistry and function which make these endothelial cells distinct from other endothelium in the body. These unique features allow precise control over the substances that leave or enter the brain.

The complex tight junctions between endothelial cells are primarily responsible for the barrier function and unlike simple tight junctions provide a high electrical resistance (up to 2000 Ω cm^2)[12]. Furthermore, there is little bulk flow of molecules through the cells because there are few pinocytotic vesicles and fenestrae. This physiological barrier protects the brain from neurotoxic substances present in the blood, e.g. glutamate and glycine which are neurotransmitters with extracellular levels in the CNS more than 1000fold lower than in the blood, or from high potassium levels. On the other hand, specific carrier systems are present in brain endothelial cells which allow control over the exchange of substances between blood and nervous system (for review of transport at the blood-brain barrier see [4,9,17]).

Several proteins have been localized to brain capillaries in vivo (for review see [27]). Some of these proteins may be blood-brain barrier specific in that sense that they are present at a much reduced or undetectable level in permeable capillaries of the body.

Astrocytes Induce the Blood-Brain Barrier in vivo

Using chick-quail transplantation experiments Stewart and Wiley[36] were the first to unambiguously demonstrate that blood-brain barrier-characteristics can be induced in endothelial cells which had invaded brain transplants. Using the chick-specific monoclonal HT7 antibody we also observed the expression of this blood-brain barrier-specific antigen in chick endothelial cells which had invaded embryonic mouse brain transplanted on the chick chorioallantoic membrane[29]. Conversely, brain capillaries became permeable after invasion of somite transplants[36]. These results indicate that organ specific characteristics of endothelial cells may be induced and maintained be the local environment. Janzer and Raff[19] have provided direct evidence that purified type I astrocytes (when transplanted into the rat anterior eye chamber or the chick chorioallantoic membrane) induce a permeability barrier in invading endothelial cells. The fact that in many brain tumors morphological irregularities of the perivascular ensheathment (enlarged perivascular space, gaps in the basal lamina, unusual or lacking glial investment) correlate with a breakdown of the blood-brain barrier (for review see [18]) suggests that astrocytes are also necessary for the maintenance of the blood-brain barrier. In addition, it supports the concept that endothelial cell differentiation can be modulated by organ and tissue environments.

Cloning of a Blood-Brain Barrier Protein

The HT7 protein is a cell surface glycoprotein and a marker for blood-brain barrier endothelium in vivo. It commences expression in embryonic brain endothelium at the time when the capillaries start to become impermeable, it is not expressed by permeable capillaries, e.g. in peripheral organs and in circumventricular organs, and it can be induced in endothelial cells which invade brain transplants[1,29,34]. Thus, it fulfills all criteria expected for a protein necessary for blood-brain barrier function. We were therefore interested in characterizing the gene encoding this protein. Sequence analysis of the cloned cDNA showed that it is a novel member of the immunoglobulin superfamily[34]. However, the function is still unknown. Homologous proteins have been identified in other species[1,35]. This protein and the analysis of the in vitro characteristics of brain endothelial cells may help to define the molecular mechanisms that are involved in blood-brain barrier induction and permeability.

Role of Growth and Permeability Factors

Growth factors for endothelial cells have been isolated and characterized from embryonic tissues and tumors[26]. Among these, vascular endothelial growth factor (VEGF) is a good candidate for a key regulatory molecule because its expression correlates with angiogenesis and endothelial cell proliferation[3,7,16,20]. For example, during embryonic brain angiogenesis VEGF is expressed in neuroectodermal cells of the subependymal layer correlating in time with the invasion of endothelial cells from the perineural plexus. It is largely switched off in the adult[7]. The high affinity tyrosine kinase receptors that bind VEGF, VEGFR-1 (flt-1[14]) and VEGFR-2 (flk-1[23,37]), are highly expressed in invading and proliferating endothelial cells during embryonic brain development but are downregulated in adult brain endothelial cells[23]. This suggests that VEGF may act as a paracrine angiogenic factor. Furthermore, a similar pattern of expression is observed in malignant brain tumors, astrocytomas and glioblastomas, in which VEGF is expressed in malignant tumor cells and the receptors are upregulated in tumor endothelial cells[24]. Therefore, its expression is also consistent with the formation of the blood-brain barrier by brain endothelial cells, i.e. capillaries are leaky in the embryonic brain and in brain tumors but are tight in the postnatal and adult brain. Interestingly, VEGF has been independently purified as a vascular permeability factor (VPF[10,11,21]) on the basis of its ability to induce macromolecular vascular permeability when injected into the hamster skin (so called Miles assay). The observation that growing vessels, e.g. vessels in tumors, are leaky suggest that VEGF/VPF may be involved in both these processes, proliferation and permeabilty. We have suggested earlier that VEGF/VPF may have a role in the fenestrations of capillaries, e.g. in the circumventricular organs (choroid plexus, area postrema) because the mRNA encoding this factor and its cognate receptors was found to be constitutively expressed in epithelial cells and endothelial cells, respectively, of these organs[6]. Therefore, we speculate that VEGF/VPF plays a major role in developmental and tumor angiogenesis in the brain and possibly elsewhere and may also be involved in hyperpermeability of vessels in tumors resulting in edematous lesions.

References

1. Albrecht U, Seulberger H, Schwarz H, Risau W (1990) Correlation of blood-brain barrier function and HT7-protein distribution in chick brain circumventricular organs. Brain Res 535: 49–61
2. Bär T (1980) The vascular system of the cerebral cortex. Adv Anat Embryol Cell Biol 59: 1–62
3. Berse B, Brown LF, Vandewater L, Dvorak HF, Senger DR (1992) Vascular-permeability factor (vascular endothelial

growth factor) gene is expressed differentially in normal-tissues, macrophages, and tumors. Mol Biol Cell 3: 211–220

4. Betz AL, Goldstein GW (1986) Specialized properties and solute transport in brain capillaries. Anniy Rev Physiol 48: 241–250

5. Bouchaud C, Bosler O (1986) The circumventricular organs of the mammalian brain with special reference to monoaminergic innervation. Int Rev Cytol 105: 283–327

6. Breier G, Albrecht U, Sterrer S, Risau W (1991) Expression of vascular endothelial growth factor during embryonic angiogenesis and endothelial cell differentiation. Development

7. Breier G, Albrecht U, Sterrer S, Risau W (1992) Expression of vascular endothelial growth-factor during embryonic angiogenesis and endothelial-cell differentiation. Development 114: 521–532

8. Brightman MW, Reese TS (1969) Junctions between intimately apposed cell membranes in the vertebrate brain. J Cell Biol 40: 648–677

9. Broadwell RD (1989) Transcytosis of macromolecules through the blood-brain barrier: a cell biological perspective and critical appraisal. Acta Neuropathol (Berl) 79: 117–128

10. Clauss M, Gerlach M, Gerlach H, Brett J, Wang F, Familletti PC, Pan YCE, Olander JV, Connolly DT, Stern D (1990) Vascular-permeability factor – a tumor-derived polypeptide that induces endothelial-cell and monocyte procoagulant activity, and promotes monocyte migration. J Exp Med 172: 1535–1545

11. Connolly DT (1991) Vascular-permeability factor – a unique regulator of blood-vessel function. J Cell Biochem 47: 219–223

12. Crone C, Olesen S-P (1982) Electrical resistance of brain microvascular endothelium. Brain Res 241: 49–55

13. Delorme P, Gayet J, Grignon G (1970) Ultrastructural study on transcapillary exchanges in the developing telencephalon of the chicken. Brain Res 22: 269–283

14. Devries C, Escobedo JA, Ueno H, Houck K, Ferrara N, Williams LT (1992) The fms-like tyrosine kinase, a receptor for vascular endothelial growth-factor. Science 255: 989–991

15. Engerman RL, Pfaffenbach D, Davis MD (1967) Cell turnover of capillaries. Lab Invest 17: 738–743

16. Ferrara N, Houck KA, Jakeman LB, Winer J, Leung DW (1991) The vascular endothelial growth-factor family of polypeptides. J Cell Biochem 47: 211–218

17. Goldstein GW, Betz AL (1986) The blood-brain barrier. Sci Am 255: 70–79

18. Greig NH (1988) In: Neuwelt EA (ed) Implications of the blood-brain barrier and its manipulation. Plenum, New York, pp 77–106

19. Janzer RC, Raff MC (1987) Astrocytes induce blood-brain barrier properties in endothelial cells. Nature 325: 253–257

20. Keck PJ, Hauser SD, Krivi G, Sanzo K, Warren T, Feder J, Connolly DT (1989) Vascular-permeability factor, an endothelial-cell mitogen related to pdgf. Science 246: 1309–1312

21. Keck PJ, Hauser SD, Krivi G, Sanzo K, Warren T, Feder J, Connolly DT (1989) Vascular-permeability factor, an endothelial-cell mitogen related to pdgf. Science 246: 1309–1312

22. Leonhardt H (1980) Ependym und circumventriculäre Organe. In: Oksche A, Vollrath L (eds) Handbuch der mikroskopischen Anatomie des Menschen. Springer, Berlin Heidelberg New York, pp 177–666

23. Millauer B, Wizigmann-Voos S, Schnürch H, Martinez R, Miller NPH, Risau W, Ullrich A (1993) High affinity VEGF binding and developmental expression suggest Flk-1 as a major regulator of vasculogenesis and angiogenesis. Cell 72: 835–846

24. Plate KH, Breier G, Weich HA, Risau W (1992) Vascular endothelial growth-factor is a potential tumor angiogenesis factor in human gliomas invivo. Nature 359: 845–848

25. Reese TS, Karnovsky MJ (1967) Fine structural localization of a blood-brain barrier to exogenous peroxidase. J Cell Biol 34: 207–217

26. Risau W (1990) Angiogenic growth factors. Progr Growth Factor Res 2: 71–79

27. Risau W (1991) Induction of blood-brain-barrier endothelial-cell differentiation. Ann NY Acad Sci 633: 405–419

28. Risau W, Hallmann R, Albrecht U (1986) Differentiation-dependent expression of protein in brain endothblium during development of the blood-brain barrier. Dev Biol 117: 537–545

29. Risau W, Hallmann R, Albrecht U, Henke-Fahle S (1986) Brain induces the expression of an early cell surface marker for blood-brain barrier specific endothelium. EMBO J 5: 3179–3183

30. Risau W, Wolburg H (1990) Development of the blood-brain barrier. Trends Neurosci 13: 174–178

31. Risau W, Wolburg H (1991) The importance of the blood-brain-barrier in fetuses and embryos-reply. TINS 14: 15–15

32. Robertson PL, Du Bois M, Bowman PD, Goldstein GW (1985) Angiogenesis in developing rat brain: an in vivo and in vitro study. Dev Brain Res 23: 219–223

33. Saunders NR, Dziegielewska KM, Mollgard K (1991) The importance of the blood-brain-barrier in fetuses and embryos. TINS 14: 14–14

34. Seulberger H, Lottspeich F, Risau W (1990) The inducible blood-brain barrier specific molecule HT7 is a novel immunoglobulin-like cell surface glycoprotein. EMBO J 9: 2151–2158

35. Seulberger H, Unger CM, Risau W (1992) HT7, neurothelin, basigin, gp42 and ox-47 – many names for one developmentally regulated immunoglobulin-like surface glycoprotein on blood-brain-barrier endothelium, epithelial tissue barriers and neurons. Neuroscience Letters 140: 93–97

36. Stewart PA, Wiley MJ (1981) Developing nervous tissue induces formation of blood-brain barrier characteristics in invading endothelial cells: a study using quail-chick transplantation chimeras. Dev Biol 84: 183–192

37. Terman Bl, Doughe-Vermazen M, Carrion ME, Dimitrov D, Armellino DC, Gospodarowicz D, Böhlen P (1992) Identification of the kdr tyrosine kinase as a receptor for vascular endothelial-cell growth-factor. Biochem Biophys Res Comm 187: 1579–1586

38. Wakai S, Hirokawa N (1978) Development of the blood-brain barrier to horseradish peroxidase in the chick embryo. Cell Tissue Res 195: 195–203

Correspondence: Werner Risau, Ph.D., Max-Planck-Institut für physiologische und klinische Forschung, W. G. Kerckhoff Institut, Abteilung Molekulare Zellbiologie, Parkstrasse 1, D-61231 Bad Nauheim, Federal Republic of Germany.

Acta Neurochir (1994) [Suppl] 60: 113–115
© Springer-Verlag 1994

Effects of Protein Kinase C Activators on Na, K-ATPase Activity in Rat Brain Microvessels

H. Johshita[1], T. Asano[2], T. Matsui[2], and T. Koide[3]

[1] Project Office, Saitama Cardiovascular Center, Takasago, Urawa, Saitama, [2] Department of Neurosurgery, Saitama Medical Center, Kamoda, Kawagoe, and [3] Chugai Pharmaceutical, Kyobashi, Chuo-ku, Tokyo, Japan

Summary

In order to study the possible role of C kinase (PKC) on sodium pump of cerebral vessels, we used diacylglycerol (diC$_8$: sn-1,2-dioctanoylglycerol) and phorbol esters (PMA: phorbol 12-myristate 13-acetate; PDA: phorbol l2,13-diacetate; 4 α-P: 4-alpha phorbol) as PKC activators, and examined their effects on Na,K-ATPase activity in rat brain microvessels (MVs). Rats were divided into non-treated (control; n = 9), four-vessel occlusion (4VO; 30–30 minutes ischemia and recirculation, n = 5), and middle cerebral artery occlusion (MCAO, n = 3) groups. MVs were passed through nylon meshes and were obtained by ultracentrifuge at 58000 g. Na,K-ATPase activity in MVs was determined by the phosphomolybdate method. DiC$_8$ enhanced Na,K-ATPase activity at 10^{-4} M in the control group, the 4VO group and the contralateral hemispheres of the MCAO group (139% ± 0.06**, 135% ± 0.2*, 133% ± 0.18, mean ± SE, * p < 0.05, ** p < 0.01, Wilcoxon rank sum) respectively, but had no effects on MVs in the ipsilateral hemispheres of MCAO group (-74% ± 0.04). This activation by diC$_8$ was inhibited by PKC inhibitors, staurosporine (3×10^{-8} M) and H7 (10^{-6} M) in the control MVs. By contrast, PMA suppressed Na, K-ATPase at 10^{-5} M in the control group (-25% ± 0.07*), but it tended to activate Na,K-ATPase activity in the ipsilateral hemispheres of the MCAO groups (33% ± 0.09). PDA and 4 α-P did not have any consistent effects at the concentration examined. The cause of difference between the effects of diC$_8$ and PMA is unclear at present, but it may stem from the mode of lipid-membrane interaction in these agents and the difference in the condition of cells as well.

Keywords: Na,K-ATPase; rat brain microvessels; protein kinase C; phorbol esters; diacylglycerol; cerebral ischemia.

Introduction

A membrane bound ionic pump, Na,K-ATPase in cerebrovascular endothelium, is considered to be important in regulating ionic homeostasis in the brain[1]. As an intracellular second messenger system, protein kinase C (PKC) activated by intrinsic lipids or phorbol esters might also be involved in the function of this ionic pump[8]. In order to investigate this possibility, we examined in vitro the effects of a sn-diacylglycerol (sn-1,2-dioctanoylglycerol: diC$_8$) as an endogenous PKC activator and active phorbol esters (phorbol 12-myristate 13-acetate: PMA, phorbol 12,13-diacetate: PDA) and an inert phorbol (4-alpha phorbol: 4 α-P) as exogenous activators of PKC on the sodium pump in brain microvessels (MVs) of rats in normal and ischemic conditions.

Material and Methods

Male Sprague-Dawley rats were anesthetized with 50mg/kg of ketamine and 10 mg/kg of xylazine for animal preparation, and were divided into non-treated (control; n - 9), four-vessel occlusion[9] (4VO; 30 minutes ischemia and 30 minutes recirculation, n = 5), and middle cerebral artery occlusion[11] (MCAO 24 hours, n = 3) experimental groups. In the 4VO group, rats underwent bilateral carotid occlusion in awake condition. Animals in which respiratory arrest occurred were excluded. In all experimental group, brain microvessels of rats (n = 3–5) were harvested[5] after heparinized saline perfusion. The hemispheres were minced with scissors in 0.25 M sucrose, 5 mM EDTA, and 50 mM Tris HCl (pH 7.3). The suspension was passed serially through nylon meshes having pore sizes of 670 μm and 335 μm, and ultracentrifuged at 58000 g for 60 min in a discontinuous sucrose density gradient. The Na,K-ATPase activity was measured by subtracting 1 mM ouabain-insensitive activity from the total activity in duplicated samples, by a determination of inorganic phosphate using the phosphomolybdate method[5]. Stocks of diC$_8$, PMA, PDA and 4 α-P were made in dimethylsulfoxide (DMSO). Vehicle concentration was less than 5% at 10^{-3} M of each agent per tube and was below at lower concentration of agents examined. The protein concentration was determined by Bio-Rad protein assay (Richmond, CA). The effects of PKC inhibitor, H-7[4] and staurosporine[10] were further studied about the ATPase activation induced by diC$_8$ at 10^{-4} M.

Results and Discussion

The Na, K-ATPase activity before drug application in MVs of the control group, the 4VO group, the ipsilateral hemispheres of the MCAO group, and the contralateral hemispheres of the MCAO group were 6.7 ± 0.91, 5.37 ± 0.91, 4.63 ± 0.95, 7.01 ± 2.13 µmol Pi/mg protein/hr (mean ± SE) respectively. In the control group, diC_8 significantly enhanced the Na,K-ATPase activity at 10^{-4} M and at 10^{-6} M while PMA showed a suppressive effect at 10^{-5} M compared with the control value. PDA and 4 α-P showed no remarkable activation of Na,K-ATPase in MVs (Fig. 1). In normal MVs, this activation induced by diC_8 at 10^{-4} M ($1.49\% \pm 0.2$, n = 3) was suppressed by a PKC inhibitor H-7 ($0.98\% \pm 0.16$ at 10^{-6} M) and staurosporine (0.67 ± 0.08 at 3×10^{-8} M) respectively. In the 4VO group, diC_8 significantly enhanced Na,K-ATPase activity at 10^{-4} M while other activators showed no constant effects (Fig. 1). In the contralateral hemispheres of MCAO group, a similar pattern of the effects of agents were observed as those in the control and the 4VO group (Fig. 2). By contrast, effect of diC_8 was rather suppressive while PMA tended to enhance Na,K-ATPase activity in the ipsilateral MCAO hemispheres (Fig. 2). This suppression resulted mainly from the

increase of the value of basic ATPase activity compared with that of control values. There is a wide discrepancy in literature regarding the effect of PKC activation on the activity of Na,K-ATPase. Either stimulation or suppression of Na,K-ATPase by various phorbol esters[8] and DAG has been reported with the rat hepatocyte[6], the rabbit diabetic nerve[3], and the rat kidney proximal tuble cells[2], indicating that PKC activation is cell specific. In the present study, the intrinsic PKC activator, diC_8 significantly enhanced Na,K-ATPase activity in rat brain microvessels in the normal condition as well as in recirculation following transient global ischemia whereas all phorbol esters examined did not. In this regard, a lipid requirement of the Na,K-ATPase has been recognized, since a variety of phospholipids activate the enzyme. This activation was observed in requirement of negatively charged lipids such as phosphatidylserine, phosphatidylinositol, or by the length of hydrocarbon of fatty acyl moiety, a determinant of a balance of hydrophobic-hydrophilic character of lipids[12]. Although involvement of such a lipid-membrane interaction cannot be excluded, the

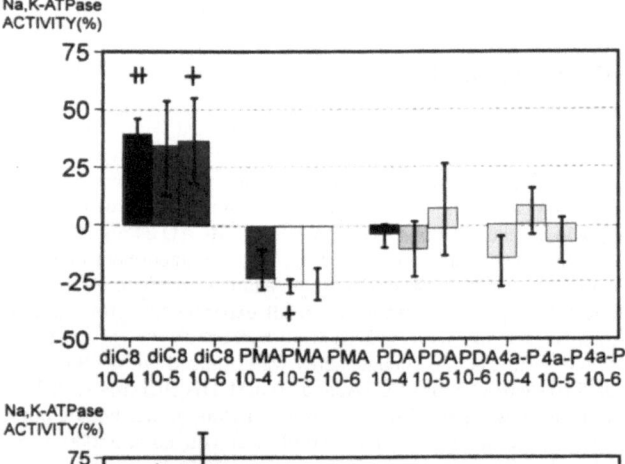

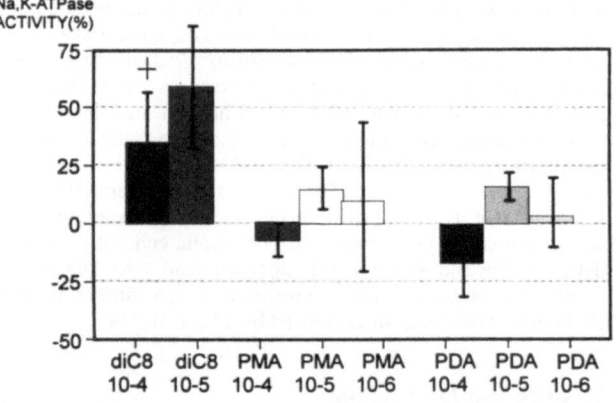

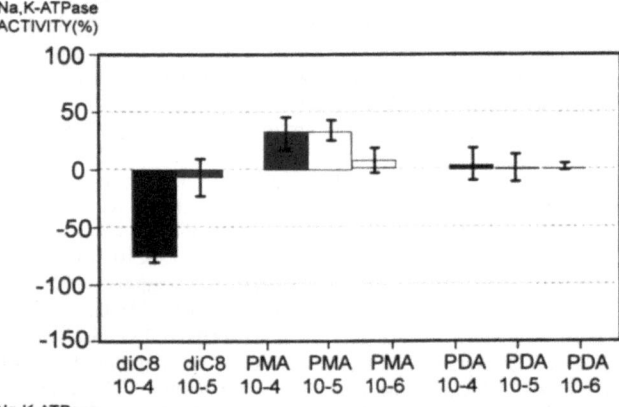

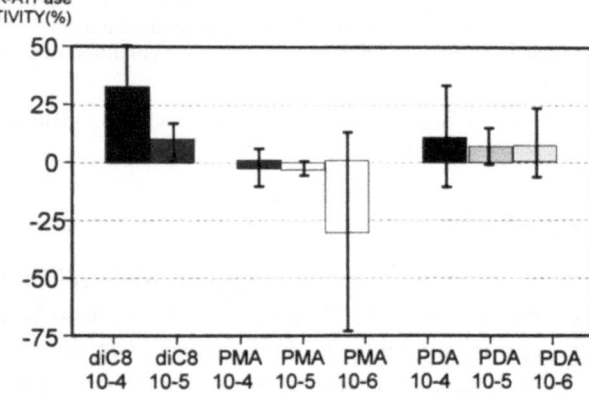

Fig. 1. Upper: Na,K-ATPase activity in MVs of normal rats, mean ± SE, $^+p < 0.05$, $^{++}p < 0.01$ Wilcoxon rank sum test. Lower: Na,K-ATPase activity in MVs of four vessel occluded rats

Fig. 2. Upper: Na,K-ATPase activity in MVs of ipsilateral hemispheres of MCAO rats. Lower: Na,K-ATPase activity in MVs of contralateral hemispheres of MCAO rats

observed enhancement of Na,K-ATPase activity by diC$_8$ is at least partially attributable to PKC activation since it was almost abolished by application of PKC inhibitors. It is of interest that PMA conversely caused suppression of Na,K-ATPase in the control group. The reason for this is unclear, but possible other effects of PMA on endothelial cells such as stimulation of generation of active oxygen species[7] might be thereby involved.

Furthermore, a reversal of the effects of diC$_8$ and PMA was observed in the ipsilateral hemispheres of MCAO group. Although no clear explanation is at hand, this may well reflect the perturbation of the cell membrane as well as the intracellular signaling induced by ischemia.

In conclusion, the results of the present study indicate that the activity of microvessel Na,K-ATPase is significantly influenced by PKC activation. Prolonged ischemia and recirculation affect this relationship in different fashions.

References

1. Betz AL (1983) Sodium transport in capillaries isolated from rat brain. J Neurochem 41: 1150–1157
2. Bertorello A, Aperia A (1988) Na$^+$,K$^+$-ATPase is an effector protein for protein kinase C in renal proximal tubule cells. Am J Physiol 254: F370–F373
3. Greene DA, Lattimer SA (1986) Protein kinase C agonists acutely normalize decreased ouabain-inhibitable respiration in diabetic rabbit nerve. Implications for (Na, K)-ATPase regulation and diabetic complications. Diabetes 35: 242–245
4. Hidaka H, Inagaki M, Kawamoto S, Sasaki Y (1884) Isoquinoline-sulfonamides, novel and potent inhibitors of cyclic nucleotide dependent protein kinase and protein kinase C. Biochemistry 23: 5036–5041
5. Koide T, Asano T, Matsushita H, Takakura K (1986) Enhancement of ATPase activity by a lipid peroxide of arachidonic acid in rat brain microvessels. J Neurochem 46: 235–242
6. Lynch CJ, Wilson PW, Blackmore PF, Exton JH (1986) The hormone-sensitive hepatic Na$^+$-pump. J Biol Chem 261: 14551
7. Matsubara T, Ziff M (1986) Superoxide anion release by human endothelial cells: synergism between a phorbol ester and a calcium ionophore. J Cell Physiol 127: 207–210
8. Nishizuka Y (1986) Studies and perspectives of protein kinase C Science 233: 305–312
9. Pulsinelli WA, Brierley JB (1979) A new model of bilateral hemispheric ischemia in the unanesthetized rat. Stroke 10: 267–272
10. Tamaoki T, Nomoto H, Isami T, Kato Y, Morimoto M, Tomita F (1986) Staurosporine, a potent inhibitor of phospholipid/Ca^{++} dependent protein kinase. Biochem Biophys Res Commun 135: 397–402
11. Tamura A, Graham DI, McCulloch J, Teasdale GM (1981) Focal cerebral ischemia in the rat: 1. description of technique and early neuropathological consequences following middle cerebral artery occlusion. J Cereb Blood Flow Metab 1: 53–60
12. Tanaka R, Sakamoto T (1969) Molecular structure in phospholipid essential to activate (Na$^+$-K$^+$-Mg^{2+})-dependent ATPase and (K$^+$-Mg^{2+})-dependent phosphatase of bovine cerebral cortex. Biochim Biophys Acta 193: 384–393

Correspondence: H. Joshita, M.D., Saitama Cardiovascular Center, 3-15-1 Takasago, Urawa, Saitama 336, Japan.

Acta Neurochir (1994) [Suppl] 60: 116–118
© Springer-Verlag 1994

Effect of Barrier Opening on Brain Edema in Human Brain Tumors

S. Sato, S. Suga, K. Yunoki, and **B. Mihara**

Institute of Brain and Blood Vessels, Mihara Memorial Hospital, Gunma, Japan

Summary

Blood-brain barrier (BBB) opening was carried out in 10 patients with cerebral lesions, and the MRI findings were evaluated following the barrier opening. An intra-arterial injection of 10% glycerol (4 ml/kg, 1 ~ 2 ml/s) was given as a hyperosmotic solution. T_2-weighted MRI was undertaken using a TOSHIBA 22A at 30 minutes after BBB opening. Barrier-opening MRI was performed 10 times in 10 patients, including 5 cases of glioblastoma multiforme, 2 cases of astrocytoma, 1 case of malignant lymphoma, 1 case of cerebral contusion and 1 case of neurinoma. The high-intensity area (HIA) was compared with that in MRI without barrier opening. Three types of changes of HIA in MRI were observed after BBB opening as follows. Type 1: Expansion of the HIA was noted in 4 of 5 cases of glioblastoma multiforme, the 1 case of malignant lymphoma and the 1 case of cerebral contusion. Type 2: Almost no change was observed in the 1 case of neurinoma. Type 3: A decrease in HIA was noted in the 2 cases of astrocytoma and in 1 case of glioblastoma multiforme.

The MRI following BBB opening evidently showed 3 types of changes according to the degree of BBB disruption. Glioblastoma multiforme or contusion with a severely disrupted BBB revealed an increase in HIA following barrier opening. Benign posterior fossa neurinoma showed no change in HIA after barrier opening. Moderate malignant tumors exhibited a decrease in HIA on barrier-opening MRI.

It was concluded that malignant tumors have a severely damaged BBB, which is readily disrupted by osmotic barrier opening.

Keywords: Blood-brain barrier; brain edema; magnetic resonance imaging.

Introduction

Osmotic opening of the blood-brain barrier (BBB) by intra-arterial injection of a hyperosmotic solution has been proposed by Neuwelt et al.[1-3] as one method for increasing the concentration of chemotherapeutic agents in malignant brain tumors. However, there may be several undetermined problems concerning the dosage, method of administration, and mechanism of barrier opening. The authors have previously conducted barrier-opening CT scans based on intra-arterial injec-

tion of mannitol and found that glioblastoma multiforme, which shows an increased contrast area after barrier opening, may be suitable for barrier-opening chemotherapy[6]. During surgery, we performed barrier-opening chemotherapy on malignant brain tumors which were suitable for such chemotherapy using barrier-opening CT scanning. Fourty-five cases of clinical BBB opening have undergone such chemotherapy and diagnosis.

The effect of BBB opening on cerebral edema is still unclear. Through the introduction of MRI, cerebral edema can be visualized more extensively than with CT. We conducted MRI on patients with various brain lesions before and after barrier opening, and evaluated the changes in high-intensity area (HIA) from T_2-weighted images. In this paper, we describe the changes in the brain edema before and after barrier opening as revealed by MRI.

Materials and Methods

Ten patients with brain tumors were studied. They included 5 cases of glioblastoma multiforme, 2 cases of astrocytoma, 1 case of malignant lymphoma, 1 case of cerebral contusion and 1 case of posterior fossa neurinoma. The method employed for barrier opening was based on MRI using glycerol as the barrier-breaking agent at an intra-arterial injection dosage of 4 ml/kg and at an administration speed of 1 ~ 1.5 ml/s. MRI (T_2-weighted image 2000/40) was performed 30 minutes after the barrier opening.

Results

Three types of MRI changes following barrier opening were observed: one group revealed expansion of the HIA (increasing type), the second group showed no changes in HIA (no-change type), and the third group had a decrease in HIA (decreasing type).

Among the patients, 4 of the 5 cases of glioblastoma multiforme, the 1 case of malignant lymphoma and the case of cerebral contusion demonstrated HIA expansion after barrier opening. The left side of Fig. 1 shows a 49-year-old male with glioblastoma multiforme in the basal ganglia and the right side illustrates the MRI following barrier opening (Fig. 1). The HIA was increased after barrier opening. As mentioned, 3 other cases of glioblastoma multiforme also belonged to the increasing type. Three of the increasing type cases were of the malignant type, including the 1 case of malignant lymphoma. The posterior fossa neurinoma exhibited no change in MRI following barrier opening. The 2 cases of astrocytoma, and 1 case of glioblastoma were of the decreasing type. The left

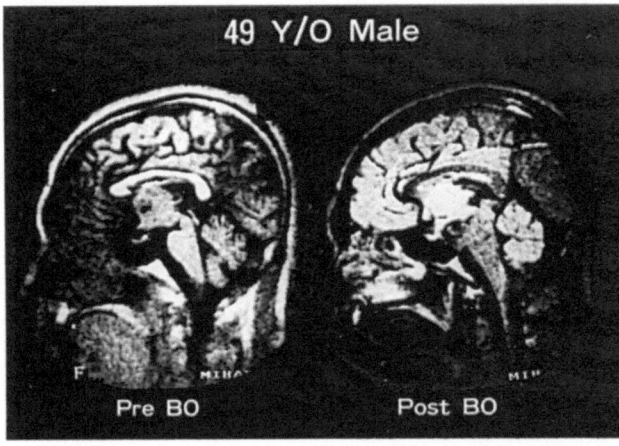

Fig. 1. 49-year-old male with glioblastoma multiforme in the basal ganglia (left side). Right side illustrates MRI following barrier opening

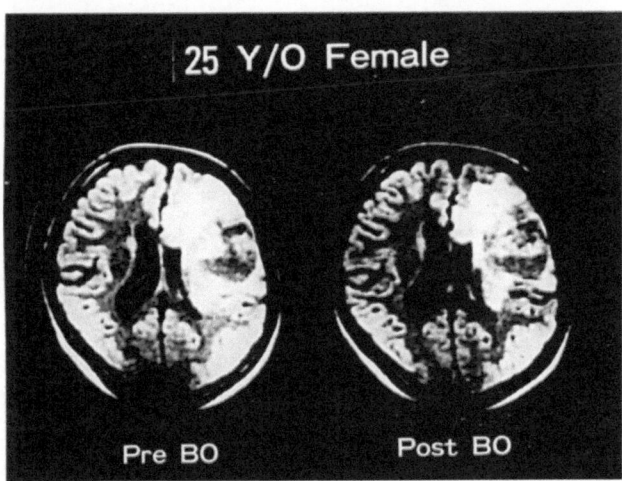

Fig. 2. 25-year-old female with right front temporal astrocytoma (left side). Right side illustrates barrier-opening MRI

side of Fig. 2 shows a 25-year-old female with astrocytoma, and the right side illustrates the barrier-opening MRI (Fig. 2). The HIA of the lesion was decreased by BBB disruption.

No change was observed following barrier opening in the 1 patient with neurinoma, and a decreasing HIA was noted in the 2 cases of astrocytoma and in 1 case of glioblastoma multiforme. An increased HIA was observed in 5 patients, of which 4 were cases of glioblastoma multiforme.

Discussion

Rapoport et al.[4] have suggested that the mechanism of BBB opening results from contact of the endothelium with the intra-vessel hyperosmotic agent, leading to osmotic shrinkage and consequent opening of the tight junction. It is unclear whether the HIA in patients with glioma represents the tumor itself or cerebral edema. In the pathological area, involving either glioma or cerebral edema, the HIA may be modified by barrier disruption. This phenomenon indicates changes in cerebral edema. In view of the differences observed in the extent and location of barrier opening in glioblastoma multiforme and astrocytoma, the stability of the BBB is regarded as an important factor. In malignant tumors, the BBB is relatively weak and barrier opening can be easily achieved with a hyperosmotic solution. In contrast, changes in mode of barrier opening in astrocytoma cases led to a decreased level of cerebral edema. whereas in glioblastoma multiforme, cerebral edema was generally increased. These observations imply that the BBB stability in astrocytoma is greater than that in glioblastoma multiforme. In other words, there are differences in tumor permeation depending on the degree of tumor malignancy, giving rise to cases in which barrier opening occurs and to those in which it does not occur.

Rapoport et al.[5] have reported 0–3 grade barrier opening in the normal brain as a result of intra-arterial injection of arabinose. However, at the speed at which we administered hyperosmotic solution for barrier opening, normal BBB modification did not occur. Barrier opening was observed only in abnormal barrier locations.

In conclusion, the degree of BBB disruption differs according to the malignancy of the tumor concerned, and tumors in which the BBB is severely damaged are indicated for barrier-opening chemotherapy.

References

1. Neuwelt EA, Maravilla KR, Frenkel EP, *et al* (1979) Osmotic blood-brain barrier disruption. Computerized tomographic monitoring of chemotherapeutic agent delivery. J Clin Invest 64: 684–688
2. Neuwelt EA, Frenkel EP (1980) Is there a therapeutic role for blood-brain barrier disruption? Ann Intern Med 93: 137–139
3. Neuwelt EA, Frenkel EP (1980) Effect of osmotic blood-brain barrier disruption of methotrexate pharmacokinetics in the dog. Neurosurgery 7: 36–43
4. Rapoport SI, Hori M, Klatzo I (1971) Reversible osmotic opening of the blood-brain barrier. Science 173: 1026–1028
5. Rapoport SI, Thompson HK, Bidinger JM (1974) Equi-osmolar opening of the blood-brain barrier in the rabbit by different contrast media. Acta Radiol 15: 21–32
6. Sato S, Toya S, Otani M (1985) Barrier opening CT scan. CT Research 7: 43–48

Correspondence: Shuzo Sato, M.D., Institute of Brain and Blood Vessels, Mihara Memorial Hospital, 366 Ootamachi, Isesaki, Gunma 372, Japan.

Acta Neurochir (1994) [Suppl] 60: 119–120
© Springer-Verlag 1994

Simple Quantitative Evaluation of Blood-Brain Barrier Disruption in Vasogenic Brain Edema

Y. Ikeda, M. Wang, and **S. Nakazawa**

Department of Neurosurgery, Nippon Medical School, Tokyo, Japan

Summary

Traumatic damage to the brain results in blood-brain barrier disruption with subsequent formation of vasogenic brain edema. The goal of this study was to evaluate a colorimetric assay of Evans blue extravasation for quantitatively studying blood-brain barrier disruption. Vasogenic brain edema was produced by a cortical freezing lesion. 52 male Wistar rats were sacrificed and cardiac perfusion was performed. A volume of dimethylformamide, twice the brain weight, was added to the brain and this was incubated for 72 hours. The supernatant was analyzed spectrophotometrically. Evans blue contents in the lesioned hemisphere was significantly increased within one hour after the lesion production relative to the normal brain and continued to increase for 24 hours (p < 0.01). This quantitative assay of Evans blue is a simple method for evaluation of blood-brain barrier disruption in vasogenic brain edema.

Keywords: Vasogenic brain edema; cold injury; Evans blue; blood-brain barrier.

Introduction

Brain edema is one of the most important clinical complications in ischemic and traumatic brain injury. The currently mostly used classification of brain edema into vasogenic and cytotoxic edema was introduced by Klatzo[2]. In vasogenic brain edema, a damage to the blood-brain barrier enables plasma proteins and electrolytes to enter the brain, mainly in the white matter. The azo dye Evans blue is known to bind quantitatively to albumin in vivo. This property has been widely used to quantify protein leakage as an indication of increased vascular permeability induced by different mediators. The extravasated Evans blue has been extracted by different solvents and the content determined by colorimetry at the absorbance maximum of 600–620 nm[3]. The goal of this study was to evaluate a colorimetric assay of Evans blue extravasation for quantitatively studying blood-brain barrier disruption in animals with a small amount of white matter.

Materials and Methods

Surgical Procedures

52 male Wistar rats, weighing 250 to 300 g each, were anesthetized intraperitoneally with chloralhydrate. A midline scalp incision and a craniectomy were made in the right parietal region. The cortical cryogenic injury was produced by application of a previously prepared metal probe cooled with dry ice to the dura of the right parietal region . The dura was left intact. The skin was closed with sutures.

The rats were sacrificed at 1 hour (n = 6), 2 hours (n = 7), 4 hours (n = 7), 6 hours (n = 9), 24 hours (n = 8) and 48 hours (n = 7) after lesion production.

Quantitative Evaluation of Blood-Brain Barrier Disruption

The rats received 2 mg/kg of a 2% solution of Evans blue in normal saline administered intravenously prior to the injury. Cardiac perfusion was performed with 200 ml of normal saline to clear the cerebral circulation of Evans blue. The brains were removed and examined for evidence of Evans blue staining. The hemisphere of interest was isolated, mechanically dissociated and weighed. A volume of dimethylformamide, twice the brain weight, was added to the brain and this was incubated for 72 hours in a 50°C water bath. The brain-dimethylformamide mixture was centrifuged at 1500 g for 10 min and the supernatant spectrophotometrically analyzed at the absorption maximum for Evans blue in dimethylformamide (635 nm). A standard curve of Evans blue in dimethylformamide was used to convert absorbance into μg Evans blue/ml dimethylformamide, which was converted into total μg Evans blue for each hemisphere.

Statistical Analysis

Data are presented as mean ± standard deviation. Student's t test was used to assess significance and p < 0.05 was considered statistically significant.

Results

Evans blue content in the ipsilateral cold lesioned hemisphere (expressed in μg/hemisphere) was 1.929 ± 0.219 in normal, 6.238 ± 1.341 at 1 hour, 6.658 ± 1.503 at 2 hours, 9.661 ± 1.736 at 4 hours, 10.785 ± 2.620 at 6 hours, 12.033 ± 3.070 at 24 hours and 8.970 ± 2.177 at 48 hours. Evans blue contents in the lesioned hemisphere were significantly increased within one hour after the lesion production relative to the normal rat brain (p < 0.01) and continued to increase for 24 hours (Fig. 1). Evans blue content in the contralateral hemisphere was 1.678 ± 0.353 in normal, 1.785 ± 0.230 at

1 hour, 1.819 ± 0.209 at 2 hours, 2.512 ± 0.609 at 4 hours, 3.363 ± 1.254 at 6 hours, 4.456 ± 1.349 at 24 hours and 3.880 ± 1.108 at 48 hours. Evans blue content in the contralateral hemisphere was significantly increased at 4 hours relative to the normal rat brain (p < 0.01) and continued to increase for 24 hours (Fig. 2).

Discussion

The importance of the extravasated plasma proteins for formation of vasogenic brain edema has been stressed[1,2]. A close interrelationship between extravasated plasma proteins and vasogenic brain edema has been suggested in a number of experimental investigations[2]. Rats and rabbits with small amount of white matter are not always good experimental animals for evaluation of vasogenic brain edema, because brain edema was strictly confined to the white matter of the affected hemisphere. The Evans blue method for macroscopic evaluation of vascular protein leakage has been widely used. In this study, we used a colorimetric absorbance method. Weissman *et al.*[5] reported that this method is a simple and excellent tool for quantitation of peritumoral brain edema. However, Uyama *et al.*[4] showed that the quantification of Evans blue by colorimetric absorbance measurement has limitations and the detection limit of the blue color makes it difficult to quantify the amount of extravasated dye. This study demonstrated that this colorimetric method may be less sensitive than the fluorescence method, but this quantative method is simple and provides an excellent tool for evaluation of blood-brain barrier disruption in cold induced brain edema.

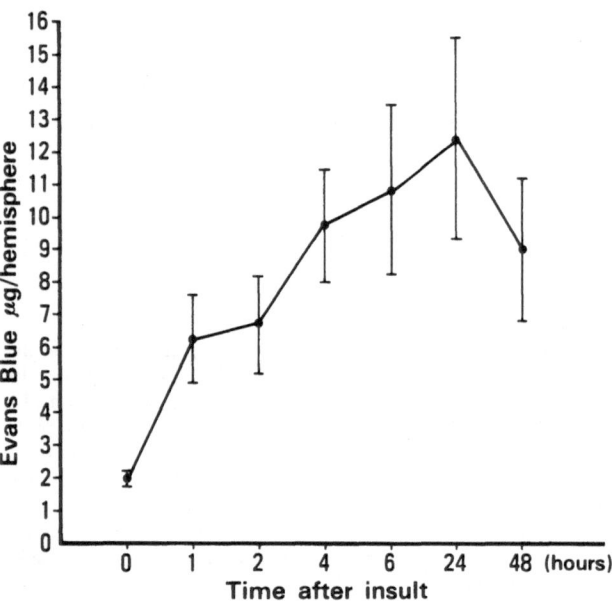

Fig. 1. Chronological changes of Evans blue content in the ipsilateral cold lesioned hemisphere. Each point represents the mean values. The vertical bars indicate standard deviations

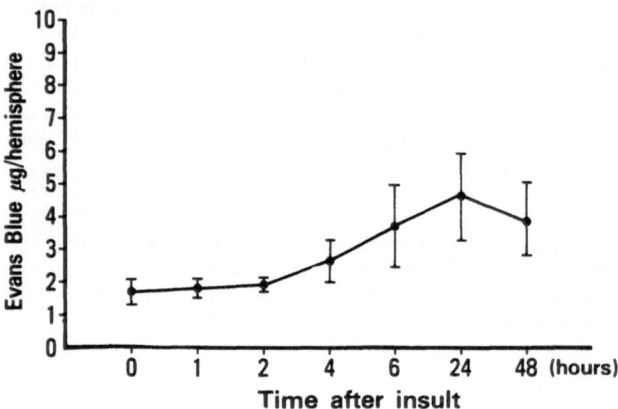

Fig. 2. Chronological changes of Evans blue content in the contralateral hemisphere. Each point represents the mean value. The vertical bars indicate standard deviation

References

1. Johansson BB (1986) Brain edema. Acta Neurochir (Wien) [Suppl] 36: 137–141
2. Klatzo I (1967) Presidential address: neuropathological aspects of brain edema. J Neuropathol Exp Neurol 26: 1–14
3. Saria A, Lundberg JM (1983) Evans blue fluorescence: quantitative and morphological evaluation of vascular permeability in animal tissues. J Neurosci Methods 8: 41–49
4. Uyama O, *et al* (1988) Quantitative evaluation of vascular permeability in the gerbil brain after transient ischemia using Evans blue fluorescence. J Cereb Blood Flow Metab 8: 282–284
5. Weissman DE, Grossman SA (1988) A model for quantitation of peritumoral brain edema. J Neurosci Meth 23: 207–210

Correspondence: Yukio Ikeda, M.D., D.M.Sc., Nippon Medical School, Department of Neurosurgery, 1-1-5, Sendagi, Bunkyo-ku, Tokyo 113, Japan.

Acta Neurochir (1994) [Suppl] 60: 121–123
© Springer-Verlag 1994

Detection of Endogenous Albumin as an Index of Blood Parenchymal Border Alteration

T. Fukuhara, M. Gotoh, M. Kawauchi, S. Asari, T. Ohmoto, K. Tsutsui[1], and **T. Shohmori**[1]

Departments of Neurological Surgery and Neurology [1], Okayama University Medical School, Okayama, Japan

Summary

We used a microdialysis technique to establish a method to detect sequential changes in disruption of the blood parenchymal border. Twelve cats were divided into two groups; one group underwent occlusion of the middle cerebral artery for 60 minutes, the other a cold injury model. Microdialysis probes were implanted bilaterally into the white matter, and dialysates were collected successively at 30 minute intervals for 6 hours in the occlusion model and 4 hours in the cold injury model. Regional cerebral blood flow was measured simultaneously using the hydrogen clearance method. The water content of the white matter was measured using specific gravity. The proteins in the dialysate were analyzed using electrophoresis with silver stain, and, with densitometric analysis, the density of the 66.2 kDa band was quantified as albumin. The ratio of this density to the preoperative density was defined as the "albumin index." On the side of the lesion in the cold injury model, this index significantly increased 3 hours after the cold injury compared with the contralateral side, and a correlation between the water content and this index was observed. The albumin index was believed to indicate the severity of disruption of the blood parenchymal border.

Keywords: Albumin; microdialysis.

Introduction

Many methods for confirming the disruption of the blood parenchymal border have been adopted in experimental models[1,3,4,12]. The common concept among most of these methods is detecting substances in the extracellular space that are usually absent. To quantify these substances, stain with Evans blue[12], autoradiograms[4] and immunohistochemical reactions[3] have been used. The advantage of these quantifications is that the extent of disruption of the blood parenchymal border is seen, but they fail to follow the sequential alteration of the disruption in the same animal. Our study aims at eliminating this disadvantage with a microdialysis technique using a high cut-off membrane.

Materials and Methods

Twelve adult cats weighing 2.3–4.0 kg were divided into two groups; one group underwent a cold injury (6 cats), the other (6 cats) underwent transient occlusion of the middle cerebral artery. The animals were anesthetized with intraperitoneal injections of ketamine (30 mg/kg), and pancuronium bromide relaxation was induced at 0.5 mg/kg/h for mechanical ventilation. The animals were ventilated with halothane at a maintenance dose of 1.0%. The left femoral vein was catheterized for intravenous injection of pancuronium bromide or Evans blue (2%, 2 ml/kg) 30 minutes before sacrifice, and the femoral artery was exposed to measure arterial blood pressure or for rapid injection of KCl (10%, 2 ml/kg) to induce a cardiac arrest. After fixing the animal in a stereotactic frame, the following experiments were carried out.

In the cold injury model, a burr hole was made over the suprasylvian gyrus on each side. A hand-made microdialysis probe and the electrode for measuring regional cerebral blood flow (rCBF) were inserted 5 mm into the white matter through the burr hole. Another burr hole was made over the frontal cortex 7 mm anterior to the microdialysis probe. After a 4-hour equilibration period, dialysates were collected after the first 30 minutes as a sample of the pre-injury state. Cold injury was then induced by applying for 1 minute a metal probe cooled with liquid nitrogen to the exposed cortex with the dura intact. Dialysates were collected every 30 minutes for 4 hours, followed by injection of Evans blue. Thirty minutes after injection of the dye, cardiac arrest was induced through rapid injection of KCl into the femoral artery. The brain was removed promptly and two coronal sections were taken. The first cut was made where the probe was inserted to confirm that the probe reached the white matter without traumatic hemorrhage. The second cut was made 3 mm anterior to the first, and was used to observe the stain of Evans blue. After observation, fractions of the white matter were dissected bilaterally to measure the water content using specific gravity.

For the reperfusion injury model, a burr hole for the insertion of the microdialysis probe was made over the middle ectosylvian gyrus. As in the cold injury model, the probes were inserted and allowed to equilibrate for 4 hours. A dialysate was collected 30 minutes after equilibration as the sample of the pre-occlusion state. The right middle cerebral artery was then occluded for 60 minutes through a transorbital approach. Dialysates were collected every 30 minutes for 6 hours: once during occlusion and once for each of the 5 hours of reperfusion. The rest of the procedure was the same as for the animals in the cold injury model.

The U-shaped dialysis probes were constructed of a 20 G and a 25

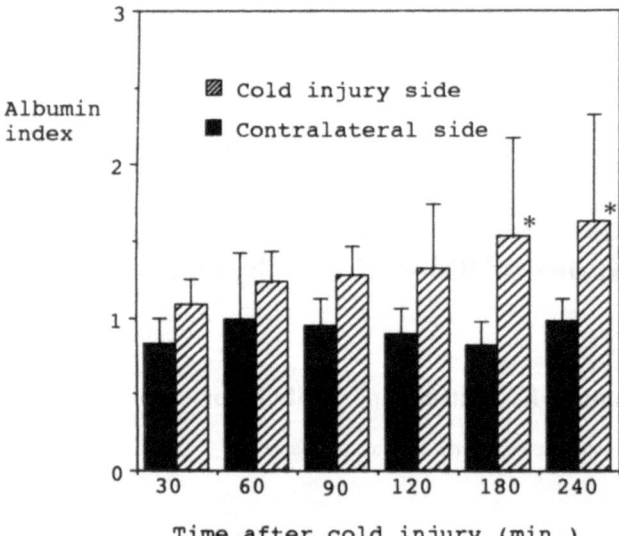

Fig. 1. Sequential change of albumin index in the cold injury model. * p < 0.05 versus contralateral side (paired t-test)

G cannula, using dialysis membrane tubing with a high molecular cut-off (Evaflux 4A, Kuraray Medica, Kitaku, Osaka, Japan). The manufacturer of the membrane has stated that Ig G, with a molecular weight of about 150 kDa, can pass through the pores of this membrane. Each probe was perfused with Ringer's solution at 2 µl/min.

To analyze the albumin in the dialysates, sodium dodecyl sulfate-polyacrylamide gel electrophoresis was carried out. The samples for electrophoresis were 15 µl of the dialysate. After electrophoresis, the gel was stained with silver (Daiichi Pure Chemicals, Tokyo, Japan). With densitometric analysis, the density of the 66.2 kDa band was quantified as albumin. The ratio of this density to the preoperative density was defined as the "albumin index". This index was analyzed in relation to rCBF and cerebral water content.

Results

In the reperfusion injury model, a decrease in rCBF of less than 60% on the occluded side compared with the preoperative value was observed during occlusion of the middle cerebral artery in all animals. Blood flow returned to the former level after reperfusion and no significant change was observed during 5 hours of reperfusion. In the cold injury model, no significant change of rCBF was observed throughout the experiment. The water content was significantly different between the side of the lesion and the contralateral side in both models. The values were 71.66 ± 1.57% on the occluded side and 68.67 ± 1.85% on the contralateral side in the reperfusion injury model, and 72.82 ± 1.96% on the injured side and 69.48 ± 0.85% on the contralateral side in the cold injury model.

In the cold injury model, the albumin index increased significantly after 3 hours compared with the

contralateral side (paired t-test, p < 0.05). In the reperfusion injury model, no significant change was observed. The sequential changes of the albumin index in the cold injury model are shown in Fig. 1. A correlation between the water content (the value of the injury side divided by that of the contralateral side) and the last albumin index before sacrifice was observed in the cold injury model (r = 0.733). Evans blue stain was confirmed only in the area of cold injury but did not spread to the area around the probe. In the reperfusion injury model, no stain was observed in any portion.

Discussion

Although the role of albumin in the formation of brain edema has long been discussed[6,10], extravasated albumin has been regarded at least as an indicator of vasogenic edema[10]. Although many methods to detect extracellular albumin have been established to analyze the function of the blood parenchymal border[1,3,4,12], a time-course study of extracellular albumin in the same animal has never been established.

The microdialysis used in our study is a widely-accepted technique that allows analysis of the sequential changes of the extracellular substances in the same animal[8]. To detect changes in the permeability of the blood parenchymal border, the microdialysis technique was originally adopted to detect injected fluorescence in the reperfusion injury model[7]. Because of its large molecular size, however, endogenous albumin was hard to detect with this technique. Therefore, we used a high cut-off membrane proven to allow albumin to pass[11], and succeeded in obtaining the sequential changes in extracellular albumin.

Our results show the different conditions of the blood parenchymal border between the two models, both of which generated brain edema. In the cold injury model, albumin-rich fluid was exuded rapidly in the extracellular space after the injury[6]. On the contrary, the albumin content did not increase in the reperfusion injury model. The disruption of the blood parenchymal border is believed to depend upon the level of rCBF during ischemia and the duration of ischemia[5]. Recent studies of vasogenic edema have gradually concentrated on the mechanism of secondary damage caused by chemical substances[2,9]. Simultaneous analysis of the extracellular fluid is essential to confirm this mechanism.

The reperfusion injury in our experiments may be insufficient to disrupt the blood parenchymal border to a level at which albumin can pass through. A longer period of reperfusion or a more severe ischemic insult may increase the albumin index in the reperfusion injury model.

References

1. Adams JC (1977) Technical considerations on the use of horseradish peroxidase as a neuronal marker. Neuroscience 2: 141–145
2. Baethmann A, Maier-Hauff K, Schürer L, Lange M, Guggenbichler C, Vogt W, Jacob K, Kempski O (1989) Release of glutamate and of free fatty acids in vasogenic brain edema. J Neurosurg 70: 578–591
3. Chui E, Wilmes F, Sotelo JE, Horie R, Fujiwara K, Suzuki R, Klatzo I (1981) Immunocytochemical studies on extravasation of serum proteins in cerebrovascular disorders. In: Cervós-Navarro J, Fritschka E (eds) Cerebral microcirculation and metabolism. Raven, New York
4. Gotoh O, Asano T, Koide T, Takakura K (1985) Ischemic brain edema following occlusion of the middle cerebral artery in the rat I: the time courses of the brain water, sodium and potassium contents and blood-brain barrier permeability to 125 I-albumin. Stroke 16: 101–109
5. Hallenbeck JM, Dutka AJ (1990) Background review and current concepts of reperfusion injury. Arch Neurol 47: 1245-1254
6. Klatzo I (1967) Neuropathological aspects of brain edema. J Neuropathol Exp Neurol 26: 1–14
7. Łazarewicz JW, Pluta R, Salińska E, Puka M (1989) Beneficial effect of nimodipine on metabolic and functional disturbances in rabbit hippocampus following complete cerebral ischemia. Stroke 20: 70–77
8. Lönnroth P, Smith U (1990) Microdialysis – a novel technique for clinical investigations. J Intern Med 227: 295–300
9. Maier-Hauff K, Baethmann AJ, Lange M, Schürer L, Unterberg A (1984) The kallikrein-kinin system as mediator in vasogenic brain edema. Part 2: studies on kinin formation in focal and perifocal brain tissue. J Neurosurg 61: 97–106
10. Menzies SA, Betz AL, Hoff JT (1993) Contributions of ions and albumin to the formation and resolution of ischemic brain edema. J Neurosurg 78: 257–266
11. Nakamura M, Itano T, Yamaguchi F, Mizobuchi M, Tokuda M, Matsui H, Etoh S, Hosokawa K, Ohmoto T, Hatase O (1990) In vivo analysis of extracellular proteins in rat brains with a newly developed intracerebral microdialysis probe. Acta Med Okayama 44: 1–8
12. Wolman M, Klatzo I, Chui E, Wilmes F, Nishimoto K, Fujiwara K, Spatz M (1981) Evaluation of the dye-protein tracers in pathophysiology of the blood-brain barrier. Acta Neuropathol (Berl) 54: 55–61

Correspondence: Toru Fukuhara, M. D., Department of Neurological Surgery, Okayama University Medical School, 2-5-1 Shikatacho, Okayama, 700, Japan.

Acta Neurochir (1994) [Suppl] 60: 124–127
© Springer-Verlag 1994

Amino Acids in Extracellular Fluid in Vasogenic Brain Edema

I. Westergren[1], **B. Nyström**[2], **A. Hamberger**[2], and **B. B. Johansson**[1]

[1] Department of Neurology, University of Lund, and [2] Institute of Neurobiology, University of Göteborg, Sweden

Summary

We have investigated if changes in brain dialysate, reflecting alterations in brain extracellular composition, can be detected during the development of vasogenic brain edema. The blood-brain barrier was opened by intracarotid infusion of 5 or 10 mg protamine sulphate. The rats were killed two hours after opening of the BBB when the brains were macroscopically edematous, after 10 mg but not 5 mg protamine sulphate. No significant alterations in the amino acid concentration in the dialysate were observed after the lower dose. 40 min after infusion of 10 mg protamine, the level of glutamate was significantly increased in the dialysate followed by that of aspartate, glycine, phosphoethanolamine and taurine 10 min later and a further delayed increase in GABA.

We conclude that the development of vasogenic brain edema is associated with significant increases in extracellular concentration of excitatory amino acids, of taurine, and of phosphoethanolamine and GABA.

Keywords: Amino acids; microdialysis; blood-brain barrier; brain edema; protamine sulphate.

Introduction

A number of studies have indicated that cerebral dialysis is useful in detecting changes in the tissue extracellular fluid[2–4,9,10]. In the present study we have investigated if changes in the extracellular environment can be detected in the dialysate during development of vasogenic edema after opening of the blood-brain barrier (BBB) with no primary tissue necrosis. We opened the BBB by intracarotid infusion of the polycation protamine sulphate which induces transient dose-dependent BBB dysfunction. Whereas 5 mg protamine infused into the carotid artery increases albumin passage through the BBB without increasing the water content of the brain, brain edema develops if the dose is doubled[13,14]. The BBB was opened 24 h after implantation of a dialysis probe and the dialysate was studied for 30 min before and 110 min after opening of the BBB.

Materials and Methods

The study is reported in detail elsewhere[16]. In short, a microdialysis probe (CMA/10, Carnegie Medicine, Sweden, Cat. no 830 9502, dialysis membrane diameter 0.5 mm, steel shaft diameter 0.6 mm) was inserted into the right parietal cortex.

The day after implantation, rats were anesthetized with methohexital i.p. and catheters were inserted in to a femoral artery and a femoral vein. After ligation of the right pterygopalatine artery and of the distal end of the right external carotid artery, a Portex PE 25 catheter connected to PE 50 tubing was inserted into the right external carotid artery towards the carotid bifurcation and secured to the artery. The skin wound was closed and the catheter was exteriorized on the back of the neck.

Anesthesia was continued with i.v. infusion of methohexital, and rectal temperature was kept between 37–38°C with a heating lamp. Heparin was given (270 IU/kg i.v.) and the probe was connected to a microinfusion pump (Carnegie Medicine, CMA 100) and perfused with sterile Krebs-Ringer bicarbonate solution at a flow rate of 2 µl/min. After collection of two 10-min samples, 2% Evans blue, 3 mg/kg, was given i.v. Ten minutes later, protamine sulphate grade X (Sigma), dissolved in 0.9% NaCl and filtered through Millex-HV Milipore 0.45 µm filter, was infused into the carotid artery within 30 s in a dose of 5 mg/100 µl or 10 mg/200 µl. Consecutive 10-min samples of the microdialysis perfusate were collected for 110 min, frozen immediately and stored at −80°C. Amino acids were determined with liquid chromatography[8,16]. Mean arterial pressure (MAP), blood gases, hematocrit (HCT) and glucose were determined before, one hour after the protamine infusion and before sacrifice.

The paired t-test was used for comparison of physiological variables within the groups; one way Anova with Scheffe's post hoc procedure was used for group comparison. For evaluation of changes in concentrations of amino acids we used Friedmann's nonparametric test.

Results

MAP, HCT and blood gases were stable throughout the experiments. Five rats given the higher dose of protamine sulphate did not survive the whole sampling time and showed signs of medullary compression indicating an increased intracranial pressure. Whereas 5 mg protamine sulphate alone did not alter the concentration of extracellular amino acids, infusion of 10 mg protamine sulphate induced significant alterations. The time course of the changes is shown in Table 1. After 40 min, glutamate was significantly increased followed by aspartate, glycine, phosphoethanolamine and taurine 10 min later and a more delayed increase in GABA. The example of an individual rat is shown in Fig. 1. In the rats that died, the increase in taurine was earlier and more pronounced than in survivors (Fig. 2).

Discussion

The concentration of amino acids in the extracellular fluid of the brain, as measured in the dialysate, did not change immediately after opening of the BBB in spite of the considerable concentration gradient between blood and extracellular fluid[4]. One explanation is that the BBB was already altered in the tissue around the probe. The trauma of inserting the probe into the brain alters BBB function as shown by the observation that radioactive inulin, which does not pass through the intact BBB, entered the dialysate to about the same extent after insertion alone as after combined insertion and intracarotid infusion of protamine sulphate. Moreover, extravasation of albumin occurred for at least three days after probe insertion (unpublished observations).

Regulation of influx and efflux of amino acids between blood and brain takes place by a number of complex mechanisms. Whereas several amino acids are transported into the brain, glutamic acid, aspartic acid, glycine and GABA have little access to the brain from the blood, and there is evidence for a net efflux of glutamate and aspartate under physiological conditions[6]. Furthermore, various uptake mechanisms modify amino acid levels in the extracellular fluid. The significant changes in concentrations of glutamate, aspartate, glycine, taurine, phosphoethanolamine and GABA that occurred concomitantly with the development of brain edema is most likely due to an increased release of amino acids from glial cells or neurons, although reduced uptake activity may also play a role.

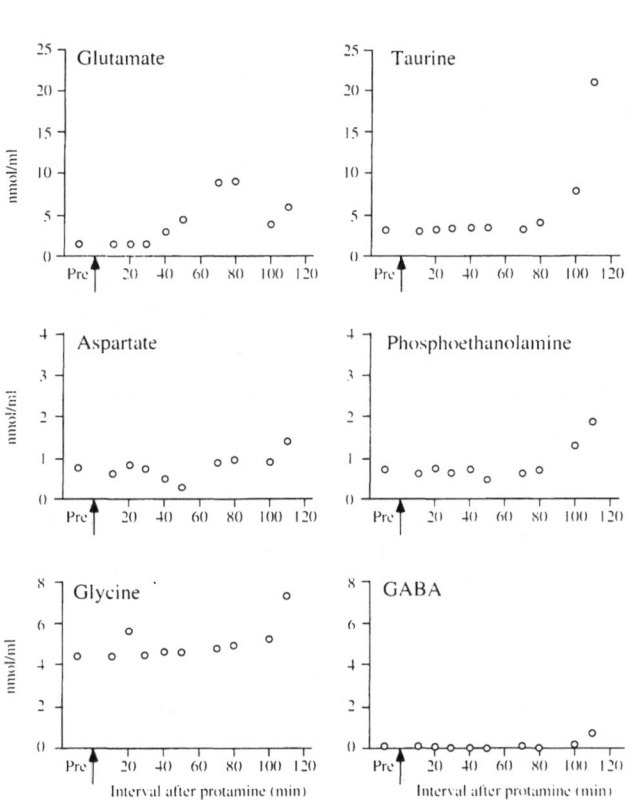

Fig. 1. Concentration of extracellular amino acids in microdialysate sampled for 110 min after infusion of 10 mg protamine sulphate into the carotid artery (arrow)

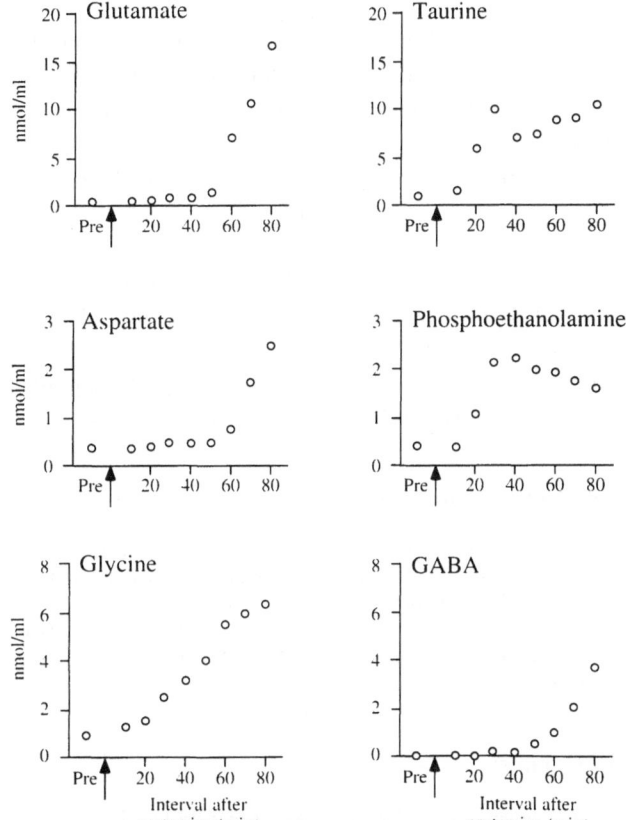

Fig. 2. Concentration of extracellular amino acids in a rat that died 80 min after intracarotid infusion of 10 mg protamine sulfate (arrow)

Table 1. *Amino Acids in the Dialysate from the Right Parietal Cortex Before (Basal Level) and After Intracarotid Infusion of 10 mg Protamine Sulfate*

Amino acid (nmol/ml)	Basal level	Interval sampled (min after protamine infusion)				
		20-30	30-40	40-50	70-80	100-110
	n=10	n=10	n=10	n=10	n=8	n=5
Aspartate	0.3±0.2	0.5±0.2	0.8±0.7	0.9±1.0	2.5±2.0*	2.3±2.5*
Asparagine	0.5±0.2	0.7±0.5	0.6±0.4	0.7±0.5	0.4±0.7	0.4±0.6
Glutamate	0.4±0.1	1.6±1.4	4.0±3.7*	5.9±5.5**	12.7±9.5*	11.9±13.3*
Serine	3.3±1.2	3.3±1.7	3.6±1.6	4.3±1.8	5.1±3.0	4.2±3.0
Glutamine	19.6±14.6	18.9±11.8	20.5±5.1	18.4±10.2	21.2±13.8	19.7±15.1
Glycine	1.7±1.2	2.7±2.3	3.0±2.1	4.4±2.2**	3.9±2.9*	4.4±2.7
Threonine	2.4±1.1	2.7±2.1	2.9±1.9	3.3±1.6	3.8±2.0°	4.0±2.2
Tyrosine	0.9±0.7	1.5±1.3	1.3±0.8	1.6±0.7	1.8±2.4	2.7±2.9
Phosphoethanolamine	0.4±0.4	0.9±0.7	1.1±1.1	1.7±1.5*	2.4±1.7*	2.2±1.9*
Taurine	1.7±1.2	4.7±0.9	4.9±4.8	7.3±6.8**	12.2±10.0**	10.0±11.4*
α-aminobutyric acid	0.2±0.2	0.4±3.0	0.2±0.3	0.1±0.3	0.4±0.3	0.3±0.3
γ-aminobutyric acid	0.05±0.07	0.16±0.17	0.46±1.0	2.26±1.93	1.13±2.3*	1.70±2.30*
Phenylalanine	0.8±0.5	1.0±0.6	1.1±0.7	1.2±0.7	1.0±0.6	1.1±0.5
Valine	1.2±1.3	2.3±1.2	2.2±1.0	3.0±0.8	1.4±1.2	2.5±1.1
Isoleucine	0.7±0.6	1.1±0.5	1.1±0.7	1.3±0.1	1.1±0.5	1.1±0.4
Leucine	0.9±0.3	1.6±1.7	1.9±1.3	1.5±0.8	1.5±1.1	1.6±0.8

Mean values ± SD
* $p<0.05$, ** $p<0.01$ for the difference to basal level (Friedmann's nonparametric test)

Taurine and phosphoethanolamine are known to be involved in the osmoregulation of the brain[7,11,12]. The increase in glutamate/aspartate was observed earlier than of taurine in the dialysate in survivors, whereas taurine tended to increase earlier in rats who developed more severe edema and died 50–80 min after infusion of protamine.

Our data support the hypothesis that glutamate may play a role in the induction and maintenance of brain edema[1,5,14–16]. However, it is possible that glutamate-independent mechanisms are, at least in part, responsible for the taurine release in severe brain edema.

Acknowledgements

The study was supported by grants from the Swedish Medical Council (Project 14X-4968), the Elsa Schmitz' Research Fund and the Rut and Erik Hardebo's Donation Fund. The technical assistance of Madeleine Jacobsson and Janina Warenholt, Laboratory of Experimental Neurology and CSF laboratory, Dept. of Neurology, is greatly appreciated.

References

1. Baethmann A, Maier-Hauff AK, Kempski O, Unterberg A, Wahl M, Schürer L (1988) Mediators of brain edema and secondary brain damage. Crit Care Med 16: 972–978
2. Benveniste H (1989) Brain microdialysis. J Neurochem 52: 1667–1679
3. Hagberg H, Anderson P, Kjellmer I, Thiringer K, Thordstein M (1987) Extracellular overflow of glutamate, aspartate, GABA and taurine in the cortex and basal ganglia of fetal lambs during hypoxia-ischemia. Neurosci Lett 78: 311–317
4. Hamberger A, Berthold CH, Karlsson B, Lehman A, Nyström B (1983) Extracellular GABA, glutamate and glutamine in vivo perfusions dialysis of the rabbit hippocampus. In: Hertz L, Kvamme E, McGeer EG, Schousboe A (eds) Glutamine, glutamate and GABA in the central nervous system. AR Liss, New York, pp 473–492
5. Kempski O, von Andrian U, Schürer L, Baethmann A (1990) Intravenous glutamate enhances edema formation after a freezing lesion. In: Advances in neurology, vol 52. Brain edema. Raven, New York, pp 219–222
6. Lefauconnier JM (1992) Transport of amino acids. In: Bradbury MWB (ed) Handbook of experimental pharmacology, vol 103. Physiology and pharmacology of the blood-brain barrier. Springer, Berlin Heidelberg New York Tokyo, pp 117–150
7. Lehmann A, Carlström C, Nagelhus EA, Ottersen OP (1991) Elevation of taurine in hippocampal extracellular fluid and cerebrospinal fluid of acutely hypoosmotic rats: contribution by influx from blood. J Neurochem 56: 690–697
8. Lindroth P, Mopper K (1979) High performance liquid chromatographic determination of subpicomole amounts of amino acids by precolumn fluorescence derivatization with o-ophthaldialdehyde. Anal Chem 51: 1667–1674
9. Sandberg M, Butcher S, Hagberg H (1986) Extracellular overflow of neuroactive amino acids during severe insulin-induced hypoglycemia: in vivo dialysis of the rat hippocampus. J Neurochem 47: 178–184
10. Tossman U, Delin A, Eriksson S, Ungerstedt U (1987) Brain cortical amino acids measured by intracerebral dialysis in portocaval shunted rats. Neurochem Res 12: 265–269
11. Thurston JH, Hauhart RE, Dirgo JA (1980) Taurine: a role in

osmotic regulation of mammalian brain and possible clinical significance. Life Sci 26: 1561–1568

12. Wade JW, Olson JP, Samson FE, Nelson SR, Pazdernik TK (1988) A possible role for taurine in osmoregulation within the brain. J Neurochem 51: 740–745

13. Westergren I, Johansson BB (1991) Dixyrazine, a phenothiazine derivate, can prevent brain oedema induced by intracarotid injection of protamine sulphate. Acta Neurochir (Wien) 113: 171–175

14. Westergren I, Johansson BB (1993a) Blockade of AMPA receptors reduces brain edema after opening of the blood-brain barrier. J Cereb Blood Flow Metab 13: 603–608

15. Westergren I, Nordborg C, Johansson BB (1993b) Glutamate enhances brain damage and albumin content in cerebrospinal fluid after intracarotid protamine infusion. Acta Neuropathol (Berl) 85: 285–290

16. Westergren I, Nyström B, Hamberger A, Johansson BB (1993) Concentrations of amino acids in extracellular fluid after opening of the blood-brain barrier by intracarotid infusion of protamine sulphate. J Neurochem, in press

Correspondence: Irena Westergren, M. D., Department of Neurology, Lund University Hospital, S-221 85 Lund, Sweden.

Acta Neurochir (1994) [Suppl] 60: 128–131

Pharmacological Reduction of Brain Edema Induced by Intracarotid Infusion of Protamine Sulphate: a Comparison Between a Free Radical Scavenger and an AMPA Receptor Antagonist

B. B. Johansson and I. Westergren

Department of Neurology, Lund University Hospital, Lund, Sweden

Summary

The blood-brain barrier (BBB) of rats was opened by infusing 10 mg protamine sulphate (200 µl in 30 s) into the right internal carotid artery. Ten minutes later, tirilazad, a 21-aminosteroid (3 mg/kg); NBQX, an AMPA receptor antagonist (5 mg/kg); or dixyrazine, a phenotiazine derivate (10 mg/kg), was administered intravenously and the rats were killed 2 h after protamine infusion. Brain specific gravity was determined in the frontal, parietal and occipital cortex and in the striatum. In separate experiments, serum albumin content was determined in the brain of rats by immunoelectrophoresis 2 h after protamine infusion with or without tirilazad pretreatment.

Specific gravity was significantly higher in all of the studied brain regions in rats given tirilazad or NBQX than in those given vehicle or dixyrazine ($p < 0.001$). A combination of tirilazad and NBQX was significantly more efficient than either drug alone in reducing edema in the occipital cortex ($p < 0.05$) and more efficient than NBQX alone in the frontal and parietal cortex ($p < 0.05$). None of the drugs reduced the albumin content in CSF; in addition, tirilazad failed to reduce albumin extravasation in the brain and CSF when given before protamine infusion. We conclude that the anti-edematous effect of tirilazad and NBQX is related to cellular events within the brain and not to a reduction of leakage over the BBB.

Keywords: Brain edema; glutamate antagonist; NBQX; tirilazad.

Introduction

21-Aminosteroids are potent inhibitors of lipid peroxidation in vitro and in vivo[2,4]. Pretreatment with the 21-aminosteroid tirilazad has been shown to reduce vascular permeability and edema after intracerebral infection of arachidonic acid, ferrous chloride or blood[5,15]. To our knowledge, the effects of lazaroids on vasogenic edema induced by opening the blood-brain barrier (BBB) without primary tissue damage have not been studied.

The aim of the present study was to see whether tirilazad could protect against the edema induced by opening the BBB with protamine sulphate[9]. We previously reported that intracarotid infusion of 5 mg protamine sulphate does not lead to brain edema, whereas 10 mg does[11]. In order to separate the possible effects on the BBB from the edema development, tirilazad was given 10 min after opening the BBB. Since pilot studies suggested that tirilazad had a benefical effect, we compared its efficacy with that of NBQX (2,3-dihydroxy-6-nitro-7-sulfamoyl-benzo(F) quinoxaline), an AMPA receptor antagonist that was earlier shown to reduce vasogenic edema induced by glutamate, hyperosmolar solutions or protamine sulphate[12,13]. Furthermore, we included a group treated with dixyrazine, a phenotiazine derivate that reduces edema when given before opening the BBB, but has so far not been tested as post-treatment[11].

Materials and Methods

Male Sprague-Dawley rats weighing 250–300 g were anesthetized with intraperitoneal methohexital sodium (Brietal) 50 mg/kg bodyweight. Catheters were inserted into a femoral vein for the application of drugs and Evans blue tracer and in a femoral artery for monitoring mean arterial pressure and sampling blood for blood gas and glucose analysis. The right external and internal carotid arteries were exposed and the pterygoid palatine artery and the distal end of the external carotid artery were ligated. A Portex PE 25 catheter connected to PE 50 tubing was inserted into the external carotid artery towards the carotid bifurcation. The catheters were exteriorized on the back of the neck.

The head was then fixed in the horizontal position, the occipital bone was exposed and a hole drilled in the sagittal midline through the suture between the interparietal and occipital bone. A CSF sampling cannula was implanted into the cisterna magna and fixed to the skull as described earlier[10]. A 50-ml CSF sample was withdrawn before and 1 h after the infusion of protamine sulphate.

After a 2- to 3-h recovery period, a 2% solution of Evans blue (3 ml/kg) was injected i.v. 10 min before a filtered solution of 10 mg protamine sulphate grade X (Sigma, USA) in 200 µl 0.9% NaCl was infused into the right carotid artery over 30 s. Tirilazad (Upjohn) 3 mg/kg, NBQX 5 mg/kg, a combination of the two drugs or dixyrazine (Esucos, USB) 10 mg/kg were administered i.v. 10 min after the protamine infusion.

Arterial blood pressure was monitored before and during the injection. Blood gases, hematocrit and blood glucose were analyzed before and 1 h after infusion of protamine sulphate.

Two hours after the intracarotid infusion of protamine sulphate, the animals were anesthetized with i.v. methohexithal and a second CSF sample was taken. The rats were then decapitated and the brains quickly removed, immersed in kerosene and sectioned into four slices. The specific gravity of tissue samples was determined in a gradient tube made of 500 ml of two different brombenzene-kerosene mixtures of specific gravity 1.005 and 1.025, respectively, the former being continuously diluted by admixture to the latter while poured into a gradient glass tube placed into a wider glass tube containing water. Stability of linearity was controlled by standard solutions of anhydrous NaCl of known specific gravity in connection with determination of every sample series[3,8]. Samples were taken from standardized areas in the frontal, parietal and occipital cortex and from the striatum, and their equilibrium depth was measured 2 minutes after insertion. The flotation point of the sample was compared with the flotation points of standardized solutions of NaCl. CSF albumin was determined by immuno-electrophoresis.

To determine if tirilazad could reduce leakage over the BBB, six rats were given 3 mg tirilazad 10 min before the protamine infusion and six rats were used as controls. Two hours after opening the BBB, the brains were perfused in situ with 0.9% NaCl. The albumin content in the right and left hemispheres was determined by immu-noelectrophoresis.

All data are expressed as mean ± SD. One way ANOVA with Scheffe's post hoc procedure was used for comparison between groups.

Results

Physiological variables (MAP, PaCO2, PaO2, pH, hematocrit and plasma glucose) did not differ between the groups. The most marked edema was seen in the parietal region. Tirilazad and NBQX significantly in-creased specific gravity in all of the tested regions ($p < 0.001$; Table 1). A combination of the two drugs was significantly more effective than either of them alone in the occipital cortex ($p < 0.05$) or NBQX alone in the frontal and parietal cortex ($p < 0.05$). The per-centage alteration in specific gravity in the parietal cortex for the various treatment groups is shown in Fig. 1. The dixyrazine group did not differ from the controls, but specific gravity was significantly lower than in rats given NBQX or tirilazad ($p < 0.05$) and in rats given a combination of NBQX and tirilazad ($p < 0.01$) (Table 2).

The serum albumin content in the right hemisphere did not differ between rats given tirilazad before BBB opening with intracarotid infusion of protamine sul-phate (2562 ± 758 µg/g wet weight) and control rats given protamine sulphate only (2776 ± 655 µg/g wet weight) corresponding to 13 and 14 times the "back-ground levels" in the left hemisphere.

Discussion

The present results confirm that AMPA receptor blockade can reduce vasogenic brain edema[13] and show that the free radical scavenger tirilazad has a similar effect. While there was no significant differ-ence between NBQX- and tirilazad-treated animals, there was a tendency for higher specific values in the tirilazad group, and a combination of the two drugs tended to be more effective than either given alone, suggesting that tirilazad and NBQX have slightly dif-ferent actions.

Neither NBQX nor tirilazad reduced protein leakage

Table 1. *Specific Gravity of the Right (Injected) and Left Brain Hemispheres 2 h After Intracarotid Injection of 10 mg/200 µl Protamine Sulphate Only (Controls) or 10 mg Protamine Followed 10 Min Later by i.v. Injection of One of the Following Substances.* Dixyrazine 10 mg/kg, NBQX 5 mg/kg, Tirilazad 3 mg/kg or NBQX (5 mg/kg) + Tirilazad (3 mg/kg)

Experimental group	n	Frontal cortex	Parietal cortex	Occipital cortex	Striatum
Controls	12	1.0340 ± 0.0035	1.0327 ± 0.0025	1.0333 ± 0.0032	1.0355 ± 0.0030
Dixyrazine	6	1.0361 ± 0.0038	1.0362 ± 0.0040	1.0372 ± 0.0054	1.0363 ± 0.0018
NBQX	10	1.0416 ± 0.0029[a,b]	1.0403 ± 0.0033[a,b]	1.0416 ± 0.0030[a,b]	1.0429 ± 0.0029[a,b]
Tirilazad	10	1.0422 ± 0.0029[a,b]	$1.0412 + 0.0031$[a,b]	1.0421 ± 0.0037[a,b]	1.0435 ± 0.0021
Tirilazad + NBQX	10	1.0443 ± 0.0021[a,bb,c]	1.0436 ± 0.0025[a,bb,c]	1.0444 ± 0.0023[a,bb,c,d]	1.0438 ± 0.033[a,bb,cc]

Mean values ± S.D.
Statistical significance between the right hemispheres: [a] $p < 0.001$ for the difference vs the controls; [bb; b] $p < 0.01$; 0.05 for the difference vs Dixyrazine; [cc; c] $p \ll 0.02$; 0.05 vs NBQX; [d] $p < 0.05$ vs Tirilazad (Anova; Scheffe's post hoc procedure).
The difference vs the left hemispheres was significant in all of the groups ($p < 0.01$–001).

Table 2. *CSF Albumin (mg/ml) Before (Basal Level) and 2 h After the Intracarotid Infusion of 10 mg Protamine Sulphate Only (Controls) or Followed 10 min Later by i.v. Injection of One of the Following Substances*. Tirilazad 3 mg/kg, NBQX 5 mg/kg, Tirilazad together with NBQX or Dixyrazine 10 mg/kg. In the pretreated group, Tirilazad was given 10 min before protamine

Experimental group	n	CSF 1	CSF 2
Controls (protamine alone)	12	0.09 ± 0.04	0.60 ± 0.23**
Tirilazad, pretreatment	6	0.12 ± 0.02	0.72 ± 0.28**
Tirilazad	10	0.08 ± 0.02	0.60 ± 0.29*
NBQX	10	0.08 ± 0.02	0.64 ± 0.31*
Tirilazad + NBQX	10	0.10 ± 0.02	0.73 ± 0.25**
Dixyrazine	6	0.08 ± 0.02	0.43 ± 0.30

Mean values ± S. D.
** $p < 0.001$, * $p < 0.05$ for difference vs basal level (paired t-test).

in the CSF, nor did pretreatment with tirilazad reduce protein entry into the brain, indicating that the anti-edematous effect of the two drugs is related to cellular events within the brain. In contrast, dixyrazine, which was earlier shown to reduce edema formation when administered before opening the BBB, had no effect when given 10 min after protamine infusion, suggesting that for this drug the protection is mediated via reduced leakage of the BBB.

Protamine produces dose-dependent and transient opening of the BBB[9]. In a recent BBB review it was stated that protamine induces permanent, non-reversible BBB opening[1]. However, the reference given was

to a single study on electrical conductance of pial vessels, where the effect was followed for less than 10 min after the end of the protamine infusion[6]. Protamine, a polycation, is thought to open the BBB by reducing the negative endothelial surface charge[7]. The molecular charge of protamine is directly related to the high content of arginine, which constitutes two-thirds or more of the protamine molecule. However, a comparison between protamine, poly-L-lysine and poly-L-arginine suggested that the effects of arginine on the BBB are not exclusively charge-related[14].

A question that remains to be answered is whether the obtained results are valid only for edema induced by protamine sulphate. In addition to its effect on protamine-induced edema, NBQX has been shown to reduce edema following hyperosmolar infusion of arabinose[13]. Whether this is also true for tirilazad remains to be shown.

Acknowledgements

NBQX was a gift from Novo Nordisk A/S, Denmark, and tirilazad from Upjohn, Kalamazoo, USA. The study was supported by a grant from the Swedish Medical Council (project 11X-4968).

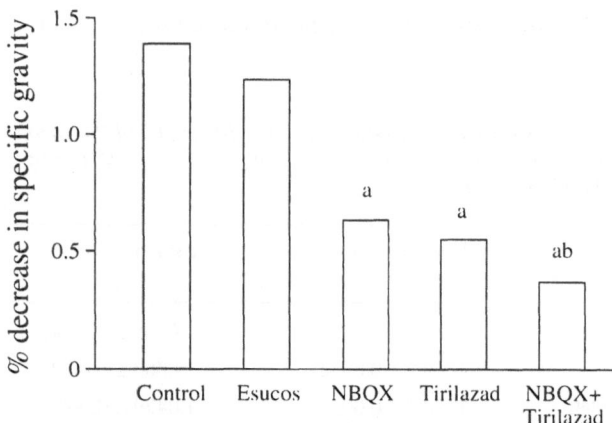

Fig. 1. Decrease (in percent) in specific gravity in the right compared with the left (control) hemisphere 2 h after the infusion of 10 mg protamine sulphate into the right carotid artery in control animals and in animals given Esucose (dixyrazine 10 mg/kg), NBQX (5 mg/kg), tirilazad (10 mg/kg) or a combination of NBQX and tirilazad 10 min after the protamine infusion. $a = p < 0.001$ compared with control animals; $b = p < 0.05$ compared with rats given NBQX

References

1. Abbott NJ, Revert P (1991) Control of brain endothelial permeability. Cerebrovasc Brain Metab Rev 3: 39–72
2. Braughler JM, Pregenzer, Chase RL, Duncan LA, Jacobsen EJ, McCall JM (1987) Novel 21-aminosteroids as potent inhibitors of iron-dependent lipid peroxidation. J Biol Chem 262: 10438–10440
3. Fujiwara K, Nitch K, Suzuki R, et al (1981) Factors in reproducibility of the gravimetric method for evaluation of edematous changes in the brain. Neurol Res 3: 345–361
4. Hall ED, Braughler JM (1989) Central nervous system trauma and stroke 2. Physiological and pharmacological evidence for

involvement of oxygen radicals and lipid peroxidation. Free Radic Biol Med 6: 303–313

5. Hall ED, Travis MA (1988) Inhibition of arachidonic acid-induced vasogenic brain edema by the non-glucocorticoid 21-aminosteroid U 74006 F. Brain Res 451: 350–352

6. Olesen S-P, Crone C (1986) Substances that rapidly augment ionic conductance of endothelium in cerebral vessels. Acta Physiol Scand 127: 233–241

7. Nagy Z, Peters H, Huttner I (1983) Charge-related alterations of the cerebral endothelium. Lab Invest 49: 662–671

8. Nelson SR, Mantz ML, Maxwell JA (1971) Use of specific gravity in the measurement of cerebral edema. J Appl Physiol 30: 268–371

9. Westergren I, Johansson BB (1990) Albumin content in brain and CSF after intracarotid infusion of protamine sulfate: a longitudinal study. Exp Neurol 107: 192–196

10. Westergren I, Johansson BB (1991) Changes in physiological parameters of rat cerebrospinal fluid during chronic sampling. Evaluation of two sampling techniques. Brain Res Bull 27: 283–286

11. Westergren I, Johansson BB (1991) Dixyrazine, a phenothiazine derivate, can prevent brain oedema induced by intracarotid injection of protamine sulphate. Acta Neurochir (Wien) 113: 171–175

12. Westergren I, Johansson BB (1992) NBQX, an AMPA antagonist, reduces glutamate-mediated brain edema. Brain Res 573: 324–326

13. Westergren I, Johansson BB (1993) Blockade of AMPA receptors reduces vasogenic brain edema. J Cereb Blood Flow Metab 13: 603–608

14. Westergren I, Johansson BB (1993) Altering the blood-brain barrier in the rat by intracarotid infusion of polycations. A comparison between protamine, poly-L-lysine and poly-L-arginine. Acta Physiol Scand, in press

15. Zuccarello M, Andersson DK (1989) Protective effect of a 21-aminosteroid on the blood-brain barrier following subarachnoid hemorrhage in rats. Stroke 20: 367–371

Correspondence: Barbro B. Johansson, M.D., Department of Neurology, Lund University Hospital, S-221 85 Lund, Sweden.

Acta Neurochir (1994) [Suppl] 60: 132–135

Visual Evoked Responses as a Monitor of Intracranial Pressure During Hyperosmolar Blood-Brain Barrier Disruption

M. K. Gumerlock, D. York, and **D. Durkis**

Department of Neurosurgery, University of Oklahoma, Health Sciences Center, Oklahoma City, Oklahoma, U.S.A.

Summary

Intracarotid mannitol (HBBBD = hyperosmotic BBB disruption) as a method of transiently increasing solute/drug delivery to brain parenchyma has been associated in animals with a 1–2% increase in brain water and an increase in cisternal ICP. To determine whether these changes are *clinically* significant, we investigated ICP changes associated with HBBBD in 33 patients with malignant brain tumor utilizing flash VERs in which the N_2 latency correlates well with ICP ($N_2 > 80$ ms linearly corresponds to ICP > 20 cm H_2O). VERs were obtained prior to, 4 and 24 h after 114 HBBBD/chemotherapy procedures. Additionally, in 10 patients (37 HBBBDs), VER monitoring was performed during the procedure. In 112/150 (75%) HBBBDs, good barrier opening was obtained (radionuclide brain scan). Postoperative mean N_2 latencies did not differ significantly from pre-HBBBD values ($N_2 = 86 \pm 3.3$ pre, $90 \pm 3.0 / 4$ h, $87 \pm 3.6 / 24$ h); there was no significant difference in N_2 latencies in those patients with good vs poor BBBD. In the 10 patients monitored during HBBBD, peak N_2 latency = 94.9 ± 1.6, however, was significantly above pre- and post-HBBBD values ($p < 0.04$). We conclude that flash VERs are a useful noninvasive measure of ICP, and that HBBBD is associated with mild transient increase in ICP which is not clinically detrimental.

Keywords: Evoked potentials; blood-brain barrier disruption; hypertonic mannitol; intracranial pressure.

Introduction

Visual sensory abnormalities are frequently described in the early stages of increased intracranial pressure (ICP). The use of visual evoked potentials or visual evoked response (VERs) to provide an indication of visual pathway disturbance is well documented[1-3]. Visual evoked potentials can be measured in response to pattern reversal stimuli or flash stimuli, either by strobe or light-emitting diode (LED). Abnormal VERs are seen with compressive lesions or demyelinative lesions along the visual pathway. They correlate with ICP and to some degree with ventricular enlargement in hydrocephalus[3]. The mechanism of ab-

normal VERs remains speculative, but has variously been ascribed to optic nerve compression, ischemic optic neuropathy secondary to papilledema, the metabolic effect of increased ICP, and interference of axoplasmic bulk flow, and mechanical stretching of the visual projection fibers. Although there may be variation among subjects, VERs are remarkably stable in a given individual. Factors known to affect VERs include temperature, blood pressure, anesthesia, age, sleep-wake stages, PO_2, and rCBF. A variety of methods to analyze the classic visual evoked potential include peak latency, amplitude, and a more complex N score[3,8]. Comparison of ICP and N_2 latency in patients with a variety of intracranial processes, including hydrocephalus, head trauma, pseudotumor, stroke, and hypertensive intracerebral hemorrhage by a least square regression analysis yielded a correlation coefficient of $r = 0.80–0.90$; $N_2 > 80$ ms corresponds linearly to ICP > 20 cm H_2O[1,6,7].

Intracarotid infusion of hyperosmolar mannitol transiently opens the BBB (HBBBD), allowing increased drug delivery to brain. A variety of animal studies have shown that HBBBD is associated with transient cerebral edema and a 1–2% increase in brain water. Neuwelt *et al.* demonstrated a significant increase in ICP after intracarotid mannitol (HBBBD) as opposed to saline infusion[5]. The purpose of this study is to demonstrate the changes in VERs as a non-invasive indicator of ICP with hyperosmolar HBBBD in a series of brain tumor patients treated with HBBBD/chemotherapy.

Materials and Methods

At the Missouri University-Columbia Health Sciences Center, patients consented to and received chemotherapeutic treatment of malignant brain neoplasia in conjunction with osmotic opening of the blood-brain barrier in accordance with the regulations of the MUHSC and its Institutional Review Board as described previously[4]. These patients underwent VER monitoring as a measure of ICP.

Flash VERs were recorded from a vertex (C_z) electrode (positive) referenced to an earlobe (negative). A ground electrode was placed on the forearm. A Cadwell Quantum 84 was used to average 8–12 flash stimuli recorded at 10^4 gain with bandpass filters of 1–30 Hz. The stimuli consisted of LED goggles placed in front of each eye with flashes at two-second intervals. The VERs were plotted and analyzed by a peak detection program using first derivative analysis. Four to six trials were averaged for the data. Comparison of common wave peaks across different subjects was done with a cluster analysis routine[6,7].

For the HBBBD/chemotherapy protocol, patients were treated under general endotracheal anesthesia with cannulation of the appropriate internal carotid or vertebral artery. Supersaturated hyperosmolar mannitol 25% was infused at a rate to replace blood flow (5–12 mls/s) for 30 seconds. Ten minutes prior to mannitol infusion, the patients were given cyclophosphamide 20 mg/kg i.v. to allow hepatic circulation required for drug activation. One minute after mannitol infusion ^{99m}Tc-DTPA 20 mCi i.v. was given and methotrexate 1–5 gm was infused intra-arterially. A nuclear brain scan was performed three hours following HBBBD/chemotherapy to document barrier opening[4].

Results

The patients ranged in age from 12–71 years with a mean age of 43 years. Initially, 33 patients underwent 114 HBBBD/chemotherapy procedures. In 85 of the 114 HBBBDs good barrier opening was obtained as determined by ^{99m}Tc-DTPA radionuclide brain scanning. Postoperative mean N_2 latencies did not differ significantly from pre-HBBBD values at 4 or 24 hours (Table 1); nor was there a significant difference between good and poor BBBDs. In 10 additional patients undergoing 37 HBBBD/chemotherapy procedures with *intra-operative* VER monitoring, the mean N_2 latency increased during the induction of anesthesia and further increased shortly after HBBBD (Fig. 1 and 2), eventually returning to baseline. The changes were statistically significant during anesthesia and HBBBD (Table 2). There was no difference between patients with good (n = 27) and poor (n = 9) barrier openings in the 36 having post-HBBBD brain scans or between

carotid (n = 23) and vertebral (n = 14) infusions. None of the patients suffered neurologic deterioration despite estimated ICP changes of up to 10 mmHg.

Discussion

While the pathophysiology of prolonged N_2 latency in patients with increased ICP remains to be delineated, this deficit may be related to prolonged hypoperfusion of the parieto-occipital lobes from chronically and/or acutely elevated ICP. This may be due either to compression of the posterior cerebral artery as with transtentorial herniation and/or direct compression of cerebral tissues in the region of the optic radiations[2]. Focal changes in interstitial fluid composition or pressure from edema may also be influential, and may be the mechanism of N_2 latency change in patients undergoing HBBBD/chemotherapy. The N_2 latency shift may also reflect an increase in cerebral blood volume[1], a cerebrovascular change that correlates with the marked increase in cerebral perfusion seen transiently with HBBBD (Gumerlock, unpublished data).

Our patient population had a mean age of 43 years; normal subjects of this age have a mean N_2 latency of approximately 84 ms. Therefore our patients demonstrate a N_2 latency (86–88) prolongation consistent

Table 2. *N_2 Latency with HBBBD*

	BBBDs n	N2 ms
Pre-	37	88.7 ± 1.3
Anesthesia	31	94.4 ± 1.6*
HBBBD	37	94.9 ± 1.6*
4 h	34	91.9 ± 1.1**
24 h	25	88.8 ± 0.8

* Differs from pre-values, p ≤ 0.0002.
** Differs from pre-values, p ≤ 0.03.

Table 1. *N_2 Latency After HBBBD*

	Total		Good BBBD		Poor BBBD	
	BBBDs n	N_2 ms	BBBDs n	N_2 ms	BBBDs n	N_2 ms
Pre-	114	86 ± 3.3	85	85 ± 3.1	29	86 ± 3.4
4 h	110	90 ± 3.0	85	88 ± 3.2	25	92 ± 2.8
24 h	94	87 ± 3.6	72	86 ± 3.3	22	88 ± 4.9

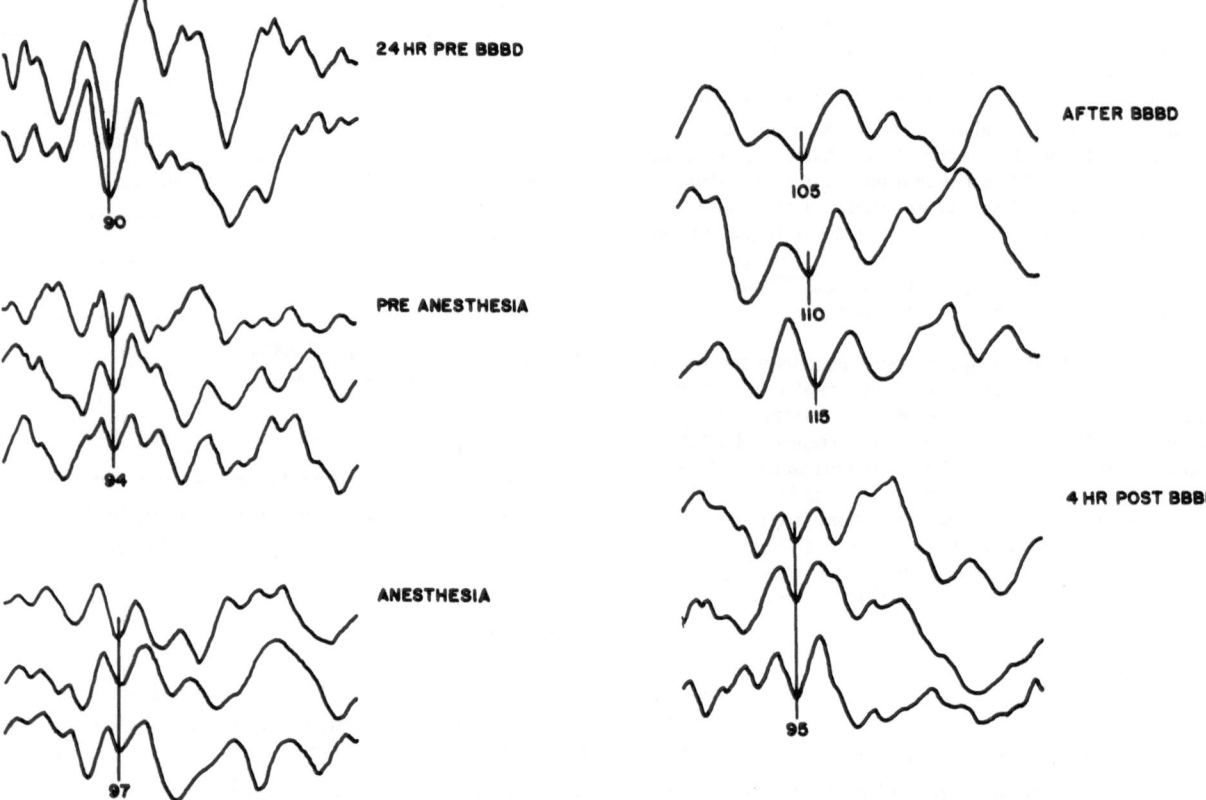

Fig. 1. VER tracings pre-, during, and post-HBBBD demonstrate N_2 latency shifts from baseline (90 ms) to peak increase with HBBBD (115 ms) and returning toward baseline at 4 h

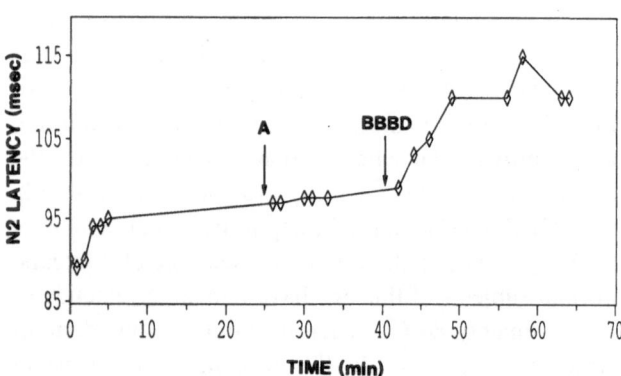

Fig. 2. Repetitive VERs during HBBBD demosntrate a change in N_2 latency with an increase during anesthesia (*A*) and after blood-brain barrier opening (*BBBD*)

with tumor and secondary increased ICP. Increased ICP is a known effect of some forms of anesthesia and, hence, our significant N_2 latency increase. A further prolonged latency following HBBBD is also anticipated based on experimental studies showing an increase in brain water of 1–2% and in cisternal CSF pressure of

2–4 mmHg. As demonstrated by Neuwelt *et al.*, the increased ICP is dissipating by 1 h[5]. Our findings are similar, with ICP improvement at 4 h and resolution at 24 h. The patients were not symptomatic beyond their neurologic baseline during the post-HBBBD time.

We have found the flash VERs more to be useful than the pattern-reversal VERs (which tend to be more stable and more sensitive). The former require no fixation and less patient cooperation and are, here, more reproducible in a neurosurgical patient population. N_2 latency reflects ICP better than P_2 latency which assesses brain maturation/development[2,3]. While our monitoring was performed in a discontinuous fashion, newer developments now allow continuous real-time monitoring. Further, latency measurements tend to be more stable than amplitude recordings[8].

We have documented that the HBBBD-induced ICP elevation is transient and unrelated to the artery which was infused. While we were expecting a difference in those patients having good vs poor disruptions, the data was not statistically significant. Rather the patients with poor HBBBDs tended to have higher baseline N_2 latencies, which may suggest that elevated ICP itself makes BBB opening more difficult. We conclude that VERs are a useful, noninvasive measure of ICP, reflecting the dynamic ICP changes which occur with hyperosmolar opening of the blood-brain barrier, changes which are not clinically detrimental.

References

1. Burrows FA, Hillier SC, McLeod ME, Iron KS, Taylor MJ (1990) Anterior fontanel pressure and visual evoked potentials in neonates and infants undergoing profound hypothermic circulatory-arrest. Anesthesiology 73: 632–636
2. Connolly MB, Jan JE, Cochrane DD (1991) Rapid recovery from cortical visual impairment following correction of prolonged shunt malfunction in congenital hydrocephalus. Arch Neurol 48: 956–957
3. Coupland SG, Cochrane DD (1987) Visual evoked potentials, intracranial pressure and ventricular size in hydrocephalus. Doc Ophthalmol 66: 321–330
4. Gumerlock MK, Belshe BD, Madsen R, Watts C (1992) Osmotic blood-brain barrier disruption and chemotherapy in the treatment of high grade malignant glioma: patient series and literature review. J Neurooncol 12: 33–46
5. Neuwelt EA, Frenkel EP, Rapoport S, et al (1980) Effect of osmotic blood-brain barrier disruption on methotrexate pharmacokinetics in the dog. Neurosurgery 7: 36–43
6. York DH, Pulliam MW, Rosenfeld JG, Watts C (1981) Relationship between visual evoked potentials and intracranial pressure. J Neurosurg 55: 909–916
7. York DH, Leagan M, Benner S, Watts C (1984) Further studies with a noninvasive method of intracranial pressure estimation. Neurosurgery 14: 456-461
8. Zaaroor M, Pratt H, Feinsod M, Schacham SE (1993) Real-time monitoring of visual evoked potentials. Isr J Med Sci 29: 17–22

Correspondence: Mary Kay Gumerlock, M.D., University of Oklahoma, Health Sciences Center, Department of Neurosurgery, P.O. Box 26901, Room 4SP200, Oklahoma City, Oklahoma 73190, U.S.A.

Acta Neurochir (1994) [Suppl] 60: 136–138
© Springer-Verlag 1994

Early Changes of Blood-Brain Barrier and Superoxide Scavenging Activity in Rat Cryogenic Brain Injury

Y. Ikeda, S. Toda, M. Wang, and S. Nakazawa

Department of Neurosurgery, Nippon Medical School, Bunkyo-ku, Tokyo, Japan

Summary

Cryogenic brain injury results in blood-brain barrier (BBB) disruption which may be mediated in part by oxygen free radicals. In this study, sequential changes of BBB and endogenous superoxide scavenging activity were investigated in brains subjected to cryogenic injury. 92 male Wistar rats were sacrificed at 15, 30 minutes, 1, 2, 3, 4, 6, 24 and 48 hours after the lesion. The extent of BBB disruption was determined by quantitative assessment of Evans blue (EB) uptake based on extraction from tissue using dimethylformamide. Determination of endogenous superoxide scavenging activity in the injured brain was performed by electron spin resonance spectrometry using a spin trapping agent. Superoxide scavenging activity was significantly decreased within one hour after the injury relative to normal rat brain ($p < 0.05$, student's t test) persisting for at least 6 hours. The EB content in the lesioned hemisphere was significantly increased within one hour after the injury relative to the normal rat brain ($p < 0.01$) and continued to increase for 24 hours. In conclusion, early changes of BBB and endogenous superoxide scavenging activity in rat cryogenic injured brain indicate that BBB may be an early target for oxygen free radicals in cryogenic brain injury.

Keywords: Cold injury; free radicals; blood-brain barrier; electron spin resonance.

Introduction

Oxygen free radicals have been implicated in a number of disparate disease processes[1,2]. In the past decade there has been particular interest for the hypothesis that brain injury and brain edema are induced by oxygen free radicals[3-5]. When oxygen free radicals exceed the capacity of the different endogenous oxygen free radical scavengers, tissue injury ensues. Increased formation of oxygen free radicals in brain injury may lead to a consumption of the endogenous oxygen free radicals scavenger compounds. Measurement of endogenous oxygen free radical scavengers can be used therefore to obtain indirect evidence of oxygen free radical generation. The present study was undertaken to investigate sequential changes of BBB and endogenous superoxide scavenging activity following cryogenic injury.

Materials and Methods

Surgical Procedure

Forty male Wistar rats were anesthetized intraperitoneally with pentobarbital (30 mg/kg). Cortical cryogenic injury was produced by application of a metal probe cooled by dry ice to the dura mater of the right parietal brain. Rats were sacrificed at 15 minutes (n = 5), 30 minutes (n = 5), 60 minutes (n = 7), 180 minutes (n = 5) and 360 minutes (n = 5) after lesion production; thirteen normal rats without cryogenic injury were included. Animals were sacrificed by intravenous injection of a pentobarbital overdose. The brain was rapidly removed, and superoxide scavenging activity was measured in small tissue samples (60–70 mg) taken from the cryogenically injured brain.

Measurement of Superoxide Scavenging Activity

The measurements were made according to the method of Hiramatsu and Kohno[4]. Fifty microliters of 2 mM hypoxanthine, 35 µl of 5.5 mM diethylenetriaminepentaacetic acid, 50 µ of xanthine oxidase (0.272 unit/ml) were put into a test tube and mixed by an automatic device. The solution was placed in a special flat cell (volume 160 µl, JEOL Ltd.) for analyzing the DMPO-superoxide spin adduct by ESR spectrometry (JES-FE1 X G). A standard curve was obtained by using 0.8 to 100 unit/ml of SOD with manganese oxide used as an internal standard. Calculations of spin numbers were carried out using the ratio of signal height intensity of 2,2-6,6-tetramethyl-4-hydroxyl-piperidine-1-oxyl which has known spin quantities. The conditions of ESR spectrometry for the measurement of superoxide scavenging activity were as follows: magnetic field, 335 ± 5 mT power, 8.0 mW; response, 0.1 s; modulation, 0.2 mT; temperature, room temperature; amplitude, 3.2×10^3; sweep time, 2 min.

Quantitative Evaluation of BBB Disruption

The animals received 2 mg/kg of a 2% solution of EB in normal saline intravenously prior to injury. Cardiac perfusion was performed with normal saline to clear cerebral blood vessels from EB. The hemisphere of interest was separated, mechanically dissociated and weighed. An amount of dimethylformamide, two times the weight of the brain was added followed by incubation for 72 h in a 50°C water bath. The brain-dimethylformamide mixture was centrifuged at 1500 g for 10 min, the supernatant was spectrophotometrically analyzed at the absorption maximum for EB in dimethylformamide (635 nm). A standard curve of EB in dimethylformamide was used to convert absorbance into g EB/ml dimethylformamide, which was converted into total μg EB per hemisphere.

Statistical Analysis

Data are presented as mean ± standard error. Student's t test was used to assess significance, with $p < 0.05$ considered statistically significant.

Results

The superoxide scavenging activity in rat brain with cryogenic injury (expressed in U/g-wet tissue) was 360.60 ± 24.71 in normal, 254.56 ± 27.09 at 15 minutes, 234.12 ± 46.26 at 30 minutes, 253.44 ± 225.37 at 60 minutes, 243.72 ± 32.74 at 180 minutes and 272.46 ± 20.26 at 360 minutes. The scavenging activity was significantly decreased at 15 minutes after the cryogenic injury relative to normal rat brain ($p < 0.05$), and it remained below normal for at least 360 minutes. The superoxide scavenging activity was 254.56 ± 27.09 in the injured brain tissue of the right hemisphere vs 401.44 ± 49.40 in non-injured brain tissue of the left hemisphere ($p < 0.05$) at 15 minutes of the cryogenic injury, 253.44 ± 25.37 vs 379.23 ± 33.44 ($p < 0.05$) at 180 minutes, and 272.46 ± 20.26 vs 375.10 ± 37.03

Table 1. *Evans Blue Contents in the Ipsilateral Cold-Lesioned Hemisphere and Contralateral Hemisphere*

Time after cold injury		Rt. hemisphere Trauma	Lt. hemisphere Control
Sham	(n = 8)	1.929 ± 0.219	1.678 ± 0.353
1 h	(n = 6)	6.238 ± 1.341*	1.785 ± 0.230
2 h	(n = 7)	6.658 ± 1.503*	1.819 ± 0.209
4 h	(n = 7)	9.661 ± 1.736*	2.512 ± 0.609*
6 h	(n = 9)	10.785 ± 2.620*	3.363 ± 1.254*
24 h	(n = 8)	12.033 ± 3.070*	4.456 ± 1.349*
48 h	(n = 7)	8.970 ± 2.177*	3.880 ± 1.108*

Values are mean ± standard deviation.
Evans blue content: μg/hemisphere.
Differences compared to sham group by student's t-test, * $p < 0.01$.

($p < 0.05$) at 360 minutes (Fig. 1). EB contents in the lesioned hemisphere were significantly increased at one hour after injury relative to the normal rat brain ($p < 0.01$) and EB continued to increase for 24 hours (Table 1).

Discussion

A potential mechanism of brain injury and brain edema is supposed to be associated with the formation of oxygen free radicals[5,8,9–11]. The pathological insult may impair the endogenous oxygen free radical scavenger systems, causing overproduction of oxygen free radicals. Averet et al.[1] observed that SOD activity is decreased at 6 hours after cryogenic injury and that it remained reduced below normal values, while catalase activity is significantly increased at 6 hours after cryogenic injury and thereafter. Glutathione peroxidase activity was also significantly increased at 96 hours post injury. In cryogenic injury, the initial trauma produces a hemorrhagic lesion with leakage of red blood cells into the extracellular space. The extravasation of red blood cells with the subsequent liberation of iron and heme compounds in head injury with contusion is critical in the formation of brain damage. Respective experiments were demonstrating the role of iron in the induction of brain edema and the formation of oxygen free radicals[6,7]. It was suggested that iron released from the shed blood due to the cryogenic insult is involved in the generation of free radicals as a major mechanism of tissue destruction. In the present study, the cryogenic brain injury produced an early decrease of the endogenous superoxide scavenging activity in the tissue affected by the insult. The decrease of superoxide scavenging activity indicates that the endogenous

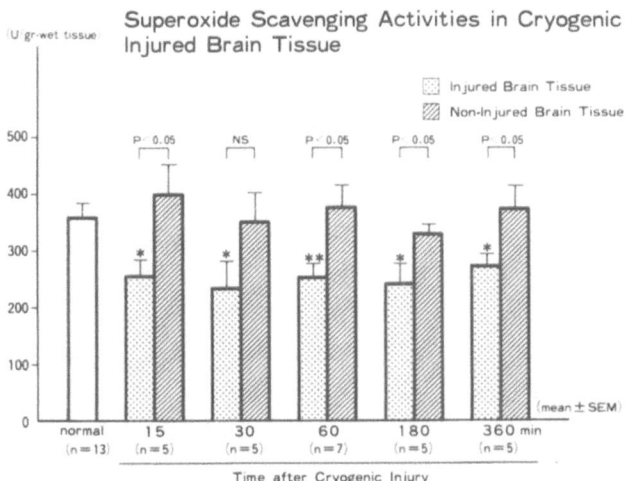

Fig. 1

Superoxide Scavenging Activities in Cryogenic Injured Brain Tissue

U/gr-wet tissue

☐ Injured Brain Tissue
▨ Non-Injured Brain Tissue

normal (n=13), 15 (n=5), 30 (n=5), 60 (n=7), 180 (n=5), 360 min (n=5)

Time after Cryogenic Injury

(mean ± SEM)

*: $P < 0.05$ vs normal, **: $P < 0.01$ vs normal.

scavenger system may be an early target in traumatic brain damage. Pharmacologic modification of this process by agents that scavenge oxygen free radicals, block their generation, or enhance the endogenous radical scavenging capability, may be considered as a rational and successful approach.

References

1. Avéret N, Coussemacq M, Cohadon F (1990) Thiobarbituric acid-reactive material content and enzymatic protection against peroxidative damage during the course of cryogenic rabbit brain edema. Neurochemical Res 15: 791–795
2. Bulkly GB (1983) The role of oxygen free radicals in human disease processes. Surgery 94: 407–411
3. Halliwell B (1987) Oxidants and human disease: some new concept. FASEB J 1: 358–364
4. Hiramatsu M, Kohno M (1987) Determination of superoxide dismutase activity by electron spin resonance spectrometry using the spin trap method. JEOL News 23: 7–9
5. Ikeda Y, Anderson JH, Long DM (1989) Oxygen free radicals in the genesis of traumatic and peritumoral brain edema. Neurosurgery 24: 769–685
6. Ikeda Y, Ikeda K, Long DM (1989) Comparative study of different iron-chelating agents in cold-induced brain edema. Neurosurgery 24: 820–824
7. Ikeda Y, Ikeda K, Long DM (1989) Protective effect of the iron chelator deferoxamine on cold-induced brain edema. J Neurosurg 71: 233–238
8. Ikeda Y, Brelsford KL, Ikeda K, Long DM (1989) Oxygen free radicals in traumatic brain edema. Neurol Res 11: 213–216
9. Ikeda Y, Long DM (1990) Review article. The molecular basis of brain injury and brain edema: the role of oxygen free radicals. Neurosurgery 27: 1–11
10. Long DM, Maxwell RE, Choi KS, Cole HO, French LA (1972) Multiple therapeutic approaches in the treatment of brain edema induced by a standard cold lesion. In: Reulen HJ, Schürmann K (eds) Steroids and brain edema. Springer, Berlin Heidelberg New York, pp 87–94
11. Long DM (1984) Microvascular changes in cold injury edema. In: Go KG, Baethmann A (eds) Recent progress in the study and therapy of brain edema. Plenum, New York, pp 45–54

Correspondence: Yukio Ikeda, M. D., D. M. Sc, Nippon Medical School, Department of Neurosurgery, 1-1-5, Sendagi, Bunkyo-ku, Tokyo 113, Japan.

Acta Neurochir (1994) [Suppl] 60: 139–141
© Springer-Verlag 1994

Evaluation of BBB Damage in an UV Irradiation Model by Endogenous Protein Tracers

J. V. Lafuente[1], **J. Cervós-Navarro**[2], and **E. Gutierrez Argandoña**[1]

[1] Department of Neuroscience, B. C. U., Leioa, Spain, and [2] Institute of Neuropathology, Free University of Berlin, Federal Republic of Germany

Summary

Autologous serum proteins have proved to be suitable tracer to evaluate vascular permeability. The dynamic behaviour of anti-HRP immunoglobulins was studied in ultraviolet (UV) irradiation induced brain edema. Cerebral cortex of 36 anaesthetized adult rats was irradiated following a 2 × 2 mm parietal craniotomy. Immunization was carried out by 3 subcutaneous injections of 10 mg HRP in 0,5 ml complete freund adjuvant (CFA), 6, 4 and 2 weeks before the injury. Control animals were immunized only with CFA; further control animals were operated and irradiated without any previous immunization. After survival times ranging from 30 min to 24 hours, postoperation animals were transcardially perfused with 4% fresh paraformaldehyde solution in phosphate buffered saline. After postfixation at 4°C, 20 μm vibratome sections were prepared for incubation with a solution of 0,05% HRP, washed and developed by the DAB reaction.

The reactions showed a remarkable exsudation and spreading of anti-HRP antibodies in the edematous brain. The antigen-antibody reaction was conspicuous in animals with shorter survival periods in the necrotic area and near the lesion (1–2 mm). After a longer survival time extravasation involved the whole hemisphere. In animals with the longest survival period labeled serum proteins were found even in the white matter of the hemisphere contralateral to the injury.

Endogenous tracer of BBB function is useful to study the spreading of brain edema in a delayed time after the edematous lesion.

Keywords: Blood-brain barrier; edema; endogenous protein tracers.

Introduction

The blood-brain barrier (BBB) permeability has been studied using different tracer substances. Their extravasation across the vascular wall is dependent on several factors like molecular size, protein binding capacity, etc. Most of the tracer substances are of exogenous origin and by its application damage of the vascular wall could occur. On the other hand, some endogenous proteins have also proved to be suitable tracers for evaluation of vascular permeability[11]. The low serum concentrations of tracer immunoglobulins is overcome by raising IgG antibodies to Horseradish Peroxidase (HRP) by immunizing the animal to HRP. Several authors proposed immunization protocols providing significant levels of circulating antibodies in an easily reproducible way[4,10]. The conjugation of these antibodies in vibratome sections with its antigen is favored due to the relative low molecular weight of HRP. Ultraviolet (UV) irradiation of the cerebral cortex is a suitable model for studying brain edema. This model in rats reproduces a cortical micronecrosis[9] similar to lesions caused by brain trauma or brain infarction[3]. The present paper reports on the dynamic behaviour of anti-HRP immunoglobulins following brain edema induced by UV-irradiation.

Material and Methods

36 anaesthetized (Ketamine hydrochloride 15 mg/100 g and Xylazine hydrochloride 2 mg/100 g) adult Wistar rats were exposed for 6 min. to an ultraviolet lamp (Osram HBO-200), placed 15 cm above the cortex, following a 2 × 2 mm parietal craniotomy. Animals were monitored and vital parameters remained constant[7].

Immunization was made by 3 subcutaneous injections of 10 mg HRP (type VI) and 0.5 ml complete freund adjuvant (CFA) 6, 4 and 2 weeks before the injury. Control animals were operated and irradiated without any previous immunization only with CFA.

After survival times of 5 min., 30 min., 1 h, 3 h, 12 h, 24 h and 1 week, the animals were transcardially perfused with 4% fresh paraformaldehyde and 0,25% glutaraldehyde solution in phosphate buffered saline. For light microscopy, 30 μ vibratome sections were incubated with a solution of 0,05% HRP after postfixation at 4°C, washed and developed by DAB reaction. A 20 μ vibratome section from each case was incubated with lcc GFAP (1 : 4000 Dakopatts, Denmark) and developed with avidin-biotin method to evaluate glial reaction. 80 μ vibratome sections were embedded in epoxy resin after HRP-incubation for ultrastructural examination.

Results

The rat brain showed a cortical micronecrosis about 1 mm in diameter and 0.2 in depth as well as sponginess of the surrounding neuropil and a glial reaction (Fig. 1). Edematous fluid accumulated mainly in the white matter. After a short survival time, the lesion center (LC) is clearly distinguished by vascular congestion contrasting to the perfused microvessels of the marginal zone (MZ). Brain sections from immunized animals, which were irradiated and subsequently incubated with HRP showed strongly positive deposits of tracer. The reaction product indicating tracer presence is clearly visible in the edema front. The tracer extravasation is especially prominent in the MZ.

In animals with a survival period up to 6 h the tracer was conspicuous in the LC and MZ. At 24 h survival time it spread to the underlying white matter of the irradiated hemisphere. In animals, which survived 48 h or more, the tracer proteins were found in the cellular

and extracellular space even in the white matter of the contralateral hemisphere.

Diffuse extravasation was found in the perivascular parenchyma of immunized rats in the MZ. The cytoplasma of neurons, of some astrocytes, the underlying white matter, and the vascular walls were HRP-positive. In the contralateral white matter only glial and endothelial cells were HRP-positive (Fig. 2).

Ultrastructurally, anti-HRP proteins were found in the endothelial layer, the basal membrane and the neuropil depending on the proximity to the necrosis and the survival time after the injury. In LC anti-HRP proteins were also found in the plasma membrane of pericytes, in intracytoplasmic lysosome-like structures as well as in the plasma membrane of neurons, astrocytes and microglial cells.

The astrocytic reaction in the lesion area was evident after 6 h survival time. Hypertrophic, GFAP positive astrocytes were found at 24 h post irradiation. Non operated / irradiated animals as well as non immunized controls did not show any reactivity.

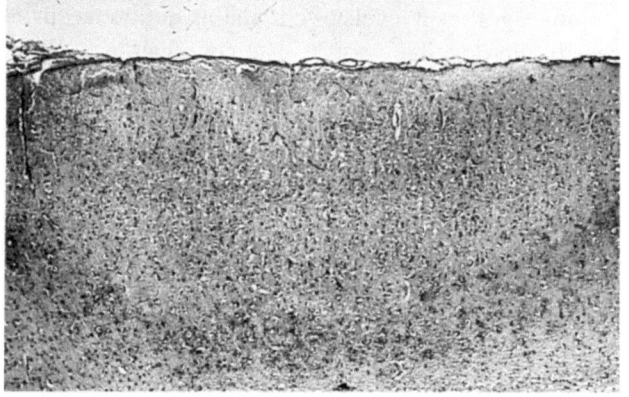

Fig. 1. Micronecrosis about 1 mm diameter and 0,2 in depth surrounded by glial reaction (HE × 80)

Discussion

Disruption of the BBB allows the penetration of antibodies to a non-brain antigen, which then can be detected by the use of immuno-enzyme techniques.

Immunoglobulins against HRP have proved to be a useful tool to study the spreading of edema under various experimental conditions[4]. Due to its endogenous nature, IgG against HRP is a physiological tracer, which reflects the extravasation of plasma proteins in a way close to reality.

Using cortical UV-irradiation as a model, a specific selectivity of tracer distribution for several exogenous substances was reported[8]. However, since the level of HRP in blood decreased continuously after its application extravasation could be correctly evaluated only a short period of time after injection of HRP. The UV-model had demonstrated its plasticity to reproduce lesions of different intensity dependent on the exposure conditions[3]. By standard conditions a direct relation between spreading of edema and survival period after irradiation was observed. Immunization after injury does not reveal any trace of anti HRP-HRP complex, the initial period of BBB disruption is crucial for the penetration by immunoglobulins.

Now, we report our results using an endogenous protein as a tracer of the BBB function. Cellular damage was more focal than diffuse, the border of the lesion was marked.

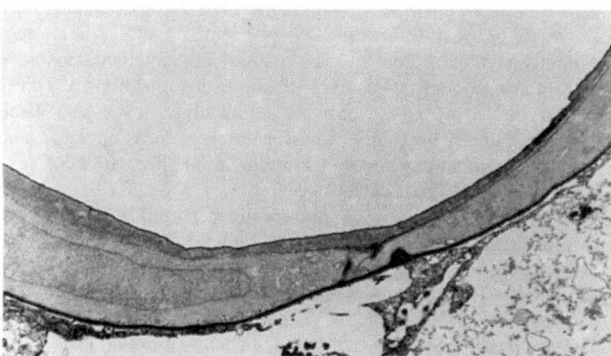

Fig. 2. Anti-HRP proteins are located ultrastructurally in the endothelial layer, the basal membrane and neuropil (× 20.000)

The tracer extravasation was prominent in the MZ and the impaired circulation in the LC is associated with thrombosis, which in agreement with Houthoff *et al.*[4] may explain the lower tracer passage. Edematous fluid spreads mainly in the subcortical white matter and is stopped at the border with the gray matter.

The time course of the spread of anti-HRP antibodies is between that of Evans blue and of HRP, taking 6 h post cortical irradiation. It is seen at 12 h in the corpus callosum and at 24 h in the contralateral hemisphere in the white matter, but never in the cortex or hippocampus[8]. Our results support the suggestion that a small unilateral lesion can induce brain edema involving remote areas.

A predominant extravasation of tracer was found in parasagittal areas. This finding agrees with others[9] about the tissue water content in this model. Several anatomical characteristics may be responsible for this, for example the venous drainage to the midline.

Some authors think that the clearance of an extravasated protein tracer occurs, at least partially, by local tracer backflow into the vessel lumen[5]. Probably several mechanisms coexist simultaneously. One of them could be the drainage through paravascular pathways to the subarachnoidal space[2,7] although no clear evidence was found in the present study. Another putative mechanism is the glial reaction: Incorporation of anti-HRP into glia was confined mainly to the perivascular astrocytes which digest anti-HRP antibodies by endocytosis: The presence of immuno-positivity in vesicular structures supports this hypothesis[6].

Our results demonstrate that endogenous tracers are useful to study the spreading of brain edema and the BBB function in a delayed time after the edematogenous lesion. It offers evidence in our model of edema spread to the contralateral hemisphere. These results suggest that the occasional passage of immunoglobulins through the BBB have important pathogenic implications for several diseases, in which edema occurs.

Acknowledgement

This research has been supported by the Basque Government Grant 9019.

References

1. Becker NH, Hirano A, Zimmermann MH (1968) Observations on the distribution of exogenous peroxidase in the rat cerebrum. J Neuropathol Exp Neurol 27: 439–452
2. Csanda E, Major O, Komoly S (1987) Presence of myelin breakdown products in cervical lymph nodes after ^{90}Yttrium implantation into brain. In: Cervós-Navarro J, Ferszt R (eds) Stroke and microcirculation. Raven, New York
3. Ferszt R, Neu S, Cervós-Navarro J (1980) Measurement of the specific gravity of the brain as a tool in brain edema research. Adv Neurol 28: 15–25
4. Houthoff HJ, Go KG, Huitema S (1981) The permeability of cerebral capillary endothelium in cold injury: comparison of an endogenous and exogenous protein tracer. Adv Neurol 29: 331–336
5. Houthoff HJ, Go KG, Gerrits PO (1982) The mechanisms of blood-brain barrier impairment by hyperosmolar perfusion. Acta Neuropathol (Berl) 56: 99–112
6. Hurley JV, Anderson MD, Sexton PT (1981) The fate of plasma protein which escapes from blood vessels of choroid plexus of the rat. An electron microscopic study. J Pathol 134: 57–70
7. Lafuente JV, Artigas J, Nakagawa Y, Aruffo C, Cervós-Navarro J, Ferszt R (1987) Paravascular pathways of CSF resorption in experimental hydrocephalus. In: Cervós-Navarro C, Ferszt (eds) Stroke and microcirculation. Raven, New York
8. Lafuente JV, Pouschmann J, Cervós-Navarro J, Sharma HS, Schreiner C, Korves M (1990) Dynamics of tracer distribution in radiation induced brain edema in rats. Acta Neurochir (Wien) [Suppl] 51: 375–377
9. Lafuente JV, Cruz-Sanchez FF, Rossi ML, Cervós-Navarro J (1992) Ultraviolet irradiation induced brain edema in rats. A microgravimetric study. Neuropath Appl Neurobiol 18: 137–144
10. Nitsch C, Goping G, Laursen H, Klatzo I (1986) The blood-brain barrier to horseradish peroxidase at the onset of bicuculine induced seizures in hypothalamus, pallidum, hippocampus, and other selected regions of the rabbit. Acta Neuropathol (Berl) 69: 1–16
11. Shinohara Y, Izumi S, Watanabe K (1981) Evaluation of BBB function and movement of protein in cold injury by antiperoxidase immunostaining method. J Cereb Blood Flow Metab 1: S393–S394

Correspondence: J. Cervós-Navarro, M.D., Institut für Neuropathologie, Freie Universität Berlin, Hindenburgdamm 30, D-12203 Berlin, Federal Republic of Germany.

Acta Neurochir (1994) [Suppl] 60: 142–144
© Springer-Verlag 1994

Blood-Brain Barrier Opening Following Transient Reflex Sympathetic Hypertension

T. Ijima[2], **Y. Kubota**, **T. Kuroiwa**[1], and **H. Sankawa**[2]

Department of Anesthesiology, School of Dentistry, Tokyo Medical and Dental University, [1] Department of Neuropathology, Medical Research Institute, Tokyo Medical and Dental University, and [2] Department of Anesthesiology, School of Medicine, Kyorin University, Japan

Summary

Numerous researchers have shown that experimentally induced hypertension opens the blood-brain barrier (BBB), whereas physiologically induced hypertension is accompanied by sympathetic activation, which exerts a protective effect on the BBB via cerebral vasoconstriction. It has not yet been established that transient reflex sympathetic hypertension can open the BBB. In this study, 14 lightly-anesthetized adult cats were subjected to electrical stimulation of a tooth and transient reflex sympathetic hypertension (duration of less than 60 s) was elicited repeatedly. Continous flowmetry of the cerebral blood flow showed that autoregulation breakthrough occurred. Light halothane anesthesia supplemented with 30 or 60% nitrous oxide, or 1.2% halothane anesthesia occasionally suppressed the elicited pressor response and prevented such breakthrough. Leakage of Evans blue (EB), which was administered before the hypertensive insult, was confirmed in the marginal and suprasylvian gyri in 5 cats. The EB positive cats reached a significantly ($p < 0.05$) higher mean arterial blood pressure (198 ± 16 mmHg) during reflex sympathetic hypertension than EB negative cats (189 ± 19 mmHg). Breakthrough occurred 16 times in EB positive, but only 8 times in EB negative cats. In conclusion, transient reflex sympathetic hypertension can elicit cerebrovascular autoregulation breakthrough and if the breakthrough occurs repeatedly it is followed by the opening of the BBB.

Keywords: Blood-brain barrier; sympathetic reflex; autoregulation.

Introduction

Hypertensive blood-brain barrier (BBB) disruption has been demonstrated in experimental models by means of injecting vasopressors or blood. In both experimental models, sympathetic nervous tone is reduced via a baroreflex. Therefore, such model would not appear to be relevant to physiologically induced hypertension. The sympathetic nervous system exerts a protective effect on the BBB by inducing cerebral vasoconstriction[3]. Gross showed that reflex sympathetic hypertension induced by sinoaortic deafferentiation also opens the BBB[2]. However, it has not yet been established that short-lasting sympathetic reflex-induced hypertension can open the BBB, overcome the protective effect of the sympathetic nervous system on cerebral blood vessels. Reflex sympathetic hypertension repeatedly was induced in cats by noxious stimulation of a tooth and the BBB permeability was evaluated by determining the cerebral leakage of Evans blue.

Methods

Fourteen adult cats were anesthetized with ketamine hydrochloride (50 mg). A femoral artery and vein were cannulated, and animals were ventilated mechanically to maintain the arterial blood CO_2 tension within the normal range. The cerebral blood flow (CBF) was monitored continously through a cranial window by thermocouple flowmetry which was calibrated by the hydrogen clearance method. A canine tooth was perforated and a wire was placed in the hole for electrical stimulation (50 Hz, 5 mA). The electroencephalogram was monitored from bipolar electrodes on the temporal and frontal bones to confirm the depth of anesthesia. Prior to electrical tooth stimulation, 5 ml of 2% w/v Evans blue (EB) was injected intravenously. The tooth was stimulated under various anesthetic conditions described below to attenuate the evoked pressor responses. This enables autoregulation breakthrough to be defined, as a weak pressor response is assumed to elicit a minimal CBF increase and vice versa. We could identify breakthrough, by comparing the CBF change that occurred under each anesthetic condition. The control condition was light halothane anesthesia (C1), for which the lowest possible concentration of halothane was administered until theta waves appeared. Then, 30% (v/v) followed by 60% (v/v) of nitrous oxide was added sequentially to the control condition (N1 and N2) respectively, after which the control condition was reestablished to confirm the pressor response to be reproducible (C2). Halothane (1.2% v/v) was administered to abolish the pressor response (H). On the completion of the experiment, the animals were sacrificed under deep halothane anesthesia and the brain was examined for cerebral leakage of EB, which indicated BBB permeability.

Data were statistically analysed with ANOVA. Difference at p < 0.05 was considered significant.

Results

Of the 14 cats, 5 brains showed focal EB leakage in the bilateral marginal and suprasylvian gyri. The animals were divided retrospectively into two groups, according to the presence or absence of EB leakage (EB-positive and EB-negative respectively).

Under control anesthesia, electrical tooth stimulation elicited a steep pressor response and the CBF increased rapidly (Fig. 1). The pressor response, however, was short-lived and the blood pressure returned to normal within 60 s. Nitrous oxide inhalation blunted this pressor response (Table 1) and suppressed the CBF increase. The EB-positive group had a significantly (p < 0.05) higher mean arterial blood pressure before and during reflex hypertension under control codition (C1) (169 ± 27 mmHg and 216 ± 16 mmHg respectively) than the EB-negative group (143 ± 9 mmHg and 189 ± 15 mmHg respectively). The mean CBFs in the EB-positive group before and during reflex hypertension were also significantly (p < 0.05) higher (109 ± 19 ml/100 g/min and 151 ml/100 g/min, repectively) than the EB-negative group (82 ± 18 ml/100 g/min and 127 ± 39 ml/100 g/min).

In order to evaluate cerebrovascular autoregulation breakthrough, the difference between the baseline and peak cerebral vascular resistance (CVR) values was calculated. A negative difference indicates cerebral vessel dilation and a value of less than –0.1 mmHg/ml/100 g/min was defined as autoregulation breakthrough. In the EB-positive group, breakthrough was seen in all anesthetic condition except for H. CVR was less than –0.1 mmHg/ml/100 g/min with C1, N1, N2, C2 in the EB-positive group, whereas it was less than –0.1

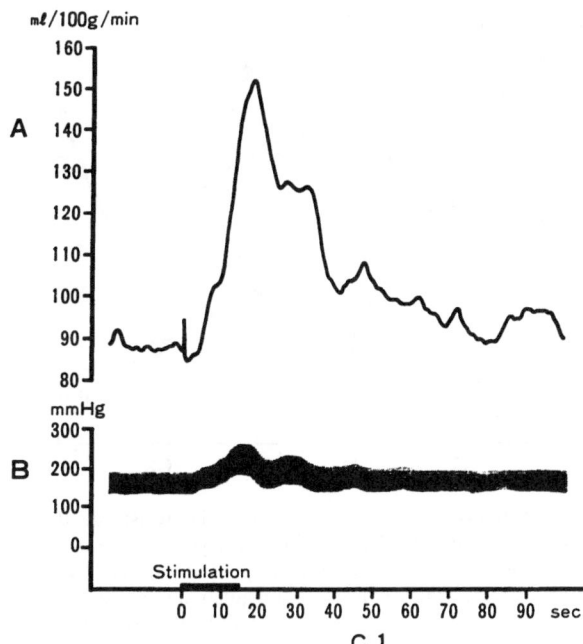

Fig. 1. Time course of cerebral blood flow (CBF) and arterial blood pressure. (A) CBF recording. (B) arterial blood pressure. Note that the steep increase in CBF is accompanied by reflex sympathetic hypertension elicited by noxious stimulation of a tooth, which indicates cerebrovascular autoregulation breakthrough

mmHg/ml/100 g/min in the EB-negative group only under the C1 and C2 conditions (Fig. 2). Thus, breakthrough occurred 16 times in the EB-positive and 8 times in the EB-negative group.

Discussion

Our results show that reflex sympathetic hypertension can open the BBB, although the duration is less than 60 s. Gross has demonstrated that reflex sympathetic hypertension opens the BBB[2]. However, the

Table 1. *Arterial Bood Pressure and Cerebral Blood Flow Before and During Reflex Sympathetic Hypertension*

		Evans blue negative				Evans blue positive			
		C1	N1	N2	C2	C1	N1	N2	C2
MABP (mmHg)	control	140 ± 14	143 ± 16	147 ± 14^a	143 ± 9	153 ± 17^b	$168 \pm 27^{a,b}$	166 ± 25^b	$169 \pm 27^{a,b}$
	peak	189 ± 19	167 ± 21^a	159 ± 17^a	189 ± 15	198 ± 16^b	194 ± 22^b	$180 \pm 26^{a,b}$	$216 \pm 16^{a,b}$
CBF (ml/100g/min)	control	82 ± 18	78 ± 25	81 ± 28	77 ± 26	109 ± 19^b	84 ± 22^a	87 ± 31^a	87 ± 41^a
	peak	127 ± 39	95 ± 35^a	89 ± 33^a	116 ± 47	151 ± 19^b	$118 \pm 29^{a,b}$	$115 \pm 49^{a,b}$	131 ± 56

MABP mean arterial blood pressure, *CBF* cerebral blood flow. Values are expressed as mean ± SD. [a] significantly different from C1, [b] significantly different from the value of Evans blue negative group. *C1, C2* control (light halothane anesthesia), *N1* supplemented anesthesia of 30% nitrous oxide to control, *N2* supplemented anesthesia of 60% nitrous oxide to control.

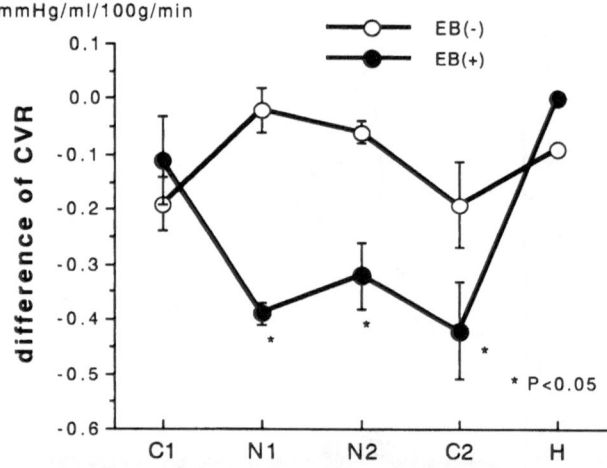

Fig. 2. Difference between cerebral vascular resistance (CVR) before and during reflex hypertension. A negative value indicates passive cerebral vasodilation during the pressor response. Note that passive dilation, which indicates autoregulation breakthrough, occurred in animals which showed Evans-blue leakage (EB-positive), compared with EB negative animals, which maintained autoregulation. C1, N1, N2, C2, and H are the different anesthetic conditions which attenuated the pressor response

induced hypertension in his model was sustained for several minutes. Short-lived reflex sympathetic hypertension is often induced physiologically by pain, fear, or other causes as part of the defence reflex and the blood pressure settles down rapidly thereafter. Therefore, our experimental model can be extrapolated to physiological conditions. It has shown that sympathetic nervous system activation prevents autoregulation breakthough and prevents passive cerebral vasodilation. The majority of the animals (9/14), showed cerebrovascular autoregulation and no BBB disruption, even though the peak arterial blood pressure exceeded 180 mmHg, which supports the concept of a protective effect mediated by the sympathetic nervous system However, BBB opening occurred in approximately 30% of the cats despite of sympathetic excitation. The main differences between EB-positive and negative cats were the mean arterial blood pressure and the incidence of breakthrough. It was difficult to define

exactly what degree cerebral vasodilation corresponded to autoregulation breakthrough. We measured the CBF continously, calculated the difference between the baseline and peak CVR values to assess autoregulation breakthrough and selected a threshold to define it. There was clear breakthrough demarcation between the retrospectively divided groups, according to the presence or absence of EB leakage. The EB-negative group had no breakthrough during nitrous oxide inhalation, whereas all the EB positive cats showed breakthrough, despite deeper anesthesia. We conclude that high arterial blood pressure leads to autoregulation breakthrough during reflex hypertension and consequently, repeated breakthrough contributes to BBB opening. This suggests that there is an arterial blood pressure threshold for BBB opening, even in the presence of a sympathetic nerve mediated protective effect.

Nitrous oxide, which we used to blunt the pressor response, increases CBF[1]. This effect may have facilitated the CBF increase that occurred during the pressor response. This effect, however, did not influence autoregulation in the EB-negative group, because we observed no abnormal CBF increases in these cats.

In conclusion, repeated reflex sympathetic hypertension opens the blood-brain barrier albeit for a short time. There seemed to be an arterial blood pressure threshold for BBB opening despite the protective effect of sympathetic nervous system activation on cerebral vessels.

References

1. Baughman VL, Hoffman WE, Miletich DJ, Albrecht RF (1990) Cerebrovascular metabolic effects of N_2O in unrestrained rats. Anesthesiology 73: 269–272
2. Gross PM, Heistad DD, Strait MR, Marcus ML, Brody MJ (1979) Cerebral vascular responses to physiological stimulation of sympathetic pathways in cats. Circ Res 44: 288–294
3. Heistad DD (1984) Protection of the blood-brain barrier during acute and chronic hypertension. Federation Proc 43: 205–209

Correspondence: Takehiko Iijima, M. D., Department of Anesthesiology, School of Medicine, Kyorin University, 20-2 Shinkawa 6-chome, Mitaka City, Tokyo 181, Japan.

Formation, Propagation and Resolution of Brain Edema

Acta Neurochir (1994) [Suppl] 60: 147–150
© Springer-Verlag 1994

Morphology of Non-Vascular Intracerebral Fluid Spaces

J. Cervós-Navarro, T. Türker, and **F. Worthmann**

Institute of Neuropathology, Free University of Berlin, Klinikum Steglitz, Berlin, Federal Republic of Germany

Summary

In the electron microscope the value for the extracellular space (ECS) in the mamallian CNS was suggested of several percent to about one third dependent on the method how it was evaluated. Since von Harreveld introduced 1965 cryofixation to estimate the extension of the ECS, the method has been never applied in brain edema research.

We carried out improved low temperature methods to measure the extracellular space of the mamallian CNS in physiological conditions. Small samples of brain tissue were cryofixed by slamm freezing on a precooled metal mirror and substituted with ethanol at −95°C over 17 hours. The embedding procedure was carried out at −22°C with LR-White under UV-irradiation. ECS was measured computer assisted with Bioquant Software. The values for the ECS of the cryofixed normal rat brain were more than twice compared to the usual transmission electron microscopy (16.3% to 7.4%, $p < 0.01$) and close to those estimated by von Harreveld (18.1–25.5%, 1965). It was interesting that the data obtained in cryofixed normal rat brain correspond to the extension measured in rat brain with irradiation edema, which was conventionally treated for EM. Greater variance of ECS in cryofixed brain (16.3% ± 3.4) demonstrate that it is far more variable than expected. This data correspond closely to the in vivo ECS. The morphological evaluation of brain edema should be revised under this premise.

Keywords: Brain cortex; extracellular spaces; cryofixation; brain edema; electron microscope.

Introduction

Light microscopic investigations of the central nervous system using any or all of the histological methods available demonstrate nerve cell bodies, myelinated and un-myelinated nerve fibers, coarser ramifications of dendrites and neuroglia cells with their processes. The remained space between the ramifications of all these elements could be light microscopically neither resolved nor measured. Various investigations were undertaken to determine the functional *extracellular space (ECS)* in the brain. As early as 1956 van Harreveld and Ochs by measurements of cortical impedance with determinations of the chloride space tended to substantiate a space of 25–30%. The need for a comparison of these measurements with the morphological characteristics of the nervous tissue, and for the correlation of these two approaches, became apparent.

The first results applying electron microscopy revealed complete filling of the extracellular spaces with cell processes closely packed together. The surface membranes of neighboring elements appeared separated by a constant distance of only 15–20 nm, and the measurements in electron micrographs of osmium tetroxide fixed tissue showed in the central nervous system a mean percentage of ECS not exceeding 5% of the total volume (Horstmann and Meves 1959).

The introduction of the electron microscope into the study of cerebral edema did not immediately clarify the situation. The predominant change was found to be swelling of the glia cell with decreased electron density, but no extracellular fluid was discovered. The suggestion was therefore offered that, unlike other tissues, edema in the brain was essentially an intracellular phenomenon and that the astrocytes were the counterparts of the extracellular spaces in other organs.

Evans and Raimondi (1961) described edema developing after subdural balloon inflation in cats and found enlarged ECS in the white matter in addition to the glial swelling previously described. Long et al. (1962) reported similar findings in human cerebral edema and in edema surrounding implanted psyllium seeds in dog cerebrum. Extracellular fluid accumulation in the white matter was observed in subsequent reports by different models of experimental edema (Gonatas et al. 1963, Cervós-Navarro 1967, Hirano 1980). In enlarged extracellular spaces of the white matter the cellular elements were separated by a clear area of low electron density which was assumed to represent distention of the ECS with fluid.

The consensus was formed, therefore, that cerebral edema was extracellular in the white matter but intracellular in the gray matter. This was not an entirely satisfactory view since many authors realized that similar observations concerning the swelling of the glial cells could be made in control animals, i.e. those supposedly free of edema. The answer to the apparent riddle was reached by means of improved perfusion fixation, epoxy embedments and the use of electron-dense tracer materials. With the newer methods, astrocytes in normal, well-preserved tissue did not show as much watery swelling as seen previously and when dense tracer material was used, it was found that the edema fluid was essentially confined to the ECS.

The Cryofixation Approach

While the enlarged ECS in edematous conditions have been clearly demonstrated in the white as well as in the grey matter, few authors have given attention to the ultrastructure of these spaces in the non edematous brain. A brain ECS of only 5% of the in vivo situation, would not be able to function as the transfering medium for the substances on their way to and from the cell. It was therefore assumed that the low values for brain extracellular space in electron microscope pictures might be the result of artifacts produced during the fixation process. Van Harreveld suggested that the handling of the tissue during fixation results in hypoxia and a rapid intracellular movement of chloride and water from the extracellular compartment. To support their contention van Harreveld and Crowell (1964) subjected the cerebellar cortex to substitution fixation. In tissue frozen immediately after circulatory arrest, an

ECS of approximately 24% was observed. In tissue frozen after 8 minutes asphyxiation, the extracellular spaces disappeared. During the last decades no further studies have been carried out with cryosubstitution on these issue. We revisited the results by applying an improved modern cryofixation method.

Material and Methods

Rats were narcotized by i.m. injection of Ketanest/Diazepam. For the preparation of brain tissue the skull was opened by a saline-cooled drill. Subsequently, brain tissue samples were taken by a scalpel and cryofixed at once. The time between preparing tissue and cryofixation did not extend more than 30 seconds. For the method of freeze-substitution a cryofixation apparatus KF 80 in combination with CSauto (Reichert-Jung) was used. Low temperature substitution is working by rapid freezing of the tissue, which is slammed on a precooled copper-mirror. Subsequent *handling of tissue was made in nitrogen atmosphere to guarantee an absolute water-free-medium.*

The water in the tissue was substituted by 100% pure ethanol at $-95°C$ over 17 hours. Afterwards ethanol was replaced by a mixture of ethanol/OsO_4 (15 : 2). Temperature was allowed to rise 2°C/h to $-60°C$. At this level the mixture was renewed to improve water/ethanol substitution. After temperature rised to $-22°C$ (2°C per hour), embedding procedure was carried out with acrylic resin embedding medium (London Resin White) under UV-irradiation.

For control, brain tissue from the contralateral hemisphere was taken and fixed in 4% glutaraldehyde for 2 hours and put into phosphate buffer solution for 24 hours. After incubation in NH_4Cl for 1 hour and in OsO_4 for 12 hours the specimens were embedded in epon.

All embedded samples were cut on an ultra-microtome to about 1 μm thick sections and stained with lead citrate and uranyl acetate. Electron micrographs were taken with a Zeiss CEM 902 electron microscope.

Electron micrographs from edematous rat brain tissue, which was treated conventionally, were chosen from former studies where brain edema was produced by UV-irradiation (Ferszt *et al.*, 1978).

ECS was tried to be measured full computerized by Quantimet 570C software, but it revealed to be insufficient. Therefore, Bioquant software in combination with a mouse was used for morpho-

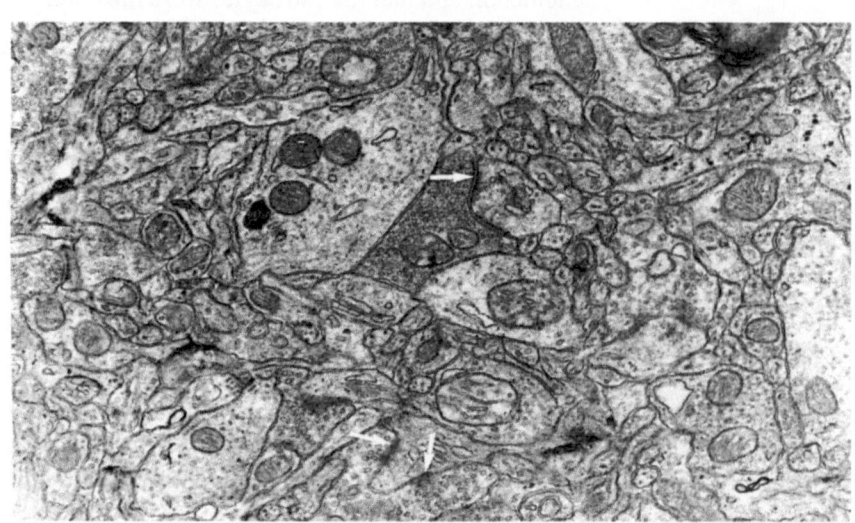

Fig.1. (Magnification 8000): Normal rat brain fixed in glutaraldehyde (4%) and embedded in epoxy resin: The gaps between the cellular structures vary between 21 nm and 26 nm. They are typically narrow in the synaptic cleft (white arrows). Confluent spaces are rare, the ECS was measured 7.4%

Table 1. *Extracellular Space (ECS) in Conventionally Treated Normal and Edematous Brain Tissue Compared to Normal Cryofixed Material: a Morphometrical Study*

Method	Area	ECS (mm²)	Total image (mm²)	ECS in%	Ø ECS in%
Normal rat brain fixed in glutaraldehyde with following epon-embedding procedure	1	161.8	2249.4	7.2	
	2	164.0	2304.9	7.1	
	3	158.7	2273.8	7.0	7.4*,a
	4	180.8	2282.4	7.9	
	5	177.1	2302.7	7.7	
Edematous rat brain fixed and embedded like above	1	360.2	2235.6	16.1	
	2	371.5	2287.4	16.1	
	3	419.8	2275.3	18.4	16.3*,b
	4	315.3	2250.1	14.0	
	5	388.9	2276.0	17.1	
Normal rat brain cryofixed and embedded in LR-White under UV-irradiation	1	370.6	2276.3	16.3	
	2	303.7	2304.5	13.2	
	3	485.0	2255.5	21.5	16.3c
	4	435.4	2281.2	19.1	
	5	255.7	2230.8	11.5	
Results of von Harreveld in mice after cryofixation (1965)	6 areas			18.1–25.5	23.6

* Student's t-test, p < 0.01. ᵃ variance = 0.03; sd = 0.18. ᵇ v = 0.5; sd = 0.7. ᶜ v = 3.4; sd = 1.8.

metrical study of ECS. 5 representative areas were chosen, the extracellular lines measured and multiplied by the width of the extracellular gap. Subsequently, this space representing ECS was related to the total area plane. The measurements were undertaken by another assistant working in the institute without knowing anything about the subject.

Statistic evaluation was performed by means of Student's t-test for unpaired groups.

Cryofixed versus Conventionally Fixed Brain Tissue

Figures 1 and 2 illustrate the electron micrographs of conventionally treated and cryofixed normal tissue.

The morphometrical examination of the different electron micrographs revealed striking differences between conventionally treated and cryofixed rat brain tissue (Table 1). ECS in *cryofixed normal* rat was measured to be the same compared to *conventionally fixed edematous* rat cortex after ultraviolet irradiation in the areas with Evans blue and HRP extravasation. The data estimated by von Harreveld, however, could not be confirmed totally.

Statistic evaluation revealed a significantly increased (p < 0.01) ECS in edematous rat brain compared to normal brain (both fixed in 4% glutaraldehyde). Mean ECS of normal rat brain cryofixed and embedded in LR-White under UV-irradiation did not differ from the values of edematous rat brain fixed in

glutaraldehyde except for an increased variance (16.3% ± 3.4 to 16.3% ± 0.5).

Fluid Movements in the ECS

In normal conditions there seems to be little or no bulk flow through neural tissue. This does not fit completely with the data obtained by cryofixation since the

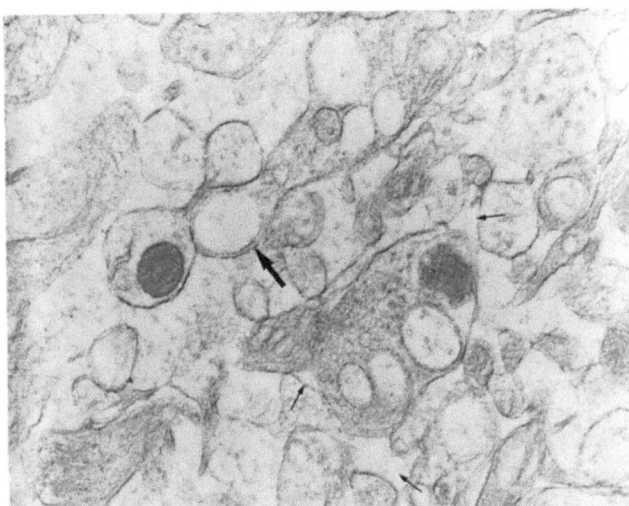

Fig. 2. (Magnification 12000): The normal rat brain after cryofixation and embedding in LR-White shows a greater variety in the width of the extracellular gap (36–47 nm, thick black arrows). Confluent spaces are more often (little black arrows)

larger extracellular spaces in the cortex should not exclude bulk flow. It seems that in the gray matter the bulk flow is very restricted due to a lack of a continuous system of parallel channels. The zig-zag-gaps facilitate diffusion but a significant convective transport cannot take place.

Furthermore, the extracellular spaces in the brain are not only filled with ions and water. *Mucopolysaccharides* (glycosaminoglycans) are present as an interstitial, i.e., extracellular, ground substance. The presence of a ground substance of a proteincarbohydrate nature occupying these spaces was suggested by Hess (1953) using histochemical staining. The glycosaminoglycans strongly determine the physiologic properties of fluid spaces in normal and edematous states. Their large molecules influence viscosity and osmotic pressure. They tend to bind water and some cations restricting the movement of water and solutes into their domain and thus impeding the bulk flow.

Marmarou *et al.* (1982) infused physiological solutions with and without additional proteins into the cerebrum of cats in order to study the mechanisms of fluid retention in brain tissue. Animals infused with the protein solution had a prolonged fluid retention as compared to those infused with mock CSF only. The authors explained this to result from the oncotic pressure of the protein solution. Therefore, other than physical forces might have been involved to retain the fluid in the tissue. One possibility could be the retention of water by proteins might be related to variable relationships between free and bound water, especially of water bound to proteins, that is dependent on physicochemical events in the extracellular spaces.

References

1. Cervós-Navarro J (1967) Brain edema due to ionizing radiation. In: Klatzo I, Seitelberger F (eds) Brain edema. Springer, New York, pp 632–638
2. Evans JE, Raimondi A (1961) Presented to the Surgical Forum. Am Coll Surg
3. Ferszt R, Neu S, Cervós-Navarro J, Sperner J (1978) The spreading of focal brain edema induced by ultraviolet irradiation. Acta Neuropathol 42: 223–229
4. Gonatas NK, Zimmermann HM, Levine S (1963) Ultrastructure of inflammation with edema in the rat brain. Am J Path 422: 455
5. Hess A (1953) The ground substance of the central nervous system revealed by histochemical staining. J Comp Neurol 98: 69–92
6. Horstmann E, Meves H (1959) Die Feinstruktur des molekularen Rindengraues und ihre physiologische Bedeutung. Zschr Zellforsch 49: 569–604
7. Long DM, Hartmann JF, French LA (1962) Presented to the Society of Neurological Surgeons
8. Marmarou A, Tanaka K, Shulman K (1982) The brain response to infusion edema: Dynamics of fluid resolution. In: Brock M, Hartmann A (eds) Treatment of cerebral edema. Springer, Berlin Heidelberg New York, pp 11–18
9. Van Harreveld A, Crowell J (1964) Electron microscopy after rapid freezing on a metal surface and substitution fixation. Anat Rec 149: 381
10. Van Harreveld A, Ochs S (1956) Cerebral impedance changes after circulatory arrest. Am J Physiol 187: 180

Correspondence: J. Cervós-Navarro, M. D., Institute of Neuropathology, Free University of Berlin, Klinikum Steglitz, Hindenburgdamm 30, D-12203 Berlin, Federal Republic of Germany.

Acta Neurochir (1994) [Suppl] 60: 151–154
© Springer-Verlag 1994

Subcortical U-Fibers Layer Preservation in Brain Edema

J. Cervós-Navarro[1], J. V. Lafuente[2], E. Gutierrez[2], and S. Kannuki[3]

[1] Institute of Neuropathology, Free University Berlin, Federal Republik of Germany, [2] Department of Neuroscience BCU, Leioa, Spain, and [3] Department of Neurological Surgery, University of Tokushima, Japan

Summary

The mechanism involved in the relative preservation of the subcortical U-fibers in the arcuate zone was studied in a post infarct edema after sagittal superior sinus occlusion.

Superior sagittal sinus (SSS) of 36 mongrel cats were occluded by polymer injection. Immediately before the occlusion Evans-blue (EB) was administered intravenously. The cats were killed 1, 2, 3, 6, 12, 24, 72 and 120 hours after sinus occlusion.

In 20 cats in which cortical veins were occluded, in addition to the SSS, EB was extravasated. In 9 of these cats, which had moderate edema, EB-staining was present only in the cortex. In 11 cats with severe edema, massive EB extravasation was observed also in the white matter. The U-fiber layer was free of EB, suggesting that the extension of edema was blocked by this zone. Our findings demonstrated that the U-fibers act not only as a resistence against extension of edema from white to gray matter, but also in a reverse direction. The characteristics of the spread of brain edema is not yet completely understood; both anatomical and biochemical peculiarities form its basis. Different morphological patterns in the astrocytic reaction as well as the U-fibres sector vascularization are important. To evaluate the role of each one of these factors in the preservation of subcortical U-fiber layer in brain edema further investigations should be done.

Keywords: Sagittal superior sinus thrombosis; subcortical U-fibers; spreading of brain edema.

Introduction

Several neurological diseases show demyelination with sparing the subcortical U-fibers. Preservation of the U-fibers can also be observed after trauma or cerebral infarcts, while deeper white matter and occasionally cerebral cortex are more seriously affected. The mechanism involved in the relative preservation of the arcuate zone (and of other regions as the corpus callosum and capsula interna) in chronic recurrent edema is unclear.

For the present study we used material from an post infarct edema after sagital superior sinus occlusion[6] to study the characteristics of the spread of brain edema in U-fiber sector as well as the gliosis in this area.

Material and Methods

Thirty-six mongrel cats weighing 2.5–4.0 kg were used. Under anaesthesia with 20 mg/kg ketamine and 1–2 mg/kg xylazine, the cats were artificially ventilated via an endotracheal tube with a Harvard animal respirator. The femoral artery and vein were catheterised, and the cats were fastened onto a stereotaxic frame. A small parietal craniotomy was made above the SSS under an operating microscope. Care was taken to avoid bleeding from the SSS. After exposure, the sinus was punctured and a 1 : 1 mixture of isobutyl-2-cyanoacrylate (IBCA) at doses of 0.1–0.2 ml in 16 cats (group A) and of 0.5 ml in 20 cats (group B) was manually injected into the SSS. The bone defect was filled with resin and the wound was closed.

Two ml/kg of 2% Evans blue (EB) were intravenously injected just before the occlusion in order to detect the maximal extension of the extravasated tracer. After recovery from anaesthesia, the animals were extubated and their neurological status carefully observed. Four cats were allowed to survive the occlusion for 1 hour, 7 for 3 hours, 7 for 4 hours, 2 for 12 hours, 5 for 24 hours, 4 for 72 hours and 4 for 120 hours. Thereafter, they were anaesthetized with the same procedure as used for surgery and sacrificed by perfusion-fixation through the heart with 500 ml of physiological saline followed by Karnovsky's solution. Three cats in comatous state died without perfusion-fixation within 12 hours of the operation.

After perfusion-fixation the brain was rapidly removed and X-rayed. The radiopaque nature of the embolus allowed to detect the extension of the occlusion. Coronal sections were placed in Karnovsky's fixative for an additional 4 hours. The material was embedded in paraffln and stained with hematoxylin-eosin, Nissl and Glial fibrillary acidic protein (GFAP). Ultrathin sections treated with uranyl acetate and lead citrate were used for EM examination.

Results

In 16 cats (group A) injected with 0.1–0.2 ml of the polymer only the SSS was occluded. In all 20 cats (group B) injected with larger amounts of the polymer

(0.5 ml), the occlusion included the cortical veins of the parasagittal region. Eleven cats of this group showed "fulminant deterioration".

The animals with an occlusion limited to the SSS (group A) showed no EB extravasation but for slight EB staining in the parasagittal cortex. EB extravasation was found in all animals of group B with occlusion of both the SSS and cortical veins. In 8 cats EB distribution was limited to the affected cortical gray matter In the other 11 cats EB extravasation spread throughout the cortex deep into the white matter. The layer of the U-fibers was mostly free of EB, even when cortex and subcortical white matter showed extravasated EB (Fig. 1).

In the group of animals with occlusion limited to the SSS only three cats showed moderate focal spongiosis of the parasagittal white matter. However, mild spongiosis was also detected in sham-operated cats. All 20 cats with occlusion of the bridge veins displayed edematous spongiosis and petechial hemorrhages to diverse degrees. Corresponding to the EB-free zone between cortex and white matter the spongiotic changes spared the U-fibers. Nine cats showed mild edema and small hemorrhagic infarctions and in 6 of them (three killed 24, one 72 and two 120 hours after the occlusion) reactive gliosis in the parasagittal grey and white matter was noted.

From the group A with occlusion limited to the SSS, two cats (24 and 72 hours after occlusion) showed slightly dilated extracellular spaces in the parasagittal subcortical white matter. The rest showed no changes at all. In the group with extensive occlusion of the veins edematous changes were prominent and their

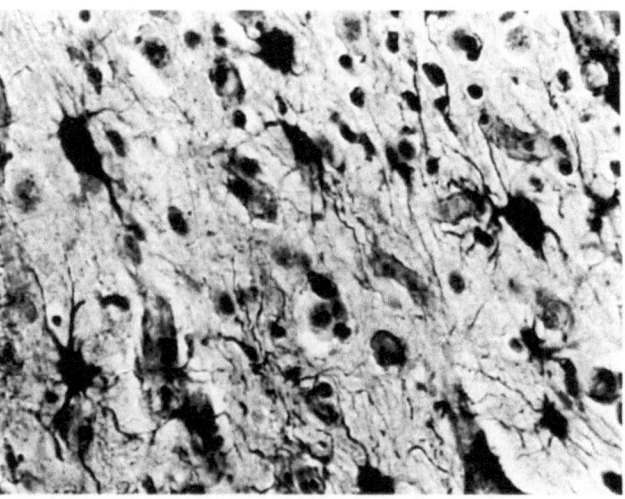

Fig. 2. Same animal, U-fibers sector. Hypertrophic reactive astrocytes with many intensely GFAP posiotive processes

intensity increased parallel to the time interval after the occlusion. The endothelium of capillaries and venules in the parasagittal cortex was damaged even in cats killed one hour after the occlusion and platelets passed through the interendothelial clefts. Markedly widened extracellular spaces filled with proteinaceous transudate were seen in cats which survived longer than 24 hours. Astrocytes with enlarged cytoplasm filled with abundant glial fibrils were frequently observed both in the cortex and the subcortical white matter of the parasagittal region. An increased number of lysosomal bodies were present in the cytoplasma of many astrocytes. The gliosis was specially conspicuous in the U-fibers sector.

At 120 h, morphological pattern changed. At longest survival studied time the mainly morphological feature is the presence of numerous, large, intensely stained astrocytes (Fig. 2) within the background of GFAP-positive fibers. The perikaryon of the astrocytes appeared mishappened, with scarce shorts prolongations surrounded by granular structures, intensely positive with GFAP. This distorted astrocytes were found all along the cortico-subcortical border. The staining profile was markedly different from those observed in matched areas or in shorter survival period, where GFAP stained astrocytes were sparse and faintly labeled in an unstained fiber background. Within the middle region of the white matter, stained astrocytes had extensive, elongated processes, and minimal glial feltwork staining. in the white matter bordering the grey matter, the interface zone, astrocytes had a few processes and existed within a prominently stained glial fiber feltwork.

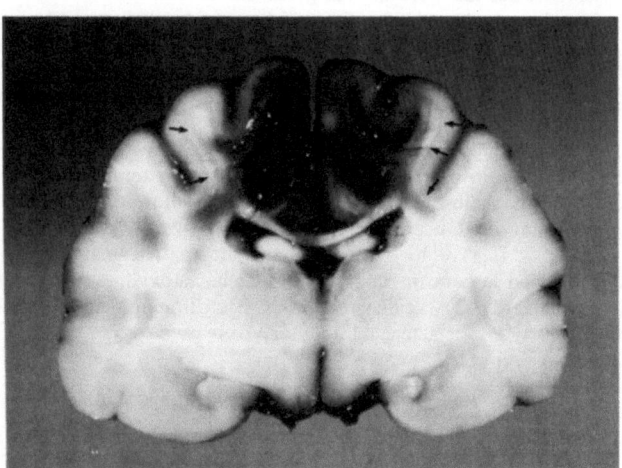

Fig. 1. Cat of group B 72 hours after occlusion extravasated EB in the cortical and subcortical white matter. EB free zone in the U-fiber layer (arrows)

Discussion

The characteristics of brain edema spread is not yet completely understood. Some anatomical factors are involved in the peculiar distribution of the extravasated fluid. Accumulation of the edematous fluid occured mainly in the white matter, but frequently the area of the subcortical U-fibers seemed to be spared.

In different experimental models of edema a band free of edema in the subcortical white matter is usually present. This sector corresponds to the short association fibers (U-fibers) which also are preserved in several neurologic diseases, which involve vascular pathology and/or secondary demyelinization.

It is known from the UV irradiation edema that the U-fibers act as resistance against edema extension from the grey matter towards the white matter. In addition to our findings we suggest that U-fibers may also act as resistance against extension of a reverse direction, namely from white matter towards grey matter. The grey matter in different pathological conditions leading to edema becomes edematous only when it is primarily injured and not as the secondary reaction to injury. The underlying mechanism for the protection of grey matter most likely is due to the U-fibers.

The mechanisms involved in the preservation by brain edema of the U fibers and other structures (corpus callosum or capsula interna) are not well understood so far. Several anatomical features such as vascular distribution, as well as glial population could be suggested.

Several authors[5] observed in vessels that cross this sector, angioarchitectonic peculiarities, such as arteriolar loops. Similar findings were reported by Cervós-Navarro et al. (1987) in senile brains. However, it does not seem probable that these vessels could provide a morphological basis for the resistance of the U-fibers to the spreading of edema. More interesting is the fact that in this zone the surface of the arteriolar band is greater than the venous one. Since brain edema starts in the venous sector of microcirculation it can be speculated that few edematous fluid is leaking in this region. Furthermore comparing the vascularization of the pre- and postcentral gyri of the former has a richer vascular pattern than the latter. The susceptibility to the development of edema is also higher in the postcentral gyrus. Is not clear yet either this refractariness of many areas to develop edema depend of a better drainage or of paucity capilaritation. The quantitative composition of the microvessels bed might determine the tendency of a tissue to edematize.

A further factor to take in consideration are the different metabolic peculiarities of the glia. In edematous areas of this model astrocytes were numerous, intensely stained and swollen; astrocytes with fewer processes were located in the interface zone nearer gray matter. The astrocytes of the white matter and of the grey matter displayed abundant processes corresponding to active process of gliosis. The subcortical astrogliosis may be related to the metabolic function of astrocytes. They are a main regulator of the glutamate metabolism within the brain. Astrocytes have distinct glutamate receptors and a higher affinity than the neurons[2,3]. There is evidences "in vitro" that astrocytes reduce the potency of glutamate as an excitotoxin[7]. In cerebral infarction glutamate-induced damage is strongly implicated[1] and the proliferation of astrocytes may ameliorate damage.

In conclusion our study shows two different morphological features, that could play a role in the preservation of the U-fibres sector from edema: the vascularization pattern and the characteristics of astrocytes in this sector. To evaluate the participation of each one in this process further investigation should be done.

Despite the lesser resistance of the white matter to extravasation of blood and spread of edema, the leaked EB in the cortex did not extend continuously into the white matter, even 72 and 120 hours after occlusion. The EB-free area along the border between cortex and white matter (U-fiber layer) persisted even when EB extravasated in the white matter. This suggests that EB beyond the U-fiber layer did not spread from the cortex, but leaked from congested veins located in the subcortical white matter. The EB-free U-fiber layer explains the delay of the spread of edema after micronecrosis limited to the superficial layers of the cortex. Both these observations suggest that the U-fiber zone acts as a barrier against extension of edema from grey into white matter.

In cats which died during the first 12 hours after the occlusion EB was restricted mostly to the parasagittal cortex although white matter has less resistance than grey matter to edema extension. Since venous drainage through collaterals is more difficult in the cortex and the vessels are more numerous here than in the white matter, the triggering events leading to the breakdown of the BBB occur in the cortex Furthermore the U-fibers act as temporary barrier to the deeper spread of the edema.

The characteristics of the spread of brain edema is not yet completely understood; both anatomical and biochemical peculiarities form its basis. Different morphological patterns in the astrocytic reaction as well as the U-fibres sector vascularization are important. To evaluate the role of each one of these factors in the preservation of subcortical U-fiber layer in brain edema could be responsible.

Acknowledgement

This research has been supported by the Basque Government Grant 9019.

References

1. Albers GV, Goldberg MP, Choi DW (1989) N-methyl-D-aspartate antagonists: ready for clinical trial in brain ischemia? Ann Neurol 25: 398–403
2. Backus KH, Kettenmann H, Schachner M (1989) Pharmacological characterization of the glutamate receptor in cultured astrocytes. J Neurosci Res 22: 274–282
3. Bridges RJ, Kesslak JP, Nieto-Sampedro M, Broderick JT, Yu J, Cotman CW (1987) A L-(3H) glutamate binding site on glia: an autoradiographic study on implanted astrocytes. Brain Res 415: 163–168
4. Cervós-Navarro J, Gertz HJ, Frydl V (1987) Cerebral blood vessel changes in old people. Mech Ageing Dev 39: 223–231
5. Duvernoy HM, Delon S, Vannson JL (1981) Cortical blood vessels of the human brain. Brain Res 415: 163–168
6. Kannuki S, Cervós-Navarro J, Matsumoto K, Nakagawa Y (1990) Experimental model in the cat for cerebral sinovenous occlusion. In: Einhäupl K, Kempski O, Baethmann A (eds) Cerebral sinus thrombosis: experimental and clinical aspects. Plenum, New York
7. Rosenberg PA, Aizemnan E (1989) Hundred-fold increase in neuronal vulnerability to glutamate toxicity in astrocyte-poor cultures of rat cerebral cortex. Neurosci Lett 103: 162–168

Correspondence: J. Cervós-Navarro, M. D., Institute of Neuropathology, Free University of Berlin, Klinikum Steglitz, Hindenburgdamm 30, D-12203 Berlin, Federal Republic of Germany.

Acta Neurochir (1994) [Suppl] 60: 155–157
© Springer-Verlag 1994

Is the Swelling in Brain Edema Isotropic or Anisotropic?

T. Kuroiwa, M. Ueki, Q. Chen, S. Ichinose[1], and **R. Okeda**

Department of Neuropathology, Medical Research Institute and [1] Laboratory of Electron Microscopy, Tokyo Medical and Dental University, Tokyo, Japan

Summary

A study was conducted to examine whether swelling of the brain due to vasogenic-type and cytotoxic-type edema is isotropic or anisotropic. Vasogenic edema was induced by cryogenic injury in cats, and coronal sections of the brain were examined at 4–5 h after injury. The swelling of the edematous white matter longitudinal to and transverse to the subcortical neuronal fibers was 2.3% and 91.1%, respectively. Ischemic edema was examined using cortical tissue specimens of cat brain subjected to either middle cerebral artery occlusion for 3 h or immersion in saline after decapitation for 3 h. The swelling parallel to the left-right axis, caudo-rostral axis and antero-posterior axis was 9.6%, 10.1% and 8.5%, respectively. Neuroglial cell swelling was prominent in the ischemic cortex. Thus swelling of the white matter in vasogenic-type edema was anisotropic, whereas that of gray matter in cytotoxic-type (ischemic) edema was isotropic. This observed difference in the biomechanical properties of brain tissue should be taken into account when the etiology of edema-mediated tissue injury, such as herniation, secondary bleeding or ischemia is investigated.

Keywords: Vasogenic brain edema; cytotoxic brain edema; isotropy; anisotropy.

Introduction

Brain edema is defined as swelling of brain tissue due to an increase of water content. However, it has never been clarified whether the swelling is isotropic or anisotropic. This issue is important when attempting to elucidate: (i) the distribution of compression injury such as perifocal ischemia, demyelination and herniation, and (ii) directional differences in the speed of spread and extent of the edema fluid[1]. In this study, we examined whether the edematous swelling of brain tissue in vasogenic-type edema and cytotoxic-type edema is isotropic or anisotropic[4].

Materials and Methods

Vasogenic-Type Edema

Cryogenic injury was created in the supramarginal gyrus of the cat brain by applying a metal plate cooled with acetone-dry ice, and Evans blue solution was injected to visualize the extent of the edema[3]. As a result of this procedure, vasogenic brain edema developed in the subcortical white matter of the lesioned cortex, in which most fibers (projection and commissural fibers) run in parallel toward the semioval center[2]. The animals were sacrificed at 4–5 h (n = 6) after injury, and the brain of each was removed and cut to obtain various coronal sections at the level of the lesion.

(i) The expansion of the edematous white matter parallel to and transverse to the subcortical fibers was evaluated morphologically in the coronal sections and compared with the corresponding area in the opposite hemisphere. Percentage volume expansion of the edematous white matter was calculated from the above morphological data.

(ii) Water content of the edematous white matter and the corresponding opposite white matter was measured by gravimetry[5]. Percentage volume expansion was calculated from the water content, and compared with the data obtained (i) to check the validity of the morphometry.

(iii) The coronal sections adjacent to those used for gravimetry were immersion-fixed with buffered formalin, embedded in paraffin and stained with cresyl violet/luxol fast blue for histological evaluation.

Cytotoxic-Type Edema

(i) Cerebra of cats (n = 4) were removed after the animals had been sacrificed with an injected overdose of pentobarbital. Cortical tissue specimens were dissected from the brain in a cooled box under high humidity, and the left-right axis, caudo-rostral axis and antero-posterior axis were marked on the surface with dyes of different colors. Each specimen was immersed in a chamber containing saline kept at 37°C and photographed intermittently for 3 h after the start of immersion. The percentage swelling parallel to the three axes was obtained by morphometry.

(ii) Morphological changes in the ischemic cortex and the corresponding water content was examined using cats (n = 4) subjected to middle cerebral artery occlusion[2]. Each animal was sacrificed with an injected overdose of pentobarbital at 3 h after the start of occlusion, and tissue samples from the ischemic and control cortices were dissected out for gravimetrical and electron microscopical examination.

Results and Discussion

Vasogenic-Type Edema

(i) Vasogenic-type edema developed in the subcortical white matter under the lesioned cortex at 4–5 h after lesioning. The swelling of the edematous white matter transverse to the fibers was 91%. and that longitudinal to the fibers 2.3% (Fig. 1). Thus white matter in which vasogenic-type edema developed showed anisotropic swelling. The percentage volume increase calculated from this morphometry was 99%.

(ii) The percentage volume increase calculated from the regional water content of the same tissue specimen as that used for the morphometry was 97%, similar in value to that obtained by morphometry.

(iii) Morphologically, most of the neuronal fibers in the subcortical white matter on the control side ran compactly in parallel toward the semioval center (Fig. 2a). In the lesioned brain, the cortex showed severe necrosis congestion and focal hemorrhage. However these changes did not extend to the underlying white matter. In the subcortical white matter,

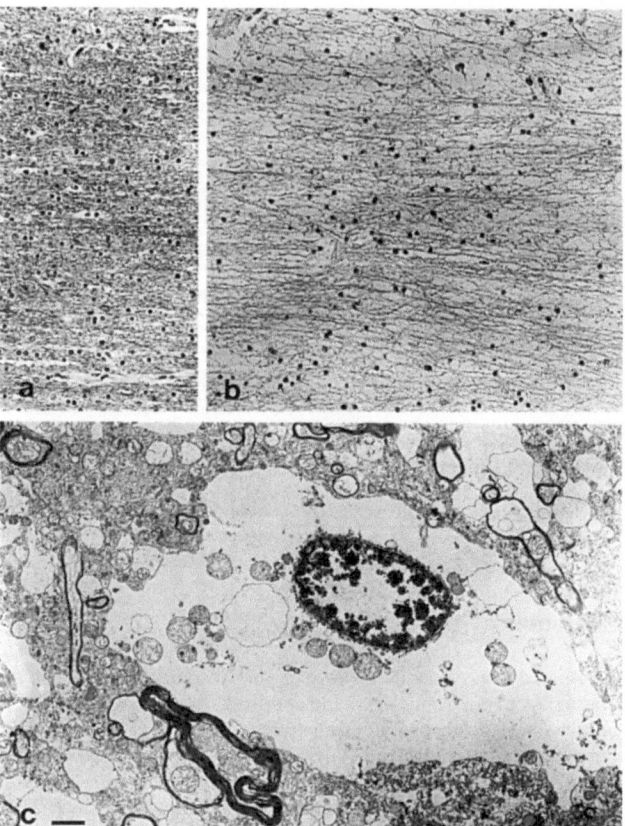

Fig. 2. Morphological change of the edematous brain tissue. In vasogenic-type edema, neuronal fiber dissociated transverse to the direction of the fibers. (a) normal white matter, (b) edematous white matter. In cytotoxic type edema induced by focal ischemia (middle cerebral artery occlusion), swelling of neuroglial cell was remarkable (c)

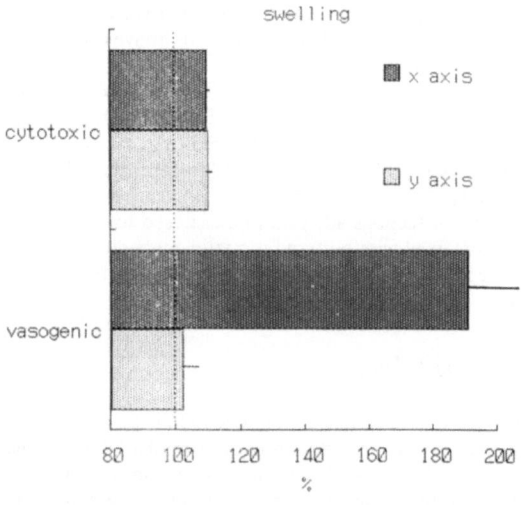

Fig. 1. Directional differences in the swelling of brain tissue in cytotoxic-type brain edema and vasogenic-type brain edema. X and y axes represent left-right axis (x axis) and caudo-rostral axis in cytotoxic type edema. In vasogenic-type edema, x and y axes represent those transverse to and parallel to the neuronal fibers in the subcortical white matter. Note the isotropic swelling in the cytotoxic-type edema and anisotropic swelling in the vasogenic-type edema

edema was prominent and most of the neuronal fibers running toward the semioval center were dissociated transversely, causing widening of the extracellular space (Fig. 2b).

Cytotoxic-Type Edema

(i) Cortical tissue specimens showed swelling after immersion in saline. At 3 h after the start of immersion, the swelling parallel to the left-right axis, caudo-rostral axis and antero-posterior axis was 9.6%, 10.1% and 8.5%, respectively. Thus swelling of the gray matter in ischemic edema was isotropic.

(ii) In the control hemisphere, the water content of the cortex was 66.4% on average. In the ischemic gray matter, the water content increased to 69.9%, corresponding to an increase in volume of approximately 11.6%. Assuming the swelling to be isotropic, this volume change corresponds to a 3.7% increase along each of the three axes. Electron microscopically, hydropic swelling of neuroglial cells was prominent, as shown in Fig. 2c.

Conclusion and Comment

The white matter underwent anisotropic swelling in vasogenic-type edema, in which edema fluid accumulates in the extracellular space. The swelling occurred predominantly in a direction transverse to the fibers and was minimal in a longitudinal direction.

This finding indicates that the spread of edema fluid is enhanced most prominently in a direction parallel to the neuronal fibers in vasogenic-type edema.

On the other hand, the swelling of the gray matter was isotropic in the cytotoxic-type edema induced by ischemic insult. Morphologically, neuroglial cells developed hydropic swelling (intracellular accumulation of edema fluid) in the ischemic cortex.

Thus the basic pattern of tissue swelling differs markedly between vasogenic-type edema and cytotoxic-type edema. The neuronal fibers run in various directions in parallel, crossing, bifurcating and intermingling throughout most of the white matter structure. Therefore, the overall swelling of the brain would not show any significant anisotropy in vasogenic-type edema. However, in many specific areas and structures of the central nervous system, such as the optic radiation, corpus callosum, subcortical white matter, internal capsule and most of the neuronal tract, neuronal fibers run in parallel. Therefore, when attempting to elucidate edema-mediated tissue injury in such structures, anisotropy in the edematous swelling of the white matter should be taken into account.

Using diffusion-weighted MRI, anisotroic spreading of edema fluid has been observed[1,6]. Ebisu et al., using cat brain subjected to cryogenic injury, observed faster spread of edema fluid in the white matter parallel to the axonal fibers. The swelling of the white matter transverse to the axonal fibers observed in the present study would be a factor enhancing anisotropy in the spread of edema fluid. However, the term "diffusion anisotropy" used in such a situation may not be appropriate, since the spread of edema fluid in cryogenic injury has been shown to occur mainly by bulk flow[7].

References

1. Ebisu T, Naruse S, Horikawa Y, et al (1991) The application of in vivo diffusion weighted magnetic resonance image to intracranial disorders. Brain Nerve 43: 677–684
2. Kuroiwa T, Seida M, Tomida S, Hiratsuka H, Okeda R, Inaba Y (1986) Discrepancies among CT, histological, and blood-brain barrier findings in early cerebral ischemia. J Neurosurg 65: 517–524
3. Kuroiwa T, Yokofujita J, Kaneko H, Okeda R (1990) Accumulation of oedema fluid in deep white matter after cerebral cold injury. Acta Neurochir (Wien) [Suppl] 51: 84–86
4. Kuroiwa T, Ueki M, Chen Q, Okeda R (1993) Swelling of the white matter is anisotropic in vasogenic brain edema. J Cereb Blood Flow Metab 13 [Suppl 1], in press
5. Marmarou A, Tanaka K, Shulman K (1982) An improved gravimetric measure of cerebral edema. J Neurosurg 56: 246–253
6. Moseley ME, Cohen Y, Kucharczyk J, Mintorovitch J, Asgari HS, Wendland MF, Tsuruda J, Norman D (1990) Diffusion-weighted MR imaging of anisotropic water diffusion in cat central nervous system. Radiology 176: 439–445
7. Reulen HJ, Graham R, Spatz M, Klatzo I (1977) Role of pressure gradients and bulk flow in dynamics of vasogenic brain oedema. J Neurosurg 46: 24–35

Correspondence: Toshihiko Kuroiwa, M. D., Yushima 1-5-45, Bunkyo-ku, Tokyo 113, Japan.

Acta Neurochir (1994) [Suppl] 60: 158–161
© Springer-Verlag 1994

Biomechanical Characteristics of Brain Edema: the Difference Between Vasogenic-Type and Cytotoxic-Type Edema

T. Kuroiwa, M. Ueki, Q. Chen, H. Suemasu[2], I. Taniguchi[1], and **R. Okeda**

Department of Neuropathology, [1] Department of Neurophysiology, Medical Research Institute, Tokyo Medical and Dental University and [2] Faculty of Engineering, Sophia University, Tokyo, Japan

Summary

Some of the basic biomechanical properties of edematous brain tissue have yet to be clarified. Therefore we measured regional tissue compliance and swelling isotropy/anisotropy in cat brain during development of vasogenic-type and cytotoxic-type edema. In vasogenic-type edema induced by cryogenic injury, the edematous white matter showed an increase of regional tissue compliance (indentation method), which paralleled the increase in the regional tissue water content (gravimetry). Swelling of the white matter due to edema was anisotropic, in which expansion transverse to the neuronal fibers caused by their dissociation was 91.1%, whereas longitudinal expansion was 2.3%. In cytotoxic-type edema induced by cerebral ischemia for 3 h, regional tissue compliance was decreased in the area suffering energy failure, which was visualized as an area of reduced succinic dehydrogenase activity. The ischemic gray matter showed isotropic swelling, and morphologically, prominent swelling of neuroglial cells. These marked differences in basic biomechanical properties between vasogenic-type and cytotoxic-type edema should be taken into account when analyzing the mechanism of edema-mediated tissue injury.

Keywords: Vasogenic-type edema; cytotoxic-type edema; anisotropy; isotropy; regional tissue compliance.

Introduction

Brain edema causes fatal complications such as cerebral herniation and brainstem hemorrhage. It is also a pathogenetic factor of demyelination and dysfunction of neuronal activity. Brain edema is defined as swelling of brain tissue due to an increase in its water content, and thus basic knowledge of the biomechanical properties of edematous as well as normal brain tissue is indispensable for analyzing the mechanism of edema-mediated tissue injury. Igor Klatzo – honorary member of the newly-born International Society of Brain Edema Research – classified brain edema into vasogenic-type and cytotoxic-type edema in 1967[1],

and this classification has been widely accepted to date. In this study, we attempted to characterize these two different types of brain edema from a biomechanical view point.

Materials and Methods

Vasogenic-Type Edema

Cryogenic injury was created in the cortex of the cat brain (n = 12) by applying a metal plate cooled with acetone-dry ice to the supramarginal gyrus, and Evans blue (EB) solution was injected to visualize the extent of the edema[3]. As a result of this procedure, vasogenic-type edema developed in the subcortical white matter of the lesioned cortex. The animals were sacrificed at 4–5 h after the lesioning, and the brain of each was removed and prepared for the following measurements.

(i) Regional tissue compliance was measured using the indentation method[4], in which a coronal block of the brain was placed in a chamber filled with silicone oil at 37°C, and the stress-strain curves obtained by compression of the regions of interest were recorded. Regional tissue water content corresponding to the site of compliance measurement was assessed by gravimetry[7].

(ii) The expansion of edematous white matter parallel to and transverse to the neuronal fibers was evaluated morphologically in the subcortical white matter of coronal sections and compared with the corresponding area in the opposite hemisphere[5]. Percentage volume expansion of the edematous white matter was calculated from the above morphological data. The water content of the edematous white matter and that of the corresponding opposite white matter were measured by gravimetry. Percentage volume expansion calculated from the water content was compared with the data obtained from the morphometry in order to check its validity. After these measurements, tissue specimens were immersion-fixed in 4% buffered formalin and prepared for histological examination.

Cytotoxic-Type Edema

(i) The change in regional tissue compliance in cytotoxic-type edema was examined in cat brain subjected to focal ischemia (n = 4). The ischemia was induced by left middle cerebral artery occlusion via the transorbital approach with/without occlusion of the left common carotid artery for 3 h[2]. The brain was removed after an injected overdose of pentobarbital, and the following parameters were examined. To determine the extent of energy failure as an indicator of the extent of ischemia, succinic dehydrogenase (SD) activity was visualized quantitatively in the coronal plane of the ischemic brain. The method used for quantitative imaging has been described elsewhere[6]. Briefly, a coronal section of the brain subjected to focal ischemia was stained with 1% triphenyltetrazolium chloride (TTC) solution kept at 35°C, and the chronological change in surface optical density (OD) was recorded by a CCD camera. The rate of formazan (product of the enzymatic reaction of SD with TTC) production was calculated from the rate of OD increase according to data obtained in a separate experiment, and the data were converted to SD activity and visualized using semi-color coding. A coronal section with the mirror surface of that used for SD activity measurement was submerged in silicone oil to prevent evaporation, and regional tissue compliance was measured by indentation technique as described above. Thereafter, tissue specimens at the site of compliance measurement were dissected out for measurement of their regional water content. Morphological changes were examined using tissue specimens obtained from the ischemic cortex.

(ii) For assessing the swelling anisotropy/isotropy of the cortical tissue subjected to ischemic insult, cortical tissue specimens from cats (n = 4) were dissected out shortly after injection of an overdose of pentobarbital, and immersed in a chamber containing saline at 37°C for 3 h[5].

Swelling of the tissue parallel to the left-right (x) axis, rostro-caudal (y) axis and antero-posterior (z) axis was measured by morphometry.

Results and Discussion

Vasogenic-Type Edema

Vasogenic edema visualized by EB staining developed in the subcortical white matter under the lesioned cortex 4–5 h after lesioning.

(i) The regional water content of the control white matter was 63.4%. In the white matter stained with EB, this increased to 82.1% (129% of the control), whereas the content in the white matter on the lesion side outside the EB- stained area was 66.5% (105% of the control). The regional tissue compliance of the edematous white matter (EB- stained area) was 366% of the control value, whereas the regional compliance of the white matter outside the edematous white matter on the lesion side was 100% of the control value. The increase in the regional tissue compliance was correlated significantly with the increase in the water content, the regression coefficient for the linear correlation being 0.83 (p < 0.001).

(ii) The swelling of the edematous white matter transverse to the neuronal fibers was 91%, and that longitudinal to the neuronal fibers 2%. The volume increase calculated from this morphometry was 99.3%. The volume increase in the edematous white matter

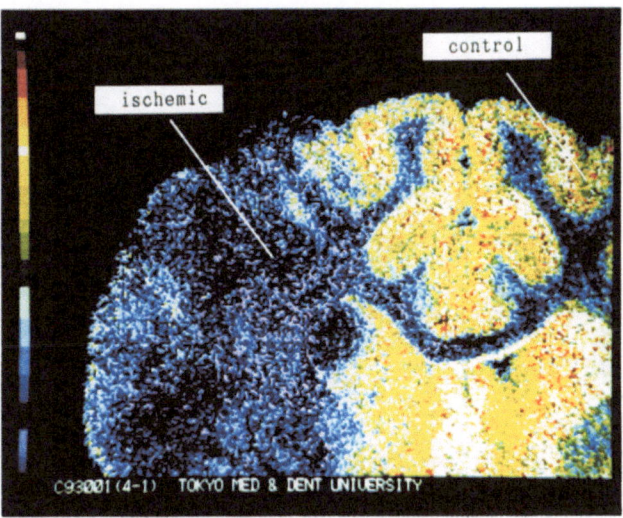

Fig. 1. Succinic Dehydrogenase activity in cat brain removed 3 h after the left middle cerebral artery (MCA) and left common carotid artery occlusion. Note the reduction of the SD activity in the MCA area which was well demarcated by the area showing normal SD activity

calculated from the regional water content was 97.3%, similar to the value obtained by morphometry.

Histological examination of the subcortical white matter on the control side at 4–5 h after lesioning showed neuronal fibers running compactly and in parallel toward the semioval center. On the lesion side, the neuronal fibers were dissociated transversely, causing widening of the inter-fiber space (extracellular space).

Cytotoxic-Type Edema

(i) SD activity was decreased significantly in the area supplied by the left middle cerebral artery at 3 h after the onset of ischemia (Fig. 1). The gray matter suffering from energy failure was well demarcated from the non-ischemic area with a few spotty areas of preserved SD activity near the border. In this ischemic gray matter, the regional tissue water content was increased (ischemic edema) and the regional tissue compliance was decreased. In a typical case shown in Figs. 1 and 2, the SD activity in an ischemic focus was reduced to 14.2% of that in the control cortex. In this area, the regional tissue compliance was reduced to 61.3% of the control value, and regional water content was increased to 113.8% of that of the control. Electron microscopical examination of the ischemic cortex revealed prominent hydropic swelling of neuroglial cells.

(ii) Cortical tissue specimens showed swelling after immersion in saline for 3 h. The swelling parallel to the

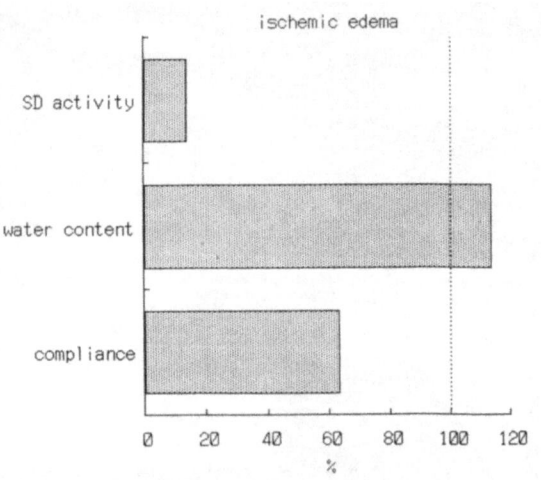

Fig. 2. The succinic dehydrogenase activity, regional tissue water content and regional tissue compliance at an ischemic focus indicated in Fig. 1. The area showed severe energy failure, increase in the regional tissue water content and decrease in the regional tissue compliance

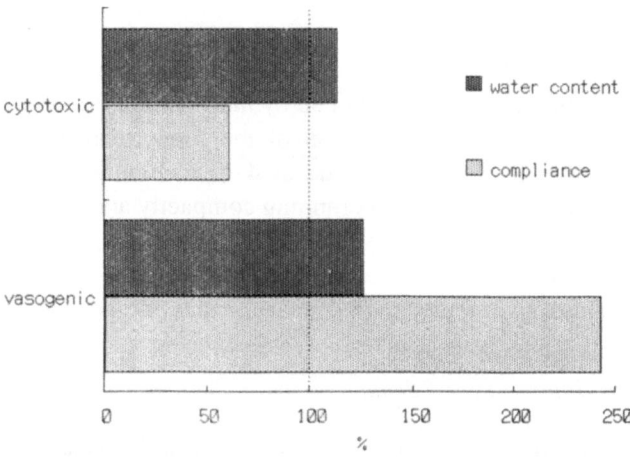

Fig. 3. Comparison of regional tissue water content and regional tissue compliance and vasogenic-type and cytotoxic-type edema. Vasogenic-type edema was examined in the white matter of cat brain 4–5 h after cryogenic injury. Cytotoxic-type edema was induced by focal cerebral ischemia for 3 h

x, y, z axes was 9.6%, 10.1% and 8.5%, respectively, thus showing isotropy (Fig. 3).

Conclusion and Comment

The present study revealed the following:

(1) Regional tissue compliance increased in parallel with the increase in water content in vasogenic-type edema induced by cryogenic injury.

(2) Swelling of the edematous white matter in vasogenic-type edeme was anisotropic (mainly transverse to the neuronal fibers), corresponding microscopically to transverse dissociation of neuronal fibers and widening of the extracellular space.

(3) In the gray matter during early ischemia visualized by SD activity imaging, ischemic edema (cytotoxic-type edema) was evident and the regional tissue compliance was decreased, corresponding ultrastructurally to swelling of the neuroglial cells.

(4) Ischemic swelling of the gray matter was considered to be isotropic (Table 1).

The compliance used in the present study is a reciprocal of the modulus of longitudinal elasticity (Young's modulus) and a quantity often used in solid mechanics and also biomechanics. It is different from the "compliance" widely used in clinical field, which is a reciprocal of a bulk modulus. The regional tissue compliance defined as the reciprocal of the longitudinal elasticity is obtained by the indentation technique. In this type of biomaterial, the value of the bulk modulus is thought to be close to that of water and insensitive to the tissue structure. The compliance measured in the present study can reflect deformability of the material and is thought to be more suitable for the evaluation of the change of the tissue structure than the compliance to the bulk change.

In his presidential address at the 42nd annual meeting of the American Association of Neuropathologists in 1967, Igor Klatzo proposed that brain edema be classified into vasogenic and cytotoxic types[1]. Since then, this classification has been widely accepted in both the clinical and experimental fields. However, the validity of the classification has been challenged by some researchers in view of the considerable variations shown by each type of edema, such as interstitial edema in hydrocephalus and ischemic edema at different stages. However, the present findings strongly indicate that the classification proposed by Klatzo is valid and essential from a biomechanical view point.

The mechanism by which the same "increase in tissue water content" would produce different tissue properties in vasogenic-type and cytotoxic-type edema is still unclear. However, the localization of the water seems to be an important factor. The edema fluid in the extracellular space (vasogenic-type edema) would make the tissue "soft" by dissociating the cellular elements, whereas intracelluar edema fluid would make the tissue "tough", the situation being comparable to the increased turgor seen in other organs such as the liver and kidney during acute cellular swelling. Recently, an attempt has been made to estimate or predict intracranial events using computer simulation technol-

Table 1. *Biomechanical Characteristics of Brain Edema*

Type	Cytotoxic	Vasogenic
Swelling	isotropic	anisotropic
Compliance	decreased	increased

ogy[9]. This approach seems to be important and promising for future research and diagnosis in the neurological field. However, the results of computer simulation depend solely on the input data, and the data necessary for making a reasonable simulation still seem to be lacking. This seems to be borne out by the fact that the marked differences in basic biomechanical features among the various types of edema have been overlooked to date.

References

1. Klatzo I (1967) Neuropathological aspects of brain edema. J Neuropathol Exp Neurol 26: 1–14
2. Kuroiwa T, Seida M, Tomida S, Hiratsuka H, Okeda R, Inaba Y (1986) Discrepancies among CT, histological,and blood-brain barrier findings in early cerebral ischemia. J Neurosurg 65: 517–524
3. Kuroiwa T, Yokofujita J, Kaneko H, Okeda R (1990) Accumulation of Oedema fluid in deep white matter after cerebral cold injury. Acta Neurochir (Wien) [Suppl] 51: 84–86
4. Kuroiwa T, Taniguchi I, Okeda R (1993) Regional tissue compliance of edematous brain after cryogenic injury in cats. In: Intracranial pressure VIII. Springer, Berlin Heidelberg New York Tokyo, in press
5. Kuroiwa T, Ueki M, Chen Q, Okeda R (1993) Is the swelling isotropic or anisotropic in brain edema. Acta Neurochir (Wien) [Suppl]
6. Kuroiwa T, Czernicki Z, Ueki M, Yamaguchi T, Okeda R (1994) Ischemic penumbra and quantitative imaging of succinic dehydrogenase activity in gerbil brain. Stroke, submitted
7. Marmarou A, Tanaka K, Shulman K (1982) An improved gravimetric measure of cerebral edema. J Neurosurg 56: 246–253
8. Nagashima T, Tamaki N, Shirakuni T, Matsumoto S, Seguchi Y, Tamura T (1985) Biomechanics of vasogenic brain edema. Application of Biot's consolidation theory and the finite element method. In: Inaba Y, Klatzo I, Spatz M (eds) Brain edema. Springer, Berlin Heidelberg New York Tokyo
9. Walsh EK, Schettini A (1976) Elastic behavior of brain tissue in vivo. Am J Physiol 230(4): 1058–1062

Correspondence: Toshihiko Kuroiwa, M. D., Yushima 1-5-45, Bunkyo-ku, Tokyo 194, Japan.

Acta Neurochir (1994) [Suppl] 60: 162–164
© Springer-Verlag 1994

Experimental Quantitative Evaluation of Transvascular Removal of Unnecessary Substances in Brain Edema Fluid

E. Kadota[1], K. Nonaka[1], M. Karasuno[1], K. Nishi[1], Y. Nakamura[2], K. Namikawa[2], Y. Okazaki[2], K. Teramura[3], and S. Hashimoto[3]

[1] Division of Pathology, Kishiwada Municipal Hospital, [2] Department of Instrument Analysis, Kinki Universitiy Faculty of Pharmaceutical Sciences, and [3] Second Department of Pathology, Kinki University School of Medicine, Osaka, Japan

Summary

We developed a model by which the transvascular removal of unnecessary substances in brain edema fluid could be measured quantitatively and chronologically. Brain stab wounds were produced in Wistar rats by insertion of paired microdialysis probes in the unilateral caudatoputamen. Homovanillic acid (HVA) was administered by microdialysis from one probe, and the HVA clearance was measured by HPLC analysis of perfusate from the other probe. Using this model, we evaluated the site of removal and whether the removal processes were affected by anesthesia or an elevated plasma concentration of the substance. As a result, 1) Probenecid did not change HVA clearance although this inhibits HVA removal via subarachnoidal vessels[4]. Therefore, HVA removal in this model was considered mainly due to intraparenchymal transvascular efflux. 2) There was no alteration in HVA removal induced by anesthesia or intravenous HVA injection. Consequently, this efflux mechanism seems to be a rather stable protective process, and seems to play a considerable role in brain microenvironmental homeostasis.

Keywords: Microdialysis; rat; homovanillic acid.

Introduction

Brain edema may be understood as a dynamic equilibrium between edema fluid production and resolution. However, only a few studies have evaluated the kinetics of brain edema fluid absorption in a chronological and quantitative manner. Microdialysis makes it possible to administer substances to the brain and to sample interstitial fluid without altering the cerebral tissue fluid volume[11]. Using microdialysis, we produced a new animal model in which the transvascular removal of unnecessary substances from brain edema fluid could be measured quantitatively over time. In addition, we evaluated 1) the site of removal and 2) whether the removal processes were affected by anesthesia or elevated plasma concentration of the substance.

Materials and Methods

Stab wounds were produced stereotactically in the brains of adult Wistar rats weighing $250 \sim 390$ g (n = 23) by inserting a pair of microdialysis probes (CMA 11, BAS Co., Ltd.) into the unilateral caudatoputamen. During the procedure, all rats were under Pentobarbital anaesthesia (25 mg/kg intraperitoneally). The probes were then fixed to the skull using dental cement. Experiments were conducted using free-moving units (CMA 125, BAS Co., Ltd.). Homovanillic acid (HVA) was used as a tracer to evaluate the efflux of unnecessary substances from the brain. HVA in Ringer's solution (500 ng/10 μl, pH 7.4) was administered by microdialysis at 1 μl/min from one probe for a 1-hour period, and HVA clearance was then measured by HPLC analysis of perfusates (perfusion rate; 1 μl/min) from the other probe. HVA clearance was measured as follows (Fig. 1). The endogenous HVA level was determined three times before exogenous HVA administration by microdialysis. Two hours after HVA administration, the HVA level was again evaluated three times. Then the mean pre- and postadministration HVA levels were used as the baseline and semilogarithmic values of the differences between the HVA levels after intracerebral administration and the baseline level were determined. The HVA half-time and fractional rate constant (k) were calculated from the semilogarithmic table thus produced using the following expressions; k = slope value/0.434, and half-time = $0.693/k$[12].

In group 1 (control group, n = 7), the HVA half-time and k values were measured 3 and 4 days after the operation.

In group 2 (n = 4), HVA clearance was measured under Probenecid administration (200 mg/kg, intraperitoneally) 4 days after probe insertion.

In group 3 (n = 6), HVA half-time and k values were studied under Pentobarbital anesthesia (25 mg/kg intraperitoneally) 3 days after the operation.

In group 4 (n = 6), 20 μg HVA was administered intravenously from the femoral vein under Pentobarbital anesthesia (25 mg/kg intraperitoneally) 4 days after the operation, and HVA half-time and k values were also measured.

Animals were sacrificed 4 days after probe insertion by perfusion with formaldehyde. All rats were examined histologically.

The HPLC analysis was performed under the following conditions; Column: 1 · 100 mm ODS 3 µm (BAS Co., Ltd.), Mobile phase: 0.1 M tartaric acid/0.1 M sodium acetate, 0.5 mM EDTA-2Na pH 3.2, 550 µM sodium 1-octanesulfonate (MW 216.28), 5% Me-CN, Flow rate: 40 µl/min, Temp: 30°C, Potential: 700 mV v.s. Ag/Cl, Filter: 0.02 Hz.

Results

In all animals, the position of probe insertion was confirmed histologically.

The HVA clearance curve was parabolic, and HVA values showed a linear distribution in the semilogarithmic tables.

In group 1, the controls, the HVA half-time was 9.86 ± 0.89 min at 3 days, and 9.82 ± 0.71 min at 4 days. The k value was 4.46 ± 0.46/h at 3 days, and 4.36 ± 0.27/h at 4 days (Fig. 2).

In group 2 with Probenecid, HVA half-time and k value were 9.15 ± 0.49 min and 4.59 ± 0.25/h, respectively. There were no statistical differences between groups 1 and 2 (Fig. 2).

In group 3 with Pentobarbital anesthesia, HVA half-time and k value were 9.60 ± 0.42 min and 4.38 ± 0.20/h, respectively. There were also no statistical differences between groups 1 and 3 (Fig. 2).

In group 4 with intravenous HVA administration, HVA half-time was 9.45 ± 0.32 min and k value was 4.42 ± 0.15/h. There was also no statistical difference between groups 1 and 4 (Fig. 2).

Discussion

It was reported that blood-brain barrier damage at the site of microdialysis probe insertion is minimal[9], and that the recovery of substances through the semipermeable membrane changes with time[2]. However, we confirmed that the microdialysis data obtained reflected not the physiological interstitial fluid but brain edema fluid, and that the data remained stable up to 4 days after probe insertion[6]. We also confirmed that exogenous HVA administration via the dialysis probe had no major effect on brain dopamine metabolism[6].

HVA removal is generally considered due to a transvascular efflux mechanism depending on an amine (or choline) carrier in the endothelium[1]. We also investigated whether intraparenchymal or subarachnoid vessels were responsible for the removal of HVA in this study. Probenecid inhibits HVA removal via the subarachnoid vessels[4] administration of the drug did not affect the HVA half-time or k value. Therefore, HVA

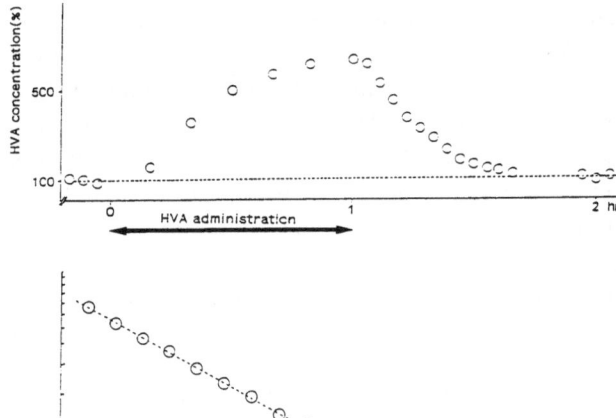

Fig. 1. Determination of HVA clearance. The endogenous HVA level was determined three times before HVA administration. 2 hours after HVA administration, the HVA level was again evaluated three times. The mean pre- and post-administration HVA levels were used as the baseline and the semilogarithmic values of the differences between the HVA levels after administration and the baseline level were determined. The HVA half-time and fractional rate constant (k) were calculated from the semilogarithmic table thus produced using the following expressions; k = slope value/ 0.434, and half-time = 0.693/k

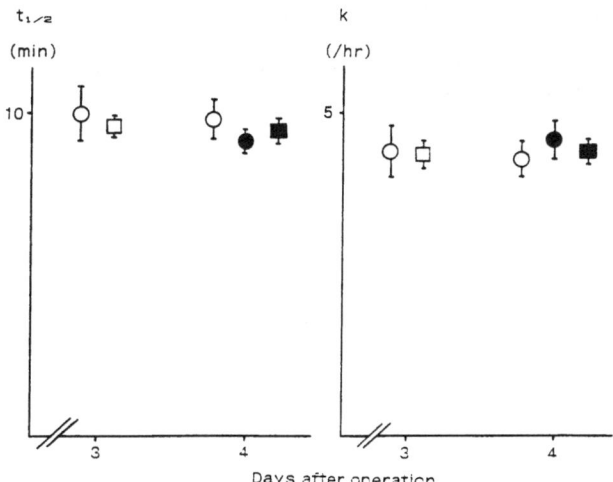

Fig. 2. The HVA half-time ($t_{1/2}$) and the k value (mean ± S.E). ○ Group 1 (Control, n = 7). ● Group 2 (with Probenecid 200 mg/kg, intraperitoneally, n = 4). □ Group 3 (with anesthesia 25 mg/kg intraperitoneally, n = 6). ■ Group 4 (with anesthesia + HVA 20 µg intravenously, n = 6)

removal in this model was considered to occur mainly via intraparenchymal transvascular efflux.

Barbiturates are assumed to have a protective effect on the neurons during ischemia or other damage[7]. However, anesthetics decrease cerebral blood flow[8]. It was reported that decrease of blood flow caused reduction in the brain efflux of unnecessary substances[10].

Therefore, we evaluated the influences of Pentobarbital anesthesia on the brain efflux mechanism. As a result, the customary dose of the drug did not change the efflux functions. There are many reports concerning the transendothelial transport of glucose. It was reported that the endothelial transport rate was determined in principle by the concentration gradient between the plasma/interstitial fluid[5]. However, there are no studies that investigated the influences of plasma/interstitial fluid concentration gradients on the brain efflux of unnecessary substances in the edema fluid. Aizenstein *et al.* reported that 20 μg HVA intravenous administration caused about a 10-fold increase of plasma HVA level for 90 minutes in rats[1]. However, this procedure proved to have no effect on the brain efflux mechanism either. Astroglia is reported to play a major role in maintaining homeostasis of the brain microenvironment[3]. However, our present data suggest that the steady transvascular efflux mechanism, which was not affected by the barbital anesthesia or the plasma/interstitial fluid concentration gradients plays a considerable protective role in the brain microenvironment during edema.

References

1. Aizenstein ML, Korf J (1978) Aspects of influx and efflux of homovanillic acid of rat cerebrospinal fluid. Brain Res 149: 129–140

2. Benveniste H, Diemer NH (1987) Cellular reactions to implantation of microdialysis tube in the rat hippocampus. Acta Neuropathol (Berl) 74: 9–14

3. Dietzel I, Heinemann U, Hofmeier G, *et al* (1982) Stimulus-induced changes in extracellular Na^+ and Cl^- concentration in relation to changes in the size of the extracellular space. Exp Brain Res 46: 73–84

4. Extein I, Korf J, Roth RH, *et al* (1973) Accumulation of 3-methoxy-4-hydroxyphenylglycol-sulfate in rabbit cerebrospinal fluid following probenecid. Brain Res 54: 403–407

5. Fishman RA (1963) Studies of transport of sugars between blood and cerebrospinal fluid in normal states and in meningeal carcinomatosis. Trans Am Neurol Assoc 88: 114–118

6. Kadota E, Nonaka K, Karasuno M, *et al* (1993) Quantitative evaluation of unnecessary substances in brain edema fluid in rats. Adv Neurotrauma Res, in press

7. Lightfoote WE, Molinari GF, Chase TN (1977) Modification of cerebral ischemic damage by anesthetics. Stroke 8: 627–628

8. Saito Y (1971) The effect of anesthetics and hypotension on cerebral cotical blood flow. Anesth 20: 504–513 (Jpn)

9. Tossman U, Ungerstedt U (1986) Microdialysis in the study of extracellular levels of amino acids in the rat brain. Acta Physiol Scand 128: 9–14

10. Ueda T, Wakisaka S, Ochiai H *et al* (1991) A study about the intracerebral accumulation of thalium 201. Jpn J Cereb Blood Flow Metab 3: 92 (Jpn)

11. Ungerstadt U, Hallstroem A (1987) In vivo microdialysis. A new approach to the analysis of neurotransmitters in the brain. Life Sci 41: 861–864

12. Venero JL, Machado A, Cano J (1991) Turnover of dopamine and serotonin and their metabolites in the striatum of aged rats. J Neurochem 56: 1940–1948

Correspondence: Eizi Kadota, M. D., Division of Pathology, Kishiwada Municipal Hospital, 4-17-1, Kamori-cho, Kishiwada City, Osaka 596, Japan.

Acta Neurochir (1994) [Suppl] 60: 165–167
© Springer-Verlag 1994

Formation and Resolution of Brain Edema Associated with Brain Tumors. A Comprehensive Theoretical Model and Clinical Analysis

T. Nagashima, N. Tamaki, M. Takada[1], and **Y. Tada**[1]

Department of Neurosurgery, Kobe University School of Medicine and [1] Department of Systems Engineering, Faculty of Engineering, Kobe University, Kobe, Japan

Summary

The purpose of this study is to quantify the relative contribution of the mechanisms in the absorption of edema fluid. The convection/diffusion and the comprehensive bulk flow model were applied for the finite element analysis of peritumoral brain edema. For clinical analysis, 90 meningiomas studied by MRI were selected. Serial CT scan and MRI were performed at 0, 2, 4, 6 hours after injection of Iopamidol or Gadpenteic acid respectively. Then the tracer distributions in the edematous brain was analyzed.

The tracer movement in the brain is well represented by the convection/diffusion equation. The absence of the preferential fluid flow directing toward the ventricle indicates that a limited role of CSF sink action into the ventricle. From capillary surface area (240 cm²/g brain), capillary hydraulic conductivity (1.8×10^{-8}ml/ cmH$_2$O/cm²/min) and the simulated average tissue pressure of 9.8 mmHg, maximum absorption rate into capillaries was estimated to be 0.003 ml/h/cm³ brain tissue. Considering the limited role of edema fluid clearance into the ventricle, the results indicate a possible role of subarachnoid CSF space for the clearance of edema fluid. The clearance of edema fluid into subarachnoid CSF space should be studied quantitatively.

Finally, unification of the convection/diffusion and the comprehensive bulk flow model will provide a more quantitative analysis of edema formation and resolution by using MRI and tracer studies.

Keywords: Brain edema; finite element method; MRI.

Introduction

Brain tumors continuously produce edema fluid, however, the edema around the tumor remains constant during a period of at least several days. This means that formation of edema fluid is dynamically balanced with its absorption. The production rate considerably varies from tumor to tumor, on the other hand, the absorption rate probably is relatively constant[5].

Several mechanisms in the resolution of edema fluid were demonstrated by clinical and experimental studies. The question is how to quantify the relative contri-

bution of these mechanisms. Methods are lacking for direct measurement of the relative contribution of each of these actions in clinical settings. When there is no method to measure directly a biological phenomenon, theoretical analysis by computer simulation provides a valuable approach for the problem. Theoretical and clinical analysis of peritumoral brain edema is reported in this paper.

Materials and Method

Theoretical Models

The convection/diffusion model: The convection/diffusion equation consists of a diffusion term and a convection term[1].

$$\frac{\partial C}{\partial t} = D_x \frac{\partial^2 C}{\partial x^2} + V_x \frac{\partial C}{\partial x} ,$$

where C is tracer concentration, V_x is convective flow vector, and D_x is diffusion coefficient of tracers.

Comprehensive bulk flow model: The comprehensive model means a unified mathematical description of the coupled behavior of various parameters; intracerebral stress, interstitial fluid flow, tissue pressure, osmotic pressure, external force, etc. The model consists of (1) equation of mechanical equilibrium, (2) equation of continuity, (3) Darcy's law, (4) Hookes' law, (5) Terzaghi's principle of effective stress, (6) Starling hypothesis. Terzaghi's principle defines the relationship between interstitial pressure and intracerebral stress. The mathematical description was reported elsewhere[4]. These equations are simple and most comprehensive at present.

Finite Element Analysis

A finite element model was constructed from a horizontal brain slice. The different material properties were assigned to the bone, CSF, gray and white matter, depending on the theoretical model used[4].

Clinical Analysis

Of 200 tumors studied by MRI, 90 meningiomas were selected for the present study. About 35% of the meningiomas showed some extent of peritumoral brain edema. The distribution and volume, associated hydrocephalus, relationship to ventricle, mass effect, tumor site were analyzed.

Serial CT scanning and MRI were performed at 0, 2, 4, 6 hours after conventional bolus injection of 100 ml of Iopamidol (330 mgI/ml) or 20 ml of Gadpenteic acid. Then the tracer distribution in the edematous brain was analyzed.

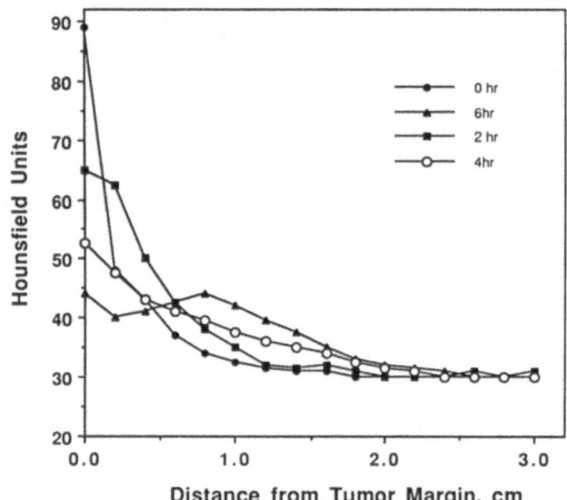

Fig. 1A. Distribution profile of tracer in brain edema. The simulation of tracer movement in the brain by the convection/diffusion model. The 2 mm/h of convective flow was assumed from the lesion to the ventricle. The diffusion coefficient of the tracer was assumed to be 2.0×10^{-10} m^{-2} s^{-1}. The tracer concentration in the brain is presented by contour line maps. The peak concentration at 10 hours was about one tenth of the initial concentration

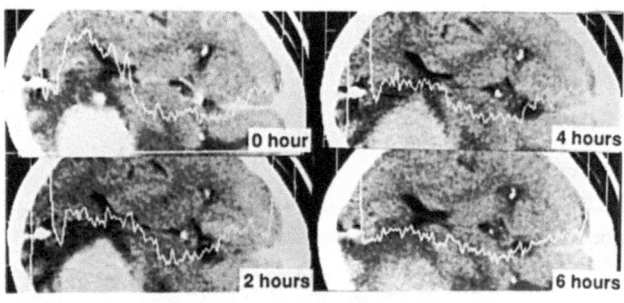

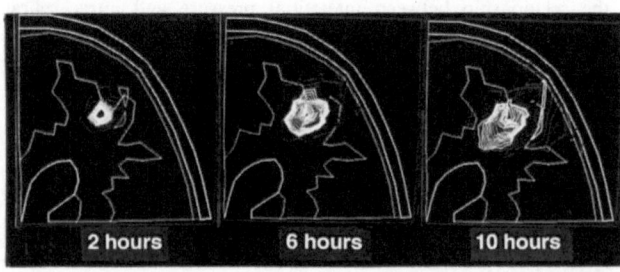

Fig. 1B. The profile of CT number from the lesion to the white matter

Results

MRI Analysis of Peritumoral Brain Edema

Expansion of white matter was heterogeneous and anisotropic. Subcortical white matter easily expands to 300% in the perpendicular plane to nerve fibers, on the other hand, almost no expansion in the parallel plane. The anisotropic expansion of white matter causes anisotropy of parenchymal hydraulic conductivity and non-linear change of it.

Ten percent of the meningiomas with peritumoral edema were associated with hydrocephalus. Coexistance of hydrocephalic and peritumoral brain edema in a same case indicates that clearance of edema fluid can be balanced with the absorption even if CSF absorption via the ventricle is obstructed.

Tracer Study Using CT Scan and MRI

The profile of CT number after contrast enhancement showed a broad peak around the tumor 6 hours after the contrast infusion (Fig. 1A), that indicate the presence of convective flow in the brain. There was no preferential movement of contrast medium directing toward the ventricle. Tracer study by MRI showed the similar profile of intensity in the brain edema. However, nonlinear relationship between the MR intensity and tracer concentration in the brain makes analysis more difficult.

Simulation by the Convection/Diffusion Model

On the assumption of edema fluid flow of 2 mm/h from the lesion to the ventricle and the diffusion coefficient of 2.0×10^{-10} m^{-2} s^{-1}, a drop of contrast medium at a point of subcortical white mater moves into the brain with time. A contour line map of tracer concentration showed a peak of tracer concentration around the point and a deviation of tracer distribution toward the ventricle(Fig. 1B). The result well represented the tracer movement observed by CT scan.

The Simulation by the Comprehensive Bulk Flow Model

A quantitative analysis of edema formation and absorption was done on the assumptions that (1) capillaries in the edematous brain has normal hydraulic conductivity, (2) ventricular pressure does not change, (3) osmotic pressure of edema fluid does not change, (4) CSF sink action does not work, (5) tissue pressure around the tumor is constantly 15 mmHg. These assumptions provides maximum edema extension and maximum fluid absorption into the capillaries in the edematous brain.

From the capillary surface area (240 cm^2/g brain), capillary hydraulic conductivity (1.8 × 10^{-8}ml/cm H$_2$O/cm^2/min)[2], and the simulated average tissue pressure of 9.8 mmHg, the calculated absorption rate into the capillaries was about 0.003 ml/ h/cm^3 brain tissue.

Discussion

The result of the two dimensional simulation by the convection/diffusion model and the clinical observations clearly showed that the tracer movement in the brain is described by the convection/diffusion equation. It supports the result of one dimensional analysis by Aaslid[1].

The estimated absorption rate of edema fluid into capillaries (0.003 ml/h/cm^3) is much smaller than the reported value of 0.02–0.05 ml/h/cm^3 which was calculated from the production rate[1,5]. This discrepancy indicates two possibilities; (1) higher capillary hydraulic conductivity of edematous brain than the reported value for normal brain, or (2) major role of CSF sink action in the absorption of edema fluid.

The simulation by the convection/diffusion model and the CT profile study indicated the absence of the preferential fluid flow directing toward the ventricle. It means that the edema fluid clearance into the ventricle has a limited role. The limited role of ventricular clearance is supported by the coexistance of hydrocephalic and peritumoral edema in a same patient. The results of clinical and simulation studies indicate a possible role of subarachnoid CSF space for the clearance of edema fluid. The clearance of edema fluid into subarachnoid CSF space should be studied quantitatively. The higher hydraulic conductivity of capillaries in the edematous brain is less possible, however it remained to be studied.

The anisotropic expansion of white matter observed by MRI indicates that anisotropy of brain hydraulic conductivity is necessary to be incorporated for three dimensional simulations. Non linear change of material properties by brain edema, for example hydraulic conductivity or brain elastic modulus, should be incorporated for the quantitative analysis. The nonlinear relationship between the material properties and tissue water content is determined only by experimental studies.

The comprehensive bulk flow model lacks the description of tracer movement. Therefore, the results of simulations are evaluated only by static MRI studies. On the other hand, convection/diffusion equation is not

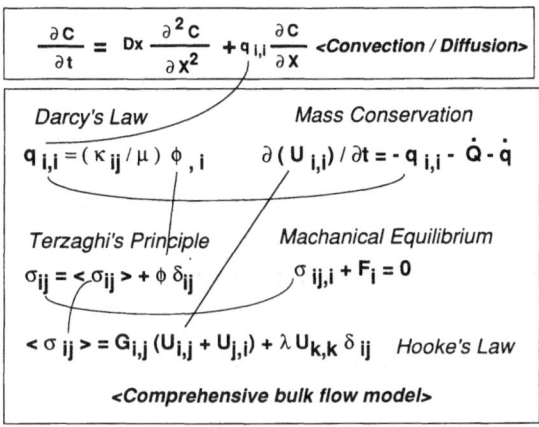

Fig. 2. The unified description of the convection/diffusion and the comprehensive bulk flow model. Where Ui,i is the divergence of the displacement of the brain, qi is the specific discharge vector of interstitial fluid, Q dot is the rate of entry of fluid from capilary bed, and q dot is metabolic water production per unit volume of the brain. The comma denoting partial differentiation. Index following comma is a conventional summation notation, implying summation over 1, 2, 3. Thus qi,i is the net volume efflux per unit volume at a point in terms of fluid mechanics. κij is parenkymal hydraulic conductivity, μ is the dynamic viscosity of the fluid. σij is the stress tensor and Fi is the volume force. <σij> is the effective stress tensor, δij is Kronecker's delta. φ is interstitial fluid pressure

supported by the law of mass conservation. Therefore, the true flow volume can not be calculated by the model. It is possible to simulate the tracer studies by convection/diffusion model in which flow vectors are given from the comprehensive bulk flow model. The unified model (Fig. 2) will provide an approach for a more quantitative analysis of edema formation and resolution by using MRI and tracer studies.

References

1. Aaslid R, Groger U, Patlak CS, Fenstermacher JD, Huber P, Reulen H-J (1990) Fluid flow rates in human peritumoral oedema. Acta Neurochir (Wien) [Suppl] 51: 152–154
2. Fenstermacher JD, Patlak CS (1976) The movement of water and solutes in the brains of mammals. In: Pappius HM, Feindel W (eds) Dynamics of brain edema. Springer, Berlin Heidelberg New York, pp 87–94
3. Ito U, Reulen H-J, Tomita H, Ikeda J, Saito J, Maehara T (1990) A computed tomography study on formation, propagation, and resolution of edema fluid in metastatic brain tumors. In: Long DM (ed) Advances in neurology, vol 52. Raven, New York, pp 459–468
4. Nagashima T, Shirakuni T, Rapoport SI (1990) A two-dimensional finite element analysis of vasogenic brain edema. Neurol Med Chir (Tokyo) 30: 1–9
5. Reulen JD, Huber P, Ito U, Groger U (1990) Peritumoral brain edema. A keynote address. In: Long DM, *et al* (ed) Advances in neurology, vol 52. Raven, New York, pp 307–458

Correspondence: Tatsuya Nagashima, M. D., Department of Neurosurgery, Kobe University School of Medicine, 7-5-1, Kusunoki-cho, Chuo-ku, Kobe 650, Japan.

Acta Neurochir (1994) [Suppl] 60: 168–170
© Springer-Verlag 1994

Medullary Cardiovascular Center and Acute Brain Swelling

M. Maeda[1], **T. Takachi**[1], **M. Nakai**[2], **A. J. Krieger**[3], and **H. N. Sapru**[3]

[1] Department of Systems Physiology, University of Occupational and Environmental Health, Kitakyushu, Japan, [2] National Cardiovascular Center, Suita, Japan, and [3] Section of Neurological Surgery, UMDNJ-New Jersey Medical School, Newark, U.S.A.

Summary

Acute brain swelling is reported to be due to acute vasodilatation of cerebral vessels. One of the causes of acute brain swelling may be disturbance of central control mechanisms of cerebral vessels. It has been reported that the anatomical location of the area which controls cerebral circulation is related to the area which controls systemic circulation. However, the role of the cardiovascular center on cerebral circulation has not been clear. The present study was, therefore, undertaken to examine the effects of chemical stimulation of the medullary cardiovascular center [nucleus tractus solitarius (NTS), ventrolateral depressor area (VLDA), and ventrolateral pressor area (VLPA)] on cerebral circulation. In anesthetized, paralyzed and artificially ventilated rats, the neurons in the NTS, VLDA, and VLPA were chemically stimulated and the cerebral blood flow (CBF) was determined using labeled microspheres. The CBF decreased significantly and the cerebrovascular resistance (CVR) increased significantly by chemical stimulation of the NTS, VLDA, and VLPA. These results suggest that the neurons within the NTS, VLDA, and VLPA control cerebral vessels vasoconstrictively. There is a possibility that the dysfunction of the NTS, VLDA, and VLPA may cause acute brain swelling.

Keywords: Acute brain swelling; cerebrovascular circulation; vasomotor center; ventrolateral medulla.

Introduction

Acute brain swelling is reported to be due to acute vasodilatation of cerebral vessels[3]. It is known that one of the causes of acute brain swelling may be disturbance of central control mechanisms of cerebral vessels[2]. It has been reported that the anatomical location of the area which controls cerebral circulation is related to the area which controls systemic circulation[6]. However, the role of cardiovascular center on cerebral circulation had not been clear. It is known that the nucleus tractus solitarius (NTS), ventrolateral medullary depressor area (VLDA) and ventrolateral medullary pressor area (VLPA) play important roles in regulation of arterial blood pressure (ABP) and heart rate (HR)[1,12,15]. Baroreceptor afferents make their primary synapus in the NTS and the pathway of the baroreflex within the brain stem is believed to be as follows: excitatory projection from the NTS to the VLDA and inhibitory GABAergic projection from the VLDA to the VLPA[13,14,16]. The NTS, VLDA, and VLPA are very important areas for vasomotor tonus. The present study was, therefore, undertaken to examine the effects of microinjection of L-glutamate into the medullary cardiovascular center (NTS, VLDA, and VLPA) on cerebral circulation: L-glutamate stimulates only neuronal cell bodies and has no effects on fibers of passage.

Materials and Methods

The methods were the same in general as those reported previously by us[7–11] and will be briefly summarized here.

General Procedures

Experiments were conducted on anesthetized Wistar rats weighing 300–350 g. On the completion of the surgical procedures, the anesthesia were maintained with alpha-cloralose (30 mg · kg^{-1}, s.c.) and urethane (300 mg · kg^{-1}, s.c.) for the NTS study; alpha-cloralose (50 mg · kg^{-1}, i.p.) and urethane (500 mg · kg^{-1}, i.p.) for the VLDA study; urethane (1.1–1.5 g · kg^{-1}, i.p.) for the VLPA study. The ABP and HR derived from the blood pressure waves were recorded on a polygraph. A small catheter was placed into the left cardiac ventricle via the right axillary artery. The trachea was cannulated and the animal was artificially ventilated by a rodent respirator (Harvard 680). The animals were immobilized by intermittent injections of d-tubocurarine (0.1–0.2 mg · kg^{-1}, i.v.). The animal was maintained in a homeothermic state at 37°C. The arterial pH, PaCO$_2$ and PaO$_2$ were frequently measured and maintained within the physiological range.

Microinjection of L-Glutamate

The animals were placed in a prone position in a stereotaxic instrument. The tooth bar was set to 11 mm below the level of the interaural line. A suboccipital craniotomy was performed and the caudal portion of the fourth ventricle was exposed. A glass single-barrel micropipette was pulled so that the outside diameter of its tip was 20–50 µm. The solutions for microinjection were freshly prepared in a 0.9% sodium chloride solution. L-glutamate was microinjected unilaterally. The coodinates were: 0.5 mm rostral, 0.5 mm lateral and 0.5 mm deep for the NTS; 0.5 mm rostral, 1.7 mm lateral and 2.0–2.4 mm deep for the VLDA; 2.0–2.5 mm rostral, 1.7 mm lateral and 2.0–2.4 mm deep for the VLPA to the calamus scriptorius.

Determination of Cerebral Blood Flow and Vascular Resistance

The regional blood flow was determined by utilizing radiolabeled microspheres (15 µm in diameter; ^{57}Co, ^{113}Sn and ^{46}Sc; New England Nuclear). One end of the silastic tube segment was connected to the catheter dipped in the blood collected from a donor rat. When the peristaltic pump was started, the blood from the donor rat infused the microspheres into the left ventricle of the recipient rat instantly. During the first 10 s, the tubing containing the microspheres was patted with a metal spatula for ensuring complete mixing of the microspheres in the blood stream. Simultaneously reference blood sample was automatically collected (via another channel in the peristaltic pump) from one of the femoral arteries. The reference sample was collected into a glass tube for 60 s and its volume was measured. The order of administration of radiolabeled microspheres was randomized. At the end of the experiments, the animals were sacrificed by hemorrhage. The right and left cerebral cortices were dissected out. The white matter and portions below the rhinal fissure were discarded. The tissue samples were then weighed. The radioactivities of microspheres in the brain and the reference blood were counted in a scintilation spectrometer. The tissue blood flow (Ft; ml · min^{-1} · (100 g)$^{-1}$) was calculated as follows; $Ft = (Ct \cdot 100 \cdot Fr) \cdot Cr^{-1} \cdot Wt^{-1}$, where Fr is the sampling rate of the reference blood (ml · min^{-1}), and Wt is the tissue weight (g), and Ct, Cr are counts in tissue and reference samples (cpm), respectively. The cerebrovascular resistance (CVR; mmHg per [ml · min^{-1} · (100 g)$^{-1}$]) was calculated by dividing the mean ABP by the CBF.

Results

Chemical Stimulation of the NTS

Unilateral chemical stimulation of the NTS produced a significant ($p < 0.05$) decrease in CBF in the whole cerebral cortex form 62 ± 28 (mean ± S.D.) to 48 ± 23 ml · min^{-1} · (100 g)$^{-1}$ and a significant ($p < 0.05$) increase in CVR in the whole cerebral cortex from 1.9 ± 1.2 (mean ± S.D.) to 2.6 ± 1.2 mmHg per [ml · min^{-1} · (100 g)$^{-1}$] (n = 6).

Chemical Stimulatlon of the VLDA

Unilateral chemical stimulation of the VLDA produced a significant ($p < 0.05$) decrease in CBF in the ipsilateral cerebral cortex form 46 ± 12 (mean ± S.E.M.) to 29 ± 7 ml · min^{-1} · (100 g)$^{-1}$ (n = 6). There was a significant ($p < 0.01$) increase in CVR in the

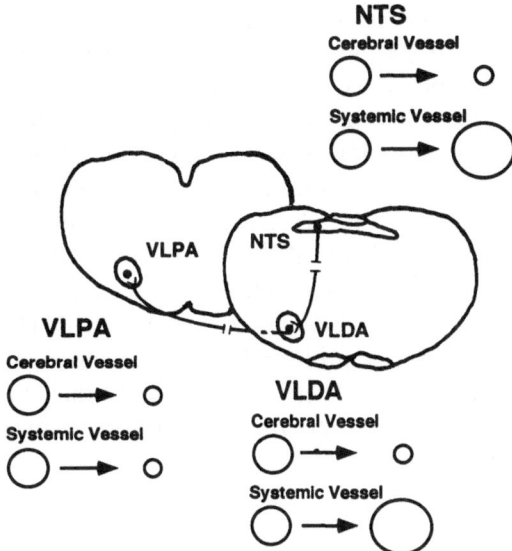

Fig. 1. The neurons within the NTS, VLDA, and VLPA control cerebral vessels vasoconstrictively. There is a possibility that the dysfunction of the NTS, VLDA, and VLPA may cause acute brain swelling

ipsilateral cerebral cortex from 2.7 ± 0.4 (mean ± S.E.M.) to 4.3 ± 0.6 mmHg per [ml · min^{-1} · (100 g)$^{-1}$] (n = 6).

Chemical Stimulation of the VLPA

The CBFs were significantly decreased and the CVRs were significantly increased in all brain regions by chemical stimulation of the VLPA. The CBFs of ipsilateral and contralateral cerebral cortices decreased significantly ($p < 0.05$) from 57 ± 14 (mean ± S.E.M.) to 41 ± 9 and from 50 ± 12 to 39 ± 9 ml · min^{-1} · (100 g)$^{-1}$, respectively (n = 8). The CVRs of ipsilateral and contralateral cerebral cortices increased significantly ($p < 0.05$) from 2.6 ± 0.6 (mean ± S.E.M.) to 3.5 ± 0.8 and from 2.7 ± 0.5 to 3.5 ± 0.8 mmHg per [ml · min^{-1} · (100 g)$^{-1}$], respectively (n = 8).

Discussion

Acute brain swelling is well known to result from augmentation of cerebrovascular blood volume due to vasoparesis of cerebral vessels and impairment of cerebrovascular tonus[3]. Neurogenic factors are reported to play a important role in development of acute brain swelling and increased intracranial pressure (ICP)[2,4–6]. Ishii[2] described that the dorsomedial nucleus in the hypothalamus controls cerebral vasculature and mechanical distortion of the hypothalamic area, through the resultant vasoparesis, leads to an acute increase in ICP: a form of brain swelling. However, many investigations had been done using the technique of electrical stimulation or lesioning. Both electrical stimulation and lesioning involve not only neuronal cell bodies but also fibers of passage. In our present study, L-glutamate, which stimulates neuronal cell bodies but not fibers, was microinjected into the NTS, VLDA, and VLPA. Our results suggest that the neurons within the NTS, VLDA, and VLPA control cerebral vessels vasoconstrictively (Fig. 1)[7-10]. There is a possibility that the dysfunction of the NTS, VLDA, and VLPA may cause acute brain swelling.

References

1. Ciriello J, Caverson MM, Polosa C (1986) Function of the ventrolateral medulla in the control of the circulation. Brain Res Rev 11: 359–391

2. Ishii S (1966) Brain-swelling; studies of structural, physiologic and biochemical alterations. In: Caveness WH, Walker AF (eds) Head injury conference proceedings. Lippincott, Philadelphia Tront, pp 276–299

3. Langffit TW, Weinstein JD, Kassell NF (1965) Cerebral vasomotor paralysis produced by intracranial hypertension. Neurology 15: 622–641

4. Maeda M, Matsuura S, Tanaka K, Katsuyama J, Nakamura T, Sakamoto H, Nishimura S (1988) Effects of electrical stimulation on intracranial pressure and systemic arterial blood pressure in cats. Part I: stimulation of brain stem. Neurol Res 10: 87–92

5. Maeda M, Matsuura S, Tanaka K, Katsuyama J, Nishimura S (1988) Effects of electrical stimulation on intracranial pressure and systemic arterial blood pressure in cats. Part II: stimulation of cerebral cortex and hypothalamus. Neurol Res 10: 93–96

6. Maeda M (1988) Changes in intracranial pressure elicited by electrical stimulation of the brainstem reticular formation in spinal cats with vagotomy. J Auton Nerv Syst 25: 155–164

7. Maeda M, Nakai M, Krieger AJ, Sapru HN (1990) Chemical stimulation of the nucleus tractus solitarii decreases cerebral blood flow in anesthetized rats. Brain Res 520: 255–261

8. Maeda M, Krieger AJ, Sapru HN (1991) Chemical stimulation of the ventrolateral medullary depressor area decreases ipsilateral cerebral blood flow in anesthetized rats. Brain Res 543: 61–68

9. Maeda M, Krieger AJ, Nakai M, Sapru HN (1991) Chemical stimulation of the rostral ventrolateral medullary pressor area decreases cerebral blood flow in anesthetized rats. Brain Res 563: 261–269

10. Maeda M, Krieger AJ, Nakai M, Sapru HN (1992) Spinal cord blood flow decreases following chemical stimulation of the rostral ventrolateral medullary pressor area in anesthetized rats. J Auton Nerv Syst 39: 151–158

11. Nakai M, Tamaki K, Yamamoto J, Shimouchi A, Maeda M (1990) A minimally invasive technique for multiple measurement of regional blood flow of the rat brain using radiolabeled microspheres. Brain Res 507: 168–171

12. Reis DJ, Granata AR, Joh TH, Ross CA, Ruggiero DA, Park DH (1984) Brain stem catecholamine mechanisms in tonic and reflex control of blood pressure. Hypertension [Suppl II] 6: II-7–II-15

13. Urbanski RW, Sapru HN (1988) Evidence for a sympathoexcitatory pathway from the nucleus tractus solitarii to the ventrolateral medullary pressor area. J Auton Nerv Syst 23: 161–174

14. Urbanski RW, Sapru HN (1988) Putative neurotransmitters involved in medullary cardiovascular regulation. J Auton Nerv Sys 25: 181–193

15. Willette RN, Barcas PP, Krieger AJ, Sapru HN (1983) Vasopressor and depressor areas in the rat medulla: identification by microinjection of L-glutamate. Neuropharmacology 22: 1071–1079

16. Willette RN, Punnen S, Krieger AJ, Sapru HN (1984) Interdependence of rostral and caudal ventrolateral medullary areas in the control of blood pressure. Brain Res 321: 169–174

Correspondence: Masanobu Maeda, M. D., Department of Systems Physiology, University of Occupational and Environmental Health, 1–1, Iseigaoka, Yahatanishi-ku, Kitakyushu 807, Japan.

Acta Neurochir (1994) [Suppl] 60: 171–173

Dysfunction of the Medullary Cardiovascular Center May Cause Acute Spinal Cord Swelling

M. Maeda[1], **M. Nakai**[2], **A. J. Krieger**[3], and **H. N. Sapru**[3]

[1] Department of Systems Physiology, University of Occupational and Environmental Health, Kitakyushu, Japan, [2] National Cardiovascular Center, Suita, Japan, and [3] Section of Neurological Surgery, UMDNJ-New Jersey Medical School, Newark, U.S.A.

Summary

Acute brain swelling is well known to be acute vasodilatation of cerebral vessels and sometimes results from brain injury. One of the causes of acute brain swelling may be disturbance of central control mechanisms of cerebral vessels. However, the presence of acute spinal cord swelling is little noticed. We present here a possibility that acute spinal cord swelling may be occur following the dysfunction of the cardiovascular center of the medulla. In urethane-anesthetized, paralyzed and artificially ventilated rats, the neurons in the rostral ventrolateral pressor area (VLPA), origin of the sympathetic nerve activities in the brain stem, were chemically stimulated by microinjection of L-glutamate and the spinal cord blood flow (SCBF) was determined using labeled microspheres. The SCBFs of cervical, thoracic, and lumbar cord decreased significantly from 27 ± 3 (mean $\pm$ S.E.M.) to 20 ± 2 (p < 0.01), from 22 ± 1 to 17 ± 2 (p < 0.05), and from 41 ± 5 to 26 ± 3 (p < 0.05) ml $\cdot$ min^{-1} $\cdot$ (100 g)$^{-1}$, respectively (n = 12). The spinal cord vascular resistances (SCVRs) of cervical, thoracic, and lumbar cord increased significantly from 3.7 ± 0.4 to 5.0 ± 0.6 (p < 0.05), from 4.2 ± 0.2 to 5.9 ± 0.7 (p < 0.05), and from 2.5 ± 0.2 to 3.8 ± 0.4 (p < 0.05) mmHg per [ml $\cdot$ min^{-1} $\cdot$ (100 g)$^{-1}$], respectively (n = 12). These results suggest that the neurons within the VLPA may play a role in the control of spinal cord circulation. There is a possibility that the dysfunction of the VLPA may cause acute spinal cord swelling.

Keywords: Acute spinal cord swelling; spinal cord circulation; vasomotor center; ventrolateral medulla.

Introduction

Acute brain swelling is well known to be acute vasodilatation of cerebral vessels and sometimes results from brain injury[3]. One of the causes of acute brain swelling may be disturbance of central control mechanisms of cerebral vessels[2]. It has been reported that the anatomical location of the area which controls cerebral circulation is related to the area which control systemic circulation[6]. It is known that the nucleus tractus solitarius (NTS), ventrolateral medullary depressor area (VLDA) and ventrolateral medullary pressor area (VLPA) play important roles as the medullary cardiovascular center[1,13,14]. We reported that the neurons within the NTS, VLDA, and VLPA control cerebral vessels vasoconstrictively and a possibility that dysfunction of the NTS, VLDA, and VLPA may cause acute brain swelling[7–9,11]. However, the presence of acute spinal cord swelling is little noticed. In the present study, we present here that the neurons in the rostral ventrolateral medullary pressor area (VLPA), origin of the sympathetic nerve activities in the brain stem, have a vasoconstrictive effects on spinal cord vessels.

Materials and Methods

The methods were the same in general as those reported previously by us[10,12] and will be briefly summarized here.

General Procedures

Experiments were conducted on a total number of 22 male Wistar rats weighing 300–350 g. The animals were anesthetized with urethane (1.1–1.5 g $\cdot$ kg^{-1}, i.p.). Mean arterial blood pressure (ABP), heart rate, derived from the blood pressure waves, and pulsatile ABP were recorded on a polygraph. A small catheter was placed into the left cardiac ventricle via the right axillary artery. The trachea was cannulated and the animal was artificially ventilated by a rodent respirator (Harvard 680). The animals were immobilized by intermittent injections of d-tubocurarine (0.1–0.2 mg $\cdot$ kg^{-1}, i.v.). The animal was maintained in a homeothermic state at 37°C. The arterial pH, PaCO$_2$ and PaO$_2$ were frequently measured and maintained within the physiological range.

Chemical Stimulation of the VLPA

The animals were placed in a prone position in a stereotaxic instrument. The tooth bar was set to 11 mm below the level of the interaural line. A suboccipital craniotomy was performed and the caudal portion of the fourth ventricle was exposed. A glass micro-pipette was pulled so that the outside diameter of its tip was 20–50 μm. The solutions for microinjection were freshly prepared in a 0.9% sodium chloride solution. L-glutamate (1.7–5.0 nmol; 100 nl) was microinjected unilaterally into the VLPA. The coodinates for the VLPA with reference to the calamus scriptorius were: 2.0–2.5 mm rostral, 1.7 mm lateral, and 2.0–2.4 mm deep.

Determination of Spinal Cord Blood Flow (SCBF) and Spinal Cord Vascular Resistance (SCVR)

The regional blood flow was determined by utilizing radiolabeled microspheres (15 μm in diameter; ^{57}Co, ^{113}Sn and ^{46}Sc; New England Nuclear). 40,000–50,000 microspheres were kept in a small segment of silastic tubing. One end of the silastic tube segment was connected to the catheter dipped in the blood collected from a donor rat. When the peristaltic pump was started, the blood from the donor rat infused the microspheres into the left ventricle of the recipient rat instantly. During the first 10 s, the tubing containing the microspheres was patted with a metal spatula for ensuring complete mixing of the microspheres in the blood stream. Simultaneously reference blood sample was automatically collected (via another channel in the peristaltic pump) from one of the femoral arteries. The reference sample was collected into a glass tube for 60 s and its volume was measured. The order of administration of radiolabeled microspheres was randomized. At the end of the experiments, the animals were sacrificed by hemorrhage. The cervical cord, thoracic cord, and lumbar cord were dissected out. The tissue samples were then weighed. The radioactivities of microspheres in the spinal cord and the reference blood were counted in a scintilation spectrometer. The tissue blood flow (Ft; ml · min^{-1} · (100 g)$^{-1}$) was calculated as follows; $Ft = (Ct \cdot 100 \cdot Fr) \cdot Cr^{-1} \cdot Wt^{-1}$, where Fr is the sampling rate of the reference blood (ml · min^{-1}), and Wt is the tissue weight (g), and Ct, Cr are counts in tissue and reference samples (cpm), respectively. The SCVR (mmHg per [ml · min^{-1} · (100 g)$^{-1}$]) was calculated by dividing the mean ABP by the SCBF.

Results

Effect of Chemical Stimulation of the VLPA on SCBF when the ABP was Maintained at Normotensive Levels

In order to avoid the influence of the hypertension induced by chemical stimulation of the VLPA on blood flow measurements, the blood flows were measured at normotension. The detailed methods were described elsewhere[10]. The local SCBFs were significantly decreased and local SCVRs were significantly increased in all spinal cord regions by unilateral chemical stimulation of the VLPA (n = 12). The SCBFs in the cervical cord before and during the chemical stimulation were 27 ± 3 (mean ± S.E.M.) and 20 ± 2 (p < 0.01); in the thoracic cord, 22 ± 1 and 17 ± 2 (p < 0.05); in the lumbar cord, 41 ± 5 and 26 ± 3 (p < 0.05) ml · min^{-1} · (100 g)$^{-1}$, respectively. The SCVRs in the cervical cord before and during the chemical stimulation were 3.7 ± 0.4 (mean ± S.E.M.) and 5.0 ± 0.6 (p < 0.05); in the thoracic cord, 4.2 ± 0.2 and 5.9 ± 0.7 (p < 0.05); in the

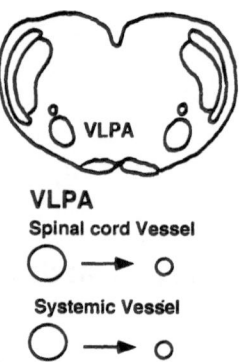

VLPA
Spinal cord Vessel

Systemic Vessel

Fig. 1. Our results suggest that the neurons within the VLPA may play a role in the control of spinal cord circulation. There is a possibility that acute spinal cord swelling may be occur following the dysfunction of the cardiovascular center of the medulla

lumbar cord, 2.5 ± 0.2 and 3.8 ± 0.4 (p < 0.05) mmHg per [ml · min^{-1} · (100g)$^{-1}$], respectively.

CO_2 Reactivity of the Spinal Cord Vessels During Chemical Stimulation of the VLPA

The reactivity of the spinal vasculature to changes in arterial $PaCO_2$ during unilateral chemical stimulation of the VLPA was tested (n = 5). When the arterial $PaCO_2$ was increased, the local SCBFs of all spinal regions increased significantly (p < 0.05). This indicates that the reactivity of spinal cord vasculature was intact during chemical stimulation of the VLPA.

Effect of Microinjection of L-Glutamate into an Area Adjacent to the VLPA on SCBF

In order to determine if the change in SCBF produced by microinjection of L-glutamate into the VLPA was due to the effect of microinjection of L-glutamate itself, L-glutamate was microinjected into an area adjacent to the VLPA on one side and SCBF was measured (n = 5). The SCBF and SCVR were not significantly altered by microinjections of L-glutamate into the area adjacent to the VLPA. This indicates that non-specific effects of L-glutamate itself, if any, were not responsible for the present results.

Discussion

Acute brain swelling is well known to result from augmentation of cerebrovascular blood volume due to vasoparesis of cerebral vessels and impairment of cerebrovascular tonus[3]. Neurogenic factors are reported to play a important role in development of acute brain swelling and increased intracranial pressure (ICP)[2,4–6]. We showed a possibility that the dysfunction of the medullary cardiovascular center (NTS, VLDA, and VLPA) may cause acute brain swelling[7–9,11]. However, the presence of acute spinal cord swelling is not brought to attention. Neurogenic factors may be important in development of acute spinal cord swelling as well as acute brain swelling. Our results suggest that the neurons within the VLPA may play a role in the control of spinal cord circulation[10]. There is a possibility that acute spinal cord swelling may be occur following the dysfunction of the cardiovascular center of the medulla (Fig. 1).

References

1. Ciriello J, Caverson MM, Polosa C (1986) Function of the ventrolateral medulla in the control of the circulation. Brain Res Rev 11: 359–391
2. Ishii S (1966) Brain-swelling; studies of structural, physiologic and biochemical alterations. In: Caveness WH, Walker AF (eds) Head injury conference proceedings. Lippincott, Philadelphia, pp 276–299
3. Langffit TW, Weinstein JD, Kassell NF (1965) Cerebral vasomotor paralysis produced by intracranial hypertension. Neurology 15: 622–641
4. Maeda M, Matsuura S, Tanaka K, Katsuyama J, Nakamura T, Sakamoto H, Nishimura S (1988) Effects of electrical stimulation on intracranial pressure and systemic arterial blood pressure in cats. Part I: stimulation of brain stem. Neurol Res 10: 87–92
5. Maeda M, Matsuura S, Tanaka K, Katsuyama J, Nishimura S (1988) Effects of electrical stimulation on intracranial pressure and systemic arterial blood pressure in cats. Part II: stimulation of cerebral cortex and hypothalamus. Neurol Res 10: 93–96
6. Maeda M (1988) Changes in intracranial pressure elicited by electrical stimulation of the brainstem reticular formation in spinal cats with vagotomy. J Auton Nerv Syst 25: 155–164
7. Maeda M, Nakai M, Krieger AJ, Sapru HN (1990) Chemical stimulation of the nucleus tractus solitarii decreases cerebral blood flow in anesthetized rats. Brain Res 520: 255–261
8. Maeda M, Krieger AJ, Sapru HN (1991) Chemical stimulation of the ventrolateral medullary depressor area decreases ipsilateral cerebral blood flow in anesthetized rats. Brain Res 543: 61–68
9. Maeda M, Krieger AJ, Nakai M, Sapru HN (1991) Chemical stimulation of the rostral ventrolateral medullary pressor area decreases cerebral blood flow in anesthetized rats. Brain Res 563: 261–269
10. Maeda M, Krieger AJ, Nakai M, Sapru HN (1992) Spinal cord blood flow decreases following chemical stimulation of the rostral ventrolateral medullary pressor area in anesthetized rats. J Auton Nerv Syst 39: 151–158
11. Maeda M, Takachi T, Nakai M, Krieger AJ, Sapru HN (1993) Medullary cardiovascular center and acute brain swelling. Acta Neurochir (Wien) [Suppl] 60: 168–170
12. Nakai M, Tamaki K, Yamamoto J, Shimouchi A, Maeda M (1990) A minimally invasive technique for multiple measurement of regional blood flow of the rat brain using radiolabeled microspheres. Brain Res 507: 168–171
13. Reis DJ, Granata AR, Joh TH, Ross CA, Ruggiero DA, Park DH (1984) Brain stem catecholamine mechanisms in tonic and reflex control of blood pressure. Hypertension [Suppl II] 6: II-7–II-15
14. Willette RN, Barcas PP, Krieger AJ, Sapru HN (1983) Vasopressor and depressor areas in the rat medulla: identification by microinjection of L-glutamate. Neuropharmacology 22: 1071–1079

Correspondence: Masanobu Maeda, M. D., Department of Systems Physiology, University of Occupational and Environmental Health, 1-1, Iseigaoka, Yahatanishi-ku, Kitakyushu 807, Japan.

Acta Neurochir (1994) [Suppl] 60: 174–175
© Springer-Verlag 1994

CSF Dynamics in Patients with Meningiomas

N. Pliushcheva and **A. Shakhnovich**

Burdenko Neurosurgery Institute, Moscow, Russia

Summary

The purpose of this work was to study CSF dynamics and brain elasticity in patients with peritumoral BE. The investigations was carried out in 14 patients with supratentorial meningiomas. The volume of tumor and the volume of BE was determined planimetrically on CT. CSF dynamics were evaluated using constant-pressure infusion-drainage tests. A significant positive correlation was observed between Ve and F. Definite increases in F in patients with high Ve may depend on the sum of the clearance of BE fluid and the amount of CSF produced. Increase in R correlates with the maximum values of Vt and Ve. This increase in R may depend on the compression of arachnoid villi by tumor and swollen brain.

Keywords: Brain edema; CSF dynamics; brain tumors; intracranial hypertension.

Introduction

The dynamics (formation, propagation and resolution) of brain edema (BE) depends on other dynamic processes (cerebral circulation and CSF dynamics) in the brain and on brain elasticity.

The purpose of this work was to study these dynamic processes in patients with peritumoral BE.

Materials and Methods

The investigation was carried out in 14 patients with supratentorial meningiomas of frontal, parietal and occipital localization (9 male and 6 female, age 22–43 years)

The volume of tumor (Vt) was calculated by planimetrically measuring each CT slice and multiplying by the 10 mm thickness of the slice.

The low-density area around the tumor visualized on CT scans expressed the extent and degree of BE. The volume of BE (Ve) was estimated on CT scans by tomodensitometry (16–22 H) and BE volume was calculated by planimetrically measuring each CT slice and multiplying by the 10 mm thickness of the slice.

The investigations were carried out only in patients having no clinical signs of tentorial herniation (normal upward gaze and normal pupillary reaction to light).

The patients were divided in two groups: those in which clinical signs of intracranial hypertension were absent (I) and those in which these signs (papilloedema) were present (II).

CSF dynamics were evaluated using constant-pressure infusion-drainage tests[3,4]. The infusion-drainage tests made it possible to determine the elasticity of the cerebrospinal system (El), the CSF outflow resistance (R), the CSF formation rate (F) and the dural sinus pressure (Pds).

Results

Our results are presented in the Table 1. As seen from this table, there was a significant difference between some parameters in the two groups, presence (II) and absence (I) of clinical signs of intracranial hypertension (papilloedema).

Vt, Ve, Pcsf, R, and F were significantly higher in the group II with intracranial hypertension than in group I. On the other hand, El was significantly lower in group II with intracranial hypertension than in group I. Pds remained normal or decreased in both clinical groups. The highest values of R (50–70 mmHg/ml/min) and F (1.0–1.2 ml/min) were discovered in patients with the highest Ve (107–135 cm).

A significant positive correlation was observed between Ve and F (r = 0.71, p < 0.01).

There was no relationship between CSF dynamics and tumor localization.

Discussion

This study demonstrates the frequent changes in CSF dynamics in patients with brain tumors and peritumoral edema.

Definite increases in F in patients with high Ve (two or three times the normal values) may depend on the sum of the clearance of Be fluid and the amount of CSF produced.

Table 1. *Quantitative Characteristics of CSF Dynamics and Venous Outflow in Patients with Brain Tumor*

Clinical groups	I		II		Normal range	
	Mean	Range	Mean	Range		
Pcsf mmHg	14.2 ± 3.2	7.2–21.3	22.3 ± 5.2**	10.2–31.3	8.0–13.5	(2)
R mmHg (ml/min)	12.3 ± 5.1	3.7–24.1	33.7 ± 15.8**	8.3–70.1	4.5–12.0	(2)
F ml/min	0.4 ± 0.1	0.2–0.6	0.7 ± 0.2**	0.3–1.2	0.2–0.6	(2)
El l/ml	0.9 ± 0.3	0.3–1.5	0.5 ± 0.1*	0.3–0.7	0.2–0.5	(4)
Pds mmHg	5.1 ± 1.7	1.9–7.1	5.2 ± 1.8	2.1–6.9	5.0–10.0	(2)
Vt ml	42 ± 11	21–64	123 ± 32**	69–201		
Ve ml	21 ± 6	11–33	32 ± 11	12–44		

* $p < 0.05$.
** $p < 0.01$.

As has been observed[1], the BE fluid clearance rate can reach 0.1 ml/min. This value is in the same order of magnitude as the CSF production rate (F).

An increase in R correlates with the highest values (group II) of Vt and Ve. This increase in R may depend on the compression of arachnoid villi by tumor and swollen brain.

An increase in El in the nonhypertensive stage of tumor growth can be considered a biologically important defense reaction which can prevent the bulk flow of Be and also dislocation of the brain and tentorial herniation.

The pathophysiological mechanism of this reaction remains vague. The decrease in Pds in some patients (usually with intracranial hypertension: group II) may be caused by direct compression of bridging veins by the tumor or swollen brain. Compression of these bridging veins may also depend on their "cuff constriction"[5], caused by an increase in Pcsf.

The changes in CBF dynamics discovered in our investigations can influence BE dynamics and must be taken into consideration in BE treatment.

References

1. Aaslid R, Groger U, Patlak CS, *et al* (1990) Fluid flow rates in human peritumoural oedema. In: Reulen H-J, Baethmann A, Fenstermacher J, *et al* (eds) Brain edema VIII. Springer, Wien New York, pp 152–154
2. Ekstedt J (1978) CSF hydrodynamics studies in man. II. Normal hydrodynamics variables related to CSF pressure and flow. J Neurol Neurosurg Psychiatry 41: 345–353
3. Portnoy HD, Croissant PD (1976) A practical method for measuring hydrodynamics of cerebrospinal fluid. Surg Neurol 5: 273–277
4. Szewzykowsky J, Slivka ST, Kunicki A, *et al* (1977) A fast-method of estimating the elastance of the intracranial system. J Neurosurg 47: 19–26
5. Wright RD (1938) Experimental observations on increased intracranial pressure. Aust N Z J Surg 7: 215–235

Correspondence: N. Pliushcheva, M.D., Ph.D., Burdenko Neurosurgery Institute, Fadeev Street 5, Moscow 125047, Russia.

Ischemic Brain Edema: Pathophysiology and Nuclear Magnetic Resonance

Acta Neurochir (1994) [Suppl] 60: 179–183
© Springer-Verlag 1994

Compartmental Analysis of Brain Edema Using Magnetic Resonance Imaging

F. A. Jolesz

Department of Radiology, Brigham and Women's Hospital, Harvard Medical School, Boston, MA, U.S.A.

Summary

The potential exists for increasing the sensitivity of magnetic resonance imaging (MRI) to white matter (WM) pathologies by identifying compartments of tissue water. We have found the physical equivalents of myelin-associated biological water compartments in normal and pathologic states by using multiexponential analysis of T2 relaxation. In addition, we have applied this multiparametric technique for the definition of various types of white matter edemas. We were able to identify some changes in physical compartments visible by MRI with simultaneous changes in biological compartments. We conclude that MRI is a very sensitive method to quantify abnormal accumulation of intracerebral water; however, it is a somewhat limited probe for identifying the biologic compartmentation of edema among the various biological compartments of the brain.

Keywords: Brain edema; MRI; water compartmentation.

Introduction

We used MRI to specifically distinguish various forms of WM edema. This goal was justified by the weakness of current clinical MRI techniques in differentiating the WM lesions of several etiologies. Since the pathophysiological differences between these conditions are primarily related to variations in the compartmentation of tissue water in the WM, we performed multiexponential analysis of T2 decay curves from in vivo MRI in a series of experiments and correlated our findings with physical and biological measures as defined by in vitro MR relaxometry, x-ray diffraction, and conventional morphological techniques[9-11,13,14,23,25,27]. This approach emphasizes the significance of the spatial distribution of edema fluid within and around the myelin membranes, in different macromolecular environments, and with different motional restriction of water molecules.

The validity of such correlations can be assessed by studying the known alterations in water compartments of WM that occur in experimental animal models of brain edema. The multiexponential components of T2 decay curves obtained from high-field MR images may improve the characterization of the various types of WM edema and the progress of various stages of edema development. If these changes are found to reflect abnormal compartmentation of tissue water in WM edema, we can relate the individual components of transverse decay curves (TDCs) to biological compartments defined with non-MR techniques. If a relationship is established, we can determine whether the various abnormal patterns of WM water distribution can be interpreted within the framework of the current pathophysiological concept of WM edema, or whether a new concept, more relevant to MRI, is necessary. We suggest that the results of these studies will not only provide an unique and improved characterization of WM lesions, but also a potential basis for interpreting clinical imaging results that will aid in the diagnosis and evaluation of WM edema.

Critical Review of MR Measurements in WM Edema

The prolongation of transverse relaxation times observed in various WM pathologies does not characterize these alterations sufficiently. The visual interpretation of images, based upon the concept that monoexponential longitudinal and transverse relaxation processes are present, is not specific enough to reveal the complex nature of macromolecular-water interactions occurring in heterogenous biological systems. The compartmentation of tissue water, the rate of exchange of protons between compartments, and the interactions between water protons and macromolecules are important contributors to tissue relaxation properties[3,4]. While the solid foundation on which MR imaging should stand is the adequate characterization of proton environments, decomposition of apparently non-exponential TDCs has not been extensively exploited for application to clinical imaging. Our current understanding of proton relaxation processes is based on a series of investigations in which the decomposition of the TDC and analysis with the two-site exchange model have been applied[21]. Improving data acquisition, sampling, and analysis to permit the resolution of biexponential or triexponential T2 relaxation significantly improved tissue characterization and made the differentiation of various pathologic processes possible in vitro[5,6,16,25]. In peripheral nerves, the myelin-associated water compartments were defined[27], and distinction between inflammatory and demyelinating white matter lesions was achieved[29]. The use of multiple transverse components for improving the specificity of current clinical MRI has already been demonstrated in vivo at lower fields[26], as well as at higher fields in our laboratory and elsewhere[7,12,15]. The implementation of this imaging technique may result not only in a deeper understanding of proton relaxation processes in vivo but also in improved specificity and therefore the clinical utility of MRI.

The Specificity of MRI for White Matter Lesions

MRI has been shown to be an extremely sensitive but not necessarily specific method for detecting lesions in cerebral WM. Its high sensitivity explains the successful application of MRI to the diagnosis of the most common demyelinating disease, multiple sclerosis (MS), or for the detection of WM edema, but it has also led to an unprecedented "diagnostic epidemic" of WM diseases of various known and unknown etiologies. As a consequence, the concept of WM pathology is now under serious reinvestigation. Progress in characterizing and classifying WM pathologies applying MRI is severely limited by the lack of an experimental correlation between MRI and conventional histopathology.

There is a growing disparity between the abundance of MRI data and the paucity of pathologic information explaining WM changes. An intense WM MRI signal may bear little relationship to the structural or staining abnormalities seen on histologic section following fixation and dehydration. Prolongation of T2 may be the result of CSF penetration into normal or ischemic WM, or be due to fluid within cystic areas of microinfarcts, or be related to demyelination. Abnormal signals from

the WM can be a reflection of a pathologic increase in tissue water and/or changes in the macromolecular water environment due to the destruction of hydrophobic myelin membranes. Both edema and demyelination will eventually prolong the relaxation processes. The pathophysiologic mechanisms cited in WM edema and demyelination do not necessarily explain the mechanisms that determine the MR characteristics of the lesion. A full understanding of the MR representation of these lesions will require a combination of measurements that reveal both the spatial distribution of water compartments and their dynamic relationship to each other.

Experimental Results

We successfully manipulated the distribution of myelin-associated biological water compartments and identified the corresponding changes in the physical compartments of water protons as revealed by MR spectroscopic and/or imaging methods.

We investigated the in vitro swelling of peripheral nerves by correlating non-perturbing x-ray diffraction and MR measurements[27]. The periodic nature of the myelin molecular organization and the presence of water spaces between the cytoplasmic and extracellular surfaces of the apposed membranes can be revealed by x-ray diffraction. The distance between the interperiod lines on x-ray diffraction is proportional to the distances between the lamellae of the myelin sheath, i.e., the periodicity. There are two main water compartments within the myelin sheath: the extracellular and the intracellular. The size of the extracellular compartment can be manipulated by incubating the nerve in different pH solutions, and the respective changes in periodicity can easily be seen with x-ray diffraction measurements as previously proved in mouse sciatic nerve studies[8]. The greater the applied pH, the greater the extracellular distance between the lamellae will become. The intracellular space between the lamellae of the myelin sheath (the intramyelinic space), however, represents a virtual space that remains unaffected by this manipulation.

We have attempted to correlate the x-ray diffraction data and the proton MR T2 relaxation times using the multiexponential decomposition technique in sciatic nerves dissected from normal rabbits[27]. All nerves were incubated for 24 hours at room temperature in bulk test solutions of NaCl. The pH of the tissue soaking solutions was adjusted to be acidic, neutral, or basic. The x-ray diffraction measurements were consistent with those performed earlier on mouse sciatic nerves[8]. The higher the pH of the solution, the longer the distance between the extracellular lines of the x-ray diffraction spectra became. There was a shortening of the overall T2 values of the nerves with increasing pH. Three TDC components were found. Both $T2_1$ and $T2_2$ showed the same trend. The long-relaxing $T2_3$ compartment, however, did not show a correlation with pH, consistent with its definition as excess tissue water. From this data it is obvious that the fractional size of the fast-relaxing compartment increases with pH. This is consistent with the swelling of the intermyelinic water compartment. Considering that the pH-induced swelling of the myelin takes place between the extracellular membrane surfaces of the myelin sheath (as has been shown with x-ray diffraction), it can be concluded that the changes in the fast-relaxing compartment represent swelling of this compartment, that is, the intermyelinic extracellular space. The second compartment would refer to the other major water compartment of the nerve, which is the extramyelinic interstitial space of the nerve. The interpretation of the respective changes in the relaxation times of these compartments is more difficult, because it is expected that the higher the degree of myelin swelling, the more water penetration to the tissue, and the longer the values of the relaxation times. On the contrary, the relaxation time of the fast-relaxing compartment (and to a certain extent, of the intermediate-relaxing compartment) showed a drop with the increased degree of swelling. The decreased values in the first compartment are most likely due to reduced mobility of the water molecules as they penetrate the extracellular surfaces of the myelin membranes where they interact with the membrane proteins, especially lipoproteins. The second compartment is in a physical exchange with the first: thus, the drop in relaxation time in the first will cause a reduction in the second. The cross-relaxation of proteins and of protein with water are also dependent on pH. Therefore, it is possible that pH may affect T2 without changing compartmental size.

The compartmentation of myelin-associated water and the TDCs from myelinated and unmyelinated peripheral nerves of the garfish were compared. Similar studies were done with the nerve cords of crayfish[17]. They concluded that the fast-relaxing component of the T2 decay curve is related to the myelin-associated water compartment. These results are consistent with our findings.

We also studied dehydration and hyperhydration states in WM using specific gravity measurements

correlated with MRI measurements[23] and TET-induced intramyelinic vacuolization as a model of cytotoxic edema using relaxation time measurements and MR imaging correlated with morphologic and physiologic data[14].

We characterized various stages of demyelinating processes using the experimental allergic encephalomyelitis (EAE) model[29]. Multiexponential T2 analysis combined with the other MR parameters distinguished the early and late changes of the experimental demyelination. We also documented the development and progression of perfusion defects produced by light-activated, IV-injected rose bengal[13]. Lesions were produced in the brains of rats and rabbits by the intravenous injection of rose bengal and subsequent irradiation of the exposed skull for 20 minutes with green light (560 nm) from a filtered xenon lamp. Brain lesions were seen on MRI less than 1 hour after their induction. The lesions increased in size during the first 24 hours, and a significant amount of accompanying edema was seen. By the fourth day, most of the edema had resolved, leaving a smaller defect. This subsequently decreased, but traces remained for as long as the animals were followed. We evaluated whether multiexponential decay was more useful than the overall T2 in characterizing the extent of edema and the time course of the lesions. Multiexponential analysis revealed that the slowly decaying component characterized the lesion more effectively than the overall T2. Although the overall T2 correlated well with overall water content, the biexponential components did not correlate, suggesting that multiexponential components represent subcompartments of tissue water, which behave in different ways during the development of cytotoxic and vasogenic edema.

Encouraged by these experimental results, we successfully used the CPMG sequence (without multiexponential decomposition) in clinical imaging[18,24]. We have developed and tested the technique of multiexponential T2 decay decomposition[25], which has been applied to several in vitro measurements. We have developed, tested, and implemented high-field 2D-FT CPMG imaging sequences suitable for clinical application[18-20].

Discussion

Our consistent hypothesis has been that analysis of non-exponential TDCs can reflect the physical compartmentation of tissue water, and that this information is also relevant to understanding the distribution of tissue water within biological compartments. Using the in vivo CPMG imaging method and displaying the parameters calculated from the biexponential analysis of TDCs, we imaged patients with multiple sclerosis and demonstrated that distinct biexponential patterns can be seen in various WM lesions[7,12]. We have now reached the stage of clinical applications and are on the verge of testing the diagnostic feasibility of the technique. What we need is further experimental verification provided by animal models in which the manipulation of water compartmentation is possible under well-controlled conditions.

To date, using MR relaxation rate measurements correlated with brain water content, different forms of brain edema (ischemic, cold- or heat-induced vasogenic, TET-induced cytotoxic, and hypoosmolality-induced osmotic) have been distinguished, suggesting that the state of water is different among these models. A linear relationship between the T1 relaxation rates and water content was found in both vasogenic and cytotoxic edema; using multicomponent T2 analysis, however, intrinsic differences were found between the two types of edema[23]. This can be attributed to the higher protein content of the vasogenic edema fluid. Multiexponential analysis unequivocally shows that there are at least two fractions of tissue water in edematous WM, fractions that do not appreciably exchange on an MR time scale. In conjunction with the morphologic data, multicomponent T2 results may be better interpreted if appropriate model systems are established. Such an interpretation has not yet been developed. It should include various morphologically and physio-chemically defined spaces (extracellular space between myelinated axons, water adsorbed at the extracellular surface of the myelin sheaths, intracellular water between the apposing cytoplasmic surfaces of myelin sheaths, intracellular water in glial cells) it should be correlated with MR data to reflect not only the distribution of water among these compartments but also the exchange of water molecules between them.

The observed T2 decay of magnetization is the complex result of contributions from different tissue types and different physiologic processes. The macromolecular content of edema fluid, the spaces resulting from the displacement of myelin membranes, the interaction between membrane surfaces and edema fluid, and the volume of fluid within the different compartments and the exchange between them, all influence the relaxation processes. The information gained from systemic examination of a series of animal models of

diverse pathogenesis should provide new insight into the pathogenesis and new approaches to the evaluation of patients in whom MR reveals significant white matter abnormalities.

References

1. Armspach JP, Gounot D, Rumbach L, Chambron J (1991) in vivo determination of multiexponential T2 relaxation in the brain of patients with multiple sclerosis. Magn Reson Imag 9: 107–114
2. Dumitresco RE, Armspach JP, Gounot D, *et al* (1986) Multiexponential analysis of T2 images. Magn Reson Imag 4: 445–448
3. Fullerton GD, Cameron IL, Hunter K, Fullerton HJ (1985) Proton magnetic resonance relaxation behavior of whole muscle with fatty inclusions. Radiology 155: 727–730
4. Fung BM, Puon PS (1981) Nuclear magnetic resonance transverse relaxation in muscle water. Biophys J 33: 27–38
5. Gersonde K, Felsberg L, Tolxdorff T, Ratzel D, Stroebel B (1984) Analysis of multiple T2 proton relaxation processes in human head and imaging on the basis of selective and assigned T2 values. Magn Reson Med 1: 463–477
6. Gersonde K, Tolxdorff T, Felsberg L (1985) Identification and characterization of tissues by T2-selective whole-body proton NMR imaging. Magn Reson Med 2: 390–401
7. Guttmann CRG, Mulkern RV, Clain A, Sandor T, Jolesz FA (1991) Discrimination of white matter lesion subtypes in multiple sclerosis using multiple T2 relaxation parameters. Society of Magnetic Resonance in Medicine Annual Meeting Vol 1, p 87
8. Inouye H, Kirschner D (1988) Membrane interactions in nerve myelin. I. Determination of surface change from effects of pH and ionic strength on period. Biophys J 53: 235–246
9. Jolesz FA, Polak JF, Adams DF, Ruenzel PW (1987) Myelinated and non-myelinated nerves: comparison of proton MR properties. Radiol 164: 89–91
10. Jolesz FA, Kirschner DA, Jakab P, Lorenzo AV (1989) Proton magnetic resonance in myelin deficient brains of mutant mice. J Neurol Sci 91: 85–96
11. Jolesz FA, Zamaroczy D, Kirschner DA, Lorenzo AV, Kleine LJ, Sandor T, Ruenzel PW (1989) Analysis of multiexponential transverse relaxation in myelin-associated water compartments. Society for Magnetic Resonance Imaging Annual Meeting, p 223
12. Jolesz FA, Mulkern RV, Wong STS, Bleier AR, Sandor T (1990) 128-echo CPMG imaging for sampling transverse decay curves of white maner lesions. Society for Magnetic Resonance Imaging Annual Meeting, p 142
13. Kleine LJ, Mulkern RV, Jolesz FA, Sandor T, Colucci VA, Zamaroczy D, Podell M (1992) Characterization of experimental cerebral infarction by multicomponent analysis of transverse magnetization. Invest Radiol 7: 422–428
14. Lorenzo AV, Jolesz FA, Wallman JK, Ruenzel PW (1989) Proton magnetic resonance studies of triethyltin-induced edema during perinatal brain development in rabbits. J Neurosurg 70: 432–440
15. Lorenzo AV, Mulkern RV, Wong STS, Colucci VM, Jolesz FA (1990) MR studies of brain edema in the developing animal. In: Reulen HJ, Baethmann A, Marmarou A, Fenstermacher J, Spatz M (eds) Brain edema VIII. Acta Neurochir (Wien) [Suppl] 51: 39–42
16. Mauss Y, Grucker D, Fornasiero D, Chambron J (1985) NMR compartmentalization of free water in the perfused rat heart. Magn Reson Med 2: 187
17. Menon RS, Rusinko MS, Allen PS (1992) Proton relaxation

studies of water compartmentalization in a model neurological system. Magn Reson Med 28: 264–274

18. Mulkern RV, Spencer RGS (1988) Diffusion imaging with paired CPMG sequences. Magn Reson Imag 6: 623–631
19. Mulkern RV, Bleier AR, Adzamli IK, Spencer RGS, Sandor T, Jolesz FA (1989) Two-site exchange revisited: a new method for extracting exchange parameters in biological systems. Biophys J 55: 221–232
20. Mulkern RV, Bleier AR, Sandor T, Jolesz FA (1990) Compatibility of the two-site exchange model and ^{1}H NMR relaxation rates. Magn Reson Med 14: 377–381
21. Mulkern RV, Wong STS, Jakab P, Bleier AR, Sandor T, Jolesz FA (1990) CPMG imaging sequences for high field in vivo transverse relaxation studies. Magn Reson Med 6: 67–79
22. Naruse S, Horikawa Y, Tanaka C, Kirakawa K, Nishikawa H, Yoshizaki K (1982) Proton nuclear magnetic resonance studies on brain edema. J Neurosurg 56: 747–752
23. Olson JE, Katz-Stein A, Reo NV, Jolesz FA (1992) Evaluation of acute brain edema using quantitative magnetic resonance imaging: effects of pretreatment with dexamethasone. Magn Reson Med 24: 64–74
24. Posner CM, Kleefield J, O'Reilly GV, Jolesz FA (1987) Neuroimaging and the lesion of multiple sclerosis. AJNR 8: 549–552
25. Sandor T, Bleier AR, Ruenzel PW, Adams DF, Jolesz FA (1988) Application of the maximum likelihood principle to separate exponential terms in T2 relaxation of nuclear magnetic resonance. Magn Reson Imag 6: 27–40
26. Staemmler M, Gersonde K (1986) Echo shape analysis and image size adjustment on the level of echoes: improvement of parameter-selective proton images. Magn Reson Med 3: 418–424
27. Zamaroczy D, Jolesz FA, Kirschner DA (1988) Correlation of x-ray diffraction and proton NMR measurements of myelin-associated fluid spaces. Society for Magnetic Resonance Imaging Annual Meeting, p 129
28. Zamaroczy D, Jolesz FA, Schluesener HJ, Sandor T (1988) Proton magnetic resonance characterization of experimental white matter disease. Invest Radiol 23: 517
29. Zamaroczy D, Schluesener HJ, Jolesz FA, Sobel RA, Colucci VM, Weiner HL, Sandor T (1991) Differentiation of experimental white matter lesions using multiparametric magnetic resonance measurements. Invest Radiol 26: 317–324

Correspondence: Ferenc Andras Jolesz, M. D., Department of Radiology, MRI Division, Brigham and Women's Hospital, 75 Francis Street, Boston, MA 02115, U.S.A.

Acta Neurochir (1994) [Suppl] 60: 184–186

A Thermal Clearance Probe for Continuous Monitoring of Cerebral Blood Flow

I. R. Chambers[1], M. S. Choksey[3], A. Clark[1], A. Green[1], A. Jenkins[2], and A. D. Mendelow[2]

[1] Regional Medical Physics Department, Newcastle General Hospital, [2] Department of Surgery (Neurosurgery), University of Newcastle upon Tyne, Newcastle upon Tyne, and [3] Royal London Hospital, London, U.K.

Summary

Cortical thermal clearance provides a practical means of continuous assessment of cerebral blood flow[3]. A probe which can be used to monitor cerebral cortical thermal clearance is described. It has proved a reliable method of predicting oligaemia, showing good correlation between measured thermal clearance and the onset of clinical signs of ischaemia. The limitations are its size, invasiveness, small sample volume and difficulty with quantification. It is a useful additional tool in the management of patients suffering from acute neurosurgical disorders.

Keywords: Thermal clearance; cerebral blood flow; ischaemia.

Introduction

Cerebral blood flow (CBF) may fall by up to 50% of normal values before the effects become clinically apparent. This reduction in blood flow leads to a reversible loss of function with neurological deficit. A system which could provide continuous information and predict the onset of ischaemia before it became clinically apparent would be very useful in the management of acute neurosurgical patients. Gibbs first demonstrated the use of thermal clearance to measure blood flow in vessels. The apparent thermal conductivity of tissue increases if it is perfused with blood. Whilst this method is difficult to quantify it has the prime advantage of providing continuous information.

Materials and Methods

Thermal clearance of tissue comprises three components: passive conductivity, convection and a metabolic term. This was first proposed by Pennes[7] as the Bio-Heat equation (Eq. 1).

$$\rho C_p \frac{\partial T}{\partial t} = K \Delta^2 T + W_b C_b (T_0 - T) + q_m, \qquad (1)$$
$$\text{\textit{Conductive} + \textit{Convective} + \textit{Metabolic}}$$

where ρ = density (g cm^{-3}), C_p = specific heat of tissue (mJ g^{-1} °C^{-1}), $\partial T/\partial t$ = rate of change of heat, K = thermal conductivity (mJ S^{-1} cm^{-1} °C^{-1}), T = temperature (°C), W_b = Blood flow (g cm^{-3} S^{-1}), C_b = Specific heat of blood (mJ g^{-1} °C^{-1}), T_0 = temperature at a remote point (°C), q_m = metabolic heat production. Subsequent authors have ignored the metabolic effect and this analysis has been reviewed by several authors and differing relationships between flow and the change in thermal conductivity have been put forward[1,8,9,11]. It is likely that the precise geometry of different probes and their thermal environment significantly influence their performance. Carslaw and Jaeger[2] showed that for a heated disc in contact with a semi-infinite solid

$$Q = 2 \cdot D \cdot \Delta T, \qquad (2)$$

where Q = applied power in mW, D = disc diameter (cm), K = thermal conductivity (mJ s^{-1} cm^{-1} °C^{-1}) and T = temperature difference between the disc and a remote point (°C) This relationship is used to calculate the apparent thermal conductivity of the underlying surface, henceforth referred to as Clearance Units (C.U.).

The flow sensor consists of two gold disks of 10 mm and 6 mm diameter (Fig. 1). The larger disc is heated by an assembly affixed to its upper surface which is electrically insulated from it by a piece of mica. Thermocouples are used to measure the temperature of the heated disc and the temperature difference between the two discs. At craniotomy, the probe was brought through the scalp via a remote stab incision and tunnelled to the bone edge, where it passed under the bone flap through a burrhole with chamfered edges, and then under the dura and onto the cortical surface. The dimensions of the probe were such that it could be placed on a gyrus, avoiding large vessels in the adjacent sulci. The design of the probe is such that it precluded loss of contact with the cortex, and movement relative to it, so that recordings were always referable to the same small volume of brain underlying the probe.

Results

Measurements were made in three different ways: in vitro, cadaveric recordings, and clinical recordings after surgery for intracranial aneurysms.

1. Studies in vitro

a) *Static:* Ten gelatine blocks were manufactured with different concentrations from 10–21% in a similar

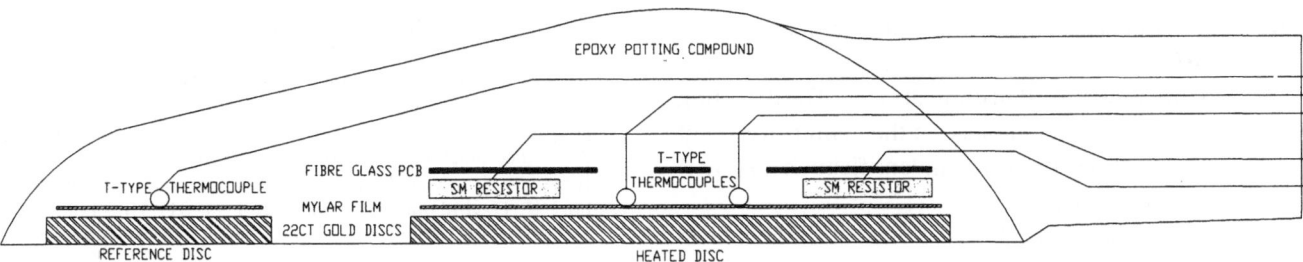

Fig. 1. Probe construction showing heated and reference disk

way to that of Grayson[6]. The linear relationship predicted by Carslaw and Jaeger was confirmed (r = 0.999).

b) Dynamic: Using a similar flow analogue to Thalasingham and Delpy[10], water was through a piece of sponge so that the flow was radially outwards. The flow rate was varied from 0.5 ml–10 ml per minute (equivalent to 5 ml–100 ml 100^{-1} min^{-1}) and a linear relationship determined between flow rate and measured heat clearance.

2. Cadaveric Recordings

Recordings were made in four cases to measure the thermal clearance of unperfused brain. Six cortical sites were measured at six different power settings. The mean thermal clearance was 24.5 ± sd 0.75 CU, and stepwise increments in power demonstrated a linear relationship between applied power and thermal clearance (r = 0.998) .

3. Clinical Measurements

Clinical recordings were made in 20 patients who had undergone clipping of aneurysms of the cerebral

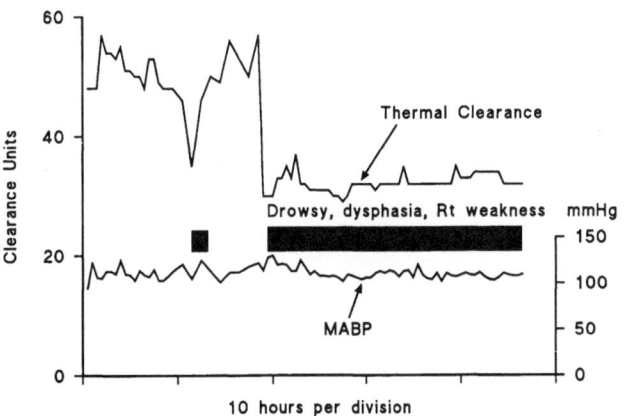

Fig. 2. Change in thermal clearance before the clinical signs of ischaemia

circulation. All went on to make good recoveries. For analysis the thermal clearance values and blood pressure were noted hourly although monitoring was continuous. Twelve patients remained clinically stable and the mean thermal clearance values stayed within a range of 48 to 59 CU with low standard deviations over long periods of monitoring (up to 72 hours). We have taken this to be the normal range for well perfused cortex. In 3 patients the initial measured clearance was low (43–47 C.U.) and all exhibited clinical symptoms of ischaemia. Over 48–72 hours their clinical state returned to normal with a concomitant rise in thermal clearance. Four patients developed severe, persistent post-operative ischaemic deficits. One patient had normal clearance values initially, which fell abruptly prior to the onset of an ischaemic deficit (Fig. 2). Two others had low thermal clearances immediately after surgery, which declined further prior to the onset of an ischaemic deficit.

However, one patient developed ischaemia despite low-normal thermal clearances. The failure of the probe to detect this was due to its placement being to far anterior. The probe must be sited where the risk of ischaemia is thought to be the greatest.

Discussion

The thermal clearance probe described conforms to the Carslaw and Jaeger equation for thermal equilibrium. The narrow range of thermal clearances obtained in the cadaveric studies confirms the assumption that the passive component of thermal conductivity of brain is constant, and that the changes are due to convection. In three instances of severe ischaemia the measured thermal clearance was low, or fell before the onset of symptoms and signs, confirming that this system is sensitive to relatively small changes in CBF. Failure to detect ischaemia in the fourth patient was due to inappropriate probe placement.

The benefit of using such a probe is its ability to

provide an indication of the moment to moment changes in CBF and to give warning of onset of ischaemia before it manifests itself in clinical signs. The detection of a falling thermal clerarance in the "clinically silent zone" could allow the early indentification of those patients at risk of ischaemia.

Thermal clearance is closely related to CBF and can be reliably monitored. Despite the limitations of small sample size, invasiveness and difficult quantification, it has a potential role in the management of acute neurosurgical patients.

References

1. Betz E, Ingvar DH, Lassen NA, Schmahl FW (1966) Regional blood flow in the cerebral cortex, measured simultaneously by heat and inert gas clearance. Acta Physiol Scand 7: 1–9
2. Carslaw HS, Jaeger JC (1959) Conduction of heat in solids. Clarendon, Oxford, p 232
3. Carter LP, Atkinson JR (1973) Cortical blood flow in controlled hypotension as measured by thermal diffusion. J Neurol Neurosurg Psychiatry 36: 906–913
4. Chato JC (1981) Reflections on the history of heat and mass transfer in bio-engineering. J Biomech Eng 103: 97–101
5. Gibbs FA (1933) A thermoelectric blood flow recorder in the form of a needle. Proc Soc Exp Biol Med 31: 141–146
6. Graham DI, Ford I, Hume Adams J, Doyle D, Teasdale GM, Lawrence AE, McLellan DR (1989) Ischaemic brain damage is still common in fatal non-missile head injury. J Neurol Neurosurg Psych 52: 346–50
7. Grayson J (1952) Internal calorimetry in the determination of thermal conductivity and blood flow. J Physiol 118: 54–72
8. Jones TH, Morawetz RB, Crowell RM, Marcoux FW, Fitzgibbon SJ, Degirolami U, Ojemann RG. (1981) Thresholds of focal cerebral ischemia in awake monkeys. J Neurosurg 54: 773–782
9. Meyer CHA, Lowe D, Meyer M, *et al* (1983) Progressive change in cerebral blood flow during the first three weeks after subarachnoid haemorrhage. Neurosurg 12: 58–76
10. Pennes HH (1948) Analysis of tissue and arterial blood temperatures in the resting human forearm. J Appl Physiol 1: 93–122
11. Perl W (1962) Heat and matter distribution in body tissues, and the determination of tissue blood flow by local clearance methods. J Theoret Biol 2: 210–235
12. Poppendieck HF, Randall R, Breeden JA, Chambers JE, Murphy JR (1966) Thermal conductivity measurements and prediction for biological fluids and tissues. Cryobiology 3: 318–327
13. Thalayasingham S, Delpy DT (1989) Thermal clearance blood flow sensor – sensitivity, linearity and flow depth discrimination. Med Biol Eng Comput 27: 394–398
14. Weinbaum S, Jiji L (1985) A new simplified bioheat equation for the effect of blood flow on local average tissue temperature. J Biomech Eng 107: 131–139

Correspondence: I. R. Chambers, M. D., Regional Medical Physics Department, Newcastle General Hospital, Westgate Road, Newcastle upon Tyne, NE4 6BE, U.K.

Acta Neurochir (1994) [Suppl] 60: 187–189
© Springer-Verlag 1994

Measurement of in vivo Cortical Thermal Clearance in Man During Complete Circulatory Standstill

M. S. Choksey[3], **I. R. Chambers**[1], **A. D. Mendelow**[2], **A. Jenkins**[2], and **R. P. Sengupta**[2]

[1] Regional Medical Physics Department, Newcastle General Hospital, [2] Department of Surgery (Neurosurgery), University of Newcastle upon Tyne, Newcastle upon Tyne, and [3] The Royal London Hospital, Whitechapel, London, U.K.

Summary

The cortical thermal clearance (CTC) was recorded continuously during surgery for a giant basilar aneurysm under cardio-pulmonary bypass. The changes observed mirrored the fall and rise in cardiac output. CTC at zero flow corresponded closely to the established cadaveric value, supporting the principle that the conductive component of thermal clearance is constant.

Keywords: Thermal clearance; cardiopulmonary bypass; giant basilar aneurysm; cortical blood flow.

Introduction

Since first proposed by Gibbs[5] thermal clearance has been used as a measure of local tissue perfusion. Mathematical analysis of the relationship between thermal clearance and local blood flow is based on the bioheat equation[3,4,5,6,7].

$$\rho C_p \frac{\partial T}{\partial t} = K \, \Delta^2 \, T. + W_b \, C_b \, (T_0 - T) + q_m \,, \quad (1)$$

Thermal Conductivity = Conductive + Convective + Metabolic

where ρ = density (g cm^{-3}), C_p = specific heat of tissue (mJ g^{-1} °C^{-1}), $\partial T / \partial t$ = rate of change of heat, K = thermal conductivity (mJ S^{-1} cm^{-1} °C^{-1}), T = temperature (°C), W_b = Blood flow (g cm^{-3} S^{-1}), C_b = Specific heat of blood (mJ g^{-1} °C^{-1}), T_0 = temperature at a remote point (°C), q_m = metabolic heat production.

It is generally assumed that the three components (conduction, convection and metabolism) can be summed to give the total thermal clearance. The metabolic component is very small provided only small temperature elevations are used, and has generally assumed to be negligible[4,6,9,11]. The conductive compo-

nent is determined largely by the water content of the tissue[10] *and is taken to be constant*. The convective component is related to the local blood flow and is variable. Thus the validity of the method rests on the assumption that the thermal clearance is the sum of fixed and variable parts.

Most workers have measured the fixed component in isolated organs, cadavers, or in protein/fat/water gels of composition similar to brain. Very rarely does the opportunity to measure cortical thermal clearance (CTC) under zero flow conditions arise in humans invivo. We report a case where thermal clearance was measured on the brain surface of a patient during clipping of a basilar aneurysm under cardio-pulmonary bypass, with a long period of circulatory standstill.

Materials and Methods

CTC was measured per-operatively using a probe designed and built in Newcastle[3]. It consists of two gold discs, 4 cm apart. One is heated, and the other is used as a reference.

The thermal clearance (K) is calculated using the Carslaw and Jaeger formula:

$$K = \frac{Q}{2 \cdot D \cdot (T_h - T_c)}$$

where K = the thermal conductivity (mJ^{-1} cm^{-1} S^{-1} °C^{-1}), D = the diameter of the heated disk in cm, T_h = temperature of the heated disc (°C), T_c = temperature of the reference disc (°C), Q = applied power in milliwatts.

The probe is connected to a control box, which performs the control, data processing, display and recording functions.

Patient Details

The patient was a 46 year old woman, who presented with headaches, unsteady gait and confusion. On ex-

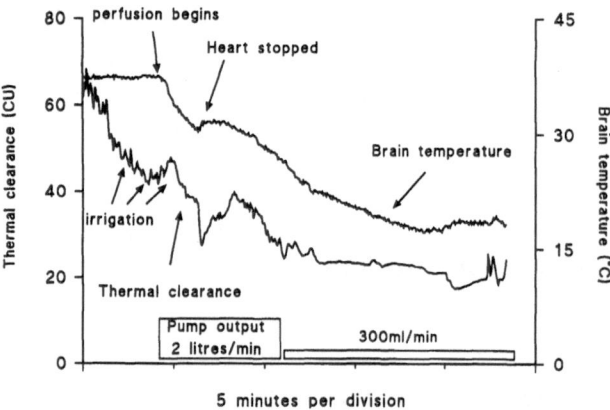

Fig. 1. Cooling curve

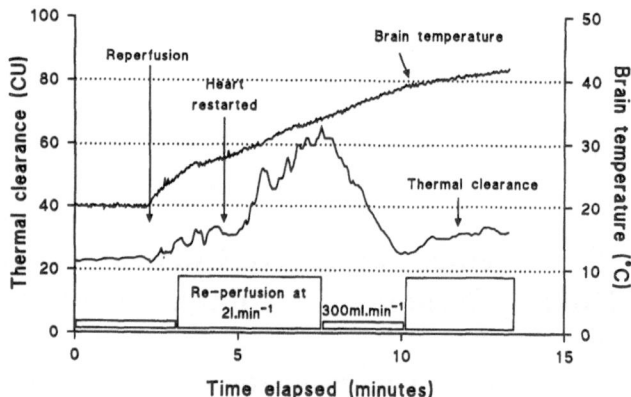

Fig. 2. Rewarming curve

amination she was lethargic, had loss of upward gaze and walked with a broad based and unsteady gait.

CT scan revealed a large partly thrombosed aneurysm, angiography showed it to be arising from the basilar artery.

After considerable discussion it was decided to treat the patient surgically, and that the aneurysm would be clipped under cardio-pulmonary bypass and circulatory standstill.

At operation, the probe was placed on the surface of the right parietal lobe, 4 cm under the craniotomy edge. It was surrounded with patties, and the area covered with gauze, to provide a stable thermal environment. The thermal clearance, the brain temperature and the cardiac pump flow during bypass were recorded continuously (Figs. 1 and 2).

As the cooling process began, and the temperature fell, the heart went into fibrillation at 30°C. The extracorporeal pump output was reduced progressively from 2 1 per minute to virtual standstill, with a trickle of 200 ml per minute. This period lasted 40 minutes, during which the aneurysm was clipped. Simultaneously, the thermal clearance was seen to fall from its initial value of 55–60 CU to 25 CU, where it remained. After the aneurysm had been clipped, the extracorporeal pump output was raised to 2 1 per minute; the thermal clearance rose correspondingly. However, it became apparent that there was considerable bleeding from the aneurysm site, so the pump output was reduced to a trickle, with a concomitant fall in thermal clearance. During the rewarming phase, at 31°C the heart began to beat spontaneously – and the cardiac output became pulsatile. At this point, even although there was no increase in mean systemic blood pressure (60 mmHg), the thermal clearance increased quite markedly.

Discussion

The use of thermal clearance as an index of local blood flow has had a long history. The search for a precise mathematical relationship between thermal clearance and flow has proved very difficult, and 60 years after its first demonstration by Gibbs, still provokes considerable controversy. However, Carter *et al.* have determined for flat, circular probes in contact with the brain surface, with increasing blood flow the thermal clearance rises linearly from a "dead brain" value, confirming the earlier findings of Grayson. If the measured thermal clearance value is to be used quantitatively, then it follows that the "zero flow" or "dead brain" value must remain constant, or very nearly so, for all such measurements to be referable to the same baseline.

Poppendick *et al.* showed that the thermal conductivity of most biological tissues could be predicted from a knowledge of the protein, water and fat content, provided that they were homogeneous (e.g. liver, kidney, brain) and did not have a laminar structure (e.g. skin) and did not contain air (e.g. lung). In cadavers we have established a narrow range of CTC (27.5 ± 1.5 CU). This value is roughly half that measured in well perfused brain[3].

However, the opportunity to observe the changes that occur in CTC in vivo, when the brain blood flow is deliberately lowered from normal to zero and then restored, is virtually unique. This recording demonstrated that – as expected – the CTC falls with decreasing temperature and cardiac output. The recorded resting value for unperfused (but still living) brain was 25 CU, correlating very well with the established cadaveric value (27.5 CU ± 1.5 SD). Finally when the cardiac output was restored, the CTC rose to its previous value.

Conclusion

This is another contribution to an accumulating large body of evidence which suggests that CTC measurement is an accurate guide to local tissue blood flow. It gives good support to the fundamental assumption that the conductive component remains virtually constant.

References

1. Carter LP, White WL, Atkinson JR (1978) Regional cortical blood flow at craniotomy. Neurosurgery 2: 223–229
2. Carslaw HS, Jaeger JC (1959) Conduction of heat in solids. Clarendon, Oxford, p 232
3. Chambers IR, Choksey MS, Clark A, Green A, Jenkins A, Mendelow AD (1992) A thermal clearance probe for continuous monitoring of cerebral blood flow. Clin Phys Physiol Meas 13:4: 311–321
4. Chato JC (1981) Reflections on the history of heat and mass transfer in bio-engineering. J Biomech Eng 103: 97–101
5. Gibbs FA (1933) A thermoelectric blood flow recorder in the form of a needle. Proc Soc Exp Biol Med 31: 141–146
6. Grayson J (1952) Internal calorimetry in the determination of thermal conductivity and blood flow. J Physiol 118: 54–72
7. Pennes HH (1948) Analysis of tissue and arterial blood temperatures in the resting human forearm. J Appl Physiol 1: 93–122
8. Perl W (1962) Heat and matter distribution in body tissues, and the determination of tissue blood flow by local clearance methods. J Theoret Biol 2: 210–235
9. Poppendick HF, Randall R, Breeden JA, Chambers JE, Murphy JR (1966) Thermal conductivity measurements and prediction for biological fluids and tissues. Cryobiology 3: 318–327
10. Weinbaum S, Jiji LM (1985) A new simplified bioheat equation for the effect of blood flow on local average tissue temperature. J Biomech Eng 107: 131–139

Correspondence: I. R. Chambers, M. D., Regional Medical Physics Department, Newcastle General Hospital, Newcastle upon Tyne, NE4 6BE, England, U.K.

Acta Neurochir (1994) [Suppl] 60: 190–192
© Springer-Verlag 1994

Cerebral Blood Flow of Rats with Water-Intoxicated Brain Edema

M. Yamaguchi[1], S. Wu[2], K. Ehara[2], T. Nagashima[2], and N. Tamaki[2]

[1] School of Allied Medical Sciences and [2] Department of Neurosurgery, School of Medicine, Kobe University, Kobe, Japan

Summary

Water intoxication brain edema was produced in rats by intraperitoneal loading of excessive amounts of distilled water (DW). In 10% and 20% groups, DW in amounts of 10 or 20% of body weight was injected, respectively. Water content of brain tissue increased proportionally to the amount of injected water, as follows: 79.8% of wet weight in control, 80.5 and 82.4 in 10% and 20% DW groups, respectively. Since cerebral blood flow (CBF) values measured by laser Doppler (LD) flowmetry were found to give a good correlation with those by hydrogen clearance method in a preliminary experiment, CBF measurement were carried out using LD flowmetry thereafter. Before the injection, CBF values were around 50 ml/min/100 g. Two hours after the water loading, CBF values in 10% and 20% DW groups were 25.6 and 20.3 ml/100 g/min, respectively. CBF values under these edematous condition decreased significantly ($p < 0.001$ by paired t-test) in proportion to the severity of the brain edema.

Keywords: Water intoxication; brain edema; CBF; laser Doppler flowmetry.

Introduction

To elucidate the effect of brain edema on cerebral blood flow (CBF), water-intoxicated brain edema was produced in rats. Measurement of CBF by LD flowmetry was performed before and after the experimental production of brain edema. Water content in the brain tissue increased in proportion to the amount of water loading, as reported previously. Decrease of CBF was demonstrated in proportion to the severity of the edema.

Materials and Methods

Wistar-strain male rats of approximately 200 to 300 g were used in this experiment. Under anesthesia with chloral hydrate (200 mg/Kg, i.p.), animals were intubated via tracheostomy and mechanically ventilated. Occasional monitoring of arterial pO_2, pCO_2, pH, Hb content, and measurement of blood pressure were carried out. When one of these parameters or more showed abnormal value(s), the rat was excluded from this experiment.

For the determination of CBF by hydrogen clearance (HC) method, a platinum electrode with a tip of 80 μ in diameter (Unique Medical Co., Japan) was fixed in a head holder. The skull bone of 5×5 mm over the parietal cortex was thinned with great care, and removed by the use of dental drill and the dura mater was opened. Animals with damage of brain surface during this process were also excluded from this experiment. Hydrogen gas (10%) was given via the intratracheal tube and CBF was calculated by wash out curve.

Determination of CBF by LD flowmetry was performed using ALF 21 laser Doppler flowmeter (Advance Co., Japan). The diameter of the probe was 1 mm. The instrument was operated at Doppler shift frequency range from 0.03 to 30 KHz.

For the correlation between HC method and LD flowmetry, the probe of LD was placed above the tip of platinum electrode under an operating microscope. The electrode was inserted 1 mm deep into the brain tissue. Both methods were used at the same time. Some animals were made hypertensive by the injection of norepinephrine and others hypotensive by withdrawal of blood. CBF data of hyper-, hypo-, and normotensive animals were compared and correlation was estimated on 27 samples as shown in Fig. 1. Significance of the correlation was checked by use of t-test.

Water content of brain tissue was determined using dry/wet weight measurements by the method of Wasterlain and Posner[5], of the tissue according to the method of Elliot and Jasper's[1]. Experimental brain edema was produced as follows: rats received intraperitoneal injection of distilled water (DW) to the amount of 10% body weight (10% DW group) or to the amount of 20% (20% DW group). Two hours after the water loading, animals were sacrificed by decapitation and the brain tissue was removed for the measurement of wet weight.

In accordance to original method of Wasterlain and Posner[5], vasopressin was injected to reduce urine output. However, no vasopressin was used in the present experiment because the substance might increase the blood pressure. As shown in the result, increase of water content was evident in the water-intoxicated groups without the injection of vasopressin.

Physiological saline instead of DW were administered in the saline group (amount of injection was 20% of body weight). To determine the effect of the brain edema on CBF, the LD flowmetry were performed before and 2 hours after the water loading.

Results

The two methods of CBF measurement were compared in Fig. 1. Correlation between the data by HC

(ml/min/100g)

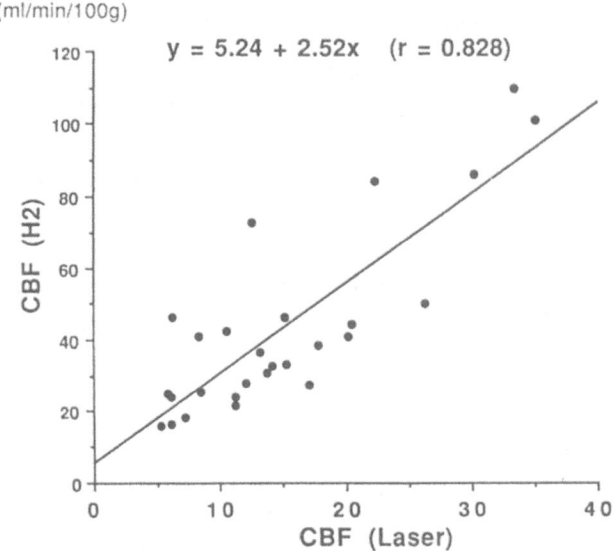

Fig. 1. Correlation between laser Doppler flow and hydrogen clearance

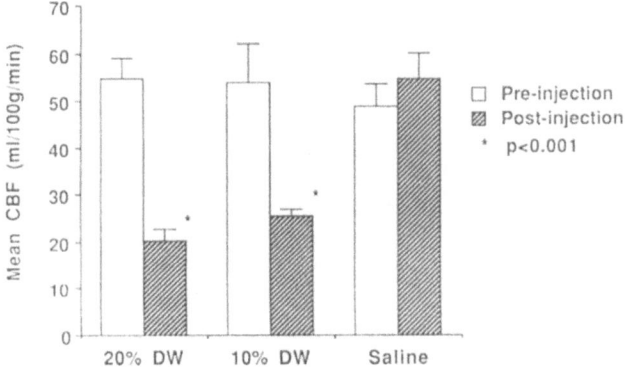

Fig. 2. Changes in the mean CBF of rats with water-intoxicated edema

and LD flowmetry was statistically significant (r = 0.828, p < 0.001 by t-test). Determinations of CBF on the edema preparations were performed by only LD flowmetry thereafter. Units of CBF measurement in LD flowmetry were converted by to those of HC method as ml/min/100 g of the brain tissue.

Water content of the normal control rat brain was 79.8 ± 1.4, (mean ± SD, number of animals = 8). In the group of 10% DW loading the water content was 80.5 ± 0.6 (n = 6). In 20% DW group, water content was 82.4 ± 0.8 (n = 6). Increased brain water content of these rats were statistically significant (p < 0.05 and 0.01 in 10% and 20% DW groups, respectively). The water content of saline group was 79.7 ± 0.7[6] and this value was not significantly different from that of the control group.

Values of CBF (ml/min/100 g) in various groups are demonstrated as the mean ± SD of 6 animals. In the 10% DW group CBF was 53.9 ± 8.3 before the water loading. After the insult, CBF increased to 25.6 ± 1.5 (42.8% of the initial value). Rats of 20% DW group showed more drastic decrease of CBF by the water intoxication brain edema: from 54.8 ± 4.3 to 20.3 ± 2.5 (6.4%). These decreases of CBF were statistically significant (paired t-test, p < 0.001 in both groups with brain edema). But the CBF values of the saline group was 49.8 ± 4.8 before the saline loading and 54.8 ± 5.5 after the injection. The CBF of the normal control was 54.8. In Fig. 2, decreased CBF values after loading of water were demonstrated in the groups of 10% DW and 20%.

Discussion

The hydrogen clearance (HC) technique provides quantitive estimates of flow. However, since the brain tissue around the platinum electrode tends to be injured and gradually develops local edema, HC may not be a suitable method for continuous recording of CBF. As Harbel et al.[2] reported, LD flowmetry is considered to be a useful method for the continuous observation of brain microcirculation. The present study showed a significant correlation between the CBF values estimated by HC method and LD flowmetry (p < 0.001). However, LD flowmetry was applied for the estimation of CBF changes under the effect of the experimental brain edema. Since hypertension apparently gave high CBF values, vasopressin was not used in this experiment because of its hypertensive effect.

The increase of water content of brain was proportional to the amount of water loading as mentioned in result. A passive shift of water by osmotic gradient may take place from blood to brain as reported by van Harreveld et al.[3]. Meinig et al. reported a decrease of CBF in water intoxication brain edema in cats[4]. They suggested that the diminution of capillary diameter due to astroglial swelling reduced CBF. Decreased CBF who correlated with the severe of water increase in the cat brain[4]. The decrease of CBF was proportional to the severity of the brain edema in our study using rats.

As reported previously[6], rats with experimental brain edema by water intoxication showed disturbed avoidance learning. Overhydration of brain tissue was considered to be responsible for the memory disturbances. The transient, but apparent dysfunction of memory process in this situation may be related to the decrease of CBF, as shown in the present experiment.

References

1. Elliot KAC, Jasper H (1949) Measurement of experimentally induced brain swelling and shrinkage. Am J Physiol 157: 122–129
2. Harbel RL, Heizer ML, Marmarou A, Ellis EF (1989) Laser-Doppler assessment of brain microcirculation: effect of systemic alteration. Am J Physiol 256: H1247–H1254
3. Harreveld A *et al* (1966) Water, electrolytes, and extracellular space in hydrated and dehydrated brains. Am J Physiol 210: 251–256
4. Meinig G, Reulen HJ, Magawly C (1973) Regional cerebral blood flow and cerebral perfusion pressure in global brain oedema induced by water intoxication. Acta Neurochir (Wien) 29: 1–13
5. Wasterlain CG, Posner JB (1969) Cerebral edema in water intoxication. I. Clinical and chemical observation. Arch Neurol 19: 71–78
6. Yamaguchi M, Kinoshita I, Sobue I (1990) Mental dysfunction in water intoxication brain edema. Advances in neurology, vol 52. Raven, New York, p 561

Correspondence: Michio Yamaguchi, M.D., School of Allied Medical Sciences, Kobe University 7-10-2 Tomogaoka, Suma, Kobe 654-01, Japan.

Acta Neurochir (1994) [Suppl] 60: 193–196
© Springer-Verlag 1994

Focal Microvascular Occlusion After Acute Subdural Haematoma in the Rat: a Mechanism for Ischaemic Damage and Brain Swelling?

H. Fujisawa[1], **W. L. Maxwell**[2], **D. I. Graham**[2], **G. M. Reasdale**[3], and **R. Bullock**[4]

[1] Department of Neurosurgery, Yamaguchi University School of Medicine, Ube, Japan, University Departments of [2] Neuropathology and
[3] Neurosurgery, Institute of Neurological Sciences, Southern General Hospital, Glasgow, Scotland, U.K., and [4] Division of Neurosurgery,
Medical College of Virginia, Virginia Commonwealth University, Richmond, U.S.A.

Summary

Using a rat model of subdural haematoma (SDH) which is associated with ischaemic damage in the ipsilateral hemisphere, we have studied the microcirculation and the time course of development of ischaemia beneath the subdural haematoma. The latter was investigated by measuring cerebral blood flow (CBF) using the hydrogen clearance technique, the former by electron microscopy (EM) and carbon black angiography. CBF fell below 5 ml/100g/min within five minutes of SDH and remained low ($\leq$ 7 ml/100g/min) for over two hours. Contralateral CBF fell to 54% of control but normalized by 1 hour. Carbon black angiography showed absent perfusion beneath the SDH. EM demonstrated: 1) normal cortical vessel traversing the haematoma; 2) occlusion of cortical vessels under SDH by clotted red cells and platelets; 3) massively enlarged perivascular spaces due to swelling of astrocytes. No vasospasm was seen. The "ischaemic core" may develop where microvascular occlusion occurs, and this microcirculatory disturbance initiates ischaemic cell damage with release of glutamate, which spreads outward from the core. This causes further damage of neurons and swelling of astrocytes, resulting in severe ischaemic necrosis, and progressive brain edema. The cause of microvessel occlusion remains unknown.

Keywords: Acute subdural haematoma; rat; ischaemic brain damage, cerebral microcirculation.

Introduction

Neuropathological studies have demonstrated that ischaemic brain damage is present in almost all patients who die after acute subdural haematoma[6,7]. However, underlying mechanisms are not yet fully understood. Cerebral blood flow (CBF) measurements in patients with an acute subdural haematoma have only rarely been made, but these have shown a profoundly reduced regional CBF (rCBF) beneath the haematoma, in most cases[1]. Using an acute subdural haematoma model in rats which consistently produces ischaemic damage beneath the haematoma[10], we have investigated the vascular pathogenesis of this condition. In the present study, we studied the early time course of ischaemia and the microcirculation under the haematoma.

Materials and Methods

Eighteen adult male Sprague-Dawley rats were studied. After tracheostomy and cannulation of femoral vessels were performed, animals were anaesthetized throughout the experiment using halothane and a nitrous oxide: oxygen mixture (70 : 30%) and ventilated attaining normocarbia and normoxia.

CBF Measurement

In order to study the time course of ischaemia, we investigated CBF changes using the hydrogen clearance technique. The animals were immobilized in a stereotactic frame and the skull was exposed. Through burr holes, two platinum electrodes were bilaterally positioned 4 mm lateral to the midline, 2 mm anterior to the coronal suture, and 1.3 mm below the surface of the parietal cortex with a lateral angle of 15°. CBF was measured before the subdural haematoma was produced and for 2 hours thereafter.

Induction of Haematoma

Another burr hole was made at the left coronal suture. A J-shaped 23-gauge needle was inserted in the subdural space and cemented into position, and 0.4 mm of non-heparinized, autologous blood was slowly injected into the subdural space.

Carbon Black Angiography

Two hours after haematoma induction, the brain of 7 rats was perfused with normal saline followed by a fine suspension of carbon black at the mean arterial blood pressure of the individual animals. The intact cranium was placed then in 40% formaldehyde, glacial acetic acid, and absolute methanol (FAM; 1:1:8, v/v/v) for 24 hours before brain removal. Thereafter, the brain was removed and immersion-fixed in FAM for 24 hours. The specimens were cut into serial coronal sections, and examined by light microscopy.

Electron Microscopic Study

Three animals were studied. Two animals were perfusion fixed with Karnovsky's fixative (2% formaldehyde, 2% glutaraldehyde in 0.1 M phosphate buffer, pH 6.2, 850 mOsm) tow hours after haematoma induction. The intact cranium was stored in Karnovsky's solution for 24 hours before brain removal. In another animal, horse radish peroxidase (HRP) was added to the autologous subdural blood in order to investigate the penetration of blood into the cortex. For this study, 30 mg HRP was dissolved in 0.25 ml of physiological saline, and 0.1 ml of this solution was added yielding 0.4 ml of subdural autologous blood. The animal was perfusion fixed with 2% lanthanum nitrate in 2.5% glutaraldehyde/2% para-formaldehyde in 0.1 mol/l sodium cacodylate buffer (pH 7.2) followed by 2% paraformaldehyde and 5% glutaraldehyde in 0.1 M cacodylate buffer (pH 7.4) containing 5% sucrose. Sections were processed and examined by EM.

Results

Adequate oxygenation, normocapnia and normo-thermia were maintained throughout the experiment (Table 1).

Changes in CBF

The CBF changes are shown in Fig. 1. CBF fell below 5 ml/100g/min within five minutes of SDH and remained low for 2 hours. Contralateral CBF fell to 54% of control, but was almost normalized by 1 hour.

Microcirculation Beneath the Haematoma

Carbon black angiography at 2 hours showed absent perfusion in the ischaemic zone beneath the haematoma. Carbon black reached to the boundary of the is-chaemic zone, while no vessels were filled with carbon black within the lesion (Fig. 2a). Vessels at the periphery of the ischaemic zone were more dilated than those in the contralateral normal cortex.

Ultrastructural Changes

Electron microscopy demonstrated: 1) Normal corti-cal vessels traversing the haematoma with no evidence

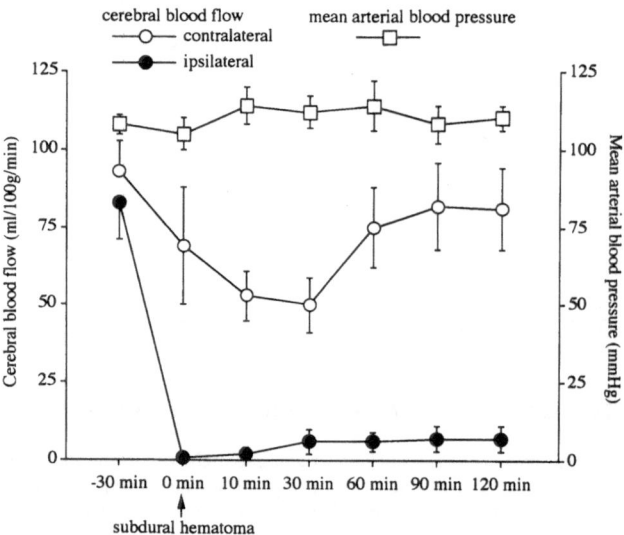

Fig. 1. Changes in cerebral blood flow in animals with subdural haematoma

of vasospasm (Fig. 2b), 2) Occlusion of cortical vessels under the SDH by clotted red cells and platelets (Fig. 2c) 3) Massively enlarged perivascular spaces due to swelling of astrocytes were seen throughout the ischaemic area, producing a vacuolated status spongio-sus in scanning EM (Fig. 2d). The HRP perfusion study indicates that the tracer penetrated freely into the cor-tex for up to 200–300 µ where it was found within the neuropil, and adjacent to the peri-vascular basal lamina (Fig. 2e). Surprisingly, lanthanum was found both within the endothelial cells of microvessels, and out-side the basal lamina, but retained by the astrocytic foot processes, indicating that the tracer passed from the intravascular space into the vessel wall, but not further into the extracellular space (Fig. 2f).

Discussion

In this study, the time course of CBF changes in the ipsi- and contralateral cortex is clearly shown. The

Table 1. *Physiological Values*

	Before haematoma	10 min	30 min	60 min	120 min
pH	7.46 ± 0.01	7.45 ± 0.02	7.52 ± 0.04	7.47 ± 0.02	7.48 ± 0.01
$PaCO_2$	38 ± 1	36 ± 2	36 ± 6	35 ± 1	34 ± 1
PaO_2	157 ± 5	151 ± 4	149 ± 4	166 ± 5	156 ± 5
Temperature	37.1 ± 0.1	37.3 ± 0.1	37.4 ± 0.6	37.3 ± 0.1	37.0 ± 0.2
MABP	105 ± 2	112 ± 4	107 ± 7	114 ± 5	108 ± 4

Values are expressed as means ± S.E.M. *MABP* mean arterial blood pressure.

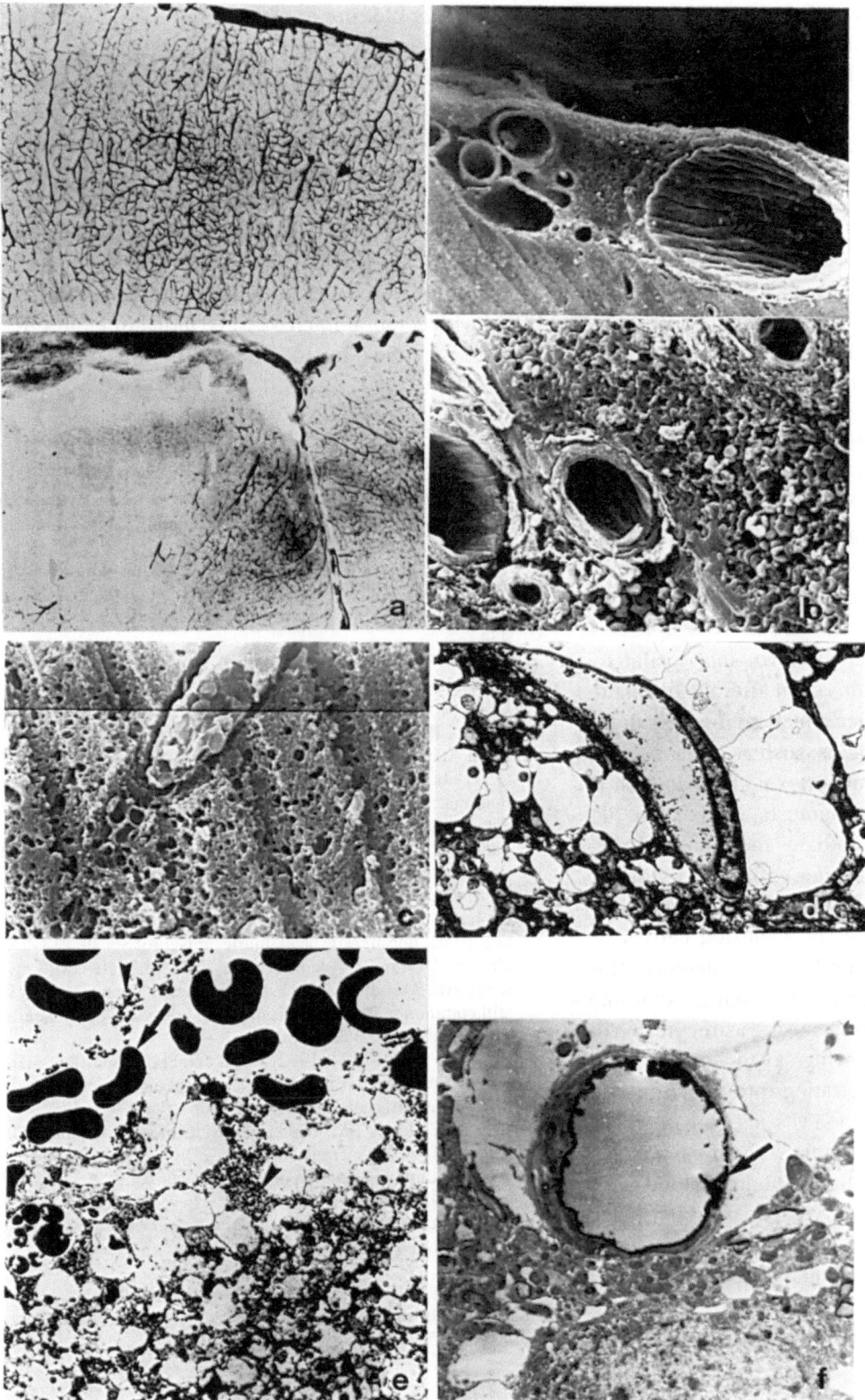

Fig. 2. (a) Carbon black angiography at 2 hours. Upper: contralateral hemisphere. Lower: ipsilateral hemisphere. (b–e) Ultrastructural changes in the lesion. (b) Cortical vessels. Upper: contralateral to the haematoma. Lower: ipsilateral to the haematoma. (c) Cortical vessels under the haematoma. (d) Massively enlarged perivascular spaces due to swelling of astrocytes. (e) HRP perfusion study. Arrow: red cells, arrow head: HRP (f) Location of lanthanum (arrow)

results are in agreement with our previous study in which ipsilateral CBF two hours after SDH was shown autoradiographically to be below 5 ml/100g/min[8]. We have also observed that intracranial pressure (ICP) after induction of the haematoma is markedly elevated (60–110 mmHg) as a cause of rapid CBF reduction to subcritical levels, although the ICP rise was brief, lasting only ± 10 minutes[10]. The ICP elevation alone is therefore unlikely as a cause of ischaemic damage. Ischaemic tissue damage was also seen in areas of cortex where the overlying haematoma mass was very thin[4]. In other experiments, a similar volume of an inert silicon mass in the subdural space produced ischaemic damage which was much smaller[9]. Duhaime, *et al.*[5] have reported that coagulated blood is toxic to the cortical structures, even in the absence of focal pressure effects, as demonstrated by removing the overlying skull bone and by opening the dura. It thus appears that clotted blood itself affects brain microvessels eliciting intravascular coagulation there, while not in the larger blood vessels.

Our previous studies have shown that ipsilateral hemispheric swelling remains even after the haematoma has been removed[8]. Since a marked increase in the level of excitatory amino acids, such as glutamate has been found in the ipsilateral cortex and a reduction in infarct size by glutamate antagonists, such as D-CPP-ene, we postulate that "excitotoxic" mechanisms exacerbate the ischaemic cell damage, enhancing cytotoxic edema and brain swelling[3,4]. It is, however, more conceivable that early microvessel occlusion causes ischaemia and the subsequent release of glutamate. It is unlikely that a release of glutamate from the clot into cortex was the primary event, causing astrocyte swelling and microvessel narrowing, even though horse radish peroxidase was penetrating into the cortex. In this study normal cortical blood vessels were observed by EM traversing the haematoma without vasospasm. Thus it is unlikely that coagulated subdural blood had a vasospastic effect on these vessels. Carbon black angiography clearly showed absent perfusion of vessels in the ischaemic lesion, and EM demonstrated occlusion of cortical vessels under SDH by clotted red cells and platelets. Although it is unclear, how the rapid

intravascular clotting occured in the lesion, we postulate that the microcirculatory disturbances led to an immediate activation of cytokines and of free radicals, such as nitric oxide, which triggered platelet adherence, endothelial pseudopodial "pit vesicle" activity, and clot formation. Although further studies are needed to test these hypotheses, similar findings have been made in humans after both SDH[1] and after contusion[2]. We have demonstrated that early focal microvessel occlusion is associated with extravascular contact of blood with cerebral microvessels, indicating the suitability of the model to study underlying mechanisms.

References

1. Bouma GJ, Muizelaar JP, Stringer WA, Choi SC, Fatouros P, Young HF (1992) Ultra-early evaluation of regional cerebral blood flow in severely head-injured patients using xenon-enhanced computerized tomography. J Neurosurg 77: 360–368
2. Bullock R, Maxwell WL, Graham DI, Teasdale GM, Adams JH (1991) Glial swelling following human cerebral contusion: an ultrastructural study. J Neurol Neurosurg Psychiatry 54: 427–434
3. Bullock R, McCulloch J, Graham DI, Lowe D, Chen MH, Teasdale GM (1990) Focal ischaemic damage is reduced by CPP-ene studies in two animal models. Stroke 21 [Suppl III]: III-32 – III-36
4. Chen MH, Bullock R, Graham DI, Miller JD, McCulloch J (1991) Ischaemic neuronal damage after acute subdural haematoma in the rat: effects of pretreatment with a glutamate antagonist. J Neurosurg 74: 944–950
5. Duhaime C, Gennarelli L, Brasko J, Ross DT (1991) Neuronal damage underlying subdural haematoma: is blood itself toxic? Abstract, Ninth Annual Meeting of Neurotrauma, New Orleans
6. Graham DI, Adams JH, Doyle D (1978) Ischaemic brain damage in fatal non-missile head injuries. J Neurol Sci 39: 213–234
7. Graham DI, Ford I, Adams JH, Doyle D, Teasdale GM, Lawrence AE, McLellan DR (1989) Ischaemic brain damages is still common in fatal non-missile head injury. J Neurol Neurosurg Psychiatry 52: 346–350
8. Kuroda Y, Bullock R (1992) Local cerebral blood flow mapping before and after removal of acute subdural haematoma in the rat. Neurosurgery 30: 687–691
9. Kuroda Y, Inglis FM, Miller JD, McCulloch J, Graham DI, Bullock R (1992) Transient glucose hypermetabolism after acute subdural haematoma in the rat. J Neurosurg 76: 471–477
10. Miller JD, Bullock R, Graham DI, Chen MH, Teasdale GM (1990) Ischaemic brain damage in a model of acute subdural haematoma. Neurosurgery 27: 433–439

Correspondence: H. Fujisawa, M. D., Department of Neurosurgery, Yamaguchi University School of Medicine, 1144 Kogushi, Ube, Yamaguchi 755, Japan.

Acta Neurochir (1994) [Suppl] 60: 197–199
© Springer-Verlag 1994

An Ischemic Opening of the Blood-Brain Barrier May Deteriorate Brain Stem Auditory Evoked Potentials Following Transient Hindbrain Ischemia in Gerbils

R. Hata[1], M. Matsumoto[1,2], K. Yamamoto[3], T. Ohtsuki[1], S. Ogawa[1], N. Handa[1], T. Kubo[3], T. Matsunaga[3], T. Nishimura[4], and T. Kamada[1]

[1] First Department of Internal Medicine, [2] Department of Neurology, [3] Department of Otolaryngology, and [4] Biomedical Research Center, Division of Tracer Kinetics University of Osaka, Osaka, Japan

Summary

To clarify the effect of vasogenic brain edema on the brainstem, the relationships between waveform changes in brainstem auditory evoked potentials (BAEP) and blood-brain barrier (BBB) disturbance following transient hindbrain ischemia were investigated. Hindbrain ischemia was induced in gerbils by bilateral occlusion of the vertebral arteries. The animals were divided into three groups subjected to 0, 5, and 30 min of bilateral vertebral occlusion (BVO-0', -5', and -30' groups; n = 4 in each group). Two hours after recirculation, Evans blue (EB) solution was injected into the saphenous vein. The brains were removed after 30 min of circulation, and all areas stained macroscopically by EB were noted and recorded. During hindbrain ischemia, BAEP disappeared within 3 min. In the BVO-5' group, BAEP reappeared and returned to normal within 10 min after reperfusion, whereas in the BVO-30' group, BAEP never returned to normal and finally disappeared within 30 min after reperfusion. In the BVO-5' group, no EB staining was visible. On the other hand, in the BVO-30' group, EB staining was seen in the medial part of the tegmentum in the midbrain in two animals, and around the vestibular nucleus in the lateral parts of the pons in three. These results demonstrate the close relationship between the reversibility of ischemia-induced changes in BAEP and BBB disturbance in the brainstem.

Keywords: Gerbils; cerebral ischemia; vasogenic brain edema; brain stem auditory evoked potentials.

Introduction

Brain edema is considered to be important because it invariably accompanies cerebral ischemia and contributes to the clinical manifestations. By using animal models of forebrain ischemia, the evolution of serum protein extravasation has been investigated extensively[2,4], since a change in blood-brain barrier (BBB) permeability is an early event that occurs before the development of brain edema. However, with regard to brainstem regions, only a few studies have been reported because of the lack of a suitable brainstem ischemia model. Recently, we have successfully established a hindbrain ischemia model in gerbils[1]. In the present study, using this model, waveform changes in brainstem auditory evoked potentials (BAEP) following transient hindbrain ischemia, and the relationship between BAEP changes and BBB disturbance were investigated.

Materials and Methods

Adult Mongolian gerbils of both sexes, weighing 60–80 g, were used. All the animals had free access to food and water until the day of the experiment. The animals were lightly anesthetized with ketamine hydrochloride (50 mg/kg i.p.). An anterior midline cervical incision was made, and the musculus longus colli was dissected to expose the vertebral arteries just before their entry into the transverse foramen of the cervical vertebra. Both vertebral arteries were loosely looped with silk sutures and the incision was closed. After this procedure, the right saphenous veins were cannulated with polyethylene catheters (PE-10, Clay Adams, Parsippany, N.J., U.S.A.) for administering Evans blue (EB). The next day, under light ether inhalation, the neck incision was reopened and the animals were orotracheally intubated. Then, the sutures around each vertebral artery were pulled with 5 g weights to occlude the circulation. During ischemia, artificial ventilation with room air was initiated immediately after apnea had been elicited, and rectal temperature was monitored continuously and maintained at 37°C by a heating lamp and a heating pad connected to a thermistor (ATB-1100, Nihon Kohden, Tokyo).

In this study, the animals were divided into three groups subjected to 0, 5, and 30 min of bilateral vertebral occlusion (BVO-0', -5', and -30' groups, respectively; n = 4 in each group). All BAEP data were recorded with commercially available instrumentation (Neu-

198

R. Hata *et al.*

ropack 4, Nihon Kohden, Tokyo) before ischemia, at 3 min of ischemia, and every 5 min up to 30 min after recirculation. Alternating polarity clicks of 0.1 ms duration were presented monaurally at a rate of 10 Hz and an intensity of 90 dB SPL. These auditory clicks were delivered to the left ear only. Each response was amplified (× 100,000), filtered at 50–3000 Hz, and averaged in runs of 200 sweeps. The positive electrode was inserted subcutaneously into the vertex, the negative electrode was inserted into the left earlobe, and the ground electrode was inserted into the abdominal wall.

All animals were killed after a postischemic period of 2.5 h. EB solution (2%, 0.2 ml) was hand-injected slowly into the saphenous vein 30 min before death. After the predetermined period, under deep pentobarbital anesthesia, the animals were fixed by transcardial perfusion with saline and 200 ml of 0.2% picric acid – 2% paraformaldehyde solution (0.1 M phosphate buffer, pH 7.4). The fixed brains were removed, and cut into 2-mm coronal sections. All areas stained macroscopically by EB were noted and recorded.

Results

Before ischemia, the BAEP consisted of four positive waves in the first 5 ms after stimulus onset (Fig. 1). During ischemia, all the BAEP waves disappeared

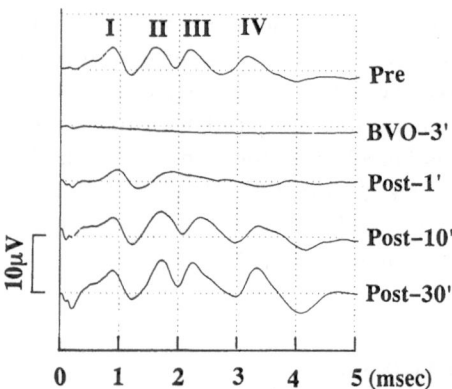

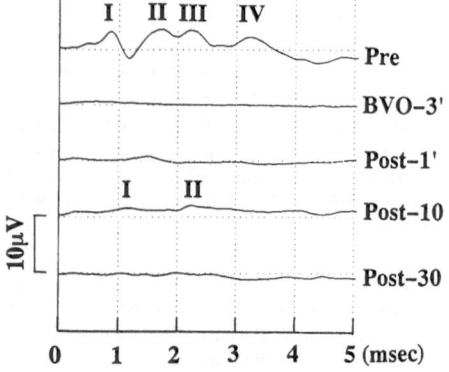

Fig. 1. Changes in brain-stem auditory evoked potentials following transient hindbrain ischemia. Upper column shows ischemia for 5 min and lower column shows ischemia for 30 min Pre: before ischemia; BVO-3′: 3 min of ischemia; Post-1′: min after reperfusion; Post-10′: 10 min after reperfusion; Post-30′: 30 min after reperfusion. For details, see text

within 3 min in all gerbils. In the BVO-5′ group, all the BAEP waves reappeared and returned to normal within 10 min after reperfusion (Fig. 1, upper column). However, in the BVO-30′ group, (deterioration of BAEP continued after reperfusion (Fig. 1, lower column). In some animals (n = 2), waves I and II reappeared 10 min after reperfusion (Fig. 1, upper column, Post-10′). However, the amplitudes of these waves were markedly reduced and their latencies were also delayed. Finally these waves disappeared again 30 min after reperfusion (Fig. 1, upper column, Post-30′) and all the animals died within 4 h because of respiratory failure.

In the BVO-5′ group, no EB staining was visible. On the other hand, in the BVO-30′ group, EB staining was seen in the lateral parts of the pons in three animals, and in the medial part of the pons and medial part of the tegmentum in the midbrain in two (Fig. 2).

Discussion

Since the introduction of the far-field scalp averaging technique for measuring BAEP by Jewett[3], BAEP has been widely used as an index of brainstem function because of the close relationship between BAEP waveforms and specific anatomical structures of the brain. This specificity allows the localisation of conduction deficits in the brainstem. Extracranial occlusion of the bilateral vertebral arteries in gerbils resulted in the disappearance of all BAEP waveforms immediately after occlusion. Previous [14C]iodoantipyrine autoradiographic measurement indicated that cerebral blood flow was reduced to less than 5 ml/100 g/min in the cerebellum, the pons, and the upper medulla, demonstrating that severe and reproducible hindbrain ischemia was induced immediately after occlusion[1]. These CBF data were consistent with the quick disappearance of BAEP waveforms. Furthermore, we demonstrated that all BAEP waveforms recovered to normal after a brief ischemic insult (5 min), but never recovered after a longer ischemic insult (30 min). In our previous study, we investigated the selective vulnerability of hindbrain neurons by applying immunohistochemical techniques, and demonstrated that after ischemia for 5 min, no ischemic lesion was detectable in the cochlear nucleus and the superior olive, which are thought to generate waves II and III, whereas after ischemia for 30 min, ischemic lesions were detected in these acoustic relay nuclei in all animals[1]. These histopathological data were consistent with the reversibility of the ischemia-induced changes in BAEP. The finding that in the BVO-30′ group, ischemia-induced

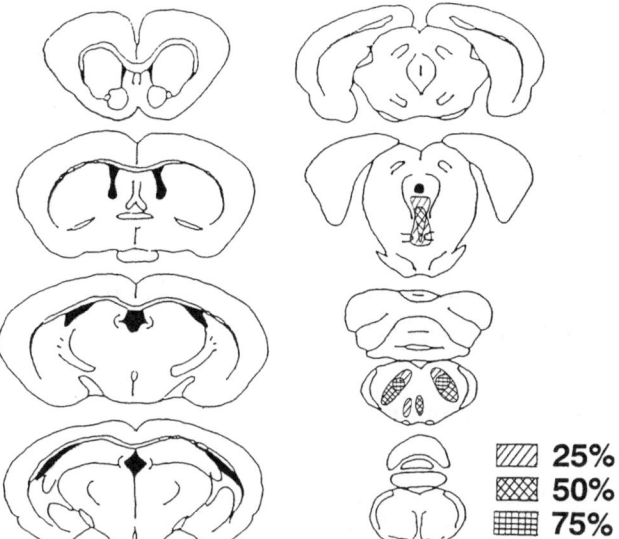

▨	**25%**
▩	**50%**
▦	**75%**

Fig. 2. Schematic distributions of Evans-blue (EB) staining in brain sections from four gerbils 2 h after 30 min of hindbrain ischemia and perfusion with EB for 30 min, illustrating the location of structure where EB staining was visible. Hatched, cross-hatched, and meshed areas indicate presence of lesions in one, two, and three animals, respectively

changes in BAEP seemed to continue and then deteriorate after reperfusion was unexpected. Although the reason for this delayed deterioration of BAEP is not clear, we speculate that brain edema might be involved. The development of brain edema often accompanies cerebral ischemia, and worsens the clinical manifestations. Therefore, in the present study, we checked serum albumin extravasation using the dye tracer technique. In the reperfusion period after cerebral ischemia, extravasation of serum proteins is a major component of vasogenic-type cerebral edema[6,7]. The dye tracer EB has been used for many years to demonstrate extravasation of albumin, in microscopy studies[8] and for the quantitation of vasogenic edema[9]. Since EB binds to albumin, enters the cell, and then dissociates from the albumin moiety[5], short-term (30 min) circulation of EB reveals BBB-leakage sites at the time of sacrifice. We demonstrated that no EB staining was present 2 h after 5 min of ischemia, whereas EB staining was visible in the brainstem 2 h after 30 min of ischemia. Interestingly, in our previous study[1], areas where EB staining was observed corresponded to the areas vulnerable to ischemia (e.g. the

lateral vestibular nucleus and oculomotor nucleus). These data suggest that ischemia-induced BBB disturbance in the brainstem may cause deterioration of BAEP following transient hindbrain ischemia in gerbils.

In conclusion, we have demonstrated that all BAEP waveforms disappeared during hindbrain ischemia, and recovered to normal after a brief ischemic insult (5 min), but never recovered after a longer ischemic insult (30 min). We also demonstrated a close relationship between the reversibility of ischemia-induced changes in BAEP and BBB disturbance in the brainstem.

References

1. Hata R, Matsumoto M, Hatakeyama T, Ohtsuki T, Handa N, Niinobe M, Mikoshiba K, Sakaki S, Nishimura T, Yanagihara T, Kamada T (1993) Differential vulnerability in the hindbrain neurons and local cerebral blood flow during bilateral vertebral occlusion in gerbils. Neuroscience, in press
2. Ito U, Ohno K, Nakamura R, Suganuma F, Inaba Y (1979) Brain edema during ischemia and after restoration of blood flow. Measurement of water, sodium, potassium content and plasma protein permeability. Stroke 10: 542–547
3. Jewett DL (1970) Volume-conducted potentials in response to auditory stimuli as detected by averaging in the cat. Electroenceph Clin Neurophysiol 28: 609–618
4. Kitagawa K, Matsumoto M, Tagaya M, Ueda H, Oku N, Kuwabara K, Ohtsuki T, Handa N, Kimura K, Kamada T (1991) Temporal profile of serum albumin extravasation following cerebral ischemia in a newly established reproducible gerbil model for vasogenic brain edema: a combined immunohistochemical and dye tracer analysis. Acta Neuropathol (Berl) 82: 164–171
5. Klatzo I, Chui E, Fujiwara K, Spatz M (1980) Resolution of vasogenic brain edema. Adv Neurol 28: 359–373
6. Klatzo I (1987) Pathophysiological aspects of brain edema. Acta Neuropathol (Berl) 72: 236–239
7. Kuroiwa T, Ting P, Martinez H, Klatzo I (1985) The biphasic opening of the blood-brain barrier to proteins following temporary middle cerebral artery occlusion. Acta Neuropathol (Berl) 68: 122–129
8. Suzuki R, Yamaguchi T, Kirino T, Orzi F, Klatzo I (1983) The effects of 5-minute ischemia in Mongolian gerbils: 1. Blood-brain barrier, cerebral blood flow, and local cerebral glucose utilization changes. Acta Neuropathol (Berl) 60: 207–216
9. Uyama O, Okamura N, Yanase M, Narita M, Kawabata K, Sugita M (1988) Quantitative evaluation of vascular permeability in the gerbil brain after transient ischemia using Evans blue fluorescence. J Cereb Blood Flow Metab 8: 282–284

Correspondence: R. Hata, M.D., First Department of Internal Medicine, Osaka University Medical School, 2-2 Yamadaoka, Suita 565, Japan.

Acta Neurochir (1994) [Suppl] 60: 200–202
© Springer-Verlag 1994

Memory Deficit Accompanying Cerebral Neurodegeneration After Stroke in Stroke-Prone Spontaneously Hypertensive Rats (SHRSP)

M. Yamaguchi[1], **K. Sugimachi**[1], **K. Nakano**[1], **M. Fujimoto**[1], **M. Takahashi**[1], **T. Chikugo**[2], and **H. Ogawa**[3]

[1] Research Department, Nihon Schering KK, Nishimiyahara, Yodogawa-ku, Osaka, [2] Department of Pathology, and [3] Department of Hygiene, Kinki University School of Medicine, Ohnohigashi, Osaka-Sayama, Japan

Summary

Memory performances of SHRSP with chronic stroke were examined on the three-panel runway task in addition to the histological evaluation and magnetic resonance imaging (MRI) of the brain. After recovery from the neurological symptoms with stroke. SHRSP were subjected to acquisition training on the memory tasks, and they exhibited both a delay and a persistent impairment of acquisition on the memory tasks, compared to the non-stroke SHRSP. T2-weighted MRI with the stroke SHRSP suggested marked edematous formation in the cortex, caudate putamen and/or thalamus, preferentially in the frontal and/or occipital cortex. The histological evaluation showed edematous degeneration such as edema, gliosis and cyst preferentially in the cortex, but no degeneration in the hippocampus. Thus, SHRSP with chronic stroke was found to exhibit impairment of learning and memory, which may be due to the cortical edematous degeneration.

Keywords: SHRSP; stroke; memory; brain edema.

Introduction

Cognitive disturbance in patients with chronic cerebrovascular diseases has been for long time a main target symptom for the drug development, but no standard therapeutic concept so far has been established due to lack of consenscious. There is no reliable animal model predictive of the clinical efficacy. It is well known that SHRSP is a suitable animal model for studying pathological and therapeutic aspects of stroke fund in human[3,6]. In this study, both memory performances and edematous degeneration in SHRSP were examined during chronic stage of stroke, in order to establish a useful animal model to evaluate the potential useful drugs for the treatment of cognitive disturbance associated with chronic stroke.

Materials and Methods

Male SHRSP (n = 22) on and after 4 weeks of age were given Funabashi SP food and drinking water containing 1% NaCl for acceleration of induction of stroke. Another group of age-matched SHRSP (n = 10) was given ordinary diet and tap water during experiments as a non-stroke control group which do not induce stroke at least up to the 10 months of age as previously reported[1]. All animals in the special diet group showed a loss of body weight and characteristic neurological symptoms including hyperkinesia between the 12 and 14 weeks of age, indicating the induction of stroke as previously reported[5]. Four animals were found to be dead immediately after the induction of stroke. Within 1–2 days after the induction of stroke, 18 animals were given ordinary food and tap water in place of the special food and salty water. Within 6–16 weeks after the induction of stroke, 10 animals recovered almost completely from the neurological symptoms, and were subjected to acquisition training on the three panel runway task, as reported previously[4]. The other 8 animals showed severe hindlimb paralysis as well as the neurological symptoms or were dead before being subjected to the acquisition training.

The three-panel runway experiments were carried out as reported previously[2] with some modifications. In the acquisition training, animals were trained with six consecutive trials per day at 2-min intervals. They could pass through only one of the three gates (correct gate) at each choice point. On the first trial of each session, the correct panel was white and the other panels were black at each choice point. Thus, the first trial reflects reference memory, in which animals were trained to discriminate the white panel from the black panel. From the second to sixth trials, all the white panels were replaced by black panels, and animal had to remember the position of correct gate from the previous trial. Thus, the 2nd to 6th trials involve working memory. The position of the correct gates were held constant within a session, but were changed from session to session. Twelve correct sequences were used, according to Furuya *et al.*[2]. The number of incorrect panel-pushes (errors) and the time from leaving the start box to receiving the food pellets (latency) were recorded in every trial. During the 30th–60th sessions, animals were subjected under anesthetized condition to T2-weighted imaging on a 4.7 T MRI system (GE Instruments). After the final 60th session, animals were sacrificed to obtain brains which were stained with hematoxylin-eosin for histological evaluation.

Results

After recovery from the neurological symptoms with stroke, SHRSP (n = 10) were subjected to acquisition training on both the reference and working memory tasks (Fig. 1). The stroke SHRSP showed a delay and an impairment of acquisition on both the memory tasks, compared to the age-matched non-stroke SHRSP (n = 10), which persisted at least for two months. On the other hand, both the stroke and non-stroke SHRSP showed almost the same latency (the time from leaving the start box to receiving the food pellets in every trial), indicating no severe motor impairment of the stroke SHRSP tested on the runway tasks (data not shown).

The stroke SHRSP were subjected to the T2-weight MRI during the 30th–60th test sessions, and the resulting brain imagings indicated a clear enhancement of the signal intensity which suggested the marked edema formation in the cortex, caudate putamen and/or thalamus, preferentially in the frontal and/or occipital cortex. The histological evaluation also showed marked edematous degeneration such as edema, gliosis and cyst preferentially in the cortex, but did not show any degeneration in the hippocampus. As shown in Table 1, the serious lesion group in the stroke SHRSP, showing marked edematous degeneration in the cortex, exhibited marked memory deficit, compared to the slight lesion group and the non-stroke control group.

Discussion

In this experiment, SHRSP with chronic stroke was found to exhibit both a delay and a persistent impairment of acquisition on the memory tasks.

Table 1. *Memory Deficit and Cortical Degeneration in SHRSP with Chronic Stroke*

	Memory deficit errors[a]	Cortical degeneration Diagnosis	Histology score[b]	MRI (%)[c]
Stroke SHRSP				
Serious lesion group (n = 5)	86 ± 17	edema, gliosis, cyst	19.5 ± 2.0	34 ± 14
Slight lesion group (n = 3)	49 ± 6	edema, gliosis	2.3 ± 0.3	1 ± 1
Non-stroke SHRSP (n = 6)	29 ± 4	non	0	0

Values are mean ± S.E.
[a] The number of errors on the runway task was summed across the 31st–40th test sessions for the 1st–6th trials.
[b] The number of degenerative regions multiplied by the severity (0 normal; 1 edema; 2 gliosis; 3 cyst) was summed across the 8 coronal brain slices.
[c] Percentage of the edematous area in whole cortical region was estimated across the 8 coronal brain slices by the T2-weighted MR imaging.

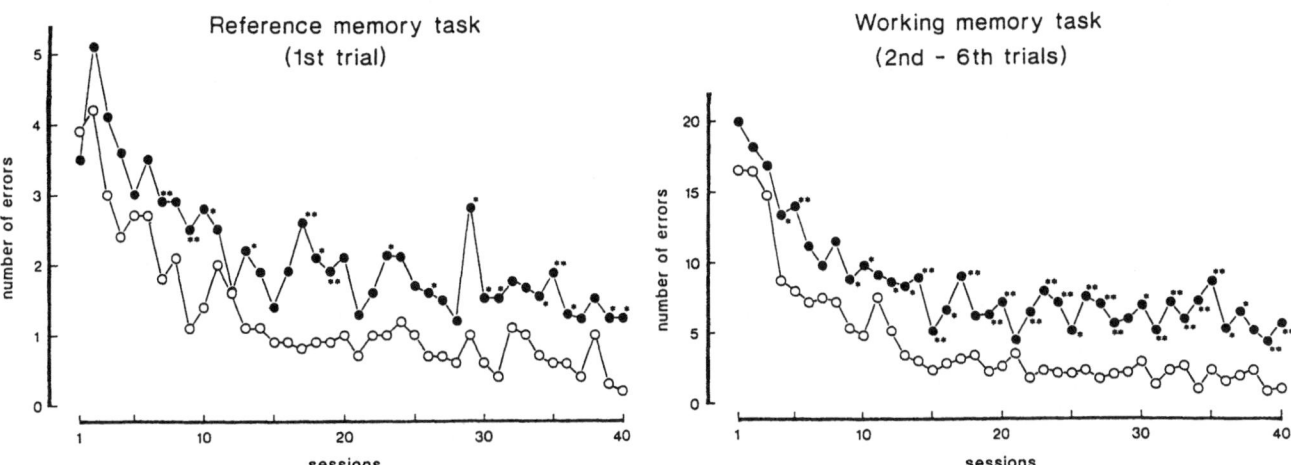

Fig. 1. Acquisition of the three panel runway task in stroke (●) and non-stroke (○) SHRSP. The left figure shows the number of errors in the first trial of each session, in which the correct panel was white and the other panels were black at each choice, in which animals were trained to discriminate the white panel from the black panel (the reference memory task). The right figure shows the number of errors from the second to sixth trials, in which all panels were replaced by black panels, and animal had to remember the position of correct gate from the previous trial (the working memory task). * p < 0. 05. **, p < 0. 01 vs non-stroke group (Student t-test)

Both the T2-weighted MRI and histological evaluation revealed marked edematous degeneration preferentially in the cortical region, but did not in the hippocampus. Thus, SHRSP with chronic stroke was found to exhibit impairment of learning and memory, which may be due to the cortical edematous degeneration. This is consistent with the cognitive disturbance in patients with chronic cerebrovascular diseases, characterized by a reduced ability to learn new information. Thus, SHRSP with chronic stroke may serve as an useful animal model to evaluate the potential nootropic drugs for the treatment of cognitive disturbance associated with chronic stroke.

In this experiment, SHRSP with chronic stroke exhibited a persistent impairment of acquisition on both the reference and working memory tasks, suggesting the impairment of the working memory as well as the impairment of acquisition on the reference memory. However, it should be noted that impairment of acquisition on reference memory may involve an impairment of acquisition on the working memory task. Our previous study with the runway task indicated that NMDA-treated rats with hippocampal CA-1 lesion may cause impairment of working memory but do not impair reference memory[4]. In our further study, the NMDA-treated rats were found to exhibit both a delay and a persistent impairment of acquisition on the working memory task but not on the reference memory task. Thus, as a conclusion, all our results on the memory performances in a rat model of chronic stroke may be derived mainly from the impairment of acquisition on the reference memory caused by the cortical edematous degeneration.

References

1. Akai T, Yamaguchi M, Tanaka T, *et al* (1992) Involvement of central serotonergic systems in enhanced startle response in SHRSP. Jpn Heart J 33: 579
2. Furuya Y, Yamamoto T, Yatsugi S, *et al* (1988) J Pharmacol 46: 183–188
3. Okamoto K. Yamori Y, Nagaoka A (1974) Establishment of the stroke-prone spontaneously hypertensive rat (SHR). Circ Res 34, 35 [Suppl 1]: 143–153
4. Sugimachi K, Izawa K, Nakamura K, *et al* (1992) Behavioral Pharmacology 3: 379–385
5. Takahashi M, Fritz-Zieroth B, Yamaguchi M, *et al* (1992) Nippon Yakurigaku Zasshi 100: 21–28
6. Yamori Y (1974) Importance of genetic factors in hypertensive cerebrovascular lesion. An evidence obtained by successive selective breeding of stroke-prone and resistant SHR. Jpn Circ J 38: 1095

Correspondence: M. Yamaguchi, M.D., Research Department, Nihon Schering KK, 2-6-64, Nichimiyahara, Yodogawa-ku, Osaka 532, Japan.

Acta Neurochir (1994) [Suppl] 60: 203–206
© Springer-Verlag 1994

Neuro-Pathophysio-Biochemical Profiles of Neonatal Asphyxia

P. Ting[1], P. Wang[2], H. Song[2], and S. Xu[1]

Departments of [1] Pediatrics and Child Health and [2] Radiology, College of Medicine, Howard University, Washington, DC, U.S.A.

Summary

Neurological and neuroelectrophysio-biochemical profiles were evaluated in newborn lambs exposed to severe temporary asphyxia. Isoelectric EEG, marked disturbances of phosphorus magnetic resonance spectrum (^{31}P-MRS), and significant brain intracellular acidosis (pHi) were noted during asphyxia. Following resuscitation, the presence of early postasphyxic blood-brain-barrier (BBB) opening was associated with a marked transient increase in intracranial pressure (ICP), a 50% neonatal mortality and a 67% incidence of severe asphyxic encephalopathy. In contrast, those lambs exposed to the same magnitude of asphyxia, but without early BBB opening experienced neither death nor severe neurological deficits. Further, these lambs showed a rapid progressive normalization of the ^{31}P-MRS and pHi, despite, the lack of EEG recovery in the first hour following resuscitation. Thus, the present study depicts that the early postasphyxic BBB disruption following temporary neonatal asphyxia is associated with poor prognosis.

Keywords: Asphyxia; blood-brain-barrier; intracranial pressure.

Introduction

Advances in perinatal medicine over the past two decades have indeed improved perinatal mortality and morbidity. However, the frequency of adverse neurological outcome following asphyxia has remained unchanged and affects approximately a third of the survivors[14]. In addition, with the increase in survivors among the very low birth weight infants, the incidence of spastic diplegia has increased recently[2]. The neuroelectrophysio-biochemical and pathological profiles associated with, and the mechanisms leading to asphyxic brain disorders are not clear. However, early transient postischemic hyperemic BBB opening has been shown to be associated with, and causally related to the progression of edema, the late necrotic BBB disruption and cerebral infarction in adult animals[3,12]. Recently, both early postasphyxic cerebral hyperemia and BBB opening were observed in newborn animals exposed to temporary asphyxia[5,9–11]. However, the sig-

nificance of the early postasphyxic BBB opening on brain edema, ICP, brain metabolism, neuroelectrophysiology and neurological function is not clear. Therefore, the present study was designed to address these pertinent issues in a temporarily asphyxiated newborn lamb model.

Materials and Methods

32 lambs less than 1 week old were fasted overnight prior to the day of an experiment. A lamb was anesthetized with 30 mg/kg IV of a-chloralose (repeated as needed up to a total of 60 mg/kg, to abolish the corneal reflexes), intubated and then placed on a respirator. Rectal temperature was kept normal between 38 and 39°C. The femoral artery and vein were cannulated. Through a midline neck incision, both common carotid arteries (CCA) were identified for later occlusion by aneurysm clips (for NMR studies, a sterile silicone rubber vascular occluder was placed around each CCA). All incisions were surgically closed; a strict aseptic technique was followed, and daily antibiotic (Flocillin) was given to lambs that survived >24 h. ICP was recorded by a miniature transducer placed in one frontal epidural space; global cerebral blood flow (CBF) was measured by standard hydrogen inhalation clearance technique, the recording platinum microelectrode was placed in the sagittal sinus. EEG was recorded from two stainless steel screws, affixed to the skull over the frontal-parietal area. All craniotomies were sealed tight with silicone rubber cement. Brain water content was determined by specific gravity (SG) method on 10 asphyxiated lambs sacrificed 1 h after resuscitation (post-R), and 7 sham-operated controls. BBB was evaluated by a 2ml/kg IV 2% Evans blue (EB) tracer given prior to asphyxia; in addition Gadolinium enhanced magnetic resonance imagings(Gd-MRI) were obtained at 1 hour, 1 and 2 day post-R. For nuclear magnetic resonance (NMR) studies, a 4.7 Tesla, 33 cm bore, multi-nuclei system was used. Both T1 and T2 weighted images (MRI) were obtained from consecutive coronal brain slices, each 2mm thick. The spatial resolution was 0.35 mm × 0.63 mm. To obtain ^{31}P-MRS, a two-turn 5cm solenoid coil was placed centrally over the vertex of a lamb's skull (with intact scalp). The spectral width was 5000 Hz, ^{31}P resonance frequency was 81 MHz, each acquisition time was 0.102 seconds, and total acquisition time for each spectrum was 5 minutes. The pHi was calculated from the chemical shift difference between Pi and PCr using Petroff's titration curve[7]. After 1 h of stabilization from surgical procedures, Pavulon 0.1 mg/kg IV was given to the lamb, and

baseline recordings were obtained. The lamb was then exposed to asphyxia by ventilation with a premixed gas mixture (5% O_2 + 10% CO_2 + 85% N_2) and occlusion of both CCA until the mean arterial blood pressure (MBP) dropped to < 25mmHg. The lamb was then immediately resuscitated by reoxygenation and release of CCA occlusions; in addition, external cardiac massage and correction of metabolic acidosis by sodium bicarbonate were instituted as medically warranted. The lambs were sacrificed by an overdose of pentobarbital at 1 h (n = 10), 24 h (n = 11) and 72 h (n = 4 for NMR) post-R. Arterial MBP, ICP, rectal temperature, arterial pH/gases, CBF, and NMR measurements were obtained before, during and after the end of asphyxia; the ^{31}P-MRS were obtained continuously during asphyxia and in the first hour following resuscitation. Neurological assessment on the level of consciousness, locomotor, sucking and swallowing functions and presence or absence of seizures was performed on 1 to 3 days post-R. Seizures were treated with phenobarbital 20 mg/kg IV.

Statistical significance (p < 0.05) was determined by t-tests (paired and unpaired) and chi-square test using the SAS program.

Results

All 25 asphyxiated lambs were successfully resuscitated, however, a 50% mortality was noted within the first 24 h post-R, in the 6 lamb with BBB opening, but in none of the 5 animals with intact BBB (Fig. 1b, p < 0.01). Asphyxiated lambs were divided into BBB + (disrupted, n = 9), and BBB- (intact, n = 16) groups; all 7 sham-operated controls had intact BBB. There were no significant differences in the MBP, duration of asphyxia and arterial blood pH/gases between the 2 groups. The respective values of BBB+ and BBB- groups obtained during asphyxia were: MBP mmHg = 18 ± 1 and 20 ± 1; pH = 7.02 ± 0.10 and 6.91 ± 0.05; pO_2 mmHg = 15 ± 1 and 17 ± 2; pCO_2 mmHg = 87 ± 3 and 86 ± 4. However, the maximum ICP observed within 20 minutes post-R was significantly higher in the BBB+ than the BBB- group (13 ± 1 and 7 ± 1 mmHg respectively, p < 0.01) (Fig. 1a). Severe asphyxic encephalopathy (stupor, quadriplegia, absent sucking-swallowing reflexes and seizures) was observed in 2 out of 3 BBB + but in none of the 9 BBB – lambs (Fig. 1c, p < 0.01). Evans-blue tracer extravasa-

tions were observed in cerebral cortex and/or basal ganglia, thalamus, hypothalamus and substance nigra. There were no significant differences in the SG values

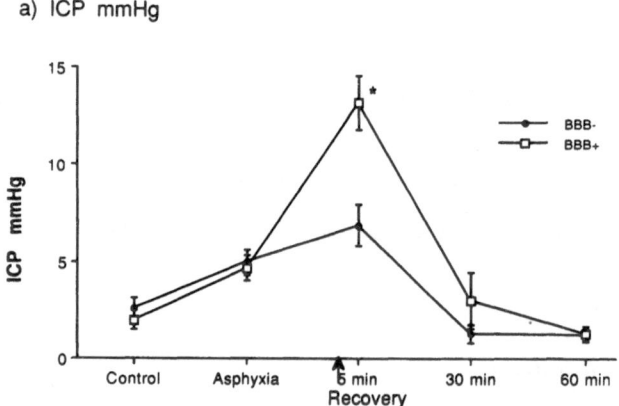

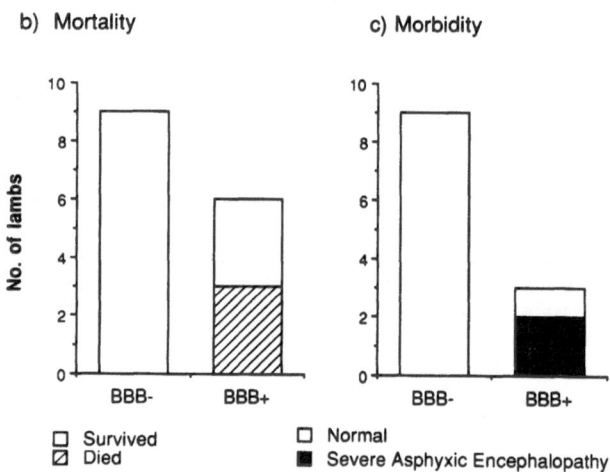

Fig. 1. Significance of early postasphxic BBB opening in temporarily asphyxiated newborn lambs. (a) ICP within the first 20 min of resuscitation was significantly higher in BBB+ than the BBB- group (p < 0.01); (b) Severe asphyxic encephalopathy was observed in 2 out 3 survivors with BBB+, but in none of the 9 animals with intact-BBB (p < 0.01). Mortality in the first 24 h following asphyxia was 50% in animals with disrupted but zero in those with intact BBB (p < 01)

Table 1. *Global Cerebral ^{31}P-MRS and pHi Data (Mean ± SE) of 4 Asphyxiated Lambs*

Variables	Baseline	Asphyxia		Recovery (min)		
		5 min	end	15	30	60
pHi	7.106 ± 0.029	6.886 ± 0.0300*	6.130 ± 0.113*	6.346 ± 0.273	7.050 ± 0.121	7.197 ± 0.120*
PCr/Pi	3.6 ± 0.3	1.4 ± 0.2*	1.0 ± 0.2*	1.6 ± 1.3	4.0 ± 1.0	3.0 ± 0.6
ΣATP	96.8 ± 13.6	102.7 ± 16.5	36.0 ± 6.0*	63.6 ± 9.9	74.7 ± 8.2	82.4 ± 8.8

Values of ATP, PCr, Pi, were peak heights for individual phosphorus-containing metabolites; ΣATP = sum of α, β, γ ATP.
* p < 0.05, compared to baseline values.

among sham-controls and asphyxiated BBB- and +group (their respective values were 1.0408 ± 0.0008, 1.0406 ± 0.0008, 1.0402 ± 0.0011). CBF was measured in 6 lambs sacrificed at 1 h post-R (no measurement during asphyxia). A significant transient hyperemia was observed at 15 minutes of post-R (Fig. 2, $p < 0.05$). EEG was recorded from 4 lambs, and it revealed slow spike waves within 1 minute of onset of asphyxia, and then became isoelectric. Upon resuscitation, incomplete EEG recovery was observed, despite the presence of an intact BBB at 1 h post-R (Fig. 2). The NMR study was done on 4 lambs, and the results were depicted in Table 1 and Fig. 2. Compared to preasphyxic baseline data, there were significant progressive decreases in brain pHi and PCr/Pi ratio starting 5 minutes of asphyxia; however, ΣATP showed an initial small insignificant increase at 5 minutes, then dropped markedly just prior to the end of asphyxia. A progressive significant improvement of the phosphorus spectra and brain pHi occurred as early as 15 minutes post-R. There was a slight but transient significant increase in the pHi to 7.197 at 1 h post-R (baseline pHi = 7.106, $p < 0.01$). Subsequent ^{31}P-MRS and pHi data of these lambs remained normal, in addition, they had normal MRI (with and without Gd), neurological function and BBB to EB.

Discussion

Exact mechanisms underlying the neonatal asphyxic encephalopathy (AE) are not clear. Following temporary cerebral asphyxia, neonates, like adults with transient cerebral ischemia, demonstrate initial postasphyxic recovery of rCBF, brain phosphorus metabolites, electrophysiological function and protein synthesis. Despite this recovery, a late unexplained deterioration ensues, with subsequent increased mortality and neurological handicaps[1,5,6,11]. The significance of early BBB opening in the first 24 h following severe temporary asphyxia on subsequent neuropathophysio-biochemical outcome is unclear. However, there is unequivocal evidence in adult animal models, that, the early postischemic BBB opening is associated with, and causally related to the progression of edema, the late necrotic BBB disruption, and cerebral infarction[3,12]. It is commonly believed that the BBB disruption is a consequence rather than the cause of ischemic asphyxic brain damage, however, this refers to the late necrotic BBB disruption, and must not be confused with the early BBB opening. In animal models one study reported that the early transient postasphyxic BBB opening in newborn piglets was associated with a significant increase in brain edema[9,10], while another

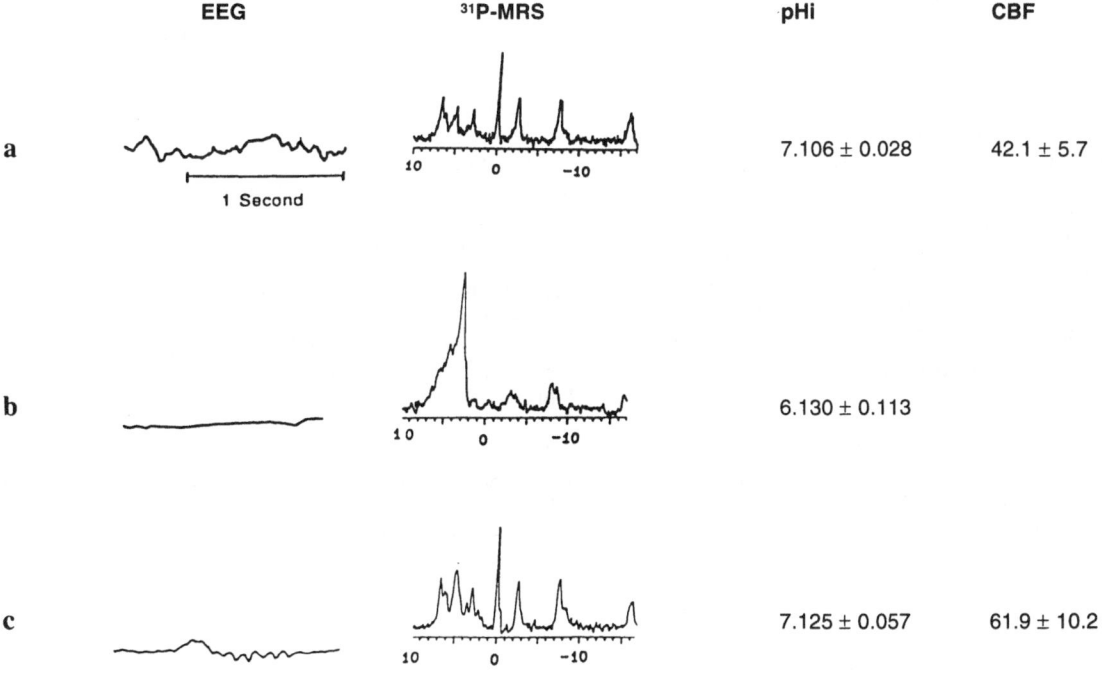

Fig. 2. Global cerebral 31 P-MRS, pHi, EEG and CBF profiles of temporarily asphyxiated lambs. (a) Baseline phosphorus peaks from right to left are β-, α-, and γ- phosphates of ATP, PCr, PDE (Phosphodiesters), Pi and PME (phosphomonoesters). (b) During asphyxia, there was marked elevation of Pi, but significant drop in PCr, ATP and pHi, and isoelectric EEG. (c) Normalization of phosphorus spectrum and pHi, significant reactive hyperemia ($p < 0.05$, paired t-test) occurred within 30 min of the resuscitation, however, the EEG recovery was minimum

study on rat pups depicted that the early postasphyxic BBB opening was associated with brain edema and neuronal necrosis[8,13]. The present study revealed that the early postasphyxic BBB opening was associated with a significant transient elevation of ICP noted shortly after resuscitation, a 50% mortality and a 67% incidence of severe AE observed 24 h after the asphyxia. The transient increase in ICP was most likely due to the increased intravascular volume (secondary to the hyperemia), rather than brain edema, because, there was no significant change in the SG value. Similar transient increased ICP was also reported elsewhere[4]. The clinical features of severe AE and increased mortality depicted in this study are very similar to those experienced by asphyxiated term human neonates[14].

Changes in EEG, brain ^{31}P-MRS and pHi during asphyxia, revealed severe energy failure and intracellular acidosis (pHi 6.1). However, a rapid progressive significant improvement in the ^{31}P-MRS and pHi was noted as early as 15 min post-R, despite, the lack of normalization of EEG. Subsequent ^{31}P-MRS and pHi of these lambs remained normal; likewise, their MRI (with or without Gd contrast), neurological function and BBB to EB tracer were normal in the first 3 days following the asphyxic insult. However, in severely asphyxiated human newborns, a normal global brain ^{31}P-MRS in the first 24 h was soon followed by a progressive deterioration of the spectrum, with subsequent increased mortality, brain atrophy and neurological handicaps[6]. Study is in progress to evaluate the effect of early postasphixic BBB opening on the phosphorus spectrum. In conclusion, the present study shows that the early postasphyxic BBB disruption in newborn lambs is associated with a significant early transient elevation of ICP, an increase in the incidence of neonatal mortality and severe asphyxic encephalopathy. However, rapid significant improvement in ^{31}P-MRS, and pHi, despite the presence of an abnormal EEG, is associated with good prognosis.

Acknowledgements

The study was funded by the United Cerebral Palsy Research and Educational Foundation Grant No. R-409-90.

References

1. Dwyer BE, Nishimura RN, Powell CL, Mailheau SL (1987) Focal protein synthesis inhibition in a model of neonatal hypoxic-ischemic brain injury. Expt Neurol 95: 277–289
2. Hagberg B, Hagberg G, Olow I, et al (1989) The changing panorama of cerebral palsy in Sweden. Acta Paediatr Scand 78: 283–290
3. Kuroiwa T, Ting P, Martinez H, Klatzo I (1985) The biphasic opening of the blood-brain barrier to proteins following temporary middle cerebral artery occlusion. Acta Neuropathol (Berl) 68: 122–129
4. McPhee AJ, Kotagal UR, Kleinman NL (1985) Cerebrovascular hemodynamics during and after recovery from acute asphyxia in the newborn dog. Pediatr Res 19: 645–650
5. Ment LR, Steward WB, Duncan CC, et al (1987) Beagle pup model of perinatal asphyxia. Stroke 18: 559–605
6. Moorcraft J, Bolas NM, Ives NK, et al (1991) Global and depth resolved phosphorus magnetic resonance spectroscopy to predict outcome after birth asphyxia. Arch Dis Child 66: 1119–1123
7. Petroff OAC, Prichard JW, Behar KL, et al (1985) Cerebral metabolism in hyper and hypocarbia: [31]P and ^{1}H nuclear magnetic resonance studies. Neurology 35: 1981–1688
8. Rice JE III, Vannucci RC, Brierley JB (1981) The influence of immaturity on hypoxic-ischemic brain damage in the rat. Ann Neurol 9: 131–141
9. Temesvari P, Joo F, Koltai M, et al (1984) Cerebroprotective effect of dexamethasone by increasing the tolerance to hypoxia and preventing brain edema in newborn piglets with experimental pneumonthorax. Neurosci Lett 49: 87–92
10. Temesvari P, Joo F, Adam G, et al (1986) Parallelism between the activation of the adenylate cyclase (AC) J in brain microvessels and the transendothelial albumin transport in newborn piglets with experimental pneumothorax (EPT) In: Kneglsein J (eds) Pharmacology of cerebral ischemia. Elsevier, New York, pp 270–274
11. Thringer K, Hrbek A, Rarlsson R, et al (1987) Postasphyxial cerebral survival in newborn sheep after treatment with oxygen free radical seavengers and a calcium antagonist. Pediatr Res 22: 62–66
12. Ting P, Masaoka H, Kuroiwa T, Wagner H, Fenton I, Klako I (1986) Influence of blood-brain barrier opening to proteins on development of post ischaemic brain injury. Neurol Res 8: 146–151
13. Vannucci RC (1985) Pathogenesis of perinatal hypoxic-ischemic brain damage. In: Thompson RA, Green JR, Johnsen SD (eds) Perinatal Neurology and neurosurgery. Spectrum, New York, pp 17–39
14. Volpe JJ (1987) Hypoxic-ischemic encephalopathy: clinical aspects. In: Volpe JJ (ed) Neurology of the newborn. Saunders, Philadelphia, pp 236–279

Correspondence: Pauline Ting, M. D., Department of Pediatrics and Child Health, Howard University, 2041 Georgia Ave N. W., Washington, DC, 20060, U.S.A.

Acta Neurochir (1994) [Suppl] 60: 207–210
© Springer-Verlag 1994

Magnetic Resonance Diffusion-Weighted Imaging: Sensitivity and Apparent Diffusion Constant in Stroke

S. C. Jones, A. D. Perez-Trepichio, M. Xue[1], A. J. Furlan, and **I. A. Awad**

Cerebrovascular Research Laboratory and [1] Magnetic Resonance Research Center, Cleveland Clinic Foundation, Cleveland, OH, U.S.A.

Summary

Magnetic resonance diffusion-weighted imaging (MR-DWI) is sensitive to the diffusibility of water and may offer characterization and anatomical localization of stroke leading to early tailored therapeutic intervention. We compared DWI, the apparent diffusion constant (ADC), and autoradiographic cerebral blood flow (CBF) in a model of focal cerebral ischemia in the rat.

Sprague-Dawley rats were embolized with a single silicone cylinder injected into the internal carotid artery. Both common carotids were permanently ligated. The animals were anesthetized (isoflurane in O_2), and paralyzed (gallamine). MR-DWI were obtained with a GE 4.7 T magnet (TE = 3 s, TR = 80 msec, b = 2393 · 10^{-3} mm^2/s, slice thickness 3 mm). DWI and CBF autoradiograms were compared visually. ADC was assessed in various regions, including ischemic cortex and a region homologous to ischemic cortex. Imaging times from stroke onset were 50 ± 6 min (mean ± SEM) for DWI, 185 ± 17 min for a second DWI. CBF was determined at 258 ± 15 min.

The specificity was 100% at both 50 min and 185 min, indicating that there were no false positives; in 3 animals ischemia was not present. However, the sensitivity analysis indicated that early DWI yields some false negatives: at 50 min the sensitivity was 60%. We attribute our result of low early sensitivity to small infarcts in relation to the slice thickness. Later, at 185 min, sensitivity was 100%. The first ADCs were higher than the second ADC values in ischemic cortex.

For infarcts larger than the slice thickness, early MR-DWI is highly sensitive for imaging evolving ischemia. Unlike other imaging methodologies (e.g. T_2 or CT Scan), MR-DWI will contribute to the diagnosis and treatment of early stages of cerebral ischemia.

Keywords: Magnetic resonance imaging; diffusion-weighted imaging; focal cerebral ischemia; edema; single embolic stroke; rats; apparent diffusion constant; cerebral blood flow.

Introduction

Early diagnosis and recognition of stroke is complicated by its symptomatology and pathophysiology. Clinical diagnosis can not provide a definite answer in the early stages. The changing pathophysiological situation during the early stages of ischemia has made predictions concerning ultimate outcome difficult. Angiography, anatomical imaging with computed tomography, single photon emission tomography, positron emission tomography, or T2-weighted magnetic resonance imaging (MRI) all have limited utility or applicability for the very early diagnosis of ischemic stroke.

Recently, MR diffusion-weighted imaging (DWI) has been used to visualize parenchymal changes in ischemic brain within one hour from vascular occlusion. This new MR modality is sensitized to the diffusion of water protons and provides an image of the apparent diffusion constant (ADC). In ischemic regions, the DWI is more intense and the ADC is less intense. We compared the distribution of DWI with cerebral blood flow (CBF) autoradiography in an experimental model of focal ischemia in the rat to define the time course of DWI intensity over the evolution of stroke and the sensitivity of stroke detection.

Methods

Sprague-Dawley rats weighing between 300–350 grams were anesthetized with 2.5% isoflurane, tracheotomized, and both femoral arteries and both femoral veins were cannulated with PE-50 polyethylene catheters. Lidocaine gel (2%) were used at incision sites. Animals were placed on a respirator, ventilated with 0.5% isoflurane, 70% N_2O, and 30% O_2, and heparinized (100 IU/kg). Body temperature was maintained at 37°C with a servo-controlled heating lamp and a rectal thermistor. Mean arterial blood pressure (MABP) was monitored from a femoral artery with a strain gauge transducer. MABP and end tidal CO_2, as determined with an infrared CO_2 analyzer, were continuously recorded on a polygraph. Animals were stabilized with PaO_2 > 90 mmHg. $PaCO_2$ was maintained between 33–37 mmHg by adjusting the respiratory rate. Isoflurane was maintained at 0.2–0.5% after surgery. An ABL-3 blood gas analyzer (Radiometer) was used to measure arterial blood gases (PaO_2, $PaCO_2$), pH and hemoglobin concentrationy [Hb].

We used a silicone cylinder embolization model[14,17] for arterial occlusion. This model minimizes tissue damage, avoids craniotomy, and possesses inherent variability, an ideal feature for investigating sensitivity-specificity. Under an operating microscope the left internal carotid was exposed and the pterygopalatine branch electrocauterized and sectioned[3]. The left common carotid was temporarily clamped for no longer than 8 min and a silicone cylinder (1 mm long, 300 μm diameter) was retrogradely infused via the external carotid into the internal carotid with 0.80 ml of normal saline solution. The total surgery time was less than 90 min. After placement into the MR unit, the animals were paralyzed (gallamine) and MABP was continuously monitored.

A General Electric CSI 4.7 Tesla unit with a 15 cm diameter bore equipped with GE Acustar self-shielded gradient coils was used for MR studies. A modified ^{1}H slotted tube resonator coil of 35 mm diameter was used[18].

Diffusion-weighted images with TR = 3 s and TE = 80 ms with 10 ms diffusion gradient pulses (G_m = 12 G/cm) and a 20 ms gradient pulse separation time, were collected for 15 min. The slice thickness was 3 mm. A T2-weighted image was then collected for 15 min with the same spin echo sequence with no diffusion gradient. A final DWI was collected.

An autoradiographic CBF study was performed after the MR sequence. The tissue equilibration technique was used with ^{14}C-iodoantipyrine (^{14}C-IAP, 100 μCi/kg)[15]. A background arterial blood sample was taken. An infusion pump containing a syringe attached to a femoral venous line was used to infuse ^{14}C-IAP in 0.8 ml saline over 45 seconds. Blood samples were taken every 2–3 sec and then counted for ^{14}C using liquid scintillation counting. After decapitation, the brain was removed and then immediately frozen in chlorodifluoromethane (–44°C). The brain was sectioned in a –20°C cryostat (20 μm thick). Every 20th section was dried on 60°C hot plate, and exposed for 5 days to X-ray film (Kodak SBS) together with 8 precalibrated ^{14}C-methyl-methacrylate standards. CBF was determined using an MCID quantitative image analysis system (Imaging Research, St. Catherines, Ontario).

Results

DWI was started within 50 ± 6 min (mean ± SEM) after stroke followed by a T2-weighted image. A final DWI was taken at 185 ± 17 min and CBF was measured after 258 ± 15 min. Four DW images and four anatomically matching CBF images for each animal were compared visually.

For the first DW image, 6 animals showed increased

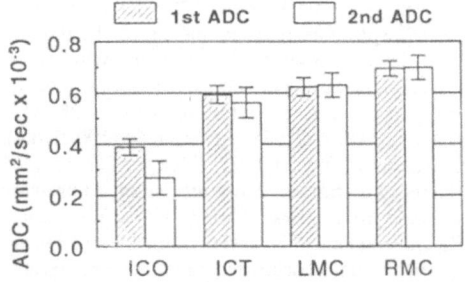

Fig. 1. ADC at 50 min and 185 min in ischemic cortex, *ICO*; homologous to ischemic cortex, *ICT*; and right and left motor cortex, *RMC* and *LMC*. The second ADC is lower in the ischemic cortex

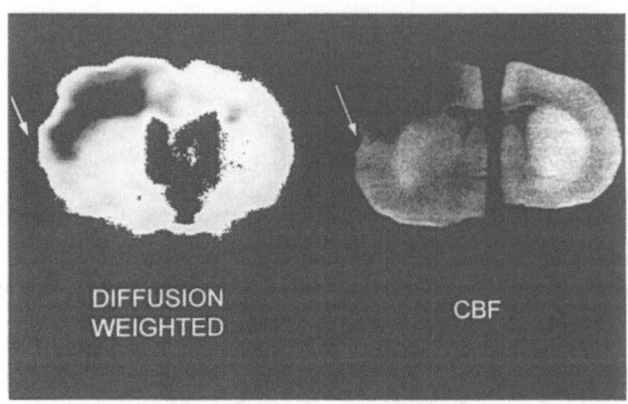

Fig. 2. Coronal images of a rat with an embolized MCl The DWI at 172 min (left) and the autoradiographic CBF image at 287 min (right) do not spatially correlate. Note that the voided area (of high DWI in this negative image, see arrow) is larger along the ventral border than the ischemic region in the CBF image

intensity in the equivalent area of low CBF, and 4 did not. For the second DW image all 10 animals showed increased signal for the corresponding areas with low CBF. In 3 animals there was no indication of stroke either in the DW images or from CBF. As shown in Table 1, the sensitivity of DWI at 50 min was 60%, and at 3 h, 100%. Specificity is 100% at both 50 min and 3 h. T2 images failed to show ischemia in all animals.

In all ischemic animals, the area of deficit from the MCA embolization in the second DWI (and ADC) images was readily visualized and corresponded with the region of ischemia in the CBF image at the same level.

The first and second ADC values from the right and left motor cortex, ischemic core and contralateral-to-ischemic core regions are shown in Fig. 1. Note that the second ADC in the ischemic region is less than that in the first ADC, but not in normal cortex. The second ADC values for the ischemic and the contralateral area were 0.27 ± 0.07 and 0.56 ± 0.06 mm²/ s · 10⁻³.

In several animals there was a region of high DW intensity at the border of the ischemic area. Figure 2 shows coronal images from a rat with an embolized MCA. The 172 min DWI image (left) does not completely correlate with the 287 min CBF autoradiographic image (right). The voided area in the DWI image (arrow) is larger at the ventral border of the ischemic region in the CBF image.

Discussion

Diffusion-weighted imaging is a relatively new MR modality that is sensitive to the translational motion of

Table 1. *Sensitivity and Specificity of DWI for Ischemic Detection*

DWI Hyperintensity	First DWI (50 ± 6 min) Focal ischemia		Second DWI (185 ± 17 min) Focal ischemia	
	yes	no	yes	no
Present	6	0	10	0
Not present	4	3	0	3
	60% sensitivity	100% specificity	100% sensitivity	100% specificity

water protons and was introduced, implemented, and refined by Le Bihan et al.[5–7]. Its role in ischemia is based on the observation that the diffusional rate of water in ischemia is much lower than that of normal brain. The ADC can be obtained from the acquired T2- and diffusion-weighted images. The ADC image provides quantitation to diffusion-weighted imaging. It is calculated, on pixel-by pixel basis, from two images, one with finite diffusion weighting and another with no diffusion weighting[5]. In agreement with our results, ADC values in normal and ischemic brain have been estimated to be $0.7–0.8 \cdot 10^{-3}$ mm^2/ s and $0.2–0.4 \cdot 10^{-3}$ mm^2/ s, respectively[11].

One study demonstrates that DWI might not be completely sensitive in the very early stages of ischemic stroke. Our results, presented in Table 1, show that DWI is not 100% sensitive at 50 min after arterial occlusion. This is in contrast to reports from other workers that DWI imaging is extremely sensitive for early stroke, although sensitivity was not measured [1,9,10,12,13]. The most likely explanation is that small regions of ischemia were not detected early due to the partial volume effect[8].

Several theories have been advanced to explain the sensitivity of DWI to cerebral ischemia[4]. The most likely reason for the sensitivity of DWI to early ischemia is that most of the freely diffusable water in normal brain is extracellular, whereas in ischemia, cytotoxic edema moves water into the cell, where its diffusion is restricted. In normal brain, the Na$^+$/K$^+$ pump maintains a large space of extracellular water, which has a higher diffusional constant than intracellular water. In ischemia, the pump is disabled and the extracellular space is decreased[16]. The evidence supporting this conjecture is based on an experiment where ouabain, which disables the Na$^+$/K$^+$ pump, was administered intraparenchymally, decreasing the ADC from $0.84 \cdot 10^{-3}$ mm^2 / s to $0.46 \cdot 10^{-3}$ mm^2 /s [2]. The initial fall of ADC after ischemia[4,11,12] certainly is con-

sistent with the initial decrease in extracellular space after arterial occlusion[16]. Thus the rapid initial decrease in ADC during early cerebral ischemia is most probably due to the increased intracellular water characteristic of cytotoxic edema. In Fig. 1 the second ADC in the ischemic cortex is slightly less than the first, suggesting that the ongoing pathologic process is changing membrane permeability or the distribution of water between intra and extracellular spaces, or that edema is progressing from the cytotoxic to the vasogenic stage.

Although the first impression from our data is that the ischemic region is equally well visualized using either DWI or CBF, the differences in the DWI and CBF noted in Fig. 2 challenge this interpretation. Instead, the area of high DWI intensity on the border of the ischemic core suggests an area of intense cytotoxic edema or other process that causes high DWI intensity, presumably associated with the ischemic penumbra.

These results indicate that very early DWI after stroke can provide delineation of areas with impaired CBF earlier and more sensitively than T2-weighted images and might lead to information concerning the ischemia penumbra.

Acknowledgements

This work was supported in part by grants from the American Heart Association – National (91013150) and the American Heart Association – Northeast Ohio Affiliate (#F244 to A D.P.-T.).

References

1. Baker LL, Kucharczyk J, Sevick RJ, Mintorovitch J, Moseley ME (1991) Recent advances in MR imaging/spectroscopy of cerebral ischemia. Am J Roentgenol 156: 1133–1143
2. Benveniste H, Hedlund LW, Johnson GA (1992) Mechanism of detection of acute cerebral ischemia in rats by diffusion-weighted magnetic resonance microscopy. Stroke 23: 746–754
3. Kaneko D, Nakamura N, Ogawa T (1985) Cerebral infarction in rats using homologous blood emboli: development of a new experimental model. Stroke 16: 76–84

4. Knight RA, Ordidge RJ, Halpern JA, Chopp M, Rodolosi LC, Peck D (1991) Temporal evolution of ischemic damage in rat brain measured by proton nuclear magnetic resonance imaging. Stroke 22: 802–808

5. Le Bihan D, Breton E, Lallemand D, Aubin ML, Vignaud J, Laval-Jeantet M (1988) Separation of diffusion and perfusion in intravoxel incoherent motion MR imaging. Radiology 168: 497–505

6. Le Bihan D, Breton E, Lallemand D, Grenier P, Cabanis E, Laval-Jeantet M (1986) MR imaging of intravoxel incoherent motions: application to diffusion and perfusion in neurologic disorders. Radiology 161: 401–407

7. Le Bihan D, Moonen CT, van Zijl PC, Pekar J, DesPres D (1991) Measuring random microscopic motion of water in tissues with MR imaging: a cat brain study. J Comput Assist Tomogr 15: 19–25

8. Mazziotta JC, Phelps ME, Plummer D, Kuhl DE (1981) Quantitation in positron emission computed tomography: 5. physical-anatomical effects. J Comput Assist Tomogr 5: 734–743

9. Minematsu K, Fisher M, Sotak CH, Davis MA, Fiandaca MS (1992) Diffusion-weighted magnetic resonance imaging: rapid and quantitative detection of focal brain ischemia. Neurology 42: 235–240

10. Mintorovitch J, Cohen Y, Chileuitt L, Shimizu H, Weinstein P, Moseley ME (1989) Early detection of regional cerebral ischemia and the effect of reperfusion on diffusion-weighted MRI in rats. In: Works in Progress – Society of Magnetic Resonance in Medicine. Society of Magnetic Resonance in Medicine, Berkeley, pp 1002

11. Mintorovitch J, Moseley ME, Chileuitt L, Shimizu H, Cohen Y, Weinstein PR (1991) Comparison of diffusion- and T2-weighted MRI for the early detection of cerebral ischemia and reperfusion in rats. Magn Reson Med 18: 39–50

12. Moseley ME, Cohen Y, Mintorovitch J, Chileuitt L, Shimizu H, Kucharczyk J, Wendland MF, Weinstein PR (1990) Early detection of regional cerebral ischemia in cats: comparison of diffusion- and T2-weighted MRI and spectroscopy. Magn Reson Med 14: 330–346

13. Moseley ME, Kucharczyk J, Mintorovitch J, Cohen Y, Kurhanewicz J, Derugin N, Asgari H, Norman D (1990) Diffusion-weighted MR imaging of acute stroke: correlation with T2-weighted and magnetic susceptibility-enhanced MR imaging in cats. AJNR 11: 423–429

14. Perez-Trepichio AD, Furlan AJ, Little JR, Jones SC (1992) Hydroxyethyl starch 200/0.5 reduces infarct volume following embolic stroke in the rat. Stroke 23: 1782–1790

15. Sakurada O, Kennedy C, Jehle J, Brown JD, Carbin GL, Sokoloff L (1978) Measurement of local cerebral blood flow with iodo[14C]antipyrine. Am J Physiol 234: H59–H66

16. Schuier FJ, Hossmann K-A (1980) Experimental brain infarcts in cats. II. Ischemic brain edema. Stroke 11: 593–601

17. Takeda T, Shima T, Okada Y, Matsumura S, Nishi Y, Uozumi T (1987) Pathophysiological studies of cerebral ischemia produced by silicone cylinder embolization in rats. J Cereb Blood Flow Metab 7 [Suppl 1]: S66

18. Xue M, Ng TC, Jones SC, Perez-Trepichio AD, Modic M (1991) 1H NMR spectroscopic imaging with very small voxel size for the detection of metabolic injury of stroke in rat brain. In: Society of Magnetic Resonance in Medicine, 10th Annual Meeting, Book of abstracts.

Correspondence: Stephen C. Jones, Ph. D., Department of Neurosciences, Cleveland Clinic Foundation, Cerebrovascular Research Laboratory, NC3-104, 9500 Euclid Avenue, Cleveland, OH 44195-5286, U.S.A.

Acta Neurochir (1994) [Suppl] 60: 211–215
© Springer-Verlag 1994

Mechanisms of Brain Injury Associated with Partial and Complete Occlusion of the MCA in Cat

Z. S. Vexler[1], **T. P. L. Roberts**[1], **N. Derugin**[1], **E. Kozniewska**[1], **A. I. Arieff**[2], and **J. Kucharczyk**[1]

[1] Neuroradiology Section, University of California, San Francisco and [2] Department of Medicine, Geriatrics Section, Veteran's Affairs Medical Center and University of California, San Francisco, CA, U.S.A.

Summary

High-speed MR diffusion/perfusion imaging was performed to assess variable degree stenosis of the MCA and the formation of cytotoxic edema in a cat model of acute ischemia. Sodium transport was estimated in synaptosomes isolated from moderately perfused or non-perfused brain tissue. Complete MCA occlusion for 50–75 min produced a major disruption of brain sodium transport, whereas continued preservation of ion homeostasis and the activation of adaptive cell volume regulatory systems was associated with longer duration of moderate severity of ischemia. Preservation of neuronal ion homeostasis might be one of the main mechanisms contributing to the relative tolerance of the brain to moderate reductions in cerebral blood flow.

Keywords: Ischemia; partial stenosis; synaptosomes; sodium transport.

Introduction

Neuronal energy failure and metabolic dysfunction following complete occlusion of the middle cerebral artery (MCA) results in brain infarction. However, incomplete ischemia has less predictable consequences. It has been suggested that maintenance of cerebral blood flow (CBF) above a critical threshold may account for reported tolerance of neurons to partial ischemia for long periods of time without irreversible brain damage. The contribution of ion homeostasis to neuronal tolerance is poorly understood.

Metabolic changes that are associated with reduced CBF have been extensively studied in different animal models by several groups[1,3,7,12,13,17,18,25,26]. The main focus of most of the studies was the energy supply, levels of high energy phosphates (ATP, PCr), intracellular pH and lactate.

The main objective of this study was to compare neuronal ion membrane transport in synaptosomes from acutely ischemic and partially perfused brain. Perfusion/diffusion-sensitive MRI was used in a cat model of acute cerebral ischemia to monitor variable degree stenosis of the MCA and the formation of cytotoxic edema, and to dynamically evaluate brain injury in relation to cell sodium transport in brain synaptosomes.

Methods

Animal models: Ten young adult female cats weighing 2.5–4.0 kg were intravenously anesthetized with 35 mg/kg sodium pentobarbital. Femoral artery and vein catheters were induced for continuous arterial blood pressure and blood gases monitoring and i.v. administration of the magnetic susceptibility contrast agent. The cats were then placed in the MRI unit in order to perform diffusion-weighted and contrast-enhanced echo-planar images (EPI). Animals were then removed from the magnet and underwent complete or incomplete suture occlusion of the MCA[20]. The suture was placed around the MCA just proximal to the origin of the lateral striate arteries. In four cats complete suture occlusion of the right MCA was induced for 50–75 min. Another group of 6 animals underwent incomplete MCA occlusion for 105–130 min. Immediately after surgery the cats were placed in the magnet, and were maintained with 1–2% isoflurane in 30% oxygen/70% nitrous oxide, and mechanically ventilated through an endotracheal tube. At the end of MR imaging the brains were quickly removed from the cranium. Regions-of-interest (ROI) were identified on the basis of observed perfusion deficits on MRI perfusion-sensitive images. Tissue samples were dissected from these regions and from the corresponding areas of the contralateral hemisphere in each individual animal and immediately placed in the isolation media for processing of synaptosomes.

Magnetic resonance imaging protocol: Imaging experiments were performed on a 2 Tesla Omega CSI system (Bruker Instruments, Fremont, CA) equipped with Acustar S-150 self shielded gradients (20 gauss/cm, 15 cm i.d.). A coronal imaging slice was selected at the level of the caudate nuclei.

Sixteen contrast-enhanced spin-echo echo-planar images (echo time 80 ms) were acquired at 2 s intervals following bolus intravenous injection of 0.25 mmol/kg of a magnetic susceptibility contrast agent, DyDTPA-BMA (Sprodiamide injection, Nycomed Salutar, Inc. and Sterling-Winthrop). Signal intensity values were obtained from cortical gray matter of the ischemic hemisphere and corresponding regions of the contralateral hemisphere. The data were expressed as normalized first-pass signal intensity/time curves for the regions-of-interest. Signal intensity versus time curves were then used to construct $\Delta R2^*$ curves, according to the formula $\Delta R2^*(t) = -\log(S(t) / S(0)) / TE$, where $S(t)$ and $S(0)$ are signal intensity in time t and 0. Calculation of relative regional cerebral blood volume was carried out based on the assumption that the peak $\Delta R2^*$ effect was proportional to the regional concentration of magnetic susceptibility contrast agent[15,21]. The ratio of peak $\Delta R2^*$ effects in ischemic to anatomically matched tissue was used as an indicator of the integrity of cerebral perfusion. Stejskal-Tanner diffusion sensitizing gradients of strengths up to 11 gauss/cm were employed to obtain corresponding diffusion-weighted echo-planar images with "b-values" in the range 0–2442 s/mm². Spatial "maps" of the apparent diffusion coefficient (ADC) were generated by pixel-by-pixel logarithmic regression analysis of a series of diffusion-weighted echo-planar images, assuming an exponential loss of signal dependent on the product of the ADC and the image "b-value"[24], $S \sim \exp(-bD)$, where D represents the ADC.

Isolation of synaptosomes: Synaptosomes were prepared as previously described from our laboratory[6], but using 7.2% and 11.5% Ficoll gradients.

Transport activity of Na+-K+-ATPase: Rubidium uptake with or without ouabain[5] was used for the estimation of potassium transport via Na+-K+-ATPase in this study. The aliquots of synaptosomes were preloaded with Na+ [(in mM) 150 NaCl, 1 MgCl₂, 5 HEPES-Tris, pH = 7.401 for 10 min. at 37°C. The suspension was then centrifuged at 18,000 g for 6 min., and the pellet was resuspended in

400–500 µl of preequilibrium media[5,6]. The reaction of the ^{86}Rb uptake was initiated with the addition of 5 µl of synaptosome suspension (~20–30 µg protein) to 95 µl uptake media [(in mM) 140 choline chloride, 5 MgCl₂, 0.2 EGTA, 1 KCl, 5 HEPES-Tris, pH = 7.40, 0.5 µCi [86]Rb 70,000 cpm] at 25°C. After 5 min, uptake was stopped by adding 2.5 ml ice-cold choline chloride (150 mM). The reaction mixture was immediately vacuum filtered through a 0.45 µm pore size cellulose acetate membrane. The zero time point Rb uptake (non specific uptake) was obtained by adding 5 µl protein suspension to 95 µl uptake media combined with 2.5 ml ice-cold choline chloride (150 mM). Each probe was done in triplicate. The same procedure was performed using 2.5 mM ouabain in the uptake media. The mean ^{86}Rb uptake with ouabain was subtracted from ^{86}Rb uptake without ouabain for each individual experiment. Protein concentration was determined using the Lowry method[14] in synaptosome suspension following each ion transport study.

Veratridine-stimulated sodium uptake: Sodium uptake with or without veratridine HCl (50 µM) was estimated in synaptosomes preloaded with potassium [(in mM) 150 KCl, 1 MgCl₂, 5 HEPES-Tris, pH = 7.40][5,6] for 10 min. at 37°C. The uptake procedure was described in the previous section. The uptake media contained (in mM) 140 choline chloride, 5 MgCl₂, 0.2 EGTA, 1 NaCl, 5 HEPES-Tris, pH = 7.40, 5.0 µCi [22]Na ~ 5 × 10⁵ cpm. Mean of the [22]Na uptake without veratridine was subtracted from uptake with veratridine for each individual experiment.

Amiloride-sensitive sodium uptake: Sodium uptake was estimated in synaptosomes with or without amiloride (1 mM)[6,9,19]. The uptake procedure was as described above. Preequilibrium media contained (in mM) 150 KCl, 1 MgCl₂, 5 HEPES-Tris, pH = 7.40. The uptake media contained (in mM) 140 choline chloride, 5 MgCl₂, 0.2 EGTA, 1 NaCl, 1 × 10⁻³ tetrodotoxin, 5 HEPES-Tris,

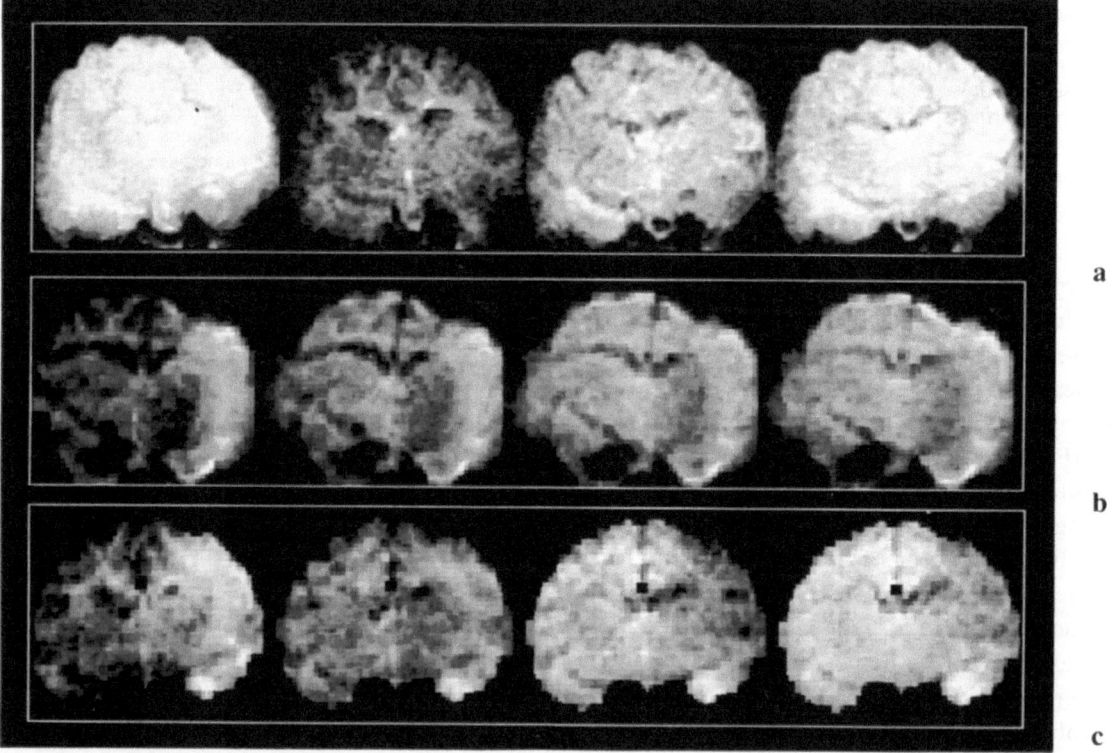

a

b

c

Fig. 1. MR imaging, of perfusion deficits associated with complete and incomplete unilateral MCA occlusion. Four of the 16 high speed echo-planar images of representative brains (pre-contrast, and three post-contrast) following bolus intravenous injection of 0.25 mmol/kg of a magnetic susceptibility contrast agent DyDTPA-BMA in normal brain (a); in the brains following unilateral complete (b) or partial occlusion of the MCA (c)

pH = 7.40, 5.0 µCi ^{22}Na ~ 5 × 10^5 cpm. The ^{22}Na uptake with amiloride was subtracted from uptake without amiloride for individual experiments.

Statistical analysis: A two-tailed t-test was used for the statistical comparisons between the groups. All data are presented as % of mean in occluded hemisphere compared to control.

Results

Detection of Perfusion Deficits and Changes in the ADC

Diffusion-weighted EPI images followed by contrast-enhanced EPI were collected in normal animals. Representative contrast-enhanced EPI are presented in Fig. 1a. Symmetrical loss of signal intensity in both hemispheres indicate normal perfusion. No significant differences in the ADC values between left and right hemispheres were found (not shown). Thirty minutes after complete MCA occlusion, there was a significant decrease of the ADC by 35% (p < 0.05) in the vascular territory of the occluded MCA. Contrast-enhanced EPI acquired immediately after the diffusion-sensitive sequence delineated perfusion deficits in the cortex and caudate nuclei ipsilateral to the occlusion. ΔR2* curves for a representative animal are shown in Fig. 1b. The peak ΔR2* was 16.9 ± 4.0% of that in corresponding area of the non-occluded hemisphere (p < 0.001).

After partial MCA stenosis, contrast-enhanced EPI showed moderate hypoperfusion of the MCA vascular territory Fig. 1c. The peak ΔR2* was 57.2 ± 7.1% of that in corresponding areas of non-occluded hemisphere. This ratio of peak ΔR2* was statistically different from both non-occluded control and complete MCA occlusion (p < 0.001). There were no significant changes in the ratio of peak ΔR2* over time. No significant changes of the ADC (92%) from control were found in this group.

Sodium Uptake in Synaptosomes

In order to assess changes in neuronal ion transport corresponding to different degrees of reduction in cerebral blood flow we isolated synaptosomes from the regions with decreased peak ΔR2* and from corresponding areas in the contralateral hemisphere. We estimated sodium transport via three different pathways: the Na$^+$-K$^+$ ATPase; veratridine-stimulated sodium channels; and Na$^+$/H$^+$ antiporter.

After 1 h of complete MCA occlusion the transport activity of Na$^+$, K$^+$-ATPase decreased by 39% in ischemic region compared with normal hemisphere

(p < 0.001). In contrast, no significant changes were found after 2 h of moderate stenosis (102%).

Sodium uptake via veratridine-stimulated Na$^+$ channels in synaptosomes isolated from ischemic tissue was 59% of that in synaptosomes from contralateral hemisphere (p < 0.01). There was a significantly higher sodium uptake via veratridine-stimulated Na$^+$ channels for partial (162%) as compared to complete occlusion (p < 0.05), but no statistical difference between partial stenosis and controls.

A tendency towards a decrease in Na$^+$ uptake via Na$^+$/H$^+$ antiporter was seen in the ischemic group (54% of control), although the values were not significantly different. However, partial stenosis resulted in an increase in Na$^+$ uptake (133% of control, p < 0.1). Changes of sodium transport through this pathway after complete occlusion and partial MCA stenosis were significantly different (p < 0.002).

Discussion

In this study, we used bolus transit contrast-enhanced high-speed MRI to assess changes in cerebral perfusion. An estimate of relative changes in cerebral perfusion in the vascular territory of partially/completely occluded MCA compared to normal was carried out by assuming that the regional concentration of magnetic susceptibility contrast agent[15,21] (DyDPTA-BMA) was proportional to relative regional cerebral blood volume. Perfusion deficits following complete MCA occlusion were also associated with a significant decrease of the ADC. A decreased ADC, which has been suggested to be a marker of cytotoxic edema[15,16,22], was the earliest MRI indication of brain edema.

A decrease in sodium transport through veratridine-stimulated sodium channels and Na$^+$, K$^+$-ATPase was found in synaptosomes isolated from ischemic brain tissue. This disruption of ion transport is likely to be the consequence of energy failure following complete arterial occlusion. Previous observations from our laboratory[10,11] and others[2,12,18,23] found almost immediate decreases in the levels of high energy phosphates in acutely ischemic brain tissue. Severe hypoperfusion may also result in failure to maintain transmembrane potential and lead to diminishing transmembrane ion gradients[27].

A degree of moderate MCA stenosis was delineated with contrast-enhanced high-speed MRI following incomplete occlusion. The degree of reduction of cerebral blood supply varied within the MCA vascular territory, being more pronounced in tissues exclusively supplied by the MCA and less pronounced in areas of collateral blood supply. No significant progression in perfusion changes over time was found. In agreement with our previous observations[20], moderate hypoperfusion was not accompanied by any significant decrease in the ADC. Moreover, a different profile of neuronal metabolic changes was found in vesicles from partially perfused brain tissue. Two hours of partial cerebral ischemia did not diminish Na uptake through sodium channels and Na$^+$, K$^+$-ATPase. These data suggest that an adequate energy supply was available for the maintenance of neuronal transmembrane potential and Na$^+$, K$^+$-ATPase in moderately hypoperfused brain tissue. Measurements of the high-energy phosphates levels in brain tissue following different degrees of reductions of CBF (using nuclear magnetic resonance spectroscopy), showed minimal changes in ATP level at CBF levels above 30–50% of normal[3,4,18]. However, depletion of PCr[18] was found after moderate reductions of the CBF.

A tendency towards activation of proton extrusion via the Na$^+$/H$^+$ antiporter, one of the key systems in cell pH and volume regulation[8, 9], was seen in synaptosomes isolated from partially perfused cortex suggesting intracellular acidosis. However, our data do not elucidate the origin or possible causes of the neuronal acidosis. The possibility of lower rates of ATP regeneration (in oxidative phosphorylation) in mitochondria, and regeneration of ATP in part by glycolysis can not be excluded. Glycolysis, in turn, is associated with lower tissue pH. At the same time, activation of proton extrusion from neurons suggests maintenance of metabolic status within the range of autoregulation.

In summary, high-speed MR diffusion/perfusion imaging can differentiate partial from total cerebral ischemia. Complete MCA occlusion for 50–75 min produced a major disruption of brain sodium transport, whereas continued preservation of ion homeostasis and the activation of adaptive cell volume regulatory systems was associated with longer duration of moderate severity of ischemia. Preservation of neuronal ion homeostasis seen in this study might be one of the main mechanisms contributing to the greater tolerance of the brain to moderate reductions in CBF.

References

1. Astrup J, Symon L, *et al* (1977) Cortical evoked potential and exracellular K$^+$ and H$^+$ at critical levels of brain ischemia. Stroke 8: 51–57

2. Behar KL, Rothman DL, Hossman KA (1989) NMR spectroscopic investigation of the recovery of energy and acid-base homeostasis in the cat brain after prolonged ischemia. J Cereb Blood Flow Metab 9: 655–665

3. Crockard HA, Gadian DG, *et al* (1987) Acute cerebral ischemia: concurrent changes in cerebral blood flow, energy metabolites, pH, and lactate measured with hydrogen clearance and ^{31}P and ^{1}H nuclear magnetic resonance spectroscopy. II. Changes during ischaemia. J Cereb Blood Flow Metab 7: 394–402

4. Eklof B, Siesjo BK (1972) The effect of bilateral carotid artery ligation upon the blood flow and the energy state of the rat. Acta Physiol Scand 86: 155–165

5. Frasers CL, Sarnacki P (1989) Na$^+$-K$^+$ ATPase pump function in male rat brain synaptosomes is different from that of females. Am J Physiol 257(Endocrinol Metab 20): E284–E289

6. Fraser CL, Sarnacki P, Arieff AI (1985) Abnormal sodium transport in synaptosomes from brain of uremic rats. J Clin Invest 75(6): 2014–2023

7. Heiss WD (1983) Flow thresholds of functional and morfological damage of brain tissue. Stroke 14(3): 329–331

8. Jean T, Frelin C, *et al* (1985) Biochemical properties of the Na$^+$/H$^+$ exchange system in rat brain synaptosomes. J Biol Chem 260: 9678–9684

9. Kanda F, Sarnacki P, Arieff AI (1992) Atrial natriuretic peptide inhibits amiloride-sensitive sodium uptake in rat brain. Am J Physiol 263: R279–R283

10. Kucharczyk J, Chew W, *et al* (1989) Nicardipine reduces ischemic brain injury. Magnetic resonance imaging/spectroscopy study in cat. Stroke 20: 268–274

11. Kucharczyk J, Moseley ME, *et al* (1989) MRS of ischemic/hypoxic brain disease. Invest Radiol 24(12): 951–954

12. Laptook AR, Corbett JT, *et al* (1988) Alterations in cerebral blood flow and phosphorylated metabolites in piglets during and after partial ischemia. Pediatr Res 23: 206–211

13. Laptook R, Corbett JT, *et al* (1992) Blood flow and metabolism during and after repeted partial brain ischemia in neonatal piglets. Stroke 23: 380–387

14. Lowry OH, Rosenberg AL, *et al* (1951) Protein measurements with Folin-phenol reagent. J Biol Chem 193: 265–275

15. Moseley ME, Cohen Y, *et al* (1990) Early detection of regional cerebral ischemia in cats: comparison of diffusion- and T2-weighted MRI and spectroscopy. Magn Reson Med 14 (2): 330–346

16. Moseley ME, Kucharczyk J, *et al* (1990) Diffusion-weighted MRI of acute stroke. Am J Neuroradiol 11: 423–429

17. Naritomi H, Sasaki M, *et al* (1988) Flow thresholds for cerebral energy disturbance and Na$^+$-pump failure as studied by in vivo ^{31}P and ^{23}Na nuclear magnetic resonance spectroscopy. J Cereb Blood Flow Metab 8: 16–23

18. Obrenovitch TP, Garofalo O, *et al* (1988) Brain tissue concentrations of ATP, phosphocreatine, lactate, and tissue pH in relation to reduced cerebral blood flow following experimental acute middle cerebral artery occlusion. J Cereb Blood Flow Metab 8: 866–874

19. Rafalowska U, Erecinska M, Wilson DF (1980) Energy metabolism in rat brain synaptosomes from nembutal-anesthetized and nonanesthetized animals. J Neurochem 34: 1380–1386

20. Roberts TPL, Vexler ZS, *et al* (1993) High-speed MR imaging of ischemic brain injury following partial stenosis of the middle cerebral artery. J Cereb Blood Flow Metab, in press

21. Rosen BR, Belliveau JM, Vevea JM, Brady TJ (1990) Perfusion imaging with NMR contrast agents. Magn Reson Med 14: 249–265

22. Sevick RJ, Kucharczyk J, *et al* (1990) Diffusion-weighted MR imaging and T2-weighted MR Imaging in acute cerebral ischaemia: comparison and correlation with histopathology. Acta Neurochir (Wien) [Suppl] 51: 210–212

23. Siesjö BK (1978) Brain energy metabolism. Wiley, New York
24. Stejskal EO, Tanner JE (1965) Spin diffusion measurements: spin-echoes in the presence of a time-dependent field gradient. J Chem Phys 42: 288–292
25. Stron, AJ, Tomlinson BE, Venables GS, Gibson G, Hardy JA (1983) The cortical ischaemic penambra associated with occlusion of the middle cerebral artery in the cat: 2. Studies of histopathology, water content, and in vitro neurotransmitter uptake. J Cereb Blood Flow Metab 3(1): 97–107
26. Tanaka K, Greenberg JH, *et al* (1985) Regional flow-metabolism couple following middle cerebral artery occlusion in cats. J Cereb Blood Flow Metab 5: 241–252
27. Thomas RC, Meech RW (1982) Hydrogen ion currents and intracellular pH in depolarize voltage-clampted snail neurons. Nature 299: 826–828

Correspondence: John Kucharczyk, Ph. D., Neuroradiology Section, University of California at San Francisco, 505 Parnassus Ave., San Francisco, CA 94143, U.S.A.

Acta Neurochir (1994) [Suppl] 60: 216–219
© Springer-Verlag 1994

Diffusion Anisotropy of Cerebral Ischemia

T. Kajima[1], **K. Azuma**[1], **K. Itoh**[1], **R. Kagawa**[2], **K. Yamane**[2], **Y. Okada**[2], and **T. Shima**[2]

[1] Department of Radiology, Hiroshima University School of Medicine and [2] Department of Neurosurgery, Chugoku Rousai Hospital, Hiroshima, Japan

Summary

Focal cerebral ischemia was produced by occlusion of the middle cerebral artery with a silicone cylinder in Wistar rats. Diffusion-weighted echo-planar images (DW-EPIs) using the motion-probing gradient (MPG) method were acquired at 1–3 hours and 24–48 hours after occlusion. Apparent diffusion coefficients (ADCs) were calculated from these images in ischemic lesions and in normal unoccluded regions. Results were as follows. 1. Ischemic lesions could be detected on the DW-EPIs at 1 hour after occulusion. 2. The ADC of water in the brain tissue was smaller than that of free water as a result of restricted diffusion. 3. Anisotropic diffusion that probably can be attributed to the myelin sheath was observed in the normal deep white matter. 4. In the ischemic lesions, the ADC decreased rapidly within 1–3 hours after occlusion and then slightly further declined after 24–48 hours. In the ischemic deep white matter, diffusion anisotropy disappeared at 24–48 hours after occlusion. Diffusion-weighted imaging may have applications in the examination of pathophysiological mechanisms in cerebral ischemia by means of evaluation of ADC and diffusion anisotropy.

Keywords: MRI; echo-planar; diffusion; ischemia.

Introduction

Incorporation of motion-probing gradient pulses (MPG) into conventional magnetic resonance imaging sequences has resulted in a new source of image contrast based upon differences in microscopic motion of water in biological tissues[3]. In vivo diffusion-weighted MR imaging (DWI) with MPG has been used recently to characterize brain tissue abnormalities caused by cerebral ischemia[2,4,5]. By changing the direction of MPG, DWI has been also used to assess anisotropically restricted diffusion which is dependent upon the orientation of biological structures[2,6,7].

In this study, diffusion-weighted echo-planar imaging (DW-EPI), which requires less than 100 ms for one image, was conducted on rats with focal cerebral ischemia, and apparent diffusion coefficients (ADCs) in each direction were examined. The purpose of this study was to investigate early ischemic changes by evaluating the microscopic motion of water.

Materials and Methods

The main intracranial cerebral artery of 20–25 week old Wistar rats (body weight of 400–500 g) was occluded by inserting a silicone cylinder via the cervical internal carotid artery. The focal cerebral ischemia thus produced was monitored by coronal DW-EPIs obtained by means of a General Electric 2. OT CSI omega system at 1–3 hours and at 24–48 hours after occlusion. DW-EPIs were acquired with MPG incorporated in the x, y, and z axis of the magnet with duration time (δ) of 10 ms, separation time (Δ) of 28 ms, and magnitude of 0–18 G/cm, resulting in b-factor of up to 2160 s/mm². ADCs were calculated from changes in signal intensity with b-factor in the 10 sequential images of these DW-EPIs. ADC analysis was accomplished in the ischemic deep white matter, amygdaloid nucleus, and thalamus and in the corresponding no-nischemic contralateral regions. The T2- weighted images (T2-WI: TR 2000 ms, TE 100 ms) were also obtained for comparison with the DW-EPIs. ADCs of a water phantom and an acetone phantom were also obtained as controls. At the end of the MRI experiments, animals were sacrificed and the brains were rapidly removed and stored in 10% buffered formalin. The brain was sectioned coronally, stained with hematoxylin-eosin and examined microscopically.

Results

Figure 1 shows T2-WI (A) and DW-EPIs (B-F) of an ischemic rat obtained at 1.5 hours after occlusion. In the DW-EPIs, MPG was incorporated in z-axis (parallel to the nerve fibers) with b-factors increasing in order of B to F. In contrast with the marked signal attenuation with increase of b-factor in the right normal side, the ischemic temporal gray matter, amygdaloid nucleus and thalamus in the left side were identified as relative high intensity areas. No significant intensity changes were observed with the conventional T2-WI produced at the same time.

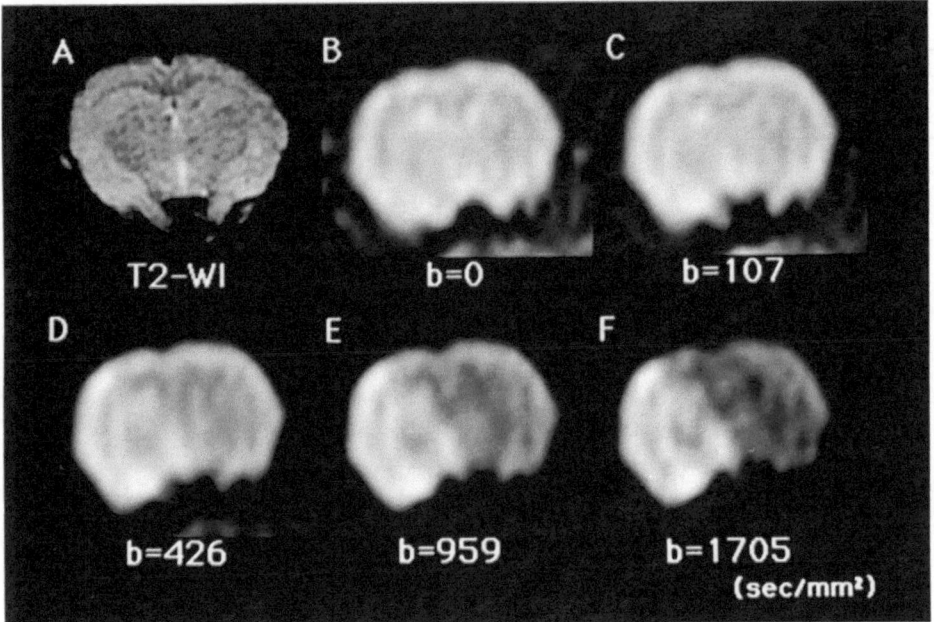

Fig. 1. (A-F) T2-WI and DW-EPIs obtained at 1.5 hours after occlusion. MPG was incorporated in Z-direction in the DW-EPIs. The ischemic lesions were identified as high intensity areas in the DW-EPIs

Changes in signal attenuation dependent on the gradient direction (anisotropic diffusion) were observed in the normal deep white matter (Fig. 2). When the gradient direction was parallel to the direction of the nerve fibers (z-direction), signal attenuation with increase of b-factor was larger than when the gradient direction was perpendicular (x-direction). ADCs were calculated from the signal attenuation curves.

Calculated ADCs are shown in the Table 1. Each

value of ADC is mean ± SD at the time in hours after occlusion. The ADCs of the phantoms were in good agreement with published values[3] without any dependency on gradient direction. The ADCs of water in the normal amygdaloid nucleus and thalamus also showed no dependency on the gradient direction. Differences of ADC dependent on the gradient direction were observed in the normal deep white matter. Anisotropic diffusion was not observed in the amygdaloid nucleus and thalamus.

ADCs decreased rapidly to about 60% of normal values within 3 hours after occlusion and then slightly further declined after 24–48 hours in the ischemic amygdaloid nucleus and thalamus. In the ischemic deep white matter, the decrease in ADC in the direction parallel to the nerve fibers showed the same tendency as that in the amygdaloid nucleus and thalamus, however, the decrease in ADC in the perpendicular direction to the nerve fibers was smaller than that in the parallel direction. Diffusion anisotropy decreased with time after occlusion and disappeared at 24–48 hours after occlusion.

Discussion

The root mean square path length of water diffusion (L) is calculated from the Einstein equation[7]:

$$L = \{2 \cdot ADC \cdot (\Delta - \sigma / 3)\}^{1/2}$$

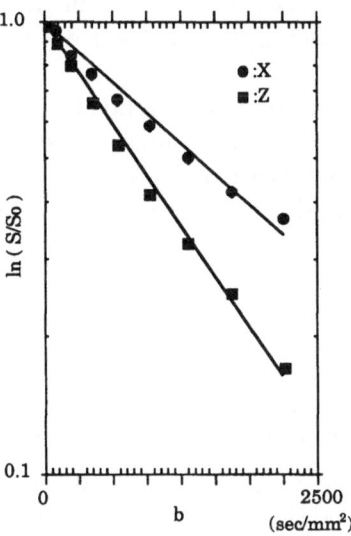

Fig. 2. Signal attenuation curve in the normal deep white matter. Anisotropic diffusion was observed

Table 1. *Calculated Values of Apparent Diffusion Coefficient (ADC)*

Subjects	Number of cases	ADC ($\times 10^{-4}$ mm²/s), direction of MPG		
		X	Y	Z
Phantoms				
water		22.9	24.5	24.6
acetone		46.6	46.6	48.9
Deep white matter				
normal	8	5.0 ± 0.3[a]	4.8 ± 0.4[a]	7.8 ± 0.8
1–3 h[b]	5	4.1 ± 0.5	3.9 ± 0.6	5.3 ± 0.8
24–48 h	6	3.8 ± 0.4	4.0 ± 0.4	4.5 ± 0.3
Amygdaloid nucleus				
normal	10	7.1 ± 0.9	7.1 ± 0.6	7.8 ± 0.4
1–3 h[b]	7	4.5 ± 1.5	4.6 ± 1.5	5.3 ± 1.4
24–48 h	4	4.3 ± 0.9	4.0 ± 0.8	4.8 ± 0.9
Thalamus				
normal	7	6.9 ± 0.5	5.7 ± 0.7	7.1 ± 1.3
1–3 h[b]	4	4.1 ± 0.7	3.9 ± 0.6	4.5 ± 0.9
24–48 h	3	3.9 ± 0.3	3.8 ± 0.4	4.3 ± 0.5

[a] ADCs in the X- and Y-direction were smaller than that in the Z-direction ($p < 0.01$) in the normal deep white matter.

[b] ADCs in the ischemic legions at 1–3 h after occlusion were smaller than that in the normal regions ($p < 0.01$).

(Δ: separation time, σ: duration time of gradient). In this experiment, $\Delta - \sigma / 3$ was 25 ms and L was 12–14 μm for free water at 37°C and 5–6 μm for water in the brain tissue. While the origin of such restricted diffusion is not totally understood, several factors like the existence of dispersed macromolecules, the cell membranes and intracellular structures[1] can be considered from judging the dimension of L. Marked restriction observed in the perpendicular direction to the nerve fibers in the normal deep white matter may suggest that the restriction is due to the myelin sheath because the diameter of myelin sheath is much shorter than the length of the axon in the long axis of the nerve fibers[2,6,7].

Various factors may be thought for the cause of decrease in the ADC in the early ischemic lesion that cannot be detected with the T2-WI. Storage of intracellular water may be a possible factor[5], however, early disruption of the blood-brain barrier (BBB) was observed using the same model of occlusion which was used in this study[8]. Exudation of plasma or ions due to change in permeability of the BBB can result in an increase of macromolecules that restricts water diffusion. Further studies are needed to understand the biological events underlying the changes of ADC in the ischemic lesions.

While no systematic report has been published describing the changes in diffusion anisotropy in the ischemic deep white matter, diffusion anisotropy decreased with time after occlusion and disappeared after 24–48 hours. The decrease in diffusion anisotropy was inferred to be the result of two-fold restriction of water diffusion: restriction due to the myelin sheath and restriction due to ischemic change. Weak diffusion anisotropy was still observed at 1–3 hours after occlusion because the decrease in ADC due to ischemic change was smaller than that due to the myelin sheath. However, at 24–48 hours after occlusion diffusion anisotropy disappeared because the decrease in ADC due to ischemic change was larger than that due to the myelin sheath.

In conclusion, DWI was useful in detection of early ischemic changes because of its contrast based on water diffusion. Observed ADCs may offer a clue to investigate the pathophysiological changes seen in ischemia.

References

1. Cooper RL, Chang DB, Young AC, Martin CJ, Ancker-Johnson BA (1974) Restricted diffusion in biophysical systems. Biophys J 14: 161–177
2. Hajnal JV, Doran M, Hall AS, Collins AG, Oatridge A, Pennock JM, Young IR, Bydder GM (1991) MR imaging of anisotropically restricted diffusion of water in the nervous system: technical, anatomic, and pathologic considerations. J Comput Assist Tomogr 15 (1): 1–18

3. Le Bihan D, Breton E, Lallemand D, Grenier P, Cabanis E, Laval-Jeantet M (1986) MR imaging of intravoxel incoherent motions: application to diffusion and perfusion in neurologic disorders. Radiology 161: 401–407

4. Mintorovitch J, Moseley ME, Chileuitt L, Shimizu H, Cohen Y, Weinstein PR (1991) Comparison of diffusion-and T2-weighted MRI for the early detection of cerebral ischemia and reperfusion in rats. Mag Res Med 18: 39–50

5. Moseley ME, Cohen Y, Mintorovitch J, Chileuitt L, Shimizu H, Kucharczyk J, Wendland MF, Weinstein PR (1990) Early detection of regional cerebral ischemia in cats: comparison of diffusion-and T2-weighted MRI and spectroscopy. Mag Res Med 14: 330–346

6. Moseley ME, KucharczykJ, Asgari HS, Norman D (1991) Anisotropy in diffusion-weighted MRI. Mag Res Med 19: 321–326

7. Sotak CH, Li L (1992) MR imaging of anisotropic and restricted diffusion by simultaneous use of spin and stimulated echoes. Mag Res Med 26 :174–183

8. Yamane K, Shima T, Okada Y, Takeda T, Uozumi T (1990) Pathophysiological studies in the rat cerebral embolization model: measurement of epidural pressure and evaluation of tissue PH and ATP. Acta Neurochir (Wien) [Suppl] 51: 223–225

Correspondence: Toshio Kajima, M. D., Department of Radiology, Hiroshima University, School of Medicine, 1-2-3 Kasumi, Minamiku, Hiroshima 734, Japan.

Acta Neurochir (1994) [Suppl] 60: 220–223
© Springer-Verlag 1994

A Sodium Magnetic Resonance Imaging Study of Acute Cerebral Ischaemia in the Gerbil

K. L. Allen, A. L. Busza[1], and **S. C. R. Williams**[2]

Institute of Neurology, [1] Royal College of Surgeons Unit of Biophysics, Institute of Child Health, and
[2] Queen Mary and Westfield College, London, U.K.

Summary

[23]Na magnetic resonance imaging has been used to investigate sodium changes during and after cerebral ischaemia in a gerbil model. The sodium signal decreased within 4 minutes of the onset of ischaemia, and subsequently increased between 4 and 8 minutes after the onset of reperfusion. These observations may be reflecting the redistribution of tissue sodium resulting from energy failure and recovery.

Keywords: Gerbil; sodium; ischaemia.

Introduction

[23]Na, which comprises virtually 100% of naturally occurring sodium, is one of the few nuclei other than proton which can be imaged at natural abundance. However, as a consequence of its lower NMR sensitivity and its much lower concentration (1/2000 that of ^{1}H), [23]Na imaging cannot approach the temporal and spatial resolution of ^{1}H imaging. In addition, the relaxation characteristics of the intra-cellular and extra-cellular compartments differ, such that intra-cellular Na is believed to be only 40% visible[1].

[23]Na imaging and spectroscopy are of great interest, and have been extensively covered in a recent review[2]. The promise of using ^{23}Na NMR to detect cellular damage and characterise changes to the neuronal sodium gradient during ischaemia, has prompted several studies using [23]Na magnetic resonance spectroscopy (MRS) to investigate such an event in the brain[3,4]. Clinically, [23]Na magnetic resonance imaging (MRI) has been used to investigate well-differentiated lesions, such as tumours, but there has been little reported to date on the application of [23]Na MRI to the study of cerebral ischaemia. Moseley *et al.* (1985) have used [23]Na MRI to investigate changes in the rat brain some hours after a focal ischaemic insult[5], but little has been done using [23]Na MRI to investigate the early, acute stages of cerebral ischaemia and reperfusion.

In this preliminary study, we have used [23]Na MRI to follow Na changes during and after cerebral ischaemia in the gerbil, with a view to seeing if the technique can be used to follow the time course of sodium changes in the initial stages of ischaemia and reperfusion.

Materials and Methods

Adult male gerbils (60–70 g, n = 6) were anaesthetised with a mixture of halothane/oxygen, and respiratory rate and body temperature (36.5–37°C, maintained by warm air) were monitored throughout the experiments. Complete cerebral ischaemia was produced by placing nylon snares around both common carotid arteries, which could be manually tightened from outside the NMR magnet[6].

Magnetic resonance imaging was performed on a standard SISCO-200 spectroscopy/imaging system in conjunction with a horizontal 4.7 T magnet, using a 2 cm diameter transmit/receive surface coil placed over the skull in the horizontal plane. [23]Na images were acquired using a refocussed gradient echo sequence incorporating a non-slice selective adiabatic 90° radiofrequency pulse (duration 1.5 ms), with imaging parameters TE = 10 ms, TR = 55 ms, field of view = 5 × 3 cm, no. averages = 256, phase encode steps = 16, imaging plane = transverse. Image acquisition time was 4 minutes. Changes in sodium intensity were calculated from regions of interest (ROI) in the cerebellum, eyes, hind brain and whole brain. The ROI's were outlined using a cursor, and the mean signal intensity within the region calculated by computer. [1]H images were acquired (as a guide to anatomical location) using a standard spin echo sequence, with imaging parameters TE = 70 ms, TR = 1000 ms, field of view = 5 × 3 cm, no. averages = 2, slice thickness 1.7 mm, phase encode steps = 256. Image acquisition time was 9 minutes.

Results

As mentioned earlier, slice selection was not used for the [23]Na images in order to maximise the signal/

noise of the inherently insensitive ^{23}Na nucleus. Panel 2 of Fig. 1 shows a typical control transaxial sodium image, showing the distribution of sodium throughout the normal brain, with the eyes clearly shown. Panels 1 and 3 of Fig. 1 are proton images acquired from the same animal, and show the area of brain which contributed most to the Na signal.

The complete sequence of events in each brain area investigated is shown in Fig. 2 (n = 6, mean ± SEM). The signal intensity is calculated as % change, with control (pre-ischaemia) values averaged and normalised at 100%. Following carotid artery occlusion, the sodium signal decreased, and this was evident by the first image, acquired over a period of 0–4 minutes after the onset of ischaemia. The ^{23}Na signal decreased between 15 and 30% depending on the anatomical region, and then remained constant for the rest of the ischaemic period (44 min). On reperfusion, the signal increased, but not immediately. The first image acquired after the onset of reperfusion was not significantly different from the ischaemia images. However, over the next 8–12 minutes the signal returned to control levels in the cerebellum and in the brain as a whole. Between the eyes and in the hindbrain there was an overshoot in the ^{23}Na signal by 12 minutes of reperfusion. The signal in the eyes increased upon reflow, but did not return to control levels for the duration of reperfusion (52 min).

Discussion

This study demonstrates that it is possible to obtain ^{23}Na images within minutes of the onset of ischaemia and reperfusion. By the end of the first image acquisition after the onset of ischaemia (4 min), the sodium signal had decreased dramatically in all areas studied. Assuming that total tissue sodium is constant, a decrease in the sodium MR signal during ischaemia implies a shift from extracellular to intracellular sodium, that is, a collapse of the transmembrane gradient. Thus, it is possible that the image changes are reflecting the development and resolution of cellular oedema. On reperfusion there is a delay of ~4 minutes before the Na signal returns to or passes control levels. This is consistent with the loss and resynthesis of ATP as seen in this model[7,8].

When acquiring sodium images, there is a compromise between using a short echo time (TE), which gives high signal/noise but low contrast, and a long TE, giving a low signal noise but higher contrast. In the case of the experiments described here a relatively long TE was used, which showed that it is possible to acquire images in a short time of sufficient quality to analyse signal intensity.

The gradient echo sequence used to obtain the images is the same type of sequence which has been used in recent studies to produce functional maps of the human brain associated with neuro-activity induced by a sen-

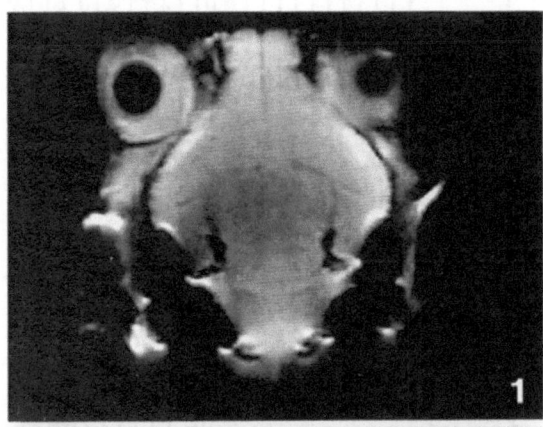

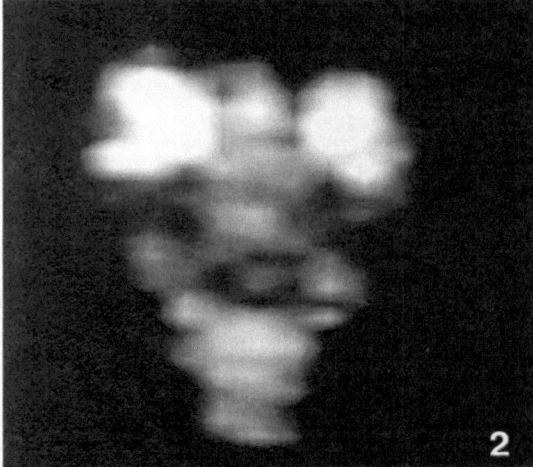

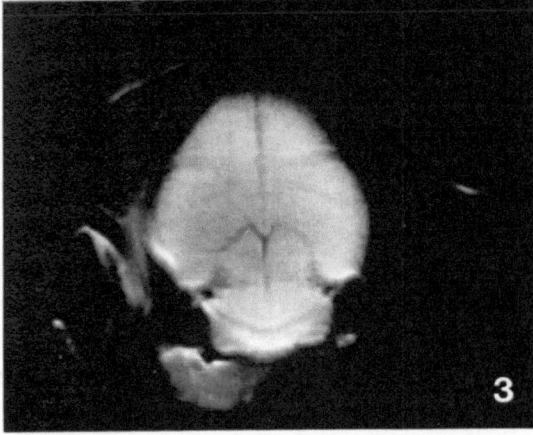

Fig. 1. (Panels 1, 3) transverse ^{1}H magnetic resonance images (slice thickness 1.7 mm), showing area of brain which contributed to the ^{23}Na signal. The ^{23}Na image is shown in panel 2

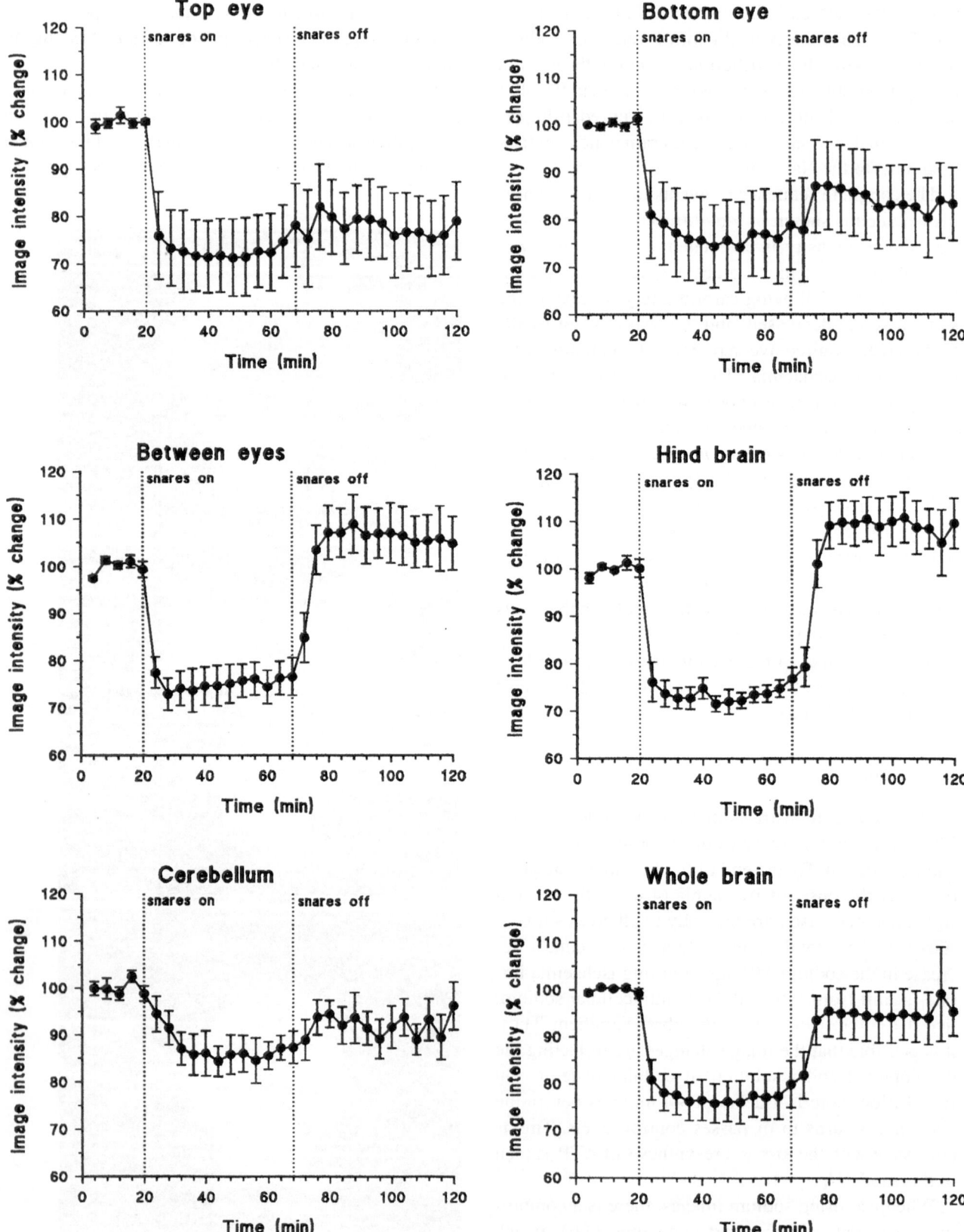

Fig. 2. Percent intensity changes in ^{23}Na signal in brain areas during and after ischaemia

sory stimulus[9]. Therefore the possibility also has to be considered that some of the changes in the images during ischaemia and reperfusion may be caused by the paramagnetic susceptibility (T_2^*) effect of deoxy-haemoglobin. Further studies will be necessary to elucidate this.

In conclusion, [23]Na MRI can detect changes in "NMR observable" sodium distribution in the brain within minutes of ischaemia and reperfusion, and may be reflecting the redistribution of tissue sodium resulting from energy failure and recovery. When allied with cerebral blood flow and/or ^{1}H and ^{31}P NMR spectroscopy measurements, we believe the technique will be a useful tool for investigating cerebral ischaemia.

Acknowledgements

We would like to thank the University of London Intercollegiate Research Service based at Queen Mary and Westfield College for providing the NMR imaging equipment used in this study. We thank Action Research and The Wellcome Trust for financial support.

References

1. Perman WH, Turski P, Houston L, Glover GH, Hayes CE (1986) Methodology of in vivo human sodium NMR imaging at 1.5 Tesla. Radiology 160: 811–820
2. Kohler SJ, Kolodny NH (1992) Sodium magnetic resonance imaging and chemical shift imaging. Progress in NMR Spectroscopy 24: 411–433
3. Eleff SM, Maruki Y, Monsein LH, Traystman RJ, Bryan N, Koehler RC (1991) Sodium, ATP, and intracellular pH transients during reversible complete ischemia of dog cerebrum. Stroke 22: 233–241
4. Naritomi H, Sasaki M, Kanashiro M, Kitani M, Sawada T (1988) Flow thresholds for cerebral energy disturbance and Na+ pump failure as studied by in vivo ^{31}P and [23]Na nuclear magnetic resonance spectroscopy. J Cereb Blood Flow Metab 8: 16–23
5. Moseley ME, Chew WM, Nishimura MC, Richards TL, Murphy-Boesch J, Young GB, Marschner TM, Pitts LH, James TL (1985) In vivo sodium-23 magnetic resonance surface coil imaging: observing experimental cerebral ischaemia in the rat. Magn Reson Imaging 3: 383–387
6. Allen KL, Busza AL, Proctor E, King MD, Williams SR, Crockard HA, Gadian DG (1993) Controllable graded cerebral ischaemia in the gerbil: Studies of cerebral blood flow and energy metabolism by hydrogen clearance and ^{31}P NMR spectroscopy. NMR in Biomedicine, in press
7. Crockard HA, Gadian DG, Frackowiak RSJ, Proctor E, Allen K, Williams SR, Ross Russell RW (1987) Acute cerebral blood flow, energy metabolites, pH, and lactate measured with hydrogen clearance and ^{31}P and ^{1}H nuclear magnetic resonance spectroscopy. II. Changes during ischaemia. J Cereb Blood Flow Metab 7: 394–402
8. Allen K, Busza AL, Crockard HA, Frackowiak RSJ, Gadian DG, Proctor E, Ross Russell RW, Williams SR (1988) Acute cerebral blood flow, energy metabolites, pH, and lactate measured with hydrogen clearance and ^{31}P and ^{1}H nuclear magnetic resonance spectroscopy. III. Changes following ischaemia. J Cereb Blood Flow Metab 8: 816–821
9. Kwong KK, Belliveau JW, Chesler DA, Goldberg IE, Weisskoff RM, Poncelet BP, Kennedy DN, Hoppel BE, Cohen MS, Turner R, Cheng H-M, Brady TJ, Rosen BR (1992) Dynamic magnetic resonance imaging of human brain activity during primary sensory stimulation. Proc Natl Acad Sci USA 89: 5675–5679

Correspondence: K. L. Allen, M. D., Department of Neurochemistry, The Institute of Neurology, Queen Square, London WC1N 3BG, U.K.

Acta Neurochir (1994) [Suppl] 60: 224–227

In vivo Differentiation of Edematous Changes After Stroke in Spontaneously Hypertensive Rats Using Diffusion Weighted MRI

M. Takahashi[1], **B. Fritz-Zieroth**[1], **T. Chikugo**[2], and **H. Ogawa**[3]

[1] Research Department, Nihon Schering, [2] Department of Pathology, and [3] Department of Hygiene,
Kinki University School of Medicine, Japan

Summary

Apparent diffusion coefficients (ADCs) of tissue water were determined in chronic brain lesions of a rat stroke model, the stroke-prone spontaneously hypertensive rat, and compared with histology. ADCs increased in the order normal < edema < gliosis < cyst. The differences between individual groups were statistically significant. The increase in ADC is thought to mainly reflect a relative increase in the extracellular space in brain tissue. ADC may be a new parameter for tissue characterization.

Keywords: Chronic stroke; brain edema; SHRSP; diffusion MRI.

Introduction

MRI is able to measure diffusion coefficients of liquids[13], and in particular that of water in tissue[16]. The observation[10] that diffusion-weighted MRI can detect lesions after stroke earlier than conventional T2 weighted images has led to renewed interest in this imaging technique. However, most of the data[5,7,8,10,17] in animals are obtained in the acute or subacute stage of stroke, when the apparent diffusion coefficients (ADC) of water is reduced.

In a previous study[15], we compared T2-weighted and enhanced T1-weighted MRI images of subacute lesions after stroke in stroke prone spontaneously hypertensive rats (SHRSP) with their histology.

In SHRSP, both infarction and hemorrhage occur, and a damage of the BBB is at least partially/temporarily associated with the observed lesions[14,15], suggesting that edema in this model can be either of cytotoxic or vasogenic[12] origin. T2-weighted images clearly showed the edematous lesions, but could not differentiate between edema, gliosis or cysts. Thus, we initiat-

ed the present study to investigate if diffusion weighted imaging could differentiate these lesions. The rationale was that the ADC of tissue water should increase with an increasing loss of cells. During the course of this study, the results of a clinical investigation were published[4]. The ADC values tended to increase with the age of the stroke, and the authors speculated that among others, gliosis might be the cause of elevated ADCs. Use of SHRSP as a stroke model offers the unique possibility to relate MRI data with histology, and to test this hypothesis.

Materials and Methods

Nine male SHRSP were provided by the Department of Hygiene, Kinki University School of Medicine. To accelerate the development of stroke, the animals were kept on Funabashi SP diet (Funabashi Co., Ltd., Chiba, Japan) containing minimal Ca^{2+} and Mg^{2+}, and the drinking water contained 1% NaCl. A sudden decrease of the body weight or increased aggressiveness, irritability or hypokinesia[1] were interpreted as occurrence of stroke, and animals were imaged between 1 week and 3 months later in order to obtain a spectrum of histological changes. As control, 7 age matched male Wistar rats were used.

The animals were anesthetized with 30–40 mg/kg sodium pentobarbital (i.p.) and 0.5 g/kg urethane (s.c.), and placed in the imager. A 4.7T animal imager (Omega CSI-II, GE NMR Instruments, Fremont, CA) equipped with self-shielded gradients in combination with a home-made bird-cage coil (5 cm diameter) was used. A T2-weighted (TR/TE = 1500/80 msec) multislice sequence giving 8, 2.5 mm thick transversal images of the brain was used to localize lesions. The slice with the severest lesion or in some cases one showing 2 independent lesions was selected for ADC measurement. A spin-echo sequence with motion-probing gradients (MPGs) of 0–5 G/cm added in phase-encode direction was used, with b values defined as usually; $b = \gamma^2 G^2 \delta^2 (\Delta - \delta / 3)$ (see [11] for details). Four to 8 images with b-values between $\cong 0$ and 1908 s/mm^2 were taken of the selected slice (the b-value of the T2-weighted image was calculated to be 19.58 s/mm^2). Other parameters were: TR = 2000 ms, TE = 90 ms, FOV = 60 mm, 256 × 128 matrix, 2 averages.

After MRI imaging, the animals were sacrificed and the brains fixed in 10% formaldehyde. Sections of the fixed brains were stained with hematoxylin/eosine and scored by a pathologist for edema, gliosis and cyst formation in the lesion detected by MRI.

Based on the histological findings, region-of-interest (ROI) were selected in the MRI images corresponding to edema (n = 11), gliosis (n = 7) or cysts (n = 5) avoiding any regions with other pathological findings as much as possible. When more than one ROI was analyzed for the same type of lesion in one animal, efforts were taken to insure the respective ROIs were selected from different independent lesions. The signal intensities (Sn) on the diffusion weighted images and (So) on the control image (b ≅ 0) were measured, and the ln (Sn/So) plotted against the b-value. ADCs were determined from the slope of these plots. Only lesions in gray matter were evaluated, to avoid possible problems with anisotropy[11]. In the control animals, ROIs were selected in the grey matter of either cortex or basal ganglia/thalamus, because the lesions evaluated occurred in these regions. ADC values were reported as mean ± SD and subjected to one-way analysis of variance with repeated measures, followed where appropriate by Scheffe multiple comparison test using commercial software (SAS/STAT® SAS Institute Inc., Cary. NC, U.S.A).

Results

Figure 1 shows typical MRI images with different diffusion weighting obtained in SHRSP with chronic stroke. While the edema shows a high signal intensity on the T2-weighted image, it shows lower intensity than normal brain tissue on the image with the highest diffusion weighting, indicating faster diffusion of tissue water in the edema than in normal brain. Note, that the signal intensity of the lesion on the T2-weighted image is homogeneously high while diffusion-weighted images allow differentiation of several sub-regions. In this area, various stages of edematous changes were found.

Figure 2 compares the ADC values for the different experimental groups. For the control group, the ADCs measured in the cortex ($0.68 \pm 0.08 \times 10^{-3}$ mm²/s) and the deep grey matter ($0.62 \pm 0.06 \times 10^{-3}$ mm²/s) were pooled because they were not significantly different.

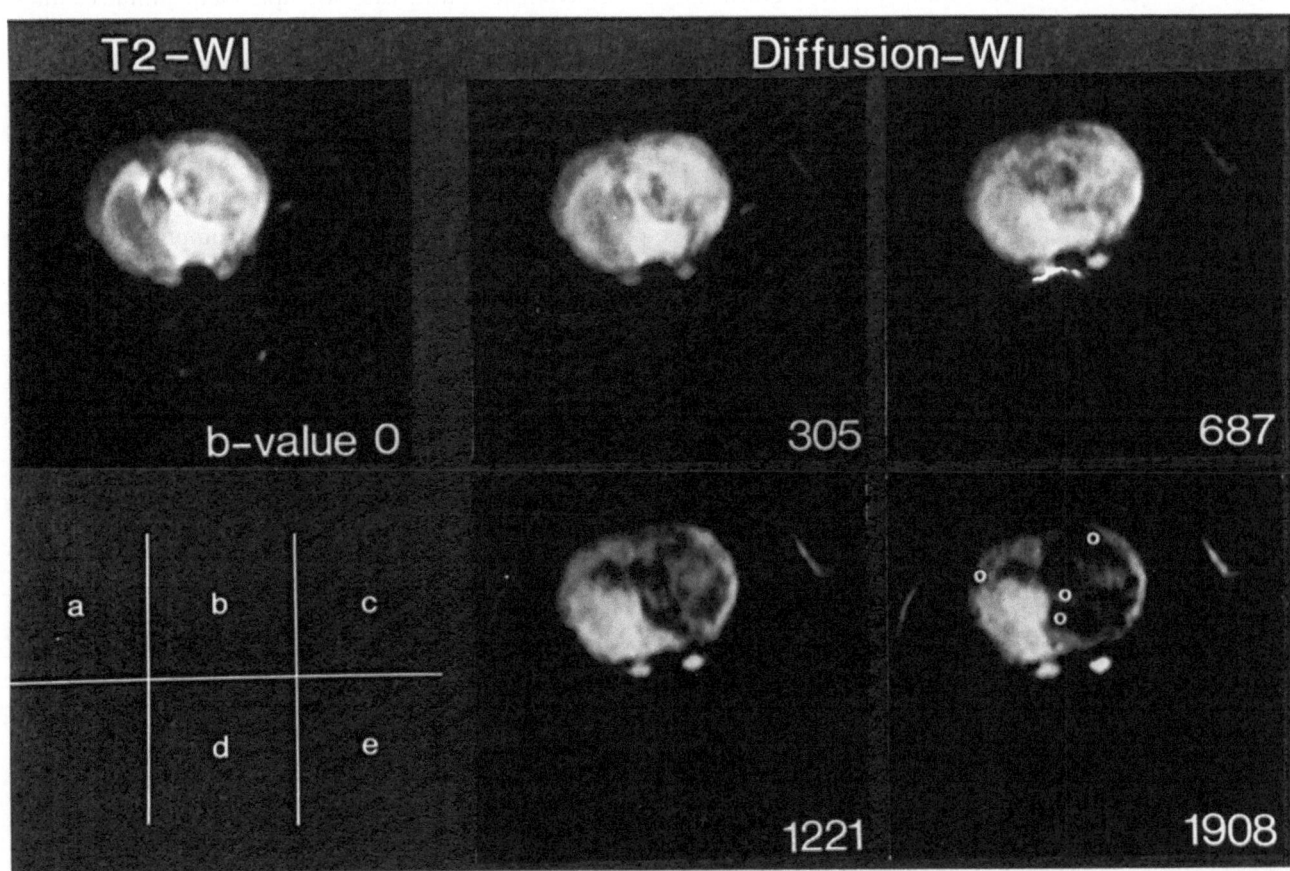

Fig. 1. MRI images with various diffusion weighting of a SHRSP 3 weeks after occurrence of stroke. *a* T2 weighted control image (b ≅ 0) shows homogenous high signal intensity in the lesion indicating increased water content. *b–e* Increasing the diffusion weighting from (*b*) b = 305 to (*e*) b = 1908 s/mm² causes rapid loss of signal from the tissue with edema. In (*d*) and (*e*), unaffected brain tissue has a higher signal intensity than the lesion. ROIs for signal intensity measurements are shown

226

M. Takahashi *et al.*

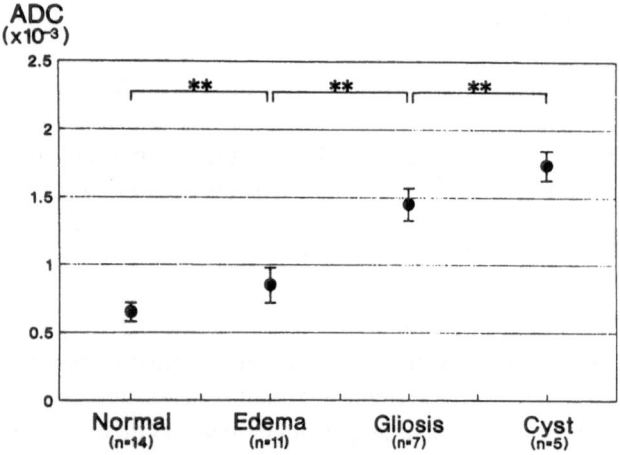

Fig. 2. The distribution of the individual ADC values is shown for the control group and the respective lesions. Mean values and standard deviations are indicated. Each group was significantly different from each other (* p < 0.01 by Scheffe multiple comparison test

However, the mean ADCs of all four groups (control: 0.65 ± 0.07; edema: 0.85 ± 0.13; gliosis: 1.45 ± 0.12; cyst: 1.74 ± 0.11 × 10⁻³ mm²/s were significantly different from each other (p < 0.01).

Discussion

Our study in a rat stroke model shows that in chronic stroke the measured ADC of tissue water is increased. The degree of increase was characteristic for the types of lesion evaluated. Our findings seem to be applicable to the clinical situation, since increased ADCs in chronic stroke have recently been reported in humans[4], where the authors attributed the increase of the ADC to gliosis or cyst formation. Our findings do not only support their interpretations, but also suggest that quantitative measurements can provide a noninvasive method to grade the histological changes. The close correlation of the ADCs with the histological results thus opens exciting aspects for monitoring stroke therapy and/or a new approach to tissue characterization.

The observed increase in the ADC of tissue water in chronic stroke contrasts with the decrease of tissue water diffusion observed in the acute stage[10]. It may be argued, that breakdown of cellular membranes occurs in subacute or chronic stroke, leading to a decrease in the restriction of diffusion of cellular water. However, the extreme thinness (= small diffusion path) of the unit membrane and the high permeability of model[2] or plant cell[6] membranes for water suggest that the plasma membrane surrounding a cell or an organelle does not cause any restriction of water exchange between intra-

and extracellular spaces by diffusion. Moonen *et al.*[9] showed that the ADC of cortex is independent of the diffusion time used, providing striking evidence that restriction of water diffusion by a single plasma membrane is negligible. In fact, a simple calculation using Einstein's equation ($d^2 = 2Dt$ for one-dimensional diffusion along the motion-probing gradient, where t is the diffusion time) shows that – assuming a time delay between the MGPs of 45 ms as used in this study and a cell diameter of 20 μm – over 75% of the intracellular water reaches the cellular surface and can leave the cell. Therefore, the measured ADCs represent a weighted average of the intracellular and extracellular diffusion coefficients, and depend on the volume ratio of intracellular and extracellular space in the measured volume.

While presently available data do not allow an independent analysis of these four parameters, the well known decrease in extracellular space in acute stroke can reasonably explain the observed decrease of the tissue ADC. Similar arguments have been used by Benveniste *et al.*[3], who attempted a quantitative interpretation of the changes in ADCs. However, in view of the significant exchange of water across the cellular membranes during the measurement discussed above, their calculations should be substantiated by more rigid methods. On the other hand, the increase in ADC in chronic stroke measured in this study implies an increase in extracellular space, an assumption which is in reasonable agreement with the histopathological findings. This increase in extracellular space is probably due to vasogenic edema, and later due to the clearing of dead cells which finally leads to formation of cysts. In order to substantiate this hypothesis, determination of the four relevant variables (intracellular space and diffusion coefficient, extracellular space and diffusion coefficient) is needed. Due to the rapid exchange of intracellular and extracellular water across the plasma membrane, diffusion experiments must use a minimal delay between the motion probing gradients. At the same time, high b-values might be necessary to achieve a sufficient suppression of the extracellular (faster diffusing) water.

Further studies, such as one mentioned above, hopefully will help to interpret ADC measurements.

References

1. Akiguchi I, Horie R, Yamori Y, Kameyama M (1979) Lethal course of stroke and therapeutic effects in stroke-prone spontaneously hypertensive rats. In: Yamori Y *et al* (eds) Prophylactic approach to hypertensive diseases. Raven, New York

2. Alberts B, Bray D, Lewis J, *et al* (1983) The cell. Garland, New York, pp 287

3. Benveniste H, Hedlund LW, Johnson A (1992) Mechanism of detection of acute cerebral ischemia in rat by diffusion-weighted magnetic resonance microscopy. Stroke 23: 746–754

4. Chien D, Kwong KK, Gress DR, *et al* (1992) MR diffusion imaging of cerebral infarction in human. AJNR 13: 1097–1105

5. Fellini DA, Hopkins AL, Selman WR, *et al* (1992) Evaluation of experimental early acute cerebral ischemia before the development of edema: use of dynamic, contrast-enhance and diffusion-weighted MR scanning. Magn Res Med 27: 189–197

6. Hills BP, Duce SL (1990) The influence of chemical and diffusive exchange on water proton transverse relaxation in plant tissue. Magn Res Imaging 8: 321–331

7. Minematsu K, Fisher M, Li L, *et al* (1992) Diffusion weighted magnetic resonance imaging of a focal ischemic lesion in rat brain. Neurology 42: 235–240

8. Minttorvitch J, Moseley ME, Chileuih L, *et al* (1991) Comparison of diffusion and T2-weighted MRI for the early detection of cerebral ischemia and reperfusion in rats. Magn Res Med 18: 39–50

9. Moonen CTW, Perkar J, DE Vleeschouwer MHM, *et al* (1991) Restricted and anisotropic displacement of water in healthy cat brain and stroke studied by NMR diffusion imaging. Magn Res Med 19: 327–332

10. Moseley ME, Cohen Y, Mintorovitch J, *et al* (1990) Early detection of regional cerebral ischemic injury in cats: evaluation of dlffusion and T2 weighted MRI and spectroscopy. Magn Res Med 14: 330–346

11. Moseley ME, Cohen Y, Kucharczyk J, *et al* (1990) Diffusion-weighted MR imaging of anisotropic water diffusion in cat central nervous system. Radiology 176: 439–445

12. Sage MR (1982) Blood-brain barrier: phenomenon of increasing importance to the imaging clinical. AJNR 3: 127–138

13. Stejskal EO, Tanner JE (1965) Spin diffusion measurements: spin-echo in the presence of a time dependent field gradient. J Chem Phys 42: 288–292

14. Tagami M, Kubota A, Sunaga T, *et al* (1983) Increase transendothelial channel transport of cerebral capillary endothelium in stroke-prone SHR. Stroke 14: 591–596

15. Takahashi M, Fritz-Zieroth B, Yamaguchi M, *et al* (1992) MR imaging of cerebral lesions accompanying stroke in stroke-prone spontaneously hypertensive rats. Folia Pharmacol Japon 100: 21–28

16. Tanner JE (1979) Self-diffusion of eater in frog muscle. Biopsy J 28: 107–116

17. Van Bruggen N, Cullen BM, King MD, *et al* (1992) T2- and diffusion-weighted magnetic resonance imaging of a focal ischemic lesion in rat brain. Stroke 23: 576–582

Correspondence: B. Fritz-Zieroth, M. D., Research Department, Nihon Schering KK, 2-6-64 Nichimiyahara, Yodogawa-ku, Osaka 532, Japan.

Acta Neurochir (1994) [Suppl] 60: 228–230
© Springer-Verlag 1994

Effect of Mannitol on Focal Cerebral Ischemia Evaluated by Magnetic Resonance Imaging

H. Kobayashi, H. Ide, T. Kodera, Y. Handa, M. Kabuto, T. Kubota, and **M. Maeda**[1]

Departments of Neurosurgery and [1] Radiology, Fukui Medical School, Fukui, Japan

Summary

We have evaluated the effect of mannitol on focal cerebral ischemia using T2-weighted magnetic resonance (MR) imaging and intravoxel incoherent motion (IVIM) MR imaging. The left middle cerebral artery (MCA) was exposed via the transorbital approach in 20 adult cats and occluded just proximal to the origin of the perforating arteries. Seven cats in treatment group received mannitol (0.5 g/kg i.v.) at 0, 6, 12 and 18 hours after MCA occlusion. The other 13 cats received saline and served as controls. Sequential MR coronal images were obtained at 2, 4, 6, and 24 hours after MCA occlusion using a GE Signa (1.5 tesla) system. IVIM MR imaging demonstrated ischemic cerebral injury as a sharply demarcated area at 2 hours after MCA occlusion in control group, while T2-weighted MR imaging failed to show clear evidence of the injury until 2–6 hours. At 24 hours after MCA occlusion, the infarcted area in the mannitol treatment group was 36.9 ± 7.7% (S.E.M.) of the left hemisphere, as compared to 57.3 ± 5.3% in control group (p < 0.05). Mannitol has beneficial effect on ischemic injury.

Keywords: Cat; focal cerebral ischemia; mannitol; magnetic resonance imaging.

Introduction

Severe ischemic injury is usually associated with brain edema. Mannitol is an osmotic agent widely used to reduce intracranial hypertension and to resolve brain edema. Magnetic resonance (MR) imaging provides fine anatomical resolution of ischemic cerebral injury[4]. Intravoxel incoherent motion (IVIM) MR imaging has been reported to be more sensitive than conventional T2-weighted imaging for the early detection of ischemic injury[1,3,8,9]. We have evaluated the effect of mannitol on focal cerebral ischemia using IVIM and T2-weighted MR imaging.

Materials and Methods

Twenty adult cats, each weighing 2.5 to 3.5 kg, were sedated with ketamine hydrochloride (10 mg/kg i.m.). Anesthesia was main-tained with sodium pentobarbital (5–10 mg/kg/h i.v.). The cat was intubated and ventilated with a Harvard respirator to maintain normocapnia. Isotonic saline solution was administered at a rate of 5 ml/h throughout the experiment. The left middle cerebral artery (MCA) was exposed via the transorbital approach and occluded just proximal to the origin of the perforating arteries[5,7]. Seven cats in treatment group received mannitol (0.5 g/kg i.v.) at 0, 6, 12, and 18 after MCA occlusion. The other 13 cats received saline and served as controls.

A 1.5 T MR imaging system (Signa; General Electric, Milwaukee) equipped with a self-shielded RF gradient coil was used. Sequential MR coronal images were obtained at 2, 4, 6 and 24 hours after MCA occlusion. Multi-slice T2-weighted spin-echo images were obtained with the following sequence: 2500/120/2 (TR/TE/excitations), 4 mm thick slice, 256 × 128 matrix. Single-slice IVIM spin echo images (1000/120/4, 5 mm thick slice, 256 × 128 matrix) were acquired by adding diffusion probing gradient pulses (duration of 45 ms, gradient separation of 11 ms) on either side of a 180° radio-frequency pulse. The animals were sacrificed after the last MR study for histological study.

Results

There was no significant difference in physiological parameters between control group and treatment group. IVIM MR images demonstrated ischemic cerebral injury as a sharply demarcated area at 2 hours after MCA occlusion in control group, while T2-weighted MR images failed to show clear evidence of injury until 2–6 hours. The high signal intensity area on the IVIM images did not increase significantly during 24 hours. In contrast, increased signal intensity area on the conventional T2-weighted images spread to the surrounding areas with time. Close anatomical correlation was found between the 24-hour T2-weighted images and the 2-hour IVIM images. The infarcted area in treatment group was significantly smaller (Fig. 1) than that in control group (Fig. 2). At 24 hours after MCA occlusion, infarction was observed in 36.9 ± 7.7%

(S.E.M.) of the hemisphere in treatment group, and in $57.3 \pm 5.3\%$ in control group ($p < 0.05$). There was a good anatomical correlation between histological investigation and MR imaging.

Discussion

MR imaging is a noninvasive technique that provides fine anatomical resolution, and exhibits greater sensitivity than other traditional methods for detecting focal cerebral ischemia[3,11]. It has been reported that IVIM MR imaging, which maps the microscopic motion of water protons, is more sensitive than conventional T2-weighted imaging for the early detection of ischemic injury in focal cerebral ischemic models[9]. Our study also demonstrated that IVIM imaging is superior to conventional T2-weighted imaging for the early identification of ischemic injury. Moseley et al.[9] detected significant hyperintensity on IVIM imaging as early as 45 minutes after induction of ischemia and a close correlation between IVIM tissue signal intensity ratios and metabolic abnormalities measured with P-31 and H-1 MR spectroscopy in a cat model of MCA and bilateral common carotid arteries occlusion.

Mannitol is a drug commonly administered in the clinic to improve cerebral blood flow; this effect has been attributed to its hyperosmolar characteristics. In our study, the infarction size in the mannitol treatment group was smaller than that in the control group. The main damaging event in cerebral ischemia is the depletion of adenosine triphosphate and phosphocreatine which accompanies the shift to anaerobic glycolysis as the dominant source of high-energy phosphates and which leads to dissipation of ion gradients, suppression of cell communication, and disruption of macromolecular assemblies[5]. Brain edema gradually develops in the periphery of the ischemic core and if edema is of sufficient magnitude to mechanically compress the capillary bed, blood flow is adversely affected and infarction of the tissues progresses[6]. Shirane et al.[10] reported that pretreatment with mannitol prevented postischemic obstruction of the microcirculation of 5 minutes of recirculation after 30 minutes of severe temporary ischemia. Hartwell et al.[2] concluded that vascular factors are more likely to explain the reduction of intracranial pressure, with mannitol than osmotic dehydration of brain.

IVIM MR imaging is a promising method to detect early cerebral ischemic lesion. Mannitol has beneficial effect on ischemic injury. This effect could be a reflec-

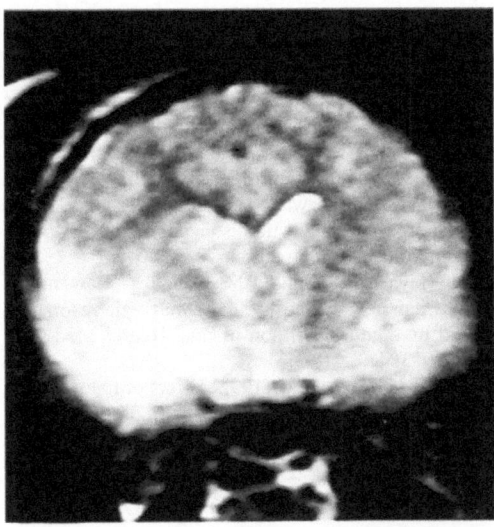

Fig. 1. T2-weighted image of coronal section at 24 hours after left middle cerebral artery occlusion in the mannitol treatment group. The left hemisphere appears on the left side of the image. High intensity area in this group is smaller than that in the control group (Fig. 2)

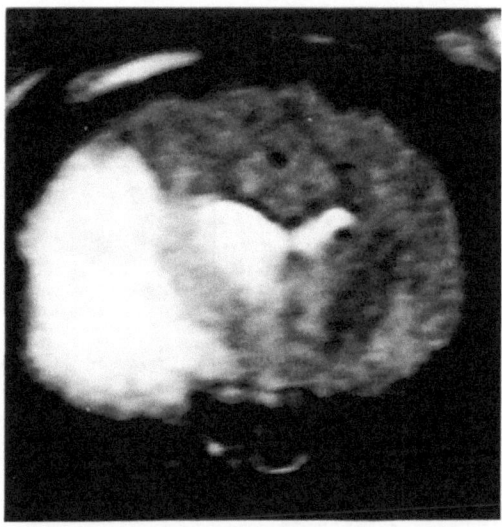

Fig. 2. T2-weighted image of coronal section at 24 hours after left middle cerebral artery occlusion in the control group. The left hemisphere appears on the left side of the image

tion of its potential to act as a scavenger of free radicals or as an agent that reduces the untoward effects of cerebral edema.

References

1. Fisher M, Sotak CH (1992) Diffusion-weighted MR imaging and ischemic stroke. AJNR 13: 1103–1105
2. Hartwell RC, Sutton LN (1993) Mannitol, intracranial pressure, and vasogenic edema. Neurosurgery 32: 444–450

3. Ide 14, Kobayashi H, Handa Y, Kubota T, Maeda M, Itoh S, Ishii Y (1993) Correlation between somatosensory evoked potentials and magnetic resonance imaging. Surg Neurol 40, in press
4. Kobayashi H, Hayashi M, Kawano H, Handa Y, Kabuto M, Ishii Y (1989) Magnetic resonance imaging of embolism from intracranial aneurysms. Surg Neurol 32: 225–230
5. Kobayashi H, Hayashi M, Kawano H, Handa Y, Kabuto M, Ide H (1990) Phosphorus 31 magnetic resonance spectroscopy of cerebral ischemia in cats. Neurosurgery 27: 240–246
6. Kobayashi H, Hayashi M, Kabuto M, Handa Y, Kawano H, Ide H (1991) Changes in Phosphorus-31 magnetic resonance spectra caused by epidural balloon in cats. Neurol Med Chir (Tokyo) 31: 379–384
7. Kobayashi H, Ide H, Hayashi M (1992) Effect of naloxone on focal cerebral ischemia in cats. Neurochirurgia 35: 69–73
8. LeBihan D, Breton E, Lallemand D, Grenier P, Cabanis E, Laval-Jeantet M (1986) MR imaging of intravoxel incoherent motions: application to diffusion and perfusion in neurologic disorders. Radiology 161: 401–407
9. Moseley ME, Cohen Y, Mintorovitch J, Chileuitt L, Shimizu H, Kucharczyk J, Wendland MF, Weinstein PR (1990) Early detection of regional cerebral ischemia in cats: comparison of diffusion- and T2-weighted MRI and spectroscopy. Magn Reson Med 14: 330–346
10. Shirane R, Weinstein PR (1992) Effect of mannitol on local cerebral blood flow after temporary complete cerebral ischemia in rats. J Neurosurg 76: 486–492
11. Yuh WTC, Crain MR, Loes DJ, Greene GM, Ryals TJ, Sato Y (1991) MR imaging of cerebral ischemia: rinding in the first 24 hours. AJNR 12: 621–629

Correspondence: H. Kobayashi, M. D., Department of Neurosurgery, Fukui Medical School, Matsuoka, Fuki 910-11, Japan.

Ischemic Brain Edema: Pathobiochemistry and Brain Edema

Acta Neurochir (1994) [Suppl] 60: 233–237
© Springer-Verlag 1994

Blood-Damaged Tissue Interaction in Experimental Brain Ischemia

J. M. Hallenbeck

Stroke Branch, National Institute of Neurological Disorders and Stroke, National Institutes of Health, Bethesda, MD, U.S.A.

Summary

This paper is a review and synthesis of work done in our laboratory by many investigators over roughly 18 years dealing with the microcirculation in a zone of acute ischemic injury. The work has been guided by a hypothesis that blood flowing through the microcirculation of an acute injury zone is capable of undergoing a multifactorial interaction at the blood-endothelial interface that can progressively impair microvascular perfusion and contribute locally to the evolution of cellular damage and death. Our work has implicated Factor VIII/von Willebrand factor, prostanoids, leukocytes, platelets, platelet-activating factor, leukotrienes, adhesion receptors, monocytes/macrophages, and cytokines in this interaction.

Keywords: Ischemia; microcirculation; stroke; inflammation.

Introduction

The work described in this review does not address brief ischemia, which leads to a maturation phenomenon and delayed neuronal death of selectively vulnerable neurons. Instead, it focuses on ischemia of sufficient density and duration to cause pannecrosis of a core of tissue. In this lesion, the core is surrounded by a zone of tissue in a meta-stable state in which the cells are potentially viable given the proper therapeutic intervention, but, failing that, tend to degenerate progressively[1].

We felt from the outset that applying the principles governing the delivery of water to the bathroom sink would not permit full understanding of the microcirculation in a zone of acute ischemic injury; that is, Ohm's Law applied to the circulation could not fully explain events in these microvessels. Initially, we borrowed from an allied field, ex-vivo organ perfusion preservation, to gain insights into microcirculatory problems in ischemia. Workers in the organ transplantation field had found that perfusion of an organ with anticoagulated whole blood led to progressive impairment of per-

fusion and something called the "early block" phenomenon[2]. Glass-wool filtration of the anticoagulated blood, however, greatly improved the blood's ability to perfuse organs ex-vivo. On the basis of this finding, we decided as a first step, to investigate the effect of glass-wool filtration of whole blood on post-ischemic reperfusion.

Blood Modification Effect on Reperfusion

Dogs treated with the anticoagulant heparin, 300 U/kg, were subjected to cerebrospinal fluid (CSF) compression ischemia for 35 min followed by 30 min of recirculation[6]. At the end of recirculation, a ^{14}C-antipyrine autoradiographic blood flow study was performed. The study revealed heterogeneous blood flows within individual neuroanatomic structures and discrete zones of extremely low blood flow, in the range of 4–6 ml/100 g/min for white matter and 6–20 ml/100 g/min for gray matter. Other dogs had their blood modified by exposure to glass-wool before the induction of ischemia. Glass-wool filters were inserted into an arteriovenous shunt from the femoral artery to the femoral vein, and several blood volumes were allowed to run through three successively inserted glass wool-filters over the course of 1 h. Reperfusion in these animals was excellent. Average flows in a sample of gray matter and white matter structures were all markedly higher in the glass-wool filtered group, and zones of extremely low flow were eliminated.

On the basis of these results, we developed a hypothesis that has continued to guide our studies over the years[6]. Blood is viewed as having two general categories of effect in a zone of acute ischemic injury. The first category is not considered to be controversial. Blood flowing above some critical rate is capable of delivering oxygen and substrates for metabolism and carrying away metabolic waste, thereby supporting normal tissue and restoring reversibly damaged tissue to normal. The other category of effect is hypothetical. Blood is viewed as a highly reactive material capable of complex, multifactorial interactions with elements of the damaged tissue, which can lead to progressive impairment of microvascular perfusion. This process staunches residual blood flow and thus interferes with the supportive and restorative properties of flowing blood. Impaired microvascular perfusion can also extend the zone of damage by injuring normal tissue and by converting an area of injured but potentially viable tissue to a zone of irreversible damage.

Factor VIII/von Willebrand Effect on Reperfusion

To continue exploring the properties of the microcirculation in ischemic brain, we again borrowed from the ex-vivo organ perfusion preservation field. Anticoagulated plasma was a less-than-ideal perfusate for these organs, and its use led to progressive perfusion difficulty with time, particularly after autotransplantation[3]. Plasma from which the cryoprecipitate had been removed, "cryoprecipitate-poor plasma", greatly improved organ perfusion. We, therefore, tested the effect of autologous cryoprecipitate on post-ischemic reperfusion. Dogs were subjected to glass-wool filtration of their blood via an arteriovenous shunt, as described earlier, and then were infused intravenously with an equal volume of either redissolved cryoprecipitate in Tris-HCI buffer or cryoprecipitate-poor plasma before undergoing CSF compression ischemia[7]. Cryoprecipitate-poor plasma did not interfere with the beneficial effects of glass-wool filtration, and the post-ischemic reperfusion was excellent. Autologous cryoprecipitate, however, nullified the beneficial effects of the glass-wool filtration and restored the defects in brain reperfusion. This finding seemed to indicate that there was some deleterious principle in cryoprecipitate. A major component of cryoprecipitate is Factor VIII/von Willebrand factor. When this protein was isolated from cryoprecipitate by Sepharose 4B gel chromatography and added back to the blood after glass-wool filtration, its effect on brain blood flow was devastating; averaged gray matter flow within single neuroanatomic structures was as low as 9 ml/100 g/min, and averaged white matter flow was as low as 5 ml/100 g/min[9].

This work demonstrated that modification of a dog's blood by exposure to glass-wool before the production

of ischemia could greatly enhance post-ischemic reperfusion and that a high-molecular-weight molecule involved in platelet adhesion and aggregation, Factor VIII/von Willebrand factor, could counteract the beneficial effects of the glass-wool filtration and contribute to post-ischemic impairment of microvascular perfusion.

Prostanoid Effect on Reperfusion

Several potent mediators of the arachidonic acid cascade appeared to provide a concrete example of blood-damaged tissue interaction. When membranes are perturbed, arachidonic acid is released from the C-2 position of phospholipids and acted upon by cyclooxygenase to form the cyclic endoperoxides, PGG_2 and PGH_2. PGH_2 then forms the substrate for two potent products with diametrically opposite effects at the blood-endothelial interface. The endothelium contains the enzyme, prostacyclin synthase, that produces prostacyclin, a vasodilator and platelet inhibitor. Circulating blood cells such as platelets and leukocytes contain thromboxane synthase, which acts on the same cyclic endoperoxide substrate to form thromboxane A_2, a potent platelet aggregator and a vasoconstrictor. We were interested to know whether the prostaglandin system played any role in the microcirculation during post-ischemic reperfusion. To study this, we blocked the cyclooxygenase enzyme with indomethacin before subjecting dogs to CSF compression ischemia, as described earlier[5,8]. In heparinized dogs that did not receive indomethacin, we observed areas of severe impairment of post-ischemic reperfusion, with blood flows in the range of 0–15 ml/100 g/min in gray matter and 0–6 ml/100 g/min in white matter. However, dogs that received, in addition, 4 mg/kg of indomethacin intravenously before the production of ischemia had very high levels of post-ischemic blood flow, averaging 102–187 ml/100 g/min in gray matter and 35–40 ml/100 g/min in white matter. Zones of greatly impaired reperfusion, as defined earlier, were absent in these animals.

Platelet and Leukocyte Effects on Reperfusion

Because some of the mediators in the arachidonic acid cascade affect platelet aggregation and others are chemotactic, we decided to investigate whether or not platelets and leukocytes accumulate in the microcirculation during post-ischemic reperfusion. These studies were conducted in a new model of ischemia[10,14,15].

Dogs were exposed to a 60-min period of localized ischemia induced by serial embolism of 20–50 μl boluses of air through a catheter placed in the internal carotid artery. The small boluses of air were given until a cortical sensory evoked response was suppressed to 10–20% of its baseline amplitude. As the response grew to exceed 20% of its baseline amplitude, more air was infused to suppress it and, in this way, a standard level of ischemia was maintained for 60 min. A 4-h period of reperfusion, with no further infusion of air, followed. After this, a ^{14}C-iodoantipyrine autoradiographic blood flow study was performed. During this study, the dogs were infused intravenously with autologous ^{111}In-labeled platelets or granulocytes, which were isolated from a sample of each animal's blood obtained during the preparation period to permit detection of ischemia-induced accumulation of these cells within the brain. Platelet or granulocyte collections in the brain tissue sections appeared as discrete dots on the autoradiograms. By 4 h after ischemia, the dots tended to cluster in zones of microcirculatory flow impairment, which implicated these cells in the process of progressive shutdown of microcirculatory flow.

Progressive Neuraxis Injury During Reperfusion

The results of the foregoing studies were all consistent with the initial concept of a blood-damaged tissue interaction, but that hypothesis also predicted that the multifactorial interaction could extend injury and lead to progression of symptoms after ischemia. The models with which we had been working did not address this possibility, so we studied the progression of neuronal damage in a model of spinal cord ischemia developed by Zivin and De Girolami[18].

In New Zealand albino rabbits, a polyethylene tube was threaded around the abdominal aorta below the renal arteries and passed through two buttons before exiting from the abdomen through a vinyl guide tube. This enabled us to compress the aorta by pulling the polyethylene tubing up tight against the guide tube. This procedure reliably produced ischemia of the lumbosacral cord and led to a painless loss of spinal cord function in awake rabbits[11]. After 25 min of ischemia, early recovery of function was observed, and between 4 and 6 h after ischemia the animals could hop with mild weakness or ataxia. This improvement was followed by a pronounced secondary deterioration of function between 8 and 24 hours of reperfusion, which left the animals able to support their weight, but unable to hop. Thromboxane B_2, a stable metabolic break-

down product of thromboxane A_2, was measured by radioimmunoassay, and elevated levels tended to coincide with both the initial and the secondary loss of neurological function. During the period of maximal recovery at 4 h post-ischemia, thromboxane B_2 levels temporarily returned to normal.

Platelet-Activating Factor and Leukotrienes in Reperfusion

In another model of spinal cord ischemia in anesthetized rabbits[12], platelet activating factor (PAF), a prothrombotic and proinflammatory mediator, was extracted from lumbosacral spinal cord subjected to 25 min of ischemia and 2 h of reperfusion. Tissue levels of PAF in the spinal cord exposed to ischemia and reperfusion were about 20-fold higher than levels in nonischemic, control spinal cord segments. The identity of the extracted substance as PAF was established by showing that the activity was susceptible to phospholipase C inhibition and that a potent PAF antagonist, BN50739, also inhibited the activity. Intraischemic administration of BN50739 improved the post-ischemic microcirculatory flow as measured by laser doppler flowmetry, and reduced edema formation relative to that occurring in animals receiving only drug vehicle during ischemia. Spinal cords exposed to 25 min of ischemia followed by reperfusion also produced increased levels of 5-HETE during the reperfusion period, implicating the leukotriene system as a possible player in the post-ischemic evolution of tissue damage[16].

Post-Injury Progression of Neuronal Damage

Although the model of spinal cord ischemia offered evidence of progressive neuronal damage, it did not have sufficient spatial and temporal reproducibility to permit detailed analysis of the progression. For this reason, we employed the cortical laser injury model, which offered excellent spatial and temporal reproducibility.

In this model, a neodymium yttrium aluminum garnet laser was used to irradiate a small area of cortex with a 20–40 watt pulse for 1 second at a distance of 1 cm so that the cortex was exposed to 20–40 joules of total energy[13]. This caused a highly reproducible injury with a necrotic core that was instantaneously coagulated. Surrounding the core was a zone of progressive tissue damage that expanded in a roughly hemispheric fashion and permitted focal measurement of blood flow, edema, neuronal damage, and polymorphonuclear leukocyte accumulation. The lesion volume roughly doubled between 4 min and 24 h after the injury. Measurement of local cerebral blood flow by laser doppler flowmetry revealed progressive deterioration of flow in a ring-shaped zone 1–2 mm wide outside the necrotic core on the brain surface. Neuronal loss was progressive and reached its peak between 8 and 24 h after the injury. In a ring of tissue on the brain surface, 4 mm outside the necrotic core where the tissue did not go on to necrosis, there was hyperemia. PAF was implicated in the blood flow changes, as a potent inhibitor of that substance, BN50739, attenuated the decrease of blood flow in the progressive injury zone surrounding the necrotic core[4]. Neutrophil accumulation was also reduced by BN50739, and neuronal counts demonstrated that neurons were preserved by the same drug in both the cortex and the hippocampus.

Potential Additional Participants in Reperfusion Injury

The work described above has implicated Factor VIII/von Willebrand factor, prostanoids, platelets, leukocytes with their free radicals, granule based enzymes and products of the arachidonic acid cascade, leukotrienes, and platelet-activating factor in reperfusion injury. The cortical laser injury studies indicate that the progression of neuronal damage is modifiable and can be attenuated by inhibiting one of the mediators, PAF, with BN50739.

Our current studies add even more complexity to the possible interactions at the blood-endothelial interface in patients with risk factors for stroke. Risk factors for stroke seem to be associated with an increase in the expression of ICAM-1, an adhesion receptor, on parenchymal brain endothelium, an increased number of ED2-positive perivascular macrophages in brain, and increased release of cytokines such as TNF-α in response to a prototypic macrophage stimulus such as lipopolysaccharide[17].

Conclusion

By inference, as blood flows into a zone of acute ischemic injury in the central nervous system, it becomes notified that there is a problem by local tissue changes and mediators that might be termed "tissue signals". Circulating cells adhere and become activated, releasing their own mediators, expressing additional adhesion receptors, and also releasing the contents of their intracellular granules. The cells can then aggregate and impair blood flow, as well as cause tissue damage by the actions of released mediators.

Our original concept of a complex, multifactoral interaction between blood and damaged tissue is rather strongly supported, and we would only modify our initial emphasis of a progressive impairment of microvascular perfusion as the primary cause of progressive damage to add that direct cytotoxicity from elaborated mediators could lead to further damage of normal or reversibly injured tissue independent of any perfusion impairment.

References

1. Astrup J, Symon L, Branston NM, Lassen NA (1977) Cortical evoked potential and extracellular K+ and H+ at critical levels of brain ischemia. Stroke 8: 51–57
2. Belzer FO, Park HY, Vetto RM (1964) Factors influencing renal blood flow during isolated perfusion. Surg Forum 15: 222–224
3. Belzer FO, Ashby BS, Huang JS, Dunphy JE (1968) Etiology of rising perfusion pressure in isolated organ perfusion. Ann Surg 168: 382–391
4. Frerichs KU, Lindsberg PJ, Hallenbeck JM, Feuerstein GZ (1990) Platelet-activating factor and progressive brain damage following focal brain injury. J Neurosurg 73: 223–233
5. Furlow TW Jr, Hallenbeck JM (1978) Indomethacin prevents impaired perfusion of the dog's brain after global ischemia. Stroke 9: 591–594
6. Hallenbeck JM (1977) Prevention of postischemic impairment of microvascular perfusion. Neurology 27: 3–10
7. Hallenbeck JM, Furlow TW Jr (1978) Influence of several plasma fractions on post-ischemic microvascular reperfusion in the central nervous system. Stroke 9: 375–382
8. Hallenbeck JM, Furlow TW Jr (1979) Prostaglandin I2 and indomethacin prevent impairment of post-ischemic brain reperfusion in the dog. Stroke 10: 629–637
9. Hallenbeck JM, Furlow TW Jr, Gralnick HR (1981) Influence of Factor VIII/von Willebrand factor protein (F VIII/vWF) and F VIII/vWF-poor cryoprecipitate on post-ischemic microvascular reperfusion in the central nervous system. Stroke 12: 93–97
10. Hallenbeck JM, Dutka AJ, Tanishima T, Kochanek PM, Kumaroo KK, Thompson CB, Obrenovitch TP, Contreras TJ (1986) Polymorphonuclear leucocyte accumulation in brain regions with low blood flow during the early postischemic period. Stroke 17: 246–253
11. Jacobs TP, Shohami E, Baze W, Burgard E, Gunderson C, Hallenbeck JM, Feuerstein G (1987) Deteriorating stroke model: histopathology, edema and eicosanoid changes following spinal cord ischemia in rabbits. Stroke 18: 741–750
12. Lindsberg PJ, Yue T-L, Frerichs KU, Hallenbeck JM, Feuerstein G (1990) Evidence for platelet-activating factor as a novel mediator in experimental stroke in rabbits. Stroke 21: 1452-1457
13. Lindsberg PJ, Frerichs KU, Burris JA, Hallenbeck JM, Feuerstein G (1991) Cortical microcirculation in a new model of focal laser-induced secondary brain damage. J Cereb Blood Flow Metab 11: 88–98
14. Obrenovitch TP, Kumaroo KK, Hallenbeck JM (1984) Autoradiographic detections of 111Indium-labeled platelets in brain tissue sections. Stroke 15: 1049–1056

15. Obrenovitch TP, Hallenbeck JM (1985) Platelet accumulation in regions of low blood flow during the postischemic period. Stroke 16: 224–234

16. Shohami E, Jacobs TP, Hallenbeck JM, Feuerstein G (1987) Increased thromboxane A_2 and 5-HETE production following spinal cord ischemia in the rabbit. Prostaglandins Leukot Med 28: 169–181

17. Siren A-L, McCarron RM, Liu Y, Heldman E, Spatz M, Feuerstein G, Hallenbeck JM (1992) Perivascular macrophage signalling of endothelium via cytokines: mechanism by which stroke-risk factors operate to increase stroke likelihood. In: Krieglstein J, Oberpichler-Schwenk H (eds) Pharmacology of cerebral ischemia. Wissenschaftliche Verlagsgesellschaft, Stuttgart

18. Zivin JA, DeGirolami U (1980) Spinal cord infarction: a highly reproducible stroke model. Stroke 11: 200–202

Correspondence: John M. Hallenbeck, M. D., Stroke Branch, NINDS, NIH, Building 36, Room 4D04, Bethesda, MD 20892, U.S.A.

Acta Neurochir (1994) [Suppl] 60: 238–241
© Springer-Verlag 1994

The Relationship Between Cerebral Ischemic Edema and Monoamines: Revisited

H. Ishii[1], **N. Bertrand**[1], **D. Stanimirovic**[1,2], **A. Strasser**[1], **B. B. Mrsulja**[2], and **M. Spatz**[1]

[1] Stroke Branch, National Institute of Neurological Disorders and Stroke, National Institutes of Health, Bethesda, MD, U.S.A., and
[2] Institute of Biochemistry, School of Medicine, Belgrade, Yugoslavia

Summary

The association of changes in the metabolic pathway of mono-amines (dopamine and 5-hydroxytryptamine) with mitochondrial enzymatic systems which are involved in the production and removal of free radicals formed during dopamine metabolism and formation of edema was investigated in bilateral brain ischemia in gerbils. The results suggest that the involvement of DA-derived free radicals in brain edema is unlikely in early reflow, because disbalance between H_2O_2-producing reactions of DA-metabolism, and mitochondrial antioxidative capacity does not occur prior to 1 hour reflow after 15 min bilateral ischemia in gerbils. However, the findings of this study reinforce the participation of 5-HT in the formation of ischemic brain edema.

Keywords: Intracerebral microdyalisis; ischemia; monoamines; free radicals.

Introduction

It is well known that cerebral ischemia is accompanied by edema (increase in tissue water content) which has both cytotoxic and vasogenic components[9]. Basically, the formation of ischemic brain edema represents an osmotic phenomenon that is triggered by biochemical dysfunction of the cellular membrane(s). The release of monoamines [especially 5-hydroxytryptamine (5-HT)] and formation of free radicals, among other factors, have been associated with the development of cerebral ischemic edema[1,12] for review and references. Hydrogen peroxide produced during degradation of dopamine (DA) by monoamine oxidase (MAO) could serve as one of the potential precursors of free radicals. Thus, we reevaluated the possible role of altered 5-HT and DA metabolism, as well as free radicals in cortical edema formation during cerebral ischemia and postischemia.

Material and Methods

Animal Preparation for Experimental Procedures

Separate groups (8–15 animals in each) of Mongolian gerbils (3-months-old, 50–70 g) were subjected to bilateral common carotid artery occlusion (15 min) under halothane (1.5%)/N_2O/O_2 (2 : 1) anesthesia, alone or with reperfusion (up to 120 min). Respective groups of sham operated animals served as controls. Pargyline (75 mg/kg b.w.) and p-chlorophenyl-alanine (PCPA) (500 mg/kg b.w.) were administered intraperitoneally, 2 hours, or repeatedly for 5 days before the induction of ischemia, respectively. Systemic blood pressure (SBP), cerebral blood flow (CBF; assessed by transcranial laser Doppler flowmetry), as well as head (temporal muscle) and rectal temperature were continuously monitored during the entire experimental period. The temperature was maintained at 37–38°C with thermostatically regulated heating lamp.

Intracerebral Microdialysis: Methodology, analysis of the monoamines and metabolites were described by us and other investigators previously ([2,6] and references).

Specific gravity was measured in cerebral cortex according to Nelson et al.[11].

Tissue analysis: Crude mitochondrial (synaptosomal) fraction was isolated from frontal cortex of brains frozen in liquid nitrogen by method of Gurd et al. (1974)[5]. Protein content in the samples was determined according to Lowry et al. (1951)[8]. MAO and glutathione reductase (GR) activities were determined fluorimetrically with kynuramine[10], and glutathione-disulfide as substrates, respectively[7]. Superoxide dismutase (SOD) activity was determined spectrophotometrically at 480 nm, as an inhibition of spontaneous epinephrine autoxidation[13]. Susceptibility of mitochondrial membranes to lipid peroxidation was estimated by the formation of thiobarbituric acid-reactive material (TBARM) during in vitro stimulation with Fe^{2+} salts ($FeSO_4$ – 0.01 mM) and ascorbic acid (0.5 mM) as described by Villacara et al.[14].

Results

Effects of Ischemia/Reperfusion Only

Bilateral cerebral ischemia for 15 min reduced CBF to < 3% and raised SBP to 145.0 ± 4.0% of the pre-ischemic level in the gerbils.

The temperature of the temporal muscle decreased by 2.3 ± 0.2°C during the same time. The initial recovery of CBF in reflow was followed by post-ischemic hypoperfusion during the observed post-ischemic period. Both SBP and temperature returned to preischemic values with reestablishment of circulation (data not shown). The ischemic insult led to a release of both DA and 5-HT which was still noticeable at 15 min of reflow (Fig. 1). The activity of MAO was markedly reduced during ischemia (Table 1), and after a transient stimulation at 3 min of reperfusion was decreased at 15 and 60 min of reflow (Table 1). Both DA and 5-HT, MAO-derived metabolites were decreased during ischemia and early reperfusion, but their content significantly increased at the later periods of reflow. The content of 3-methoxytyramine (3-MT), the MAO-independent DA metabolite increased (about 4-fold) during ischemia, and remained significantly enhanced in dialysate at 30 min of postischemia (data not shown). The marked reduction in the specific gravity of cortex seen at 15 min of ischemia, continuously decreased during the recirculation period (Fig. 2).

Mitochondrial Changes in GR, SOD, and TBARM

Ischemia alone reduced the activity of GR. Subsequently, in the early period of reperfusion SOD activity was increased, whereas no significant changes were observed in the mitochondrial susceptibility to lipid peroxidation in vitro. However, an increased formation of TBARM was seen at 15 min of reflow, prior to the reduction of antioxidative enzymes (Table 1).

Effects of Pargyline

Pretreatment of the gerbils with pargyline, the inhibitor of MAO, abolished the initial recovery of CBF (data not shown). The release of both monoamines was not only potentiated, but also prolonged, whereas the levels of their MAO-metabolites were decreased in the dialysate (Fig. 1). Pargyline augmentened and length-

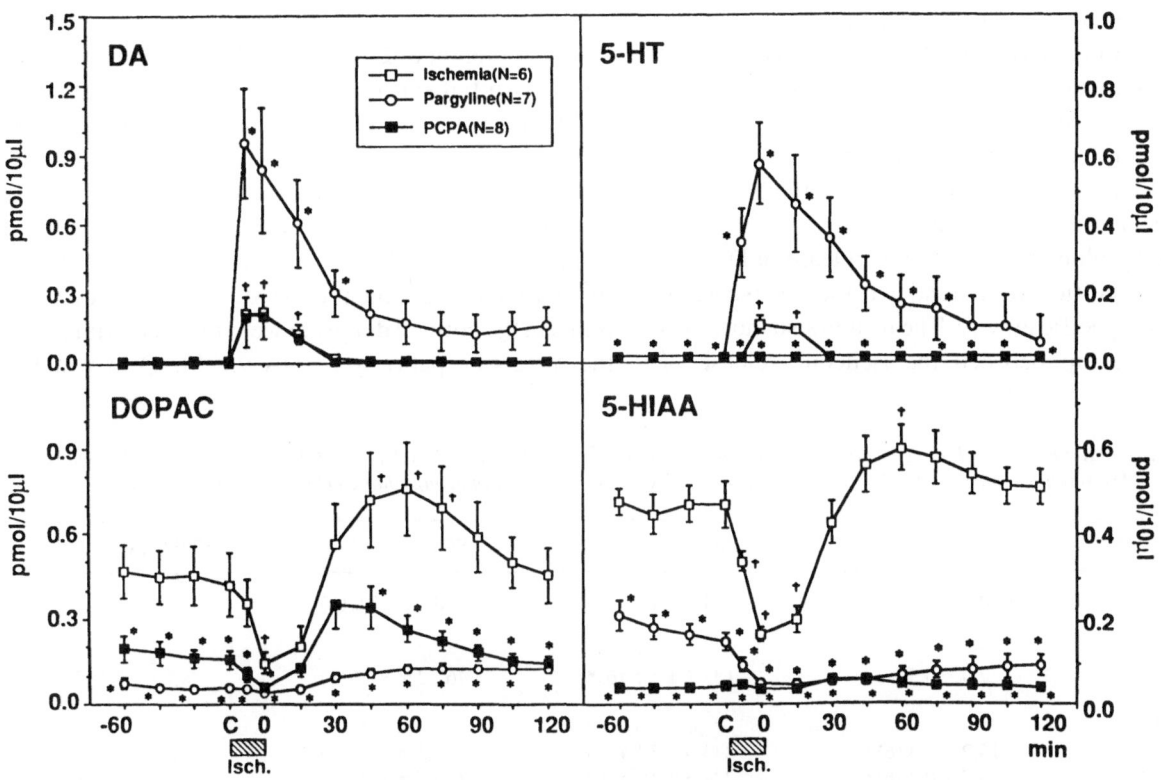

Fig. 1. Effect of pargyline or p-chlorophenylalanine (PCPA) on ischemia/reperfusion-induced DA, DOPAC, 5-HT and 5-HIAA changes in the cerebrocortical extracellular space of gerbils as measured by intracellular microdialysis. Each point represents mean ± SEM values obtained from 6–8 animals. + Indicates significant difference between the control (C) and experimental values in untreated gerbils (p < 0.01, ANOVA followed by Fisher's protected least squares difference test); * indicates significant difference between untreated and pargyline or PCPA treated gerbils (p < 0.01, ANOVA).

ened the production of 3-MT, the methylated metabolite of DA[6]. Moreover, a significantly greater accumulation of water was seen in the cortex of pargyline-pretreated than in saline-treated ischemic gerbils (Fig. 2).

Effects of PCPA

The ischemic release of 5-HT, but not that of DA, into the cortical extracellular space, was completely abolished in the PCPA-pretreated animals (Fig. 1). Moreover, a depletion of 5-HIAA, and reduction of DOPAC (Fig. 1), and HVA, but not 3-MT (not shown) was observed in cortical dialysate of PCPA-pretreated ischemic animals (Fig. 1). A significantly lesser reduction of specific gravity (except for 15 min of reflow) was observed in the cortex of PCPA-pretreated gerbils as compared to saline-treated animals in ischemia and recirculation (Fig. 2).

Discussion

The results of this study demonstrate that the transient bilateral cerebral ischemia causes a release of monoamines and changes in the cortical tissue water content as indicated by the increased extracellular content of the amines (DA and 5-HT) and decrease levels of their deaminated metabolites, as well as decreased specific gravity of the cortex. The association of the monoamine release and the accumulation of tissue water is strengthened by the observed effects of pargyline (the inhibitor of MAO activity), which augmented and prolonged the ischemic release of DA and 5-HT, and significantly potentiated the postischemic reduction in specific gravity. The results also indicate a close relationship between the ischemic release of

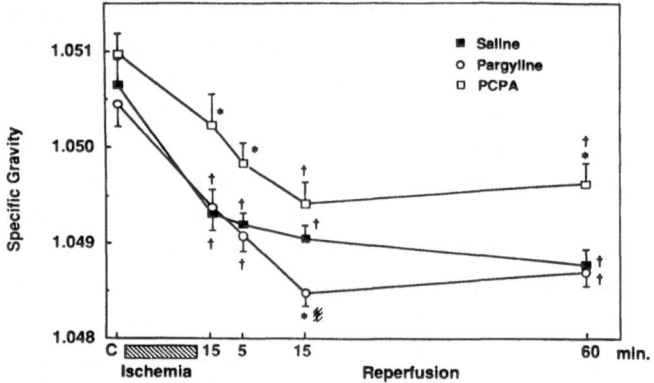

Fig. 2. Effect of pargyline or p-chlorophenylalanine (PCPA) on cerebrocortical specific gravity during ischemia/reperfusion in gerbils. + Indicates significant difference between the control (C) and experimental values in untreated gerbils (p < 0.01, ANOVA followed by Fisher's protected least squares difference test); * indicates significant difference between untreated and pargyline or PCPA treated gerbils (p < 0.01, ANOVA)

5-HT and tissue accumulation of water, since preischemic treatment of gerbils with PCPA, the inhibitor of 5-HT synthesis depleted 5-HT, but not DA, from cortical extracellular dialysate, and significantly prevented the reduction of specific gravity in the cortex. Moreover, our preliminary results suggest that partial depletion of DA by preischemic treatment with a-methyl-p-tyrosine (AMPT), an inhibitor of catecholamine biosynthesis, does not affect the ischemic/postischemic reduction in specific gravity.

The contribution of free radicals derived from DA metabolism to the early formation of ischemic brain edema is unlikely, since the mitochondrial MAO activity as well as the content of MAO-derived DA metabolites in the cortical dialysate were reduced. Moreover, the metabolic pathway of DA via MAO in which a

Table 1. *Ischemia Reperfusion – Induced Changes in Mitochondrial Monoamine Oxidase (MAO), Glutathione Reductase (GR), Superoxide Dismutase (SOD) Activities, and in vitro Lipid Peroxidation (TBARM) in the Cortex of Mongolian Gerbils*

	MAO[a]	GR[b]	SOD[c]	TBARM[d]
Sham	10.2 ± 0.4 (6)	234.4 ± 12.8 (5)	460.8 ± 27.6 (6)	92.2 ± 7.0 (6)
Ischemia				
15 min	7.9 ± 0.5 (7)*	155.2 ± 5.5 (6)*	467.2 ± 34.4 (8)	93.1 ± 7.2 (7)
Reperfusion				
3 min	15.2 ± 1.3 (6)*	243.5 ± 12.4 (6)	577.3 ± 37.8 (7)	102.6 ± 7.5 (6)
15 min	5.7 ± 0.4 (6)*	221.3 ± 10.5 (6)	473.2 ± 41.9 (8)	144.8 ± 9.3 (6)*
60 min	8.1 ± 0.6 (6)*	184.3 ± 12.1 (5)*	372.2 ± 23.7 (7)*	156.7 ± 10.4 (8)*

Data are given as means ± SEM for the number of animals indicated in the parenthesis. [a] nmol 4-HOQ/mg protein/min; [b] nmol NADPH/mg protein/h; [c] units; [d] nmol MDA/mg protein/h.
* Indicates significant difference (p < 0.01, ANOVA) from respective sham group.

substantial amount of H_2O_2 is produced is linked to mitochondrial antioxidative system including GR and SOD, that are either activated or unchanged up to 15 min of reflow.

The concept of 5-HT involvement in the development of ischemic brain edema is not new, since it has been supported by many studies conducted by us and other investigators for about two decades. However, previous investigations implicating 5-HT release in the formation of ischemic cerebral edema were based on pharmacologic monoamine turnover studies in the tissue performed mainly during postischemia[12]. Recent availability of microdialysis technique permitted direct measurement of monoamine neurotransmitter release into extracellular space and comparison of these findings with the events occurring in the tissue during ischemia and reflow.

The initial increase in 5-HT content in the extracellular space observed at 7.5 min of bilateral ischemia in gerbils, correlates well with the first appearance of tissue swelling between 5–6 min of ischemia, as well as with the reported marked reduction of Na^+, K^+-ATPase activity, an increase in Na^+ and decrease in K^+ in the tissue at 6 min of ischemia[4] in the same model of ischemia. Thus, these studies not only confirm the involvement of 5-HT in the formation of ischemic brain edema, but strongly suggest that the ischemically released 5-HT represents one of the main factors that contributes to the primary alteration of cellular water homeostasis in ischemia. Furthermore, the observed persistence of ischemic 5-HT release into extracellular space in reflow and progressive accumulation of tissue water are in agreement with our earlier studies demonstrating correlation between kinetic changes in 5-HT_{1A+B}, 5-HT_{1B}, and 5-HT_2-receptor binding sites, increased 5-HT turnover, spontaneous release of 5-HT from synaptosomes and tissue swelling in the same model of cerebral ischemia[3, 4, 12].

References

1. Chan PK, Schmidley WJ, Fishman AR, Longar MS (1984) Brain injury, edema and vascular permeability changes induced by oxygen-derived free radicals. Neurology 34: 315–320
2. Chang CJ, Ishii H, Yamamoto H, Yamamoto T, Spatz M (1993) Effects of cerebral ischemia on regional dopamine release and D_1 and D_2 receptors. J Neurochem 60: 1483–1490
3. Cvejic V, Kumami K, and Spatz M (1990) Effects of cerebral ischemia on synaptosomal uptake and release of ^3H-5-hydroxytryptamine in adult and young Mongolian gerbils. Metab Brain Dis 5: 1–6
4. Djuricic B, Micic DV, Mrsulja BB (1984) Phase recognition of edema caused by ischemia. In: Go KA, Baethmann A (eds) Recent progress in the study and therapy of brain edema. Plenum, New York, pp 491–498
5. Gurd, WJ, Jones RL, Mahler RH, Moore J (1974) Characterization, isolation and partial characterization of rat brain synaptic membrane. J Neurochem 22: 281–290
6. Ishii H, Stanimirovic DB, Chang CJ, Mrsulja BB, Spatz M (1993) Dopamine metabolism and free-radical related mitochondrial injury during transient brain ischemia in gerbils. Neurochem Res 18: 1193–1201
7. Lowry OH, Passonneau JV (1974) A flexible system of enzymatic analysis. Academic Press, New York
8. Lowry OH, Rosebrough NJ, Farr AL, Randall RJ (1951) Protein measurement with the folin phenol reagent. J Biol Chem 193: 952–958
9. Klatzo I (1967) Neuropathological aspects of brain edema. J Neuropath Exp Neurol 26: 1–14
10. Krajl M (1963) A rapid microfluorometric determination of monoamine oxidase. Biochem Pharmacol 14: 1684–1685
11. Nelson SR, Manz ML, Maxwell JA (1971) Use of specific gravity in the measurement of cerebral edema. J Appl Physiol 30: 268–271
12. Spatz M, Mrsulja BB (1990) Monoamines and cerebral ischemia. In: Schurr A, Rigor BM (eds) Cerebral ischemia and resuscitation. CRC, Boca Raton, Florida, pp 179–189
13. Sun M, Zigman S (1978) An improved spectrophotometric assay for superoxide dismutase based on epinephrine autoxidation. Anal Biochem 90: 81–89
14. Villacara A, Kumami K, Yamamoto T, Mrsulja BB, Spatz M (1989) Ischemic modification of cerebrocortical membranes: 5-hydroxytryptamine receptors, fluidity, and inducible in vitro lipid peroxidation. J Neurochem 53: 595–601

Correspondence: Maria Spatz, M. D., Bldg. 36, Room 4D04, Stroke Branch, NINDS, NIH, 9000 Rockville Pike, Bethesda, MD 20892, U.S.A.

Acta Neurochir (1994) [Suppl] 60: 242–245

Free Fatty Acid Liberation and Cellular Swelling During Cerebral Ischemia: the Role of Excitatory Amino Acids

Y. Katayama, T. Kawamata, T. Maeda, and **T. Tsubokawa**

Department of Neurological Surgery, Nihon University School of Medicine, Tokyo, Japan

Summary

We previously demonstrated the occurrence of cellular swelling during cerebral ischemia in vivo using microdialysis. ^{14}C-Sucrose was pre-perfused into the extracellular space (ECS) as an ECS marker, and cellular swelling was detected as an increase in extracellular concentration of ^{14}C-sucrose. In the present study, we examined in rats the time courses of the increase in FFA in relation to the occurrence of cellular swelling, and the role of excitatory amino acids (EAAs) in these events. A pair of microdialysis probes were placed in the hippocampus bilaterally. One probe was perfused with modified Ringer solution containing kynurenic acid (KYN, 10 mM), a broad-spectrum EAA antagonist, and the other with Ringer solution as a control. At 30 minutes after the initiation of perfusion, ischemia was induced by decapitation. The brain was subjected to microwave irradiation at 0–8 minutes after decapitation, and the FFA levels in the dorsal hippocampus were measured by high performance liquid chromatography. The time course of cellular swelling was determined in a separate group of animals using the previously described method. It was found that arachidonic and stearic acids began to demonstrate a rapid increase during the period from 1 to 2 minutes following ischemia induction. The levels of oleic and palmitic acids demonstrated similar but less marked increase. The rate of increase became less rapid after 4 minutes, suggesting a transient rapid increase superimposed on a background slow increase. The rapid increase roughly corresponded in timing to the occurrence of the early cellular swelling. The cellular swelling and concomitant rapid increase in FFAs during the early period of ischemia were both inhibited by KYN. Mutual relationships between these two events and the role of EAAs are discussed.

Keywords: Cerebral ischemia; excitatory amino acid; free fatty acid; phospholipase; hippocampus.

Introduction

The levels of free fatty acids (FFA) increase rapidly from 0.5 to 4 minutes after induction of cerebral ischemia[1,3,10,11]. Little is known concerning the precise mechanism of this early and rapid increase in FFAs. One of the most remarkable events occurring in the brain at 1–3 minutes after onset of ischemia is sudden and massive ionic fluxes and associated cellular swelling[2,6]. Concomitantly with this event, various neurotransmitters are released probably through the depolarization of nerve terminals[5]. In such circumstances, phospholipase A_2 and C would be activated through a mechanism coupled to Ca^{2+} entry and a guanosine triphosphate-binding protein. It appears possible therefore that the massive ionic fluxes and resultant neurotransmitter release may be responsible for the early rapid increase in FFAs.

We have previously demonstrated the occurrence of cellular swelling during cerebral ischemia in vivo using microdialysis[8]. ^{14}C-Sucrose was pre-perfused into the extracellular space (ECS) as an ECS marker, and cellular swelling was detected as an increase in extracellular concentration of ^{14}C-sucrose. In the present study, we examined the time courses of the increase in FFA in relation to the occurrence of cellular swelling. Furthermore, we tested whether or not the cellular swelling and the increase in FFAs could be inhibited by kynurenic acid (KYN), a broad-spectrum EAA antagonist, administered by microdialysis. The cellular swelling and the changes in FFA levels observed in the hippocampus were compared with those in the contralateral hippocampus of the same animal into which vehicle alone was perfused. This technique can sensitively detect pharmacological effects of KYN on other ischemia-induced neurochemical as well as morphological changes[4,7].

Materials and Method

Young adult Sprague-Dawley rats were anesthetized with a mixture of N_2O, O_2 and enflurane. The rectal temperature was maintained at 37–38°C. A pair of microdialysis probes (O.D., 500 µ;

effective length, 3 mm; cut off, 20,000 MW) were positioned vertically in the hippocampus (CA1 and dentate areas; 3.8–4.8 mm posterior to the bregma, 2.0 mm lateral to the midline, and 3.5 mm below the surface of the dura) bilaterally. The temperature of the perfusate for microdialysis was maintained at 37°C.

Detection of Cellular Swelling

The probes were initially perfused with modified Ringer solution containing ^{14}C-sucrose (10 mM) and 3 mM Ca^{2+} at a rate of 5.0 µl/ min. At 20 min after initiation of the ^{14}C-sucrose perfusion, the perfusate was switched to Ringer solution without ^{14}C-sucrose, and measurements of the ^{14}C-sucrose concentrations of each dialysate ($[^{14}C$-sucrose$]_d$) were initiated. The dialysate fractions were collected at 1-min intervals. $[^{14}C$-sucrose$]_d$ was expressed as relative radioactivity compared to radioactivity of the dialysate obtained immediately before ischemia induction. The osmolarity of each perfusate was maintained constant by adjusting the concentration of sodium chloride. Cerebral ischemia was induced by decapitation at 15–25 min following the termination of the ^{14}C-sucrose perfusion. In one of the 2 probes, the effects of in situ administration of KYN, a broad-spectrum EAA antagonist, were tested. Based on previous dose response studies[3,6], the smallest perfusate concentration of KYN which demonstrated near maximum effects were employed (10 mM). The microdialysis through the other probe served as the control.

Measurement of Free Fatty Acids

In separate group of animals, KYN was similarly administered through one of the 2 probes throughout the experiment. At 30 minutes after the initiation of KYN perfusion, ischemia was induced by decapitation. The brain was subjected to microwave irradiation at 1 or 4 minutes before, and 1, 2, 4 or 8 minutes after decapitation. The dorsal hippocampus was removed and homogenized for measurement of the FFA levels using a high performance liquid chromatograph (Beckman, System Gold) fitted with Ultrasphere C-18 Column (Beckman, 4.6 mm I.D. × 15 cm L.), a fluorescence detector (Japan Spectroscopic Co., Type 821-FP; excitation wave length, 56 nm; emission, 412 nm) and a Beckman 507 Autosampler. Fluorescent labeling of FFAs was carried out by the method of Nimura and Kinoshita[8]. The chloroform extract of the samples was dissolved in 50 µl 9-anthryldiazomethane (ADAM; Funakoshi Yakuhin, Tokyo) solution (0.5 mg/ml methanol). The conditions of analysis were as follows: mobile phase, 94% methanol; flow rate, 1.0 ml/min; temperature, 60°C; sample injection volume, 20 µl; analysis time, 25 min. The FFA peaks were identified by comparing their retention time with those of FFA standards processed by identical procedures. The FFA amounts were expressed as µg/g wet weight.

$[^{14}C$-sucrose$]_d$ and the FFA levels in the hippocampus perfused with KYN (test side) and those in the hippocampus of the other side perfused with Ringer solution alone (control side) were compared by employing the paired t–test and Wilcoxon matched-pair rank sum test.

Results

A sudden increase in $[^{14}C$-sucrose$]_d$ was usually observed during the period from 1 to 3 minutes following ischemia induction (Fig. 1A, solid dots). $[^{14}C$-sucrose$]_d$ remained to be elevated for 2–4 min. The increase in $[^{14}C$-sucrose$]_d$ was significantly delayed by microdialysis with perfusate containing KYN (Fig.

1A, open circles; p < 0.01 at 2–4 minutes). Thus, the latency of the fraction which displayed the largest increase as compared to the immediately preceding fraction was prolonged.

It was found that the levels of arachidonic and stearic acids began to demonstrate a rapid increase during the period from 1 to 2 minutes following ischemia induction (Fig. 1B, solid dots). The increase continued further, although the rate of increase became less rapid after 4 minutes, suggesting a transient rapid increase superimposed on a background slow increase. The rapid increase roughly coincided with the occurrence of the early cellular swelling. The levels of oleic and palmitic acids also demonstrated a rapid increase

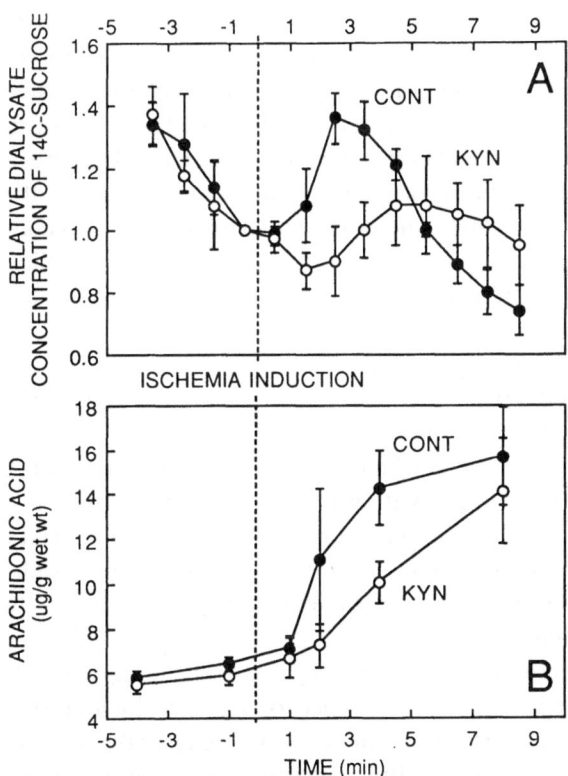

Fig. 1. (A) Changes in $[^{14}C$-sucrose$]_d$ in the dorsal hippocampus during cerebral ischemia, showing the effect of in situ administration of kynurenic acid (*KYN*, 10 mM) through the microdialysis probe. $[^{14}C$-sucrose$]_d$ was expressed as relative radioactivity compared to radioactivity of the dialysate obtained immediately before ischemia induction. (B) Changes in arachidonic acid level in the dorsal hippocampus during cerebral ischemia, showing the effect of in situ administration of KYN (10 mM) through the microdialysis probe. The results for each time point are from 5 animals (means ± S.D.). In each animal, two probes (control and test) were placed symmetrically in the hippocampus; one was perfused with KYN (open circles) and the other was perfused without KYN (solid dots). Ischemia was induced at time zero. p < 0.01 during the period between 2 and 4 minutes

beginning during the period from 1 to 2 minutes following ischemia induction. The rapid increase in these FFAs was, however, less pronounced. The rapid increases in FFA levels were profoundly inhibited by KYN (Fig. 1B, open circles; $p < 0.01$ at 2 and 4 min). In contrast, the background increases in the FFAs appeared unaffected by KYN.

Discussion

The present study demonstrated that [^{14}C-sucrose]$_d$ suddenly increases at 1–3 min following ischemia induction. Since sucrose is not taken up by either cells or capillaries, the absolute amount of ^{14}C-sucrose in the ECS must be unchanged. This increase, therefore, appears to represent a relative decrease in water volume in the ECS resulting from a movement of water into the cells, i.e., cellular swelling[2,8]. We have previously demonstrated a decrease in recovery rate of the microdialysis system in vivo in association with the massive ionic fluxes during cerebral ischemia[5]. The decrease in the recovery rate is presumably caused by a decrease in the effective surface area for microdialysis due to the ECS shrinkage. Therefore, actual increase in ECS concentration of ^{14}C-sucrose due to the cellular swelling may be larger than the magnitude estimated by the observed increase in [^{14}C-sucrose]$_d$.

The onset of a rapid increase in FFAs during the period from 1 to 2 minutes is in agreement with the observations reported by Abe *et al.*[1]. This suggests that certain event(s) which suddenly trigger a rapid FFA liberation occur with a latent period of 1–2 minutes. The period in which the rapid increase in FFAs takes place, corresponds to the timing of the cellular swelling, suggesting that these two events are causally related.

The early cellular swelling occurring during cerebral ischemia is a consequence of massive ionic fluxes across the plasma membrane which causes the osmotic pressure of intracellular impermeable anions no longer to be counterbalanced. There appear to be several mechanisms by which the ionic fluxes and the FFA release are mutually related. Phospholipase C could be activated by various neurotransmitters which are released as a result of such ionic fluxes (e.g.,[9]). The ionic fluxes involve an increase in the intracellular Ca^{2+} and thereby activate phospholipase A_2. The phospholipase C activation also results in an increase in inositol triphosphate-mediated Ca^{2+} mobilization from the intracellular store. Furthermore, the massive ionic fluxes facilitate adenosine triphosphate depletion and lactate

accumulation[3]. This would compromise energy-dependent FFA reacylation, enhancing the increase in FFA levels. At the same time, the released FFAs could be one of the cause of the ionic fluxes through changes in ion permeability of the plasma membrane.

The profound effect of KYN demonstrated in the present study suggests that EAAs are critically involved in both of these concomitant events. Our previous experiments revealed that a sudden increase in extracellular K^+, which represents an example of the consequences of sudden ionic shifts during cerebral ischemia, can be delayed by KYN administered by the same procedure[6]. The observed inhibition of the early cellular swelling can be attributed to such a delay in developing massive ionic fluxes. The delay in ionic shifts may be responsible for the inhibition of the rapid FFA liberation. At the same time, the attenuated FFA liberation may contribute to the delay in ionic shifts.

In summary, the present study demonstrated that the early cellular swelling and the rapid FFA liberation occur concomitantly after ischemia induction, and suggested that these two events are mutually related. EAAs which are also released concomitantly[4] appear to play a vital role in producing these events during the early period of ischemia. This work was supported by a research grant for Cardiovascular Disease (2A-2) from the Ministry of Health and Welfare of Japan.

References

1. Abe K, Kogure K, Yamamoto H, Imazawa M, Miyamoto K (1987) Mechanism of arachidonic acid liberation during ischemia in gerbil cerebral cortex. J Neurochem 48: 503–509
2. Hansen AJ, Olsen CE (1980) Brain extracellular space during spreading depression and ischemia. Acta Physiol Scand 108: 355–365
3. Ikeda M, Yoshida S, Busto R, Santiso M, Ginsberg MD (1986) Polyphospho -inosities as a probable source of brain free fatty acids accumulated at the onset of ischemia. J Neurochem 47: 123–132
4. Katayama Y, Kawamata T, Kano T, Tsubokawa T (1992) Excitatory amino acid antagonist administered via microdialysis attenuates lactate accumulation during cerebral ischemia and subsequent hippocampal damage. Brain Res 584: 329–333
5. Katayama Y, Kawamata T, Tamura T, Becker DP, Tsubokawa T (1991) Calcium-dependent glutamate release concomitant with massive potassium flux during cerebral ischemia in vivo. Brain Res 558: 136–140
6. Katayama Y, Tamura T, Becker DP, Tsubokawa T (1991) Calcium-dependent component of massive increase in extracellular potassium during cerebral ischemia as demonstrated by microdialysis in vivo. Brain Res 558: 136–140
7. Katayama Y, Tamura T, Becker DP, Tsubokawa T (1991) Inhibition of rapid potassium flux during cerebral ischemia in vivo with excitatory arnino acid antagonist. Brain Res 568: 294–298

8. Katayama Y, Tamura T, Becker DP, Tsubokawa T (1992) Early cellular swelling during cerebral ischemia in vivo is mediated by excitatory amino acids released from nerve terrninal. Brain Res 577: 121–126

9. Nimura N, Kinoshita T (1980) Fluorescent labeling of fatty acids with 9-anthryl -diazomethane (ADAM) for high performance liquid chromatography. Anal Lett 13 (A3): 191

10. Umemura A, Mabe H, Nagai H, Sugino F (1992) Action of phospholipase A2 and C on free fatty acid release during complete ischemia in rat neocortex. Effect of phospholipase C inhibitor and N-methyl-D-asparate antagonist. J Neurosurg 76: 648–651

11. Yasuda H, Kishiro K, Izumi N, Nakanishi M (1985) Biphasic liberation of arachidonic and stearic acids during cerebral ischemia. J Neurochem 45: 168–172

Correspondence: Yoichi Katayama, M. D., Ph. D., Department of Neurological Surgery, Nihon University School of Medicine, Tokyo 173, Japan.

Acta Neurochir (1994) [Suppl] 60: 246–249
© Springer-Verlag 1994

Severe Brain Edema Associated with Cumulative Effects of Hyponatremic Encephalopathy and Ischemic Hypoxia

Z. S. Vexler[1], T. P. L. Roberts[1], J. Kucharczyk[1], and A. I. Arieff[2]

[1] Neuroradiology Section, University of California, San Francisco and [2] Department of Medicine, Geriatrics Section, Veteran's Affairs Medical Center and University of California, San Francisco, CA, U.S.A.

Summary

Hyponatremia in cats produced brain edema, detectable by both magnetic resonance imaging (MRI) and increased brain water, with a compensatory decrease of brain sodium. Sodium transport was measured in synaptosomes from hyponatremic cat cerebral cortex. The sodium efflux via Na^+-K^+-ATPase was significantly higher (144%) than control, while sodium influx via the Na^+/H^+ antiporter was significantly decreased (74%). Both responses tend to decrease brain intracellular sodium and thus, brain cell osmolality. Ischemia following unilateral middle cerebral artery occlusion also resulted in brain edema. However, the efflux of sodium via both Na^+- K^+-ATPase and sodium channels actually decreased, both maladaptive responses. Furthermore, when ischemia was superimposed upon hyponatremia, all of the cerebral adaptive changes which had been induced by hyponatremia alone were rendered ineffective. This resulted in further elevations of brain water and sodium. Hyponatremia superimposed upon ischemia thus worsens the brain edema associated with ischemia alone. Thus, ischemia impairs the ability of the brain to adapt to hyponatremia, probably by eliminating the compensatory mechanisms of brain sodium transport initiated by hyponatremia.

Keywords: Ischemia; hyponatremia; synaptosomes; sodium.

Introduction

Hyponatremia is among the most common metabolic abnormalities seen in a general hospital population[1,8,24]. Among patients with hyponatremia, about 1% develop encephalopathy, and of these, about 20% die or suffer permanent brain damage[1,5]. Death in such patients is largely the result of brain edema which, if the brain is unable to adapt, often leads to cerebral herniation.

It has been postulated from the clinical observations[1,5,24] that ability of the brain to adapt to hyponatremia-induced edema can be seriously impaired by either hypoxia, ischemia or other pathophysiologic factors. Brain edema is the eventual cause of irreversible brain damage in patients with hyponatremic encephalopathy. The objective of this study was to evaluate effects of ischemic hypoxia on the ability of the brain to adapt to acute hyponatremia.

Methods

Animal models: Experiments were performed in two groups of female cats: normal controls (6 animals) and acute hyponatremia for three hours (8 animals). Additionally, unilateral middle cerebral artery (MCA) occlusion was induced in each group. Thus, four subgroups were created: normal controls; hyponatremic normoxic animals; normonatremic cats with focal ischemia (ischemic hypoxia); and, hyponatremic cats in which focal ischemia for 90 min was induced following 90 min, of induction of hyponatremia.

Animals weighing 2.5–4.0 kg were intravenously anesthetized with 35 mg/kg sodium pentobarbital. Femoral artery and vein catheters were placed for arterial blood gases and blood pressure monitoring and administration of the contrast agent. During the MRI experiment cats were then maintained with 1–2% isoflurane in 30% oxygen/70% nitrous oxide, and mechanically ventilated through an endotracheal tube. Acute hyponatremia was induced in cats by subcutaneous injection of 1 unit of arginine vasopressin in oil and intraperitoneal injection of 12.5% body weight of 140 mM, followed by two subcutaneous injections of 0.5 units of arginine vasopressin in water each 30 min. Serum sodium was monitored on the 30 min basis. Serum sodium was 123 ± 3 mM after 3 h, which is 30 mM below the control value of 153 ± 1 mM. Unilateral ligation of the MCA for 90 min was induced as previously described[12,17]. After ligation of the MCA, the cats were immediately placed in the magnet. At the end of the imaging, the cats were sacrificed, and brain tissue samples were quickly dissected from ischemic tissue in right hemisphere (areas were defined by contrast-enhanced MRI) and from the corresponding areas of the contralateral hemisphere in each individual animal. Most of the tissue was immediately placed in the isolation media for processing synaptosomes. The rest of the tissue from chosen regions was used for the water and electrolytes analysis.

Magnetic resonance imaging (MRI) protocol: Imaging experiments were performed on a 2 Tesla Omega CSI system equipped with Acustar S-150 self shielded gradients (20 gauss/cm, 15 cm i.d.). A home built 9 cm i.d. volume birdcage RF coil was used in all studies. Stejskal-Tanner diffusion sensitizing gradients[23] of strengths up to 11 gauss/cm were employed to obtain diffusion-weighted images with "b-values" in the range 0–2442 s/mm². Spatial "maps" of the apparent diffusion coefficient (ADC) were generated by pixel-by-pixel logarithmic regression analysis of the diffusion-weighted images, assuming an exponential loss of signal dependent on the product of the ADC and the image "b-value"[14], $S \sim \exp(-bD)$, where D represents the ADC. Sixteen contrast-enhanced high speed echo-planar images (echo time 80 ms) were acquired at 2 s intervals following bolus intravenous injection of 0.25 mmol/kg of a magnetic susceptibility contrast agent, Dy-DTPA-BMA (Sprodiamide injection, Nycomed Salutar, Inc. and Sterling Winthrop). The signal intensity variations before, during and after the injection of contrast agent were compared in the regions-of-interest in the occluded MCA in the right hemisphere and corresponding areas in non occluded right hemisphere.

Isolation of synaptosomes: Synaptosomes were isolated from brain regions which were selected on the basis of observed perfusion deficits on MRI perfusion-sensitive imaging[13,20], and from the corresponding areas of the contralateral hemisphere in each individual animal. Synaptosomes were prepared as previously described from our laboratory[7]. Isolation media contained 320 mM sucrose, 0.2 mM K-EDTA, 5 mM Tris-HCl, pH = 7.40. We modified the previously described methods[7] by using 7.2% and 11.5% Ficoll gradients.

Transport activity of Na^+-K^+-ATPase: Rubidium (Rb) uptake with or without ouabain (2.5 mM) was used for the estimation of potassium transport via Na^+-K^+-ATPase in synaptosomes preloaded for 10 min. at 37° C with sodium [in mM) 150 NaCl, 1 $MgCl_2$, 5 HEPES-Tris, pH = 7.40].The uptake media contained (in mM) 140 choline chloride, 5 $MgCl_2$, 0.2 EGTA, 1 KCl, 5 HEPES-Tris, pH = 7.40, 0.5 µCi ^{86}Rb 70,000 cpm. The mean of the ^{86}Rb uptake with ouabain was subtracted from the uptake without ouabain for each individual experiment. Protein was estimated by the Lowry method[15].

Veratridine-stimulated sodium uptake: Sodium uptake with or without veratridine HCl (50 µM) was estimated in synaptosomes preloaded for 10 min at 37° C with potassium [(in mM) 150 KCl, 1 $MgCl_2$, 5 HEPES-Tris, pH = 7.40][7]. The uptake media contained (in mM) 140 choline chloride, 5 $MgCl_2$, 0.2 EGTA, 1 NaCl, 5 HEPES-Tris, pH = 7.40, 5.0 µCi ^{22}Na ~ 5 × 10⁵ cpm. The mean of the ^{22}Na uptake without veratridine was subtracted from uptake with veratridine for each individual experiment.

Amiloride-sensitive sodium uptake: Sodium uptake was estimated in synaptosomes with or without amiloride (1 mM)[11]. Preequilibrium media contained (in mM) 150 KCl, 1 $MgCl_2$, 5 HEPES-Tris, pH = 7.40. The uptake media contained (in mM) 140 choline chloride, 5 $MgCl_2$, 0.2 EGTA, 1 NaCl, 1 × 10⁻³ tetrodotoxin, 5 HEPES-Tris, pH = 7.40, 5.0 µCi ^{22}Na ~ 5 × 10⁵ cpm. The ^{22}Na uptake with amiloride was subtracted from uptake without amiloride for each individual experiment.

Water and tissue sodium content: Water and sodium content were assessed in brain gray matter as previously described[2].

Statistical analysis: Quantitative data are expressed as means ± SE or as a % of mean in occluded hemisphere compared to control. For the comparisons between groups, data was analyzed by ANOVA.

Results

MRI Study

We performed diffusion-weighted MRI experiments to assess brain edema using the ADC as previously described[10]. In non-ischemic cerebral tissue, hy-ponatremia alone caused an 11.4% decrease of the ADC compared to the control mean of 0.79 ± 0.02 × 10⁻⁵ cm²/s (p < 0.03). Ischemia alone resulted in a 20.3% decrease of ADC values in ischemic (MCA-occluded) brain tissue. In ischemic/hyponatremic brain, the ADC was 21.5% lower than in normal brain (p < 0.01).

In normonatremic cats, the water content in the ischemic vascular territory after 90 min of MCA occlusion was 103% of the normal value (389 ± 10 ml/100 gm dry tissue) (p > 0.05). Following 3 hours of hyponatremia without ischemia, there was an increase of brain water content to 112%, which was significantly greater than the values in both control (p < 0.001) and ischemic cat brain (p < 0.025). Combined hyponatremia and ischemia caused a further elevation in brain water (122%), which was significantly greater than the value in normal controls (p < 0.001), ischemic cats (p < 0.001), and hyponatremic animals without ischemia (p < 0.001).

Brain sodium content was 121% of the control value (262 ± 14 mmol/kg dry wt.) (p < 0.05) in ischemic areas. In contrast, hyponatremia without ischemia produced a significant adaptive decrease of brain sodium by 11% (p < 0.04). However, the adaptive decrease of brain sodium was not adequate to prevent a significant increase of brain water. Combined hyponatremia and ischemia was also associated with a reversal of the decreased brain sodium observed with hyponatremia alone. This condition was also associated with a significant increase in brain edema.

Sodium Transport in Synaptosomes

We then investigated how ischemia impaired the ability of the hyponatremic brain to transport sodium. Synaptosomes have proven useful in the investigation of cellular (neuronal) mechanisms of brain damage associated with different disorders of the central nervous system[11,19].

Hyponatremia alone resulted in an increase of Na^+-K^+-ATPase transport activity compared to normal (144%, p < 0.001). However, in synaptosomes isolated from normonatremic ischemic tissue, the Na^+-K^+-ATPase activity was decreased to 56% of normal, significantly less than either the control or hyponatremic values (p < 0.001). When ischemia was superimposed upon hyponatremia, the Na^+-K^+-ATPase activity was only 66% of normal, significantly lower than the value in cats with either hyponatremia alone, or in synaptosomes from normal cat brain (p < 0.001).

Ischemia itself resulted in a significant decrease of sodium uptake via veratridine-stimulated sodium channels (57% of normal, p < 0.002). In hyponatremic animals, veratridine-stimulated synaptosomal sodium uptake tended to be lower than in control, although it was not significantly different (91%). However, it was significantly higher than control after ischemia (p < 0.002). In animals with hyponatremia plus ischemia, veratridine-stimulated sodium uptake was also significantly higher than in cats with ischemia alone (152%, p < 0.002).

Amiloride-sensitive sodium uptake was significantly less than the control value in animals which had hyponatremia without ischemia (79% of normal, p < 0.05). However, when ischemia was added to the hyponatremia, the amiloride-sensitive sodium uptake was not significantly different from either the control or hyponatremic values.

Discussion

In this study we evaluated brain edema in cat using non-invasive high speed echo-planar diffusion MR imaging and measuring the content of cerebral cortex (gray matter) water. Both methods delineated brain edema associated with ischemia, hyponatremia or combinations of those.

We found progressive elevation of brain water content associated with ischemia, hyponatremia and hyponatremia combined with ischemia when compared to either hyponatremia or ischemia alone. We also found a significant decrease of the ADC in each of experimental groups versus control. The maximum decrease in the ADC was found in the areas where there was no blood flow, i.e. in ischemic areas, regardless of the serum sodium level. At the same time, water content only tended to be elevated in ischemic gray matter in comparison with control. The brain water was much higher after the combination of hyponatremia and ischemia when compared to either normal or hyponatremic brain. In contrast, hyponatremia itself caused major shift of water into the brain as a result of osmotic disequilibrium between blood and tissue, which was significantly higher than in cats with ischemia alone. Hyponatremia itself resulted in a moderate reduction of the ADC. These two experimental facts together provide information not only about development of cytotoxic edema[17,20,21], but also indicate the dynamics of water movement in brain following hyponatremia. The results suggest activation of mechanisms participating in regulatory volume decrease

following hyponatremia, including cell membrane pumps. Changes in brain (gray matter) sodium content support this last point. Brain sodium appeared to be lower after acute hyponatremia, indicating activation of mechanisms to extrude sodium from the brain. Loss of brain sodium as an initial response to hyponatremia was also reported previously[3,16]. In contrast, ischemia, either alone or combined with hyponatremia, resulted in elevation of brain sodium. These data suggest that ischemia eliminates the brain adaptive mechanisms to hyponatremia.

To test this hypothesis we evaluated activity of neuronal (synaptosomal) membranes to transport osmolitically active cations (sodium and potassium). The main findings with regard to ion transport activity of Na^+,K^+-ATPase were activation of sodium extrusion induced by hyponatremia, and a decrease in sodium efflux via Na^+,K^+-ATPase induced by ischemia. Activation of sodium extrusion via Na^+,K^+-ATPase seems to be one of the main mechanisms contributing to brain adaptation to hyponatremic encephalopathy. Although we did not do any measurements of high-energy phosphates in this study, the maintenance of functional activity, and even activation, of Na^+,K^+-ATPase in our experiments indirectly indicates availability of ATP (no external ATP was added) in the hyponatremic state. In contrast, ischemia resulted in major decreases of Na^+,K^+-ATPase activity. This very likely occurred because of lack of intracellular ATP during ischemia[13,22].

Following hyponatremia, a significant decrease of sodium uptake in synaptosomes from hyponatremic animals was found in sodium transport via the Na^+/H^+-exchanger. This pathway appears to be important for the regulation of cell pH, intracellular sodium concentration and cell volume in most mammalian systems[10,11]. Initial regulatory volume decrease demands extrusion of intracellular sodium and/or limitation of influx of sodium. Decreases in sodium influx found in our experiments serve to compensate tissue-blood osmotic disequilibrium. However, the combination of hyponatremia and ischemia resulted in an inversion of the decreased sodium influx via Na^+-K^+-ATPase associated with hyponatremia alone.

Summarizing, ischemia not only eliminated the increased Na^+-K^+-ATPase transport activity initiated by hyponatremia, but actually decreased the transport activity to levels below normal and minimized the adaptive decrease in sodium influx via Na^+/H^+-exchanger. In aggregate, those factors together caused a net inflow of sodium into the brain, resulted in increased cerebral edema and significantly exacerbated brain injury induced by hyponatremia. Hyponatremia superimposed upon ischemia worsens the degree of brain edema associated with ischemia alone. Relatively short periods of ischemia or incomplete ischemia generally can be tolerated with the possibility of no irreversible brain damage[9,18,20,22] in spite an obvious metabolic disturbance[13,18]. However, exacerbation by metabolic abnormalities associated with hyponatremia may prevent tissue from recovering from ischemia, leading to delayed brain damage.

Thus, the severe brain edema associated with the combined effects of hyponatremic encephalopathy and ischemic hypoxia might be an important cause of the permanent brain damage seen in such patients[1,5].

References

1. Arieff AI, Ayus JC, Fraser CL (1992) Death or permanent brain damage in healthy children with hyponatremia. BMJ 304: 1218–1222
2. Arieff AI, Kleeman CR, et al (1973) Studies on mechanisms of cerebral edema in diabetic comas. Effects of hyperglycemia and rapid lowering of plasma glucose in normal rabbits. J Clin Invest 52 (3): 571–583
3. Arieff AI, Llach F, et al (1976) Neurological manifestations and morbidity of hyponatremia: correlation with brain water and electrolytes. Medicine (Baltimore) 55(2): 121–129
4. Ayus JC, Olivero JJ, Frommer JP (1982) Rapid correction of severe hyponatremia with intravenous hypertonic saline solution. Am J Med 72: 43–48
5. Ayus JC, Wheeler JM, Arieff AI (1992) Postoperative hyponatremic encephalopathy in menstruant women. Ann Intern Med 117: 891–897
6. Fraser CL, Sarnacki P (1989) Na$^+$-K$^+$ ATPase pump function in male rat brain synaptosomes is different from that of females. Am J Physiol 257 (Endocrinol Metab 20): E284–E289
7. Fraser L, Sarnacki P, Arieff AI (1985) Abnormal sodium transport in synaptosomes from brain of uremic rats. J Clin Invest 75(6): 2014–2023
8. Griggs RC, Arieff AI (1992) Hypoxia and the central nervous system. In: Gripps RC, Arieff AI (eds) Neurologic manifestations of systemic disorders. Little, Brown, Boston, pp 39–54
9. Heiss WD (1983) Flow thresholds of functional and morphological damage of brain tissue. Stroke 14(3): 329–331
10. Jean T, Frelin C, et al (1985) Biochemical properties of the Na$^+$/H$^+$ exchange system in rat brain synaptosomes. J Biol Chem 260: 9678–9684
11. Kandai F, Sarnacki P, Arieff AI (1992) Atrial natriuretic peptide inhibits amiloride-sensitive sodium uptake in rat brain. Am J Physiol 263: R279–R283
12. Kucharczyk J, Mintorovitch J, Moseley ME, Asgari H, Sevick R, Derugin N, Norman D (1991) Ischemic brain damage: reduction by sodium-calcium channel modulator RS-87476. Radiol 179 (1): 221–227
13. Kucharczyk J, Moseley ME, et al (1989) MRS of ischemic/hypoxic brain disease. Invest Radiol 24(12): 951–954
14. LeBihan D (1990) Magnetic resonance imaging of perfusion. Magn Reson Med 14: 283–292
15. Lowry OH, Rosenberg AL, et al (1951) Protein measurements with Folin-phenol reagent. J Biol Chem 193: 265–275
16. Melton JE, Patlak CS, et al (1987) Volume regulatory loss of Na, Cl, and K from rat brain during acute hyponatremia. Am J Physiol 252 (Renal Fluid Electrolyte Physiol 21): F661–F669
17. Moseley ME, Kucharczyk J, et al (1990) Diffusion-weighted MRI of acute stroke. Am J Neuroradiol 11: 423–429
18. Obrenovitch TP, Garofalo O, et al (1988) Brain tissue concentrations of ATP, phosphsreatine, lactate, and tissue pH in relation to reduced cerebral blood flow following experimental acute middle cerebral artery occlusion. J Cereb Blood Flow Metab 8: 866–874
19. Rafalowska U, Erecinska M, Wilson DF (1980) Energy metabolism in rat brain synaptosomes from nembutal-anesthetized and nonanesthetized animals. J Neurochem 34: 1380–1386
20. Roberts TPL, Vexler ZS, et al (1993) High-speed MR imaging of ischemic brain injury following partial stenosis of the middle cerebral artery. J Cereb Blood Flow Metabolism: in press
21. Sevick RJ, Kucharczyk J, et al (1990) Diffusion-weighted MR imaging and T2-weighted MR imaging in acute cerebral ischaemia: comparison and correlation with histopathology. Acta Neurochir (Wien) [Suppl]51: 210–212
22. Siesjö BK (1978) Brain energy metabolism. Wiley, New York
23. Stejskal EO, Tanner JE (1965) Spin diffusion measurements: spin-echoes in the presence of a time-dependent field gradient. J Chem Phys 42: 288–292
24. Tien R, Arieff AI, et al (1992) Hyponatremic brain damage: Is central pontine myelinolysis common? Am J Med 92 (5): 513–522

Correspondence: Allen I. Arieff, M.D., Department of Medicine, V.A. Medical Center (111G), 4150 Clement Street, San Francisco, CA 94121, U.S.A.

Acta Neurochir (1994) [Suppl] 60: 250–252

Endogenous Superoxide Dismutase Activity in Reperfusion Injuries

T. Fukuhara, M. Gotoh, M. Kawauchi, S. Asari, and **T. Ohmoto**

Department of Neurological Surgery, Okayama University Medical School, Okayama, Japan

Summary

To elucidate the relationship between reperfusion injuries and free radicals, we monitored the endogenous superoxide dismutase (SOD) activity by intracerebral microdialysis. Six cats underwent a transient occlusion of the middle cerebral artery for 60 minutes after microdialysis probes were implanted bilaterally into the white matter under the ectosylvian gyrus. Dialysates were collected at 30 minute intervals over the course of 5 hours after reperfusion. The SOD activity of the dialysates was measured with electron spin resonance spectrometry. Regional cerebral blood flow was measured simultaneously and the water content of the white matter was assayed at the end of the experiment. After reperfusion, SOD activity increased significantly in the first 30 minutes compared with the preoperative value, and decreased over 4–4.5 hours and 4.5–5 hours in the occluded side. The water content in the occluded side was significantly higher than that in the contralateral side. The highest SOD activity during reperfusion and the water content in the occluded side seemed to correlate, although not significantly. A leakage of intracellular SOD or a reactive increase of SOD activity in response to the reperfusion injury are possible mechanisms of increase in extracellular SOD.

Keywords: Microdialysis; reperfusion; superoxide dismutase.

Introduction

Free radicals are assumed to play an important role in reperfusion injury after brain ischemia[2]. Exogenous superoxide dismutase (SOD) has been administered to scavenge free radicals, and some investigators have reported its protective effect against reperfusion injury[1]. Although the induction of SOD mRNA during reperfusion injury has been reported[7,11], changes of endogenous SOD have yet to be determined. We analyzed sequential changes in SOD activity in the extracellular space of the brain during reperfusion.

Materials and Methods

Six cats were anesthetized with halothane inhalation at a maintenance dose of 1.0% and immobilized with pancuronium bromide at 0.5 mg/kg/h. After tracheostomy, the left femoral vein and artery were catheterized and the skull was fixed in a stereotactic frame. A burr hole was made over the middle ectosylvian gyrus on each side. The dura was incised and both, a microdialysis probe and an electrode to measure regional cerebral blood flow (rCBF) were inserted into the subcortical white matter. The probe was made of dialysis membrane tubing with a high molecular cut-off (Evaflux 4A, Kuraray Medica, Kitaku, Osaka, Japan). With this type of probe, the recovery rate of Cu, Zn-SOD in vitro was determined to be $12.88 \pm 3.18\%$ (mean value of five probes). The recovery rate in vivo was determined immediately before implantation by immersing the probe into cat serum. Each probe was perfused with Ringer's solution at 2 μl/min. After implantation the probes were allowed an equilibration time of 4 hours.

The right middle cerebral artery of each cat was occluded with a temporary clip utilizing a transorbital approach.

The clip was removed after 60 minutes, and dialysates were sampled at 30 minute intervals during a course of 5 hours of reperfusion. Evans blue (2%, 2 ml/kg) was then injected into the femoral vein. Thirty minutes later, cardiac arrest was induced by rapid injection of KCL (10%, 2 ml/kg) into the femoral artery. The brain was removed promptly and white matter tissue near the microdialysis probe was dissected. The water content was determined using specific gravity. The SOD activity of the dialysate was assayed with an electron spin resonance spectrometer (JES-FE1XG, JEOL Ltd., Tokyo, Japan) using spin trapping[3]. To obtain a standard baseline, SOD solutions were prepared at concentrations of 0.5, 1, 5 and 10 units/ml with human erythrocyte Cu, Zn-SOD (Sigma, St. Louis, Missouri). To obtain a relative signal intensity, manganese oxide was used as internal standard. SOD activity in the extracellular space was calculated by dividing the activity of the dialysate by the recovery rate of the probe.

Results

During occlusion of the middle cerebral artery, rCBF decreased to less than 60% in all animals as compared with the preoperative value. During reperfusion, however, no significant changes were observed on either side. The water content was significantly different between the occluded side $(71.66 \pm 1.57\%)$ and the contralateral side $(68.67 \pm 1.85\%)$ ($p < 0.002$, paired t-test). Evans blue staining was not observed.

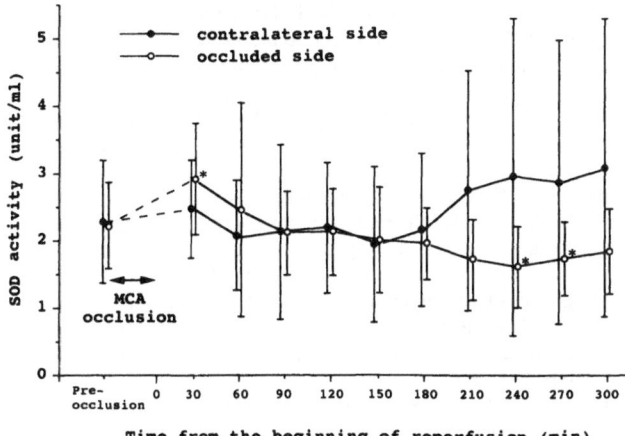

Fig. 1. Sequential changes of SOD activity during reperfusion. * Significantly different from the pre-occlusion value at p < 0.05 by paired t-test

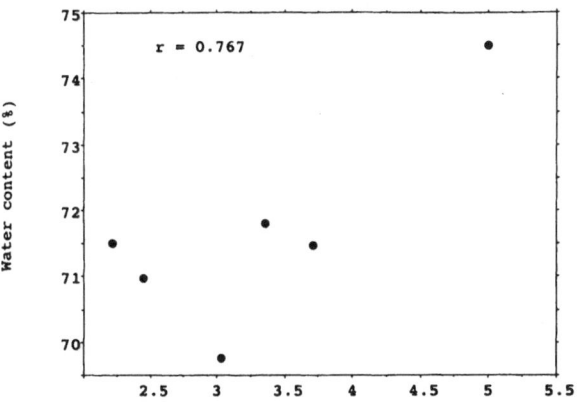

Fig. 2. Scattergram showing a correlation between the highest SOD activity during reperfusion and the brain water content after 1 hour of occlusion followed by 5 hours of reperfusion of the occluded side

On the occluded side, significant changes in SOD activity were observed during reperfusion as compared with the preoperative value (Fig. 1). SOD activity increased in the first 30 minutes of reperfusion and decreased at 4–4.5 and 4.5–5 hours (p < 0.05, paired t-test). There was no correlation between rCBF and SOD activity. The highest SOD activity obtained during reperfusion and the water content of the occluded side seemed to correlate (r = 0.767), but not significantly (Fig. 2).

Discussion

The SOD activity determined in this study may reflect not only SOD but also low molecular antioxidants, such as ascorbic acid or glutathione[6]. These substances may pass through the microdialysis membrane at different rates and may have some effects on the activity measurement. Consequently, results should be expressed as SOD-like activity. The activity of homogenized brain tissue, from which low molecular substances were removed by dialysis, has been evaluated by using the spin trapping method[10]. The studies confirmed that these substances had no significant effect for up to 4 hours. Based on this finding, the activity currently observed may largely reflect SOD-derived activity.

An increase in SOD activity after acute ischemic stroke has been reported[5,9] with different interpretations. Some authors believe that the increase in SOD activity originated from leakage[9]; others assume that it is indicative of a protective reaction against free radical injury[5]. Our results showed that the increase occurred during the initial stage of reperfusion but not during occlusion. If SOD leakage from injured cells was a major factor, an increase should be observed already during occlusion. Studies showing that SOD-mRNA expression is increased by reperfusion injury[7,11] may support the concept of a protective reaction. Neither interpretation, however, explains the decrease in SOD activity observed in this experiment. Some studies report a decrease in SOD activity after reperfusion of ischemic brain tissue[8,12]. Therefore, a mechanism lowering SOD activity may exist. Another study noted that accumulation of hydrogen peroxide irreversibly[4] inactivates SOD, which also may be a factor.

The presently found correlation between the highest SOD activity during reperfusion and the cerebral water content suggests that endogenous SOD activity is an indicator of the severity of reperfusion injury. Thus, SOD activity may increase in association with the severity of the reperfusion injury. Administration of exogenous SOD has been reported to attenuate the increased permeability of the blood-brain barrier[1]. Based on this finding, an increase in SOD activity should be expected to reduce brain edema. Such a protective effect, however, was not observed in this study, possibly because the endogenous increase alone may be insufficient. Further studies are necessary to elucidate the role of endogenous SOD activity.

References

1. Armstead WM, Mirro R, Thelin OP, Shibata M, Zuckerman SL, Shanklin DR, Busija DW, Leffler CW (1992) Polyethylene glycol superoxide dismutase and catalase attenuate increased blood-brain barrier permeability after ischemia in piglets. Stroke 23: 755–762

2. Hallenbeck JM, Dutka AJ (1990) Background review and current concepts of reperfusion injury. Arch Neurol 47: 1245–1254

3. Hiramatsu M, Kohno M, Edamatsu R, Mitsuta K, Mori A (1992) Increased superoxide dismutase activity in aged human cerebrospinal fluid and rat brain determined by electron spin resonance spectrometry using the spin trapping method. J Neurochem 58: 1160–1164

4. Hodgson EK, Fridovich I (1975) The interaction of bovine erythrocyte superoxide dismutase with hydrogen peroxide: inactivation of the enzyme. Biochemistry 14: 5294–5303

5. Horáková L, Uraz V, Ondrejicková O, Lukovic L, Juránek I (1991) Effect of stobadine on brain lipid peroxidation induced by incomplete ischemia and subsequent reperfusion. Biomed Biochem Acta 50: 1019–1025

6. Landolt H, Lutz TW, Langemann H, Stäuble D, Mendelowitsch A, Gratzl O, Honegger CG (1992) Extracellular antioxidants and amino acids in the cortex of the rat: monitoring by microdialysis of early ischemic changes. J Cereb Blood Flow Metab 12: 96–102

7. Matsuyama T, Michishita H, Nakamura H, Tsuchiyama M, Shimizu S, Watanabe K, Sugita M (1993) Induction of copper-zinc superoxide dismutase in gerbil hippocampus after ischemia. J Cereb Blood Flow Metab 13: 135–144

8. Nadasy GL, Mela-Riker L, Reivich M, Kovach AG (1989) Parallel changes in brain tissue blood flow and mitochondrial function during and after 30 minutes of bilateral forebrain ischemia in the gerbil. Acta Physiol Hung 74: 267–276

9. Strand T, Marklund SL (1992) Release of superoxide dismutase into cerebrospinal fluid as a marker of brain lesion in acute cerebral infarction. Stroke 23: 515–518

10. Takagi K, Kanemitsu H, Tomukai N, Oka H, Tamura A, Kohno M, Mitsuda K, Yoshida S, Sano K (1992) Changes of superoxide dismutase activity and ascorbic acid in focal cerebral ischaemia in rats. Neurol Res 14: 26–30

11. Uyama O, Matsuyama T, Michishita H, Nakamura H, Sugita M (1992) Protective effects of human recombinant superoxide dismutase on transient ischemic injury of CA1 neurons in gerbils. Stroke 23: 75–81

12. Yang Y-J, Tang W-X, Tian H-C, Yu P-L (1991) A new model of global postischemic reperfusion in rabbit. Mol Chem Neuropathol 14: 11–23

Correspondence: Toru Fukuhara, M. D., Department of Neurological Surgery, Okayama University Medical School, 2-5-1 Shikatacho, Okayama, 700, Japan.

Acta Neurochir (1994) [Suppl] 60: 253–256
© Springer-Verlag 1994

Endogenous Opioid System Activity Following Temporary Focal Cerebral Ischemia

P. Ting, S. Xu, and **S. Krumins**

Departments of Pediatrics and Child Health, College of Medicine, Howard University, Washington, DC, U.S.A.

Summary

We studied changes in opioid receptors (μ, δ, κ) concentrations during temporary middle cerebral artery occlusion (MCAO) in cats by sequential displacement of unselective opioid antagonist, [^{3}H]-diprenorphine with highly selective ligands for μ, δ and κ, subsites. Following threshold cerebral ischemia (rCBF < 10ml/100 g/min) there was a 2 to 3 fold increase in the 3 opioid receptor subtype concentrations at 10 min following the release of MCAO. Further, 56% of the cats depicted early postischemic hyperemia BBB opening, at 1 h and 3 h following the release of occlusion, with significant subsequent progression of brain edema. We believe that the enhanced brain opioid activity may be relevant to the neuronal damage caused by the early postischemic BBB opening.

Keywords: Cerebral ischemia; blood-brain barrier; opioid receptors.

Introduction

Scientific and biomedical technical advances in the past 2 decades have indeed made significant progress in both identification, and understanding of the complex cerebral vascular-cellular-molecular events associated with cerebral ischemia. However, the exact mechanisms responsible for the irreversible ischemic brain damage are not clear. Among the various postischemic factors suggested, an early transient postischemic hyperemic blood-brain barrier (BBB) disruption has been shown to be associated with, and causally related to progression of brain edema, late necrotic BBB opening and cerebral infarction; but a cellular basis for this irreversible brain injury has not been explored[11,12,20]. Among many cellular mediators proposed, enhanced endogenous opioid system activity has been suggested to mediate ischemic neurological dysfunction and neuronal damage, while treatment with opioid antagonists, ameliorate these adverse effects[5,14].

We therefore, proposed to study the possible role of enhanced brain opioid system activity in mediating neuropathology following the early transient postischemic hyperemic BBB disruption in cats subjected to temporary MCAO.

Materials and Methods

Male mongrel cats weighing 2.5 to 3.5 kg were studied after overnight fasting, but were allowed free access to water. Following initial induction with Ketamine (20 mg/kg) IM, a cat was intubated and placed on a respirator. Femoral artery and vein were cannulated to monitor arterial blood pressure (BP), pH, PCO_2, and PO_2. Anesthesia was maintained with α-chloralose (60 mg/kg IV). With the aid of a heating pad, rectal temperature was kept between 37–38°C. The cat's head was then fixed in a stereotaxic apparatus and a left middle cerebral artery (MCA) was exposed via a transorbital approach. The proximal MCA was occluded for an hour, and then released. Regional cerebral blood flow (rCBF) was measured by standard hydrogen clearance technique described elsewhere[12]. The recording platinum microelectrodes were inserted stereotaxically into those coordinates (Snyder's Atlas) of caudate and cortex supplied by the MCA. The rCBF is calculated using the formula $F = \lambda\ (0.693/T) \times 100$, where F = flow in ml/100 g/min, λ brain blood partition coefficient (for H_2, $\lambda = 1$), 0.693 = $\log_e 2$, and T is the effective half time in minutes for desaturation. To evaluate BBB opening to albumin, 2 ml/kg of 2% Evans blue (EB) IV was injected either prior to MCAO for cats sacrificed at the end of 1 h occlusion, or immediately after the release of occlusion for cats sacrificed at 10, 60, and 180 min later (early transient postischemic BBB opening). In sham-operated controls, EB was given at 1 h after the surgical procedures and then, sacrificed at 6–7 h after the surgery. Passage of the tracer into the brain tissue was assessed macroscopically in the brain sections, in the vicinity of the electrode recording tips. For brain water content evaluation, the specific gravity (SG) method[19] was used on regional tissue samples obtained from close to the recording electrode tips. Brain opioid receptor subtypes (μ, δ, κ) were determined by radioligand binding technique[6,22]. Briefly, brain tissue samples of 50–70 mg were obtained from regions close to the microelectrode needle tips and frozen immediately on dry ice and stored at –70°C. Different receptor subtypes were studied by sequential displacement of the "overall" binding of the unselective opioid antagonist, < 10^{-9}M [^{3}H]-diprenorphine (DPN), with highly selective ligand for various subsites: 10^{-7}M [D-Ala^2, MePhe^4, Gly(ol)^5]-enkephalin (DAGO = μ agonist), 10^{-6} M, ICI 174,864 (ICI = delta antagonist), 10^{-5}M U-50488H (U-50 = kappa agonist) and 10^{-5} Naloxone (NAL) to define non-specific binding. The receptor concentration was expressed as fmol/mg protein in the sample.

Following an hour of stabilization from surgery, rCBF was determined before, during and after the release of 1 h MCAO. Arterial BP and rectal temperature were recorded continuously until sacrifice with an overdose of IV pentobarbital. After sacrifice, the brain was rapidly removed and a coronal section was made through the H$_2$ electrode entrance marks, the anterior 5mm block was used for SG, and the posterior 5 mm section was frozen immediately for opiod receptor assays.

Student's test (paired and unpaired) was used for statistical analysis, and significance was established at $p < 0.05$.

Results

Six cats were used as sham-operated controls, and 20 were subjected to temporary MCAO. The arterial BP and pH/gases remained fairly stable throughout the experiment, and the changes observed were insignificant from baseline pre-ischemic values (Table 1). However, during ischemia, the ipsilateral cortical rCBF dropped from a baseline level of 58 ± 7 to 8 ± 2 ml/100 g/min ($p < 0.05$), and upon recirculation, a significant increase in rCBF to 120 ± 30 ml/100 g/min was noted at 10 min (Table 1, Fig. 1a). Distinct areas of EB extravasation were observed in caudate (superior and lateral) and cortex (ectosylvian) ipsilateral to MCAO, but in not of the sham-controls. The early post-ischemic BBB opening was noted in 6 out of 11 cats sacrificed at 1 h and 3 h following the release of occlusion. Compared to sham-operated controls (SG = 1.0439 ± 0.0003), the ipsilateral cortical SG values were significantly lower ($p < 0.05$) at the end of 1 h MCAO (1.0423 ± 0.0002) and during recirculation at

10 min (1.0425 ± 0.0003), 1 h (1.0425 ± 0.0007) and 3 h (1.0410 ± 0.0007); but in caudate, the significance was reached only at 3 h postischemia (1.0398 ±

a) Cortical CBF Following Temporary MCAO

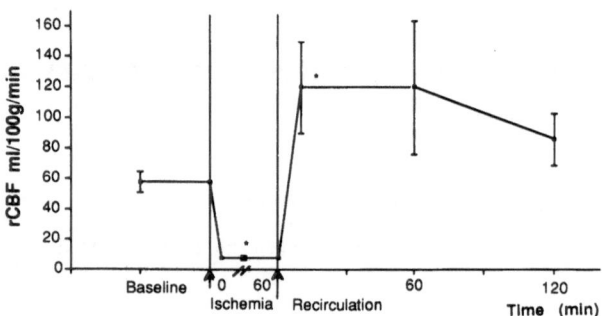

b) Specific Gravity Following Temporary MCAO

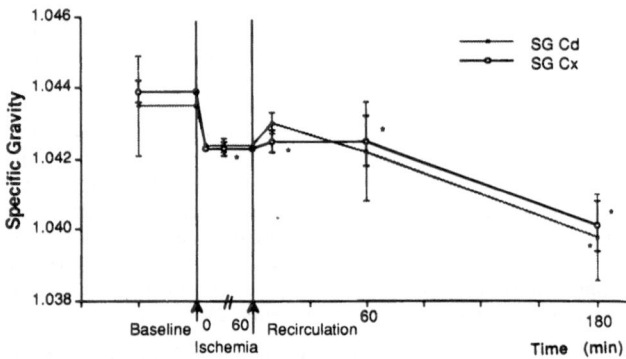

Fig. 1. rCBF and brain SG data of cats subjected to temporary MCAO. (a) Ipsilateral cortical CBF dropped to 8 ± 2 ml/100 g/min during ischemia, and upon recirculation, a significant reactive hyperemia was noted (* $p < 0.05$, paired t-test); (b) there was a progressive decrease in ipsilateral brain SG when compared to the sham-control (* $p < 0.05$)

Table 1. *Physiological Variables (Mean ± SE) of Cats Exposed to Temporary MCAO*

Variables	Baseline	MCAO	Recirculation (min)		
			10	60	180
MBP (mmHg)	119.5 ± 4.4	108.8 ± 7.8	113.2 ± 5.3	112.6 ± 7.2	123.9 ± 8.5
pH	7.35 ± 0.01	7.37 ± 0.02	7.38 ± 0.02	7.31 ± 0.02	7.37 ± 0.02
pO2 (mmHg)	124.1 ± 7.7	146.7 ± 12.6	123.8 ± 8.6	144.6 ± 17.3	157.8 ± 23.6
pCO2 (mmHg)	31 ± 1	27 ± 1	26 ± 2	26 ± 1	25 ± 2
rCBF[a] (ml/100g/min)	58 ± 7	8 ± 2[c]	120 ± 30[c]	120 ± 44	86 ± 1 7
SG[b]					
Caudate	1.0435 ± 0.0007	1.0424 ± 0.0002	1.0430 ± 0.0003	1.0422 ± 0.0014	1.0398 ± 0.0012[c]
Cortex	1.0439 ± 0.0003	1.0423 ± 0.0002[c]	1.0425 ± 0.0003[c]	1.0425 ± 0.0007[c]	1.0410 ± 0.0007[c]

[a] rCBF was recorded from cortex ipsilateral to MCAO.
[b] Baseline SG values were obtained from sham-operated controls.
[c] $p < 0.05$ when compared with baseline values.

0.0012) (Table 1, Fig. 1b). The opioid receptor subtypes (μ, δ, κ) concentrations (fmol/kg protein) in ipsilateral caudate and cortex of sham-operated controls (n = 6), and of cats subjected to temporary MCAO were depicted in Fig. 2. Compared to the levels of sham-controls (caudate = 131 ± 48 (μ), 41 ± 16 (δ), 116 ± 49 (κ); cortex = 55 ± 25 (μ), 25 ± 8 (δ), 56 ± 18 (κ), the ischemic cortex depicted a significant transient increase in all 3 receptor subtypes at 10 min (n = 4) following the release of MCAO (μ = 179 ± 66, δ = 101 ± 12, κ = 132 ± 11); whereas, in ischemic caudate, only μ binding was significantly elevated (294 ± 32). The remaining observations on changes in the receptor concentrations during ischemia, and recirculation were insignificant, except, at 1 h (n = 5) following the release of occlusion, there was a significant decrease of κ binding in cortex (10 ± 3).

Discussion

The present study on cats subjected to severe temporary MCAO (rCBF < 10ml/100 g/min) revealed a significant *increase* in concentrations of brain μ, δ, κ receptors during the *critical* early post-ischemic hyperemic period. In addition, 56% of these cats showed early post-ischemic BBB opening, and progressive brain edema. Published work using the same ischemic animal model has established an association and a causal relationship between the early post-ischemic hyperemic BBB opening, and progression of brain edema and cerebral infarction[11,12,20]. The cause of the enhanced brain opioid activity is not known. Despite the controversies over the effects of brain opioid activity and opioid antagonists on ischemic neurological and neuropathological outcome[5,14], we believe that the conflicts can be resolved by a systematic sequential study of brain opioid activity, especially at critical time points during and following cerebral ischemia. Among a few studies on the brain opioid system, one reported an 80% increase in β-endorphins in ischemic cortex of gerbils subjected to 9 h unilateral carotid occlusion[10]; while other studies showed a decrease in κ receptor binding in the ischemic cortex of cats exposed to 9 h of MCAO[3], and decreased dynorphin-A immunoactivity in hippocampus of gerbils 1 h following the release of forebrain ischemia[7].

Mechanisms by which the perturbed brain opioid system produces ischemic and post-ischemic neuronal injury are not known. However, the following opioid-mediated microcirculatory and cellular effects may be pertinent. Opioids are vasoactive, Met-enkephalin,

dynorphin and β-endorphin decreased rCBF and impaired autoregulation in normal rats, but, these effects were reversed by naloxone[15]. In addition, an intracarotid injection of a stable enkephalin, produced an 87% decrease in ischemic rCBF in cats subjected to MCAO[16]. Further, endorphins have been shown to produce epileptic discharges with a concomitant marked increase in cerebral glucose utilization (without a proportional increase in CBF) and selective increase in glutamate release[1,8,9,18,21]. Recently, dynorphin has been shown to act as a potent agonist at the N-methyl-D-aspartate (NMDA) receptor[4,13]. Another relevant factor is that opioids have been reported to have chemotactic properties, and that, β-EP has been shown to stimulate polymorphonuclear leucocytes' superoxide production via stereoselective naloxone-sensitive opiate receptors[17].

In summary, enhanced brain endogenous opioid system activity may contribute to irreversible ischemic and post-ischemic neuronal damage by: 1) further reduction of rCBF in ischemic and penumbral regions,

a) Caudate

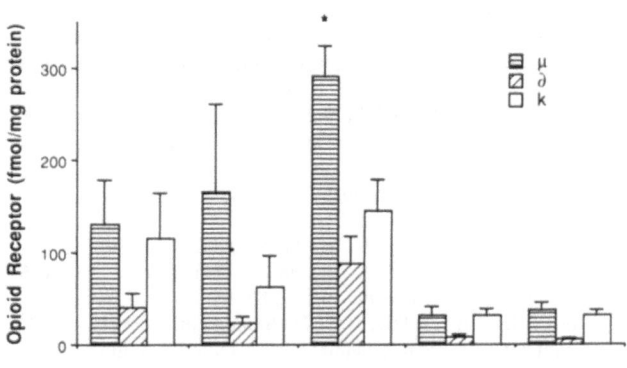

b) Cortex

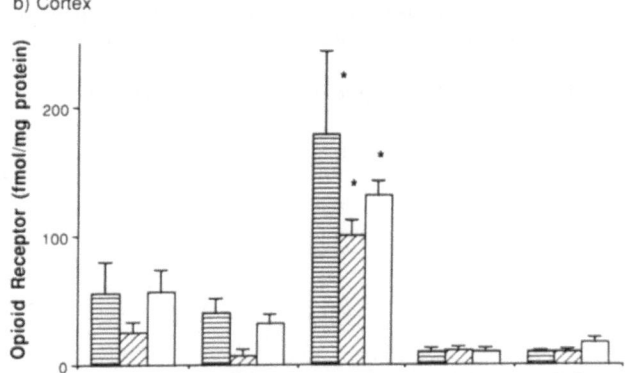

Fig. 2. (a, b) Ipsilateral brain opioid receptor subtypes μ, δ, κ concentrations (fmol/mg protein) of sham-operated controls and of cats subjected to temporary MCA occlusion. * Significantly different from the sham-controls (p < 0.05)

and by aggravation of postasphyxic hypo or hyperperfusion; 2) enhancement of glucose metabolism and neuronal activity, and anaerobic glycolysis due to uncoupling of blood flow and metabolism, 3) increased glutamate release, and 4) stimulation of superoxide production.

The significance of the transient enhanced brain opioid activity is not clear and awaits further study using opioid antagonists. However, due to the above opioid-mediated cerebral microcirculatory and cellular interactions, perturbations in the brain opioid activity may be relevant to ischemic neuronal injury.

Acknowledgement

The study is supported by NIH, NIGMS SO6-GM08244-06. The authors are indebted to the support provided by Howard University Veterinary Services and Ms. Angela Moody.

References

1. Arneric SP, Meeley MP, Reis DJ (1986) Met-enkephalin selectively increases the spontaneous release of amino acid neurotransmitters from rat striatal slices. Neurosci Lett 66: 73–78
2. Aukland K, Bruce BF, Berliner RW (1964) Measurement of local blood flow with hydrogen gas. Circ Res 14: 164–187
3. Baskin DS, Kuroda H, Hosobuchi Y, Lee NM (1985) Treatment of stroke opiate antagonists – effects of exogenous antagonists and dynorphin 1–13. Neuropeptides 5: 307–310
4. Contreras PC, Ragan DM, Bremer ME, *et al* (1991) Evaluation of U-50,488H analogs for neuroprotective activity in the gerbil. Brain Res 546: 79–82
5. Faden AI (1986) Neuropeptides and central nervous system injury. Arch Neurol 32: 501–504
6. Fedynyshyn JP, Lee GK, NM (1989) Characterization of high affinity opioid binding sites in rat periaqueductal gray P_2 membrane. Eur J Pharmacol 159: 83–88
7. Fried RL, Nowak TS, Jr (1987) Opioid peptide levels in gerbil brain after transient ischemia: lasting depletion of hippocampal dynorphin. Stroke 18: 765–770
8. Goodman RR, Snyder SH, Kuhar MJ, Young III WS (1980) Differentiation of delta and mu opiate receptor localizations by light microscopic autoradiography. Proc Natl Acad Sci USA 77: 6239–6243
9. Hommer DW, Pert A (1983) The action of opiates in the rat substantia nigra: an electrophysiological analysis. Peptides 4: 603–608
10. Hosobuchi Y, Baskin DS, Woo SK (1982) Reversal of induced ischemic neurologic deficit in gerbils by the opiate antagonist naloxone. Science 215: 69–71
11. Klatzo I, Seida M, Wagner GH, Ting P (1986) Features of ichemic brain oedema. In: Krieglstein J (ed) Pharmacology of cerebral ischemia. Elsevier, Amsterdam, pp 23–30
12. Kuroiwa T, Ting P, Martinez H, KIatzo I (1985) The biphasic opening of the blood-brain barrier to proteins following temporary middle cerebral artery occlusion. Acta Neuropathol (Berl) 68: 122–129
13. Massardier D, Hunt PR (1989) A direct non-opiate interaction of dynorphin-(1-13) with the N-methyl-E-aspartate (NMDA) receptor. Eur J Pharmacol 170: 125–126
14. Olinger CP, Adams HP Jr, Brott TG, *et al* (1990) High-dose intravenous naloxone for the treatment of acute ischemic stroke. Stroke 21: 721–725
15. Sandor P, DeJong W, DeWied D (1986) Endorphinergic mechanism in cerebral blood flow autoregulation Brain Res 386: 122–129
16. Sandor P, Gotoh F, Tomita M, Tanahashi N, Gogolak 1(1986) Effects of a stable enkephaline analogue, (DiMet2, Pro5)-enkephalinamide, and naloxone on cortical blood flow and cerebral blood volume in experimental brain ischemia in anesthetized cats. J Cereb Blood Flow Metab 6: 553–558
17. Sharp BM, Tsukayama DT, Gekker G, *et al* (1987) β-endorphin stimulates human polymorphonuclear leukocyte superoxide production via a stereoselective opiate receptor. J Pharmacol Exp Ther 242: 279-282
18. Strejckova A, Jakoubek B, Kraus M, Mares P (1985) Changes in the activity of rat cortical and hippocampal neurones after the inotophoretic administration of beta-endorphin, glutamate and GABA. Physiol Bohemoslov 34: 567–573
19. Tengvar C, Hultstrom D, Olsson Y (1983) An improved Percoll density gradient for measurements of experimental brain edema. Acta Neuropathol (Berl) 61: 201–06
20. Ting P, Masaoka H, Kuroiwa T, Wagner H, Fenton I, Klatzo I (1986) Influence of blood-brain barrier opening to proteins on development of post ischaemic brain injury. Neurol Res 8: 146–151
21. Urca G, Frenk H, Leibeskind JC, *et al* (1977) Morphine and enkephalin: analgesic and epileptic properties. Science 197: 83–86
22. Wood MS, Traynor JR (1989) [³H]Diprenorphine binding to κ-sites in guinea-pig and rat brain: evidence for apparent heterogeneity. J Neurochem 53: 173–178

Correspondence: Pauline Ting, M. D., Department of Pediatrics and Child Health, Howard University, 2041 Georgia Ave N.W., Washington, DC 20060, U.S.A.

Acta Neurochir (1994) [Suppl] 60: 257–260
© Springer-Verlag 1994

P-Glycoprotein Expression in Brain Capillary Endothelial Cells After Focal Ischemia in Rat

K. Samoto, K. Ikezaki, N. Yokoyama, and **M. Fukui**

Department of Neurosurgery, Neurological Institute, Faculty of Medicine, Kyushu University, Fukuoka, Japan

Summary

We investigated the time kinetics of P-glycoprotein (P-gp), a membrane bound drug efflux pump for many anti-cancer drugs in multidrug resistant cells, using a rat ischemic brain model. Frozen sections of the brain were studied immunohistochemically with anti-Factor VIII antibody for endothelial cells, with anti-glial fibrillary acidic protein (GFAP) antibody for reactive astrocytes, and with MC6-4 monoclonal antibody for P-gp. A putative blood-brain barrier (BBB) marker, γ-glutamyl transpeptidase (γ-GTP), and the progression of the brain edema were also studied. P-gp positive endothelial cells disappeared in the ischemic lesion by post-ischemic Day 3. Factor VIII-positive regenerating capillaries were first observed on Day 3 without P-gp expression when the brain edema reached a maximum. P-gp positive endothelial cells began to reappear on Day 5, and were detected in all endothelial cells by Day 8. The time kinetics of P-gp expression in the endothelial cells showed a similar pattern as that of γ-GTP, and its induction is associated with GFAP-positive reactive astrocytes. These results suggest that P-gp might play an important role in maintaining the BBB function in conjunction with glial cells.

Keywords: P-glycoprotein; brain; endothelial cells; edema.

Introduction

Multidrug resistant cells consistently overexpress 170 kD membrane glycoprotein known as P-glycoprotein[7] and has the function of decreasing the intracellular levels of anti-cancer drugs by pumping them out of the cells[3]. Moreover, the expression of P-gp has also been shown in specialized epithelial cells with either secretory or excretory functions[12]. P-gp has also been detected in endothelial cells at BBB sites[4]. Recent studies indicated the efflux of vincristine from cultured mouse brain capillary endothelial cells in which P-gp was found to exist[11]. Thus, P-gp expression in the capillary endothelial cells of the brain may indicate a physiological role for P-gp in regulating the entry of certain molecules into the central nervous system. In addition, we realized that the systemic distribution of P-gp is very similar to that of y-glutamyl transpeptidase (γ-GTP)[1], which is a putative marker of the BBB function. In brain capillaries, γ-GTP controls the transfer of some amino acid across the BBB[9]. The purpose of the present study is to investigate the time kinetics of P-gp expression in the capillary endothelial cells of ischemic lesions comparing with the γ-GTP activity and the development of brain edema using a rat MCA occlusion model.

Materials and Methods

Male Sprague-Dawley rats weighing about 200 g were anesthetized with intraperitoneal pentobarbital sodium. After exposing the infratemporal fossa, a small craniectomy was made and the trunk of the MCA was ligated and divided to ensure complete occlusion[10]. The animals were sacrificed following post MCA occlusion survival times of 1, 2, 3, 5, 8, and 14 days. The brains were then removed and fixed with 4% paraformaldehyde, and proceeded to frozen section (6 μm). The sections underwent hematoxylin and eosin (HE) staining for histological evaluation, or subsequent immunohistochemistry and γ-GTP histochemistry.

The monoclonal antibody MC6-4 (a generous gift of Prof. S. Akiyama, Kagoshima Univ.) was used for P-gp immunohistochemistry. The monoclonal antibody from clone G-A-S and the rabbit polyclonal antibody were used for GFAP and Factor VIII (von Willebrand factor) staining, respectively. γ-GTP activity in the endothelial cells was examined histochemically by the modified technique of Glenner[6].

The brain edema was measured by the dry-weight method. Three small tissue specimens were taken from the cortex of both normal and injured hemispheres. They were dried up at 100°C for 24 h in a constant temperature oven. Their weights were measured before and after the drying, and the water content percentage was calculated from their differences.

Ischemic lesions were subdivided into three histological zones[5]. Histologically, the central zone was characterized by a necrosis, infiltration of inflammatory cells, newly formed capillaries, and cavitation. The reactive zone, which surrounds the central zone, was characterized by ischemic neuronal changes, infiltration of inflammatory cells, newly formed capillaries, and astrogliosis. In the marginal zone, which was located peripherally to the reactive zone, ischemic neuronal changes could be seen, but neither any changes in the capillary endothelial cells nor infiltration of inflammatory cells could be detected.

Results

In the contralateral hemisphere, Factor VIII, P-gp and γ-GTP were detected in all endothelial cells of the parenchymal capillaries and no significant changes were observed throughout the course. In the marginal zone, they are also normally detected throughout the course.

In the ischemic lesion, Factor VIII was detected in residual capillary endothelial cells equally to that in the normal side from Day 1 to Day 3. On Day 3, Factor VIII-positive newly formed capillaries began to be identified in the reactive zone, that had prominent nuclei and thickened endothelial cells (Fig. 1A). Abundant newly formed capillaries can be seen in the central zone from Day 5 to Day 14. All these capillaries were

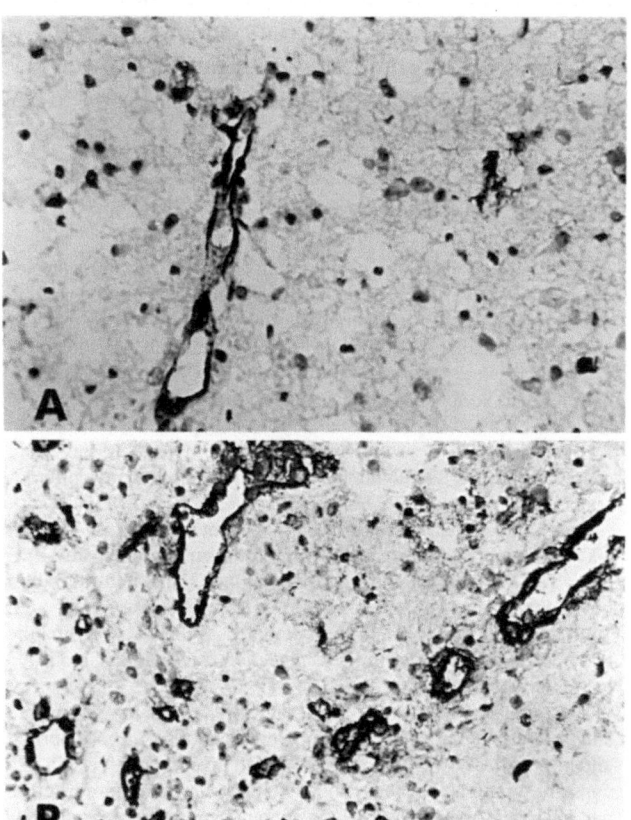

Fig. 1. Immunohistochemical staining of Factor VIII. Factor VIII positive newly formed capillaries appeared in the reactive zone on Day 3 (A). Abundant newly formed capillaries were seen in the central zone on Day 5 (B). × 250

also positive to Factor VIII in their endothelial cells (Fig. 1B).

The P-gp was found only in the capillary endothelial cells, not in the hemispheric large arteries in the contralateral normal brain (Fig. 2A). In the ischemic lesion, P-gp expression at the preexistent capillaries was detected at almost normal levels on Day 1 and Day 2, but then totally disappeared on Day 3 in the central and reactive zones (Fig. 2B). Factor VIII-positive capillaries appeared in the reactive zone, but they did not have P-gp immunoreactivity on Day 3. On Day 5, P-gp expression was still undetectable in the central zone (Fig. 2C) but a few P-gp-positive endothelial cells were recognized in newly formed capillaries in the reactive zone. On Days 8 and 14, all capillary endothelial cells in the ischemic lesion expressed P-gp (Fig. 2D). This reappearance of P-gp in endothelial cells tended to be delayed when the ischemic lesion was large. The time course of the γ-GTP activity in the ischemic lesion was almost the same as that of P-gp. The consequences of Factor VIII and P-gp immunohistochemistry and g-GTP histochemistry are summarized in Table 1.

On Day 2, GFAP-positive reactive astrocytes appeared in the marginal and reactive zone. By 8 days after MCA occlusion, GFAP-positive cells had increased gradually in number in the reactive zone and extended along the vessels toward the central zone. On Day 8, a few reactive astrocytes are recognized in the central zone where P-gp and y-GTP were positive in the newly formed capillaries. On Day 14, activated astrocytes were seen in the marginal wall of the cavity in the ischemic lesion.

Water content in the ischemic cortex gradually increased from Day 1 to Day 3. In the intact hemisphere, there was no significant change in the water content throughout the time course (mean 78.4%). The water content in the ischemic cortex reached a maximum 3 days after MCA occlusion (82.3%, $p < 0.05$). Fourteen days after MCA occlusion, there was no significant difference in the water content between the ischemic and normal cortecies (78.8 and 78.5%, respectively) (Fig. 3).

Discussion

In the present study, the endothelial P-gp decreased and disappeared along with the progress of the infarction by Day 3 as the brain edema advanced. Although, P-gp was immunoreactive in the preexisting capillary endothelial cells on Days 1 and 2, adenosine triphos-

Table 1. *Summary of Factor VTII, P-gp and g-GTP Expression in Newly Formed Capillary Endothelial Cells in the Central and Reactive Zones*

	Days after MCA occlusion			
	1–2	3	5	8–14
Factor VIII	n.d.	+/n.d.	+/+	+
P-gp	n.d.	–/n.d.	±/–	+
γ-GTP	n.d.	–/n.d.	±/–	+

+ homogeneously stained, ± weakly stained, – negative. *n.d.* No newly formed capillaries were detected. On Days 3 and 5, the ischemic lesion was subdivided into the reactive zone/central zone.

phate (ATP) dependent P-gp might not function in the ischemic lesion, since a remarkable reduction of ATP would occur soon after the onset of the ischemia. P-gp gradually reappeared from Day 5 along with the improvement of brain edema. The factors that influence the brain edema could be numerous, however, P-gp might be one of the factors that restricts the brain

edema, by keeping the osmotic pressure of the brain tissue by regulating the entry of certain toxic normal metabolites and foreign agents into the central nervous system.

It is also interesting to note that the time kinetics of P-gp expression in endothelial cells was very similar to that of γ-GTP. A previous study has demonstrated the role of astrocytes in the induction of γ-GTP in endothelial cells using a mouse frozen injury brain model[2]. It suggested the necessity of the activation of astrocytes and its association with newly formed vessels before γ-GTP expression. Other in vitro studies have demonstrated the role of astrocytes in the induction of γ-GTP[8].

It took 2 or 3 days more until the newly formed capillaries obtained P-gp and γ-GTP in their endothelium in which Factor VIII had already been expressed. On Day 5, a few GFAP-positive astrocytes appeared in the central zone and elongated their foot processes along with the capillary extension. GFAP was, however, not always observed around the capil-

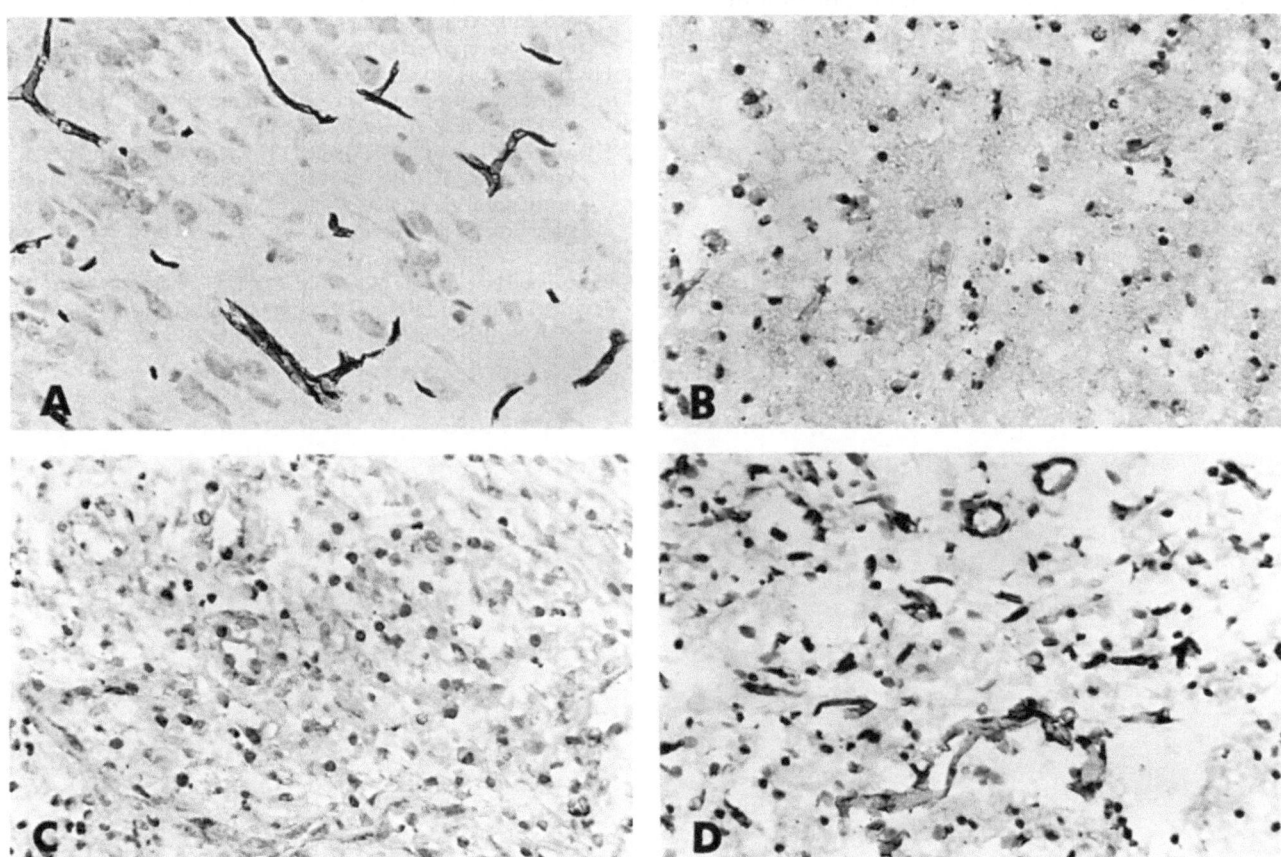

Fig. 2. Immunohistochemical staining of P-gp. In the contralateral hemisphere, all normal endothelial cells were positive to P-gp on Day 5 (A). In the reactive zone, residual capillaries were almost negative to P-gp on Day 3 (B). Newly formed capillaries were negative to P-gp in the central zone on Day 5 (C). All capillaries exhibited P-gp on Day 8 (D). × 250

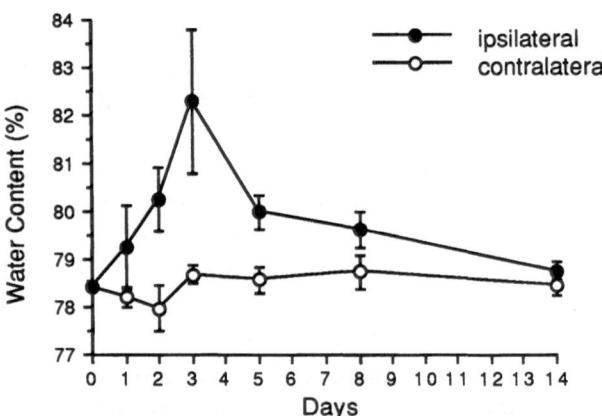

Fig. 3. Water content in brain tissue after MCA occlusion. Values are expressed as means ± SE (n = 3)

laries in the central zone where P-gp and γ-GTP were positive in endothelial cells. These findings suggested that astrocytes might participate in P-gp and γ-GTP induction in capillary endothelial cells either by some humoral factors and/or by direct contact with the endothelial cells.

The present investigation has shown only preliminary data, but suggests that P-gp expression in brain capillary endothelial cells might play an important role in one of the BBB function, in conjunction with glial cells. It is probable that P-gp could be a new BBB marker.

References

1. Albert Z, Orlowska J, Orlowski M, Szewczuk A (1964) Histochemical and biochemical investigations of gamma-glutamyl transpeptidase in the tissue of man and laboratory rodents. Acta Histochem 18: 78–89

2. Cancilla PA, Berliner JA, Bready JV (1990) Astrocytes and the blood-brain barrier. Kinetics of astrocyte activation after injury and induction effects on endothelium. In: Johansson BB, Owman C, Widner H (eds) Pathophysiology of the blood-brain barrier. Elsevier, Amsterdam, pp 31–39

3. Chen Cj, Chin JE, Ueda K, *et al* (1986) Internal duprication and homology with bacterial transport proteins in the mdr1 (P-glycoporotein) gene from multidrug-resistant human cells. Cell 47: 381–389

4. Cordon-Cardo C, O'Brien JP, Casals D, *et al* (1989) Multidrug-resistance gene (P-glycoprotein) is expressed by endothelial cells at blood-brain barrier sites. Proc Natl Acad Sci USA 86: 695–698

5. Garcia JH, Kamijyo Y (1974) Cerebral infarction: evolution of histopathological changes after occlusion of a middle cerebral artery in primates. J Neurophathol Exp Neurol 33: 408–421

6. Glenner GG, Fork JE, McMillan PJ (1962) Histochemical demonstration of a gamma-glutamyl transpeptidase-like activity. J Histochem Cytochem 10: 481–489

7. Juliano RL, Ling V (1976) A surfase glycoprotein modurating drug permeability in chinese hamster ovary cell mutants. Biochim Biophys Acta 455: 152–162

8. Maxwell K, Berliner JA, Cancilla PA (1987) Induction of γ-glutamyl transpeptidase in cultured cerebral endothelial cells by a product released by astrocytes. Brain Res 410: 309–314

9. Orlowski M, Sessa G, Green JP (1974) γ-Glutamyl transpeptidase in brain capillalies: possible site of a blood-brain barrier for amino acids. Science 184: 66–68

10. Tamura A, Graham DI, McCulloch J, Teasdale GM (1981) Focal cerebral ischaemia in the rat: 1. Description of technique and early neuropathological consequences following middle cerebral artery occlusion. J Cereb Blood Flow Metab 1: 53–60

11. Tatsuta T, Naito M, Oh-hara T, Sugawara I, Tsuruo T (1992) Functional involvement of P-glycoprotein in blood-brain barrier. J Biol Chem 267: 20383–20391

12. Thiebaut F, Tsuruo T, Hamada H, Gottesman MM, Pastan I, Willingham MC (1987) Cellular localization of the multidrug-resistance gene product P-glycoprotein in normal human tissues. Proc Natl Acad Sci USA 84: 7735–7738

Correspondence: Ken Samoto, M. D., Department of Neurosurgery, Neurological Institute, Kyushu University Faculty of Medicine, Fukuoka 812, Japan.

Acta Neurochir (1994) [Suppl] 60: 261–264
© Springer-Verlag 1994

Blood-Borne Macromolecule Induces FGF Receptor Gene Expression After Focal Ischemia

K. Yamada, T. Sakaguchi, T. Yuguchi, E. Kohmura, H. Otsuki, T. Koyama, and **T. Hayakawa**

Department of Neurosurgery, Osaka University Medical School, Osaka, Japan

Summary

We have detected fibroblast growth factor receptor (FGFR) gene expression in the focal ischemia model. The FGFR gene expression in neurons can be explained by neuronal network disturbances, but the mechanism of astroglial gene expression remains uncertain. We speculated that blood-borne edema fluid may activate gene expression of astroglias. To prove this hypothesis, we compared the pattern's of gene expression of FGFR and distribution of edema fluid by using serial tissue sections of the middle cerebral artery (MCA) ischemia.

The left MCA of twenty-four male Wistar rats were occluded, and sacrificed 1, 3, 4, 7 and 14 days later by transcardiac perfusion and fixation. The tissues were sliced thinly to 14 μm sections. Part of the tissue sections was used for in situ hybridization for rat FGFR with [35S]labeled RNA probes. The other part of the sections was used for immunostaining for albumin, immunoglobulin G (IgG) and IgM.

The FGFR mRNA expression was evident in the lesion-side hemisphere. In the cortex, neurons mainly expressed FGFR gene in the cortex, whereas astroglias and capillary endothelium expressed FGFR in the corpus callosum and internal capsule. The albumin distributed cortex and white matter of the lesion-side and it extended to the contralateral side. The IgG distributed mainly in the lesion-side white matter, and in part extended to the contralateral side. The IgM only distribute to the infarcted area. When we compared topographical distribution of FGFR in the white matter and pattern of albumin, IgG and IgM distribution, pattern of IgG distribution correlated well to the area of FGFR expression.

The results indicate that blood-borne macromolecules distribute to the periinfarcted brain tissue depending on their molecular size and the FGFR gene expression might be activated by blood-borne macromolecules with molecular weights similar to IgG.

Keywords: Fibroblast growth factor receptor; middle cerebral artery; focal ischemia; blood-borne macromolecules.

Introduction

We have detected fibroblast growth factor receptor (FGFR) gene expression in the focal ischemia model, and part of the result was reported elsewhere[3]. Upregulation of FGFR gene expression in neurons can be well explained by neuronal network disturbances[2]. Mechanism of astroglial FGFR gene upregulation, however, remains uncertain. We speculate that blood-borne edema fluid may activate gene expression of astroglias. We therefore compared in situ hybridization autoradiographs of FGFR and adjacent tissue sections immunostained for albumin, IgG and IgM.

Materials and Methods

Ischemia model: Twenty-two male Wistar rats weighing approximately 300 g were anesthetized by an intraperitoneal injection of ketamine hydrochloride (150 mg/kg) and a local injection of lidocaine under the skin. The left middle cerebral artery (MCA) was approached by a small opening of the skull and dura. The arachnoid around the artery was opened, and the MCA was cauterized at the level of the olfactory tract. The rats were deeply anesthetized with chloral hydrate and sacrificed with transcardiac perfusion with saline and fixature of 4% paraformaldehyde in 0.1 M sodium phosphate buffer (pH 7.4). The brains were removed and postfixed with same fixatives, dehydrated and embedded in paraffin. The blocks were sliced into thin 14 μm coronal sections and used for in situ hybridization and immunostaining.

In situ hybridization: Preparation of RNA probe for FGFR has been reported elsewhere[4]. In brief, partial complementary DNA (cDNA) fragment (300 base pairs) of rat FGFR (flg-type) was used as a template. Antisense and sense RNA probes were created by linearization of cDNA with appropriate restriction endonucleases, and then reacting with appropriate RNA polymerases. In situ hybridization techniques were based on those of Wilkinson et al.[5]. Tissue sections were deparaffinized and hydrated. The [35S] labeled RNA probes (antisense or sense) in hybridization buffer were placed on the sections, covered with siliconized coverslips and hybridized in a humid chamber overnight at 55°C. After hybridization, sections were rinsed, dehydrated and exposed to the X-ray film. The film was developed and tissue sections were counterstained with thionin.

Detection ot edema fluid extension: To detect blood-borne macromolecules, tissue sections which were adjacent to the sections used for in situ hybridization were used. Those sections were immunostained for albumin, immunoglobulin G (IgG) and immunoglobulin M (IgM). Anti-rat albumin, biotinylated anti-rat IgG,

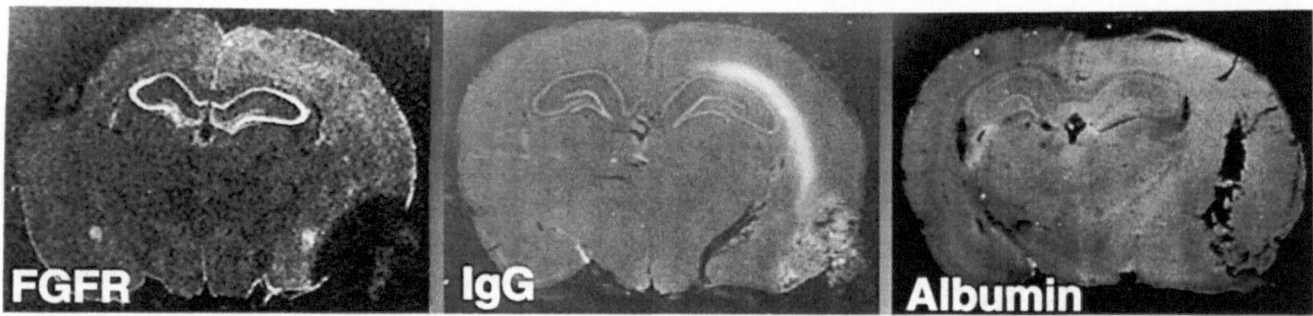

Fig. 1. MCA occlusion (1 day). In situ hybridization macroautoradiograph for fibroblast growth factor receptor (FGFR) and immuno-staining for albumin and immunoglobulin G (IgG) at one day after occlusion of the middle cerebral artery (MCA)

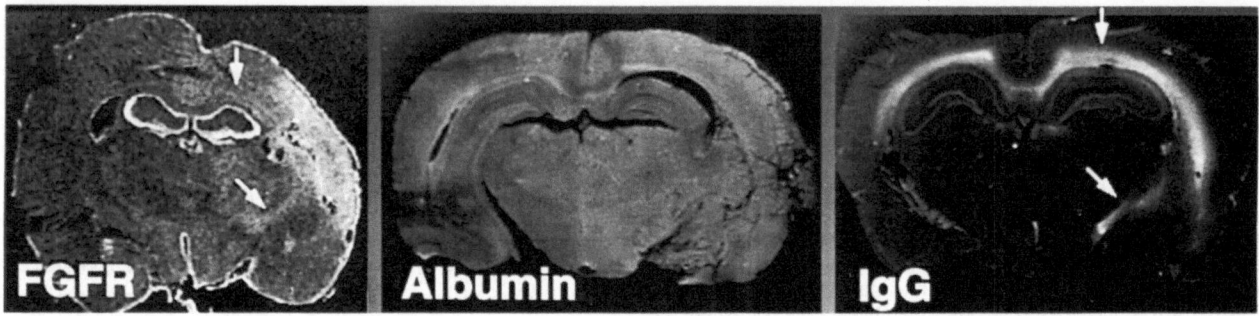

Fig. 2. MCA occlusion (3 days). In situ hybridization autoradiograph for FGFR and immunostaining for albumin and IgG at 3 days after MCA occlusion

and anti-rat IgM antibodies, and ABC-Elite kit (Vectastain) was used for visualization.

Results

FGF receptor expression: One day after ischemia, the FGFR expression was upregulated in the lesion side cortex. Expression of FGFR was obvious in the cingulate cortex of the lesion side when compared to the contralateral side (Fig. 1). Microscopic observations revealed that the FGFR expression was detectable in the neurons of the cortical layer 5. No definite expression was evident in the white matter.

Three days after ischemia, the expression became limited to the area of ischemia. The cingulate cortex became less intense in the FGFR expression. The corpus callosum and internal capsule of the lesion-side showed a higher level of FGFR expression (Fig. 2: arrows). Upregulation of FGFR in those white matter area lasted for 2 weeks, and microscopic observations discovered that most of the signals was located in the cell bodies (Fig. 3).

Extension of the edema fluid: One day after ischemia, albumin distributed cortex, internal capsule and corpus callosum of the lesion side. Distribution in the cortex is rather wide and diffuse. Cingulate cortex of the lesion side was stained for albumin when compared to the contralateral side (Fig. 1). In three days, albumin distribution was extended to the thalamus of the lesion side, and into parts of the contralateral cingulate cortex and corpus callosum (Fig. 3).

The IgG distribution was different from albumin distribution. One day after ischemia, the IgG was only distributed in the corpus callosum of the lesion side (Fig. 1). The distribution extended to the internal capsule of the lesion side and corpus callosum of the contralateral side in 3 days (Fig. 2). The IgG distribution in the internal capsule and corpus callosum of the lesion side showed a high correlation to the FGFR expression (Fig. 3: arrows). However, it never extended to the cortex or thalamus. Distribution of IgG in the white matter lasted for 2 weeks (Fig. 3). Microscopical observation showed that IgG was mainly distributed in the exracellular space, but it was taken up by some cells in the white matter (Fig. 3).

Discussion

In this report we have studied FCFR gene expression in the focal ischemia model and analyzed upregulation mechanism of the FGFR expression. We started this study due to the fact that slowly progressing thalamic degeneration after cortical infaction can be prevented by intracisternal injection of FGF[6]. Neurons in the ventral posterolateral (VPL) nucleus of the lesion-side thalamus became shrunken and disappeared due to retrograde degeneration. The FGF injected intracisternally prevented this shrinkage. However, it is not clear whether FGF directly affects the VPL neurons or indirectly through non-neural cells. We therefore studied the expression of FGFR in this model and found that FGFR was not expressed in the thalamic neurons but expressed in the cortical neurons and non-neural cells in the corpus callosum and internal capsule. We believe that the FGF acts indirectly on the thalamic neurons through activation of the non-neural cells and modification of the local environment.

Diffuse upregulation of FGFR in the lesion side cortex during early stage of ischemia is of interest. For the rat model of MCA occlusion, the area of ischemia is limited to the area of cell necrosis, and the penumbra zone is minimal. Yet, the area of FGFR gene upregulation is extended to the entire hemisphere. We have shown with Fink-Heimer's silver staining that degenerated nerve terminals were diffusely distributed in the cortex and thalamus of the lesion side[1]. With acetylcholine esterase histochemistry, we have also shown that cholinergic deafferentation was observed in the widespread area of the lesion side cortex[2]. Those areas of deafferentation were well correlated to the areas of FGFR mRNA upregulation. We therefore suggest that upregulation of FGFR gene expression in the cortex is caused by neuronal deafferentation.

The other possibility for diffuse cortical deafferentation is spreading depression. We have detected in the previous experiments, area of expression of *fos* mRNA was rather diffuse in the lesion side cortex extending to the area of cingulate cortex and the pattern of fos expression was similar to the pattern of FGFR expression (Yuguchi *et al.*, unpublished observation). The fos mRNA expression is known to relate spreading depression, which may play an important role for diffuse cortical FGFR expression.

Because albumin was distributed rather widely to the lesion side cortex, albumin can be a candidate for causing diffuse cortical expression of FGFR. However, albumin was also distributed to the thalamus, but FGFR was not expressed in the thalamus. It is therefore unlikely that albumin induced gene expression of FGFR.

The question then arises what is the mechanism of receptor expression in the corpus callosum and internal

capsule. We speculated that blood-borne edema fluid may induce gene expression in those areas. We therefore stained albumin and immunoglobumins. As a result, IgG distribution showed a high correlation to the area fo FGFR gene upregulation. Furthermore, microscopical observation revealed that non-neural cells, presumably astroglias took IgG up to their cytoplasm. This suggests that blood-borne materials with molecular weight similar to IgG might be taken up by astroglias, and regulate the metabolic activity of astroglias.

To prove this working hypothesis, we topically applied FGF to the rat cortex and observed cell reaction in

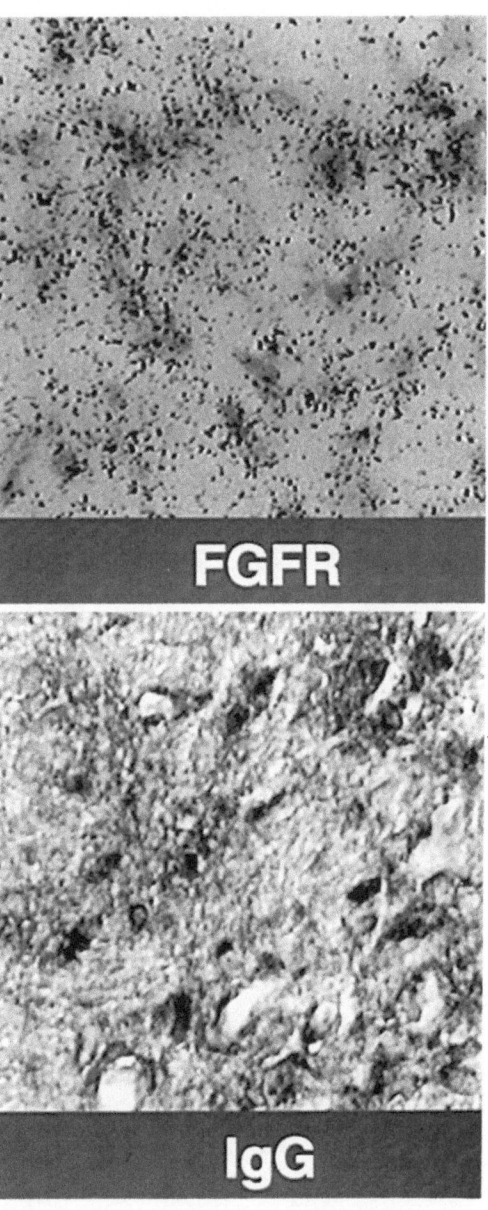

Fig. 3. Periischemic white matter (2 weeks). Microautoradiograph for FGFR and microscopical appearance of immunostaining for IgG 2 weeks after MCA occlusion

the adjacent tissue (Yuguchi *et al.*, unpublished observation). The FGF indeed activate non-neural cells in the corpus callosum and internal capsule, and they became BUdR-positive cells when BUdR was given intravenously. The facts indicate that FGF has actions to non-neural cells, induced cell mitosis and support neuronal survival after injury. We also detected that FGF prolonged expression period of c-fos mRNA in the cortical neurons. The data supports the hypothesis that blood-borne materials indeed activate neurons and non-neural cells and support survival of injured neurons and modify the environments.

References

1. Kataoka K, Hayakawa T, Yamada K, Mushiroi T, Kuroda R, Mogami H (1989) Neuronal network disturbance after focal ischemia in rats. Stroke 20: 1226–1235

2. Kataoka K, Hayakawa T, Kuroda R, Yuguchi T, Yamada K (1991) Cholinergic deafferentation after focal infarct in rats. Stroke 22: 1291–1296

3. Sakaguchi T, Wanaka A, Yamada K, Kohmura E, Taneda M, Tohyama M, Hayakawa T (1991) Expression of bFGF receptor mRNA in the central nervous system with ischemic injury. Adv Neurotrauma Res 3: 53–58

4. Wanaka A, Johnson EM Jr, Milbrandt J (1990) Localization of FGF receptor mRNA in the adult rat central nervous system by in situ hybridization. Neuron 5: 267–281

5. Wilkinson DG, Peters G, Dickson C, McMahon AP (1988) Expression of the FGF-related proto-oncogene int-2 during gastrulation and neurulation in the mouse. EMBO J 7: 691–695

6. Yamada K, Kinoshita A, Kohmura E, Sakaguchi T, Taguchi J, Kataoka K, Hayakawa T (1991) Basic fibroblast growth factor prevents thalamic degeneration after cortical infarction. J Cereb Blood Flow Metab 11: 472–478

Correspondence: Kazuo Yamada, M. D., Department of Neurosurgery, Osaka University Medical School, 2-2 Yamadaoka, Suita, Osaka 564, Japan.

Acta Neurochir (1994) [Suppl] 60: 265–267
© Springer-Verlag 1994

Expression of Basic Fibroblast Growth Factor in Astrocytes at the Site of Cerebral Lesions and Edematous Areas Under Chronic Hypertension

E. Yamada, H. Kataoka, C.-H. Chue, and **F. Hazama**

Department of Pathology, Shiga University of Medical Science, Seta Tsukinowa-cho, Otsu, Japan

Summary

In an attempt to obtain information about changes of the basic fibroblast growth factor (bFGF) in the brain under chronic hypertensive condition, we studied immunohistochemically the distribution of bFGF in the brain of stroke-prone spontaneously hypertensive rats (SHRSP). In the control normotensive rats, only weak immunoreactivity for bFGF was demonstrated in nerve cells and some astrocytes. In SHRSP marked immunoreactivity was demonstrated in the densely packed reactive cells, particularly astrocytes, in and around cerebral lesions. Slightly increased reaction for bFGF was found in the nerve cells around lesions. Astrocytes in the white matter also showed immunoreactivity for bFGF, the localization of which corresponded very well with the site of edema. This finding indicates the possibility that brain edema expresses bFGF in astrocytes and possibly in nerve cells.

Keywords: Hypertension; brain edema; basic fibroblast growth factor; astrocyte.

Introduction

In previous studies of the brain of stroke-prone spontaneously hypertensive rats (SHRSP), we reported that proliferation[8], hypertrophy and increased oxidoreductase enzyme activities[7] were observed in astrocytes not only in and around the cerebrovascular lesions but also in the edematous subcortical white matter. Recently it has been reported that the immunoreactivity for basic fibroblast growth factor (bFGF) was increased in reactive astrocytes at the site of focal brain wounds[6]. Since in vitro and in vivo studies demonstrated that bFGF had effects on the survival of neurons[1,9], and proliferation of neuroglial cells[4] and endothelial cells[5], such increased bFGF at the site of focal brain wound was supposed to contribute to the cascade of cellular changes following brain injury.

In order to obtain information about the participation of bFGF in the hypertensive cerebral changes, we histochemically studied the immunoreactivity of bFGF in the brain, particularly in astrocytes, of SHRSP. For the purpose to correlate the changes of bFGF in astrocytes with cerebral edema and to identify the species of bFGF-positive cells, immunoreactions for fibrinogen, bromodeoxiuridine (BrdU) and glial fibrillary acidic protein (GFAP) were also carried out.

Materials and Methods

Thirty male SHRSP aged 12, 16, 18 and 24 weeks and age-matched 16 male Wistar rats of Kyoto strain (WKY) were intraperitoneally injected with BrdU (25 mg/kg) for 7 times at 8 hours intervals. Under pentobarbital anesthesia animals were perfused through the left ventricles with 4% paraformaldehyde in 0.1 M phosphate buffer (pH 7.2). The brains were removed and embedded in paraffin. Serial paraffin sections of 2.5 μm thickness were cut and stained with hematoxylin and eosin. For electron microscopical cytochemistry, brains were rinsed in 20% sucrose for 48 hours, frozen in liquid nitrogen and serially cut at 6 μm with a cryostat. Immunostainings for bFGF, fibrinogen BrdU and GFAP were carried out using avidin-biotin-peroxidase method.

Results and Discussion

In the brain of SHRSP, various cerebral lesions such as petechiae, spongy changes, cyst formation with or without necrosis and severe edema with white matter degeneration were observed. In these cerebral lesions, numerous reactive cells with strong immunoreaction for bFGF (Fig. 1a) were located, around arterioles showing hyaline or fibrinoid degeneration. Some of these reactive cells also showed positive reaction for GFAP (Fig. 1b) and fibrinogen. These findings suggest that astrocytes incorporate extravasated plasma components, change to the reactive type, and express bFGF.

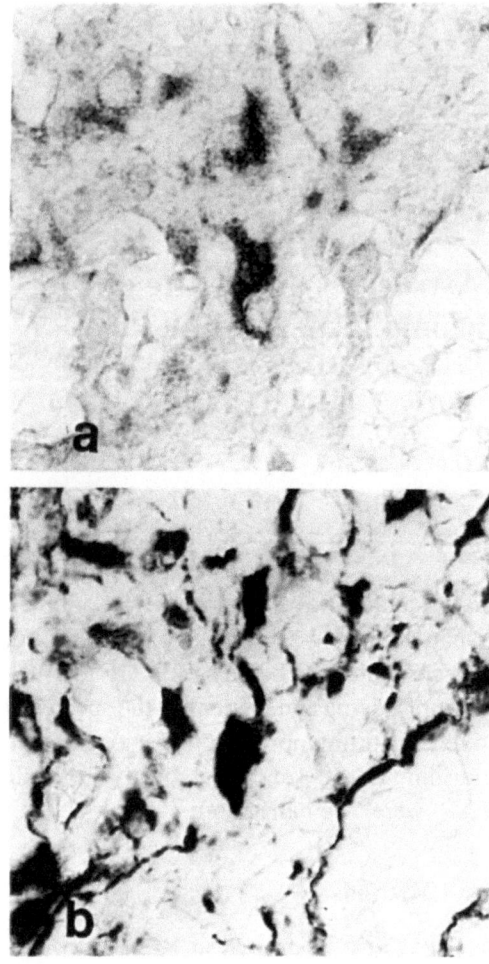

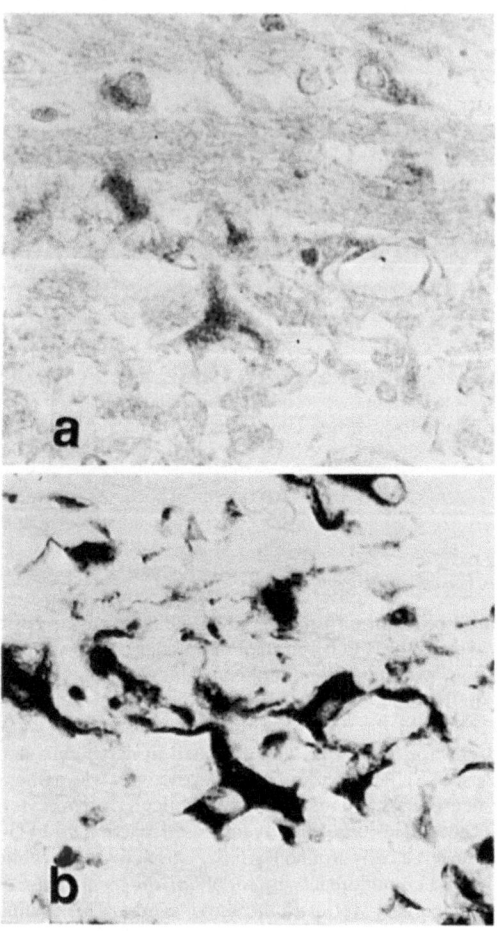

Fig. 1. In cerebral cortex of SHRSP, numerous glial cells show intense immunoreactivity against bFGF (a) and GFAP (b), indicating that these cells are mainly astrocytes. bFGF (a) and GFAP (b) immunostaining × 400

Fig. 2. Astrocytes (b) show intense immunoreactivity against bFGF (a) in the edematous and degenerated subcortical white matter of SHRSP. bFGF (a) and GFAP (b) immunostaining × 400

Many glial cells in the edematous portion around the cerebral lesions and in the edematous subcortical white matter also revealed strong positive bFGF immunoreactivity (Fig. 2a). Even in young SHRSP of 12 weeks of age, bFGF-positive cells were demonstrated in the edematous subcortical white matter. These bFGF-positive cells were identified mainly as astrocytes by positive reaction for GFAP (Fig. 2b). The distributions of bFGF-positive cells, GFAP-positive cells and the areas of edematous changes corresponded closely to each other. Some astrocytes with positive bFGF reactivity showed also positive reaction for fibrinogen in the cytoplasm. These findings suggest that edema fluid itself is enough to evoke bFGF expression in astrocytes.

Nerve cells in the cortex of control animals showed weak bFGF immunoreactivity while in SHRSP more intence immunostaining was demonstrated in the nerve

cells in the cerebral lesions and in the edematous portion around the lesions.

In the central nervous system, several kinds of cells, including neurons, astrocytes, microglia and endothelial cells, are reported to synthesize bFGF[2,10]. The question is which kind of cells are responsible for the increased immunoreactivity in and around the cerebral lesions in SHRSP. In the present study, mainly astrocytes showed strong immunoreactivity in the brain of SHRSP. It has been reported that bound bFGF internalizes into vesicles in the cytoplasm, and localizes to the perinuclear cytoplasm in astrocytes[11]. Basic FGF in the extravasated edema fluid could, therefore contribute the immunoreactivity in the astrocytes. However, bFGF immunoreactivity was electron microscopically observed along the ribosomes of rER and in the cytosol of the reactive astrocytes, indicating that bFGF was synthesized in the cells themselves.

Basic FGF has trophic effects on neurons[1,9] as well as mitogenic activity for astrocytes resulting in astrocytosis in injured brains[3,4]. In and around the cerebral lesions of SHRSP, most neurons remained intact or were only mildly affected unlike in the infarction. Glial proliferation and strong expression of GFAP were also observed in these regions. These findings suggest that bFGF is expressed in astrocytes by edema, and that it plays some role in the development of hypertensive cerebral changes.

References

1. Anderson KJ, Dam D, Lee S, Cotman CW (1988) Basic fibroblast growth factor prevents death of lesioned cholinergic neurons in vivo. Nature 332: 360–361
2. Araujo DM, Cotman CW (1992) Basic FGF in astroglial, microglial, and neuronal cultures: characterization of binding sites and modulation of release by lymphokines and trophic factors. J Neurosci 12: 1668–1678
3. Eclancher F, Perraud F, Faltin J, Labourdette G, Sensenbrenner M (1990) Reactive astrogliosis after basic fibroblast growth factor (bFGF) injection in injured neonatal rat brain. Glia 3: 502–509
4. Engele J, Bohn MC (1992) Effects of acidic and basic fibroblast growth factors (aFGF, bFGF) on glial precursor cell prolifera-

tion: age dependency and brain region specificity. Dev Biol 152: 363–372
5. Ferrara N, Ousley F, Gospodarowicz D (1988) Bovine brain astrocytes express basic fibroblast growth factor, a neurotropic and angiogenic mitogen. Brain Res 462: 223–232
6. Finkelstein SP, Apostolides PJ, Caday CG, Prosser J, Philips MF, Klagsbrum M (1988) Increased Basic fibroblast growth factor (bFGF) immunoreactivity at the site of focal brain wounds. Brain Res 460: 253–259
7. Hanakita J, Hazama F, Amano S, Yamada E, Handa H (1980) Histochemical study on oxidative enzyme activity in the brain, particularly of astrocytes, in spontaneously hypertensive rats. Neuropathol Apl Neurobiol 6: 471–482
8. Hazama F, Haebara H, Amano S, Ozaki T (1977) Autoradiographic investigation of cell proliferation in the brain of spontaneously hypertensive rats. Acta Neuropathol (Berl) 37: 231–236
9. Morrison RS, Sharma A, Vellis JD, Bradshhaw RA (1986) Basic fibroblast growth factor supports the survival of cerebral cortical neurons in primary culture. Proc Natl Acad Sci USA 83: 7537–7541
10. Pettmann B, Labourdette G, Weibel M, Sensenbrenner M (1986) The brain fibroblast growth factor (FGF) is localized in neurons. Neurosci Lett 68: 175–180
11. Walicke PA, Baird A (1991) Internalization and processing of basic fibroblast growth factor by neurons and astrocytes. J Neurosci 11: 2249–2258

Correspondence: Eiji Yamada, M. D., Department of Pathology, Shiga University of Medical Science, Seta-Tsukinowa-cho, Otsu 520-21, Japan.

Acta Neurochir (1994) [Suppl] 60: 268–270

Cytochrome Oxidase and Hexokinase Activities in an Infusion Edema Model with Preserved Blood Flow

T. Kawamata, Y. Katayama, K. Kinoshita, A. Yoshino, H. Hirota, and **T. Tsubokawa**

Department of Neurological Surgery, Nihon University School of Medicine, Tokyo, Japan

Summary

Despite numerous investigations, the mechanisms underlying the neurological deficits observed in association with interstitial edema remain unclear. A recent study has demonstrated that the cerebral blood flow (CBF) in edematous white matter is unchanged if the blood flow values are corrected for dilution. In contrast, the cerebral glucose metabolism (CMRgl) has been found to be increased. In order to examine the effects of interstitial edema on the oxidative metabolism, we measured the cytochrome oxidase (CYO) activity, a marker of mitochondrial respiration, and the hexokinase (HK) activity, a marker of glycolysis, together with CBF and CMRgl employing the iodoantipyrine and deoxyglucose autoradiography in an infusion edema model in rats. In agreement with the previous study, CBF was not significantly changed in the edematous hemisphere. No significant alterations in CMRgl and HK activity were noted. In contrast, there was a significant decrease in CYO activity in the edematous hemisphere (-17%; $p < 0.01$), which was correlated to the edema. These findings suggest that interstitial edema causes a decreased mitochondrial respiratory function despite a maintained circulation. This may be explained by postulating a decreased oxygen delivery and/or accumulation of lactate, both of which have been shown to interfere with mitochondrial respiratory function.

Keywords: Interstitial edema; oxygen metabolism; glycolysis; cerebral blood flow; cytochrome oxidase activity.

Introduction

Interstitial edema is defined as an increased water content in the extracellular space, as typically observed in the white matter of hydrocephalus[2]. Although it is clear that interstitial edema clinically induces neurological deficits, such as memory disturbance, urinary incontinence or gait disturbance[1], the mechanisms underlying these symptoms remain unclear. A recent study demonstrated that the cerebral blood flow (CBF) in edematous white matter is unchanged if the blood flow values are corrected for dilution. In contrast, the cerebral glucose metabolism (CMRgl) has been found

to be increased[9]. This suggests that interstitial edema may disturb aerobic metabolism, resulting in activation of anaerobic glycolysis. In the present study, in order to examine the effects of interstitial edema on the oxidative metabolism, we measured the cytochrome oxidase (CYO) activity, a marker of mitochondrial respiration[10], and the hexokinase (HK) activity, a marker of glycolysis[3], together with CBF and CMRgl employing iodoantipyrine and 2-deoxyglucose autoradiography[7,8] in an infusion edema model in rats.

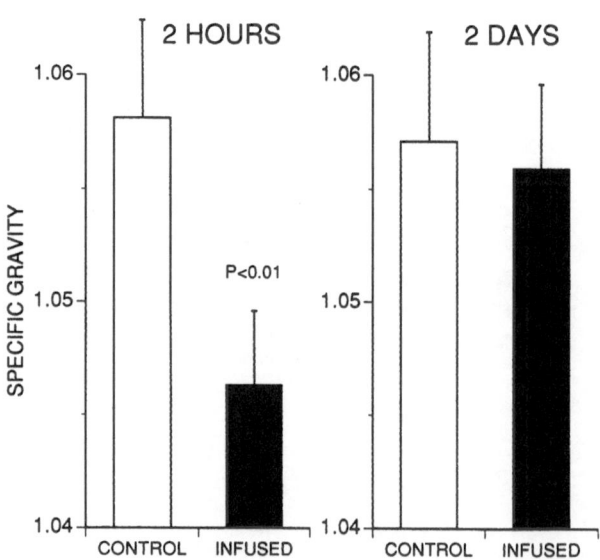

Fig. 1. Specific gravities of the control (open columns) and infused brain tissue (closed columns). The specific gravity was significantly decreased in the infused brain at 2 hours after the infusion ($p < 0.01$) (left). In contrast, no significant difference was noted at 2 days after the infusion (right), indicating that the edema had improved during this period

Methods

Young adult Sprague-Dawley rats (250–300 g) were anesthetized with a mixture of 66% nitrous oxide, 33% oxygen and 1% halothane. The rectal temperature was maintained at 37–38°C with a heating pad. A large craniectomy for decompression was performed on the fronto-parietal skull to avoid increase of intracranial pressure due to the infusion edema, which could decrease the cerebral blood flow resulting in the derangement of cerebral metabolism. A pair of 31G needles was placed bilaterally in the frontal centrum semiovale (A = 1.7 mm, L = 2.0 mm, D = 3.1 mm). Through one needle, modified Ringer's solution (pH = 7.2, osmolarity = 308 mOsm) was infused at a rate of 0.25 µl/min for 120 min (total, 30 µl). In the contralateral hemisphere, no solution was infused, and this part of the brain served as a control. Edema was quantified by the specific gravity method[4] at 2 hours and 2 days after the infusion. For CYO and HK histochemical staining, the brain was perfused transcardially with 4% paraformaldehyde solution at 2 days after the infusion. The CYO and HK activities were determined densitometrically from the histochemical staining, which is known to correlate directly with the biochemically quantitated CYO and HK activities[3,10]. The CBF and CMRgl values were estimated autoradiographically[7,8], at 2 hours after the infusion. The CBF, CMRgl, and CYO and HK activities in the edematous hemisphere were corrected for dilution and compared with those in the control hemisphere[9].

Results

At 2 hours after the infusion, the specific gravity was significantly decreased in the infused brain (infused/control = 1.0463 ± 0058 / 1.0581 ± 0.0044) (p < 0.01). At 2 days after the infusion, however, no significant difference was observed between the specific gravity in the control and infused brains (infused/control = 1.0559 ± 0.0037 / 1.0571 ± 0.0048) (Fig. 1). The value of CBF was not significantly changed in the edematous hemisphere (infused/control = 157.9 ± 15.6/160.4 ± 17.6 ml/100 g/min). The CMRgl was slightly increased in the infused brain; however, the change was not statistically significant (infused/control = 105.4 ± 5.9 / 95.7 ± 6.0 µmol/100 g/min). No significant alteration was noted in the HK activity (infusion/control = 0.801 ± 0.051 / 0.799 ± 0.043) (Fig. 2). In contrast, as shown in Fig. 2, there was a significant decrease in CYO activity in the edematous hemisphere (infusion/control = 0.794 ± 0.050 / 0.951 ± 0.053) (–17%; p < 0.01).

Discussion

One of the mechanisms underlying the neurological deficits observed in brain edema has been considered to be a decrease in regional CBF, resulting in impairment of axonal transmission in the edematous white matter[6]. However, in an infused edema model, Sutton *et al.* demonstrated that the CBF was not decreased, if

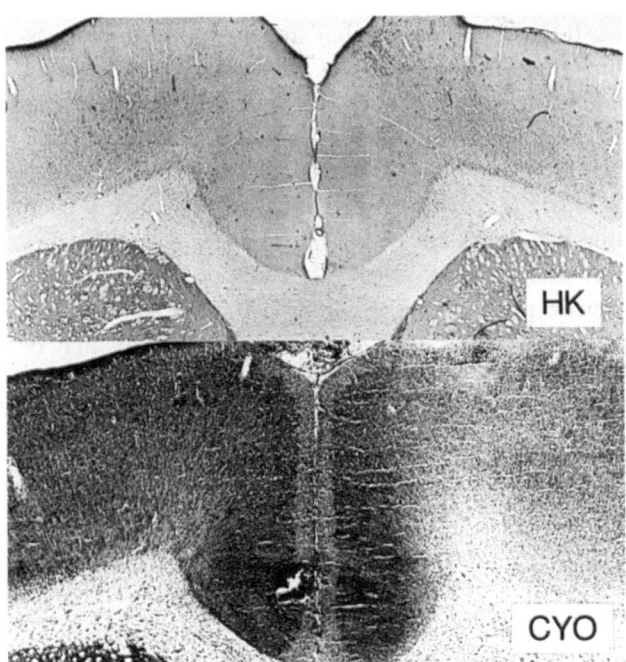

Fig. 2. Upper: Representative HK histochemical staining, demonstrating that the HK activity was not significantly changed in the edematous hemisphere (right side). Lower: Representative CYO histochemical staining, demonstrating that the CYO activity was significantly decreased in the edematous hemisphere (right side) as compared to the control hemisphere (left side)

the CBF value calculated from the autoradiography was corrected for the dilution caused by the edema formation[9]. They also demonstrated increased level of lactate and CMRgl in the edematous brain in the same model. These results may be explained by postulating an increase in anaerobic metabolism to produce high-energy phosphates despite maintained circulation.

Oxygen and glucose are utilized as energy sources to ensure maintenance of neuronal cell functions. For the delivery of these energy sources, two major routes exist between the capillary vessels and neuronal cells. One is diffusional delivery through the extracellular space, and the other is active transport via glial cells. Glucose is delivered by both these routes, while oxygen is delivered mainly by diffusion. Sutton *et al.*[9] suggested that enlargement of the extracellular space more strongly disturbs the oxygen diffusion than the glucose transport. According to this hypothesis, edema formation could impair the oxidative metabolism, resulting in activation of anaerobic glycolysis.

In the present study, although the CMRgl and HK activity were not statistically significantly changed in the edematous brain, they did show a tendency to increase, and were certainly not decreased. In contrast, the oxidative metabolism, as evaluated by measures the CYO activity, was markedly decreased in the edematous brain. These findings are consistent with the hypothesis that edema formation disturbs the oxygen diffusion in the extracellular space.

In summary, the results of the present study indicate that interstitial edema causes a unique metabolic dysfunction, involving decreased mitochondrial respiration despite maintained cerebral circulation, which can be regarded as a state of relative hypoxia. This may be explained by postulating a decreased oxygen delivery and/or accumulation of lactate, both of which have been shown to interfere with mitochondrial respiratory function.

References

1. Adams RD, Fisher CM, Hakim S, Ojemann RG, Sweet WH (1965) Symptomatic occults hydrocephalus with "normal" cerebrospinal fluid pressure: a treatable syndrome. N Engl J Med 273: 117–126
2. Go GK (1981) The classification of brain edema. In: Becks JWF (ed) Brain edema. Wiley, New York, pp 3–9
3. Lawrence GM, Trayer IP (1984) Histochemical and immunohistochemical localization of hexokinase isoenzymes in rat kidney. Histochem J 16: 697–708
4. Marmarou A, Poll W, Shulman K, Poll W (1978) A simple gravimetric technique for measurement of cerebral edema. J Neurosurg 49: 530–537
5. Ohta K, Marmarou A, Povlishock JT (1990) An immunocytochemical study of protein clearance in brain infusion edema. Acta Neuropathol (Berl) 81: 162–177
6. Polly M (1985) Blood-brain barrier; cerebral edema. Wilkins, Rengachary (eds) Neurosurgery. McGraw-Hill, New York, pp 322–331
7. Sakurada O, Kennedy C, Jehle J, *et al* (1978) Measurement of local cerebral blood flow with iodo[^{14}C] antipyrine. Am J Physiol 234: H59–H66
8. Sokoloff L, Reivich M, Kennedy C, Des Rosiers MH, Patlak CS, Pettigrew KD, Sakurada O, Shinohara M (1977) The [^{14}C]-deoxyglucose method for the measurement of local cerebral glucose utilization: theory, procedure, and normal values in the conscious and anesthetized albino rat. J Neurochem 28: 897–916
9. Sutton LN, Greenberg J, Welsh F (1990) Blood flow and metabolism in vasogenic oedema. Acta Neurochir (Wien) [Suppl] 51: 397–400
10. Wong-Riley MTT (1989) Cytochrome oxidase: an endogenous metabolic marker for neuronal activity. TINS 12(3): 94–101

Correspondence: Tatsuro Kawamata, M. D., Department of Neurological Surgery, Nihon University School of Medicine, Oyaguchi-kamimachi, Itabashi-ku, Tokyo 173, Japan.

Acta Neurochir (1994) [Suppl] 60: 271–273
© Springer-Verlag 1994

Blood-Brain Barrier, Cerebral Blood Flow, and Brain Edema in Spontaneously Hypertensive Rats with Chronic Focal Ischemia

K. Shima, K. Ohashi, H. Umezawa, H. Chigasaki, and **S. Okuyama**

Department of Neurosurgery, National Defense Medical College, Tokorozawa and Research Center, Taisho Pharmaceutical, Omiya, Saitama, Japan

Summary

This study was conducted to explore the participation of blood-brain barrier (BBB) permeability and cerebral blood flow (CBF) on the development of ischemic brain edema in rats with chronic arterial hypertension. Young spontaneously hypertensive rats were used, and focal ischemia was produced by occluding the distal middle cerebral artery (MCA). On day 7 after MCA occlusion, BBB permeability and CBF were measured by autoradiographic methods using ^{14}C-alpha-amino-isobutyric acid (AIB) and ^{14}C-iodoantipyrine. BBB permeability (transfer constant for AIB) was significantly higher in the ischemic center and periphery. The CBF of the ischemic cortex showed a graded reduction from the ischemic center to the surrounding area. The ischemic brain regions showed significantly decreased specific gravity. We conclude that SHRSP may be more vulnerable to BBB disruption after ischemia.

Keywords: Cerebral ischemia; blood-brain barrier; cerebral blood flow; brain edema.

Introduction

Recent studies have indicate that chronic hypertension plays a significant role in the development of ischemic brain edema. In the presymptomatic stage of sustained hypertension in stroke-prone spontaneously hypertensive rats (SHRSP), there is an intermittent opening of blood-brain barrier (BBB) to serum proteins[6]. SHRSP develop much larger cortical infarctions following middle cerebral artery (MCA) occlusion than normotensive rats[1]. It is thought that this susceptibility to infarction in SHRSP may be linked to hypertensive vascular features of SHRSP[2]. Some studies have shown that the SHRSP strain suffers more severe ischemia and edema formation following global ischemia induced by bilateral common carotid artery occlusion[5]. However, there have so far been no studies on the role of chronic hypertension in the development of brain edema following focal ischemia.

The present study was designed to determine the regional alterations of BBB permeability and cerebral blood flow (CBF) on the development of ischemic brain edema in SHRSP.

Materials and Methods

Male SHRSP at the age of 11 weeks were anesthetized, and focal ischemia was produced by the permanent MCA occlusion. The left MCA was coagulated distal to striate branches and 0.7–1.0 mm dorsal to the rhinal fissure via a retroorbital approach[1,9]. On day 7 following MCA occlusion, each animal was anesthetized with halothane, and blood pressure, rectal temperature and blood gases were monitored. Three sets of experiments were done.

BBB permeability was measured by autoradiographic method using ^{14}C-alpha-amino-isobutyric acid (AIB). After confirming the acceptable physiological state of each animal, 100 μCi of AIB dissolved in 1 ml of saline, after pH adjustment to 7.2 to 7.4, was given intravenously at a constant rate. Timed blood samples were taken for analysis of ^{14}C activity. Ten min after the injection of the tracer, the rat was decapitated, the brain rapidly removed, frozen, and cut into 20 μm-thick sections for autoradiographic studies. BBB permeability was expressed as a unidirectional blood-to-brain transfer constant (Ki) for AIB, which can be calculated according to the following equation[4]

$$ Ki = \frac{Am - (Vp \times CT)}{\int_0^T Cp\,(t)\,dt} $$

where Am is the tracer concentration in the tissue, Vp is the plasma volume retained in the tissue, CT is the terminal tracer concentration in the plasma and Cp is the tracer concentration in the plasma. Residual plasma volume was determined as the ^{14}C-sucrose space at 30 s after the injection of the tracer (100 μCi/kg).

CBF was measured by autoradiographic method using ^{14}C-iodoantipyrine (^{14}C-IAP) according to Sakurada et al.[10]. Each animal received an intravenous injection of 100 μCi/kg of ^{14}C-iodoantipyrine. Blood samples were taken every 5 s to assess ^{14}C activity. The animal was decapitated 30 s after the injection of the tracer. Autoradiographs were prepared in the same manner as for BBB permeability.

Brain water content was determined by the gravimetric method[7] using a bromobenzene-kerosene linear gradient columm (r > 0.99).

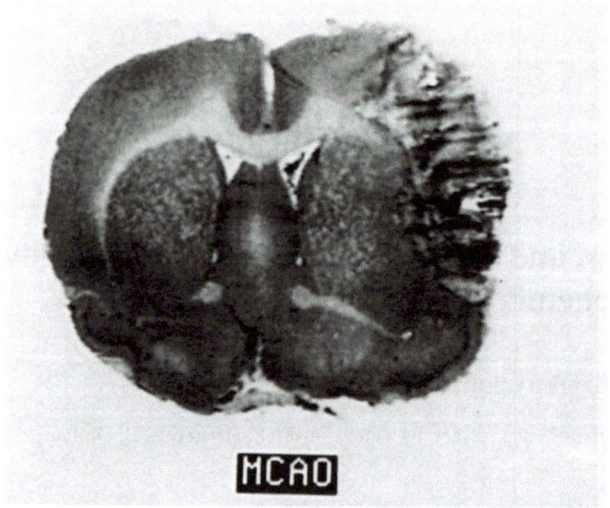

a

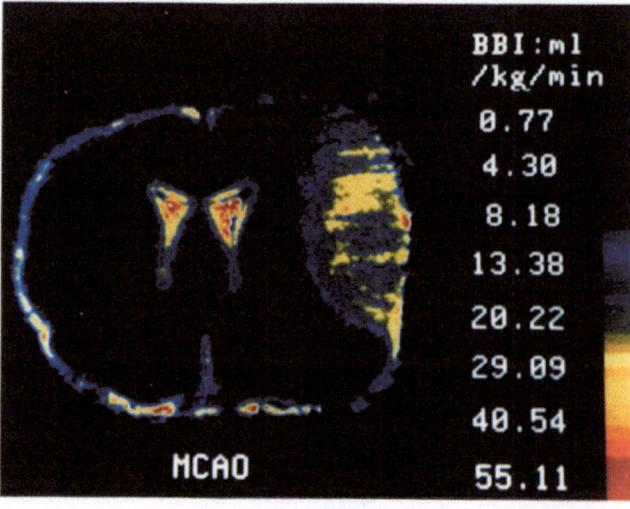

b

Fig. 1. Histological brain section to show cortical infarction (a) and corresponding ¹⁴C-alpha-amino-isobutyric acid autoradiographic image of the same section (b)

Samples 1 mm in size were dissected from each brain area and put into the column.

Data were analysed by analysis of variance (ANOVA) and statistical differences between groups were determined by Dunnett's test. All values are expressed as mean ± SD.

Results

The physiological variables at the start of autoradiographic studies were similar in all groups. The infarcted area in each animal was limited to the ipsilateral cerebral cortex and did not extend to the striatum[9,11].

Regional BBB permeability was significantly higher in the ischemic center (19 times) and periphery (14 times) compared to the corresponding contralateral region (Fig. 1, Table 1).

The CBF of ischemic cortex showed a graded reduction. In the central area, the CBF decreased markedly,

to around 6%, compared with both the contralateral side and the sham control group. The CBF in the peripheral and surrounding areas decreased less markedly, to 15–18% and 53–65%, respectively (Table 2).

The ischemic brain regions showed significantly decreased specific gravity values, compared with those in the nonischemic side (Table 1).

Discussion

The present results demonstrate that a prominent increase in the BBB permeability to AIB occur in the ischemic center and border until the seventh day after MCA occlusion in SHRSP. An increase in the BBB permeability in normotensive rats is also recognized until the seventh day after MCA occlusion, but the values assessed with albumin is only 2 times higher

Table 1. *Regional Blood-to-Brain Tansfer Constant (Ki, n = 5), Plasma Volume (Vp; n = 5) and Specific Gravity (n = 8) in SHRSP 7 Days After MCA Occlusion*

Brain region	Ki (ml/kg/min)	Vp (μl/g)	Specific gravity
Frontal cortex			
Contralateral side	0.36 ± 0.19	11.55 ± 0.90	1.045 ± 0.001
Center of infarct	7.00 ± 1.53**	9.98 ± 3.19	1.028 ± 0.007**
Periphery	4.93 ± 1.08*	37.60 ±12.95	1.036 ± 0.003**
Surrounding	1.94 ± 0.56	11.03 ± 2.84	1.041 ± 0.002

Values are expressed as means ± SD.
* $p < 0.05$, ** $p < 0.01$ vs contralateral side.

Table 2. *Regional CBF in SHRSP 7 Days After MCA Occlusion*

Brain region	Sham-operated (5 rats)	MCA-occluded (8 rats)	Percent change
Frontal cortex			
Contralateral side	171 ± 57	172 ± 33	100
Center of infarct		10 ± 7*	6
Periphery		27 ± 14*	16
Surrounding		92 ± 21*	53

Values are ml/100g/min (means ± SD).
Percent change: percent changes in CBF values compared to the contralateral side. * $p < 0.001$ vs contralateral side.

than those in the sham-operated group[3]. Recent studies have shown no significant difference between AIB and albumin on the BBB permeability in the ischemic cortex after MCA occlusion in normotensive rats, and the BBB permeability to AIB remained 2–2.5 times higher values for 14 days, compared with non-ischemic cortex[8]. It is difficult to compare the absolute values measured in different experimental protocol, but the BBB alterations in normotensive animals were less pronounced than those in SHRSP. In the existence of the BBB opening, the hydrostatic pressure gradient between intracranial vessels and brain tissue such as chronic hypertension in SHRSP may influence the development of brain edema[12].

Thus, our results indicate that, in chronic focal ischemia, the postischemic low CBF and BBB disruption in the ischemic rim may be related to the development of ischemic brain edema. SHRSP may be more vulnerable to BBB disruption after focal ischemia.

References

1. Coyle P, Jokelainen PT (1983) Differential outcome to middle cerebral artery occlusion in spontaneously hypertensive stroke-prone rats (SHRSP) and Wistar Kyoto (WKY) rats. Stroke 14: 605–611
2. Coyle P (1986) Different susceptibility to cerebral infarction in spontaneously hypertensive (SHR) and normotensive Sprague-Dawley rats. Stroke 17: 520–525
3. Gotoh O, Asano T, Koide T, Takakura K (1985) Ischemic brain edema following occlusion of the middle cerebral artery in the rat. 1: the time course of the brain water, sodium and potassium contents and blood-brain barrier permeability to ^{125}I-albumin. Stroke 16: 101–109
4. Gross PM (1987) The microcirculation of rat circumventricular organs and pituitary gland. Brain Res Bull 18: 73–85
5. Katayama Y, Terashi A, Sugimoto S, Inamura K, Suzuki S, Sekiguchi F, Akashi A (1984) Experimental cerebral ischemia after bilateral common carotid artery ligation in SHRSP, SHRSR and Wistar rats. Brain Nerve (Tokyo) 36: 1069–1075
6. Klatzo I (1986) Relationship between water and protein transfer across the barrier. In: Mchedlishvili G, Cervós-Navarro J, Hossmann K-A, Klatzo I (eds) Brain edema. A pathogenetic analysis. Akadémiai Kiadó, Budapest, pp 127–128
7. Marmarou A, Poll W, Shulman K, Bhagavan H (1978) A simple gravimetric technique for measurement of cerebral oedema. J Neurosurg 49: 530–537
8. Menzies SA, Hoff JT, Betz AL (1990) Acta Neurochir (Wien) [Suppl] 51: 220–222
9. Okuyama S, Shimamura H, Karasawa K, Araki H, Kimura M, Aihara H (1991) Protective effects of minaprine in infarction produced by occluding of the middle cerebral artery in stroke-prone spontaneously hypertensive rats. Gen Pharmac 22: 143–150
10. Sakurada O, Kennedy C, Jehle J, Brown JD, Carbin GL, Sokoloff L (1978) Measurement of local cerebral blood flow with iodo-^{14}C-antipyrine. Am J Physiol 234: H59–H66
11. Shima K, Ohashi K, Umezawa H, Chigasaki H, Krasawa Y, Okuyama S, Araki H, Otomo S (1990) Acta Neurochir (Wien) [Suppl] 51: 242–244
12. Umezawa H, Shima K, Chigasaki H, Ishii S (1993) Local cerebral blood flow and glucose metabolism in hydrostatic brain edema. Neurol Med Chir (Tokyo) 33: 139–145

Correspondence: Katsuji Shima, M.D., Department of Neurosurgery, National Defense Medical College, 3-2 Namiki, Tokorozawa, Saitama 359, Japan.

Acta Neurochir (1994) [Suppl] 60: 274–276
© Springer-Verlag 1994

Abnormalities of the Blood-Brain Barrier in Global Cerebral Ischemia in Rats Due to Experimental Cardiac Arrest

M. J. Mossakowski, A. S. Lossinsky, R. Pluta, and **H. M. Wisniewski**

Medical Research Centre, Warsaw, Poland

Summary

The aim of the study was to characterize the impact of global cerebral ischemia resulting from cardiac arrest on the BBB permeability. Survival time of experimental animals after 3.5, 5 or 10 min ischemia range from 3.5–10 min to 24 h. Vascular permeability was evaluated with horseradish peroxidase (HRP).

BBB disturbances were of biphasic nature. In the first phase, appearing immediately after ischemia and persisting till 1st postischemic hour, HRP extravasation involved mainly venous site of microcirculation and was limited to the cerebral and cerebellar cortex and the central periventricular structures. The second phase covering the period between 6 and 24 h after resuscitation was characterized by random HRP leakage in all the CNS structures. HRP penetrated through increased microvesicular and canalicular endothelial systems, through interendothelial junctions and via disintegrated endothelial cells.

Distribution and perivenous localization of BBB changes in early phase suggests their connection with venostasis resulting from cardiac arrest. The second phase seems to be pathogenetically related with the consequences of the ischemic process.

Keywords: Global cerebral ischemia; cardiac arrest; BBB-damage.

Introduction

A great number of studies have been devoted to the influence of cerebral ischemia on the vascular permeability in the brain. What inclined us to come back to this problem, was the availability of a relatively nontraumatic model of global cerebral ischemia resulting from experimentally induced cardiac arrest. The model described by Korpachev *et al.*[4] permits exceptionally long survival of experimental animals and as such seems to be particularly useful for studies on the late sequelae of the ischemic incident. However, in contrast to other models in use, this is not a model of isolated cerebral ischemia but rather of experimental clinical death with all consequences resulting from generalized ischemia of all body organs[9]. The data concerning vascular permeability in this particular model are controversial. Mossakowski and Krajewski[7] demonstrated BBB abnormalities appearing in early postischemic period. So did Gannushkina *et al.*[1]. On the contrary, Kapuscinski[2] was not able to detect any changes in vascular permeability with quantitative isotope methods. Aiming to elucidate these controversies we have undertaken more detailed studies on the impact of this type of global cerebral ischemia on the nature and dynamics of BBB changes.

Material and Methods

The experiments were performed on adult Albino rats in which clinical death was induced according to the method described by Korpachev *et al.*[4]. Compression of the heart vascular bundle by a special hook inserted into the thorax led within 2–3.5 min to cardiac arrest and cessation of respiratory function lasting until resuscitation was undertaken. In our experiments this was done either after 3.5, 5 or 10 min of complete cessation of the brain bioelectric activity. Following resuscitation which included external heart massage and artificial respiration, the rats survived for 3.5, 5 and 10 min, as well for 1, 3, 6, 12 and 24 h.

Horseradish peroxidase (HRP, Sigma type VI) was used as an indicator of the BBB changes. It was infused into the femoral vein (50 mg HRP/rat) and allowed to circulate for 30 min. Animals were sacrificed by transcardiac perfusion with a solution containing 2% formalin and 1% glutaraldehyde in 0.1 M sodium cacodylate, pH 7.3 at 22°C[8]. For demonstration of HRP vibratome tissue slices from cerebral hemispheres, brainstem and cerebellum were incubated in a solution of 3,3'-diaminobenzidine-tetra HCl[5] and than processed for electron microscopic study in a routine way. Chopped tissue blocks for light microscope evaluation of HRP extravasation were prepared according to Lossinsky *et al.*[6].

Results

As assessed by light microscopy, HRP extravasation was scanty. Arterioles, capillaries and venules surrounded by HRP products were randomly distributed.

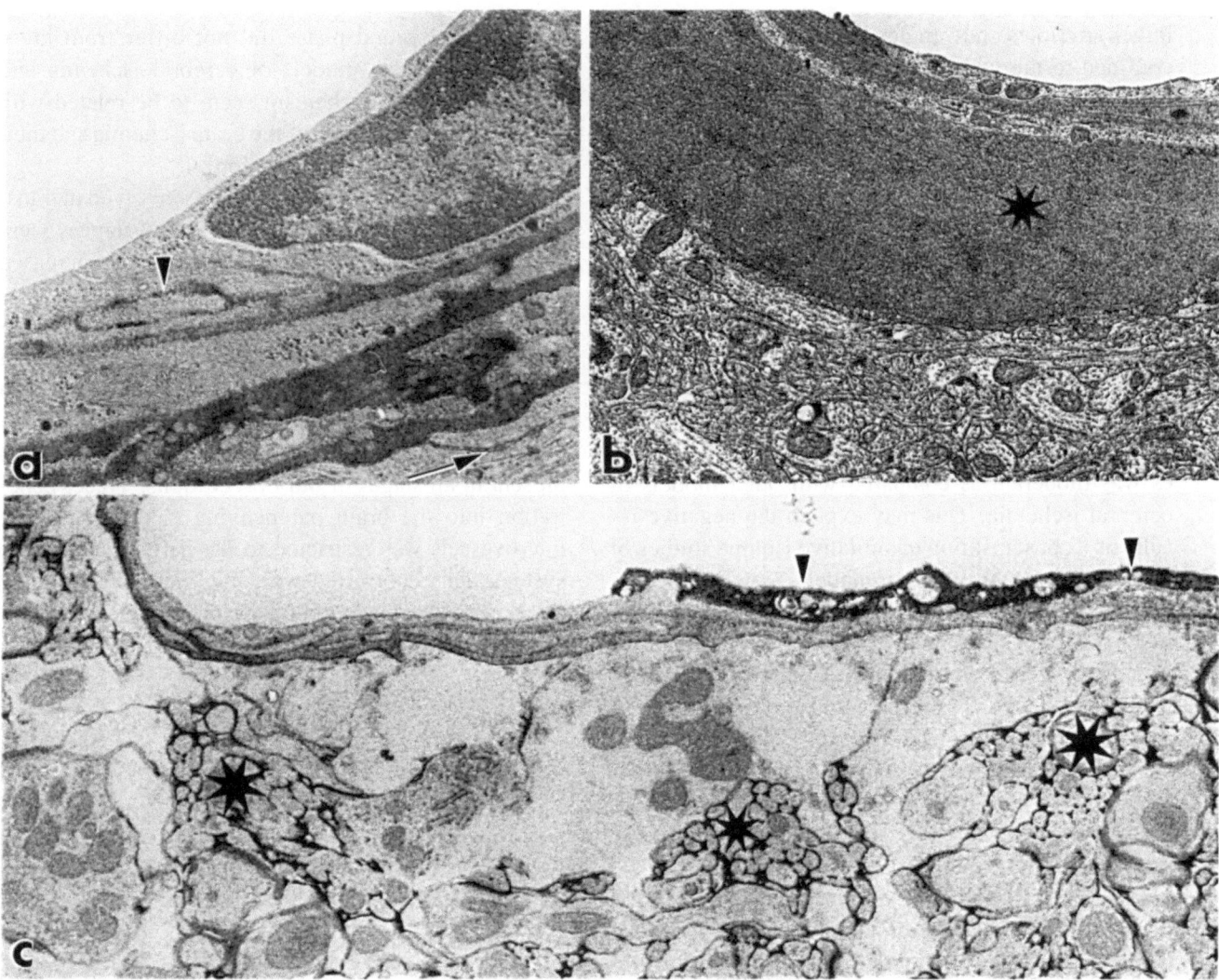

Fig. 1. HRP extravasation into brain tissue: (a) 5 min ischemia, 1 h survival. HRP accumulated within the interendothelial junction (arrowhead) and in the basal membrane systems, slightly penetrating to the brain extracellular space (arrow). × 18000; (b) 10 min ischemia, 10 min survival. Massive accumulation of HRP within perivascular space (asterisk). × 8000; (c) 5 min ischemia, 5 min survival. Fragment of the microvessel with endothelial HRP accumulation (arrow heads) and diffuse spread into the surrounding tissue

Tracer extravasations at the point of vascular branchings were common. Biphasic appearance of HRP extravasation was noticed. Extravascular tracer leakage appeared immediately after resuscitation and was present at 1 h after ischemia. No BBB abnormalities were noticed at the 3rd postischemic hour. Their features reappeared 6 h after resuscitation and persisted till 24 h. HRP extravasations surrounding mostly small veins in the first stage were limited to the cerebral and cerebellar cortex and the central periventricular structures, while those appearing in the second phase were randomly distributed all over the central nervous system.

Electron microscopic observations showed remarkable changes in the structure of microvessel endotheli-um. They appeared as deep invaginations of luminal endothelial plasmolemma, in some instances reaching almost the outer surface of the basal membrane. Increased number of intracytoplasmic microvesicles was noticed. So was formation of intracytoplasmic canalicular structures. Interendothelial junctions were widened. In many instances, particularly in the early postischemic phase, severe disintegration of endothelial cells became apparent.

HRP products were present within the walls of microvessels. They appeared within endothelial surface pits, intracytoplasmic microvesicles and canalicular structures, in distended intercellular junctions as well as in basal membrane systems (Fig. 1a). In some instances HRP filled the whole endothelial cytoplasm. In larger arterioles and venules HRP product was mostly confined to the widened perivascular space (Fig. 1b) with only slight penetration to the extracellular spaces. HRP leaking through damaged capillaries was spreading diffusely to the surrouding tissue (Fig. 1c). Perivascular astrocytic processes were swollen during the whole period of observation in the vicinity of both leaking and not-leaking vessels (Fig. 1c).

Discussion

Our studies indicate that global cerebral ischemia due to experimentally induced cardiac arrest results in generalized BBB disturbances. The intensity of changes was moderate as compared with other models of cerebral ischemia. This may explain the negative results of Kapuscinski[2] in quantitative isotope studies of vascular permeability in our model.

The intensity of BBB changes appeared to be largely independent of the duration of the ischemic incident, except for the alterations noticed immediately after resuscitation. In this stage the number of microvessels, mostly venules, showing features of tracer extravasation seemed to be greater in animals with the longer period of cardiac arrest. This may be connected with the severe damage of endothelial cells observed in this stage of the experiment.

Like in most models of cerebral ischemia the BBB disturbances in our experiments were biphasic. In that respect, however, our model revealed some distinguishing features, concerning the early phase of barrier alterations. HRP extravasation in this phase was almost totally limited to superficial layers of the cerebral and cerebellar cortex and central periventricular structures. It was connected with the venous side of microcirculation and its severity was dependent on the duration of the ischemic incident. This nature of the early BBB alterations may suggest that they are connected with severe venostasis forming an important element of cardiac arrest pathology[10]. In that context pathogenetic mechanism of the early barrier disturbances may be somewhat different from that generally accepted for other models of cerebral ischemia. Klatzo[3] considers that early stage of the ischemic BBB damage results from rapid increase of blood pressure appearing immediately after release of vascular occlusion. Postischemic hyperperfusion was also observed in our model[8]. However its appearance was not as rapid as in experimental models connected with ligation of the supplying vessels. Distribution and nature of the BBB altera-

tions in the second phase did not differ from those described in most models of cerebral ischemia and their pathogenic mechanism seem to be related with metabolic consequences of the brain ischemia and their influence on the microcirculation[3].

Features of all known mechanisms involved in transvascular transfer of macromolecular substances were observed in our material. These included increased pinocytic activity, formation of intracytoplasmic canalicular structures as well as widening of interendothelial junctions[5,11]. Severe damage of endothelial cells as a possible factor facilitating tracer penetration is worth mentioning. This phenomenon, although most remarkable in the earliest postischemic period, was present in all stages of our observations. Tracer penetration into the brain parenchyma in case of larger microvessels was restricted to the perivascular space system. This contrasted with the wide spread of the tracer into the surrounding brain parenchyma from damaged capillary vessels.

References

1. Gannushkina IY, Weinrauder-Semkow H, Matuskina EA, Gurvich AM, Mossakowski MJ (1990) Postresuscitation changes of blood-brain barrier permeability and their possible pathogenetic importance. Pat Fizjol Exp Ter 3: 13–16
2. Kapuscinski A (1988) Blood-brain barrier in a model of clinical death in rats. Neuropatol Pol 26: 175–183
3. Klatzo I (1983) Disturbances in blood-brain barrier permeability in cerebrovascular disorders. In: Hossmann K-A, Klatzo I (eds) Cerebrovascular transport mechnisms. Acta Neuropathol (Berl) [Suppl] 8: 81–88
4. Korpachev WG, Lysenkov SP, Tiel LZ (1982) Modelling of clinical death and postresuscitation disease in rats. Patol Fizjol Eksp Ter 3: 78–80
5. Lossinsky AS, Garcia JH, Iwanowski L, Lightfoote WE Jr (1979) New ultrastructural evidence for a protein transport system in endothelial cells of gerbil brains. Acta Neuropathol (Berl) 47: 105–110
6. Lossinsky AS, Vorbrodt AW, Wisniewski HM, Moretz RC (1981) A simple screening procedure for evaluating central nervous tissue sections showing structural and cytochemical alterations of the blood-brain barrier. Stain Technol 56: 279–282
7. Mossakowski MJ, Krajewski S (1988) Antineuronal antibodies in blood sera of rats subjected to global cerebral ischemia. Neuropatol Pol 26: 37–43
8. Pluta R, Lossinsky AS, Mossakowski MJ, Faso L, Wisniewski HM (1991) Reassessment of a new model of complete cerebral ischemia in rats. Method of induction of clinical death, pathophysiology and cerebrovascular pathology. Acta Neuropathol (Berl) 83: 1–11
9. Safar P (1986) Cerebral resuscitation after cardiac arrest: a review. Circulation 74 [Suppl 4] : 138–153

Correspondence: M. J. Mossakowski, M. D., Medical Research Centre, PASci, Dworkowa Street 3, 00-784 Warsaw, Poland.

Ischemic Brain Edema: Treatment

Acta Neurochir (1994) [Suppl] 60: 279–281
© Springer-Verlag 1994

Do N-methyl-D-aspartate Antagonists Have Disproportionately Greater Effects on Brain Swelling than on Ischemic Damage in Focal Cerebral Infarction?

C. K. Park, J. McCulloch, D. S. Jung, J. K. Kang, and **C. R. Choi**

Department of Neurosurgery, Catholic University Medical College, Seoul, Korea

Summary

Using a frozen section technique, we have assessed the effects of N-methyl-D-aspartate(NMDA) antagonist upon brain swelling caused by ischemic brain edema in a rat model of focal cerebral infarction. Although pretreatment with the competitive NMDA antagonist, D-CPPene or the noncompetitive NMDA antagonist, CNS 1102 reduced both the volumes of infarction and ischemic edema in the cerebral hemisphere, mean reduction in brain edema was proportionately similar to decrease in infarct volume in the same animals (correlation coefficienct, r = 0.82, p < 0.001). There was, therefore, no evidence of disproportionately greater effects with NMDA antagonist upon brain edema.

Keywords: Ischemic brain edema; glutamate; N-methyl-D-aspartate antagonists.

Introduction

Excitatory amino acids play a major role in the potentiation of neuronal ischemic damage. Numerous studies in animal models of focal cerebral ischemia have demonstrated that the antagonists of the N-methyl-D-aspartate (NMDA) subtype of glutamate receptor can attenuate ischemic brain damage[3,5,12,13,17]. Brain swelling caused by brain edema after an ischemic episode remains a major clinical problem, regardless of the mechanism of the brain edema (cytotoxic or vasogenic). In views of the proposed role of excitatory amino acid neurotransmitters in ischemic brain edema[1,14], NMDA antagonists could be a pharmacologic therapy for brain swelling additional to the benefit of their antiischemic effects. To study the effects of NMDA antagonists upon brain edema in 24-hour focal ischemia model in rats, we used a novel competitive NMDA antagonist, D-(E)-4-(3-phosphonoprop-2-enyl) piperazine-2-carboxylic acid (D-CPPene)[5] and a novel non-competitive NMDA antagonist, N-(1-Naphthyl)-N-(3-ethyphenyl)-N-methyl guanidine hydro-

chloride (CNS 1102)[6], the neuroprotective effects of which have been established[10,11].

Materials and Methods

We studied eighteen adult male Sprague-Dawley rats. The animals were anesthetized using halothane and a nitrous oxygen mixture (70 : 30%) during the surgical procedure, less than 30 min in duration. The mask was placed over the face, and spontaneous respiration was maintained. Cannulation of femoral vessels was first carried out, and the animals were maintained normothermic by a homeothermic system (Havard Apparatus, Kent, U.K.). The arterial catheter was removed and anesthesia was discontinued immediately after completion of the surgical procedures, but the venous catheter was kept by a swivel system to administer the drugs throughout the experiments.

Middle Cerebral Artery (CA) Occlusion

All animals underwent the occlusion of the left MCA via a subtemporal approach without removal of zygomatic arch or temporal muscle. Under high magnification of surgical microscope, the left MCA was coagulated with a micro-bipolar coagulator from the olfactory tract to the most proximal portion of the MCA and divided.

Assessment of Cerebral Infarction and Edema Volumes

Twenty-four hours following MCA occlusion, the rats were sacrificed. Brains were removed from the calvaria immediately after sacrifice, and frozen in a cryostat (−25°C). Coronal sections (20 μm thick) were cut with a cryostat and stained with hematoxylin-eosin. Infarcted area and the extent of brain swelling were readily delineated (Fig. 1). Eight coronal sections which corresponded to planes of the preselected forebrain were selected among the stained slices, and the volume of ischemic damage was computed from the areas of ischemic damage measured at the different coronal planes and their anteroposterior coordinates, as described previously[9,13]. To measure the volume of brain swelling, the total volume of the nonischemic hemisphere in each brain was subtracted from the total volume of the ischemic one.

Experimental Design

The animals were divided into three groups: vehicle-administered group (n = 6); D-CPPene-treated group (n = 6); CNS 1102-

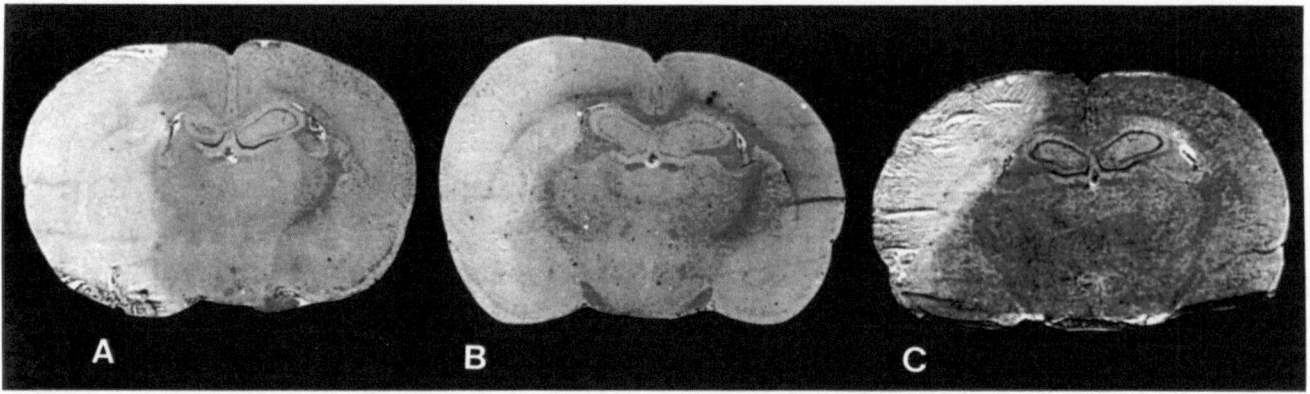

Fig. 1. Hematoxylin-eosin stained coronary sections of the brain at the level of the anterior hypothalamus (5.15 mm anterior to interaural line) 24 hours after the middle cerebral artery occlusion. The infarction area is well delineated, and the amounts of ischemic damage and brain swelling in the control (A) are much larger than those in the D-CPPene-treated (B) or the CNS 1102-treated animals (C)

treated group (n = 6). The drugs were administered 15 min before MCA occlusion. The drugs-treated animals received a bolus dose and then a constant infusion (D-CPPene: 4.5 mg/kg iv + 3 mg/kg/h iv; CNS 1102 : 0.25 mg/kg iv + 0.17 mg/kg/h iv). In the vehicle-administered group, only saline was infused in the same fashion.

Statistical Analysis

The relationship between brain swelling and infarction was assessed by linear regression analysis and Pearson's correlation coefficient values.

Results and Discussion

Adequate oxygenation, normocapnia and normothermia were maintained throughout the surgical procedure. Animals treated with D-CPPene induced small (4.5% compared to the controls) and transient reduction of MABP immediately following an intravenos loading bolus.

Pretreatment with either D-CPPene or CNS 1102 significantly reduced the volume of infarction in the cerebral hemisphere by 36% (p < 0.001) or by 26% (p < 0.01) respectively. The volume of brain swelling, obtained by subtracting the nonischemic hemispheric volume from the ischemic volume, was also significantly reduced both with D-CPPene (37% less than the controls; p < 0.001) and with CNS 1102 (28% less; p < 0.05). The assessment of the relationship between the volumes of cerebral infarction and edema revealed an excellent correlation (r = 0.82, p < 0.001) with almost every animal close to the overall regression line irrespective of whether they received vehicle or the NMDA antagonists (Fig. 2). There was no evidence that the NMDA antagonists treatment altered the size of the contralateral hemisphere (Table 1).

Our results with D-CPPene and CNS 1102 are consistent with the previous reports that NMDA receptor antagonists are capable of reducing ischemic brain damage. Particularly, the positive results with the competitive NMDA antagonist, D-CPPene can be added to the few investigations[15,16] demonstrating anti-ischemic effects of competitive NMDA antagonists, administered systemically.

Brain swelling, caused by ischemic brain edema, may play an important role in the pathogenesis of focal cerebral ischemic lesions. Numerous studies have revealed that NMDA antagonists attenuated ischemic brain damage but little attention has been given to the effect of NMDA antagonist on ischemic brain edema[7]. In the

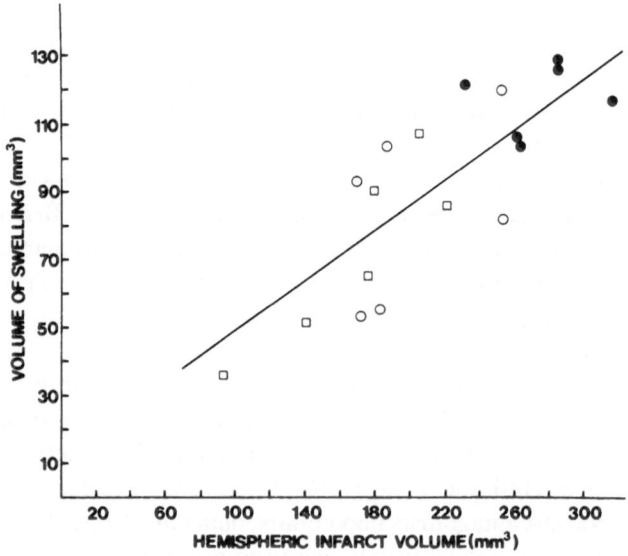

Fig. 2. Scatter diagram of hemispheric swelling (mm³) as a function of infarct volume (mm³) in vehicle-treated (filled circles), D-CPPene-treated (open squares) and CNS 1102-treated animals (open circles). The regression line is calculated for vehicle, D-CPPene and CNS 1102-treated animals (r = 0.82, p < 0.001)

Table 1. *Effect of NMDA Antagonists on the Volumes of Brain Swelling and the Contralateral Cerebral Hemisphere at 24 h Following Left MCA Occlusion*

	Group		
	Vehicle	D-CPPene	CNS 1102
Hemispheric volume, contralateral	560.8 ± 26.1	580.5 ± 22.8	581.7 ± 12.4
Brain swelling	116.0 ± 6.0	72.5 ± 11.0**	84.4 ± 10.8*

Data are presented as mm^3 (mean ± SEM).
* $p < 0.05$; ** $p < 0.001$ for comparision of vehicle and drug-treated animals (Student's t-test).

present study, the NMDA antagonists attenuated brain swelling in direct proportion to the reduction of ischemic brain damage. Both D-CPPene and CNS 1102 have been reported to reduce ischemic brain damage even in short term (less than 6 h) experimental models of focal cerebral ischemia[2,8], when brain edema did not reach its maximal intensity. The interpretation of the present data would be that the primary effect is the reduction in infarction with less brain swelling as a consequence.

These studies suggest that NMDA antagonists do not have disproportionately greater effects upon ischemic brain edema despite the proposed role of NMDA neurotransmitters in ischemic brain edema.

Acknowledgement

The investigations were supported in part by Non Directed Research Fond, Korea Research Foundation, 1991, and by Catholic Medical Center, Korea. The address of Professor J. McCulloch is University of Glasgow, Glasgow, U.K. We thank Miss Y. J. Chung in the preparation of the manuscript.

References

1. Baethmann A, Oettinger W, Rothenfuer W, Kempski 0, Unterberg A, Geiger R (1980) Brain edema factors: current state with particular reference to plasma constituents and glutamate. In: Cervós-Navarro J, Ferszt R (eds) Advances in neurology, vol 28. Raven, New York, pp 171–195
2. Bullock R, Graham DI, Chen MH, Lowe D, McCulloch J (1990) Focal cerebral ischemia in the cat: Pretreatment with a competitive NMDA receptor antagonist, D-CPPene. J Cereb Blood Flow Metab 10: 668–674
3. Germano IM, Pitts LH, Meldrum BS, Bartkowski HM, Simon RP (1987) Kynurenate inhibition of cell excitation decreases stroke size and deficits. Ann Neurol 22: 730–734
4. Gotti B, Duverger D, Bertin J, Carter C, Dupont R, Frost J, Gaudillier B, MacKenzie ET, Rousseau J, Scatton B, Wick A (1988) Ifenprodil and SL 82.0715 as cerebral anti-ischemic agents. I. Evidence for efficacy in models of focal cerebral ischemia. J Pharmacol Exp Ther 247: 1211–1221
5. Lowe DA, Neijt HC, Aebischer B (1990) D-CPPene (SDZ EAA 494), a potent and competitive N-methyl-D-aspartate (NMDA) antagonist: effect on spontaneous activity and NMDA-induced depolarizations in the rat neocortical slice preparation, compared with other CPP derivatives and MK-801. Neurosci Lett 113: 315–321
6. McBurney RN, Radd NL, Hamilton PN, Taylor EM, Cotter RE, Fischer JB, Goldin SM, Keana JFW, Wolcott TJ, Kirk CJ (1991) Neuroprotective efficacy of CNS 1102, a novel noncompetitive NMDA antagonist. J Cereb Blood Flow Metab 11 [Suppl 2]: S219
7. McCulloch J, Bullock R Teasdale GM (1991) Excitatory amino acid antagonists: opportunities for the treatment of ischemic brain damage in man. In: Meldrum BS (ed) Excitatory amino acid antagonists. Blackwell, Edinburgh, pp 287–327
8. Minematsu K, Fisher M, Li L, Kanapp AG, Cotter RE, McBurney RN, Sotak CH (1993) Effects of a novel NMDA antagonist on experimental stroke rapidly and quantitatively assessed by diffusion-weighted MRI. Neurology 43: 397–403
9. Osborne KA, Shigeno T, Ford I, McCulloch J, Teasdale GM, Graham DI (1987) Quantitative assessment of early brain damage in a rat model of focal cerebral ischaemia. J Neurol Neurosurg Psychiatry 50: 402–410
10. Park CK, McBurney RN, Holt WF, Cotter RE, McCulloch J, Kang JK, Choi CR (1993) The dose-dependency of the antiischemic efficacy and of the side effects of a novel NMDA antagonist, CNS 1102. J Cereb Blood Flow Metab 13 [Suppl 3]: S641
11. Park CK, McCulloch J, Kang JK, Choi CR (1992) Efficacy of D-CPPene, a competitive N-methyl-D-aspartate antagonist in focal cerebral ischemia in the rat. Neursci Lett 147: 41–44
12. Park CK, Nehls DG, Graham DI, Teasdale GM, McCulloch J (1988) Focal cerebral ischaemia in the cat: treatment with the glutamate antagonist MK-801 after induction of ischaemia. J Cereb Blood Flow Metab 8: 757–762
13. Park CK, Nehls DG, Graham DI, Teasdale GM, McCulloch J (1988) The glutamate antagonist MK-801 reduces focal ischaemic brain damage in the rat. Ann Neurol 24: 543–551
14. Rothman SM, Olney JW (1986) Glutamate and pathophysiology of hypoxic-ischemic brain damage. Ann Neurol 19: 105–111
15. Simon R, Shiraishi K (1990) N-methyl-D-aspartate antagonist reduces stroke size and regional glucose metabolism. Ann Neurol 27: 606–611
16. Simon RP, Swan JH, Griffiths T, Meldrum BS (1984) Blockade of N-methyl-D-aspartate receptors may protect against ischemic damage in the brain. Science 226: 850–883
17. Steinberg GK, George CP, DeLaPaz RD, Shibata DK, Gross T (1988) Dextromethorphan protects against cerebral injury following transient focal ischemia in rabbits. Stroke 19: 1112–1118

Correspondence: C. K. Park, M. D., Department of Neurosurgery, Kangnam St. Mary's Hospital, 505, Banpo-dong Seocho-ku, Seoul 137-040, Korea.

Acta Neurochir (1994) [Suppl] 60: 282–284
© Springer-Verlag 1994

The Effect of Age on Cerebral Oedema, Cerebral Infarction and Neuroprotective Potential in Experimental Occlusive Stroke

M. Davis, A. D. Mendelow, R. H. Perry, I. R. Chambers, and **O. F. W. James**

Departments of Geriatric Medicine and Surgery, University of Newcastle upon Tyne, England, U.K.

Summary

A model of occlusive stroke in the aging brain has been developed and used to evaluate the effects of age upon cerebral infarction, cerebral oedema and neuroprotective potential. Focal ischaemia following left middle cerebral artery occlusion has been compared in aged (30 month) and adult (< 17 month) rats, with histological assessment of infarct volume and analysis of specific gravity as an index of cerebral oedema. Aging was associated with a significant increase in cerebral infarct size. The mean infarct volume in aged rats was $40.5\% \pm 2.6\%$ of the hemisphere volume, compared to $30.9\% \pm 0.7\%$ in adults ($p < 0.01$). Pre-treatment with the competitive N-Methyl-D-Aspartate (NMDA) receptor antagonist 3-(2-Carboxy Piperazin-4-yl)Propyl-l-Phosphonate (D-CPP-ene) reduced infarct volumes in both age groups to $33.0\% \pm 1.8\%$ and $20.7\% \pm 3.2\%$ in aged and adult animals, respectively ($p < 0.05$). There was significantly less oedema of the cerebral cortex in D-CPP-ene pre-treated rats; mean cortical specific gravity 4 hours post-infarction was 1.0381 ± 0.0013 in untreated aged rats and 1.0391 ± 0.0014 in untreated adults, compared to 1.0458 ± 0.0031 in treated aged rats and 1.0442 ± 0.0014 in treated adults ($p < 0.05$). At 24 hours post-infarction, D-CPP-ene pre-treated aged rats had a mean cortical specific gravity of 1.0403 ± 0.0006 compared to 1.0361 ± 0.0014 in untreated aged animals ($p < 0.05$). This study has demonstrated an age-related increase in cerebral infarct size, but has shown that the aging brain is amenable to neuroprotection by NMDA receptor antagonism.

Keywords: Aging; cerebral ischaemia; NMDA.

Introduction

Stroke is a major cause of death and disability in the elderly . However, most experimental stroke research, including the evaluation of neuroprotective strategies, has incorporated the use of young animals. The N-Methyl-D-Aspartate (NMDA) receptor antagonists have been reported to yield large reductions in focal ischaemic damage in young animal models of stroke[1]. Competitive antagonists of the NMDA receptor, including 3-(2-Carboxy Piperazin-4-yl)Propyl-1-Phos-phonate (D-CPP-ene), are thought to limit glutamate-mediated neurotoxicity by competing with the excitotoxin at the neurotransmitter recognition site, thereby blocking the calcium influx that is thought to be associated with cell death[5]. However, there is a significant decline in the density, sensitivity and function of cortical NMDA receptors with age[2,9]. This suggests that there may be limited scope for neuroprotection in elderly stroke patients. In this study, the effect of age on cerebral infarct size, cerebral oedema and neuroprotective potential has been evaluated, using a model of occlusive stroke in the aging rat brain.

Materials and Methods

The investigations were performed in adult (< 17 month) and aged (30 month) male Wistar rats. Anaesthesia was induced and maintained hy inhalation of halothane in nitrous oxide/oxygen (70%/30%) via intubation and ventilation. The femoral vessels were cannulated to allow sampling for glucose, haematocrit and arterial gases, administration of drugs and fluids and continuous monitoring of arterial blood pressure. Core temperature was controlled using a rectal thermistor probe and a heating pad. Adult and aged rats were randomly allocated into treated and untreated groups. Treated animals received D-CPP-ene (15 mg/kg) 15 minutes prior to infarction followed by an infusion (0.17 mg/kg/min) for the duration of the experiment. Untreated animals received a similar regime using the vehicle (0.9% saline). A microcraniectomy was performed and the left middle cerebral artery was permanently occluded proximal to the lenticulo-striate branch, using microbipolar diathermy, to yield a focal ischaemic lesion involving the cortex and caudate nucleus[8]. Brains were perfusion fixed 6 hours post infarction, using 40% formaldehyde/glacial acetic acid/methanol (FAM 1 : 1 : 8). Stained sections were analysed by light microscopy and on those corresponding to 8 pre-determined stereotactic levels, the ischaemic areas were measured by a video plan image analyser. The infarct volume was calculated by integrating the ischaemic areas at each coronal section, over their antero-posterior coordinates[6]. Specific gravity was measured as an index of cerebral oedema, using calibrated gravimetric columns[7].

A coronal slice (from an identical site corresponding to the bregma in each brain) was used to cut paired 1 mm cubes of tissue from the cerebral cortex, caudate nucleus and white matter of each hemisphere. Samples taken from the cerebellar cortex acted as a control. The level of descent of the samples in the gravimetric columns was recorded at 1 minute. The specific gravity of the specimen was calculated using a linear regression analysis derived from calibration of the column with potassium sulphate droplets of known specific gravity. Measurements were made at 4 and 24 hours post-infarction. In the 24 hour experiments, the pretreatment D-CPP-ene (4.5 mg/kg intravenously) was followed by intraperitoneal injections of 4.5 mg/kg 3 hourly post-infarction. Untreated animals received similar volumes of saline. Procedures were performed aseptically, a topical antibiotic and 2% marcaine were applied to incisions, with suturing prior to recovery. Animals were placed in an incubator post-operatively with free access to food and water and regular monitoring of core temperature and oxygen saturation. Data has been presented as means ± their standard errors and comparisons made with non-parametric Mann-Whitney tests.

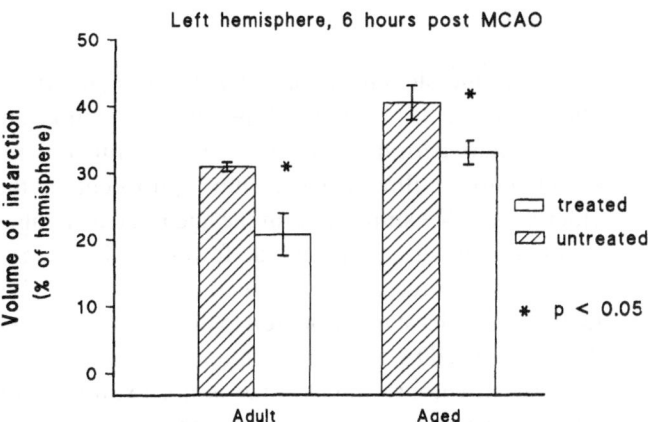

Fig. 1. The effects of age and D-CPP-ene on infarct volume. In untreated groups, aged rats had larger infarcts ($p < 0.01$). D-CPP-ene reduced infarct size in both age groups ($p < 0.05$)

Results

Aging was associated with a significant increase in cerebral infarct size. The mean infarct volume, (expressed as a percentage of the total volume of the left hemisphere), was 40.5% ± 2.6% in aged, untreated rats (n = 9), compared with 30.9% ± 0.7% in untreated adults (n = 10, $p < 0.01$, Fig. 1). This age related increase in infarct size, was predominantly a reflection of more extensive cortical infarction in the older animals. Pre-treatment with D-CPP-ene significantly reduced the volume of cerebral infarction in the aged animals, from a mean value of 40.5% ± 2.6%, to a mean volume of 33.0% ± 1.8% (n = 10, $p < 0.05$). This was largely attributable to a neuroprotective effect in the cerebral cortex, with treated, aged animals having less ischaemic cortical damage than their untreated counterparts, at all of the 8 stereotactic levels that were analysed. D-CPP-ene had a more marked neuroprotective effect in the adult animals, reducing the mean infarct volume from 30.9% ± 0.7% (n = 10), to 20.7% ± 3.2% (n = 8, $p < 0.05$, Fig. 1), representing a 33% reduction in infarct size, as compared to an 18.5% reduction in the aged brains. These differences in infarct volume were not a reflection of significant changes in physiological variables, since these parameters were similar in all groups. All brains were well perfusion fixed. Classical ischaemic changes were seen in all infarcted areas and the pattern of ischaemic damage corresponded to the typical distribution of the left middle cerebral artery. The specific gravities of samples from the cerebral cortex, caudate nucleus and white matter of the left hemisphere, were significantly lower than those of corresponding samples from the right, non-lesioned hemisphere in all rats ($p < 0.05$). Pre-treatment with D-

CPP-ene resulted in significantly less cortical oedema in the lesioned hemispheres of both adult and aged animals, at 4 hours post-infarction. Treated aged rats (n = 7) had a mean cortical specific gravity of 1.0458 ± 0.0031 compared with 1.0381 ± 0.0013 in untreated aged rats (n = 9, $p < 0.05$). Treated adults (n = 8) had a mean value of 1.0442 ± 0.0014 compared with 1.0391 ± 0.0014 in untreated adults (n = 8, $p < 0.05$, Fig. 2). This neuroprotective effect was still evident at twenty-four hours, with a mean cortical specific gravity of 1.0403 ± 0.0006 in treated aged rats, compared with a mean value of 1.0361 ± 0.0014 in untreated aged animals (n = 6 in each group, $p < 0.05$).

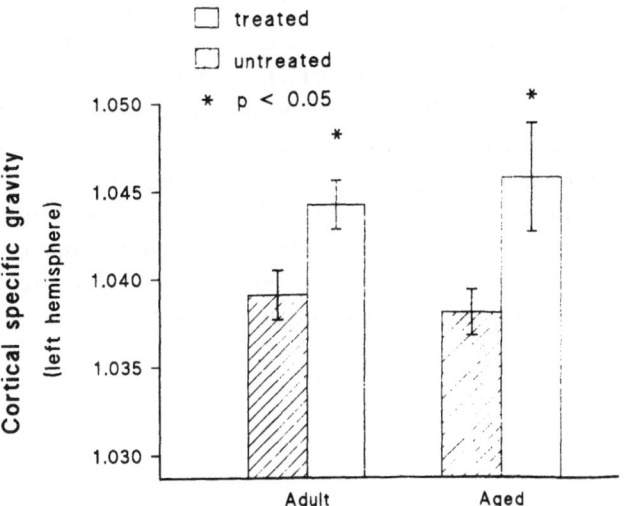

Fig. 2. The effect of D-CPP-ene on cortical SG (4 hours post-infarction). Treatment reduced oedema in both age groups ($p < 0.05$)

Discussion

A reproducible, physiologically stable model of occlusive stroke has been developed in the aging brain. This should permit more relevant experimental research into an illness that is most prevalent in the elderly. Notable similarities in the age related changes in rat and human brains include fewer and less sensitive dopamine receptors, a reduction in neurotrophic factors, a decrease in the neuron to glial cell ratio and a decline in NMDA receptor function[2]. The age related increase in cerebral infarct size is not explicable by factors such as arterial carbon dioxide or blood pressure levels, since although they may influence cerebral blood flow, they were similar in the two age groups. The explanation may lie in a relative deficiency of the collateral circulation with age. Documented morphological changes in the cerebral vasculature of aging rats include thinning of the endothelium[3], increases in collagen and elastin within vessel walls and impaired reactivity to vasodilator stimuli, thought to result from declining endothelial function[4]. These factors may contribute to a reduced capacity in aged cerebral vessels to respond to ischaemia and hypoxia. However, NMDA receptor antagonism is neuroprotective in focal ischaemic lesions of the aging brain, with D-CPP-ene pretreatment leading to significantly less oedema and infarction of the cerebral cortex, in the absence of any significant effect on physiological variables. The prominence of the cortical protection reflects the distribution of NMDA receptors and has been noted previously in young brains. The potential for benefit becomes less marked with age. This may be attributable to the reported decline in NMDA receptors with age, although with such a reduction in receptor function, it is surprising that a significant benefit from D-CPP-ene, was demonstrable. D-CPP-ene may have an alternative site of action, other than its role as a calcium channel blocker via the NMDA receptor, such as a direct effect on the vasculature and cerebral blood flow. The results of this study are encouraging in that there is scope for neuroprotection in the aging brain. Although the effect is less marked with age, the unprotected aged brain is liable to sustain a larger and potentially more disabling infarct and modest degrees of neuronal salvage may ultimately be of clinical use in this age group. A potential role for NMDA receptor antagonists in ischaemic aged brains has been identified and a study of the value of a post-infarction treatment regime in this age group would be useful, as it would be more representative of the clinical situation in stroke patients.

Acknowledgements

This study was supported by The Northern Regional Research Committee, The Sir James Knott Trust and Sandoz Ltd., Berne.

References

1. Bullock R, Graham DI, Chen MH, Lowe D, McCulloch J (1990) Focal cerebral ischaemia in the cat: pre-treatment with a competetive NMDA receptor antagonist SDZ-EAA-494. J Cereb Blood Flow Metab 10: 668–675
2. Gonzales RA, Brown LM, Jones TW, Trent RD, Westbrook SL, Leslie SW (1992) N-Methyl-D-aspartate mediated responses decrease with age in Fischer 344 rat brain. Neurobiol Aging 12: 219–225
3. Hajdu MA, Siems JE, Baumbach GL (1989) Effects of aging on mechanics of cerebral arterioles. Faseb J 3: A486
4. Hongo K, Nakagomi T, Kassell NF, Sasaki T, Lehman M, Vollmer DG, Tsukahara T, Ogawa H, Torner J (1987) Effects of aging and hypertension on endothelium-dependent vascular relaxation in rat carotid artery. Stroke 19: 892–897
5. Meldrum BS (1990) Protection against ischaemic neuronal damage by drugs acting on excitatory neurotransmission. Cerebrovasc Brain Metab Rev 2: 27–57
6. Osborne KA, Shigeno T, Balarsky AM, Ford I, McCulloch J, Teasdale GM, Graham DI (1987) Quantitative assessment of early brain damage in a rat model of focal cerebral ischaemia. J Neurol Neurosurg Psychiatry 50: 402–410
7. Shigeno T, Brock M, Shigeno S, Fritschka E, Cervos-Navarro J (1982) The determination of brain water content: microgravimetry versus drying-weighing method. J Neurosurg 57: 99–107
8. Tamura A, Graham DI, McCulloch J, Teasdale GM (1981) Focal cerebral ischaemia in the rat I. Description of technique and early neuropathological consequences following middle cerebral artery occlusion. J Cereb Blood Flow Metab 1: 53–60
9. Wenk GL, Walker LC, Price DL, Cork LC (1991) Loss of NMDA, but not GABA-A binding in the brains of aged rats and monkeys. Neurobiol Aging 12: 93–98

Correspondence: M. Davis, M. D., c/o A. D. Mendelow, The Regional Neurosciences Centre, Newcastle General Hospital, Westgate Road, Newcastle upon Tyne, England, U.K.

Acta Neurochir (1994) [Suppl] 60: 285–288
© Springer-Verlag 1994

Temporal Profiles of Ca^{2+}/Calmodulin-Dependent and -Independent Nitric Oxide Synthase Activity in the Rat Brain Microvessels Following Cerebral Ischemia

T. Nagafuji[1], **M. Sugiyama**[1], and **T. Matsui**[2]

[1] Fuji-Gotemba Research Laboratories, Chugai Pharmaceutical, Komakado, Gotemba, Shizuoka and [2] Department of Neurosurgery, Saitama Medical Center/School, Kamoda, Kawagoe, Saitama, Japan

Summary

The present study was aimed at determining chronological alterations of Ca^{2+}/calmodulin (CaM)-dependent and -independent nitric oxide synthase (NOS) activities in brain microvessels (MV) isolated from the affected hemisphere following an occlusion of the middle cerebral artery (MCAo) in rats. It was shown that significant enhancements of Ca^{2+}/CaM-independent NOS activity to 922% and 920% of control level were manifested at 4 h and 24 h, respectively, which returned to the control level at 48 h after MCAo. Regarding Ca^{2+}/CaM-dependent NOS, on the other hand, it was shown that the activity was invariably increased to 374% and 743% of control level at 48 h and 1 week following MCAo, respectively. Thus, the present study provided the first evidence that two distinct types of NOS activities were increased with different temporal patterns after MCAo. These heterogeneous alterations of NOS activities may be of critical importance for the induction of brain damage following cerebral ischemia.

Keywords: Focal cerebral ischemia; nitric oxide synthase inhibitor; brain edema; cerebral infarction.

Introduction

Although hitherto accumulated evidence suggested that increases in the intracellular cytosolic calcium concentration, acidosis and productions of free radicals are involved in the manifestation of ischemic brain damage[21], no definitive cellular mechanisms have been drawn. As regards to the calcium hypothesis, particular attention has been paid to the excitatory amino acids within the neuron. Excessive releases of excitatory amino acids during and shortly after cerebral ischemia result in neuronal necrosis that is mediated predominantly by the N-methyl-D-aspartate (NMDA) class of glutamate receptors. This neurotoxicity depends on influx of extracellular calcium via ligand-gated or other cation channels[21].

One enzyme activated by stimulations of NMDA receptors is Ca^{2+}/calmodulin (CaM)-dependent nitric oxide synthase (NOS) in neurons[5,8]. From our series of experiments, N^{ω}-nitro-L-arginine (L-NNA), a NOS inhibitor, was shown to mitigate brain edema observed after 48 h permanent[20] or during recirculation after temporary occlusion[15] of the middle cerebral artery (MCAo) in rats, respectively. In discussing these mechanisms of action, it should be recalled that distribution patterns of NOS and those of NMDA receptors within the rat brain differ markedly[5,7,8,18]. These data together with the fact that brain edema observed at both 48 h permanent and during recirculation after temporary MCAo can be designated as vasogenic type[9,11], led us to speculate that ameliorative effects of L-NNA may be elicited by preferential inhibitions of NOS activity in the cerebral endothelium rather than the neurons.

To test our working hypothesis, the present study is designed to examine the temporal changes of Ca^{2+}/CaM-dependent and -independent NOS activities in the brain microvessels (MV) obtained from the affected hemisphere following MCAo in rats.

Materials and Methods

Surgical Operations

Adult male Sprague-Dawley rats (278–342 g) were anesthetized with 2% halothane for induction and 1% halothane for maintenance, in a nitrous oxide/oxygen mixture (70 : 30%). Briefly, a tiny craniectomy was made with a dental drill just anterior to the foramen of the mandibular nerve as described previously[20]. The inner margin of the olfactory nerve and the main trunk of the MCA could be seen through the thin dura matter. Then, the left MCA trunk proximal to the lenticulostriate artery (arteries) was exposed and Zen-clip was applied to the MCA trunk. Throughout the surgical manipulations, the temperature of both the right temporal muscle and the rectum were kept constant at around 37°C.

Isolation of Brain Microvessels

Following MCAo, animals were sacrificed by transcardial perfusion with saline containing 10 I.U./ml of heparin under anesthesia with ether at the time of 4 h, 24 h, 48 h and 1 week, respectively. The left hemisphere thus obtained was minced with scissors in a solution containing 10 volumes of 0.25 M sucrose, 5 mM EDTA and 50 mM Tris-HCl (pH 7.4). The suspension was passed serially through nylon mesh with pore sizes of 670 μm and 335 μm. The filtrate was centrifuged at 3,500 g for 10 min and the resultant precipitate was resuspended in the same buffer. Each 2.5 ml was layered onto a discontinuous density gradient consisting of 1.0 M, 1.3 M and 1.5 M sucrose (3 ml each). Then, the gradient was centrifuged at 58,000 g for 60 min and the fraction at the bottom of the tube was collected[13].

NOS Activity Assay

NOS activity was measured according to the method described by Bredt and Snyder[4] with a slight modification. The reaction medium contained MV (10–45 μg/ml), 100 nM [^{3}H]arginine, 1 mM EDTA, 1 mM NADPH, 10 μM tetrahydrobiopterine, 100 μM FAD, 100 μM FMN, 50 mM Tris-HCl (pH 7.4). Incubation with/without 1.25 mM CaCl$_2$ and 10 μg/ml CaM lasted for 10 min at 37°C. The reaction was terminated by addition of ice-cold 50 mM Tris-HCl (pH 5.5) containing 1 mM EDTA. The reaction mixtures were then applied to a cation exchange resins-prefilled column (Dowex AGSO WX-8, 2.0 ml, Na$^+$ form) and [^{3}H]citrulline formed in elution with an aliquot of distilled water was quantified by a liquid scintillation counter. Ca^{2+}/CaM-dependent NOS activity was calculated by subtracting the enzyme activity observed in Ca^{2+}/CaM-absent medium from that observed in Ca^{2+}/CaM-present medium. Protein concentrations were determined using Bio-Rad Microassay kits.

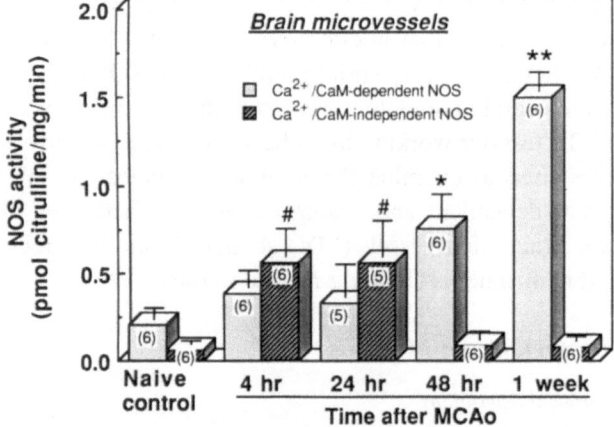

Fig 1. Temporal profiles of Ca^{2+}, CaM-dependent and -independent NOS activity in brain microvessels obtained from the affected hemisphere following MCAo in rats. Each histogram represents the mean ± S.E.M. for the number of animals indicated in parentheses. * p < 0.05 vs. naive control (Ca^{2+}, CaM-dependent NOS); ** p < 0.01 vs. naive control (Ca^{2+}, CaM-dependent NOS); # p < 0.05 vs. naive control (Ca^{2+}, CaM-independent NOS) (ANOVA, followed by Dunnett test)

Results

In the normal control group, either Ca^{2+}/CaM-dependent or -independent NOS activity in the cerebral MV obtained from the left hemisphere was found to be 0.202 ± 0.063 or 0.061 ± 0.023 pmol [^{3}H]citrulline/mg protein/min (mean ± S.E.M., n = 6, in each), respectively. Figure 1 illustrates the time-course analysis of these two NOS activities following MCAo. For the comparison of NOS activities between ischemic and non-ischemic hemispheres, the contralateral right hemisphere was not utilized as a normal control since the impact of "diaschisis" seems likely to affect the final outcome[19]. Under the conditions employed, it was shown that significant enhancements of Ca^{2+}/CaM-independent NOS activity to 922% and 920% of control level were manifested at 4 h and 24 h after MCAo, respectively. These alterations of the activity returned to the control level at 48 h after MCAo. Ca^{2+}/CaM-dependent NOS activity, on the other hand, was continuously increased to 374% and 743% of control level at 48 h and 1 week following MCAo, respectively.

Discussion

From the present study, two major findings can be drawn. First, the cerebral MV possess two distinct types of NOS, i.e. Ca^{2+}/CaM-dependent NOS [constitutive NOS (cNOS)] and Ca^{2+}/CaM-independent NOS [inducible NOS (iNOS)][16]. Second, NOS activity in the brain MV was shown to be enhanced following cerebral ischemia. Regarding these data, of particular interest is the finding that distinct forms of NOS activities were increased with different temporal patterns after MCAo. In the following, we will mainly focus on the functional relevance of the present finding.

It seems very appropriate to begin by comparing the temporal evolution of ischemic brain edema with those data obtained from the present study after MCAo. Brain MV is known to play a crucial role as the functional site of the blood-brain barrier which regulates the permeability of intrinsic/extrinsic substances[3]. Furthermore, a great deal of available evidence indicates that significant increases in the water and sodium contents within the brain develop as early as three hours after MCAo, concomitant with decreases in the potassium contents[11,21]. Thus, the initial increase of iNOS activity observed in the present study seems to correspond to the development of ischemic brain edema. Namely, increased iNOS activity in the cerebral MV precedes the maximum brain edema formation[11]. Further supporting evidence to our idea comes from the result that dexamethasone, a compound which inhibits the expression of iNOS but not cNOS in vascular endothelial cells[16], ameliorates the ischemic brain edema[14]. Taken together, it seems very likely that an activation of iNOS, particularly in the initial stage of ischemia, plays a pathological role for the development of ischemic brain edema.

Although the factors involved in causing initial Ca^{2+}/CaM-independent iNOS activations following MCAo remain unknown, a possible candidate could be cytokines and/or endotoxin formation by immunological stimuli. It has been known that murine brain endothelial cells produce NO in response to cytokines and endotoxin[12]. Moreover, very recently it has been reported that microglial cells in the neocortex are activated by ischemic attack at several hours following MCAo[17]. In this regard, the finding is noteworthy that in vitro applications of cytokines or endotoxin to glial/microglial cells induce an iNOS expression via de novo synthesis mechanisms[10,12] and that produced NO and reactive nitrogen oxides mediate neuronal cell injury[2,6]. Thus, there are ample reasons why iNOS rather than cNOS within endothelium and glial/microglial cells, both of which are known to be resistant to ischemic insults[3,21], contributes to the development of brain edema following ischemia.

Cytotoxicity of NO or its oxides derivatives including peroxynitrite, hydroxy radicals, and nitrogen dioxide has been firmly established: these substances inhibit enzyme activities associated with mitochondrial respiration, glycolysis, or DNA synthesis[1,8,16]. In any event, it can be envisaged that the amelioration of ischemic brain edema and infarction by L-NNA[15,20] are at least partly due to interference with the cytotoxic activities and/or metabolites of NO. Finally, it should be emphasized that the here described Ca^{2+}-independent cellular mechanisms give an explanation why calcium antagonists have been proven to be ineffective against ischemic brain edema[21].

Since consistent enhancements of Ca^{2+}/CaM-dependent cNOS activity were detected at 48 h and 1 week following ischemia, the functional significance of the data will be discussed. Although no clear interpretation is at hand at the present moment, it seems likely that an increased activity of cNOS in the late stage of cerebral ischemia may be related to the resolution mechanisms underlying ischemic brain edema[11]. The literature reveals that cNOS synthesizes low or intermediate amounts of NO for short periods of time upon stimulation[16]. These continuous releases of NO

seem to serve for the maintenance of physiological functions such as regulating vascular tone and arterial blood pressure[16]. Further studies on this issue will definitely be required. In conclusion, the present study presents the first evidence that microvascular NOS in the brain is activated by focal cerebral ischemia. These results, together with our previous report that a NOS inhibitor mitigates ischemic brain edema and subsequent cerebral infarction[15,20], render it very likely that blockade of iNOS expression/activity in the brain MV becomes a novel pharmacotherapeutic strategy for the treatment of cerebral stroke.

Acknowledgements

The authors would like to thank Dr. T. Koide (Business Development Department, Chugai Pharmaceutical Co. Ltd.) for his discussion with us on this manuscript.

References

1. Beckman JS (1991) The double-edged role of nitric oxide in brain function and superoxide-mediated injury. J Devel Physiol 15: 53–59
2. Boje KM, Arora PK (1992) Microglial-produced nitric oxide and reactive nitrogen oxides mediate neuronal cell death. Brain Res 587: 250–256
3. Bradbury MWB (1992) Physiology and pharmacology of the blood-brain barrier. Springer, Berlin Heidelberg New York Tokyo
4. Bredt DS, Snyder SH (1990) Isolation of nitric oxide synthetase, a calmodulin-requiring enzyme. Proc Natl Acad Sci USA 87: 682–685
5. Bredt DS, Snyder SH (1990) Nitric oxide, a novel neuronal messenger. Neuron 8: 3–11
6. Chao CC, Hu S, Molitor TW, Shaskan EG, Peterson PK (1992) Activated microglia mediate neuronal cell injury via nitric oxide mechanism. J Immunol 149: 2736–2741
7. Cotman CW, Monaghan DT, Ottersen OP, Storm-Mathisen J (1987) Anatomical organization of excitatory amino acid receptors and their pathways. Trends Neurosci 10: 273–280
8. Dawson TM, Dawson VL, Snyder SH (1992) A novel neuronal messenger molecule in brain: the free radical, nitric oxide. Ann Neurol 32: 297–331
9. Flynn CJ, Farooqui AA, Horrocks LA (1989) Ischemia and hypoxia. In: Siegel G, Agranoff B, Alberts RW, Molinoff P (eds) Basic neurochemistry, 4th Ed. Raven, New York, pp 787–788
10. Galea E, Feinstein DL, Reis DJ (1992) Induction of calcium-independent nitric oxide synthase activity in primary rat glial cultures. Proc Natl Acad Sci USA 89: 10945–10949
11. Gotoh O, Asano T, Koide T, Takakura K (1985) Ischemic brain edema following occlusion of the middle cerebral artery in the rat. I: the time courses of brain water, sodium and potassium contents and blood-brain barrier permeability to ¹²⁵I-albumin. Stroke 16: 101–109
12. Kiboum RG, Belloni P (1990) Endothelial cell production of nitrogen oxides in response to interferon γ in combination with tumor necrosis factor, interleukin-1, or endotoxin. J Natl Cancer Inst 82: 772–776
13. Koide T, Asano T, Matushita H, Takaura K (1986) Enhancement of ATPase activity by a lipid peroxide of arachidonic acid in rat brain microvessels. J Neurochem 46: 235–242

14. Koide T, Wieloch TW, Siesjö BK (1986) Chronic dexamethasone pretreatment aggravates ischemic neuronal necrosis. J Cereb Blood Flow Metab 6: 395–404

15. Matsui T, Nagafuji T, Auer RN, Koide T, Tsutsumi K, Asano T (1993) Beneficial effect of nitric oxide synthase inhibitor on reversible and permanent focal cerebral ischemia in rats. XVIth International Symposium on Cerebral Blood Flow and Metabolism [Suppl], in press

16. Moncada S, Palmer RMJ, Higgs EA (1991) Nitric oxide: physiology, pathophysiology, and pharmacology. Pharmacol Rev 43: 109–142

17. Morioka T, Kalehua AN, Streit WJ (1993) Characterization of microglial reaction after middle cerebral artery occlusion in rat brain. J Comp Neurol 327: 123–132

18. Moriyoshi K, Masu M, Ishii T, Shigemoto R, Mizuno N, Nakanishi S (1991) Molecular cloning and characterization of the rat NMDA receptor. Nature 354: 31–37

19. Nagafuji T, Koide T, Takato M (1992) Neurochemical correlates of selective neuronal loss following cerebral ischemia: role of decreased Na⁺, K⁺-ATPase activity. Brain Res 571: 265–271

20. Nagafuji T, Matsui T, Koide T, Asano T (1992) Blockade of nitric oxide formation by N^{ω}-nitro-L-arginine mitigates ischemic brain edema and subsequent cerebral infarction in rats. Neurosci Lett 147: 159–162

21. Siesjö BK (1992) Pathophysiology and treatment of focal cerebral ischemia. Part II: mechanisms of damage and treatment. J Neurosurg 77: 337–354

22. Simmons M, Murphy S (1992) Induction of nitric oxide synthase in glial cells. J Neurochem 59: 897–905

Correspondence: T. Nagafuji, M. D., Fuji-Gotemba Research Laboratories, Chugai Pharmaceutical Co. Ltd., 1-135 Komakado, Gotemba, Shizuoka 412, Japan.

Acta Neurochir (1994) [Suppl] 60: 289–292
© Springer-Verlag 1994

Effect of a New Calcium Antagonist (SM-6586) on Experimental Cerebral Ischemia

F. Kashiwagi, Y. Katayama, H. Igarashi, S. Iida, H. Muramatsu, and **A. Terashi**

Second Department of Internal Medicine, Nippon Medical School, Tokyo, Japan

Summary

SM-6586 (SM) is a new derivative of dihydropyridine with potent calcium blocking activity and inhibitory activity of the Na^+/H^+ and Na^+/Ca^{++} exchange transport. The effect of SM on survival rate, brain edema and metabolites was evaluated using two different models in spontaneously hypertensive rat (SHR). Global ischemia was induced by bilateral common carotid artery ligation (BLCL) and focal ischemia was induced by middle cerebral artery occlusion. The survival rate after BLCL was higher in the SM-treated group. The brain water content was lower, the ATP level was higher and lactate level was lower in the SM-treated group compared to the control group. In focal ischemia models, the SM-treated group showed a reduction of T1 relaxation time. The brain water content was significantly decreased in the SM-treated group. These results indicate that SM was effective in ameliorating the ischemic insult in global and focal cerebral ischemia models.

Keyword: Calcium antagonist; cerebral ischemia; brain edema; brain metabolism; SHR.

Introduction

It is well known that intracellular calcium overload causes cell damage in cerebral ischemia[1]. It has been reported that many calcium antagonists have protective effect on ischemic brain damage[2]. SM-6586 (Methyl 1,4-dihydro-2,6-dimethyl-3-[3-(N-benzyl-N-methylaminomethyl)-1,2,4-oxadiazolyl-5-yi]-4-(3-nitrophenyl)pyridine-5-carboxylate), a new derivative of dihydropyridine, has a potent Ca^{++} blocking action via the voltage sensitive calcium channel and the inhibitory action of Na^+/H^+ and Na^+/Ca^{++} exchange transport. In this study, the effect of SM-6586 on survival rate, brain edema and brain metabolites using two different ischemia models in spontaneously hypertensive rat (SHR) was evaluated.

Materials and Methods

1. Global Ischemia Model

Sixteen week old, male spontaneously hypertensive rats were used in this experiment. All surgical procedures were performed under anesthesia of inhalation of 1.5% halothane, 30% oxygen and 70% nitrous oxide. Cerebral ischemia was induced by bilateral common carotid artery ligation (BLCL). The rats received either 30

Table 1. *Physiological Variables After BLCL*

	Vehicle (7)	30 µg/kg (7)	100 µg/kg (7)
MABP (mmHg) before	157.1 ± 9.9	156.7 ± 9.8	157.1 ± 9.9
MABP (mmHg) after	166.4 ± 14.9	165.0 ± 7.7	166.4 ± 14.9
PH before	7.42 ± 0.33	7.42 ± 0.04	7.40 ± 0.06
PH after	7.43 ± 0.06	7.41 ± 0.06	7.43 ± 0.07
PaO2 (mmHg) before	100.9 ± 11.6	93.6 ± 5.4	103.6 ± 5.8
PaO2 (mmHg) after	127.5 ± 6.1*	111.7 ± 10.1*	127.1 ± 8.3*
PaCO2 (mmHg) before	29.8 ± 3.9	31.5 ± 4.1	29.9 ± 4.2
PaCO2 (mmHg) after	23.8 ± 11.6	23.2 ± 5.8	24.9 ± 5.8

Data represent means ± S.D. *MABP* mean systolic arterial blood pressure; *before* 5 min before ischaemia; *after* 3 h after ischaemia. * $p < 0.05$ significantly different from before ischaemia.

Table 2. *Brain Metabolism After BLCL*

	ATP	Lactate	Pyruvate	L/P
Control	0.17 ± 0.13	25.96 ± 3.45	0.26 ± 0.07	103.6 ± 19.0
SM (30 µg/kg)	0.20 ± 0.11	$14.16 \pm 13.69^*$	0.34 ± 0.07	$49.96 \pm 55.6^*$
SM (100 µg/kg)	0.10 ± 0.07	20.11 ± 10.47	0.29 ± 0.07	77.75 ± 49.7

Data represent means ± S.D. *ATP,* lactate and pyruvate values are µ mol/g brain; *L/P* lactate/pyruvate ratio. * $p < 0.05$ significantly different from the vehicle group.

or 100 µg/kg of SM-6586 30 min before ischemia via intraperitoneal injection. The survival rate was observed through 6 h, and the brain water content and metabolite concentration (ATP, lactate, pyruvate) were determined 3 hrs after ischemia. The brain water content was measured by the freeze dry method. Metabolites were measured by enzymatic methods.

2. Focal Ischemia Model

Focal cerebral ischemia was induced by MCA (middle cerebral artery) occlusion. The rats received 150 µg/kg of SM-6586 intraperitoneally 30 min before and immediately after occlusion and 300 µg/kg 23 h after ischemia. At 24 h after ischemia, T1 relaxation time was investigated using MRI (JEOL GX-270, 6.34 T). The brain water content was measured at 24 h of ischemia.

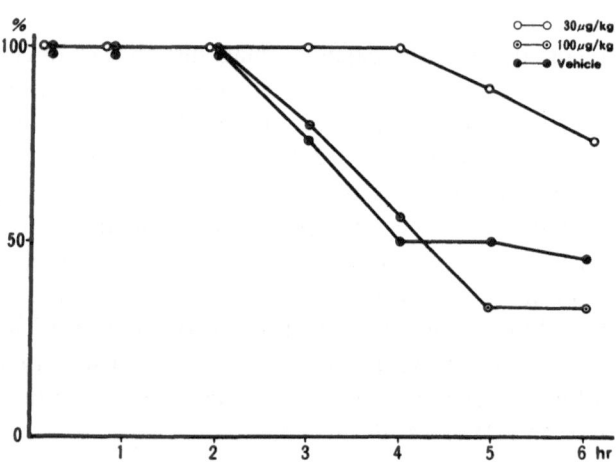

Fig. 1. Survival rate after BLCL. Data expressed as %. Note higher survival rate in the 30 mg/kg SM-treated group compared to the vehicle group

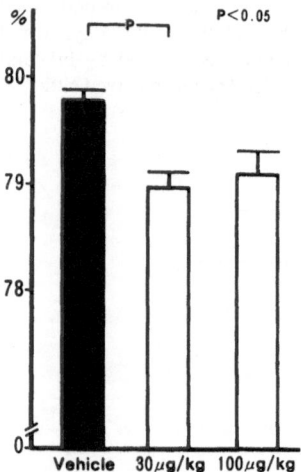

Fig. 2. Brain water content after BLCL. Data represent means ± S.D. The brain water content is lower in the 30 mg/kg SM-treated group than in the vehicle group. $p < 0.05$, significantly different from the vehicle group

Results

1. Global Ischemia Model

Physiological parameters including mean arterial blood pressure (MABP), arterial pH, pCO2 and PaO2 are summarized in Table 1.

The survival rate shown in Fig. 1. shows that the group administered 30 µg/kg had a higher survival rate compared both with the controls and the 100 µg/kg SM-treated group until 6 h after ischemia. However 100 µg/kg of SM-treated group did not show higher survival rate compared to the control group.

The brain water content was significantly decreased in the 30 µg/kg SM-treated group compared to the control groups (Fig. 2). However there was no significant difference between the 100 µg/kg SM-treated group and the control group. The values of the brain metabolites after BLCL are shown in Table 2. There were no significant differences in ATP levels between both the SM-treated group and the control group. The levels of lactate and lactate/pyruvate ratio were lower in the 30 µg/kg SM-treated group compared to the control group. However there was no significant difference between the 100 µg/kg SM-treated group and the control group.

2. Focal Ischemia Model

After 24 h of MCA occlusion, the brain water content was significantly reduced in the SM-treated group compared to the control group (Fig. 3).

The T1 relaxation time: at 24 h of ischemia, SM-treated group showed remarkable reduction of T1 relaxation time compared to the ischemic hemisphere of the control group (Fig. 4).

Discussion

In this study, there were no significant differences betweeen the SM-treated and the control group in physiological parameters after BLCL. It is suggested that the 30 and 100 µg/kg of SM administered had no effect on systemic blood pressure after BLCL in rats. The 30 µg/kg SM-treated group showed a higher survival rate and a remarkable reduction of brain edema. Intracellular calcium overload has been recognized as one of the detrimental factors leading to cell damage during and following brain ischemia. The use of a calcium antagonist may have the potential to reduce brain edema, through the inhibitory action of many calcium mediated lipases and proteases[3] resulting in cell damages. Moreover this novel calcium antagonist SM may reduce the accumulation of sodium due to its inhibitory action of Na^+/H^+ exchange transport.

In brain ATP levels showed no significant difference between the SM-treated group and the control group; but brain lactate levels and the L/P ratio were lower in the SM-treated group than in the control group. It has been reported that calcium antagonists can ameliorate the disturbance of brain metabolism by increasing cerebral blood flow due to vasodilation[4], inhibiting platelet aggregation[5] and by a direct protective effect on the neuron. However, higher doses of SM (100 µg/kg) did not further reduce brain edema or improve brain metabolism. The higher dose in the SM-treated group may not improve the brain metabolism due to a further reduction of cerebral blood flow in the ischemic area, caused by the intracranial steal phenomenon.

In the focal ischemia model, brain edema was reduced and metabolism was improved in the SM-treated group. The T1 relaxation time was also reduced in the SM-treated group. It is well known that the prolongation of T1 and T2 relaxation time is reflected in an increase of brain water.

Excessive influx of calcium may occur through: voltage sensitive calcium channel (VSCC), Na^+/Ca^{++} exchanger, NMDA-operated calcium channel and discharge from endoplasmic reticulum during ischemia. It has been reported that the existing calcium antagonists do not have the inhibitory effect on intracellular calcium influx via Na^+/Ca^{++} exchange transport. It is suggested that SM can reduce excessive

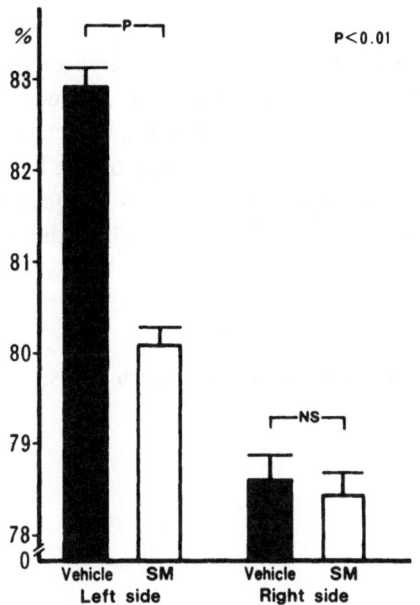

Fig. 3. Brain water content after MCA occlusion. Data represent means ± S.D. Left side is ischemic and right side is non-ischemic hemisphere. At the ischemic hemisphere (left), the brain water content is significantly decreased in the SM-treated group. p < 0.01

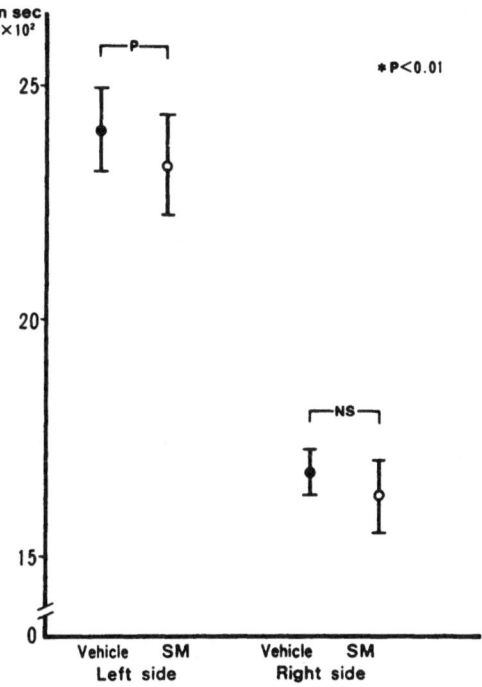

Fig. 4. T1 relaxation time after MCA occlusion. Data represent means ± S.D. Left side is ischemic and right side is non-ischemic hemisphere. At the ischemic hemisphere, T1 relaxation time is reduced in the SM-treated group. p < 0.01

calcium overload not only through VSCC but also Na^+/Ca^{++} exchanger resulting in the amelioration of ischemic brain damage.

Conclusion

These results indicate that a new derivative of the dihydropyridine Ca^{++} antagonist (SM-6586) decreased mortality, and ameliorated brain edema and metabolism. This may reflect its inhibitory action on the Na^+/H^+ and Na^+/Ca^{++} exchange transport. It is suggested that SM-6586 could be effective in acutely induced cerebral ischemia.

References

1. Faber JL (1981) The role of calcium in cell death. Life Sci 29: 1289–2295

2. Meyer FB, *et al* (1986) Effect of nimodipine on intracellular brain pH, cortical blood flow and EEG in experimental focal cerebral ischemia. J Neurosurg 64: 617–626

3. Abe K, *et al* (1988) Prevention of ischemic and postischemic brain edema by a novel calcium antagonist (PN200-110) J Cereb Blood Flow Metab 8: 436–439

4. Harper AM, *et al* (1981) Effect of calcium antagonist nimodipine, on cerebral blood flow and metabolism in primate. J Cereb Blood Flow Metab 1: 349–356

5. Aragno R, *et al* (1976) The exclusion of the cerebral circulation; effect on in vivo platelet aggregation. Throm Res 9: 319–327

Correspondence: F. Kashiwagi, M. D., Second Department of Internal Medicine, 355 Iidabashi, Chiyoda-ku, Tokyo 102, Japan.

Acta Neurochir (1994) [Suppl] 60: 293–295
© Springer-Verlag 1994

The Effect of BAY K-8644 on Cytotoxic Edema Induced by Total Ischemia of Rat Brain

M. Haida, Y. Shinohara, M. Yamamoto, T. Nagayama, and **D. Kurita**

Department of Neurology, Tokai University School of Medicine, Kanagawa, Japan

Summary

The calcium channel activator BAY K-8644, a dihydropyridine (DHP) derivative, has been shown to possess neurochemical and behavioral activities, but its effect on ischemic brain damage has remained unknown. This report describes the effect of the drug on the progression of cytotoxic edema induced by total ischemia of the brain, evaluated by measuring the time constant, k, of elongation of the ^{1}H-NMR relaxation time (T_2) after brain biopsy. Twenty-six male Wistar rats were divided into four groups, (a) control (saline) group (n = 10), (b) BAY K-8644 vehicle group (n = 4), (c) BAY K-8644 0.03 mg/kg group (n = 6) and (d) BAY K-8644 0.3 mg/kg group (n = 6). The k value of group (d), 18.2 ± 5.8 min (mean ± SD), was significantly higher compared with those of groups (a) 10.3 ± 1.6, (b) 11.8 ± 1.5 and (c) 9.8 ± 3.3 min (p < 0.01 by ANOVA). These results indicate that BAY K-8644 delayed the progression of cytotoxic edema induced by total ischemia of the rat brain.

Keywords: NMR; cytotoxic edema; BAY K-8644; brain ischemia.

Introduction

The Ca channel antagonists have a protective effect on development of brain edema after ischemic insults[6,7]. On the other hand, the dihydropyridine (DHP) derivative, BAY K-8644 (methyl-1,4-dihydro-2,6-dimethyl-3-nitro-4-(2-trifluoromethylphenyl)-pyridine-5-carboxylate), which acts as a Ca channel agonist[4,5], has behavioral and neurochemical effects on mice, though its effect on ischemic brain damage remains to be elucidated. The behavioral effect of this drug suggests that the drug can pass through the blood-brain barrier (BBB). If the activating action of this drug on the voltage-dependent Ca channel is present in the ischemic brain, an adverse effect such as acceleration of brain edema would be expected. The ^{1}H-NMR relaxation time (T_2) is especially sensitive to brain edema, and we have developed a ^{1}H-NMR method to evaluate the rate of development of cytotoxic edema[6]. In this study, we examined the effects of BAY K-8644 on the development of cytotoxic edema after total ischemia of rat brain by using NMR relaxation time measurements.

Materials and Methods

Twenty-six male Wistar rats (8- to 10-week-old, weighing 260 to 280 g) were divided into four groups, (a) control (saline) group (n = 10), (b) BAY K-8644 vehicle group (n = 4), (c) BAY K-8644 0.03 mg/kg group (n = 6) and (d) BAY K-8644 0.3 mg/kg group (n = 6). Under light anesthesia induced by intraperitoneal urethane (500 mg/kg body weight) and chloralose (50 mg/kg body weight), a small burr hole (5 mm ID) was drilled in the right frontoparietal region of the rat skull, and the dura and arachnoid were carefully incised. A brain tissue sample, about 2 mm thickness, and consisting mainly of gray matter, was punched out with a sharp-edged Pyrex glass tube (1.8 mm ID), which was then sealed tightly with a waterproof material, Hematoseal (a compound used for hematocrit measurement which was confirmed not to affect the NMR signals), and placed in an NMR spectrometer with a special sample holder. This sample can be regarded as a highly simplified model of brain ischemia. This method has two advantageous points: Firstly, the total water content in the sample is constant during measurements, which means the effect of vasogenic factor can be eliminated and we can examine mainly the effect of cytotoxic edema. Secondly, we can evaluate development of cytotoxic edema of the brain numerically, and can make a quantitative evaluation of brain edema by using this method.

A 90 MHz NMR spectrometer (FX 90A, JEOL, Japan) and Hahn's spin echo method (TR = 8 s, TE = 40, 60, 80 and 120 s, AV = 4 times) were used to measure T_2. NMR measurements were performed every 2 minutes up to one hour at 25°C. An exponential function was fitted to the obtained time course of $T_2(t)$ by using the least-squares method to calculate a time constant, k. Details of this method had been reported elsewhere[6]. Normal rat brain tissue has two components of T_2, T_2 fast, T_{2f}, and T_2 slow, T_{2s}, and they have been assigned to intracellular and extracellular space of rat brain tissue, respectively. Thus, the elongation of T_{2f} in the sample can be explained as reflecting the development of cytotoxic edema induced by ischemia following brain biopsy. The time constant, k, can then be regarded as a measure of the time taken to develop cytotoxic edema of the brain due to total ischemia.

The drugs were slowly injected intravenously 2 minutes before brain biopsy. The obtained k values for the four groups were statistically tested using the multiple comparison method (ANOVA).

Results

Table 1 and Fig. 1 show the obtained k values of the four groups. There was no significant difference in k between the saline control group and BAY K-8644 vehicle group. A slight shortening of k was seen in low-dose of BAY K-8644 (0.03 mg/kg) group compared with its vehicle group, but there was no statistically significant difference. Statistically significant differences in k were seen between the hig-dose BAY K-8644 (0.3 mg/kg) group and the saline control or its vehicle group (p < 0.01).

Discussion

The elongation of k value can be interpreted as being due to an effect of the administered drug on the development of cytotoxic edema. Since a high dose of BAY K-8644 (0.3 mg/kg) elongates k, the development of cytotoxic edema was delayed. Similar elongation of k has been observed using several voltage-dependent Ca channel antagonists[7]. This result indicates that the membrane depolarization and influx of Ca ions through the voltage dependent Ca channels, which can be reduced by introduction of Ca channel antagonists, have important roles in the development of cytotoxic edema following ischemic insult. Ca channel activators such as BAY K-8644 turn into Ca channel blockers if the membrane is depolarized[3]. This fact explains why administration of the high dose of Bay K-8644 elongates the k value in this model. The effect of the Ca channel antagonist, nimodipine (10 mg/kg), and the activator, BAY K-8644 (1.25 mg/kg), on rat local cerebral glucose utilization have been examined[2], and

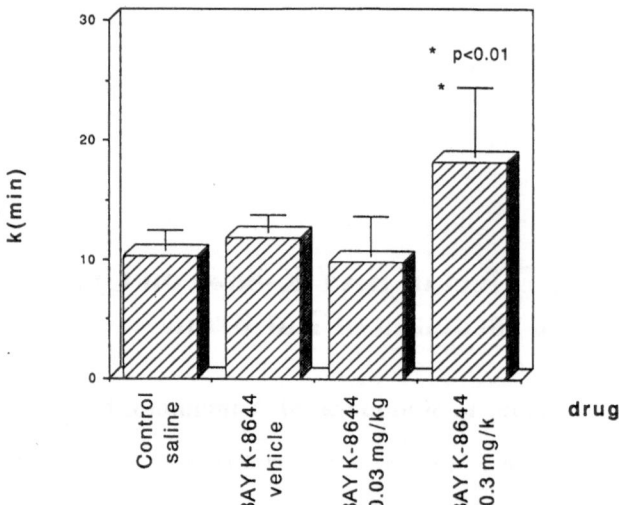

Fig. 1. Effect of BAY K-8644 on the time constant, k

the results showed that both drugs reduced local cerebral glucose metabolism in rat. This also indicates that the action of BAY K-8644 is not necessarily solely that of an agonist.

On the other hand, the low-dose of BAY K-8644 slightly shortened the k value, which may be explained by its Ca channel agonist action. In conclusion, the BAY K-8644 may have a different action on the voltage-dependent Ca channels, depending on its dosage. At a high dose such as 0.3 mg/kg, BAY K-8644 acts as Ca channel antagonist in the case of depolarization due to ischemia. At a low dose such as 0.03 mg/kg, BAY K-8644 shows a slight Ca channel-activating action, though this effect was not statistically significant.

References

1. Haida M, Yamamoto M, Matsumura H, Shinohara Y, Fukuzaki M (1987) Intracellular and extracellular spaces of normal adult rat brain determined from proton nuclear magnetic resonance relaxation times. J Cereb Blood Flow Metab 7: 552–556
2. Hovath E, Warmuth I, Zilles K, Traber J (1990) Effect of the 1,4-dihidropyridine-sensitive L-type calcium channel antagonist nimodipine and calcium channel activator Bay K 8644 on local cerebral glucose utilization in the rat. Stroke 21 [Suppl IV]: IV-126–IV-129
3. Sanguinetti MC, Krafte DS, Kass RS (1986) Voltage-dependent modulation of Ca channel current in heart cells by Bay K 9644. J Gen Physiol 88: 369–392
4. Schramm M, Thomas G, Towart R, Franckowiak G (1983) Activation of calcium channels by novel 1,4-dihydropyridines: a new mechanism for positive inotropics or smooth muscle stimulants. Arzneim-Forsch Drug Res 33: 1268–1272
5. Schramm M, Thomas G, Towart R, Franckowiak G (1983) Novel dihydripyridines with positive inotropic action through activation of Ca^{2+} channels. Nature 303: 535–537

Table 1. *Effect of BAY K-8644 on the Time Constant, k*

Drug	k (min) (mean ± SD)	n
Control (saline)	10.3 ± 1.60	10
BAK Y-8644 vehicle	11.8 ± 1.47	4
BAY K-8644 0.03 mg/kg	9.83 ± 3.32	6
BAY K-8644 0.3 mg/kg	18.2 ± 5.80	6

6. Shinohara Y, Yamamoto M, Haida M, Yazaki K, Kurita D (1990) Effect of glutamate and its antagonist on shift of water from extracellular to intracellular space after cerebral ischemia. Acta Neurochir (Wien) [Suppl 51]: 198–200
7. Shinohara Y, Haida M, Yamamoto M, Yazaki K (1991) Effects of glutamate and calcium antagonists on the progression of cytotoxic edema after total ischemia of rat brain. J Cereb Blood Flow Metab 11: S306

Correspondence: Yukito Shinohara, M. D., Department of Neurology, Tokai University School of Medicine, Boseidai, Isehara, Kanagawa, 259-11, Japan.

Acta Neurochir (1994) [Suppl] 60: 296–299
© Springer-Verlag 1994

LTC$_4$/LTB$_4$ Alterations in Rat Forebrain Ischemia and Reperfusion and Effects of AA-861, CV-3988

Y. Namura, H. Shio, and **J. Kimura**

Department of Neurology, Kyoto University School of Medicine, Kyoto, Japan

Summary

LTC$_4$, which enhances vascular permeability and promotes tissue edema, and LTB$_4$, which is a potent chemotactic and activating factor for leukocytes, were measured in rat brain after ischemia and several time periods of reperfusion. Forebrain ischemia was induced by 4-vessel occlusion. LTC$_4$ / LTB$_4$ in the brain were measured by RIA. We also studied the effects of a 5-lipoxygenase inhibitor, AA-861 and a PAF antagonist, CV-3988 on LTC$_4$ / LTB$_4$ concentrations. LTC$_4$ in brain tissue increased during 30 min forebrain ischemia (p < 0.001). After reperfusion, LTC$_4$ increased further, but at 15 min reperfusion LTC$_4$ returned to the control level. Tissue levels of LTB$_4$ in the brain increased during 30 min ischemia and remained high until 5 min after reperfusion (p < 0.01) returning at 15 min reperfusion to the control level. AA-861 inhibited elevation of LTC$_4$ / LTB$_4$ in the reperfusion phase, but was not effective during ischemia. CV-3988 had a similar effect. LTC$_4$ and LTB$_4$ may be involved in the pathogenesis of ischemia brain edema and leukocyte infiltration. Further, PAF and LTs have many similarities of their pathophysiological properties, and may interact therefore in pathologic process.

Keywords: LTC$_4$; LTB$_4$; 5-lipoxygenase inhibitor; PAF antagonist.

Introduction

LT (leukotriene) is a metabolite of AA (arachidonic acid), which accumulates in the brain during ischemia. LTC$_4$ and LTD$_4$ are the largest fraction of leukotrienes which is metabolized from LTA$_4$. It has been shown that LTC$_4$ and LTD$_4$ enhance vascular permeability and that they are potent vasoconstricting agents. In ischemia LTC$_4$ and LTD$_4$ are thought to exacerbate cell damage and to promote tissue edema. LTB$_4$, which is metabolized from LTA$_4$, is one of the most potent chemotactic leukocyte activating factors. Leukocytes have been implicated in the pathogenesis of cerebral ischemia[12]. In this experiment, we measured alterations of LTC$_4$ and LTB$_4$ concentrations in brain tissue after 4-vessel occlusion and reperfusion in rats. We studied also effects of the 5-lipoxygenase inhibitor, AA-861 after ischemia and reperfusion. PAF (platelet activating factor) might also contribute to the enhancement of ischemic cell damage[14]. PAF and LTs have many common pathophysiological properties. We studied therefore a specific PAF antagonist, CV-3988 on changes of LTC$_4$ and LTB$_4$ concentrations in the brain after ischemia and reperfusion.

Materials and Methods

Male Wistar rats were subjected to 30 minutes of global forebrain ischemia induced by 4-vessel occlusion. The procedure consists of bilateral temporary clamping of the common carotid arteries in rats with permanent occlusion of the vertebral arteries. Rats that did not become unresponsive within 60 seconds after clasp tightening were excluded from the study. Following the predetermined period of ischemia and reperfusion, the rats were sacrificed by decapitation and the heads were frozen in situ by liquid nitrogen. The brains were removed by chiselling them under liquid nitrogen. The brain samples were stored at –80°C until analysis. The frozen cerebral hemispheres were weighed and homogenized in ethanol and then centrifuged at 1000 g for 10 minutes at –20°C. The supernatant extracts were dried under N$_2$ and diluted with phosphate buffer (pH 7.4), adjusted to pH 2.3 with 1.0 NHCL and passed through a Sep-Pac C18 cartridge (Waters Assoc., Milford, Massachusetts). The PG fraction was eluted with ethanol and dried using a rotary evaporator. The eluent was studied by RIA for LTC$_4$ and LTB$_4$. ^{3}H-Kits (New England Nuclear, Boston, Massachusetts) were used for LTC$_4$ and LTB$_4$ assay. AA-861 (2-(12-hydroxydodeca-5,10-diynyl)-3,5,6-trimethyl-1,4-benzoquinone), and CV-3988 (rac-3-(N-n-octadecylcarbamoyloxy)-2-methoxypropyl 2-thiazolioethyl phosphate) were supplied by Takeda Chemical Industries Ltd. (Osaka, Japan). AA-861 (6 mg/kg) and CV-3988 (1 mg/kg) were given intraperitoneally 10 min before ischemia. Values are reported as mean ± SE. Statistical differences were calculated by Student's t-test.

Results

LTC$_4$

The LTC$_4$-concentration in the brain increased during 30 min forebrain ischemia ($p < 0.001$). After reperfusion LTC$_4$ increased further reaching maximum at 1 min. At 2.5 min LTC$_4$ started to decrease slowly, returning at 15 min to the preischemic or control level. Pretreatment with AA-861 did not change the increase of LTC$_4$ during 30 min ischemia, but at 1 and 2.5 min reperfusion elevation of LTC$_4$ was significantly inhibited in the brain as compared to untreated animals ($p < 0.01$). Correspondingly treatment with CV-3988 had no effect on LTC$_4$ accumulation during 30 min ischemia, but inhibited its increase at 1 min postischemic reperfusion ($p < 0.1$) (Fig. 1).

LTB$_4$

LTB$_4$ was also increased in the brain during 30 min ischemia as compared with the preischemic control level ($p < 0.01$). At 1 and 2.5 min after reperfusion LTB$_4$ remained high but did not further rise contrary to LTC$_4$. After 2.5 min LTB$_4$ decreased gradually, and returned at 15 min reperfusion to the preischemic level. Treatment with AA-861 or CV-3988 had no effect on LTB$_4$, concentrations after 30 min ischemia, while at 1 or 2.5 min reperfusion both compounds lowered the LTB$_4$ level ($p < 0.05$) (Fig. 2).

Discussion

AA (arachidonic acid) is the precursor of eicosanoids, including thromboxane, prostaglandins, and leukotrienes (LTs). It has been suggested that TXB$_2$ and PGI$_2$ are the major metabolites among the other PGs which affect cerebral blood flow (CBF) and influence neuronal damage in ischemia and/or reperfusion. Gaudet et al.[7] first described changes of thromboxane, prostacyclin and other prostanoids in gerbil brain after 5 min global ischemia and reperfusion. The resulting thromboxane-prostacyclin imbalance in ischemia and reperfusion was thought to aggravate cell damage.

However, Moufarrij et al.[11] reported that infarction was more extensive if a TX synthetase inhibitor was given in cats with MCA occlusion. This might be attributable to enhancement of the lipoxygenase pathway with overproduction of LTs and other lipoxygenase products.

LTC$_4$ and LTD$_4$ markedly potentiate tissue exudation and contraction of smooth muscle. They appeared to be involved in the pathogenesis of inflammation and allergic reaction. Later accumulation of LTC$_4$ and LTD$_4$ was detected in ischemic brain and following reperfusion[10], just as other PGs and eicosanoids. It has been suspected that LTs increase vascular permeability, such as at the blood-brain barrier (BBB) and that LTs mediate brain edema formation. In rats with MCA occlusion hydroxyacids were significantly enhanced in

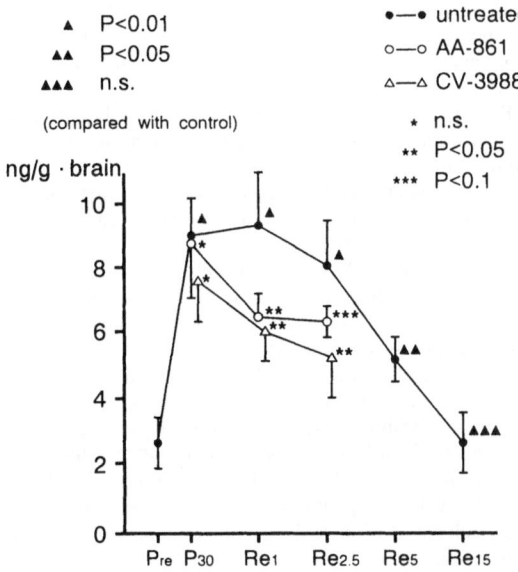

Fig. 1. Alterations of LTC$_4$ concentration of rat brain after forebrain ischemia and at different intervals of postischemic reperfusion. Effect of AA-861 and CV-3988 on LTC$_4$ concentrations are included

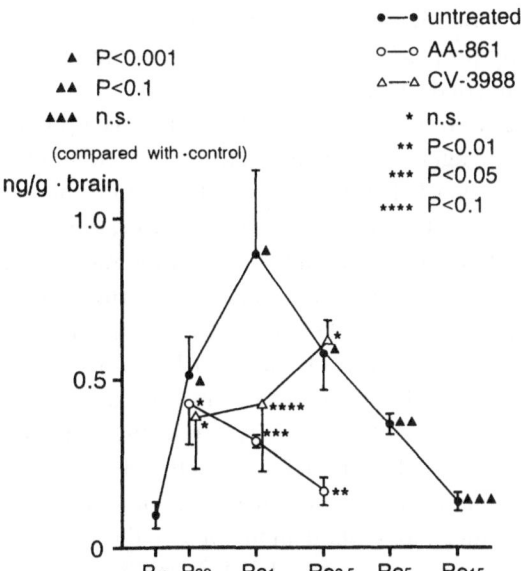

Fig. 2. Alterations of LTB$_4$ concentration of rat brain after forebrain ischemia and at different periods of reperfusion. Effects of AA-861 and CV-3988 are included

brain microvessels after 1 and 3 days ischemia[1]. Black *et al.*[5] found an increased uptake of Evans blue after injection of LTC$_4$ into rat brain.

In the four-vessel occlusion model, side effects of anesthesia and by convulsion can be avoided. Early hyperemia and late hypometabolism and neuronal death were observed in this model[13]. In our experiments, LTC$_4$ increased further after reperfusion and reached a maximum at 1 min followed by slow decrease. The LTC$_4$ concentration change might be correlated with early hyperemia. After treatment by AA-861, the LTC$_4$ concentration was not changed during the ischemic period, while further elevation of LTC$_4$ was prevented during the reperfusion period. This might be an important point for the treatment of brain ischemia. In other models of experimental ischemia AA-861 and other 5-lipoxygenase inhibitor were also reported to suppress an increase of LTC$_4$ during reperfusion[8,9].

After administration of the PAF antagonist CV-3988, we obtained smaller but similar effects as with AA-861. PAF was recently recognized as a putative candidate to exacerbate ischemia. The PAF antagonist, BN 52021 was found to improve clinical signs and mitochondrial respiration in brain tissue after 10 min global ischemia and reperfusion in gerbils[14]. In these experiments BN 52021 improved also the cerebral water and sodium content. It was further reported that BN 52021 decreases accumulation of free fatty acids in mouse brain during ischemia[4]. PAF induces aggregation of platelets and leukocytes, vasoconstriction, and production of superoxide anions. PAF and LTs thus have many properties in common and may interact in pathologic conditions.

Another lipoxygenase metabolite, LTB$_4$, is a potent chemotactic agent for human neutrophils. Leukocytes have been regarded as an important factor in ischemic brain damage. Clinically, leukocyte infiltration was detected by using the Indium-111 labelling method in areas of human cerebral infarction[12]. Experimentally, polymorphonuclear leukocyte (PMNL) infiltration was observed in infarcted tissue after MCA occlusion in spontaneously hypertensive rats in correlation with myeloperoxydase (MPO) activity[2]. In addition, Dutka *et al.*[6] described beneficial effects of mechlorethamine-induced depletion of PMNL prior to cerebral ischemia by air embolism in dogs. In these experiments, both CBF and cortical SSEP amplitudes were improved by neutropenia. Few studies have been made on LTB$_4$ accumulation in cerebral ischemia. In this study, a significant increase of LTB$_4$ concentra-

tions was found in brain tissue of rats with global cerebral ischemia and reperfusion. However, alterations of LTB$_4$ by ischemia and reperfusion were less prominent as of LTC$_4$. Recently Xu *et al.*[15] reported detection of LTB$_4$ in rat spinal cord injury. Barone *et al.*[3] demonstrated increased LTB$_4$ receptor binding in rat focal ischemia and reperfusion. Yet, the mechanisms of LTB$_4$ elevation and relationships between activation of leukocytes and LTB$_4$ or other mediators must be further investigated. As seen the PAF antagonist, CV-3988 suppressed accumulation of LTB$_4$ in the reperfusion phase. PAF induces also leukocyte activation which may result in a potent release of free radicals by the cells. Free radicals can be formed through the lipoxygenase pathway of AA. Thus interactions of LTs, PAF, and free radicals should be further studied.

References

1. Asano T, Gotoh O, Koide T, Takakura K (1985) Ischemic brain edema following occlusion of the middle cerebral artery in the rat. II: alteration of the eicosanoid synthesis profile of brain microvessels. Stroke 16: 110–113
2. Barone FC, Hillegass LM, Price WJ, White RF, Lee EV, Feuerstein GZ, Sarau HM, Clark RK, Griswold DE (1991) Polymorphonuclear leukocyte infiltration into cerebral focal ischemic tissue: myeloperoxidase activity assay and histologic verification. J Neurosci Res 29: 336–345
3. Barone FC, Schmidt DB, Hillegass LM, Price WJ, White RF, Feuerstein GZ, Clark RK, Lee EV, Griswold DE, Sarau HM (1992) Reperfusion increases neutrophils and leukotriene B4 receptor binding in rat focal ischemia. Stroke 23: 1337–1348
4. Birkle DL, Kurian P, Braquet P, Bazan NG (1988) Platelet-activating factor antagonist BN52021 decreases accumulation of free polyunsaturated fatty acid in mouse brain during ischemia and electro-convulsive shock. J Neurochem 51: 1900–1905
5. Black KL, Hoff JT (1985) Leukotrienes increase blood-brain barrier permeability following intraparenchymal injections in rats. Ann Neurol 18: 349–351
6. Dutka AJ, Kochanek PM, Hallenbeck JM (1989) Influence of granulocytopenia on canine cerebral ischemia induced by air embolism. Stroke 20: 390–395
7. Gaudet RJ, Alam I, Levine L (1980) Accumulation of cyclooxygenase products of arachidonic acid metabolism in gerbil brain during reperfusion after bilateral common carotid artery occlusion. J Neurochem 35: 653–658
8. Mabe H, Nagai H, Suzuka T (1990) Role of brain tissue leukotriene in brain oedema following cerebral ischaemia: effect of a 5-lipoxygenase inhibitor, AA-861. Neurol Res 12: 165–168
9. Minamisawa H, Terashi A, Katayama Y, Kanda Y, Shimizu J, Shiratori T, Inamura K, Kaseki H, Yoshino Y (1988) Brain eicosanoid levels in spontaneously hypertensive rats after ischemia with reperfusion: leukotriene C$_4$ as a possible cause of cerebral edema. Stroke 19: 372–377
10. Moskowitz MA, Kiwak KJ, Hekimian K, Levine L (1984) Synthesis of compounds with properties of leukotriens C$_4$ and D$_4$ in gerbil brains after ischemia and reperfusion. Science 224: 886–889

11. Moufarrij NA, Little JR, Skrinska V, Lucas FV, Latchaw JP, Slugg R M, Lesser RP (1984) Thromboxane synthetase inhibition in acute focal cerebral ischemia in cats. J Neurosurg 61: 1107–1112

12. Pozzilli C, Lenzi GL, Argentino C, Carolei A, Rasura M, Signore A, Bozzao L, Pozzilli P (1985) Imaging of leukocytic infiltration in human cerebral infarcts. Stroke 16: 251–255

13. Pulsinelli WA, Levy DE, Duffy TE (1982) Regional cerebral blood flow and glucose metabolism following transient forebrain ischemia. Ann Neurol 11: 499–509

14. Spinnewyn B, Blavet N, Clostre F, Bazan N, Braquet P (1987) lnvolvement of platelet-activating factor (PAF) in cerebral post-ischemic phase in Mongolian gerbils. Prostaglandins 34: 337–349

15. Xu J, Hsu CY, Liu TH, Hogan EL, Perot PL Jr, Tai H-H (1990) Leuko-triene B4 release and polymorphonuclear cell infiltration in spinal cord injury. J Neurochem 55: 907–912

Correspondence: Yasuhiro Namura, M. D., Department of Neurology, School of Medicine, Kyoto University, 54 Shogoin-Kawaharacho, Sakyoku, Kyoto 606, Japan.

Acta Neurochir (1994) [Suppl] 60: 300–302

Blocking of Interleukin-1 Activity is a Beneficial Approach to Ischemia Brain Edema Formation

Y. Yamasaki[1], **H. Shozuhara**[1], **H. Onodera**[1], and **K. Kogure**[2]

[1] Department of Neurology, Institute of Brain Disease, Tohoku University School of Medicine, Sendai, and
[2] Institute of Pathology, Saitama, Japan

Summary

We examined the therapeutic value of an interleukin-1 inhibitor on brain edema formation using a transient focal ischemia model in rats. Rats were given an interleukin-1 blocker, or interleukin-1 release inhibitor immediately after reperfusion. In rats treated with interleukin-1 inhibitor, ischemic brain edema 1 day after reperfusion was significantly decreased compared to that of saline-treated control rats. The simultaneous application of an IL-1 release inhibitor and a lipoxygenase inhibitor showed an additive beneficial effect on brain edema formation. These findings suggest that blocking IL-1 activity ameliorates brain edema and attenuates the neuronal damage induced by focal transient ischemia in rats.

Keywords: Ischemia; edema, interleukin-1.

Introduction

Recently, no-reflow phenomenon by leukocytes aggregation[4], increased LTB4 binding sites after ischemia-reperfusion[1], amelioration of ischemic damages by anti-neutrophil antibody treatment[6] and anti-adhesive antibody treatment[2] have been reported. We have also suggested the contribution of interleukin-1 (IL-1) on the pathogenesis in ischemic brain edema formation[8]. IL-1 is a known chemotactant for leukocytes and induces the adhesive molecule on the endothelium to entrap the activated leukocytes. These findings suggest that the inflammatory reaction plays an important role in ischemia-reperfusion tissue injury. However, few studies have focused on the therapeutic value of the IL-1 inhibitor for post-ischemic brain edema. We examined the therapeutic value of a IL-1 blocker on the ischemic edema formation.

Materials and Methods

Adult male Wistar rats, each weighing 300–330 g, (purchased from Japan S.L.C., Inc.) were allowed free access to food and water. Transient focal cerebral ischemia was induced using a slightly modified version of Longa's method[3], as described previously[8]. Briefly, silicon coated 4-0 nylon thread was inserted into the right internal carotid artery to occlude the origin of the right middle cerebral artery (MCA). One hour after MCA occlusion, the thread was removed to allow complete reperfusion of the ischemic area. One day after reperfusion, animals were decapitated and brain samples were dissected into the cerebral cortex perfused by MCA (MCA area), the dorsal area of the caudate putamen (DCP), and the ventral area of the caudate putamen (VCP). Water contents of these samples were measured by the wet and dry method.

To assess the efficacy of the interleukin-1 blocker (Zinc proto-porphyrin, ZnPP, purchased from Aldrich), an interleukin-1 release inhibitor (IL-1 inhibitor, kindly donated by Sandoz, Switzerland), and 5-lipoxygenase inhibitor (AA-861, synthesized in our lab), these were administered intraperitoneally immediately after reperfusion.

Results and Discussion

A significant increase in the brain water content was observed in MCA, DCP and VCP (86.1 ± 0.5, 86.5 ± 0.5, and 83.9 ± 0.3, respectively) 1 day after ischemia. Treatment with an IL-1 blocker, ZnPP, decreased the water content dose dependently, and ZnPP at doses of 30 and 100 mg/kg significantly reduced the brain water content in all three areas observed (83.3 ± 0.7, 83.3 ± 1.2, and 80.1 ± 0.6 in MCA, DCP and VCP area, respectively, at a dose of 100 mg/kg). An IL-1 inhibitor also dose dependently ameliorated the water content in all three areas as did ZnPP (Fig. 1a). The administration of lipoxygenase inhibitors, AA-861 or TMK866, protected the ischemic brain damages dose-dependently in MCA, DCP, and VCP areas. Furthermore, the simultaneous administration of IL-1 inhibitor and AA861 showed an additive protective effect (Fig. 1b).

On the other hand, a calcium channel blocker, nicardipine, also decreased the water content in the

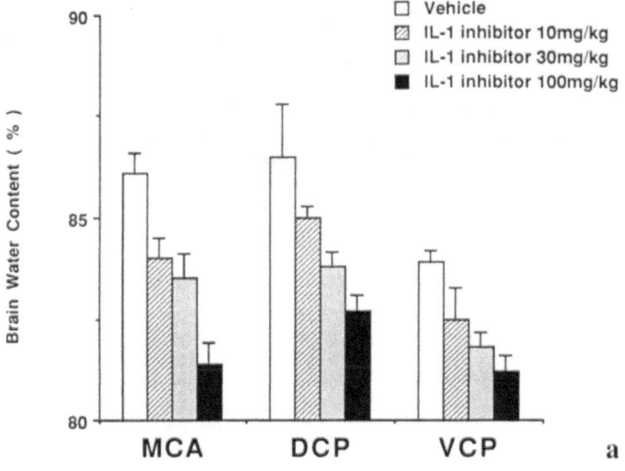

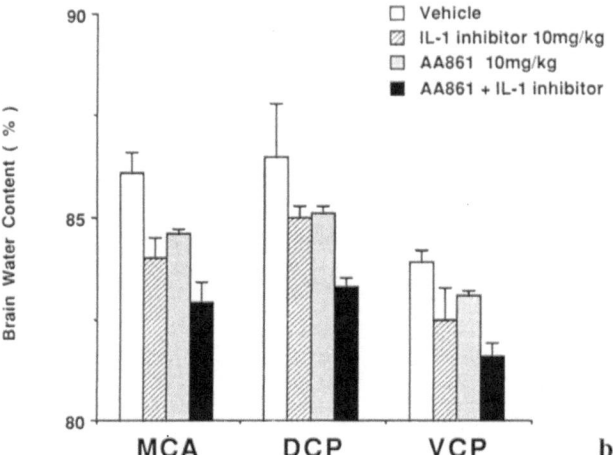

Fig 1. (a) Effect of IL-1 inhibitor on ischemic brain edema formation. (b) Additive effects of IL-1 inhibitor and AA861 on ischemic
brain edema formation. Both were administered immediately after
reperfusion and water contents were measured 24 h after ischemia.
MCA cerebral cortex perfused by middle cerebral artery, *DCP*
dorsal area of caudate putamen, *VCP* ventral area of caudate putamen. Note the dose-dependent protective effect (a) and an additive
protective effect (b)

MCA area, but not in the DCP and VCP areas. The
infarct areas and neuronal damages estimated by H.E.
staining also revealed the ameliorative effect of ZnPP
and IX207 in the cortex and in the caudate putamen
(data not shown). Since we recently confirmed that
interleukin-1 is a pathogenetic mediator in ischemic
brain edema formation, we studied the effects of IL-1
blockers and lipoxygenase inhibitors. Intraperitoneal
application of ZnPP, and IL-1 inhibitor could reduce
the brain water content. This suggested that ZnPP is a
potent IL-1 blocker without having anti-inflammatory
effects[5] and IL-1 inhibitor is a potent IL-1 synthesis

and release inhibitor. IL-1 has been suggested to be
released from ameboid microglia and/or reactive astrocytes, in addition to invaded macrophages and neutrophils. IL-1 mPNA was also increased at an early
stage after reperfusion. Furthermore, IL-1 is accepted
as a chemotactant to polymorphonuclear cells and has
been suggested to play an important role for inducing
adhesive molecules to entrap inflammatory cells to
the endothelium. Therefore, blocking of IL-1 activity
may inhibit the entrapping and invasion of leukocytes
to the endothelium, subsequently inhibiting the infiltration of leukocytes into ischemic areas.

Our findings showed that AA-861 could reduce
brain edema formation. Subsequent reports have demonstrated that the destruction of membranes by lipid
peroxidation, and an explosive activation of arachidonate cascades and levels of lipoxygenase products,
including 12-hydroxyeicosatetraenoic acid and leukotriene B4, have been reported to occur in ischemia-
reperfusion injury[7].

Interestingly, smilutaneous application of an IL-1
release inhibitor and a lipoxygenase inhibitor showed
an additive beneficial effect on brain edema. This
indicates that both IL-1 and LTB4 are important
chemotactic factors in the infiltration of leukocytes
into ischemic areas for brain edema formation.

In conclusion, the IL-1 inhibitor was found to ameliorate brain edema and attenuate the neuronal damages
induced by focal transient ischemia. Furthermore,
combination therapy using an IL-1 release inhibitor
and a lipoxygenase inhibitor may be useful for the
treatment of stroke patients.

References

1. Barone FC, Schmidt DB, Hillegass LM, Price WJ, White
   RF, Feuerstein GZ, Clark RK, Lee EV, Griswold DE, Sarau
   HM (1992) Reperfusion increases neutrophils and leukotriene B4 receptor binding in rat focal ischemia. Stroke 23:
   1337–1348
2. Clark WM, Madden KP, Rothlein R, Zivin JA (1991) Reduction of central nervous system ischemic injury by monoclonal
   antibody to intracellular adhesion molecule. J Neurosurg 75:
   623–627
3. Longa EZ, Weinstein PR, Carlson S, Cummins R (1989) Reversible middle cerebral artery occlusion without craniectomy
   in rats. Stroke 20: 84–91
4. Mori E, Zoppo GJ, Chambers JD, Copeland BR, Arfors KE
   (1992) Inhibition of polymorphonuclear leukocyte adherence
   suppresses no-reflow after focal cerebral ischemia in baboons.
   Stroke 23: 712–718
5. Nagai H, Kitagaki K, Kuwabata K, Koda A (1992) Anti-
   inflammatory properties of zinc protoporphyrin disodium
   (ZnPP-2Na). Agents Action 37: 273–283

6. Shiga Y, Onodera H, Kogure K, Yamasaki Y, Yashima Y, Shouzuhara H, Sendo F)1991) Neutrophil as a mediator of ischemic edema formation in the brain. Neurosci Lett 125: 110–112

7. Xu J, Hsu CY, Liu TH, Hogan EL, Perot PL, Tai HH (1990) Leukotriene B4 release and polymorphonuclear cell infiltration in spinal cord ischemia. J Neurochem 55: 907–912

8. Yamasaki Y, Suzuki T, Yamaya H, Matsuura N, Onodera H, Kogure K (1992) Possible involvement of interleukin-1 in ischemic brain edema formation. Neurosci Lett 142: 45–47

Correspondence: Y. Yamasaki, M. D., Department of Neurology, Institute of Brain Disease, Tohoku University School of Medicine, 2-1 Seiryo-machi, Aoba-ku, Sendai 980, Japan.

Acta Neurochir (1994) [Suppl] 60: 303–306
© Springer-Verlag 1994

Complete Cerebral Ischemia, Prostacyclin Deficiency, and Therapeutic Possibilities

R. Pluta

Department of Neuropathology, Medical Research Centre, Polish Academy of Sciences, Warsaw, Poland

Summary

This report summarizes the results of our studies on the effects of prostacyclin (PGI_2) on the outcome of global cerebral ischemia (GCI). GCI was produced for 15 and 20 min. In vivo dialysis of the hippocampus was used to determine the changes in extracellular concentrations of calcium ($Ca^{+2}e$) and blood-brain barrier (BBB) permeability. Moreover, EEG and general physiological parameters were recorded. This was combined with morphological observations. PGI_2 was infused continuously i.v. at a rate of 2 µg/kg/min. Rabbits with untreated GCI served as reference. Treatment with PGI_2 significantly enhanced EEG recovery and normalization during recirculation, and reduced both the decrease in $Ca^{+2}e$, and the BBB leakage. The number of ischemic neurons in the PGI_2-treated rabbits was significantly lower than in the non-treated ones. PGI_2 reduced brain edema. These data suggest that PGI_2 may protect against postischemic brain damage, in part by inhibiting excessive calcium influx to neurons and in part by tightening of BBB.

Keywords: Brain ischemia; prostacyclin; calcium; blood-brain barrier.

Introduction

Altered prostanoid synthesis following GCI represents one aspect of a broader hypothesis relating tissue damage in the cerebrum during ischemia to a multifactorial sequence of events resulting in a general increase in microcirculatory resistance during recirculation. This sequence is termed the blood vessel damaged tissue interaction. The platelet-vessel wall interaction represents one part of this process potentially important for postischemic recirculation. Platelets accumulate in the injured brain vessels after short- and long-term survival following GCI[10]. Postischemic obstruction by platelets with absence of circulation through small brain vessels has been demonstrated in several models of ischemia[10]. This suggested that microcirculatory reocclusions may play a significant role in the production of an infarction, and that vessels and blood may have a lower threshold to ischemic injury than neurons and neuroglial cells. Thromboxan A_2 (TXA_2) and PGI_2 production at the platelet-vessel wall interface occurs through selective metabolism of the cyclic endoperoxide PGH_2. Platelet thromboxane synthetase converts PGH_2 to TXA_2 during platelet aggregation. TXA_2 is a potent stimulant of platelet aggregation and vasoconstriction and is overproduced during recirculation after GCI[1]. Endothelial PGI_2 synthetase converts PGH_2 to the vasodilatory PGI_2, which strongly inhibits platelet aggregation. Diminished level of PGI_2 was noted during reperfusion with deficiency of PGI_2 in brain microcirculation[1]. It seemed reasonable to propose that in normal conditions a "balance" between these mediators might be of fundamental importance in the regulation of blood-vessel homeostasis. Neither anticoagulant drugs nor thrombolytic therapy were shown to be effective in patients and animals with GCI. These findings indicated that treatment of GCI with PGI_2 may compensate the endogenous loss of controlling activity of platelet aggregability and vascular resistance. Great interest has recently emerged in the literature on the potential efficacy of exogenous PGI_2 in treating GCI. In this paper we present some of the pharmacotherapeutic effects of PGI_2 in animals following GCI[4–9].

Materials and Methods

Rabbits (n = 90; 2.5–3.0 kg) were implanted with transhippocampal microdialysis fibre one day before the experiment[9]. On the next day they were anesthetized, tracheostomized, relaxed and artificially ventilated[9]. Complete 15- or 20-min cerebral ischemia was induced by the occlusion of the brachiocephalic trunk, the left subclavian and bilateral internal thoracic artery combined with the hypotension. Dialysis perfusion of hippocampus was performed before GCI, during the 15 min of its duration and for 2 h thereafter. The changes of $Ca^{+2}e$ were measured using ^{45}Ca as a probe. BBB permeability was measured with Na-fluorescein in the dialysate. During the experiments general physiological parameters and EEG were recorded. This was combined with morphological examination by light and electron microscopy (brain, heart, adrenal glands – 3 or 6 h survival). PGI_2 (Wellcome Foundation Ltd., London and Sigma) was applied to all treated rabbits. The vehicle control was run parallel.

Results

Pharmacology of Exogenous Prostacyclin in the Treatment of Global Cerebral Ischemia and on the Ischemia-Associated Changes

I. Dosage and Administration

Prostacyclin was administered i.v. continuously 3 min before, during and for 15 min after ischemia or for the last 3 min of ischemia and for 40 min after it in a dose 2 μg/kg/min.

II. Pharmacotherapeutic Effects

Brain

– Enhanced both EEG recovery and normalization of the recordings during recirculation.
– Reduced the decrease in extracellular calcium concentration with some acceleration of recovery during recirculation. Reduced the blood-brain barrier leakage.
– Prevented hemorrhages.
– Reduced edema (cytotoxic and vasogenic types).
– Reduced number of constricted cerebral blood vessels.
– Reduced the spectrum of neuronal changes.
– Decreased the number of pathologically changed neurons.
– Prevented ultrastructural changes in neuronal cytoplasm.
– Prevented ultrastructural changes in neuroglial cytoplasm.
– Improved the early neurological state following ischemia.

Heart

– Prevented changes typical in EKG for myocardial infarct.
– Reduced the intensity of hemorrhages in cardiac muscle.
– Reduced bradycardia during and immediatly after ischemia.

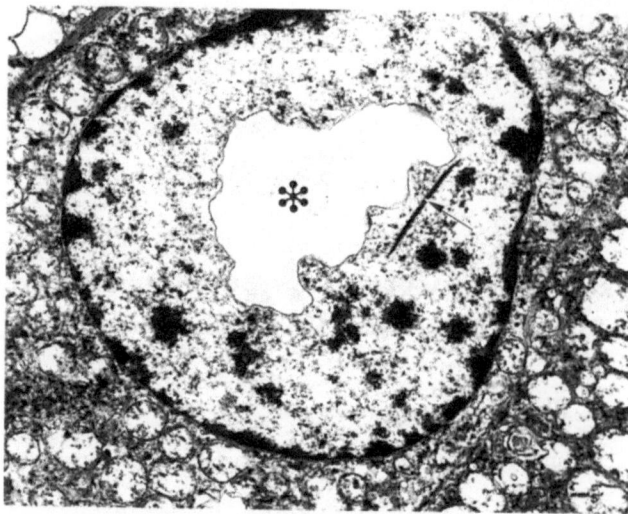

Fig. 1. Changes in adrenal cortex cell nucleus (vesicular structure (*), and filamentous inclusion (arrow)) after 15-min GCI with PGI_2. 6 h survival. 10,800 ×

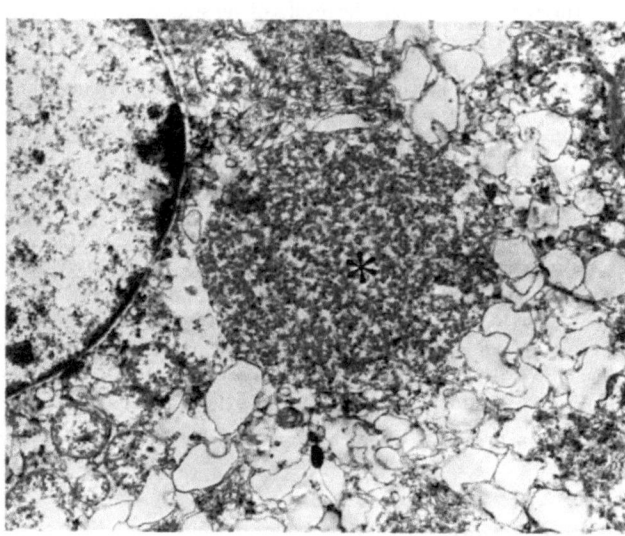

Fig. 2. Necrosis of adrenal cortex cell cytoplasm (*) after 15-min GCI with PGI_2. 3 h survival. 12,500 ×

– Reduced hypertension in the early period of ischemia.
– Reduced hypotension in the early period after ischemia.
– Acted antiarythmically during and following ischemia.

III. No Effects

Brain

– Did not ameliorate the ultrastructural changes in neuronal cell nuclei (clumping and marginal shift of

chromatin, vesicular structures, intranuclear membranaceous or filamentous inclusions)
- Did not ameliorate the ultrastructural changes in neuroglial cell nuclei (the changes were the same as in neurons).

Adrenal Glands

- Did not influence the ultrastructural changes in adrenal medulla and cortex cell nuclei (the changes were the same as in neurons, Fig. 1).

IV. Side Effects

Heart

- Developed hypotension.
- Developed bradycardia.

Adrenal Glands

- Produced severe destruction of adrenal cortex cell cytoplasmic organization (myelin figures, necrosis of cytoplasm structure, Fig. 2).

Discussion

There are contradictory data in the literature concerning improvement of the neurologic state and survival of ischemic patients and animals treated with PGI₂[2,3,9]. The present study demonstrates that i.v. treatment with PGI₂ accelerates the postischemic EEG recovery and improves the morphologic status of neurons in the early periods after GCI. The mechanisms of PGI₂-evoked brain protection may either be vascular or parenchymal, and the former appears to dominate. This appears to be the case with the reduction of the decrease of $Ca^{+2}e$, which may be related to potent cerebral vasodilatatory effects of i.v. applied PGI₂. Protection of the BBB by i.v., applied PGI₂ is another example of the strong vasotropic action of PGI₂ and further supports the importance of vascular sites in the brain protection by PGI₂. Some of our findings (e.g. the acceleration of EEG recovery) clearly point at early beneficial effects of PGI₂ treatment on postischemic brain damage. Our other data, which include such effects of PGI₂ as the diminution of calcium influx to the neurons during and after GCI and protection of the BBB during recirculation, may be regarded as directed towards parenchyma. It seems that there is a direct causal relation between the effects of PGI₂ on calcium homeostasis and BBB permeability and the final

effect of this treatment; protection against and even more likely facilitation of postischemic recovery. The facilitation of early postischemic recovery contrasts with the ineffectiveness of the PGI₂ towards nuclear damage. In this way, the vicious circle evoking ischemic changes would be temporarily broken with regard to cytoplasm, but not to cell nuclei. In this situation nuclear damage may be responsible for the lack of effect[3] or disappearance[2] of early beneficial effect of treatment with PGI₂. It may be hypothetized that ischemic changes in the nuclei, possibly reflecting or being a cause of disturbances of their transcriptional functions, are the primary cause of severe secondary metabolic disorder in the whole cell, further leading to far-reaching functional changes resulting in the so-called "maturation phenomenon". It would be unrealistic to expect that a single type of treatment would be beneficial in GCI, as various pathological factors are involved during and after GCI, and therapy with PGI₂ may not be a "causative" treatment for all of them. It is tempting to set up a trial to test the effects of some combined treatments (e.g. PGI₂ + nimodipine) in animals with GCI. Other options include some new unknown substances which will act directly on nuclei during and/or after GCI.

References

1. Gaudet RJ, Levine L (1979) Transient cerebral ischemia and brain prostaglandins. Biochem Biophys Res Comm 86: 893–901
2. Huczyński J, Kostka-Trąbka E, Sotowska W, Bieroń K, Grodzińska L, Dembińska-Kieć A, Pykosz-Mazur E, Pęczak E, Gryglewski RJ (1985) Double-blind controlled trial of the therapeutic effects of prostacyclin in patients with completed ischemic stroke. Stroke 16: 810–814
3. Kerckhoff van den W, Hossmann KA, Hossmann V (1983) No effect of prostacyclin on blood flow, regulation of blood flow and blood coagulation following global cerebral ischemia. Stroke 14: 724–730
4. Pluta R (1985) Influence of prostacyclin on early morphological changes in the rabbit brain after complete 20-min ischemia. J Neurol Sci 70: 305–316
5. Pluta R (1986) The influence of prostacyclin on the recovery of bioelectric cerebral activity after complete ischemia. Acta Neurol Scand 73: 44–54
6. Pluta R, Dydyk L (1987) Failure of prostacyclin effect on the reversal of early ultrastructural changes in cell nuclei of adrenal cortex after complete 15-min cerebral ischemia in rabbit. Exp Pathol 32: 163–168
7. Pluta R (1988) The effects of prostacyclin on early ultrastructural changes in the neuron nuclei of the motor cortex in rabbits after complete 20-min cerebral ischemia. Exp Mol Pathol 48: 161–173
8. Pluta R (1991) The effects of prostacyclin on early ultrastructural changes in the cytoplasm and nuclei of neuroglial cells following complete cerebral ischemia. J Hirnforsch 32: 175–184

9. Pluta R, Salińska E, Łazarewicz JW (1991) Prostacyclin atten-
    uates in the rabbit hippocampus early consequences of transient
    complete cerebral ischemia. Acta Neurol Scand 83: 370–377
10. Sampaolo S, Cervós-Navarro J, Djouchadar D, Figols J (1987)
    Clinical and experimental evidence of microthrombosis in
    cerebral ischemia. In: Hartmann A, Kuschinsky W (eds) Cer-
    ebral ischemia and hemorrheology. Springer, Berlin Heidel-
    berg New York, pp 386–393

Correspondence: Ryszard Pluta, M.D., Department of Neuropathology, Medical Research Centre, Polish Academy of Sciences,
Dworkowa St. 3, 00-784 Warsaw, Poland.

Acta Neurochir (1994) [Suppl] 60: 307–309

Histochemical Demonstration of Free Radicals (H$_2$O$_2$) in Ischemic Brain Edema and Protective Effects of Human Recombinant Superoxide Dismutase on Ischemic Neuronal Damage

H. Morooka[1], N. Hirotsune[2], T. Wani[2], and T. Ohmoto[2]

[1] Department of Neurological Surgery, Bizen City Hospital and [2] Okayama University Medical School, Okayama, Japan

Summary

A new histofluorescence method by HPAA (p-hydroxyphenyl acetic acid) for free radicals in the brain tissue was devised to study neuronal damage induced by ischemia. Cerebral ischemia was produced in rats by injection of plastic microspheres and arachidonic acid (AA) into the right carotid artery.

The concentration of malondialdehyde (MDA; free radical) in cerebral cortex of aminotriazol (an H$_2$O$_2$-dependent inhibitor of catalase) treated rats 2 h after stroke was 6.33 times the level before infarction, while the concentration of MDA in h-r SOD (free radical-scavenging enzyme) treated rats 2 h after stroke was significantly lower than in untreated rats.

The histochemical findings demonstrated marked H$_2$O$_2$ production around blood vessels occluded by microspheres in the cerebral cortex of the aminotriazole treated rats 2 h after stroke together with disruption of the BBB. Light microscopical findings demonstrated extensive edematous changes in the aminotriazole treated rats 2 h after stroke, while pathological damage in SOD treated rat brains was absent or minimal.

We conclude that free radicals are formed during ischemia, and that AA appears to be a major source of activated oxygen radicals. The findings indicate that SOD is protective against ischemia-induced neuronal damage.

Keywords: Free radical; histochemistry; superoxide dismutase; cerebral ischemia.

Introduction

O$_2^-$, H$_2$O$_2$ and OH$^.$ are highly reactive molecules which could damage cellular structures and subcellular organelles. Previous studies have demonstrated that free radical formation during ischemia plays an important role in microcirculatory damage[1-3]. The present study was designed to elucidate histochemical changes due to oxygen-derived free radicals (H$_2$O$_2$) in ischemic brain edema and to examine protection by human recombinant superoxide dismutase (h-r SOD) against ischemia-induced neuronal damage.

Materials and Methods

The experimental animals used were male Wistar strain rats weighing approx. 300–350 g. Cerebral ischemia was produced by a right intracarotid injection of plastic microspheres (35 µm ø) together with arachidonic acid (AA, 5 mM, 25 µl). The brain was fixed by freezing with liquid nitrogen followed by biochemical analysis. The tissue malonyldialdehyde (MDA) concentration was measured by the thiobarbituric acid reaction[4], the water content by lyophilization. The histochemical method employed to detect H$_2$O$_2$ was based on the conversion of p-hydroxyphenyl acetic acid (HPAA) to a highly fluorescent compound (λ_{ex} = 317 nm, λ_{em} = 414 nm) in the presence of peroxidase[5]. A frozen 12 µm section was cut by a cyrostat and placed on a slide coated with the gelatin, containing HPAA (2.5 mg/ml).

The reaction was started by adding horseradish peroxidase (1 mg/ml). The slide was incubated at 37°C for 10 min. For examination of vascular permeability, the animals were given 2% Evans blue (1 mg/kg) into the femoral vein 5 min before sacrifice. The experiment was performed in three groups, the first group (n = 20) being untreated, the second group (n = 15) receiving 3-amino-1,2,4-triazole (1 g/kg i.p.), an H$_2$O$_2$-dependent inhibitor of catalase 1 h before embolization, and the third group (n = 15) administered with h-r SOD (8 × 10^4 units/kg) intravenously 1 min before embolization.

Results

The water content of brain tissue of the embolized side in untreated rats increased gradually after stroke. 1 h after stroke 79.6 ± 1.1% (SD) was obtained which was significantly higher than in controls, i.e., 77.0 ± 1.0% (p < 0.01). The water content of embolized hemispheres in aminotriazole treated rats 1 h after stroke was 82.5% ± 1.5, which was significantly higher than that of the untreated rats (p < 0.01). On the other side, the water content of the embolized hemispheres in SOD treated rats 1 h after the stroke was 77.1 ± 1.0, which was not significantly higher than that of controls without embolization.

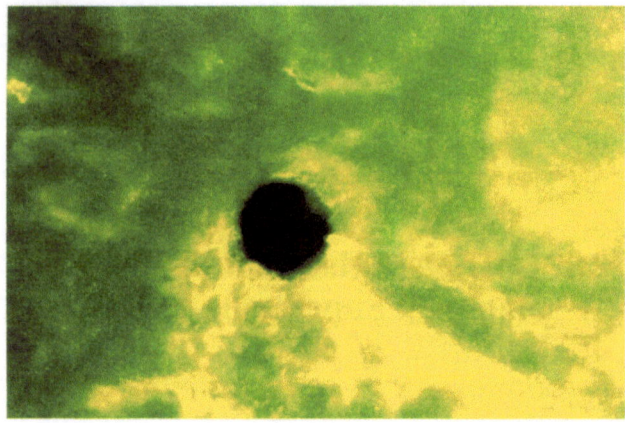

Fig. 1. Free radical (H_2O_2) fluorescence of cerebral cortex in a aminotriazole treated rat 2 h after stroke

The concentration of MDA (free radical) in brain tissue of the embolic side in untreated rats showed a peak at 2 h after infarction of 1.5 times the value before infarction ($p < 0.001$). The concentration of MDA in brain tissue of the embolic side in aminotriazol treated rats 2 h after the infarction was 6.33 times the level before infarction ($p < 0.001$).

The concentration of MDA in brain tissue of the embolic side in SOD treated rats at 2 h was 20 ± 5 n moles/g tissue, which is significantly lower than that of untreated rats, i.e. 380 ± 100 n moles/g tissue ($p < 0.001$).

The histochemical findings demonstrated marked H_2O_2 production around blood vessels occupied by microspheres in cerebral cortex of the aminotriazole treated rats 2 h after stroke (Fig. 1). The light microscopic findings indicate extensive edematous changes in the aminotriazole treated rats, while pathological damage in SOD treated rat brain 2 h after the stroke was absent or minimal (Fig. 2).

Further disturbances of blood-brain barrier function (BBB) shown by leakage of Evans blue were noted in the aminotriazole treated rat 2 h after the stroke.

Discussion

In cerebral infarction caused by intracarotid injection of AA in rats receiving aminotriazole (an H_2O_2 dependent inhibitor of catalase[6]) marked edema was found, disruption of the BBB, and extensive H_2O_2 production around blood vessels. In all rats receiving h-r SOD, edematous changes in the area of infarction and around vessels, were slight or totally absent, as compared to rats with infarction not receiving treatment. Our results strongly suggest a role of oxygen-

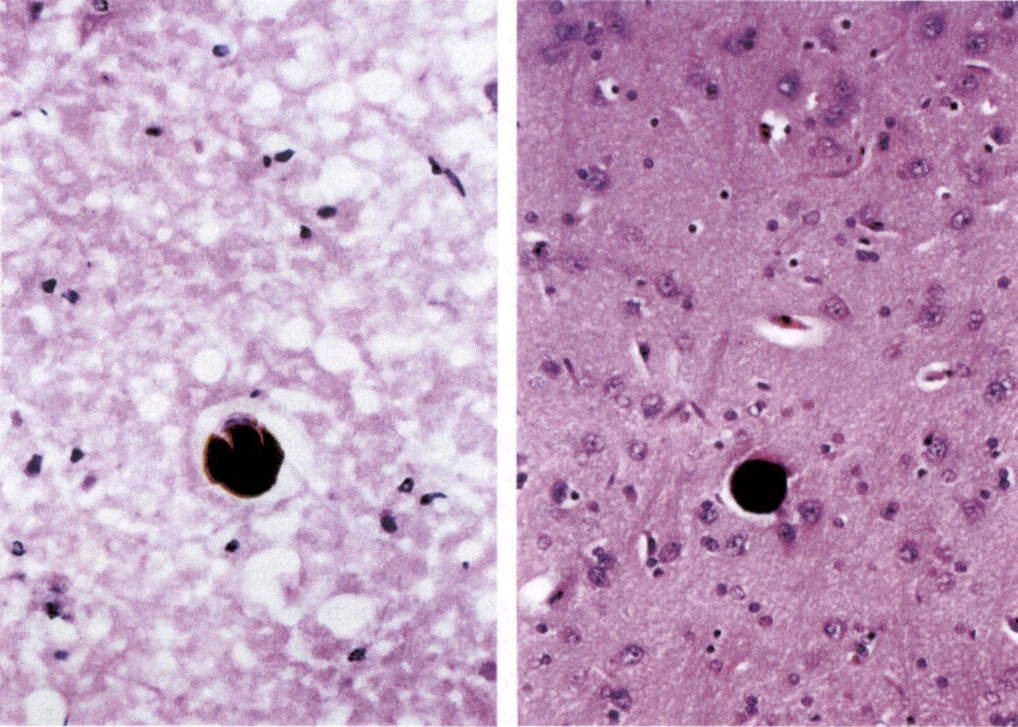

Fig. 2. Cerebral cortex of aminotriazol treated rat 2 h after stroke, showing marked edematous changes (left panel). Cerebral cortex of the SOD treated rat 2 h after stroke, showing only minimal changes (right panel). H.& E. × 400

derived free radicals which were particularly abundant in post-ischemic brain tissue.

AA is converted by cyclooxygenase to prostaglandis and lipid peroxides, or by lipoxygenase to hydroxy fatty acids and leukotrienes[7]. AA is reported to modulate K$^+$ currents associated with alterations of membrane integrity[8,9].

Superoxide radicals and other activated oxygen species may contribute to changes of cerebral brain vessel endothelium, as characterized by an increase of cytoplasmic inclusions of surface pits, and mitochondrial injury together with intravascular neutrophilic granulocyte accumulation and platelet aggregation[1,8].

It has been proposed that free radicals are mainly formed during reperfusion of previously ischemic tissues, involving xanthine oxidase[10] or ubisemiquinone in mitochondria[11]. However, free radicals may also be formed during incomplete ischemia or even in complete ischemia or anoxia, as demonstrated by the generation of H$_2$O$_2$ in the current experiments. Accumulation of free radicals either in ischemic conditions or by enzymatic reactions may lead to nonspecific release of neurotransmitters[1,12]. Previous experiments of our group have shown that histamine release is accelerated by free radicals, which may contribute to the increase of macromolecular permeability of the BBB and to vasogenic edema[13].

References

1. Morooka H, Nishimoto A (1985) The role of adrenergic activity and histamine in ischemic brain edema. In: Inaba Y, Klatzo I, Spatz M (eds) Brain edema. Springer, Berlin Heidelberg New York Tokyo, pp 302–309

2. Armstead WM, Mirro R, Busija DW, Leffler CW (1988) Post-ischemic generation of superoxide anion by newborn pig brain. Am J Physiol 255: 401–403

3. Pellegrini-Giampietro DE, Cherici G, Alesiani M, Caria V, Moroni F (1990) Excitatory amino acid release and free radical formation may cooperate in the genesis of ischemia-induced neuronal damage. J Neurosci 10 (3): 1035–1041

4. Ohkawa H, Ohishi N, Yagi K (1979) Assay for lipid peroxide in animal tissue by thiobarbituric acid reaction. Anal Biochem 95: 351–358

5. Guilbault GG, Brignac PJ, Juneau M (1954) New substrates for the fluorometric determination of oxidative enzymes. Anal Chem 40 (8): 1256–1263

6. Sinet PM, Heikkila RE, Cohen G (1980) Hydrogen peroxide production by rat brain in vivo. J Neurochem 34 (6): 1421–1428

7. Kukreja RC, Kontos HA, Hess ML, Ellis EF (1986) PGH synthase and lipoxygenase generate superoxide in the presence of NADH or NADPH Circ Res 59: 612–619

8. Chan PH, Fishman RA, Caronna J, Schmidley JW, Prioleau G, Lee J (1983) Induction of brain edema following intracerebral injection of arachidonic acid. Ann Neurol 13: 625–632

9. Ehrengruber MU, Deranleau DA, Kempf C, Zahler P, Lanzrein M (1993) Arachidonic acid and other unsaturated fatty acids alter membrane potential in PC 12 and bovine adrenal chromaffin cells. J Neurochem 60 (1): 282–288

10. McCord JM (1985) Oxygen -derived free radicals in post-ischemic injury. N Engl J Med 312: 159–163

11. Del Maestro RF (1980) An approach to free radicals in medicine and biology. Acta Physiol Scand [Suppl] 492: 153–168

12. Pellegrini-Giampietro DE, Cherici G, Alesiani M, Garia V, Moroni F (1988) Excitatory amino acid release from rat hippocampal slices as a consequence of free-radical formation. J Neurochem 51: 1960–1963

13. Morooka H, Sasayama H, Sakai K, Namba S, Nishimoto A (1989) The role of catecholamine-induced lipid peroxidation and histamine in ischemic brain edema. In: Hoff JT, Betz AL (eds) Intracranial pressure VII. Springer, Berlin Heidelberg New York, pp 818–820

Correspondence: H. Morooka, M. D., Department of Neurosurgery, Bizen City Hospital, 2245 Inbe, Bizen City, Okayama 705, Japan.

Acta Neurochir (1994) [Suppl] 60: 310–313

Tirilazad Reduces Brain Edema After Middle Cerebral Artery Ligation in Hypertensive Rats

A. Karki[*], **I. Westergren, H. Widner,** and **B. Johansson**

Department of Neurology, University Hospital, Lund, Sweden

Summary

Tirilazad (3 mg/kg i.v.) or vehicle was given 10 min, 3 and 12 hours after ligation of the right middle cerebral artery in male spontaneously hypertensive rats (n = 12 in each group). Brain specific gravity was determined in 23 predetermined cortical regions covering the core and penumbra areas 24 hours after the occlusion. The specific gravity was significantly higher in tirilazad-treated rats than in controls (p < 0.0001). When individual regions of the two groups were compared, the difference was significant in 9 of the 23 samples predominantly, but not exclusively, in the penumbra zone.

Keywords: Tirilazad; brain edema; focal ischemia; brain specific gravity.

Introduction

Evidence is accumulating that oxygen free radicals are important mediators of brain tissue injury[8,16] and vasogenic edema[4] in focal cerebral ischemia. Tirilazad mesylate, a 21-aminosteroid free radical scavenger, inhibits iron catalyzed lipid peroxidation in vitro[2] and reduces postischemic lipid peroxidation, as assessed by the preservation of brain vitamin E in transient forebrain ischemia in the gerbil[9]. The aim of the present study was to evaluate if tirilazad could reduce brain edema after permanent occlusion of the middle cerebral artery in the spontaneously hypertensive rat (SHR). The brain specific gravity was determined 24 hours after the ligation of the middle cerebral artery in 23 predetermined cortical areas covering the core and the penumbra.

Materials and Methods

Twenty-four adult, non-fasting, male, spontaneously hypertensive rats (SHR) (Møllegaard Centre, Denmark) weighing between

---

[*] On leave of absence from Institute of Medicine, Tribhuvan University Teaching Hospital, Kathmandu, Nepal

280–390 g were used. The animals were anesthetized with methohexital (Brietal; Eli Lilly & Co., USA), 50 mg/kg i.p. and with additional doses of 1 mg/kg i.v., as required. The rats were breathing spontaneously throughout. Catheters were placed in the right femoral vein for drug and Evan's blue administration, and in the right femoral artery for continuous monitoring of mean arterial pressure (MAP) and blood sampling. Middle cerebral artery occlusion (MCAO) was performed as described by Tamura *et al.*[17]. Briefly, a craniotomy was made and the right middle cerebral artery was ligated with a 10-0 monofilament nylon thread at its crossing with the olfactory tract. Rectal temperature, PaO_2, $PaCO_2$, pH, hematocrit and blood glucose levels were determined 1 hour before and immediately after the MCAO. Twelve rats were treated with tirilazad mesylate (Upjohn Co., USA) at a dose of 3 mg/kg i.v. and the other twelve with vehicle, 10 min, 3 hours and 12 hours after the MCAO. Evan's blue (3 ml/kg) was given i.v. at the end of the operation to delineate the infarct and penumbral zones. After 24 hours, the rats were reanesthetized and the brains were rapidly removed and immersed in kerosene, to reduce evaporation. The brains were sectioned into 5 coronal slices. As depicted in Fig. 1, five tissue samples were taken from the right prefrontal, frontal, parietal, and parieto-occipital slice, and 3 samples from the occipital slice. Two samples, corresponding to the "b" and "c" sites, were taken from each slice on the left side. The specific gravity of the samples was determined as described by Nelson *et al.*[12] and modified by Fujiwara *et al.*[6]. Briefly, the samples were placed in a column containing a gradient mixture of bromobenzene-kerosene and the equilibration depth of each sample was measured after two minutes. Data were analysed using a oneway ANOVA, with Scheffé's post hoc procedure. All other values are given as mean ± 95% confidence intervals (CI).

Results

MAP before and after the MCAO was 180 ± 13 (95% CI) and 158 ± 20 mmHg in controls and 170 ± 13 and 165 ± 16 mmHg in tirilazad-treated rats. Preoperative and postoperative blood glucose levels were 6.5 and 6.0 mmol/l and 6.8 and 6.6 mmol/l in the former and latter groups. The PaO_2, $PaCO_2$, pH, hematocrit, and temperature values were within physiological limits during the entire operative period. All operated rats survived until the time of sacrifice and developed macroscopically visible infarcts in the right cortex.

The cortical specific gravity in the left control hemisphere did not differ between the two groups and corresponded to earlier published values (1.0469 ± 0.002). In the right infarcted hemisphere, the specific gravity was significantly reduced in all 23 samples in both groups with the lowest values in the infarct core, the "c" samples of the prefrontal, frontal, parietal and parieto-occipital slices (Fig. 1A). In corresponding samples, the specific gravity was consistently higher in rats given tirilazad than in control rats (p < 0.0001). When individual regions of the two groups were compared, the difference was significant in 9 of the 23 samples (Figs. 1 and 2), mainly in the penumbra zone but also in some of the most severely edematous regions. Thus, in tirilazad-treated rats the specific gravity was < 1.0300 in two regions only, compared to 5 in the control group (Fig. 1).

Discussion

Earlier studies on the effect of tirilazad in ischemic brain edema are conflicting. In a model of perinatal hypoxia-ischemia in the rat, Bågenholm et al.[3] found no effect on the edema although pre- as well as posttreatment with the drug reduced brain damage. In contrast, Young et al.[20] observed a reduction of ischemic brain edema after ligation of the middle cerebral artery in normotensive Sprague-Dawley rats.

In a recent report, tirilazad mesylate reduced the infarct size after intermittent, but not permanent occlusion of the middle cerebral artery in SHR, nor did the drug reduce the infarct size after permanent ligation in normotensive Wistar rats[19]. We are unaware of anyother study dealing with the effect of tirilazad on ischemic brain edema in SHR. The specific gravity method[6,12] has the advantage that it allows a detailed regional determination of the edema. Picozzi et al. have evaluated the method in ischemic edema[15]. They found that the decrease in specific gravity was highly correlated to the increase in water content and that cerebral blood volume changes did not significantly influence the measurements. Our data show that treatment with tirilazad mesylate 10 min, 3 hours and 12 hours after the occlusion reduces the formation of brain edema as indicated by increased specific gravity. The largest effect was observed in the penumbral zone (Fig. 1).

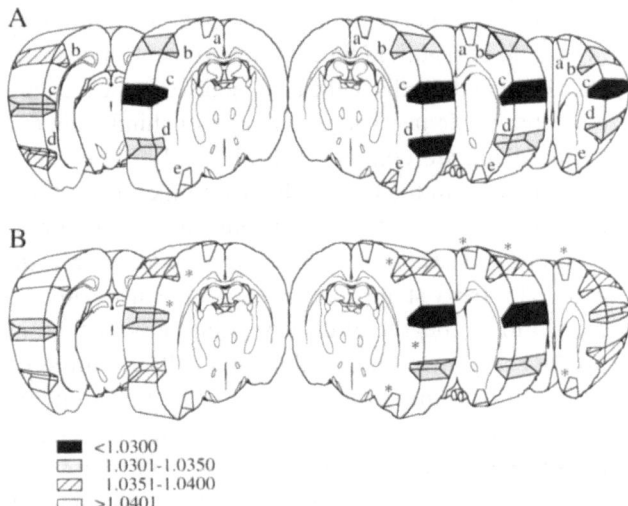

■ <1.0300
☐ 1.0301–1.0350
▨ 1.0351–1.0400
☐ >1.0401

Fig. 1. The brain specific gravity 24 hours after ligation of the right middle cerebral artery in spontaneously hypertensive rats illustrated in a schematic drawing showing where cortical tissue samples were taken. Mean specific gravity values of each site are arbitrarily divided into four groups: < 1.0300; 1.0301–1.0350; 1.0351–1.0400, > 1.0401, and shaded as indicated by key. (A) 12 control vehicle treated rats. (B) 12 rats given tirilazad mesylate (3 mg/kg i.v.) at 10 min, 3 and 12 hours after occlusion. Asterisk (*) marks the sample sites where post hoc ANOVA analysis of specific gravity values were significantly different (95% significance level) between two groups

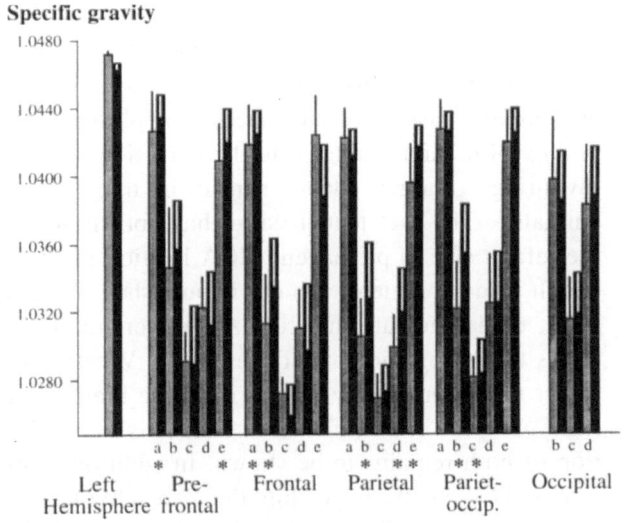

Fig. 2. Bar diagram of means ± 95% confidence intervals, of specific gravity measurements of brain tissue samples taken from areas as depicted in Fig. 1, with same labels. Shaded columns represents vehicle treated groups and solid black columns represents tirilazad treated groups. Asterisks (*) represent sample sites with significantly higher specific gravity as determined by one way ANOVA, and Scheffé's post hoc test, at 95% significance level

Successful intervention against ischemic brain edema would be a major clinical improvement since many stroke-related deaths in the acute phase can be attributed to brain edema. The mechanisms behind ischemic brain edema are complex[10]. In the early stage, water will move from the extracellular into the intracellular compartment due to insufficient membrane pump function and accumulation of metabolites within the cells. This early stage of intracellular edema is directly related to the degree of energy failure and metabolite production. Since brain endothelial cells are more resistant to ischemia than nerve cells, the BBB remains intact as to passage of larger molecules like serum proteins during the early stage. After a time that varies with the degree in duration of ischemia, plasma constituents that normally cannot enter the brain start leaking into the brain. Whereas it has earlier been thought that the early ischemic phase determines the fate of neurons, more recent studies indicate that vasogenic edema may be of main importance for tissue survival[18].

Hypertension might aggravate the early intracellular as well as the vasogenic component of ischemic brain edema. Because of compromised collateral circulation ischemia will be more severe distal to a ligation in hypertensive animals[5,7]. Thus, the brain specific gravity was significantly reduced two hours after a 30 min occlusion in SHR but not in normotensive Wistar-Kyoto rats[14]. When the blood-brain barrier is altered, passage of substances from plasma to the brain will be highly related to the perfusion pressure favouring a larger edema spread in hypertensive animals[11]. The fact that tirilazad has not influenced the infarct size in permanent MCA ligation in SHR[9], which is in agreement with our unpublished observations, could indicate that tirilazad preferentially reduces the vasogenic part of the edema. Whether the effect is predominantly on the ischemic brain tissue itself or on the brain endothelial cells or is a combination of both remains to be shown. In addition to the free radical synthesis within the parenchyma, free radicals form in the brain endothelial cells during ischemia[1].

Our results show that tirilazad can also reduce brain edema after permanent ligation in hypertensive rats. Even if no reduction in infarct size occurs, this observation has importance for two reasons. Firstly, high intracranial pressure due to brain edema is an important cause of death in patients who die in the acute stage. Secondly, brain edema can induce permanent cellular lesions outside the infarcted area[13].

Acknowledgements

This work was supported by grants from the Swedish Medical Research Council (14X–4968), King Gustaf V and Queen Victoria's Foundation, Thorsten och Elsa Segerfalk's Foundation for Medical Research and Thuring's Foundation.

References

1. Asano T, Gotoh O, Koide T, Takakura K (1985) Ischemic brain edema following occlusion of the middle cerebral artery in the rat. II: alteration of the eicosanoid synthesis profile of brain microvessels. Stroke 16: 110–113
2. Braughler JM, Pregenzer JF, Chase RL, Duncan LA, Jacobsen EJ, McCall JM (1987) Novel 21-aminosteroids as potent inhibitors of iron-dependent lipid peroxidation. J Biol Chem 262: 10438–10440
3. Bågenholm R, Andiné P, Hagberg H, Kjellmer I (1991) Effect of 21-aminosteroid U74006F on brain damage and edema following perinatal hypoxia-ischemia in the rat. J Cereb Blood Flow Metab 11 [Suppl 2]: S 134
4. Chan PH, Fishman RA, Wesley MA, Longar S (1990) Pathogenesis of vasogenic edema in focal cerebral ischemia: role of superoxide radical. Adv Neurol 52: 177–183
5. Coyle P (1984) Outcomes to middle cerebral artery occlusion in hypertensive and normotensive rats. Hypertension 6 [Suppl I]: 169–174
6. Fujiwara K, Nitch K, Suzuki R, Klatzo I (1981) Factors in reproducibility of the gravimetric method for evaluation of edematous changes in the brain. Neurol Res 3: 345–361
7. Grabowski M, Nordborg C, Brundin P, Johansson BB (1988) Middle cerebral artery occlusion in the hypertensive and normotensive rat: a study of histopathology and behaviour. J Hypertens 6: 405–411
8. Hall ED, Braughler JM (1989) Central nervous system trauma and stroke. II. Physiological and pharmacological evidence for involvement of oxygen radicals and lipid peroxidation. Free Radic Biol Med 6: 303–313
9. Hall ED, Pazara KE, Braughler JM (1991) Effects of tirilazad mesylate on postischemic brain lipid peroxidation and recovery of extracellular calcium in gerbils. Stroke 22: 361–366
10. Johansson BB (1986) Brain edema. Acta Neurochir (Wien) [Suppl] 36: 137–141
11. Klatzo I, Wisniewski H, Steinwall O, Streicher E (1976) Dynamics of cold injury edema. In: Klatzo I, Seitelberger F (eds) Brain edema. Springer, Berlin Heidelberg New York, pp 554–563
12. Nelson SR, Mantz ML, Maxwell JA (1971) Use of specific gravity in the measurement of cerebral edema. J Appl Physiol 30: 268–271
13. Nordborg C, Sokrab TOD, Johansson BB (1991) The relationship between plasma protein extravasation and remote tissue changes after experimental brain infarction. Acta Neuropathol (Berl) 82: 118–126
14. Olsson A-L, Westergren I, Johansson B (1989) Brain edema after middle cerebral artery occlusion. A comparison between normotensive and spontaneously hypertensive rats. Acta Neurol Scand 80: 12–16
15. Picozzi P, Todd NV, Crockard AH (1985) The role of cerebral blood volume changes in brain specific-gravity measurements. J Neurosurg 62: 704–710
16. Siesjö BK (1992) Pathophysiology and treatment of focal cerebral ischemia. Part II: mechanisms of damage and treatment. J Neurosurg 77: 337–354

17. Tamura A, Graham DI, McCulloch J, Teasdale GM (1981) Focal cerebral ischemia in the rat. 1. Description of technique and early neuropathological consequences following middle cerebral artery occlusion. J Cereb Blood Flow Metab 1: 53–60

18. Ting P, Masaoka H, Kuroiwa T, Wagner HG, Fenton I, Klatzo I (1988) Influence of the blood-brain barrier opening to proteins on development of post-ischemic brain injury. Neurol Res 8: 146–151

19. Xue D, Slivka A, Buchan AM (1992) Tirilazad reduces cortical infarction after transient but not after permanent focal cerebral ischemia in rats. Stroke 23: 984–899

20. Young W, Wojak JC, DeCrescito V (1988) 21 aminosteroids reduces ion shifts and edema in the rat middle cerebral artery occlusion model of regional ischemia. Stroke 19: 1013–1019

Correspondence: Babro Johansson, M. D., Ph. D., Department of Neurology, University Hospital, S-221 85 Lund, Sweden.

Acta Neurochir (1994) [Suppl] 60: 314–317

Riboflavin Reduces Edema in Focal Cerebral Ischemia

A. L. Betz[2], X. D. Ren[1], S. R. Ennis[1], and D. E. Hultquist[3]

Departments of Surgery (Neurosurgery), [1] Pediatrics, [2] Neurology, and [3] Biological Chemistry, University of Michigan, Ann Arbor, MI, U.S.A.

Summary

Oxidized iron has been proposed as a mediator of the free radical-induced damage that occurs during cerebral ischemia. Dihydroriboflavin, a compound produced from riboflavin (B_2) by NADPH-dependent flavin reductase, rapidly reduces oxidized iron. Since treatment with riboflavin offers protection from ischemic injury in other tissues, we tested the effect of pretreatment with B_2 on brain edema formation during focal ischemia.

Two different models of middle cerebral artery occlusion (MCAO) in rats were tested: transcranial electrocautery and intra-carotid occlusion with a nylon thread. Groups of 6–8 animals were treated with 7.5 mg of B_2/kg or saline vehicle 1 h before MCAO and brain water content was determined after 4 h of ischemia. Pretreatment with B_2 reduced total hemisphere edema formation from 0.37 ± 0.05 to 0.19 ± 0.05 ml/g dry wt. (48% protection, $p < 0.01$) following transcranial MCAO. Edema was greater following MCAO with the intra-carotid thread (0.54 ± 0.05 ml/g) but protection by B_2 was less (21%).

We conclude that pretreatment with B_2 reduces ischemic brain injury, perhaps by reacting with oxidized iron. However, the larger stroke produced by the thread MCAO method makes it more difficult to observe protection following brief ischemia in this model.

Keywords: Ischemia; middle cerebral artery occlusion; riboflavin; vitamin B_2.

Introduction

Oxygen free radicals have been implicated in the brain injury that occurs during focal ischemia[4,9] and oxidized iron has been proposed as a mediator of the damage produced by free radicals[2]. Dihydroriboflavin, a compound produced from riboflavin (vitamin B_2) through the action of an NADPH-dependent flavin reductase, rapidly reduces hemeproteins containing higher oxidation states of iron[3]. Since treatment with riboflavin (B_2) offers protection from ischemic injury in other tissues[3], one goal of this study was to determine whether B_2 can reduce brain edema formation during focal ischemia. A second goal of this study was to compare the sensitivity for detecting cerebral protection of two different methods of producing focal ischemia in rats, proximal middle cerebral artery occlusion (MCAO) by trans-cranial electrocautery[1] and the recently-described intra-carotid thread MCAO[5].

Materials and Methods

Groups of 6–8 adult male Sprague Dawley rats were pretreated 1 h prior to MCAO. B_2 was given at doses of 7.5 mg/kg or 1.5 mg/kg (50 ml/kg or 10 ml/kg of a solution containing 150 mg/l of B_2 in 0.9% NaCl). Vehicle-treated controls received the same volume of saline. Additional control groups were treated with 7.5 mg/kg or 1.5 mg/kg of flavin mononucleotide (FMN), the phosphorylated form of B_2 that penetrates the blood-brain barrier (BBB) poorly[7,8].

Animals were anesthetized with Ketamine (50 mg/kg) and xylazine (10 mg/kg) given i.m. In some, focal cerebral ischemia was produced by MCAO according to the method of Bederson et al.[1] as described previously[6]. In the other half, MCAO was produced by passing a 4-0 nylon suture into the internal carotid artery and advancing it past the origin of the MCA until it lodged in the proximal anterior cerebral artery as described by Longa et al.[5]. Following MCAO, the animal's body temperature was maintained at 37°C with a heating blanket until fully recovered from anesthesia.

About 3.5 h after MCAO, rats were reanesthetized and a catheter was placed in the femoral artery for monitoring blood pressure and sampling arterial blood. Exactly 4 h after MCAO, the animals were killed by decapitation, the brains were removed, and the cortical mantels from the ischemic and non-ischemic cerebral hemispheres were dissected into 3 concentric samples using 7 and 10 mm cork borers as described previously[6]. Brain samples were placed in pre-weighed crucibles, re-weighed, dried for 48 h at 100°C and weighed once again. The water content (ml/g dry wt) of each sample was calculated as the difference between the wet and dry weights divided by the dry weight. The water content of the entire cortex was calculated using the sums of the weights of the 3 different zones. The difference in the water content between the corresponding samples from the ischemic and non-ischemic cortex (Δ water content) was used as a measure of the amount of edema formation.

Data are reported as means $\pm$ SE. Results from drug-treated animals were compared to those of the corresponding vehicle-treated control group using a one-tailed Student's t test.

Table 1. *Effect of Riboflavin and Flavin Mononucleotide on Ischemic Brain Edema*

	n	Difference in water content between ischemic and non-ischemic cortex (ml/g dry wt.)		
		Center zone	Intermediate zone	Outer zone
Trans-cranial MCAO				
vehicle	6	0.692 ± 0.059	0.344 ± 0.126	0.176 ± 0.068
7.5 mg riboflavin/kg	6	0.506 ± 0.026^{c}	0.041 ± 0.020^{b}	0.058 ± 0.028
vehicle	7	0.473 ± 0.073	0.136 ± 0.053	0.033 ± 0.047
7.5 mg FMN/kg	6	0.432 ± 0.091	0.092 ± 0.079	0.038 ± 0.043
vehicle	6	0.634 ± 0.030	0.212 ± 0.068	0.054 ± 0.032
1.5 mg riboflavin/kg	6	0.647 ± 0.047	0.139 ± 0.076	0.001 ± 0.027
vehicle	6	0.620 ± 0.031	0.230 ± 0.062	0.060 ± 0.063
1.5 mg FMN/kg	6	0.577 ± 0.051	0.165 ± 0.068	0.098 ± 0.043
Thread MCAO				
vehicle	8	0.733 ± 0.063	0.717 ± 0.041	0.334 ± 0.078
7.5 mg riboflavin/kg	8	0.769 ± 0.018	0.579 ± 0.062^{a}	0.149 ± 0.073^{a}
vehicle	6	0.753 ± 0.013	0.598 ± 0.060	0.299 ± 0.038
7.5 mg FMN/kg	6	0.678 ± 0.055	0.484 ± 0.085	0.249 ± 0.058

Values are means $\pm$ SE.
[a] $p < 0.05$, [b] $p < 0.02$, [c] $p < 0.01$ compared to vehicle-treated control group.

Results

Mean values of physiological parameters were in the normal range for all groups. The only values that differed between trans-cranial and thread occlusion models were the mean arterial blood pressures which were slightly lower in the thread occlusion group (89 ± 2 vs. 97 ± 2 mmHg, $p < 0.01$).

Following MCAO by the trans-cranial method, brain edema was most severe in the center of the infarct and less severe in the intermediate and outer zones (Table 1). Pre-treatment with 7.5 mg of B_2/kg significantly reduced brain edema formation in the center and intermediate zones. However, the reduction was not significant in the outer zone where the amount of edema was small (Fig. 1). Overall, B_2 decreased total hemisphere edema formation from 0.37 ± 0.08 ml/g to 0.19 ± 0.02 ml/g, a decrease of 48% ($p < 0.01$). Treatment with a lower dose of B_2 (1.5 mg/kg) did not significantly reduce brain edema formation, nor did treatment with FMN at either 7.5 or 1.5 mg/kg (Table 1).

Following MCAO by the thread method, the total edema accumulation was greater than that seen following trans-cranial MCAO (e.g. 0.54 ± 0.05 ml/g in thread and 0.37 ± 0.08 ml/g in transcranial control groups, $p < 0.05$). Further evidence for a much larger stroke following thread MCAO was found in the large amount of edema present in the intermediate sample which was nearly equal to the edema seen in the center zone (Table 1, Fig. 1). With this model, a significant reduction of edema formation was noted in the intermediate and outer zones following 7.5 mg of B_2/kg (Fig. 1). Overall, however, total cortical edema formation was reduced by only 21% (from 0.54 ± 0.05 to 0.43 ± 0.05 ml/g) and this difference was not quite significant ($p < 0.1$). FMN had no significant effect on the formation of ischemic edema in this model.

Discussion

NADPH-dependent flavin reductase is an enzyme that was first identified in erythrocytes based upon its ability to reduce methemoglobin in the presence of methylene blue[10]. More recently, the enzyme has been shown to reduce riboflavin and the resulting dihydroriboflavin can, in turn, reduce methemoglobin[11] as well as hemeproteins which contain iron in the higher oxidation states (e.g., Fe(IV) and Fe(V))[3]. In the process, dihydroriboflavin is converted back to riboflavin and is available to react again with flavin reductase in a redox cycle.

Since hemeproteins that contain oxidized iron may contribute to cellular injury in a variety of pathological conditions, Hultquist *et al.*[3] hypothesized that flavins may protect cells from oxidative injury by reducing oxidized hemeproteins. B_2 is a logical choice for exogenous therapy since it readily enters cells, while FMN should not be very effective because it doesn't. Our initial studies demonstrated that B_2 therapy was effective in reducing cellular injury in 3 different models:

A

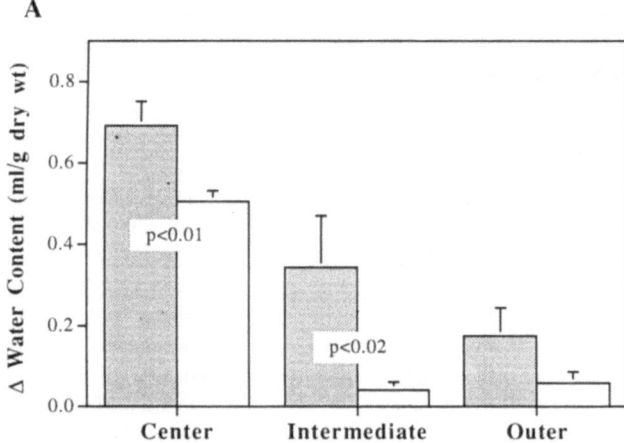

B

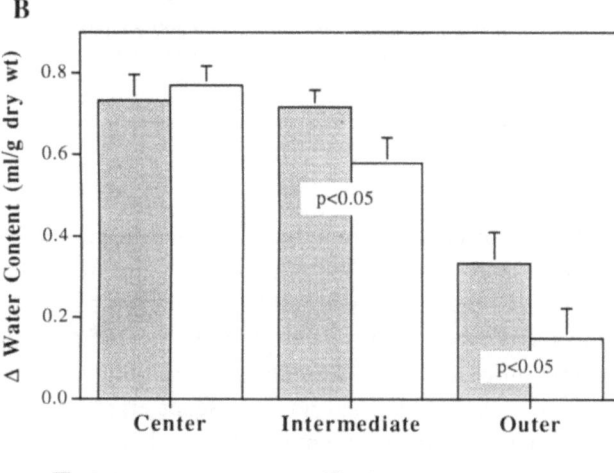

☐ Saline (50 ml/kg)   ☐ Riboflavin (7.5 mg/kg)

Fig. 1. Effect of riboflavin on brain edema formation following MCAO. Saline vehicle (shaded bars) or 7.5 mg/kg of riboflavin (open bars) was administered to groups of 6–8 animals 1 h prior to MCAO produced by either trans-cranial electrocautery (A) or an intracarotid thread (B). Brain water content of the ischemic and non-ischemic cerebral cortex was measured 4 h after MCAO in the center, intermediate, and outer zones of the infarct. The difference between the ischemic and non-ischemic samples (Δ water content) was used as a measure of brain edema formation. Values are means ± SE

1) reoxygenation-induced injury in isolated, saline-perfused rabbit hearts, 2) lung injury induced by complement-activated neutrophils in rats, and 3) focal cerebral ischemia in rats[3].

This study extends our previous preliminary report of cerebral protection by B_2. FMN was used as a control for non-specific systemic effects of flavins since FMN does not cross the BBB[7,8]. Our results show that B_2 was effective in reducing brain edema at a dose of 7.5 mg/kg but not 1.5 mg/kg. Because of the poor solubility of B_2, these doses required the administration of large volumes of saline (10–50 ml/kg). However, the vehicle

was well tolerated in both B_2 and saline-treated animals since the physiological parameters were normal. As anticipated, FMN, even at the high dose, was without effect. This is consistent with the fact that the BBB is intact after 4 hrs of focal ischemia[6]. Thus, B_2 must have either interacted with brain cells to reduce cellular injury or reduced ischemia by improving CBF.

This study also compared the ability to detect cerebral protection in 2 different models of rat MCAO. Passage of an intra-carotid nylon thread resulted in a greater amount of cortical brain edema after 4 h of ischemia, most likely because the MCA was occluded at its origin, thereby stopping flow in proximal MCA branches that are not occluded by the trans-cranial approach. Not only did the edema appear greater in the center of the infarct, but severe edema encompassed a larger amount of the cortex. This suggests that the penumbra is smaller in the thread model and, therefore, the volume of tissue that can be protected in permanent ischemia may be smaller. Indeed, the protection that was seen following B_2 administration to animals subjected to thread MCAO was smaller and involved the more peripheral tissue. Thus, the thread MCAO model may not be optimal for testing the effects of potential cerebroprotective agents because the stroke is large and the salvageable penumbra is small. The thread model, however, is likely to provide a convenient and easily-reproducible model of reperfusion injury since blood flow can be re-established by withdrawal of the thread[5].

Acknowledgements

This work was supported by grants from the National Institutes of Health (NS-2387 and AG-07046) and the Office of the Vice President for Research, University of Michigan.

References

1. Bederson JB, Pitts LH, Tsuji M, Nishimura MC, Davis RL, Bartkowski H (1986) Rat middle cerebral artery occlusion: evaluation of the model and development of a neurologic examination. Stroke 17: 472–476
2. Halliwell B, Gutteridge JMC (1984) Oxygen toxicity, oxygen radicals, transition metals and disease. Biochem J 219: 1–14
3. Hultquist DE, Xu F, Quandt KS, Shlafer M, Mack CP, Till GO, Seekamp A, Betz AL, Ennis SR (1993) Evidence that NADPH-dependent methemoglobin reductase and administered riboflavin protect tissues from oxidative injury. Am J Hematol 42: 13–18
4. Ikeda Y, Long DM (1990) The molecular basis of brain injury and brain edema: the role of oxygen free radicals. Neurosurgery 27: 1–11
5. Longa EZ, Weinstein PR, Carlson S, Cummins R (1989) Reversible middle cerebral artery occlusion without craniectomy in rats. Stroke 20: 84–91

6. Martz D, Beer M, Betz AL (1990) Dimethylthiourea reduces ischemic brain edema without affecting cerebral blood flow. J Cereb Blood Flow Metab 10: 352–357

7. Nagatsu T, Nagatsu-Ishibashi I, Okuda J, Yagi K (1967) Incorporation of peripherally administered riboflavine into flavine nucleotides in the brain. J Neurochem 14: 207–210

8. Spector R (1980) Riboflavin homeostasis in the central nervous system. J Neurochem 35: 202–209

9. Traystman RJ, Kirsch JR, Koehler RC (1991) Oxygen radical mechanisms of brain injury following ischemia and reperfusion. J Appl Physiol 71: 1185–1195

10. Warburg O, Kubowitz F, Christian W (1930) Über die katalytische Wirkung von Methylenblau in lebenden Zellen. Biochem Z 227: 245–271

11. Yubisui T, Takeshita M, Yoneyama Y (1980) Reduction of methemoglobin through flavin at the physiological concentration by NADPH-flavin reductase of human erythrocytes. J Biochem (Tokyo) 87: 1715–1720

Correspondence: A. Lorris Betz, M.D., Ph.D., D3227 Medical Professional Building, University of Michigan, Ann Arbor, MI 48109-0718, U.S.A.

Acta Neurochir (1994) [Suppl] 60: 318–320
© Springer-Verlag 1994

Effect of YM737, a New Glutathione Analogue, on Ischemic Brain Edema

O. Gotoh, M. Yamamoto[1], A. Tamura[1], and K. Sano[1]

Department of Neurosurgery, Toshiba Hospital and [1] Department of Neursurgery, Teikyo University School of Medicine, Tokyo, Japan

Summary

We investigated the effect of YM737, a monoester of glutathione (GSH), on brain edema and GSH content after occlusion of a middle cerebral artery (MCA) in the rat. The drug possesses stronger radical scavenging activity than GSH itself, and is more effectively transported into cells. Hemispheric water, sodium, and potassium contents were determined at 2.5 and 24 hours after MCA occlusion. The animals received either YM737 or GSH immediately after occlusion. Cerebral GSH content was measured by HPLC after 2.5 hours of ischemia. The increases in water and sodium contents at 2.5 and 24 hours after MCA occlusion were significantly suppressed by YM737. GSH content decreased by 53% in the caudate, and by 22% in the cortex after ischemia. YM737 significantly ameliorated the GSH decrease in the caudate, while administration of GSH showed little effects on the ischemia-induced changes in water, sodium, and GSH contents. The result suggests the role of free radicals in the pathogenesis of ischemic brain edema.

Keywords: Brain edema; cerebral ischemia; free radical; glutathione.

Introduction

Recent studies have implicated oxygen free radicals in ischemic brain edema. However, free radicals are extremely short-lived, and produced only in minute quantities. Therefore, it is technically difficult to measure their concentrations directly in vivo, making the free radical hypothesis controversial[3]. In the present study, we examined the hypothesis by measuring changes in glutathione (GSH), one of endogenous radical scavengers, in focal cerebral ischemia in the rat. We also studied the effect of YM737[N-(N-r-L-glutamyl-L-cysteinyl)glycine 1-isopropyl ester sulfate monohydrate] on the development of brain edema. YM737, a monoester of CSH, possesses stronger radical scavenging activity than GSH itself, and is more effectively transported into cells[1].

Materials and Methods

Male Sprague-Dawley rats, weighing approximately 300 grams, were anesthetized with 2% halothane. The proximal portion of the left middle cerebral artery (MCA) was permanently occluded by a microsurgical technique[2].

1. Measurements of Brain Water, Sodium, and Potassium Contents

Water content was determined by comparing wet and dry weights of cerebral hemispheres[2]. The wet weight (W) of each hemisphere was measured on a chemical balance within 90 seconds after decapitation. The dry weight (D) was obtained after the brain was desiccated in an oven at 105°C for 4 days. The water content of each hemisphere was calculated as follows:

$$\text{water content (\%)} = W - D / W \times 100$$

Table 1. *Effect of GSH on the Brain Water, Sodium, and Potassium Contents After 2.5 Hours of MCA Occlusion*

	Ipsilateral			Contralateral		
	H_2O (%)	Na (mEq/kg)	K (mEq/kg)	H_2O (%)	Na (mEq/kg)	K (mEq/kg)
Normal (8)	78.76 ± 0.07	249 ± 4	458 ± 3	78.66 ± 0.06	227 ± 4	471 ± 9
MCA-occluded – saline (8)	79.47 ± 0.06[a]	247 ± 5	469 ± 11	78.71 ± 0.08	228 ± 6	469 ± 13
MCA-occluded – GSH (8)	79.27 ± 0.06[b]	260 ± 7	463 ± 4	78.53 ± 0.04	238 ± 4	483 ± 5

Data are means ± SE. GSH was intraperitoneally given at the dose of 300 mg/kg three times after induction of ischemia.
[a] $p < 0.001$: normal vs MCA-occluded saline groups; [b] $p < 0.05$: saline- vs GSH-treated groups.

The cation contents of each hemisphere were measured by flame photometry, and expressed as mEq/kg, dry weight. The rats received either YM737 or GSH (300 mg/kg, i.p.), immediately, 1, and 2 hours after MCA occlusion. Control rats were given the same volume of saline. The animals were decapitated 150 minutes after vascular occlusion. In other series, rats received intravenous infusion of YM737 at the rate of 300 mg/kg/h for 3 hours after induction of ischemia, or at the rate of 50 mg/kg/h for 24 hours. Those animals were decapitated at 24 hours after MCA occlusion.

2. Measurement of Cerebral GSH Content

Rats were given either YM737 or GSH (300 mg/kg, i.p.), immediately, 1, and 2 hours after induction of ischemia. The rats were decapitated at 2.5 hours, and the sensori-motor cortex and caudate nucleus were removed. The GSH content was determined by high-performance liquid chromatography.

All data are presented as mean ± SE. Student's t test was used for the analysis of differences between each group. A p-value less than 0.05 was considered significant.

Results and Discussion

In the saline-control rats, water and sodium were observed to increase in the affected hemispheres at 2.5 and 24 hours after MCA occlusion, while potassium content showed a tendency to decrease. Administration of YM737 significantly reduced the increases in water

Table 2. *Effect of Administration of YM737 on GSH Concentration in the Caudate Nucleus and the Cortex After 2.5 Hours of MCA Occlusion*

	Caudate	Cortex
Normal (7)	1.58 ± 0.04	1.34 ± 0.03
MCA-occluded		
Saline (7)	0.74 ± 0.10[b]	1.04 ± 0.10[a]
YM737 (7)	1.14 ± 0.12[c]	1.12 ± 0.10[a]

Values are μmol/g, expressed as means ± SE.
[a] p < 0.05, [b] p < 0.001: comparison between normal and MCA-occluded saline groups; [c] p < 0.05: comparison between saline- and YM736-treated groups.

Table 3. *Effect of Administration of Exogenous GSH on the GSH Content in the Caudate After 2.5 Hours of MCA Occlusion*

	GSH concentration	
	Ipsilateral	Contralateral
Normal (8)	1.76 ± 0.05	1.76 ± 0.07
MCA-occluded		
Saline (8)	1.21 ± 0.10[a]	1.55 ± 0.09
GSH (8)	1.30 ± 0.09[a]	1.68 ± 0.08

Values are μmol/g, expressed as means ± SE.
[a] p < 0.01: difference from normal controls. There is no significant difference between saline- and GSH-treated rats.

and sodium at 2.5 hours (Fig. 1), but GSH showed little effect on the increases (Table 1). Increased water and sodium contents after 24 hours of MCA occlusion were suppressed by the intravenous infusion of YM737 at 300 mg/kg/h for 3 hours, but not by the infusion of the drug at 50 mg/kg/h for 24 hours (Fig. 2). Such a difference in the drug effect suggests that pharmacological intervention be initiated in the early stage of ischemia at full doses before irreversible damage occurs.

Following 2.5 hours of MCA occlusion, GSH concentration decreased by 53% in the caudate, and by 22% in the cortex (Table 2). YM737 significantly ameliorated the GSH decrease in the caudate (Table 2), while exogenous GSH showed no effect on the decrease (Table 3). YM737 caused no change of the GSH content in the contralateral caudate, though the concentration of

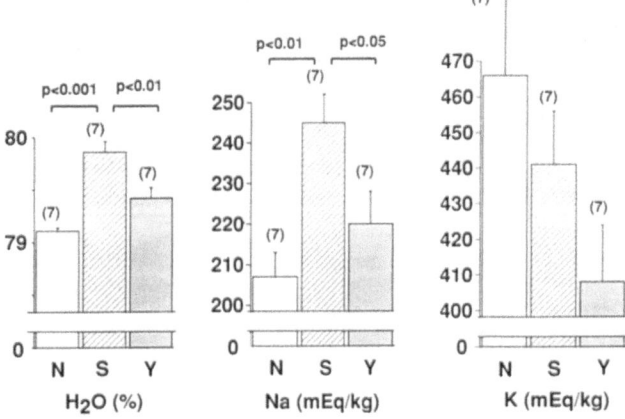

Fig. 1. Effect of intraperitoneally administered YM737 on the brain water, sodium, and potassium contents after 2.5 hours of MCA occlusion. *N* normal, *S* MCAO-saline, *Y* MCAO-YM737

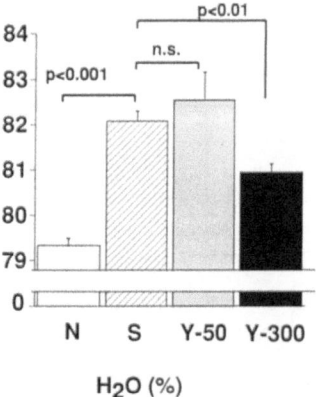

Fig. 2. Effect of intravenously administered YM737 on the brain water content after 24 hours of MCA occlusion. *N* normal, *Y-50* MCAO-YM737 (50 mg/kg/h for 24 h), *Y-300* MCAO-YM737 (300 mg/kg/h for 24 h)

YM737 in the contralateral caudate was twice as high as that in the ipsilateral caudate(10.9 vs 5.0 µg/g). It can be assumed that YM737 mitigated ischemic edema, directly by scavenging radicals, or indirectly by lessening the decrease in GSH as an endogenous scavenger. In conclusion, YM737, a monoester of GSH with potent radical scavenging activity and high permeability through cell membranes, ameliorated edema development and CSH decrease in focal cerebral ischemia in the rat. The result suggests the role of free radicals in the pathogenesis of ischemic brain edema.

References

1. Anderson ME, Porrie F, Puri RN, Meister A (1985) Glutathione monoethyl ester: preparation, uptake by tissues, and conversion to glutathione. Arch Biochem Biophys 239: 538–548
2. Gotoh O, Asano T, Koide T, Takakura K (1985) Ischemic brain edema following occlusion of the middle cerebral artery in the rat. Stroke 16: 101–109
3. Schmidley JW (1990) Free radicals in central nervous system ischemia. Stroke 21: 1086–1090

Correspondence: Osamu Gotoh, M.D., Department of Neurosurgery, Toshiba Hospital, 6-3-22 Higashi-Ohi, Shinagawa-ku, Tokyo 140, Japan.

Acta Neurochir (1994) [Suppl] 60: 321–324
© Springer-Verlag 1994

Comparison of the Effects of Glycerol, Mannitol, and Urea on Ischemic Hippocampal Damage in Gerbils

K. Otsubo, Y. Katayama, F. Kashiwagi, H. Muramatsu, and **A. Terashi**

Second Department of Internal Medicine, Nippon Medical School, Tokyo, Japan

Summary

The effects of glycerol and mannitol, as well as urea, on delayed neuronal death (DND) in the gerbil hippocampus were investigated. 20% solution of glycerol, mannitol and urea were prepared, and 6.5 ml/kg of each agent, or saline, was administered to male Mongolian gerbils intraperitoneally 30 min before ischemia. The animals were subjected to transient forebrain ischemia for 5 min. Seven days after the ischemic insult, the brains were fixed and stained for histopathological analysis. The number of normal neurons (neuronal density, ND) in a 1 mm linear length of hippocampal CA1 region was counted. ND of sham-operated group (n = 6) was 275.3 ± 16.7 (mean $\pm$ SD). ND in the saline-treated group (n = 6) was 14.8 ± 5.0. ND of groups treated with glycerol (n = 6), mannitol (n = 6) and urea (n = 4) was 68.2 ± 56.7 ($p < 0.01$), 52.8 ± 54.4 ($p < 0.01$) and 12.0 ± 2.5 (NS), respectively. The present study demonstrates that glycerol and mannitol have some protective effects against DND in the gerbil hippocampus, whereas urea has no effect.

Keywords: Glycerol; mannitol; urea; delayed neuronal death.

Introduction

Transient forebrain ischemia in animals produces delayed neuronal death (DND) in the hippocampal CA1 region[10]. An analogous phenomenon is known to occur also in the human hippocampus[15], and the lesion in CA1 causes amnesia[21]. Therefore, it is important to evaluate whether clinically useful agents protect CA1 neurons from ischemia. Glycerol and mannitol are clinically used as hyper-osmolar agents for the treatment of brain edema. The effects of these agents on acute cerebral infarction have been well documented, however, the effects of these agents to DND in the hippocampus following brief transient ischemia have not yet been studied. In the present study, the effects of glycerol and mannitol, as well as urea, on DND in the gerbil hippocampal CA1 region were examined.

Methods

Twenty-eight male Mongolian gerbils weighing 60–70 g were used. The experimental groups were as follows: (A) Sham-operated group (n = 6); (B) Ischemia-reperfusion group pretreated with saline (n = 6); (C) Ischemia-reperfusion group pretreated with glycerol (n = 6); (D) Ischemia-reperfusion group pretreated with mannitol (n = 6); (E) Ischemia-reperfusion group pretreated with urea (n = 4). The animals were fasted overnight and anesthetized with a mixture of 1.5% halothane, 70% N_2O, and 3% O_2. Afterward the bilateral common carotid arteries were exposed and halothane was discontinued. Saline, 20% glycerol, 20% mannitol and 20% urea were administered, by the dose of 6.5 ml/kg each, intraperitoneally to the gerbils 30 min before ischemia. They were subjected to transient forebrain ischemia for 5 min by bilateral common carotid artery occlusion. Sham operation was carried out in six animals as normal control. Skull and rectal temperatures were maintained at 36.5 ± 0.5 °C from 30 min prior to occlusion through 60 min post ischemia. Seven days after the ischemic insult, the brains were perfusion fixed with a 4% formaldehyde solution, and 6 μm-thick brain coronal sections were stained with hematoxylin and eosin for histopathological analysis. The number of normal neurons (neuronal density, ND) in a 1 mm linear length of the hippocampal CA1 region was counted under a light microscope. Statistical comparisons were made with a two-tailed Mann-Whitney U-test.

Results

No necrotic CA1 pyramidal cells were observed in the sham-operated group (Fig. 1A). The hippocampal CA1 regions in the ischemic animals are shown in Fig. 1B–D. In the saline-treated animals (Fig. 1B), normal CA1 neurons were hardly recognized and ischemic neuronal injury characterized by cell shrinkage, vacuolation, cytoplasmic eosinophilia and nuclear pyknosis was apparent. In the groups treated with glycerol (Fig. 1C) and mannitol (Fig. 1D), some normal neurons were observed, but in the urea-treated group (Fig. 1E), few normal pyramidal cells survived.

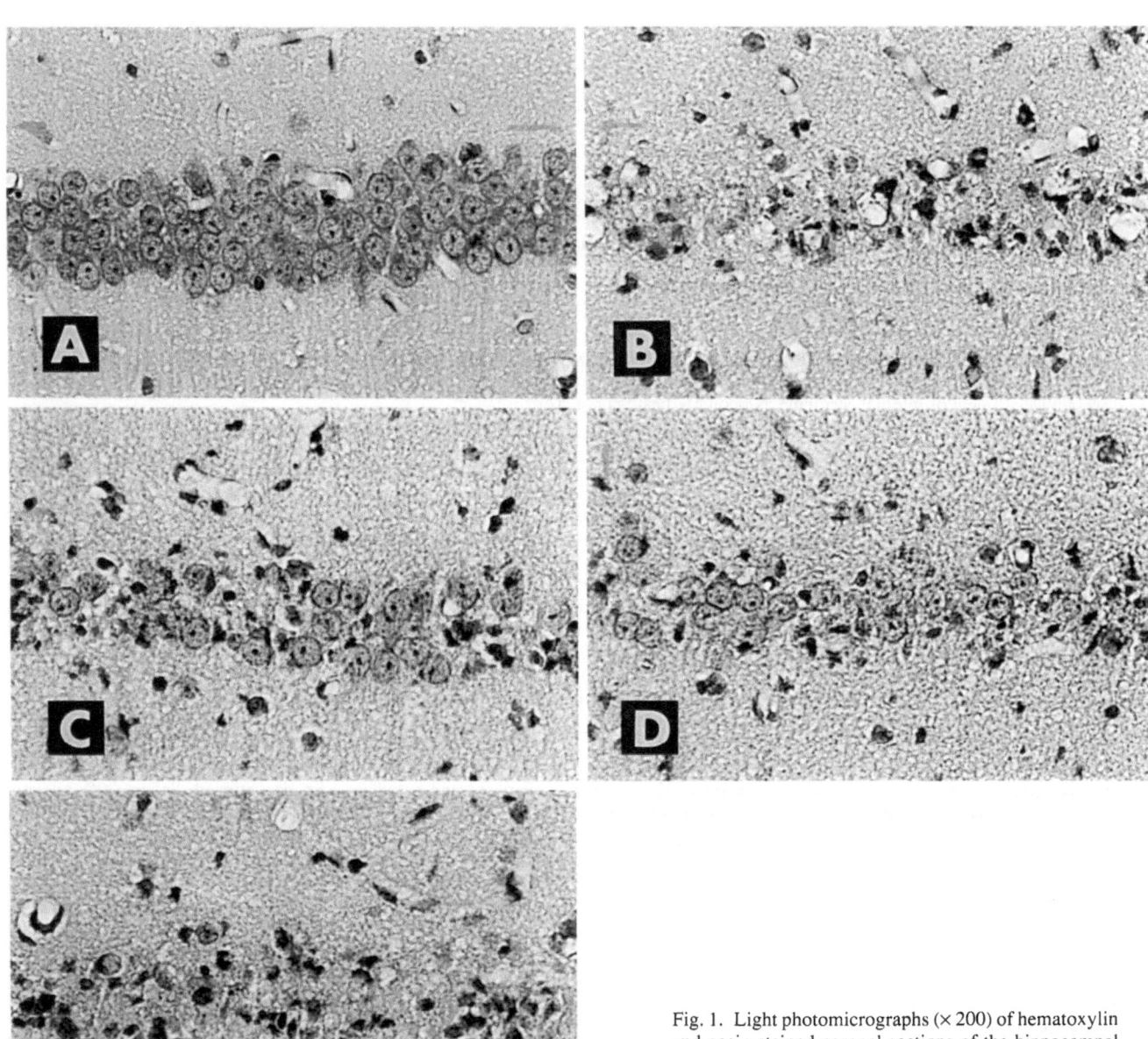

Fig. 1. Light photomicrographs ($\times$ 200) of hematoxylin and eosin-stained coronal sections of the hippocampal CA1 region in gerbils. (A) Sham-operated normal control. (B–E) Ischemia-reperfusion groups pretreated with saline (B), glycerol (C), mannitol (D) and urea (E). Note preservation of some neurons in glycerol and mannitol-treated groups

ND in the sham-operated group was 275.3 $\pm$ 16.7 (mean $\pm$ SD). In the ischemic animals treated with saline, ND was markedly decreased to 14.8 $\pm$ 5.0* (*p < 0.0001, compared with sham-operated control). The groups treated with glycerol, mannitol and urea showed 68.2 $\pm$ 56.7**, 52.8 $\pm$ 54.4** and 12.0 $\pm$ 2.5, respectively (**p < 0.01, compared with saline-treated group). There was no statistically significant difference in ND between the saline and the urea-treated groups (Fig. 2).

Discussion

Glycerol and mannitol have been used clinically as hyperosmolar agents for the treatment of brain edema. The present study showed that glycerol and mannitol had some protective effects against DND in the gerbil hippocampal CA1 region, while, urea had no effect to protect CA1 neurons. The influence of glycerol and mannitol on excitatory amino acid release or calcium influx mechanism, which are believed to be the major

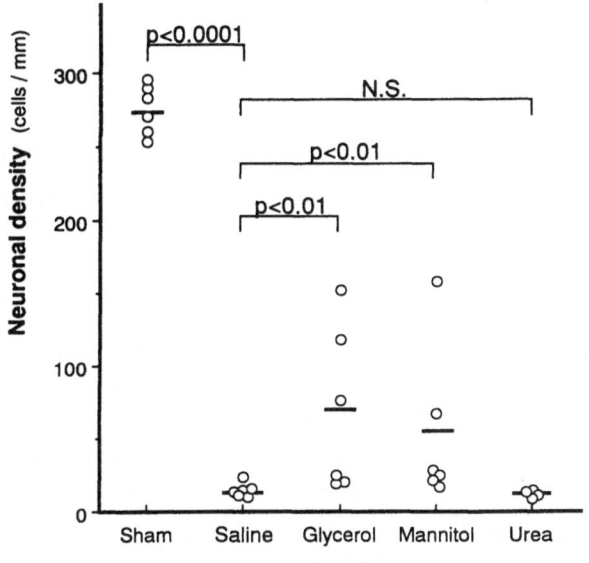

Fig. 2. Neuronal density (cell counts per 1mm linear length in the hippocampal CA1 region) in each treatment group. Statistical comparisons were made with a two-tailed Mann-Whitney U-test. *N.S.* not significant.

causes of ischemic neuronal cell death, has never been reported. Consequently, it is difficult to explain why these agents had protective effect on DND. Reduction of cerebral edema[11,13] and scavenging action of free radicals[9,18] are some known effects common to glycerol and mannitol.

A previous study demonstrated that 5 min of transient forebrain ischemia in rats resulted in no apparent regional edema during recirculation[12]. In the gerbil model, there has been no precise study of the time course of brain edema during an early recirculation period following 5 min of ischemia. It is known that severe ischemia is accompanied by a rapid decrease in extracellular sodium and chloride ion contents[5,7], and this is generally interpreted as a reflection of a cellular influx of sodium and chloride ions. This ion shift is accompanied by influx of water causing cellular edema. This process is triggered by depolarization which may involve glutamate neurotoxicity, and this finally leads to the influx of calcium and delayed death of the neuron[5]. It remains to be demonstrated that this process produces DND. If this process occurred in this model, glycerol and mannitol might have participated in the reduction of the cellular swelling and prevented the neuronal cell death.

Previous studies indicated that glycerol reduced the serum lipid peroxide level after transient cerebral ischemia in rats[9] and mannitol showed a hydroxyradi-

cal scavenging action[18]. In contrast, urea, which is metabolically inert, is not a free radical scavenger. Many reports have been described that free radical scavengers and lipid peroxidation inhibitors prevented DND in CA1[3,4,6,20]. In this study, therefore, glycerol and mannitol might ameliorate DND owing to their antioxidative ability.

In addition, circulatory effects of glycerol[14] and mannitol[2] may have contributed to the therapeutic effect. It has been shown that glycerol and mannitol improve microcirculation and improve the delayed hypoperfusion during the early reperfusion phase.

Although glycerol and mannitol significantly reduced neuronal damage in the hippocampal CA1 region, the improvement was not pronounced. This suggests that the effects of glycerol and mannitol are not enough to completely prevent the production and progression of DND under present experimental conditions.

An analogous damage to neurons in the hippocampal CA1 region following transient ischemia occurs also in the human brain[15]. Another study showed that the bilateral lesion limited in CA1 caused amnesia[21]. Because DND appears slowly after an ischemic episode, there may be a therapeutic window for clinical protection. Previously, some agents, such as N-methyl-D-aspartate receptor antagonists[17,19], calcium channel blockers[1,8] and free radical scavengers[3,4,6,20], have been considered to have protective effects on DND in experimental animals. The present study demonstrated that glycerol and mannitol had also protective effects on ischemic neuronal damage in the hippocampal CA1. Thus, these results may open a new category of protection, and may be clinically useful.

Conclusions

The present study demonstrates that glycerol and mannitol have protective effects against DND in the gerbil hippocampus, whereas urea has no effect. The pharmacological actions of glycerol and mannitol may be free radical scavenging along with improvement of microcirculation.

References

1. Alps BJ, Calder C, Hass WK, Wilson AD (1988) Comparative protective effects of nicardipine, flunarizine, lidoflazine and nimodipine against ischaemic injury in the hippocampus of the Mongolian gerbil. Br J Pharmacol 93: 877–883
2. Burke AM, Quest DO, Chien S, Cesare C (1981) The effect of mannitol on blood viscosity. J Neurosurg 55: 550–553

3. Clemens JA, Ho PPK, Panetta JA (1991) LY178002 reduces rat brain damage after transient global forebrain ischemia. Stroke 22:1048–1052

4. Hall ED, Pazara KE, Braughler JM (1988) 21-Aminosteroid lipid peroxidation inhibitor U74006F protects against cerebral ischemia in gerbils. Stroke 19: 997–1002

5. Hansen AJ, Zeuthen T (1981) Extracellular ion concentrations during spreading depression and ischemia in the rat brain cortex. Acta Physiol Scand 113: 437–445

6. Hara H, Kato H, Kogure K (1990) Protective effect of α - tocopherol on ischemic neuronal damage in the gerbil hippocampus. Brain Res 510: 335–338

7. Hossman K-A, Sakaki S, Zimmerman V (1977) Cation activities in reversible ischemia of the cat brain. Stroke 8: 77–81

8. Izumiyama K, Kogure K (1988) Prevention of delayed neuronal death in gerbil hippocampus by ion channel blockers. Stroke 19: 1003–1007

9. Kashiwagi F, Katayama Y, Suzuki S, Shimizu J, Nagazumi A, Terashi A (1988) Effect of glycerol administration on experimental cerebral ischemia. Part 1. Studies on lipid peroxides, prostaglandins, brain edema and brain metabolites. Brain Nerve 40: 179–185

10. Kirino T (1982) Delayed neuronal death in the gerbil hippocampus following ischemia. Brain Res 239: 57–69

11. Little JR (1978) Modification of acute focal ischemia by treatment with mannitol. Stroke 9: 4–9

12. Mellergård P, Bengtsson F, Smith M-L Riesenfeld Z, Siesjö BK (1989) Time course of early brain edema following reversible forebrain ischemia in rats. Stroke 20: 1565–1570

13. Meyer JS, Charney JZ, Rivera VM, Mathew NT (1971) Treatment with glycerol of cerebral edema due to acute cerebral infarction. Lancet 2: 993–997

14. Meyer JS, Fukuuchi Y, Shimazu K, Ohuchi T, Ericsson AD (1972) Effect of intravenous infusion of glycerol on hemispheric blood flow and metabolism in patients with acute cerebral infarction. Stroke 3:168–180

15. Petito CK, Feldmann E, Pulsinelli WA, Plum F (1987) Delayed hippocampal damage in humans following cardiorespiratory arrest. Neurology 37: 1281–1286

16. Rothmann SM, Olney JW (1986) Glutamate and pathophysiology of hypoxic-ischemic brain damage. Ann Neurol 19: 105–111

17. Simon RP, Swan JH, Griffiths T, Meldrum BS (1984) Blockade of N-methyl-D-aspartate receptors may protect against ischemic damage in the brain. Science 226: 850–852

18. Suzuki J, Imaizumi S, Kayama T, Yoshimoto T (1985) Chemiluminescence in hypoxic brain – the second report: cerebral protective effect of mannitol, vitamine E and glucocorticoid. Stroke 16: 695–700

19. Swan JH, Meldrum BS (1990) Protection by NMDA antagonists against selective cell loss following transient ischaemia. J Cereb Blood Flow Metab 10: 343–351

20. Uyama O, Matsuyama T, Michishita H, Nakamura H, Sugita M (1992) Protective effects of human recombinant superoxide dismutase on transient ischemic injury of CA1 neurons in gerbils. Stroke 23: 75–81

21. Zola-Morgan S, Squire LR, Amaral DG (1986) Human amnesia and the medial temporal region: Enduring memory impairment following a bilateral lesion limited to field CA1 of the hippocampus. J Neurosci 6: 2950–2967

Correspondence: K. Otsubo, M. D., Second Department of Internal Medicine, Nippon Medical School, 3-5-5 Iidabashi, Tokyo 102, Japan.

Acta Neurochir (1994) [Suppl] 60: 325–328

The Effect of Alkalizing Agents on Experimental Focal Cerebral Ischemia

H. Kuyama, T. Kitaoka, K. Fujita, and **S. Nagao**

Department of Neurological Surgery, Kagawa Medical School, Miki-cho, Kita-gun, Kagawa Prefecture, Japan

Summary

We investigated the immediate effect of tris (hydroxymethyl) aminomethane (THAM) and $NaHCO_3$ on focal cerebral ischemia produced by occlusion of the left middle cerebral artery (MCA) in cats. The animals were divided into three groups. In the control group, physiological saline was infused continuously. The THAM and $NaHCO_3$ groups received continuous administration of 0.3 mol THAM and 7% $NaHCO_3$, respectively, to normalize arterial pH. Local CBF measured in the marginal and suprasylvian gyri decreased less than 30 ml/100 g/min after the MCA occlusion and there were no significant differences among the three groups. Extracellular pH of the marginal gyrus (peri-infarct zone) decreased from 7.21 to 6.86 in the control group. However, extracellular pH did not show significant changes in the THAM and $NaHCO_3$ groups. Intracellular pH of the infarct area decreased from 7.23 to 6.13 in the control group within 6 hours after occlusion. THAM had a tendency to normalize intracellular pH compared with that in the control and $NaHCO_3$ groups. THAM significantly ($p < 0.05$) decreased water content of the gray matter in the marginal gyrus at 6 hours after occlusion and the infarct size compared with those in the control and $NaHCO_3$ groups. Therefore, normalization of systemic and perifocal acidosis with THAM is effective for reducing cortical edema and infarct size in the early stage of focal cerebral ischemia probably due to the improvement of intracellular acidosis.

Keywords: Focal cerebral ischemia; acidosis; THAM, $NaHCO_3$.

Introduction

Metabolic acidosis in cerebral ischemia is considered deleterious to cell function and neurologic recovery[8]. Amelioration of systemic and focal cerebral acidosis by hyperventilation and/or alkalizing agents may reduce ischemic brain damage. Muizelaar *et al.*[6] reported hyperventilation could not sustain alkalinization in the CSF, whereas tris (hydroxymethyl) aminomethane (THAM) could in patients with traumatic brain injury. However, the effect of THAM on focal cerebral ischemia remains unclear. We investigated the effect of THAM and $NaHCO_3$ on focal cerebral ischemia produced by occlusion of the middle cerebral artery in cats.

Materials and Methods

Fifty-five adult cats were anesthetized with ketamine hydrochloride (30 mg/kg) and artificially ventilated with room air. Muscle paralysis was obtained with intravenous pancuronium bromide (2 mg/kg). Catheters were placed in the femoral artery for pressure monitoring and blood sample withdrawal and in the femoral vein for drug infusion. PaO_2, $PaCO_2$, arterial pH and mean arterial blood pressure were maintained within their normal ranges. Focal cerebral ischemia was produced by coagulation of the left middle cerebral artery (MCA) using the transorbital approach.

Water content and local cerebral blood flow (lCBF) were measured by specific gravimetric method and hydrogen clearance method, respectively. Extracellular pH was measured by glass electrode (Kemikarukiki Co., CM-151) on the cerebral cortex (marginal gyrus). Intracellular pH was calculated with chemical shift of inorganic phosphate measured by magnetic resonance spectroscopy (^{31}p-MRS) using BEM-250/80 NMR system 2 Tesla (Otsuka Electronics, Japan). The surface coil, 2 cm in diameter, was placed on the skull of the left cerebral hemisphere. The measurement was repeated every hour for 6 hours after the MCA occlusion.

Six hours after focal ischemia the animals were sacrificed by KCl injection and the brains were coronally sectioned with 5 mm-thick slices at the level of the optic chiasm. The samples of gray and white matter were taken from ectosylvian, suprasylvian and marginal gyri for measurement of water content. The coronal slices of the brain were also immersed in a 2% solution of 2, 3, 5-triphenyltetrazolium chloride (TTC) in normal saline at 37°C for 30 min[2]. The surface of the TTC-stained sections was photographed and the area of infarction was expressed as a percentage of the whole coronal section.

The animals were divided into three groups. In the control group, physiological saline was infused continuously (2.0 ml/kg/hour). The THAM group received continuous administration of 0.3 mol THAM (2.0 ml/100 g/hour). In $NaHCO_3$ group 7% $NaHCO_3$ (0.7 ml/kg/hour) and physiological saline (1.3 ml/kg/hour) were infused continuously.

Results

Mean arterial blood pressure, PaO_2 and $PaCO_2$ did not show significant changes in the three groups. In the

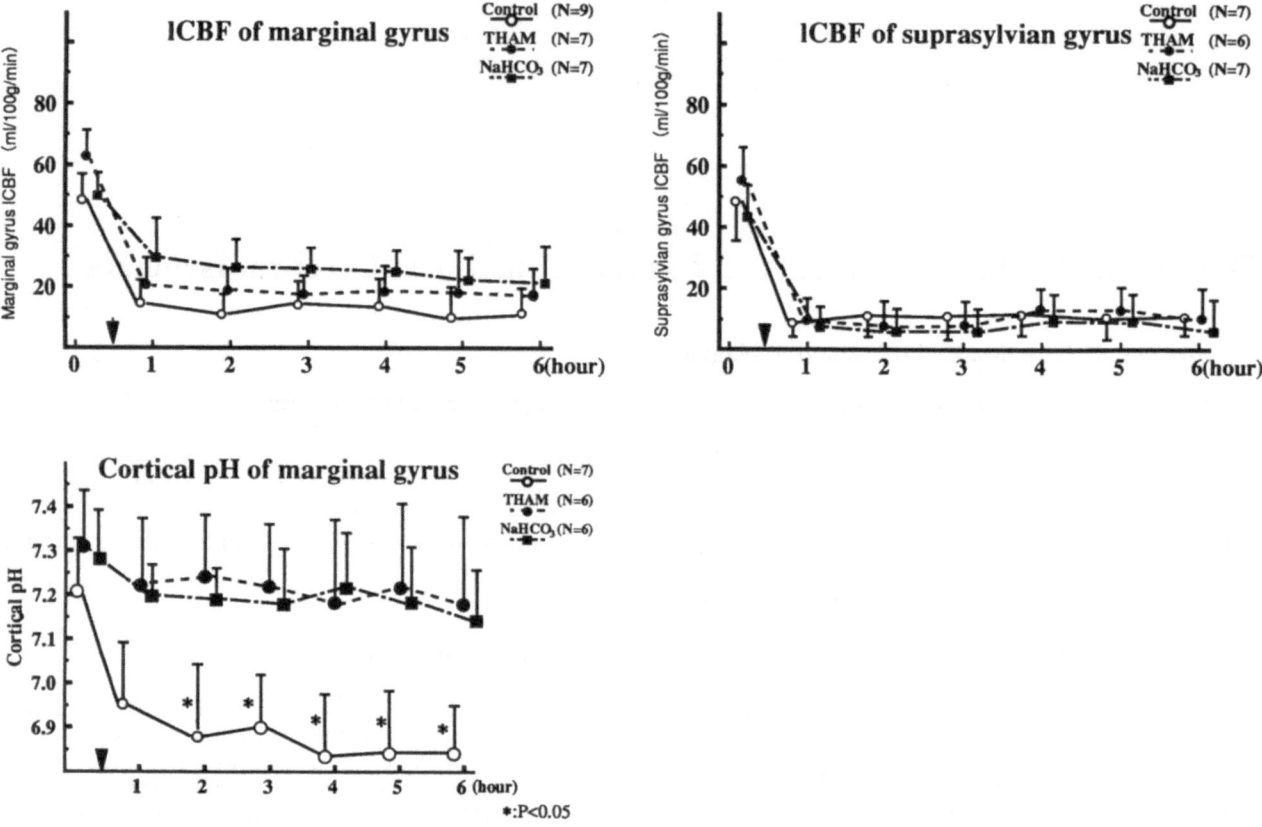

Fig. 1. Upper: Changes in lCBF. lCBF measured in the marginal gyrus (left) and suprasylvian gyrus (right) showed a prompt decrease after MCA occlusion; there were no significant differences among the three groups. Lower: Changes in extracellular pH. Extracellular pH of the marginal gyrus decreased in the control group. However, in the THAM and NaHCO₃ groups extracellular pH did not show significant changes

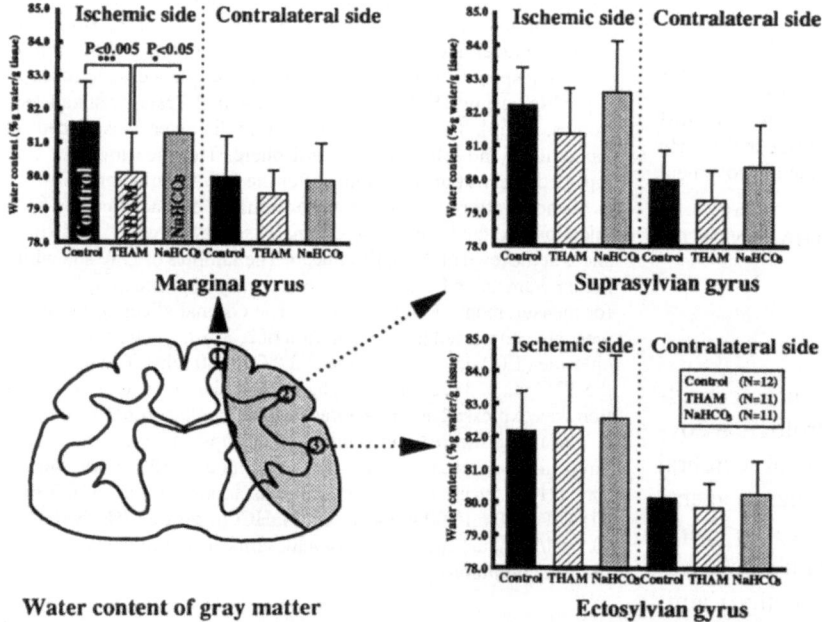

Fig. 2. Water content of gray matter at 6 hours after MCA occlusion. Lower left: Schematic representation of a coronal brain slice at the level of the optic chiasm. Sites of tissue samples for water content analysis are shown. Graph shows water content of gray matter. In the marginal gyrus, water content of gray matter in THAM group significantly decreased compared with that in control and NaHCO₃ groups. * p < 0.05, *** p < 0.005

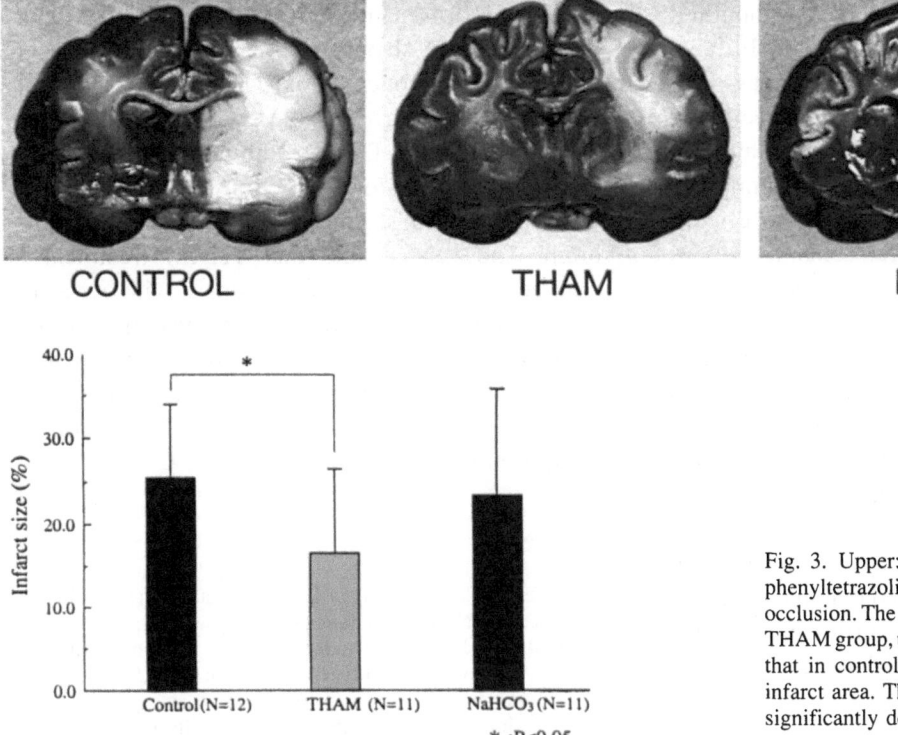

CONTROL   THAM   NaHCO$_3$

Fig. 3. Upper: Coronal brain slices stained with tri-phenyltetrazolium chloride (TTC) 6 hours after MCA occlusion. The infarct area was not stained with TTC. In THAM group, the infarct area decreased compared with that in control and NaHCO$_3$ groups. Lower: Size of infarct area. The size of infarct area in THAM group significantly decreased compared with that in control group

THAM and NaHCO$_3$ groups, arterial pH was within the normal range, whereas in the control group, arterial pH gradually decreased from 7.42 ± 0.04 to 7.30 ± 0.09 six hours after the MCA occlusion. LCBF measured in the marginal and suprasylvian gyri showed a prompt decrease below 30 ml/100 g/min after the MCA occlusion and there were no significant differences among the three groups (Fig. 1).

Extracellular pH of the marginal gyrus decreased from 7.21 ± 0.11 to 6.86 ± 0.10 with 6 hour-ischemia in the control group. However in the THAM and NaHCO$_3$ groups extracellular pH did not show significant changes (Fig. 1). Intracellular pH over the ischemic brain gradually decreased from 7.23 ± 0.06 to 6.13 ± 0.61 with the MCA occlusion. Intracellular pH in the THAM and NaHCO$_3$ groups was 6.23 ± 0.68 and 6.16 ± 0.53, respectively, 6 hours after the MCA occlusion. Although these values were not significantly different, THAM had a tendency to normalize intracellular pH.

In the THAM group, water content of the gray matter in the marginal gyrus located in peri-infarct zones 6 hours after the MCA occlusion significantly decreased compared with that in the control and NaHCO$_3$ groups (Fig. 2). The size of the infarct area in the THAM group was $16.6 \pm 10.0\%$, which was signifi-

cantly decreased compared with that in the control group ($25.6 \pm 8.4\%$) (Fig. 3).

Discussion

Cerebral ischemia leads to an immediate cessation of oxidative phosphorylation, resulting in a rapid depletion of mitochondrial ATP. Lactate accumulates as a consequence of the anaerobic metabolism. There is an increased intracellular hydrogen ion concentration because of the excessive lactate accumulation, resulting in intracellular acidosis. Staub et al.[9] reported that acidification below pH 6.8 led to an immediate swelling of glial cells. The degree of cell swelling was related to the decrease in pH and the duration of exposure. Cell viability decreased in relation to the severity of acidosis at pH 5.6 and below. They also supposed the mechanism of the acidosis induced glial swelling. H$^+$ ions which are accumulating in the extracellular compartment are buffered by bicarbonate, resulting in the formation of CO$_2$ and water. CO$_2$ diffuses into the intracellular compartment resulting in intracellular acidification by formation of H$_2$CO$_3$. The H$^+$ and HCO$_3^-$ ions are exchanged against extracellular Na$^+$ and Cl$^-$ ions. Consequently Na$^+$ and Cl$^-$ are accumulating in the cell. In addition, protonated lactic acid can freely penetrate the cell membrane due to its lipophilicity, which raise the intracellular osmotic load. Hillered *et al.*[4] reported that mitochondrial respiration was inhibited by a decrease in pH which may lead to a permanent suppression of ATP production. Therefore, amelioration of cerebral acidosis may reduce ischemic brain damage. The present study examined the effect of alkalizing agents on focal cerebral ischemia.

Ischemia due to MCA occlusion encompasses a densely ischemic core and a less densely ischemic penumbral zone. The gyral topography of the ischemia following the MCA occlusion in cats was reported that ectosylvian and marginal gyri were ischemic core and penumbral zone respectively[10]. In this study THAM significantly decreased water content of the gray matter in the marginal gyrus, which suggested THAM had an effect in the area of peri-infarct penumbra, but not in the ischemic core. It was supposed that in the ischemic core irreversible cell damage occurred and the effective dosage of THAM did not reach the core due to the markedly decreased lCBF.

In this study, there were no significant differences in lCBF among the three groups. Extracellular pH did not change significantly in the THAM and $NaHCO_3$ groups, whereas that in the control group decreased. These data suggest that alkalizing agents improve extracellular pH, but have no effect on cerebral circulation in cerebral ischemia. THAM has a tendency to normalize intracellular pH and significantly decreased water content of the gray matter and the infarct size. Thirty percent of THAM is the non-ionized form[7], which can cross the plasmatic membrane and directly effect intracellular acidosis. Moreover, THAM acts partially as an osmotic diuretic, since THAM is eliminated in its ionized form with an equimolar amount of bicarbonate[3]. Thus THAM is effective for reducing brain edema and infarct size. It is also likely that a decrease in infarct area may be a result of improved cerebral metabolism by THAM, because our preliminary experiment showed THAM infusion decreased lactate content of the marginal gyrus. $NaHCO_3$ infusion was reported to cause paradoxical intracellular acidosis, which is attributed to increased intracellular CO_2[1]. This may account for the ineffectiveness of NaHCO3 for focal ischemia.

Meyer *et al.*[5] reported that THAM infusion had no effect on metabolism and hemispheric blood flow in patients with cerebral ischemia. In their study initial treatment with THAM was begun more than 3 days after trauma onset and THAM dosage might have been less compared with our study. THAM is more effective in the early stage of cerebral ischemia, since prolonged ischemia leads to irreversible cell damage. Yoshida *et al.*[11] showed that THAM administration minimized oxidative energy store depletion and decreased production of brain lactate in experimentally injured brain. They also found that THAM treatment decreased brain edema, which is consist with our results.

In conclusion, normalization of systemic and perifocal acidosis with THAM is effective for reducing cortical edema and infarct size in the early stage of focal cerebral ischemia, probably due to the improvement of intracellular acidosis.

References

1. Arieff AI, Leach W, Park R, Lazarowitz VC (1982) Systemic effects of $NaHCO_3$ in dogs. Am J Physiol 242: F586–FS91
2. Bederson JB, Pitts LH, Germano SM, Nishimura MC, Davis RL, Bartkowsky HM (1986) Evaluation of 2,3,5-triphenyltetrazolium chloride as a stain for detection and quantification of experimental cerebral infarction in rats. Stroke 17: 1304–1308
3. Goetz RH, Selmonosky A, State D (1961) Anuria of hypovolemic shock relieved by tris (hydroxymethyl) aminomethane (THAM). Surg Gynecol Obstet 112: 724–728
4. Hillered L, Ernster L, Siesjö BK (1984) Influence of in vitro lactic acidosis and hypercapnia on respiratory activity of isolated rat brain mitochondria. J Cereb Blood Flow Metab 4: 430–437
5. Meyer JS, Fukuuchi Y, Shimazu K, Ohuchi T, Ericsson AD (1972) Abnormal hemispheric blood flow and metabolism in cerebrovascular disease. II. Therapeutic trials with 5% CO_2 inhalation, hyperventilation and intravenous infusion of THAM and mannitol. Stroke 3: 157–167
6. Muizelaar JP, Marmarou A, Ward JD, Kontos HA, Choi SC, Becker DP, Gruemer H, Young HF (1991) Adverse effect of prolonged hyperventilation in patients with severe head injury: a randomized clinical trial. J Neurosurg 75:731-739
7. Rosner MJ, Becker DP (1984) Experimental brain injury: successful therapy with the weak base, tromethamine with an overview of CNS acidosis. J Neurosurg 60: 961–971
8. Siesjö BK (1988) Acidosis and ischemic brain damage. Neurochemical Pathology 9: 31–88
9. Staub F, Baethmann A, Peters J, Kempsi O (1990) Effects of lactacidosis on volume and viability of glial cells. Acta Neurochir (Wien) [Suppl] 51: 3–6
10. Strong AJ, Venables GS, Gibson G (1983) The cortical ischemic penumbra associated with occlusion of the middle cerebral artery in the cat: 1. Topography of changes in blood flow, potassiumion activity, and EEG. J Cereb Blood Flow Metab 3: 86–96
11. Yoshida K, Marmarou A (1991) Effects of trometamine and hyperventilation on brain injury in the cat. J Neurosurg 74: 87–96

Correspondence: H. Kuyama, M.D., Department of Neurological Surgery, Kagawa Medical School, 1750-1 Miki-cho, Kita-gun, Kagawa Prefecture, 761-07, Japan.

Acta Neurochir (1994) [Suppl] 60: 329–331
© Springer-Verlag 1994

Effect of Acetazolamide on Early Ischemic Cerebral Edema in Gerbils

Z. Czernicki[1], **T. Kuroiwa**[2], **K. Ohno**[3], **S. Endo**[4], and **U. Ito**[5]

[1] Department of Neurosurgery, Medical Research, Polish Academy of Sciences, Warsaw, Poland,
[2] Medical Research Institute, [3] Animal Research Centre, [4] Department of Neurosurgery, Tokyo Medical and Dental University, and
[5] Department of Neurosurgery, Musashino Red Cross Hospital, Tokyo, Japan

Summary

Acetazolamide was given in the early stage of ischemic cerebral edema produced by unilateral permanent carotid occlusion in gerbils. The animals were studied 1, 4, and 6 hours after ischemia. The tissues were examined for water and electrolyte concentrations and ischemic areas were visualized by 2,3,5-triphenyltetrazolium chloride (TTC) and H-E staining.

Acetazolamide injected just after occlusion showed a positive effect in reducing edematous changes. Later administration of the drug had neither positive nor harmful effect on the ischemic brains. Thus, acetazolamide seems to be useful for cerebrovascular response studies in the early stage of a brain lesion.

Keywords: Acetazolamide; ischemic cerebral edema; water content; electrolytes.

Introduction

Acetazolamide (trade name, Diamox) is widely used in vascular reserve studies using CBF measurement techniques and transcranial Doppler sonography[8,9]. Diamox has been used only in the chronic stage of the disease because the drug's effect on early cerebral ischemia is unclear as well as the possibility of complications related to vasodilation caused by Diamox[5]. This study aims to examine the effect of Diamox on early ischemic cerebral edema in gerbils.

Material and Methods

Cerebral ischemia was produced in gerbils using a permanent unilateral carotid artery occlusion model. The animals of either sex, each weighing 60–80 g, were anesthetized with ether and underwent occlusion of the left common artery with a Yasargil clip. Only so-called positive animals were evaluated; selection was done according to Ohno's criteria[4]. The animals were divided into the following groups: (1) control (sham operation), (2) 1 hour ischemia, (3) 4 hours ischemia, (4) 6 hours ischemia. In each group there were 5 or 6 treated and 5 or 6 untreated animals. The animals were treated with Diamox, 20 mg/kg b.w. injected intraperitoneally. The drug was always administered 1 hour before decapitation. After decapitation the brains were removed and cut into three portions (Fig. 1). Water content was determined by means of wet/dry tissue weight method and electrolyte concentration was determined using flamephotometry. The dry tissue samples were dissolved in 0.75 N nitric acid solution after incubation at room temperature for 3 days. The tubes containing samples were kept plugged to prevent evaporation[7]. TTC staining was applied to 3 mm coronal brain sections to visualize the ischemic area; H-E staining was used for routine histological examination (Fig. 1). The slices were placed in Dulbecco tissue culture medium and incubated with 1% TTC for 20 minutes at 35°C.

Results

The results are shown in Tables 1–3. Water content and sodium concentration were lower than in No-Diamox group, but potassium concentration was significantly higher. All changes were statistically significant in the 1 hour occlusion group. TTC staining enabled visualization of ischemic changes in the

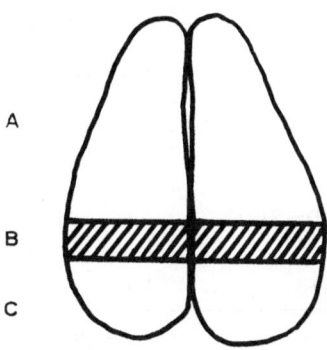

Fig. 1. *A* Parts used for water and electrolyte determinations. *B* Parts used for 2,3,5-triphenylotetrazolium (TTC) staining. *C* Parts used for histological examination, H-E staining

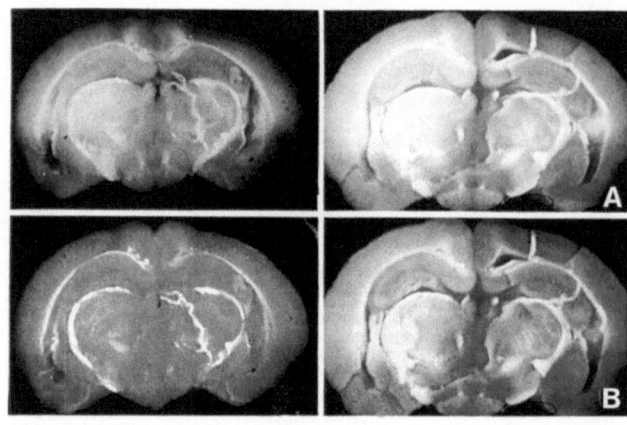

Fig. 2. Left: 4-hour ischemia animal. (A) Ischemic area visualized after 6 min of staining. (B) Homogenous coloration after 20 min of staining. Right: 6-hour ischemia animal. (A) Ischemic area after 6 min of staining. (B) Ischemic area after 20 min of staining

4 hour ischemia and 6-hour ischemia animals. In 4-hour ischemia group the ischemic area was visualized only in the early phase staining while in 6-hour ischemia group the ischemic lesion was also well defined at the end of the staining procedure (Fig. 2).

TTC staining revealed no differences between treat-

ed and untreated animals; H-E stained brains showed no differences either (Fig. 2).

Discussion

Long *et al.*[3] suggested the beneficial diuretic effect of Diamox in the treatment of vasogenic brain edema. The drug decreases extracellular pH and is used in vascular reserve studies[8,9]. Bickler *et al.*[1] found that in rabbits Diamox decreases extracellular pH but has no effect on intracellular pH. High energy compounds were not effected and no lactate acidosis was observed. On the other hand, Regli *et al.*[5] and Bremer *et al.*[2] suggested a deleterious effect of acetozalamide on ischemic cerebral edema. Both papers, however, studied prolonged therapy and harmful effects were observed at least 20 hours after the treatment was started. Our study shows that acetozalamide administered just after carotid occlusion has a positive effect on ischemic brain edema. It decreased water content and sodium concentration, while potassium increased. The values obtained were close to control values. In animals treated later, i.e. 3 or 5 hour after clippling, the effect was not statistically significant

Table 1. *Water Content %*

	Control	1 h	4 h	6 h
No diamox	77.6 ± 0.22	79.3 ± 0.33	80.2 ± 0.42	81.6 ± 0.48
Diamox	76.4 ± 0.29	77.3 ± 0.37*	79.8 ± 0.39	80.5 ± 0.31

* Statistically significant compared with no-diamox group.

Table 2. *Sodium in mEq/kg Dry Tissue*

	Control	1 h	4 h	6 h
No diamox	235 ± 25	306 ± 28	418 ± 28	463 ± 32
Diamox	212 ± 20	252 ± 31*	408 ± 27	430 ± 28

* Statistically significant compared with no-diamox group.

Table 3. *Potassium in mEq/kg Dry Tissue*

	Control	1 h	4 h	6 h
No diamox	492 ± 21	378 ± 19	321 ± 17	287 ± 18
Diamox	479 ± 25	441 ± 20*	332 ± 21	305 ± 23

* Statistically significant compared with no-diamox group.

positive, but not harmful either. Therefore, a single dose of acetazolamide is indicated for patients with acute brain lesions.

References

1. Bickler PE, Litt L, Banville DL, Severinghaus JW (1988) Effects of acetazolamide on cerebral acid-base balance. J Appl Physiol 65: 422–427
2. Bremer AM, Yamada K, West ChR (1980) Ischemic cerebral edema in primates: effects of acetazolamide, phenytoin, sorbitol, dexamethasone and methylprednisolone on brain water and ectrolytes. Neurosurgery 6 (2): 149–154
3. Long DM, Maxwell R, Choi KS (1976) A new therapy regimen for brain edema. Springer, Berlin Heidelberg New York, pp 291–300
4. Ohno K, Ito U, Inaba Y (1984) Regional cerebral blood flow and stroke index after left carotid artery ligation in the conscious gerbil. Brain Res 297: 151–157
5. Piepgras A, Schmiedek P, Leinsinger G, Haberl RL, Kirsch CM, Einhäupl KM (1990) A simple test to assess cerebrovascular reserve capacity using transcranial Doppler sonography and acetazolamide. Stroke 21 (9): 1306–1311
6. Regli F, Yamaguchi T, Waltz AG (1971) Effects of acetazolamide on cerebral ischemia and infarction after experimental occlusion of middle cerebral artery. Stroke 2: 456–460
7. Reulen HJ, Aigner P, Brendel W, et al (1966) Elektrolytveränderungen in tiefer Hypothermie I. Die Wirkung akuter Auskühlung bis 0°C und Wiedererwärmung. Pflugers Arch: 288: 197–219
8. Sullivan HG, Kingsbury TB, Morgan MS, Jeffoat RD, Allison JD, Goode JJ, McDonnell DE (1987) The rCBF response to diamox in normal subjects and cerebrovascular disease patients. J Neurosurg 647: 525–534
9. Vorstrup S, Eugell HC, Lindewald H, Lassen NA (1984) Hemodynamically significant stenosis of the internal carotid artery treated with endarterectomy. J Neurosurg 60: 1070–1075

Correspondence: Zbigniew Czernicki, M. D., Ph. D., Department of Neurosurgery, Medical Research Centre, Polish Academy of Science, Barska Street 16, 02-315 Warsaw, Poland.

Acta Neurochir (1994) [Suppl] 60: 332–334

Neurotropin Treatment of Brain Edema Accompanying Acute Middle Cerebral Artery Infarction

J. De Reuck, D. Decoo, P. Boon, and **C. Van Der Linden**

Department of Neurology, University Hospital, Ghent, Belgium

Summary

The present study is a second analysis of a randomized double-blind controlled trial on the efficacy of neurotropin on brain edema in a subgroup of patients with acute middle cerebral artery infarct treated within 24 hours. Neurotropin is a biological extract that specifically inhibits the release of bradykinin. The mortality rate was significantly lower in the neurotropin than in the placebo group. In the surviving patients the neurological deficit decreased to a significantly greater extent by neurotropin therapy after 15 days. The CTscan findings in the brain of the neurotropin-treated patients demonstrated a significant reduction in the size of the infarct and of the edematous area.

Patients with middle cerebral artery infarct, which is prone to give rise to fatal intracranial hypertension, may derive great benefit from treatment of brain edema with neurotropin.

Keywords: Middle cerebral artery infarct; vasogenic edema; neurotropin.

Introduction

Brain edema after acute ischemic insult contributes to the high morbidity and mortality rate in stroke patients[1,9]. Brain edema evolves with time and peaks between 24 and 48 hours after onset[6]. Ischemic brain edema is primarily vasogenic and all mediators involved are inflammatory in nature[11]. Bradykinin has been implicated in the pathogenesis of brain edema because it stimulates phospholipases A_2 and C, leading to arachidonic acid release and eicosanoid formation[2].

Neurotropin is a non-protein extract of cutaneous tissue from rabbits previously inoculated with vaccinia virus[5]. It inhibits the release of bradykinin in a dose-dependent matter without altering the release of histamine, serotonin and prostaglandin E_2[8]. Neurotropin has been shown to significantly reduce brain swelling in an experimental model of brain infarction in mice[10].

A randomized double-blind controlled clinical trial was performed to investigate the efficacy and the usefulness of neurotropin in the treatment of brain edema accompanying acute ischemic stroke. The present study is a second analysis of this trial in a subgroup of patients with acute middle cerebral artery infarct, treated with neurotropin.

Patients and Methods

A randomized double-blind controlled trial was performed from August 1988 to February 1991 at the Neurological Department of the University Hospital of Ghent. The study was approved by the Ethical Committee and the informed consent of the patients or their relatives was obtained.

Patients with ischemic stroke clinically diagnosed within 24 hours after the onset of symptoms were randomized to treatment. The general design of the trial, the exclusion criteria, the clinical work-up and the CTscan analysis are described in detail in a previous paper.

On admission, all of the patients received a bolus of six 3.6-mg ampuoles neurotropin or placebo intravenously, followed by 10 ampoules of the same solution in a 5% glucose infusion daily for 10 days.

Of 220 randomized patients, 109 with an infarct in the area of the middle cerebral artery were included in the present analysis. The clinical outcome was assessed on day 15 and was based on a global impression rating by the treating physicians (complete recovery, partial recovery, unchanged, aggravated or dead) and with the Toronto Stroke Scale[7] at repeated times.

The CTscans of the brain performed on day 3 and day 11 were compared with respect to the size and degree of hypodensity and the edematous zone. For statistical evaluation, Friedman's repeated measures analysis, the Mann-Whitney U-, Student's t-, χ^2 (with Yate's correction) and log-rank tests were used.

Results

The neurotropin- and placebo-treated groups of patients were comparable for age, sex, and etiology and

Table 1. *Matching of the Relevant Variables of the Treatment Groups*

	All patients		Surviving patients	
	Neurotropin (n = 61)	Placebo (n = 48)	Neurotropin (n = 54)	Placebo (n = 36)
Age (years ± s.e.m.)	69.8 ± 1.6	73.5 ± 1.7	69.5 ± 1.4	72.7 ± 1.4
Sex				
Male	52.4%	40.0%	55.5%	44.5%
Female	47.6%	60.0%	44.5%	55.5%
Etiology				
Thrombo-embolic	65.0%	54.3%	60.3%	61.7%
Cardiac embolic	35.0%	45.7%	39.7%	38.3%
Stroke pattern				
Complete stroke	79.6%	85.1%	79.2%	80.0%
Stroke in evolution	8.5%	4.2%	7.5%	5.7%
Reversible stroke	11.9%	10.7%	13.3%	14.3%

pattern of stroke (Table 1). The clinical outcome in both groups at day 15 is shown in Table 2. The percentage of patients who improved reached 70.5% in the neurotropin group, compared with 35.4% in the placebo group (X^2-test, $p < 0.01$).

The neurological deficit scores on the Toronto Stroke Scale for the surviving patients of both groups are compared in Fig. 1. The neurological deficit score was significantly (U-test, $p = 0.05$) lower in the neurotropin than in the placebo group on day 15. A significant reduction between day 0 and 5 was observed only in the neurotropin-treated group (Friedman's analysis, $p < 0.05$).

Mortality was significantly (X^2-test, $p < 0.05$) lower in the neurotropin group (7 patients or 11.5%) than in the placebo (14 patients or 29.1%) group after 3 weeks. The difference reached even greater significance after 5 weeks, with a mortality rate of 35.4% (17 patients) in

Table 2. *Comparison of the Number of Patients with Different Clinical Outcomes in Both Study Groups*

	Neurological deficit	
	Neurotropin (n = 61)	Placebo (n = 48)
Dead	7 (11.5%)	12 (25.0%)
Aggravated	5 (8.2%)	5 (10.4%)
Unchanged	6 (9.8%)	14 (29.2%)
Partial recovery	34 (55.7%)	14 (29.2%)
Complete recovery	9 (14.8%)	3 (6.2%)
Global number of improved patients (p < 0.01)	43 (70.5%)	17 (35.4%)

the placebo- and 11.5% (7 patients) in the neurotropin-treated group. The survival rates for both groups are compared in Fig. 2. For a 2-month period after stroke, survival probability was significantly higher in the neurotropin group (log-rank test, $p < 0.004$).

Comparison of CTscans on day 3 and day 11 showed a very significant reduction in mean infarct and edema size from 6.40 (SD ± 6.50) cm² to 5.11 (SD ± 5.43) cm² in the neurotropin group (paired t-test, $p < 0.002$) and no change in size, from 5.56 (SD ± 5.87) cm² to 5.56 (SD ± 4.96) cm², in the placebo group.

The degree of hypodensity of the infarct and the edematous area on CTscan did not change significantly in either group between day 3 and day 11.

Discussion

The present study of a subgroup of patients with an acute middle cerebral artery infarct demonstrates a significant beneficial effect of neurotropin on the outcome of the patients. The mortality rate after 5 weeks was strikingly lower in the neurotropin- than in the placebo-treated group. The clinical outcome of the surviving patients was also better in the neurotropin group after 15 days.

The CTscan findings revealed that neurotropin may reduce the size of the infarct and the edematous area and without significantly influencing the degree of hypodensity. These results suggest that the clinical improvement and the reduced mortality rate in the neurotropin-treated patients was due to a decrease of brain edema. They also confirm the findings of our previous study on postmortem particle-induced X-ray analysis (PIXE) of trace and minor elements and of

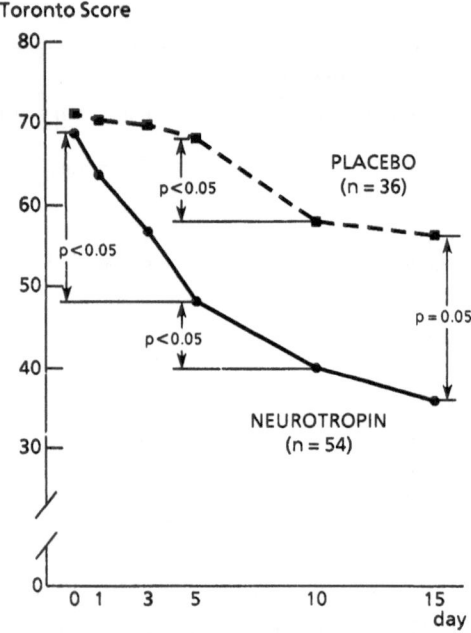

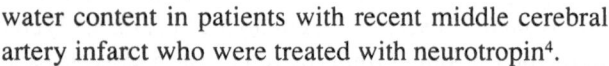

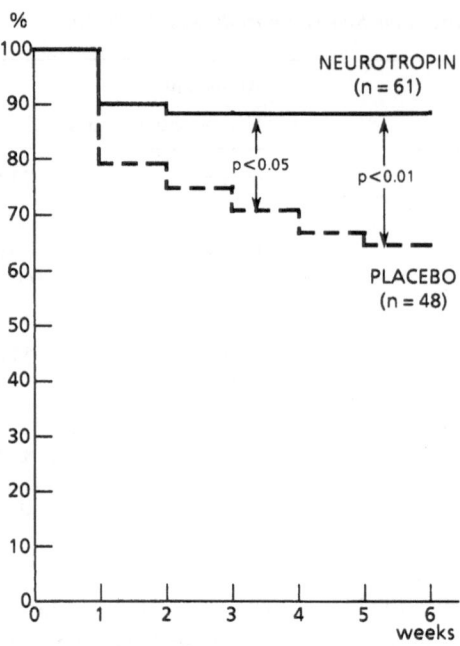

Fig. 1. Comparison of the neurological deficit scores on the Toronto Stroke Scale in the surviving neutropin- and placebo-treated patients during the 15-day study period

Fig. 2. Comparison of survival rates in the neutropin- and placebo-treated groups until 6 weeks after stroke

water content in patients with recent middle cerebral artery infarct who were treated with neurotropin[4].

Our results in this subgroup of patients with middle cerebral artery infarct appear to be more convincing than those previously reported in the entire group. A possible explanation may be differences in the degree of edema and fatal intracranial hypertension between middle cerebral artery and other infarcts[3].

References

1. Bounds JV, Wiebers DO, Whisnant JP, *et al* (1981) Mechanism and timing of death from cerebral infarction. Stroke 12: 474–477
2. Conklin BR, Burch RM, Steranka JR, *et al* (1988) Distinct bradykinin receptors mediate stimulation of prostaglandin synthesis by endothelial cells and fibroblasts. J Pharmacol Exp Ther 244: 646–649
3. De Reuck J, Sieben G, De Coster W, Vander Eecken H (1981) Fatal intracranial hypertension in patients with cerebral infarction. In: Meyer J, Lechner O, Reivich M, Ott E, Aranibar A (eds) Cerebral vascular disease. Excerpta Medica, Amsterdam, pp 324–328
4. De Reuck J, Hebbrecht G, Maenhaut W (1993) Influence of neurotropin on the water content and regional distribution of two minor and six trace elements in human brains affected by recent middle cerebral artery infarcts. Cerebrovasc Dis, in press
5. Imai Y, Saito K, Maeda S, *et al* (1984) Inhibition of the release of bradykinin-like substances into the perfusate of rat hind paw by neurotropin. Japan J Pharmacol 36: 104–106
6. Klatzo I (1967) Neuropathological aspects of brain edema. J Neuropathol Exp Neurol 26: 1–13
7. Norris JW, Hachinski VC (1982) Comment on "study design of stroke treatments." Stroke 113: 527–528
8. Ohara H, Namimatzu A, Fukahara K, *et al* (1988) Release of inflammatory mediators by noxious stimuli: effect of neurotropin on the release. Europ J Pharmacol 157: 93–99
9. Ropper AH, Shafran B (1984) Brain edema after stroke, clinical syndrome and intracranial pressure. Arch Neurol 41: 26–30
10. Sprumont P, Caprano G, Lintermans J (1993) Morphometrical quantification of brain oedema related to experimental multiple micro-infarcts in mice: assessment of neurotropin effect. Meth Find Exp Clin Pharmacol 15, in press
11. Wahl M, Unterberg A, Baethmann A, *et al* (1988) Mediators of blood-brain barrier dysfunction and formation of vasogenic brain edema. J Cereb Blood Flow Metab 8: 621–634

Correspondence: J. De Reuck, M. D., Department of Neurology, University Hospital, De Pintelaan 185, B-9000 Ghent, Belgium.

Acta Neurochir (1994) [Suppl] 60: 335–337

Anti-Ischemia Activity of HU-211, a Non-Psychotropic Synthetic Cannabinoid

M. Vered, A. Bar-Joseph, L. Belayev, Y. Berkovich, and **A. Biegon**

Pharmos Ltd., Kiryat Weizmann, Rehovot, Israel

Summary

The novel tricyclic-terpenoid type cannabinoid, HU-211, was tested in gerbils and rats for protection against the effects of cerebral ischemia. Our transient ischemic models in gerbils and rats are based on the protection afforded against the lethal effects of global ischemia. HU-211 gave over 30% protection, by i.v. administration. The optimal dose ranges of HU-211 were between 5 and 10 mg/kg i.v. In gerbils we used a transient ischemia model induced by occlusion (10 min) of the bilateral common carotid arteries (BCCA). HU-211, 8 mg/kg i.v., gave a significantly ($p \leq 0.05$) better in vivo performed than a control group over three days following ischemia. Histology performed in gerbil model also resulted in significantly ($p < 0.001$) diminished degeneration area of CA1 in the hippocampus after treatment of HU-211. In the rat model, after four vessel occlusion (4VO) (20 min), HU-211 treatment significantly ($p < 0.01$) improved neurobehavior scoring. These results show that the new synthetic cannabinoid can protect against hippocampal neuron damage due to selective brain injury induced by cerebral global ischemia in gerbils or rats.

Keywords: Cannabinoid; neuroprotection; cerebral ischemia; neurological score.

Introduction

Recent evidence from animal models has helped to substantiate the idea that neuropathology, associated with stroke and cerebral ischemia, results from large increases of extracellular glutamate and aspartate[1]. HU-211 has a major potential in the treatment of stroke, by its ability to block glutamate receptors. HU-211 is a novel structure of the cannabinoid family which lacks psychotropic activity[2]. The highly lipophilic molecule readily crosses the blood-brain barrier and has potent central effects when given by peripheral routes. Thus, it seems reasonable to assume that systemic administration of HU-211 reduces excitotoxicity in the brain, mediated by overactivity of glutamate receptors. Our findings demonstrated that HU-211 is effective in improving both edema and clinical status after head injury. HU-211 is effectively protecting neural damage caused by: (1) neurotoxins, such as NMDA[1], (2) brain ischemia in the gerbil or rat; and (3) head injury in the rat[3]. The mechanism underlying the neuroprotective action of HU-211 is not completely understood.

Material and Methods

Male Mongolian gerbils (60–80 g) were supplied by Tumble-Brook Farm (MA, U.S.A.); male Sprague-Dawley rats (250 g) were supplied by Anilab (Hulda, Israel), and received food and water ad libitum. The HU-211 was synthesized and provided by R. Mechoulam from the Faculty of Pharmacology at the Hebrew University of Jerusalem, Israel.

Drug Preparation and Administration

a. Submicron emulsion preparation. The procedure of PHARMOS was used to prepare cannabinoids in sub-micronized emulsion (SME) for i.v. administration[4].

b. Hydroxypropyl-β-cyclodextrin (HPCD) preparation. HU-211 was dissolved in a minimum volume of absolute ethanol and the solution was added dropwise to HPCD powder. The ethanol was dried out (1 h at 45°C) and water was added and mixed by stirring or sonication. The HU-211/HPCD solution was filtered through 0.2 μ and placed in sterilized vials ready to use.

Global Ischemia Studies

a. Bilateral common carotid arteries (BCCA) in gerbils were isolated and 3–0 silk suture material was positioned loosely around them. The tips of each loop were tied together, and the suture material was placed beside the trachea. The ventral neck incision was then sutured. On the following day, the animals were lightly anesthetized with ether, the neck skin wounds were opened and the BCCA were occluded for 10 minutes using small artery clips. The drug (HU-211 0.2% w/v in SME) was administered i.v. 30 min post-ischemia insult. Injection volume (1.5 ml/kg) was used for vehicle control.

b. In 4VO rats the vertebral arteries were occluded on the first day, according to Pulsinelli et al.[5]. On the second day the common carotid arteries (CCA) were occluded. At that stage, the animal was anesthetized lightly (ether), the skin opened and CCA closed for 20 min with arterial clips. HU-211 was administered 15 min before occlusion of CCA, 8 mg/kg i.v. (4 ml/kg) in 45% HPCD; the same volume was used for vehicle control.

Clinical Evaluation

a. Gerbils. Three to 5 hours post-ischemia, animals were observed for their clinical appearance using the Rudolphi method[6]. This was done every 24 hours, and the score was averaged over the entire 72-hour period.

b. Rats. The neurological deficit was evaluated daily for three days after 4VO or sham operation, using a modification of Wahl's protocol[7].

Results

Global Ischemia in Gerbils

After 10 min BCCA occlusion in gerbils, HU-211, that was administered 30 min post-ischemia (8 mg/kg i.v.) significantly (p < 0.05) reduced the average score from 7.5 in untreated group to 1.5 in the treated group. Vehicle alone was not significantly different from the untreated group (Fig. 1).

Histopathological evaluation of CA1 after 10 min of ischemia demonstrated a substantial loss of pyramidal neurons (not shown); in the medial, middle and the lateral parts only 25% survived. The HU-211 treated

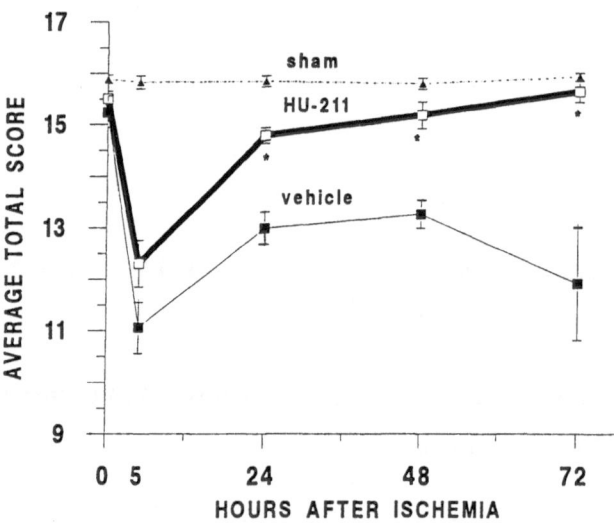

Fig. 2. Neurobehavioral scores (normal score = 16) of rats after 4VO, and treated by HU-211 (8 mg/kg i.v.). Line graph of mean ± SEM of data expressed a progress in neurological score starting right after occlusion for four consecutive days. Each line graph represents a group study (n = 10); untreated group, vehicle (HPCD, 4 ml/kg i.v.) and treated group. * Statistically significantly higher than vehicle group (p < 0.01, Kruskal-Wallis ANOVA)

group showed approx. 50% cell survival in all three areas, and that was significantly above the two control groups.

Global Ischemia in Rats

Neurological score in rats subjected to 20 min 4VO is given in Fig. 2. After seventy-two hours of recovery, we observed a reduction of neurological score to approx. 35% in the vehicle group, but full recovery in the HU-211 treated group. The rats treated with HU-211 (8 mg/kg i.v.) reached the level of the sham group. In the sham and drug treated groups one out of eight animals died, but six out of eight in the vehicle group (on days 4, 6, 8, 12 and 13).

In a parallel study we performed a follow-up of body temperature throughout the recovery time (not shown). The body temperature of the treated group was the same as in the two control groups.

Discussion

The pharmacological profile of the non-psychotropic cannabinoid HU-211 in gerbils and rats includes neuroprotection. The results demonstrated that administration of HU-211 (8 mg/kg i.v.) 30-min post-occlusion (gerbils) or at zero time (rats) markedly improved

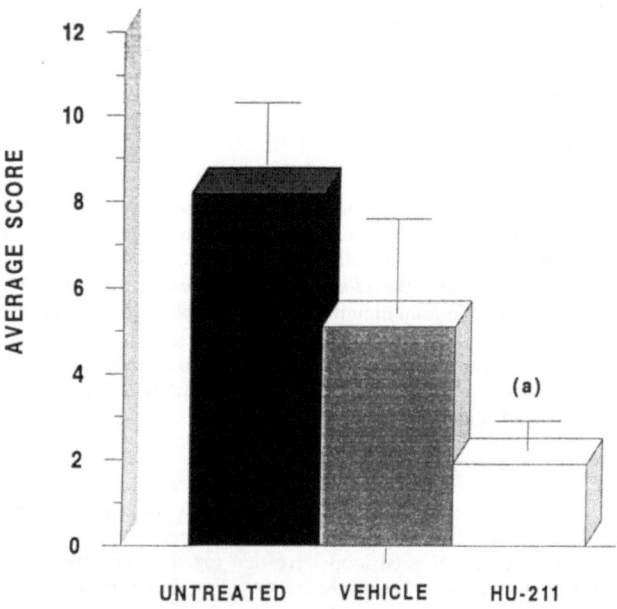

Fig. 1. Bar graph of average neurological scoring protection of gerbils, by HU-211 (8 mg/kg i.v.); normal = 0; maximum = 12. The scoring was measured during 4 days following transient global ischemia (10 min). Each graph (mean ± SEM) presents a group study (n = 10); untreated group, vehicle (SME, 4 ml/kg i.v.) and treated group. (*a*) p < 0.05 HU-211 vs. control (Wilcoxon Rank Sum Test)

the neurobehavior in both species and reduced loss of hippocampal neurons in gerbil model. The mechanism by which HU-211 protects the ischemic tissue is not yet established. HU-211, whose half-life is about 3 hours, has been shown to block NMDA-receptors in Xenopus oocytes injected with mammalian brain RNA. In this electrophysiological study, HU-211 acts as a non-competitive antagonist in response to NMDA[8].

In conclusion, HU-211 acts as a neuroprotective agent in forebrain ischemia in rats, as well as bilateral carotid artery occlusion in gerbils. In both species, HU-211 exerts a cerebroprotective effect at zero time or 30 min post-ischemia.

References

1. Feigenbaum JJ, Bergmann F, Richmond SA, Mechoulam R, Nadler V, Kloog Y, Sokolovsky M (1989) Nonpsychotropic-cannabinoid acts as a functional N-methyl-D-aspartate receptor blocker. Proc Natl Acad Sci USA 86: 9584–9587

2. Little PJ, Compton DR, Mechoulam R, Martin BR (1989) Stereochemical effects of 11-OH-Δ^8-THC-dimethylheptyl in mice and dogs. Pharmacol Biochem Behav 32: 661–666

3. Shohami E, Novikov M, Mechoulam R (1993) A non-psychotropic cannabinoid, HU-211, has cerebroprotective effects after closed head injury in the rat. J Neurotrauma, in press

4. Levy MY, Langerman L, Gottschalk-Sabag S, Benita S (1989) Side-effect evaluation of a new diazepam formulation: venous sequela reduction following intravenous (iv) injection of a diazepam emulsion in rabbits. Pharm Res 6 (6): 510–516

5. Pulsinelli WA, Brierley JB, Plum F (1982) Temporal profile of neuronal damage in a model of transient forebrain ischemia. Ann Neurol 11: 491–498

6. Toth J, Schubert L, Rudolphi P, Kreutzberg K (1987) Ischemia-induced neuronal cell death, calcium accumulation, and glial response in the hippocampus of the Mongolian gerbil and protection by propentofylline. J Cereb Blood Flow Metab 7: 745–751

7. Wahl F, Allix M, Plotkine M, Boulu RG (1992) Neurological and behavioral outcomes of focal cerebral ischemia in rats. Stroke 23 (2): 267–272

8. Dascal N (1993) Screening of the pharmacological properties of putative NMDA antagonists of the HU group in the Xenopus expression system. Personal communication

Correspondence: M. Vered, M. D., Pharmos Ltd., Kiryat Weizmann, Rehovot 76326, Israel.

Tumor and Brain Edema: Pathophysiology, Laboratory Studies

Acta Neurochir (1994) [Suppl] 60: 341–343
© Springer-Verlag 1994

Simultaneous Measurement of Bidirectional Capillary Permeability, Vascular Volume, Extracellular Space, and rCBF in Experimental Gliomas and Surrounding Edema

P. C. Warnke, F. J. Hans, A. Korst, C. Mall, and **C. B. Ostertag**

Department of Neurosurgery, University of Freiburg, Federal Republic of Germany

Summary

Measurements of bidirectional capillary permeability in the ASV tumor model revealed a heterogeneous distibution of transport rates for the small water-soluble molecules which are believed to be the driving force in perifocal edema formation. Mean K1 values were 8.1 ± 5.5 µl/g/min in whole tumor and 18.89 ± 12.3 µl/g/min in highly permeable areas. Tumor blood flow ranged between normal white matter and cortex, whereas the plasma vascular space in the tumors was increased as compared to the normal brain. The size of the extracellular space was depending on the regional capillary permeability, underlining the vasogenic nature of peritumoral edema which was capable of being studied using in vivo methods.

Keywords: Capillary permeability; extracellular space; brain edema.

Introduction

Perifocal edema in brain tumors is due to increased capillary permeability, as has been clearly shown in numerous models[2]. The driving force for the increased flow of water into the extracellular space has been discussed controversially[7]. Nevertheless, it has been shown that it is the increased capillary permeability for water-soluble molecules which accounts for the increased water flow resulting in peritumoral edema[6]. Unfortunately, only a few papers report on the measurement of bidirectional capillary permeability, multiplying the time by the concentration and calculating the distribution space of water-soluble markers in order to calculate edema formation rates. If other pertinent physiological parameters, such as rCBF, vascular volume, distribution space, and extraction fraction, could be measured as well, more insight into the mechanisms of edema formation could be gained[7]. In addition, the impact of brain edema on microcirculation and drug delivery should be studied. Furthermore, therapeutic modification of edema by steroids can be examined quantitatively. Following up the through study of Nakagawa[7], we therefore measured bidirectional capillary permeability for meglumine iothalamate, vasular volume, size of extracellular space and rCBF in an experimental canine glioma model employing in vivo methods. The objective of our study was to measure the aforementioned parameters in vivo to gain a data base in order to test different pharmacological regimens[8] for the treatment of edema in patients.

Material and Methods

Tumor Model

Brain tumors were induced in 3-day-old beagle puppies (n = 6) by intracerebral inoculation of 0.01 ml avian sarcoma virus (ASV) containing 10^5 focus-forming units/ml. At 6 to10 weeks after injection, the dogs were anesthetized as described by Groothuis[3] and CT scans were performed to detect the tumors. At this point in time the dogs were still asymptomatic. Retrospectively, one animal had to be excluded from the analysis as histology showed a sarcoma whereas the remaining animals came down with anaplastic gliomas.

Measurements of Capillary Permeability, Vascular Volume and Extracellular Space

Bidirectional capillary permeability was measured as described by Groothuis[3] using a CT scanner and a two-compartment model. Variables measured were the influx transport constant for meglumine iothalamate K1 (µl/g/min), the efflux transport constant k2 (l/min) and the plasma vascular space Vp (ml/100 g). As it has been shown previously that meglumine iothalamate distributes only in the extracellular space and does not enter cells, the ratio K1/k2 in this model can be used as a lower limit estimate of the steady-state distribution space, i.e. the extracellular space[5].

Measurement of rCBF

rCBF was measured in the same animals using stable xenon CT under stereotactic conditions to assure identical slice position as compared to the permeability measurements. The dogs were intubated and allowed to breathe a mixture of 35% xenon/30% oxygen/ 35% room air. Blood gases and blood pressure were monitored throughout the experiment and stayed within the physiological range. For multiple capillary passages, the ratio of K1/F (1-hematocrit), where F is blood flow in ml/100 g/min, equals the net extraction fraction under circumstances of merely unidirectional flow.

Results

The mean K1 value for all tumors was 8.1 ± 5.5 μl/g/ min as compared to 1.1 ± 0.4 in normal white matter. As indicated already by the standard deviation and as can be seen in Table 1, K1 values varied widely among tumors. So did all other physiological parameters, as summarized in Table 1. Nevertheless regional analysis of the variables revealed interesting details which deserve further comment. All tumors, whether regarding whole-tumor values or separate tumor regions, showed in all aspects significant differences from the normal brain. Blood flow ranged 49.1 ± 16.4 ml/100 g/min between normal white matter (31.2 ± 5.37 ml/100 g/ min) and normal cortex (80.1 ± 12.54 ml/100 g/min). Blood flow also varied greatly within the same tumors. The mean plasma vascular space was slightly above that of the normal brain (1.85 ± 1.33 ml/100 g), but some tumors showed highly increased vascular volumes (No. 2). Relatively consistent were the values of the extracellular space, with a mean of 0.15 ± 0.04 μl/g. Analysis of whole-tumor values versus values in areas with maximum permeabilities as judged from the image pixels with high K1 values showed a relationship between K1 values and size of the extracellular space as shown in Fig. 1. Mean K1 in these areas was 18.89 ± 12.3 μl/g/min, and the extracellular space increased to 1.61 ± 1.1 μl/g. In addition, Vp increased two-fold to 3.58 ± 2.2 ml/100 g. None of the measured physiological parameters was correlated to tumor size.

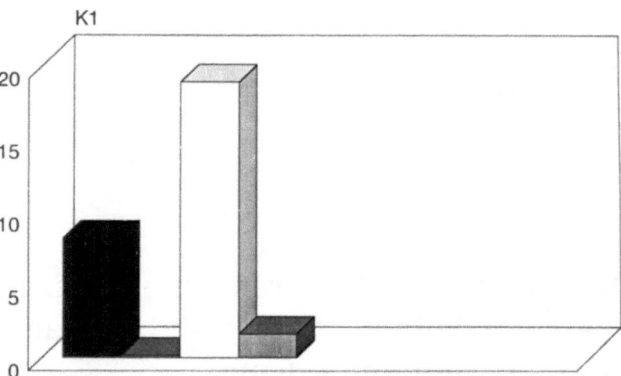

Fig. 1. Capillary permeability values (K1) in whole tumor (black area) and in highly permeable areas (white area). To the right of each are grey areas showing the corresponding sizes of the extracellular space of the same dimension (μl/g). Note the correlation of K1 and extracellular space

Furthermore, there was no cross-correlation between blood flow and permeability.

The net extraction fraction ranged between 0.15 and 0.016 among individual tumors and showed no correlation with size, permeability or blood flow.

Discussion

It is shown that, in the ASV-model, in vivo measurements of a variety of physiological variables by in vivo methods are possible, thus implying that this should also be possible in patients. The K1 values were on the lower range of those reported[3], but comparable to the K1 values measured in human gliomas by this method[1,4]. Blood flow values were similar to those reported for this specific tumor model. The size of the extracellalar space was almost ten-fold higher than described by Nakagawa using the RG-2 model and albumin as a marker[7]. As the K1 values in the models

Table 1. *Physiological Parameters of ASV-Gliomas*

No.	K1 (μl/g/min)	k2 (min⁻¹)	rCBF (ml/100 g/min)	Vp (ml/100 g)	E$_s$ (μl/g)
1	6.9	41.2	23.3	1.81	0.16
2	15.7	92.5	45.5	4.24	0.17
3	11.9	53.5	50.7	2.78	0.22
4	2.5	20.6	64.1	0.59	0.12
5	3.6	31.1	62.3	1.37	0.11
Mean:	8.1 ± 5.5	47.7 ± 27.8	49.1 ± 16.4	2.15 ± 1.4	0.15 ± 0.04

Summary of the physiological variables measured in five ASV tumors using computerized tomography. Note the widespread variation of all variables.

are rather similar, this difference in the 30-minute experiments can be explained by the different molecular size of albumin (MW 68500) and meglumine iothalamate (MW 677). Taking into consideration that the square root of the molecular weight of a compound relates to it's K1 value, as established by Fenstermacher and Rapoport[2], our values corroborate this theoretical relationship. Our methodology can now be used to measure the effects of various drugs on edema formation in brain tumor patients. This will hopefully further elucidate the findings obtained by positron emission tomography, which have attributed the positive effects of steroid treatment to the decrease in K1. However, without being able to measure k2 due to the short half-life of the positron emitter, the product of the time multiplied by the concentration and the size of the extracellular space could not be measured, and there crucial to the assessment of the effects of dexamethasone[7]. These questions can be answered in brain tumor patients by using the CT compartment model and iodinated compounds.

References

1. Aaslid R, Gröger U, Patlak CS, Fenstermacher JD, Huber P, Reulen HJ (1990) Fluid flow rates in human peritumoural oedema. Acta Neurochir (Wien) [Suppl] 51: 152–154

2. Fenstermacher JD Rapoport SI (1985) The blood-brain barrier In: Renkin EM, Michel CC (eds) Handbook of physiology: the microcirculation. American Physiological Society, Bethesda, pp 969–1000
3. Groothuis DR, Lapin GD, Vriesendorp FJ, Mikhael MA, Patlak CS (1991) A method to quantitatively measure transcapillary transport of iodinated compounds in canine brain tumors with computed tomography. J Cereb Blood Flow Metab 11: 939–948
4. Groothuis DR, Vriesendorp FJ, Kupfer B, Warnke PC, Lapin GD, Kuruvilla A, Vick NA, Mikhael MA, Patlak CS (1991) Quantitative measurements of capillary transport in human brain tumors by computed tomography. Ann Neurol 30: 581–588
5. Groothuis DR, Fross R, Molnar P, Pasternak J, Blasberg R, Patlak C (1983) Applicability of multiple-time-point measurements for in vivo determination of blood-to-tissue transfer constants of iothalamate meglumine (Conray) and AIB in experimental gliomasc J Cereb Blood Flow Metab 3 [Suppl 1]: 93–94
6. Hossmann KA, Bothe HW, Bodsch W, Paschen W (1983) Pathophysiological aspects of blood-brain barrier disturbances in experimental brain tumors and brain abscesses. Acta Neuropathol (Berl) [Suppl] 8: 89–102
7. Nakagawa H, Groothuis DR, Owens ES, Fenstermacher JD, Patlak CS, Blasberg RG (1987) Dexamethasone effects on [125-I]albumin distribution in experimental RG-2 gliomas and adjacent brain. J Cereb Blood Flow Metab 7: 687–701
8. Reichman HR, Farrell CL, Del Maestro RF (1986) Effects of steroids and nonsteroid anti-inflammatory agents on vascular permeability in a rat glioma model. J Neurosurg 65: 233–237

Correspondence: Peter Warnke, M.D., Department of Neurosurgery, University of Freiburg, Hugstetter Strasse 55, D-79106 Freiburg, Federal Republic of Germany.

Acta Neurochir (1994) [Suppl] 60: 344–346
© Springer-Verlag 1994

Quantitative Diffusion MR Imaging of Cerebral Tumor and Edema

M. Eis, T. Els, M. Hoehn-Berlage, and **K.-A. Hossmann**

Department of Experimental Neurology, Max-Planck-Institute for Neurological Research, Cologne, Federal Republic of Germany

Summary

The detection of brain tumors using standard techniques of qualitative, relaxation-weighted magnetic resonance imaging (MRI) requires the application of contrast agents. We investigated whether or not it is possible to use diffusion-weighted MRI to localize tumors without contrast enhancement. Three different experimental rat brain tumors were studied: F98 glioma, RN6 Schwannoma and E376 neuroblastoma.

We found a marked hypointensity in the region of the tumor and edema in heavily diffusion-weighted images, which corresponded well with the histological presentation. Quantitative maps of the apparent diffusion coefficient (ADC) allowed a better localization of the tumor than that obtained by regional presentation of T_2 times, particularly under conditions in which peritumoral edema was absent. The ADC differences of the three tumor types were statistically not significant. Based upon regions-of-interest evaluations, tumor could be distinguished from peritumoral edema and normal brain tissue. However, a sharp demarcation between tumor and peritumoral edema was not possible, and this is attributed to a similar enlargement of interstitial space.

It was concluded that diffusion-weighted MRI possesses a high potential for the detection of brain tumors but does not allow precise demarcation of the tumor border.

Keywords: Tumor; edema; magnetic resonance imaging; diffusion imaging.

Introduction

Diffusion imaging provides unique information about early pathological changes of the brain. Previous studies of acute brain infarcts have shown a decrease in the apparent diffusion coefficient (ADC)[3,4], which has been attributed to the shrinkage of the interstitial space[1] occurring as a consequence of cell membrane depolarization[8]. In brain tumors and peritumoral brain tissue, interstitial space widens because of extracellular accumulation of vasogenic edema[6], resulting in an increase in the ADC.

In a clinical setting, localization of the tumor is often difficult using standard techniques of qualitative, re-laxation-weighted magnetic resonance imaging (MRI). On T_2-weighted images, the peritumoral edema can be detected by its strong hyperintensity but the tumor itself can only be visualized after application of a paramagnetic contrast agent using T_1-weighted imaging[7].

We therefore investigated whether or not it is possible to use diffusion-weighted images to localize the tumor without contrast agents. Furthermore, quantitative ADC images were computed to find out if the differences between tumors, peritumoral edema and normal brain tissue are statistically significant.

Materials and Methods

Experimental brain tumors were induced in male Fischer rats (160–300 g body weight) by stereotaxic allotransplantation of cloned rat F98 glioma (n = 9), RN6 Schwannoma (n = 6) and E376 neuroblastoma cells (n = 7) into the nucleus caudatus of the right hemisphere. MR measurements were performed 11 to 13 days after tumor inoculation. Prior to imaging, the animals were anesthetized with 0.8% halothane in a mixture of 70% N_2O and 30% O_2, and allowed to breathe spontaneously. Body temperature and respiration were continuously monitored and kept within physiological range. At the end of the MR study, the rat brains were fixated by perfusion with 4% paraformaldehyde. For histological examination, sections (10 μm thick) of the brains were stained with hematoxylin-eosin or cresyl violet for demarcation of the tumor. Peritumoral edema was visualized using immunohistochemistry of extravasated serum immunoglobulins2.

A Biospec 4.7 T system (Bruker, Karlsruhe, Germany), equipped with a homogeneous Alderman-Grant-resonator and fitted with an actively shielded gradient insert (maximum gradient strength 110 mT/m), was used for the MR studies. Diffusion-weighted imaging was performed using a Stejskal-Tanner type pulsed gradient spin echo sequence[9]. Slice thickness was 1.1 mm, field of view 5.0 cm or 4.0 cm and image matrix 128 × 256. Recovery times and the number of averages varied between 400 ms/16 and 2300 ms/2, and echo times varied between 27 ms and 33 ms. Quantitative maps of the apparent diffusion coefficient were calculated on the basis of 6 images recorded under identical conditions but with different diffusion weighting ($b_{max} = 1600$ s/mm^2). All ADC values were corrected for interference of the imaging gradients with the diffusion sensitizing gradients[5].

Apparent diffusion coefficients were determined in regions of interest (ROI) in tumor and edema as well as in the contralateral cortex and caudate-putamen (CP). The positions of the ROI were defined on the basis of the histological slices. The values found for each investigated structure were averaged for each animal. The ability to differentiate among tumor, edema and healthy brain tissue on the basis of the ADC was statistically evaluated using Scheffe's F-test and a significance level of $p < 0.01$.

Results

Heavily diffusion-weighted images showed a marked hypointensity in the region of tumor and peritumoral edema: the ratio between signal intensities of the central tumor area and the corresponding contralateral side was reduced to 0.72 ± 0.10 (mean $\pm$ SD, b-factor: 700 s/mm^2, TR: 400 ms, TE: 28 ms). Sharp demarcation between tumor and peritumoral edema on diffusion-weighted images was not possible, even when short recovery times were used to increase contrast by additional T_1-weighting. However, quantitative ADC maps allowed a better localization of the tumor-edema area than that obtained using a regional presentation of T_2 times. The region of increased ADC values corresponded to the area of the tumor and the peritumoral edema as visualized by histology and extravasated serum albumins, respectively (Fig. 1).

ADC values (given in μm^2/s) were 798 ± 81 for F98, 756 ± 111 for E367, 763 ± 38 for RN6, 902 ± 152 for

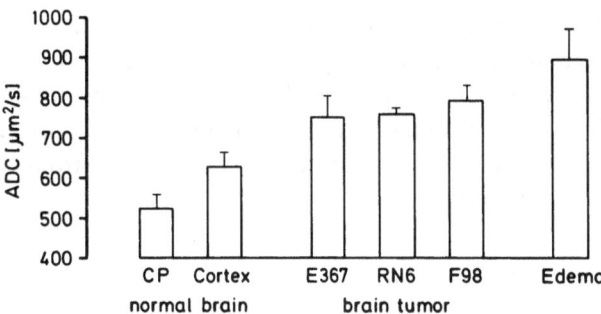

Fig. 2. ADC values measured in healthy brain, central areas of three different brain tumors (neuroblastoma E367, schwannoma RN6, and glioma F98) and peritumoral edema, respectively. Data values are given in μm^2/s and represent the means of all animals; error bars indicate the standard deviation. Differences among caudate-putamen (*CP*), cortex, tumor and edema (all possible combinations) are statistically significant ($p < 0.01$)

edema, 630 ± 75 for cortex, and 524 ± 72 for caudate-putamen (Fig. 2). The ADC differences among tumor types were statistically not significant. The mean tumor value of 775 ± 82 μm^2/s was significantly different from peritumoral edema, normal cortex and caudate-putamen ($p < 0.01$). Differences between edema and healthy brain tissue as well as between cortex and caudate-putamen were also statistically significant ($p < 0.01$).

Discussion

We found statistically significant differences in the apparent diffusion coefficients of tumor, peritumoral edema, intact cortex and caudate-putamen, which finding is in line with the previously documented enlargement of extracellular space in tumor and peritumoral edema[6]. The inverse changes of ADC in acute infarcts and tumors support the theory that ADC is a function of the relative volume of the extracellular compartment[1].

We were also able to show that both tumor and peritumoral edema can be visualized by a substantial decrease in signal intensity in native diffusion-weighted images. However, a sharp demarcation of the tumor was not possible either in diffusion-weighted images or on ADC maps. This is explained by the similar enlargement of the extracellular compartment induced by the spread of edema fluid from the tumor into the surrounding tissue.

These observations demonstrate that diffusion-weighted MRI and ADC maps possess high potentials for the reliable detection of brain tumors because of their ability to visualize tumors without contrast enhancement. This is of great advantage in comparison to

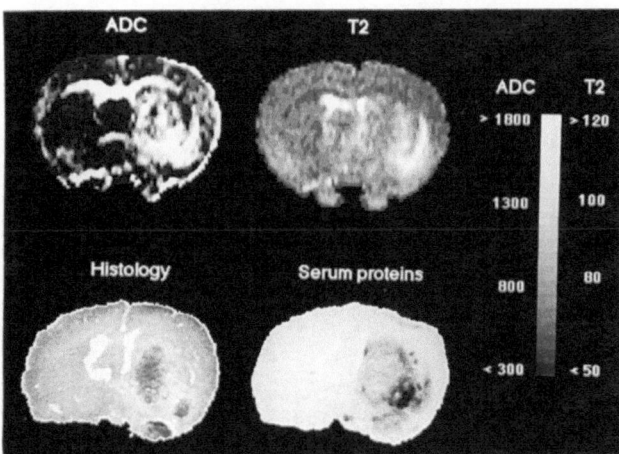

Fig. 1. Comparison of the regional distribution of the MR-parameters ADC (upper left) and T_2 (upper right) with histology (lower left) and extravasation of serum proteins (lower right) for a coronal brain slice of an animal with implanted F98 glioma. The region of increased ADC values corresponds to the area of the tumor and the peritumoral edema as visualized by histology and extravasated serum albumins. On the T_2-image, the tumor cannot be detected; only the peritumoral edema is visible due to its prolonged relaxation time

T$_2$-weighted imaging, which may fail to detect the pathological process if peritumoral edema is absent. However, for the precise demarcation of the tumor from the surrounding tissue, additional T$_1$-weighted imaging after application of a contrast agent is strongly recommended[7,10].

Acknowledgement

Preparation of the histological slices by Mrs. Stephanie Krause is gratefully acknowledged. This project was supported in part by a grant from the Deutsche Forschungsgemeinschaft (SFB 194/B1).

References

1. Benveniste H, Hedlund LW, Johnson GA (1992) Mechanism of detection of acute cerebral ischemia in rats by diffusion-weighted magnetic resonance microscopy. Stroke 23: 746–754
2. Hossmann K-A, Hürter T, Oschlies U (1983) The effect of dexamethasone on serum protein extravasation and edema development in experimental brain tumors of cat. Acta Neuropathol 60: 223–231
3. Moseley ME, Kucharczyk J, Mintorovitch J, Cohen Y, Kurhanewicz J, Derugin N, Asgari H, Norman D (1990) Diffusion-weighted MR imaging of acute stroke: correlation with T$_2$-weighted and magnetic susceptibility-enhanced MR imaging in cats. Am J Neuroradiol 11: 423–429
4. Moseley ME, Cohen Y, Mintorovitch J, Chileuitt L, Shimizu H, Kucharczyk J, Wendland MF, Weinstein PR (1990) Early detection of regional cerebral ischemia in cats: comparison of diffusion- and T$_2$-weighted MRI and spectroscopy. Magn Reson Med 14: 330–346
5. Neeman M, Freyer JP, Sillerud LO (1990) Pulsed-gradient spin-echo diffusion studies in NMR imaging. Effects of the imaging gradients on the determination of diffusion coefficients. J Magn Reson 90: 303–312
6. Okada Y, Kloiber O, Hossmann K-A (1990) Regional metabolism in experimental brain tumors in cats: relationship with acid/base, water and electrolyte homeostasis. J Neurosurg 77: 917–926
7. Schoerner W, Laniado M, Kornmesser W, Felix R (1989) Comparison of multi-echo and contrast enhanced MR scans: image contrast and delineation of intracranial brain tumors. Neuroradiology 31: 140–147
8. Schuier FJ, Hossmann K-A (1980) Experimental brain infarcts in cats II. Ischemic brain edema. Stroke 11: 593–601
9. Stejskal EO, Tanner JE (1965) Spin diffusion measurements: spin echoes in the presence of a time-dependent field gradient. J Chem Phys 42: 228–292
10. Wilmes LJ, Hoehn-Berlage M, Els T, Bockhorst K, Eis M, Bonnekoh P, Hossmann K-A (1993) In vivo relaxometry of three brain tumors in the rat: effect of Mn-TPPS, a tumor-selective contrast agent. JMRI 3: 5–12

Correspondence: Mathias Hoehn-Berlage, M. D., Max-Planck-Institut für neurologische Forschung, Gleueler Strasse 50, D-50931 Köln, Federal Republic of Germany.

Acta Neurochir (1994) [Suppl] 60: 347–349

Localization of Experimental Brain Tumors in MRI by Gadolinium Porphyrin

K. Bockhorst, T. Els, K. Kohno, and **M. Hoehn-Berlage**

Department of Experimental Neurology, May-Planck-Institute for Neurological Research, Cologne, Federal Republic of Germany

Summary

The contrast between edema and F98 glioma in rat brain was distinctly enhanced in T_2-weighted MRI (TE 130 ms, TR 3 s) by intraperitoneal injection of the synthetic gadolinium-porphyrin complex, GdTPPS. The T_1 relaxation time of the gliomas was selectively shortened by about 50% from 1339 ± 109 ms to 628 ± 106 ms, and the T_2 relaxation time was shortened by about 35% from 86 ± 6 ms to 57 ± 5 ms. The relaxation times of normal tissues under investigation (cortex, corpus callosum, temporal muscle, ventricles) were unaltered. Therefore, GdTPPS-application causes F98 gliomas to appear hyperintense in T_1-weighted MRI and hypointense in T_2-weighted MRI.

Keywords: Paramagnetic contrast agent; metalloporphyrin T_2-weighted imaging; glioma.

Introduction

In recent years, MRI has become an important tool for the detection of cerebral tumor and peritumorous edema. Its reliability for the detection of brain tumors has been substantially improved by the introduction of paramagnetic contrast agents such as gadolinium-di-ethylene-triamine-pentaacetic acid (GdDTPA). These substances shorten the longitudinal relaxation time T_1 in tissues such as experimental gliomas which do not have a blood-brain barrier (BBB). Thus, gliomas appear hyperintense in T_1-weighted MRI following GdDTPA application. Since GdDTPA does not bind to glioma cells it rapidly diffuses into peritumorous tissue making tumors appear larger in T_1-weighted MRI than they really are[2].

Peritumorous edema, as visualized by T_2-weighted imaging, frequently serves as indirect marker of cerebral tumors. However, since the difference in T_2 relaxation times for edema and glioma is only small the demarcation between these two tissues is difficult.

The paramagnetic metalloporphyrin complex, MnTPPS, selectively binds to tumor cells. It is not washed out into the peritumorous interstitial space, and its efficiency as a contrast agent for the detection of gliomas has been prevlously demonstrated[4,6,11]. However, MnTPPS has only a minor influence on T_2 and is therefore not suited for application in T_2-weighted MRI. Here, replacement of manganese by gadolinium in the porphyrin complex is shown to produce a tumor selective contrast agent suited for both tumor detection in T_1-weighted MRI and separation of tumor from edema in T_2-weighted MRI.

Materials and Methods

Tetra-(p-sulfonatophenyl)porphyrin (TPPS) was synthesized by condensation of benzaldehyde and pyrrole in propionic acid (140°C, 1 h). Gadolinium was inserted into the porphyrin by reaction of gadolinium acetylacetonate with TPPS in imidazol in a dry nitrogen atmosphere (240°C, 1 h)[7]. The crude product was purified by recrystallisation from > 95% aqueous ethanol. It was then dried by heating 24 h at 100°C and subsequently by storing over KOH in vacuum. The tumors were produced by stereotactic inoculation of 10,000 viable F98 glioma cells into the right caudate nucleus of Fischer rats under pentobarbital anesthesia[8]. 10 days after the inoculation, the tumors had reached a diameter of about 5 mm. At this time the anesthetized animals ($N_2O : O_2 / 7 : 3, 0.5–1.5\%$ halothane) were fixed in a head holder and scanned by MRI.

The NMR studies were performed on a 4.7 T Bruker Biospec MRI/MRS system (Bruker, Karlsruhe, Germany), equipped with a homogeneous Alderman-Grant resonator and fitted with an actively shielded gradient insert (maximum gradient strength 100 mT/m). Images were recorded using a modified CPMG spin-echo sequence, which allowed the determination of T_1, T_2 and proton density at the same time[3,11]. Eight echoes were recorded at a TR of 0.6 s and 3.0 s (TE 13 ms) for calculation of T_1, while 32 echoes were recorded at a TR of 3 s (TE 13 ms) for determination of T_2. A slice thickness of 1 mm and field of view of 6 cm was used for all scans.

The GdTPPS was dissolved in 1 ml sodium bicarbonate (8.4%, pH 7–8) and, after recording of control images, injected intraperitoneally. T_1 and T_2 measurements were made 1, 2, 24 and 48 hours after injection. Immediately after the final measurement, the animals were perfused with formalin in situ, the brains excised and stored in formalin. Histological sections were stained with cresyl violet or hematoxylin/eosin for detection of neoplastic tissue, and with serum-specific antibodies for detection of edematous tissue.

Results

Figure 1 shows coronal T_1- and T_2-weighted images of a rat brain with an F98 glioma recorded before (top) and after (bottom) the injection of GdTPPS. The tumor was not visible in the T_1-weighted images (top left) before contrast agent application, while T_2-weighted images (top right) clearly display pathological tissue although tumor and edema were not differentiated clearly.

After the injection of GdTPPS the signal intensity of glioma is clearly enhanced in the T_1-weighted images (bottom left), and the tumor becomes hypointense compared to the edema in the T_2-weighted images (bottom right) with a clear demarcation between tumor, cyst and edema.

Figure 2 shows the change in the T_1 (mean ± S.D.) and T_2 (mean ± S.D.) relaxation times after injection of

GdTPPS. In the F98 glioma (full circles) both, T_1 (top) and T_2 (bottom), decrease within one hour of GdTPPS application. Thereafter, changes of T_1 and T_2 are not significant compared to the values, measured directly after GdTPPS-application. In the contralateral striatum (open circles) neither T_1 nor T_2 are altered by GdTPPS.

Table 1 shows the mean T_1- and T_2-values of four rats, determined before and after the injection of

Table 1. *T_1 and T_2 in Glioma Before and 2 d After GdTPPS- or MnTPPS-Application (0.25 mmol/kg bw)*

	Native	MnTPPS*	GdTPPS
T_1 (ms)	1315 ± 199	894 ± 100	628 ± 106
T_2 (ms)	86 ± 4	81 ± 3	57 ± 5

* Values taken from [11].

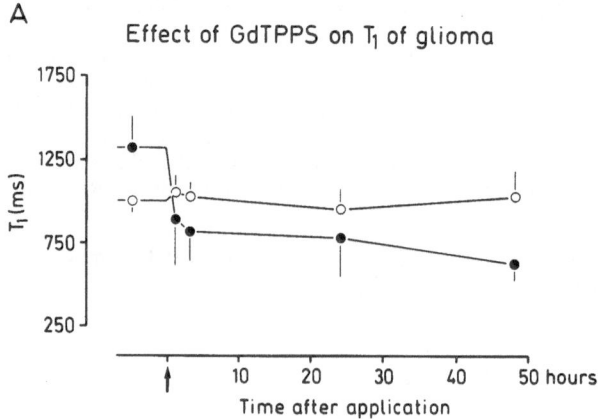

A

Effect of GdTPPS on T_1 of glioma

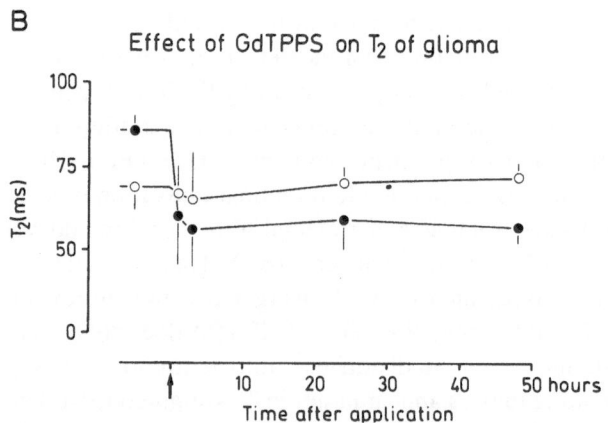

B

Effect of GdTPPS on T_2 of glioma

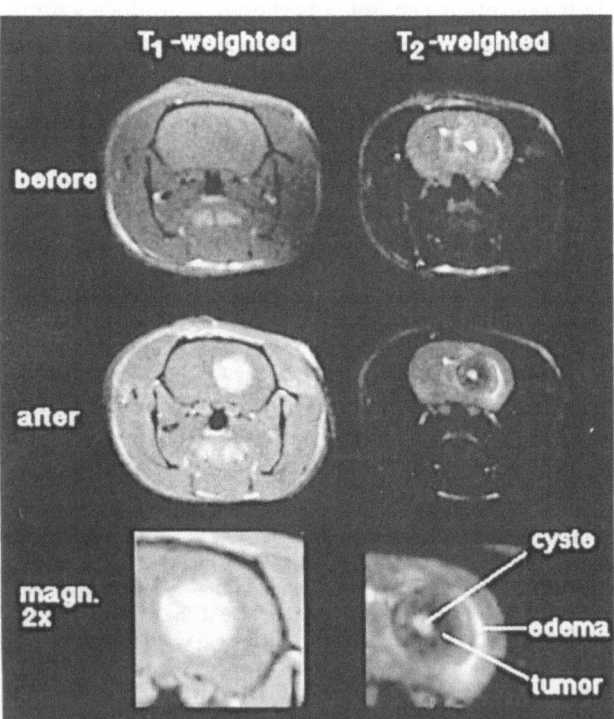

Fig. 1. GdTPPS enhancemant of glioma. Coronal MRI of rat brain glioma, recorded before and after i.p. injection of 0.25 mmol GdTPPS/kg bw. Without injection of GdTPPS tumor and edema cannot be demarcated clearly on T_1- (top left) or T_2-weighted imaging (top right). However, after the GdTPPS injection the tumor becomes hyperintense on T_1- (bottom left) and hypointense in T_2-weighted imaging (bottom right), which clearly improves the delineation of tumor and edema

Fig. 2. T_1 and T_2 (mean ± S.D.) of glioma and contralateral striatum before and after i.p. injection of 0.25 mmol GdTPPS/kg bw. The time of injection is marked by an arrow on the x-axis. Within one hour after the injection both T_1 (A) and T_2 (B) of glioma (full circles) decrease. Changes of T_1 and T_2 during the following observation period are not significant, compared to the value recorded directly after the application. GdTPPS has no influence on T_1 or T_2 of the contralateral striatum (open circles)

GdTPPS. For comparison, the relaxation times measured after injection of an equivalent amount of MnTPPS are taken from an earlier study[11]. T_1 is shortened to 50% of control by GdTPPS, compared to 70% of control by MnTPPS. GdTPPS shortens T_2 to about 65% of control, while MnTPPS does not produce any distinct effect on T_2.

Discussion

In 1987, Marzola and Cannistraro proposed the use of GdTPPS as tumor-selective NMR contrast agent[10]. In the same year, Lyon *et al.* suggested, GdTPPS might not be stable enough in vivo[9]. However, in 1988, Gersonde *et al.* reported the successful administration of GdTPPS as contrast agent[5].

In agreement with our earlier findings[5] we do not find evidence suggesting GdTPPS to be substantially more toxic than MnTPPS. GdTPPS, used in the present study, is synthesized from gadolinium(III)acetylacetonate. Acetylacetonate derivatives of lanthanide porphyrins are known to be more stable than halides etc.[1]. In the other studies, where GdTPPS was reported as unstable in serum[9], probably gadolinium halides were used for synthesis.

Porphyrin complexes with the latter lanthanides (Tb-Lu) as central cations are even more stable than GdTPPS[7]. Since some of these ions have larger magnetic moments than gadolinium (Ho, Dy), their effect on T_2 is expected to be even greater and further improve the differentiation of edema and tumor.

Acknowledgement

This project was supported by a grant from the Deutsche Forschungsgemeinschaft (SFB 194/B1).

References

1. Adler AD, Newman W, Mulvaney M, Paone J (1979) Lanthanide binding by porphyrins and the mechanisms of porphyrin metal binding. In: Longo RF (ed) Porphyrin chemistry advances. New York, 265 pp
2. Baba T, Moriguchi M, Natori Y, Katsuki C, Inoue T, Fukui M (1990) Magnetic-resonance-imaging of experimental rat-brain tumors – histopathological evaluation. Surg Neurol 34 (N6): 378–382
3. Eis M, Hoehn-Berlage M, Bockhorst K, Hossmann K-A (1991) A time efficient method for combined T_1/T_2 relaxation time measurements. Evaluation for multi-parameter tissue characterization. In: Book of abstracts: Eighth Annual Meeting of the Society of Magnetic Resonance in Medicine. SMRM, Berkeley, 701 pp
4. Ernestus R-I, Hoehn-Berlage M, Bockhorst K, Kloiber O, Wilmes L, Bonnekoh P, Hossmann K-A (1990) Differentiation of tumor and edema in experimental tumors of the cat brain. Application of the tumor-specific NMR contrast agent manganese (III) tetraphenylporphinesulfonate. In: Book of abstracts: Tenth Annual Meeting of the Society of Magnetic Resonance in Medicine. SMRM, Berkeley, 701 pp
5. Gersonde K, Bockhorst K, Hoehn-Berlage M, Staemmler M, (1988) Manganese and gadolinium porphyrin complexes with high relaxivity and tumor selectivity useful for NMR imaging and photodynamic therapy. In: Work in progress: Eighth Annual Meeting of the Society of Magnetic Resonance in Medicine. Berkeley, p 10
6. Hoehn-Berlage M, Norris D, Bockhorst K, Ernestus R-I, Kloiber O, Bonnekoh P, Hossmann K-A (1992) T_1 Snapshot flash measurement of rat brain glioma: kinetics of the tumor-specific contrast agent manganese (III) tetraphenylporphine sulfonate. Magn Reson Med 27: 201–213
7. Horrocks WD, Hove EG (1978) Water-soluble lanthanide porphyrins: shift reagents for aqueous solution. JACS 100: 4386–4392
8. Hossmann K-A, Mies G, Paschen W, Szabo L, Dolan E, Wechsler W (1986) Regional metabolism of experimental brain tumors. Acta Neuropathol 69: 139–147
9. Lyon RC, Faustino PJ, Cohen JS, Katz A, Mornex F, Colcher D, Baglin C, Koenig SH, Hambright P (1987) Tissue distribution and stability of metalloporphyrin MRI contrast agents. Magn Reson Med 4: 24–33
10. Marzola P, Cannistraro S (1989) Gd3+ TPPS: a potential paramagnetic contrast agent in NMR imaging. Physiol Chem Phys Med NMR 19: 279–282
11. Wilmes LJ, Hoehn-Berlage M, Els T, Bockhorst K, Eis M, Bonnekoh P, Hossmann K-A (1993) In vivo relaxometry of three differnet experimental brain tumors in the rat. JMRI 3: 5–12

Correspondence: Mathias Hoehn-Berlage, M. D., Max-Planck-Institut für neurologische Forschung, Gleueler Strasse 50, D-50931 Köln, Federal Republic of Germany.

Acta Neurochir (1994) [Suppl] 60: 350–352
© Springer-Verlag 1994

Proton MR Spectroscopy of Experimental Brain Tumors in vivo

M. L. Gyngell, T. Els, M. Hoehn-Berlage, and **K.-A. Hossmann**

Department of Experimental Neurology, Max-Planck-Institut for Neurological Research, Cologne, Federal Republic of Germany

Summary

F98 gliomas, E367 neuroblastomas, and RN6 Schwannomas in rat brain were studied non-invasively in vivo by localized proton MR spectroscopy (MRS). The spectra obtained from homotopic brain contralateral to the tumors were qualitatively indistinguishable from those of normal rat brain in vivo and showed resonance lines assigned to N-acetylaspartate, glutamate, total creatine (creatine and phosphocreatine), choline, glucose, and myo-inositol. The tumor spectra displayed marked differences compared to those obtained from contralateral brain. There were increases in choline, myo-inositol and lipids, which are presumably associated with increased membrane turnover. The presence of lactate indicated anaerobic glycolysis. Other differences included the absence of signals from NAA resulting from the destruction or displacement of neuronal tissue by the tumor. There was also a loss of total creatine. Although the spectra of all three tumor types were distinct from contralateral brain, there were no obvious differences between the different tumor types.

Keywords: Proton MRS in vivo; experimental tumos; rat brain.

Introduction

Magnetic resonance spectroscopy (MRS) provides a unique opportunity to obtain biochemical information non-invasively in vivo. Significant differences are observed when comparing the spectra of diseased tissue with those of normal tissue[1,2]. The emergence of reliable localization methods for MRS enables the acquisition of spectroscopic data from precisely defined regions of interest[3]. The wide availability of whole-body MRI systems and the high sensitivity of proton spectroscopy has led to its early application in the study (and diagnosis) of disease in humans. Applications in the brain have included studies of brain tumors, and metabolic disorders, as well as functional investigations. However, in vivo MRS still needs to be validated through the study of disease models under controlled experimental conditions so that the spectral patterns can be reliably related to the metabolic state of the tissue. In order to do this, the various resonance lines in the proton spectrum need to be unambiguously and correctly assigned to particular biochemical substrates. Possible sources of misinterpretation include incorrect assignment of overlapping resonance and tissue heterogeneity within the selected volume.

An earlier study on implanted F98 gliomas using localized proton MRS in vivo and high-resolution MRS of tissue extracts showed major differences between the spectra of tumors and contralateral brain both in vivo and in vitro. It also showed that the tumors contained glycine, which could be misinterpreted as myo-inositol in the in vivo spectra[4].

The purpose of this study was to use in vivo proton MRS to compare three different experimental brain tumor types originating from completely different cell lines in order to look for potential inter-tumor-type differences.

Materials and Methods

Male Fischer rats of 200 to 350 g body weight were used (n = 20). Ten thousand viable cells of F98 glioma (n = 6), E367 neuroblastoma (n = 7), or RN6 Schwannoma (n = 7) were stereotactically inoculated into the right caudate nucleus under pentobarbital anaesthesia[5]. The resulting tumors were allowed to grow for 10 to 15 days. The NMR studies were performed with the animals anaesthetized with 0.8 to 1.5% halothane in a 7:3 $N_2O:O_2$ mixture and artificially ventilated. Catheters were inserted into the femoral artery to monitor blood pressure and to obtain blood samples, and into the femoral vein to administer pancuronium bromide for muscle relaxation. The body temperature was recorded by a rectal thermometer and maintained at 37°C by a warm water blanket. The animals were positioned prone and fixed in a stereotactic head holder for accurate and reproducible positioning within the NMR spectrometer. The studies were made using a 4.7 T Bruker Biospec MRI/MRS system fitted with actively shielded gradient coils (Bruker, Karlsruhe, Germany). A surface receiver coil was placed over the skull while the RF excitation was made with a 12-cm diameter Helmholtz transmitter coil.

The position and extent of the tumor were determined from T_2-weighted MRI. On the basis of the images, volumes of interest (VOI) for spectroscopy were defined within the tumor or in homotopic contralateral brain. Volumes of 27 μl (3 × 3 × 3 mm) or 64 μl (4 × 4 × 4 mm) were used depending on tumor size, and the same size was used for each pair of tumor and contralateral measurements. All ^{1}H MRS measurements were made using the STEAM sequence[3].

Results

The spectra of the brain regions contralateral to the tumors were indistinguishable from the spectra of normal rat brain (see Fig. 1a). The major resonances are assigned to N-acetylaspartate (NAA) at 2.0 ppm, glutamate (Glu) at 2.4 and 3.75 ppm, total creatine (Cr + PCr) at 3.03 and 3.9 ppm, choline (Cho) at 3.19 to 3.22 ppm, glucose (Glc) at 3.4 ppm, and myo-inositol (Ins) at 3.5 ppm. Signals due to lactate (Lac) at 1.3 pm were below the detectable limit in the contralateral regions.

The tumor spectra (Fig. 1b–d) were completely different from those of the contralateral brain regions. The NAA signal was universally below the detectable limit. The signal from total creatine was significantly reduced to between 30 and 50% of that contralaterally. There were increases in resonances attributed to choline, and to myo-inositol and glycine (3.5 ppm).

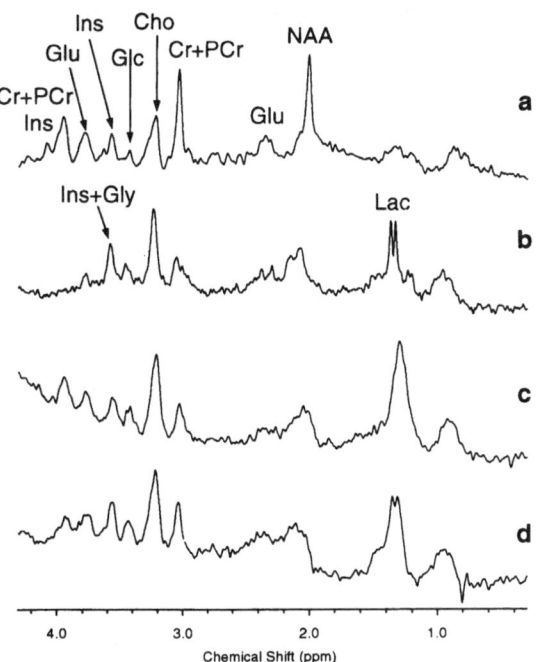

Fig. 1. Typical localized ^{1}H MRS spectra of (a) brain contralateral to an implanted tumor, (b) F98 glioma, (c) E367 neuroblastoma, and (d) RN6 Schwannoma in vivo. Resonances are assigned to N-acetylaspartate (*NAA*), glutamate (*Glu*), creatine and phosphocreatine (*Cr + PCr*), choline (*Cho*), glucose (*Glc*), myo-inositol (*Ins*), glycine (*Gly*), and lactate (*Lac*)

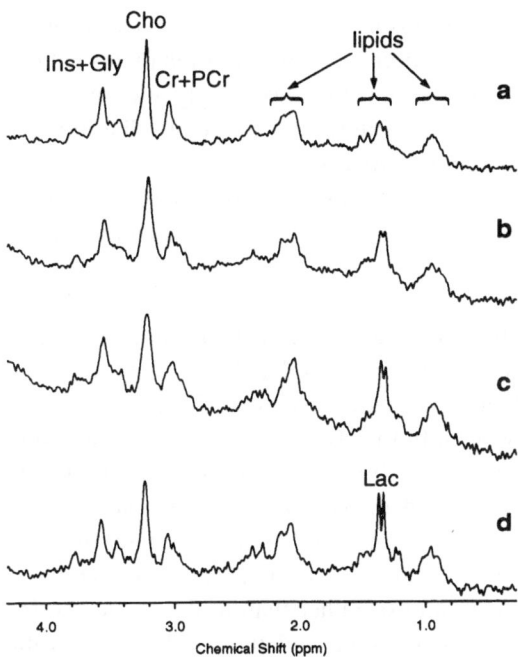

Fig. 2. (a–d) Localized ^{1}H MRS spectra of four different animals with F98 gliomas, showing strong differences in the 1.3-ppm resonance of lactater

Mobile lipid components at 0.9, 1.3 and 2.1 ppm were also elevated. In contrast to the other components, the coincident lipid and lactate signals at 1.3 ppm displayed considerable variability. Figure 2 demonstrates the variability of the 1.3 ppm signal in the spectra from four different F98 gliomas, and the characteristic lactate doublet is clearly visible.

Discussion

NAA is thought to be located exclusively in neurons, and it is therefore assumed to be a neuronal marker. Its absence in the tumor spectra indicates destruction or displacement of vital neurons by the tumors. This also implies that there was no normal brain tissue within the selected volumes of interest, and it is a confirmation of the excellent spatial selectivity achievable by localized ^{1}H MRS. In the case of the E367 neuroblastoma spectra, the absence of a detectable NAA signal may at first seem surprising, since the cell clone originates from neurons. However, it is in agreement with NAA specific immunohistochemical staining[6], where neuroblastomas showed only a weak NAA immunoreaction compared to normal brain (T. Els, 1993, private communication).

An earlier high-resolution ^{1}H MRS study of tissue extracts in solution showed that F98 gliomas contain significant amounts of glycine, which is indistinguishable from myo-inositol in in vivo studies. This is due to the inherently poorer spectral resolution that is achievable in vivo. The accumulation of glycine (and other amino acids) is considered indicative of a disruption of normal metabolic processes.

The increases in choline, inositol and lipids probably reflect increased cell membrane turnover and cell proliferation.

The presence of polyunsaturated fatty acids and other mobile lipids at 1.3 ppm compromises the identification of the lactate doublet. In Fig. 1b and d the lactate doublet is clearly resolved, whereas in Fig. 1c it is not resolved. In Fig. 1c the 1.3-ppm signal is probably due mainly to lipids, but the presence of lactate cannot be excluded. The narrow line width of the lactate doublet in Fig. 1b in comparison with the other resonances suggests that the lactate is in a collection of fluid, namely a cyst, rather than in the tissue. Differences in vascularization and the resulting anaerobic glycolysis are likely to be the causes of the variability of the lactate resonance. The spectral differences observed within a tumor type may be associated with differences in tumor development.

In conclusion, in agreement with studies of patients, proton MRS in vivo can distinguish between normal brain tissue and cerebral tumors. However, ^{1}H MRS in vivo cannot alone differentiate among different tumor types.

Acknowledgement

The financial support of the Deutsche Forschungsgemeinschaft (SFB 194/Bl) is gratefully acknowledged.

References

1. Bruhn H, Frahm J, Gyngell ML, Merboldt KD, Hänicke W, Sauter R, Hamburger C (1989) Noninvasive differentiation of tumors using localized ^{1}H NMR spectroscopy in vivo: initial experience in patients with cerebral tumors. Radiology 172: 541–548
2. Demeaerel P, Johannik K, Van Hecke P, Van Ongeval C, Verrellen S, Marchal G, Wilms G, Plets C, Goffin J, Van Calenbergh F, Beart AL (1991) Localized ^{1}H NMR spectroscopy in fifty cases of newly diagnosed intracranial tumors. J Comput Assist Tomogr 15: 67–76
3. Frahm J, Michaelis T, Merboldt KD, Bruhn H, Gyngell ML, Hänicke W (1990) Improvements in localized proton NMR spectroscopy of human brain. Water suppression, short echo times, and 1 ml resolution. J Magn Reson 90: 464–473
4. Gyngell ML, Hoehn-Berlage M, Kloiber O, Michaelis T, Ernestus RI, Hörstermann D, Frahm J (1992) Localized proton NMR spectroscopy of experimental gliomas in rat brain in vivo. NMR Biomed 5: 335–340
5. Hossmann K-A, Mies G, Paschen W, Szabo L, Dolan E (1986) Regional metabolism of experimental tumors. Acta Neuropathol (Berl) 69: 139–147
6. Moffet JR, Aryan Nambodiri MA, Cangro CB, Neale JH (1991) Immunohistochemical localization of N-acetylaspartate in rat brain. Neuro Report 2: 131–134

Correspondence: Michael L. Gyngell, Ph. D., Max-Planck-Institut für neurologische Forschung, Gleueler Strasse 50, D-50931 Köln, Federal Republic of Germany.

Acta Neurochir (1994) [Suppl] 60: 353–355
© Springer-Verlag 1994

MRI Contrast Enhancement by Mn-TPPS in Experimental Rat Brain Tumor with Peripheral Benzodiazepine Receptors

K. Ikezaki, T. Nomura, M. Takahashi[1], B. F. Zieroth[1], and **M. Fukui**

Department of Neurosurgery, Neurological Institute, Faculty of Medicine, Kyushu University and
[1] Research Department, Diagnostic Laboratory, Nihon Schering KK, Japan

Summary

Recent studies have disclosed that Mn-TPPS, a paramagnetic metalloporphyrin, may be a tumor-specific contrast media for magnetic resonance (MR) imaging. We investigated whether or not Mn-TPPS enhancement of the glioma could be mediated by peripheral benzodiazepine receptor. Using a transplanted rat C6 glioma model, Mn-TPPS enhancement was performed with or without pretreatment by peripheral or central benzodiazepine-specific receptor ligands. Signal intensity analysis disclosed that the enhancement was not inhibited by these ligands. Post-contrast replacement studies showed that neither of these ligands reduced Mn-TPPS enhancement. Although the tissue concentration of Mn-TPPS was significantly higher in the glioma than in the contralateral brain, PK11195 pretreatment did not replace the intratumoral Mn-TPPS. This data suggested that the tumor-specific enhancement of Mn-TPPS was not mediated by peripheral type benzodiazepine receptor.

Keywords: Brain neoplasm; benzodiazepine receptors; porphyrin; magnetic resonance imaging.

Introduction

The application of gadolinium DTPA facilitates the detection of brain tumor MR imaging. The limitations of this method, however, become apparent in glial and other infiltrative tumors due to the failure to identify tumor cells that reside beyond the borders of the imaging abnormality[3]. The need for a tumor cell-specific nuclear magnetic contrast agent led to the development of manganese(III)tetra-(4-sulfanatophenyl) porphyrin (Mn-TPPS), a synthetic metal porphyrin complex[6]. Bockhorst *et al.*[2] speculated that tight binding of Mn-TPPS to tumor cells might be mediated by PBZ receptors at the mitochondrial membrane based on their preliminary experiment using a rat brain tumor model.

In the central nervous system, PBZ receptors are specifically expressed in brain tumors, as shown in experimental and clinical studies[1,4,5,7]. Although the endogenous ligand for this receptor has not yet been identified, porphyrins have been described as one candidate for a PBZ receptor ligand[9]. In this study, we investigated whether Mn-TPPS contrast enhancement of the brain tumor was mediated by PBZ receptors.

Materials and Methods

PK11195, 1-(2-chlorophenyl)-N-methyl-N-(1-methylpropyl)-3-isoquinoline carboxamide, a specific ligand for the PBZ receptor, and clonazepam, a specific ligand for central type receptors, were dissolved in ethanol:DMSO (1:1, vehicle). Mn-TPPS was dissolved in normal saline.

Animal Experiment

Exponentially growing C6 glioma cells ($2 \times 10^5/5$ μl) were implanted stereotactically into the right hemisphere of male Wistar rats under pentobarbital anesthesia. After 10 days, rats were anesthetized with pentobarbital and urethane and separated into 4 groups. Groups 1 to 3 were pretreated with intraperitoneal injection of vehicle, 5 mg/kg of PK11195, or clonazepam, respectively. After 15 min, 0.16 mmol/kg of Mn-TPPS was given intravenously. T1-weighted MR images were taken on an animal imager (GE, Omega CSI-II, 4.7 T) using the spin echo method (TR/TE = 600/24.5 ms). MR images of Group 4 were taken 17 h after intravenous Mn-TPPS injection. PK11195 (5 mg/kg) was then administered intraperitoneally. After 15 min, MR images were taken again for the replacement study.

Each signal intensity (SI) value was related to the SI of a simultaneously measured external standard, consisting of a cylindrical plastic tube that contained water. The SI was likewise determined by a region-of-interest technique in the external standard, the tumor tissue, the contralateral brain, the tongue, and the facial muscle.

Tissue Mn-TPPS Measurement

After image analysis, the removed brain was histologically evaluated. The tumor tissue, the contralateral cortex, and the kidney were carefully dissected and removed for the tissue concentration analysis of Mn-TPPS by ICP-AES[8] with 257.6 nm (SPS1200A, Seiko Electric Co., Ltd.). Standard solutions of Mn (Wako) were diluted with H_2O.

Statistical analysis was performed by pared Student's t-test and ANOVA analysis.

Results

Image Analysis

Figure 1 (A, B, and E) shows MR images of a tumor before and at various times after Mn-TPPS application. In glioma tissue, the tongue, and the muscle, Mn-TPPS enhancement was maximal at 5 min after Mn-TPPS injection and stabilized after 15 min for 30 min. C6 glioma showed the highest SI among the tissues examined after Mn-TPPS injection. The mean enhancement ratios (post-injection SI at time t / pre-injection SI) in the tongue, muscle, and C6 glioma (n = 3 each) were approximately 2.0, 1.7, and 1.6, respectively. However, the normal brain was not enhanced by Mn-TPPS. Post-contrast tumor versus normal brain SI ratio was 1.8 (p = 0.0001). The enhancement was prolonged to over 17 h although the contrast ratio decreased by 1.4 fold. In vehicle, PK1195 and clonazepam pretreated groups, there were no significant differences in glio-

ma/contralateral cortex SI ratios, being 1.8 ± 0.2, 1.7 ± 0.2, and 1.6 ± 0.2, respectively (Fig. 1C and D). Furthermore, PK11195 did not reduce Mn-TPPS enhancement 17 h after Mn-TPPS injection (Fig. 1E and F).

Mn-TPPS Concentration

Tissue Mn-TPPS concentrations are summarized in Table 1. The selective accumulation of Mn-TPPS was observed in the tumor tissue and reached a level approximately 4 times higher than that of the contralateral brain. This tumor-selective accumulation of Mn-TPPS was not, however, inhibited by preteatment with either PK11195 or clonazepam. Although the Mn-TPPS concentration decreased to approximately 60% of the initial value after 17 h (93.3 ± 11.4), tumor tissue maintained a 3.5-fold higher accumulation of Mn-TPPS than the contralateral brain (26.4 ± 1.0). In the kidney, Mn-TPPS concentration decreased (0.54 ± 0.06) to 5% of the control 17 h after Mn-TPPS injection. Intravenous administration of PK11195, however, did not replace this accumulation of Mn-TPPS in either the tumor tissue or the kidney.

Discussion

The present study disclosed that Mn-TPPS did not pass through the BBB into the normal brain. However, the blood-tumor barrier was disrupted, and extravasated Mn-TPPS remained in the tumor tissue. The prolonged enhancement in tumor tissue suggested that selective Mn-TPPS binding or incorporation into the tumor cells. We investigated whether or not this enhancement was mediated by PBZ receptor. The in vivo receptor competition study disclosed that this enhancement was not inhibited by the PBZ receptor-specific ligand. Further, PK11195 did not replace the pre-existing Mn-TPPS. Considering that membrane-bound

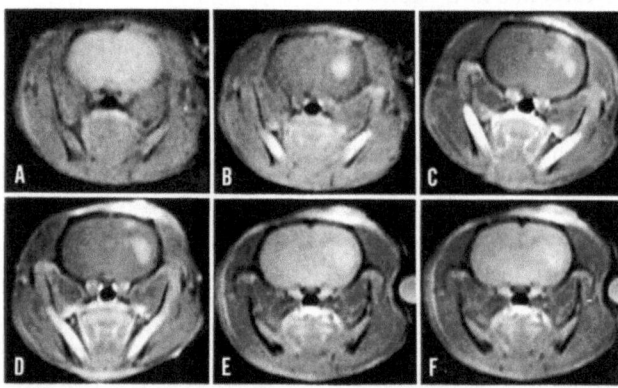

Fig. 1. T1-weighted MR images. (A) Pre-contrast, (B) post-contrast 15 min, (C) pretreated with PK11195, post-contrast 15 min, (D) pretreated with clonazepam, post-contrast 15 min, (E) post-contrast 17 h, (F) PK11195 replacement study, post-contrast 17 h

Table 1. *Tissue Concentration of Mn-TPPS*

Pretreatment	Tumor	Normal	T/N ratio	Kidney
Vehicle	161.9 ± 104.4	41.9 ± 3.6	3.79 ± 2.4	14.5 ± 2.1
PK11195	139.2 ± 28.1	36.9 ± 1.7	3.80 ± 0.9	11.1 ± 2.0
Clonazepam	149.0 ± 45.2	39.0 ± 5.6	3.80 ± 1.0	11.7 ± 2.4

Rats were pretreated with vehicle, PKl 1195, or clonazepam 15 min before Mn-TPPS administration. Tissues were removed approximately 15 min after Mn-TPPS injection. Normal: Contralateral brain, T/N ratio: Mn-TPPS concentration ratio tumor vs. contralateral brain. Data is expressed as mean ± SD (n = 3 for each experiment).

receptor turn-over usually occurs within a couple of hours, the long-lasting enhancement by Mn-TPPS may not be mediated by PBZ receptors.

The tissue Mn-TPPS concentration study also disclosed tumor-specific incorporation of Mn-TPPS. Although the proximal tubules of the kidney also posses a high concentration of PBZ receptors, tissue accumulation of Mn-TPPS decreased to 5% of the initial amount after 17 h. There was an apparent difference between the tumor tissue and the kidney for the prolonged enhancement and tissue Mn-concentration. This discrepancy also supports the assumption that selective accumulation of Mn-TPPS in tumor tissue is not mediated by PBZ receptors.

Many porphyrins are known to concentrate in tumor tissue both in vivo and in vitro, but the mechanisms of porphyrin-tumor affinity are unknown. Rat glioma cells may have a substantial affinity for the boronated porphyrin, and that affinity is largely cytoplasmic. These observations might provide a great opportunity when considering photosensitization therapy. However, we concluded that Mn-TPPS enhancement of the brain tumor was not simply mediated by PBZ receptors but predominantly by destruction of the BBB. The prolonged Mn-TPPS enhancement may be due to: 1) low diffusion rate in extracellular space, and/or 2) non-receptor mediated incorporation of Mn-TPPS into the tumor cells.

Acknowledgments

This work was partly supported by a grant from the Japanese Ministry of Health and Welfare.

References

1. Black KL, Ikezaki K, Santori EM, Becker DP, Vinters HV (1990) Specific high-affinity binding of peripheral benzodiazepine receptor ligands to brain tumors in rat and man. Cancer 65: 93–97
2. Bockhorst K, Hoehn-Berlage M, Kocher M, Hossmann K-A (1990) MRI contrast enhancement by MnTPPS of experimental brain tumours in rats. Acta Neurochir (Wien) [Suppl] 51: 134–136
3. Earnest F IV, Kelly PJ, Scheithauer BW, Kall BA, Cascino TL, Ehman RL, Forbes GS, Axley PL (1988) Cerebral astrocytomas: histopathologic correlation of MR and CT contrast enhancement with stereotactic biopsy. Radiology 166: 823–827
4. Ikezaki K, Black KL, Toga AW, Santori EM, Becker DP, Smith ML (1990) Imaging peripheral benzodiazepine receptors in brain tumors in rats: in vitro binding characteristics. J Cereb Blood Flow Metab 10: 580–587
5. Ikezaki K, Black KL, Tabuchi K (1991) Peripheral-type benzodiazepine receptors in brain tumors: in vitro binding characteristics and implication for new image analysis. In: Tabuchi K (ed) Biological aspects of brain tumors. Springer, Berlin Heidelberg New York Tokyo, pp 484–493
6. Patronas NJ, Cohen JS, Knop RH, Dwyer AJ, Colcher D, Lundy J, Mornex F, Hambright P, Sohn M, Myers CE (1986) Metalloporphyrin contrast agents for magnetic resonance imaging of human tumours in mice. Cancer Treat Rev 70: 391–395
7. Starosta-Rubinstein S, Ciliax BJ, Penney JB, McKeever P, Young AB (1987) Imaging of a glioma using peripheral benzodiazepine receptor ligands. Proc Natl Acad Sci USA 84: 891–895
8. Subramanian KS, Meranger JC (1982) Simultaneous determination of 20 elements in some human kidney and liver autopsy samples by inductively coupled plasma atomic emission. Sci Total Environ 24: 147–154
9. Verma A, Snyder SH (1988) Characterization of porphyrin interactions with peripheral-type benzodiazepine receptors. Mol Pharmacol 34: 800–805

Correspondence: Kiyonobu Ikezaki, M. D., Department of Neurosurgery, Neurological Institute, Faculty of Medicine, Kyushu University 60, Fukuoka 812, Japan.

Acta Neurochir (1994) [Suppl] 60: 356–358
© Springer-Verlag 1994

MRI Contrast Enhancement by Gd-DTPA-Monoclonal Antibody in 9L Glioma Rats

A. Matsumura, Y. Shibata, K. Nakagawa, and **T. Nose**

Department of Neurosurgery, Institute of Clinical Medical Science, University of Tsukuba, Tsukuba, Ibaraki, Japan

Summary

To achieve a tissue-specific enhancement in diagnosis of brain tumor, a magnetic resonance imaging (MRI) study was performed using conjugate of Gd-DTPA and monoclonal antibody (MoAb) against 9L glioma cells. Fisher 344 strain rats were used for this study. MoAb against 9L glioma cells was conjugated with Gd-DTPA according to the method of Hnatowich et al. (1983) and used for the MRI study. The gadolinium (Gd) concentration in the Gd-MoAb injected to the rats was 0.01–0.03 mmol/kg. The enhancement effect increased gradually and persisted for 24 hours after the injection. This was longer than Gd-DTPA, which showed a peak of enhancement effect within 30 minutes after injection and was washed out within 120 min. This result was compatible with scintigraphy studies using ^{125}I labeled anti 9L monoclonal antibody, in which the accumulation of the ^{125}I antibody increased at 24, 48 and 72 hours after the injection. By using tumor-specific contrast agents such as Gd-MoAb, it may be possible to differentiate among tumor, perifocal edema and radiation injury.

Keywords: Magnetic resonance imaging; brain tumor; monoclonal antibody; gadolinium.

Introduction

Gadopentate dimegulumine (Gd-DTPA) is widely used as a contrast agent for magnetic resonance imaging (MRI). Gd-DTPA is useful in delineating the margin of the tumor on T1-weighted images.

However, it is known that Gd-DTPA crosses the damaged blood-brain barrier (BBB), and thus infarction area and radiation necrosis are also enhanced by Gd-DTPA. Therefore, the developement of tissue-specific contrast agents deserves considerable attention due to their specificity to the tumor. The use of monoclonal antibody as an MRI contrast agent in subcutaneous tumor in nude mice has been successful[1]. In the present study, we demonstrate the tumor-specific enhancement of intracerebrally transplanted 9L glioma

tumor in Fisher 344 rats and discuss the usefullness and limitations of this contrast agent.

Materials and Methods

Monoclonal antibody (MoAb) against 9L glioma cells was produced by using hybridoma of Balb/c mouse spleen cells and P3U1 myeloma cells.

For the scintigraphy study, ^{125}I labeled monoclonal antibody to 9L was used. Nonspecific IgG was also labeled by ^{125}I and served as a control. Scintigraphy was performed in Fisher rats with subcutaneous 9L tumor.

The Gd-MoAb was conjugated according to the method of Hnatowich et al. (1983)[2]. DTPA anhydride and 9L monoclonal antibody were first conjugated, and then, in order to remove the excess DTPA, gel fitration through a Sephadex G-50 was performed. Next, Gd-acetate was added to label DTPA-MoAb with Gd. The final molar ratio was 1 : 1.

For the experimental brain tumor model, Fisher 344 strain rats were used. $1 \times 10^4/10 \mu l$ 9L glioma cells were inoculated into the brain parenchyma at a depth of 5 mm from the surface through a burr hole near the coronal suture using a stereotactic manipulator with a 27G fine needle.

MRI was performed 14 days after the innoculation under general anesthesia by pentobarbital with spontaneous breathing. MRI was performed by a BMT24/40 superconducting MRI system (Bruker Co., Ltd., 2.4T) by using spin echo images for T1-weighted (Tr/Te = 512/28) and T2-weighted (Tr/Te = 2040/80) images. The enhancement effect of Gd-MoAb was compared with that of Gd-DTPA.

After MRI analysis, the rats were sacrificed using an overdose of pentobarbital. Tumor, brain, skin, liver, kidney and serum were sampled and immediately kept in a deep freezer until being used for the analysis. Gd concentrations in the tissue were measured by an ICP-analyzer (Seiko Electric Co., Ltd., SPS 1200A).

Results

[125]I MoAb Scintigraphy Study

After injection of ^{125}I-MoAb, the rats were serially evaluated by scintigraphy. As shown in Fig. 1, [125]I-

MoAb to 9 L cells (left side) accumulated in the tumor (black arrows) and in the liver after 48 hours, while nonspecific IgG (right side) showed no accumulation in the tumor (white arrows).

MRI Study

The gadolinium concentration in the Gd-MoAb injected to the rats was 0.01–0.03 mmol/kg. The enhancement effect of Gd-MoAb increased gradually and persisted for 24 hours after the injection (Fig. 2), although the change in intensity was rather weak compared to that of Gd-DTPA. Gd-DTPA injected at 0.1 mmol/kg showed a peak of enhancement effect in the tumor within 30 minutes after injection, and there was almost no enhancement effect after 120 minutes.

Tissue Gd Concentration

The tissue concentration of Gd in the Gd-DTPA group rapidly increased in the serum, kidney and tumor, while there was only very low accumulation in the normal brain. The peak concentration of gadolinium in the tumor was 18 ppm. The clearance of Gd-DTPA from all tissues was fairly rapid. Unlike Gd-DTPA, Gd-MoAb did not accumulate in the kidney, but mainly in the liver. The gadolinium concentration in the tumor was below 0.03 µg/g tissue throughout the entire time course in the majority of cases. The accumulation of gadolinium in the liver slowly increased after injection and reached about 40 ppm after 48 hours.

Discussion

Monoclonal antibody to 9L glioma cells exhibited a specific binding activity to intracerebrally transplanted experimental tumor cells evaluated by both scintigraphy and MRI.

The enhancement effect of Gd-MoAb on MRI continued for 24 hours after injection. This result was compatible with a scintigraphy using ^{125}I labeled anti-9L monoclonal antibody study in which accumulation of the ^{125}I-MoAb in the tumor relatively increased at 24, 48 and 72 hours after the injection. The results indicate tissue-specific binding activity and this may be useful in differential diagnosis of various types of tumor and/or damaged blood-brain barrier (e.g. cerebral infarction and radiation necrosis).

The increase of signal intensity of MRI was rather weaken with Gd-MoAb than with Gd-DTPA. This is thought to be due to the amount of Gd given to the rats.

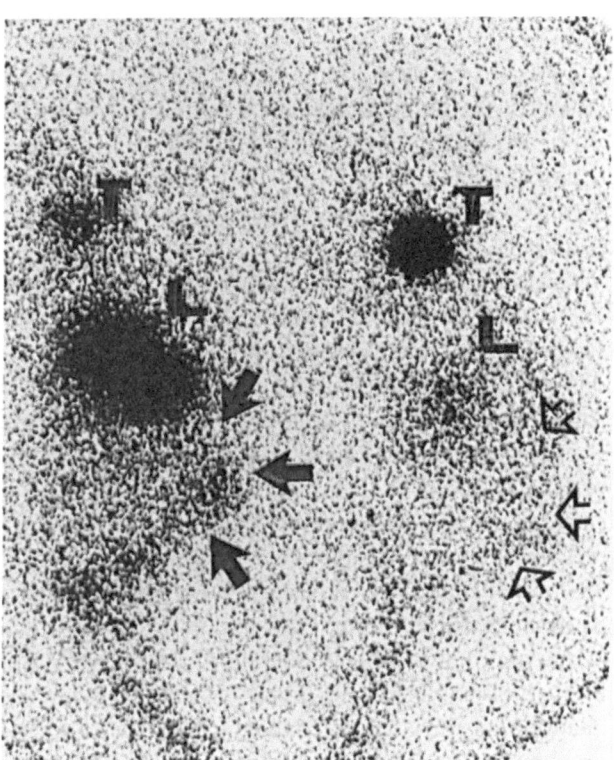

Fig. 1. [125]I scintigraphy of 9L monoclonal antibody (left side) and non specific IgG (right side) in 9L tumor-bearing rats. Note the accumulation in the tumor (*T*), with monoclonal antibody (black arrows), but no accumulation with IgG (white arrows). *L* liver, *T* thyroid

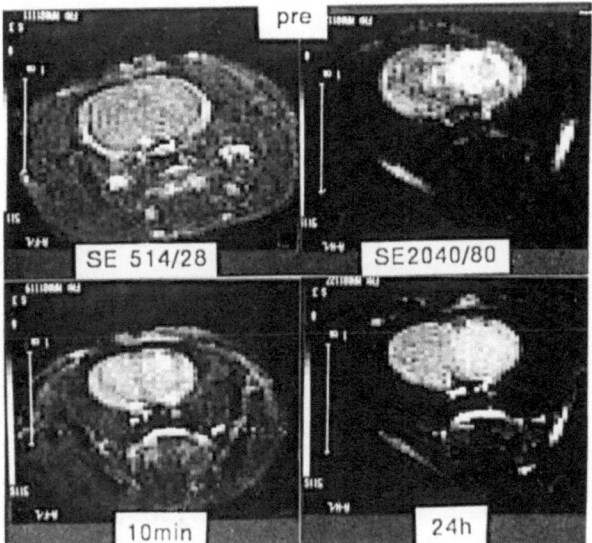

Fig. 2. MRI showing the 9L tumor. Upper row showing pre-contrast T1 image (left) and T2 image (right). After injection of Gd-MoAb, the tumor is enhanced aftr 10 min, and also after 24 hours

Gd-DTPA is used at 0.1 mmol/kg in clinical settings, and so this concentration was used in this study. Gd-MoAb could be administrated at only 0.01–0.03 mmol/kg in the present study due to the technical limitations of injecting a large amount of Gd-MoAb to the rats. Another reason for the faint and delayed enhancement effect is the high molecular weight of Gd-MoAb. While the molecular weight of Gd-DTPA is only 743, that of Gd-MoAb is about 160,000, which is 200 times that of Gd-DTPA, and therefore its uptake in the tumor is limited.

Gd-MoAb enhancement has been succesful in various subcutaneous tumors using nude mice[1]. Another study using normal hamsters and CEA failed to demonstrate a contrast enhancement[4]. As demonstrated from our data, for in vivo use of Gd-MoAb in subjects with normal immunological systems, the accumulation of Gd-MoAb in the liver makes it difficult to achieve sufficient accumulation of Gd-MoAb in the tumor itself. To overcome this problem, several improvements has to be developed. The intra-arterial administration of monoclonal antibody[3], the use of only Fab fragments in order to decrease the molecular weight[5], and the use of polylysine for more effective labeling of gadolinium[5] are all under investigation. If these improvements are succesful, tumor-specific contrast agents such as Gd-MoAb may become useful in the differential diagnosis of various types of tumor and also in the delineation of perifocal edema and/or radiation injury.

Acknowledgement

This work was supported by University of Tsukuba Research Projects. The authors express their gratitude to Schering Japan Co., Ltd. for their assistance.

References

1. Göhr-Rosenthal S, Schmitt-Willich, Ebert W, Gries H, Vogler H, Weinmann HJ, Conrad J (1991) An immunoselective contrast medium for MRI: detection of colorectal tumor transplants in mice with a Gd-labeled monoclonal antibody. Abstract of 10th SMRM, p 356
2. Hnatowich DJ, Childs RL, Lanteigne D, Najafi A (1983) The preparation of DTPA-coupled antibodies radiolabeled with metallic radionuclides: an improved method. J Immunol Meth 65: 147–157
3. Ingvar C, Norrgren K, Strand SE, Brodin T, Jönsson PE, Sjögren HO (1991) Tumor uptake of monoclonal antibody after regional intraarterial injection. Acta Oncol 30: 65–69
4. Unger EC, Totty WG, Neufeld DM, Otsuka FL, Murphy WA, Welch MS, Connet JM, Philpott GW (1985) Magnetic resonance imaging using gadolinium-labeled monoclonal antibody. Invest Radiol 20: 693–700
5. Wang TST, Fawwaz RA, Alderson PO (1992) Reduced hepatic accumulation of radiolabeled monoclonal antibodies with indium-111-thioether-poly-L-lysine-DTPA-monoclonal antibody-TP41.2F (ab')2. J Nucl Med 33: 570–574

Correspondence: A. Matsumura, M.D., Department of Neurosurgery, Institute of Clinical Medical Science, University of Tsukuba, Ibaraki 305, Japan.

Tumor and Brain Edema: Pathophysiology, Clinical Studies

Acta Neurochir (1994) [Suppl] 60: 361–364
© Springer-Verlag 1994

Peritumoral Edema in Meningioma: a Contrast Enhanced CT Study

U. Ito, H. Tomita, O. Tone, H. Masaoka, and **B. Tominaga**

Department of Neurosurgery, Musashino Red Cross Hospital, Tokyo

Summary

The propagation of extravasated contrast medium around 6 supratentorial meningiomas with peritumoral white matter of low density (PWL) of Lanksch II~III was investigated by repeated CT scanning at 4 h intervals, following a 1 h drip infusion of 200 ml of Iopamidol. The volume of the expanding peritumoral contrast enhancement was calculated according to a method previously described[5-8]. By calculating the increase in volume from the first to the second scan, and from the second to third, we derived the rate of edema formation as well as the resolution rate of edema in the PWL. The surface area of the entire tumor (TS) and area of tumor surface facing the PWL (LS) were calculated by summating the surface areas of all CT slices, each area of which was derived from the measured length of the entire circumference of the tumor and circumference of the tumor facing the PWL, respectively, multiplied by the slice thickness of 0.5 cm. The volume of PWL, edema formation rate of entire tumor, and tumor volume $\times$ LS/TS were well correlated with each other. We concluded that the severity of peritumoral edema in meningiomas depends on the size of the tumor and the extent of tumor surface contact with the PWL.

Keywords: Brain edema; meningioma; CT enhancement; steroid therapy.

Introduction

The characteristics of peritumoral edema induced by exudation of plasma filtrate through leaky tumor capillaries have been well described in different types of brain tumors[2,4-10]. However, not all meningiomas exhibit peritumoral edema. As to the edema formation of meningiomas, such various factors as tumor location[3,11], tumor size and clinical evolution[2,9,11], sinus involvement of the tumor[11], tumor vascularity[1,3], presence of secretory activity in the tumor cells, and histopathological subtype[2,3,9] have been described. Recently, the relevance of breached leptomeninx and cortical tissue separating the extramedullary growing tumor from the white matter has been reported[2,9].

In previous studies[5-8], propagation of the extravasated contrast medium around various intra-axial brain tumors have been investigated by repeated CT scans, following a 1 h drip infusion of 200 ml of Iopamidol. By measuring the increase in volume of the expanding peritumoral contrast enhancement, we calculated the rate of formation and resolution of peritumoral brain edema. In the present study, applying this method to patients with meningiomas, the pathomechanism responsible for the peritumoral brain edema associated with meningiomas is investigated.

Materials and Methods

Using a GE 9200 CT scanner, we obtained serial axial scans of 0.5 cm in thickness in 5 patients with infratentorial tumor and in 9 patients with supratentorial tumor. The first scan was taken at the end of a 1 h continuous drip infusion of 200 ml of Iopamidol (66% meglumin amidotrizoate: 300 mgI/ml, molecular weight 809.14); and the second and third were made at 4 h intervals (t_1) thereafter. Following this procedure, all 14 patients received i.v. dexamethazone (8 mg/day) for 4 days; and the same study was then repeated. In the 6 supratentorial tumors with peritumoral white matter of low density (PWL) of Lanksch grade II and III[4-6], the contrast-enhanced area expanded around the tumor border into the adjacent PWL, as we had previously observed in various intramedullary brain tumors[5-8]. The volume of the peritumoral enhancement was calculated by summating the volumes of all slices, the enhanced area of each of which was planimetrically measured and multiplied by the 0.5 cm thickness of the slice. The increase in volume of the peritumoral enhancement from the first to second scan (ΔV_1) and from the second to third (ΔV_2) was calculated for each tumor, and the formation rate (ΔV_F) of edema from 1 cm^3 of tumor and resolution rate (ΔV_R) of edema in 1 cm^3 of PWL were calculated from the two formulas below based on the following assumptions: 1) the concentration of the contrast medium in the extravasated edema fluid was similar to that in the plasma, 2) the contrast medium was distributed only in the extracellular space and moved with a similar speed as other solutes in the edema fluid, and 3) the extracellular space in the peritumoral edematous white matter was about 30% of the total volume[6,7]:

$$\Delta V_F = \frac{2\Delta V_1 - \Delta V_2}{V_1 \times t_1} \times 0.3 \; \text{ml/cm}^3 \; \text{tumor/h}$$

$$\Delta V_R = \frac{\Delta V_1 - \Delta V_2}{V_1 \times t_1} \times 0.3 \; \text{ml/cm}^3 \; \text{white matter/h}$$

The statistical analysis was performed by the paired t-test (Table 1). Informed consent was obtained from all patients for the repetition of the CT examination following the drip infusion.

The volume of tumor as well as that of PWL of the 6 patients was calculated in the same way by summating the volumes of tumor and PWL for all CT slices, each of which was derived by multiplication of the planimetrically measured area of tumor and PWL, respectively, by the slice thickness of 0.5 cm.

The area of the entire tumor surface (TS) and area of tumor surface facing the PWL (LS) of the 6 patients was calculated by summating the surface areas of all CT slices, each of which was derived by multiplication of the length of the entire circumference of the tumor amd circumference of the tumor facing the PWL, respectively, by the slice thickness of 0.5 cm.

Results

The infratentorial tumors without PWL were either petrosal or tentorial in origin, and most of the tumor mass was located in the basal cistern and had made partial contact with the cerebellum and brain stem with a thin but distinct CSF space intervening between them. The same structure was also recognized by the operative microscope around the supratentorial meningiomas without PWL. No peritumoral expansion of the contrast enhancement was observed in any of the 3 infratentorial and 2 supratentorial tumors without PWL after a continuous 1 h drip infusion of 200 ml of Iopamidol. Among the tumors of different histological subtypes, no special tendency in edema size was observed, except for one malignant meningioma that showed large PWL in proportion to the tumor size.

All supratentorial meningiomas with PWL of Lanksch II and III showed peritumoral expansion of the contrast enhancement into the adjacent PWL (Fig. 1). Corresponding to the tumor surface that faced the PWL (LS) seen in the CT, strong adhesion between tumor capsule and brain with breached leptomeninx and cortical tissue between them was recognized by the operation microscope.

Formation rate (ΔV_F) and resolution rate (QV_R) of edema in 6 supratentorial tumors are listed in the Table. After steroid treatment, the average rate of edema formation was reduced to $63.3 \pm 17.6\%$ ($p < 0.05$), but the resolution rate did not change significantly. On the same 6 supratentorial tumors, volume of PWL, edema formation rate of entire tumor ($\Delta V_F \times$ tumor volume), and tumor volume $\times$ LS/TS were assessed for possible correlations (Fig. 2). Following strong positive correlations were obtained; volume of PWL vs. tumor volume $\times$ LS/TS ($Y = 4.0 \, X - 12.7$, $r = 0.90$, $p < 0.025$), edema formation rate of entire tumor vs. tumor volume $\times$ LS/TS ($Y = 0.076 \, X = 0.54$, $r = 0.98$, $p < 0.005$), and volume of PWL vs. edema formation rate of entire tumor ($Y = 55.6 \, X = 47.2$, $r = 0.82$, $p < 0.05$). Therefore, we can conclude that the edema formation rate of the entire tumor depends on the size of the tumor multiplied by the ratio of the tumor surface in direct contact with the PWL (LS) to the total tumor surface (TS), and thus determine the size of the PWL.

Table 1. *PWL Volume, Tumor Volume, Formation Resolution Rate of Edema*

Patient			PWL volume (cm³)	Tumor volume (cm³)	Formation rate of edema (ΔV_F) (ml/cm³ tumor/h) Steroid			Resolution rate of edema (ΔV_R) (ml/cm³ white matter/h) Steroid		
					−	+	(%)	−	+	(%)
H.H.	70	M	11.5	28.1	0.026	0.010	(38.5)	0.035	0.016	(45.7)
A.J.	49	F	37.5	32.1	0.047	0.021	(44.7)	0.028	0.046	(164.3)
N.K.	56	M	83.1	42.4	0.039	0.030	(76.9)	0.031	0.036	(116.1)
T.R.	69	M	57.6	52.7	0.036	0.028	(77.8)	0.031	0.029	(93.6)
U.H.	61	F	70.2	77.6	0.034	0.022	(64.7)	0.012	0.022	(83.3)
M.Y.	63	M	8.3	116.9	0.013	0.010	(76.9)	0.038	0.035	(92.1)
Average ± SD					0.033 ± 0.012	0.020 ± 0.009	(63.3)* ± (17.6)	0.029 ± 0.009	0.031 ± 0.011	(115.9) ± (50.8)

PWL peritumoral white matter of low density. * $p < 0.05$.

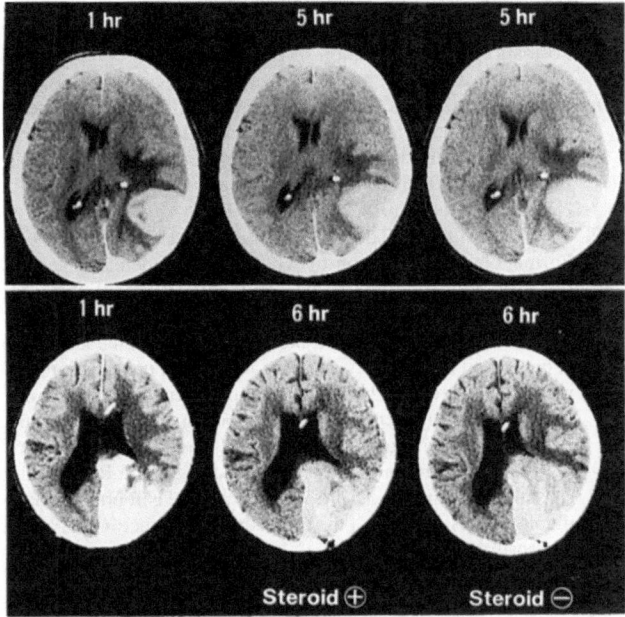

Fig. 1. Enhanced tumor and peritumoral expansion of contrast enhancement 4 or 5 hours after a 1 h continuous drip infusion of 200 of Iopamidol, before and after steroid treatment. Upper: meningothelial meningioma. Lower: malignant meningioma

Discussions

PWL size has been reported to be positively correlated with tumor size of meningiomas[2,9,11]. In the present study, we could not find a significant correlation between them by measurement on CT slices. However, we have found the volume of the PWL and formation rate of edema in the entire tumor are both well correlated with the value of tumor volume multiplied by the percentage ratio of the tumor surface facing the PWL to the total tumor surface. In other words, the severity of peritumoral edema in meningiomas is well correlated with tumor volume and area of the tumor surface facing the PWL[2,9]. By the operative microscope, this tumor surface (SL) is observed to be strongly attached to the brain surface where leptomeninx and cortical tissue are breached, and received an arterial feeding supply from the leptomeningeal arterial branches. These findings are compatible with angiographic data indicating that the vascular supply from the intrinsic cerebral arteries influences the severity of the peritumoral edema of meningiomas[1,3].

An alternative source for peritumoral edema that is typically vasogenic in type would be the peritumoral brain tissue. Different kinds of mediators such as leukotrienes, arachidonate metabolites from macrophages, and tumor plasminogen activator (TPA), as well as some immunological reactions and ischemia with sup-

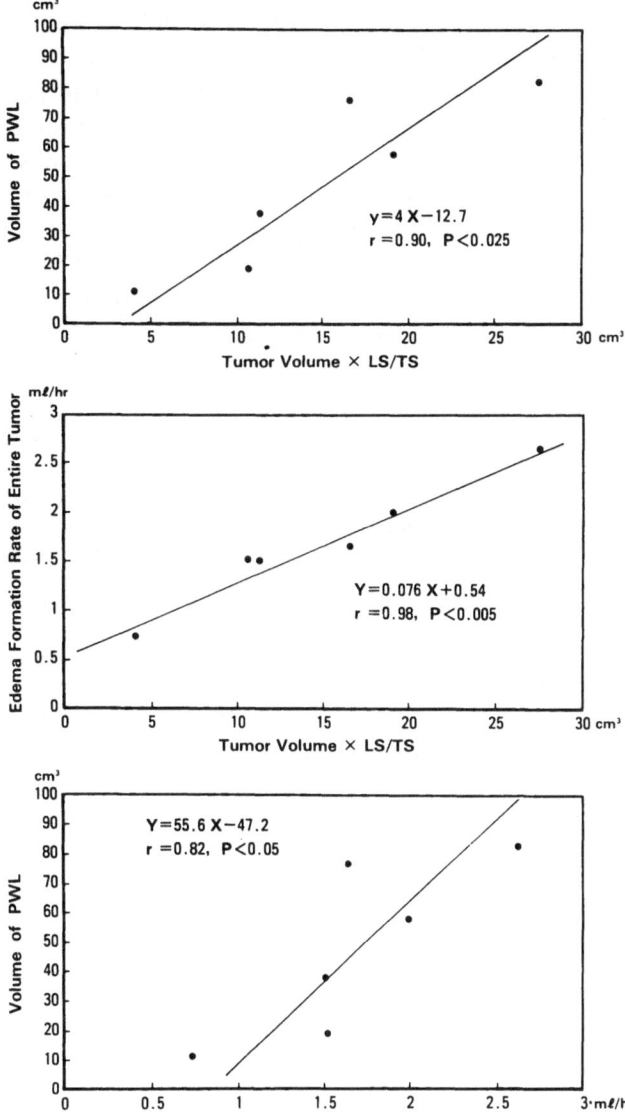

Fig. 2. Correlation and correlation coefficients (r) among volume of PWL, edema formation rate of entire tumor ($\Delta V_F \times$ tumor volume), and tumor volume $\times$ LS/TS. PWL peritumoral white matter of low density (cm³), LS area of tumor surface facing PWL (cm³), TS area of entire tumor surface (cm³)

pressed oxygen metabolism, have been reported to induce edema in the peritumoral brain tissue[11]. However, we never see contrast enhancement, which is a cardinal indicator of BBB breakdown, immediately after a 1 h continuous infusion of 200 ml of Iopamidol. As the spread of the contrast medium is suppressed by the steroid treatment used to reduce the size of the PWL, simple peritumoral diffusion of the contrast medium is not indicated[5–8].

One malignant meningioma showed an extraordinarily large volume of PWL relative to the size of the tumor. However, we encountered difficulty in investigating the correlation of edema size and histological subtype, basically because the subtype was not homogeneous in each tumor, and because only small portions of the tumor was biopsied removing other tumor portion crushed by the ultrasonic mess[2,3,9]. The major therapeutic effect of steroid treatment on the peritumoral edema of the meningioma is, as in the case of other intramedullary brain tumors[4-8], a reduction in the rate of edema formation (Fig. 1), probably due to a suppression of vascular permeability of the tumor tissue.

References

1. Casasco A, Mani J, Alachkar F, Jahara M, Theron J (1986) L'oedeme peritumoral dans les meningiomes intracraniens. Correlation angiographique et tomodensitometrique. Neurochirurgie 32: 296–303
2. Go KG, Wilmink JT, Molenaar WM (1988) Peritumoral brain edema associated with meningiomas. Neurosurgery 23: 175–179
3. Inamura T, Nishio S, Takeshita I, Fujiwara S, Fukui M (1992) Peritumoral brain edema in meningiomasinfluence of vascular supply on its development. Neurosurgery 31: 179–185
4. Ito U, Reulen H-J, Huber P (1986) Spatial and quantitative distribution of human peritumoural brain oedema in computerized tomography. Acta Neurochir (Wien) 81: 53–60
5. Ito U, Reulen H-J, Tomita H, Ikeda J, Saito J, Maehara T (1988) Formation and propagation of brain oedema in computerized tomography. Acta Neurochir (Wien) 81: 53–60
6. Ito U, Tomita H, Reulen H-J, Maehara T, Kohmo Y, Ito Y (1989) A CT study on formation, propagation and resolution of brain oedema fluid in human peritumoral oedema. In: Hoff JT, Betz AL (eds) Intracranial pressure VII. Springer, Berlin Heidelberg New York Tokyo, pp 959–965
7. Ito U, Tomita H, Tone O, Shishido T, Hayashi H (1990) Formation and resolution of white matter oedema in various types of brain tumours. Acta Neurochir (Wien) [Suppl] 51: 149–151
8. Ito U, Reulen H-J, Tomita H, Ikeda J, Saito J, Maehara T (1990) A computed tomography study on formation, propagation, and resolution of edema fluid in metastatic brain tumors. In: Long DM (ed) Brain edema; pathogenesis, imaging and therapy. Advances in neurology, vol 52. Raven, New York, pp 456–468
9. Ohno K, Matushima Y, Aoyagi J, Ikeda J, Suzuki R, Ichimura M, Takei M, Hirakawa K (1992) Peritumoral cerebral edema in meningiomas: the role of the tumor-brain interface. Clin Neurol Neurosurg 94: 291–295
10. Reulen H-J, Graber S, Huber P, Ito U (1988) Factors affecting the extension of peritumoral brain ooedema. A CT-study. Acta Neurochir (Wien) 95: 19–24
11. Stevens JM, Ruiz JS, Kendall BE (1983) Observations on peritumoural oedema in meningioma: part II. Mechanisms of oedema production. Neuroradioloy 25: 125–131

Correspondence: Umeo Ito, M.D., Department of Neurosurgery, Musashino Red Cross Hospital, 1-26-1, Kyonan-cho, Musashino-shi, Tokyo 180, Japan.

Acta Neurochir (1994) [Suppl] 60: 365–368

A Study on Peritumoral Brain Edema Around Meningiomas by MRI and Contrast CT

K. G. Go, R. L. Kamman, J. T. Wilmink, and **E. L. Mooyaart**

Departments of Neurosurgery, Neuroradiology and MRI, University of Groningen, The Netherlands

Summary

In the present study upon 9 meningiomas, the volume of peritumoral brain edema was calculated by integration of the cross-sectional edematous areas on serial MRI slices. It was zero in 3 cases and ranged from 11 to 176.4 ml in the other cases. There was disruption of the cortex in all cases, ranging from slight to severe. T_1 and T_2 measurements were made at the level of maximum edema extension (at 1.5 T), using a mixed sequence. From the T_2 values tissue water content in % was calculated using equations, which had been obtained by correlating water content with relaxation rates. These had been measured on human brain autopsy specimens subjected to hydration or dehydration. Assuming that spread of contrast agent marks the advancement of the front of edema produced by the tumor, calculation of the excess of edema production per unit tumor volume V_{eff}/V_N was performed on the basis of CT-studies made before, and at 1.5, 3 and 6 hr after contrast infusion. Of 6 tumors with edema (mean water content of peritumoral white matter of 91% and mean edema volume of 69.2 ml) the V_{eff}/V_N was 0.18–1.08 ml/h/cm³ tumor volume, whereas 3 tumors without associated edema had a V_{eff}/V_N of 0.03–0.12 ml/h/cm³. In meningiomas cortex penetration seemed to be a separate factor determining the spread of edema.

Keywords: Brain edema; meningiomas; brain water content; contrast exudation.

Introduction

As meningiomas have an extracerebral origin, peritumoral edema which arises in the tumor, can spread into the adjacent brain only when it has crossed the leptomeninges, subarachnoid space and cerebral cortex. These are the layers which separate the tumor from the cerebral white matter in which the edema tends to accumulate. Indeed it could be demonstrated, that the amount of edema correlated with the incidence of cortical penetration by the tumor, with its size and histological subtype, but not with its location[1]. In this study the edema is assessed more accurately by magnetic resonance imaging (MRI). The spread of edema fluid from the tumor was studied by computerized tomography (CT) in conjunction with the infusion of contrast agent, using the movement of the contrast agent into the surrounding brain as a marker of the spread of edema fluid[2,3,5].

Material and Methods

The study was conducted on 8 patients with meningiomas (3 frontal, 5 parietal), in one of whom there were 2 parietal tumors separated by the falx (Fig. 1). The patients did not use glucocorticosteroids at the time of the study.

The volume of peritumoral edema was measured by MRI using T_1-weighted (IR-TR: 2000, TI: 400, TE: 20) and T_2-weighted (SE-TR: 2000, TE: 50) images in 2 directions. On serial slices (5 mm thick) the volume of edema was calculated by integration of cross-sectional surfaces of the edematous areas.

Tissue water content was calculated from tissue T_1 and T_2 on the basis of an equation previously obtained: $R_i = b/WC + a$ for the relation between relaxation rate R_i (i = 1 for $R_1 = 1/T_1$ and i = 2 for $R_2 = 1/T_2$) and water content WC[4]. For application to human tissue, T_1 and T_2 measurements were performed (mixed-sequence at 1.5 T (Philips Gyroscan) with TR-SE = 1000, TR-IR = 1400, TI = 400, TE = 30 (8 echos), # averaging = 2) on fresh autopsy samples of human cortex and white matter, both native and after alteration of their water content by immersion in distilled water, in distilled water/saline, or in 20% mannitol solution for 24 h. Subsequently their % water content was determined from the wet and dry weights. By regression analysis the relations between T_2 and WC were obtained (Fig. 2). T_1 and T_2 maps were made of the patients, using the same mixed sequence at the level of maximum tumor size. By applying the equations on readings of the T_2 maps, tissue water content was obtained.

The kinetics of edema fluid production was studied from the spreading of contrast agent. Assuming that contrast agent is carried by the edema fluid as it spreads from the tumor into the adjacent brain tissue, the increase in size of the contrast-stained area on CT-scans may be considered to mark the moving front of edema.

The kinetics comprise competition of production versus absorption. Production may be expressed as:

$$V_p = \int_0^t J_p(t) \cdot dt = \int_0^t p \cdot V_N \cdot dt = p \cdot V_N \, \Delta t \qquad (1)$$

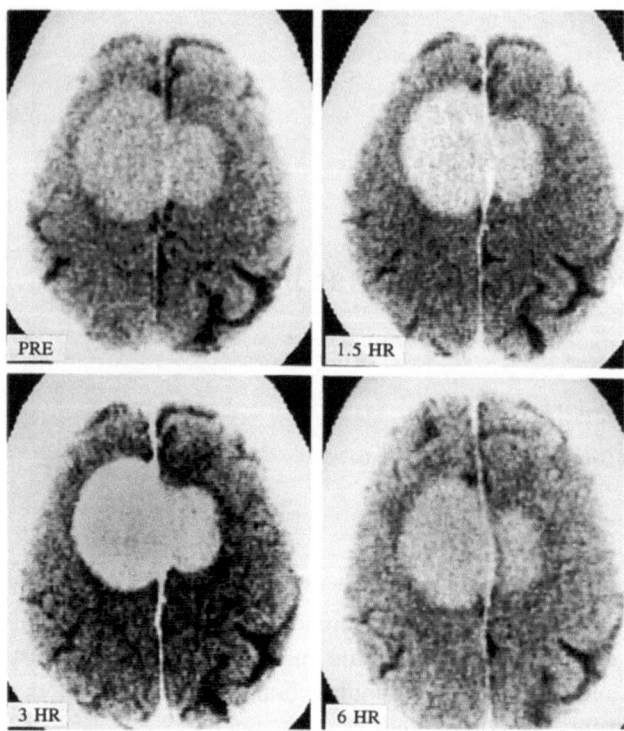

Fig. 1. Sequential CT-scans of case no. 5 made (*PRE*) pre-infusion, (*1.5 HR*) 1.5 hours, (*3 HR*) 3 hours, and (*6 HR*) 6 hours after starting intravenous contrast infusion in patient with tumors on either side of the falx. There is minimal change in the cross-sectional surface of the contrast-enhanced area. Note also disruption of the cortex, and absence of edema

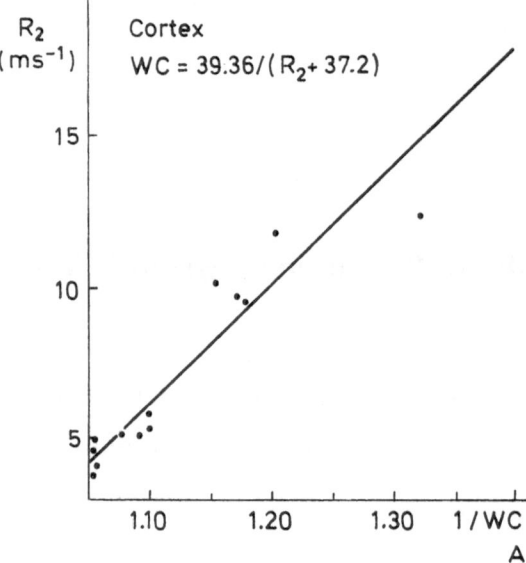

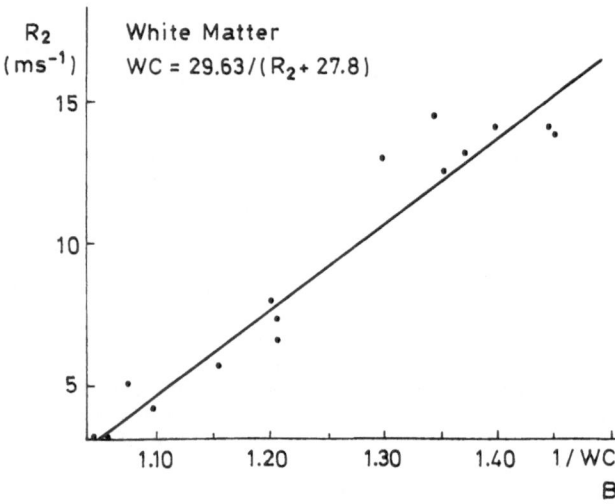

Fig. 2. Regression line for the relation between water content WC and relaxation rates (A) R_2 for cortex and (B) R_2 for white matter

V_p: total edema fluid production (ml) by the tumor (ml/s/cm³ tumor). J_p: rate of edema fluid production (ml/s). V_N: initial volume at time t_0 before contrast infusion.

The spreading edema fluid occupies a volume ΔV of brain, or rather the extracellular space of the measured volume, the effective volume V_{eff}.

As the edema fluid occupies a volume of the adjacent brain tissue, it is liable to absorption by the tissue. Therefore, in fact $V_{eff} = V_p - V_a$, with V_a: the volume of edema fluid (ml) absorbed in a period Δt within the volume V_{eff}.

$$V_a = \int_0^{t_f} J_a(t) \cdot dt \qquad (2)$$

with J_a: absorption rate (ml/s).

During the production of edema fluid volume V_p within a period Δt there is a continuous absorption of a final amount of V_a. In the steady-state (equilibruim) no further increase in occupied volume will be noticed. Then

$$\frac{dV_{eff}}{dt} = 0 \rightarrow \frac{dV_p}{dt} = \frac{dV_a}{dt} \qquad (3)$$

and

$$p \cdot V_N = a \cdot V_{eff} \rightarrow p = \frac{V_{eff}}{V_n} \cdot a \qquad (4)$$

V_{eff}/V_N: the excess of edema fluid production per unit of time per unit of tumor volume is a direct measure for differences in edema production between tumors, and can be calculated from the increases of the circumference of the contrast-stained areas on the CT-

scan. CT-slices (9 mm thick) were made at 4 time points: before, and at 1.5 h, 3 h and 6 h after starting the infusion of contrast agent (Fig. 1 and 3). The contrast agent (200 ml of Rayvist) was intravenously infused at a constant rate for 3 h. At every time point the sagittal and frontal diameters (cm) as well as the surface area (cm²) of the tumor and contrast-stained area were measured.

Results and Discussion

Application of the equations, relating WC to T_2 on the T_2-readings of normal cortex and normal white matter, respectively, gave a mean water content of 82.53 ± 1.19% for normal cortex, and 74.72 ± 1.85% for normal white matter. (The water content calculated from the T_1-values by means of the T_1/WC relations resulted in unrealistically high values, presumably because of postmortem changes in the autopsy specimens.)

Table 1

Patient	Tumor volume (ml)	Penetration of cortex	Volume of edema (ml)	Water content (%)	V_{eff}/V_N (ml/h/cm³)
1.	59.4	+	0	77.3	0.03
2.	18.1	+	39.6	88.4	0.42
3.	2.0	++	57.2	91.5	0.30
4.	29.7	++	176.4	94.1	1.08
5. right	21.3	+	0	75.9	0.12
left	6.7	+	0	73.5	0.12
6.	140.0	+	48.2	90.1	0.18
7.	38.8	±	11.0	94.8	0.18
8.	37.2	++	82.3	87.3	0.18

From the measurements of edema volume by integration of the cross-sectional surfaces of edematous areas on the MRI-scans, it appeared that in 3 tumors there was no peritumoral edema, i.e. volume of edema was 0, while the water content of peritumoral white matter was normal. In the other 6 cases there was edema to a greater or lesser extent, with a water content ranging from 87.3% to 94.8% and a volume of edema which ranged from 11.0 to 176.4 ml (Table 1). But the volume of edematous tissue did not always correlate with its water content. Notably in patient # 7 (Table 1) there was only 11 ml of edematous tissue, while its water content was the highest of the series (94.8%).

The excess of edema fluid production V_{eff}/V_N at steady-state was in the range of 0.03 to 0.12 ml/h/cm³ in 3 cases, with normal (73.5 and 75.9%) or slightly increased (77.3%) peritumoral tissue water content. In the other 8 cases the excess of edema fluid production ranged from 0.18 to 1.08 ml/h/cm³, while peritumoral water content was considerably increased (87.3 to 94.8%) (Table 1). The patient # 4 with the largest volume of edema (176.4 ml) also appeared to be the one with the highest production excess (1.08 ml/h/cm³), while there was no peritumoral edema in the patients with a low production excess, because apparently the minimal excess of edema formation in these cases does not significantly differ from the normal rate of extracellular fluid formation.

In all cases there was penetration of the cortical layer in various degrees. The degree of cortex penetration did not correlate with the excess of edema fluid production. Although the cases with the most severe degree of cortex penetration did show high values of peritumoral water content, the case (no. 7) with only 11 ml of edema and a high peritumoral water content

(94.8%) had only slight cortical disruption. For comparison, another case (no. 8) with similar tumor volume (37.2 ml) and equal excess edema fluid production (0.18 ml/h/cm³), but severe cortex penetration, had a

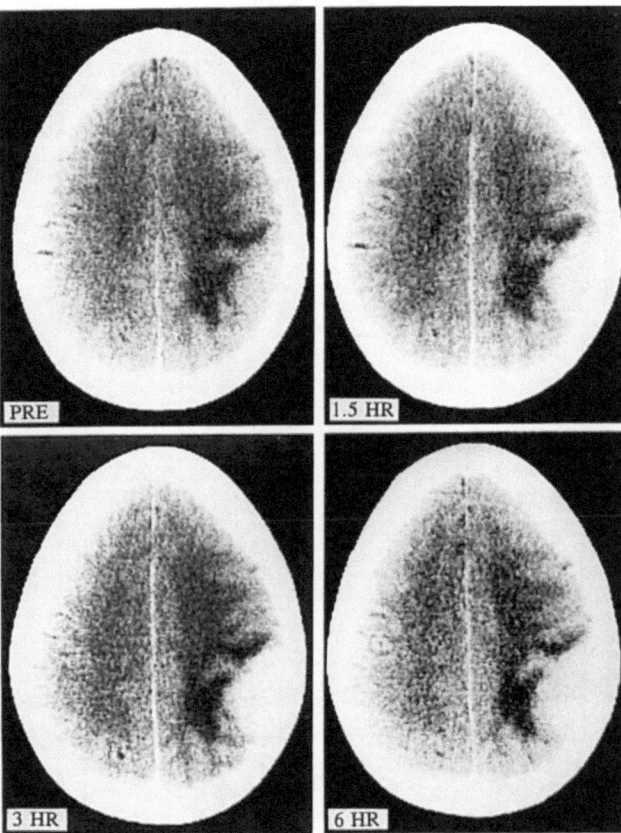

Fig. 3. Sequential CT-scans of case no. 2 made (*PRE*) pre-infusion, (*1.5 HR*) 1.5 hours, (*3 HR*) 3 hours, and (*6 HR*) 6 hours after starting intravenous contrast infusion in patient with right frontal tumor. There is a gradual increase in the cross-sectional surface of the contrast-stained area. Note also severe disruption of the cortex and severe edema

much larger volume of edema (82.3 ml) but a lower water content (87.3 ml). Apparently cortical penetration constitutes a separate factor determining the spread of peritumoral edema in meningiomas.

References

1. Go KG, Wilmink JT, Molenaar WM (1988) Peritumoral brain edema associated with meningiomas. Neurosurgery 23: 175–179
2. Ito U, Reulen HJ, Huber P (1986) Spatial and quantitative distribution of human peritumoural oedema in computerized tomography. Acta Neurochir (Wien) 81: 53–60
3. Ito U, Reulen HJ, Tomita H, Ikeda J, Saito J, Maehara T (1988) Formation and propagation of brain oedema fluid around human brain metastases. A CT study. Acta Neurochir (Wien) 90: 35–41
4. Kamman RL, Go KG, Brouwer W, Berendsen HJC (1988) Nuclear magnetic resonance relaxation in experimental brain edema: effects of water concentration, protein concentration, and temperature. Magn Reson Med 6: 265–274
5. Reulen HJ, Graber S, Huber P, Ito U (1988) Factors affecting the extension of peritumoural brain oedema. A CT-study. Acta Neurochir (Wien) 95: 19–24

Correspondence: K. G. Go, M. D., Neurosurgery, Academisch Ziekenhuis Groningen, Oostersingel 59, Postbus 30.001, NL-9700 Groningen, The Netherlands.

Acta Neurochir (1994) [Suppl] 60: 369–372
© Springer-Verlag 1994

Peritumoral Brain Edema and Cortical Damage by Meningioma

M. Ide, M. Jimbo, O. Kubo, M. Yamamoto, E. Takeyama, and **H. Imanaga**

Department of Neurosurgery, Tokyo Women's Medical College Dai-ni Hospital, Tokyo, Japan

Summary

Forty supratentorial meningiomas were analyzed to identify factors causing peritumoral brain edema. Parasagittal, sphenoid ridge, and olfactory groove meningiomas induced edema more frequently than those in other locations. Meningothelial meningiomas were more invasive than other types and were associated with more peritumoral edema. Brain edema correlated significantly with tumor size and histological evidence of leptomeningeal and cortical damage from the tumor. Larger tumors destroy the leptomeninges and cerebral cortex, allowing direct transmission of edema fluid into the white matter, resulting in vasogenic edema.

Keywords: Peritumoral brain edema; meningioma; tumor-margin; cortical invasion.

Introduction

The incidence of peritumoral brain edema in meningiomas 46 to 92%[9], despite the tumor's generally benign histology. Many factors related to edema development have been studied but its pathogenesis remains to be elucidated. Recent reports have closely linked cerebral cortical destruction by the tumor to peritumoral brain edema[5]. However, edema development and signs of cortical damage were investigated only by high-resolution computed tomographic (CT) scan, and no histological verification of actual cortical damage was provided[9]. We studied the histological features of the marginal portion of the meningioma and the boundary between the tumor and brain tissue adhering to its margin.

Materials and Methods

We examined 40 supratentorial meningiomas excised between 1983 and 1992, for which either adequate CT or magnetic resonance (MR) images, or both, were available. Multiple and recurrent meningiomas were excluded. We neuroradiologically assessed: 1) tumor size (maximal diameter based on three dimensional measurement), 2) tumor location, and 3) presence or absence of peritumoral brain edema. There were seven male and 33 female patients, 25 to 77 years old (mean ± SD, 56 ± 12). Among the 40 cases, 23 (57.5%) had brain edema and 17 (42.5%) did not.

Excised tumors were histologically classified into subtypes according to WHO criteria. Immunohistochemical examination was performed by the peroxidase-antiperoxidase method with antibody against glial fibrillary acidic protein (GFAP) (DAKO, Copenhagen) to identify GFAP-positive cells in the tumor margin. Excised tumor surfaces were examined as widely as possible by light microscopy to identify the brain-tumor interface. Three cases underwent only intracapsular removal. The tumor surface facing the brain was identifiable in the remaining 37 cases, in 17 of whom (46%) brain tissue adhered to the tumor margin. The brain-tumor interface was histologically examined, and the relationship between cortical destruction and the development of brain edema was analyzed.

Results

Table 1 summarizes tumor size measurements and the presence of brain edema. There was a statistical correlation ($p < 0.001$, Kruskal and Wallis test). The larger the tumor, the more frequent was peritumoral edema.

Parasagittal, sphenoid ridge, and olfactory groove meningiomas induced edema more frequently than

Table 1. *Size of Meningiomas and Presence/Absence of Brain Edema (40 Cases)*

Edema	≤ 2.0	2.0 <	3.0 <	4.0 <	5.0 <	6.0 <	7.0 < (cm)
+	1	0	3	9	4	4	2
−	4	6	2	3	1	1	0

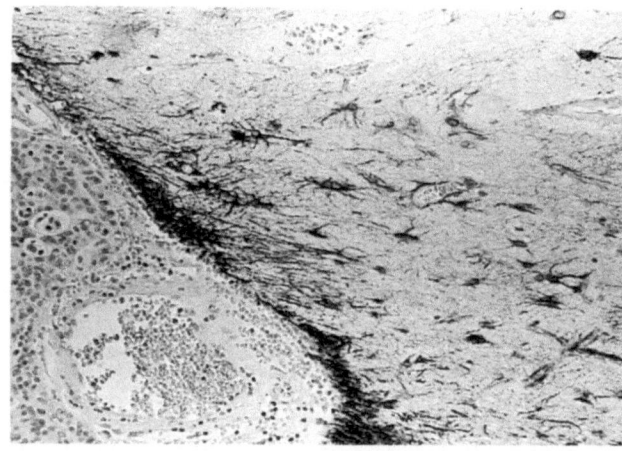 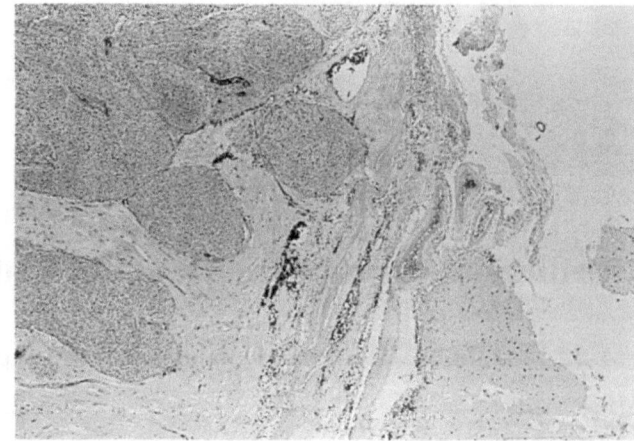

Fig. 1. Tumor-brain interface. Left: Finger-like tumor invasion is seen. Arachnoid and pial membrane can not be clearly identified. (Hematoxylin and eosin stain × 40, reproduced with permission from Ide *et al.*[6]) Right: Brain tissue shows gliotic change with loss of neurons. (immunoperoxidase stain for glial fibrillary acidic protein, hematoxylin counter stain × 200)

those in other locations (p < 0.05, chi-square test) (Table 2).

Histologically, brain tissue adhered to the tumor in the 17 cases (46%). There was no intact arachnoid membrane between the tumor and brain tissue and thus no subarachnoid space (Fig. 1). In nine of the 17 cases, tumors were invasive and interdigitation was seen. The adherent brain tissue showed gliotic change with loss of neurons, and cystic change. Fifteen (88%) of the 17 cases with brain tissue adhesion showed peritumoral brain edema on CT or MR images, while only six (30%) of the 20 without adhesion had brain edema. There was a statistically significant correlation between the presence of adhesion around the tumor margin and peritumoral brain edema (p < 0.001, chi-square test).

Intraoperatively, larger tumors were found to be more adherent to surrounding brain tissue, making dissection more difficult. Exceptionally, two large convexity meningiomas (> 5.0 cm in diameter) had loose adhesions and dissection was very easy. Tumor removal revealed the cerebral surface to be completely covered with well-preserved arachnoid membrane, and cerebral cortical integrity was preserved. In these two cases, no edema was seen on CT and MR images and a peritumoral fluid cleft was seen on MR images (Fig. 2). In contrast, a small tumor had brain edema with histological evidence of cortical destruction despite being less than 2 cm in diameter.

Table 3 shows the relationship between the histological subtype of meningioma and brain edema. Thirteen of the 17 tumors with adherent brain tissue were meningothelial meningiomas. These tumors were more invasive and more frequently associated with peritumoral brain edema than other subtypes (p < 0.05, chi-square test).

Table 2. *Location of meningiomas and Presence/Absence of Brain Edema (40 Cases)*

Edema	Parasagittal Falx	Convexity	Sphenoid ridge	Middle fossa	Olfactory groove	Tuberculum sellae	Lateral ventricle
+	8	5	6	0	3	0	1
−	2	7	2	2	0	3	1

Table 3. *Histological Subtypes and Presence/Absence of Brain Edema (40 Cases)*

Edema	Meningothelial	Fibrous	Transitional	Angiomatous
+	17	4	2	0
−	5	7	4	1

Discussion

The source of edema fluid in meningioma has not been clarified. Vasogenic edema might result from damage to the blood-brain barrier (BBB) of cortical vessels. Alternatively, humoral edema-promoting factors derived from the meningioma might destroy the BBB[1,12], but there are no firm data. Brain edema associated with gliomas and metastatic brain tumors is caused by plasma leakage from tumor vessels where BBB is damaged or absent[7,15], as shown on postcontrast CT scan by abnormal enhancement. Ito *et al.* calculated formation and propagation rates of edema fluid in metastatic brain tumors by CT scan with intravenous administration of contrast material[7]. In meningiomas, electron microscopy has revealed little evidence of BBB damage in edematous white matter, while fenestrated endothelium, pinocytotic vesicles and punctate-type junctions, suggesting increased permeability, are present in blood vessels of the meningioma[4]. Some reports emphasize the secretory activity of meningioma cells[11]. These findings point to meningioma tissue as the source of edema fluid.

We found a correlation between peritumoral brain edema and histological evidence of leptomeningeal and cortical destruction by the tumor. A small meningioma is separated from the cerebral cortex by the arachnoid membrane and pia mater, and can easily be dissected without damaging the brain. At this stage, brain edema is extremely rare. However, as tumor growth leads to destruction of the arachnoid membrane and pia mater, the tumor comes into contact with the cortex, compressing and adhering to it, and causing cortical degeneration. Few studies have investigated the interface between the meningioma and surrounding brain tissue[2,3,10]. Nakasu *et al.* examined this interface in autopsy cases on which surgery had not been performed[10]. They found that meningiomas occasionally invade the brain despite their generally benign nature, destroying the leptomeninges and even the cerebral cortex, making direct contact with white matter. As our study dealt with surgical cases, incidental attachment of brain tissue to the excised tumor may have occurred. However, interface histology included destruction of the leptomeninges and cerebral cortex, and numerous interdigitations. Thus, incidental adhesion due to technical factors is unlikely. In such situations, flawless dissection between tumor and brain is virtually impossible.

The association of brain edema with destruction of the leptomeninges and cortex by meningioma may be

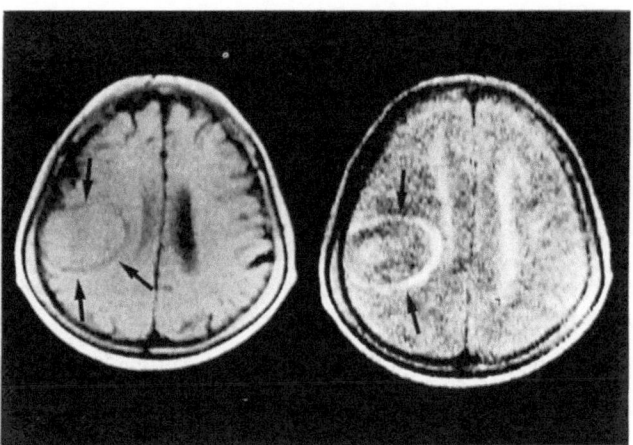

Fig. 2. Magnetic resonance images of a large meningioma, exceeding 5 cm in diameter, without peritumoral brain edema. Peritumoral fluid cleft (arrows) can be seen as a hypointense rim on T_1-weighted (left) and a hyperintense rim on T_2-weighted images (right)

explained as follows. The arachnoid membrane is normally impermeable to fluid. Pia mater permeability to macromolecular substances like edema fluid protein is low. The cerebral cortex is composed of dense networks of neuronal and glial processes and is thus quite resistant to the storage and spread of vasogenic edema fluid[5], while the white matter is a potential space for edema fluid accumulation as described by Klatzo[8]. Large tumors damage the leptomeninges and cortex, making direct contact with and thereby allowing transmission of edema fluid into the white matter, resulting in vasogenic edema.

Acknowledgment

The authors are very grateful to Bierta E. Barfod, M.D., University of Washington School of Medicine, for her assistance in the preparation of this manuscript.

References

1. Baethmann A, Oettinger W, Rothenfuber W, Kempski O, Unterberg A, Geiger R (1980) Brain oedema factors. Current state with particular reference to plasma constituents and glutamate. In: Cervós-Navarro J, Ferszt R (eds) Brain oedema. Advances in neurology, vol 28. Raven, New York
2. Böcker DK, Meurer H, Gullotta F (1985) Recurring intracranial meningiomas. J Neurosurg Sci 29: 11–17
3. Crompton MR, Gautier-Smith PC (1970) The prediction of recurrence in meningiomas. J Neurol Neurosurg Psychiatry 33: 80–87
4. Gilbert JJ, Paulseth JE, Coates RK, Malott D (1983) Cerebral edema associated with meningiomas. Neurosurgery 12: 599–605
5. Go KG, Wilmink JT, Molenaar WM (1988) Peritumoral brain edema associated with meningiomas. Neurosurgery 23: 175–179

6. Ide M, Jimbo M, Kubo O, Yamamoto M, Imanaga H (1991) Peritumoral brain edema associated with meningioma. Neurol Med Chir 32: 65–71

7. Ito U, Reulen HJ, Tomita H, Ikeda J, Saito J, Maehara T (1988) Formation and propagation of brain edema fluid around human brain metastases. Acta Neurochir (Wien) 90: 35–41

8. Klatzo I (1967) Neuropathological aspects of brain edema. J Neuropathol Exp Neurol 26: 1–14

9. Lindley JG, Challa VR, Kelly DL Jr (1991) Meningiomas and brain edema. In: Al-Mefty O (ed) Meningiomas. Raven, New York

10. Nakasu S, Hirano A, Llena JF, Shimura T, Handa J (1989) Interface between the meningioma and the brain. Surg Neurol 32: 206–212

11. Philippon J, Foncin JF, Glob R, Srour A, Poisson M, Pertuiset BF (1984) Cerebral edema associated with meningiomas. Possible role of secretory-excretory phenomenon. Neurosurgery 14: 295–301

12. Shinonaga M, Chang CC, Suzuki N, Sato M, Kuwabara T (1988) Immunohistochemical evaluation of macrophage infiltrates in brain tumors. J Neurosurg 68: 259–265

13. Smith HP, Challa VR, Moody DM, Kelly DL (1981) Biological features of meningiomas that determine the production of cerebral edema. Neurosurgery 8: 428–433

14. Stevens JM, Ruiz JS, Kendall BE (1983) Observation on peritumoural brain oedema in meningioma. Part II: mechanism of oedema production. Neuroradiology 25: 125–131

15. Yamada K, Hayakawa J, Ushio Y, Arita N (1981) Regional blood flow and capillary permeability in the ethylnitrosourea-induced rat glioma. J Neurosurg 55: 922–928

Correspondence: Mitsunobu Ide, M.D., Department of Neurosurgery, Tokyo Women's Medical College, Dai-ni Hospital, 2-1-10 Nishiogu, Arakawa-ku, Tokyo 116, Japan.

Acta Neurochir (1994) [Suppl] 60: 373–374
© Springer-Verlag 1994

Formation and Resolution of Human Peritumoral Brain Edema

U. Gröger[1], **P. Huber**[1], and **H. J. Reulen**[2]

[1] Departments of Neurosurgery and Neuroradiology, University of Berne, Switzerland, and [2] Department of Neurosurgery, Ludwig-Maximilians-University Munich, Federal Republic of Germany

Summary

In 16 patients with 21 metastastic brain tumors and 9 patients with a malignant glioma, tumor volume, volume of the edematous tissue, edema production, speed of edema propagation and edema resolution were examined by using the CT. Edema production was determined according to a technique described previously[3] and ranged between 0.09 and 1.63 ml/h in metastases and between 0.42 and 3.49 ml/h in gliomas. The speed of edema propagation ranged from 0.2–2.2 mm/h. Edema resolution can take place within the tissue (i.e. reabsorption into blood) as well by drainage into the entricular or subarachnoid CSF. In a few small metastases with a small perifocal edema (without contact to the ventricule or the subarahnoid space) the amount of edema resolution within the tissue could be determined and averaged 0.0086 ml/h/cm³. This probably represents the reabsorption of edema fluid into capillaries within the edematous tissue. If this value is used to calculate the edema reabsorption in larger tumors, the resulting data are considerable lower than the respective edema production rate of that tumor. This indicates, that in larger tumors the main fraction of the edema fluid is draining into the ventricular and/or subarachnoid CSF.

Keywords: CT-study; peritumoral edema; edema formation; edema resolution.

Introduction

Until recently the pathophysiology of peritumoral brain edema had to be investigated mainly in experimental models. With the access to modern imaging techniques we now have a tool to examine this entity also in patients with brain tumors[1,2,5]. Several of these studies were presented on former meetings. The intention of the present study was to get a better understanding about the dynamics of the fluid turnover in human peritumoral edema.

In particular we tried to determine the various components of the fluid turnover, that means the tumor volume, the volume of the edematous tissue, the edema production and the edema resolution. The knowledge of the various components offers a better insight into the mechanisms, which in some patients may lead towards a decompensation of intracranial volume and intracranial pressure.

Material and Methods

9 patients with malignant gliomas and 16 patients with 21 metastatic brain tumors were investigated. Five patients had metastases in both hemispheres so that both foci could be examined. All patients showed a well circumscribed peritumoral edema of Grade II–III according to Lanksch[1]. The studies were done with a GE 9800 CT scan and axial slices of 3 or 5 mm thickness were taken to show the whole cranio-caudal extension of the tumor and the edema as well. Following the initial CT scan a bolus of contrast medium was administered i.v. and the second CT study was performed 10 minutes thereafter. Following the bolus a drip-infusion of 100 ml contrast medium was administered during one hour and the third CT study was performed 3 hours after the second study.

The area of contrast enhancement which represents the tumor and the area of low density which represents the edema were measured planimetrically on the 10 min and 180 min axial CT slices. The volume of the tumor and the volume of the edematous tissue was then determined with the volume summation method by adding the values of the various slices.

Edema production was determined according to a technique published previously together with Ito[2]. This technique is based on the assumption that the CT scan after three hours shows the expansion of the contrast labelled edema fluid. The radius of the edema fluid spreading as well as the volume of the newly formed edema fluid can then be calculated – by using several assumptions – from the difference between the scan of 180 min minus the scan after 10 min.

The determination of the edema resorption during the passage through the tissue is based upon the following concept: since there is continous edema formation and the total edema volume remains practically constant during a period of 24 hours, the same amount of fluid must be cleared, either within the tissue and/or by drainage into the CSF. If small tumors with a limited edema halo are selected, where no contact exists with the ventricular ependyma, resolution should take place preferentially by reabsorption within the tissue, i.e. by reabsorption into blood. The edema resorption (R) represents the volume of fluid, which is reabsorbed in a certain tissue volume in a time unit and can be calculated, because the edema production (V) and the total edema tissue volume are known.

$$R = V / h : \text{total edema volume}$$

In contrast, the volume of edema fluid which is cleared into the ventricular system or the subarachnoid space is not directly measurable at the moment. However, it can be calculated as difference between edema production rate and tissue resorption.

Results and Discussion

The values of the tumor volume in gliomas ranged from 5.62 to 76.73 cm³ and in metastases from 0.27 to 28.31 cm³; the volume of the edematous tissue ranged from 29.12. to 119.71 cm³ in gliomas and from 3.89 to 137.40 cm³ in metastases.

The values for the speed of edema propagation varied only slightly between the two groups. For gliomas it ranged between 0.9 and 2.2 mm/h and for metastases we found values between 0.2 and 1.7 mm/h. The edema production rate ranged between 0.09 and 1.63 ml/h in metastases and between 0.42 and 3.49 ml/h gliomas. The large inter-individual differences can at least partly be explained with the before-mentioned different tumor volumes. A close relationship between the tumor diameter and the edema formation rate was found in the investigated patients.

Seven small metastases fullfilled the criteria of having a limited peritumoral edema halo. The average value for the tissue resorption of edema fluid in these 7 cases was 0.0086 ml/h/cm³. This value still may be slightly overestimated since the edema in some patients bordered the cortex and there may have been some drainage of edema fluid into the subarachnoid CSF.

On the other hand, the calculated average tissue resorption of metastases with broad ventricular and/or cortical contact was 0.021 ml/h/cm³, a value which is 2.4 times higher than that of the 7 metastases without ventricular contact. It may be speculated that this difference of 0.012 ml/h/cm³ represents the fraction which is drained into the ventricular and/or subarachnoid CSF. This now allows to describe the dynamic edema fluid turnover with two typical examples.

a)   In a small metastasis the edema fluid production is 3.4 ml/day. The surrounding edematous tissue volume amounts to 35 ml. Assuming a tissue resorption of 0.0086 ml/h/cm³ the edema fluid can be reabsorbed exclusively within the edematous tissue volume (Fig. 1).

b)   The second example is a large metastasis with a edema fluid production of 48 ml/day and an edematous tissue volume of 90 ml. In this case only 8.6 ml of edema fluid per day can be explained by tissue resorption (Fig. 2). This means that in such a case the majority of fluid resolution occurs by CSF drainage, a find-

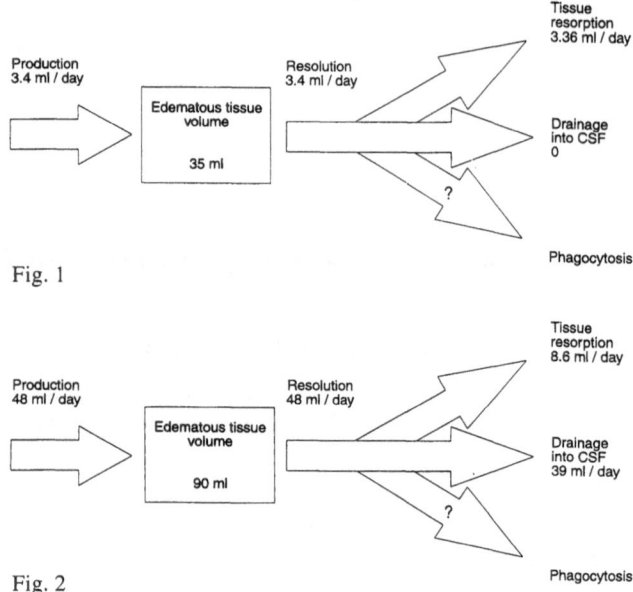

Fig. 1

Fig. 2

ing which is consistent with recent experimental data from Ohata *et al.*[3]. In summary, in small tumors the resulting edema production can be reabsorbed within the surrounding tissue. If the tumor grows, the edema production increases, and at a certain point exceeds the capacity of tissue resorption. This excess fluid then is drained into the CSF, when the edema border progresses toward the ventricle or the cortex and reaches a CSF surface. In larger tumors this second mechanism must be the prevailing mechanism. This may explain why in certain patients a compression of the lateral ventricle may lead to a decompensation of this delicately balanced fluid turnover with a rapid and dramatic increase in intracranial volume and pressure.

References

1. Lanksch WR (1982) The diagnosis of brain oedema by computed tomography. In: Hartmann A, Brock M (eds) Treatment of cerebral oedema. Springer, Berlin Heidelberg New York, pp 43–80
2. Ito U, Reulen HJ, Tomita H, Ikeda J, Saito J, Maehara J (1988) Formation and propagation of edema fluid around human brain metastases. A CT-study. Acta Neurochir (Wien) 81: 53–60
3. Ohata K, Marmarou A, Povlishock HT (1990) An immuno-cytochemical study of protein clearance in brain infusion edema. Acta Neuropathol (Berl) 81: 162–177
4. Reulen HJ, Graham R, Spatz M, Klatzo J (1977) Role of pressure gradients and bulk flow in dynamics of vasogenic brain edema. I. Neurosurgery 46: 24–35
5. Tsuyumn M, Reulen HJ (1981) Oruikeazm -g,: measurement of edema clearance into ventricular CSF. Acta Neurochir (Wien) 57: 1–13

Correspondence: U. Gröger, M. D., Department of Neurosurgery, University of Bern, Inselspital, CH-3010 Bern, Switzerland.

Acta Neurochir (1994) [Suppl] 60: 375–377
© Springer-Verlag 1994

Peri-Tumoural Hypoxia in Human Brain: Peroperative Measurement of the Tissue Oxygen Tension Around Malignant Brain Tumours

G. S. Cruickshank and **R. Rampling**

Institute of Neurological Sciences, Glasgow, Scotland, U.K.

Summary

Malignant brain tumours contain focal hypoxic areas that may increase their resistance to chemotherapy and radiotherapy. Following surgical excision, the peri-tumoural area will contain residual viable tumour cells, and this area is therefore the logical site for subsequent therapy. The new bioreductive agents are metabolized under hypoxic conditions to produce a cytotoxic species. Peroperative peri-tumoural micro-polarographic measurments have been made to establish the oxygen environment of this region and to determine whether the hypoxic conditions might allow for bioreductive drug activation. The micro-polarographic method is described and results are presented for "normal" white matter (8 patients) to allow comparison with peri-tumoural brain (8 patients) before and after removal of the tumour. The results suggest that peri-tumoural brain (median pO_2 10.8 mmHg, 18% pO_2 < 2.5 mmHg) is markedly hypoxic in comparison with the "normal" brain (median pO_2 15.3 mmHg, less than 2% < 2.5 mmHg), and that surgery improves peri-tumoural oxygenation towards that of the "normal" white matter. It is concluded that the hypoxic peri-tumoural area can provide the conditions under which bioreductive agents may be activated.

Keywords: Bioreductive; hypoxia; brain; polarography.

Introduction

Malignant brain tumours are more resistant to radiotherapy and chemotherapy and have a faster rate of recurrence than most other human tumours[1]. There has been much interest shown in the role of focal hypoxia as an important cause of this resistance[5]. Hypoxic resistance has been demonstrated in many types of human tumours, but its presence in malignant brain tumours has been disputed[9]. Recently, using high-resolution dynamic micro-polarography, we have been able to confirm significant radiobiological hypoxia in malignant intracranial tumours[3]. Radiobiological hypoxia has been shown to result in half-maximal radio-resistance at a level of 2.5 mmHg[2]. A number of compounds known as hypoxic cell sensitizers, which include the drug Misonidazole, have been used to try and decrease the hypoxic cell fraction. Recently it has been suggested that the bioreductive agents that are metabolized under hypoxic conditions to produce cytotoxic free radical compounds lethal to tumours may actually be able to exploit the hypoxic environment with a somewhat higher threshold level of 10–15 mmHg[10]. After surgery, the peri-tumoural area contains residual viable tumour cells. In this study, we have focussed our attention on the oxygenation of peri-tumoural tissue, as this is the logical site for further therapy. The oxygen status of the peri-tumoural area is therefore of great interest in describing whether these drugs may be activated in this region.

Materials and Methods

In tissues, at any point the pO_2 varies between 100 and 0 mmHg[7]. Multiple high-resolution spatially separate samples of pO_2 allow an accurate estimate of the statistical distribution of the oxygen supplied to tissues characteristic of the micro-environment of that tissue. We have used a 300 μm polarographic needle with a 12 μm oxygen tip to record tissue pO_2 levels at the microvascular level. The needle records 90% of available oxygen in 500 ms. This allows a near instantaneous measurement of oxygen tension before local changes in pO_2 can occur. The sterile pre-calibrated needle was introduced into the tissues under direct visual observation and tracked through the tissue in steps of 1 mm with 0.3 mm back-step to reduce tissue distortion. The position and tracking of the needle could be followed by real-time ultrasound at 7 MHz. This made possible a point-to-point relationship of pO_2 measurements to tissue position. Usually, six tracks of 32 readings at 700 μm intervals could be recorded in 6–8 minutes, giving approximately 192 values. The number of values required to estimate the tissue distribution was evaluated by comparing the variation in the cumulative median value until the 5% level was reached. The graph for values in tumours reflects this tissue-particular heterogeneity by requiring nearly double the number of readings. Pre- and post-calibration compensate for calibration drift and allow interpolation of the results. In addition, probe poisoning could also be excluded. Artefacts due to brain movement were estimated and eliminated by altering the individual step lengths.

Patients were anaesthetized with either nitrous oxide and isoflurane or intravenous Propofol with fractional inspired oxygen of 40%. The arterial pCO_2 was controlled to between 3.5 to 4.5 kPa. Polarography was performed after dural opening and a 15 minute equilibration period. Eight patients (Table 1) were evaluated for "normal" human white matter pO_2 distribution. Eight patients with malignant brain tumours were evaluated to examine the tissue pO_2 distribution in peritumoural white matter.

Results and Discussion

To establish for comparison the oxygenation pattern of normal white matter is ethically difficult. It was, however, possible to evaluate white matter tissue oxygenation in patients undergoing surgery for intractable epilepsy. Data has also been collected for those patients whose underlying pathology was focally limited and unlikely to exert significant general cerebral impact. The average values for eight patients are shown in Fig. 1. The pO_2 frequency distribution for "normal brain" is shown less than 2% of the values below 2.5 mmHg and a median value of approximately 15.3 mmHg. Such pO_2 levels are generally lower than those reported by previous authors[8], especially in man. Firstly we have to remember that we are recording from white matter, which has one-tenth the oxygen supply of grey matter and a quarter or less the capillary density[6]. Secondly, we are using a dynamic system so that both the smoothing of results seen by the diffusion barrier of static systems and the destructive nature of those systems are avoided[4]. Thirdly, high-resolution polarography makes it possible to discern the full spectrum of oxygen values within the diffusion field. Finally, it must be remembered that this pO_2 distribution represents the normal heterogeneity of oxygen supply seen with measurements made within the oxygen diffusion field, i.e. less than the intercapillary distance. Of course, the ability to maintain low oxygen levels under these conditions is a function of the dynamic nature of the blood supply of oxygen.

In peri-tumoural white matter at 2 cm from the margin of the tumour (Fig. 2a), a marked skewing of the distribution to the left can be seen, reflecting the high percentage of low pO_2 values.

Low pO_2 values in the peri-tumoural area may well relate to the impaired brain function commonly observed, but they are not explained by peri-tumoural edema.

These results suggest that the peri-tumoural mantle of white matter and residual tumour cells contain a

large proportion of radiobiologically significant hypoxic areas. This, of course, would represent a significant obstacle to radiation treatment.

Conversely, with bioreductive agents, it is apparent that such high levels of hypoxia might well allow such compounds to exploit the environment and improve their degree of activation.

It is important to consider the size of this peritumoural hypoxic effect. An estimate was gained from assessment of continuous pO_2 tracks through tissue. pO_2 values from 4 tracks of the polarographic needle

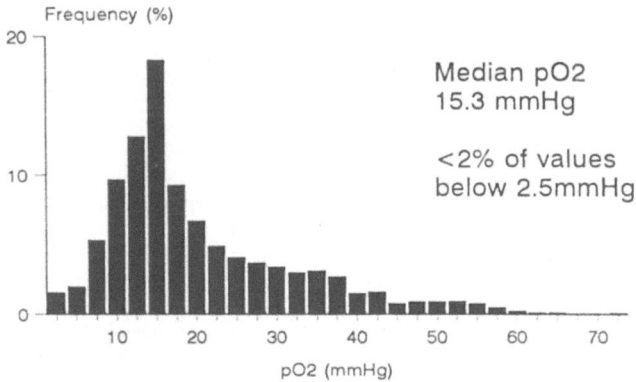

Fig. 1. Average values of "normal" white matter from 8 patients

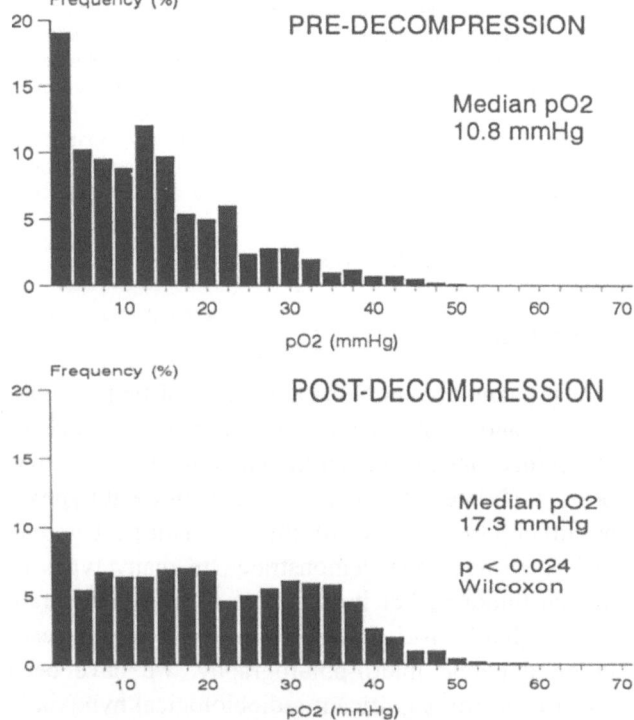

Fig. 2. (Top) Peri-tumoural white matter at 2 cm from tumour edge. (Bottom) Peri-tumoural white matter from same location as in (top) following surgical decompression of the tumour

Table 1. *Details Concerning Patients from which "Normal" White Matter pO₂ Recordings were made*

Patient	Age	Sex	White matter site	Operative indication	Pathology	Complications
1	45	F	r. frontal	l. frontal SOL[a]	AVM	none
2	32	F	l. frontal	r. frontal SOL[a]	AVM	none
3	51	F	r. frontal	l. frontal SOL[a]	venous haemangioma	none
4	54	M	r. frontal	l. frontal SOL[a]	infarct	none
5	30	M	l. temporal	epilepsy	sclerosis < 0.5 cm	none
6	40	M	l. temporal	epilepsy	NAD	none
7	24	F	r. temporal	epilepsy	NAD	none
8	37	F	r. temporal	epilepsy	NAD	none

SOL Space occupying lesion. [a] Recording from contralateral side.

across the wall of a malignant cystic tumour were made with the needle penetrating about 10 mm into the peritumoural region. Extrapolation of the gradient of pO₂ in the peritumoural white matter in a linear fashion makes it possible to estimate that, although the median value for the "normal brain" is achieved at about 3.0 cm, it is likely that the gradient is exponential, achieving the "normal" median value at between 5 and 7 cm.

After surgically removing the tumour, there has been a statistically significant improvement in the distribution of oxygen levels as assessed by the Wilcoxon Signed Ranks test. The change in the pattern of the distribution is very clearly seen, with a shift to higher pO₂ levels. The fact that this improvement can occur within the time span of an operation implies that it is more likely to be due to decompression of vascular structures with improved vascular patency than to resolution of edema. It would be reasonable to expect that this improvement in oxygenation would continue postoperatively, and indeed it is known that these areas do recover with a concomitant improvement in brain function. There was clearly a reduction in the degree of hypoxia, and this supports the concept that tumour decompression may be valuable in enhancing the effect of radiotherapy. It was also clear that simple biopsy alone without tissue debulking leaves the peri-tumoural area still compromised.

Although the degree of hypoxia does decrease after surgical decompression, there is still a significant residual level. This will have implications for treatment with radiotherapy, but may enable activation of the bioreductive agents which will be metabolized to the active cytotoxic component under these conditions.

Because of the difference in thresholds, the radiobiologically insensitive (pO₂ < 2.5 mmHg) hypoxic tumour volume in the peritumoural area will always be less and included within the volume of tissue with a higher threshold (10–15 mmHg), within which bioreductive agents such as SR4233 will be activated.

References

1. Bloom HJG (1982) Intracranial tumours: response and resistance to therapeutic endeavours 1970–1980. Int J Radiation Oncol Biol Phys 8: 1083–1113
2. Bush RS, Jenkin RDT, Allt WEC, Beale FA, Bean H, Dembo AJ, Pringle JF (1978) Definitive evidence for hypoxic cells influencing cure in cancer therapy. Br J Cancer 37 [Suppl 3]: 302–306
3. Cruickshank GS, Rampling R, Cowans W (1992) Direct measurement of the pO₂ distribution in malignant brain tumours. In: Vaupel P, *et al* (eds) Proceedings of the International Society on Oxygen Transport to Tissue XV. Adv Exp Med Biol, in press
4. Fleckenstein W, Schaffler A, Heinrich R, Petersen C, Gunderoth-Palmowski M, Nollert G (1990) On the difference between muscle pO₂ measurements obtained with hypodermic needle probes and with multiwire surface probes (Parts 1 and 2). In: Ehrly AM *et al* (eds) Clinical oxygen pressure measurement II. Blackwell, Ueberreuter, Berlin, pp 256–278
5. Gray LH, Conger AD, Ebert M, Hornsey S, Scott OCA (1953) Concentration of oxygen dissolved in tissues at time of irradiation as a factor in radiotherapy. Br J Radiol 26: 638–648
6. Grote J (1960) In: Schmidt RF, Thews G (eds) Physiologie des Menschen, 23rd edn. Springer, Berlin Göttingen Heidelberg, p 635
7. Lubbers DW (1969) The meaning of the tissue oxygen distribution curve and its measurement by means of Pt electrodes. Prog Resp Res 3: 112
8. Schultheiss R, Leuwer R, Leniger-Follert E, Wassmann H, Wullenwever R (1988) Tissue pO₂ of human brain cortex-method, basic results and effects of pentoxifylline. Angiology 38: 221–225
9. Whittle IR (1992) The biology of glioma. In: Teasdale GM, Miller JD (eds) Current neurosurgery. Churchill Livingstone, Edinburgh, pp 255–284
10. Workman P, Walton MI (1991) Enzyme-directed bioreductive drug development. In: Adams GE, Breccia A, Fielden EM, Wardman P (eds) Selective activation of drugs by redox processes. Plenum, New York, pp 173–191

Correspondence: G. S. Cruickshank, MBBS, PhD, FRCS (Ed/Lond), FRCS (SN), Department of Neurosurgery, Institute of Neurological Sciences, Southern General Hospital Trust, 1345 Govan Road, Glasgow G51 4TF, Scotland, U.K.

Acta Neurochir (1994) [Suppl] 60: 378–380
© Springer-Verlag 1994

Does Tumour Related Oedema Contribute to the Hypoxic Fraction of Human Brain Tumours?

G. S. Cruickshank and **R. Rampling**

Institute of Neurological Sciences, Glasgow, Scotland, U.K.

Summary

Focal hypoxia has been demonstrated and is known to contribute to the resistance of malignant brain tumours to radiation and chemotherapy. Using dynamic needle micro-polarogaphy and tissue morphometry on biopsy specimens, the relationship between the effect of oedema on tissue structure and tissue pO_2 was investigated in 24 patients undergoing craniotomy for tumour decompression. An inverse correlation ($r = -0.84$) was found for intercapillary distance and pO_2 levels in peritumoural white matter, but this was less marked ($r = -0.22$) in tumour, probably as a result of sampling difficulties from tissue heterogeneity. Comparison of maximum pO_2 levels in oedematous peritumoural white matter with those in tumour suggests that peritumoural oedema is unlikely to contribute to tumour hypoxia.

Keywords: Brain; hypoxia; oedema; polarography.

Introduction

Tumour hypoxia is an important factor in determining the resistance of tumours to the cytotoxic effects of radiotherapy and chemotherapy[1]. Using dynamic needle micro-polarography in human brain in patients undergoing surgery for excision of tumours, we have been able to demonstrate high levels of radiobiologically significant hypoxia[2]. An important clinical feature of malignant brain tumours is the degree of peritumoural oedema. This study examines the relationship between tissue pO_2 distribution and the degree of tumour-induced oedema. The relationship between the extent of peritumoural oedema and tumour hypoxia was examined.

Materials and Methods

Micropolarography

There is a diffusion gradient for oxygen between capillary and tissue such that at any one point the tissue oxygen can vary between 100 mmHg and zero, depending on the perpendicular distance from the capillary and the oxygen uptake by the tissue cells[3]. Oedema might be expected to distort the dimensions of this field in such a way that the oxygen diffusion field might be insufficient to prevent intercapillary hypoxia. Single measurements of pO_2 cannot truly reflect the dynamic nature of this oxygen field gradient. However, by using multiple high-resolution spatially separate micro-polarographic measurements of pO_2, it is possible to achieve an accurate statistical description of the oxygen supply to tissues[4]. A precalibrated 300 μm polarographic needle with a 12 μm oxygen-sensitive tip was inserted into the brain under direct vision. The needle was advanced under computer control, with its movement directed and monitored using an ultrasound probe at 7 MHz. Further details of the technique are described elsewhere[6,7]. pO_2 measurements were made over 6 22 mm tracks, each of which were consisting of 32 readings 700 μm apart. This provided at least 192 readings in each tissue described in each patient. In this study we have recorded pO_2 values in malignant brain tumours and in the peritumoural area at 2 cm from the margin of the tumour.

Morphometry

The effect of oedema on the dimensions of the tissue microenvironment were examined by morphometric assessment of paraffin sections from core samples using standard techniques[5]. *Capillary densitiy* was expressed as the number per square millimeter from microscopic counts of multiple measuring fields of 700×500 μm. *Intercapillary distance* was determined from multiple field measurements in which the nearest neighbour distances between capillaries (highlighted by endothelial immunostaining) were recorded. Cumulative measurements of individual fields were determined until the variation of the mean achieved 5%.

Patients

Twenty-four patients aged between 16 and 70 years with a CT diagnosis of malignant tumour were recruited. General anaesthesia was maintained with either nitrous oxide and isoflurane or intravenous Propofol at fractional inspired oxygen of 40%. Arterial pCO_2 was controlled to between 3.5 and 4.5 kPa. All patients were treated with dexamethasone for at least 48 hours prior to surgery. These patients underwent routine craniotomy and decompression of their tumours. Twelve patients had malignant gliomas and 12 metastases (lung 4, breast 4, melanoma 3, and sarcoma 1). During craniotomy, needle recording of pO_2 values were made, after which core samples were removed for histopathological assessment and morphometry.

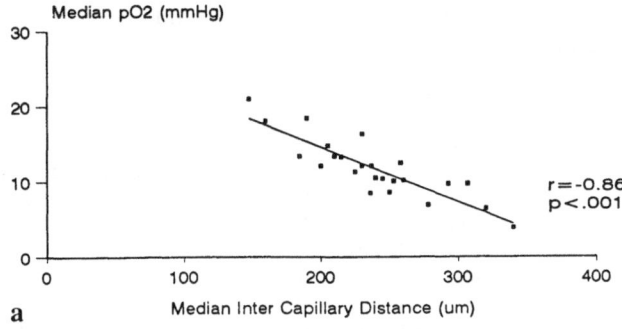

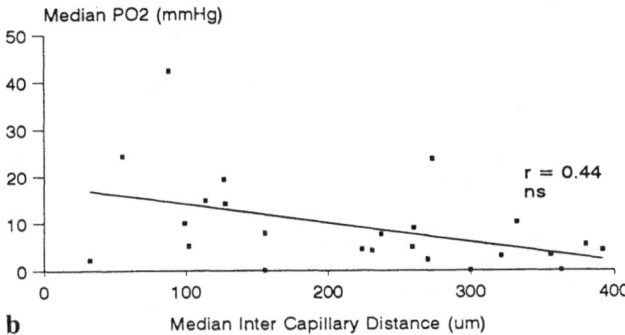

Fig. 1. Comparison of median micropolarographic pO_2 measurements with morphometric estimates of intercapillary distance in oedematous white matter (a) and in malignant brain tumours (b) in 24 patients

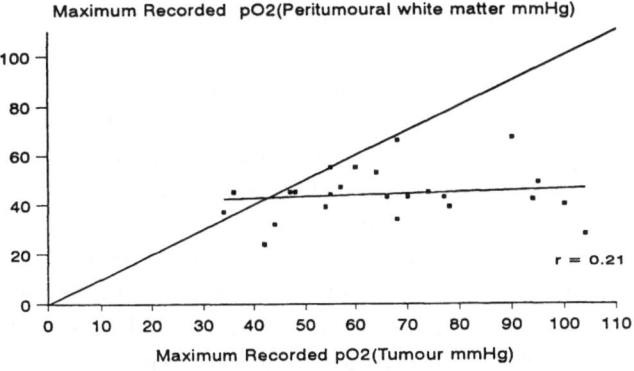

Fig. 2. Maximum recorded tissue pO_2 levels in oedematous peritumoral white matter compared with maximum pO_2 levels recorded in brain tumors

Results and Discussion

The peroperative arterial oxygen tension was recorded prior to tissue oxygen measurement and correlated with tumour pO_2 measurements. As expected, tumour pO_2 was generally lower than the arterial level. However, the variation in tumour pO_2 was not explained by the variation in arterial pO_2. This indicates that variations in arterial pO_2 at the levels obtained during surgery were not limiting factors influencing tissue pO_2 in this study.

The oxygen distribution in peritumoural brain and in tumour were compared. There was profound skewing of the distribution profile for both tissues, reflecting the high percentage of low pO_2 values.

This non-parametric distribution of values necessitates the use of the median value as a description of the distribution. In peritumoural tissue (Fig. 1a), there was a clear negative association between median pO_2 and median intercapillary distance expressed in micrometers, such that the decreased pO_2 was well explained by the increased intercapillary distance. In tumours (Fig. 1b), such an association was not nearly so evident. There was a larger variation in intercapillary distances reflecting the well known tissue heterogeneity of these tumours. As such, intrinsic difficulties with

sampling errors rather than counting errors may be important. White matter is homogeneous in comparison to tomour tissue, so that white matter tissue sampling is easily representative of polarographic tracks. In comparison, the variability of tumour structure results in a high likelihood of a biopsy being unrepresentative, particularly at the microvascular level. Comparable results to those above were seen with estimates of capillary density. There was a good correlation between pO_2 and capillary density in oedematous white matter, but results varied widely in tumours, reflecting the problem of relating pO_2 measurements to the exact microscopic environment demonstrated histologically.

It was therefore possible to explain the influence of oedema on tissue hypoxia based on alterations in tissue structure in white matter. A different approach was required to establish whether the same relationship can be extended to include the heterogeneous pattern found in tumours.

Using whole-brain SPECT scans in patients preoperatively, it was possible to relate tumour blood supply using HMPAO to functional tumour volume using Thallium 201. During surgery, an ultrasound probe was used to guide the polarographic needle across a region of poorly vascularized tumour into well vascularized tumour, thus making it possible to follow the change in recorded pO_2. The improved oxygenation correlates very clearly with the pattern of HMPOA supply to the tumour.

This data supports the contention that tissue pO_2 levels relate to the position of the recording electrode within an oxygen field, determined by the density and patency of the blood vessels supplying it. A further problem in brain tumours is the presence of necrosis as a source of hypoxic volume. Comparison of pO_2 frequency distributions for glioblastoma and anaplastic astrocytoma, both highly malignant brain tumours, shows a very similar pattern. However, anaplastic astrocytoma are by definition devoid of microscopic necrosis. Thus, such a high degree of hypoxia must be to a large extent a function of the tissue blood supply.

The influence of peritumoural oedema on measurements of pO_2 in tumours was compared by relating tissue pO_2 in the oedematous peritumoural white matter to pO_2 levels recorded in tumours. There appeared to be only a very weak association between the two series. The pO_2 levels of tumours were comparable, if not better than, that in white matter of the brain at higher levels. To test this relationship, the maximum pO_2 levels (Fig. 2) in the peritumoural tissue were compared with the maximum pO_2 levels of tumours. The values clustered beneath the equivalence line in such a manner that, for a given level of pO_2 in the tumour, the brain level was lower. It thus appears unlikely, that peritumoural oedema or its related mass effect would exert much influence on tumour pO_2 levels.

Tumour hypoxia is a function of the heterogeneity of local blood supply, both spatial and temporal, and not simply necrosis. Peritumoural hypoxia is, however, related to oedema in terms of its local impact on capillary beds. Finally, tumour-related oedema and tumour expansion result in local mass effects on capillaries which appear to result in profound peritumoural hypoxia.

References

1. Bloom HJG (1982) Intracranial tumours: response and resistance to therapeutic endeavours 1970–1980. Int J Radiation Oncol Biol Phys 8: 1083–1113
2. Cruickshank GS, Rampling R, Cowans W (1992) Direct measurements of the pO_2 distribution in human malignant brain tumours. In: Vaupel P *et al* (eds) Proceedings of the International Society on Oxygen Transport to Tissue XV. Adv Exp Med Biol, in press
3. Lubbers DW (1981) Grundlagen und Bedeutung der lokalen Sauerstoffdruckmessung und des pO_2-Histogramms für die Beurteilung der Sauerstoffversorgung der Organe und des Organismus. In: Ehrly AM (ed) Messung des Gewebesauerstoffdruckes bei Patienten. Witzstrock, Baden-Baden, pp 11–12
4. Lubbers DW (1969) The meaning of the tissue oxygen distribution curve and its measurement by means of Pt electrodes. Prog Resp Res 3. Karger, Basel, p 112
5. Ruthman A (1970) Methods in cell research. Bell, London, pp 211–214
6. Schramm U, Fleckenstein W, Weber C (1990) Morphological assessment of skeletal muscle injury caused by pO_2 measurements with hypodermic needle probes. In: Ehrly AM *et al* (eds) Clinical oxygen pressure measurement II. Blackwell, Ueberreuter, Berlin, pp 38–50
7. Vaupel P, Schlenger K, Knoop C, Hockel M (1991) Oxygenation of human tumours: evaluation of tissue oxygen distribution in breast cancers by computerized O_2 measurements. Canc Res 51: 3316–3322

Correspondence: G. S. Cruickshank, MBBS, Phd, FRCS (Ed/Lon), FRCS (SN), Department of Neurosurgery, Institute of Neurological Sciences, Southern General Hospital Trust, 1345 Govan Road, Glasgow G51 4TF, Scotland, U.K.

Acta Neurochir (1994) [Suppl] 60: 381–383

Peritumoral Brain Edema Associated with Pediatric Brain Tumors: Characteristics of Peritumoral Edema in Developing Brain

A. Kawamura, T. Nagashima, K. Fujita, and **N. Tamaki**

Department of Neurosurgery, Kobe University School of Medicine, Kobe, Japan

Summary

The incidence and clinical significance of peritumoral brain edema in pediatric patients is not well understood. The purpose of this study is to clarify the clinical significance of peritumoral brain edema in pediatric patients. Seventy seven pediatric patients (under 15 year old) with brain tumor were studied by MRI. The volume of peritumoral edema and brain tumor were measured by integration of the cross-sectional area on serial MRI. The severity of brain edema was expressed by the ratio of edema volume to tumor volume. The results were compared with that of 408 adult patients with brain tumor.

Incidence of the brain edema associated with supratentorial tumors is lower in pediatric group than in adult. Severity of brain edema is also less extent in the pediatric group. Forty two percent of supratentorial gliomas in pediatric group showed peritumoral edema. On the other hand, 63% of adult group showed peritumoral brain edema. In general, pediatric brain tumors rarely associated with marked peritumoral brain edema which observed in the adult group. However, intraventricular or paraventricular tumors with obstructed hydrocephalus showed remarkable brain edema in pediatric group.

Low incidence and less severity of peritumoral edema are caused not only by histological bias of tumors, but also by the biological nature of developing brain.

Keywords: Brain edema; brain tumor; MRI.

Introduction

Developing brain is characterized by higher water content, larger extracellular space, and higher metabolic rate. Experimental and clinical studies have demonstrated that developing brain shows different response to traumatic and ischemic brain damages. However the incidence and clinical significance of peritumoral edema in pediatric patients is not well known. The purpose of this study is to clarify the characteristics of peritumoral brain edema in developing brain.

Material and Method

Magnetic resonance imaging of 77 pediatric cases with brain tumors (under 15 year old) at our institute were analyzed retrospectively. Of those, 36 cases were supratentorial tumors and 41 cases were infratentorial tumors. Of the supratentorial tumors, 26 were intra-axial tumors and 10 were extra-axial tumors. All of the infratentorial tumors were intra-axial tumors. The volume of peritumoral edema and brain tumor were measured by integration of the cross-sectional area on serial MRI. The area of brain edema was measured by subtracting the area of Gd-enhanced tumor from the high intensity area on T_2 image. The severity of brain edema was expressed by the ratio of edema volume to tumor volume. The severity of brain edema was classified as none (0 ~ 10%), slight (10–30%), mild (30 ~ 50%) and severe (over 50%). The results were compared with that of 408 adult patients with brain tumor.

Results

In the pediatric group, 31% of supratentorial tumor showed peritumoral brain edema (Fig. 1). The majority showed slight to mild brain edema. Only 4 cases showed severe peritumoral edema. Those tumors with severe brain edema located in the ventricle and were accompanied with obstructed hydrocephalus. Only 20% of extra-axial tumors showed brain edema in the pediatric group. In infratentorial tumors, peritumoral edema were found in 45% of the cases (Fig. 1). But the severity of the edema was mild in infratentorial tumors.

Histological study revealed that 42% of gliomas, 27% of germ cell tumors, and 13% of craniopharyngioma showed peritumoral edema in the pediatric group. Sixty four percent of cerebellar astrocytomas, 34% of medulloblastomas and 75% of brain stem gliomas showed peritumoral brain edema (Fig. 2). The incidence of brain edema in infratentorial tumors was higher than that of supratentorial tumors. Two primi-

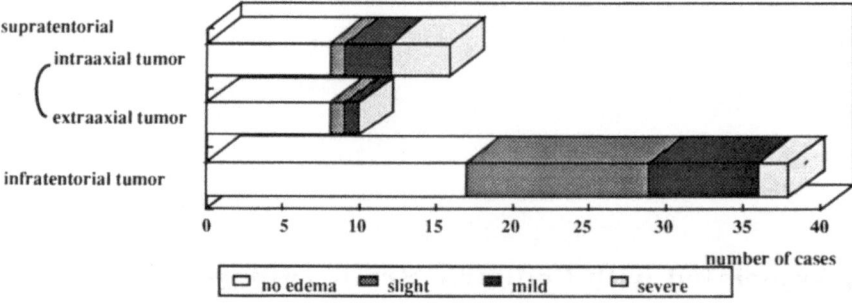

Fig. 1. Incidence of peritumoral brain edema in the pediatric group. The severity of brain edema was analyzed in relation to the locations

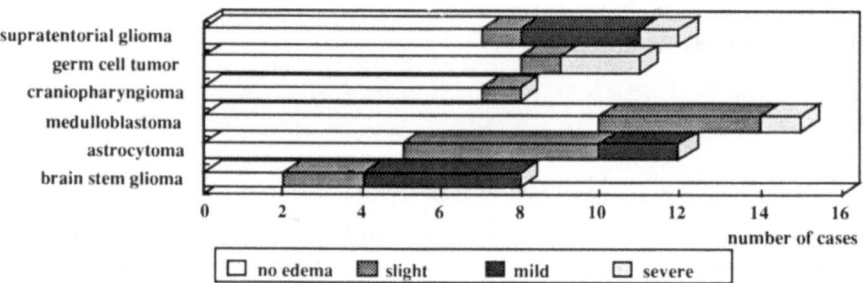

Fig. 2. Incidence of peritumoral brain edema in the pediatric group. The severity of brain edema was analyzed in relation to the histological classification

tive ectodermal tumors (PNET), one meningioma, and two neurofibroma of von Recklinghausen's disease showed slight brain edema. Although PNETs are malignant intra-axial tumor, all of them showed only slight peritumoral brain edema in contrast to the malignant tumors in the adult group. Supratentorial glioma in the pediatric group rarely associated with marked peritumoral edema that was observed in adult cases. Only one meningioma in our pediatric series was associated with von Recklinghausen's disease. This convexity meningioma was very large and showed exceptionally severe brain edema and hydrocephalic edema around the opposite side ventricle.

In general, marked brain edema extending into the white matter which was observed in the adult group was not observed in the pediatric group.

Discussion

Peritumoral edema is caused by leakage of serum from the vasculature in the brain tumors. The capillaries in the tumors are newly formed and lacks of normal blood brain barrier[2,4]. The extravasated edema fluid migrates into the white matter and absorbed into the ventricle or from the parenchymal capillaries. The developing brain is characterized by higher cerebral blood flow and higher cerebral metabolic rate. Cerebrospinal absorption capacity of developing brain may differ from that of adult brain. The different biological characteristics of developing brain produces different reaction to brain edema induced by brain tumors. Some studies showed more extent brain edema in traumatized geriatric brains. A experimental study demonstrated the tolerance of developing brain to the cold injury brain edema[1,3].

The extent of brain edema is demonstrated as high intensity area in MRI-T_2 image which correspondens to the accumulation of water in extracellular space[4]. However, it is rather difficult to measure exactly the volume of brain edema, because tumor margin is obscure in some cases. The calculation error in our study was 10%. Even if the error factor is considered, the incidence and severity of brain edema around the pediatric brain tumors was lower than that of adult brain tumors.

The marked peritumoral brain edema in the adult group was observed around meningioma, metastatic tumors, and malignant gliomas. The low incidence of those tumors in the pediatric group can reflect the lower incidence of brain edema. However, lower incidence and less severity of brain edema around pediatric supratentorial glioma than the adult gliomas indicates different characeristics of developing brain. The differences may be attributed to the lower blood pressures, higher metabolic rate, and higher cerebrospinal fluid absorption capacity in developing brain. In pediatric brain tumors, remarkable brain edema were recognized only in cases with marked obstructive hydrocephalus. It may indicate the effect of cerebrospinal absorption capacity on the resolution of brain edema.

In conclusion, lower incidence of brain edema associated with pediatric brain tumors might attributed not only to histological bias, but also the biological nature of developing brain.

References

1. Cohadon F, Desbordes P (1984) Brain edema, brain water, and aging. In: Inaba Y, Klatzo I, Spatz M (eds) Brain edema. Springer, Berlin Heidelberg New York Tokoy, pp 331–335
2. Kunishio K, Maeshiro T, Matsuhisa T, Mishima N, Tsuno K, Shigematsu H, Matsumoto K, Furuta T, Nishimoto A (1992) Neurol Surg 20: 39–44
3. Streicher E, Wisniewski H, Klatzo I (1969) Resistance of immature brain to experimental cerebral edema. Neurology: 833–836
4. Yamaki T, Hashi K, Kubota T (1990) Brain tumor and associated edema. Kyukyuigaku 14: 303–310

Correspondence: Atsufumi Kawamura, M. D., Department of Neurosurgery, Kobe University School Medicine, 7-5-1 Kusunoki-cho, Chuo-ku, Kobe 650, Japan.

Acta Neurochir (1994) [Suppl] 60: 384–386
© Springer-Verlag 1994

Perivascular Edema Fluid Pathway in Astrocytic Tumors

S. Kida[1], **D. W. Ellison, P. V. Steart, F. Iannotti,** and **R. O. Weller**

[1] Department of Neurosurgery, School of Medicine, University of Kanazawa, Japan and Department of Neuropathology and Clinical Neuroscience, School of Medicine, University of Southampton, U.K.

Summary

Perivascular spaces are anatomical routes for the bulk flow drainage of fluid from the gray matter to the subarachnoid space in normal rat brain. Perivascular cells are the resident scavengers in perivascular spaces. Following focal brain damage, perivascular cells upregulate MHC Class II antigens associated with uptake of edema fluid. Similar cells can be defined in damaged human brain. In the present investigation, the distribution of MHC Class II upregulated perivascular cells was measured in 30 astrocytic tumors and adjacent edematous tissues by immunocytochemistry using the following antibodies: HLA-DR (MHC Class II), PGM1 and MAC387 (macrophages). Perivascular cells were PGM1+/MAC387- and were located in perivascular spaces along blood vessels of all sizes. MHC Class II+ perivascular cells were distributed mainly in the tumors but in some cases (4 of 10 in astrocytomas, 4 of 10 in anaplastic astrocytomas, and 7 of 10 in glioblastomas) they were also found in adjacent edematous brain. The extensive MHC Class II expression on perivascular cells suggests that perivascular cells play a scavenging role in the perivascular spaces in human brain. The results of the present study indicate the similarity between perivascular spaces in human and rat brains and emphasize the significance of perivascular spaces as anatomical routes for edema fluid drainage from human brain tissue.

Keywords: Perivascular space; perivascular cell; brain edema; astrocytic tumor.

Introduction

Tracer studies in small mammals and anatomical studies in man suggest that perivascular spaces (PVS) are the drainage pathways for fluid, particularly from the gray matter of the brain[1,4]. It appears that fluid and soluble substances drain from the PVS into paravascular CSF compartments in the subarachnoid spaces[6]. In the rat brain, PVS contain an immunophenotypically distinct population of histiocytic cells, i.e. perivascular cells (PVC)[2]. Such cells avidly phagocytose particulate material injected into the PVS. Furthermore, PVC show a widespread rapid response following focal brain damage. Part of such a reaction is the upregulation of MHC Class II associated with the spread of cerebral edema[2]. There is now good evidence that PVC also exist in normal human brain and that they express MHC Class II[3]. The present study tests the hypothesis that a defined population of PVC in the human brain responds to tumor cells and the edema those cells provoke in a similar way to the PVC in rat brain. In this study, we have used formalin-fixed paraffin-embedded tissues in which the expression of constitutive MHC Class II is no longer detectable, but in which upregulated MHC Class II can be detected. This method not only makes possible the study of larger section areas and provided better histological resolution, it also functions as a discriminating factor in estimating the upregulation of MHC Class II.

Materials and Methods

Thirty astrocytic tumors (10 astrocytomas, 10 anaplastic astrocytomas and 10 glioblastomas) with adjacent edematous brain tissue were selected for the study. All tumors were supratentorial. Materials were obtained when the first surgery was performed.

Specimens were fixed in 10% buffered formalin for 24–48 hours, embedded in paraffin and serially sectioned at 3 μm. Sections were stained with HE, reticulin and immunohistochemistry. Five specimens of normal white matter and cortex, remote from tumor sites, were selected and processed in the same way as the tumor tissue.

Immunohistochemistry was performed using an avidin-biotin-peroxidase complex (ABC) technique. The following monoclonal antibodies were used at their optimal dilutions with 0.05M Tris/0.01 M phosphate-buffered saline (PBS): HLA-DR beta-chain (Dako, U.K.) (1 : 4000) for MHC Class II, PGM1 (Dako, U.K.) (1 : 200) and MAC387 (Dako, U.K.) (1:4000) for macrophages. A rabbit polyclonal antibody raised against glial fibrillary acidic protein (GFAP) in our laboratory was used at a dilution of 1 : 2000. Primary antibodies were applied to the sections for 30 min at room temperature. Specific antibody binding sites were visualized with 3,3′-diaminobenzidine (DAB) (Sigma, U.K.). Control sections for antibody binding were exposed to non-immune rabbit serum.

Results and Discussion

The histologically normal brain, remote from tumor tissue, showed no staining with the antibodies against MHC Class II, PGM1 and MAC387. These observations provided a baseline against which the reactions of PVC and microglia in tumor tissue and edematous peritumoral area could be assessed.

In the tumor area, MHC Class II+ cells were seen in 5 of 10 astrocytomas, 5 of 10 anaplastic astrocytomas and 9 of 10 glioblastomas. PVC showed a flattened morphology closely opposed to the outer aspect of vessel walls. By using the panel of 3 antibodies, it was possible to distinguish PVC from microglia, macrophages and blood monocytes (Table 1). This distinction between PVC and the other cells was most clearly seen in glioblastomas, in which the degree of PVC and microglial activation and infiltration of macrophages were most prominent. PGM1 labels differentiated macrophages, whereas MAC387 labels blood monocytes and hematogenous macrophages of recent recruitment, suggesting that PVC are mature tissue histiocytes and not recently derived from blood monocytes. The location of PVC within PVS can be demonstrated by reticulin stains, which identify the basement membrane of the glia limitans and the outer aspect of the blood vessel; the PVC are interposed between these two layers of reticulin. PVC are thus distinguished from vascular endothelial cells and from parenchymal microglia which lie outside the PVS. In the present study, the upregulation of MHC Class II is considered to be a general marker for activated cells in pathological conditions rather than a marker for antigen-presenting cells, as we found no spatial correlation between MHC Class II+ cells and lymphocyte infiltration. Therefore, MHC Class II upregulation in this study may reflect cellular uptake of soluble materials associated with tumors.

In peritumoral edematous tissue, both reactive PVC and microglia were labeled by PGM1 antibody. MHC Class II+ cells were detected in 4 of 5 astrocytomas, 4 of 5 anaplastic astrocytomas and 7 of 9 glioblastomas, all of the which contained MHC Class II+ cells within the tumors themselves. For the most part, MHC Class II labeling was confined to PVC from blood vessels of all sizes in the gray matter and occasional MHC Class II+ microglia was seen in the white matter (Table 2).

There was a gradual decrease in the number of MHC Class II+ cells as distance from the tumor increased.

The present study has defined PVC in astrocytic tumors as a distinct cell population occupying PVS

Table 1. *Idendification of PVC by Immunocytochemistry*

	MHC II	PGM1	MAC387
PVC	+	+	−
Microglia	+	+	+
Macrophages	+	+	+

Table 2. *Distribution of MHC Class II Positive Cells*

	Tumor	Edema	Normal tissue
PVC	++	+	−
Microglia	++	+/−	−
Macrophages	++	+/−	−

Number of labeled cells per mm²: ++ 11–25 cells; + 5–10 cells; +/− 1–4 cells; − none.

around blood vessels. MHC Class II upregulated PVC were widely distributed not only throughout tumor tissue but in surrounding edematous brain. A similar distribution of MHC Class II+ PVC associated with cerebral edema has been observed in the rat brain following surgical brain damage[2]. We have also reported that the anatomical structure of PVS in the human brain is similar to that in the rat brain[5] and that PVS could also act as fluid drainage pathways. This analogy has been further strengthened by the presence of PVC in human brain. Thus, in the present study, the upregulation of PGM1 and MHC Class II by PVC in edematous peritumoral tissue suggests that PVC are upregulated by edema fluid and may play a similar scavenging role in PVS as that seen in the rat[2]. Such humoral factors draining from tumors may also upregulate MHC Class II on microglia in edematous white matter, through which a substantial percentage of edema fluid seems to drain into the cerebral ventricles[1,5].

In conclusion, it appears that PVC in the human brain play a scavenging role which is closely linked to the drainage fo fluid along the PVS. Such PVS may not only provide anatomical routes for the resolution of edema fluid, but also fascilitate the reactions of brain resident histiocytic cells remote from tumors.

References

1. Kida S, Weller RO (1993) Morphological basis for fluid transport in and around ependymal, arachnoidal and glial cells. In: Raimondi AJ (ed) Principles of pediatric neurosurgery: intracranial cyst lesions. Springer, Wien New York

2. Kida S, Steart PV, Zhang ET, Weller RO (1994) Perivascular cells act as scavengers in the cerebral perivascular spaces and remain distinct from pericytes, microglia and macrophages. Acta Neuropathol (Berl), in press
3. Sasaki A Nakazato Y (1992) The identity of cells expressing MHC class II antigens in normal and pathological human brain Neuropathol Appl Neurobiol 18: 13–26
4. Weller RO, Kida S, Zhang ET (1992) Pathways of fluid drainage from the brain: morphological aspects and immunological significance in rat and man. Brain Pathol 2: 277–284
5. Zhang ET, Inmann CBE, Weller RO (1990) Interrelationships of the pia mater and perivascular (Virchow-Robin) spaces in the human cerebrum. J Anat 170: 111–123
6. Zhang ET, Richards HK, Kida S, Weller RO (1992) Directional and compartmentalized drainage of interstitial fluid and cerebrospinal fluid from the rat brain. Acta Neuropathol (Berl) 83: 233–239

Correspondence: S. Kida, M.D., Department of Neurosurgery, School of Medicine, University of Kanazawa, 13-1 Takara-machi, Kanazawa 920, Japan.

Acta Neurochir (1994) [Suppl] 60: 387–389
© Springer-Verlag 1994

The Plasma and CSF Vasopressin Levels in Brain Tumors with Brain Edema

V. D. Tenedieva, P. V. Lyamin, and **V. P. Nepomnyaschi**

Burdenko Neurosurgical Institute, Moscow, Russia

Summary

Vasopressin (VP) levels were evaluated by radioimmunoassay (RIA) in the arterial (A), peripheral (Vp) and jugular (Vj) vein blood and in CSF in 102 patients with brain tumors. In 60 cases the patients' state was complicated by brain edema (BE) and hemodynamic disturbances (HDD). The obtained data revealed significantly higher VP levels: 1) in A, Vp and CSF in patients with BE (Group A) in comparison with patients without BE (Group B), 2) in Vj in pateints with HDD only (Group Bc) and 3) in Vp in patients with HDD and BE (Group Ac) in comparison with Group Bc ($p < 0.05$). There were marked extremely high VP levels in Vj in patients with severe haemorrhage, tachycardia and high blood pressure (BP) and in CSF in patients with tachycardia, high BP and cardiac arrest ($p < 0.05$ correspondingly in each of the cases). Our results on a clinical basis confirmed CSF VP influence on BE development. We also confirmed the neurohumoral (through blood) and neurotransmitter (possibly through CSF and/or vasopressinergic pathways) VP influences on cardiocvascular regulation mechanisms. We content that this is a pathogenetic basis for application of VP direct or indirect antagonists for preventing and treating brain edema in neurosurgical patients.

Keywords: Vasopressin; CSF; brain edema.

Introduction

There are three neuropeptides (NP) functions in body; 1) releasing factors, 2) transmitters and modulators and 3) hormons. NP and blood-brain-barrier (BBB) interaction is a basis for humoral brain-body axis[2,5]. The aim of this study was to investigate the degree of VP influence in BE genesis and to estimate (indirectly) its neurotransmitter and its neurohormone influence on a clinical basis.

Materials and Methods

102 patients (51 men and 51 women aged 3–69 years) after brain tumor neurosurgical operations were studied. Among the 59 patients with BE, 57 died after surgery. Among 41 patients without BE, 17 died. BE was verified by computer tomography and clinical neurological investigations. Table 1 shows the detailed distribution of the hemodynamic complications (HDC) in examined patients. Plasma and CSF VP levels were simultaneously measured by radioimmunoassays in arterial (A), peripheral (Vp) and jugular (Vj) vein blood and CSF for 1–10 days after neurosurgery. To determine the VP levels, Buhlmann Laboratories AG (Switzerland) kits were used. The arterial blood pressure (BP) was measured by sphygmomanometry. Taking into account the possibility of reaction cascade and "harmful circuit" development in examined patients we performed a post hoc analysis of VP levels which were measured in the postoperative period (POP) and compared these measures with complications that took place during the operation period (OP). The analysis was made for VP levels and complications in POP also. The obtained data were analysed in patients without HDD with BE (Group A) and without BE (Group B) and with HDD also with and without BE (Group Ac and Bc correspondingly).

Statistics

Results were expressed as means and standard errors (SEM), and means were compared using Student's paired and unpaired t-test. Linear regression values were also calculated.

Results

The absolute VP values (Fig. 1) had a tendency to be lower in all survived patients with and without BE in comparison with patients who died. However, there are considerably higher Vj VP levels in survivors without BE ($p < 0.05$) and CSF VP levels in non-survivors with BE ($p < 0.05$). There were significant VP differences in A and Vj in non-survivors with BE in comparison with survivors with and without BE ($p < 0.05$), excluding VP levels in Vj in survivors without BE.

The VP levels depending on the HDD (in addition to presence of BE) revealed (Fig. 2 upper part) significantly higher VP levels: 1) in A, Vp and CSF in patients of Group A in comparison with Group B ($p < 0.05$) and 2) in CSF in patients of Group Ac in comparison with Group Bc and 3) vice versa in Vj in patients of Group Bc in comparison with Group Ac ($p < 0.05$). The analysis of VP levels in some HD

Table 1. *Hemodynamic Complications in Examined Patients with Brain Tumors*

	Operation period	Postoperative period
Haemorrhage	56 (56)	8 (8)
Cardiac arrest	8 (8)	4 (4)
Instability blood pressure	66 (64)	110 (60)
Heart rate changes:		
Tachycardia > 80 bpm	36 (35)	151 (68)
Bradycardia > 60 bpm	12 (12)	11 (9)
Ventricular arrythmia	4 (4)	4 (1)

Incidents (number of patients).

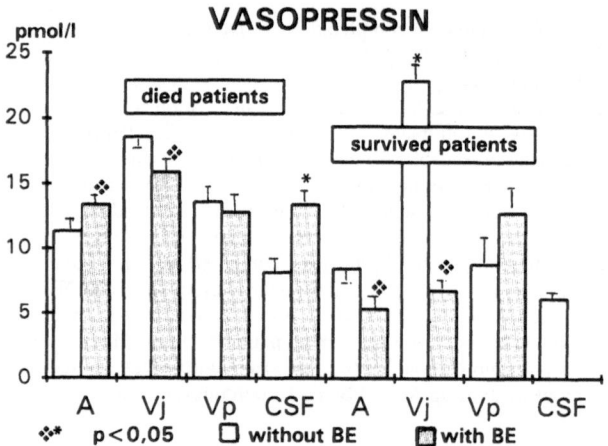

Fig. 1. Plasma (A, Vj, Vp) and CSF vasopressin levels in brain tumor patients with and without BE in dependence of outcome

symptoms revealed an extremely high VP levels in Vj in patients (Group Bc) with severe haemorrhage, tachycardia during OP and with high BP episodes in POP in comparison with patients of Group Ac (p < 0.05). In contrast, significantly higher VP levels were revealed in CSF in patients (Group Ac) with tachycardia, cardiac arrest and high BP during OP (p < 0.05) (Fig. 2 lower part).

No correlations were found between VP levels in plasma (A and Vj) and CSF in patients with BE, however, there were positive correlations in patients without BE (r = 0.4–0.52, p < 0.05). In spite of the considerable increase of the absolute values both plasma and CSF VP levels, CSF/plasma (A, Vj, Vp) VP ratios were significantly higher in patients with BE, especially when HDD was observed (Table 2).

Discussion

A great number of factors may contribute to changes in VP level and one of them is haemorrhage[3,6,10]. Our results show a considerable increase of plasma and CSF VP levels especially in patients with haemorrhage. Our studies also confirm by clinical data the well known CSF VP role of water accumulation in brain[1,8,9] and thus in the pathogenesis of BE.

The comparison of results in patients' groups depending on presence or absence of BE and HDD (Fig. 2 upper part) showed a relatively adequate reaction of VP hypothalamic system on neurosurgery in patients of Group B (which had no BE and had as a stress factor only the operation) and in patients of Group Bc (which had HDD plus the operative stress). Significantly increased CSF VP levels in patients of Group A may serve as an indicator of excessive VP secretion and release from many locations of the brain (not only hypothalamic nuclei)[7,10]. This fact by itself in combination with BE in examined patients supports the pathogenetic VP role in BE development.

On the other hand, significant increase in VP levels may probably be assumed to be a consequence of disturbances in body metabolism (reception and degradation processes) in patients of Group A. Namely metabolic disturbances could cause the lowering VP clearance particularly in POP. As we consider, Vp and CSF VP high levels in patients of Group A (with BE) taken together indicate both the lower VP clearance from blood and the permeability of BBB changes. These facts could create an additional "harmful pathway". The above

Table 2. *The CSF/Plasma Vasopressin Ratios*

Groups of patients	A (with BE)	B (without BE)	Ac (with BE and HDD)	Bc (without BE with HDD)
CSF/A	1.09	0.42	0.98	0.74
CSF/Vj	0.83	0.16	0.84	0.34
CSF/Vp	0.53	0.26	1.01	0.61

A arterial, *Vj* and *Vp* jugular and peripheral vein blood.

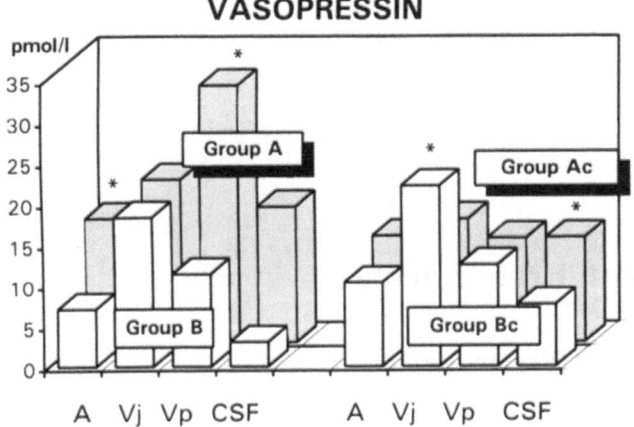

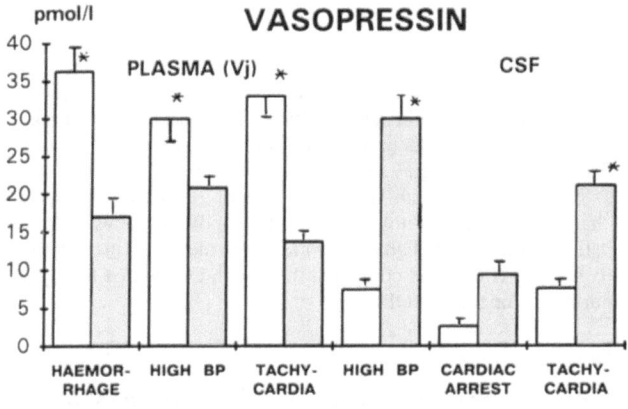

Fig. 2. Plasma (A, Vj, Vp) and CSF vasopressin levels in brain tumor patients in dependence of brain edema and hemodynamic disturbances (Groups A, B, Ac, Bc) (upper part), and some hemodynamic complications (lower part)

mentioned applies equally to patients of Group Ac too. However patients in more serious conditions of Group Ac were characterized not only by BE but also by different hard HD disturbances. As we may propose, the relative suppresion of VP secretion and releasing in brain (through the absolute VP values were significantly increased beyond the normal) in patients of Group Ac were due to more severe metabolic changes.

The mapping of central vasopressinergic pathways and receptors within the hypothalamus and brainstem of rat and human and studies of cardiovascular regulation mechanisms by VP revealed that the effects of this neuropeptide depend on site of injection in the cerebrospinal space or brain nuclei[4,11,12]. These data are the basis which may explain complex of diverse HDD that were manifest in examined patients by lowering or increasing BP, cardiac arrest, tachy- and bradycardia etc. Thus our results support neurotransmitter and neurohormonal VP roles in cardiovascular functions on a

clinical basis. Increased CSF/plasma VP level ratios and also the absence of correlation between VP levels in plasma and CSF in our study suggests the probable existence of two independent VP systems in the brain; one of which is connected with secretion of this peptide by hypothalamic nuclei and its release into the blood, and the other one with extrahypothalamic neurons and VP secretion directly into the cerebrospinal fluid.

It should be noted in summary that our results show a more reliable appraisal of hypothalamic releasing VP may be obtained by VP measurement in Vj in comparison with A and Vp. This accessory conclusion of our studies may be useful in clinical conditions when there is no possibility of direct measurement of VP levels in pituitary portal venous blood (in contrast to experimental investigation).

Our data serve as a pathogenetic basis for prescribing *direct antagonists of VP (its V1-receptors in brain) or indirect antagonists (Ca+2 channels' inhibitors)* in patients undergoing surgical removal of brain tumor.

References

1. Doczi T, Szerdahrlyi P, Gulya K, Kiss J (1982) Brain water accumulation after the central administration of vasopressin. Neurosurgery 11: 402–407
2. Ermisch A, Ruhle H-J, Landgraf R, Hess J (1985) Blood-brain barrier and peptides. J Cereb Blood Flow Metab 5: 348–350
3. Forsling ML (1976) Anti-diuretic hormone. Eden, Montreal
4. Gunter O, Kovacs GL, Szabo G, Telegly G (1986) Opposite effects of intraventricular and intracisternal administration of vasopresine on pressure in rats. Peptides 7: 539–540
5. Jezova D, Johansson BB, Oprsalova Z, Vigas M (1989) Changes in blood-brain barrier function modify the neuroendocrine response to circulating substances. Neuroendocrinology 49: 428–433
6. Laycock JF, Penn W, Shirley DG, Walter SJ (1976) The role of vasopressin in blood pressure regulation immediately following acute haemorrhage in the rat. J Physiol 296: 267–275
7. Luerssen TG, Shelton RL, Robertson GL (1977) Evidence for separate origin of plasma and cerebrospinal fluid vasopressin. Clin Res 25: 14A
8. Raichle ME, Grubb RL (1978) Regulation of brain water permeability by cntrally-released vasopressin. Brain Res 142: 191–194
9. Rosenberg CA, Estrada E, Kyner WT (1988) The effect of arginine vasopressin and V1-receptor antagonist on brain water in cat. Neurosci Lett 95: 241–245
10. Sorenson PS (1986) Studies of vasopressin in the human cerebrospinal fluid. Acta Neurol Scand 74: 81–102
11. Unger T, Romeis D, Demmert G, Ganten D, Lang R, Luft F (1986) Brain vasopressin and plasma AVP differentially modulate baroreceptor reflex. Hypotention 8 [Suppl 11]: 157–162
12. Zerbe RL, Kirtland S, Faden AI, Feuerstein G (1983) Central cardiovascular effects of mammalian neurohypophyseal peptides in conscious rats. Peptides 4: 627–630

Correspondence: V. D. Tenedieva, M. D., Burdenko Neurosurgery Institute, Fadeev Street 5, Moscow 125047, Russia.

Acta Neurochir (1994) [Suppl] 60: 390–394
© Springer-Verlag 1994

The CSF Aldosterone in Brain Tumors with Brain Edema

V. D. Tenedieva, V. P. Kulikovsky, P. V. Lyamin, and **V. P. Nepomnyaschi**

Burdenko Neurosurgical Institute, Moscow, Russia

Summary

The study of renin-angiotensin-aldosterone (RAA) and vaso-pressin (VP) systems in neurosurgical patients with brain tumors and brain edema (BE) had revealed an excessive activity of these systems with secondary hyperaldosteronism especially with BE that proves the pathogenetic role of these systems. Measurement of Aldosterone (Ald) in CSF may serve as a diagnostic test to help manage the patient's clinical condition. Mechanisms of Ald penetration in CSF assumed to be the result of blood-brain-barrier (BBB) destruction (especially in astrocytomas) and/or the mediation by neuropeptides (for example increasing activity of VP V1-receptors). Results serve as a basis for application of the neuropeptide and hormone antagonists and inhibitors on all stages of cascade reactions taking part in the water and sodium retention.

Keywords: Aldosterone; CSF; brain edema; astrocytomas.

Introduction

The presence of renin angiotensin system (RAS) in the brain has been demonstrated[3], moreover attempts were undertaken[11] to use the Ald antagonists (spirono-lactones) in BE treatment. However, the exact role of Ald in BE pathogenesis is not yet clear. The goal of this study was: 1) to analyse the blood-born Ald in BE pathogenesis taking into account the morphological structure of brain tumors (astrocytomas and nonastro-cytomas) and 2) to estimate the role of Ald and RAS components in development of BE and cardiovascular disturbances which in turn can lead to fatal outcome.

Materials and Methods

102 patients (51 men and 51 women aged 3–69 years) after brain tumor surgery were studied. Among the 60 patients with BE 57 died after surgery, and among 42 without BE 17 died. BE was verified by computer tomography and clinical neurological investigations. Hemodynamic complications (HDC) are presented in Table 1. Plasma and CSF hormones were simultaneously measured by radioimmunoassays in arterial (A), peripheral (Vp) and jugular (Vj) vein blood and the CSF for 1–10 days after neurosurgery. To determine the Ald levels and plasma renin activity (PRA). International CIS (France) kits were used. To determine the VP and angiotensin II (AII), Buhlmann Laboratories AG (Switzerland) kits were used.

Taking into account the possibility of reaction cascade and "harmful circuit" development we applied post hoc analysis for the patient hormone level changes which were measured in postoperative period (POP) and compared them with events that look place during the operation (OP).

Statistics

Results were expressed as means and standard errors (S.E.M.), and means were compared using Student's paired and unpaired t-test. Linear regression values were also calculated.

Results

Figure 1 shows the levels of neuropeptides and hormones at different HDC. In patients with astrocytomas (22 persons) and BE (10 persons), as seen on Fig. 2, Ald levels in plasma (A and Vj) and in CSF were significantly increased. VP CSF levels were significantly increased also in BE. In nonastrocytomas cases VP levels were markedly increased in Vj in patients without BE and in CSF in patients with BE. CSF Ald content in patients with nonastrocytomas and BE was lower ($p < 0.05$) than in those with astrocytomas and BE. There were no significant differences in CSF Ald levels in groups nonastrocytomas patients with and without BE.

Discussion

The results of this study indicate the high level of activity of all components RAA and Vp systems particularly in patients with BE. This fact is in correspondence with the known mechanisms by which these systems are stimulated (different structure affections of

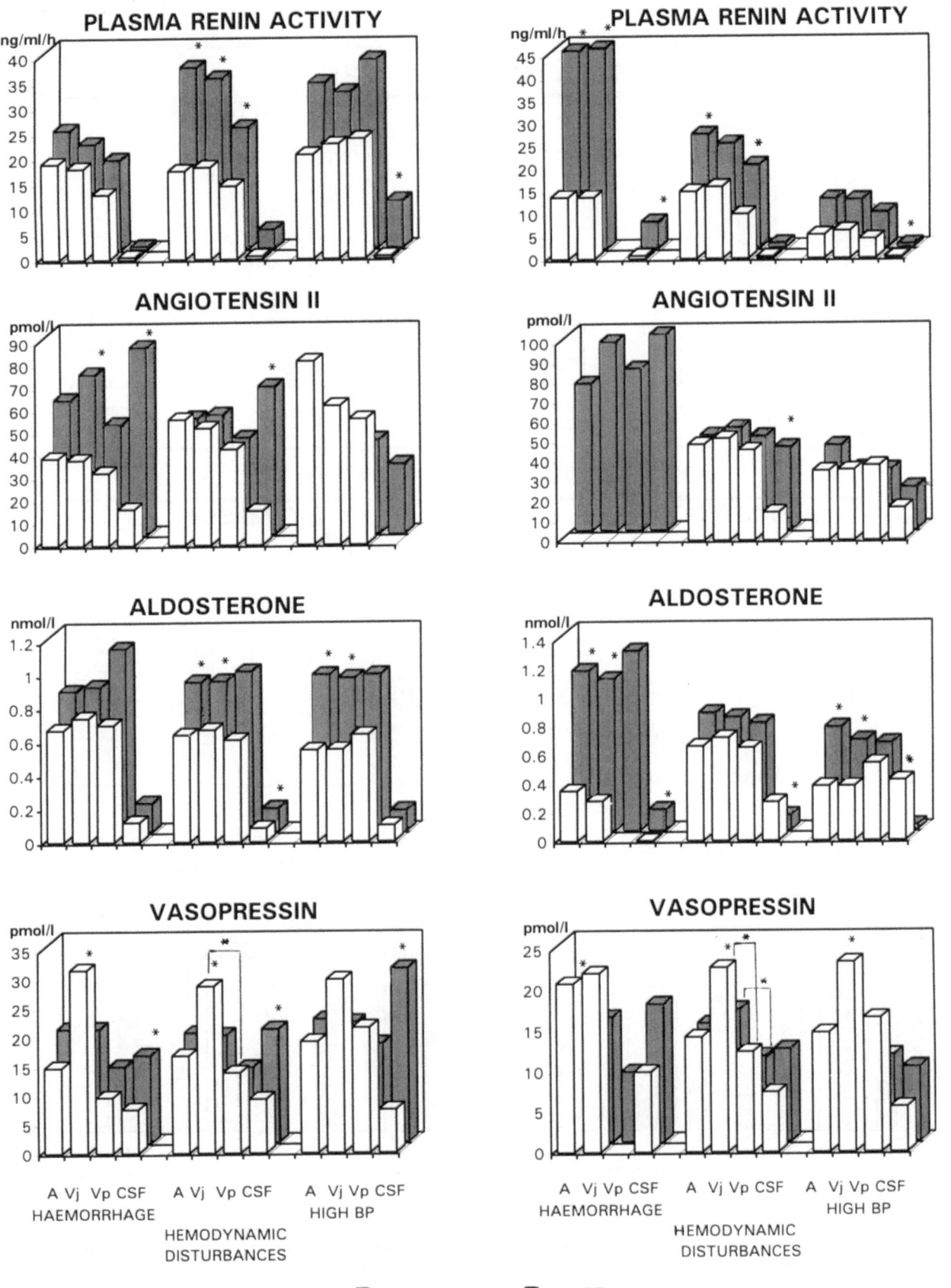

Fig. 1. PRA and AII, Ald, VP levels in brain tumor patients with and without BE depending on some clinical events (haemorrhage, hemodynamic disturbances) during OP (left side) and POP (right side). * indicates p < 0.05

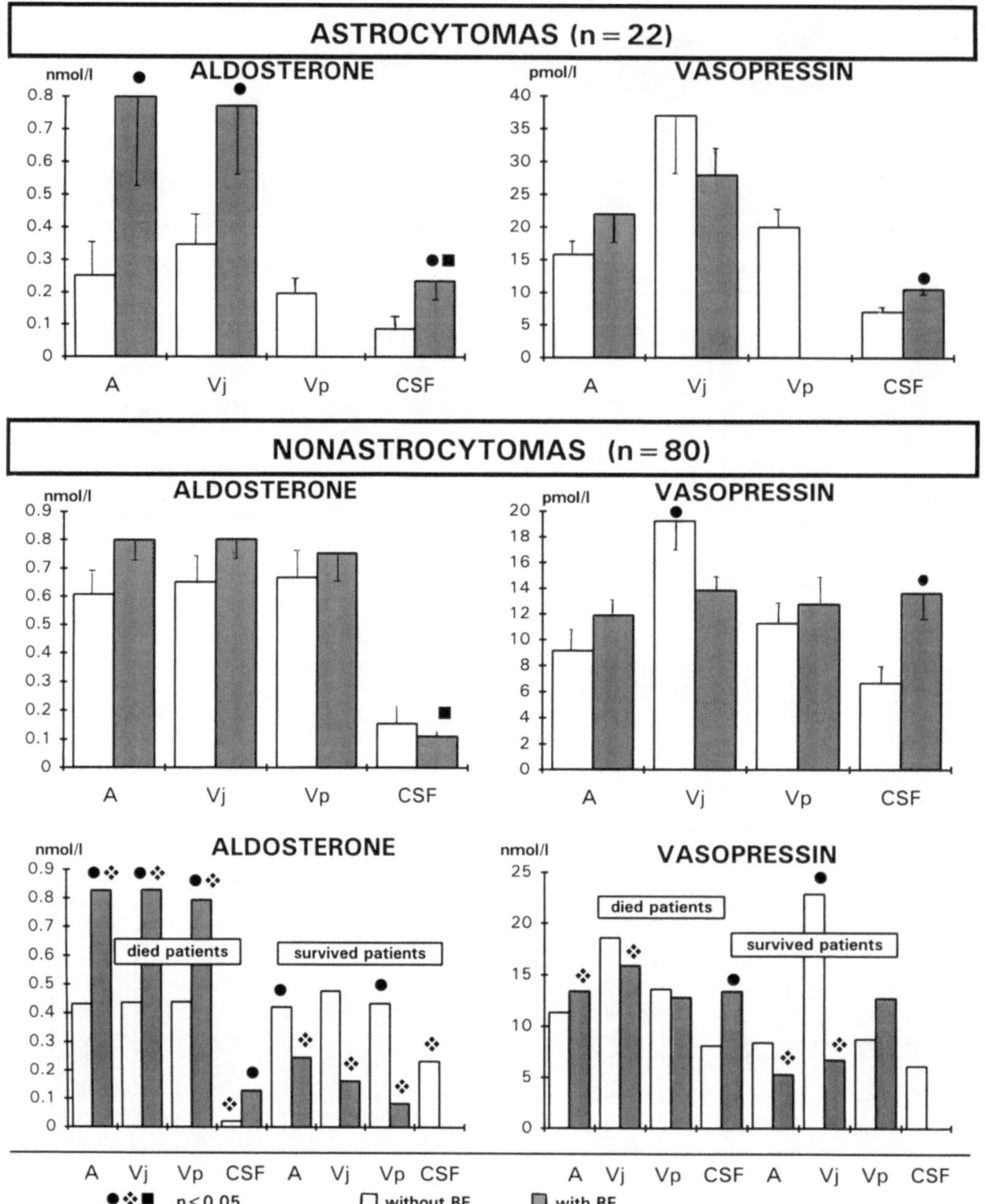

Fig. 2. Plasma (A, Vj, Vp) and CSF aldosterone and vasopressin levels in patients with astrocytomas and nonastrocytomas depending on presence or absence of brain edema and outcome (survived and died). * indicates p < 0.05

Table 1. *Hemodynamic Complications in Examined Patients with Brain Tumors*

	Operation period	Postoperative period
Haemorrhage	56 (56)	8 (8)
Cardiac arrest	8 (8)	4 (4)
Instability blood pressure	66 (64)	110 (60)
Heart range changes:		
Tachycardia > 80 bpm	36 (35)	151 (68)
Bradycardia > 60 bpm	12 (12)	11 (9)
Ventricular arrythmia	4 (4)	4 (1)

Number of incidents (number of patients).

brain itself, neurosurgery, haemorrhage and its sequences)[1,14].

Our results indicate extremely high CSF AII levels. However it is difficult to estimate whether the origin of high AII levels is due to a brain-born source or due to an increasing penetration across the BBB. Correlation coefficients between AII and PRA in plasma were 0.9 with BE ($p < 0.05$) and 0.7–0.8 without BE ($p < 0.05$). Correlation coefficients between AII (A, Vj) and AII in CSF were 0.8 and 0.81 correspondingly ($p < 0.05$) with BE and between Vp and CSF AII was 0.59 ($p < 0.05$) without BE. We believe such extremely high level of RAS activity in those patients examines to be a specific feature for neurosurgery and undoubtely it contributes to the BE development. This specificity is probably due to the increased releasing renin-releasing factor under the influence of increased potassium concentration[13]. The hemodynamic disturbances (HDD) together with increased endogenous CSF and plasma AII and VP levels confirm the pathogenetic role of these neuropeptides in HDD due to their neurohormonal and neurotransmitter influence. These data are in agreement with results published elsewhere[8,14]. The second-

ary hyperaldosteronism with high Ald levels in plasma and CSF are considered to be the additional factor in pathogenesis of HDD. The negative inotropic effect of Ald is well known[10].

The detectable amounts of Ald in CSF may be considered as one of the most important findings in our study. It is known that the steroid hormones traverse the BBB by lipid mediation[9]. To explain this fact two mechanisms in neurosurgical brain tumor patients are to be considered. The first one is probable transient mediation by neuropeptides including VP by means of its specific V1-receptors on luminal membrane of brain capillaries endothelium tight junctions[2]. There is considerable evidence that interaction between neurons and BBB is a basis of humoral brain-body axis. This mechanism is confirmed by positive correlations (Table 2) between VP levels in CSF and Ald in A, Vj and CSF and negative ones between Vp levels in CSF and Ald in Vp in patients with BE. Correlations between CSF VP levels and plasma and CSF Ald levels were absent in patients without BE.

The second mechanism is the BBB damage (free drainage between plasma and CSF). In the case of astrocytomas our results assumed this mechanism to be most likely. We posit that the astrocytes physically create BBB surrounding the brain capillaries endothelium thus limiting the penetration into the brain of substances circulating in the blood[4,5]. The absence of BBB tight junctions on endothelium brain capillaries in astrocytomas[7] by transformation of astrocytes, as it was proved by BBB ultrastructure examination, may create fenestrations like in circumventricular organs. In addition, the declined Ald clearance from CSF cannot be excluded.

Therefore the presence of Ald in CSF in patients in critical condition may serve as a diagnostic test not only in neurosurgical patients with BE but in other clinical cases[6]. Analysis of Ald contents in A, Vj, Vp and CSF in POP has revealed an interesting fact: the

Table 2. *Coefficients of Correlation (All Values are Significant) Between Plasma and CSF Vasopressin and Aldosterone Levels in Examined Patients*

		Vasopressin without BE				Vasopressin with BE			
		A	Vj	Vp	CSF	A	Vj	Vp	CSF
Aldosterone	A	0.3	0.4	0.5	0.0	0.0	0.2	0.6	0.5
	Vj	0.3	0.4	0.5	0.0	0.0	0.2	0.6	0.5
	Vp	0.3	0.5	0.0	0.0	0.7	0.8	0.6	− 0.5
	CSF	0.0	0.0	0.0	0.0	0.0	0.0	0.6	0.3

content of Ald in CSF could be detected until its levels in plasma exceeded 1.0 nmol/l (normal values are ranged 0.033–0.346 nmol/l in supine).

What is the role of blood-born Ald in CSF? It is well known recently that Na, K, Cl co-transport stimulated by sodium[12] and K+"siphoning"[5] takes place in human astrocytes. Tas *et al.*[12] speculate that a large increase of potassium concentration in extracellular space may result in the accumulation of osmotically active particles inside the cell with concomitant water movement. Our results assume CSF Ald to be involved in the above mentioned processes and thus promotes the development of brain edema.

It may be proposed that the astrocytes transformation and consequent CF Ald increase in astrocytomas are one more argument in favor of the CSF Ald role in the pathogenesis of brain edema.

Our results may serve as a basis for pharmacologic intervention which will preclude an excessive activation of the systems studied in this report. It might be possible to influence the cascade reactions on any stages by applying: 1) medication of the renin substrate antagonists (for example enalkiren type), 2) angiotensin-converting enzyme (ACE) inhibitors and 3) direct or indirect (for example Ca^{++} channel blockers) VP antagonists.

References

1. Carlson DE, Gann DS (1987) Response of adrenocorticotropin and vasopressin to hemorrhage after lesions of caudal ventrolateral medulla in rats. Brain Res 406: 385–390
2. Ermisch A, Ruhle H-J, Landgraf R, Hess J (1985) Blood-brain barrier and peptides. J Cereb Blood Flow Metab 5: 350–357
3. Ganten D, Lang RE, Lehmann E, Unger T (1984) Brain angiotensin: on the way to becoming a well-studied neuropeptide system. Biochem Pharmacol 33: 3523–3528
4. Goldstein GW (1988) Endothelial cell-astrocyte interactions. In: Strand FL (ed) Forth colloquium in biological sciences: blood-brain transfer. Annals of the New York Academy of Sciences, Vol 529. New York, pp 31–39
5. Iadecola C (1992) Intrinsic and extrinsic neural regulation of the cerebral circulation. In: Schmiedek P, Einhaupl K, Kirsch CM (eds) Stimulated cerebral blood flow. Springer, Berlin Heidelberg New York Tokyo, pp 19–36
6. Johansson BB (1980) The blood-brain barrier in acute and chronic hypertension. In: Eisenberg HM, Suddith RJ (eds) The cerebral microvasculature. Plenum, New York, pp 211–216
7. Long DM (1970) Capillary ultrastructure of blood-brain barrier in human malignant brain tumors. J Neurosurg 32: 127–144
8. Matsuguchi H, Sharabi FM, Gordon FJ, Johnson AK, Schmid PG (1982) Blood pressure and heart rate responses to microinjection of vaopressin into the nucleus tractus solitarius region of the rat. Neuropharmacology 21: 687–693
9. Pardridge WM (1981) Transport of protein-bound hormones into tissue in vivo. Endocr Rev 2: 103–123
10. Sayers G, Solomon N (1960) Work performance of a rat heartlung preparation: standardization and influence of corticosteroids. Endocrinology 66: 719–722
11. Schmiedek P, Baethmann A, Schneider E, Brendel W, Enzenbach R, Marguth F (1972) Use of spirolactone in the treatment of cerebral edema in neurosurgical patients. In: Brendel W, Coraboeuf E, Ebling FJG, Friis T, Greenwood G, Henderson IW, Hokfelt B, Lauler D, Witzmann R (eds) Extrarenal activity of aldosterone and its antagonists. Excerpta medica, Amsterdam, pp 234–237
12. Tas PWL, Massa PT, Koschel K (1986) Preliminary characterization of an Na+, K+, Cl' co-transport activity in cultured human astrocytes. Neurosci Lett 70: 369–373
13. Urban JH, Brownfield MS, Levine JE, Van der Kar LD (1992) Distribution of a renin-releasing factor in the central nervous system of the rat. Neuroendocrinology 55: 574–582
14. Zerbe RI, Feuerstein G, Meyer DK, Kopin IJ (1982) Cardiovascular, sympathetic and renin angiotensin system responses to hemorrhage in vasopressin deficient rats. Endocrinology 111: 608–613

Correspondence: V. D. Tenedieva, M.D., Burdenko Neurosurgery Institute, Fadeev Street 5, Moscow 125047, Russia.

Tumor and Brain Edema: Treatment

Acta Neurochir (1994) [Suppl] 60: 397–399
© Springer-Verlag 1994

The Effects of Topical Dexamethasone on Experimental Brain Tumors and Peritumoral Brain Edema

Y. Ikeda[1], **B. S. Carson**[2], and **D. M. Long**[2]

Departments of Neurosurgery, [1] Nippon Medical School, Tokyo, Japan, and [2] The Johns Hopkins Hospital, Baltimore, MD, U.S.A.

Summary

To determine if topical dexamethasone administered to brain tumor beds would not only control peritumoral edema and suppress tumor growth but also prevent systemic steroid complications, we studied experimental brain tumors produced in 102 rabbits by implanted VX2 carcinoma cells. We separated 58 animals into three groups: 1) untreated rabbits (n = 15), 2) systemic dexamethasone-treated (4 mg/kg/day) rabbits (n = 18), and 3) topical dexamethasone-treated (2.5 µl/h, osmotic pump) rabbits (n = 25). We administered systemic or topical dexamethasone from the third day or from the seventh day after tumor implantation, and sacrificed the animals on the 13th day. We compared survival in these three groups with that of another 44 rabbits, beginning treatment on the seventh day. We measured brain water content in the white matter of the sacrificed rabbits by the specific gravity method. We measured the length and width of the brain tumors of all the rabbits and estimated tumor volume. Systemic and topical dexamethasone administered from the third day produced statistically significant inhibition of tumor volume as well as a mean reduction in peritumoral brain edema in most tested sites. Systemic and topical dexamethasone treatment resulted in a statistically significant increase in survival relative to the untreated group. These results suggest that topical dexamethasone is efficacious in a brain tumor model and its administration to brain tumor beds constitutes a new therapeutic modality.

Keywords: Brain edema; dexamethasone; osmotic pump; VX2 carcinoma.

Introduction

Peritumoral brain edema remains a substantial cause of mortality and morbidity in patients with brain tumors[1]. For the past thirty years, systemically administered steroids have proved to be one of the few therapies effective in treating this problem; however, the use of steroids is limited by many systemic side effects[4]. In this study, we investigated the therapeutic effects of topical dexamethasone on brain tumors to determine if the beneficial effects of systemic dexamethasone could be achieved concomitant with a reduc-

tion in harmful side effects. We used a rabbit brain tumor model established by Carson et al.[3]. This tumor model forms a single intracerebral tumor mass days after surgical implantation, with rapid growth of the tumor between the ninth and the 13th day following tumor cell implantation. The therapeutic effects of systemic and topical dexamethasone on peritumoral brain edema and brain tumor growth were evaluated both before and after formation of the intracerebral tumor mass in the present study.

Materials and Methods

Tumor Implantation

Adult New Zealand white rabbits weighing 3.5–4.5kg were used for this procedure. They were anesthetized with ketamine hydrochloride (25 mg/kg) and acepromazine maleate (2.5 mg/kg). Experimental brain tumors were produced by injecting a 25-µl suspension of 3×10^5 viable VX2 carcinoma cells into the right front-parietal lobe through a 1-mm diameter cranial burr hole.

Implantation of the Osmotic Pump

A 2-cm para-saggital incision was made. The periosteum was reflected and a 23-gauge burr hole was made 4 mm lateral and 3 mm posterior to the bregma. A Teflon catheter was inserted to a depth of 7 mm and glued in place. The Alzet osmotic pump was then implanted subcutaneously and connected to the Teflon catheter with P.E.60 tubing.

Peritumoral Brain Edema Study

Fifty-eight rabbits were separated into three groups: 1) untreated tumor-bearing rabbits (n = 15), 2) systemic dexamethasone-treated (4 mg/kg/day) tumor-bearing rabbits (n = 18), and 3) topical dexamethasone-treated (2.5 µl/h, 2ML4 Alzet osmotic pump) tumor-bearing rabbits (n = 25).

Systemic or topical dexamethasone was administered from day 3 or day 7 after tumor implantation, with sacrifice at day 13. Brain water content was measured by the specific gravity (SG) method. Brain samples, approximately 1 mm³, were excised and placed onto a kerosene/monobromobenzene column. The eight sampling areas are illustrated in Fig. 1.

Measurement of Tumor Volume

Tumor volume was estimated by measuring tumor dimensions along the long axis of the tumor and across the tumor at the widest point. Tumor volume was calculated by the equation[2]:

tumor volume (cu mm) = length (mm) x width $(mm)^2/2$.

Survival Study

Forty-four rabbits were separated into three groups: l) untreated tumor-bearing rabbits (n = 16), 2) systemic dexamethasone-treated (4 mg/kg/day) tumor-bearing rabbits (n = 14), and 3) topical dexamethasone-treated (2.5 µl/h, 2ML4 Alzet osmotic pump) tumor-bearing rabbits (n = 14). Treatment was started on day 7 after tumor implanation and continued until death.

Statistical Analysis

Statistical comparisons were performed using the Student's t-test for independent variables, and a 95% confidence level was considered statistically significant. All data are given as the mean ± standard error.

Results and Discussion

Peritumoral Brain Edema Study

In the third-day treatment groups, mean white matter edema in all sites was found to be significantly different from the mean edema in the corresponding sites in the control rabbits at the 5% level. In the seventh-day topical treatment group, mean white matter edema in one site in the contralateral hemisphere was found to be

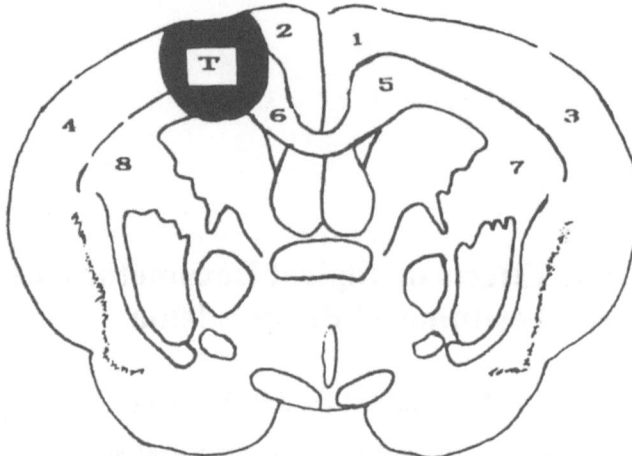

Fig. 1. The eight sampling areas used for measurement of water content by the specific gravity method in brains subjected to transplantation of VX2 carcinoma cells

significantly different from the mean edema in the corresponding site in the control rabbits at the 5% level. In the seventhy-day systemic treatment group, mean white matter edema in three sites, two in the contralateral and one in the ipsilateral hemisphere, was found to be significantly different from the mean edema in the corresponding sites in the control rabbits at the 5% level. In no group was grey matter edema found to be significantly different from the corresponding sites in the control rabbits at the 5% level. The brain edema data are reported in Table 1.

Table 1. *Specific Gravity Values of Eight Sampling Areas in Untreated Tumor-Bearing Rabbits and in Those Treated with Local or Systemic Dexamethasone*

	Area							
	1	3	5	7	2	4	6	8
Untreated (n = 15)	1.0469 ± 0.0017	1.0465 ± 0.0018	1.0367 ± 0.0046	1.0432 ± 0.0033	1.0461 ± 0.0032	1.0465 ± 0.0020	1.0377 ± 0.0038	1.0394 ± 0.0038
3 Local[a] (n = 15) p-value	1.0467 ± 0.0010 none	1.0467 ± 0.0011 0.315	1.0441 ± 0.0021 < 0.001	1.0455 ± 0.0012 0.006	1.0461 ± 0.0018 0.475	1.0465 ± 0.0009 0.497	1.0394 ± 0.0036 0.047	1.0443 ± 0.0019 < 0.001
3 System[a] (n = 11) p-value	1.0462 ± 0.0010 none	1.0465 ± 0.0009 0.451	1.0440 ± 0.0026 < 0.001	1.0449 ± 0.0015 0.027	1.0455 ± 0.0019 none	1.0465 ± 0.0014 none	1.0411 ± 0.0032 < 0.001	1.0441 ± 0.0020 < 0.001
7 Local[b] (n = 10) p- value	1.0466 ± 0.0017 none	1.0463 ± 0.0015 none	1.0393 ± 0.0048 0.019	1.0440 ± 0.0029 0.187	1.0458 ± 0.0019 none	1.0465 ± 0.0015 none	1.0378 ± 0.0028 0.441	1.0403 ± 0.0031 0.207
7 System[b] (n = 7) p-value	1.0458 ± 0.0007 none	1.0461 ± 0.0007 none	1.0437 ± 0.0006 < 0.001	1.0447 ± 0.0004 0.048	1.0455 ± 0.0015 none	1.0458 ± 0.0008 none	1.0385 ± 0.0029 0.225	1.0431 ± 0.0013 < 0.001

[a] Treatment from 3 days after tumor implantation sacrificed at 13th day.
[b] Treatment from 7 days after tumor implantation sacrificed at 13th day.

Tumor Volume

Systemic and topical dexamethasone administered from day 3 with sacrifice on day 13 produced significant inhibition of tumor volume (mm^3) relative to the untreated group (31.72 ± 9.66, 26.46 ± 6.51, 135.78 ± 28.79, respectively) ($p < 0.05$). Systemic and topical dexamethasone administered from day 7 with sacrifice on day 13 tended to inhibit tumor volumes, but there was no statistical significance relative to the untreated group (76.18 ± 27.22, 84.25 ± 20.14, 135.78 ± 28.79, respectively). The differences in final tumor volume between the three groups were not significant (137.35 ± 15.40, 143.79 ± 10.96, 179.25 ± 8.86, respectively).

Survival Study

Systemic and topical dexamethasone treatment both resulted in a significant increase in survival relative to the untreated group (17.57 ± 0.58, 17.92 ± 0.79, 13.25 ± 1.00, days, respectively) ($p < 0.05$).

References

1. Bartkowski HM (1984) Peritumoral edema. Prog Exp Tumor Res 27: 179–190
2. Bullard DE, Schold SC, Bigner SH, Bigner DD (1981) Growth and chemotherapeutic response in athymic mice of tumors arising from human glioma-derived cell lines. J Neuropath Exp Neurol 40: 410–427
3. Carson BS, Anderson JH, Grossman SA, Hilton J, White CL, Colvin OM, Clark AW, Grochow LB, Kahn A, Murray KJ (1982) Improved rabbit brain tumor model amenable to diagnostic radiographic procedures. Neurosurgery 2: 603–608
4. Weissman DE, Dufer D, Vogel V, Abeloff MD (1987) Corticosteroid toxicity in neurooncology patients. J Neurooncol 5: 125–128

Correspondence: Yukio Ikeda, M. D., D. M. Sc., Department of Neurosurgery, Nippon Medical School, 1-1-5, Sendagi, Bunkyo-ku, Tokyo 113, Japan.

Acta Neurochir (1994) [Suppl] 60: 400–402
© Springer-Verlag 1994

Effect of Histamine on the Blood-Tumor Barrier in Transplanted Rat Brain Tumors

T. Nomura, K. Ikezaki, K. Matsukado, and **M. Fukui**

Department of Neurosurgery, Neurological Institute, Faculty of Medicine, Kyushu University, Fukuoka, Japan

Summary

We studied the effect of intracarotid administration of histamine on the blood-tumor barrier permeability and also on the blood-brain barrier permeability in transplanted rat C6 glioma. There was no definite Evans blue (EB) extravasation either in normal or tumor tissue after intracarotid saline infusion. In contrast, histamine at doses of 1 and 10 µg/kg/min produced slight to moderate EB extravasation in the tumor without any significant extravasation in the normal brain tissue. Intravenously administered H_1 and H_2 receptor antagonists (5 mg/kg each) reduced the histamine (10 µg/kg/min) induced extravasation of EB in the tumor tissue. These results indicated that brain tumor vessels responded to histamine in a different fashion from normal brain capillaries. Histamine could thus be utilized for selective drug delivery to brain tumors without affecting normal brain tissue.

Keywords: Histamine; permeability; brain neoplasm; C6 glioma.

Introduction

The regulatory action of neurotransmitters to the blood-tumor barrier (BTB) has not been studied. Histamine has been proposed as a vasoactive neurotransmitter due to the histaminergic innervation of the cerebral micro-vasculature and its vasodilatory effect on normal pial vessels[1-3]. Furthermore, intracarotid administration of histamine has been reported to extravasate various tracers into normal brain tissue[4,5]. Experiments were undertaken to determine the effect of intracarotid administration of histamine on the BTB and also on the blood-brain barrier (BBB) using a transplanted rat brain tumor model.

Materials and Methods

C6 glioma cells were subcultured in Ham's F-10 medium containing 10% fetal bovine serum. Tumor cells (5×10^5 cells/5 µl) were stereotactically implanted into the right basal ganglia of male Wistar rats (weighing 190–250 g) under intraperitoneal pentobarbi-

tal anesthesia (50 mg/kg). Eleven to 13 days after tumor inoculation, the rats were anesthetized with pentobarbital for a tracheostomy and catheter placement in the femoral arteries, veins and internal carotid artery. Each animal was mounted onto a stereotactic apparatus, immobilized with pancuronium (0.6 mg/kg, i.v.), and mechanically ventilated with room air. During the experiment, anesthesia was maintained with halothane (0.2–0.5%) and the arterial blood pressure was monitored. End-tidal CO_2 and recta temperature were also monitored continuously and adjusted regularly throughout the experiment. After the arterial blood pressure became stable, arterial blood samples were taken for the blood gas analysis (Techno Medica GASTAT-1). Histamine (Sigma) was dissolved in 0.9% saline and administered intracarotidly by constant infusion at a rate of 1.5 ml/h. Histamine at doses of 1 and 10 µg/kg/min was infused through the internal carotid artery for 10 minutes. EB (2%) was injected (5 mg/kg, i.v., bolus) immediately after histamine infusion. In place of histamine, intracarotid saline infusion was performed for the control experiment. Histamine-receptor antagonists (pyrilamine: H_1 blocker, cimetidine: H_2 blocker, 5 mg/kg i.v. each) were administered 10 minutes before histamine infusion for the blocking experiment. Animals were perfused with Ringer's solution through the aorta and decapitated 20 minutes after EB injection. The extravasation of EB was visually investigated in the removed brain and graded. The specimens were then sectioned, stained with hematoxylin and eosin, and microscopically examined.

Results

No significant changes were observed in physiological parameters, such as arterial blood pressure, body temprature, and arterial blood gas. Neither intracarotid administration of histamine (1 and 10 µg/kg/min) nor intravenous injection of histamine receptor antagonists (pyrilamine and cimetidine, 5 mg/kg each) changed systemic blood pressure.

Intracarotid infusion of saline did not show any extravasation of EB either in the tumor or in the normal brain. Histamine at a dose of 1 µg/kg/min (n = 4) produced a slight homogenous EB extravasation in the tumor, with no significant extravasation in the normal brain (Fig. 1). At 10 µg/kg/min (n = 4), a

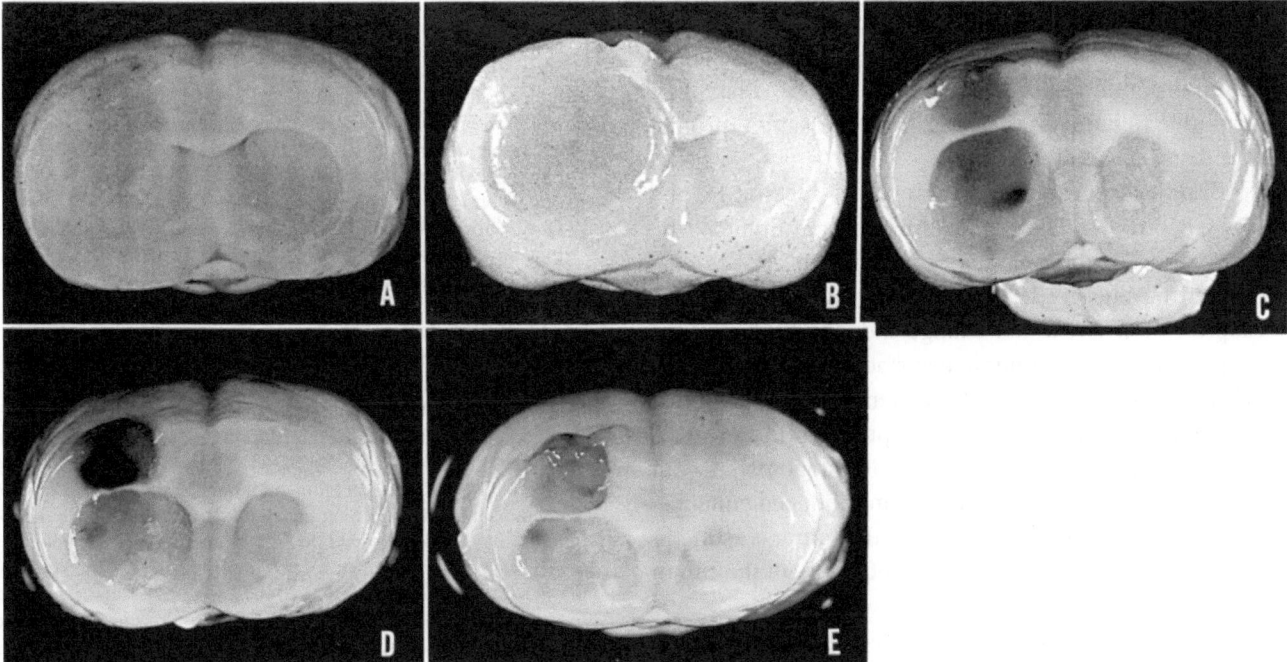

Fig. 1. Coronal brain sections and EB extravasation. (A) Intracarotid saline infusion: no definite EB extravasation either in normal or tumor tissue. (B) Histamine (1 μg/kg/min): slight extravasation in the tumor. (C) Histamine (10 g/kg/min): moderate extravasation in the tumor. (D) Pretreatment with H₁ receptor antagonists before histamine infusion (10 μg/kg/min): slight extravasation of EB in the tumor. (E) Pretreatment with H₂ receptor antagonists before histamine infusion (10 μg/kg/min): slight extravasation of EB in the tumor

Table 1. *Grading of Evans Blue Dye Extravasation in the Tumor*

	None	Slight	Moderate	n
Control	4			4
Histamine (μg/kg/min)				
1	1	3		4
10			4	4
10 + H₁ blocker		3	3	
10 + H₂ blocker		3	3	

moderate homogenous extravasation of EB in the tumor was observed, with no significant extravasation in the normal brain. This histamine-induced EB extravasation in the tumor was partly reduced by the histamine-receptor antagonists pyrilamine and cimetidine (Table 1). A slight heterogenous extravasation of EB in the tumor was observed when the rats were pretreated with histamine-receptor antagonist 10 minutes before histamine infusion at 10 μg/kg/min. Cimetidine exhibited a greater reduction of EB extravasation than pyrilamine (Fig. 1). The area of EB extravasation by histamine was histologically consistent with the tumor tissue.

Discussion

Various neurotransmitters have been known to resulate cerebral blood vessels in normal brain tissue. Although there have been some reports on the effects of neurotransmitters on blood flow in brain tumor models, no study of neurotransmitters on BTB permeability has been reported to date. Histamine is regarded as one type of neurotransmitter in the normal brain. There have been, however, contradictory reports concerning the effect of histamine on BBB permeability in normal brain capillaries[4-6]. These different effects of an intracarotid infusion of histamine on the BBB permeability might be explained by methodological reasons, such as different doses of histamine and/or species differences as discussed elsewhere[6]. In our experiments, EB did not extravasate into either the normal or the tumor tissue after intracarotid saline infusion. In contrast, the intracarotid administration of histamine resulted in selective EB extravasation in the tumor tissue. These results indicated that brain tumor vessels responded to histamine in a different fashion from normal brain capillaries. Although EB extravasation was not seen in the tumor after saline infusion in our study, the disruption of the BTB in the experimental brain tumor model has previously been reported in various literature[10-12].

High-molecular compounds such as EB-albumin conjugate could not extravasate into the tumor tissue under normal conditions in our model, whereas histamine was able to induce selective EB extravasation through the disrupted BTB.

Brain vessels are known to possess specific H_1 and H_2 receptors[7,8]. The permeability change induced by intra-arterial administration of histamine has been assumed be due predominantly to activation of H_2 receptors[4,5], whereas the change induced by histamine application to the abluminal surface was mediated via H_1 receptors[9]. In our study, histamine-induced selective enhancement of the BTB permeability was reduced by both H_1 and H_2 receptor antagonists in our study. This indicated that histamine-induced enhancement of the BTB permeability in the tumor was mediated in some part by histamine receptors in the tumor vessels.

Thus, it is possible to utilize histamine-induced selective enhancement of the BTB permeability for selective drug delivery to brain tumors without affecting the normal brain. It is considered useful to use histamine in the management of malignant brain tumors using monoclonal antibodies, anti-cancer agents, or even larger molecules for gene therapy.

References

1. El-Ackard TM, Brody MJ (1975) Fluorescence histochemical localization of non-mast cell histamine. In: Boissier JR, Hippius H, Pichot P (eds) Neuropsychopharmacology. Excerpta Medica 359: 551–559
2. Schwartz JC, Pollard H, Quach TT (1980) Histamine as a neurotransmitter in mammalian brain: neurochemical evidence. J Neurochem 35: 26–33
3. Steinbusch HWM, Verhofstad AAJ (1986) Immunocytochemical demonstration of noradrenaline, serotonin and histamine and some observations on the innervation of the intracerebral blood vessels In: Owman C, Hardebo JE (eds) Neural regulation of brain circulation. Elsevier, Amsterdam, pp 181–194
4. Dux E, Joo F (1982) Effects of histamine on brain capillaries. Fine structural and immunohistochemical studies after intracarotid infusion. Exp Brain Res 47: 252–258
5. Gross PM, Teasdale GM, Graham DI, Angerson WJ, Harper AM (1982) Intra-arterial histamine increases blood-brain transport in rats. Am J Phisiol 243: H307–H317
6. Wahl M, Unterberg A, Baethmann A, Schilling L (1988) Mediators of blood-brain barrier dysfunction and formation of vasogenic brain edema. J Cereb Blood Flow Metabol 8: 621–634
7. Edvinsson L, Owman C, Sjoberg N (1976) Autonomic nerves, mast cells, and amine receptors in human brain vessels. A histochemical and pharmacological study. Brain Res 115: 377–393
8. Peroutka SJ, Moskowitz MA, Reinhard JF Jr, Snyder SH (1980) Neurotransmitter receptor binding in bovine cerebral microvessels. Science 208: 610–612
9. Olesen SP (1987) Leakiness of rat brain microvessels to fluorescent probes following craniotomy. Acta Physiol Scand 130: 63–68
10. Inoue T, Fukui M, Nishio S, Kitamura K, Nagara H (1987) Hyperosmotic blood-brain-barrier disruption in brains of rats with an intracerebrally transplanted RG-C6 tumor. J Neurosurg 66: 256–263
11. Shapiro WR, Hiesiger EM, Cooney GA, et al (1990) Temporal effects of dexamethasone on blood-to-brain and blood-to-tumor transport of C14-alpha-aminoisobutyric acid in rat C6 glioma. J Neuro Oncol 8: 197–204
12. Yamada K, Ushio Y, Hayakawa T, Kato A, Yamada N, Mogami H (1982) Quantative autoradiographic measurements of blood-brain-barrier permeability in the rat glioma model. J Neurosurg 57: 394–398

Correspondence: Kiyonobu Ikezaki, M. D., Department of Neurosurgery, Faculty of Medicine, Kyushu University 60, Fukuoka 812, Japan.

Acta Neurochir (1994) [Suppl] 60: 403–405
© Springer-Verlag 1994

Selective Increase in Blood-Tumor Barrier Permeability by Calcium Antagonists in Transplanted Rat Brain Tumors

K. Matsukado, T. Nomura, K. Ikezaki, and **M. Fukui**

Department of Neurosurgery, Neurological Institute, Kyushu University, Faculty of Medicine, Fukuoka, Japan

Summary

To clarify the altered response of calcium antagonists on pathological vessels, we investigated the effect of intracarotid infusion of nifedipine on the blood-brain barrier (BBB) permeability using a rat glioma model. Animals were treated with 0, 0.1, 1, 5, and 10 µg/kg/min of intracarotid continuous infusion of nifedipine. 2% Evans blue (EB, 2 ml/kg) was injected intravenously immediately after nifedipine infusion. BBB and blood-tumor barrier (BTB) permeability were evaluated by direct visual and histological observation. During the entire experiment, systemic parameters such as arterial blood pressure and blood analysis were not changed significantly. There was a dose-dependent increase of EB permeability selectively in the tumor tissue without affecting the normal brain. These results indicate that tumor vessels may show an altered response to calcium antagonists. Intracarotid administration of calcium antagonists contribute to a selective enhancement of drug delivery to malignant brain tumors without affecting the normal brain.

Keywords: Calcium antagonist; brain neoplasm; blood-brain barrier; permeability.

Introduction

The normal BBB restricts the entry of various large molecules or ionic compounds into the brain tissue. The blood vessels of brain tumors are structurally altered and the capillary permeability in brain tumors differs from that of normal brain tissue[6]. However, malignant brain tumors are frequently refractory to chemotherapy. Limited effectiveness of chemotherapy may be attributed to insufficient drug delivery to the tumor. To enhance the effect of chemotherapy on malignant brain tumors, several attempts have been undertaken to increase drug delivery to the tumor[1,5].

A variety of calcium antagonists have been reported to relax vascular smooth muscle and to increase regional cerebral blood flow (rCBF)[2,3,7,8]. Intravenous administration of calcium antagonists increased permeability in the normal brain under induced systemic hypertension[4]. The present study was designed to examine the effects of the intracarotid administration of nifedipine, a dihydropyridinic calcium antagonist, on BTB permeability using a intracerebrally transplanted rat C6 glioma model.

Materials and Methods

As nifedipine is very sensitive to light of normal wavelengths, drug dispension and experiments were performed under sodium light, and powder and solutions were covered with silver foil.

Tumor Inoculation

Tumor cells (5×10^5 in 5-µl solution) were stereotactically implanted into the right cerebral hemisphere under intraperitoneal pentobarbital sodium anesthesia (Nembutal: 50 mg/kg). The stereotactic coordinates used were 4 mm lateral and 2 mm anterior to the bregma and 5 mm vertically deep to the dural surface.

Animal Preparation

Male Wistar rats weighting 180–210 g were used for experiments 9 to 15 days after tumor inoculation. Under intraperitoneal pentobarbital anesthesia, tracheostomy and catheter placement were performed as previously described in detail[1]. Each rat was mounted onto a stereotactic apparatus, immobilized with pancuronium (0.6 mg/kg, i.v.), and mechanically ventilated with room air. During the experiment, anesthesia was maintained with 0.2–0.5% halothane, and arterial blood pressure was monitored. End-tidal CO_2 and rectal temperature were also monitored continuously and adjusted regularly throughout the experiment. After arterial blood pressure stabilized, arterial blood samples were taken for the blood gas analysis.

Permeability Analysis

The animals were treated with 0.1, 1, 5, or 10 µg/kg/min of intracarotid continuous infusion of nifedipine (n = 5 in each group) for 15 min. 20% mannitol or 2.5% ethanol in normal saline (vehicle) were also infused intracarotidly to compare the BBB permeability with that of nifedipine. Immediately after nifedipine infusion, 2%

EB (2 ml/kg) was intravenously injected. The animals were decapitated 10 min after EB injection. The brain was removed and then sliced coronally at the maximum diameter of the tumor. EB extravasation was evaluated by direct visual observation for BBB and BTB permeability analysis. The removed brain was further evaluated histologically for EB permeability.

Results

Throughout the experiment, systemic paraments such as arterial blood pressure and blood analysis were not significantly changed.

The intensity of tumor staining with EB was graded as follows: Grade 0, no staining; Grade 1+, faint localized staining within the tumor; Grade 2+, moderate localized staining; Grade 3+, severe localized staining; and Grade 4+, extensive staining of the ipsilateral hemisphere[9]. The vehicle did not induce any visible extravasation of EB either in the tumor tissue or in the normal brain. Intracarotid administration of nifedipine, however, produced a dose-dependent increase of EB extravasation within the tumor tissue (Table 1). Nifedipine at 0.1 and 1 µg/kg/min showed slight but restricted EB extravasation within the tumor. A selective and moderate increase of EB extravasation was seen in the tumor tissue at 5 µg/kg/min and a significant increase was seen at 10 µg/kg/min. On the other hand, 20% mannitol displayed a strong and diffuse EB extravasation in the normal brain (Fig. 1). EB extravasation within the tumor was far less than that of the normal brain.

Table 1. *Nifedipine and Permeability Grading*

Nifedipine (µg/kg/min)	Grade				
	0	1+	2+	3+	4+
0	4	1			
0.1	2	3			
1		2	3		
5		1	3	1	
10				3	2
20% mannitol				1	4

Data indicate the number of rats in each grading.

Discussion

The present study has demonstrated that the intracarotid administration of nifedipine selectively increased BTB permeability in transplanted rat glioma. Recently we also observed a selective increase of rCBF by nifedipine in a rat brain tumor model (unpublished data). These results indicate that the intracarotid administration of calcium antagonists may contribute to a selective delivery of both hydrophilic and lipophilic moieties to malignant brain tumor without affecting the normal brain. Using an altered response of the tumor vessels to various vasoactive agents, it might be possible to enhance the anti-tumor strategy in the central nervous system with minimal risk to the normal brain. Because nifedipine increased BTB permeability

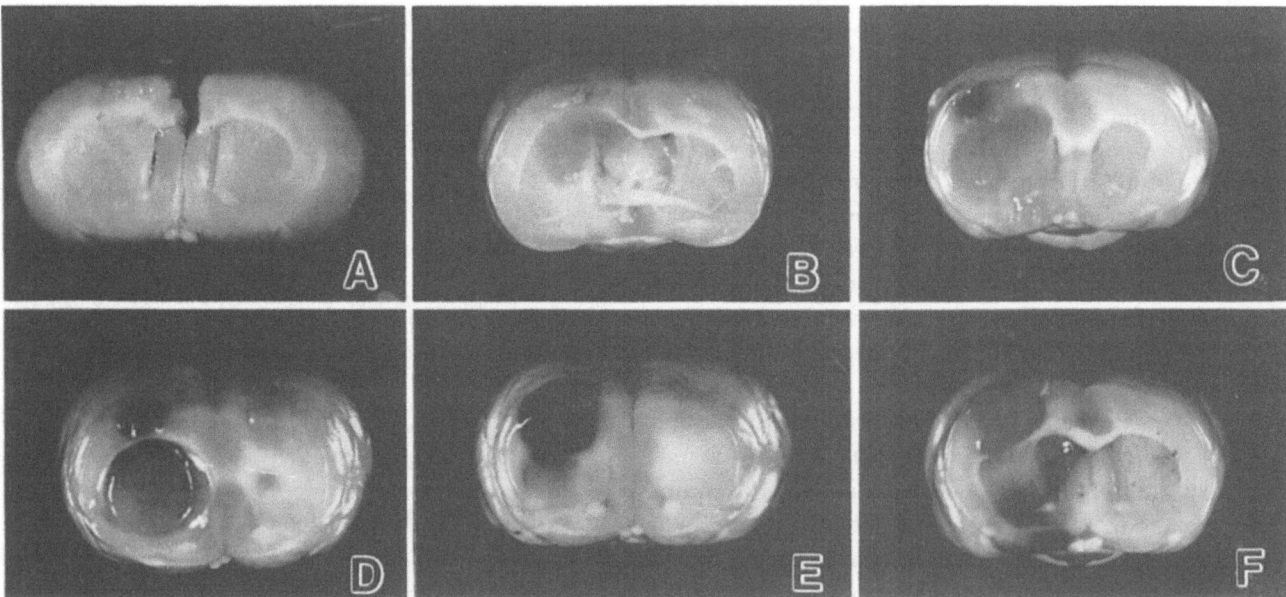

Fig. 1. Direct visual observation of BTB permeability. (A) Vehicle, (B) Nifedipine at 0.1, (C) 1, (D) 5, (E) 10 µg/kg/min, and (F) 20% mannitol

to a large molecule such as EB albumin conjugate, it may be possible to utilize calcium antagonists for selective and enhanced delivery of tumor-specific therapeutic moieties such as monoclonal antibody and antitumor genes in addition to chemotherapeutic agents.

Acknowledgment

This work was supported in part by grants-in-aid from the Japanese Ministry of Education, Science and Culture.

References

1. Baba T, Fukui M, Sakata S, Tashima T, Takeshita T, Nakamura T, Inoue T (1989) Selective enhancement of intratumoural blood flow in malignant gliomas: experimental study in rats by intracarotid administration of adenosine or adenosine triphosphate. Acta Neurochir (Wien) 101: 66–74

2. Brandt L, Andersson KE, Bengtsson B, Edvinsson L, Ljunggren B, MacKenzie ET (1979) Effects of nifedipine on pial arteriolar calibre: an in vivo study. Surg Neurol 12: 349–352

3. Grabowski M, Johansson BB (1985) Nifedipine and nimodipine: effect on blood pressure and regional cerebral blood flow in conscious normotensive and hypertensive rats. J Cardiovasc Phamacol 7: 1127–1133

4. Höllerhage HG, Gaab MR, Zumkeller M, Walter GF (1988) The influence of nimodipine on cerebral blood flow autoregulation and blood-brain barrier. Neurosurgery 69: 919–923

5. Inoue T, Fukui M, Nishio S, Kitamura K, Nagara H (1987) Hyperosmotic blood-brain barrier disruption in brains of rats with an intracerebrally transplanted RG-C6 tumor. J Neurosurg 66: 256–263

6. Long DM (1970) Capillary ultrastructure and the blood-brain barrier in human malignant brain tumors. J Neurosurg 32: 127–144

7. Mohamed AA, McCulloch J, Mendelow AD, Teasdale GM, Harper AM (1984) Effect of the calcium antagonist nimodipine on local cerebral blood flow: relationship to arterial blood pressure. J Cereb Blood Flow Metab 4: 206–211

8. Steen PA, Newberg LA, Mide JH, Michenfelder JD (1983) Nimodipine improves cerebral blood flow and neurologic recovery after complete cerebral ischemia in the dog. J Cereb Blood Flow Metab 3: 38–43

9. Tomiwa T, Hazama F, Mikawa H (1982) Reversible osmotic opening of the blood-brain barrier. Prevention of tissue damage with filtration of the perfusate. Acta Pathol Jpn 32: 427–435

Correspondence: Koichiro Matsukado, M. D., Department of Neurosurgery, Neurological Institute, Faculty of Medicine Kyushu University 60, Fukuoka 812, Japan.

Acta Neurochir (1994) [Suppl] 60: 406–409

Induced Hyperthermia in Brain Tissue in vivo

A. J. A. Terzis, G. Nowak, E. Mueller, O. Rentzsch, and **H. Arnold**

Department of Neurosurgery, Medical University of Lübeck, Federal Republic of Germany

Summary

Concerning hypothermia treatment, knowledge of time-temperature and of temperature distributions within tumor volumes is essential in order to obtain the maximal therapeutic effect. New techniques are being developed to overcome these difficulties. Two different heat sources, a contact Nd:YAG laser system and an automatically controlled high-frequency current system were investigated on 15 rabbits. Changes of the intracerebral temperature were registered at 4 different distances from the energy source. The intracerebral temperature was increased to 42.5°C at a distance of 5 mm to the heat source and maintained at this level for a period of 60 min. The contact Nd:YAG laser system reached 42.5°C at 3 W of output power. Using higher laser output power, brain tissue herniation (brain edema) through the burrhole was observed. The automatically controlled high-frequency current system reached 42.5°C at 18.75 W of output current. A very small herniation of brain tissue could be observed using higher output current. Both heat sources presented an exponential decrease of the temperature profile depending on the distance. The tissue heat clearance was compensated for by intermittent laser or high-frequency current application. Both systems proved efficient for inducing hyperthermia as needed for antitumoral therapy.

Keywords: Hyperthermia; Nd:YAG laser; high-frequency current; brain tissue.

Introduction

Hyperthermia is a potentially useful means of cancer treatment, either by itself or in combination with radiation and/or chemotherapy[3]. The treatment of cancer with adjunctive hyperthermia is based on a strong biological rationale. It is known that a marked increase in sensitization to radiation is achieved at temperatures of about 40–41°C. Above this threshold the sensitivity approximately doubles for each degree of increase. This may occur due to two facts. First, tumors, particularly large masses, are often poorly vascularized and their vessels lack normal vasoregulatory mechanisms that result in increased blood flow under conditions of heat stress. Therefore, they may be unable to cool themselves by blood perfusion as efficiently as the surrounding normal tissue, resulting in the attainment of comparably higher temperatures in the interior of the tumor. Second, the sensitization is probably greater for malignant cells than for normal tissue. There is considerable evidence that the mechanism of hyperthermia sensitization may be due to denaturation or inactivation of enzymes and other proteins which are involved in the repair processes of radiation-induced DNA strand breaks[1]. Hyperthermic exposure above the critical temperature produces prolonged inactivation of these enzymes.

However, since both normal tissue destruction and tumor destruction are very sharp functions of thermal exposure, a difference of only a fraction of one degree centigrade could mean the difference between tumor recurrence and tumor destruction, or between acceptable and unacceptable damage to normal tissue.

Therefore, knowledge of the temperature distribution within brain tissue and tumor volume is crucial. It is critically important that as little normal tissue as possible be destroyed, allowing regeneration of the region with continued function. On the other hand, all regions of the tumor should receive a tumoricidal hyperthermia dose with no areas of inadequate treatment. This is often difficult and may be equipment specific. In addition, a thermal map of heat energy deposition in tumor and normal tissue provides precise and accurate control of therapy.

In this study, a contact Nd:YAG laser system and an automatically controlled high-frequency current are used to induce hyperthermia in rabbit brain in vivo. The purpose is to analyze both systems as local heating sources and, in addition, to obtain a thermal map of heat energy deposition in perfused normal brain tissue.

Materials and Methods

Heat Sources

The laser system used was a Nd:YAG laser (Medilas 2, MBB Medizintechnik) with a wavelength of 1.064 µm and an output power of up to 100 W. It was armed with a helium-neon pilot laser having a constant output power of 2 mW and a wavelength of 0.633 µm. An interstitial sapphire contact tip (MBB Medizintechnik) connected to a gastro-transmission system was used. The sapphire was a monocrystalline form of aluminum oxide (AL_2O_3) and did not change the wavenlength of the laser beam. The probe delivered the laser beam quite effectively and quantitatively into the tissue.

The automatically controlled high-frequency current (TUR-Unit type 27800, Firma Karl Storz) was composed of a generator with an adjustable voltage and current output at a frequency of 450 kHz, and a control unit to regulate the constant current during the complete heating time. This mechanism protected the tissue against burning. The electrode was based on two parallel isolated wires with a diameter of 0.7 mm each and a distance of 5 mm between them.

Thermal Map of Heat Energy

Temperature measurements were done using thermocouples which we fabricated of copper and nickel wires having a diameter of 50 µm. The wires were integrated in a glass capillary for mechanical stability and thermal isolation. Specially designed aluminum plates were constructed in order to have constant distances between the thermocouples during all measurements. The distances of 2.5 mm, 5.0 mm, 7.5 mm and 10 mm from the heat probe were used for the measurements. Temperatures were monitored by a Multitech computer system model Accel 900 programmable datalogger at preset time intervals of 5 s. For both systems the dependence of the temperature profiles on the distance to the heating source and on the heating time was recorded.

In order to determine the optimal parameters for energy application, studies were performed using 2 different energy power levels (3 W and 12 W) for the contact Nd:YAG laser system and 3 different energy levels (16.85 W, 18.75 W and 22.5 W) for the automatically controlled high-frequency current.

The tissue temperature was increased to 42.5°C and maintained for a period of 60 min by means of repeated exposure to the heat source. In addition, the temperature decrease in relation to time was measured.

Animal Model

15 male rabbits weighing 3000–3500 g were anesthetized with 0.8 mg/kg ketamine hydrochloride and 0.2 mg/kg xylazine hydrochloride i.m. and then placed in a stereotaxic device. Through a midline sagittal incision, the scalp was sub-periostally elevated and dissected in order to expose the right frontal and parietal regions of the skull. Fronto-parietal craniectomies (2 × 1 cm) were performed. This procedure was accomplished without bleeding from pial-cortical vessels. The rabbits then underwent intracerebral implantation of 4 temperature probes and either a laser or a high-frequency current probe.

Results

Energy Application

Measurements were performed at different energy levels for the two systems in order to have a temperature rise of 5.5°C (from 37°C to 42.5°C) at a distance of

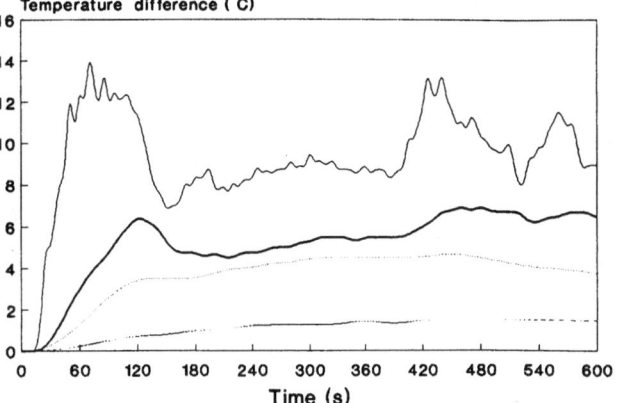

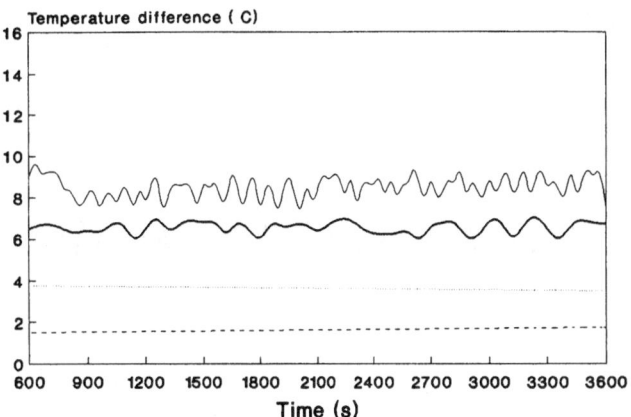

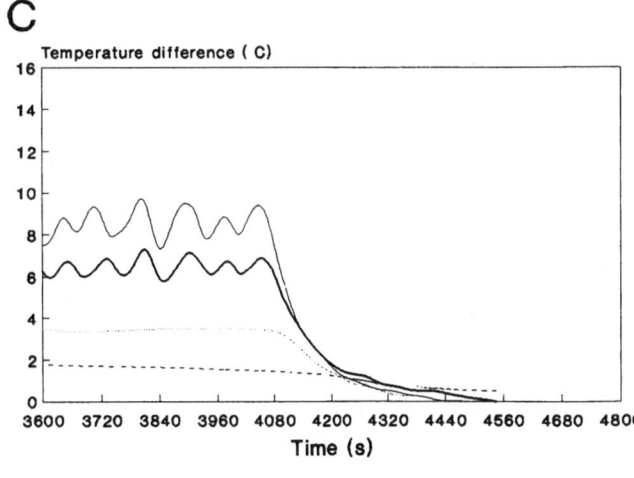

Fig. 1. Themal map for a 60-min exposure time using the contact Nd:YAG laser system: (A) 0–600 s, (B) 600–3600 s (C) 3600–4800 s

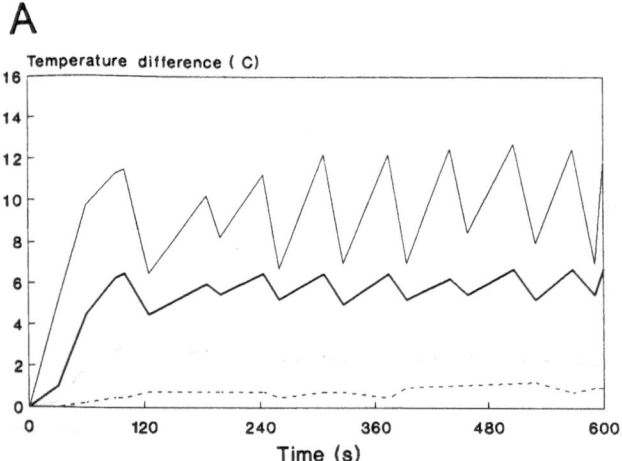

A

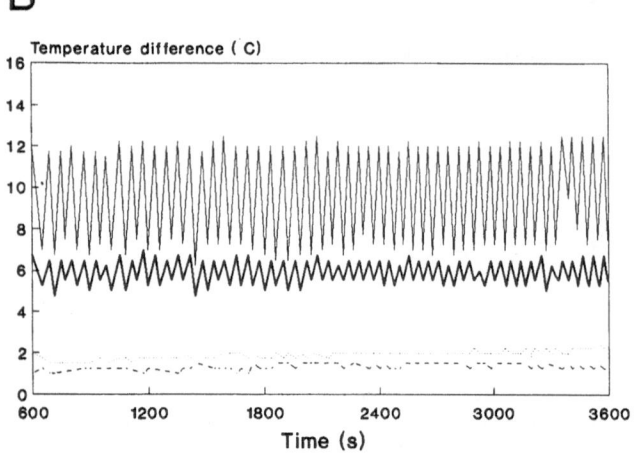

B

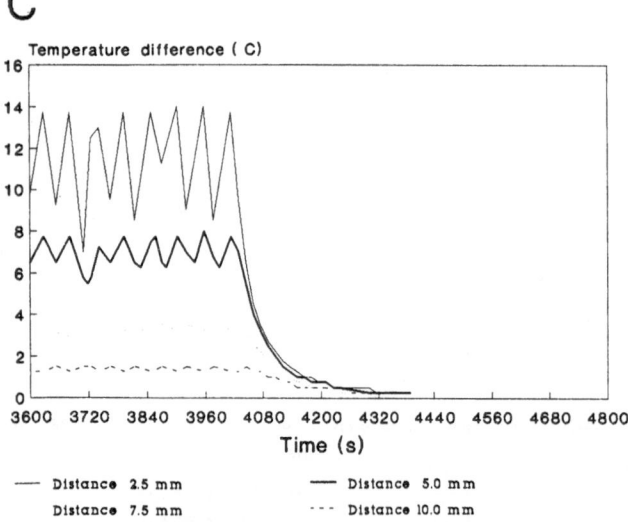

C

Distance 2.5 mm   Distance 5.0 mm
Distance 7.5 mm   Distance 10.0 mm

Fig. 2. Thermal map for a 60-min exposure time using the automatically high-frequency current system: (A) 0–600 s, (B) 600–3600 s, (C) 3600–4800 s

5 mm from the heat source probe. On activation of the heating source systems, the heat energy needed to be delivered rapidly in a continuous mode in order to increase the tissue temperature to 42.5°C. Once the tissue temperature reached 42.5°C, the heating source systems delivered the heat energy more slowly using a pulse mode.

The contact Nd:YAG laser system effected a temperature increase of 5.5°C at an output of 3 W. By using higher output power (12 W), the temperature was maintained constant for 20 min by 0.7 s of energy exposure. After this period, a critical brain tissue herniation (brain edema) through the burrhole was observed. In addition, the development of a coagulum was seen.

The automatically controlled high-frequency current caused a temperature increase of 5.5°C at an output of 18.75 W. Due to technical limitations, it was not possible to maintain a temperature increase of 5.5°C for 60 min using lower output power (16.85 W). By using higher output power (22.5 W), a minimal brain tissue herniation through the burrhole was observed. The development of a coagulum was also seen. Thus the energy levels found to be suitable for further experiments were a 3-W output for the contact Nd:YAG laser system and a 18.75-W output for the automatically controlled high-frequency current.

Thermal Map of Heat Energy

The tissue temperature was increased to 42.5°C and maintained for a period of 60 min by means of repeated exposure to the heat source. Measurements were done depending on the distance of the thermocouples from the heat source (2.5 mm, 5.0 mm, 7.5 mm and 10 mm). In addition, the temperature decrease in relation to time was also measured (Figs. 1 and 2).

Using the contact Nd.YAG laser system, the temperature reached a ΔT of 5.5°C at a distance of 5 mm from the heat source after 49 s, 95 s, 235 s, 556 s and 1112 s of exposure to a continuous mode at 3 W of output power for 5 rabbits, respectively. The long heating time observed in 1 rabbit (1112 s) was due to a severe venous bleeding resulting from the introduction of the thermocouples into the brain. The temperature was maintained constant for 60 min by 10 s of energy exposure every 20 s or 15 s (Figs. 1A and B). The thermocouples placed at 2.5 mm, 7.5 mm and 10 mm from the heat source measured increases of 8.7°C, 3.0°C and 1.5°C, respectively. After the laser had been switched off, an accentuated decrease in temperature was registered from all thermocouples (Fig. 1C).

Using the automatically controlled high-frequency current system, the temperature reached a ΔT of 5.5°C at a distance of 5 mm from the heat source after 60 s, 84 s, 75 s, 101 s and 289 s of exposure to a continuous mode at 18.75 W of output power for 5 rabbits, respectively. The temperature was maintained constant for 60 min by energy exposure for different lengths of time (Figs. 2A and B). The pulse mode was reduced from 50 s to 35 s during the experiment for 1 rabbit, from 50 s to 30 s for 1 rabbit, from 50 s to 25 s for 1 rabbit, and from 50 s to 20 s for 2 rabbits. The thermocouples placed at 2.5 mm, 7.5 mm and 10 mm from the heat source measured increases of 9.4°C, 2.9°C and 1.1°C, respectively. After the current was switched off, an accentuated decrease in temperature was registered from all thermocouples (Fig. 2C).

Discussion

Hyperthermia has been used in an experimental setting for the treatment of refractory, recurrent primary, and metastatic human brain tumors[3]. Saleman and Samaras[4] noted that temperatures above 44–45°C are not tolerated by the brain, regardless of the species studied. This correlates to a 6.5°C rise over the baseline for the rabbit. The objective of this work was to quantify responses of the normal brain tissue to local hyperthermia treatment at a temperature of 42.5°C using interstitial laser energy and high-frequency current fields. Interstitial laser irradiation offers the theoretical advantage of transmitting laser energy directly into tumor tissue, thereby minimizing undesired effects on surrounding normal structures. This approach is based on the fact that thermal effects are confined to the region where the optical energy is absorbed.

The high-frequency current produces a small amount of brain damage only within the electrical field and around its probe. In addition, the control unit measures the rising electrical resistance of the tissue, and stops immediately when the tissue temperature rises and vaporization of cytoplasm occurs, thus protecting the tissue against burning.

Both heat sources feature small-size probes which constantly distribute heat into the tissue. The tissue temperature rapidly increased and decreased, after the heat sources were switched on and off, respectively.

The thermal properties of viable tissue are determined by three principles: the ability to transport heat by thermal conduction, the ability to store heat, and the ability to transfer heat by the vasculature. These tissue parameters determine the steady-state temperature distribution in irradiated tissue[5]. We confirmed these observations in our studies in vivo, where differences in heating time and heat clearance were very variable due to individual tissue conditions. The substantial heating time-differences observed among the rabbits for both heating sources was probably due to different degree of venous bleeding following insertion of the thermocouples into the brain, and to the differences in the ability of the normal brain to transfer heat through the vessels. As the experiment proceeded, the heat energy needed to maintain constant hyperthermia decreased. This was probably caused by reduced blood flow through the tissue. No differences in heating time or pulse mode were observed in earlier studies performed in vitro[2].

The formation of brain edema was due to thermal injury. Edema production was dependent of the rate of energy delivery per volume of tissue and time. Brain edema was partly avoided by using a low output power and a long exposure time.

The development of a coagulum was attributed to the seepage of blood into the lesion that resulted from the insertion of the probes, and due to local tissue coagulation secondary to thermal injury.

Induced hyperthermia may provide one form of cancer treatment, either when used alone or as an adjuvant to chemotherapy, immunotherapy, radiotherapy, or photodynamic therapy. Further studies will be undertaken to investigate the influence of induced hyperthermia on neoplastic brain tissue.

References

1. Lett JT, Clark EP (1977) Effects of hyperthermia on the rejoining of radiation-inducd DNA strand breaks. In: Streffer C (ed) Proc 2nd Int Symp Cancer Therapy by Hyperthermia and Radiation. Urban and Schwarzenberg, Baltimore
2. Nowak G, Rentzsch O, Terzis AJA, Arnold H (1990) Induced hyperthermia in brain tissue: comparison contact Nd.YAG laser system and automatically controlled high-frequency current. Acta Neurochir (Wien) 102: 76–81
3. Page RC, Ricca GF, Dohan FC (1990) Hyperthermia for the treatment of brain tumors. Adv Exp Med Biol 267: 145–153
4. Saleman M, Samaras GM (1981) Hyperthermia for brain tumors: biophysical rationale. Neurosurgery 9: 327–335
5. Svaasand LO, Boerslid T, Oeveraasen M (19875) Thermal and optical properties of living tissue: application to laser-induced hyperthemia. Lasers Surg Med 5: 589–602

Correspondence: A. Jorge A. Terzis, M.D., Department of Neurosurgery, Medical University of Lübeck, Ratzeburger Allee 160, D-23562 Lübeck, Federal Republic of Germany.

Acta Neurochir (1994) [Suppl] 60: 410–412
© Springer-Verlag 1994

Partial Deuteration and Blood-Brain Barrier (BBB) Permeability

Y. Nakagawa[1], **H. Hatanaka**[2], **M. Moritani**[2], **K. Kitamura**[3], **K. Matsumoto**[3], and **M. Kobayashi**[4]

[1] Department of Neurosurgery, National Kagawa Children's Hospital, Kagawa, [2] Department of Neurosurgery, Teikyo University, Teikyo, [3] Department of Neurosurgery, Tokushima University, Tokushima, and [4] Research Reactor Institute, Kyoto University, Kyoto, Japan

Summary

Boron neutron capture therapy (BNCT) is one method of radiosurgery used for malignant brain tumor. The theory of this method is based on the nuclear reaction that occurs when boron-10 is radiated with and absorbs neutrons. When treating deep-seated brain tumor, partial deuteration of body water is used to improve the penetration of neutrons into the tissue. In order to investigate the change in the BBB function under partial deuteration, we measured brain water content and the permeability of protein. Wister rats were given 99% heavy water (approx. 10% of body weight/day) as drinking water. On the day of the experiment, all animals received 125-I labeled albumin as a tracer of protein. Light water in the control group or heavy water (D_2O) in the experimental group was given by drip infusion for 60 min. Heavy water was measured by infrared spectroscopic analysis. Water content was measured by the freeze-dry method. The radioactivity of 125-I was determined with a gamma scintilation counter. Mean values of D_2O concentration in the tissue were 17.9–36.5%. Water content increased in the cortex in all animals. The ratio of 125-I in the brain tissue to blood showed significant differences between the control group and animals deuterated to more than 30% of body water. Brain tissue deuterated over 10% showed a mild leakage of water, indicating the early stage of brain edema. Deuteration higher than 30% caused a leakage of protein, which might indicate the leakage of boron compound from the vessels into the normal brain tissue.

Keywords: Brain tumor; brain edema; boron neutron capture therapy; heavy water.

Introduction

Boron neutron capture therapy (BNCT) using thermal neutrons is an effective and valuable treatment for malignant brain tumors (Hatanaka 1991). The theory of BNCT is based on the nuclear reaction that occurs when boron-10, a non-radioactive isotope, is irradiated with thermal neutrons. However, it is widely believed that thermal neutrons cannot ideally treat patients with deep-seated brain tumors because of the rapid attenuation of neutron flux in the brain tissue (Barth 1990). On the other hand, it is also well known that the partial

deuteration of body water can increase the penetration of neutrons into the tissue (Nakagawa 1992). However, if deuteration of brain tissue alters the BBB permeability, the boron compound will leak out from the vessels into the normal brain tissue, thus possibly minimizing the advantages of BNCT. In our previous studies, we reported that there are two different conditions for the opening of the blood-brain barrier. Typical findings were observed in an experimental ischemic model and in kaolin-induced hydrocephalus in rats (Nakagawa 1985, 1987) In the early stage of brain edema, very small molecules pass through the tight junctions between the endothelial cells of the small vessels. When disruption of the blood-brain barrier progressed, the leakage of large molecules was demonstrated. If deuteration results in large molecular substances such as HRP or albumin leaking out from the vessels, we have to assume the leakage of the boron compound. In order to investigate the change in BBB function under partial deuteration, we measured brain water content and the permeability of the protein.

Materials and Methods

Adult wistar rats of either sex weighing 150–180 g were used in this study. Animals were divided into 6 groups: one control and 5 experimental groups. Each animals was given light water as drinking water except for a few days before the experiment. The animals in group 2 were given heavy water by gastric tube for one day before the experiment, those in group 3 for two days, those in group 4, for three days, and those in group 5 for four days. The dosage of heavy water was 0.1 ml/g body weight/day. On the day of investigation, under general anesthesia by pentobarbital (5 mg/kg), a polyethylene catheter was inserted into the femoral vein and all animals were given 125-I labeled albumin as a tracer of protein. Light water in the control group or heavy water (0.1 ml/g body weight) in the experimental groups was given by drip infusion through the femoral vein for 60 minutes. One hour after drip infusion of light or heavy water,

411

the animals were sacrificed by decapitation and samples were taken from three different areas of the brain (cortex, brain stem and cerebellum) and from the kidney. Heavy water content in the blood, cortex and kidney was measured by infrared spectroscopic analysis. Water content of the brain samples was measured by the freeze-dry method. Radioactivity of 125-I in the blood and brain samples was determined using a gamma scintillation counter. Passage of the tracers from the blood into the brain tissue was calculated as the following distribution ratio index:

$$Index = \frac{Radioactivity/brain}{Radioactivity/blood} \times 1000.$$

Sodium fluorescein (0.2 ml of a 10% solution in saline) and Evans blue (0.2 ml of a 10% solution) were given 5–10 minutes before sacrifice to some animals and the site and degree of extravasation of the dye was grossly evaluated.

Results

Tracer Study

Faint fluorescein staining was observed in the animals of groups 3 and 4. In the animals of group 5, more intense fluorescein staining was observed. However, there was usually no extravasation of Evans blue.

Concentration of Heavy Water

The changes in heavy water (D$_2$O) content in the blood, brain cortex and kidney are shown in Fig. 1. The concentration of D$_2$O gradually increased in each group. However, there were no remarkable differences

among specimens in the same group. Mean values of D$_2$O concentration in blood were 9.1% in group 1, 19.3% in group 2, 24.4% in group 3, 32.6% in group 4 and 37.2% in group 5.

Brain Tissue Water Content

Water content significantly increased in the cortex in groups 2, 3, 4 and 5 (Fig. 3). However, there was no

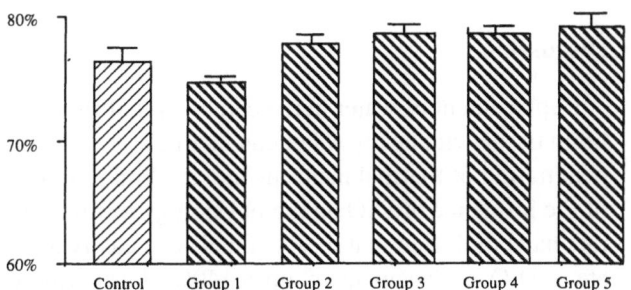
Fig. 3. Water content in cortex

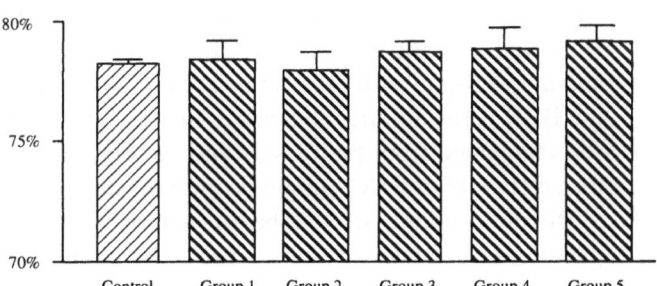
Fig. 4. Water content in cerebellum

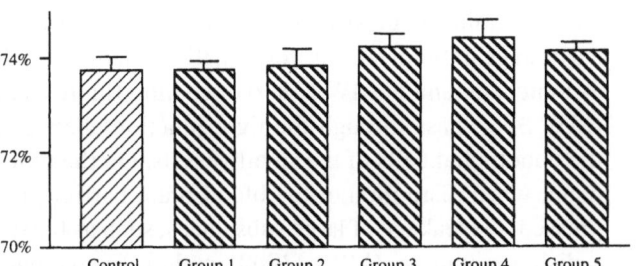
Fig. 5. Water content in brain stem

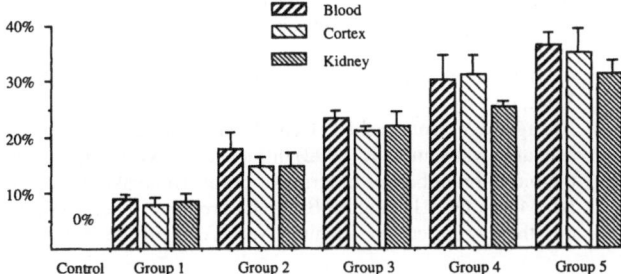
Fig. 1. D$_2$O concentration

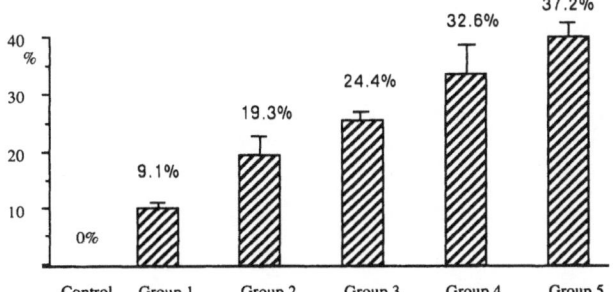
Fig. 2. D$_2$O concentration in blood

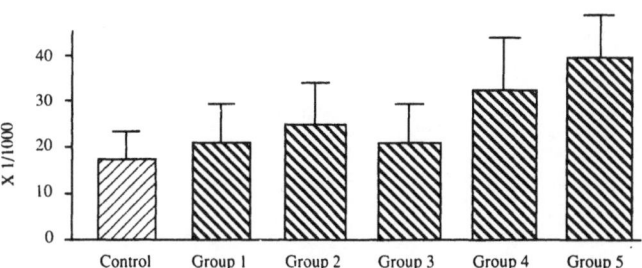
Fig. 6. Ratio of 125-I in brain tissue and blood

remarkable changes in the cerebelum and brain stem (Figs. 4 and 5).

BBB Permeability to 125-I-BSA

Changes in the BBB permeability to 125-I-BSA in the five control rats and the experimental groups are shown Fig. 6. The increase in the uptake of 125-I-BSA in groups 1, 2 and 3 was not significant. However, the ratio of 125-I in groups 4 and 5 significantly increased.

Discussion

Replacement of water in the cerebrospinal fluid and brain tissue with heavy water can facilitate an efficient penetration of thermal neutrons into the brain. Studies of the influence of D_2O on the physiology of mice and rats have indicated that these animals can easily tolerate a D_2O replacement of up to 40%. Furthermore, recent study concerning conducted toxicity of D_2O has demonstrated that there is no difference in acute toxicity in terms of LD50 in rats between H_2O and D_2O. However, the influence of deuteration against BBB is still not well known. If deuteration of brain tissue alters the BBB permeabilities and causes the leakage of boron compound from the vessels into the brain tissue, the selectivity of BNCT might be minimized.

Disruption of the BBB has been discussed in various pathological states and a discrepancy between BBB permeability to small and large substances has been studied in many experimental models. Typical findings were observed in an experimental ischemic model and in kaolin-induced hydrocephalus in rats. In the early stage of brain edema, very small substances, such as ionic lanthanum (MW 138.9) or sodium fluorescein (MW 376), pass through the tight junctions between the endothelial cells of the small vessels. On the other hand, when disruption of the blood-brain barrier progressed, the leakage of large substances, such as horseradish peroxidase (MW 43000) and/or Evans blue (combined with serum albumin), was demonstrated by

electron microscope. In the present study, we found similar changes of BBB permeabilities for partially deuterated brain in rats. Deuterated brain tissue of over 20% showed mild leakage of small molecules and water, indicating the early stage of brain edema. Over 30% deuteration caused leakage of protein, which might indicate the leakage of boron compound from the vessels into the extracellular space around the normal brain tissue. These findings indicated that the safety level of deuteration is below 30%.

References

1. Barth RF, Soloway AH, Fairchild RG (1990) Boron neutron capture therapy for cancer. Sci Am Oct: 69–73
2. Hatanaka H (1991) Boron neutron capture therapy for tumors. In: Karim AB, Laws ER (eds) Glioma. Springer, Berlin Heidelberg New York Tokyo, pp 233–249
3. Kadosawa T, Takeuchi A, Nakaichi T, Matsumoto T, Nozaki T, Aizawa O, *et al* (1987) Basic research in neutron capture therapy. 1. Preliminary study on the use of D_2O for boron neutron capture therapy. Annu Rep Joint Stud Musashi Inst Technol React 12: 9–1
4. Kobayashi M (1990) Decrease in deuterium content of heavy water in contact with air. Annual reports of the Research Reactor Institute, vol 23. Kyoto University, Kyoto, pp 188–193
5. Nakagawa Y, Cervós-Navarro J, Artigas J (1985) Trace study on a paracellular route in experimental hydrocephalus. Acta Neuropathol (Berl) 65: 247–254
6. Nakagawa Y, Fujimoto N, Matsumoto K, Cervós-Navarro J (1987) Blood-brain barrier permeability in acute cerebral ischemia after MCA occlusion and reperfusion in the rat. In: Cervós-Navarro J, Ferszt (eds) Stroke and microcirculation, pp 271–276
7. Nakagawa Y, Fujimoto N, Matsumoto K, Cervós-Navarro J (1990) Morphological changes in acute cerebral ischemia after occlusion and reperfusion in the rat. In: Long Kadosawa DM (eds) Advances in neurology. Brain edema, vol 52. Raven, New York, pp 21–27
8. Nakagawa Y, Kobayashi T, Ueno Y, Hatanaka H, Moritani M, Mukai K, Matsumoto K (1992) Influence of BNCT radiation on the blood-brain barrier in terms of boron-10 uptake. In: Allen BJ, Moore DE, Harrington BV (eds) Progress in neutron capture therapy for cancer. Plenum, New York, pp 525–529

Correspondence: Y. Nakagawa, M.D., Department of Neurosurgery, National Kagawa Children's Hospital, 2603 Zentsuji-Cho, Zentsuji, Kagawa 765, Japan.

Acta Neurochir (1994) [Suppl] 60: 413–415
© Springer-Verlag 1994

Quantitation of Peritumoural Oedema and the Effect of Steroids Using NMR-Relaxation Time Imaging and Blood-Brain Barrier Analysis

C. Andersen[1,3], **J. Astrup**[1], and **C. Gyldensted**[2]

Departments of [1] Neurosurgery and [2] Neuroradiology, Århus University Hospital, and [3] MR-Centre, Skejby Hospital, Århus, Denmark

Summary

A new method of in vivo quantitation of peritumoural brain oedema using NMR relaxation time imaging (T_1-maps) and Gd-DTPA-enhanced Blood-Tumour-Barrier (BTB) analysis is presented. The method is based on image pixel histogram analysis and a fast imaging method combined with arterial [Gd-DTPA]-measurement. The method was applied in 26 brain tumour patients, studied prior to – and 1, 3 and 7 days after initiation of dexamathasone. Oedema resorption rate following dexamethasone treatment was almost equal in glioblastomas and metastases, mean T_1 being reduced by ca. 12% in 7 days. In meningiomas no significant changes in peritumoural oedema could be detected. Simultaneous BTB-analysis was obtained in 4 patients showing correlation between oedema resorption in vivo and reduction of BTB transport rate constant K_i after 7 days of steroid treatment.

This method is a powerful tool in quantitation and monitoring of brain oedema in vivo, as both steroid influence on oedema resorption and on BTB-defect can be monitored simultaneously.

Keywords: Peritumoural brain oedema; MRI; brain tumours; glucocorticoids.

Introduction

Peritumoural brain oedema is generated by extravasation of plasma water and macromolecules through a defective blood-brain-barrier (BBB) within the tumour. Glucocorticoids are known to reduce the tissue water content in the oedema area, while the exact mechanism is unknown. In order to monitor changes in peritumoural brain oedema (PTBOe) during steroid treatment, a model using NMR relaxation time imaging was developed. The relationship between relaxation times and tissue water content is well established[1,2], and therefore relaxation times seem highly suitable for in vivo, noninvasive studies of brain oedema. In this study calculated T_1-images were used to monitor the effect of steroids in different types of brain tumours as a function of time, and a new method to quantitate Blood-to-Tumour-Barrier (BTB) defects in tumours using Fast-Field-Echo MRI is introduced.

Patients and Methods

The study included 26 patients (8 metastases, 8 meningiomas, 10 glioblastomas). A Philips Gyroscan S15 HP 1.5 T MR-scanner was used. The patients were scanned before treatment and 1, 3 and 7 days after initiation af dexamethasone (DEX) (0.23–0.64 mg/kg b.w.). We used calculated T_1-images (T_1-maps) of the same tumour slice. Quantitation of oedema was done by a new image histogram evaluation technique, in which the histogram of the pixels in an oedema area is divided in 2 parts: The highest water content being the 'super-oedema' and a low-grade oedema area (Fig. 1a), i.e. the oedema area with the highest water content is monitored as a function of time.

Modeling of Blood-Tumour-Barrier Defects

In 3 metastasis and 1 glioblastoma patients simultaneous BTB-defect quantitation was done, giving the blood-to-tumour transfer rate constant K_i. We employ a single capillary, two compartment model[4]. In summary, the cerebral concentration $C_{br}(t)$ of Gadolinium-DTPA (Gd-DTPA) at time t after an intravenous injection is given by

$$C_{br}(t) = C_{br}(0) + K_i v \int_0^t e^{k_i(\tau - t)} C_a(\tau) d\tau \qquad (1)$$

($C_{br}(0)$: Cerebral Gd-DTPA-concentration at time t = 0, C_a: Arterial concentration of Gd-DTPA, $K_i = EF/v$, E is the extraction fraction of Gd-DTPA, F the plasma flow and v the apparant volume of distribution of Gd-DTPA in brain per unit mass). Detection of $C_{br}(t)$ is done by reading increasing MRI signal intensity in consecutive slices obtained by a fast imaging technique (duration: 10 min.), knowing the equation for the image sequence employed, (in our case Fast Field Echo[3]) as a function of time after a bolus injection of Gd-DTPA. We assume that the relaxation rates $R_{1,2} = 1/T_{1,2}$ change in time proportionally to cerebral [Gd-DTPA], i.e.

$$R_i(t) = R_i^0 + R(t), \quad i = 1,2 \qquad (2)$$

thus establishing the connection between equation 1 and a quantity measurable by MRI. Arterial [Gd-DTPA] is measured for every image in the sequence. Convolution between arterial [Gd-DTPA]

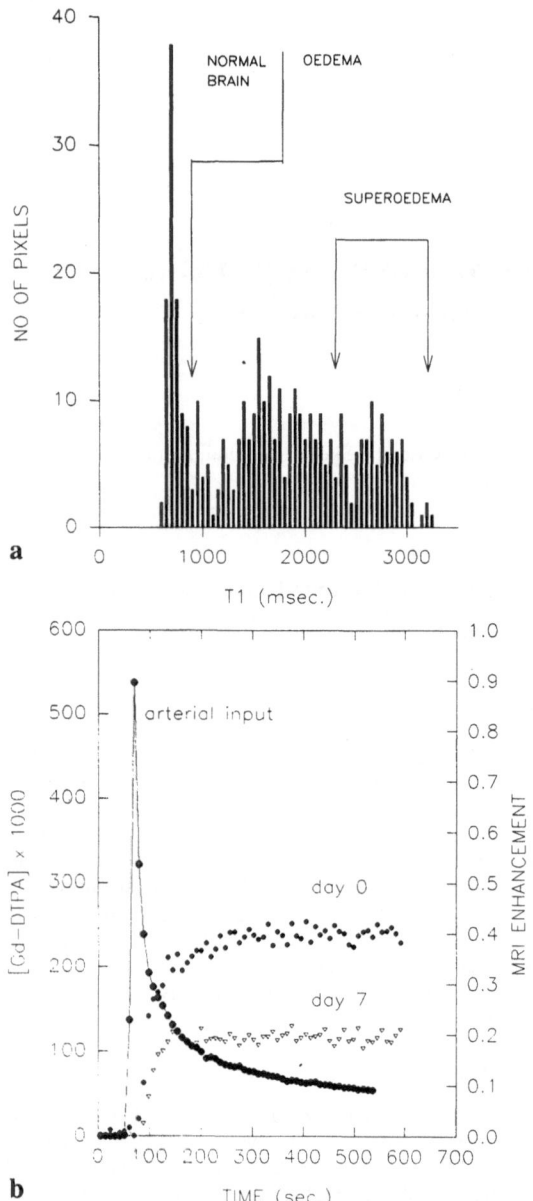

a

b

Fig. 1. (a) Image pixel histogram of oedematous brain: There are 3 populations of pixels: 1: Normal white matter, 2: Low grade oedema, 3: Super-oedema, with the highest water content. In a normal brain the histogram would be gaussian distributed with $\bar{T}_1 = 630$ msec. (b) Output from BTB-analysis in a metastasis patient, showing arterial input curve of Gd-DTPA, and 2 MRI detection curves, the upper before steroid treatment and the lower after 7 days of treatment

curve and the $C_{br}(t)$-curve is done using linear interpolation between the measured arterial input values[5], giving the transfer rate constant K_i.

Results

The lower threshold for oedema was defined as T_1 above 850 milliseconds (msec.) by studying 35 normal volunteers (mean T_1 ($\bar{T}_1$) + 3 SD. of normal white matter). Steroid treatment did not affect T_1 in the normal hemisphere of the tumour patients. The relative change in 'super-oedema' area in each scan is shown in Fig. 2 (a–e) for glioblastomas, metastases and meningiomas respectively. Glioblastomas and metastases were almost equal in the oedema resorption rate: The mean change in super-oedema area was 57% after 7 days of treatment in glioblastomas ($p < 0.005$), 67% in metastases ($p < 0.01$) and 1% in meningiomas (not significant). The change in $\bar{T}_1$ of the entire oedema area was 13% for glioblastomas ($p < 0.003$), 11% for metastases ($p < 0.02$) and 0.9% for meningiomas. The change in $\bar{T}_1$ and 'super-oedema' area in meningiomas was not significant; i.e. an effect of steroids in meningiomas could not be detected. BTB-defect analysis showed reductions in transfer rate constant K_i after steroid treatment (Table 1), that correlated with oedema resorption.

Discussion

The fact that tissue water content is the most important factor determining the relaxation time (RT) T_1 in NMR, makes RT imaging highly suitable in noninvasive, in vivo monitoring of brain oedema. The present study combines RT imaging with a pixel histogram evaluation technique and analysis of the transfer rate constant over the blood-to-tumour barrier for Gd-DTPA. It is shown that changes in water content following steroid treatment can be measured with a high sensitivity and that after 7 days of treatment $\bar{T}_1$ changes by ca. 12%. The area with the highest water content ('super-oedema') is the most sensitive to changes in water content. A reduction of more than 50% in 7 days was measured, except in meningiomes, where no significant change was detectable, suggesting a different pathophysiology in meningioma oedema. Glucocorticoid receptors are found in large numbers in meningi-

Table 1. *Correlation Between Oedema Reduction and Change in Blood-to-Tumour Transfer Rate Constant K_i After Steroid Treatment*

Patient	K_i day 0 $\times 10^{-3}$ s^{-1}	K_i day 7 $\times 10^{-3}$ s^{-1}	Diff. %	Oedema change %
Metast.	2.68	2.21	−17.5	−14.7
Metast.	4.75	3.77	−20.6	−15.3
Glioblast.	3.81	3.11	−18.4	−0.9
Metast.	4.21	3.71	−11.9	
Mean			−17.1	−12.7

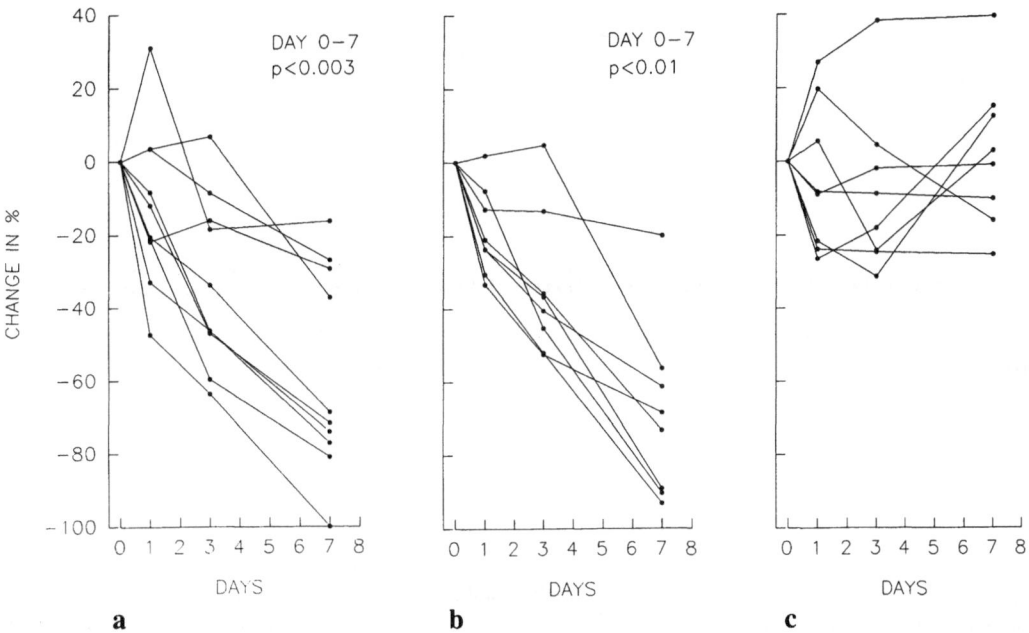

Fig. 2. (a) Relative changes in super-oedema area in 10 glioblastoma patients before and after 7 days of steroid treatment. (b) Relative change in 8 metastasis patients, oedema resorption rate was not different from glioblastomas. (c) No significant change after 7 days of treatment in meningiomas

omas[6], and our finding suggests that oedema resorption in meningiomas may be quantitatively independent of the receptor. Simultaneous monitoring of the blood-to-tumour transfer rate constant K_i showed that steroids reduced the BTB permeability by an amount comparable to the oedema reduction. In other words, steroids are not capable of closing the leaky BTB completely, but merely reduces permeability to such an extent that resorption is larger than production.

References

1. Kamman RL, Go KG, Berendsen HJC (1988) Nuclear magnetic resonance in experimental brain oedema: effects of water concentration, protein concentration, and temperature. Magn Reson Med 6: 265–274

2. MacDonald HL, Bell BA, Smith MA, *et al* (1986) Correlation of human NMR T_1 values measured in vivo and brain water content. Br J Radiol 59: 355–357

3. Van der Meulen P, Groen JP, Tinus AMC, Bruntink G (1988) Fast field echo imaging: an overview and contrast calculations. Magn Reson Imaging 6: 355–368

4. Paulson OB, Hertz MM (1983) Tracer kinetics and physiologic modeling. In: Levin S, Lambrecht RM, Rescigno A (eds) Lecture notes in biomathematics. Springer, Berlin Heidelberg New York

5. Press WH, Flannery BP, Teukolsky SA, *et al* (1988) Numerical recipes in C. The art of scientific computing. Cambridge University Press, Cambridge

6. Yu ZY, Wrange O, Boethius J, *et al* (1981) A study of glucocorticoid receptors in intracranial tumors. J Neurosurg 55: 757–760

Correspondence: C. Andersen, M. D., Department of Neurosurgery, Århus University Hospital, DK-8000 Århus, Denmark.

Acta Neurochir (1994) [Suppl] 60: 416–418

Resolution of Peritumoral Brain Edema Following Excision of Meningioma

T. Shirotani, K. Shima, and **H. Chigasaki**

Department of Neurosurgery, National Defense Medical College, Saitama, Japan

Summary

We studied the resolution of peritumoral brain edema after meningioma excision. In twenty-nine patients with meningioma, the total volume of tumory and the peritumoral edema were measured planimetrically by serial CT scans and MRI with or without contrast enhancement. Four different patterns of postoperative resolution of hypodense volume on CT were observed: Group A: a large hypodensity rapidly decreased and disappeared, which may be related to the clearance of the real peritumoral edema in meningioma. Group B: a small hypodensity gradually disappeared. Group C: the hypodensity remained unchanged, which may result from the damaged brain tissue. Group D: the hypodensity progressively decreased but persisted, which may represent both the peritumoral edema and damaged brain tissue. We have calculated the resolution rate of edema fluid using the clearance curve of Groups A and D. The average resolution rate of edema fluid during the passage through 1 cm³ of the peritumoral white matter was 0.0493 ml/day. We speculate that 50% of edematous white matter, which presented as hypodensity on a CT scan, may be resolved in 4 days after total removal, and that 90% may be resolved in 14 days.

Keywords: Brain edema; meningioma CT scan; MRI.

Introduction

Computed tomography (CT) and magnetic resonance imaging (MRI) are easily used for evaluating the areas surrounding meningiomas macroscopically. Several reports have demonstrated the resolution of human peritumoral edema[2,4,5]. Experimental studies have shown that resolution of brain edema, when tissue pressure gradients subside, was sustained throughout the experiments but reduced. To test this hypothesis, we analyzed the time course of peritumoral brain edema following total excision of brain tumor and after the dissipation of tissue pressure gradients.

Patients and Methods

Twenty-nine patients with a peritumoral hypodensity, examined repeatedly by CT scans or MRIs following total removal of menin-giomas without intraoperative or postoperative complications, were retrospectively analyzed. The tumor area on enhanced CT, the peritumoral hypodense area on CT without enhancement, and the hyperintense area on T2-weighted MRI were determined planimetrically on each CT or MRI slice by an image analysis system (SPICCAII, Nippon Avionics, Tokyo, Japan). If we assume that tumor and peritumoral hypodensity are conical-shaped, the conical volume may be calculated by the following formulae:

$$V_T = \Sigma V_{Ti}, V_H = \Sigma V_{Hi}$$

$$V_{Ti} = 1/3\, d\, (A_{T(i+1)}^{3/2} - A_{Ti}^{3/2}) / (A_{T(i+1)}^{1/2} - A_{Ti}^{1/2})$$

$$V_{Hi} = 1/3\, d\, \{(A_{T(i+1)} + A_{H(i+1)})^{3/2} - (A_{Ti} + A_{Hi})^{3/2}\} / \{(A_{T(i+1)} + A_{H(i+1)})^{1/2} - (A_{Ti} + A_{Hi})^{1/2}\} - V_{Ti}$$

$$A_{T(i+1)} > A_{Ti}, A_{T(i+1)} + A_{H(i+1)} > A_{Ti} + A_{Hi}$$

V_T: tumor volume, V_H: hypodense volume, V_{Ti}: tumor volume or dead space volume between section i and section (i + 1), V_{Hi}: hypodense volume between section i and section (i + 1), A_{Ti}: tumor area or dead space area on section i, A_{Hi}: hypodense area on section i, d: distance between section i and section (i + 1).

We assumed that the hypodense volumes on CT scan and the hyperintense volumes on MRI are identical, because the relationship between the two values shows a strong and significant correlation (r = 0.9795, p < 0.001) with almost direct proportions between each value (Y = 0.97 X + 2.94).

Results

Clearance of Peritumoral Hypodensity

Figure 1 shows typical cases with different clearance patterns of peritumoral hypodensity. We classified them into 4 groups. The findings in Group A may be related to the clearance of edema fluid from the edematous white matter. Damaged brain caused by multiple mechanisms may manifest the Group C findings. We considered that a gradually decreasing volume of peritumoral edematous white matter and unchanged volume of damaged brain may represent the findings in Group D.

Resolution Rate of Edema Fluid

Six cases in Group B were excluded from evaluation of edema fluid resolution, because precise measurement of hypodensity was not possible. In 3 cases in

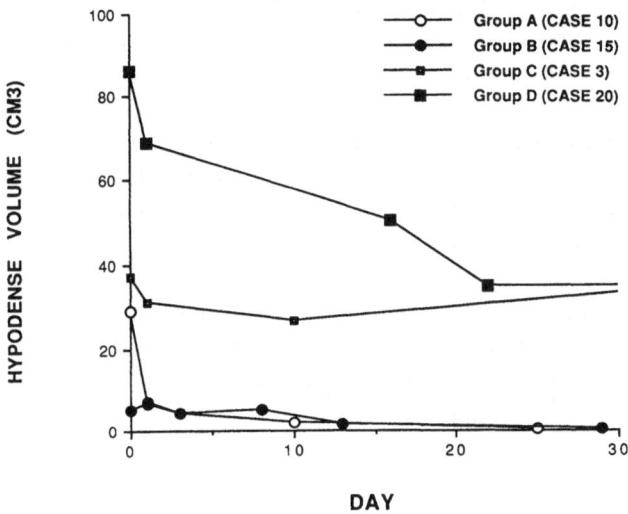

Fig. 1. Clearance of hypodensity. Time course of hypodense volume on CT in four typical cases. Group A: hypodensity of more than 10 cm³ rapidly decreased and disappeared (3 cases). Group B: hypodensity of less than 10 cm³ gradually decreased and disappeared (6 cases). Group C: no remarkable change of hypodensity was observed (7 cases). Group D: hypodensity progressively decreased, but a hypodense lesion in the brain parenchyma persisted (10 cases)

Group A, the hypodense volume may represent the edematous white matter volume (V_E). However, the possibility remains that the hypodensity in Group D contains edematous white matter and damaged brain. We estimated that the hypodense volume in the last evaluation (V_H-last) on and after the thirtieth day following tumor removal is similar to the damaged brain volume (V_D) (Fig. 2A). Only 4 cases examined by CT or MRI on and after the thirtieth day were analyzed in Group D. The extracellular space in the peritumoral edematous white matter was assumed to be about 30%[2]. We presumed that the resolution rate of edema fluid during passage through 1 cm³ of peritumoral edematous white matter is constant (k_2), and calculated the resolution rate by the following formula:

$$dVet/dt = k_2 \, V_Et, \quad dV_Et/dt = k_2/0.3 \times V_Et$$
$$V_Et = V_{E0} \exp(k_2/0.3 \times t) = V_{E0} \exp(k_1 \times t)$$

k_1: resolution rate of edematous white matter (ml/day/cm³ white matter), k_2: resolution rate of edema fluid (ml/day/cm³ white matter), t: days after operation, V_Et: volume of edematous white matter (cm³) at t days after

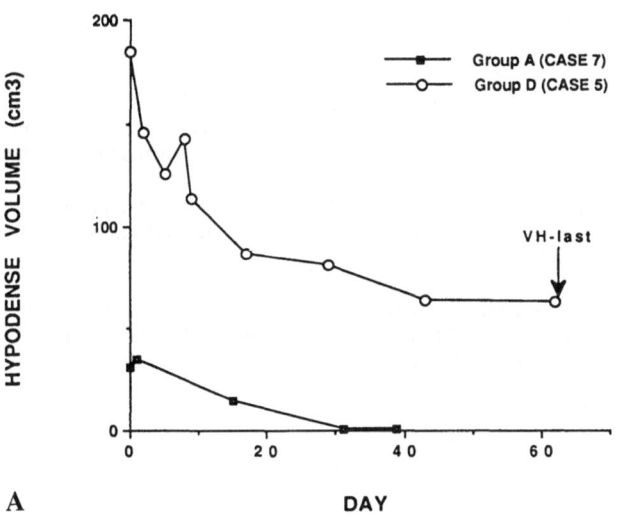

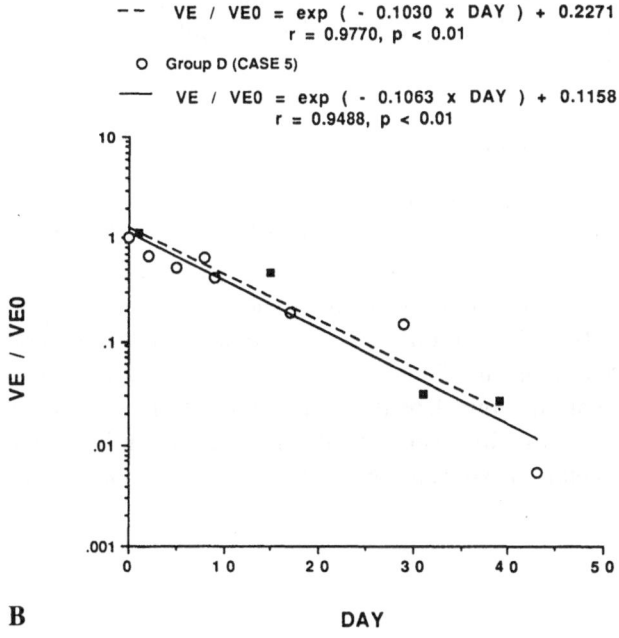

Fig. 2. (A) Clearance of hypodensity. Time course of hypodense volume. The hypodensity in case 7 rapidly decreased and almost disappeared. The hypodensity in case 7 represents edematous white matter volume (V_E). The hypodense volume (V_H) in case 5 continuously decreased but persisted even at the last evaluation. The hypodensity in case 5 represents edematous white matter that decreased (V_E) and the damaged brain that remained after tumor removal (V_D). We estimate that the hypodense volume in the last evaluation ($V_{H\text{-last}}$) is similar to the damaged brain volume. V_E is calculated by the following formula; $V_E = V_H - V_D = V_H - V_{H\text{-last}}$. (B) Changes of edematous white matter volume ratio (V_E/V_{E0}). There is a logarithmic correlation between the V_E/V_{E0} and time in case 7 ($V_E/V_{E0} = \exp(-0.1030 \times day) + 0.2271$, $r = 0.9770$, $p < 0.01$) and case 5 ($V_E/V_{E0} = \exp(-0.1063 \times day) + 0.1158$, $r = 0.9488$, $p < 0.01$). We estimate that the resolution rate of edematous white matter is -0.1030 ml/day/cm³ white matter in case 7, and -0.1063 ml/day/cm³ white matter in case 5. V_{E0}: volume of edematous white matter before operation

Table 1. *Resolution Rate of Edematous White Matter and Edema Fluid*

Resolution rate (ml/day/cm³)		k_1	k_2
Group A	case 7	−0.1030	−0.0309
	case 10	−0.3358	−0.1007
	case 14	−0.2224	−0.0667
Group B	case 5	−0.1063	−0.0319
	case 11	−0.1319	−0.0396
	case 18	−0.1007	−0.0302
	case 20	−0.1509	−0.0453
	Average	−0.1644	−0.0493

k_1 resolution rate of edematous white matter for 1 cm³ edematous white matter. k_2 resolution rate of edema fluid for 1 cm³ edematous white matter.

operation, Vet: volume of edema fluid (cm³) at t days after operation, V_{E0}: volume of edematous white matter before operation (cm³).

The resolution rate of edematous white matter and edema fluid are shown in Table 1. The resolution rate of edematous white matter averaged in 7 cases was −0.1644 ml/day/cm³ white matter, and the resolution rate of edema fluid was −0.04983 ml/day/cm³ white matter.

Discussion

Ito *et al.* studied the formation and resolution of edema fluid in different types of tumor and found that the average resolution rate of edema fluid during passage through 1 cm³ of peritumoral white matter was 0.03 ± 0.003 ml/h[2]. Stevens *et al.* analyzed postoperative scans in 52 patients with meningiomas and found that the low density disappeared completely in 10 patients[4]. In seven patients of this group, it had completely disappeared by 2 to 3 weeks postoperatively,

and in the remaining patients, low density has disappeared in scans done 2 months to one year postoperatively. We showed that the resolution rate of edema fluid was 0.0493 ml/day/cm³ white matter, which is considerably lower than Ito's data. The resolution of brain edema is accomplished in part by a process of bulk flow. A pressure gradient available for bulk flow clearance is provided by the increased difference in regional brain tissue pressure that is caused by the formation of edema fluid, as demonstrated experimentally by the infusion edema model of Marmarou[3]. Following the total removal of meningioma, the formation of edema fluid may stop, tissue pressure may decrease, and the resolution rate may decrease. Using the resolution rate of edematous white matter calculated in our study, we estimated that 50% of the hypodense volume may be resolved in 4 days after total removal, and 90% may be resolved in 14 days. This is in agreement with Stevens's data.

References

1. Fenske A, Samii M, Reulen HJ, Hey O (1973) Extracellular space and electrolyte distribution in cortex and white matter of dog brain in cold induced oedema. Acta Neurochir (Wien) 28: 81–94
2. Ito U, Tomita H, Tone O, Shishido T, Hayashi H (1990) Formation and resolusion of white matter oedema in various types of brain tumors. Acta Neurochir (Wien) [Suppl] 51: 149–151
3. Marmarou A, Tanaka K, Shulman K (1982) The brain response to infusion edema: Dynamics of fluid resolusion. In: Hartmann A, Brooks M (eds) Treatment of cerebral edema. Springer, Berlin Heidelberg New York, pp 11–18
4. Stevens JM, Ruiz JS, Kendall BE (1983) Observation on peritumoral edema in meningioma: part I. Distribution, spreed and resolution of vasogenic edema seen on computed tomography. Neuroradiology 25: 71–80
5. Trittmacher S, Traupe H, Schmid A (1988) Pre- and postoperative changes in brain tissue surrounding a meningioma. Neurosurgery 22: 882–885

Correspondence: Toshiki Shirotani, M.D., Department of Neurosurgery, National Defense Medical College, 3-2 Namiki, Tokorozawa, Saitama 359, Japan.

Traumatic Brain Edema:
Pathophysiology, Laboratory Studies

Acta Neurochir (1994) [Suppl] 60: 421–424
© Springer-Verlag 1994

Traumatic Brain Edema: an Overview

A. Marmarou

Division of Neurosurgery, Medical College of Virgina, Richmond, VA, U.S.A.

Summary

This article provides a brief summary of concepts describing the formation and resolution of traumatic brain edema. Recent laboratory and clinical data are reviewed targeted toward resolving the contribution of edema to the swelling process. These data, indicate that blood volume is reduced in areas of ischemia following traumatic injury and edema volume is increased. Thus, edema is the major contributor to the swelling process in diffuse injury. As clinical MRI studies have not revealed barrier compromise in the presence of swelling, it is considered that other forms of edema, primarily ischemic and neurotoxic, make a substantial contribution to the edema volume.

Keywords: Vasogenic edema; traumatic edema; MRI; resolution.

Introduction

The contribution of edema and blood volume toward brain swelling and intracranial pressure (ICP) rise particularly in cases of traumatic brain injury remains a critical problem. In head injured patients, raised ICP is the single most frequent cause of death in head injured patients[2,30]. In earlier studies Narayan et al.[34] reported a 51% mortality rate in patients whom exceeded 20 mmHg compared to a 16% mortality rate in patients with normal ICP throughout the period of observation. Later studies by Miller et al.[32] indicated that high ICP was initially present in two thirds of patients without intracranial space occupying lesions who were comatose upon admission. In this group, 50% of patients who developed ICP or in whom elevated ICP persisted died despite intense therapy. A more recent analysis by the American Traumatic Coma Data Bank indicates that probability of mortality and morbidity increases with time at ICP elevated above 20 mmHg[28]. Thus raised ICP with accompanying neurologic deficit continues as a salient feature of traumatic head injury, yet the mechanisms leading to brain swelling and subsequent ICP rise remain unclear. It is the purpose of this article to provide a brief overview of current concepts regarding formation and resolution of traumatic brain edema as well as present more recent findings to more clearly identify the role that edema plays in the swelling process.

Brain Swelling: Edema or Vascular Engorgement?

It remains uncertain if the primary disruption of volume homeostasis and subsequent development of ICP is caused by brain edema or vascular engorgement. Primary traumatic swelling usually involves both hemispheres and has been reported to occur in 16% of all patients[50] and in 28% of pediatric head injuries[51]. In a more recent study of the traumatic coma data bank, 31% of patients showed evidence of brain swelling[10]. Earlier studies relating CBF to CT scan data in head injured patients reported that the increased CBF found could not quantitatively account for the rise in attenuation number. Moreover, it was argued that vascular congestion could not provide the increase in volume necessary to collapse the cerebral ventricles[38]. Despite these reports, and in the absence of methods for measurement of tissue water, the contribution of blood volume and edema in head injured patients was still unresolved.

Laboratory Studies of Edema and Vascular Engorgement

Attempts have been made to resolve this issue in the laboratory and most studies in acute swelling have utilized the balloon compression model. This model produces rapid swelling upon deflation and simulates the rapid increase in brain volume following surgical decompression of a supratentorial extracerebral hematoma. The main causes of brain swelling with this model have been attributed to development of edema[9,16,17,27,31,39,42,43,49] or hyperemia[6,15,26]. However this model of brain swelling is unlike the diffuse injury produced by mechanical impact. In a recent study by our laboratory (see Kita and Marmarou this volume), we utilized a new impact acceleration model which produces a delayed and pronounced diffuse axonal injury similar to the DAI observed in Man)[12]. Using this model, cerebral blood volume and gravimetric water content was measured in animals subjected to trauma, trauma plus hypotension, and trauma plus hypoxia combined with hypotension two hours post injury. In these studies, water content was observed to increase and cerebral blood volume to decrease in all groups with the greatest change seen in injured animals with combined hypotension and hypoxia. When the increase in water per gram tissue was compared to the decrease in blood volume per gram tissue, it was found that the edema far exceeded the reduction in blood volume suggesting that the CSF compartment was reduced. These studies are the first to demonstrate a consistent decrease in blood volume with traumatic injury coupled with a dramatic increase in brain water.

Clinical Studies of Cerebral Blood Volume in Traumatic Injury

Clinical studies employing measurements of blood volume in head injured patients have been rare. Kuhl and co-workers, using technetium-labelled red blood cells and CT scans, showed that diffuse traumatic swelling in patients may occur in the presence of decreased cerebral blood volume[25].

The most recent data comes Muizelaar and co-workers who have utilized a unique CT method for measurement of blood volume and who have studied changes in both cerebral blood flow[3,4,5] and blood volume. Although work is still in the preliminary stage, these investigators have shown, for the first time that cerebral blood volume is reduced in patients with ischemia[44].

Fortunately, MRI brain water was measured by Marmarou in several of these patients and tissue water was increased by up to 3.5%[29]. Taking the laboratory and clinical studies in concert, these preliminary data provide compelling evidence that edema plays a major role in the swelling process following trauma and that cerebral blood volume is actually reduced.

Components of Traumatic Brain Edema

Traumatic brain edema has usually been distinguished from other forms of edema by its origin, namely the leakage of plasma bome substances from the vasculature as a result of a breakdown of the blood brain barrier[23]. However, in light of more recent studies, it may be incorrect to strictly assign the term "vasogenic" to traumatic edema as other forms of edema are now considered to play an important role in the swelling process. Cellular edema associated with ischemia has been designated cytotoxic according to the original classification by Klatzo. Cellular swelling that occurs in the absence of ischemia is considered neurotoxic and may result as a consequence of ionic disruption associated with traumatic injury[18]. Thus, at least three forms of edema, vasogenic, ischemic and neurotoxic may be contributing to the increased tissue water following trauma. The contribution of each type in traumatic brain injury remains unknown.

Development of Traumatic Neurotoxic Edema

Information on ionic flux change in traumatic brain injury are limited. A significant increase of potassium in cortex was reported by Takahashi[45] using ion sensitive electrodes in a closed head injury model. Katayama confirmed the trauma induced increase in potassium using microdialysis technique[18], Nilsson studied extracellular potassium and calcium shifts in a dural impact model and reported prolonged reduction of calcium suggesting massive influx into the cellular compartment where neuronal injury was observed[35]. However, although several lines of evidence relate the release of excitatory amino acids to astrocytic swelling[19-21], and acting as mediators of brain edema[1] no animal studies have been conducted relating the ionic dysfunction to the brain swelling process in diffuse brain injury. Nevertheless, in the absence of ischemia or barrier compromise, it is reasonable to suspect neurotoxic edema playing a major role in the swelling process.

Resolution of Vasogenic Brain Edema

A considerable effort over the years has been directed toward clarifying mechanisms of resolution and Go has provided an excellent review of these studies[13]. In brief, the three possible mechanisms which have been considered responsible for resolution of vasogenic edema are: (a) migration to CSF by bulk flow in the presence of pressure gradients[40,41] (b) glial and neuronal uptake of the protein compounds[22,24,47], and (c) reverse vesicular transport from the ECS to the blood via transendothelial passage[46]. Studies utilizing the infusion model of edema have shown that the primary route for clearance of a proteinaceous fluid is the CSF pathway[36]. Our studies support the earlier findings of Reulen who demonstrated the movement of edema toward the ventricle in the presence of pressure gradients[40,41]. In addition to this route, we observed in our model that during the infusion when the fluid is pressure driven, the fluid traveled to cortex by three paths; the ECS of the neuropil, the expanded pericapillary space around microvessels and the expanded perivascular space. The possibility that materials are eliminated from the perineuronal environment by drainage along perivascular channels into the CSF in the subarachnoid space has been reported[7,8,11,14]. Our study would suggest that the perivascular spaces of venules and veins are the pathway for the efflux of solutes from the brain parenchyma to the CSF space however further work is necessary to confirm this route.

A striking observation in these infusion studies was that neither astrocytes nor neurons were observed to take part in the clearance of the albumin which is in direct contrast to previous reports which indicated that cellular uptake of protein compounds was the major mechanism for clearance of vasogenic brain edema[22,24]. This has led us to the hypothesis that concomitant brain trauma or ischemia is a necessary trigger of the glial response.

In summary, studies in our laboratory indicate that in normal brain infused with an extracellular edema, the primary path of resolution is via the cerebrospinal fluid front moving toward the ventricle and the cortical surfaces to reach the CSF. This occurs in the late stage of clearance where tissue pressure gradients subside and may include transport of solutes by newly formed extracellular fluid derived from the cerebral microcirculation[33,48].

Does this occur solely by a process of bulk flow? Studies by Cserr using intracerebral injections of radiolabeled tracers of different molecular weight (69,000 D albumin, 4000 D polyethylene glycol, and 900 D polyethylene glycol) were all cleared at a similar rate and with a half time of approximately 12 hours[8]. In contrast, recent studies by our laboratory utilizing fluorescein labeled dextran of high (71,200) and low (4400) molecular weight observed 15 minutes post infusion indicated that movement occurs by diffusion[37] Reulen et al.[41] also observed differences in movement

of tracers utilizing the ventriculo-cisternal perfusion technique. The issue remains controversial but important to the development of more effective therapies to enhance the clearance process.

Conclusion

A great deal of progress has been made in elucidating mechanisms responsible for formation and resolution of traumatic brain edema. Whereas in the past, a clear distinction was made between traumatic edema, considered mostly vasogenic, and cytotoxic edema. It appears now that all three types of edema, vasogenic, ischemic, neurotoxic, are involved following diffuse head injury. Future studies must concentrate on identifying the contribution of each form to the brain swelling process as it is becoming increasingly evident that brain swelling in patients with early ischemia is water and not blood volume.

References

1. Baethmann A, Maier-Hauff K, Kempski O, et al (1988) Mediators of brain edema and secondary brain damage. Crit Care Med 16: 972–978
2. Becker DP, Miller JD, Ward JD, et al (1977)The outcome from severe head injury with early diagnosis and intensive management. J Neurosurg 47: 491–502
3. Bouma GJ, Muizelaar JP, Choi SC, et al (1991) Cerebral circulation and metabolism after severe traumatic brain injury: the elusive role of ischemia. J Neurosurg 75: 685–693
4. Bouma GJ, Muizelaar JP, Stringer WA, Choi SC, Fatouros P, Young HF (1992) Ultra-early evaluation of regional cerebral blood flow in severely head-injured patients using xenon-enhanced computerizedtomography. J Neurosurg 77: 360–368
5. Bouma GJ, Muizelaar JP, Schuurman R, et al (1992) Cerebral blood volume in acute head injury: relationship to CBF and ICP. In: Avezaath CJJ (ed) Intracranial pressure VIII. Springer, Berlin Heidelberg New York
6. Clubb RJ, Maxwell RE, Chou S (1980) Experimental brain injury in the dog. The pharmacological effects of pentobarbital and sodium nitroprusside. J Neurosurg 52: 189–196
7. Cserr HF, Cooper DN, Milhorat TH (1977) Flow of cerebral interstitial fluid as indicated by the removal of extracellular markers from rat caudate nucleus. Exp Eye Res [Suppl] 25: 461–473
8. Cserr HF, Cooper DN, Suri PK, Patlak CS (1981) Efflux of radiolabeled polyethylene glycols and albumin from rat brain. Am J Physiol 240: F329–F328
9. Czernicki Z, Kozniewska E (1977) Disturbances in the blood-brain barrier and cerebral blood flow after rapid brain decompression in the cat. Acta Neurochir (Wien) 36: 181–187
10. Eisenberg H, Gary H, Jane J, et al (1988) CT scan findings in a series of 595 patients with severe closed head injury: a report from the NIH traumatic Coma data bank. The Annual Meeting of the American Association of Neurological Surgeons, Toronto, Ontario, April 24–28
11. Flexner LB (1933) Some problem of the origin, circulation and absorpion of the cerebrospinal fluid. Q Rev Biol 8: 397–422

12. Foda M, Marmaro, A (1993) A new model of diffuse brain injury in rats. Part II: morphological characterization. J Neurosurg, accepted
13. Go KG (1991) Cerebral pathophysiology. Elsevier, New York
14. His W (1865) Über ein perivasculares Canalsystem in den Lymphsystemen. Z Wiss Zool 15: 127–141
15. Ishii S (1966) Brain swelling. Studies of structural physiological and biochemical alterations. In: Caveness WH, Walker AF (eds) Head injury conference proceedings. Lippincott, Philadelphia, pp 276–299
16. Jakobsson KE, Thuomas KA, Bersttrom K, et al (1990) Rebound of ICP after brain compression. An MRI study in dogs. Acta Neurochir (Wien) 104: 126–135
17. Jakobsson DE, Lofgren J, Zwetnow NN, et al (1990) Cerebral blood flow in experimental intracranial mass lesions. Part 2: the postdecompression phase. Neurol Res 12: 153–157
18. Katayama Y, Becker DP, Tamura T, Havda DA (1990) Massive increases in extracellular potassium and the indiscriminate release of glutamate following concussive brain injury. J Neurosurg 73: 889–900
19. Kettenmann H, Schechner M (1985) Pharmacological properties of gamma-aminobutyric acid-glutamate-, and aspartate-induced depolarizations in cultured astrocytes. J Neurosci 5: 3295–3301
20. Kimelberg H (1987) Amisotonic media and glutamate-induced ion transport and volume responses in primary astrocyte cultures. J Physiol (Paris) 82: 294–303
21. Kimelberg H, Pang S, Treble D (1989) Excitatory amino acid-stimulated uptake of 22Na+ in primary cultures. J Neurosci 9: 1141–1149
22. Klatzo I, Miquel J (1960) Ovservations in pinocytosis in nervous tissue. J Neuropathol Exp Neurol 19: 475–481
23. Klatzo I (1967) Neuropathological aspects of brain edema. J Neuropathol Exp Neurol 26: 1–14
24. Klatzo I, Chui E, Fujiwara K, Spatz M (1980) Resolution of vasogenic brain edema. Adv Neurol 28: 359–373
25. Kuhl D, Alavi A, Hoffman E, et al (1980) Local cerebral blood volume in head-injured patients. Determination by emission computed tomography of 99mTc-labeled red cells. J Neurosurg 52: 309–320
26. Langfitt TW, Kassell NF, Weinstein JD (1965) Cerebral blood flow with intracranial hypertension. Neurology 15: 761–773
27. Lowell HM, Bloor BM (1971) The effect of increased intracranial pressure on cerebrovascular hemodynamics. J Neurosurg 34: 760–769
28. Marmarou A, Anderson R, Ward J, et al (1988) The traumatic coma data bank: monitoring of ICP. The Seventh International Symposium on Intracranial Pressure and Brain Injury, Ann Arbor, Michigan
29. Marmarou A, Fatouros P, Bandoh K, Zerate B, Laine F, Deyo D, Brockenbrough P, Neuroscience ICU nursing staff, Young HF (1991) The contribution of brain edema to brain swelling. ICP and craniospinal dynamics, in press
30. Marshall LF, Smith RW, Shapiro, HM (1979) The outcome with aggressive treatment in severe head injuries. Part I: the significance of intracranial pressure monitoring. J Neurosurg 50: 20–25
31. Miller JD, Stanek AE, Langfitt TW (1973) Cerebral blood flow regulation during experimental brain compression. J Neurosurg 39: 186–196
32. Miller JD, Becker DP, Ward JD, et al (1977) Significance of intracranial hypertension in severe head injury. J Neurosurg 47: 503–516
33. Milhorat TH, Davis DA, Hammock MK (1983) Cerebrospinal fluid as reflection of internal milieu of brain. In: Wood JH (ed) Neurobiology of cerebrospinal fluid, Vol 2. Plenum, New York, pp 1–23
34. Narayan RK, Greenberg RP, Miller DJ, et al (1981) Improved confidence of outcome prediction in severe head injury. J Neurosurg 54: 751–762
35. Nilsson P, Hillered L, Olsson Y, Sheardown MJ, Hansen AJ (1992) Regional changes in interstitial K+ and Ca2+ levels following cortical compression contusion trauma in rats. J Cereb Blood Flow Metab 13: 183–192
36. Ohata K, Marmarou A, Povlishock JT (1990) An immunocytochemical study of protein clearance in brain infusion edema. Acta Neuropathol (Berl) 81: 162–177
37. Ohata K, Marmarou A (1992) Clearance of brain edema and macromolecules through cortical extracellular space. J Neurosurg 77: 387–396
38. Penn RD, Clasen RA (1982) Traumatic brain swelling and edema. In: Cooper PR (ed) Head injury. Williams and Wilkins, Baltimore, pp 233–255
39. Ponten U, Thuomas KA, Bergstrom K, et al (1986) Evaluation of intracranial pressure rebound after evacuation of intracranial expanding lesions. An experimental study in dogs. Acta Radiolog [Suppl] 369: 360–364
40. Reulen HJ, Graham R, Spatz M, Klatzo I (1977) Role of pressure gradients and bulk flow in dynamics of vasogenic brain edema. J Neurosurg 46: 24–36
41. Reulen HJ, Tsuyumu M, Tack A, Fenske AR, Prioleau GR (1978) Clearance of edema fluid into cerebrospinal fluid. A mechanism for resolution of vasogenic edema. J Neurosurg 48: 754–764
42. Sakamoto H, Tanaka K, Nakamura S, et al (1986) Direct observation of autonomic nerve activity in experimental acute brain swelling. In: Miller JD, Teasdale GM, Roman JQ, Galbraith SL, Mendelow AD (eds) Intracranial pressure VI. Springer, Berlin Heidelberg New York, pp 166–173
43. Schettini A, Lippman HR, Walsh EK (1989) Attenuation of decompressive hypoperfusion and cerebral edema by superoxide dismutase. J Neurosurg 71: 578–587
44. Schroeder M, Muizelaar JP, Fatouros P, Young H (1993) Local cerebral blood volume (CBV) after severe head injury in patients with local cerebral ischemia. Abstract AAN
45. Takahashi H, Manaka S, Sano K (1981) Changes in extracellular potassium concentration in cortex and brain stem during the acute phase of experimental closed head injury. J Neurosurg 55: 708–717
46. Vorbrodt AW, Lossinsky AS, Wisniewski HM, Suzuki R, Yamaguchi T, Masaoka H, Klatzo I (1985) Ultrastructural observations on the transvascular route of protein removal in vasogenic brain edema. Acta Neuropathol (Berl) 66: 265–273
47. Wolman M, Klatzo I, Chui E, Wilmes F, Nishimoto K, Fujiwara K, Spatz M (1981) Evaluation of the dye-protein tracers in pathophysiology of the blood-brain barrier. Acta Neuropathol (Berl) 54: 55–61
48. Wood JH (1980) Physiology, pharmacology and dynamics of cerebrospinal fluid. In: Wood JH (ed) Neurobiology of cerebrospinal fluid, Vol 1. Plenum, New York, pp 1–16
49. Yamaguchi M, Shirakata S, Yamasaki S, et al (1976) Ischemic brain edema and compression brain edema. Water content, blood-brain barrier and circulation. Stroke 7: 77–83
50. Zimmerman R, Bilaniuk L, Gennarelli T (1978) Computed tomography of shearing injuries of the cerebral white matter. Radiology 127: 393–396
51. Zimmerman R, Bilaniuk L, Bruce D, et al (1978) Computed tomography of pediatric head trauma: acute general cerebral swelling. Radiology 126: 403–408

Correspondence: Anthony Marmarou, M. D., Division of Neurosurgery, Medical College of Virginia, Box 508, MCV Station, Richmond, VA, 23298, U.S.A.

Acta Neurochir (1994) [Suppl] 60: 425–427
© Springer-Verlag 1994

Growth Kinetics of a Primary Brain Tissue Necrosis from a Focal Lesion*

J. Eriskat[1], **L. Schürer**[2], **O. Kempski**[3], and **A. Baethmann**[1]

[1] Institute for Surgical Research, [2] Department of Neurosurgery, Klinikum Großhadern, Ludwig-Maximilians-Universität, München, and
[3] Institute for Neurosurgical Pathophysiology, University Mainz, Mainz, Federal Republic of Germany

Summary

Secondary brain damage, such as brain edema or impairment of the cerebral microcirculation may evolve from tissue necrosis of the brain induced by trauma or ischemia. This laboratory has provided novel information on the secondary increase of a primary brain tissue necrosis resulting from a focal lesion. We have presently investigated more closely the growth kinetics of this process during 24 h after trauma. Rats were subjected to a standardized focal freezing injury of the brain. Area and volume of the resulting necrosis were quantitatively assessed by morphometry after different periods of survival (i.e., 5 min, 3, 6,12, 18 and 24 h after trauma). The maximal area of necrosis increased by 45% (p < 0.001) during the posttraumatic observation period. Growth of necrosis after trauma was not limited to the early period, but continued between 12 and 24 h, amounting then to 29% (p < 0.05). The volume of necrosis calculated on the basis of histological serial sections was also observed to increase by 45%. The current findings confirm that a primary brain tissue lesion induced by a standard cryogenic injury, studied as model of a contusion focus in severe head injury, is subjected to secondary growth within a period of 24 h after trauma, longer periods of survival were not investigated yet. Quantification of lesion growth makes possible not only to study underlying mechanisms, but also of whether this process can be therapeutically inhibited.

Keywords: Secondary brain damage; cold injury; quantitative histology; secondary lesion growth.

Introduction

Brain edema, intracranial hemorrhage, or impairment of the cerebral microcirculation are pertinent intracranial manifestations of secondary brain damage in severe head injury[1]. A novel as well as intriguing aspect is that a primary necrosis of brain parenchyma evolving from a focal cerebral insult may be subjected to secondary growth. Experiments utilizing different methods of brain injury have consistently confirmed an increase in size of the resulting tissue necrosis within 24 h, amounting to 50% in rats[2,4,7], or even to 300% in rabbits[8]. It is not clear yet, however, whether the phenomenon is reflecting a delayed but irreversible primary process which is resistant to treatment, or is representing a manifestation of secondary brain damage, thuslserving as a potential target of therapeutic inhibition. Confirmation of the latter would require more detailed information on the time course as well as extent of lesion growth on a quantitative basis – the point of the present experimental investigations.

Materials and Methods

In male Sprague-Dawley rats (278 ± 39 g b.w., n = 59) anesthesia was introduced by ether after premedication with 0.25 mg atropin subcutaneously to reduce salivation. Anesthesia was maintained by i.p. chloralhydrate (360 mg/kg b.w.). The animals were breathing room air enriched with oxygen. The skull was positioned then in a stereotactic frame. For exposure of the left parietal cerebral cortex a craniotomy of 3 mm Ø was made with a dental drill under continuous cooling and employment of an operating microscope. The dura remained intact and was frequently moistened with isotonic saline. The freezing insult was induced by a metal probe (Ø 1 mm) attached to a cylinder and cooled to –68°C by a dry ice/acetone mixture. The cooling probe was fixed to a stereotactic device powered by a computer-controlled stepper motor for maximum precision of placement and duration of the freezing insult. The focal cold injury was inflicted according to Klatzo et al.[3] through the cranial burr hole during a freezing period of the exposed cortex of 10 sec duration. Morphological analysis of the resulting lesion demonstrated confinement of the insult to the cerebral cortex. Body temperature was monitored and maintained during anesthesia between 37.5°C and 38.0°C by a servo-controlled heating pad. The animals were assigned to one of the six experimental groups with different survival periods after trauma (i.e. 5 min, 3, 6,12,18 and 24 h, respectively) for examination of progression of the lesion during the posttraumatic course. Animals sacrificed immediately after trauma were rapidly thoracotomized for perfusion fixation of the brain through the left cardiac ventricle, starting with isotonic saline for 30 sec followed by 2% buffered paraformaldehyde (pH 7.4) for 20 min. Animals of other

* Supported by Deutsche Forschungsgemeinschaft Ba 452/6–7.

experimental groups surviving longer periods were allowed to recover from anesthesia after surgical closure of the scalp. Anesthesia was induced again after respective survival periods for perfusion fixation of the brain as described above. Following in situ fixation of the brain, the animals were decapitated, and the head was placed in 10% formalin over night at 4°C. The brain was subsequently removed carefully and stored again in 10% formalin prior to preparation for histology following dehydration and embedding in paraffin. Coronal serial sections of the brain of 3–5 µm thickness were made at 100 µm intervals throughout the lesion area. The sections were stained with cresyl violet according to NISSL. The area of necrosis was measured planimetrically in all sections containing lesion. The maximal lesion area was determined from these data, and the areas of necrosis of the total brain were utilized for calculation of the necrosis volume by employment of a 'basic volume estimator' according to[6]. Data are given as mean ± standard error of the mean. Differences betwoen groups were tested by the Kruskal-Wallis test with Conover's multiple comparison according to[5].

Results

Kinetics of the spread of necrosis and the calculated necrosis volume are summarized in Table 1. The maximum area of necrosis increased within 3 h by approximately 20%, i.e. from 0.75 ± 0.06 mm^2 to 0.90 ± 0.05 mm^2, while growth of necrosis was not observed between 3 and 12 h after trauma. Thereafter, the area of necrosis was increasing further between 12 and 24 h, from 0.85 ± 0.05 mm^2 to 1.09 ± 0.08 mm^2 ($p < 0.05$). The maximal area of necrosis altogether was increasing by 45% within 24 h after trauma ($p < 0.001$). The volume of necrosis calculated on the basis of the serial brain sections was increasing from 0.78 ± 0.09 mm^3 at 5 min after trauma to 1.09 ± 0.08 mm^3 at 24 h after trauma. Hence, the increase in necrosis volume amounted also to 45% within the posttraumatic observation period. Due to the greater statistical scatter of individual data, the differences between results obtained at 5 min and at 24 h after trauma are statistically not significant.

Discussion

The present findings confirm that a primary brain tissue necrosis induced by a standard insult is subjected to secondary growth within 24 h after trauma. Two periods of spread of the brain tissue necrosis can be clearly distinguished, an early phase of up to 3 h after trauma, and a late period between 12 and 24 h after trauma. The progressive destruction of brain parenchyma, as assessed by histological morphometry provided the basis of the determination of lesion growth. The primary lesion by purpose was limited to the gray matter of cerebral cortex, making possible thereby a more accurate distinction between irreversibly damaged and surviving perifocal brain parenchyma in histological preparations. In addition, growth of the area of necrosis in the histological sections was tested by assessment of the density of viable nerve cells in perifocal brain tissue areas (data not shown). The findings obtained thereby confirm that lesion growth following the primary insult must be attributed to a progressive extinction of cells and not to mere expansion of the necrotic area by tissue swelling.

As to the mechanisms underlying growth of the primary lesion, a variety of factors might be involved. These are, among others, impairment of the cerebral microcirculation resulting from damage of the arterial blood supply, as well as activation of necrobiotic processes attributable to the release and accumulation of neurotoxic mediator compounds. Whatsoever, identification of secondary growth of a primary brain tissue lesion at a quantitative level as presently confirmed, provides a basis not only for elucidation of underlying mechanisms but also for the development of specific methods of therapeutical inhibition. Respective findings would have clinical significance.

Table 1. *Maximal Area and Volume of Necrosis of Brain Parenchyma in Rats with Focal Cold Injury*

Survival	n	Maximal lesion area [mm^2]	Lesion volume [mm^3]
5 min	9	0.75 ± 0.06[c]	0.78 ± 0.09
3 h	8	0.90 ± 0.05	1.02 ± 0.09
6 h	10	0.85 ± 0.05[b]	0.97 ± 0.11
12 h	7	0.85 ± 0.05[b]	1.04 ± 0.10
18 h	10	0.94 ± 0.06[a]	1.07 ± 0.07
24 h	15	1.09 ± 0.08	1.13 ± 0.09

[a] $p < 0.05$ vs. 5 min; [b] $p < 0.01$ vs. 24 h; [c] $p < 0.001$ vs. 24 h.

Acknowledgements

The technical and secretarial assistance of Ann Hedblom, Helga Kleylein and Monika Stucky is gratefully acknowledged.

References

1. Baethmann A, Maier-Hauff K, Kempski O, Unterberg A, Wahl M, Schürer L (1988) Mediators of brain edema and secondary brain damage. Crit Care Med 16: 972–978
2. Eggert HR, Kiessling M, Kleihues P (1985) Time course and spatial distribution of neodymium: yttrium-aluminium-garnet (Nd: YAG) laser-induced lesions in the rat brain. Neurosurgery 16: 443–448
3. Klatzo I, Piraux A, Laskowski EJ (1958) The relationship between edema, blood-brain barrier and tissue elements in a local brain injury. J Neuropathol Exp Neurol 17: 548–564
4. Lindsberg PJ, Frerichs KU, Burris JA, Hallenbeck JM, Feuerstein G (1991) Cortical microcirculation in a new model of focal laser-induced secondary brain damage. J Cereb Blood Flow Metab 88–98
5. Theodorsson-Norheim E (1986) Kruskal-Wallis test: BASIC computer program to perform nonparametric one-way analysis of variance and multiple comparisons on ranks of several independent samples. Comp Meth Prog Biomed 23: 57–62
6. Uylings HBM, van Eden CG, Hofman MA (1986) Morphometry of size/volume variables and comparison of their bivariate relations in the nervous system under different conditions. J Neurosci Meth 18: 19–37
7. Vonhof S (1993) Autoradiographische Untersuchungen des Gehirns über das sekundäre Nekrosewachstum nach primärem Trauma. Thesis, Ludwig-Maximilians-Universität, München, in preparation
8. Wyrwich W (1993) Die Entwicklung einer sekundären Hirngewebsnekrose nach primärem Trauma. Thesis, Ludwig-Maxmilians-Universität, München, in preparation

Correspondence: Alexander Baethmann, M. D., Institute for Surgical Research, Klinikum Großhadern, Ludwig-Maximilians-Universität, Marchioninistrasse 15, D-81377 München, FRG.

Acta Neurochir (1994) [Suppl] 60: 428–430
© Springer-Verlag 1994

Focal Brain Injury: Histological Evidence of Delayed Inflammatory Response in a New Rodent Model of Focal Cortical Injury

P. Mathew, D. I. Graham, R. Bullock, W. Maxwell, J. McCulloch, and **G. Teasdale**

Institute of Neurological Sciences, University of Glasgow, Southern General Hospital, Glasgow, Scotland, U.K.

Summary

Cortical contusions are one of the most common characteristics in head injury and are regarded by many as the hallmark of significant injury. No experimental study has clarified the roles of mechanical forces, haemorrhage and ischaemia in the process of progressive acute brain damage and later neurobehavioural dysfunction. We have devised a new, simple reproducible rodent model of focal cortical injury which employs a 'pure' mechanical/physical force applied through the intact dura. Using this model we have investigated the time course and pattern of changes in neurons, glia and microvasculature. With the exception of haemorrhage, this model closely reproduces the light- and electron microscopy features of human contusion. In the absence of perivascular haemorrhage we have demonstrated delayed perivascular protein leakage and polymorpho-nuclear-leukocyte infiltration of the damaged cortex. We postulate that a component of the delayed blood brain barrier breakdown demonstrated in human focal head injury (which may contribute to swelling and brain damage) is due to an acute inflammatory response, the magnitude of which is dependent on the amount of tissue injury.

Keywords: Contusion; head injury; animal model; acute inflammation.

Introduction

Cortical contusions are one of the most common and characteristic lesions of head injury, regarded by many as the hallmark of significant damage. Contusions are important in the processes of progressive brain damage in the acute stage and may cause many late sequelae, such as change in personality and memory, and post-traumatic epilepsy. Despite their importance, the mechanisms causing contusion and in particular leading to their dynamic evolution and maturation are not established. So far, no experimental study has clarified the roles of mechanical forces, haemorrhage and ischaemia. Research into the mechanisms of contusions is limited by the lack of a reliable animal model of focal cortical injury. We have devised a new, simple, reproducible model of focal cortical injury, which employs a "pure" mechanical force applied through the intact dura. This model closely reproduces the light- and ultrastructural features of human contusion. Using this model, we have investigated the time course and pattern of the changes in neurons, glia and microvasculature.

Methods

Fifty-two adult male Sprague-Dawley rats were studied. Focal cortical lesions were produced by applying a localised suction impact (minus 700 mmHg) to the exposed (left parietal) dura. The tip of the suction device (internal diameter 5 mm) was applied to the intact dura, and maintained in position with a stereotactic frame, (Fig. 1). The negative pressure, generated by allowing a spring loaded plunger to spring back when a release screw was rapidly released, was maintained for 5 seconds. A three-way tap in the system was then opened to air, to allow rapid dissipation of negative pressure (Fig. 2). Animals were randomly allocated to lesioned and control groups of six and five rats respectively, for each survival time of four hours, 24 hours, three days and seven days, followed by perfusion fixation. In addition, two control and two lesioned animals, again randomly allocated, were studied at five minutes and 30 minutes survival and processed for EM. Blood brain barrier damage was assessed ultrastructurally in animal perfusion fixed with lanthanum nitrate at five minutes and one hour.

Results

There were no significant differences in mean arterial blood pressure, pO_2, pCO_2, temperature and pH between the sham operated and injured animals.

Light Microscopy

Sham operated animals did not show any abnormality. Each of the injured animals had a lesion centred over the sensory motor cortex. The AP length of the lesion was 4 mm. A minimal amount of subarachnoid haemorrhage was found in each case. A "saucer-shaped" area of pallor of staining, extending to involve cortical layers I and II, with a sharply defined boundary, was present in each animal. After 24 hours the subarachnoid haemorrhage blood had dispersed, but the area of pallor appeared larger with an AP length of 7.5 mm and involvement of layer three of the cortex. By day three the lesion size was becoming less well demarcated microscopically and reducing in size.

At four hours, within the zone of pallor of staining macroscopically, there was marked vacuolization of the neuropil; the neurons demonstrated pyknosis and shrinkage to an irregular triangular shape with associated surface encrustations. The nucleus showed loss of the nucleolus and was surrounded by an eosinophilic cytoplasm. Haemorrhage within the lesion was not seen.

At 24 hours, there was progression to a more obviously necrotic appearance which was well defined and surrounded by a zone of pyknotic shrunken neurons with associated enlarged, reactive astrocytes. Perivascular protein leakage was present in layer II and part of layer III with associated disruption of the surrounding neuropil. Polymorphs were seen scattered throughout the lesion.

At three days there was a marked monocyte/macrophage response, which persisted at seven days. At this stage an astrocyte response at the margin of the lesion was still present.

Electron Microscopy

At five minutes a very marked neuronal dendritic response was present with disruption of the neuropil, in cortical layers I, II and part of III. At four hours there was selective neuronal cell necrosis. The astrocyte response was well developed at four hours, with development of swelling of perivascular and perineuronal processes permeating the neuropil. Swollen astrocyte cell bodies were also present. In animals perfusion

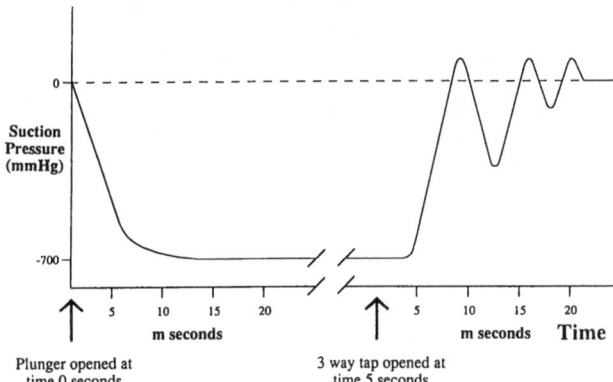

Fig. 2. Pressure characteristics of suction impact

fixed with lanthanum there was no evidence of breakdown of blood-brain barrier at either five minutes or one hour survival.

Discussion

Mechanical injury to the cortex can replicate the histopathology of contusional head injury without acute vascular damage. In this model early traumatic disruption of the blood-brain barrier and haemorrhage did not occur. Swelling in this acute phase appears to be the result of cytotoxic accumulation of fluid in cells[1]. Previous models of spinal cord- and head injuries have demonstrated early accumulation of polymorphonuclear granulozytes in the lesion during the first eight hours after trauma[2]. However, this may have been in part due to direct disruption of the vascular endothelium and subsequent haemorrhage in these models[2,3].

The presence of a polymorphonuclear reaction and perivascular protein leakage in our model suggests that a component of the delayed swelling associated with a traumatic focal lesion is a delayed inflammatory response to local tissue injury which may contribute to the formation of perilesional oedema. This blood-brain barrier breakdown may be promoted/initiated by polymorphonuclear leukocytes via the release of various factors (e.g. hydrolytic enzymes, lipid mediator production, or oxygen free radical production)[4,5].

References

1. Bullock R, Maxwell WL, Graham DI, Teasdale GM, Adams JH (1991) Glial swelling following human cerebral contusion: an ultrastructural study. J Neurol Neurosurg Psychiatry 54: 427–434
2. Means ED, Anderson DK (1983) Neuronophagia by leukocytes in experimental spinal cord injury. J Neuropath Exp Neurol 42: 707–719

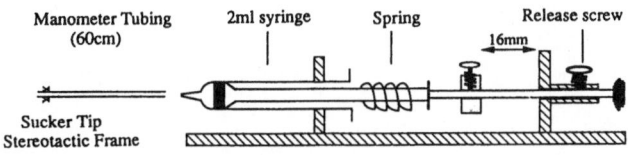

Fig. 1. Suction apparatus: entire apparatus filled with normal saline and primed to deliver a constant negative pressure

3. Schoettle RJ, Kochanek PM, Margargree MJ, Uhl MW, Nemoto EM (1990) Early polymorphonuclear accumulation correlated with the development of post-traumatic cerebral oedema in rats. J Neurotrauma 7: 207–217
4. Shasby DM, Shasby SS, Peach MJ (1983) Granulocytes and phorbol myristate acetate increase permeability to albumin of cultured endothelial monolayers and isolated perfused lungs. Am Ref Respir Dis 127: 72–76
5. Kochanek PM, Hallenbeck JM (1992) Polymorphonuclear leukocytes and monocytes/macrophages in the pathogenesis of cerebral ischaemia and stroke. Stroke 23: 1367–1379

Correspondence: Mr. P. Mathew, Institute of Neurological Sciences, University of Glasgow, Southern General Hospital, Glasgow, G51 4TF, Scotland, U.K.

Acta Neurochir (1994) [Suppl] 60: 431–433

Development of Traumatic Brain Edema in Old versus Young Rats

A. Unterberg, G. H. Schneider, J. Gottschalk, and **W. R. Lanksch**

Department of Neurosurgery, Rudolf Virchow Medical Center, Free University of Berlin, Berlin, Federal Republic of Germany

Summary

Age is an important factor of mortality and morbidity following traumatic brain injury. The causes for the adverse effect of old age remain obscure. The aim of this study was to clarify whether age affects the development of posttraumatic brain edema.

In Wistar rats, a cortical freezing lesion was applied to the parietal region in ketamine-xylazine anesthesia. 18 young rats (4–6 months) were compared to 15 old animals (36–40 months). In the early peritraumatic and late posttraumatic period blood pressure was monitored. 24 hours after trauma, the brains were removed and hemispheric swelling, water- und electrolyte-contents were measured. In addition, the brains of 3 animals of each group were histologically evaluated.

In the old age group, 3 animals died during the 24 hours observation period (mortality 20%), whereas all young rats survived ($p < 0.01$). The cortical freezing lesion resulted in a hemispheric swelling of $6.9 \pm 0.5\%$ in young, and $10.4 \pm 0.8\%$ in old animals ($p < 0.001$). Accordingly, the increase of cerebral water content due to the lesion was significantly more pronounced in the group of old rats, i.e. 2.05% in old versus 1.50% in young animals ($p < 0.01$).

The increase of swelling and edema in the old age group could not be attributed to arterial hypertension. On the contrary, mean arterial blood pressure was significantly lower in old animals. Histological examinations did not reveal significant differences between the two groups.

Edema generation following a standardized cryogenic lesion is markedly enhanced in old versus young rats. This might be one factor among others for higher mortality and morbidity following traumatic brain injury in old versus young individuals.

Keywords: Age; edema formation; rats; cold injury.

Introduction

Morbidity, mortality as well as outcome following a traumatic brain injury are markedly influenced by age, both experimentally and clinically[1,3,4,6,7]. The causes for the adverse effect of age are still not clear. They seem to be based upon an alteration in the pathophysiological response of the aging central nervous system to severe trauma rather than on an increased incidence of age-related non-neurological complications[6]. It has thus been proposed to further study physiology and pathophysiology of aging in the nervous system including the effects of age on ICP, on the generation of cerebral edema, and the response of the aging brain to mechanical stress[6].

The aim of this study was therefore to examine the effect of age on the development of vasogenic brain edema following a standardized cryogenic cortical lesion.

Methods

The experiments were performed in Wistar rats weighing 420 ± 10 g. The animals were anesthetized with ketamine and xylazine. A circular craniectomy was performed over the right parietal region for induction of cold injury by means of a copper-cylinder (diameter 5 mm), cooled with a dry-ice-acetone mixture[2]. This cylinder was stereotactically attached onto the intact dura for 30 seconds. Blood pressure was monitored via a cannula in the femoral artery in the early peritraumatic as well as the late posttraumatic period.

The old age group consisted of 15 animals (36–40 months) and was compared to a group of 18 young rats (4–6 months).

24 hours after trauma the animals were sacrificed and the brains were quickly removed. In 15 young and 11 old rats the cerebral hemispheres were separated, weighed for determination of hemispheric swelling, dried and weighed again. Hemispheric swelling was determined as the difference of weight of the traumatized and contralateral hemispheres[5]. Cerebral water content of both hemispheres was obtained by dry-wet-weight measurement. Tissue electrolyte contents (solium and potassium) were thereafter determined in desiccated brain specimen.

In addition, the brains of 3 animals of each group were removed 24 hours after trauma and fixed in 10% neutral formaldehyde for histological evaluation.

Results

The cortical freezing lesion was survived by all young rats, whereas 3 animals of the old age group died during the 24 hours observation period. This corresponds to a mortality of 20% in the old age group

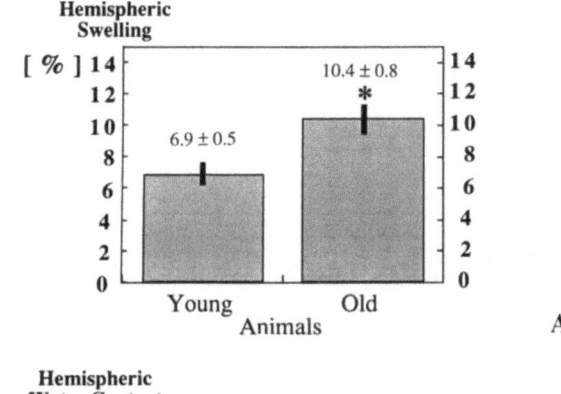

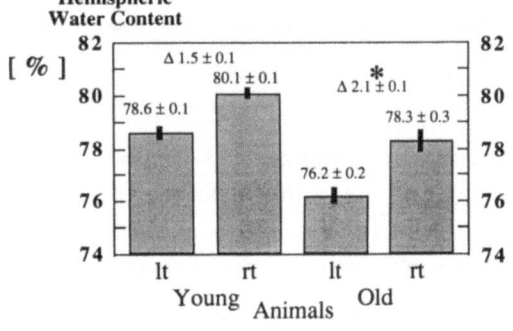

Fig. 1. (A) Hemispheric swelling following a cryogenic lesion in young and old rats. Swelling in markedly enhanced in the old age group. (B) Cerebral water content of young and old rats following a cortical freezing lesion (in ml/100 g f.w.). The increase in water content due to the lesion is more pronounced in old animals, while cerebral water content is generally lower in old animals

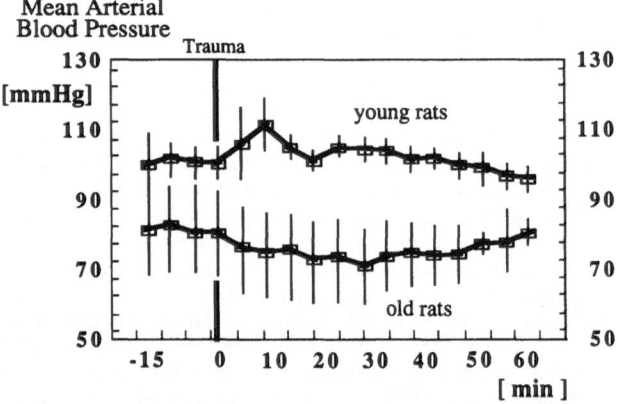

Fig. 2. Blood pressure in young and old animals before, during and following the cryogenic injury. Mean arterial blood pressure is significantly lower in the old age group

($p < 0.01$). There were no obvious reasons for the observed mortality, especially intracerebral hemorrhage was not found.

Hemispheric swelling due to the cortical freezing lesion was $6.9 \pm 0.5\%$ in young, and $10.4 \pm 0.8\%$ in old animals ($p < 0.001$) (Fig. 1A). Accordingly, the increase of cerebral water content due to the perifocal edema was significantly more pronounced in the group of old rats. In the elderly animals this difference was 2.1%, whereas it was 1.50% in young animals ($p < 0.01$) (Fig. 1B). In general, brain water content was significantly lower in old rats, e.g. $78.6 \pm 1\%$ in the contralateral control hemisphere of young, and $76.2 \pm 0.2\%$ in the control hemisphere of old animals (Fig. 1B).

Monitoring of mean arterial blood pressure in both groups revealed that older animals had a significantly decreased blood pressure compared to the young group (Fig. 2). Also, in the end of the experiments, blood pressure was significantly lower in the old age group.

Histological evaluation did not reveal significant differences between the two age groups. Areas of necrosis and perifocal edema were comparable in old and young animals.

Discussion

Our study demontrates for the first time that edema formation following a standardized cryogenic lesion is significantly enhanced in old versus young rats. This might be one factor among others for higher mortality and morbidity after a severe head injury in old versus young individuals.

It is a well-known clinical experience that elderly patients fare worse compared to younger individuals[1,6]. Also, experimental studies were able to demonstrate that aging is associated with an increased mortality rate and greater neurological deficits following traumatic brain injury[3,4]. The adverse effect of age following cerebral injury is not restricted to trauma, but also holds true for ischemia and subarachnoid hemorrhage[7]. Since edema is a common feature of all acute cerebral injuries, enhanced edema formation might be one factor among others.

The question remains to be solved why edema formation is more pronounced in old animals. The increase of swelling and edema could not be attributed to arterial hypertension. On the contrary, blood pressure was significantly lower in old animals. It might therefore be argued that hypotension favours the cytotoxic component of edema. Further experiments shall clarify whether blood-brain barrier damage, i.e. permeability is increased in old animals.

Acknowledgement

The excellent technical and secretarial assistance of Ms. J. Kopetzki and Ms. A. Riede is highly appreciated.

References

1. Jamjom A, Nelson R, Stranjalis G, Wood S, Chissel H, Kane N, Cummins B (1992) Outcome following surgical evacuation of traumatic intracranial haematomas in the elderly. Br J Neurosurg 6: 37–32
2. Klatzo I, Piraux A, Laskowski EJ (1958) The relationship between edema, blood-brain barrier and tissue elements in local brain injury. J Neuropath Exp Neurol 17: 548–564
3. Hamm RJ, Jenkins LW, Lyeth BG, White-Gbadebo DM, Hayes RL (1991) The effect of age on outcome following traumatic brain injury in rats. J Neurosurg 75: 916–921
4. Hamm RJ, White-Gbadebo DM, Jenkins LW, Hayes RL (1992) The effect of age on motor and cognitive deficits after traumatic brain injury in rats. Neurosurgery 31: 1072–1078
5. Unterberg A, Schmidt W, Dautermann C, Baethmann A (1990) The effect of various steroid treat,ment regimens of cold-induced brain swelling. Acta Neurochir (Wien) [Suppl] 51: 104–106
6. Vollmer DG, Torner JT, Eisenberg HM, Foulkes MA, Marmarou A, Marshall LF (1991) Age and outcome following traumatic coma: why do older patients fare worse? J Neurosurg 75: 537–549
7. Yao H, Sadomisha S, Ooboshi H, Sato Y, Uchimaara H, Fujishima M (1991) Age-related vulnerability to cerebral ischemia in spontaneously hypertensive rats. Stroke 22: 1414–1418

Correspondence: A. Unterberg, M. D., Department of Neurosurgery, Rudolf Virchow Medical Center, Free University of Berlin, Augustenburger Platz 1, D-13353 Berlin, Federal Republic of Germany.

Acta Neurochir (1994) [Suppl] 60: 434–436
© Springer-Verlag 1994

Differences in Wound Healing Pattern Between Mature and Immature Brain Behavior of Extravasated Serum Protein

M. Suzuki, O. Motohashi, A. Nishino, K. Umezawa, N. Shida, and **T. Yoshimoto**

Division of Neurosurgery, Institute of Bran Diseases, Tohoku University School of Medicine, Sendai, Japan

Summary

To clarify the fate of extravasated serum protein and tissue reaction following blood brain-barrier breakdown in mature and immature brain, we produced cryogenic injury in the cortices of adult and post natal day 2 rats, and immunohistochemical examination with GFAP, vimentin and albumin, endogenous tracer of serum proteins, antibodies. Mature and immature brain showed the same histological changes by day 3. However, fibrotic scar, cyst formation and GFAP, vimentin-positive astrocytosis were main features in mature brain, fusion of adjoining cortical plate without scar and astrocytosis was typical in the immature brain. In adult rats, presence of albumin was observed near the lesion on day 1, evidently extending to the contralateral hemisphere on day 7 and localized again around scar and cyst on day 14. In the P2 rats, albumin was present in both hemispheres on day 1, but was localized to molecular layer periventricular region and choroid plexus on day 7. No albumin was detected on day 14. These results suggest that rapid spreading and clearance of extravasated serum proteins may take place in the immature brain and this may be deeply involved in the characteristic wound healing pattern.

Keywords: Aging; blood brain barrier; albumin; wound healing.

Introduction

The stage of development in the central nervous system (CNS) is known to influence the prognosis of injury such as trauma and cerebral ischemia. We previously demonstrated differences in wound healing pattern between mature and immature brain following cryogenic injury in rat cortices[5,6]. There were several histological features of wound healing in neonate rats following cryogenic injury, i.e., fusion of the necrotic zone resembled human micropolygyria, lack of fibrous scar formation and less astrocytic reaction. One of the sequelae after CNS injury is brain edema because of BBB impairment, and it could influence the histological changes and the result of the CNS injury. Serum components in the edema fluid is also thought to not only increase intracranial pressure, but exaggerate wound healing process of CNS tissue. The development of BBB and the distribution of serum protein in the developing brain have been studied intensively[1,3,4]. However, there is no investigation to clarify the chronological changes of the condition of BBB and the distribution of extravasated serum protein. Therefore, we examined the behavior of extravasated serum albumin, the most rich serum component and an ideal endogenous tracer, by immunohistochemical techniques in the same cryogenic injury model, and compared the behavior of immature and mature brain up to four weeks. Comparative studies between the distribution of albumin and histological changes, especially astrocytic reaction, were also carried out.

Materials and Methods

Sprague-Dawley rats were used. Cryogenic injury, reaching the deep cerebral cortex, was caused by a copper cylinder cooled by liquid nitrogen according to the previous studies[5,6]. The probe was contacted for 90 seconds for adult animals and 5 seconds for P2 animals. The lesion was produced at postnatal day 2 and 12 weeks (adult). Nembutal were used for adult animals as an anesthetic and hypothermia for P2 animals. Animals were sacrificed by perfusion with 4% paraformaldehyde under ether inhalation at 1, 3, 5, 7, 14 and 28 days post injury (PI). The brains were removed and 6-μm paraffin sections were made. Following deparaffinization, blocking of endogenous peroxidase activity and non-specific binding, the sections were incubated with the appropriate primary antibodies at 4°C for 12 hours, and then processed along with avidin-biotin complex method. The primary antisera used were rat albumin (Cappel polyclonal, 1 : 1000), glial fibrillary acidic protein (GFAP) (DAKO polyclonal, 1 : 1000) and vimentin (DAKO monoclonal, 1 : 100). Sections were also stained by hematoxylin-eosin (HE).

Results

Histological Changes

A well-delineated coagulative lesion was consistently produced on the cerebral cortices in both adult and P2 rats on day 3 PI (Fig. 1). Although infiltration by macrophages-monocytes and proliferation of capillaries were observed, these reactions were less intense in P2 animals. On day 7 PI, adult animals showed the lesion filled with young fibroblasts, capillaries and inflammatory cells, on the other hand, resolution of inflammatory reactions was considerable and subsequent fusion of adjoining cortical plates had already took place with the formation of a microsulcus in P2 animals (Fig. 1). On day 14 PI, organization of thick fibrotic scar and cyst formation were noted in adults animals, and there was no scar formation in P2 animals (Fig .1) . Immunohistochemical examination revealed that GFAP-positive reactive astrocytes were present around the lesion, more dense distribution in the deeper part, on day 7 PI in the adult animals. On day 28 PI, GFAP-positive astrocytes still existed beneath the fi-

brotic scar. The distribution of vimentin-positive astrocytes was relatively narrow. Although, in P2 animals, germ cell layer was stained with GFAP and vimentin up to day 28 PI, and the intensity decreased gradually, there was no obvious GFAP and/or vimentin-positive astrocytes without those in the vicinity of the microsulcus on day 28 PI.

Behavior of Albumin

In adult animals, albumin reaction products were observed in the adjacent grey and white matter of the lesion, and partly extending to the contralateral hemisphere through the corpus callosum on day 1 PI. These distribution pattern evolved by day 5 to 7 PI. On day 14 PI, reaction products were confined again to the gray matter around the cyst and to the bilateral subcortical white matter. On day 28 PI, the intensity decreased further. In P2 animals, reaction products widely spread through the bilateral cortices and the deep structures even on day 1 PI. On day 3 PI, those were confined to the subcortical white matter, decreased the intensity on

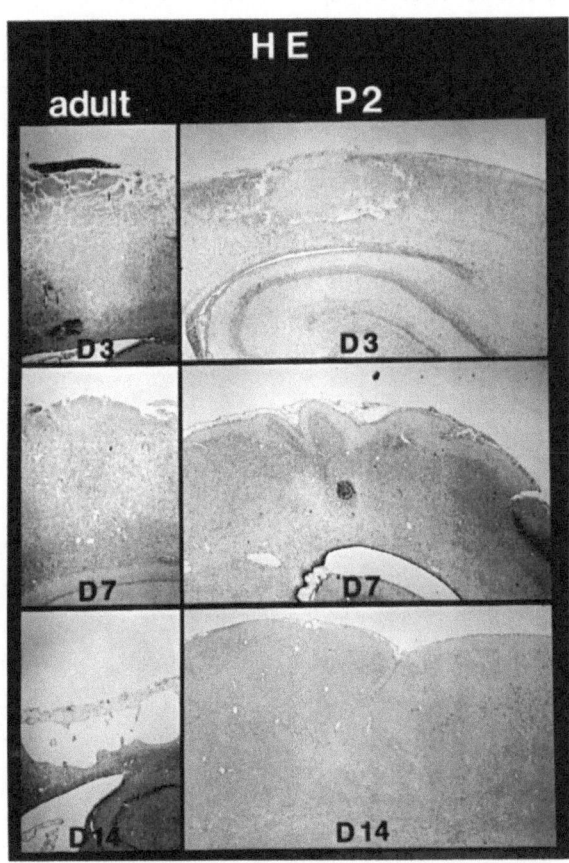

Fig. 1. Hematoxylin-eosin preparation. Left: adult rats; right: P2 rats. Upper: day 3 post injury; middle: day 7; lower: day 14

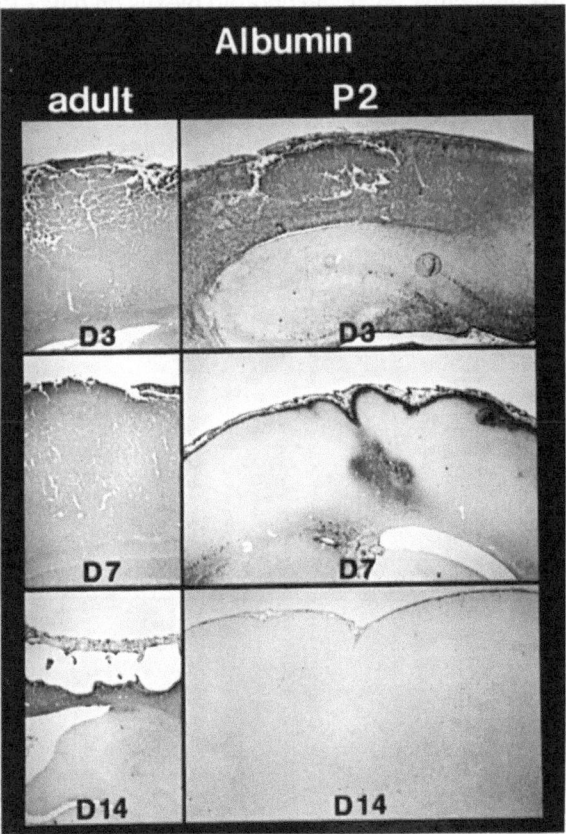

Fig. 2. Albumin immunohistochemistry. Left: adult rats; right: P2 rats. Upper: day 3 post injury; middle: day 7; lower: day 14

day 7 PI, and diminished on day 14 PI (Fig. 2). The characteristic features of P2 animals in the edema resolution stage was rapid disappearance of reaction product in the vicinity of the lesion, and accumulation of those in the molecular layer, periventricular region and choroid plexus on day 7 PI.

Discussion

In vertebrates, BBB is well established and serum proteins are usually withheld from entering the CNS[1,3,4]. This investigation disclosed a great difference between mature and immature CNS in spreading and resolution of extravasated albumin, once BBB was disintegrated. Rapid spreading of albumin into brain parenchyma in P2 animals might be a reflection of a morphological features at this stage. Because some neurons are still migrating, extracellular space is wide, nerve connections including synapses and neuritogenesis are scanty, and myelination has not been initiated[6]. Consequently, edema fluid containing albumin can freely spread into the parenchyma without resistance. Although further studies are necessary to clarify the mechanism of accumulation of albumin in the molecular layer and choroid plexus on day 7 PI, those findings suggest the enhancement of elimination of albumin from brain parenchyma into cerebro-spinal fluid (CSF) and retrograde transport of albumin from CSF to blood respectively. Various substances and mediators are assumed to exist in the edema fluid. In this context, rapid resolution of serum albumin might be one of the explanations for the lesser activation of astrocytes and the lack of fibrotic scar in the immature brain.

References

1. Dziegielewska KM, Hinds LA, Mollgard K, Reynolds ML, Saunders NR (1988) Blood-brain, blood-cerebrospinal fluid and cerebrospinal fluid-brain barriers in a marsupial during development. J Physiol 403: 367–388
2. Liu HM, Sturner WQ (1988) Extravasation of plasma proteins in brain trauma. Forensic Sci Int 38: 285–295
3. Mollgard K, Dziegielewska KM, Saunders NR, Zakut H, Soreq H (1988) Synthesis and localization of plasma proteins in the developing human brain. Dev Biol 128: 207–221
4. Risau W, Wolburg H (1990) Development of the blood-brain barrier. TINS 13: 174–178
5. Suzuki M, Choi BH (1990) The behavior of the extracellular matrix and the basal lamina during the repair of cryogenic injury in the adult rat cerebral cortex. Acta Neuropathol (Berl) 80: 355–361
6. Suzuki M, Choi BH (1991) Repair and reconstruction of the cortical plate following closed cryogenic injury to the neonatal rat cerebrum. Acta Neuropathol (Berl) 82: 93–101

Correspondence: Michiyasu Suzuki, M. D., Division of Neurosurgery, Institute of Brain Diseases, Tohoku University School of Medicine, Seiryo-machi 1-1, Aoba-ku, Sendai 980, Japan.

Acta Neurochir (1994) [Suppl] 60: 437–439

Dynamics of Posttraumatic Brain Swelling Following a Cryogenic Injury in Rats

G.-H. Schneider, S. Hennig, W. R. Lanksch, and A. Unterberg

Department of Neurosurgery, Rudolf Virchow Medical Center, Free University of Berlin, Berlin, Federal Republic of Germany

Summary

For evaluation of novel therapeutic regimens against brain edema in experimental models, it is important to know the temporal profile of edema formation. Since development of brain swelling and edema following a cryogenic injury is poorly documented in rats, these parameters have been investigated in this study.

27 Sprague-Dawley rats were used. Three rats (controls) were sham-operated. Their brains were removed 15 min after sham-operation. In all other animals a right parietal cryogenic lesion was applied in ketamine-xylazine anesthesia. Systemic blood pressure was monitored during the perioperative period. Brains were removed in defined intervals following cryogenic injury (0.25, 0.5, 1, 3, 6, 12, 24, and 72 h; n = 3 each). Hemispheres were then separated and weighed for determination of brain swelling, dried and weighed again. Cerebral water content was calculated as the difference of hemispheric wet- and dryweight.

Posttraumatic hemispheric swelling reached its maximum (7.9 ± 0.4%) as early as 12 h post trauma. During the first hour after injury, brain swelling showed a steep, linear increase to 3.9%, i.e. swelling amounted to 50% of its maximum within one hour. 6 h post trauma swelling was 5.7%. After 12 h brain swelling started to decline. 72 h post cryogenic injury, 6.1% hemispheric swelling were found. Development of brain swelling was paralleled by a linear increase in brain water content of the lesioned hemisphere. The maximum of hemispheric water content was seen 24 h post trauma. In non-lesioned hemispheres only a marginal transient increase in cerebral water content was observed.

Thus, in rat models the rapid development of brain swelling and edema has to be considered when planning therapeutic protocols.

Keywords: Cold injury; brain swelling; cerebral edema; time course; rats.

Introduction

A rational treatment for traumatic brain swelling and edema, a major determinant of mortality and outcome following cerebral injury, still remains to be found. In evaluation of novel therapeutic regimens against brain swelling the cortical freezing lesion is often used modelling disruption of the blood-brain barrier following cerebral trauma. When studying the effect of new therapeutic agents it is paramount to know the time course of posttraumatic brain swelling and edema. In cats, Pappius found that hemispherical swelling following a cryogenic cortical lesion amounted to its maximum 48 hours following trauma[6]. Time course of hemispheric swelling and edema in rats, however, is poorly documented so far. The present study was therefore designed to investigate the development of brain swelling and edema following a cortical freezing lesion in rats as a basis for further therapeutical studies.

Methods

The experiments were performed with 27 male Sprague-Dawley rats (506 ± 9 g). In ketamine-xylazine anesthesia a craniectomy was made over the right parietal region and a standard cortical freezing lesion was applied to the cortex through the intact dura[1,3,7]. For the lesion a copper probe 5 mm in diameter cooled to –60°C by a dry-ice/acetone mixture was used which was lowered stereotactically onto the exposed dura for 30 seconds. During the perioperative period systemic blood pressure was monitored via a cannula in the tail artery.

Three rats (controls) were sham-operated and had their brains removed 15 min after exposing the dura. All other animals were sacrificed in defined intervals following the cryogenic injury (0.25, 0.5, 1, 3, 6, 12, 24, and 72 h post trauma; n = 3 each). After removal the brains were meticulously separated with the help of an operating microscope. Therafter hemispheres were weighed for determination of brain swelling, dried for 48 h at 100°C and weighed again. Cerebral water content was calculated as the difference of hemispheric wet- and dryweight.

Results

The time course of hemispheric swelling due to the cryogenic injury is given in Fig. 1 (rats = filled circles). Swelling of the traumatized hemisphere was already measurable within half an hour following the cortical freezing lesion (2.9 ± 0.1%) and continued to increase until reaching its maximum (7.9 ± 0.4%) as early as 12 h post injury. During the first hour after injury brain swelling showed a steep, linear increase to 3.9% (y = 3.6x + 0.5, r = 0.96, p < 0.01), i.e. swelling amounted to 50% of its maximum within one hour. 6 h post trauma swelling was 5.7 ± 0.9%. Resolution of brain swelling started 12 h after the trauma and could be approximated by linear regression: y = −0.03x + 8.3, r = 0.99, p < 0.001. 72 h post cryogenic injury, 6.2 ± 0.7% hemispheric swelling was found.

Development of brain swelling was paralleled by a linear increase in brain water content of the traumatized hemispheres during the first 6 h (y = 0.2x + 78.6, r = 0.91, p < 0.001; Fig. 2). In contrast to brain swelling the maximum of posttraumatic hemispheric water con-

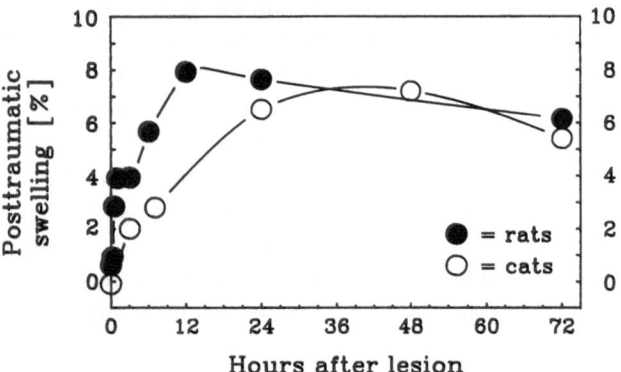

Fig. 1. Development of posttraumatic brain swelling in rats and cats. Formation of posttraumatic hemispheric swelling in rats is much faster than in cats reaching its maximum already 12 h after trauma while cats show peak swelling 48 h post cryogenic injury. (Cat data taken from Pappius and McCann, Arch Neurol 20: 207–216, 1969)

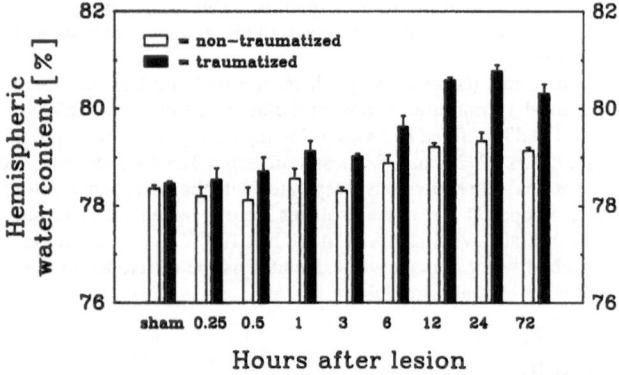

Fig. 2. Time course of hemispheric water content following a right parietal cryogenic injury in rats. Development of posttraumatic brain swelling is paralleled by a linear increase of water content of the lesioned hemisphere during the first 6 h. The highest posttraumatic water content is found after 24 h. In non-lesioned hemispheres a marginal rise in water content was observed after 6 h and later

tent was seen after 24 h (80.8 ± 0.1 ml/100 g f.w., p < 0.01; Fig. 2). In non-lesioned hemispheres a marginal transient rise in cerebral water content was observed after 6 h and later (Fig. 2).

Discussion

In contrast to Pappius' findings in cats[6] hemispheric swelling following a cortical freezing lesion reached its maximum much faster in rats (Fig. 1; rats = filled circles, cats = open circles). Time course of cold-induced brain swelling is obviously species-specific. While posttraumatic swelling in rats was already 50% of its maximum one hour after trauma, cold-induced brain swelling in cats reached only 30% of its maximum after 3 h. Peak brain swelling in rats was observed as early as 12 h following the cryogenic injury, while in cats the maximum of posttraumatic swelling was found after 48 h (Fig. 1). In concordance with our findings of fast edema development in rats, Csanda observed the highest Evans blue extravasation as an indicator of the extent of blood-brain barrier damage 2 h after cryogenic injury in rats. After 8 h repair of the blood-brain barrier disruption had already started[2].

A possible reason for the different time course of formation of traumatic hemispheric swelling and edema between rats and cats could be the difference in white/gray-matter ratio and tissue architecture below the cortical lesion. Kuroiwa studying the distribution of vasogenic edema fluid in cats found a regional difference in edema accumulation depending on local structural characteristics[4]. Since spreading of brain edema is also determined by the size of blood-brain barrier damage[5], the observed species-specific dynamics of brain swelling may as well be due to the different ratios of lesion probe surface to brain volume of rats and cats.

Taken together, the rapid development of brain swelling and edema in rats has to be considered when planning experimental protocols.

Acknowledgement

The excellent technical assistance of Mrs. J. Kopetzki is gratefully acknowledged.

References

1. Berenberg P v, Unterberg A, Schneider G-H, *et al* (1994) Treatment of traumatic brain edema by multiple doses of mannitol. Acta Neurochir (Wien) [Suppl] 60: 531–533
2. Csanda E, Komoly S, Major O (1985) The effect of lymphatic blockage on resolution of vasogenic brain edema. In: Inaba Y,

Klatzo I, Spatz M (eds) Brain edema VI. Springer, Berlin Heidelberg New York Tokyo

3. Klatzo I, Piraux A, Laskowski EJ (1958) The relationship between edema, blood-brain barrier and tissue elements in local brain injury. J Neuropath Exp Neurol 17: 548–564

4. Kuroiwa T, Yokofujita J, Kaneko H, *et al* (1990) Accumulation of oedema fluid in deep white matter after cerebral cold injury. Acta Neurochir (Wien) [Suppl] 51: 84–86

5. Pappius HM, Gulati DR (1963) Water and electrolyte content of cerebral tissues in expermentally induced edema. Acta Neuropathol (Berl) 2: 451–460

6. Pappius HM, McCann WP (1969) Effects of steroids on cerebral edema in cats. Arch Neurol 20: 207–216

7. Schneider G-H, Unterberg A, Lanksch WR (1993) Effect of the 21-aminosteroid U-74389F on brain edema following a cryogenic lesion in rats. Adv Neurosurg 21: 299–301

Correspondence: G.-H. Schneider, M. D., Department of Neurosurgery, Rudolf Virchow Medical Center, Free University of Berlin, Augustenburger Platz 1, D-13353 Berlin, Federal Republic of Germany.

Acta Neurochir (1994) [Suppl] 60: 440–442
© Springer-Verlag 1994

Blood-Brain Barrier, Cerebral Blood Flow, and Cerebral Plasma Volume Immediately After Head Injury in the Rat

H. Nawashiro, K. Shima, and **H. Chigasaki**

Department of Neurosurgery, National Defense Medical College, Namiki, Tokorozawa, Saitama, Japan

Summary

The purpose of the present study was to determine blood-brain barrier (BBB) permeability, regional cerebral blood flow (rCBF) and regional cerebral plasma volume (rCPV) in the period immediately after head injury, and thereby to evaluate the effects of vascular factors in the pathophysiology of traumatic brain injury.

Male Sprague-Dawley rats (350–450 g) anesthetized with 1.0–1.5% halothane were subjected to an impact acceleration closed head injury at the moderate level. BBB permeability (n = 5), rCBF (n = 8) and rCPV (n = 9) were measured by quantitative autoradiographic techniques using ^{14}C-alpha-aminoisobutyric acid (AIB), ^{14}C-iodoantipyrine and ^{14}C-sucrose, respectively. Intravenous administration of each radiotracer was simultaneous with the traumatic impact.

At 10 min after injury, BBB permeability, the transfer constant for AIB, was less than 0.1 ml/kg/min for all regions except for those with a relatively leaky BBB. At 30 s after injury, a significant and heterogeneous increase in rCBF was observed at 9 subcortical regions (p < 0.05). RCPV increased significantly in the frontal cortex, parietal cortex, thalamus, and hypothalamus (p < 0.05).

In our closed head injury model without severe hypertension, BBB disruption did not occur immediately after trauma. Vascular responses in the period immediately after trauma may result from the derangement of cerebral autoregulatuion.

Keywords: Head injury; blood-brain barrier; cerebral blood flow; plasma volume.

Introduction

Cerebrovascular alterations immediately after traumatic brain injury may play an important role in the development of secondary brain damage and may be one of the major factors governing the outcome. In the clinical setting, it is impossible to measure cerebral blood flow (CBF) soon after injury, and it is difficult to assess to what extent the cerebrovascular alterations are primary or secondary. The purpose of the present study was to determine blood-brain barrier (BBB) permeability, regional CBF, cerebral vascular resistance (CVR), and regional cerebral plasma volume (rCPV) in the period immediately after trauma, and thereby to evaluate the effects of vascular factors in the pathophysiology of closed head injury. In the present study, a new experimental model for closed head injury introduced by Marmarou *et al.*[1] was used.

Materials and Methods

Male Sprague-Dawley rats weighing 350 to 450 g about 12 weeks of age), maintained on a normal diet, were used in the present study. Each animal was anesthetized with a mixture of 1% halothane, air, and oxygen delivered by a closely fitting face mask. The femoral arteries and veins were catheterized for arterial blood sampling and continuous monitoring of arterial blood pressure, and for intravenous injection of radioisotopes. After reflection of the scalp, a stainless steel disc, 10 mm in diameter, was secured on the skull of the area midway between the bregma and lambda. Following catheterization, the animals were immobilized by a loose-fitting plaster cast around the hindquarters to facilitate blood sampling for autoradiographic study. The physiological condition of each rat was assessed by measurements of rectal temperature, blood pressure, hematocrit and arterial blood gases. Soon after discontinuing halothane inspiration, each animal was mounted in a prone position on a foam bed and was placed under hollow plexiglass tubing. A blunt head injury was delivered by dropping a brass weight (450 g) through the tubing from a height of 1 meter, which was adjusted to represent a moderate injury level. Following head injury, autoradiographic studies were carried out. BBB permeability was measured autoradiographically using ^{14}C-alpha-aminoisobutyric acid[2] (^{14}C-AIB, New England Nuclear Corp., Boston; specific activity 1.48-2.22 Gbq/mmol). ^{14}C-AIB (3.7 Mbq/one rat), dissolved in 1 ml of saline and adjusted to pH 7.4, was administered simultaneously with the traumatic impact. Timed arterial blood samples were rapidly taken and plasma radioactivity was measured by a liquid scintillation counter. At 10 min, each rat was decapitated. The brain was rapidly removed, frozen and cut into 20-μm-thick sections. Serial histology slices were picked up and processed for light microscopic analysis to match autoradiograms with the histology. Each of the sections were mounted on glass coverslips, dried on a hot plate and placed in an X-ray cassette with calibrated ^{14}C-methylmethacrylate standards (Amersham Corp., Arlington Heights, IL) against X-ray film (SB-S, Kodak) for at least 7 days. The optical densities of cerebral structures were determined using a microdensitometer (PDS-15, Sakura). Traumatic changes of BBB permeability were also assessed by intravenous injection of Evans blue (2% in saline) at 5 min pre-injury. RCBF was measured using ^{14}C-iodoantipyrine (^{14}C-IAP, specific activity 1.48–2.22 Gbq/ mmol, New England Nuclear Corp., Boston) as described by Sakurada et al.[6]. Simultaneous with the delivery of the impact, ^{14}C-IAP (3.7 Mbq/kg) infusion was started. The time for isotope injection and arterial sampling was 30 s. At 30 s the rat was decapitated, and the brain was rapidly removed, frozen and cut into 20-μm-thick sections. Each of the sections were mounted on glass coverslips, dried on a hot plate and placed in an X-ray cassette with calibrated ^{14}C-methylmethacrylate standards (Amersham Corp., Arlington Heights, IL) against X-ray film (SB-5, Kodak) for at least 7 days. The optical densities of cerebral structures were determined using a microdensitometer (PDS-15, Sakura). RCBF was calculated using the equation derived by Kety and developed by Sakurada et al.[6] in 24 anatomically discrete brain regions. CVR was calculated by dividing mean arterial blood pressure (MABP) by CBF and was expressed as mmHg/ml/100 g brain/min. RCPV was determined as the ^{14}C-sucrose space[5], (dpm/g brain)/(dpm/ml plasma), soon after intravenous ^{14}C-sucrose injection. Three point seven Mbq of ^{14}C-sucrose was administered simultaneously with the traumatic impact. Samples of arterial blood were withdrawn to determine the average tracer concentration and hematocrit. Each rat was decapitated at 30 s. The brain was removed quickly and then frozen and cut in the same manner as described for the CBF studies.

Results and Discussion

The transfer constant for AIB was less than 0.1 ml/ kg/min for all brain regions except those with a relatively leaky BBB. There was no extravasation of Evans' blue in the brains of all animals examined at 30 s or 10 min after trauma. RCBF data in the groups are presented in Fig. 1. All brain regions except the parietal cortex which was underneath the impact site, showed a varied increase in rCBF, up to 115–226% of control values. Compared with mean values of sham-operated control animals (n = 4), trauma resulted in significant rCBF increases of 88% in amygdala (p < 0.01), 63% in

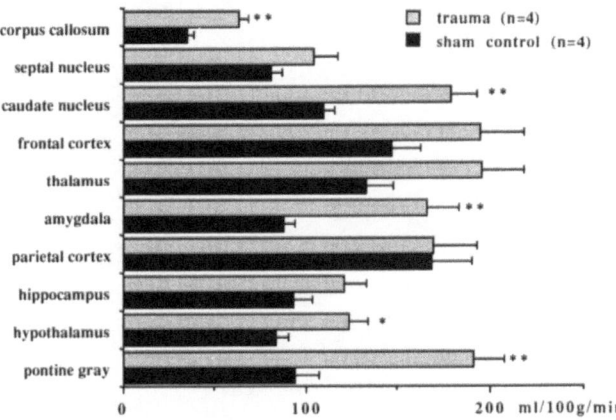

Fig. 1. Regional cerebral blood flow values (means±S.E.M.) at 10 regions examined. Signiflcant (* p < 0.05, ** p < 0.01 by unpaired Student's t - test) increases immediately after trauma were observed at 5 regions compared to control values

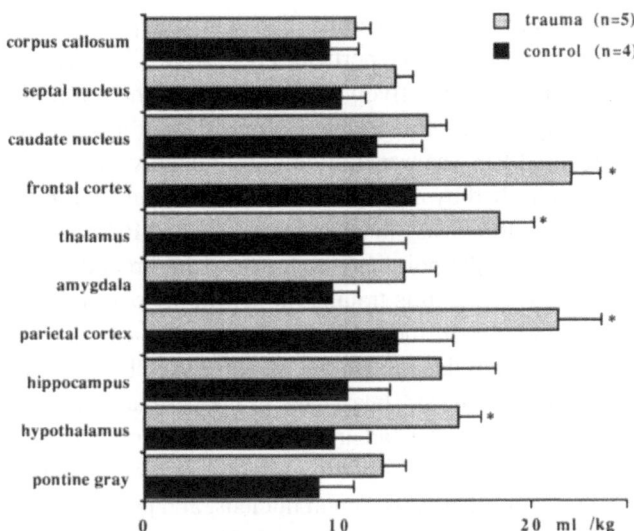

Fig. 2. Regional cerebral plasma volume (residual ^{14}C-sucrose space) at 10 regions examined. Significant (* p < 0.05 by unpaired Student's t - test) increases immediately after trauma were observed in the frontal cortex, parietal cortex, thalamus, and hypothalamus

caudate-putamen (p < 0.01), 105% in pontine gray (p < 0.01), 48% in hypothalamus (p < 0.05), 84% in mamillary body (p < 0.05), 126% in cerebellar cortex (p < 0.05), and 83–91% in white matter (p < 0.01). RCVR was decreased significantly in all brain regions except the parietal cortex and hippocampus undemeath the impact site. All brain regions showed some increase in rCPV at 30 s after trauma, and the increase was particularly significant in the frontal cortex, parietal cortex, thalamus, and hypothalamus (p < 0.05) (Fig. 2).

The blood-brain barrier and cerebral autoregulation are two important mechanisms for preserving brain homeostasis. In the present study, we investigated immediate cerebrovascular responses to trauma using a newly developed impact-acceleration injury model. The blood pressure surge association with this model was not as dramatic as that seen in fluid-precussion injury[3]. The BBB remains intact for AIB and Evans' blue, but pressure auto-regulation was impaired once resulted in widespread hyperperfusion immediately after trauma.

The present study demonstrated that widespread but heterogeneous alterations in rCBF occurred immediately after closed head injury in the rat with a pronounced increase in pontine gray, hypothalamus, amygdala, caudate nucleus, and white matter. This uneven increase in rCBF could be a reflection of the increased energy demands coupling with increased neuronal activity. However the evidence of decreased local cerebral glucose utilization at 40 min post-injury[8] suggests that luxury perfusion occurred. The increase in rCPV was relatively proportional to the capillary density in the normal physiological condition[4]. The elevated BP was transmitted unmitigatedly into every capillary bed by a traumatically induced vasoparalysis and resulting vascular engorgement. MABP was not elevated to the point at which a pressure breakthrough would occur in this trauma model. In the parietal cortex, underneath the impact site, the increase of CBF was minimal despite a large increase in rCPV. Vascular congestion may take place in the parietal cortex.

In our closed head injury model, regional hyperemia occurred 15 min after trauma in the substantia nigra, hypothalamus, septal nucleus, and pontine gray, while there were decreases of rCBF in many other regions examined[7]. The different time course of rCBF may reflect the regional variety of neuronal activity, including vasomotor activity. The heterogeneous increases in rCBF immediately after trauma may represent a specific pathology of concussive brain injury. The mechanisms for the varied regional and temporal increases in CBF are unknown and require further investigation.

References

1. Abd-elfattah Foda MA, Marmarou A (1993) A new model of diffuse brain injury in rats: morphological characterization. J Neurosurg: in press
2. Blasberg RG, Fenstermacher JD, Patlak CS (1983) Transport of a-aminoisobutyric acid across brain capillary and cellular membranes. J Cereb Blood Flow Metab 3: 8–32
3. DeWitt DS, Jenkins LW, Wei EP, *et al* (1986) Effects of fluid-percussion brain injury on regional cerebral blood flow and pial arterial diameter. J Neurosurg 64: 787–794
4. Klein B, Kuschinsky W, Schröck H, Vetterlein F (1986) Interdependency of local capillary density, blood flow, and metabolism in rat brains. Am J Physiol 251: H1333–H1340
5. Ohno K, Pettigrew KD, Rapoport SI (1978) Lower limits of cerebrovascular permeability to nonelectrolytes in the conscious rat. Am J Physiol 235: H299–H307
6. Sakuradas O, Kennedy C, Jehle J, *et al* (1978) Measurement of local cerebral blood flow with iodo-[^{14}C] antipyrine. Am J Physiol 234: H59–H66
7. Shima K, Koshimae N, Wakoh N, Chigasaki H, Marmarou A (1990) Cerebral circulation following expe ental head injury. Advances in Experimental Neurotrauma Research 2: 53–57
8. Shima K, Ohashi K, Wakoh N, Chigasaki H, Marmarou A (1991) Cerebral blood flow and metabolism following experimental head injury. Advances in Experimental Neurotrauma Research 3: 84–87

Correspondence: Hiroshi Nawashiro, M. D., Department of Neurosurgery, National Defense Medical College, 3–2, Namiki, Tokorozawa, Saitama 359, Japan.

Acta Neurochir (1994) [Suppl] 60: 443–445
© Springer-Verlag 1994

Long Term Exposure to Heat Reduces Edema Formation After Closed Head Injury in the Rat

E. Shohami[1], **M. Novikov**[1], and **M. Horowitz**[2]

Departments of [1] Pharmacology and [2] Physiology, Hadassah Schools of Pharmacy, Medicine and Dental Medicine, Hebrew University, Jerusalem, Israel

Summary

Cerebral edema is one of the major consequences of head trauma (HT); its evolution may cause secondary ischemia and neuronal damage. In a closed head injury model in rats, we have shown BBB disruption and edema formation during the post traumatic period. We have previously shown that chronic exposure to moderate heat improves clinical outcome of rats subjected to HT. Long term exposure to heat results in the achievement of a stable acclimated state, characterized by a lower metabolic rate and improved heat tolerance. In the present study, we investigated the effect of chronic exposure to heat on edema formation after HT. Rats were held at 24°C (CON) or 34°C (ACC) for one month. Injury was then induced under ether anesthesia by a weight drop device. Four or 48 hours later, they were sacrificed for evaluation of BBB integrity (Evans blue, EB, extravasation) or edema formation (specific gravity, SG, or percent water). We found that EB uptake by the contused hemisphere was 6 fold lower in the ACC rats as compared to CON (p < 0.001). Furthermore, edema measured at 48 h by both SG and percent water methods was significantly lower in the acclimated rats (p < 0.01). We suggest that heat acclimation offers protection to rats subjected to head injury, possibly by reduction of plasma proteins extravasation.

Keywords: BBB; head trauma; heat acclimation; brain edema.

Introduction

Traumatic injury to the brain triggers a cascade of events leading to delayed tissue edema, necrosis and impaired function. These effects are associated with the accumulation of harmful mediators as a result of activation of various biochemical pathways[7]. Such mechanisms are probably coupled to disruption of energy supply (secondary ischemia), impaired energy metabolism and imbalanced ionic homeostasis. Studies on the ischemic cardiac muscle of rats show that chronic exposure to high ambient temperature (heat acclimation), provides protection during ischemia and reperfusion, associated with better ATP preservation[3].

Heat acclimation status is accompanied by lowered basal metabolic rate[9].

In the past several years we have developed and studied in our laboratory a rat model of closed head injury[5]. During the post traumatic period, we showed edema formation following opening of the blood brain barrier (BBB). The edema peaked at 24–48 h and spontaneously dissolved by 7–10 days. We also showed a severe neurological deficit, which gradually improved during the post traumatic period[6]. In view of the findings on the beneficial effects of heat acclimation on the ischemic heart, we hypothesized, that heat acclimation may offer protection against head trauma insult. In this investigation the evolution of edema during the post traumatic period and the BBB function after closed head injury in acclimated rats were examined and compared to those, in non-acclimated rats.

Materials and Methods

Male rats (Sabra strain) weighing 200–220 g were used. The animals were divided into two groups: 1) Control rats (CON) were held at an ambient temperature of 24 ± 1°C. 2) Heat acclimated (ACC) were exposed to 34 ± 1°C for one month. This period was found to be sufficient for achievement of a stable acclimated state, characterized by a lower basal metabolic rate, lower heart beat rate and improved heat tolerance. A considerable knowledge on this acclimation model has been accumulated in our laboratory[2,3].

Head trauma (HT) was induced under ether anesthesia, confirmed by loss of corneal and pupillary reflexes, as previously described[5]. In brief, trauma was induced by a well calibrated weight drop device which falls over the exposed skull covering the left cerebral hemisphere 1–2 mm lateral to the midline, in the mid-coronal plane. The ACC rats were returned to the climatic chamber after the impact. Sham rats kept at 24 or 34°C, were anesthetized, their skulls were exposed but no trauma was induced. They were decapitated at the same time as the traumatized rats and served as controls.

E. Shohami *et al.*

Specific gravity (SG) measurements: The SG of brain tissue was determined according to the method of Marmarou *et al.*[4]. Cortical tissue segments (~ 20 mg), were placed on top of linear gradient columns of bromobenzene and kerosene, as described in detail[5]. *Water content:* Another approach to measure edema, which does not distinguish between cytotoxic and vasogenic edema is to determine the tissue water content. Water percentage was calculated from wet and dry weight measurements. Tissue was taken from the same animals for both SG and water content measurements.

The integrity of the BBB was evaluated according to the method of Uyama *et al.*[8]. Evans blue dye (2%), was given iv and allowed to circulate for at least 60 min. Before sacrifice, the chest wall of the rat was opened under ether anesthesia and the animals were perfused with saline to remove the intravascularly dye. The brain was removed, the hemispheres separated, both hemispheres were weighed, homogenized and dye was extracted and quantified using a Perkin-Elmer LS-5 fluorospectrophotometer, (excitation at 620 nm and emission at 680 nm).

Clinical evaluation was also performed using a set of criteria, testing reflexes and motor function. Neurological severity score (NSS), which gives 1 point for failure in performing task[6], is zero for sham rats, and maximal at 1h after HT and declines spontaneously thereafter. The NSS of severely injured rats ranges between 15–20 point. Rats gaining 20–24 points will die within a few hours after HT.

Statistics: Values are expressed as mean ± SEM of at least n = 8 rats per group. One-way ANOVA was performed to compare the SG, percent water or amount of dye in the various experimental groups, followed by Student's t-test. Values of NSS are expressed as medians and range, and non-parametric Wilcoxon's rank test was performed for comparisons between groups.

Results and Discussion

The initial NSS, measured at 1 hour after HT was similar in both ACC and CON rats (17.5 range: 16–20 and 16.6 range: 14–17). This indicates that both groups were subjected to similar trauma severity. Figure 1 summarizes the NSS of CON and ACC rats, evaluated 48 h after HT. These results show that during the post-traumatic period the spontaneous recovery of the ACC rats was significantly better, as shown by the lower NSS (p < 0.001).

Figure 2 depicts water content of both hemispheres of CON and ACC rats. It is clear that while the non-contused (right) hemisphere maintained its normal water content, the contused hemisphere had a significantly higher water level after HT. The increase of water content in ACC was significantly lower than in CON. SG measurements revealed similar effect of acclimation to heat, and its decrease as a result of HT was attenuated. (SGs 1.0364 ± 0.0007 and 1.0389 ± 0.0013 for CON and ACC respectively (p = 0.0057) as compared to 1.0445 ± 0.0005 for sham). Thus the extent of edema is remarkably lower in ACC rats. It should be noted that basal water content did not change during the acclimation regimen. Vasogenic edema is associated with increased permeability of the BBB, we

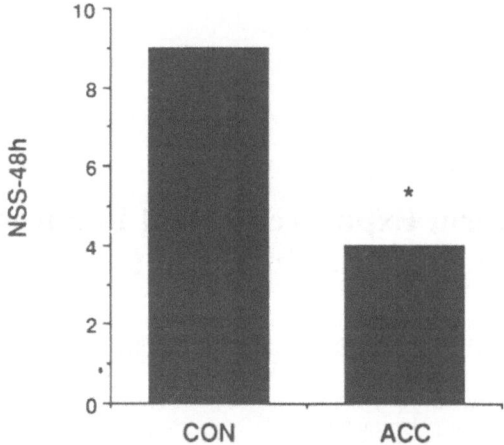

Fig. 1. The NSS – neurological severity score of CON and ACC, evaluated 48 h after HT. Sham rats have a score of 0, and the maximal score is 20. * p < 0.01 as compared to CON

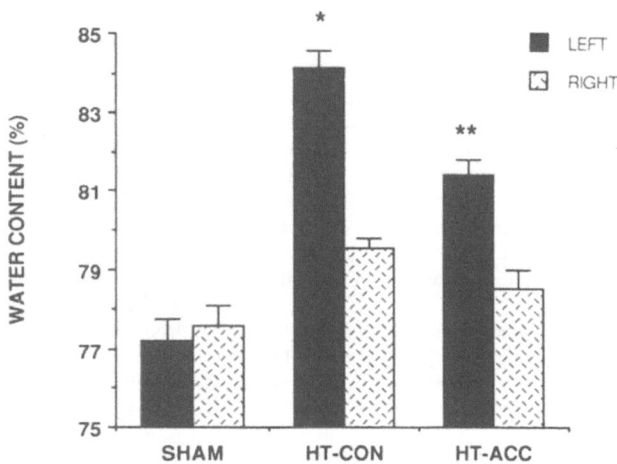

Fig. 2. Water content (%) in Sham, CON and ACC rats 48 h after HT. Wet and dry weight were measured for the contused (left) and contralateral (right) hemispheres. * p < 0.01 as compared to SHAM. ** p < 0.01 as compared to CON

therefore evaluated its integrity 4 h after HT. At that time we found the maximal impairment of BBB function[6]. The uptake of EB by contused hemispheres of ACC rats was significantly lower than that of CON rats, (50 ± 23 and 315 ± 61 ng/g tissue, p = 0.0083). Evidently, ACC rats suffered less BBB disruption as a result of the trauma than CON.

Our results demonstrate for the first time, that heat acclimation protects the BBB and reduces edema formation after head injury. The molecular or cellular events that occur during the acclimation process, leading to lower susceptibility of the BBB to injury is still not known. It is note worthy, however, that rats subjected to chronic heat, similar to our experimental conditions, showed lower basal rate of PGE_2 production by the hypothalamus[1] and lower plasma corticosterone levels, which were not elevated after LPS administration (Hadas, personal communication). Thus, we speculate that some endogenous mechanisms may be regulated during chronic exposure to high ambient temperatures so that upon stressful conditions, like those occurring after head injury, will provide neuroprotection.

Acknowledgement

The authors are indebted to Ms. Judith Shlaier for her excellent technical work.

References

1. Hadas H, Sod-Moriahi UA, Kaplanski J (1989) The effect of lipopolysaccharide on fever and thermoregulation in rats. In: Mercer JB (ed) Thermal physiology 1989. Elsevier, New York, pp 395–400
2. Horowitz M (1976) Acclimation of rats to mild heat: body water distribution and adaptability of submaxillary salivary gland. Pflugers Arch 366: 173–176
3. Levi E, Vivi A, Hasin Y, Vivi M, Navon G, Horowitz M (1993) Heat acclimation improves cardiac mechanical and metabolic performance during ischemia and reperfusion. J Appl Physiol, in press
4. Marmarou A, Tanaka K, Shulman K (1982) An improved gravimetric measure of cerebral edema. J Neurosurg 56: 246–253
5. Shapirac Y, Shohami E, Sidi A, Soffer D, Freeman S, Cotev S (1988) Experimental closed head injury in rats. Mechanical, pathophysiology and neurologic properties. Crit Care Med 16: 258–265
6. Shapira Y, Setton D, Artru AA, Shohami E (1993) Blood brain barrier permeability, cerebral edema and neurological function following closed head injury in rats. J Anasth Analg, in press
7. Siesjo BK (1988) Mechanisms of ischemic brain damage. Crit Care Med 16: 954–963
8. Uyama O, Okamura N, Yamase M, Narita M, Kawabata K, Sugita M (1988) Quantitative evaluation of vascular permeability in the gerbil brain after transient ischemia using Evans blue fluorescence. J Cereb Blood Flow Metab 8: 282–294
9. Yousef MK, Johnson OH (1975) Thyroid activity in desert rodents: a mechanism for lower metabolic rate. Am J Physiol 229: 427–431

Correspondence: E. Shohami, M. D., Department of Pharmacology, Hadassah School of Pharmacy, POB 1172, Jerusalem 91010, Israel.

Acta Neurochir (1994) [Suppl] 60: 446–448
© Springer-Verlag 1994

Metabolic Changes Following Cortical Contusion: Relationships to Edema and Morphological Changes

R. L. Sutton[1], **D. A. Hovda**[2], **P. D. Adelson**[2], **E. C. Benzel**[1], and **D. P. Becker**[2]

Divisions of Neurosurgery, [1] University of New Mexico, Albuquerque, NM, and [2] University of California, Los Angeles, CA, U.S.A.

Summary

Rats with contusion injury to the right cortex exhibited significant formation of edema 6 and 24 hours after injury which resolved by 8 days and was replaced by cavitation necrosis. The contusions produced hyperglycolysis and ischemia in the impacted cortical tissue and underlying hippocampus immediately through 30 minutes post-injury. Glucose utilization was depressed throughout the contused cortex and in ipsilateral subcortical regions, as was blood flow, at chronic (1 and 10 days) periods after injury.

Keywords: Cerebral edema; cortical contusion; glycolysis; ischemia; necrosis.

Introduction

Localized trauma to the cerebral cortex results in pronounced changes in both cerebral edema and cellular morphology. These changes have a temporal course affecting different regions of the brain depending on the severity and localization of the injury. Previous work[2,4,8,9] has demonstrated that following diffuse traumatic injury produced by fluid percussion, the cerebral cortex exhibits dynamic metabolic changes in response to disrupted neurochemistry. However, it is not known if similar or identical changes occur following cerebral contusion.

Here, we report our preliminary findings on the extent and duration of cerebral edema, metabolic dysfunction, and cortical necrosis following cortical contusion as produced with a pneumatic injury device[7]. The extent of cortical edema and necrosis was determined using quantitative planimetry. Metabolic measures included those for local cerebral glucose utilization (LCMRGLC) and local cerebral blood flow (LCBF). Animals were assessed at acute and chronic time points following the insult, with results indicating that the cortical contusion injury produces morphological and metabolic dysfunctions that differ from those produced by diffuse concussive injury.

Materials and Methods

Subjects: Forty male Sprague Dawley rats (300–350 gm) were subjected to an experimental cortical contusion and studied autoradiographically for measurements of cerebral metabolism and/or histologically to evaluate the extent of cerebral edema and necrosis.

Injury induction: All animals were placed under either ethrane (1–2%, 33% oxygen, 66% nitrous oxide) or sodium pentobarbital (50 mg/kg, i.p.) anesthesia and positioned in a stereotaxic. A craniotomy was made over the right sensorimotor cortex and a cortical contusion was induced with a pneumatic impactor (2 mm compression, 2.25 m/s velocity), as previously described[7]. Following injury, the craniotomy was covered with gelfoam and the skin sutured closed. Sham operated controls did not receive craniotomy or contusion. Animals were placed under close observation for 6 hours until they had recovered from anesthesia prior to being returned to their home cage.

Metabolic analysis of LCMRGLC and LCBF: Animals were prepared under general gas anesthesia (see above) for [^{14}C]2-deoxy-D-glucose[6] or [^{14}C]iodoantipyrine[5] autoradiography for the analysis of LCMRGLC (μmol/100 g/min) or of LCBF (ml/100 g/min), respectively. Optical density measurements were taken in 23 structures utilizing an image analysis system (JAVA; Jandel Scientific). Animals were studied immediately, 30 min, 6 hours, 1 and 10 days following surgery.

Histology and edema assessment: Additional animals sacrificed at 6 hours, 1 or 8 days after surgery were overdosed with sodium pentobarbital (100 mg/kg, i.p.) and perfused with 3% formaldehyde. Histological sections stained with thionin and acid fuchsin were examined microscopically, and the extent of cortical swelling/edema and necrosis at 11 anterior-posterior planes was quantified using planimetry methods.

Results

All injured animals developed consistently similar contusions which were localized to the right sensori-

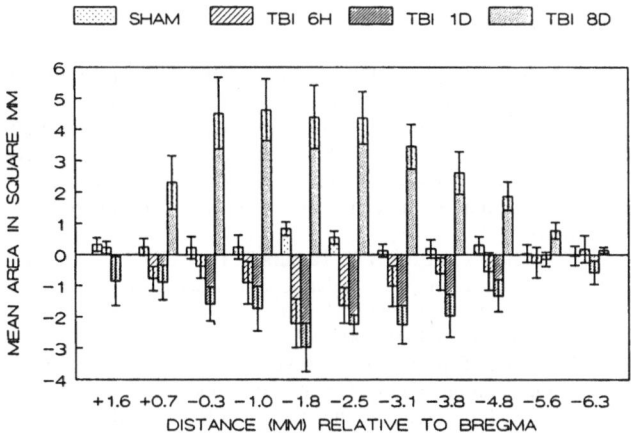

SHAM   TBI 6H   TBI 1D   TBI 8D

MEAN AREA IN SQUARE MM

DISTANCE (MM) RELATIVE TO BREGMA

+1.6 +0.7 −0.3 −1.0 −1.8 −2.5 −3.1 −3.8 −4.8 −5.6 −6.3

Fig. 1. Mean areas of left minus right (L-R) cortical mantle at 11 anterior-posterior levels. Sham operates had very little L-R asymmetry and the total area of left and right cortex did not significantly differ in these animals. Rats with contusion injury to the right sensorimotor cortex had negative values for the L-R cortical measures at 6 hours (H) and 1 day (D) after injury, reflecting the formation of swelling/edema in the injured cortex. By 8 days (D) after injury, the L-R cortical measures in contused rats were positive, reflecting the formation of cavitation necrosis in the contused cortex

motor and parietal cortex. Within the first 30 seconds following the injury, the cerebral cortex usually swelled up within the cranial defect. During the first 24 hours post-injury, the right cortical mantle exhibited an edematous response which showed a spatial distribution extending over most of the anterior-posterior length of the cortex. However, this cortical swelling was most pronounced within the tissue nearest the site of trauma (see Fig. 1). Tissue under the impact site was spongiform in appearance, and acid fuchsin-positive neurons were most numerous in these regions at 6 hours and 1 day after trauma. By 8 days after the insult edema had resolved, acid fuchsin-positive neurons were not detectable in the contused cortex, and a significant cortical cavitation necrosis had formed (see Fig. 1).

Metabolic measurements revealed that tissue adjacent to the contusion site had high rates of LCMRGLC (range = 122.5 to 169.2) during the first few minutes after trauma (see Fig. 2). In addition, the underlying hippocampus exhibited high LCMRGLC during this acute period, with the CA1 and CA3 regions reaching levels of 156 and 163, respectively. At acute time points, the trauma-induced increases of LCMRGLC remained localized to regions directly adjacent or under the site of injury.

As time after injury increased, a greater region of the cerebral cortex, including the hippocampus, exhibited

a metabolic depression with LCMRGLC reaching levels as low as 29.9. At 10 days after injury metabolic depression encompassed a large portion of the ipsilateral cerebral cortex and also extended into the contralateral cortex and the ipsilateral thalamic nuclei.

Measurements of LCBF revealed that the contused cortex is exposed to a pronounced level of ischemia, primarily at the impact site, immediately after the insult (see Fig. 2) This low level of LCBF (as low as 4.9) was also seen within the hippocampus ipsilateral to injury, and was chronic (up to 10 days), particularly in regions exhibiting neuronal death. In areas of the cerebral cortex remote from the impact site LCBF maintained relatively low levels (39.5), coinciding with low rates of glucose metabolism at chronic times after injury.

Discussion

The results from this study indicate that following a local cortical impact injury cerebral edema, necrosis, and disruptions of metabolism have a dynamic time course. The sequela of events is both similar and dissimilar to that seen in diffuse experimental trauma as produced by the fluid percussion device. Like fluid percussion, cortical contusion produced edema and changes in cellular pH within injured tissue. Unlike

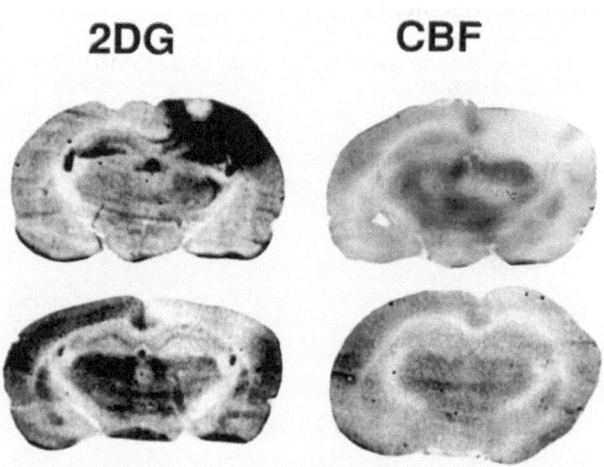

2DG   CBF

Fig. 2. Representative autoradiographs showing LCMRGLC or LCBF of cortically contused rats immediately (upper) and 1 days after (lower) trauma. Note the focal increase in cortical and hippocampal LCMRGLC immediately after insult and the widespread depression of glucose metabolism by 1 day. LCBF dropped to ischemic levels in impacted cortex and in some regions of the underlying hippocampus immediately after insult, and remained depressed in regions exhibiting low LCMRGLC at 1 day after injury

fluid percussion injury, morphological consequences after cortical contusion are much more marked as evident by the extent of neuronal loss and cavitation. Cortical contusion also resulted in a core region of cerebral ischemia localized to the impact site and the underlying hippocampus, which is not observed following fluid percussion injury. Finally, although both types of traumatic brain injury models produce dynamic changes in cerebral metabolism of glucose, the acute injury-induced increase in glucose utilization after contusion is more focal, being restricted to the ipsilateral hippocampus and the penumbra of the cortical contusion. Within 30 minutes of contusion, the increase in glucose utilization is replaced by a more diffuse metabolic depression that endures for at least 10 days.

The mechanisms underlying edema formation, metabolic dysfunction, and neuronal death following contusion injury may include the increase in excitatory amino acids resulting in a massive flux of ions across cellular membranes[1]. However, these or other proposed mechanisms must include the added component of cerebral ischemia which occurs concurrently with cortical contusion. The current work begins to define potential cellular pathophysiological differences between local and diffuse traumatic brain injury. The results imply that the sequela of cellular changes following these two types of trauma demand a different conceptual approach in terms of therapeutic treatments.

Acknowledgements

This work was supported by U.S. Army Grant DAMD-17-91-Z-1006 and NS30308.

References

1. Hovda DA, Becker DP, Katayama Y (1992) Secondary injury and acidosis. J Neurotrauma 9 [Suppl 1]: S47–S60
2. Hovda DA, Yoshino A, Kawamata T, Katayama Y, Becker DP (1991) Diffuse prolonged depression of cerebral oxidative metabolism following concussive brain injury in the rat: a cytochrome oxidase histochemistry study. Brain Res 567: 1–10
3. Katayama Y, Becker DP, Tamura T, Hovda DA (1990) Massive increases in extracellular potassium and the indiscriminate release of glutamate following concussive brain injury. J Neurosurg 73: 889–900
4. Kawamata T, Katayama Y, Hovda DA, Yoshino A, Becker DP (1992) Administration of excitatory amino acid antagonists via microdialysis attenuates the increase in glucose utilization seen following concussive brain injury. J Cereb Blood Flow Metab 12: 12–24
5. Sakurada O, Kennedy C, Jehle J, Brown JD, Carbin GL, Sokoloff L (1978) Measurement of local cerebral blood flow with iodo[^{14}C]antipyrine. Am J Physiol 234: H59–H66
6. Sokoloff L, Reivich M, Kennedy C, Des Rosiers MH, Patlak CS, Pettigrew KD, Sakurada O, Shinohara M (1977) The [^{14}C]deoxyglucose method for the measurement of local cerebral glucose utilization: theory, procedure, and normal values in the conscious and anesthetized albino rat. J Neurochem 28: 897–916
7. Sutton RL, Lescaudron L, Stein DG (1993) Unilateral cortical contusion injury in the rat: vascular disruption and temporal development of cortical necrosis. J Neurotrauma, in press
8. Yoshino A, Hovda DA, Katayama Y, Kawamata T, Becker DP (1992) Hippocampal CA3 lesion prevents post-concussive metabolic dysfunction in CA1. J Cereb Blood Flow Metab 12: 996–1006
9. Yoshino A, Hovda DA, Kawamata T, Katayama Y, Becker DP (1991) Dynamic changes in local cerebral glucose utilization following cerebral concussion in rats: evidence of a hyper- and subsequent hypometabolic state. Brain Res 561: 106–119

Correspondence: Richard L. Sutton, Ph. D., Neurosurgical Research, Department of Surgery, Hennepin County Medical Center, 701 Park Avenue South, Minneapolis, MN 55415-1829, U.S.A.

Acta Neurochir (1994) [Suppl] 60: 449–451
© Springer-Verlag 1994

Retraction Induced Brain Edema

S. Harada and **T. Nakamura**

Department of Neurosurgery, Saiseikai Kanagawa-ken Hospital, Tomiya-cho, Kanagawa-ku, Yokohama City, Japan

Summary

A local cerebral retraction apparatus which simulates cerebral retraction was devised in an effort to clarify the relationship among the retraction pressure, somatosensory evoked potential (SEP) and cerebral blood flow (CBF) by measuring these parameters both simultaneously and chronologically. Twenty seven cats were divided into three goups according to the retraction pressure (10, 30 and 50 mmHg, respectively). Each group underwent 30 minutes' retraction. At a retraction pressure of 10 mmHg, the reduction rate of CBF was low and the recovery of SEP was excellent. With an excessive pressure of 50 mmHg, both SEP and CBF reduced to 60% of the control value during the retraction and the recovery of SEP was extremely poor. Marked hyperemia of the brain surface was seen immediately after the release of retracton in more than 50% of the animals. At 30 mmHg, the recovery of SEP was moderately disturbed but, nevertheless, with a satisfactory value of more than 60% of the control. When the N_1 component of SEP was abolished, the residual CBF showed approximately 60% of the control value, which seemed relatively high as compared to cases where cerebral artery were occluded. Extravasation of Evans blue which is an indication of vasogenic brain edema due to disruption of the blood-brain barrier, occurred extensively in the cerebral cortex involving a deep-lying white matter and an increase of the retraction pressure, corresponding to a poor recovery of neuronal function.

Keywords: Cerebral blood flow; vasogenic brain edema; somatosensory evoked potential; cat.

Introduction

Cerebral retraction can't be avoided for operative manipulations in neurosurgery, for example, removal of brain tumor, evacuation of hematoma, microsurgical dissection, etc. But an excessive retraction often results in a poor neurological outcome due to parenchymal damage. Although many investigations have been carried out on the relationship between retraction force and neuronal damage, it isn't clear what degree of ichemia of the cerebral surface is enough to affect the neuronal function. The objective of this study is to clarify the relationships among retraction pressure, neuronal function and cerebral blood flow. Observations were also made to check the degree of vasogenic brain edema caused by this iatrogenic brain damage due to retraction, which we call "retraction induced brain edema".

Materials and Method

Prior to the experiment, a local cerebral retraction or compression apparatus which simulates brain retraction was devised. This apparatus incorporates three important parts of thermister couple probe, silver ball electrodes and strain gauge. Measurements or monitoring of somatosensory evoked potential (SEP), cerebral blood flow (CBF), and retraction pressure (RP), can be performed by means of this apparatus both simultaneously and chronologically (Fig.1). Twenty seven cats were anesthetized for surgical procedures with chloralose and urethane, and ιne left primary sensory cortex was exposed. The animals were divided into three groups according to the magnitude of retraction pressure exerted, i.e. 10 (group A), 30 (group B) and 50 mmHg (group C). Each group underwent 30 minutes' retraction, and observations were made 2 hours after the release of retraction. Both the amplitude of SEP and CBF were evaluated by the relative change of values during the retraction and after the release of the retraction compared with control values before retraction. At the end of the experiment, 10% Evans blue was injected via the thoracic aorta into the bilateral carotid artery to investigate the integrity of the blood brain barrier in each group. Animals were sacrificed humanely and brains were removed. After fixation with formalin, the brain surface was observed and the brains were cut into coronal sections to check the width and depth of stained area.

Results

In group A, a mild reduction in the amplitude of SEP was seen during the retraction and its recovery was complete by 2 hours after the release of retraction. Also CBF reduced slightly to 80 or 90% of control value during retraction. In group B, the reduction in the amplitude of SEP was greater than that in group A during the retraction and its recovery was moderately

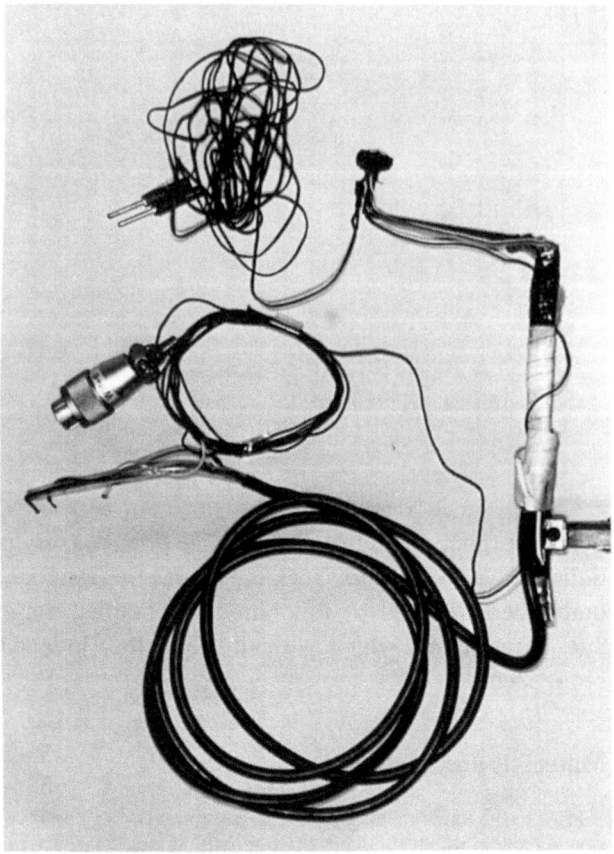

Fig. 1. Photograph of retraction apparatus

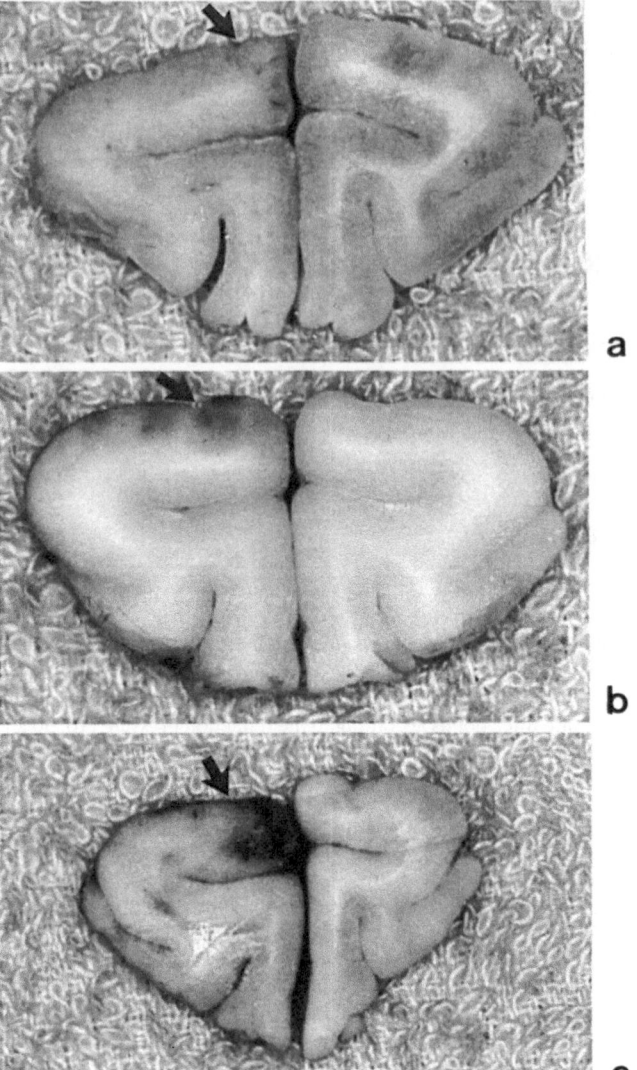

Fig. 2. Coronalsection of each group; arrow is pointing out the retracted area after the injection of Evans blue. (a) 10, (b) 30, (c) 50 mmHg, respectively. Arrow: compressed area

disturbed. CBF reduction during the retraction ranged from 60 to 80% of control value.With an excessive pressure of 50 mmHg, both SEP and CBF reduced below 60% of control value during the retraction and the recovery was poor. Marked hyperemia was seen over the surface of the brain immediately (within 30 minutes) after the release of the retraction in 5 of 9 cats. The average recovery rates of SEP 2 hours after the release of the retraction in group A, B, C, were 85.3 ± 12.4, 61.9 ± 9.8, and 23.2 ± 2 0.3% of control, respectively. And there were statistically significant differences among the groups in this respect (p < 0.01). As to the relative CBF change during the retraction,the average residual CBF in group A, B, C, were 87.0 ± 6.7, 75.7 ± 12.0, and 71.4 ± 7.3% of control value, respectively. There was no statistically significant difference between the 30 and 50 mmHg groups although the 30 mmHg group had a lower value than 10 mmHg group (p < 0.05). Extravasation of Evans blue, an indication of vasogenic brain edema due to disruption of the blood-brain barrier, it was minimal and seen on the retracted area and a part of the gray matter in

group A. In group B, extravasation was noted to extend beyond the retracted area, involving slightly the white matter (cortico-medullary junction). Marked extension of extravasation to the cortex and the deeplying white matter was seen in group C, corresponding to a poor recovery of SEP (Fig. 2).

Discussion

Cerebral retraction during operative manipulations in neurosurgery can cause neuronal dysfunction. To avoid this postoperative complication of retraction induced brain damage, some important points at issue, i.e. the manner of retracting the brain[5], innovation of spatula[3], drug administrations[2] for brain protection have been discussed. And from another point of view, the relationships between cerebral blood flow and neuronal activity were investigated by Branston et al.[1], who estimated critical cerebral blood flow at which SEPs are abolished to be 12 ml/min/100 g in case of middle cerebral artery occlusion. Izumi[4] demonstrated that the critical cortical blood flow as measured by hydrogen clearance method at the time of disappearance of N_1 component of SEP was 18 ml/min/100 g using acute intracranial presure raising model in cats. In our study, the critical level of cerebral blood flow was relatively higher compared with these reports. The difference in CBF value was presumably attributable to complexed insult of brain damage brought about by cerebral retraction, that could cause not only ischemia on the retracted area but also mechanical damage including distortion of neuronal fibers, tearing of the pia-arachnoid structure. Consequently, both ischemic and mechanical insults of brain retraction affect electrical activity of the cerebral cortex. So meticulous care should be taken during cerebral retraction not to reduce cerebral blood flow so that a satisfactory recovery of neuronal function could be achieved. With increasing retraction pressure, the extent of extravasation of dye increased both in depth and width, which signified disruption of the blood-brain barrier with consequent vasogenic brain edema. As a matter of fact, poor recovery of neuronal function was proven to result from morphological change of cerebral parenchyma, "retraction induced brain edema". Further study is required to ascertain as to whether this edema itself would be reversible. In conclusion, cerebral retraction is a partially ischemic insult causing vasogenic brain edema and neuronal dysfunction. The critical retraction pressure for reversible neuronal dysfunction is below 30 mmHg, as long as the brain is retracted no longer than 30 minutes.

References

1. Branston NM, et al (1974) Relationship between the cortical evoked potential and local cortical blood flow following acute middle cerebral artery occlusion in the baboon. Exp Neurol 45: 195–208
2. Harada S, Kawase T, Toya S (1990) The effects of glycerol and fluosol on local cerebral compression. Advances in Neurotrauma Research 2: 29–35
3. Harada S, et al (1993) A new versatile spatula with a rounded tip for corticotomies. Jpn J Neurosurg (Tokyo) 2: 70–73
4. Izumi J (1989) Change and reversibility of somatosensory evoked potentials and cerebral blood flow in experimentally increased intracranial pressure. Keio Igaku 66: 89–103
5. Yokoh A, Sugita K, Kobayashi S (1983) Intermittent versus continuous brain retraction: an experimental study. J Neurosurg 58: 918–923

Correspondence: Shunichi Harada, M. D., Department of Neurosurgery, Saiseikai Kanagawa-ken Hospital, 6-6 Tomiya-cho, Kanagawa-ku, Yokohama City, Japan.

Acta Neurochir (1994) [Suppl] 60: 452–455
© Springer-Verlag 1994

The Cause of Acute Brain Swelling After the Closed Head Injury in Rats

H. Kita and **A. Marmarou**

Division of Neurosurgery, Medical College of Virginia, Richmond, VA, U.S.A.

Summary

The major component of acute brain swelling was determined using a new closed head injury (CHI) model in rats. Twenty seven Sprague-Dawley rats were separated into four groups (Sham, CHI, CHI combined with hypotension and CHI combined with hypoxia and hypotension). Hypoxia (pO_2 of 40 mmHg) and hypotension (mean arterial blood pressure of 30 to 40 mmHg) were induced immediately after head injury and were maintained for 30 minutes. These experiments were terminated at two hours after CHI by transcalvarial freezing with liquid nitrogen. Blood pressure, intracranial pressure (ICP) and physiological parameters were monitored. Regional cerebral blood volume and water content were measured quantitatively. Rats with CHI, and with CHI and hypotension, had mild increase in ICP. Otherwise, rats with CHI, hypoxia and hypotension showed a significant increase in ICP (36.2 ± 5.6 mmHg). Water content showed an increase of 1.6% in the estimated total brain and 2.4% in the cerebral cortex in those rats. Cerebral blood volume decreased by 61.4% in the total brain and 57.3% in the cortex. There was a reduction in the cerebral hematocrit of 2.4% in the total brain and 4.7% in the cortex. The main component of brain swelling in this head injury model was brain edema. Cerebral blood volume and hematocrit were reduced in the remarkable edematous brain.

Keywords: Brain swelling; head injury; brain edema; cerebral blood volume.

Introduction

Acute head-injured patients frequently have increased intracranial pressure (ICP) without intracranial mass lesions and the outcome of such patients is more severe than those patients without brain swelling. The cause of brain swelling has been attributed by some workers to vascular congestion[15], while other studies demonstrated that brain swelling may occur in the presence of a decreased cerebral blood volume[10]. Diffuse hemispheric swelling has also been reported to be associated with an early episode of either hypoxia or hypotension[4]. Therefore, hypoxia and/or hypotension were combined with a closed head injury model to determine the contribution of edema and blood volume to brain swelling under these conditions.

Materials and Methods

For this study, twenty-seven adult male Sprague-Dawley rats, weighing 350 to 375 gram, were divided into four groups: Group I: sham operation (n = 6); Group II: closed head injury (CHI) alone (n = 7); Group III: CHI and hypotension (n = 7); and Group IV: CHI, hypoxia and hypotension (n = 7).

Rats were initially anesthetized with halothane, then intubated and artificially ventilated with a gas mixture containing N_2O (70%), O_2 (30%), and halothane. Head injury was produced by dropping a sectioned brass weight of 450 gram from a height of two meters onto the center of round stainless steel disc mounted on the skull. After CHI, rats were artificially ventilated and the head was fixed by placing it in a stereotaxic frame. Rats that developed a skull fracture were excluded from the study. In Group III and IV, hypotension (mean arterial blood pressure of 30 to 40 mmHg) was induced by withdrawing arterial blood immediately after CHI and was sustained for 30 minutes. In Group IV, hypoxia (pO_2 of 40 mmHg) was obtained by a reduction in oxygen content and was maintained for 30 minutes. ICP was measured using a ventricular needle. Cerebral blood volume (BV) and hematocrit (Hct) were measured from the distributions of ^{51}Cr-tagged red cells and ^{125}I-labeled bovine serum albumin (BSA), using a modification of the method described previously[2]. The packed red blood cells (0.4 ml) were labeled for 30 minutes at 37°C with 7.4 MBq of ^{51}Cr-sodium chromate solution (Amersham) and subsequently washed three times with PBS to remove extracellular chromate. Just before use, ^{51}Cr-tagged red cell was suspended in PBS with 1.11 MBq of ^{125}I-labeled BSA (ICN Radiochemicals). At two hours after CHI, an aliquot of the suspension was injected into a femoral vein. After three minutes, terminal samples of blood (300 μl) were collected and rats were then sacrificed by transcalvarial freezing with liquid nitrogen[11]. The brains were removed and regional samples were dissected in a cryostat at –12°C and weighed. Clear visible blood vessels were removed. ^{125}I and ^{51}Cr activity in brain and blood samples were measured by a gamma counter (AUTO-GAMMA® Model 5550, Packard Instrument Company) with two energy windows (15–80 keV, 240–400 keV). Cerebral RCV, PV, BV and Hct were calculated as follows:

$$\text{Cerebral RCV } (\mu l/g) = \frac{^{51}\text{Cr cpm/g brain tissue} \times \text{large vessel Hct}}{^{51}\text{Cr cpm/}\mu l \text{ whole blood}}.$$

$$\text{Cerebral PV } (\mu l/g) = \frac{^{125}\text{I cpm/g brain tissue}}{^{125}\text{I cpm/}\mu l \text{ plasma}} .$$

$$\text{Cerebral BV } (\mu l/g) = \text{Cerebral RCV} + \text{Cerebral PV} .$$

$$\text{Cerebral Hct } (\%) = \frac{\text{Cerebral RCV}}{\text{Cerebral BV}} \times 100 .$$

After the measurement of radioactivity, brain samples were then analyzed for water content (WC) by gravimetric techniques[6,13]. Blood-brain barrier (BBB) permeability was studied in Groups I and IV using a separate group of rats by Evans blue (EB) dye infused intravenously at 30 minutes before sacrifice. Statistics were analyzed by computing a two-tailed unpaired t-test, comparing control and the experimental groups. Values of all graphs are shown by the means ± SE.

Results

Pre-injured physiological data did not differ among each of the groups. In groups III and IV, the rats were induced with controlled hypotension (MABP = 32.9 ± 1.0 and 31.1 ± 0.5 mmHg at 15 minutes after CHI, 32.4 ± 0.8 and 31.1 ± 0.7 at 30 minutes (p < 0.01), respectively). In all groups, MABP showed no significant change just prior to sacrifice. In Group IV, pO_2 decreased to 45.1 ± 3.2 mmHg at 15 minutes after CHI. Before sacrifice, rats in each group had no significant difference at pCO_2.

Rats in Group I, had a stable ICP and measured 4.0 ± 0.4 mmHg at the end of study. In Group II, rats had increase in ICP (5.6 ± 0.4, 5.5 ± 0.3 mmHg) at 90 and 120 minutes after CHI (p < 0.01, 0.05) over that of control. Rats in groups III and IV, maintained a significantly higher ICP, 45 minutes after CHI (p < 0.01). Especially in Group IV, ICP demonstrated a rapid increase during 30 to 60 minutes after CHI and 36.2 ± 5.6 mmHg at the time of sacrifice (Fig. 1).

In Group I, WC of striatum, cortex and thalamus were 79.20 ± 0.19, 79.49 ± 0.09 and 78.88 ± 0.07%. The water content of cortex was significantly higher than that of the thalamus (p < 0.01). The estimated percentage of WC in this group was 79.21 ± 0.07%. In Group II, striatum showed significantly lower WC (78.33 ± 0.17%, p < 0.01). Other regions showed the same reduced water content. Cortex in Group III and IV yielded significantly higher WCs (80.77 ± 0.50, 81.94 ± 0.41%) than that of Group 1. The estimated WC of total brain increased in Group IV (Fig. 2).

In Group I, Cerebral BV was 11.33 ± 0.57 in the striatum, 15.10 ± 0.64 in the cortex, 12.73 ± 0.52 in the thalamus and 13.63 ± 0.56 μl/g in the estimated total brain. BV of cortex was higher than that of striatum and thalamus (p < 0.01 compared with striatum, p < 0.05 compared with thalamus). In Group II, BV of all regions showed a lower value than Group I, but with no significant differences. In Group III, BV was lower than in Groups I and II. Only in the striatum of Group III (8.98 ± 0.81 μl/g) showed a significant difference as compared with Group I (p < 0.05). In Group IV, all regions had significantly lower values of BV. The striatum was 5.90 ± 0.70, the cortex 8.65 ± 1.84, the thalamus 8.89 ± 0.73 and the estimated total cerebrum 8.37 ± 1.22. The BV of cortex especially decreased over that of the other regions (Fig. 2).

Hct in Group I, was 28.6 ± 1.0 in the striatum, 32.5 ± 0.7 in the cortex, 29.7 ± 0.6 and 31.0 ± 0.7% in the estimated total cerebrum. The Hct of cortex showed the highest value as compared with the striatum and thalamus (p < 0.05). In the other groups, cerebral Hcts showed no significant differences as compared with Group I except in the cortex of Group IV (27.7 ± 1.9%, p < 0.05). The striatum and thalamus of Group III had high values. The estimated total cerebral Hct of Group IV was low. But, these regions had no significant differences.

In Group III and IV, changes of WC and of cerebral BV were calculated in each region of each rats. WC was subtracted by the mean values of each region in Group I to calculate increases of WC. To calculate decreases of cerebral BV, the same subtraction was also used. The relationship between the increase of WC and the decrease of cerebral BV revealed a linear correlation. The increase of WC was greater than the decrease of cerebral BV.

No extravasation of EB could be detected in any of the groups macroscopically. In Group IV, brains were swollen with smaller ventricles and massive subarachnoid hemorrhage around the basal cistern and brain stem areas.

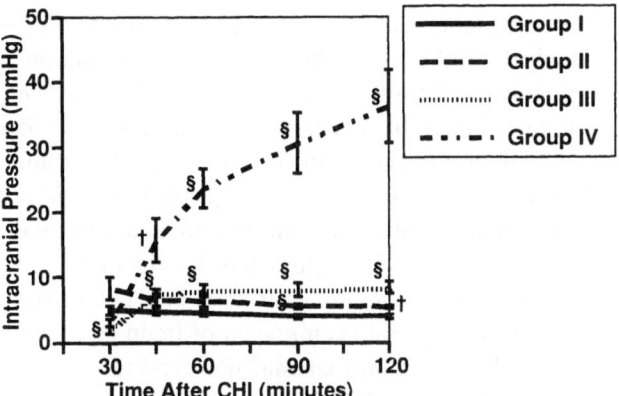

Fig. 1. Time course of intracranial pressure (ICP). In Group III and IV, ICP rose to significantly high values 45 minutes after CHI. Especially in Group IV, ICP revealed rapid increase during 30 to 60 minutes after CHI. Values are means i SE. § p < 0.01 (vs. Group I), † p < 0.05 (vs. Group I).

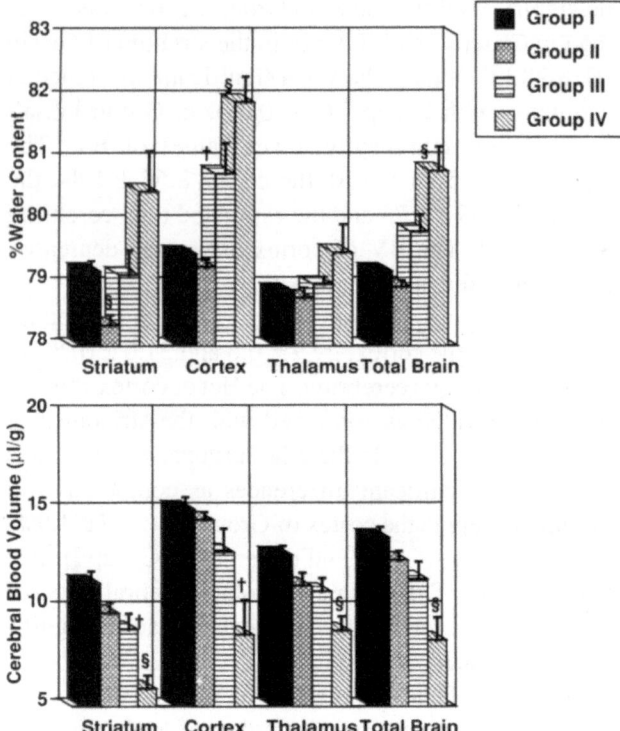

Fig. 2. % Water contents (WC) and Cerebral blood volume (BV). In Group II, striatum showed significant low WC. Cortex in Group III and IV revealed significant high WCs. The estimated WC of total brain increased in the Group IV. In Group II, BVs of all regions were lower than Group I, but not significantly. In Group III, BVs were lower than Group I and II. Only striatum of Group III showed a significant difference. But, in Group IV, all regions had significantly low values of BV. The BV of cortex was especially decreased. Values are means ± SE. §: p < 0.01 (vs. Group I) p < 0.05 (vs. Group I).

Discussion

Brain Swelling Model

Brain edema is defined as a type of brain swelling in which the brain is expanded by the accumulation of tissue water. Conversely, another form of brain swelling may occur by hyperemia or vasoparalysis that results in an increase in the cerebral vascular volume. The discrimination of these two types of brain swelling is important because each must be treated differently. This study was designed to determine the major component of brain swelling after head injury. Our study showed that the main component of brain swelling in this model was brain edema, not cerebral BV, and cerebral BV was reduced by severe brain edema.

Data from the NIH traumatic coma data bank (TCDB) showed that 61% of patients with diffuse brain swelling had early hypoxia and 34% of patients without such swelling showed hypoxia. The relation-ship between early hypotension and diffuse brain swelling was similar to that of early hypoxia (57% and 35%)[1]. A review of patients within the TCDB data shows that 17.6% (13 of 74) of patients with diffuse brain swelling have signs of both hypoxia and hypotension, and only 6.6% (40 of 610) had hypoxia and hypotension without brain swelling. It has been reported that hypoxia and hypotension had the most adverse effect after head injury[7,8]. In preliminary studies, hypotension, and hypoxia combined with hypotension were also studied in the absence of trauma. But, these groups showed no increase in ICP and no presence of brain edema at two hours post injury. Therefore, the combination of head injury, hypoxia and hypotension renders the brain vulnerable to massive swelling and ICP rise.

Brain Edema

Recent contradictions about the classification of brain edema are being discussed and the actual cause of brain edema after head injury is still unclear. It has been suggested that ischemia may be the main mechanism[7]. Our model was similar to that of postischemic reperfusion. In groups involving CHI alone, hypotension alone, and hypoxia combined with hypotension, each group did not show any measurable edema. These insults were not sufficient to produce brain edema at such an acute phase. The combination of CHI and hypotension did yield some edema. When hypoxia was added to the combination, brain edema was greatly aggravated. This data suggests that the formation of edema in our model is dependent on the severity of secondary insults, and metabolic crisis.

Cerebral Blood Volume and Hematocrit

It was important that isotopes were trapped in situ at the time of death in the measurement of cerebral BV. Therefore, we sacrificed in situ by liquid nitrogen using the transcalvarial freezing method. In fact, our preliminary data showed that the estimated cerebral BV of rats sacrificed by the freezing method was about 1.6 times of that by the guillotine. Our data of cerebral BV sacrificed by guillotine was similar to values in a previous study[2]. Some intravascular blood was lost in this method of sacrifice. Our cerebral Hct of the sham group were quite close to previous data[2]. In our non sham groups, our cerebral Hct data showed that some regions of groups had significantly lower Hct especially the striatum and cortex that showed severe brain edema in Group IV. Previous investigators have reported cerebral Hcts lower than that of large vessels[2]. Fågræus found that Hcts decreased when the experimental tube diameter was reduced from 1.10–0.05 mm[5]. The phenomenon called "plasma skimming" also happens in the capillaries[9]. "Plasma skimming" is one cause of the decreased Hct in the microcirculation. When capillaries are compressed partially, a striking fall in the Hct occurs[12]. This supports cerebral Hct reduction when arterioles and capillaries are compressed by brain edema. Regions that showed the more severe edema had lower cerebral Hct. The edema at reperfused period after ischemia was shown to be associated with compression of capillary channels and with dilatation of vessels of arteriolar caliber[14]. Arteriolar caliber may depend on the severity of edema. In severe edematous brain, like our brain swelling model, we suspect that vessels were compressed by the edema, collapsed and vascular volume was lost.

Causes of Increased ICP

In Group III and IV, ICP was increased continuously after the removal of the second insult such as hypoxia and hypotension. The ICP chart in Group IV showed a rapid increase of ICP occurring just after the recovery from second insult such as hypoxia and hypotension. It was suggested that at this time one cause of increased ICP was increased cerebral BV after reperfusion[3] and another was the edema formation, although we did not measure the time course of cerebral BV and WC. Data showed high WC and low cerebral BV at two hours after CHI. Rats of Group IV did not show ventriculomegaly due to subarachnoid hemorrhage. We conclude that the major component of brain swelling in this model was brain edema.

References

1. Aldrich EF, Eisenberg HM, Saydjari C, Luerssen TG, Foulkes MA, Jane JA, Marshall LF, Marmarou A, Young HF (1992) Diffuse brain swelling in severely head-injured children. A report from the NIH traumatic coma data bank. J Neurosurg 76: 450–454

2. Cremer JE, Seville MP (1983) Regional brain blood flow, blood volume, and haematocrit values in the adult rat. J Cereb Blood Flow Metab 3: 254–256
3. Crumrine RC, LaManna JC (1991) Regional cerebral metabolites, blood flow, plasma volume, and mean transit time in total cerebral ischemia in the rat. J Cereb Blood Flow Metab 11: 272–282
4. Eisenberg HM, Gary Jr. HE, Aldrich EF, Saydjari C, Turner B, Foulkes MA, Jane JA, Marmarou A, Marshall LF, Young HF (1990) Initial CT findings in 753 patients with severe head injury: a report from the NIH Traumatic Coma Data Bank. J Neurosurg 73: 688–698
5. Fahraeus R, Lindqvist T (1931) The viscosity of blood in narrow capillary tubes. Am J Physiol 96: 562–568
6. Groeger U, Marmarou A (1989) The importance of protein content in the oedema fluid for the resolution of brain oedema. Acta Neurochir (Wien) 101: 134–140
7. Ishige N, Pitts LH, Berry I, Carlson SG, Nishimura MC, Moseley ME, Weinstein PR (1987) The effect of hypoxia on traumatic head injury in rats: alternations in neurologic function, brain edema, and cerebral blood flow. J Cereb Blood Flow Metab 7: 759–767
8. Ishige N, Pitts LH, Berry I, Nishimura MC, James TL (1988) The effects of hypovolemic hypotension on high-energy phosphate metabolism of traumatized brain in rats. J Neurosurg 68: 129–136
9. Krogh A (1921) Studies on the physiology of capillaries. II. The reactions to local stimuli of the blood-vessels in the skin and web of the frog. J Physiol 55: 412–422
10. Kuhl DE, Alavi A, Hoffman EJ, Phelps ME, Zimmerman RA, Obrist WD, Bruce DA, Greenberg JH, Uzzell B, (1980) Local cerebral blood volume in head-injured patients. Determination by emission computed tomography of ^{99m}Tc-labeled red cells. J Neurosurg 52: 309–320
11. Pontén U, Ratcheson RA, Salford LG, Siesjo BK (1973) Optimal freezing conditions for cerebral metabolites in rats. J Neurochem 21: 1127–1138
12. Svanes K, Zweifach BW (1968) Variations in small blood vessel hematocrits produced in hypothermic rats by microocclusion. Microvasc Res 1: 210–220
13. Van den Brink WA, Marmarou A, Avezaat CJJ (1990) Brain oedema in experimental closed head injury in the rat. Acta Neurochir (Wien) [Suppl] 51: 261–262
14. Yamaguchi M, Shirakata S, Yamasaki S, Matsumoto S (1976) Ischemic brain edema and compression brain edema. Water content, blood-brain barrier and circulation. Stroke 7: 77–83
15. Zimmerman RA, Bilaniuk LT, Bruce D, Dolinskas C, Obrist W, Kuhl D (1978) Computed tomography ot pediatric head trauma: acute general cerebral swelling. Radiology 126: 403–408

Correspondence: Anthony Marmarou, Ph.D., Division of Neurosurgery, Medical College of Virginia, MCV Station, P.O. Box 508, Richmond, Virginia 23298, U.S.A.

Acta Neurochir (1994) [Suppl] 60: 456–458
© Springer-Verlag 1994

Quantitative Analysis of Blood-Brain Barrier Damage in Two Models of Experimental Head Injury in the Rat

W. A. van den Brink[1,2], **B. O. Santos**[2], **A. Marmarou**[2], and **C. J. J. Avezaat**[1]

[1] Dijkzigt University Hospital of Rotterdam, The Netherlands, and [2] Medical College of Virginia, Richmond, VA, U.S.A.

Summary

The integrity of the blood-brain barrier was studied in a new model of closed head injury, and in an established model of fluid percussion injury, in the rat. Brain injury in this new model is induced by impact and acceleration of the protected rat skull. Severe hypertension is not a characteristic of this new model as compared to the tremendous surge following direct dural percussion. This is important because of the well known sensitivity of the cerebral microvasculature for acute hypertension. Using a radioactive tracer technique the dysfunction of the barrier was quantified. It is shown that the BBB is temporarily damaged due to trauma, subsequent arterial pressure surge, as seen in the percussed animals, deteriorates the dysfunction of the barrier even further. This study indicates that vascular damage is a key event following head injury. Yet the concomitant basic pathophysiological sequelae of different models must be considered when studying barrier damage and cerebral edema following brain injury. Time window studies of the barrier indicate that the barrier seals within a few hours following severe concussive head injury, and in the absence of a hypertensive surge.

Keywords: Head injury; blood brain barrier; hypertension.

Introduction

The disruption of the blood-brain barrier (BBB) and development of brain edema concomitant with traumatic brain injury has been well established[7,8]. In experimental fluid percussion injury (FPI) studies, at least two factors contribute to this disruption; the mechanical deformation of tissue and cerebral vessels immediately upon impact, and the hypertensive surge that accompanies this injury. In fact, models of pharmacologically induced hypertension in the range of the surge observed with (FPI) produce similar defects in the barrier[1,9]. The objective of this study was twofold: First to compare the arterial pressure response and BBB disruption in the established FPI model with the new closed head Injury (CHI) model. This new model

developed by Marmarou traumatizes the rat brain by means of impact-acceleration to the closed calvarium, and does not cause a severe hypertensive surge[10,11]. Secondly to study the "time-window" after head injury in which the BBB remains opened. The BBB was studied using iodine-125 radio active labelled bovine albumin (RISA), as a tracer of a blood borne high molecular weight substance, to show a general BBB-dysfunction. The use of this technique made it possible to perform a quantitative analysis of barrier damage, whereas the majority of the BBB studies in trauma have been purely descriptive.

Materials and Methods

Trauma Models

The CHI model in the rat produces brain trauma by 450 gram/ 2 meter weight drop on a stainless steel disc secured to the intact skull. Limited movement and thus impact, compression and acceleration are provided by supporting the skull with foam.

The FPI model in the rat produces brain trauma by direct 3.2 to 3.4 atmosphere fluid bolus percussion to the exposed dura under a midsagittal craniectomy. Mechanics, physiology and details of this model are well described[1,2].

RISA Protocols

Four protocols of animal preparation were used. In the first two groups a cumulative permeability Index (CPI) study was performed. The rats were injected with 30 µCi radioactive I-125 labelled albumin prior to CHI. Survivors of injury were sacrificed after 4 (group 1) and 15 (group 2) hours respectively (5 injured, 5 sham-controls in each group). In the third protocol the rats were first injured. The surviving rats were then injected with I-125 labelled albumin 30 minutes before sacrifice 2 hours post injury (group 3: 3 injured, 4 sham-controls), to calculate a BBB-permeability index (BBB-PI). Finally the FPI injured rats (group 4: 3 injured, 3 sham-controls) were studied after 15 hours of CPI study, to be comparable to the 15 hour survival of the CHI group. Sham-controls were treated exactly the same as injured animals, save for the actual delivery of brain injury. After saline wash-out perfusion under deep pentobarbital anaesthesia the brains were extracted and arachnoid was removed carefully. This to avoid contamination with the tracer of the subarachnoid hemorrhage often encountered. The brain was frosted to facilitate standardized cutting in 5 rostro-caudal supra tentorial and two infra tentorial segments. Permeability index for each segment was calculated by the following formula: PI = brain-CPM/g divided by blood-CPM/g; all multiplied by 1000. Data are displayed as mean ± standard deviation. The student t-test for unpaired variables was used; $p < 0.05$ was considered significant.

Results

Both models result in a 50% mortality indicating severe and comparable brain damage in both groups. One of the CHI injured rats suffered a skull fracture and did not survive injury. Surviving animals were not artificially ventilated or supported by other means than keeping the airway free. These data are consistent with mortality and fracture rates in previous CHI studies[10–12]. The disruption of the BBB is reflected in the rise in pemmeability index (PI) for I^{-125} RISA. In all slices the CPI of injured animals is consistently higher than CPI of the comparable slices of the sham animals (see Table 1).

CPI after 4 hours (group 1) is significantly higher in the three caudal tissue samples. Main increase in the supratentorial compartment is in the brain slices 3, 4, and 5. These are located under the helmet. Unexpectedly high is the vascular permeability in the brain stem and the cerebellum. The increase in CPI 15 hours after CHI (group 2) is much smaller than 4 hours after trauma. Still the differences are statistically signficant in 4 segments. 15 hours after FPI (group 4) there is CPI inaease that is markedly higher than 15 hours after CHI, with significant barrier damage in all but one brain segment. Again the increase is most outspoken just beneath the area of trauma impact, in slice 3. In the third group an actual "time window study" was performed as compared to the cumulative permeability studies, the barrier is still open two hours after injury in all slices. The total-brain BBB-PI is significantly higher in the injured group.

Discussion

Severe brain concussion is the single cause of BBB dysfunction in the CHI model of head injury. The mild transient of hypertension, followed by a period of sustained hypotension seen in CHI is in contrast to the documented surge of arterial pressure following FPI[1,2]. Early qualitative BBB studies have determined that an arterial pressure surge, even in absence of craniocere-

Table 1. *Permeability Index*

Slice no.	Closed head injury 4 hours			
	sham		CHI	
1	1.59	(0.45)	2.36	(0.50)
2	1.34	(0.36)	2.90	(0.97)
3	1.50	(0.56)	3.42	(1.63)
4	1.50	(0.78)	4.47	(1.88)
5	1.66	(0.64)	7.04	(2.96)
Stem	1.89	(0.76)	10.34	(4.99)
Cerebellum	1.48	(0.76)	14.50	(12.35)

Slice no.	Closed head injury 15 hours			
	sham		CHI	
1	1.39	(1.36)	2.01	(0.41)
2	0.88	(0.48)	1.42	(0.46)
3	0.77	(0.36)	1.53	(0.31)
4	0.73	(0.29)	1.81	(0.58)
5	0.99	(0.33)	2.21	(0.78)
Stem	1.37	(0.63)	2.96	(0.87)
Cerebellum	1.21	(0.39)	2.44	(0.50)

Slice no.	Fluid percussion injury 15 hours			
	sham		CHI	
1	1.88	(0.74)	3.77	(0.76)
2	1.42	(0.38)	5.46	(2.90)
3	1.36	(0.65)	10.52	(4.87)
4	1.25	(0.86)	14.73	(2.58)
5	1.32	(0.20)	11.97	(2.09)
Stem	1.27	(0.10)	5.89	(0.65)
Cerebellum	0.67	(0.07)	6.50	(0.77)

Slice no.	Closed head injury 2 hours			
	sham		CHI	
1	2.88	(1.48)	4.61	(0.83)
2	2.84	(1.23)	4.42	(0.90)
3	3.60	(1.02)	4.32	(0.64)
4	3.42	(1.04)	4.71	(0.69)
5	3.57	(0.83)	4.78	(0.51)
Stem	2.62	(1.11)	5.63	(1.48)
Cerebellum	3.12	(1.14)	5.17	(0.82)

Brain CPM/g divided by blood CPM/g for the experimental group.

bral trauma, compromises the barrier[3,5]. This quantitative radioactive tracer study of the BBB, has confirmed our hypothesis of considerable barrier damage caused by brain trauma alone. Furthermore data from the fluid percussed rats implicate that the dramatic arterial pressure surge in this model of brain injury, plays an important role in the opening of the barrier. The increase in PI for albumin after brain injury is consistent. In both models the main increase is seen in the brain slices in the areas where most of the brain deformation is to be expected.

Comparison of the CPI's, after CHI at different times after trauma demonstrate that the cumulative radioactivity concentration after 15 hours is considerably less than after 4 hours. This finding implicates that the protein that was extravasated initially is being cleared from the brain parenchyma already, without influx of new protein bound tracer. Obviously the barrier seals again at some point in time after trauma. The time window study of barrier damage indicates that the BBB is still damaged two hours after CHI. The barrier dysfunction as studied with a large albumin molecule is general, in the sense that when albumin can freely cross the barrier, the permeability can be considered severe enough to allow an indiscriminate flow of plasma components into the surrounding tissue. Plasma components in the neuropil are further contributing to the brain damage, the main response to it being cerebral edema, initially being of vascular origin. The osmotic pressure gradients that develops after introduction of blood borne high molecular weight substances in the brain parenchyma is a well known factor in the process of edema formation[4,6]. FPI does cause significant edema formation[7,8]; and the characteristics on edema formation in this new model of CHI were reported earlier[10,11]. Since in these studies an ongoing formation of edema was found after the sealing of the barrier, part of the edema formation after several hours is likely to be of cellular origin. In the light of this study a major part of the edema formation after FPI will be attributable to the arterial pressure surge in this model.

Conclusion

Experimental traumatic brain injury is followed by a temporal dysfunction of the BBB, enabling indiscriminate passage of plasma constituents into the neuropil. Barrier damage is exacerbated when it occurs in a experimental setup in which the initial trauma is followed by a severe blood pressure surge, as in the fluid percussion model. The new closed head injury model in the rat is characterized by a relative absence of pressure rise. Also in this scenario the barrier is compromised, giving rise to a cascade of possible pathological events, the indicator of these being cerebral edema. The duration of barrier opening is at least two

hours after CHI. It seals however within 15 hours. Thus there is a short but significant post trauma interval of barrier compomise. Future studies of edema in traumatic brain injury will be directed to identify the time of barrier closure more precisely.

Acknowledgements

This study was funded by: NIH Grant 2POINS12587 and NIDRR HI33B80029. Further support was obtained from the Richard Roland Reynolds Research Laboratories and the "Stichting L. van Stolk Fonds" from The Netherlands.

References

1. Dixon CE, Lighthall JW, Anderson TE (1988) Physiologic, histopathologic, and cineradiographic characterization of a new fluid percussion model of experimental brain injury in the rat. J Neurotrauma 5: 99–104
2. Dixon CE, Lyeth BG, Povlishock JT, Hamm RJ, Marmarou A, Young HF, Hayes RL (1987) A fluid percussion model of experimental brain injury in the rat. J Neurosurg 67: 110–119
3. Hardebo JE, Nilsson B (1981) Opening of the blood-brain barrier by acute elevation of intracarotid pressure. Acta Physiol Scand 111: 43–49
4. Ikeda Y, Long DM (1990) The molecular basis of brain injury and brain edema: the role of oxygen free radicals. Neurosurgery 27: 1–11
5. Johansson B, Linder LE (1974) Blood-brain barrier dysfunction in acute arterial hypertension induced by clamping of the thoracic aorta. Acta Neurol Scand 50: 360–365
6. Kuroiwa T, Shibutani M, Tajima T, Hirasawa H, Okeda R (1990) Hydrostatic pressure versus osmotic pressure in the development of vasogenic brain edema induced by cold injury. In: Long D (ed) Advances in neurology. Raven, New York, pp 11–19
7. Marmarou A, Shima K (1990) Comparative studies of edema produced by fluid percussion injury with lateral and central modes of injury in cats. In: Long D (ed) Advances in neurology. Raven, New York pp 233–236
8. McIntosh TK, Soares H, Thomas M, Cloherty K (1990) Development of regional cerebral oedema after lateral fluid-percussion brain injury in the rat. Acta Neurochir (Wien) [Suppl] 51: 263–264
9. Zhang XM, Ellis EF (1990) Superoxide dismutase reduces permeability and edema induced by hypertension in rats. Am J Physiol
10. van den Brink WA, Marmarou A, Avezaat CJJ (1990) Brain oedema in experimental closed head injury in the rat. Acta Neurochir (Wien) [Suppl] 51: 261–262
11. van den Brink WA, Santos BO, Marmarou A, Avezaat CJJ (1991) Temporary blood brain barrier damage and continued edema formation in experimental closed head injury in the rat. ICP 8
12. van den Brink WA, Slomka W, Marmarou A (1991) Neurophysiological evidence of preservation of brain stem function in a new model of closed head injury in the rat. ICP 8

Correspondence: A. Marmarou, Ph. D., Medical College of Virginia, Division of Neurosurgery, MCV Station, P.O. Box 509, Richmond, VA 23298, U.S.A.

Acta Neurochir (1994) [Suppl] 60: 459–461
© Springer-Verlag 1994

Transdural Cortical Stabbing Facilitates the Drainage of Edema Fluid Out of Cold-Injured Brain

T.-L. Chiou, Y.-H. Chiang, W. S. Song, and **S.-S. Lin**

Division of Neurosurgery, Tri-Service General Hospital and National Defense Medical Center, Taipei, Republic of China

Summary

Recent experimental results indicate that cerebral glia lining and glia limitans may be barriers for plasma protein extravasated from injured cerebral microvessels flowing into the adjacent subarachnoid space. Therefore, it has been hypothesized that a transdural cortical stabbing which opens both the pia lining and glia limitans may facilitate drainage of edema fluid into the subarachnoid space and minimize brain edema. This hypothesis was tested in Sprague-Dawley rats with a transdural cold-injury on the right parietal cortex. The animals were sacrificed 24 hours later. One hour before being sacrificed 0.6 ml of 2% Evans blue was intravenously injected to determining the Evans blue distribution area. For measuring the inulin retention volume in the brain, ^{14}C-inulin (10 µCi) in 1 ml of saline was injected intravenously at 10 min before sacrifice. The extent of brain edema was assessed by measuring the water content, the inulin retention volume, and the distribution area of Evans blue in the brain. Our results showed that the transdural cortical stabbing did not alter the water content of the cerebral hemisphere with cold lesion. However, it did effectively diminish the inulin retention volume by 26% as well as the distribution area of Evans blue by 22% in the cerebral hemisphere with cold lesion. In conclusion, a transdural cortical stabbing on the injured cortex may be beneficial for vasogenic brain edema.

Keywords: Brain edema; cortical drainage.

Introduction

Edema fluid extravasated from injured cerebral microvessels often accumulates in the surrounding brain parenchyma and causes a severe local cerebral edema. Edema fluid flows slowly along the interstitium of white matter into either the cerebral ventricles or the adjacent subarachnoid space[2,3]. However, the cortical surface of glia limitans and the pia lining may limit the passage of plasma protein from brain parenchyma into the subarachnoid space[1]. Thus, we hypothesized that a transdural cortical stabbing which destroys both the glia limitans and the pia lining may facilitate the drainage of edema fluid into the adjacent subarachnoid space, thereby ameliorating brain edema. We tested this hypothesis with the rat cortical injury model of vasogenic edema.

Materials and Methods

Sprague-Dawley rats weighing 300–400 g were used for these experiments. The rats were anesthetized with a mixture of halothane and oxygen. The scalp was incised and about 5 mm × 5 mm of the right parietal bone was removed. A freezing probe 3 mm in diameter, prechilled with liquid nitrogen, was applied to the exposed dura mater of the right parietal cortex for 30 s. For the control group, the skull defect was occluded with bone wax and the scalp incision was closed. For the experimental group, immediately after the freezing, a needle 1 mm in diameter was advanced via the dura mater into the center of the cold-injured cerebral cortex to a depth of 1.5 mm; the skull defect was not occluded with bone wax, but the scalp incision was closed. Measurements of cerebral water content, Evans blue distribution area, and inulin retention volume were made 24 h after trauma. But prior to that, rats from both groups were re-anesthetized for femoral artery and vein catheterization. The rats were then wrapped in loosely fitting plaster bandages, taped to stainless steel bricks, and allowed to recover from anesthesia. At 23 h, and 23 h 50 min after the lesioning, 0.6 ml of 2% Evans blue solution and 1 ml of ^{14}C-inulin saline solution (10 µCi) were intravenously injected, respectively. Twenty-four hours after trauma, the rats were sacrificed by injecting 3 ml of 15% potassium chloride solution. Immediately before sacrifice the rats, a 100 µl blood sample was obtained from the femoral artery. The brains were quickly removed from the skull, and the center of lesioned forebrain was coronally sliced into a 3 mm thick section, which was subsequently split into right and left halves, and then weighed separately.

For determining their water contents, the brain samples were dried at 105°C for 48 h. The water content was calculated using the formula:

$$\text{water content} = (\text{wet weight} - \text{dry weight}) / \text{wet weight}$$

For the measurements of inulin retention volume, the radioactivity of ^{14}C-inulin in the brain (Ab) and plasma samples (Cp) were determined by β counting. Inulin retention volume (IRV) was defined as IRV (ml/100 g) = Ab / Cp.

The rostral and caudal cutting surfaces adjacent to the removed brain slice were photographed and used for measuring the Evans blue distribution area. The blue discolored area (Ae) and the total brain section area (At) were calculated using a computerized image analyzer system. Evans blue distribution area (EBDA) was derived from the formula:

$$EBDA\ (\%) = Ae / At \times 100$$

Paired and unpaired Student's t-tests were used for statistical analysis of the data. The value for significant difference was defined as p < 0.05.

Results and Discussion

Transdural cortical stabbing on the injured cortex had little effect on blood pressure. The mean blood pressures for the control and the experimental groups were 115 ± 5 and 111 ± 3 mmHg, respectively; there was no significant difference. The water content of the control group was 0.788 ± 0.002 on the non-lesioned side and 0.81 ± 0.005 on the lesioned side; the difference was significant. For the cortical stabbing group, the water content of the non-lesioned side was 0.789 ± 0.002, the same as that of the control group. On the lesioned side, the water content was 0.803 ± 0.004 for the cortical stabbing group, less than that of the control group (0.81 ± 0.005) but not a statistically significant difference.

The Evans blue distribution areas on the rostral cutting surface was 7.41 ± 0.79% for the control group and 6.05 ± 0.49% for the cortical stabbing group (Fig. 1). The latter was 18% smaller than the former, but statistically insignificant (p > 0.05). However, on the caudal cutting surface, the Evans blue distribution areas were 7.17 ± 0.64% and 5.28 ±

0.53% for the control and experimental groups, respectively. The latter was significantly less than the former (p < 0.05) by 26% (Fig. 1). On the non-lesioned side, the inulin retention volume was 1.45 ± 0.11 ml/100 g for the control group and 1.71 ± 0.12 ml/100 g for the cortical stabbing group. (Fig. 2); the data were not significantly different. On the lesioned side, the inulin retention volume of the cortical stabbing group (2.38 ± 0.31 ml/100 g) was significantly less (p < 0.05) than that of the control group (3.16 ± 0.32 ml/100 g) by 25%.

Our results revealed that transdural cortical stabbing on the lesioned cerebral cortex led to significant decreases in both inulin retention volume and Evans blue distribution area. These findings indicate that a small opening of the glia limitans and pia lining on the cerebral cortex with cold lesion may facilitate the drainage of edema fluid into its adjacent subarachnoid space as well as the epidural space, and lessen the retention and distribution of edema fluid in the brain parenchyma.

The results also indicate that for evaluating brain edema, the measurement of inulin retention space is more sensitive than that of water content. This is because the difference in lesioned and non-lesioned brain was 2–3% by measuring water content, but more than 100% by determining inulin retention volume in the brain.

Acknowledgements

The authors gratefully acknowledge the technical support of Mrs. Fen-Ying Wu as well as the assistance of Ms. Hui-Ju Chong with the preparation of manuscript. This work was supported by National Defense Research Grants of 1991 War Injury and 1992 War Injury 15705-1.

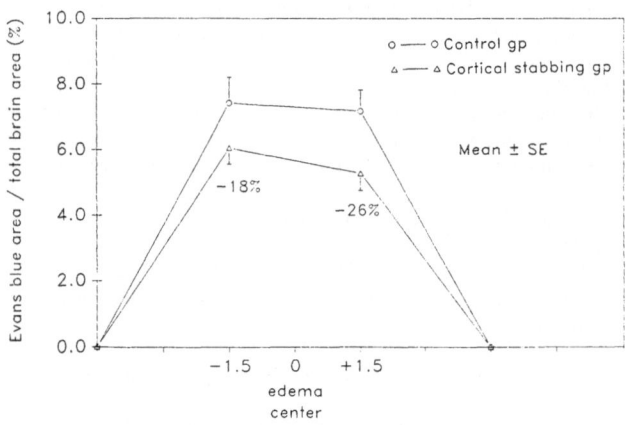

Fig. 1. Evans blue distribution area

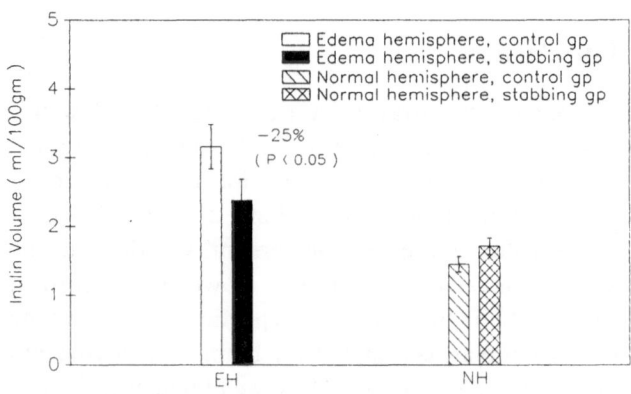

Fig. 2. Inulin retention volume. *EH* edema hemisphere, *NH* normal hemisphere

References

1. Ohata K, Marmarou A (1992) Clearance of brain edema and marcromolecules through the cortical extracellular space. J Neurosurg 77: 387–396
2. Nakato H, Yen M-M, Tajima A, Lin S-Z, Pettigrew K, Blasbery R, Fenstermacher J (1990) Dexamethasone effects on the distribation of water and albumin in cold-injury cerebral edema. In: Long D, *et al* (eds) Advances in neurology. Raven, New York, pp 335–342
3. Reulen H (1976) Vasogenic brain edema. Br J Anaesth 48: 741–754

Correspondence: Shinn-Zong Lin, M. D., Ph. D., Division of Neurosurgery, Tri-Service Gerneral Hospital, 8, Set. 3, Ting-Chou Rd., Taipei, Republic of China.

Traumatic Brain Edema: Pathophysiology, Clinical Studies

Acta Neurochir (1994) [Suppl] 60: 465–467
© Springer-Verlag 1994

Massive Astrocytic Swelling in Response to Extracellular Glutamate – a Possible Mechanism for Post-Traumatic Brain Swelling?

W. L. Maxwell[1], R. Bullock[2], H. Landholt[2], and H. Fujisawa[2]

Departments of [1] Anatomy and [2] Neurosurgery, University of Glasgow, Scotland, U.K.

Summary

Little attention has been paid to the responses of astrocytes in the brain to the application of neurotoxic excitatory transmitters such as glutamate. We have developed a simple model to study the responses of cells within the cerebral cortex to neurotoxic levels of glutamate. After short periods of perfusion with glutamate, perivascular and interstitial astrocytes swell and become electron lucent. The astrocyte swelling extends, with increasing periods of perfusion, up to 400 um into the adjacent, intact neuropil. After 90 minutes of glutamate perfusion, intermediate filament bundles and glycogen granules occur within the astrocyte cytoplasm. We obtained no evidence for compromised blood flow. We suggest that astrocyte swelling serves to limit the diffusion of glutamate from the site of the lesion.

Keywords: Astrocytic swelling; glutamate; post-traumatic brain.

Introduction

The most compelling evidence in support of the excitotoxic hypothesis comes from many pharmacological neuroprotective studies which have demonstrated that focal ischaemic lesions following middle cerebral artery occlusion may be markedly reduced by both presynaptic and postsynaptic glutaminergic blockade[1-3]. Recent in vitro studies comparing astrocyte rich and astrocyte poor cultures have demonstrated that astrocytes may be a major factor limiting glutamate toxicity[4,5]. In these culture experiments, the presence of astrocytes reduced the magnitude of glutamate neurotoxicity by a factor ranging between 30 to 100 times[4,6,7]. In addition, cell culture studies have indicated a direct response by astrocytes to exicitatory amino acid transmitters, e.g. glutamate, where a transient swelling of astrocytes has been demonstrated[8]. However, there is no published data concerning the response by astrocytes to the application of glutamate in vivo.

We have devised a simple model to evaluate and quantitify glutamate neurotoxicity in intact brain and used this model to study the early histopathological responses by astrocytes after perfusion of glutamate via a microdialysis probe placed in the cerebral cortex of adult rats. This material has been examined ultrastructurally to characterise the response by astrocytes to extracellular glutamate.

Material and Methods

We have studied ultrastructurally eight adult male Sprague-Dawley rats with weights ranging from 350–390 g (mean 370 g). All experiments were performed under 2.5% Halothane in 2 : 1 nitrous oxide/oxygen mixture. Tracheostomy and mechanical ventilation was then performed. Blood gases were monitored frequently and maintained within the physiological range (Table 1).

Table 1

	Control	Glutamate
PaO_2	166 ± 4	166.5 ± 4
$PaCO_2$	36.9 ± 0.1	37.25 ± 0.1
MABP	92 ± 2	93.5 ± 2
Rectal temp.	35 ± 1	36.5 ± 1
pH	7.45 ± 0.01	7.44 ± 0.01

Core normothermia, monitored by a rectal temperature probe, was maintained using a heating pad.

Using an operating microscope, the parietal part of the right side of the skull was exposed and a burr hole made through which was passed a microdialysis probe (Carnegie Medicine CMA 12, Biotech Instruments, Luton, UK). The probe was angled at 15° to the sagittal plane and lowered 3.5 mm into the cortex after the dura had been opened and the pia incised. The microdialysis probe was precalibrated in vitro and perfused throughout the procedure with mock

CSF (pH range 7.3–7.4) at a rate of 1.5 µl/min using a Carnegie Medicine CMA 100 infusion pump (Biotech Industries, Luton, UK). The 3.0 × 0.5 mm diameter polycarbonate dialysis membrane had a practical molecular weight cut off of approximately 500D.

Regional cerebral blood flow was measured autoradiographically using the iodo-antipyrene (IAP)[9] method.

After a 90 minute stabilisation period in the cortex, the perfusate was changed from mock CSF to a glutamate solution made up from mock CSF by adding 1 M monosodium glutamate. Glutamate concentrations in the perfusion fluid and the dialysate were checked by HPLC after collection of 45 µl samples harvested at 30 minute intervals throughout the experiment. The difference between these two concentrations was assumed to be the concentration of the perfusate delivered to the brain.

The glutamate solution was perfused through the microdialysis probe for 15 (n = 1), 30 (n = 1), 45 (n = 1) and 90 (n = 2). In order to take into account the high osmolarity of the 1 M glutamate solution, a 1 M sodium chloride solution was used to provide a control group (n = 3).

After glutamate perfusion was discontinued, the animal was monitored and ventilated for a further two and a half hours. The dialysis probe was then withdrawn and the animals perfusion fixed through the left ventricle with 2% glutaraldehyde, 2% paraformaldehyde in 0.1 M phosphate buffer (pH 6.2, 850 mOs). The animals was then decapitated and the head immersion fixed in the same solution for 24 hours before removal of the brain. A coronal slice 1.5 mm thick was taken from the cerebral cortex into which the microdialysis probe had been inserted. All material was post-fixed in 1% osmium tetroxide in 0.1 M phosphate buffer for one hour and routinely processed for resin embedding. A mesa containing the probe area but extending to intact neuropil on either side of the probe path was made, and thin sections cut for routine examination in the transmission electron microscope (TEM).

Results

No difference in values for regional cerebral blood flow occurred between control and experimental animals.

After perfusion with glutamate for 15 minutes there was pronounced astrocytic swelling/lucency extending from the core of the injection site for a distance of 150–200 µm. Within the relatively intact neuropil bordering the lesion, the most prominent feature was the occurrence of numerous lucent astrocytic cell bodies and perivascular astrocyte foot processes in relation to capillaries within the neuropil. In control material astrocytic swelling occurred in the site of insertion of the dialysis probe but did not extend into the surrounding neuropil.

Astrocytic swelling extended for a distance of 400 µm from the boundary zone of the lesion core into the relatively intact neuropil after perfusion with glutamate for 45 or 90 minutes. Within this area astrocytic structure was characterised by a centrally placed nucleus within a lucent cytoplasm, which contained no other feature than structurally normal mitochondria (Fig. 1). There was therefore a dramatic loss of normal cytoplasmic organelles, except mitochondria, from these

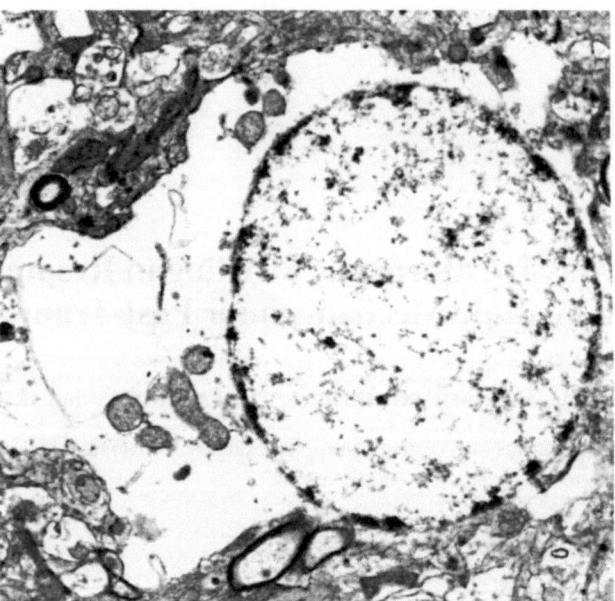

Fig. 1. A lucent astrocyte cell body obtained after 15 minutes perfusion with glutamate. The cell cytoplasm lacks all normal organelles apart from structurally normal mitochondria. x 8600

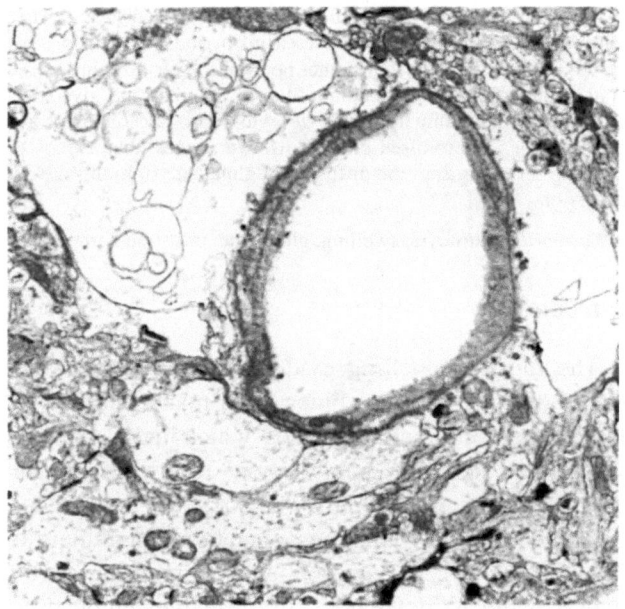

Fig. 2. A transverse section of a cerebral capillary surrounded by swollen perivascular astrocyte foot processes after 90 minutes perfusion with glutamate. The micrograph is taken from within the "intact" neuropil at the boundary of the lesion. x 5900

cells. Swollen perivascular astrocyte processes surrounded blood vessels (Fig. 2). In the core of the lesion membrane organisation was lost and the swollen, perivascular region was suggestive of vascular occlusion. But in the boundary zone and extending into the intact neuropil for a distance of 3–400 µm, although astrocyte foot processes were enlarged, blood vessels were patent. The astrocyte "sheet reaction"[10] was well established 90 minutes perfusion with glutamate as aggregates of glycogen particles occurred frequently in the cytoplasm of astrocytes.

Discussion

We have demonstrated in an in vivo model that rapid astrocytic swelling occurs after the slow delivery of glutamate, via a microdialysis probe, into the cerebral cortex of the adult rat. This ultrastructural demonstration serves to confirm and extend findings reported from recent in vitro[8] studies which demonstrated that transient swelling of astrocytes occurs without cell death after exposure to glutamate. However, there is a distinction to be made between in vitro studies and our own studies in that the concentrations of glutamate necessary to elicit cell swelling are far higher in vivo. This disparity can only remain controversial at the present time. Our findings do, however, support the concept that astrocytes are a major factor in limiting gluatamate toxicity in vivo[4,6,11].

We have provided ultrastructural evidence in support of the concept of a widespread astrocyte response and circumstantial evidence, in the core of the lesion, for a compromise in blood flow by vascular occlusion in the affected part of the brain. The rapid development of astrocytic swelling is clearly an acute response to injury to the brain which does not previously appear to have been noted in vivo. We suggest that acute astrocytic swelling may serve a neuroprotective role by increasing the tortuosity of the extracellular space and thereby limit diffusion of toxic substances, among them glutamate, from the site of injury.

References

1. McCulloch J, Bullock R, Teasdale GM (1991) Excitatory amino acid antagonists: opportunities for the treatment of ischaemic brain damage. In: Meldrum B (ed) Excitatory amino acid antagonists. Blackwell, Oxford, pp 287–326
2. Meldrum B (1990) Protection against ischaemic neuronal damage by drugs acting on excitatory neurotransmission. Cerbrovasc Brain Metabol Rev 2: 27–57
3. Meldrum B, Swan JH, Leach MJ, et al (1992) Brain Res 593: 1–6
4. Choi DW (1991) Excitotoxicity. In; Meldrum B (ed) Excitatory amino acid antagonists. Blackwell, Oxford, pp 216–237
5. Garthwaite G, Garthwaite J (1984) Differential sensitivity of rat cerebellar cells in vitro to the neurotoxic effects of excitatory amino acid analogues. Neurosci Lett 48: 361–367
6. Rosenberg PA, Aizenman E (1989) Hundred fold increase in neuronal vulnerability to glutamate toxicity in astrocyte poor cultures of rat cerebral cortex. Neurosci Lett 103: 162–168
7. Rosenberg PA, Amin S, Leitner M (1992) Glutamate uptake disguises neurotoxic potency of glutamate antagonists in cerebral cortex in dissociated cell culture. J Neurosci 12: 56–61
8. Nobel LJ, Hall JJ, Chen S, Chan PH (1992) Morphologic changes in cultured astrocytes after exposure to glutamate. J Neurotrauma 9: 255–267
9. Sakurada O, Kennedy C, Jehle J, et al (1978) Measurement of local cerebral blood flow with iod [14C] antipyrene. Am J Physiol 234: H59–H66
10. Torvik A, Soreide AJ (1972) Nerve cell regeneration after axon lesions in newborn rabbits. Light and electron microscopic study. J Neuropath Exp Neurol 31: 683–695
11. Schwartz R, Kohler C, Mangano RM, Neophytides AN (1981) Glutamate induced neuronal degeneration: studies on the role of glutamate re-uptake. In: Di Chiara G, Gessa GL (eds) Glutamate as a neurotransmitter, pp 403–412

Correspondence: W. L. Maxwell, Department of Anatomy, University of Glasgow, Glasgow G12 8QQ, U.K.

Acta Neurochir (1994) [Suppl] 60: 468–471
© Springer-Verlag 1994

Testing of Cerebral Autoregulation in Head Injury by Waveform Analysis of Blood Flow Velocity and Cerebral Perfusion Pressure

M. Czosnyka, E. Guazzo, V. Iyer, P. Kirkpatrick, P. Smielewski, H. Whitehouse, and **J. D. Pickard**

Neurosurgical Unit, University of Cambridge, Addenbrooke's Hospital, U.K., and Warsaw University of Technology, Poland

Summary

Thirty five head injured patients were subjected to day-by-day monitoring of intracranial pressure (ICP), arterial blood pressure (ABP) and blood flow velocity (FV) in middle cerebral artery. Parameters describing cerebral autoregulation: flow velocity time average, systolic, diastolic values, pulsatility index (PI), estimate of cerebrovascular resistance (CVR = CPP/FV) etc. were analyzed as functions of CPP (ABP-ICP). The results show that for CPP below 55 mmHg the evidence of exhausted autoregulation can be observed in FV, CVR, PI. The method is helpful in assessment ot the optimal level of CPP in order to reduce the probability of ischaemic brain insults.

Keywords: Cerebrovascular haemodynamics; autoregulation; ultrasonography; head injury; computer analysis.

Introduction

Transcranial Doppler (TCD) ultrasonography is a useful non-invasive technique for monitoring of cerebral blood flow velocity (FV) in the basal cerebral arteries[1]. However, following head injury FV may not be proportional to CBF as a result of changes in arterial diameter and local abnormalities in CBF may not necessarily be reflected by changes in FV in the insonated artery. Nevertheless, TCD velocimetry provides continuous information on the cerebral blood supply and its variations. Such fluctuations usually indicate changes in cerebral perfusion pressure (CPP), $PaCO_2$, and blood viscosity, or local or global changes in cerebrovascular resistance secondary to vasoparalysis, spasm, or infarct. Is it possible to distinguish between different factors that affect the FV waveform?

The absolute value of time averaged FV may be elevated because of hyperaemia, spasm, or increased CPP, $PaCO_2$ or temperature. The pulsatility indices may be affected by low CPP[3] but also by high pulsation of arterial blood pressure (ABP) or intracranial pressure (ICP), bradycardia or hypocarbia. Hence, raw FV values reflect a complex array of factors and often cannot be interpreted unambiguously.

Direct monitoring of CPP removes uncertainties related to the assessment of cerebral autoregulation[1,2,8]. Any variations of FV caused by changes in CPP may be interpreted in terms of whether CBF is remaining stable. Our method presumes that in head injured patients spontaneous waves of CPP are almost always present, and therefore an analysis of even small changes in FV waveform caused by endogenous variations in CPP permits the continuous assessment of cerebral autoregulation. Although the results are usually easy to interpret in individual cases, is it possible to reveal any statistically significant patterns in a large population of head injured patients?

Materials and Methods

Thirty five head injured patients were included in the study. All patients were sedated, paralysed and ventilated with $PaCO_2$ maintained between 3.3 and 4.5 kPa. ICP was monitored using a Camino transducer placed subdurally (n = 27) or by a subdural drain connected to an external pressure transducer (n = 8). ABP was measured using an arterial line. TCD blood flow velocity in the middle cerebral artery was recorded daily for periods of 10 to 80 min. The electrical signals from the analog outputs of pressure monitors and TCD ultrasonograph (SciMed, Bristol, UK) were sampled (50 Hz), digitized (12 bit analog-to-digital converter DT 2814, USA), calibrated in appropriate units and stored in computer memory using customized software (W. Zabolotny, Warsaw University of Technology, Poland). The signals were then analyzed (TCMR, M. Czosnyka) to reveal changes in FV and ICP waveforms caused by variations of CPP; the following parameters were calculated:

– flow velocity time average (FV), systolic (FVs) and diastolic (FVd);

– the amplitudes of the fundamental harmonic components of the FV (FVa), ICP (ICPa) and ABP (ABPa) waveforms;
– and cerebral perfusion pressures: time average (CPP), systolic (CPPs) and diastolic (CPPd).

The additional parameters: pulsatility index (PI = (FVs – FVd) / FV), standardized pulsatility index (SPI = PI/ABPa), moving correlation coefficient between mean ICP and amplitude of its pulse wave (RAP), flow velocity diastolic to time average ratio (FVd / FV), estimate of cerebrovascular resistance (CPP / FV) calculated for time average (CVR), systolic (CVRs) and diastolic (CVRd) values of FV and CPP, respectively, were also computed. FVd / FV ratio reflects the proportion of continuous to total (i.e. pulsatile plus continuous) blood inflow. Continuous assessment of autoregulation was performed using multifactorial analysis of the FV waveform parameters with varying CPP. All recordings were analyzed individually; one minute time averages of selected parameters were submitted to statistical analysis (Statgraphic+, Manugistic, USA) and their dependence on CPP was evaluated using analysis of variance.

Results

The parameters recorded during a plateau wave of ICP are shown in Fig. 1. In such individual patients diastolic flow velocity started to fall much earlier than systolic flow velocity; pulsatility index steadily increased, CVR decreased and then reached the minimum at a very low CPP (below 35 mmHg). At the same critically low CPP, the RAP coefficient fell significantly.

The results of the statistical analysis of 2800 one minute averages of selected parameters are presented in Fig. 2. Mean values of FV time average, systolic and diastolic, are shown (along with 95% confidence limit) in plot A. The threshold for autoregulation of FV time average and diastolic was detected (as the point below which the FV time average decreased significantly below 90% of baseline recorded between 60 and 70 mmHg) at CPP 55 mmHg (see vertical line A on Fig. 2B). However, below 55 mmHg FV time average and diastolic were significantly correlated with CPP (r = 0.41 and 0.60 respectively; p < 0.0001). Flow velocity systolic did not exhibit a consistent threshold even at very low CPP's. All the flow velocities started to increase with CPP above 90 mmHg (r > 0.35, p < 0.001). CVR time average, systolic and diastolic (Fig. 2C) showed a good correlation with CPP within the middle range of pertusion pressures (for CVR diastolic from CPP 85 to 55 mmHg, r = 0.378, p < 0.0001; for CVR systolic from 95 to 20 mmHg, r = 0.6766. p < 0.00001). CVR diastolic reached the minimum and was not significantly correlated to CPP between 55 to 30 mmHg and started to rise significantly with CPP falling below 30 mmHg (r = –0.53, p < 0.0001). CVR systolic did not exhibit any breakpoints for low CPP.

The relationships between the amplitudes of ICP and FV pulse waveforms versus CPP showed a gradual increase with falling CPP down to 35 mmHg (r = –0.37 for ICPa and r = –0.24 for FVa; p < 0.001). Below this level they fell sharply with CPP (r = 0.38 for ICPa and r = 0.22 for FVa; p < 0.01). The averaged pulsatility index (Fig. 2.) showed considerable variation with higher CPP and then a significant increase when CPP fell below 55 mmHg. The standardized pulsatility index rose gradually for CPP's below 95 mmHg, but below 55 mmHg this gradient became much steeper. The correlation coefficient for the inverse regression model of SPI and PI versus CPP is higher for SPI (r = 0.58) than for PI (r = –0.40); the significance of the difference between these coefficients is less than 0.01. The parameters, RAP and the ratio of FV diastolic to time average showed a very similar relationship to CPP with both falling significantly when

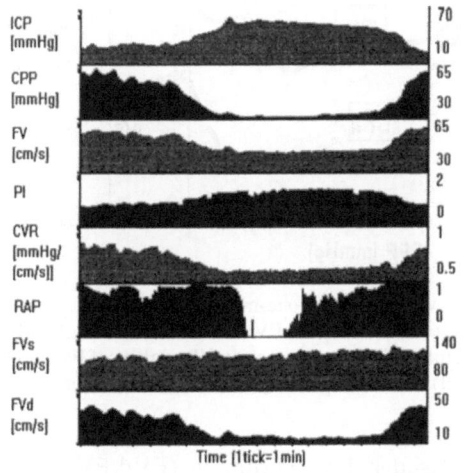

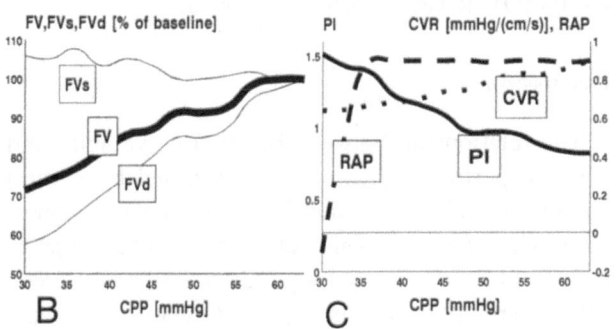

Fig. 1. (A) Time plot of parameters. *ICP* intracranial pressure, *CPP* cerebral perfusion pressure, *FV, FVs, FVd* flow velocity time average, systolic, diastolic, *PI* pulsatility index, *CVR* cerebrovascular resistance, *RAP* correlation coefficient between amplitude of pulse wave and mean ICP. (B) Relationships between mean FVs, FV and FVd versus CPP. (C) Relationships between PI, CVR and RAP versus CPP recorded during 12 minute period of elevated ICP (plateau wave)

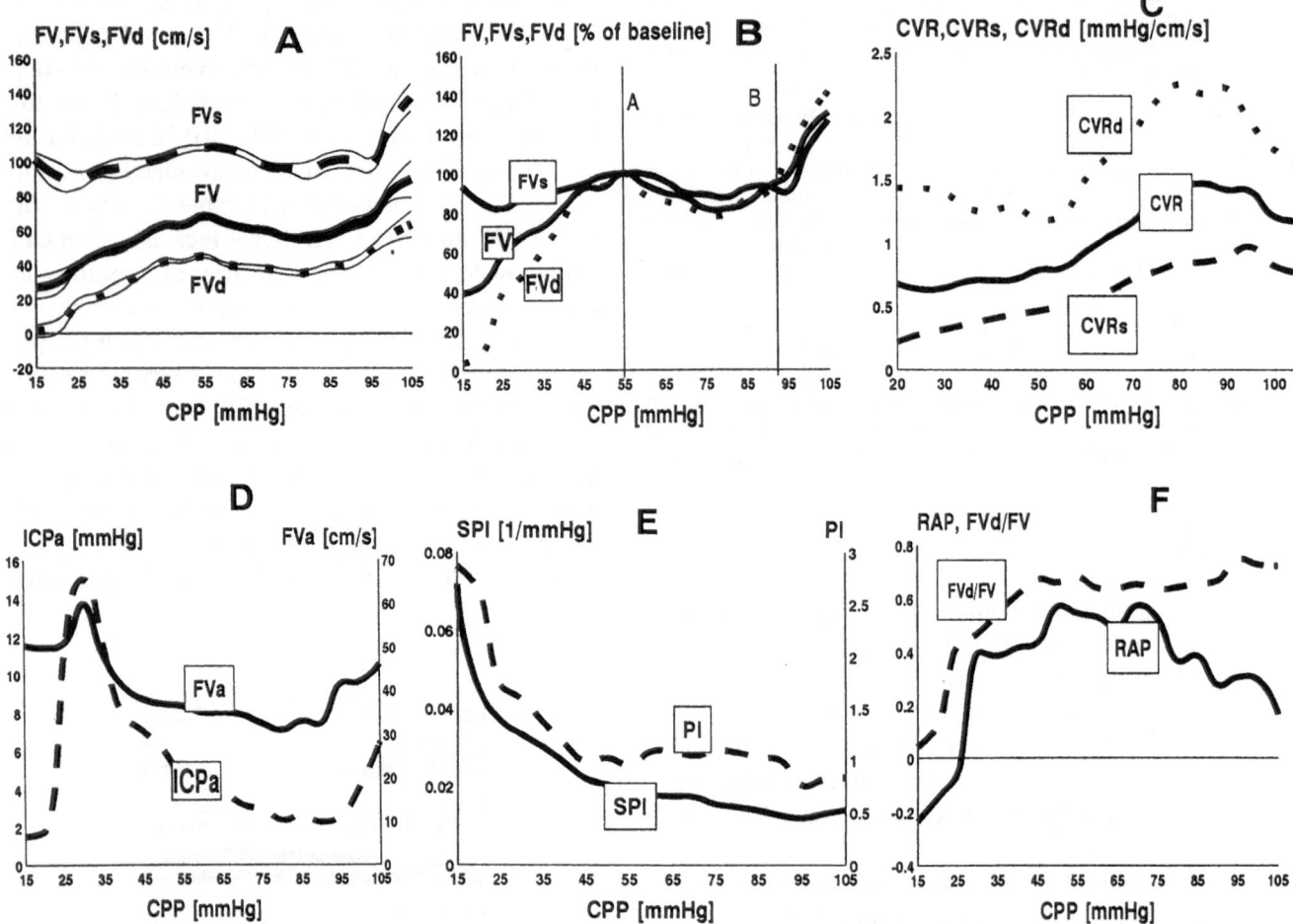

Fig. 2. Mean values of 2800 one-minute averaged parameters expressed as functions of CPP (x-axis): (A) FVs, FV, FV – absolute values; (B) percentage changes in FV, FVd, FVs (100% is the baseline recorded for CPP from 55 to 75 mmHg); (C) CVR calculated for diastolic (CVRd), systolic (CVRs) and mean (CVR) values of FV and CPP waveforms; (D) amplitudes of ICP (ICPa) and FV (FVa) waveforms; (E) pulsatility (PI) and standardized pulsatility (SPI) indices; (F) RAP and FV diastolic to FV time average ratio

CPP decreased below 35 mmHg (FVd/FV: r = 0.88, p < 0.0001; RAP: r = 0.48, p < 0.001).

Discussion

The relationship between blood flow velocity and cerebral perfusion pressure is interpreted by several authors as a reliable method for assessment of autoregulation in experimental models[2,8,9] and in clinical practice. Dahl *et al.*[6] showed a good correlation between SPECT and TCD studies in patients with cerebrovascular diseases. Aaslid *et al.*[1] demonstrated that, during step decreases in CPP of 20 mmHg, the change of the lumen of the basal artery was minimal and hence dynamic study of FV versus varying CPP may reveal the force of the autoregulatory reflex. Giller[7] and Czosnyka *et al.*[4] presented the transient hyperaemic response of FV after a decrease in CPP produced by

carotid artery compression lasting 5–7 seconds as evidence of intact autoregulation. Chan *et al.*[3] showed in head injured patients that increasing TCD pulsatility index correlated well with decreases of global CBF in head injured patients (measured indirectly by jugular bulb oxygen saturation).

Monitoring of FV and CPP gives an opportunity to assess the cerebrovascular reserve continuously. Our results with the averaging of more than 140 independent periods of monitoring in 35 patients strongly support our methodology. Time averaged flow velocity starts to decrease below a CPP of 55 mmHg- lower than the averaged limit of autoregulation reported by Chan *et al.*[3]. It is remarkable that while flow velocity diastolic decreases significantly with decreasing CPP below this limit, flow velocity systolic remains constant (Fig. 2B) causing an increase in the systolic-diastolic amplitude of FV waveform and increase in pulsatility index (Fig. 2D, E). The discrepancy between the behaviour of FV systolic and diastolic reflects the difference in conditions of blood transport tor systolic and diastolic CPP (for CPP time average of 55 mmHg, mean systolic CPP was 85 and diastolic CPP was 35 mmHg). FV diastolic expressed as a function of diastolic CPP exhibits an autoregulation breakpoint at 34 mmHg. FV systolic does not show any breakpoint because in our material the lowest systolic CPP was 35 mmHg. Is then 34 mmHg a real breakpoint of autoregulation? There is a wide transient zone reported[8] within which autoregulation is still intact but unable to keep CBF at a constant level. The first sign of entering this zone is the decrease in diastolic FV when CPP diastolic falls below this critical level. As PI Increases, CBF starts to decrease slightly[3] and the amplitude of FV waveform starts to increase[8]. This transient zone is as wide as the CPP peak to peak pressure. This finding supports the methodology of continuous monitoring of autoregulation using the comparison of moving regression coefficients between FV systolic, diastolic and CPP: When the gradient of FV diastolic versus CPP is much higher then for FV systolic, the transient zone between intact and impaired autoregulation has been entered already. If both gradients are equal with strict correlation between flow velocity and CPP, autoregulation is exhausted. Similar gradients with poor or negative correlation between FV and CPP indicate a good autoregulatory reserve. The upper breakpoint of autoregulation was detected at a much lower CPP level (95 mmHg) than it has been reported experimentally. This point requires further study in the context of management of head injury where there is a high incidence of hyperaemia.

The material presented provides an excellent opportunity to compare the different indices of FV pulsatility. The pulse waveform of blood flow velocity depends not only on the state of the cerebrovascular bed but also on the absolute value of the pulse wave of arterial blood pressure (ABPa). Good correlation between ABPa and CPP ($r = 0.643$ $p < 0.0001$) makes standardized pulsatility index correlate better with CPP than Gosling PI. However, it is obvious that any kind of pulsatility indices evaluated using joint changes of FV systolic and diastolic is less accurate than separate analysis of systole and diastole FV.

The correlation between FV findings and the results of ICP pulse waveform analysis reveals that FV and ICP waveform amplitudes increase with decreasing CPP and both reach breakpoints at a CPP time average of 35 mmHg. The increase in ICPa is better correlated to CPP than changes in FVa which starts to rise with decreasing CPP only below 50 mmHg. This seems to be in accordance with both experimental findings[8] and modelling predictions[5]. The correlation coefficient between pulse wave amplitude and mean ICP level (RAP) starts to decrease below 55 mmHg and abruptly switches to negative values for CPP below 35 mmHg. This confirms our earlier interpretation about this coefficient as an index of exhausted autoregulatory reserve. The specific similarity between the course of RAP and ratio of diastolic flow velocity to time average FV (Fig. 2F) shows that an increase of pulsatile blood inflow can be interpreted as a symptom of diminishing autoregulatory reserve. Blood is supplied more easily in pulsations requiring less energy expenditure than in continuous stream. From this point an increase in blood flow pulsatility can be interpreted as defensive reflex aimed against disturbances in brain blood supply when CPP is too low.

References

1. Aaslid R, Lindegaard KF, Sorteberg W, et al (1989) Cerebral autoregulation dynamics in human. Stroke 20: 45–52
2. Barzo P, Doczi T, Csete K, et al (1991) Measurements of regional cerebral blood flow and blood flow velocity in experimental intracranial hypertension: infusion via cisterna magna in rabbits. Neurosurgery 28: 821–25
3. Chan KH, Miller DJ, Dearden M, et al (1992) The effect of changes in cerebral perfusion pressure upon middle cerebral artery blood flow velocity and jugular bulb venous oxygen saturation after severe brain trauma. J Neurosurg 77: 55–61
4. Czosnyka M, Pickard JD, Whitehouse H, et al (1992) The hyperaemic response to a transient reduction in cerebral perfusion pressure. A modelling study. Acta Neurochir (Wien) 115: 90–97
5. Czosnyka M, Harris NG, Pickard JD, Piechnik S (1993) CO_2 reactivity as a function of perfusion pressure – a modelling study. Acta Neurochir (Wien) 121: 159–165
6. Dahl A, Lindegaard KF, Russell D, Nyberg-Hansen R, Rootwelt K, Sorteberg W, Nornes H (1992) A comparison of transcranial doppler and cerebral blood flow studies to assess cerebral vascoreactivity. Stroke. 23: 15–19
7. Giller CA (1991) A bedside test for cerebral autoregulation using transcranial Doppler ultrasound. Acta Neurochir (Wien) 108: 7–14
8. Nelson RJ, Czosnyka M, JD Pickard, et al (1992) Experimental aspects of cerebrospinal haemodynamics: the relationship between blood flow velocity waveform and cerebral autoregulation. Neurosurgery 31: 705–710
9. Richards G, Czosnyka M, Pickard JD (1993) Correlation between basilar artery flow velocity and cortical laser doppler flux during haemorrhagic hypotension in the rabbit: the importance of estimating laser doppler biological zero. Am J Physiol: submitted for publication

Correspondence: Marek Czosnyka, M. D., Neurosurgical Unit, Block A, Level 4, Addenbrooke's Hospital, Hills Road, Cambridge CB2, 2QQ, U.K.

Acta Neurochir (1994) [Suppl] 60: 472–474
© Springer-Verlag 1994

Gadolinium DTPA-Enhanced Magnetic Resonance Imaging of Cerebral Contusions

H. Kushi[1], **Y. Katayama**[1], **T. Shibuya**[1], **T. Tsubokawa**[1], and **T. Kuroha**[2]

Departments of [1] Neurological Surgery and [2] Radiology, Nihon University School of Medicine, Tokyo, Japan

Summary

The morphological characteristics of cerebral contusions in head trauma patients suggest that an increase in cerebrovascular permeability is responsible for the contusion edema which develops within 1–3 days posttrauma. In the present study, 10 patients with cerebral contusions (mean age, 38 years old; 8 males and 2 females) were examined by gadolinium (Gd)-DTPA enhanced magnetic resonance imaging (MRI) at 1–2 days after trauma. Gd-DTPA (0.3 mmol/kg) was infused intravenously over a period of 30 min. MRIs were taken before, and at 2 and 4 hours after initiation of the Gd-DTPA administration. It was found that contusion edema areas were frequently enhanced by Gd-DTPA at 2 hours. The enhancement diminished at 4 hours. These findings appear to be inconsistent with the results of previously reported similar studies in which enhancement was detected at 6–9 days posttrauma but not during the period earlier than 6 days. This discrepancy may be attributable to the presence of a high blood concentration of Gd-DTPA for a longer period of time and a delay in the time at which MRIs were taken in the present study. The present data indicate that an increased cerebrovascular permeability occurs at as early as 1–2 days posttrauma, and suggest that contusion edema which progresses during the initial 1–3 days may be at least partially vasogenic in nature.

Keywords: Cerebral contusion; head trauma; vasogenic edema; magnetic resonance imaging.

Introduction

Cerebral contusions in head trauma patients cause progressive contusion edema, sometimes leading to clinical deterioration during the initial 1–3 days posttrauma[4]. An increase in cerebrovascular permeability has been shown to develop in experimental models of cerebral contusion (e.g.,[10]) In addition, the morphological characteristics of contusion edema in head trauma patients[3,4] suggest that an increased cerebrovascular permeability underlies its progressive development.

It has been reported, however, that Gd-DTPA enhanced magnetic resonance imaging (MRI) revealed no evidence of an increased cerebrovascular permeability in patients with cerebral contusions during the initial few days posttrauma[5]. Furthermore, an immunohistochemical study of the post-mortem brain of such patients failed to reveal any plasma protein leakage around the contused brain areas during this early period[9]. Such findings appear to contradict the widely held view that an increased cerebrovascular permeability is responsible for the development of contusion edema.

It has been reported that an increase in cerebrovascular permeability can be detected more clearly by intravenous slow infusion of contrast medium and delayed computerized tomography (CT) scanning[2]. In the present study, we examined the changes in cerebrovascular permeability in patients with cerebral contusions during the initial 1–2 days posttrauma, employing Gd-DTPA enhanced MRI with a similar technique.

Patients and Methods

A total of 10 patients with cerebral contusions were included in the present study. The 10 patients consisted of 8 males and 2 females and their mean age was 38 years old (range, 17–58). Six patients had sustained a head trauma due to traffic accidents and 4 due to falls. On admission, 4 patients exhibited disturbance of consciousness and scored less than 8, while the remaining 6 patients were scored 9–14 on the Glasgow Coma Scale (GCS). Plain craniograms revealed linear fractures in 8 patients. No fracture was observed in the remaining 2 patients. CT scans on admission demonstrated cerebral contusions in 9 patients (frontal lobe in 4; temporal lobe in 5). One of them was complicated with shearing injury. In another patient, contusional hemorrhage occurred in the left temporal lobe after removal of an acute epidural hematoma on the right side. The cerebral contusions in all of the 10 patients were not restricted to the cortex but also extended into the white matter.

CT scans (Shimadzu 200T) and MRI (Toshiba 200 FX, 1.5 Tesla; or Hitachi NRH 500, 0.5 Tesla) were performed in all patients. CT scanning was carried out 3 times, i.e., within 24 hours, at 36–48 hours, and at approximately 1 week posttrauma. MRIs of horizontal and coronary sections without enhancement and Gd-DTPA enhanced MRIs were taken within 1–2 days post-trauma. Gd-DTPA at a dose of 0.3 mmol/kg[7] was then infused intravenously over a period of 30 min. Gd-DTPA enhanced MRIs were taken at 2 and 4 hours after initiation of the Gd-DTPA infusion.

Results

MRIs without enhancement clearly depicted contusion edema extending along the white matter fibers on T2-weighted images, as reported previously. The technique employed in the present study for Gd-DTPA enhanced MRI disclosed that cerebral contusions can be enhanced at as early as 1–2 days posttrauma. In MRIs taken at 2 hours after Gd-DTPA administration, enhancement on T1-weighted images was observed either in areas of contusion necrosis, in areas of contusion edema or both. Enhancement was also noted in the damaged area of the corpus callosum due to shearing injury. After 4 hours, the enhancement on MRI diminished markedly.

Enhanced CT scans taken within 1–2 days posttrauma with conventional technique revealed only localized enhancement in some cases. The enhancement observed by CT scans during this period was dissmilar to that observed by Gd-DTPA enhanced MRI. At 1 week posttrauma, CT scans demonstrated enhancement of contusion edema areas in most cases, which resembled to the enhancement observed by MRI earlier.

A representative case is illustrated in Fig. 1. This 25-year-old male sustained a head trauma due to a fall. He was admitted to hospital with disturbance of consciousness and scored 9 on the GCS. Contusional hemorrhage was observed in the left frontal lobe on CT scans. In MRIs taken at 48 hours posttrauma, T2-weighted images without enhancement revealed contusion edema extending along the white matter fibers (Fig. 1). T1-weighted images taken at 2 hours after Gd-DTPA administration demonstrated enhancement of contusion edema areas in the white matter surrounding the cerebral contusion (cf. Fig. 2A and B, inidicated by

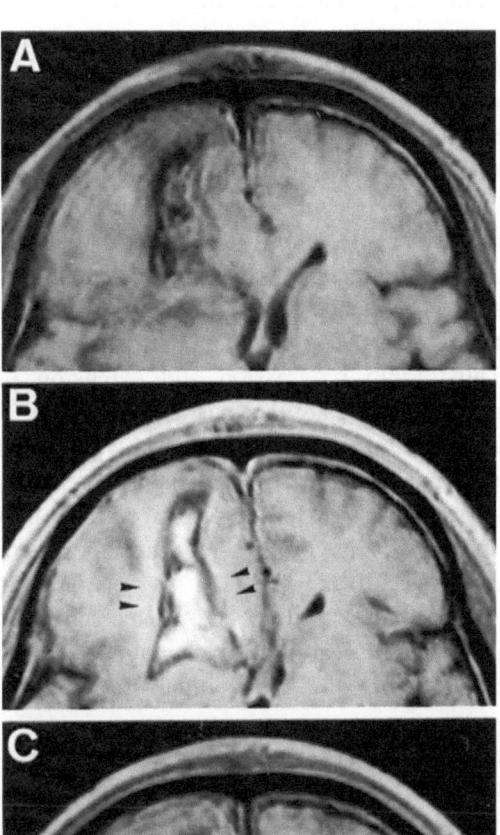
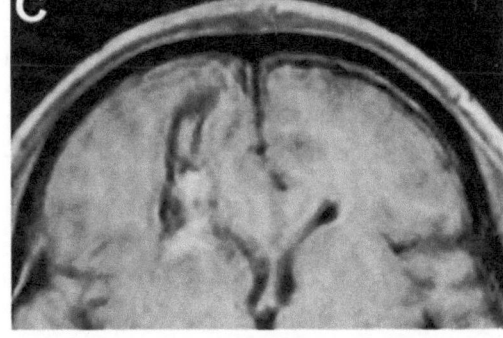

Fig. 2. MRIs of the same patient as shown in Fig. 1 (48–52 hours posttrauma). (A) T1-weighted images without enhancement. (B) T1-weighted images taken at 2 hours after Gd-DTPA administration. Note the enhancement of the contusion edema areas in the white matter surrounding the cerebral contusion (arrow). (C) T1-weighted images taken at 4 hours after Gd-DTPA administration. Note that the enhancement has diminished

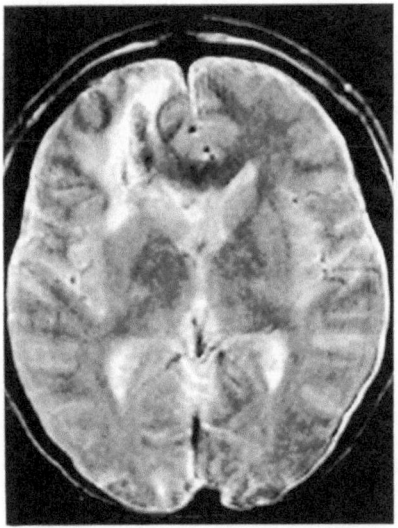

Fig. 1. MRI of a 25-year-old male (48 hours posttrauma). T2-weighted images without enhancement. Note the contusion edema extending along the white matter fibers

arrows). The enhancement was diminished at 4 hours after initiation of Gd-DTPA administration (cf. Fig. 2B and C).

Discussion

Since Gd-DTPA, like plasma protein, does not normally cross the blood-brain barrier, enhancement with Gd-DTPA, if observed in association with cerebral contusions, implies an increased cerebrovascular permeability. In a previous study[5], Gd-DTPA (0.2 mmol/kg) was administered intravenously by bolus injection and MRI was undertaken soon after the administration. Such a technique failed to detect any increase in cerebrovascular permeability associated with cerebral contusions during the initial few days posttrauma. Enhancement with Gd-DTPA has been reported to become detectable at 6–9 days posttrauma, which we confirmed by enhanced CT scans in the present study. This is clearly inconsistent with previous findings for experimental models of cerebral contusion[10] and those obtained in head trauma patients in the present study.

The tissue compliance of the brain is essentially small at the early stage of development of vasogenic edema. Thus, rapid progression and extension of edema require a large capillary perfusion pressure[6,8]. When the extracellular space is enlarged by persistent vasogenic edema[1], however, the tissue compliance increases and even a slight elevation of capillary perfusion pressure can cause progression and extension of vasogenic edema[6,8]. It may be that, by means of the conventional technique for Gd-DTPA enhanced MRI, the increase in cerebrovascular permeability becomes detectable only after the extracellular space is enlarged.

In order to detect vasogenic edema earlier by Gd-DTPA enhanced MRI, it appears necessary to maintain a sufficiently high blood concentration of Gd-DTPA for a longer period of time and to allow for delays caused by the slow penetration of Gd-DTPA into edema areas. The present results demonstrate that an increased cerebrovascular permeability occurs in patients with cerebral contusions at as early as 1–2 days posttrauma. This is consistent with previous findings for experimental models of cerebral contusion (e.g.,[10]) and suggests that contusion edema which progresses during the initial 1–3 days may be at least partially vasogenic in nature.

References

1. Hirano A (1969) The structure of brain edema. In: Bourne GH (ed) The structure and function of nervous system. Academic Press, New York, pp 89–135
2. Ito U, Reulen H-J, Tomita H, *et al* (1988) Formation and porpagation of brain oedema fluid around human brain metastasis. A CT study. Acta Neurochir (Wien) 90: 35–41
3. Katayama Y, Tsubokawa T, Kinoshita K, *et al* (1992) Intraparenchymal fluid-blood levels in traumatic intracerebral hematomas. Neuroradiology 34: 381–383
4. Katayama Y, Tsubokawa T, Miyazaki S, *et al* (1990) Oedema fluid formation within contused brain tissue as a cause of medically uncontrollable elevation of intracranial pressure in head trauma patients. Acta Neurochir (Wien) [Suppl] 51: 308–310
5. Lang DA, Hadley DM, Teasdale GT, *et al* (1991) Gadolinium DTPA enhanced magnetic resonance imaging in acute head injury. Acta Neurochir (Wien) 109: 5–11
6. Marmarou A, Takagi H, Shulman K (1980) Biomechanics of brain edema and effects on local cerebral blood flow. In: Advances in neurology, Vol 28. Raven, New York, pp 345–358
7. Niendorf HP, Haustein J, Louton T, *et al* (1991) Safety and tolerance after intravenous administration of 0.3 mmol/Kg Gd-DPTA. Results of randomized, controlled clinical trial. Invest Radiol [Suppl] 26: 221–223
8. Reulen HJ, Graham R, Spatz M, *et al* (1977) Role of pressure gradients and bulk flow in dynamics of vasogenic brain edema. J Neurosurg 46: 24–35
9. Todd NV, Graham DI (1990) Blood-brain barrier damage in traumatic brain contusion. Acta Neurochir (Wien) [Suppl] 51: 296–299
10. Tornheim PA, Prioleau GR, McLaurin RA, *et al* (1984) Acute responses to experimental blunt head trauma topography of cerebral cortical edema. J Neurosurg 60: 473–480

Correspondence: Hidehiko Kushi, M. D., Department of Neurological Surgery, Nihon University School of Medicine, Tokyo 173, Japan.

Acta Neurochir (1994) [Suppl] 60: 475–478
© Springer-Verlag 1994

Neurochemical Monitoring and On-Line pH Measurements Using Brain Microdialysis in Patients in Intensive Care

H. Landolt, H. Langemann, A. Mendelowitsch, and **O. Gratzl**

Neurosurgical Clinic, Kantonspital, Aarau, Switzerland

Summary

We will report on our preliminary findings using microdialysis to monitor three patients In intensive care with either severe head injury (SHI) or severe subarachnoid hemorrhage (SAH) for up to 72 hours. In addition, basal levels in uninjured brain were assessed during an extra-intracranial bypass operation. Samples were collected hourly or half-hourly (flow rate 2 μl/min, perfusion medium 0.9% saline). Parameters measured were the antioxidants ascorbic acid, uric acid, glutathione and cysteine. In 2 patients, the pH of the dialysate (pH$_D$) was also measured on-line with a specially constructed flow-through meter, and glucose and lactate levels were assessed in the dialysate. In patient 1 (SHI), there was practically no cerebral perfusion pressure because of high ICP; cysteine and lactate levels were very high and glucose not measurable. In patient 2 (SAH) a hypoxic episode was accompanied by increased uric acid and decreased glucose. In patient 3 (SHI), the pH$_D$ reflected normalisation of blood gases after hyperventilation. Results indicate that parameters are in the range known from experimental studies, and can be correlated with clinical situations. The pH$_D$ as valuable indicator of metabolic changes is also feasible bedside.

Keywords: Microdialysis; human; hydrogen ion concentration; antioxidants.

Introduction

Monitoring in patients in neurosurgical intensive care is limited to EEG, ICP, blood velocity and recently also oxygen saturation measurement. The only approach to measuring cerebral metabolism continously is by the assessment of arterial-venous difference between lactate and oxygen from the jugular bulb. However, recently microdialysis has been used to continuously monitor neurochemical and metabolic dynamics in neurosurgical patients (parameters measured include lactate-pyruvate ratio and amino acids)[6]. Our reports preliminary findings used cerebral microdialysis for monitoring patients in intensive care with severe head injury (SHI) or severe subarachnoid hemorrhage (SAH), who were treated with either craniectomy, routine ICP-measurement or open ventricular drainage. The aim was to show the feasibility of the method, and to correlate levels of the antioxidants ascorbic acid, uric acid, cysteine, glutathione, and glucose, lactate and on-line pH (pH$_D$) in the dialysates with traditional parameters and clinical course. In previous experimental studies we found significant changes in all these parameters during focal cerebral ischemia in the rat[4].

Methods

The microdialysis probe (CMA 20, 4 mm membrane, 0.5 mm diameter, sterilised with ethylene dioxide at 60°C) was implanted into the cortex of the patient after a 2 mm incision of the dura, either through a craniectomy for intracranial hematoma, or through a burr hole for routine ICP measurement. Perfusion with 0.9% saline was started before implantation. Flow rate was 2 μl/min and fractions were collected every 30 or 60 min. All apparatus was obtained from Carnegie Medecine (Stockholm, Sweden). The samples were frozen as soon as possible. Ascorbic acid, uric acid, cysteine and glutathione were determined in the dialysates (20 μl) using high pressure liquid chromatography with electrochemical detection[2]. The recoveries at room temperature of the probes used were subsequently determined in vitro for the parameters measured.

The following special procedures were performed in different cases, according to the permission of the local Ethical Committee:

1. In a preliminary measurement, a probe was inserted into the temporal cortex during an extra-intracranial bypass operation in general anesthesia, and fractions were collected for 2 h. The aim was to assess basal levels in uninjured human brain.

2. Fractions were collected hourly or half-hourly for between 8 and 72 h in 3 patients with severe head injury or severe subarachnoid hemorrhage.

3. In patient 1, flow was reduced to 1 μl/min for an hour, in an attempt to assess true extracellular levels[3].

4. In patients 2 and 3, the PH$_D$ was continuously monitored using a specially constructed pH meter (Hamilton, Bonaduz, Switzerland), mounted between the probe and the fraction collector[5].

5. In patients 1 and 2, glucose and lactate were also measured using enzymatic determination; glucose colorimetrically in 15–20 µl using a kit (Trinder from Sigma), lactate fluorometrically in 30–40 µl using lactate dehydrogenase.

Results

In the patient with the extra-intracranial bypass operation, all four antioxidants were detectable in the dialysate. Levels, which were high 30 min after insertion, quickly approached basal values. After 1.5 h these were: ascorbic acid, 7.98 µM, uric acid, 0.72 µM, cysteine, 0.065 µM, glutathione, 0.825 µM, levels comparable to those which we previously obtained in rats[4].

Patient 1 (male, age 47 years) had an emergency craniectomy for a large subdural hematoma. The probe was inserted into the underlying cortex after evacuation of the hematoma. The clinical data and concentrations in the dialysate are shown in Table 1. In spite of treatment, intracranial pressure was so high that there was only minimal cerebral perfusion pressure. Cysteine levels were massively elevated compared with those of other patients. Ascorbic acid, uric acid and glutathione levels were not remarkably different. The flow was reduced to 1 µl/min for one hour, in an attempt to extrapolate to real extracellular levels at virtual no-flow. The extrapolated values obtained were as follows: ascorbic acid: 10.5 µM, uric acid: 5 µM; cysteine: 24 µM; glutathione: 0.5 µM. The patient died 6 h later (Fig. 1).

Patient 2 (female, 60 years, Glasgow Coma Scale 5 on admission) received an open ventricular drainage after subarachnoid hemorrhage (Hunt-Hess grade IV). Microdialysis samples were collected for 72 hours and PH_D measured subsequently for 13 hours. The concentrations of neurochemical parameters and clinical data are given in Fig. 2A and B.

During a temporarily extubated period with insufficient spontaneous respiration (drop of arterial blood oxygen) there was an early marked decrease in glucose, followed by a 4-fold increase in uric acid. The on-line pH_D during ventilation was stable between 6.4 and 6.5, a value similar to that obtained in experimental animals[5].

Patient 3 (female, 29 years, intubated and sedated on admission) had a severe head injury with diffuse contusions and was monitored neurochemically using on-line pH during 10 hours. At the beginning the patient was hyperventilated, and the pH_D followed the normalisation of blood gases from 6.7 (blood pH 7.55, CO_2 3.2 kPa) to 6.45 (blood pH 7.4, CO_2 5.0 kPa). After this the patient could be extubated.

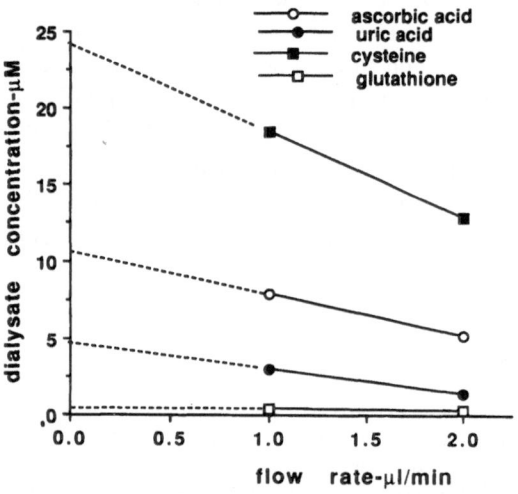

Fig. 1. Patient 1: The relation of flow rate of the perfusion medium (0.9% NaCl) to concentrations of neurochemical parameters in the dialysate. The dotted lines indicate extrapolation to virtual no-flow. For clinical data see Table 1

Table 1. *Clinical Data and Concentrations of Neurochemical Parameters in Patient 1 During Monitoring in Intensive Care*

Clinical data					Dialysate concentrations – µM (flow 2 µl/min)					
Hours	GCS	ICP	CPP	Plasma gluc	Asc	UA	Cys	GSH	Gluc	Lac
0	3	50	31							
0.5	3	45	31	12.3	2.33	1.02	3.64	0.27	< 7	858
1.5	3	43	41							
2.5	3	48	35	6.8						
3.5	3	51	36							
4.5	3	49	36		5.21	1.4	1 2.84	0.27	< 7	866

GCS Glasgow coma scale; *ICP* intercranial pressure in mmHg; *CPP* cerebral perfusion pressure in mmHg; *Plasma gluc* concentration of glucose in plasma in mM; *Asc* ascorbic acid; *UA* uric acid; *Cys* cysteine; *GSH* glutathione; *Gluc* glucose; *Lac* lactate.

Table 2. *The in vitro Relative Recoveries of the Probes Used, Measured After the End of Neurochemical Monitoring.* Values are in percent

%	Ascorbic acid	Uric acid	Glutathione	Cysteine	Glucose	Lactate
Bypass	19	29	7	7		
Patient 1	15	20	11	10	29.5	29
Patient 2		7			2	2
Patient 3	10	13	1.4	1.6		

The in vitro relative recoveries of all probes are given in Table 2. The probe used in patient 2 was accidently sterilised with ethylene dioxide at a high temperature (100°C). There was a problem with stability of glutathione and cysteine when collection time was more than half a hour. Bacteriological smears of all 4 probes after removal were negative.

Discussion

At the moment clinical cerebral microdialysis represents the only method to monitor multiple neurochemical parameters in intensive care. Until now no direct chemical information was available from the extracellular space. Our findings show that neurochemical monitoring and on-line pH_D measurements are technically feasible in patients in neurosurgical intensive care. The acceptance by nursing personnel and physicians was high. No major technical troubles occurred. Bacteriology of the probe after removal was negative in all 3 cases. The longest duration of monitoring was 72 hours.

First observations also indicate correlations of the parameters with clinical situations:

In patient 1 (Table 1) the increase in cysteine correlated with the experimental findings during focal ischemia (1992). There was a 2-fold increase of lactate and a disappearance of glucose, as we found during focal ischemia in rats (unpublished results).

In patient 2 (Fig. 2A, B), although plasma glucose was in normal range, glucose in the dialysate started to fall rapidly before extubation. This was followed by a hypoxic episode. Subsequently uric acid rose, and decreased after artificial ventilation was restored. The Glasgow Coma Score deteriorated after extubation. The pH_D was similar to that found in animal experiments[5]. It is remarkable that glucose in the dialysate fell before the beginning of the clinical deterioration. This finding might therefore be related to the cause of clinical impairment, rather than being the consequence of it. The hypoxic event would then be consequent to the clinical worsening. These observations suggest that glucose in the dialysate would be a valuable early neurochemical parameter. In patient 3 pH_D followed blood gas values during hyperventilation and normoventilation. This could correlate with changes of carbondioxide or pH in the extracellular space[5]. Further

A)

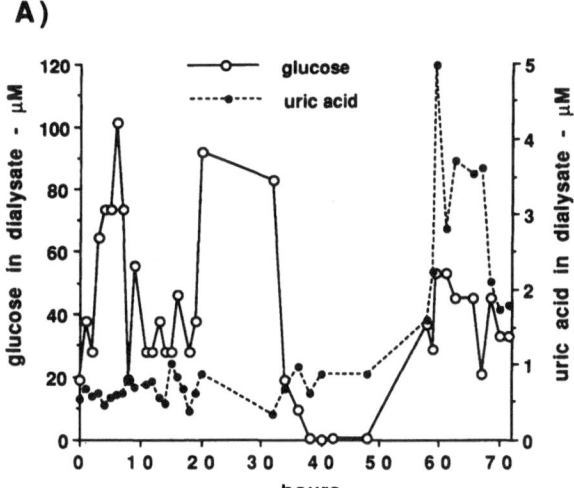

B)

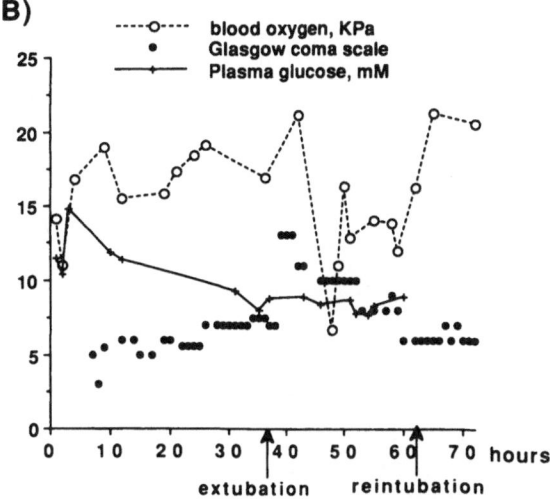

Fig. 2. Patient 2: Concentrations of uric acid and glucose in the dialysate (A) and clinical data (B) during monitoring with microdialysis

observations are necessary to assess the value as an early bedside parameter for critical cerebral disturbances. An advantage of microdialysis is that several parameters can be analysed at each time point represented by a single sample. On-line pH measurement allows continuous bedside evaluation of the acid-base balance of the dialysate. The future strategy for clinical cerebral monitoring with microdialysis could include both measurements of still intact brain to prevent secondary lesions[1] as well as measurements in areas of therapeutic interest (contusions, hemorrhage, edema)[6]. In the latter case it could help to assess the efficacy of different therapeutic tools.

References

1. Graham DI, Ford I, Adams JH, Doyle D, Teasdale GM, Lawrence AE, McLellan DR (1989) Ischaemic brain damage is still common in fatal non-missile head injury. J Neurol Neurosurg Psychiatry 52: 346–350
2. Honegger CG, Langemann H, Krenger W, Kempf A (1989) Liquid chromatographic determination of common water soluble antioxidants in biological samples. J Chromatogr 487: 453–468
3. Jacobson J, Sandberg M, Hamberger A (1985) Mass transfer in brain dialysis devices – a new method for the estimation of extracellular amino acids concentration. J Neurosci Meth 15: 263–268
4. Landolt H, Lutz TW, Langemann H, Stäuble D, Mendelowitsch A, Gratzl O (1992) Extracellular antioxidants and amino acids in the cortex of the rat: monitoring by microdialysis of early ischemic changes. J Cereb Blood Flow Metab 12: 96–102
5. Landolt H, Langemann H, Gratzl O (1993) On-line monitoring of cerebral pH using microdialysis. J Neurosurg June 1993
6. Persson L, Hillered L (1992) Chemical monitoring of neurosurgical intensive care patients using intracerebral microdialysis. J Neurosurg 76: 72–80

Correspondence: H. Landolt, M. D., Neurosurgical Clinic, Kantonspital, CH-5001 Aarau, Switzerland.

Acta Neurochir (1994) [Suppl] 60: 479–481
© Springer-Verlag 1994

Regional Cerebral Blood Flow Trends in Head Injured Patients with Focal Contusions and Cerebral Edema

M. J. Alexander, N. A. Martin, R. Khanna, M. Caron, and **D. P. Becker**

Division of Neurosurgery, U.C.L.A. Medical Center, Los Angeles, CA, U.S.A.

Summary

Focal contusions following head injury may be associated with focal or diffuse cerebral edema. Early global hyperemia and perifocal hyperemia may play a role in cerebral edema, although causal relationships have yet to be clearly been defined. We studied 27 patients with head injury (admission GCS 3–12) resulting in focal contusions (without evidence of subarachnoid, intraventricular or intraparenchymal hemorrhage by CT). Patients were studied with ICP monitors, head CTs, and intravenous ^{133}Xenon regional cerebral perfusion studies serially over several days post injury. Low cortical blood flow and a low mean CBF_{15} flow were evident on the day of the injury. Additionally, F1 analysis indicated significantly ($p < 0.05$) greater cortical blood flow in the surrounding brain (mean 60 cc/100 g/min) compared to the contusion area (mean 43 cc/100 g/min) on the day of trauma. Mean regional CBF remained below normal in the contused areas ($CBF_{15} < 35$ cc/100 g/min), however the cortical flow increased in the first few days post-injury (peak F1 = 95 cc/100 g/min on day 3) then decreased to sub-normal levels. The mean CBF in the surrounding brain was low on the day of injury ($CBF_{15} = 29$ cc/100 g/min), although higher than the contused area, and increased to a peak of 45 cc/1009/min on day 3 posttrauma. Cortical flow in the surrounding brain, however, exhibited a different trend. The mean F1 was low on the day of trauma and significantly higher one day after trauma (mean 105 cc/100 g/min). Only 15 of the 27 patients with focal contusions had evidence of cerebral edema. Eleven of these exhibited focal edema and 4 exhibited diffuse edema. Focal edema developed over the first few days posttrauma as seen in followup CT, whereas patients with diffuse edema exhibited edema on the admission CT. Initial oligemia in the contused areas was associated with a subsequent hyperemic rim about the contusion. Focal hyperemia was associated with focal edema in 41% of the patients, whereas diffuse edema appeared to be independent of the hyperemic response in contusions.

Keywords: Cerebral contusion; regional cerebral blood flow; cerebral edema; cortical blood flow.

Introduction

Cerebral hyperemia following head injury has been implicated as a factor in the development of intracranial hypertension[4]. The fact that many patients with hyperemia do not develop edema and many without hyperemia may develop edema, indicates there may not be a direct cause and effect involved, rather the two may occur coincidentally. These early studies focused, however, on global or hemispheric cerebral blood flow (CBF) as opposed to regional blood flow trends. Enevoldsen and coworkers[1] found that by doing regional CBF compartmental analysis, tissue peak hyperemia could be seen in areas of severe contusion, though $rCBF_{10}$ values showed poor topographical correlation. Since contusions are often cortical in nature, $rCBF_{10}$ or $rCBF_{15}$ values may not accurately reflect regional flow changes in contusion areas, in as much as these numbers reflect both a cortical and subcortical component. Therefore, in the present study, regional F1 values were analyzed to focus on the fast flow compartment (gray matter) in patients with cerebral contusions, and to determine whether a mechanism of hyperemia-induced edema was evident or not.

Materials and Methods

Patient characteristics. This was a retrospective analysis of 27 patients with acute head injury who were evaluated initially in the U.C.L.A. Medical Center Emergency Department and admitted to the Neurosurgical service from October 1989 to March 1993. Initial head CT studies indicated focal contusions with or without concomitant cerebral edema. Patients were excluded who had evidence of subarachnoid, intraventricular, or gross intraparenchymal hemorrhage by CT. Patients ranged in age from 18 to 84 (mean 34.9). Admission Glasgow Coma Scale scores ranged 3 to 12 (mean 7.8). There were 22 males and 5 females.

Regional cerebral blood flow measurements. Regional CBF was measured by the intravenous ^{133}Xenon method[4] from 10 extracranial detectors, five placed over each hemisphere with a portable unit (Cortexplorer 10, Simonsen Medical AJS, Randers, Denmark). Each patient was studied serially over several days following injury. ^{133}Xenon clearance curves were subjected to

2-compartment analysis, yielding regional CBF_{15}, the mean blood flow of both fast and slow-clearing compartments, and regional F1, an estimate of the fast-clearing compartment, primarily gray matter. Regions of CBF study included temporal, parietal, and posterior frontal lobes. Regional blood flow mapping was then compared to the most recent head CT to delineate contusion and non-contusion areas of interest.

Patient management protocol. Six patients had craniotomies for decompressive procedures. Seventeen patients had ICP monitoring by either a ventriculostomy (Becker EDMS II drainage system, Pudenz-Schulte Medical, Goleta, California) or a fiberoptic ICP monitor (Model 110-4B, Camino Laboratories, San Diego, California). Intracranial hypertension was treated aggressively by CSF drainage, hyperventilation, mannitol, and sedation as needed for ICP in excess of 20 mm Hg. Intial CBF studies were done within 24 hours of injury and serial studies were performed over the next several days.

Results

Regional cortical bood flow. Low cortical flow and a low mean CBF^{15} flow were evident on the day of injury (Figs. 1 and 2). Regional F1 analysis indicated significantly ($p < 0.05$) greater cortical blood flow in the surrounding brain (mean 105 cc/100 g/min) compared to the contusion area (mean 52 cc/100 g/min) on the day after trauma. Cortical flow in the contusion areas increased in the first few days post-injury (peak F1 = 95 cc/100 g/min on day 3) then diminished to subnormal levels. The cortical flow in the surrounding brain exhibited a different trend. The mean F1 was low on the day of injury as mentioned, but subsequently increased more quickly than the contused areas. The highest mean F1 for contused areas was on day 3 posttrauma, and the highest mean F1 for non-contused areas was on day 1 posttrauma.

Regional CBF_{15} trends. The mean regional CBF (Fig. 2) remained below normal in the contused areas

($CBF_{15} < 35$ cc/100 g/min). The mean regional CBF in the surrounding brain was low on the day of injury ($CBF_{15} = 31$ cc/100 g/min), although higher than the contused area, and increased to a peak of 45 cc/100 g/min on day 2 posttrauma.

Cerebral edema. Fifteen of the 27 patients had evidence of either focal or diffuse cerebral edema by CT. Focal edema was seen in 11 patients and diffuse edema was seen in 4 patients. All 4 cases of diffuse edema were seen on the admission CT. In contrast, the 11 patients with localized edema had no evidence of edema on admission CT, but developed local edema in the contusion areas a few days following trauma. Seven patients had intracranial pressures greater than 20 mmHg, including all 4 diffuse edema patients and 3 of the 11 focal edema patients. Mean ICP values were highest on day 5 posttrauma (16.4 mmHg).

Relationship between edema and CBF. Patients with diffuse edema on the admission CT demonstrated lower global and regional blood flows for days 0–3 than the patients with contusions in the absence of edema (mean CBF_{15} for diffuse edema patients in days 0–3 was 28.2 cc/100 g/min, mean CBF_{15} for other patients in days 0–3 was 35.1 cc/100 g/min, $p < 0.05$). This difference in global CBF values resolved by days 7–10, correlating with a diminution in edema by CT and lower ICP values (mean for edema group in days 7–10 was 10.3 mmHg). Focal edema was present in 11 of the 27 patients (41%). Though the peak of hyperemia in contusion area cortical flow for focal edema patients (mean 101 cc/100 g/min) was higher than those patients who did not demonstrate focal edema (mean 92 cc/100 g/min), the difference was not statistically significant ($p = 0.104$).

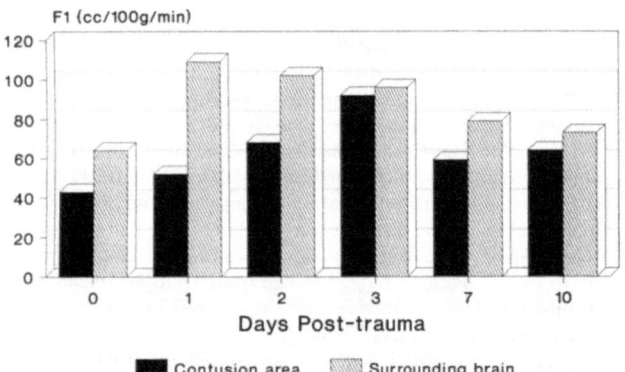

Fig. 1. Regional cortical blood flow in patients with contusions represented as F1 values in cc/100 g/min. Note the difference in hyperemic peaks: day 3 for contusion areas and day 1 for surrounding brain

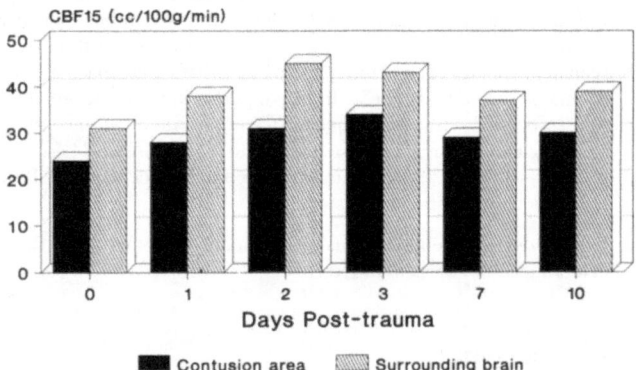

Fig. 2. Regional CBF_{15} values in patients with contusions in cc/100 g/min. Blood flow values are consistently lower in the contusion areas compared to surrounding brain. Maximal blood flow values in both areas occur day 2–3

Discussion

Obrist and coworkers have reported a high incidence of intracranial hypertension (77%) in trauma patients with hyperemia[4]. Whether this represents a causal relationship, however, or an epiphenomenon has yet to be defined. Tornheim and coworkers[6] have used and experimental model of head-injury in cats to examine acute regional cerebral blood flow changes and presence of edema within an hour of injury. These studies revealed hypoperfusion bilaterally in the injured brains without evidence of edema. Similarly, our data shows initial hypoperfusion on the day of injury. Hyperemia in this study group did not present until at least 24 hours after insult. Therefore in patients with diffuse edema on admission, it is unlikely that cerebral hyperemia is the etiology.

Similar to previous experimental and clinical reports[2,4,6–8], we found decreased regional blood flow in the contusion areas compared to surrounding brain. This difference was most noticable on the day following injury in the cortical blood flow values. By doing serial exams, however, we noticed a gradually increasing contusion cortical flow until day three posttrauma. In fact, by day 3, cortical flow in contused and surrounding brain are virtually the same. Metabolism in contusion areas has been demonstrated to be lower than that of surrounding brain[3], thus cortical blood flow in the contused areas on day 3 in our patients most likely represents focal luxury perfusion. This hyperperfusion is also evident in the non-contused areas, however appears to occur earlier following injury, occuring at post-injury day 1.

The subset of patients with focal edema in contusion areas appears to present a more convincing argument for causality between hyperemia and edema. Although not all patients with regional hyperemia developed focal edema (in fact only 41% did), there was an excellent correlation between areas which demonstrated a hyperemic peak in contusion sites and regions of delayed (2–5 days posttrauma) focal edema by CT as evidenced by focal areas of increased water content by Hounsfield units. This relationship also appeared temporally related.

Although clinical studies do not allow for in depth analysis of the entire posttraumatic pathologic process, these studies do indicate at least two patterns of posttraumatic cerebral edema in patients with contusions: a diffuse more acute edema and a regional subacute edema. Of these two, rCBF studies indicate hyperemia is more likely to associated with the latter.

References

1. Enevoldsen EM, Jensen FT (1977) Compartmental analysis of regional cerebral blood flow in patients with acute severe head injuries. J Neurosurg 47: 699–712
2. Madsen FF (1990) Regional cerebral blood flow after a localized cerebral contusion in pigs. Acta Neurochir (Wien) 105: 150–157
3. Nilsson P, Hillered L, Ponten U, Ungerstedt U (1990) Changes in cortical extracellular levels of energy-related metabolites and amino acids following concussive brain injury in rats. J Cereb Blood Flow Metab 10: 631–637
4. Obrist WD, Langfitt TW, Jaggi JL, Cruz J, Genarelli TA (1984) Cerebral blood flow and metabolism in comatose patients with acute head injury. Relationship to intracranial hypertension. J Neurosurg 61: 241–253
5. Ozawa Y, Nakamura T, Sunami K, Kubota M, Ito C, Murai H, Yamaura A, Makino H (1991) Study of regional cerebral blood flow in experimental head injury changes following cerebral contusion and during spreading depression. Neurol Med Chir 31: 685–690
6. Tornheim PA, McDermott F, Shiguma M (1990) Effect of experimental blunt head injury on acute regional cerebral blood flow and edema. Adv Neurol 52: 377–384
7. Wyper DJ, Bullock SR, Patterson J, Maxwell W, Hadley D, Teasdale GM (1991) Traumatic cerebral contusions cause severely reduced perifocal CBF and ischaemic damage. J Cereb Blood Flow Metab 11: S831
8. Yamakami I, Yamaura A, Isobe K (1993) Types of traumatic brain injury and regional cerebral blood flow assessed by ^{99m}Tc-HMPAO SPECT. Neurol Med Chir 33: 7–12

Correspondence: Neil A. Martin, M. D., Division of Neurosurgery, U.C.L.A. Medical Center, CHS 74-140, 10833 Le Conte Avenue, Los Angeles, CA 90024-7039, U.S.A.

Acta Neurochir (1994) [Suppl] 60: 482–484
© Springer-Verlag 1994

The Effect of Blood Volume Replacement on the Mortality of Head-Injured Patient

T. Sakano, S. Yamayoshi, K. Higashi, H. Ikeuchi, Y. Abe, Y. Kinoshita, M. Kishikawa, and **K. Katsurada**

Department of Emergency Medicine, Osaka Prefectural Hospital, Osaka, Japan

Summary

In 77 head-injured and transfused patients, the amount of blood volume replacement (BVR) and patient outcome were retrospectively analyzed. They were divided into four groups of intracranial lesion by initial CT; acute subdural hematoma (SDH) with or without other lesions, traumatic subarachnoid hemorrhage only, epidural hematoma only and all other lesions. Result shows SDH is the most vulnerable to massive transfusion and BVR more than 5000 ml was fatal. Patients with other lesions have high possibility of survival even if BVR amounts to 7000ml. It is concluded, for patients resuscitated with excessive amount of transfusion (> 5000 ml), follow up CT and some vigorous treatment such as administration of hypertonic solutions should be scheduled.

Keywords: Head injury; transfusion; brain edema; prognosis.

Introduction

While institution of comprehensive trauma center systems have dramatically improved survival of the critically injured patient, head injury remains the single largest contributor to deaths[2]. One of the important factors which contributes to high mortality is difficulty in the management of head-injured patient with massive hemorrhage. Clinically, it is occasionally encountered that patient become brain-dead after vigorous and successful treatment to exsanguinating hemorrhage. To date, it is well known that blood volume replacement (BVR) increases mortality of head injury[4], but the type of vulnerable intracranial lesion, or the amount of BVR which affects mortality has not been reported. This retrospective study was designed to clarify these issues.

Materials and Methods

251 patients with head injury were admitted to our emergency unit from January 1989 through July 1992. Criteria for inclusion in this study were: (1) blunt and closed head injury with or without associated injuries; (2) age between 15 to 65 and (3) receiving blood transfusion immediately after injury. Patients who died of associated injuries or uncontrollable hemorrhage was excluded. A total of 77 patients met these criteria and their clinical records were reviewed. They were transfused to keep hematocrit around 30% and crystalloid solutions and plasma were also administered. Patients had various kinds of associated injuries and CT scan was taken after confirmation of circulatory stabilization. Almost all of initial CT were taken within 2 hours after admission. They were divided by the findings of initial CT. Group A (n = 35); acute subdural hematoma (SDH) with or without other lesions, Group B (n = 13); traumatic subarachnoid hemorrhage (SAH) only, Group C (n = 7); epidural hematoma (EDH) only, and Group D (n = 22); all other lesions. This group includes contusion, intracerebral hematoma (ICH), intraventricular hemorrhage (IVH), combined lesions of SAH and EDH, and any combination of these lesions. In each group, relationship between the amount of BVR within 48 hours after injury, minimum base excess (MBE) during volume resuscitation and prognosis at every initial Glasgow coma scale (GCS) were retrospectively analyzed.

Results

In Group A, all patients with GCS less than 4 died irrespective of the amount of BVR and they were excluded from analysis. In the other patients, all 7 patients transfused between 1400 ml and 4200 ml survived and all 5 transfused more than 4400 ml died (Fig. 1): the cause of death was brain death in 4 and sepsis in 1. In Group B, one patient transfused 14800 ml died due to brain death. 6 patients received between 3000 ml and 9800 ml all survived (Fig. 2). In Group C, one patient transfused 10400 ml survived. All 6 transfused from 400 ml to 1800 ml survived except one. In Group D, 2 patients given around 9000 ml died due to brain death. Their initial CT findings were SAH with IVH in one and contusion with IVH in the other. 4 received between 5200 ml and 7400 ml all survived. All 16 transfused less than

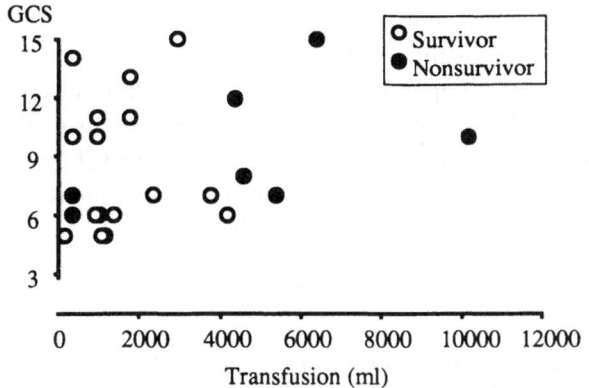

Fig. 1. The effect of BVR on outcome at every GCS in patients with lesions accompanied by SDH

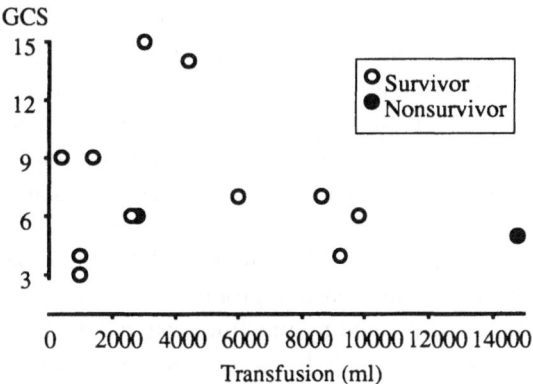

Fig. 2. The effect of BVR on outcome at every GCS in patients with SAH only

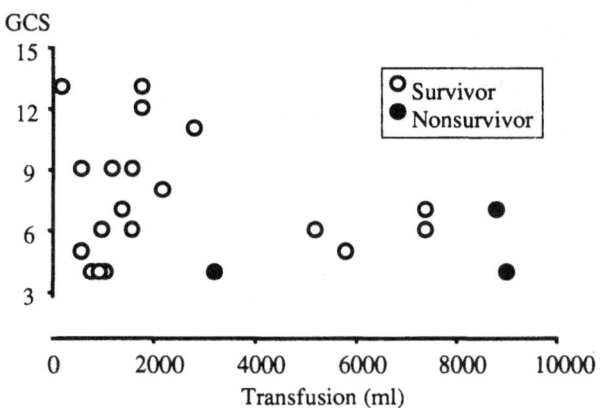

Fig. 3. The effect of BVR on outcome at every GCS in patients with all other lesions

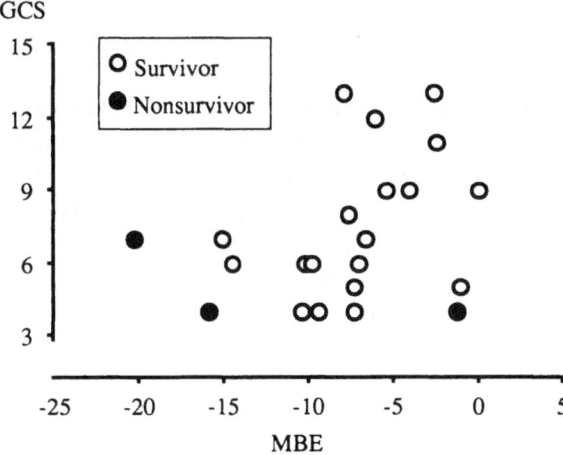

Fig. 4. Relationship between MBE and outcome at every GCS in patients with all other lesions

3200 ml survived except one (Fig. 3). MBE was related to outcome except in Group A. Figure 4 shows relationship between MBE and outcome in Group D.

Discussion

While volume resuscitation always take precedence in the management of head-injured patient with hypovolemic insult, increased risk of death is reported[4]. One possible mechanism involved is exacerbation of brain edema and the other is deincrease of intracranial bleeding due to coagulopathy. In animal models of hemorrhagic shock with brain injury, BVR is proved to increase brain edema and intracranial pressure[3]. Hematologic derangements become critical, when a patient receives massive transfusion of twice his estimated circulating blood volume[1], although this evaluation may be modified by the coagulopathy induced by brain injury itself. In our study, the cause of death of patients received massive transfusion was brain death in all except one, suggesting exacerbation of edema and/or aggravation of intracranial hemorrhage.

The most susceptible type of intracranial lesion was SDH, and BVR more than 5000 ml, which does not seem to induce coagulopathy, was fatal. Generally, most of SDH is accompanied by the brain tissue injury and remarkable edema, as being reflected by midline shift or hemispheric swelling, is present. Intracranial pressure is also increased near to the critical level in many patients, resulting in minimum intracranial buffering capacity of the brain to increase of volume. For these reasons, edema may be easily aggravated by BVR and an addition of even slight edema may be fatal.

Contusion is also speculated to be fragile to BVR, because tissue injury is associated with increased vascular permiability. Although, one patient died after massive transfusion had contusion, the number of patients is too small to induce conclusion.

It was surprising that all 4 patients massively transfused from 6000 ml up to 9800 ml survived in SAH only group, although one patient given 14800 ml, which is more than twice circulating volume, died. This may be explained by the more important role of coagulopathy rather than edema formation, because injury is not to the brain itself but to vessels.

In the current study, SDH only and SDH with other lesions were included in one group, because most of SDH was accompanied by the other lesions. ICH and contusion was also included in one group due to the difficulty in differentiation between the two lesions. Although traumatic coma data bank classification is becoming popular for categorization of lesions, standard classification was used because of the difficulty in the accurate estimation of high or mixed density volume.

MBE is an index of tissue hypoxia yielded by hypotension and hypoxia. Patient outcome was not as related to MBE in SDH group and it may be due to the high mortality with a rather small amount of transfusion.

There is some limitations in this study. First, lesions were categorized by initial CT only and delayed onset of other lesions was ignored. Second, the effect of crystalloid and colloid infusion or change of hematology was not evaluated. In spite of these problems, this study offers important therapeutic considerations concerning BVR and patient outcome. For patients resuscitated with massive transfusion, a follow up CT shoud be taken and vigorous treatment such as administration of hypertonic solutions should be considered.

References

1. Collins JA (1974) Problems associated with the massive transfusion of stored blood. Surgery 75: 274–295
2. Gennarelli TA, Champion HR, Sacco WJ, *et al* (1989) Mortality of patients with head injury and extracranial injury treated in trauma centers. J Trauma 29: 1193–1202
3. Gunner W, Jonasson O, Merlotti G, *et al* (1988) Head injury and hemorrhagic shock: studies of blood brain barrier and intracranial pressure after resuscitation with normal saline solution, 3% saline solution and dextran-40. Surgery 103: 398–407
4. Siegel JH, Gens D, Mamantov T (1989) The effect of associated injuries and blood volume replacement on death, rehabilitation needs and disability in blunt traumatic brain injury (BTBI). J Trauma 29: 1731

Correspondence: T. Sakano, M. D., Department of Emergency Medicine, Osaka Prefectural Hospital, 1-56, Bandai Higashi 3-chome, Sumiyoshi-ku, Osaka 558, Japan.

Acta Neurochir (1994) [Suppl] 60: 485–487

Monitoring of Rectal, Epidural, and Intraventricular Temperature in Neurosurgical Patients

P. Mellergård

Department of Neurosurgery, Lund University Hospital, Lund, Sweden

Summary

We will report our accumulated experience in monitoring of brain temperatures in neurosurgical patients. The intraventricular temperature was monitored with a thermocouple designed for the purpose. This thermocouple was introduced through a plastic catheter, which was also used for monitoring intracranial pressure. The rectal and epidural temperature was simultaneously measured, with commercially available thermocouples. Human brain temperature is higher than the central core temperature, and there is also a temperature gradient within the brain, with the central parts being warmer than the surface. The relationship between rectal, epidural and intraventricular temperatures is maintained during anaesthesia. We have also shown that it is possible to lower the temperature of the human brain.

Keywords: Brain; temperature; hypothermia.

Introduction

Recent reports that the ischemic brain is very sensitive to minor changes in brain temperature[1,6] have been following by a surge of interest in the effects of brain temperature upon different pathophysiological events. A number of groups have now shown that a small increase in cerebral temperature has a significantly negative effect on the outcome following an ischemic event, while mild hypothermia is protective. Particularly interesting are the increasing number of experimental reports showing that post-ischemic mild hypothermia has a cerebro-protective effect (for brief review, see [5]).

Optimal use of hypothermia as a therapeutic tool requires a method for continuous monitoring of brain temperature. We have, therefore, developed such a method, and have thus also been able to provide some information regarding human brain temperature. Surprisingly, very little information has previously been availabe on this subject.

Methods

A copper-constantan thermocouple were made, covered and isolated with a thin Teflon layer, with an outer diameter of 0.5×0.8 mm[2,3]. This thermocouple could be introduced into the plastic intraventricular catheter routinely for monitoring of intracranial pressure (ICP) in neurosurgical patients (the catheter then being positioned in the anterior born of the right ventricle). For epidural and rectal measurements commercially available thermocouples were used (Ellab, Copenhagen). The epidural catheter were introduced through the same burr-hole as the intraventricular catheter, with the tip positioned 1–2 cm lateral to the burrhole. The rectal thermocouple were positioned with its tip 3–4 cm inside of the anal sphincter. All probes were connected to an expanded voltmeter, designed for temperature measurements in humans (Ellab, Copenhagen). The thermocouples were carefully checked in a thermostatically controlled water-bath in between measurements. The temperature resolution were 0.1°C, and the different thermocouples always gave the same reading over a temperature interval between 30°–43°C.

Except during the measurements made in the operation theatre, all monitoring of temperature were made in the intensive care unit, with continuous monitoring of ICP, electrocardiograms, blood pressure, consciousness level, etc. The decision to introduce an intraventricular catheter were made according to the routines of our clinic, the diagnosis varying from severe head trauma to monitoring of ICP during posterior fossa surgery. All patients had their hair shaved off at the time of implantation of the ICP-catheter, and all had bandages around their heads during all measurements. The room temperature was 20–22°C.

Results

Brain-Body Temperature Gradient

As measured in 27 patients, the ventricular temperature (T_{ventr}) was higher than rectal temperature during more than 90% of the time. The largest difference observed between rectal and intraventricular temperature (ΔT_1) was 2.3°C. Usually the difference was much lower, however. For all measurements, ΔT_1 averaged 0.33°C. We monitored esophageal temperature in two

Table 1

	T_{ventr}	ΔT_1	ΔT_2
RLS 1–2	37.2 ± 0.12	0.40 ± 0.07	0.46 ± 0.06
RLS 3–5	38.2 ± 0.09	0.30 ± 0.04	0.45 ± 0.07
RLS 6–8	38.2 ± 0.18	0.27 ± 0.11	0.62 ± 0.16

°C, mean ± SEM.

of our patients. In agreement with the literature, esophageal temperature was always lower (0.2–0.8°C) than rectal temperature in these two patients. Thus, during most of the monitoring, there was a brain-body temperature gradient in our patients.

Intracerebral Temperature Gradient

Epidural temperature were monitored simultaneously with intraventricular and rectal temperature in ten patients. The epidural temperature was always lower than the ventricular temperature. The largest difference observed between epidural and intraventricular temperature (ΔT_2) was 1.0°C. However, the average difference was smaller, ΔT_2 being 0.55°C.

In two, additional, patients undergoing stereotactically guided thalamotomy, the epidural temperature and the temperature around the stereotactic probe was measured, while the probe was slowly and stepwise advanced towards the target point. The temperature increased with the depth of the probe, the largest difference between the epidural space and central brain being 0.9°C.

Brain Temperature and Consciousness

The cerebral temperature is dependent on the metabolic activity of the brain. It could therefore be expected that comatose patients should have a lower brain temperature, and/or a smaller ΔT_1, compared to patients that are fully awake. On the other hand, patients that are unconscious due to severe brain damage often exhibit serious disturbances in cerebral blood flow. This may influence brain temperature in the opposite direction, as the blood normally serves a function of removing heat from the brain.

Twenty-seven patients were divided into three groups (of similar size), based on their grading on the Reaction Level Scale 85 (RLS). (RLS 1-2 roughly corresponding to Glasgow Coma Scale ≥ 12, RLS 3–5 corresponding to GLS ≅ 7–11, and RLS 6–8 corresponding to GLS ≤ 6). As can be seen in Table 1, there

was a tendency that ΔT_1 decreased with decreasing consciousness. However, there was no significant correlation between consciousness level and brain temperature.

Brain Temperature During Major Intracranial Operations

In nine patients the brain temperature were measured during anaesthesia for major intracranial operations[4]. The intraventricular temperature followed closely the rectal temperature, but was almost always slightly higher. The difference between ventricular and rectal temperateure was usually only 0.2–0.4°C, with a uniform pattern in all patients. The mean ΔT_1 for all measurements in all patients during anaesthesia were 0.30°C. The epidural temperature closely followed, but was always lower than ventricular temperature.

The lifting of the boneflap, or opening of the dura, did not significantly increase the temperature gradients the epidural space and the ventricle. Once opened, the craniotomy as such did not significantly influence ventricular temperature, and there were no continuous fall in intraventricular temperature following the craniotomy. In most patients intraventricular temperature rather showed a tendency to increase (0.1–0.2°C) during the operation.

Postoperatively both rectal and intraventricular temperature rose. Intraventricular temperature usually showed a relatively larger increase compared to rectal temperature, thus also increasing $\Delta T_{v–r}$. In three patients this increase in temperature were quite impressive, with intraventricular temperature rising 2.5–4°C during the first six hours following termination of anaesthesia, thereafter decreasing to pre-operative levels.

Brain Cooling

In ten seriously injured and unconscious patients we made different attempts to lower the brain temperature[5]. In five patients we tried to cool solely the head, either by blocks of frozen liquid (Cool Pac©) wrapped around the head, or by means of a special "cooling helmet". This helmet, was made of fabric enclosing the whole head and part of the neck, and had a system of thin plastic channels on the inside, through which water could be circulated. The water was supplied from a thermo-stat bath, continuously circulated through an electromechanical pump, and with the temperature of the water bath chosen such that the temperature beneath the helmet was around 15°C. A steep temperature gradient versus the head was thus maintained during the whole cooling period. With both these methods there were very little or no effect on the intraventricular temperature during the cooling.

Solitary nasopharyngeal cooling was tried in three patients. The tip of a Foley catheter was cut oblique, and was placed in the nasopharynx with the technique used for application of a posterior tamponade in case of heavy nasal haemorrhage. Humified oxygen (from the "connection" in the wall of the ICU), with a flow rate of 5 litres/min, was led through in approximately 4 m long copper spiral placed in a large bucket filled with ice and water, and from there through an insulated tubing to the Foley catheter. The nasopharyngeal temperature decreased to at least 30–31°C during the cooling period, as measured by a thermocouple placed in nasopharynx during the cooling. However, there was no, or very little (maximally 0.2°C) reduction of intraventricular temperature in any of the three patients exposed to nasopharyngeal cooling for two hours, or more. In two patients rectal and epidural temperature actually increased slightly.

Five patients were treated with 1 gram paracetamol given orally or per rectum, in some cases repeatedly. All these patients had a rectal temperature exceeding 38.5°C when given the drug. In all patients the treatment led to a decrease in rectal temperature, varying between 0.6–1.2°C. However, the reduction in brain temperature was larger, sometimes 2°C. Thus, when rectal temperature fell, the intraventricular temperature usually fell to a larger degree, decreasing the temperature gradient between the lateral ventricle and rectum (ΔT_1) that was usually observed. The changes in intraventricular temperature, (whether the initial decrease, or the increase seen when the effect of paracetamol wore off), usually started before the corresponding changes in rectal temperature. Thus, the administration of paracetamol was the single most effective method to lower brain temperature.

However, in order to achieve a lasting reduction of human brain temperature to 34–35°C, we had to use a combination of head cooling, paracetamol, and intensive whole body cooling (with classical methods including alcohol washing and ice in the axillae and groins).

Conclusion

Our experience with monitoring of human brain temperature thus tell us the following: It is possilbe to monitor human brain temperature in a safe and reliable way. As in experimental animals, human brain temperature is higher than central core temperature. There is also a temperature gradient within the brain, with the central parts being warmer than the surface, or cortical parts. The relationship between rectal, epidural and intraventricular temperatures is maintained during anaesthesia. We have also shown that it is possible to lower the temperature of the human brain. In order to efficiently achieve this, it is necessary to combine direct head cooling with intensive whole body, surface cooling, preferably in combination with administration of paracetamol. However, the beneficial effect of mild hypothermia, during pathological events in humans, remains to be proven.

References

1. Busto R, Dietrich W, Globus M, Valdes I, Scheinberg P, Ginsberg M (1987) Small differences in intraischemic brain temperature critically determine the extent of ischemic neuronal injury. J Cereb Blood Flow Metab 7: 729–738
2. Mellergård P, Nordström CH (1990) A method for monitoring intracerebral temperature in neurosurgical patients. Neurosurgery 27: 654–657
3. Mellergård P, Nordström CH (1991) Brain temperature in neurosurgical patients. Neurosurgery 28: 709–713
4. Mellergård P, Nordström CH, Messeter B (1992) Human brain temperature during anaesthesia for intracranial operations. J Neurosurg Anesth 4: 85–91
5. Mellergård P (1992) Changes in human brain temperature in response to different means of brain cooling. Neurosurg 31: 671–677
6. Minamisawa H, Smith M, Siesjö B (1990) The effect of mild hyperthermia and hypothermia on brain damage following 5, 10 and 15 min of forebrain ischemia. Ann Neurol 28: 26–33

Correspondence: Pekka Mellergård, M. D., Department of Neurosurgery, Lund University Hospital, S-221 85 Lund, Sweden.

Traumatic Brain Edema: Treatment

Acta Neurochir (1994) [Suppl] 60: 491–493

A High-Field Magnetic Resonance Imaging Study of Experimental Vasogenic Brain Edema and its Response to AVS: 1,2-Bis (Nicotinamido)-Propane

K. Kamada, K. Houkin, Y. Iwasaki, and **H. Abe**

Department of Neurosurgery, Hokkaido University School of Medicine, Hokkaido, Japan

Summary

We clearly represented brain structures of rats and permitted a rapid assessment of water gradient of the brain edema by cortical freezing utilizing a high-field (7T) proton magnetic resonance imaging (MRI). The typical time course of vasogenic edema and the efficacy of AVS;1,2-bis (nicotinamide)-propane upon the edema were presented. Twelve rats with edema induced by cortical freezing were divided into two groups; one group of animals received 0.5 ml of physiological saline with 100 mg (/kg) AVS every eight hours intraperitoneally. The other group of untreated animals received only saline. One three, six, 12, and 24 hours after lesion production, the profiles of edema fluid spreading and the maximum signal intensity (MI) of some regions of interest (ROI) were assessed by T2 weighted images (TE = 70 ms, TR = 3500 ms). One hour after lesion production in the untreated group, a low heterogeneous intensity area was seen mainly in the primarily injured cortex. Two hours later, the margin of the lesion gradually increased in intensity and MI of ROIs around the lesion also gradually increased. Twenty-four h after lesion production edema extended contralaterally via corpus callosum.

AVS reduced edema fluid spreading beginning from about six hours after lesion production. The MIs of the AVS treated group were significantly lower than in the untreated group ($p < 0.01$). We conclude that sequential observation of edema using MRI is a quite practical technique for evaluation of the efficacy of any therapeutic agent.

Keywords: Brain edema; proton magnetic resonance imaging; AVS (1,2-bis (nicotinamide)-propane).

Introduction

Vasogenic brain edema occurs in association with pathological processes of the CNS such as tumors or infarctions, where the blood-brain barrier (BBB) is damaged. In experimental studies of vasogenic edema, the cold-injury model has been widely accepted to produce constant lesions. Therefore, many investigators used this model to evaluate many aspects of edema and the efficacy of some drugs upon edema by various means[2-6]. In edema, increased brain water content linearly correlates with alterations in density and relaxation times of proton[2,5,6]. We present a high-resolution proton magnetic resonance imaging (MRI) of rat brain structures which allows to precisely estimate the time course and signal intensity of edema area. Also, the efficacy of AVS; 1,2-bis (nicotinamide)-propane upon the edema is presented.

Materials and Methods

Vasogenic edema was produced by cortical freezing in twelve female Wistar rats weighing 280–320 grams[3]. During the experiment, the body temperature and blood pressure were monitored and kept constant. One-half of the animals received 0.5 ml of physiological saline with 100 mg (kg) AVS every eight hours intraperitoneally. Untreated animals received only saline.

High-resolution images of rat brains were obtained by a 300 MHz imaging spectrometer with 183 mm horizontal bore magnet (Spectroscopy Imaging Systems, USA) and a home-made modified saddle-shaped coil. After the pilot sagittal images were obtained, T2-weighted spin-echo image, repetition time (TR) = 3500 msec, echo time (TE) = 70 s, were acquired at one, 3, 6, 12 and 24 hours after lesioning. The profiles of edema fluid spreading and the maximum signal intensity (MI) of some regions of interest (ROI) in the brain were analyzed using equipped image processing software. On the same plane, MI of the right putamen (MIr) was measured for reference ROI and the normalized MI of each ROI (MIn) was calculated from the ratio of these data:

$$MIn = MI / MIr$$

Results

Image Appearances in MRI (Fig. 1)

At one hour after lesion production, a mainly low heterogeneous signal intensity area was seen in the injured cortex and white matter. Two hours later, the margin of the lesion gradually increased in intensiy and edema fluid spread downward toward the corpus callosum.

In the untreated animals the whole lesion area reached a high intensity and a extension of edema fluid into the opposite hemisphere was observed after six hours following lesion production. The maximum of these changes was observed at 24 hours. In the AVS treated group, the image appearance markedly differed. The edema extension into the opposite site was markedly suppressed six hours after lesion production.

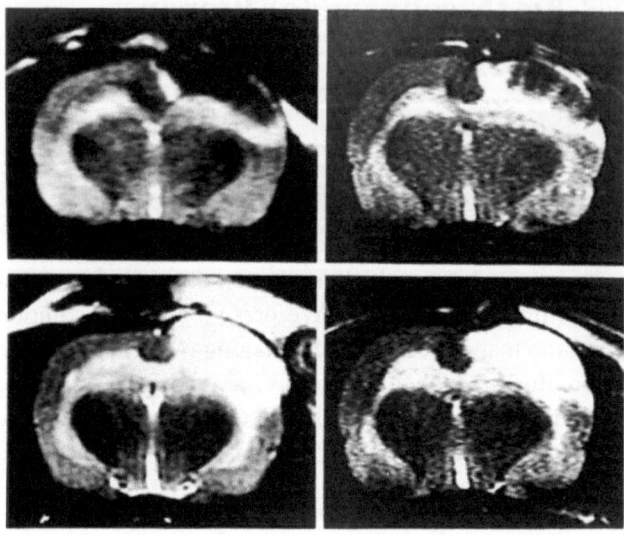

Fig. 1. Coronal spin echo images of a rat brain without (left) and with (right) AVS treatment at 1 h in (above) and 24 h (below) after lesioning. The edema extension into the opposite site was markedly suppressed 24 hours after lesion production

Time Course of MIn of Each ROI

Whereas the T2-weighted images reflected the anatomical extent of pathological changes, signal intensity was measured to provide an objective assessment of the severity of the edema. The results are given in Fig. 2. The MIn of ROI1 (lesion), 2 and 3 (surrounding regions) increased continuously. In contrast, in ROI 4, an area distant from the lesion, no change was observed. The MIn of ROI 1, 2 and 3 in animals with AVS treatment were significantly lower than in untreated rats and AVS reduced edema formation as well as its extension area (*: $p < 0.01$).

Discussion

Proton MRI technique is able to reveal the regional brain water content which is reflected by proton NMR relaxation times or density[5]. AVS, which was used in this study, is one of the free radical scavengers which may possess cerebral protective action and suppress the influx of water and sodium across the BBB through modulation of the arachidonate metabolism[1]. Therefore it is expected to be effective upon vasogenic edema. Using these MRI techniques, we were able to reveal the typical time course of edema spreading in vivo. Pretreatment with AVS resulted in a suppression of the rate of edema spreading after lesion production.

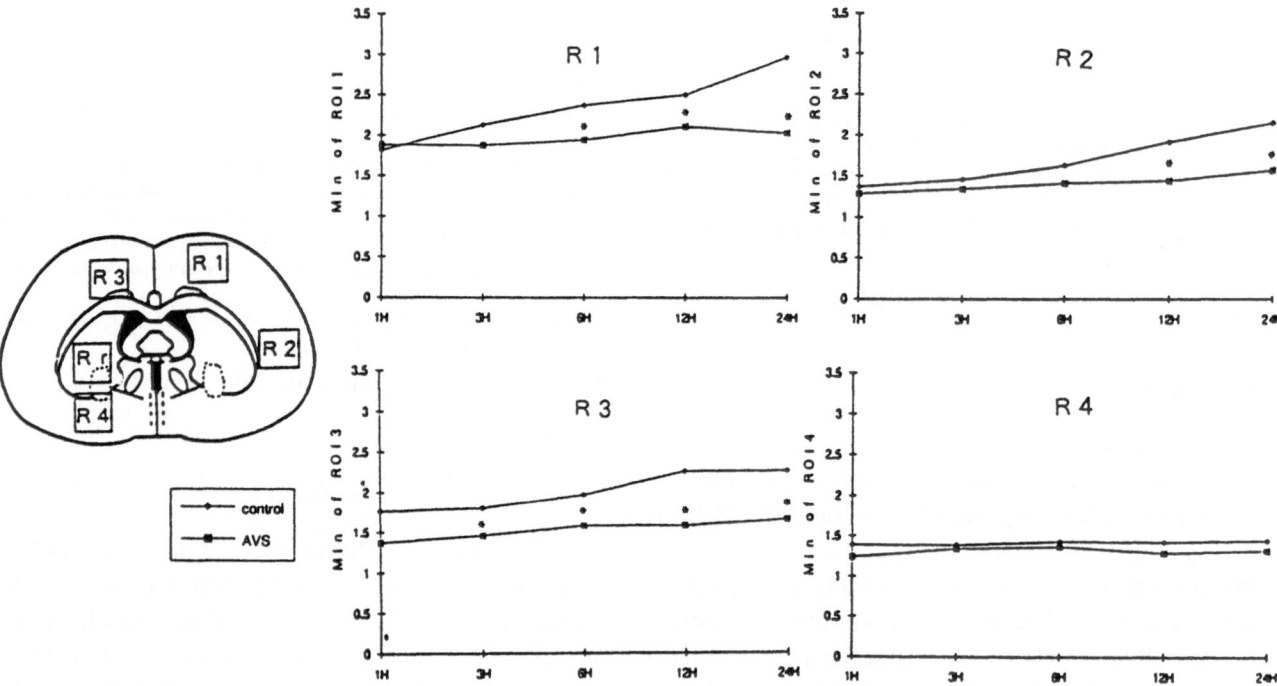

Fig. 2. Illustration of the slice taken edema for measurement and regions of interest and the time course of MIn of each ROI

The application of MRI techniques should be extremely useful not only in experimental edema, but also in human vasogenic edema; e.g. with tumors and cerebrovascular disease. It probably could be used assessing the efficacy of new therapeutic agents including AVS.

References

1. Asano T, Johshita H, Koide T, Takakura K (1984) Amelioration of ischemic cerebral oedema by a free radical scavenger, AVS:1,2-bis (nicotinamido)-propane. An experimental study using a regional ischaemia model in cats. Neurol Res 6: 163–168
2. Branes D, McDonald WI (1988) A magnetic resonance imaging study of experimental cerebral edema and its response to dexamethasone. Magn Reson Med 7: 125–131
3. Combs DJ, Ott L, Mcaninch PS, Dempsey RJ, Young B (1989) The effect of total parenteral nutrition on vasogenic edema development following cold injury in rats. J Neurosurg 70: 623–627
4. Maxwell RE, Long DM, French LA (1971) The effects of glucosteroids on experimental cold-induced brain edema. Gross morphological alterations and vascular permeability changes. J Neurosurg 34: 477–487
5. Olson JE, Karz-Stein A, Reo NV, Jolesz FA (1992) Evaluation of acute brain edema using quantative megnetic resonance imaging: Effects of pretreatment with dexamethasone. Magn Reson Med 24: 64–74
6. Shinohara Y, Ohsuga H, Ohsuga S, Takizawa S, Haida M (1990) Routed protein migration after protein extravasation and water leakage caused by cold injury. Acta Neurochir (Wien) [Suppl] 51: 90–92

Correspondence: Kyousuke Kamada, M.D., Department of Neurosurgery, Hokkaido University, School of Medicine, North-15, West-7 Kita-Ku, Sapporo 060, Japan.

Acta Neurochir (1994) [Suppl] 60: 494–498
© Springer-Verlag 1994

7.2% NaCl/10% Dextran 60 versus 20% Mannitol for Treatment of Intracranial Hypertension

S. Berger[1], L. Schürer[2], R. Härtl[1], T. Deisböck[1], C. Dautermann[2], R. Murr[3], K. Messmer[1], and A. Baethmann[1]

[1] Institute for Surgical Research, [2] Department of Neurosurgery, and [3] Institute of Anesthesiology, Klinikum Großhadern, Ludwig-Maximilians-University, München, Federal Republic of Germany

Summary

Severe head injury is frequently associated with extracranial injuries causing hemorrhagic hypotension. Volume replacement with isotonic fluids not only is therapeutically of limited efficacy but may aggravate posttraumatic brain edema. On the other side, hypertonic/hyperoncotic saline/dextran solution (HHS) shown to restore cardiovascular function in hemorrhagic shock instantaneously, was found to decrease intracranial pressure in experimental head injury. Currently the therapeutic efficacy of HHS and mannitol on ICP was compared at 24 hrs after a focal cerebral lesion and inflation of an epidural balloon in rabbits. Both solutions given at an equimolar dose rapidly lowered the ICP. After the first injection, ICP reduction was longer maintained with mannitol (189 ± 27 min) as compared to HHS (98 ± 14 min), while no difference in duration of lowering ICP was found after the second injection. Due to its blood pressure effects, HHS afforded a higher cerebral perfusion pressure than mannitol. In animals with HHS, the water content of the traumatized hemisphere was increased while the contralateral hemisphere was dehydrated. With mannitol, no differences in water content were found between the injured and uninjured hemisphere. The efficiency of HHS in hemorrhagic shock and intracranial hypertension render the fluid mixture particularly promising in patients with polytrauma in combination with head injury.

Keywords: Experimental head injury; hypertonic/hyperoncotic saline/dextran; mannitol; ICP.

Introduction

Severe head injury is a leading prognostic factor in 50–60% of trauma fatalities[7,28]. The most important extracranial complication responsible for poor outcome in head injury is arterial hypotension[22]. An unfavourable role of hypotension in cerebral trauma has been demonstrated during the entire posttraumatic course of the patient: i.e. on admission to hospital[15], in the operation theatre[23], during intrahospital transport[1], and in the intensive care unit[22]. The combination of systemic hypotension with intracranial hypertension endangers maintenance of the cerebral perfusion pressure thereby promoting the development of secondary brain damage[20]. Primary care therefore must reestablish and maintain an appropriate cerebral perfusion based on competent systemic circulation. In severe hemorrhagic shock this goal cannot be reached by conventional therapy with large volumes of isotonic fluids since institution of an adequate infusion rate is difficult[31] and, because high water load may aggravate development of cerebral edema[8].

Small volume resuscitation by hypertonic/hyperoncotic saline/dextran solution (HHS) is a novel method to circumvent this therapy dilemma. Infusion of 4 ml/kg b.w. of HHS within 2 min was shown to instantaneously restore systemic blood pressure and cardiac output in severe hemorrhagic shock instantaneously[12]. Since hypertonic saline alone reduces intracranial pressure in head injury[6,36] we postulated that hypertonic/hyperoncotic saline/dextran might be of special benefit in polytraumatized patients with head injury. In respective animal investigations Schürer *et al.*[27] did not observe adverse effects of HHS on the cerebral perfusion or O_2-supply in rabbits resuscitated from hemorrhagic shock with or without head injury. Härtl *et al.*[11] demonstrated reduction of intracranial hypertension by saline/dextran in rabbits with experimental head injury. This aspect was presently studied in comparison with a standard treatment of raised ICP using mannitol. For this purpose, rabbits were subjected to a focal cerebral lesion 24 h before administration of the respective treatment to assess effectiveness of HHS and of mannitol under conditions when vasogenic edema is maximally developed.

Materials and Methods

Instrumentation: 12 New Zealand albino rabbits (3.2 ± 0.1 kg) were anesthetized by intravenous ketamine/xylazine. After fixation of the skull in a stereotactic frame a circular trephination (Ø 7 mm) of the left temporal bone was made, leaving the dura mater intact. A focal cold lesion of the left cerebral hemisphere was induced then by attachment of a cooled (liquid nitrogen) metal probe (Ø 6 mm) to the dura for 5 min. An epidural balloon was inserted above the lesion and the bone defect was closed by dental cement. 21 h later the animals were reanesthetized with thiopental. After tracheotomy for mechanical ventilation, anesthesia was maintained by α-chloralose via infusion to a femoral vein catheter. A femoral artery catheter was inserted for blood pressure monitoring and withdrawal of blood samples. Hematocrit, blood gases, plasma-electrolytes, -osmolality, -colloidosmotic pressure, and blood cell counts were studied at intervals. Body temperature was maintained at 39°C by a servo-controlled heating pad. ICP was measured manometrically in the subdural space.

Protocol: 23 h after trauma the ICP was raised to 17 mmHg by inflation of the epidural balloon (max. volume 0.6 ml). During a subsequent observation period of 30 min ICP remained at 15–20 mmHg without changing the balloon volume. At 24 h after induction of trauma, either 4 ml/kg b.w. of 7.2% NaCl/10% dextran 60 (HHS) or 9 ml/kg b.w. of 20% mannitol were infused into an ear vein within 2 min. The osmotic load was equal in both groups amounting to 9.6 mosmol/kg. Both solutions rapidly decreased ICP, reaching a minimum at 20 min after injection. When the ICP was increasing again above 15 mmHg the same treatment was repeated. Saturated KCl was injected then for termination of the experiment when the ICP exceeded 15 mmHg after the third treatment episode or if ICP remained below 15 mmHg for > 240 min. The cerebrum was rapidly removed from the skull and frozen to –70°C. For gravimetric determination of the tissue water content, the cerebral hemispheres were separated and dried at 80°C for 24 h, Cerebral Na+- and K+-contents were analysed by flame photometry.

Statistics: Unpaired data were compared by the Kruskal-Wallis-test, paired data by the Friedmann-Quade-test, both in the computer version of Theodorssen-Norheim. p < 0.05 was considered significant.

Results

Systemic parameters: No differences between groups were observed during the control period. Administration of HHS was found to increase, whereas

Table 1

	MAP mmHg	Hct %	Na+ mM/l	Osm mosm/l	COP mmHg	pO2 mmHg	pCO2 mmHg	pH -lg[H+]
Control								
HHS	88.8 ± 5.7	35.8 ± 1.3	144.4 ± 1.7	298.2 ± 1.7	17.9 ± 1.8	147.2 ± 7.0	32.4 ± 1.7	7.411 ± 0.02
Mannitol	89.2 ± 8.5	39.6 ± 0.7	145.4 ± 4.3	297.2 ± 2.5	17.1 ± 1.3	149.9 ± 5.9	34.2 ± 2.3	7.408 ± 0.01
10' after inj. 1								
HHS	93.0 ± 2.6	29.6 ± 1.1	56.8 ± 2.9	320.0 ± 26	18.6 ± 1.6	140.8 ± 10.1	36.1 ± 1.9	7.345 ± 0.02
Mannitol	88.4 ± 8.1	34.5 ± 1.1	135.3 ± 2.8	326.0 ± 4.6	15.9 ± 1.4	115.6 ± 11.8	45.4 ± 4.5	7.287 ± 0.05
60' after inj. 1								
HHS	98.5 ± 3.6	33.0 ± 1.0	147.9 ± 2.3	310.8 ± 1.7	19.7 ± 1.6	151.5 ± 4.5	33.5 ± 0.9	7.383 ± 0.02
Mannitol	91.2 ± 7.7	39.5 ± 1.2	138.1 ± 3.3	309.8 ± 3.7	16.3 ± 2.1	148.1 ± 7.9	33.2 ± 1.3	7.410 ± 0.00
10' after inj. 2								
HHS	98.8 ± 3.9	27.6 ± 8.9	157.7 ± 7.5	328.5 ± 2.2	20.5 ± 0.9	150.9 ± 6.3	34.1 ± 0.9	7.353 ± 0.02
Mannitol	87.8 ± 6.6	34.9 ± 1.8	146.1 ± 5.3	327.0 ± 5.7	17.5 ± 1.7	121.8 ± 9.1	36.5 ± 3.4	7.352 ± 0.03
60' after inj. 2								
HHS	102.7 ± 3.4	29.7 ± 1.0	158.0 ± 3.7	322.0 ± 2.7	19.1 ± 2.0	155.7 ± 4.8	31.3 ± 0.9	7.401 ± 0.02
Mannitol	80.6 ± 7.0	39.2 ± 0.8	146.3 ± 2.8	313.0 ± 3.5	18.1 ± 0.3	134.1 ± 5.2	33.5 ± 2.5	7.414 ± 0.02
10' after inj. 3								
HHS	101.0 ± 3.6	25.9 ± 1.1	169.1 ± 4.9	341.1 ± 4.3	23.1 ± 3.1	150.3 ± 7.6	33.7 ± 0.9	7.365 ± 0.01
Mannitol	83.7 ± 6.6	34.2 ± 1,4	131.9 ± 4.2	327.0 ± 9.0	16.1 ± 2.0	137.0 ± 1.5	35.7 ± 2.9	7.384 ± 0.02
60' after inj. 3								
HHS	100.8 ± 7.5	28.9 ± 1.4	160.1 ± 2.8	333.0 ± 8.1	20.7 ± 2.4	151.8 ± 7.9	32.7 ± 1.0	7.383 ± 0.01
Mannitol	73.3 ± 7.7	38.3 ± 0.9	150.3 ± 2.9	319.0 ± 5.6	18.6 ± 1.3	132.9 ± 5.6	30.0 ± 3.2	7.443 ± 0.02

Blood pressure, hematocrit, plasma-Na+, -osmolality, -colloidosmotic pressure, and blood gases after subsequent injections of either 4 ml/kg b.w. of HHS or 9 ml/kg b.w. of 20% mannitol. *MAP* mean arterial pressure. *Hct* hematocrit. *Na+* plasma sodium concentration. *Osm* plasma osmolality. *COP* colloid osmotic pressure of the plasma. *pO2, pCO2* and *pH:* arterial blood gas parameters. *10', 60' after inj. 1, 2, 3* = time in minutes after first, second and third injection of the respective hypertonic solution.

mannitol to decrease blood pressure. This resulted in a significantly different cerebral perfusion pressure after the second injection period. Hct decreased with either injection by approximately 5%. This effect was completely reversed with mannitol within 60 min, while Hct remained decreased with HHS. The colloid-osmotic pressure increased moderately with HHS, no changes were seen with mannitol. Plasma osmolality increased by 15–25 mosmol/l after either injection but tended to normalize thereafter, albeit at a higher level with HHS. A maximum of 341 ± 4 mosmol/l was obtained after the 3rd injection of HHS while osmolality peaked at 327 ± 6 mosmol in animals with mannitol injections. Plasma-Na^+ was increased by HHS as compared to a moderate reduction following the 1st and 3rd administration of mannitol resulting from diuresis. The first injection of mannitol led to a temporary decrease of the blood pH accompanied by an increased pCO_2 and decreased pO_2 (Table 1).

ICP and cerebral water content: The inflation of the epidural balloon by fluid injection led to an ICP-level of 17 mmHg with a considerable pressure amplitude. Either solution decreased promptly the ICP-level and -amplitude. At 20 min after injection an ICP-minimum of 4–6.5 mmHg was obtained. Thereafter, ICP increased again in most animals of both groups, attributable to persistence of the space occupation. After infusion of HHS, an ICP-level of 15 mmHg was reached within 98 ± 14 min, whereas the same level was obtained within 189 ± 27 min after infusion of mannitol. The initial ICP response was identical when the treatment was repeated. Duration of ICP-reduction was then 142 ± 21 min by HHS while 107 ± 11 min by mannitol. In two animals with mannitol and one with HHS ICP remained normal for > 240 min making further injections unnecessary. The cerebral water content measured by the dry/wet weight method of the cerebral hemispheres, was more increased in the traumatized hemisphere of animals with HHS ($p < 0.05$). Conversely, the water content of the uninjured hemispheres was more decreased by HHS than by mannitol. The total brain water content was $78.76 \pm 0.29\%$ in animals with HHS as compared to $78.99 \pm 0.23\%$ in animals with mannitol (Fig. 1).

Discussion

"Scoop and run" with critically injured trauma victims has been under debate for decades. This discussion was revived by the concept of small volume resuscitation by HHS[32,18]. Immediate or rapid resuscitation

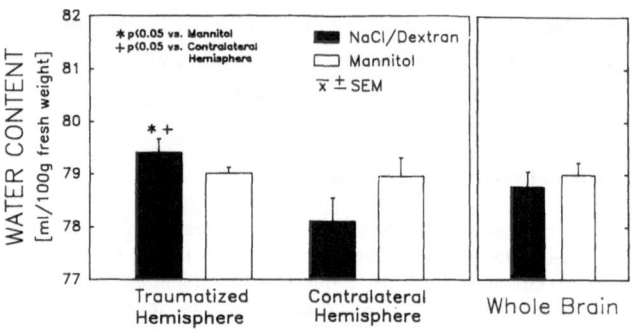

Fig. 1. Tissue water content of the traumatized (cold injury) left hemispheres and the contralateral nonexposed hemispheres (right) of rabbits receiving either HHS, or hypertonic mannitol

from severe hemorrhagic shock is hardly possible by i.v. administration of isotonic fluids[31,14]. Since the introduction of both, HHS[32] and intraosseous infusion devices[10,4] thereby avoiding futile search for venous access to infuse large volumes of fluid, better pre-hospital stabilization can be accomplished now. Some authors[9,17] demand control of bleeding from large arterial vessels as a prerequisite for shock treatment with hypertonic saline, since uncontrolled hemorrhage might be enhanced by the acute blood pressure increase following hypertonic saline administration[3,16]. In a model of hemorrhage driven by blood pressure, Prist *et al.*[24] have shown that HHS improves survival if administered alone, and increases recovery from hemorrhage if combined with Ringer's lactate. Uncontrolled bleeding therefore is no contraindication for HHS in patients with head injury and shock. Nevertheless, the combination of head injury and hemorrhagic shock is the most lethal pattern of injury seen in emergency rooms[7]. The old conflict of neurosurgeons and trauma surgeons notwithstanding on whether the patient s life is more in danger by either delayed evacuation of space occupying intracranial lesions (hematomas and contusions) or by delayed control of e.g. abdominal bleeding, maintenance of the cerebral perfusion pressure is indispensable. Therefore, methods of treatment are required which stabilize the patient for peripheral surgery or neurosurgery. Recent findings that HHS effectively reduces posttraumatic intracranial hypertension[11] and increases survival in head injured patients with polytrauma[34] are clinically promising. Employment of hypertonic saline/dextran solutions in trauma patients has been criticised to induce a dangerous increases in plasma-osmolality and -sodium concentration[5]. Brain damage, such as pontine myelinolysis occuring in ICU patients during the final course of internal diseases associated with hyperosmolality[21], has never been shown in patients receiving emergency treatment with hypertonic saline or HHS[33,34]. Induction of intracranial bleedings by this treatment has neither been confirmed as suggested by some authors[2,19] e.g. following disruption of bridging veins with rapid shrinkage of the brain. On the other hand, therapeutical benefits of hypertonic saline were demonstrated in cases of otherwise untreatable intracranial hypertension[36]. Fisher et al.[6] showed lower ICP in head injured children with administration of 3% NaCl. While Shapira et al.[30] did not find therapeutical differences with regard to the tonicity of the fluid regimen on brain edema induced by a weight drop device in rats, Shackford et al.[29] demonstrated superior reduction of ICP and improvement of CBF after administration of hypertonic compared to isotonic fluid (500 mosmol/l sodium lactate) in a porcine model of vasogenic brain edema. The results are in accordance with findings of Wisner et al.[35] on reduction of cerebral water content of rats by resuscitation from fluid-percussion injury with hypertonic saline (6.5%). Schmoker et al.[25] observing reduction of ICP and improvement of cerebral oxygen delivery in hemorrhagic shock in pigs by 500 mosmol/l sodium lactate, could not demonstrate a correlation between the total load of sodium or fluid and ICP or Glasgow Outcome Scale in patients with head injury and peripheral injuries receiving Ringer's lactate[26]. Therefore, it is apparently not the total amount of sodium which is infused, but the sodium concentration of the infusion fluid which is of importance. Nevertheless, the high osmolality of HHS limits its administration to a few injections in the early posttraumatic phase, which must be followed by conventional fluid therapy. HHS is unlikely to replace mannitol for chronic (several days) therapy of an elevated ICP, although alternate infusions of both solutions might be advantageous. The different effects of mannitol and HHS on hemispheric water content or vasogenic brain edema, respectively, found in the present study might indicate different modes of action of the solutions. Kaufmann and Cardoso[13] reported accumulation of mannitol in perifocal brain tissue following multiple injections of mannitol. Our group is currently evaluating extravasation and tissue accumulation in the brain of either mannitol or dextran, and potential effects thereof on the rebound phenomenon of ICP.

References

1. Andrews PJD, Piper IR, Dearden NM, Miller JD (1990) Secondary insults during intrahospital transport of head-injured patients. Lancet 335: 327–330
2. Arvidsson S, Häggendal E, Winsö I (1981) Effect on cerebral blood flow of infusion of hyperosmolar saline during cerebral vasodilation in the dog. Acta Anaesth Scand 25: 153
3. Bickell WH, Shaftan GW, Manox KL (1989) Intravenous fluid administration and uncontrolled hemorrhage. J Trauma 29: 409
4. Chávez-Negrete A, Cruz SM, Munari AF, Perches A, Argüero R (1991) Treatment of hemorrhagic shock with intraosseous or intravenous infusion of hypertonic saline dextran solulion. Eur Surg Res 23: 123–129
5. Dawidson I (1990) Hypertonic saline for resuscitation: a word of caution. Crit Care Med 18: 245
6. Fisher B, Thomas D, Peterson B (1992) Hypertonic saline lowers raised intracranial pressure in children after head trauma. J Neurosurg Anesthesiol 4: 4–10
7. Gennarelli T, Champion H, Sacco W, Copes WS, Alves WM (1989) Mortality of patiens with head injury and extracranial injury treated in trauma centers. J Trauma 29: 1193–1201
8. Gunnar W, Jonasson O, Merlotti G, Stone J, Barrett J (1987) Head injury and hemorrhagic shock: studies of the blood brain barrier and intracranial pressure after resuscitation with normal saline solution, 3% saline solution, and dextran-40. Surgery 103: 398–407
9. Gross D, Landau EH, Klin B, Krausz MM (1990) Hypertonic saline treatment of uncontrolled hemorrhagic shock. Surg Gyn Obstet 170: 106–112
10. Halvorsen L, Bay BK, Perron PR, Gunther RA, Holcroft JW, Blaisdell FW, Kramer GC (1990) Evaluation of an intraosseous infusion device for the resuscitation of hypovolemic shock. J Trauma 30: 652–658
11. Härtl R, Schürer L, Dautermann C, Röhrich F, Berger S, Murr R, Messmer K, Baethmann A (1993) Effect of hypertonic-hyperoncotic solutions (HHS) on increased intrascranial pressure after a focal brain lesion and inflation of an epidural balloon. In: Avezaat C, et al (eds) ICP and craniospinal dynamics, in press
12. Holcroft JW, Vassar MJ, Turner JE, Derlet RW, Kramer GC (1987) 3% NaCl and 7.5% NaCl/Dextran 70 m the resuscitation of severly injured patients. Ann Surg 206: 279–287
13. Kaufmann AU, Cardoso MS (1992) Aggravation of vasogenic cerebral edema by multiple-dose mannitol. J Neurosurg 77: 584–589
14. Kaweski SU, Sise MJ, Virgilio RW (1990) The effect of prehospital fluids on survival in trauma patients. J Trauma 30: 1215–1218
15. Kohi YM, Mendelow AD, Teasdale GM, Allardice GM (1984) Extracranial insults and outcome in patients with acute head injury – relationship to the Glasgow coma scale. Injury 16: 25–29
16. Kowalenko T, Stern S, Dronen S, Wang X (1992) Improved outcome with hypotensive resuscitation of uncontrolled hemorrhagic shock in a swine model. J Trauma 33: 349–353
17. Krausz MM, Bar-Ziv M, Rabinovici R, Gross D (1992) "Scoop and run" or stabilize hemorrhagic shock with normal saline or small-volume hypertonic saline? J Trauma 33: 6–10
18. Kreimeier U, Messmer K (1988) Small volume resuscitation. In: Kox W, Gamble J (eds) Fluid resuscitation, Baillière's clinical anesthesiology, Vol 2. Baillière Tindall, London, pp 545–577
19. Luttrell CN, Finberg L, Drawdy LP (1959) Hemorrhagic encephalopathy induced by hypernatremia. II. Experimental observations on hyperosmolarity in cats. Arch Neurol 1: 153
20. Marmarou A, Andersson RL, Ward JD, Choi SC, Young HF, Eisenberg HM, Foulkes MA, Marshall LF, Jane JA (1991) Impact of ICP instability and hypotension on outcome of patients with severe head injury. J Neurosurg 75: S59–S66
21. Mattar JA (1989) Hypertonic and hyperoncotic solutions in patients. Crit Care Med 17: 297

22. Piek J, Chesnut RM, Marshall LF, van Berjum-Clark M, Klauber MR, Blunt BA, Eisenberg H, Jane JA, Marmarou A, Foulkes MA (1992) Extracranial complications of severe head injury. J Neurosurg 77: 901–907

23. Pietropaoli JA, Rogers FB, Shackford SR, Wald SL, Schmoker JD, Zhuang J (1992) The deleterious effects of intraoperative hypotension on outcome m patients with severe head injuries. J Trauma 33: 403–407

24. Prist R, Rocha e Silva M, Velasco IT, Loureiro MI (1992) Pressure-driven hemorrhage: a new experimental design for the study of crystalloid and small-volume hypertonic resusatation in anesthetized dogs. Circ Shock 36: 13–20

25. Schmoker JD, Zhuang J, Shackford SR (1991) Hypertonic fluid resuscitation improves cerebral oxygen delivery and reduces intracranial pressure after hemorrhagic shock. J Trauma 31: 1607–1613

26. Schmoker JD, Shackford SR, Wald SL, Pietropaoli JA (1992) An analysis of the relationship between fluid and sodium administration and intracranial pressure after head injury. J Trauma 33: 476–481

27. Schürer L, Dautermann C, Härtl R, Murr R, Berger S, Röhrich F, Messmer K, Baethmann A (1992) Therapie des hämorrhagischen Schocks mit kleinen Volumina hyperton-hyperonkotischer NaCl-Dextranlösung – Auswirkungen auf das Gehirn. Anästhesiol Intensivmed Notfallmed Schmerzther 27: 209–217

28. Shackford SR, Mackersie RC, Dacis JW, Wolf PL, Hoyt DB (1989) Epidemiology and pathology of trauma deaths occurring at a level 1 trauma center in a regionalized system: the importance of secondary brain damage. J Trauma 29: 1392–1397

29. Shackford SR, Zhuang J, Schmoker J (1992) Intravenous fluid tonicity: effect on intracranial pressure, cerebral blood flow, and cerebral oxygen delivery in focal brain injury. J Neurosurg 76: 91–98

30. Shapira Y, Artru AA, Cotev S, Muggia-Sulam M, Freund HR (1992) Bram edema and neurologic status following head trauma in the rat. Anesthesiology 77: 79–85

31. Smith GJ, Bodai BI, Hill AS, Frey CF (1985a) Prehospital stabilization of critically injured patients: a failed concept. J Trauma 25: 65–70

32. Smith GJ, Kramer GC, Perron P, Nakayama S, Gunther RA, Holcroft JW (1985b) A comparison of several hypertonic solutions for resuscitation of bled sheep. J Surg Res 39: 517–528

33. Vassar MJ, Perry CA, Holcroft JW (1990) Analysis of potential risks associated with 7.5% sodium chloride resuscitation of traumatic shock. Arch Surg 125: 1309–1315

34. Vassar MJ, Perry CA, Gannaway WL, Holcroft J (1991) 7.5% sodium chloride/dextran for resuscitation of trauma patients undergoing helicopter transport. Arch Surg 126: 1065–1072

35. Wisner DH, Schuster L, Quinn C (1990) Hypertonic saline resuscitation of head injury: effects on cerebral water content. J Trauma 30: 75–78

36. Worthley LIG, Cooper DJ, Jones N (1988) Treatment of resistant intracranial hypertension with hypertonic saline. J Neurosurg 68: 78–481

Correspondence: Steffen Berger, M. D., Institute for Surgical Research, Klinikum Großhadern, Ludwig-Maximilians-University, Marchioninistrasse 15, D-81377 München, Federal Republic of Germany.

Acta Neurochir (1994) [Suppl] 60: 499–501
© Springer-Verlag 1994

31P-Magnetic Resonance Spectroscopic Study on the Effect of Glycerol on Cold-Induced Brain Edema

T. Kamezawa, T. Asakura, K. Yatsushiro, M. Niiro, S. Kasamo, and **T. Fujimoto**[1]

Department of Neurosurgery, [1] South Japan Health Science Center, Kagoshima University School of Medicine, Sakuragaoka, Kagoshima, Japan

Summary

The aim of the present study was to determine the effects of a hyperosmotic agent, 10% glycerol, on both brain energy metabolism and intracellular pH (pHi) in experimental vasogenic brain edema. Vasogenic brain edema was induced by cold injury applied to bilateral parietal portions in 13 mongrel dogs (7 glycerol, 6 control) while, 3 dogs were used as control. Before and at 24 hours after the injury, sequential phosphorous-31 magnetic resonance spectroscopy (31P-MRS) was performed for 2 hours in order to determine phosphocreatine (PCr), β-adenosine triphosphate (β-ATP), inorganic phosphate (Pi) levels and pHi. At 24 hours following cold injury, both PCr/Pi and ATP/Pi ratios significantly decreased from 7.75 to 3.97 and from 2.26 to 1.25, respectively. Furthermore, a moderate decrease in pHi of 7.16 to 7.01 was significantly demonstrated during the same experimental period. Administration of glycerol for 30 minutes significantly increased PCr/Pi from 3.97 to 5.06 and ATP/Pi from 1.25 to 1.72, respectively. Also, glycerol administration caused a significant increase in pHi from 7.01 to 7.11. This study indicates that cryogenic injury, in which formation and expansion of vasogenic brain edema a known to occur, results in disturbed brain energy metabolism and in intracellular acidosis; moreover, the administration of glycerol can ameliorate either or both of these derangements.

Keywords: Brain energy metabolism; cold-induced brain edema; glycerol; phosphorous-31 magnetic resonance spectroscopy.

Introduction

Vasogenic brain edema, which often causes increased ICP, is frequently encountered in a clinical setting. In spite of numerous studies[1,2,4,7–9], it still remains unclear whether or not vasogenic edema causes a derangement of brain energy metabolism. Recently, with the advent of 31P-MRS, it has become possible to monitor high-energy metabolites and pHi continuously under in vivo experimental conditions in animals. In the present study, we have undertaken experiments to measure the changes in brain energy metabolism and pHi in cold-induced brain injury employing 31P-MRS. Regarding the treatment of the edema produced by this model, hyperosmotic agents have been established to have beneficial effects owing to an osmotic dehydration[3,5,11]. Therefore, the effects of administering a hyperosmotic solution, 10% glycerol, on brain metabolism and pHi were also examined.

Materials and Methods

Sixteen adult mongrel dogs, each weighing 9 to 12 kg, were used for this study. Prior to each operation or MRS study, anaesthesia was initially induced with ketamine hydrochloride and was maintained by the intravenous administration of ketamine hydrochloride and pancuronium bromide. The animals were endotracheally intubated and artificially ventilated to maintain a $PaCO_2$ value of 39.2 ± 1.8 mmHg and PaO_2 value of 101.9 ± 4.1 mmHg, respectively. A craniectomy was initially performed and two cold lesions were produced by contact with a copper cylinder cooled in liquid nitrogen and placed on the dural surfaces of the bilateral parietal portions for 15 seconds in 13 dogs, with 3 dogs not being subjected to these lesions (the control group). At 24 hours after the injury, sequential 31P-MRS was performed for 2 hours in order to determine PCr, β-ATP, Pi levels and pHi, and other physiological data, such as serum osmotic pressure and hematocrit, were simultaneously obtained.

Following the first measurement, 4 g/kg of 10% glycerol was infused intravenously for 30 minutes in 7 animals (the glycerol group), and saline was infused in 6 others (the control group). MRS (Super 200 2.0 Tesla, ASAHI) was used to measure PCr/Pi and ATP/Pi ratios for the peak of each spectrum. A 50-mm two-turn surface coil which measured qualitative factors at 34.2 MHz was placed over the dural surface. The peak of β-ATP was chosen for the measurement of ATP concentration because α - and γ-ATP peaks contained other signals. PHi was determined from the chemical shift (δ) of the Pi peak relative to the PCr peak by the following equation[6]:

$$pHi = 6\,77 + \log \frac{\delta - 3.29}{5.58 - \delta} \ .$$

We determined the normal values of these parameters from the normal group. These values were analyzed by Student's t-test. Significance between the experimental groups was assigned at a level of p < 0.05.

Results

Physiological parameters obtained are shown in Table 1. Following the administration of glycerol, serum osmotic pressure significantly increased (from 303.6 ± 7.4 to 404.7 ± 6.6 mOsm/1, p < 0.001), but hematocrit significantly decreased (from 36.6 ± 1.2 to 32.0 ± 3.6%, p < 0.01).

The time course of PCr/Pi ratio is shown in Fig. 1. The ratio fell to about 50% of the normal value at 24 hours after cold injury (p < 0.0001). In the glycerol

group, this ratio significantly increased compared to that at 24 hours postinjury (p < 0.001). In the control group, on the other hand, the ratio significantly decreased (p < 0.05).

In Fig. 1, the time course of ATP/Pi ratio is also shown. Again, at 24 hours postinjury, the ratio fell approximately to 50% of the normal value (p < 0.0001). Although the ratio significantly increased after injury in the glycerol group (p < 0.05), no changes in the ratio were evident in the control group.

Regarding intracellular pH (pHi), the values significantly decreased from 7.16 ± 0.05 to 7.01 ± 0.07 (p < 0.01) at 24 hours following injury, as shown in Fig. 2. Glycerol infusion significantly attenuated the decrease in pHi (p < 0.05). In contrast, there were no significant changes in pHi in the control group.

Discussion

Metabolic changes associated with cryogenic injury and subsequent formation of vasogenic brain edema have been extensively studied by many workers[1,2,4,7–9]. However, it still remains unclear whether vasogenic edema results in a derangement of brain energy metabolism. The present investigation was designed to measure noninvasively the changes in high-energy metabolites and pHi associated with the formation of vasogenic brain edema by cold injury. Furthermore, our cold injury model has an advantage that edematous lesions were acquired without hemorrhagic lesions.

Regarding the treatment of vasogenic edema, the administration of glycerol is known to raise the serum osmolarity and reduce ICP[3,5,11]. In addition to this osmotic dehydrating effect, glycerol increases both cerebral blood flow and glucose utilization, and also enters the glycolytic pathway as a substrate[3,10]. Yatsushiro, our coworker[11], reported that T2 relaxation times were significantly reduced for at least 2 hours after administration of 10% glycerol in a cold injury model. The ameliorative effects of glycerol against brain edema were also examined employing 31P-MRS in this study.

The present study revealed that PCr/Pi and ATP/Pi ratios, and pHi were significantly decreased at 24 hours following cold injury. Moreover, this study demonstrated that the administration of glycerol almost completely abated these pathological alterations.

It is concluded that the formation and expansion of vasogenic brain edema results in disturbed brain energy metabolism and in intracellular acidosis. Further,

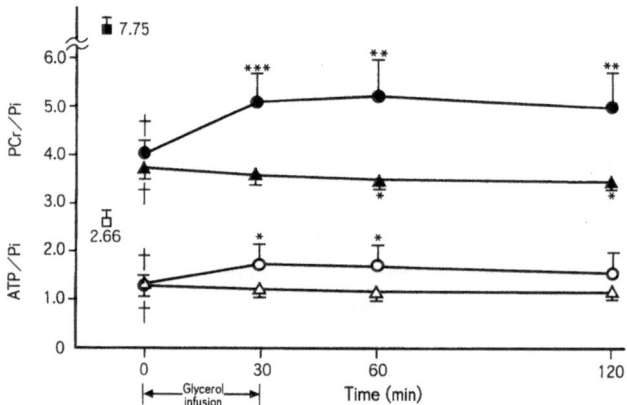

Fig. 1. The time course of the changes in PCr/Pi and ATP/Pi ratios. PCr/Pi ratio: ■ normal group (n = 3); ● glycerol group (n = 7); ▲ control group (n = 6). ATP/Pi ratio: □ normal group; ○ glycerol group; △ control group. Significant differences: + p < 0.0001 vs normal, * p < 0.05, ** p < 0.01, *** p < 0.001 vs. time 0

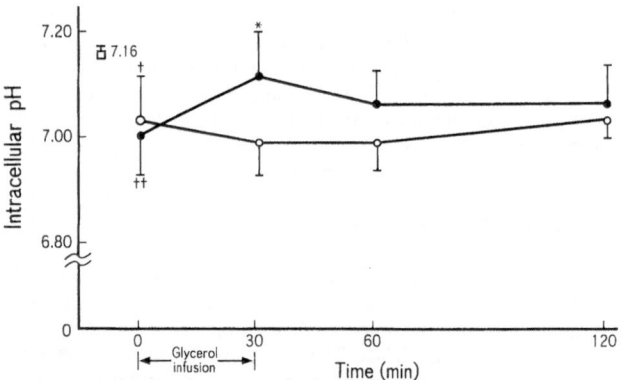

Fig. 2. The time course of the changes in intracellular pH. □ normal dogs (n = 3); ● Cl dogs with the infusion of glycerol (n = 7); ○ control grop (n = 6). Significant differences; + p < 0.05, ++ p < 0.01 vs. normal, * p < 0.05 vs. time 0. Cl: cold injury. Time 0: 24 h after Cl

Table 1. *Serum Osmotic Pressure and Hematocrit in Control and Clycerol Goups*

Factor	Control	30 min postinfusion	60 min postinfusion	120 min postinfusion
Control group				
SOP	298.2 ± 9.2	299.8 ± 6.1	302.0 ± 8.1	301.3 ± 8.2
Hematocrit (%)	37.1 ± 1.1	37.3 ± 1.3	37.7 ± 1.7	38.7 ± 1.3
Glycerol group				
SOP	303.6 ± 7.4	404.7 ± 6.6[b]	389.9 ± 11.3[b]	382.3 ± 15.5[b]
Hematocrit (%)	36.6 ± 1.2	32.0 ± 3.6[a]	36.0 ± 5.5	38.6 ± 13.5

SOP Serum osmotic pressure (mOsm/l). Values are means ± standard deviation. Significant differences = [a] $p < 0.01$, [b] $p < 0.001$ when compared with control values.

the administration of glycerol has ameliorative effects on metabolic derangements associated with vasogenic edema, although the effects of increased ICP cannot be clearly separated from the direct effects of edema. It can be envisaged that the amelioration of energy failure by intravenous glycerol may represent a mitigation of cerebral microcirculatory disturbances caused by the expansion of brain edema.

References

1. Frei HJ, Wallenfang T, Poll W, *et al* (1973) Regional cerebral blood flow and regional metabolism in cold induced oedema. Acta Neurochir (Wien) 29: 15–28
2. Imataka K, Handa H, Ishikawa M, *et al* (1987) The sequential changes of energy metabolism following cold-induced brain edema. Brain Nerve 39: 447–453
3. Meyer JS, Itoh Y, Okamoto S, *et al* (1975) Circulatory and metabolic effects of glycerol infusion in patients with recent cerebral infarction. Circulation 51: 701–712
4. Mies G (1983) Regional evaluation of blood flow and metabolism in experimental brain tumor of rats. In: Ishii S, Nagai H, Brock M (eds) Intracranial pressure V. Springer, Berlin Heidelberg New York Tokyo, pp 429-435
5. Niiro M, Asakura T, Yatsushiro K, *et al* (1990) Magnetic resonance studies in human brain oedema following administration of hyperosmotic agents. Acta Neurochir (Wien) [Suppl] 51: 131–133
6. Petroff OAC, Prichard JW, Behar KL, *et al* (1985) Cerebral intracellular pH by 31P nuclear magnetic resonance spectroscopy. Neurology 35: 781–788
7. Reulen HJ, Medzihradsky F, Enzenbach R, *et al* (1969) Electrolytes, fluids, and energy metabolism in human cerebral edema. Arch Neurol 21: 517–525
8. Schmiedek P, Baethmann A, Sippel G, *et al* (1974) Energy state and glycolysis in human cerebral edema. J Neurosurg 40: 351–364
9. Sutton LN (1980) Metabolic and electrophysiologic consequences of vasogenic edema. In: Cervos-Navarro J, Ferszt R (eds) Advances in neurology, Vol 28. Brain edema. Raven, New York, pp 241–254
10. Thurston JH, Hauhart RE, Dirgo JA (1981) Effects of a single therapeutic dose of glycerol on cerebral metabolism in the brains of young mice. J Neurochem 36: 830–838
11. Yatsushiro K, Niiro M, Asakura T, *et al* (1990) Magnetic resonance study of brain oedema induced by cold injury. Acta Neurochir (Wien) [Suppl] 51: 113–115

Correspondence: T. Kamezawa, M.D., Department of Neurosurgery, Kagoshima University School of Medicine, 8-35-1, Sakuragaoka, Kagoshima, 890, Japan.

Acta Neurochir (1994) [Suppl] 60: 502–504

Treatment of Vasogenic Brain Edema with Arginine Vasopressin Receptor Antagonist – an Experimental Study

S. Nagao, M. Kagawa, I. Bemana, T. Kuniyoshi, T. Ogawa, Y. Honma, and **H. Kuyama**

Department of Neurological Surgery, Kagawa Medical School, Kagawa, Japan

Summary

We determined the effect of a centrally administered V_1 receptor antagonist of arginine vasopressin on the brain water content in an animal model of vasogenic brain edema. Using adult rats, a cold injury was induced in the left hemisphere of the brain by applying a frozen copper rod. 50 ng of V_1 receptor antagonist was administered into the left lateral ventricle 10 minutes prior to and/or 1 hour after injury. Twenty four hours after the cold injury, the brain water and sodium contents and plasma osmolality were measured. The V_1 receptor antagonist significantly suppressed the increase of the brain water and sodium contents in the cortical structure adjacent to the lesion without any changes in plasma osmolality. Our results demonstrate the effectiveness of a V_1 receptor antagonist of vasopressin on vasogenic brain edema.

Keywords: Brain edema; arginine vasopressin; V_1 receptor antagonist; cold injury.

Introduction

Significant increases in arginine vasopressin (AVP) in cerebrospinal fluid (CSF) were reported in patients with intracranial hemorrhage[10], ischemic stroke[10], and craniocerebral trauma[10]. Intraventricularly administered AVP also increased the brain water content in the normal brain of monkeys[5] and cold-injured brain of cats[6]. These findings suggest that centrally released AVP plays an important role in regulation of the brain water content in normal and pathological conditions.

We undertook this study to examine the effect of intraventricular administration of a V_1 receptor antagonist of AVP before and after cold injury on the brain water content in rats.

Materials and Methods

General Procedure

The experiments were performed in male Sprague-Dawley rats with body weight of 250 to 350 g. During operation and before decapitation, the animals were anesthetized by an intraperitoneal injection of pentobarbital sodium (30 mg/100 g body weight). The V_1 receptor antagonist (50 ng in 1 μl of Ringer solution) was administered into the left lateral ventricle by stereotaxy. A cold injury was induced of the left cerebral hemisphere by applying a copper rod (5 mm in diameter) cooled by liquid nitrogen to the exposed skull bone. Twenty-four hours after injury, the animals were sacrificed by decapitation. The brain was removed immediately and both the injured and contralateral hemispheres were divided into cortical and deep structures. The brain water content was measured using the dry-wet-weight method. Brain tissue sodium and potassium were extracted with 2N nitric acid for 48 hours and analyzed using atmoic absorption spectrophotometry. Before decapitation, a blood sample was taken from the femoral artery for measurement of plasma osmolality, sodium, and potassium. Statistical analysis were performed using the Student's t-test.

Experiment 1. Cold injury of 20 seconds was induced at ten minutes after intraventricular adiminstration of the antagonist. Twenty rats receiving intraventricular Ringer's solution served as controls.

Experiment 2. Cold induced edema was produced by 1 minute application of the cooled copper rod. 50 ng of the V_1 receptor antagonist was administered into the ventricle at 1 hour after cold injury.

Results

Experiment 1. The tissue water and sodium contents of bilateral cortical structures were significantly increased concomitantly with a decrease in potassium. In the deep cerebral structures, no significant changes in brain water and electrolyte contents were observed. The bilateral changes of the cortical structures by trauma were significantly suppressed by the intraventricular administration of AVP V_1 receptor antagonist (Fig. 1), which did not affect plasma osmolality.

Experiment 2. The water contents of the bilateral cortical and deep cerebral structures were significantly increased by cold injury of 1 min duration. Cortical

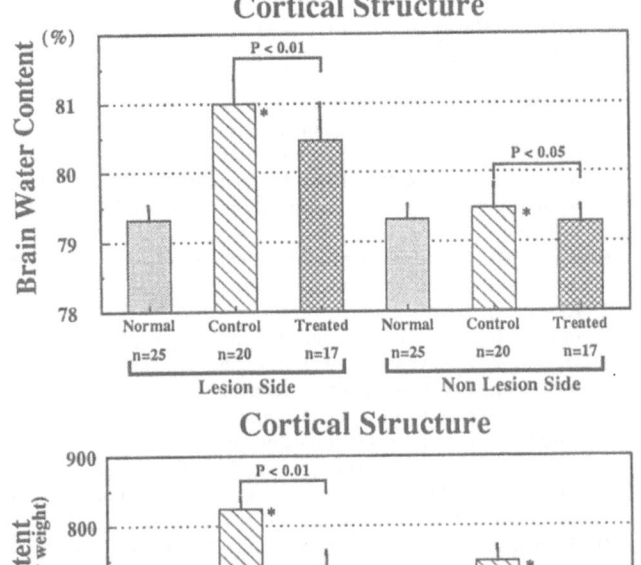

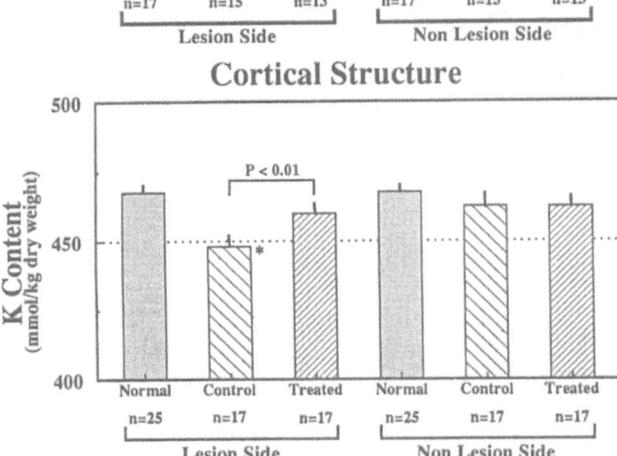

Fig. 1. Changes in brain water, sodium, and potassium contents of bilateral cortical structures. Animals received a V_1 receptor antagonist 10 minutes prior to injury. * Values significantly different from the normal group by $p < 0.05$. Treated animals with central administration of a V_1 receptor antagonist (50 ng) prior to cold injury. Values are mean ± standard error

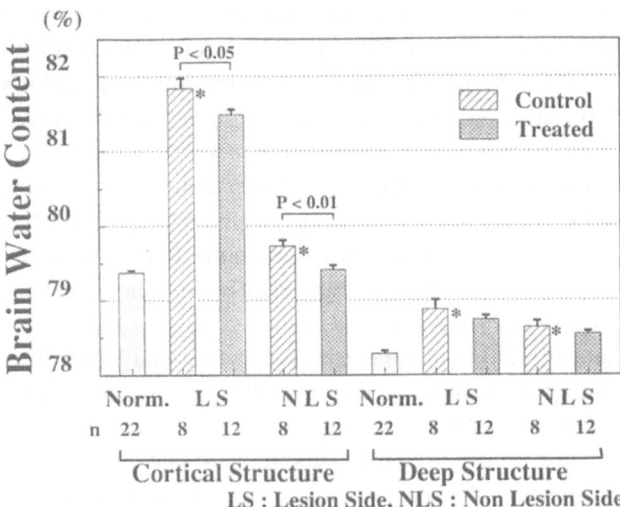

Fig. 2. Changes in brain water content of cortical and deep brain structures of both cerebral hemispheres. Animals received a V_1 antagonist 1 hour after injury. * Values significantyl different from the normal group by $p < 0.05$. Treated: animals with central administration of V_1 receptor antagonist (50 ng) after cold injury. Values are mean ± standard error

Discussion

In normal brain of rat[1] and cold injured brain of cat[6], intraventricular administration of AVP was reported to increase the brain water content by changing brain capillary permeability to water. In patients with brain edema and intracranial hypertension, a high concentration of AVP in the CSF was associated with poor clinical conditions. Thus centrally released AVP has been considered to have an influences on the water permeability of the brain.

There are two types of AVP receptors. The V_1 receptor found in vascular smooth muscle and hepatocytes links to the phosphatidylinositol system with calcium mobilization, while the V_2 receptor found in the renal medulla links to generation of cyclic AMP. In brain capillary endothelium, the vasopressin-induced rise of intracellular calcium is completely abolished by V_1 receptor antagonists[2].

Intracranial injection of AVP was reported to cause an increase in brain water content, which could be inhibited by V_1 receptor antagonist[7]. Further, increased brain water and sodium contents in the brain following collagenase-induced cerebral hemorrhage were reduced by central administration of a V_1 receptor antagonist[8]. Based on radiolabeled ligand binding studies, the AVP-receptor of isolated cerebral microvessel is of the V_1 type[4]. These reports support that the increase in

edema was more pronounced, however, than edema in the deep structures. Intraventricular administration of the V_1 receptor antagonist 1 hour after cold injury resulted also in a significant inhibition of edema formation in the bilateral cortical structures, while no effect was obtained on edema of the deep structures (Fig. 2).

brain water and sodium contents in vasogenic brain edema may be mediated by the V_1 receptor for AVP.

We have previously reported that intraventricularly administered V_1 and V_2 receptor antagonists given before cold injury, suppress the increase in brain water in rats, and that the V_1 receptor antagonist is more effective in the reduction of vasogenic brain edema[3]. In the present experiments, intraventricular administration of a V_1 receptor antagonist before and after cold injury significantly reduced cortical edema and sodium accumulation in the tissue, suggesting that V_1 receptor antagonists suppress edema development in the central nervous system.

In conclusion, our results indicate that centrally administered V_1 receptor antagonists of AVP may have significance for the treatment of brain edema.

References

1. Doczi T, Szerdahelyi P, Gulya K, Kiss J (1982) Brain water accumulation after the central administration of vasopressin. Neurosurgery 11: 402–407
2. Hess J, Jensen CV, Diemer NH (1991) The vasopressin receptor of the blood-brain barrier in the rat hippocampus is linked to calcium signaling. Neurosci Lett 132: 8–10
3. Kagawa M, Nagao S, Kuniyoshi T, Ogawa T, Itoh T, Honma Y, Itano T, Hatase O (1991) Treatment of brain edema with arginine vasopressin receptor antagonist. Advances in neuro-trauma research, Vol 3, pp 75–79
4. Pearlmutter AF, Szkrybalo M, Kim Y, Harik SI (1988) Arginine vasopressin receptors in pig cerebral microvessels, cerebral cortex and hippocampus. Neurosci Lett 87: 121–126
5. Raichle ME, Grubb RL (1978) Regulation of brain water permeability by centrally-released vasopressin. Brain Res 143: 191–194
6. Reeder RF, Nattie EE, North WG (1986) Effect of vasopressin on cold-induced brain edema. J Neurosurg 64: 941–950
7. Rosenberg GA, Estrada E, Kyner WT (1990) Vasopressin-induced brain edema is mediated by the V_1 receptor. In: Long D, *et al* (eds) Advances in neurology. Raven, New York
8. Rosenberg GA, Scremin O, Estrada E, Kyner WT (1992) Arginine vasopressin V_1-antagonist and atrial natriuretic peptide reduce hemorrhagic brain edema in rats. Stroke 23: 1767–1774
9. Sorensen PS, Gjerris F, Hammer M (1985) Cerebrospinal fluid and plasma vasopressin during short-time induced intracranial hypertension. Acta Neurochir (Wien) 77: 46–51
10. Sorensen PS, Gjerris A, Hammer M (1985) Cerebrospinal fluid vasopressin in neurological and psychiatric disorders. J Neurol Neurosurg Psychiatry 48: 50–57
11. Wang BC, Share L, Crofton JT, Kimura T (1982) Effect of intracerebroventricular infusion of hypertonic solutions on plasma and cerebrospinal fluid vasopressin concentrations. Neuroendocrinology 34: 215–221

Correspondence: Seigo Nagao, M.D., Department of Neurological Surgery, Kagawa Medical School, 1750-1 Ikenobe, Miki-cho, Kita-gun, Kagawa, 761-07, Japan.

Acta Neurochir (1994) [Suppl] 60: 505–507
© Springer-Verlag 1994

Attenuation of Hemispheric Swelling Associated with Acute Subdural Hematomas by Excitatory Amino Acid Antagonist in Rats

K. Kinoshita, Y. Katayama, T. Kano, T. Hirayama, and **T. Tsubokawa**

Department of Neurological Surgery, Nihon University School of Medicine, Tokyo, Japan

Summary

Acute subdural hematoma (ASDH) gives rise to a mass effect not only by itself but also through unilateral hemispheric swelling. The present study tested the hypothesis that hemispheric swelling is mediated by mechanisms which involve excitatory amino acids (EAAs). After removal of the subdural clot, introduced by homologous blood (0.1–0.2 ml), the % brain water was determined from the formula: ((wet weight – dry weight) / wet weight) × 100. The % brain water of the left hemisphere was significantly greater than that of the right hemisphere during the initial 6 hours after induction of ASDH in animals injected with 0.2 ml blood. A less marked but significant increase was observed in the animals injected with 0.1 ml blood. Systemic pretreatment with kynurenic acid (KYN; 800 mg/kg, i.p.), a broad-spectrum EAA antagonist, attenuated the increase in % brain water in the animals injected with 0.2 ml blood. In order to determine the changes in cerebral metabolism induced by the model of ASDH employed in the present study, we measured the cortical cytochrome oxidase (CYO) activity, a marker of mitochondrial respiration, in a separate group of animals. The CYO activity estimated densitometrically from the histochemical staining was not significantly altered in the animals injected with either 0.1 or 0.2 ml blood, suggesting absence of ischemia. These results indicated that the hemispheric swelling associated with thin ASDHs may be partially mediated by mechanisms other than ischemia, in which EAAs appear to be involved.

Keywords: Acute subdural hematoma; brain swelling; excitatory amino acids.

Introduction

Acute subdural hematomas (ASDHs) cause a mass effect not only by the SDH volume but also through umlateral hemispheric swelling. While various mechanisms have been proposed[7,10,13], no universally accepted explanation for the frequent occurrence of unilateral hemispheric swelling in association with ASDHs has yet been established. Recently, Bullock and his coworkers[1,2] demonstrated that excitatory amino acids (EAAs) plays a crucial role in inducing ischemic neu-

ronal damage in an experimental model of ASDHs. Our previous studies have indicated that EAA neurotransmitters are involved in the mechanisms of the ionic fluxes and cellular swelling that occur traumatic as well as ischemic brain insults[8,9]. If EAAs are released in association with ASDHs, they may also contribute to hemispheric swelling, since EAAs produce massive ionic fluxes through the neuronal and glial cell membranes[3,11,14]. The present study tested the hypothesis that hemispheric swelling is mediated by mechanisms involving EAAs.

Materials and Methods

Young Sprague-Dawley rats (weighing 200–250 g) were used for the experiments. The rectal temperature was maintained at 37–38°C. The animals were anesthetized with 1% halothane and 2 : 1 N_2O in O_2. Each animal was then fixed in a stereotaxic instrument, and small skull holes were made in the left parietal region. Although the procedures for inducing ASDH were basically similar to those described by Bullock et al.[12], the amount of the blood injected into the subdural space in the present study was smaller. A curved No. 23 blunt-tipped needle was inserted into the left subdural space. The needle was cemented into position using rapid-setting cyanoacrylate glue. Nonheparinized venous blood was withdrawn from the femoral vein and 0.1–0.2 ml of the blood was injected into the subdural space over a period of 2–4 minutes. In the sham-control group, the same volume of 0.9% saline was infused. The anesthesia was then discontinued and the animals were returned to their cages. All animals were pretreated (i.p.) with either kynurenic acid (KYN; 800 mg/kg), a broad-spectrum EAA antagonist[5], or vehicle alone at 30 minutes prior to the induction of ASDH.

The animals were anesthetized with an overdose of pentobarbital sodium (i.p.) and decapitated at 0–12 hours after the induction of ASDH. The brains were immediately removed and separated into left and right hemispheres. Due to the small volume of injected blood, the ASDHs produced in the present study were relatively thin as compared to those in previous reports. After removal of the subdural clot, the % brain water was determined from the formula[6]:

$$\% \text{ brain water} = ((\text{wet weight} - \text{dry weight}) / \text{wet weight}) \times 100$$

In order to characterize the changes in cerebral metabolism in the model of ASDH employed in the present study, the cytochrome oxidase (CYO) activity, a marker of mitochondrial respiration[4], was examined in the cerebral cortex in a separate group of animals. These animals were anesthetized with an overdose of pentobarbital sodium (i.p.) at 12 hours after induction of ASDH, and then perfused transcardially first with heparinized saline and subsequently with 4% paraformaldehyde in phosphate buffer (0.1 M, pH 7.4). The brains were removed, postfixed in the same fixative overnight and cytoprotected with 40% sucrose in phosphate buffer. Frozen sections at a thickness of 25 μm were obtained from the region of the S1, and collected in phosphate buffer and tested for their CYO activity. The CYO activity in the cortex was analyzed by photometric semi-quantitation. The changes in CYO activity is known to reflect the metabolic activity during the previous 12–24 hours.

Results

When 0.2 ml blood was injected, the survival rate of the animals with ASDH decreased progressively as time elapsed (Fig. 1). In contrast, all animals survived for 12 hours when 0.1 ml blood was injected. The brains removed at predetermined times demonstrated varying amounts of blood spreading into the subarachnoid space in addition to subdural clots. The data from

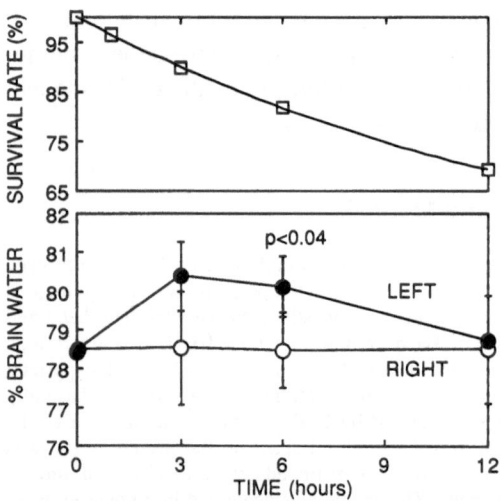

Fig. 1. Survival rate (top) and % brain water (bottom) versus time after induction of ASDH in animals injected with 0.2 ml blood. Statistical comparison: paired t-test

animals which predominantly revealed subdural clots were analyzed in the present study. The clots appeared, however, too thin to produce a mass effect by themselves.

The % brain water of the left hemisphere was significantly greater as compared to that of the right hemisphere during the initial 6 hours after induction of ASDH in the animals injected with 0.2 ml blood (Table 1; Figs. 1 and 2). A less marked but significant increase was observed in the animals injected with 0.1 ml blood (Table 1). Substantially no change was observed in the sham-controls. The effects of pretreatment with KYN were tested at 6 hours after induction of ASDH, at which time most animals still survived. KYN clearly attenuated the increase in % brain water in the animals injected with 0.2 ml blood (Fig. 2). The effect of KYN in the animals injected with 0.1 ml blood was unclear, since the increase in % brain water was too small. The CYO activity evaluated at 12 hours after induction of ASDH was not significantly altered in the animals injected with either 0.1 or 0.2 ml blood (Table 1).

Discussion

It is a common clinical experience for unilateral hemispheric swelling not always to be associated with voluminous ASDHs. It could occur in association even with thin ASDHs, suggesting that this phenomenon is not merely a consequence of compression ischemia. The marked difference in survival rates observed between animals injected with 0.1 ml blood and those injected with 0.2 ml blood in the present study, indicates that the threshold volume of subdural clots for the induction of the pathophysiology unique to ASDHs in rats lies somewhere between 0.1 and 0.2 ml.

The present findings demonstrate that a subdural clot volume of around the threshold produces a significant increase in % brain water in the ipsilateral hemisphere, which indicates the occurrence of hemispheric swelling. The absence of any significant alteration in

Table 1. *% Brain Water, Survival Rate and Cytochrome Oxidase Activity*

Group	% Brain water at 6 hours (left / right)	Survival rate at 12 hours (%)	Cytochrome oxidase activity at 12 hours (% left / right)
0.1 ml blood	79.2 ± 0.2 / 78.8 ± 0.2*	100.0	117.0 ± 8.2
0.2 ml blood	80.4 ± 0.5 / 78.5 ± 1.0**	69.2	89.2 ± 7.4

Mean ± SD. ASDHs were induced on the left side. * p < 0.05, ** p < 0.04.

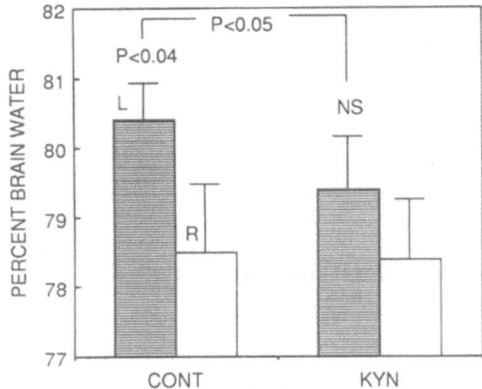

Fig. 2. % brain water of the hemispheres in animals injected with 0.2 ml at 6 hours after induction of ASDH. *R* right; *L* left. *CONT* pretreated with vehicle alone; *KYN* pretreated with 800 mg/kg kynurenic acid. Statistical comparison: paired t-test for that between R and L: unpaired t-test for that between CONT and KYN

CYO activity in the cortex again implies that ischemia is not a predominant contributor to such a change, at least in animals injected with 0.1 ml blood. While all these animals survived for 12 hours, they exhibited a significant increase in % brain water but no significant change in cortical CYO activity.

The effects of KYN observed in the present study suggest that the hemispheric swelling associated with thin ASDHs may be mediated by mechanisms involving EAAs. Further studies are needed to determine the mechanisms of EAA release. The present results imply that administration of EAA antagonists may be useful in the management of ASDHs.

References

1. Bullock R, Butcher SP, Chen M-H, Kendall L, McCulloch J (1991) Correlation of the extracellular glutamate concentration with extent of blood flow reduction after subdural hematoma in the rat. J Neurosurg 74: 794–802
2. Chen M-H, Bullock R, Graham DI, Miller JD, McCulloch J (1991) Ischemic neuronal damage after acute subdural hematoma in the rat: effects of pretreatment with a glutamate antagonist. J Neurosurg 74: 944–950
3. Choi DW, Koh J-Y, Peters S (1988) Pharmacology of glutamate neurotoxicity in cortical cell culture: attenuation by NMDA antagonist. J Neurosci 8: 185–196
4. Dimlich RVW, Showers MJ, Shipley MT (1990) Densitometric analysis of cytochrome oxidase in ischemic rat brain. Brain Res 516: 181–191
5. Gill R, Woodruff GN (1990) The neuroprotective actions of kynurenic and MK-801 in gerbils are synergistic and not related to hypothermia. Eur J Parmacol 176: 143–149
6. Gotoh O, Koide T, Takakura K (1985) Ischemic brain edema following occlusion of the middle cerebral artery in the rat: the time course of the brain water sodiume and potassium contents and blood-brain barrier permiability to ^{125}I-albumin. Stroke 16: 101–109
7. Gutierrez FA, Raimondi AJ (1975) Acute subdural hematoma in infancy and childhood. Childs Brain 1: 269–290
8. Katayama Y, Becker DP, Tamura T, Hovda DA (1990) Massive increase in extracellular potassium and indiscriminate glutamate release after concussive brain injury. J Neurosurg 73: 889–900
9. Katayama Y, Tamura T, Becker DP, Tsubokawa T (1992) Early cellular swelling during cerebral ischemia in vivo is mediated by excitatory amino acids released from nerve terminals. Brain Res 577: 121–126
10. Lobato RD, Sarabia R, Cordobes F, Rivas JJ, Adrados A, Cabrera A, Gomez P, Madera A, Lamas E (1988) Posttraumatic cerebral hemisheric swelling. Analysis of 55 cases studied with computerized tomography. J Neurosurg 68: 417–423
11. Mayer ML, Westbrool GL (1987) Cellular mechanisms underlying excitotoxicity. TINS 10: 59–61
12. Miller JD, Bullock R, Graham DI, Chen M-H, Teasdale GM (1990) Ischemic brain damage in a model of acute subdural hematoma. Neurosurgery 27: 433–439
13. Thomas A, Gennarelli TA, Spielman GM, Longfitt TW, Gildenberg PL, Harrington T, Jane A, Marshall LF, Miller JD, Pitts LH (1982) Influence of the type of intracranial lesion on outcome frome severe head injury. J Neurosurg 56:26–32
14. Wong-Riley MTT, Merzenich MM, Leake PA (1978) Changes in endogenous reactivity to DAB induced by neuronal inactivity. Brain Res 141: 185–192

Correspondence: Kosaku Kinoshita, M. D., Department of Neurological Surgery, Nihon University School of Medicine, Tokyo 173, Japan.

Acta Neurochir (1994) [Suppl] 60: 508–510
© Springer-Verlag 1994

Effect of Platelet-Activating Factor Antagonist on Brain Injury in Rats

T. Tokutomi, M. Sigemori, T. Kikuchi, and **M. Hirohata**

Department of Neurosurgery, Kurume University School of Medicine, Asahi-machi, Kurume, Japan

Summary

The ability of a platelet-activating factor (PAF) antagonist to reduce infarct size has been reported in a animal model of focal brain ischemia. The authors studied the effect of PAF antagonist (TCV-309) on cold brain injury in rats. Twenty-four hours after injury, water content was determined by both drying-weighing and spesific gravimetric techniques, and ischemic brain damage was assesed with 2,3,5-triphenyltetrazolium chloride in multiple coronal sections. Pretreatment with TCV-309 (lmg/kg) significantly reduced the water content (p < 0.01) and volume of ischemic damage (p < 0.001) produced by the cold brain injury. These results indicate that PAF antagonist can ameliorate secondary brain tissue damage following brain injury.

Keywords: PAF antagonist; brain ischemia; brain edema cold injury.

Introduction

Platelet-activating factor (PAF) is a potent mediator of anaphylaxis and shock[2,13] and there is also evidence for the involvement of PAF in pathophysiology of brain edema and cerebral blood flow reduction[2,8]. Moreover, PAF antagonists have also been reported to reduce brain edema and damage in ischemic models[7,11,12]. As secondary brain injury, such as brain edema and ischemia, adversely affects recovery following head trauma, we studied the effect of a PAF antagonist on post-traumatic cerebral edema and ischemia, using a rat model of cold brain injury.

Materials and Methods

Thirty-seven male Wister rats weighing 300 to 350 g each were anesthetized with intraperitonial pentobarbital sodium (40 mg/kg). The scalp and periosteum was incised and exposed. A metal tube probe (outer diameter = 4 mm) filled with liquid nitrogen was then applied to the exposed intact right parietal bone for 30 seconds.

Rats were divided into two groups and received the PAF antagonist TCV-309 (lmg/kg; manufactured by Takeda Chemical Indus-tries Ltd., Osaka, Japan) or a vehicle administered intravenously immediately prior to injury. TCV-309 is a specific and potent PAF antagonist with no agonistic action in vivo[13].

Twenty-four hours after injury, 24 rats were anesthetized again and decapitated. The forebrain was removed after section at the midbrain and the hemispheres were divided carefully in the midline for determining the brain water content. Freeze-drying procedures and the specific gravimetric technique were used for measurement of water content in cerebral tissue. Water content measured by the drying method was determined by weighing both fresh samples and samples dried for 72 h and expressed as percentages ((wet weight-dry weight)/wet weight)[5]. Liquid gradient columns were prepared for the specific gravity method according to the technique described by Marmarou et al.[9].

The other 14 rats were anesthetized again 24 hours after injury, and a lethal intracardiac perfusion of 2,3,5-triphenyltetrazolium chloride (TTC) was performed. The TTC-stained brains were re-moved and fixed in 10% phosfate-buffered formalin for 3 days. Two mm-thick coronal sections were then cut and the size of poorly stained area on each slices were measured to determine the volume of the ischemic area.

Another six rats were used for normal control of brain water content. Statistical comparisons between the two groups were performed using Student's t-test.

Results

Water Content

Results of the water content in these groups are summarized in Table 1. Pretreatment with TCV-309 significantly reduced the water content measured by both drying-weighing and specific gravimetric techniques.

TTC Staining

The extent of ischemic area was visualized as an area of poorly stained tissue around the cold lesion, as opposed to viable tissue, which stained brick red. The extent of the ischemic area around the cold lesion was more severe in the control brain (Fig. 1). The calculated volume of ischemic brain damage was significantly decreased in the TCV-309-treated group (165.9 ± 25.7 mm^3 in the control group vs. 36.8 ± 19.3 in the TCV-309-treated group; p < .001, Fig. 2).

Discussion

Ischemic lesions identified by TTC staining have been shown to the an identical to the lesions detected utilizing hematoxylin and eosin-stained sections[1]. Perfusion staining of the cerebral tissue in vivo with TTC is therefore a reliable method for detemination of ischemic brain damage in rats[6]. Our results show that PAF antagonist effectively reduces not only brain edema but also secondary ischemic damage produced by cold brain injury.

Human endothelial cells stimulated by PAF retract and separate, while stress fibers disappear or become less regular, causing bleb formation, and these effects are inhibited by PAF antagonists[3]. In a preliminary study, we examined the morphological changes of blood vessels by electron microscopy in the same model of cold brain injury, and we defeeted a marked increase of endothelial microvilli in microvessels from the peri-lesion at 6 and 24 hours after injury. Pretreatment with TCV-309 apparently reduced the development of microvilli. Microvillus formation in the cerebral microvasculature has been observed after a variety of insults[4,10]. Although the pathogenesis of the morphologic changes of the endotherial cell remains to be clarified, the major possible target of PAF antagonist is the endothelium.

As microvascular damage plays a very important role in a development of brain edema and ischemia following brain injury, an effective PAF antagonist may protect the microvessels and prevent the formation of brain edema and ischemia.

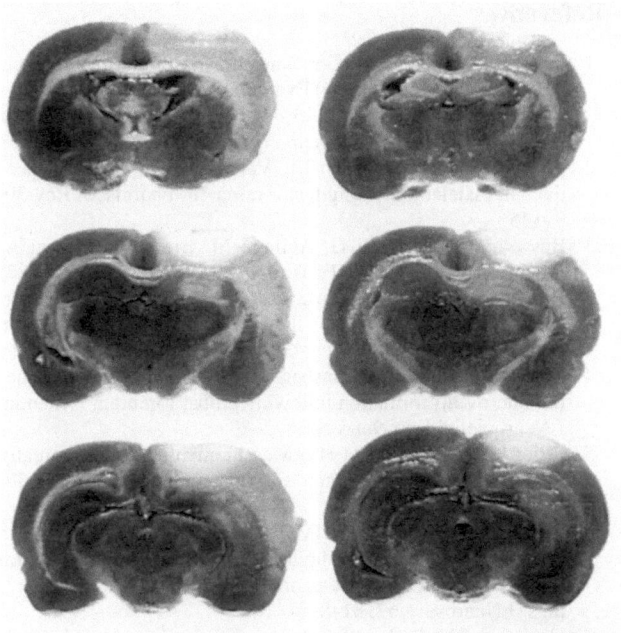

Fig. 1. Photomicrographs of brain sections stained in vivo with 2,3,5-triphenyltetrazolium chloride from control (left) and TCV-309-treated rats (right) 24 hours after cold injury. The zone of pallor staining which indicates ischemic damage was smaller in the TCV-309-treated brain than that in the control brain

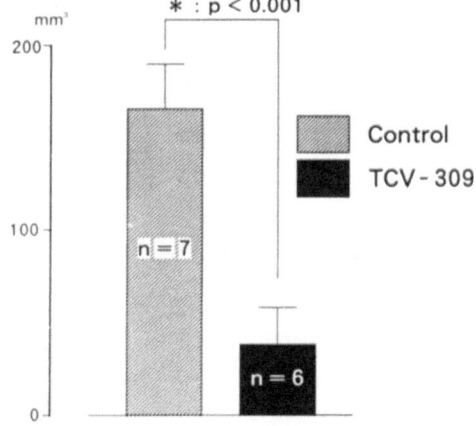

Fig. 2. Mean volume $\pm$ SD of ischemic damage after cold injury in control rats and TCV-309-treated rats. The difference was significant (p < 0.001)

Table 1. *Water Content Measured by Freeze-Drying and Specific Gravity in Control and TCV-309 Treated Groups and in Normal Group (Without Cold Injury)*

Group	Freeze-drying (%)		Specific gravity	
	Lesion hemisphere	Contralateral hemisphere	Lesion hemisphere	Contralateral hemisphere
Control n = 6	80.90 ± 0.50	79.67 ± 0.61	1.0364 ± 0.0007	1.0385 ± 0.0006
TCV 309 n = 6	79.53 ± 0.76[a]	78.42 ± 0.35[a]	1.0384 ± 0.0005[b]	1.0412 ± 0.0005[a]
Normal n = 3	77.35 ± 0.35		1.0433 ± 0.0002	

Values are mean $\pm$ SD. [a] Significantly different from control: p < 0.01; [b] significantly different from control: p < 0.05.

References

1. Bederson JB, Pitts LH, Germano SM, Nishimura MC, Davis RL, Bartkowski HM (1986) Evaluation of 2,3,5-triphenyltetrazolium chloride as a stain for detection and quantification of experimental cerebral infarction in rats. Stroke 17: 1304–1308
2. Braquet P, Touqui L, Shen TY, Vargaftic BB (1987) Perspectives in platelet-activating factor research. Pharmacol Rev 39: 9–145
3. Bussolino F, Camussi G, Aglietta M, Braquet P, Bosia A, Pescarmona G, Sanvio F, D'urso N, Marchisio PC (1987) Human endothelial cells are target for platelet-activating factor. 1. Platelet-activating factor induces changes in cytoskeleton structures. J Immunol 139: 2439–2446
4. Dietrich WD, Busto R, Ginsberg MD (1984) Cerebral endothelial microvilli: formation following global forebrain ischemia. J Neuropathol Exp Neurol 43: 72–83
5. Elliot KAC, Jasper H (1949) Measurement of experimentally induced brain swelling and shrinkage. Am J Physiol 157: 122–129
6. Germano IM, Bartkowski HM, Cassel ME, Pitts LH (1987) The therapeutic value of nimodipine in experimental focal cerebral ischemia. Neurological outcome and histopathological findings. J Neurosurg 67: 81–87
7. Kochaneh PM, Dutka AJ, Kumaroo KK, Hallenbeck JM (1987) Platelet activating factor receptor blockade enhances recovery after multifocal brain ischemia. Life Sci 41: 2639–2644
8. Kochanek PM, Nemoto EM, Melick JA, Evans RW, Burke DF (1988) Cerbrovascular and cerebrometabolic effects of intracarotid infused platelet-activating factor in rats. J Cereb Blood Flow Metab 8: 546–551
9. Marmarou A, Poll W, Shulman K, Bhagavan H (1978) A simple gravimetric technique for measurement of cerebral edema. J Neurosurg 49: 530–537
10. Maxwell W, Irvine A, Watt C, Graham DI, Adams JH, Gennarelli TA (1991) The microvascular response to stretch injury in the adult guinea pig visual system. J Neurotrauma 8: 271–279
11. Panetta T, Marcheselli VL, Braquet P, Spinnewyn B, Bazan NG (1987) Effects of a platelet activating factor antagonist (BN 52021) on free fatty acids, diacylglycerols, polyphosphoinositides and blood flow in the gerbil brain: inhibition of ischemia-reperfusion induced cerebral injury. Biochem Biophys Res Commun 149: 580–587
12. Spinnewyn B, Blavet N, Clostre F, Bazen N, Braquet P (1987) Involvement of platelet-activating factor (PAF) in cerebral post-ischemic phase in mongolian gerbils. Prostaglandins 34: 337–349
13. Terashita Z, Kawamura M, Takatani M, Tsushima S, Imura Y, Nishikawa K (1992) Beneficial effects of TCV-309, a novel potent and selective platelet activating factor antagonist in endotoxin and anaphylactic shock in rodents. J Pharmacol Exp Ther 260: 748–755

Correspondence: Takashi Tokutomi, M. D., Department of Neurosurgery, Kurume University School of Medicine, 67 Asahi-machi, Kurume-shi, Fukuoka-ken 830, Japan.

Acta Neurochir (1994) [Suppl] 60: 511–515
© Springer-Verlag 1994

Naloxone Reduces Alterations in Evoked Potentials and Edema in Trauma to the Rat Spinal Cord

T. Winkler[1], H. S. Sharma[2], E. Stålberg[1], Y. Olsson[2], and F. Nyberg[3]

[1] Department of Clinical Neurophysiology, [2] Laboratory of Neuropathology, University Hospital, and
[3] Department of Pharmacology, Biomedical Centrum, University of Uppsala, Uppsala, Sweden

Summary

The influence of naloxone (an opioid receptor antagonist) on spinal cord conduction and edema formation as a result of trauma to the cord was investigated in a rat model. The spinal cord injury (SCI) was inflicted in urethane anesthetized animals by a longitudinal incision into the right dorsal horn of the T10-11 segments, about 2 mm deep and 5 mm long. Spinal cord evoked potentials (SCEP) were recorded epidurally from the T9 (rostral) and T12 (caudal) segments after stimulation of the ipsilateral tibial and sural nerves at the ankle. The edema was measured by determining water content of the cord at s h after injury. In rats not given naloxone SCI resulted in an immediate long-lasting depression of the rostral maximal negative peak (MNP) amplitude (about 60%) and a significant increase in the latency of the rostral maximal positive peak (MPP). Pretreatment with naloxone inhibited the immediate post-injury decrease of the rostral MNP and some of the increase of MPP latency. The water content in thc traumatized spinal cord was reduced by 3% in naloxone treated animals compared with untreated injured controls. Our results indicate that endogenous opioid peptides participate in changes of spinal cord conduction after trauma and influence edema formation probably via multiple opioid receptors.

Keywords: Maloxone; spinal cord evoked potentials; edema.

Introduction

In addition to classical neurotransmitters the neuropeptides have important roles in the communication network of the nervous system. The dorsal horn of mammalian spinal cord is very rich in various opioid peptides such as dynorphin, and enkephalins. Additionally, other peptides like substance P, somatostatin, neurotensin and oxytocin are present in particularly high concentrations[3]. However, their pathophysiological importance in trauma and other neurological diseases of the cord is still not clarified.

Micro-iontophoretic application of opioid peptides depresses the single unit discharge in extracellular single unit recordings[1]. This effect of peptides is blocked by naloxone, a selective opioid receptors antagonist[5]. Thus, peptides can influence the electrical activity of the nervous system via specific receptors. However, the involvement of neuropeptides in various bio-electrical events occurring immediately after a trauma to the spinal cord is not known in all its details.

We are investigating the pathophysiological consequences of focal spinal cord injuries using a rat model, particularly the relation between early changes of spinal cord evoked potentials (SCEP) and formation of edema. Our model consists of a unilateral longitudinal incision into the right dorsal horn leaving the major ascending and descending tracts in the white matter mainly intact[9]. An alteration in electrical activity occuring in this model will thus mainly depend on local neuronal activities. Additionally, a release of neurochemicals from injured cell bodies and their terminals may affect tracts of ascending fibres leading to local inhibition of cord conduction. Earlier investigations have indicated that serotonin and prostaglandins influence early SCEP changes following trauma indicating an invovement of neurochemicals in changes of bio-electrical activities[11,15].

Previous biochemical determinations in our model of spinal cord injury have demonstrated marked alterations in the contents of dynorphin, enzephalin and substance P in the segments located around the primary injury[10,12,13]. The significance of altered peptide concentrations in spinal cord injury is unclear but the possibility exists that neuropeptides may modulate SCEP changes and somehow contribute to secondary injuries such as edema. The present investigation was carried out to find out whether opioid receptors may be involved in early SCEP changes and edema formation following spinal cord injury using a pharmacological approach.

Materials and Methods

Animals

The experiments were carried out on 17 Sprague-Dawley male rats weighing between 350–400 g. The animals were housed at a temperature of 22 ± 1°C with a 12 h light and 12 h dark schedule. Food and drinking water were supplied ad libitum.

Spinal Cord Injury

Under urethane anaesthesia (1.5 g/kg intraperitoneally) laminectomy was performed over the T10-11 segments of the spinal cord. A 3 mm deep and 5 mm long unilateral incision was made 2 mm to the right of the midline using a scalpel blade (Fig. 1). The lesion was mainly in the right dorsal horn and its deepest part was located approximately in Rexed's lamina VII[8,9]. This experimental condition is approved by the Ethical Committee of Uppsala University.

Spinal Cord Evoked Potentials (SCEP)

The right tibial and sural nerves were stimulated simultaneously and supramaximally at a frequency of 10 Hz with needles (Terumo injection needles 0.6 × 25 mm) placed behind the achilles tendon (cathode) and below the plantar surface (anode). The duration of the stimulus was 0.1 ms. The unipolar electrodes were made from copper wire winded around a polythene tubing (PE 10). The electrodes were inserted in the epidural space and directed towards the right side of the cord both rostrally and caudally and advanced to the level of T9 and T12 (Fig. 1). The reference needle electrodes were placed in the paravertebral muscles. For each recording two hundred responses were collected and averaged by an EMG equipment (Medelec MS 20 "Mystro", filter 3 Hz–5 kHz, sweep duration 20 ms). The results were analyzed manually. Amplitudes and latencies of the maximal positive peak (MPP) and the maximal negative peak (MNP) were recorded[11,15]. In calculations, latencies and amplitudes measured 30 min before injury was used as references (100%).

Spinal Cord Edema

Five hours after trauma the injured segment of the spinal cord was removed and immediately weighed. Then the tissue was dried in an oven maintained at 90°C for 72 h. The water content was determined from the difference between wet and dry weights[9].

Administration of Naloxone

Naloxone hydrochloride (an opioid receptor antagonist, Sigma Chemical Co., U.S.A.) was dissolved in 0.9% saline and administered at a dose of 10 mg/kg intraperitoneally 30 min prior to injury[6]. SCEP (n = 6) and edema were recorded in the animals as above.

Statistical Analysis

The data were computed using commercial software (Excel, Microsoft Corp, Statview II and SE+, Abacus Concepts Inc.) on a personal computer (Macintosh Plus, Apple Computer Inc.). Amplitudes and latencies of MPP and MNP in percent of values at –30 min were used in statistical calculations with ANOVA and Fisher PLSD test. p-values less than 0.05 were considered significant. In each group of untreated pretreated animals paired comparisons were made for each variable from –2 min and then at each recording time.

Results

Untreated Animals

A representative record of the effect of spinal cord injury on SCEP in one untreated animal is illustrated in Fig. 2. There is an immediate abolition of rostral MNP amplitude after injury with minor late recovery. Mean rostral MNP amplitude showed a decrease by 60% immediately after injury which persisted up to 120 min after injury (Fig. 2). The increase of mean latencies of rostral MPP from the pre-injury level was seen from 120 min after injury and onwards (Fig. 2).

Naloxone-Pretreated Animals

Preteatment of naloxone markedly attenuated the MNP amplitude changes in rostral recordings after trauma (Fig. 2). The mean rostral MNP amplitudes did not show any decrease after injury compared to –2 min (Fig. 2). The MNP amplitude was significantly higher from the untreated traumatized group at 2 to 120 min (Fig. 2). The MNP latencies showed some increase in later recordings (Fig. 2).

Spinal Cord Edema

In untreated rats (n = 5) there was a 4% increase of water content corresponding to a volume increase of 16% at 300 min after injury (Table 1). Pretreatment with naloxone inhibited this accumulation of water in the traumatized spinal cord.

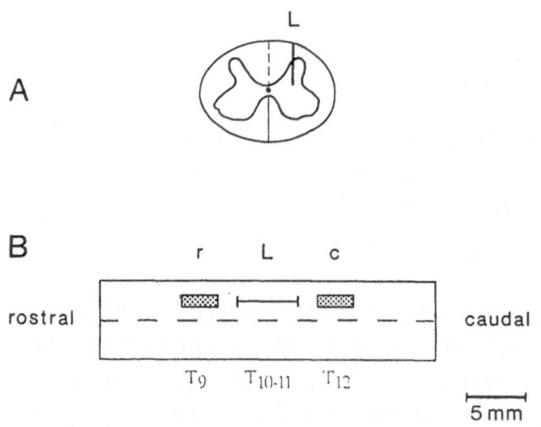

Fig. 1. Schematic diagram showing placement of rostral (*r*) and caudal (*c*) recording electrodes around lesion (*L*) (B). The extent of lesion is shown in a cross section of the spinal cord (A)

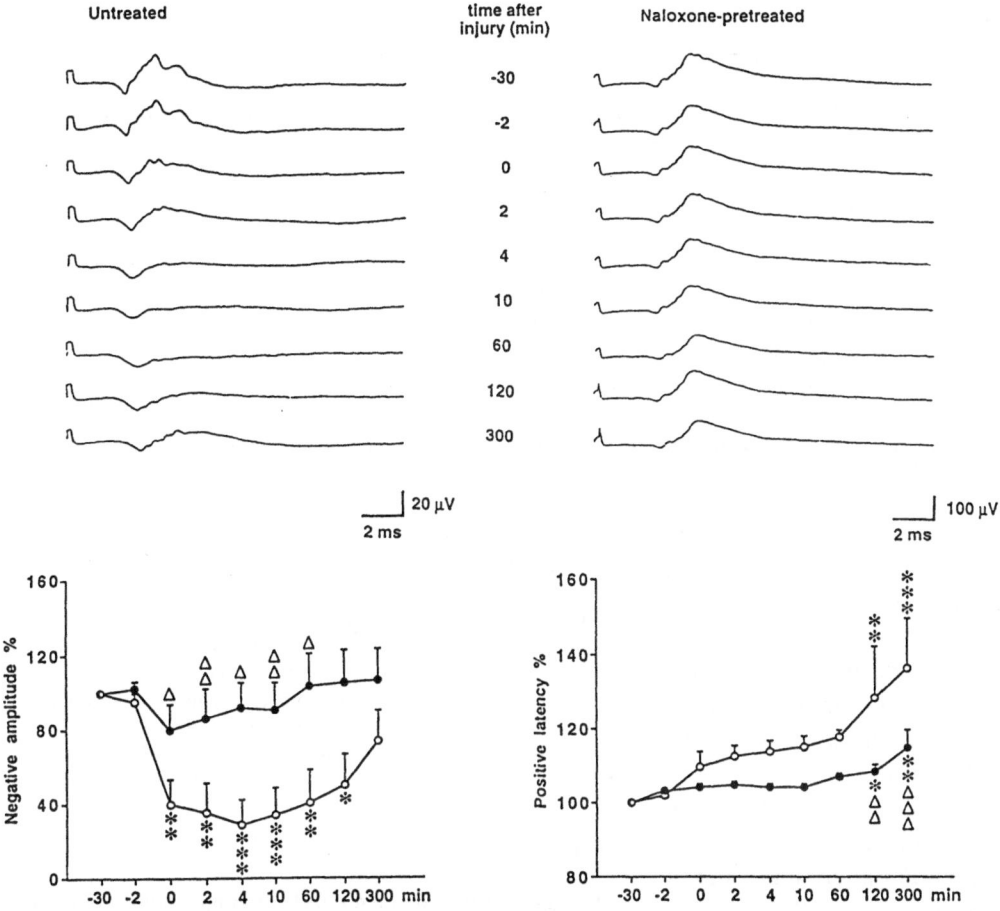

Fig. 2. Individual SCEP recorded from the rostal electrode in one untreated and one naloxone pretreated rat after stimulation of the right tibial and sural nerves at the ankle (above) . Mean SCEP MNP amplitude and MPP latencies recorded from rostral electrode are displayed below. After a longitudinal lesion in the right dorsal horn between T10–11 segments the untreated animal showed a total loss of MNP amplitude (upward deflection). On the other hand, the naloxone treated animal shows apparently no change in the MNP amplitude. Untreated rats show a pronounced depression of the rostral MNP amplitude lasting 120 min (ANOVA, Fisher PLSD, * $p < 0.05$, ** $p < 0.01$, *** $p < 0.001$), which was mainly absent in the naloxone treated group. The difference between the drug-treated and untreated group was significant (ANOVA, Fisher PLSD, Δ $p < 0.05$, $\Delta\Delta$ $p < 0.01$). Mean SCEP MPP latencies showed increase in later parts of the recording in both groups (ANOVA, Fisher PLSD, * $p < 0.05$, ** $p < 0.01$, *** $p < 0.001$). The difference between the drug-treated and untreated group was significant (ANOVA, Fisher PLSD, $\Delta\Delta$ $p < 0.01$, $\Delta\Delta\Delta$ $p < 0.001$). Naloxone (10 mg/kg, i.p.) was injected between –35 and –0 min

Discussion

The present study clearly shows that endogenous opioids are involved in the alterations of spinal cord conduction following a focal trauma to the cord. This effect seems to be mediated by specific receptors since pretreatment with naloxone in high doses markedly attenuated the changes in SCEP amplitude. Recording of such potentials by epidural electrodes is reliable because SCEPs are quite stable in repeated recordings performed before injury. Individual variations of SCEP present in different groups of rats may be due to different biological responses to injury rather than to the extent of injury[11,15].

Depression of MNP amplitude in untreated animals immediately after lesion suggests that nerve cell conduction may entered a "shock phase"[2,4]. This may be due to the injury or to a release of different neurochemicals around the site of lesion. Diffusion of peptides from the injury zone and local neuronal shock may both block the conduction of the long dorsal tracts and be reflected as a decrease in SCEP amplitudes.

Since naloxone attenuated the SCEP alterations after trauma it seems likely that these changes are mediated by peptides via blocade of opioid receptors. This may inhibit the depressor effects of peptides and may attenuate local neuronal shock after trauma. An increase in SCEP latencies after trauma may well be caused by

Table 1. *Influence of Naloxone on Edema Physiological Variables in 5 h Spinal Cord Traumatized Rats*

Parameters measured	Control	5 h Spinal cord injury	
	(n = 5)	Untreated (n = 5)	Naloxone treated (n = 6)
Edema			
1. Water content %	69.34 ± 0.31	73.56 ± 0.34***	71.21 ± 0.23[a],**
Physiological variables			
1. MABP torr	110 ± 5	98 ± 8**	100 ± 6[a],**
2. Arterial pH	7.38 ± 0.04	7.37 ± 0.03	7.37 ± 0.6
3. PaO_2 torr	80.56 ± 0.23	81.65 ± 0.76	80.89 ± 0.98
4. $PaCO_2$ torr	34.67 ± 0.38	33.68 ± 0.56	33.39 ± 0.78

Values are mean ± SD, *** $p < 0.001$, ** $p < 0.01$, significantly different from control group, [a] $p < 0.05$, significantly different from untreated group, Student's unpaired t-test, *MABP* mean arterial blood pressure.

damage to myelin. The protective effects of naloxone on latency parameters may be due to a reduced myeline damage after trauma.

An increase of MPP amplitude in rostral recordings in a few untreated animals of our study may be due to a decrease in synaptic inhibition[14]. An increase of the rostral MPP amplitudes may represent 'positive injury potentials'. Such potentials may be generated by the the positive 'killed end' signals recorded by an electrode distal to it; and/or alterations in ionic permeability in the extra- and intracellular spaces around the lesion[2,4]. These positive potentials were not seen in naloxone treated injured animals indicating that release of opioid peptides within the injured spinal cord may somehow influence the generation of MPP.

Our results further show that blockade of opioid receptors reduced the edema formation after spinal cord injury. The mechanisms of naloxone induced reduction in edema after spinal cord injury is not clear. Improved local blood flow and reduced microvascular permeability disturbances may be implicated[7]. We used a very high dose of naloxone in the present study which not only block the opioid μ-, but also δ- and κ-oipoid receptors[5,6]. Thus we do not know which type of receptors that is particularly important in mediating SCEP changes and edema formation after spinal cord injury. Further studies using specific antagonists to the various receptors are needed to clarify these points.

In summary, it appears that changes of negative amplitude and positive latency of SCEP can be used for assessing spinal cord dysfunction. These parameters may also be used to predict the degree of edema which may occur at a later stage after injury.

Acknowledgements

This investigation was supported by grants from Swedish Medical Research Council Nos. 3020, 135, 9459, 10357; Trygg-Hansa, Söderbergs stiftelse, Hedlunds stiftelse, Thyrings stiftelse, RTP, Wallenius Line, Åhléns and University Grants Commission, New Delhi, India. The skilful technical assistance of Gunilla Tibbling, Madeleine Thörnwall, Madeleine Jarild and Åke Pettersson is acknowledged with thanks.

References

1. Duggan AW, North RA (1984) Electrophysiology of opioids. Pharmacol Rev 35: 220–281
2. Fehlings MG, Tator CH, Linden RD (1989) The relationships among the severity of spinal cord injury, motor and somatosensory evoked potentials and spinal cord blood flow. Electroenceph Clin Neurophysiol 74: 241–259
3. Gibson SJ, Polak JM (1986) Neurochemistry of the spinal cord. In: Polak JM, Noorden SV (eds) Immunocytochemistry. Wright, Bristol, pp 360–389
4. Homa S, Tamaki T (1984) Fundamentals and clinical applications of spinal cord monitoring. Saikon, Tokyo
5. Hylden JLK, Nahin RL, Traub RJ, Dubner R (1991) Effects of spinal kappa-opioid receptor agonists on the responsiveness of nociceptive superficial dorsal horn neurons. Pain 44: 187–193
6. Jones SL, Sedivec MJ, Light AR (1990) Effects of ionophoresed opioids on physiologically characterized laminae I and II dorsal horn neurons in the cat spinal cord. Brain Res 532: 160–174
7. Means ED, Anderson DK (1987) The pathophysiology of acute spinal cord injury. In: Davidoff RA (ed) Handbook of the spinal cord, Vol 5. Marcel Dekker, New York, pp 19–61
8. Olsson Y, Sharma H S, Pettersson CÅV, Cervós-Navarro J (1992) Endogenous release of neurochemicals may increase vascular permeability, induce edema and influence cell changes in trauma to the spinal cord. Prog Brain Res 91: 197–203
9. Sharma HS, Olsson Y (1990) Edema formation and cellular alterations following spinal cord injury in the rat and their modification with p-chlorophenylalanine. Acta Neuropathol (Berl) 79: 604–610
10. Sharma HS, Nyberg F, Olsson Y, Dey PK (1990) Alteration of substance P after trauma to the spinal cord. An experimental study in the rat. Neuroscience 38: 205–212
11. Sharma HS, Winkler T, Stålberg E, Olsson Y, Dey PK (1991) Evaluation of traumatic spinal cord edema using evoked poten-

tials recorded from the spinal epidural space. An experimental study in the rat. J Neurol Sci 102: 150–162

12. Sharma HS, Nyberg F, Olsson Y (1992) Dynorphin A content in the rat brain and spinal cord after a localized trauma to the spinal cord and its modification with p-chlorophenylalanine. An experimental study using radioimmunoassay technique. Neurosci Res 14: 195–203

13. Sharma HS, Nyberg F, Thörnwall M, Olsson Y (1993) Met-Enkephalin-Arg[6]-Phe[7] in spinal cord and brain following traumatic injury to the spinal cord: Influence of p-chlorophenyla-lanine. An experimental study in the rat using radioimmu-noassay technique. Neuropharmacology, in press

14. Tator CH, Fehlings MG (1991) Review of the secondary injury theory of acute spinal cord trauma with emphasis on vascular mechanisms. J Neurosurg 75: 15–26

15. Winkler T, Sharma HS, Stålberg E, Olsson Y (1993) Indometh-acin, an inhibitor of prostaglandin synthesis attenuates altera-tion in spinal cord evoked potentials and edema formation after trauma to the spinal cord. An experimental study in the rat. Neuroscience 52: 1057–1067

Correspondence: Hari Shaker Sharma, Ph. D., Laboratory of Neuropathology, University Hospital, S-751 85 Uppsala, Sweden.

Acta Neurochir (1994) [Suppl] 60: 516–518
© Springer-Verlag 1994

21-Aminosteroid U-74389F Reduces Vasogenic Brain Edema

G.-H. Schneider, A. Unterberg, and **W. R. Lanksch**

Department of Neurosurgery, Rudolf Virchow Medical Center, Free University of Berlin, Berlin, Federal Republic of Germany

Summary

21-Aminosteroids have been shown to attenuate neuronal damage and to improve neurological outcome after experimental ischemia. The aim of this study was to determine whether brain edema induced by a cryogenic injury can be influenced by the 21-aminosteroid U-74389F.

A cortical freezing lesion was applied to the right parietal region of Sprague-Dawley rats under ketamine-xylazine anesthesia. Systemic blood pressure was monitored in the peritraumatic period. Four different doses of U-74389F (A–D) were studied for their effect on post-traumatic brain swelling and edema. Respective control groups received only the solvent, citric acid buffer.

(A) 3 mg/kg b.w. i.p. (total dose) 30 min before, 1 and 12 h; post trauma (p.t.); (B) 9 mg/kg b.w. i.v. 30 min before, 1 and 12 h p.t.; (C) 25 mg/kg b.w. i.v. 30 min before, 1, 6, and 12 h p.t.; (D) 50 mg/ kg b.w. i.v. 15 min before, 15 and 30 min as well as 1, 2, 6, and 12 h p.t. 24 h after trauma, brains were removed and hemispheric swelling and water content were determined from the difference between wet and dry weight.

Application of the 21-aminosteroid U-74389F moderately reduced post-traumatic brain swelling in all treatment groups: (A) 5%, (B) 9%, (C) 12%, and (D) 14%.

In parallel with this, the increase in water content of the traumatized hemisphere was marginally lowered by U-74389F in all groups; in (C) e.g. from $1.9 \pm 0.1\%$ to $1.7 \pm 0.1\%$, $p = 0.07$.

These two findings taken together indicate that the 21-aminosteroid U-74389F moderately reduces post-traumatic swelling and edema.

Keywords: Cold injury; brain swelling; cerebral edema; lazaroids; rats.

Introduction

Mortality and neurological outcome following severe head injury are markedly influenced by posttraumatic brain swelling[6].

During the past decades steroids have been used extensively to ameliorate post-traumatic brain swelling and edema[8,11]. In contrast to results in experimental studies evidence regarding the effectiveness of steroids in clinical trials of head injury have been unconvincing.

Nevertheless, a trend towards using ever higher doses of steroids has arisen in the treatment of central nervous system injury due to the fact that high doses of steroids display inhibitory effects on lipid peroxidation[1]. Free radical generation is supposed to be an important factor in central nervous system injury[2,5]. Recently, a new class of steroids with strong antiperoxidative properties, the 21-aminosteroids or lazaroids, was developed for treatment of acute central nervous system lesions. Lazaroids have been demonstrated to exert neuroprotective effects in various experimental models of ischemia and subarachnoid hemorrhage[7,13].

This study was undertaken to determine whether brain swelling and edema induced by a cortical freezing lesion is attenuated by the 21-aminosteroid U-74389F.

Methods

For the study, 64 male Sprague Dawley rats weighing 470 ± 14 g were used. Under ketamine-xylazine anesthesia a burr hole was made over the right parietal region. Using a copper probe with a diameter of 5 mm, a standard freezing lesion was applied to the right parietal hemisphere by stereotactically lowering the probe onto the intact dura for 30 seconds[4]. The probe was cooled to $-60°C$ by a mixture of dry ice and acetone. Systemic blood pressure was monitored in the peritraumatic period via a cannula in the femoral artery.

Four different doses (A–D) of the aminosteroid U-74389F were studied for their effect on posttraumatic brain swelling and edema. Respective control groups received a corresponding volume of just the solvent, citric acid buffer (pH = 3.0). Group A was injected intraperitoneally, while groups B, C, and D received the drug intravenously via a catheter in the jugular vein.

(A) 3 mg/kg b.w. i.p. (total dose) 30 min before, 1 and 12 h post trauma (p.t.); n = 10.

(B) 9 mg/kg b.w. i.v. 30 min before, 1 and 12 h p.t.; n = 6.

(C) 25 mg/kg b.w. i.v. 30 min before, 1, 6, and 12 h p.t.; n = 10.

(D) 50 mg/kg b.w. i.v. 15 min before, 15 and 30 min as well as 1, 2, 6, and 12 h p.t.; n = 6.

24 h after trauma the animals were sacrificed and the brains were quickly removed. For determination of hemispheric swelling the

Table 1. *Post-traumatic Brain Swelling and Hemispheric Increase in Water Content*

	Hemispheric swelling [%]		Post-traumatic increase in hemispheric water content [%]	
	U-74389F	Vehicle	U-74389F	Vehicle
Group A – 3 mg/kg i.p.	8.1 ± 0.8	8.5 ± 0.7	1.5 ± 0.1	1.8 ± 0.1
Group B – 9 mg/kg i.v.	7.9 ± 0.9	8.7 ± 0.8	1.5 ± 0.9	1.6 ± 0.1
Group C – 25 mg/kg i.v.	9.2 ± 0.8	10.4 ± 0.8	1.7± 0.1	1.9 ± 0.1
Group D – 50 mg/kg i.v.	9.3 ± 0.6	10.8 ± 0.9	1.6 ± 0.1	1.7 ± 0.1

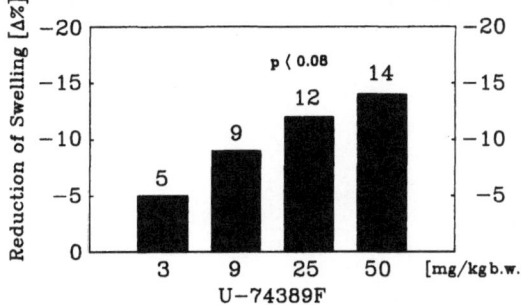

Fig. 1. Moderate reduction of post-traumatic hemispheric swelling by four different doses of the 21-aminosteroid U-74389F

cerebral hemispheres were meticulously separated using a microscope and weighed. Thereafter hemispheres were dried at 100°C for 48 h and dessicated for another 24 h. Cerebral water content was then calculated as the difference between hemispheric wet and dry weight.

The study was conducted in a blinded manner to eliminate bias when separating the hemispheres.

Results

Treatment with the aminosteroid U-74389F moderately reduced post-traumatic hemispheric swelling in all groups. Hemispheric swelling for group A–D due to the cortical freezing lesion in comparison to that in vehicle-treated animals is given in Table 1. Post-traumatic brain swelling was reduced by 5% in group A (3 mg/kg i.p.), by 9% in group B (9 mg/kg i.v.), by 12% ($p < 0.08$) in group C (25 mg/kg i.v.), and by 14% in group D (50 mg/kg i.v.) versus the respective control groups (Fig. 1). Accordingly, the increase of cerebral water content due to the perifocal edema was slightly lowered in all U-74389F treated groups (A–D); in group C, for example, it was lowered from 1.9 ± 0.1% to 1.7 ± 0.1%, $p = 0.07$ (Table 1).

Discussion

A rational treatment for edema, a common feature of acute cerebral injuries, still remains to be found. Generation of oxygen free radicals following trauma or ischemia[12] may be one starting point for novel therapeutic strategies. As lipid peroxidation is one proposed mechanism of edema generation in the cold injury model[9,3], one might expect lazaroids to influence post-traumatic brain swelling following a cryogenic injury. Indeed, post-traumatic hemispheric swelling could be moderately, but not significantly reduced by administration of any of the four doses of the 21-aminosteroid U-74389F. Application of lazaroids seemingly leads to a dose-dependent reduction of post-traumatic brain swelling. The rise in hemispheric water content due to the cortical freezing lesion was only marginally reduced in all four treatment groups.

This rather moderate effect of U-74389F may be due to a minor role of free radicals in edema generation[10] or due to a minor role played by ischemic factors in the cold injury model. Beneficial effects of lazaroids are obviously more pronounced in models of experimental ischemia.

Ongoing clinical trials in head injury and subarachnoid hemmorrhage will soon demonstrate whether lazaroids will fullfil the hopes elicited by experimental studies.

Acknowledgement

The excellent technical assistance of J. Kopetzki and S. Hennig is highly appreciated.

References

1. Anderson DK, Saunders RD, Demediuk P, *et al* (1986) Lipid hydrolysis and peroxidation in injured spinal cord: partial protection with methylprednisolone or vitamine E and selenium. CNS Trauma 2: 257–268

2. Hall ED, Braughler JM (1989) Central nervous system trauma and stroke. II. Physiological and pharmacological evidence for involvement of oxygen radicals and lipid peroxidation. Free Radic Biol Med 6: 303–313

3. Hall ED, Travis MA (1988) Inhibition of arachidonic acid-induced vasogenic brain edema by the non-glucocorticoid 21-aminosteroid U74006F. Brain Res 451: 350–352

4. Klatzo I, Piraux A, Laskowski EJ (1958) The relationship between edema, blood-brain-barrier and tissue elements in local brain injury. J Neuropath Exp Neurol 17: 548–564

5. Kogure K, Watson BD, Busto R (1982) Potentation of lipid peroxides by ischemia in the rat brain. Neurochem Res 7: 437–454

6. Marshall LF, Eisenberg HM, Jane JA, *et al* (1991) The outcome of severe closed head injury. J Neurosurg 75: S28–S36

7. Perkins WJ, Newbwerg Milde L, Milde JH, *et al* (1991) Pretreatment with U74006F improves neurologic outcome following complete cerebral ischemia in dogs. Stroke 22: 902–909

8. Reulen HJ, Schürmann K (eds) (1972) Steroids and brain oedema. Springer, Berlin Heidelberg New York

9. Suzuki O, Yagi K (1974) Formation of lipoperoxide in brain edema induced by cold injury. Experientia 30: 248

10. Unterberg A, Wahl M, Baethmann A (1988) Effects of free radicals on permeability and vasomotor response of cerebral vessels. Acta Neuropathol (Berl) 76: 238–244

11. Unterberg A, Schmidt W, Dautermann, *et al* (1990) The effect of various steroid treatment regimens on cold-induced brain swelling. Acta Neurochir (Wien) [Suppl] 51: 104–106

12. Wahl M, Unterberg A, Baethmann A, *et al* (1988) Mediators of blood-brain barrier dysfuction and formation of vasogenic brain edema. J Cereb Blood Flow Metab 8 (5): 621–634

13. Zuccarello M, Anderson DK (1989) Protective effect of a 21-aminosteroid on the blood-brain barrier following subarachnoid hemorrhage in rats. Stroke 20: 367–371

Correspondence: G.-H. Schneider, M. D., Department of Neurosurgery, Rudolf Virchow Medical Center, Free University of Berlin, Augustenburger Platz 1, D-13353 Berlin, Federal Republic of Germany.

Acta Neurochir (1994) [Suppl] 60: 519–520
© Springer-Verlag 1994

Effect of Torasemide on Intracranial Pressure, Mean Systemic Arterial Pressure, and Cerebral Perfusion Pressure in Experimental Brain Edema of the Rat

C. A. Plangger

Department of Neurosurgery, University of Innsbruck, Innsbruck, Austria

Summary

Torasemide, a puridine-3-sulfonylurea derivative, is a lipophilic diuretic compound. It interacts with the $Na^+2Cl^-K^+$ carrier and at higher concentrations with the chloride channels. The lipophilic nature of torasemide may determine its access to glial and neuronal cells across the blood brain barrier. The present study was performed to establish whether torasemide could modify intracranial pressure and cytotoxic brain edema in functionally nephrectomized Wistar rats (150–240 g). Brain edema of the cytotoxic type was induced by infusion of 100 ml aqua bidest/kg body weight. After the end of the infusion 100 mg torasemide/kg body weight or a corresponding volume of isotonic saline were injected followed by continuous continued recording of ICP and systemic arterial pressure for at least 3 hours. Torasemide prevented the rise in intracranial pressure seen in control animals. The ICP values for the torasemide-treated animals were lower at all points. Following intravenous injection of 100 mg torasemide / kg body weight at 50, 60, 70, 90 and 120 minutes a significant decrease of intracranial pressure compared to that in controls was observed. Torasemide may be a useful adjunct in the therapy of brain edema and increased intracranial pressure.

Keywords: Intracranial pressure; rat brain; cytotoxic brain edema; torasemide.

Introduction

Torasemide, a puridine-3-sulfonylurea derivative, is a lipophilic diuretic compound. It interacts with the $Na^+2Cl^-K^+$ carrier and at higher concentrations with the chloride channels[3]. The chemical structure of torasemide places this compound between blockers of the $Na^+2Cl^-K^+$ co-transport, such as furosemide and blockers of the chloride channels, such as diphenylamine-2-carboxylate. The affinity of torasemide for the $Na^+2Cl^-K^+$ carrier is far greater than affinity for the chloride channel[7]. The lipophilic nature of torasemide may determine its access to glial and neuronal cells across the blood brain barrier. The present study was performed to establish whether torasemide could modify intracranial pressure in rats with cytotoxic brain edema.

Materials and Methods

Male Wistar rats (150–250 g, Ivanovas, Kisslegg, Germany) were anesthetized by intraperitoneal injection of Inactin (120 mg/kg body weight, Byk-Gulden, Konstanz, Germany), placed on a heatable operating table and maintained at a body temperature of $37 \pm 1°C$. The animals were tracheotomized and a catheter inserted in the right jugular vein. The rats were functionally nephrectomized by ligating the ureter and the veins and arteries of the kidneys. Systemic arterial pressure was continously recorded by a catheter in the right femoral artery. A small burr hole was made on the left frontal part of the cranium and a Wick-catheter introduced for determination of brain tissue pressure[5]. The Wick-catheter was sealed using an acryl glue (Histoacryl, B. Braun, Melsungen, Germany). The Wick-catheter measures pressure through a column of fluid which establishes a connection between the brain tissue and the transducer. Brain tissue pressure and ventricular fluid pressure respond similarly to changes in intracranial pressure[4]. The brain tissue pressure was continously measured using an electromagnetic Statham transducer connected to an electromanometer and recorded on a two-channel recorder. 100 ml aqua bidest/kg body weight were infused at a rate of 0.5 ml/min through the vena jugularis catheter, thus inducing brain edema of the cytotoxic type[6]. After the end of the infusion 100 mg torasemide/kg body weight corresponding to a volume of 10 ml/kg body weight or 10 ml isotonic saline/kg body weight were injected followed by continuous continued recording of ICP and systemic arterial pressure for at least 3 hours. Statistical evaluation of the effect of torasemide on ICP, mean systemic arterial pressure and cerebral perfusion pressure, defined as the difference between mean systemic arterial pressure and ICP, was performed by Student's two-tailed t-test for unpaired data.

Results

Infusion of 100 ml aqua bidest/kg body weight led to a marked increase of brain tissue pressure from 2.80 mmHg ± s.d. 0.87 mmHg (n = 11) to 15.15 mmHg ±

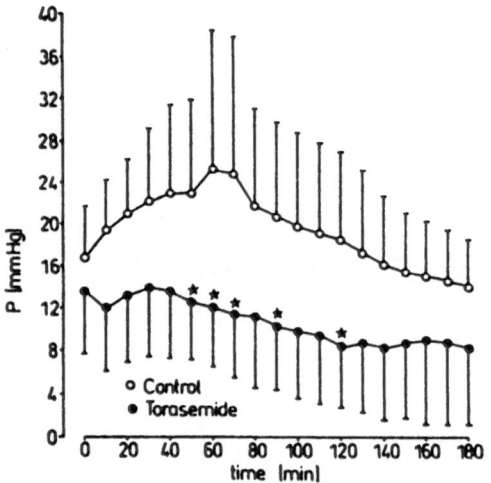

Fig. 1. Intracranial pressure (ICP) following infusion of 100 ml aqua bidest/kg body weight and subsequent application of either 100 mg torasemide/kg body weight corresponding to a volume of 10 ml/kg body weight (closed symbols, n = 6) or 10 ml 150 mM NaCl solution / kg body weight (open symbols, n = 5). Arithmetic means ± s.d. * Significantly different from control, p < 0.05, two-tailed Student's t-test for unpaired data

s.d. 5.53 mmHg (n = 11). Mean systemic arterial pressure increased during this period from 82.88 mmHg ± s.d. 10.54 mmHg (n = 11) to 115.94 mmHg ± s.d. 24.64 mmHg. Cerebral perfusion pressure (CCP) increased from 80.08 mmHg ± s.d. 10.20 mmHg to 100.78 mmHg ± s.d. 21.84 mmHg.

Figure 1 shows the ICP values in animals treated with torasemide and in untreated control animals. 50, 60, 70, 90 and 120 minutes after injection of 100 mg torasemide / kg body weight the intracranial pressure was significantly lower in the torasemide-treated animals using a two-tailed Student's t-test for unpaired data.

Torasemide induced a fall in mean systemic arterial pressure lasting for 40 minutes, but without reaching statistical significance.

From 30 until 150 minutes the cerebral perfusion pressure was higher in the torasemide-treated animals due to a lower ICP, but also due to agenerally higher mean systemic arterial pressure. These differences of cerebral perfusion pressure did not reach statistical significance by the two-tailed Student's t-test.

Discussion

Torasemide has potent diuretic activity in rats and dogs. In both species urinary volume and electrolyte

excretion increased linearly with the logarithm of the dose, thus resembling the profile of a high-ceiling diuretic. The minimum effective dose by the oral route is 0.2 mg/kg in the rat, with the maximal effect obtained with a dose of about 10 mg/kg[2]. Despite the same natriuretic effect, potassium losses with torasemide were significantly less than those with furosemide. On a weight basis torasemide is 9–40 times more potent than furosemide in the rat. After i.v. injection of torasemide, the onset of diuresis was observed within 5 to 10 minutes and peak effect within 20–40 minutes in the rat[1]. The lipophilic nature of torasemide may determine its access to glial and neuronal cells across the blood brain barrier. In a H_2O/ CH_2Cl_2 mixture at pH 7,4 33% of torasemide is recovered in the lipid phase[7]. For comparison, 0% was found for furosemide.

In our model of brain edema of the cytotoxic type induced by hypo-osmolar hyperhydration torasemide prevented the rise in intracranial pressure seen in control animals receiving a corresponding volume of isotonic saline. Torasemide may be a useful adjunct in the therapy of brain edema and increased intracranial pressure.

References

1. Ghys A, Denef J, Delarge J, Georges A (1985) Renal effects of the high ceiling diuretic torasemide in rats and dogs. Arzneimittelforsch Drug Res 35 (II): 1527–1531
2. Ghys A, Denef J, De Suray JM, et al (1985) Pharmacological properties of the new potent diuretic torasemide in rats and dogs. Arzneimittelforsch Drug Res 35 (II): 1520–1526
3. Greger R (1980) Inhibition of active NaCl reabsorption in the thick ascending limb of the loop of Henle by torasemide. Arzneimittelforsch Drug Res 38 (I): 151–152
4. Jannotti F, Hoff JT, Schielke GP (1984) Brain tissue pressure: physiological observations in anesthetized cats. J Neurosurg 60: 1219–1225
5. Poll W, Brock M, Markakis E, Winkelmüller W, Dietz H (1972) Brain tissue pressure. In: Brock M, Dietz H (eds) Intracranial pressure. Springer, Berlin Heidelberg New York, pp 188–194
6. Weed LH, Mc Kibben PS (1919) Pressure changes in the cerebrospinal fluid following intravenous injections of solutions of various concentrations. Am J Physiol 48: 512–530
7. Wittner M, Di Stefano A, Schlatter E, Delarge J, Greger R (1986) Torasemide inhibits NaCl reabsorption in the thick ascending limb of the loop of Henle. Pflugers Arch 407: 611–614

Correspondence: Clemens A. Plangger, M. D., Department of Neurosurgery, University of Innsbruck, Anichstrasse 35, A-6020 Innsbruck, Austria.

Acta Neurochir (1994) [Suppl] 60: 521–523
© Springer-Verlag 1994

Administration of an Omega-Conopeptide One Hour Following Traumatic Brain Injury Reduces ^{45}Calcium Accumulation

D. A. Hovda, K. Fu, H. Badie, A. Samii, P. Pinanong, and **D. P. Becker**

Division of Neurosurgery and the Brain Research Institute UCLA School of Medicine, Los Angeles, CA, U.S.A.

Summary

The omega-conopeptide, SNX-111 (NEUREX Corporation) was administered to rats 1 hour following a lateral fluid percussion brain injury to determine if the drug could reduced the extent and duration of trauma-induced calcium accumulation. Administration at doses of 3 or 5 mg/kg (i.v.) markedly reduced the extent of calcium accumulation as determined using ^{45}calcium autoradiography primarily within the cerebral cortex and hippocampus. The reduction of calcium accumulation was particularly event within the parietal cortex beginning as early as 6 hours and lasting out to 48 hours following injury. Although not as effective as in the cerebral cortex, SNX-111 did exhibit a reduction of calcium accumulation within the dorsal hippocampus especially at 24 and 48 hours after the insult. These preliminary results demonstrate that SNX-111 can reduce the injury-induced accumulation of calcium even when administered 1 hour after the insult and offers this compound as a potential therapeutic treatment for traumatic brain injury.

Keywords: Calcium; brain injury; concussion.

Introduction

Resent work had indicted that conopeptides have the capacity to be a potent blocker of calcium flux across the cell membrane and can provide substantial protection to neurons when administered after cerebral ischemia. Specifically a new compound (SNX-111; NEUREX Corporation) has provided substantial enthusiasm in this regard due to its specificity for blocking the N-type calcium channel. Although this compound appears to provide substantial protection in models of cerebral ischemia, little is known regarding its efficacy following traumatic brain injury where ischemia is not a prominent feature[4,5].

Even with relatively preserved cerebral blood flow, experimental traumatic brain injury results in a marked accumulation of calcium[1]. The injury-induced calcium accumulation contributes to a massive ionic flux[3] which is thought to be responsible for the metabolic dysfunction exhibited following traumatic brain injury. These pathophysiological events may underlie the development of a state of cellular vulnerability which can threaten neuronal survivability. Consequently, the reduction of this injury-induced ionic flux may provide a therapeutic benefit both in terms of reducing cell loss and enhancing recovery of function. However, it remains to be determined if SNX-111 can reduce the trauma-induced accumulation of calcium. Therefore, the following study was specifically designed to address this question.

Materials and Methods

Subjects: Sprague-Dawley male rats (n = 79, wt = 223–367 g) were randomly divided into sham operated (n = 28) or injured (n = 51) groups.

Surgery: All animals were placed under general anesthesia (1.5–2.0% enflurane, 33% O_2, 66% NO_2) and then positioned in a stereotaxic apparatus. A 2×2 mm craniotomy was positioned 1 mm posterior to bregma and 6 mm lateral (left) of the midline. A rigid hollow injury screw (O.D. 6.5 mm; I.D. 4.5 mm) was inserted on top of the exposed dura and secured to the skull with dental cement.

Fluid percussion injury: After surgery, animals were removed from the stereotaxic frame and the injury screw was connected to the fluid percussion device by a 20 cm pressure resistive tube (I.D. 2 mm). The device was set to deliver a 3.9–5.0 atm fluid pulse resulting in a 2.4–2.6 atm pressure wave at the dural surface. The fluid pulse was delivered 60 seconds after the removal of inhalation anesthesia. with the level of unconsciousness determined by the assessing the presence of a paw-pinch response. Following injury, animals exhibited a brief period of apnea mean = 21 s) and unconsciousness (mean = 1 min 30 s). At the first sign of consciousness, animals were again placed under general anesthesia, the injury cap was removed and the wound sutured closed.

SNX-111 administration and ^{45}Calcium autoradiography: One hour after injury, rats were placed under general anesthesia (see above) and the superficial portion of the femoral vein was exposed. SNX-111 was delivered (i.v.) at a dose of 0 (saline), 3 or 5 mg/kg.

[45]Calcium (1 µCi/kg) was then injected (i.v.) at 6 (n = 9), 12 (n = 11), 24 (n = 11), 48 (n = 10) and 96 (n = 10) hours after drug injection and was allowed to circulate for 5 hours. Animals were then administered a lethal dose of pentobarbital, the brain quickly removed being placed immediately in powdered dry ice. Frozen coronal 20 µm sections were then processed for autoradiography and nissl staining.

Data analysis: Using an image analysis system (JAVA: Jandel Scientific) 20 structures were measured bilaterally in terms of optical density. To calculate relative optical density, the ipsilateral value (left) was subtracted from the contralateral value (right) and divided by their sum.

Results

The relative optical density measurements indicated that in saline treated animals, fluid percussion brain injury resulted in a marked accumulation of calcium. This accumulation was particularly evident within the ipsilateral cerebral cortex and hippocampus and persisted up to 48 hours following the insult. In contrast, animals who were administered SNX-111 exhibited a marked reduction of calcium accumulation following fluid percussion injury.

This drug-induced reduction of calcium accumulation was particularly evident within the ipsilateral cerebral cortex. Although the effect of SNX-111 was evident as early as 6 hours after injury it maintained its ability to retard calcium accumulation out to 48 hours. The most pronounced effect of the drug occurred 24 hours post-injury when the relative densitometry scores within the parietal cortex indicated a significant difference ($p < 0.01$) between saline treated (0.312) and SNX-l 11 treated animals (3 mg = 0.048: 5 mg = 0.061). Finally, this effect on calcium accumulation by SNX-111 was more marked in the high dose, particularly at the later time points (see Fig. 1).

Although effects of SNX-111 on calcium accumulation were seen within the dorsal hippocampus these effects were not as prevalent as within the cerebral cortex. Only at the 24 and 48 hours post-injury was there any substantial evidence of the reduction of the injury-induced calcium accumulation (see Fig. 1). This is likely due to the fact that in our model of fluid percussion brain injury the majority of calcium accumulation occurs within the cortex which bears the major brunt of the insult.

Discussion

These preliminary results indicate that SNX-111 is very effective in reducing the extent and duration of calcium accumulation following traumatic brain injury. Its effectiveness is most evident in regions of the

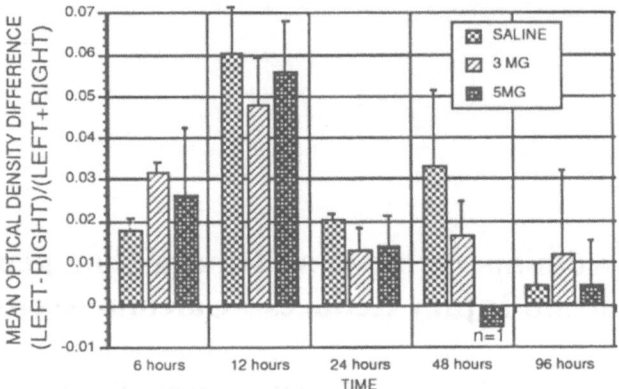

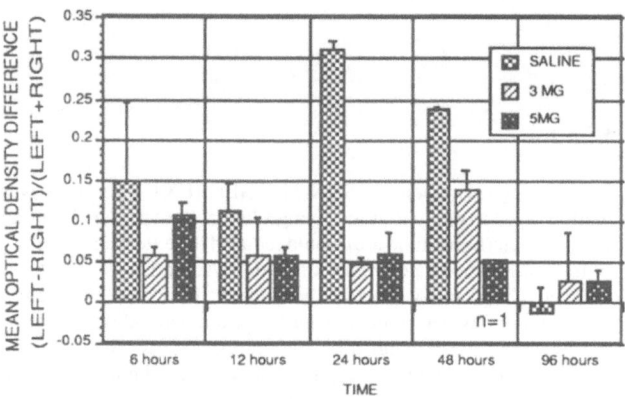

Fig. 1. Mean (± SEM) relative optical density measurements in the parietal cortex (top) and dorsal hippocampus (bottom) at different time after fluid percussion injury for animals treated with saline or SNX-111. The effect of the compound was most apparent within the parietal cortex and lasted for as long as 2 days. Note the difference in scale on the ordinate between the two graphs

brain which exhibit the greatest amount of calcium flux. Furthermore, its effectiveness lasts long after administration with the reduction of calcium accumulation persisting for up to 48 hours.

Although not producing marked cell death, the calcium accumulation following fluid percussion brain injury may play a significant role in the post traumatic metabolic dysfunction or the enhancement of cellular vulnerability to a second insult[2,3]. Furthermore, the extended length of time calcium can accumulate following traumatic brain injury demands new approaches in therapeutic treatments which are not only effect in reducing calcium accumulation but maintain a prevention of this flux for several hours if not days. According to our current data, SNX-111 is a pharmacological agent which appears to satisfy both of these requirements.

To determine if SNX-111 has a therapeutic effect on functional outcome following fluid percussion injury, behavioral studies are required. These studies are currently in the design phase and will incorporate both sensorimotor and cognitive functioning.

Acknowledgements

We thank Ms. Nabila Balady for her technical assistance. This work was supported by a grant from the NEUREX Corporation and by NS30308).

References

1. Fineman I, Hovda DA, Smith M, Yoshino A, Becker DP (1993)Concussive brain injury is associated with a prolonged accumulation of calcium: a ^{45}calcium autoradiographic study. Brain Res, in press

2. Hovda DA, Becker DP, Katayama Y (1992) Secondary injury and acidosis. J Neurotrauma 9 [Suppl 1]: S47–S60
3. Katayama Y, Becker DP, Tamura T, Hovda DA (1990) Massive increases in extracellular potassium and the indiscriminate release of glutamate following concussive brain injury. J Neurosurg 73: 889–900
4. Yamakami I, McIntosh TK (1989) Effects of traumatic brain injury on regional cerebral blood flow in rats as measured with radiolabeled microspheres. J Cereb Blood Flow Metab 9: 117–124
5. Yamakami I, McIntosh TK (1991) Alterations in regional cerebral blood flow following brain injury in the rat. J Cereb Blood Flow Metab 11: 655–660

Correspondence: David A. Hovda, Ph. D., Division of Neurosurgery, 74-140 CHS, UCLA School of Medicine, Los Angeles, CA 90024-6901, U.S.A.

Acta Neurochir (1994) [Suppl] 60: 524–527

Excitatory Amino Acid Release from Contused Brain Tissue into Surrounding Brain Areas

H. Tanaka, Y. Katayama, T. Kawamata, and **T. Tsubokawa**

Department of Neurological Surgery, Nihon University School of Medicine, Tokyo, Japan

Summary

The EAA release from contused brain tissue and its effect on the extracellular EAA levels in brain areas surrounding the contusion were investigated with microdialysis technique in the rat. A significant increase in extracellular EAA levels was observed in the contused brain tissue. The EAA increase was significantly greater in the contused brain tissue than in the isolated but non-contused brain tissue. It was further demonstrated that EAAs were released from non-contused brain areas 1–2 mm distant from contused brain tissue. No such EAA release from surrounding brain areas was demonstrated when the cavity was filled with isolated but non-contused brain tissue. The increase in EAAs was attenuated by KYN administered through microdialysis, suggesting that the EAA release from the surrounding brain areas appears to be a consequence that is secondary to the EAA release from the contused brain tissue. Such a diffusion-reaction process is probably mediated by the neurotransmitter actions of EAAs. The results of the present study are of clinical importance, since surgical removal of contused brain tissue and administration of EAA antagonists may serve to protect the surrounding brain areas from EAA neurotoxity.

Keywords: Cerebral contusion; excitatory amino acid; microdialysis.

Introduction

Cerebral contusions demonstrate morphological evolution and progressive edema formation which lead to clinical deterioration during the initial 12–24 hours postinjury. Most vertebrate neurons contain millimolar concentrations of excitatory amino acids (EAAs), as neurotransmitters and as cytosolic intermediary metabolites. Disruption of these cells may therefore liberate EAAs into the extracellular space[1]. Our previous studies have indicated that EAA transmitters are involved in the mechanisms of the early ionic fluxes, cellular swelling, hypermetabolism and subsequent metabolic derangements which follow concussive brain injury (e.g.,[4,10]). If EAAs are released from con-

tused brain tissue, they may contribute to progressive edema formation and secondary damage[2,4,8,10]. Little is currently known, however, regarding the role of EAAs in contusional brain injury.

In the present study, we examined the EAA release from contused brain tissue and its effect on the EAA levels in brain areas surrounding the contusion. Cerebral contusions are composed of several components including areas of necrosis, hemorrhage, infarct and edema. In order to investigate the EAA release from the contused brain tissue in isolation, we employed an experimental model of isolated cerebral contusion[9].

Materials and Methods

Young Wistar rats weighing 250–300 g were anesthetized with a mixture of 33% oxygen, 66% nitrous oxide and 1% halothane. With the rectal temperature kept at 37.0–38.0°C with a heating pad, the animals were placed in a stereotaxic frame, the calvarium was exposed and a craniotomy was performed. A precisely determined volume of brain tissue was removed from the frontal cortex using a specially designed cannula (O.D. = 3 mm, I.D. = 2 mm). The removed brain tissue was contused with an ultrasonic homogenizer (2.0 W for 2 s) and was then returned to the cavity from which the brain tissue had been obtained.

One microdialysis probe (O.D. = 500 μm; effective length = 3.0 mm; cut off = 20,000 MW) was placed within the isolated and contused brain tissue, and an other probe was placed within the intact brain of the contralateral frontal cortex as a control (control-1). In a separate group of animals, the brain tissue was removed from the frontal cortices of both sides. In these animals, one of the cavities was filled with isolated and contused brain tissue, and the other was filled with gelatine of equivalent volume (control-2). In some animals, the cavity was filled with isolated but non-contused brain tissue. Additional pair of microdialysis probes were placed bilaterally in the parietal cortex 1.5 mm or 3 mm posterior to the edge of the cavities.

Measurements of the dialysate concentration of glutamate ([Glu]$_d$) were initiated within 5 min after induction of contusion. Measurements of [Glu]$_d$ in the intact brain areas were begun simultaneously by placing microdialysis probes through the tracks formed by pre-penetration. Modified Ringer solution (pH = 7.4, osmolarity = 308 mOsm), containing 1 mM Ca²⁺, was employed as a perfusate for microdialysis. The dialysate was perfused using a 1000 μl Hamilton microsyringe and a microperfusion pump at a rate of 5.0 μl/min, and the temperature was maintained at 37.0°C with a perfusion warmer.

[Glu]$_d$ was measured by high-performance liquid chromatography. Precolumn derivatization of the amino acids with ophthaldialdehyde was employed to form highly fluorescent reaction products which were separated using a Beckman HPLC fitted with a Beckman Ultrasphere C-18 Column (5 μ, 4.6 mm i.d. × 250 mm) and a fluorescence detector (Japan Spectroscopic Co. Ltd. Type 821-FP; excitation wave length, 340 nm; emission, 440 nm). The concentrations of amino acids were normalized against a known concentration of homoserine (25 μM) which was added to the dialysate as an internal standard.

In order to test the effects of released EAAs on the [Glu]$_d$ levels of the surrounding brain areas, kynurenic acid (KYN, 10 mM), a broad-spectrum EAA antagonist, was perfused through the microdialysis probe placed in the surrounding brain areas. The dose of KYN was determined from previous studies which had demonstrated pharmacological effects of KYN administered by microdialysis. The osmolarity was maintained constant by removing equimolar sodium chloride from the perfusate.

Results

Initial penetration of the brain tissue by the microdialysis probe induced a large increase in [Glu]$_d$. In contrast, re-insertion of probes through tracks previously made produced a less marked increase in [Glu]$_d$, and the [Glu]$_d$ rapidly reached a semi-stabilized level.

No glutamate was detected in dialysates obtained from gelatine.

EAA release from isolated but non-contused brain tissue. A high [Glu]$_d$ was observed within the isolated but non-contused brain tissue. The [Glu]$_d$ level reached 10 times the [Glu]$_d$ level measured simultaneously within the normal brain tissue. Its value then gradually decreased. The [Glu]$_d$ remained 2–3 times higher than the [Glu]$_d$ level within the normal brain tissue at 1 hour after production of lesions.

Similarly, a high [Glu]$_d$ was observed within the isolated and contused brain tissue immediately after probe insertion. The [Glu]$_d$ level was approximately 10 times the [Glu]$_d$ level measured simultaneously within the normal brain tissue. As compared to the isolated but non-contused brain tissue, however, the contused brain tissue continued to demonstrate a high [Glu]$_d$ level for a longer time period. The [Glu]$_d$ remained 4–5 times higher than the [Glu]$_d$ level within the normal brain tissue at 1 hour after production of lesions. The [Glu]$_d$ level within the brain areas surrounding the cavity filled with contused brain tissue was found to be elevated. Dialysate collected for 1 hour from brain areas 1.5 mm distant from the cavity filled with contused brain tissue revealed an elevated [Glu]$_d$ level, which was approximately 3-fold measured simultaneously within the normal brain tissue. No such increase

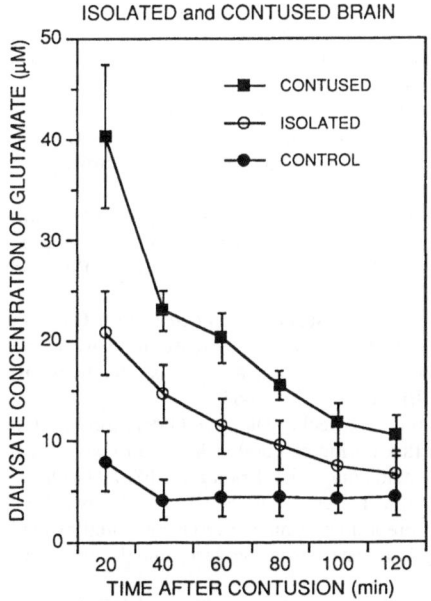
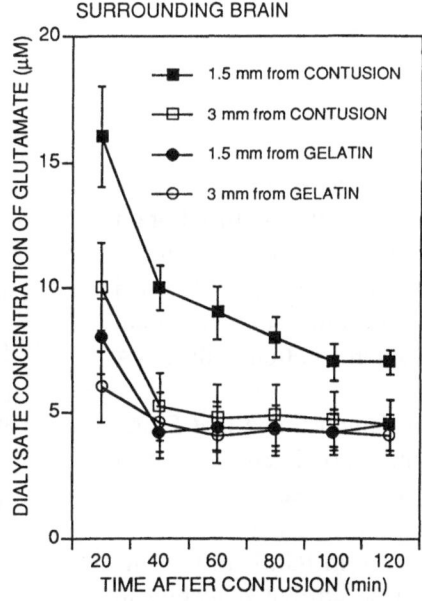

Fig. 1. Right: Changes in glutamate concentration of the dialysates obtained from isolated and contused brain tissue (closed squares) or isolated but non-contused brain tissue (open circles) returned to the cavity, and from the intact brain (closed circles). Left: Changes in glutamate concentration of the dialysates obtained from brain areas surrounding the cavity in which isolated and contused brain tissue (closed or open squares) or gelform of equivalent volume (closed or open circles) was placed. Data for each time points were from 5 rats. Mean ± S.D.

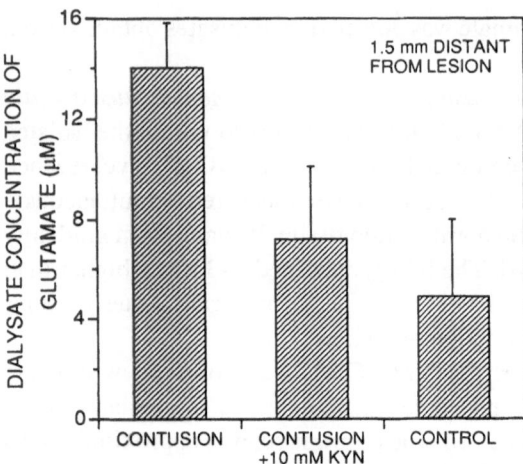

Fig. 2. Effects of kynurenic acid (*KYN*) administered through microdialysis probe on glutamate concentration in the dialystae obtained from brain areas surrounding the cavity (1.5 cm distant from the edge of the cavity) in which isolated and contused brain tissue was placed. *CONTUSION* dialysate obtained with perfusate containing no KYN. *CONTUSION + KYN* dialysate obtained with perfusate containing 10 mM KYN. Glutamate concentration in the dialysate obtained from the intact brain is shown for comparison. Values represent concentration in dialysate collected for 1 hour. Data for each column were from 5 rats. Mean ± S.D. $p < 0.01$ for CONTUSION vs. CONTUSION + 10 mM KYN

in $[Glu]_d$ was demonstrated when the cavity was filled with isolated but non-contused brain tissue. In situ administration of KYN significantly inhibited the increase in $[Glu]_d$ within the brain areas surrounding the cavity filled with contused brain tissue.

Discussion

The results of the present study demonstrated a significant increase in extracellular EAA levels in the contused brain tissue. The magnitude of the EAA increase was significantly greater in the contused brain tissue than in the isolated but non-contused brain tissue. Since isolation of the brain tissue can be regarded as equivalent to complete ischemia, contused brain tissue appears to release EAAs in addition to the EAA release caused by ischemia.

The present study further demonstrated that EAAs were released from non-contused brain areas adjacent to contused brain tissue. The attenuation of the increase in EAAs observed after in situ administration of KYN suggests that the increase in $[Glu]_d$ in the surrounding brain areas cannot be attributed merely to diffusion of EAAs from the contused brain tissue. Thus, the EAA release from the surrounding brain areas appears to be a consequence that is secondary to the EAA release from the contused brain tissue. Such a

diffusion-reaction process is probably mediated by the neurotransmitter actions of EAAs[7].

We have demonstrated previously an increase in glucose metabolism and subsequent depression of cytochrome oxidase activity without lethal cell damage in the surrounding brain areas in the same model of isolated contusion[9]. Similar observations have been reported previously employing other models of cerebral contusion[12]. These findings suggest that contused brain tissue exerts considerable metabolic influences on the surrounding brain areas. The EAA release from the contused brain may play some role in the expression of such metabolic influences, since EAAs produce massive ionic fluxes through the cell membrane[3,7,8,11]. If the ionic fluxes are sufficiently large, the nerve terminals may be depolarized, and further release of EAA neurotransmitter may result[5,6]. The results of the present study are of clinical importance, since surgical removal of contused brain tissue and administration of EAA antagonists may serve to protect the surrounding brain areas from EAA neurotoxicity.

References

1. Faden AI, Demdeiuk P, Panter SS, *et al* (1989) The role of excitatory amino acids and NMDA receptors in traumatic brain injury. Science 244: 798–800
2. Bullock R, Butcher SP, Chen M-H, *et al* (1991) Correlation of the extracellular glutamate concentration with extent of blood flow reduction after subdural hematoma in the rat. J Neurosurg 74: 794–802
3. Choi DW, Koh J-Y, Peters S (1988) Pharmacology of glutamate neurotoxicity in cortical cell culture: attenuation by NMDA antagonist. J Neurosci 8: 185–196
4. Katayama Y, Becker DP, Tamura T, *et al* (1990) Massive increase in extracellular potassium and indiscriminate glutamate release after concussive brain injury. J Neurosurg 73: 889–900
5. Katayama Y, Kawamata T, Tamura T, *et al* (1991) Calcium-dependent glutamate release cocomitant with massive potassium flux during cerebral ischemia in vivo. Brain Res 558: 136–140
6. Katayama Y, Tamura T, Becker DP, *et al* (1991) Calcium-dependent component of massive increase in extracellular potassium during cerebral ischemia as demonstrated by microdilaysis in vivo. Brain Res 567: 57–63
7. Katayama Y, Tamura T, Becker DP, *et al* (1991) Inhibition of rapid potassium flux during cerebral ischemia in vivo with an excitatory amino acid antagonist. Brain Res 568: 294–298
8. Katayama Y, Tamura T, Becker DP, *et al* (1992) Early cellular swelling during cerebral ischemia in vivo is mediated by excitatory amino acids released from nerve terminals. Brain Res 577: 121–126
9. Katayama Y, Yoshino A, Kawamata T, *et al* (1991) Metabolic effects of contused brain tissue implanted in the brain as demonstrated by ^{14}C-deoxyglucose autoradiography, NADH fluorescence and cytochrome oxidase histochemistry. 9th Neurotrauma Symposium, November 9–10, 1991, LA, U.S.A.

10. Kawamata T, Katayama Y, Hovda DA, *et al* (1992) Administration of excitatory amino acid antagonists via microdialysis prevents the increase in glucose utilization seen immediately following concussive brain injury. J Cereb Blood Flow Metab 12: 12-24
11. Mayer ML, Westbrool GL (1987) Cellular mechanisms underlying excitotoxicity. TINS 10: 59–61
12. Sumani K, Nakamura T, Ozawa Y, *et al* (1989) Hypermetabolic state following experimental head injury. Neurosurg Rev 12: 400–411

Correspondence: Hiroaki Tanaka, M.D., Department of Neurological Surgery, Nihon University School of Medicine, Tokyo 173, Japan.

Acta Neurochir (1994) [Suppl] 60: 528–530
© Springer-Verlag 1994

Combined Treatment with Nicardipine, Phenobarbital, and Methylprednisolone Ameliorates Vasogenic Brain Edema

S.-Z. Lin, T.-L. Chiou, Y.-H. Chiang, and W.-S. Song

Division of Neurosurgery, Tri-Service General Hospital and National Defense Medical Center, Taipei, Republic of China

Summary

Free radicals formed around the edematous areas of the brain can cause lipoperoxidation of the cellular membrane, followed by calcium influx into the cell through calcium channels. These secondary insults may aggravate vasogenic brain edema. Since phenobarbital is a free radical scavenger; methylprednisolone has an antilipoperoxidation effect; and nicardipine is a calcium channel blocker, we hypothesized that combined treatment with phenobarbital, methylprednisolone, and nicardipine would be beneficial in vasogenic brain edema. This hypothesis was tested in Sprague-Dawley rats with a transdural cold-injury on the right parietal cortex. The animals were randomly divided into two groups. Animals in the treatment group were injected intraperitoneally with phenobarbital (4 mg/kg), methylprednisolone (50 mg/kg), and nicardipine (10 µg/kg) at 5 min and 8 hours after the cold-injury. The control animals were injected with saline. These animals were sacrificed 24 hours after the injury. The extent of brain edema was assessed by measuring the water content, the inulin distribution volume, and the distribution area of Evans blue in the brain. Our results showed that the water content of the edematous hemisphere was similar in the control and the treatment groups. However, Evans blue distribution area and inulin distribution volume of the treatment group were less than those of the control group by 12% and 31%, respectively. In conclusion, the combined treatment with phenobarbital, methylprednisolone and nicardipine is beneficial in vasogenic brain edema.

Keywords: Brain edema; calcium blockade; free radicals; methylprednisolone.

Introduction

The injuries of cerebral microvascular endothelia and the concomitant damage of neurons are the two major causes for the development of severe brain edema after a brain injury[1-3].

These endothelial and neuronal damages involve the processes of breakdown of cellular membrane by the abnormally activated phospholipase into arachidonic acid, which is then turned into thromboxanes via cyclooxygenase[1,3]. During these cellular breakdown precesses, free radicals are produced, causing membrane peroxidation and opening calcium channels[1,2]. Thus, cellular death may ensue.

Accordingly, we hypothesized that combined treatment with free radical scavengers and calcium channel blockers was beneficial in vasogenic brain edema. This hypothesis was tested in this study.

Materials and Methods

Sprague-Dawley rats weighing 300–400 g were used for these experiments. The rats were anesthetized with a mixture of halothane and oxygen gases. The scalp was incised, and a section of the right parietal bone about 5 mm × 5 mm in size was removed. A freezing probe with 3mm in diameter prechilled with liquid nitrogen was applied to the exposed dura matter of the right parietal cortex for 30 s. The skull defect was replaced with bone wax and the wound was sutured. These lesioned rats were randomly divided into the treatment and control groups. The treatment animals were injected intraperitoneally with nicardipine (10 µg/kg), phenobarbital (4 mg/kg), and methylprednisolone (50 mg/kg) at 5 min and 8 h after the cold injury. The control group was treated with saline.

The next morning, the rats were anesthetized with halothane before undergoing femoral artery and vein catheterization. The rats were then wrapped in loosely fitting plaster bandages, taped to stainless steel bricks, and allowed to recover from anesthesia before study.

At 23 h and at 23 h 50 min after the freezing, 0.6 ml of 2% Evans blue solution and 1 ml of ^{14}C-inulin saline solution (10 uCi), respectively, were injected intravenously. At 24 h after the lesioning, the rats were sacrificed by injecting 3 ml of 15% potassium chloride. The brain was removed from the skull, and the center of lesioned forebrain was coronally sliced into a 3 mm thick section, which was subsequently split into right and left halves, and then weighed separately. Immediately before the sacrifice of the rats, a blood sample of about 100 µl was obtained from the femoral artery. For the purpose of determining water content, the brain sections were placed in an oven set at 105°C for 48 h. The water content was calculated from the data using the following formula: water content = (wet weight – dry weight) / wet weight.

For the measurements of inulin distribution volume, the radio-activity of ^{14}C-inulin in the brain (Ab) and plasma samples (Cp) were determined by β counting. Inulin distribution volume (IDV) was defined as IDV (ml / 100 g) = Ab / Cp.

The rostral and caudal cutting surfaces adjacent to the removed brain slice were photographed. The photos were used for the measurements of Evans blue distribution area (EBDA), which was derived from the following: EBDA (%) = Ae / At × 100 where Ae was the blue-colored area on the cutting surface and At was the total brain section area.

Paired and unpaired t-tests were used for data comparisons.

Results and Discussion

Mean blood pressure was 126 ± 10 mmHg for the treatment group and 122 ± 2 mmHg for the control group. Pulse rates were 395 ± 30 and 394 ± 13 beat/min for the treatment and control group, respectively. These results indicated that the combination of drugs (nicardipine, methylprednisolone, and phenobarbital) did not alter the blood pressure and pulse rate.

The water content measurements of the non-lesioned side were 0.793 ± 0.002 and 0.788 ± 0.002 for the control and treatment groups, respectively. The two values were not significantly different (p < 0.05). On the lesioned side, the water content values were also similar in the control and the treatment groups (0.810 ± 0.005 vs. 0.806 ± 0.002, p > 0.05).

Evans blue distribution area was an indicator for the degree of extravasation of plasma protein into the brain. On the rostral cutting surface, the Evans blue distribution areas were 7.71 ± 0.68% for the control group, and 7.21 ± 0.56% for the treatment group (Fig. 1). The latter was 13% less than the former, but statistically insignificant. On the caudal cutting surface, the Evans blue distribution areas were 6.74 ±

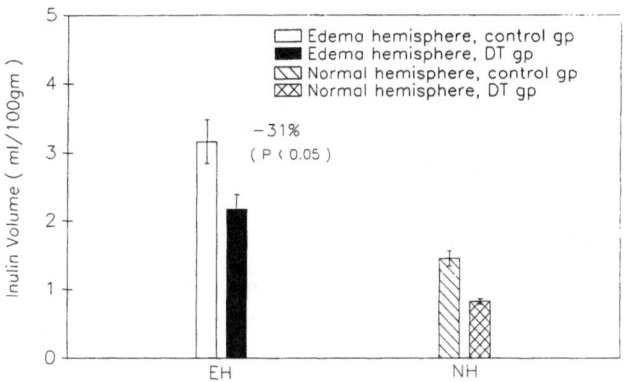

Fig. 2. Inulin distribution volume. *EH* edema hemisphere, *NH* normal hemisphere

0.43, and 6.42 ± 0.63% for the control and the treatment groups, respectively (Fig. 1). Although the comparison of the two sets of data revealed no significant difference, the Evans blue distribution area of the treatment group was 11% less than that of the control group.

Inulin has a large molecular weight and cannot across the intact blood-brain barrier into the brain interstitium. However, if the blood-brain barrier is damaged, inulin will become extravasated into the brain parenchyma, thereby incresing the inulin distribution volume in the brain.

This theory was supported by the present results. For the control group, the inulin distribution volume was 1.45 ± 0.11 ml/100 g for the non lesioned side, and 3.16 ± 0.32 ml/100 g for the lesioned side (Fig. 2). The latter was at least one-fold greater than the former, and the difference between the two sides was very significant (p < 0.001). This was also true for the treatment group (0.82 ± 0.04 vs. 2.17 ± 0.21 ml/100 g).

For the non-lesioned side, the inulin distribution volume was significantly less in the treatment group than the control group (Fig. 2), indicating that the three drugs used in the present study caused a decrease in cerebral microvascular plasma volume. A similar decrease (−31%) in inulin distribution volume in the lesioned side was also noted for the treatment group compared to that for the control group (2.17 ± 0.21 vs. 3.16 ± 0.32 ml/100 g). This decrease in inulin distribution volume of the treatment group may be due to the decreases in both the extravasated inulin in the damaged brain region and the retained inulin in the normal cerebral microvessels.

In conclusion, the combined treatment with nicardipine, methylprednisolone, and phenobarbital may be beneficial for vasogenic brain edema.

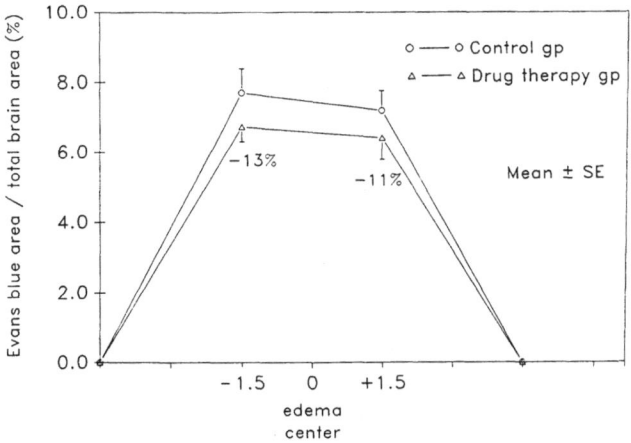

Fig. 1. Evans blue distribution area

Acknowledgement

The authors gratefully acknowledge the technical support of Mrs. Fen-Ying Wu as well as the assistance of Ms. Hui-Ju Chong with the preparation of this manuscript. This work was supported by National Defense Research Grants of 1991 war injury and 1992 war injury 15705-1.

References

1. Baethmann A (1978) pathophysiological and pathochemical aspects of cerebral edema. Neurosurg Rev 85–100

2. Hewitt K, Corbett D (1992) Combined treatment with MK-801 and nicardipine reduces global ischemic damage in the gerbil

3. Nakato H, Yen M-M, Tajima A, Lin S-Z, Pettigrew K, Blasberg R, Fenstermacher J (1990) Dexamethasone effects on the distribution of water and albumin in cold-injury cerebral edema. In: Long D, *et al* (eds) Advances in neurology. Raven, New York, pp 335–342

Correspondence: Shinn-Zong Lin, M. D., Ph. D., Division of Neurosurgery, Tri-Service General Hospital, 8, set. 3, Ting-Chou Rd. Taipei, Republic of China.

Acta Neurochir (1994) [Suppl] 60: 531–533
© Springer-Verlag 1994

Treatment of Traumatic Brain Edema by Multiple Doses of Mannitol

P. von Berenberg, A. Unterberg, G. H. Schneider, and **W. R. Lanksch**

Department of Neurosurgery, Rudolf Virchow Medical Center, Free University of Berlin, Berlin, Federal Republic of Germany

Summary

Mannitol is frequently used to reduce elevated intracranial pressure often associated with brain edema. In cases of a damaged blood-brain barrier, however, mannitol might aggravate vasogenic cerebral edema, as has recently been stressed. The aim of this study was to investigate whether multiple doses of mannitol administered during development of vasogenic brain edema following a cryogenic cortical injury affect hemispheric swelling and edema.

Sprague-Dawley rats were anesthetized with ketamine and xylazine. A cortical freezing lesion was applied to the right parietal region. A first series of eight rats received four doses of 20% mannitol (0.4 g/kg within 10 minutes) thirty minutes, 3, 6 and 9 hours after trauma. Twelve hours after cryogenic injury, the brains were removed for determination of hemispheric swelling and cerebral water content. Eight control rats were infused with saline only.

In a second series nine rats received eight doses of 20% mannitol 30 minutes, 3, 6, 9, 12, 15, 18 and 21 hours after trauma. In this series, the brains were removed 24 hours after freezing. Again respective control animals were infused with saline only.

Hemispheric swelling was $7.2 \pm 0.5\%$ after four doses of mannitol compared to $7.6 \pm 0,5\%$ in control animals (n.s.). Following eight doses of mannitol hemispheric swelling was $8.9 \pm 0.4\%$ compared to $10.1 \pm 0.4\%$ in control rats ($p < 0.05$). Accordingly, the water content of traumatized hemispheres was lower following repeated mannitol treatment (80.5 versus 80.8%). Water content in control hemispheres was not affected by mannitol.

Taken together, these results indicate that multiple doses of mannitol do not aggravate total hemispheric swelling, nor global water content following induction of vasogenic edema. In contrast, repeated administration of mannitol reduces traumatic swelling, thus favouring repeated mannitol application even when the blood-brain barrier is defective.

Keywords: Cerebral edema; mannitol; cold injury; rats.

Introduction

Mannitol is successfully used in the management of raised intracranial pressure associated with intracranial masses or cerebral edema. Though there is consent that a single dose of mannitol is beneficial for transient reduction of cerebral bulk[1,6,7], Kaufmann and Cardoso (1992) provided evidence that multiple doses of man-

nitol might aggravate vasogenic cerebral edema[2]. They observed an increase in water content in edematous regions following a cryogenic injury with repeated doses of mannitol, associated with a reversal of the osmotic concentration gradient between edematous brain and plasma. They concluded that the long-term use of mannitol should be reevaluated[2].

The aim of this study was, therefore, to investigate whether multiple doses of mannitol administered during the development of vasogenic brain edema following a cryogenic injury affect total hemispheric swelling and edema.

Methods

Thirty-six male Sprague-Dawley rats (551 ± 26 g) were used for the study. A cortical freezing lesion was applied to the right parietal region in ketamine-xylazine-anesthesia[3,10]. The standard lesion was made by a probe 5 mm in diameter cooled to –60°C by a dry ice/acetone-mixture and lowered to the exposed dura for 30 seconds. Systemic blood pressure was monitored via a cannula in the tail artery.

Two series of experiments with multiple intravenous doses of mannitol (0.4 g/kg per injection) were performed.

Subset A (n = 9) received four doses of mannitol 20% at 30 minutes, 3, 6 and 9 hours post trauma. Subset B (n = 9) was treated with eight doses at 30 minutes, 3, 6, 9, 12, 15, 18 and 21 hours post trauma. Respective control animals in both series (n = 9 each) received corresponding volumes of normal saline at the same time intervals.

In series A, the brains were removed 12 hours after trauma, in series B 24 hours following cold injury. Thereafter, the hemispheres were meticulously separated using a conventional operating microscope and weighed for determination of brain swelling. Hemispheres were then dried for 48 hours at 100°C and weighed again. Cerebral water content was calculated as the difference of hemispheric wet- and dry-weight.

Results

After four doses of mannitol (subset A), hemispheric swelling was slightly lower compared to control ani-

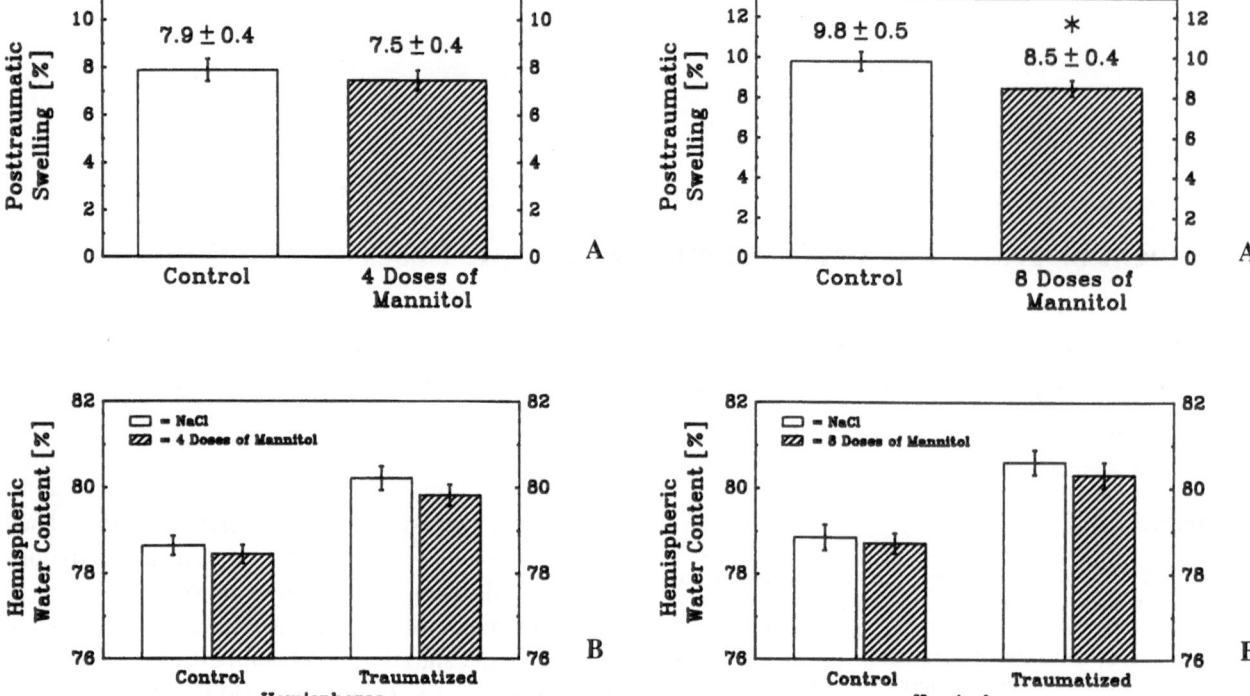

Fig. 1. (A) Posttraumatic hemispheric swelling following four doses of mannitol, survival 12 hours. Swelling is marginally reduced by treatment. (B) Hemispheric water content following four dosis of mannitol, or saline respectively. 12 hours survival. Cerebral water content is only slightly lowered by treatment

Fig. 2. (A) Posttraumatic hemispheric swelling following eight doses of mannitol. 24 hours survival. Significant reduction of swelling by treatment. (B) Hemispheric water content following eight doses of mannitol, or saline respectively. 24 hours survival. Hemispheric water content is moderately lowered in traumatized hemispheres by mannitol treatment, and unaffected in control hemispheres

mals (7.5 ± 0.4% versus 7.9 ± 0.5%, n.s.; Fig. 1). Concomitantly, the water content of the lesioned hemisphere was marginally lower in mannitol treated rats, 79.8 ± 0.3 versus 80.2 ± 0.3% (Fig. 1). The water content of the control hemispheres was nearly identical (mannitol 78.5%, saline 78.6%).

Following eight doses of mannitol (0.4 g/kg each) hemispheric swelling was significantly reduced from 10.1 ± 0.4% in controls to 8.5 ± 0.4% in mannitol-infused rats (p < 0.05; Fig. 2). Accordingly, the water content of traumatized hemispheres was lower in mannitol treated animals (80.5 ± 0.3% versus 80.8 ± 0.2%; Fig. 2). Mannitol-treatment did not affect the water content of the non-lesioned hemispheres (Fig. 2).

Discussion

The effect of mannitol on cerebral water content, edema, increased intracranial pressure, cerebral blood flow, viscosity etc. has extensively been studied during the last 20 years[1,2,4–7,9,10]. Two main mechanisms of action of mannitol on cerebral compliance are discussed. First, mannitol increases CBF by a transient hypervolemia and reduced blood viscosity. This induces a compensatory vasoconstriction, thus reducing cerebral blood volume[6]. Second, osmotic dehydration of the brain is a widely accepted mechanism of action[2,7,8]. Additional beneficial effects of mannitol, e.g. reduction of ischemic damage, have been reported and partly attributed to a free radical scavenging action[5,9].

Repeated mannitol administration is widely used in many centers and a basis of various therapeutic regimens for severe head injury. It has been argued, however, that administration of multiple doses of mannitol might aggravate vasogenic edema and cerebral swelling, if the blood-brain barrier is damaged[2].

Our findings, that repeated mannitol administration does not aggravate total hemispheric swelling, but rather diminishes swelling and edema in a model with gross blood-brain barrier damage are in contrast to the observations of Kaufmann and Cardoso[2], though both studies cannot be directly compared. While local accumulation of mannitol and local increase in tissue water content in the edema zone may be true, total hemispheric swelling and total hemispheric water content are reduced. This raises the question of the mode of action of mannitol once again. Since the water content of the non-lesioned hemisphere was not affected by multiple doses of mannitol, a possible osmotic dehydration is unlikely.

It is suggested that mannitol interferes with other mechanisms, e.g. acts as a free radical scavenger[5], thus specifically interfering with edema promoting mediators, or that mannitol reduces the growth of the primary brain tissue necrosis from the focal injury. Findings that mannitol reduces the area of infarction following middle cerebral arterial occlusion support this hypothesis[9].

In conclusion, our study favours repeated mannitol application, e.g. in the treatment of reduced intracranial compliance following severe head injury, even if the blood-brain barrier is grossly damaged.

Acknowledgement

The technical and secretarial assistance of Ms. J. Kopetzki and Ms. A. Riede is gratefully acknowledged.

References

1. Hartwell RC, Sutton LN (1993) Mannitol, intracranial pressure, and vasogenic edema. Neurosurgery 32: 444–450
2. Kaufmann AM, Cardoso ER (1992) Aggravation of vasogenic cerebral edema by multiple-dose mannitol. J Neurosurg 77: 584–589
3. Klatzo I, Piraux A, Laskowski EJ (1958) The relationship between edema, blood-brain barrier and tissue elements in local brain injury. J Neuropath Exp Neurol 17: 548–564
4. Long DM, Maxwell RE, Choy KS, Cole HO, French LA (1972) Multiple therapeutic approaches in the treatment of brain edema induced by a standard cold lesion. In: Reulen HJ, Schurmann K (eds) Steroids and brain edema. Springer, Berlin Heidelberg New York, pp 87–96
5. Macgovern GJ, Bolling SF, Casale AS (1984) The mechanism of mannitol in reducing ischemic injury: hyper-osmolarity or hydroxyl scavenger. Circulation 70 [Suppl I]: 191–195
6. Muizelaar JP, Wei EP, Kontos HA, Becker DP (1983) Mannitol causes compensatory cerebral vasoconstriction and vasodilation in response to blood viscosity changes. J Neurosurg 59: 822–828
7. Nath F, Galbraith S (1986) The effect of mannitol on cerebral white matter water content. J Neurosurg 65: 41–43
8. Pappius HM, Dayes LA (1965) Hypertonic urea. Arch Neurol 13: 295–402
9. Rickels E, Gaab MR, Heissler H, Dietz H (199 The effect of mannitol and nimodipine-treatment in a rat model of temporary focal ischemia. Zentralbl Neurochir 54: 3–12
10. Unterberg A, Schmidt W, Dautermann C, Baethmann A (1990) The effect of various steroid treatment regimens on cold-induced brain swelling. Acta Neurochir [Suppl] 51: 104–106

Correspondence: P. von Berenberg, M. D., Department of Neurosurgery, Rudolf Virchow Medical Center, Free University of Berlin, Augustenburger Platz 1, Berlin, Federal Republic of Germany.

Acta Neurochir (1994) [Suppl] 60: 534–537
© Springer-Verlag 1994

Therapeutical Efficacy of a Novel Chloride Transport Blocker and an IP₃-Analogue in Vasogenic Brain Edema

S. Berger, F. Staub, M. Stoffel J. Eriskat, L. Schürer, and **A. Baethmann**

Institute for Surgical Research, Department of Neurosurgery, Klinikum Großhadern, Ludwig-Maximilians-University, München, Federal Republic of Germany

Summary

The efficacy of torasemide, a novel chloride-channel blocker, and of PP56, an IP₃ analogue, was currently examined in experimental brain edema. Following trephination in anesthesia rats were subjected to a focal cold injury of the left cerebral hemisphere. Animals of 4 experimental groups receiving either torasemide (i.v. at 30 min before and 6 h after lesion) or PP56 (continuous infusion beginning at 30 min before until 24 h after lesion) at two dose levels were compared with controls administered with i.v. saline. 24 h after trauma the brain was removed from the skull, and the hemispheres were separated in the median plane for gravimetric assessment of hemispheric swelling. Hct, blood gases and body temperature remained constant in all groups. Blood pressure was found to increase in a dose-dependent manner in animals with torasemide. No significant reduction of brain swelling was found in animals with low-dose torasemide ($8.51 \pm 0.63\%$) or low- (7.91 ± 0.60) and high-dose PP56 ($6.85 \pm 1.05\%$) as compared to the untreated controls. Brain swelling, however, was significantly attenuated by high-dose torasemide to $7.04 \pm 0.36\%$, as compared to $8.89 \pm 0.29\%$ of the untreated group ($p < 0.005$). It is currently studied whether torasemide reduces brain swelling when given after the insult.

Keywords: Experimental head injury; torasemide; PP56; vasogenic brain edema.

Introduction

Brain edema leading to intracranial hypertension is of major concern in cerebral disorders, such as head injury, stroke, or brain tumors. Although cerebral tissue water content can be lowered by hypertonic solutions, e.g. by mannitol, hypertonic saline or glycerol, the treatment does not prevent further formation of edema. Methods of treatment might be more beneficial which ameliorate in addition breakdown of the blood-brain barrier and extravasation of edema fluid. This might be accomplished by inhibition of the formation or release, respectively of mediator compounds which enhance development and persistence of brain edema.

The present experiments were conducted to study the therapeutical efficacy of two novel pharmacological agents on vasogenic brain edema in rats subjected to an acute cerebral lesion. Torasemide is a newly developed loop diuretic. It is an inhibitor of the $Na^+/K^+/Cl^-$-cotransporter and of Cl^--channels in kidney epithelial cells[15]. Therapeutical effects of torasemide were observed in intracranial hypertension from water intoxication in vivo, as well as in cytotoxic glial swelling from acidosis in vitro[10,14]. PP56 (D-myo-inositol-1,2,6-trisphosphate) is an isomer of the intracellular second messenger IP₃. The compound, however, is probably no competitive antagonist of IP₃, as it does not affect release of Ca^{++} in cells[5]. Antiedematous properties of PP56 were found in peripheral organs in experimental inflammation or burns[6]. Further, PP56 is antagonizing vasoconstriction of neuropeptide Y (NPY), a potent constrictor of rat or human cerebral arteries[4]. PP56 was currently studied with regard to reduction of vasogenic brain edema from an acute cerebral lesion. For that purpose, the cryo-lesion model was substantially refined in rats with improved reproducibility of the insult by employment of a computer-controlled stepper motor during induction of the lesion.

Materials and Methods

Preparation: 54 male Sprague-Dawley rats of 250–350 g b.w. were anesthetized by ether followed by i.p. chloralhydrate. The animals were spontaneously breathing room air supplied with additional oxygen. A catheter was introduced into the tail artery for monitoring of blood pressure and to obtain blood samples for determination of blood gases and hematocrit. Another catheter in the right jugular vein served for infusions of fluid and drugs. Body temperature was maintained at 38.5°C by a servo-controlled heating pad. The animal's head was fixed in a stereotactic holder for preparation of a circular trephination (Ø 6 mm) of the left temporal skull leaving the dura mater intact. A metal probe (Ø 4.5 mm) was attached to a stereotactic device powered by a computer-controlled stepper motor. The probe cooled to –68°C by a dry ice/acetone mixture was placed onto the exposed brain surface for 15 s.

Protocol: Administration of treatment started prior to induction of the lesion in both groups; treatment was made by a blinded investigator. Untreated animals with trauma (n = 12) were continuously infused with 0.5 ml saline/h. Animals with PP56 (Perstorp Pharma, Lund, Schweden) received either 40 mg/kg b.w. (low-dose; n = 10) or 120 mg/kg b.w. (high-dose; n = 10) dissolved in physiological saline (0.5 ml/h) which was continuously administered i.v. from 1 h prior to until 24 h after lesion. Animals with torasemide (Boehringer Mannheim, Germany) received either 1 mg/kg b.w. (low-dose; n = 10) or 10 mg/kg b.w. (high-dose; n = 10) at 30 min prior to and 6 h after trauma. 24 h after trauma the animals were reanesthetized and sacrificed by exsanguination. The cerebrum was rapidly removed and separated in the median plane. The hemispheres were placed into an air-tight glass vial for gravimetrical determination of fresh weight of the tissue. Water content of blood plasma or the brain specimen was obtained by the dry/wet weight method (drying at 110°C for 24 h). Na^+- and K^+-contents of dried tissue specimen were determined by flame spectrophotometry. Hemispheric brain swelling was determined as increase in weight of the traumatized hemisphere over that of the contralateral hemisphere. Data are given as mean ± SEM. Unpaired data were analyzed by the Kruskal-Wallis test, paired data by the Friedman-Quade test. $p < 0.05$ was considered significant.

Results

Systemic parameters: No differences of the systemic blood pressure were found between the various groups during the control period. Following treatment, however, a dose-dependent increase of MAP was found ($p < 0.05$) in animals receiving torasemide within 90 min after first administration of the drug (Fig. 1), whereas PP56 had no effect. PP56 and torasemide at either high or low dose were without other effects, e.g. on blood gases, hemoglobin or hematocrit. The body weight was found to decrease by 8% within 24 h after trauma in all groups (Table 1).

Cerebral water and electrolytes: In animals with focal lesion the water content of the traumatized hemisphere was increased by approximately 1% as compared to the contralateral hemisphere ($p < 0.001$). In untreated controls, the water content of the injured (left) hemisphere was 81.25 ± 0.14 g/100 g f.w., while 79.93 ± 0.08 g/100 g f.w. were found in the control hemisphere. Low-dose PP56 had no effect on the cerebral water content of either hemispheres. High-dose PP56 had a tendency to lower the water content of the exposed and of the non-exposed hemisphere, however, without attaining statistical significance. Low-dose torasemide had no influence on the formation of brain edema in the traumatized hemisphere. On the other side, high-dose torasemide was found to reduce the water content of the traumatized hemisphere to 80.82 ± 0.13 g/100 g f.w. ($p = 0.09$ vs. control, i.e. not significant). The tissue electrolytes were in accordance with

Table 1

Time from trauma	Group	MAP [mmHg]	Hct [%]	pH -Ig[H+]	B.W. [g]
Prior to trauma	Control	77.2 ± 3.5	43.6 ± 0.9	7.33 ± 0.01	312 ± 11
	Tora l.d.	77.9 ± 3.7	43.5 ± 1.0	7.35 ± 0.01	328 ± 11
	Tora h.d.	80.1 ± 3.7	42.3 ± 0.6	7.34 ± 0.01	312 ± 9
	PP56 l.d.	67.0 ± 3.1	43.8 ± 0.9	7.31 ± 0.02	335 ± 14
	PP56 h.d.	73.5 ± 4.3	43.3 ± 0.8	7.32 ± 0.01	309 ± 16
60 min after trauma	Control	72.4 ± 3.3	44.4 ± 0.9	7.34 ± 0.01	–
	Tora l.d.	85.1 ± 3.8*	42 8 ± 1.0	7.37 ± 0.01	–
	Tora h.d.	90.9 ± 4.0**	43.6 ± 0.5	7.37 ± 0.01	–
	PP56 l.d.	67.0 ± 3.7	46.7 ± 0.9	7.36 ± 0.01	–
	PP56 h.d.	69.7 ± 3.9	42.9 ± 1.0	7.33 ± 0.01	–
24 h after trauma	Control	–	43.1 ± 1.3	–	289 ± 11
	Tora l.d.	–	42.2 ± 0.9	–	302 ± 10
	Tora h.d.	–	42.4 ± 1.2	–	282 ± 10
	PP56 l.d.	–	42.9 ± 0.7	–	310 ± 13
	PP56 h.d.	–	42.9 ± 1.0	–	280 ± 15

Blood pressure, hematocrit, arterial blood pH, and body weight in animals subjected to a focal cold lesion of the brain under different treatment protocols prior to trauma and thereafter. * $p < 0.05$ vs. control, ** $p < 0.01$ vs. control. *Tora l. (h.) d.* Torasemide low (high) dose; *PP56 l. (h.) d.* PP56 low (high) dose; *B.W.* body weight.

extravasation of a plasma-like edema fluid from disruption of the blood-brain barrier. Hence, the Na^+ content was increased, whereas K^+ was decreased in the traumatized hemispheres of experimental animals with and without treatment.

Hemispheric brain swelling: In untreated controls, the freezing lesion led to an increase in weight of 8.89 ± 0.29% over that of the contralateral hemisphere. Hemispheric swelling was somewhat attenuated by PP56 resulting in an increase in weight of 7.91 ± 0.60% (p = 0.0646) in the low-dose group, or of 6.85 ± 1.05% in the high-dose group. High-dose torasemide was affording significant inhibition of hemispheric swelling from the lesion to 7.04 ± 0.36% (p < 0.005), whereas low-dose treatment had no effect. Hemispheric swelling amounted in the latter group to 8.51 ± 0.63% (Fig. 2).

Discussion

Torasemide: Loop diuretics, such as furosemide which is related to the currently studied agent, have been found to reduce the cerebral water content, CT hypodensity of injured cerebral regions, or intracranial pressure in animals with a focal brain lesion[3]. Loop diuretics may enhance bulk movement of fluid across the blood-brain barrier into blood[13], which is therapeutically utilized to improve dehydration by hyperosmotic solutions[12]. The similarities between furosemide and torasemide notwithstanding, it must be noted that torasemide is more lipophilic than furosemide, allowing for its entrance into the brain through the intact blood-brain barrier. Specific mechanisms of torasemide may relate with inhibition of the $Na^+/K^+/Cl^-$-cotransporter and Cl^-/HCO_3^--antiporter at the blood-brain barrier and other interfaces in brain tissue, resulting among others in antagonization of CSF-production[8]. Although pretreatment with torasemide was currently found to attenuate vasogenic edema formation at 24 h after trauma, it is not known yet whether the agent is also effective if administered post insult. Investigations of the latter point are presently carried out in this laboratory.

PP56: Aside from inhibition of edema formation in peripheral tissue, PP56 has properties which may be relevant for the brain. It is an antagonist of neuropeptide Y (NPY), a polypeptide which is widely distributed in central nervous tissue[2]. NPY is a potent vasoconstrictor of cerebral arteries. Its systemic administration was found to lower cerebral blood flow in rats and rabbit[4]. Abel *et al.*[1] consider that NPY has a role in

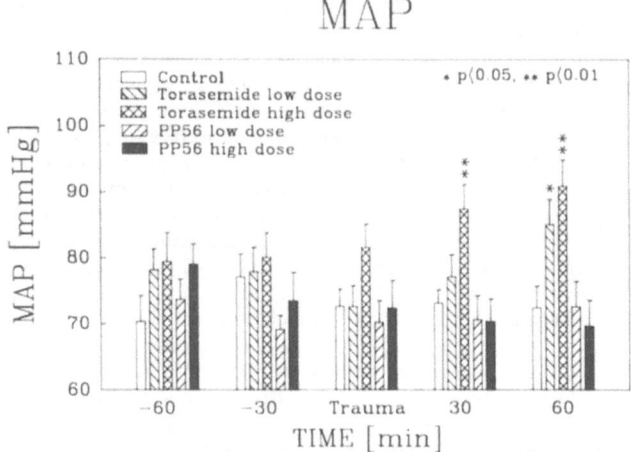

Fig. 1. Arterial blood pressure prior to lesion and up to 60 min thereafter in untreated controls and following various treatments. * p < 0.05 vs. control, ** p < 0.01 vs. control

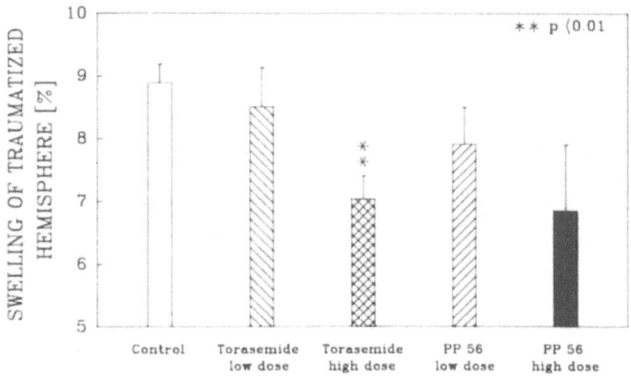

Fig. 2. Swelling of traumatized hemispheres as increase in weight in % over that of the contralateral hemisphere, in untreated controls with trauma and animals with torasemide or PP56 at high or low dose treatment

vasospasm from subarachnoid hemorrhage. This, however, is not supported by studies of Pluta *et al.*[11] on relationships between plasma- and CSF-levels of NPY and appearance of delayed vasospasm from SAH. On the other side, significant release of NPY has been demonstrated in experimental head injury in rats[9].

Observations that PP56 is blocking NPY-vasoconstriction in vitro and in vivo[7] might be noteworthy in this context. It is, nevertheless, difficult to reconcile with inhibition of vasogenic brain edema, since edema formation is markedly influenced by the regional tissue perfusion. Inhibitory effects of NPY on the tissue perfusion, however, may be counteracted by enhancement of necrosis formation from severe blood flow disturbances. It is then conceivable that enhancement of brain tissue necrosis by the release of NPY facilitates edema formation. The limited efflcacy of PP56 on vasogenic brain edema might thus be explained by these antagonistic, admittedly hypothetical mechanisms, namely inhibition of the blood flow reduction resulting from release of NPY and, thereby, of necrosis formation. Further information is certainly required on the function of NPY in secondary brain damage, such as vasogenic brain edema from ischemia and trauma.

Taken together, the present findings on novel methods of brain edema treatment using an experimental screening procedure not only are promising as far as enhancement of treatment specificity is concerned, but also pertinent as to the information obtained on mechanisms underlying edema formation, as e.g. activation of cotransport mechanisms across the blood-brain barrier.

References

1. Abel PW, Han C, Noe BD, McDonald JK (1988) Neuropeptide Y: vasoconstrictor effects and possible role in cerebral vasospasm after experimental subarachnoid hemorrhage. Brain Res 463: 250–258
2. Adrian TE, Allen JM, Bloom SR, Ghatewi MA, Rossor MN, Roberts GW, Croe TJ, Tatemoto K, Polak JM (1983) Neuropeptide Y distribution in the human brain. Nature 306: 584–586
3. Albright AL, Latchaw RE, Robinson AG (1984) Intracranial and systemic effects of osmotic and oncotic therapy in experimental cerebral edema. J Neurosurg 60: 481–489
4. Allen JM, Todd N, Crockard HA, Schon F, Yeats JC, Bloom SR (1984) Presence of neuropeptide Y in human circle of Willis and its possible role in cerebral vasospasm. Lancet 2: 550–552
5. Authi KS, Gustafsson T, Crawford N (1989) The effects of inositol-1,2,6-trisphosphate (PP56) on the calcium sequestration properties of human platelets. Thromb Haemos 62: 250–256
6. Claxson A, Morris C, Blake D, Sirén M, Halliwell B, Gustafsson T, Löfkvist B, Bergelin I (1990) The anti-inflammatory effects of D-myo-inositol-1,2,6-trisphosphate (PP56) on animal models of inflammation. Agents Actions 29: 68–70
7. Edvinsson L, Adamsson M, Jansen I (1990) Neuropeptide Y antagonistic properties of D-myo-inositol-1,2,6-trisphosphate in guinea pig basilar arteries. Neuropeptides 17: 99–105
8. Javaheri S (1991) Role of NaCl cotransport in cerebrospinal fluid production: effect of loop diuretics. J Appl Physiol 71: 795–800
9. McIntosh TK, Ferriero D (1992) Changes in neuropeptide Y after experimental traumatic brain injury in the rat. J Cereb Blood Flow Metab 12: 697–702
10. Plangger C (1992) Effect of torasemide on intracranial pressure, mean systemic arterial pressure and cerebral perfusion pressure in experimental brain edema of the rat. Acta Neurol Scand 86: 252–255
11. Pluta RM, Deka-Staarosta A, Zauner A, Morgan JK, Muraszko KM, Oldfield EH (1992) Neuropeptide Y in the primate model of subarachnoid hemorrhage. J Neurosurg 77: 417–423
12. Roberts PA, Pollay M, Engles C, Pendleton B, Reynolds E, Stevens FA (1978) Effects on intracranial pressure of furosemide combined with varying doses and administration rates of mannitol. J Neurosurg 66: 440–446
13. Schettini A, Stahurski B, Young HF (1982) Osmotic and oncotic loop diuresis in brain surgery. J Neurosurg 56: 679–684
14. Staub F, Peters J, Kempski O, Schneider GH, Schürer L, Baethmann A (1993) Swelling of glial cells in lactacidosis and by glutamate: significance of Cl–transport. Brain Res 610: 69–74
15. Wittner M, Di Stefano A, Schlatter E, Delarge J, Greger R (1986) Torasemide inhibits NaCl reabsorption in the thick ascending limb of the loop of Henle. Pflugers Arch 407: 611–614

Correspondence: Steffen Berger, M.D., Institute for Surgical Research, Klinikum Großhadern, Ludwig-Maximilians-University, Marchioninistrasse 15, D-81377 München, Federal Republic of Germany.

Acta Neurochir (1994) [Suppl] 60: 538–540
© Springer-Verlag 1994

Pharmacokinetic Study of Mannitol in Subjects with Increased ICP

T. Kobayashi, T. Ichikawa, R. Kondo, Y. Yoshiyama, F. Tomonaga, and **T. Ohwada**

School of Pharmaceutical Sciences, Kitasato University and Emergency and Critical Care Medicine, Kitasato University Hospital, Tokyo, Japan

Summary

Six human subjects received mannitol as an introvenous dip (DIV) infusion. In each subject, the concentration of mannitol and serum osmolality were rapidly elevated during mannitol administration and reached the maximum at the end of mannitol administration. Then, it exponentially decreased. There was a strong positive correlation between mannitol concentration and extrinsic serum osmolality. The integrated values of mannitol concentration difference between the central (Cc) and the peripheral compartment (Pc) correlated with the changes of ICP reduction during mannitol administration. The results indicate that the changes of ICP reduction depend on the mannitol concentration difference between Cc and Pc.

Keywords: Mannitol; pharmacokinetics; ICP; human.

Introduction

Mannitol is a well known hypertonic agent for the treatment of intracranial hypertension[5]. However, the development of an optimal dosage regimen has been limited by the absence of information regarding the relationship among mannitolpharmacokinetics and pharmacodynamics[3,4]. In an attempt to clarify the optimal dosage regimen of mannitol to obtain the most pronounced and longest reduction of raised ICP, serial determinations of serum concentration and ICP were performed in six human subjects receiving mannitol as an DIV infusion.

Subjects and Methods

Six human subjects (3 men and 3 women) aged average 43 years were administered mannitol as DIV infusion for the treatment of intracranial hypertension. Dose was 20% mannitol solution, 150 ml or 300 ml. Small baloons filled with sterile water were inserted epidurallay for monitoring ICP. In each subject, mannitol concentration (mg/ml), serum osmolality (mOsm/kg) were sequentially examined for 120 minutes. Blood samples were taken 5, 10, 20, 30, 60 and 120 minutes after the drug was given. The samples were stored frozen at –20°C until the time of assay. Serum concentrations of mannitol were measured by gas-liquid chromatography.

Results

Mannitol Concentration

In each subject, the serum concentration of mannitol rapidly increased during mannitol administration and reached the maximum at the end of mannitol administration. Then, it exponentially declined.

ICP Changes

In each subject, ICP slowly decreased after beginning of mannitol infusion. Then, the ICP gradually returned to the initial ICP level.

Serum Osmolality

Effective serum osmolality reflecting the effect of mannitol was calculated by subtracting the calculated osmolality from actual serum osmolality. In each subject, it reached a maximum at the end of mannitol, administration respectively, and then exponentially declined in parallel with the mannitol concentration curve.

Discussion

Changes of mannitol concentration showed a bi-exponential curve and best fitted to a two-compartment

Fig. 1. The time course of ICP (●– – – –●), mannitol concentration (●– – – ●), serum osmolality (●– - - ●) and sodium concentration (●——●)

Fig. 2. The time course of integrated values of mannitol concentration difference between the central and the peripheral compartment in relation to the change of ICP. (——) Integrated values, (▲- - -▲) change of ICP

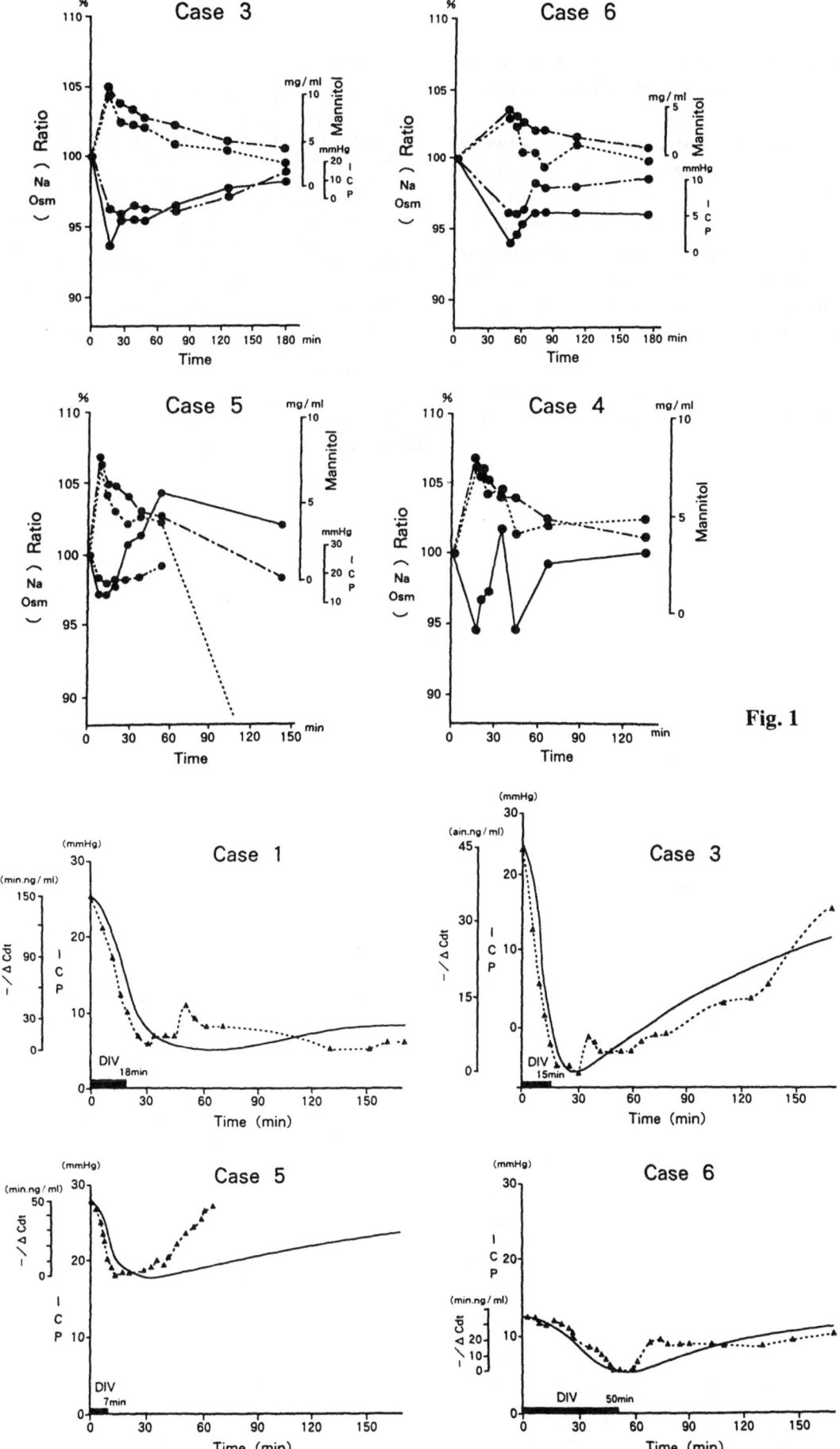

Fig. 1

Fig. 2

model. The pharmacokinetic parameters of mannitol in this study were similar to those which have been reported in humans[1,2].

There was a strong positive correlation between mannitol concentration and extrinsic serum osmolality. The integrated values of mannitol concentration difference between the central (Cc) and the peripheral compartment (Pc) correlated with the changes of ICP reduction during mannitol administration. The results indicate that the changes of ICP reduction depend on the mannitol concentration difference between Cc and Pc.

In clinical practice, the optimal dosage regimen may play an important role for producing effective ICP reduction.

References

1. Cloyd JC, Snyder BD, Cleeremans B, Bundlie SR (1986) Mannitol pharmacokinetics and serum osmolality in dogs and humans. J Pharmacol Exp Ther 236: 301–306
2. Domingnez R, Corcoran AC, Page IH (1947) Mannitol: kinetics of distribution, excretion, and utilization in human beings. J Lab Clin Med 32: 1192–1202
3. Goldwasser P, Fotino S (1984) Acute renal failure following massive mannitol infusion. Arch Intern Med 144: 2214–2216
4. Kullberg G, Sundbarg G (1976) Reduction of raised intracranial pressure recordings. In: Beks JWF, Bosch DA, Brock M (eds) Intracranial pressure 3. Springer, Berlin Heidelberg New York
5. Shenkin HA, Goluboff B, Haft H (1962) The use of mannitol for the reduction of intracranialpressure in intracranial surgery. J Neurosurg 19: 897–901

Correspondence: Teruaki Kobayashi, M. D., School of Pharmaceutical Sciences, Kitasato University, 9-1, Shirokene 5 Chome, Minato-ku, Tokyo 108, Japan.

Acta Neurochir (1994) [Suppl] 60: 541–543
© Springer-Verlag 1994

Systemic Management of Cerebral Edema Based on a New Concept in Severe Head Injury Patients

N. Hayashi, T. Hirayama, A. Udagawa, W. Daimon, and **M. Ohata**

Nihon University Memorial Critical Care and Emergency Center, Oyaguchi Kami-machi, Itabashi-ku, Tokyo, Japan

Summary

Cerebral hypothermia treatment of critical brain injury patients was studied based on the management and control of cerebral thermo-pooling, synaptic excitation, hypermetabolic demand, and the systemic critical condition of the metabolic reserve. The initial pathophysiological changes after trauma included a progressive increase in brain tissue temperature. Such cerebral thermo-pooling, which reached a maximum of 43.8°C, can change or damage the vascular proteins directly.

The brain tissue temperature was influenced by four factors: 1. the cerebral metabolism, 2. the systemic excess energy metabolism, 3. the CPP that carries the systemic energy to the brain tissue, and 4. the cerebral blood flow that leads to washout of brain tissue temperature. Mild cerebral hypothermia (32–33°C) managed by the whole body compartment cooling technique in the critical conditions of diffuse brain injury patients (GCS < 4) produced a good recovery in 8 of 10 patients. Continuous monitoring of the jugular venous oxygen saturation and BTT/TMT was effective for evaluating cerebral ischemia and oxygen metabolic disturbances even during cerebral hypothermia treatment.

Keywords: Diffuse brain injury; cerebral hypothermia; brain tissue temperature; intracranial hypertension.

Introduction

Intracranial hypertension following severe brain injury continues to remain a serious problem affecting borth morbidity and mortality. The major causes of such seriously elevated levels of intracranial pressure are mainly related to either vascular engorgement (brain swelling) or brain edema. In recent studies, we demonstrated that the cerebral perfusion pressure, systemic oxygen metabolism and thermoregulation can modify the changes of vascular tone, permeability and cell membrane damage in ischemic injury, resulting in brain edema and swelling[5]. In this paper, we present a computed management method for critical brain injury patients based on a new concept of systemic cerebral metabolic-thermoregulation.

Materials and Methods

The clinical studies were divided into two groups. In group A, the effects of systemic thermoregulation, oxygen metabolism, and hemodynamic changes on the development of cerebral edema were examined in 52 critical patients (GCS < 8). In group B, the effect of mild cerebral hypothermia (32–33°C) for preventing brain edema and thermal dysregulation was examined in 10 cerebral trauma patients with dilated pupils (GCS < 4).

All of the patients underwent monitoring of their intracranial pressure (ICP), jugular venous bulb O_2 saturation (SJO$_2$, systemic mixed venous O_2 saturation (SVO$_2$), systemic arterial O_2 saturation (SaO$_2$), cardiac output (CO), brain tissue temperature (BTT), tympanic membrane temperature (TMT), bladder temperature (BT), pulmonary arterial pressure, and end-tidal CO_2. The cardiac output was measured 4 times a day by the thermodilution technique to calibrate the respiratory K for calculation of the CO value based on Fick's theory. The oxygen delivery (DO$_2$), systemic oxygen consumption (VO$_2$), oxygen extraction ratio (O$_2$ER), cerebral perfusion pressure (CPP), systemic vascular resistance (SVR), systemic oxygen metabolic reserve $\{SO_2MR = [(VO_2 - \text{initial } VO_2) / \text{initial } VO_2] / [(DO_2 - \text{initial } DO_2) / \text{initial } DO_2]\}$ and dynamic cerebral oxygen metabolic reserve $\{CO_2MR - [(SaO_2 - SJO_2) - \text{initial } (SaO_2 - SJO_2)] / \text{initial } (SaO_2 - SVO_2) / [(CPP - \text{initial } CPP) / \text{initial } CPP]\}$[4], cerebral thermal index (CTI = BTT / TMT), and cerebral thermal distribution (CTD = TMT / BT) were computed automatically, and the data were presented as a graphic or table on the bedside work station display of an EMTEK computer management system.

Results and Discussion

The computed monitoring system provided a new technique for the management of the cerebral blood flow, cerebral oxygen metabolism, systemic hemodynamic-metabolic changes, and their correlations at each bed side in real time.

1. Cerebral Thermodysregulation

The initial pathophysiological changes after head injury included progressive brain edema followed by thermodysregulation in the brain tissue. The brain tissue temperature rose to 40–43.8°C at 17–30 hours after surgery and trauma. Such specific pathophysiological changes, viz, *cerebral thermo-pooling*, were triggered by hypotension and activated with increasing CPP and SO_2MR. The specific thermo-pooling phenomenon could not be controlled successfully by previous ICP management methods. Only elevation of the MABP (> 65 mmHg) could reduce the cerebral thermo-pooling by brain temperature washout with an increased cerebral blood flow. The typical cerebral thermo-pooling phenomenon was observed in patients with severe diffuse brain injury associated with shock (Fig. 1). Elevation of the brain tissue temperature above 42°C could change the neuronal and vascular proteins directly. A new post-reperfusion injury mechanism was identified in these studies.

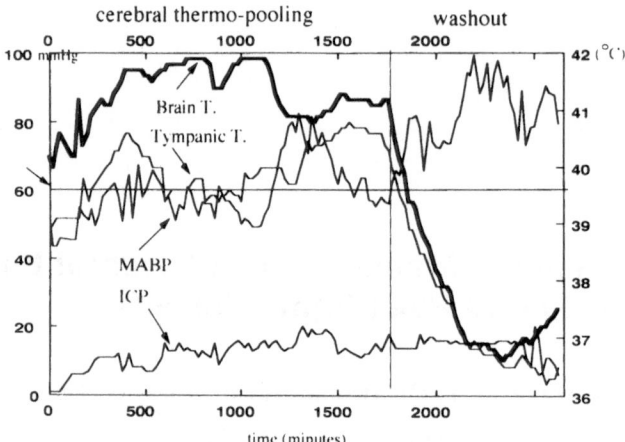

Fig. 1. Cerebral thermo-pooling after diffuse brain injury with primary brain stem injury. The brain tissue temperature increased to 41.8°C during the initial 30 hours after injury. This elevated brain tissue tempe;ature was washed out by increasing the mean arterial blood pressure to a level in excess of 65 mmHg

2. Cerebral Thermo-Pooling and Oxygen Metabolism

Our preliminary studies demonstrated that the brain tissue temperature and CTI changed in inverse correlation with the cerebral blood flow. The changes in BTT/TMT ratio and SJO_2 were therefore, indirectly evaluated as conditions reflecting the cerebral blood flow and cerebral oxygen metabolism, respectively.

At the early stage with thermo-pooling, SJO_2 changed to various levels. Simultaneous monitoring of SJO_2 and CTI indicated that the basic patterns of cerebral blood flow and oxygen metabolism in critical head injury patients progressed the in following stages: I, ischemia and hypoxia (progressive increase of SJO_2 and elevation of CTI); II, hypermetabolism (progressive lowering of SJO_2 and mild reduction of CTI); III, ischemia with hypometabolism or no metabolic changes (acute elevation of CTI with or without elevation of SJO_2); IV, temporary recovery (lowering of CTI and SJO_2); and V, progressive metabolic distur-

bances with delayed cerebral ischemia in cases with poor prognosis, manifested as acute elevation of SJO_2 with delayed elevation of CTI (Fig. 2). Such indirect evaluations of the cerebral blood flow and metabolism were also useful during cerebral hypothermia treatment.

3. Cerebral Mild Hypothermia Treatment of Critical Head Injury Patients

Cerebral hypothermia was produced by whole-body cooling throughout a compartment cooling jacket which was developed at our Critical Care Center using a Branketrol II-model (Cincinnati-Sub Zero Products, Inc., 12011 Mosteller, Cincinnati, Ohio 45241). The major negative biological responses were hypopotassemia (2–3 mEqt/dl), thrombocytopenia (10–11 × 10^4/mm³), and a lower CO value (1.3–2.5 L/min). The

Table 1. *Prognosis of Patients with Acute Subdural Hematoma in a Condition with Non Reactive Bilateral Dilated Pupils*

Autors	Year	No. of cases	Prognosis (no. of cases) (%)		
			GR/MD	SD/VS	D
Becker *et al.*[1]	1977	23	5 (22%)	3 (13%)	15 (65%)
Bricolo *et al.*[2]	1980	59	–	–	45 (76%)
Seeling *et al.*[6]	1981	37	–	–	28 (76%)
Hayashi *et al.*	1993	8	5/1 (75%)	0 (0%)	2 (25%)
Coma with anisocolia		2	2/0	0	0
Total		10	7/1 (80%)	0 (0%)	2 (20%)

GR good recovery; *MD* mild disability; *SD* severe disability; *VS* vegetable state; *D* death.

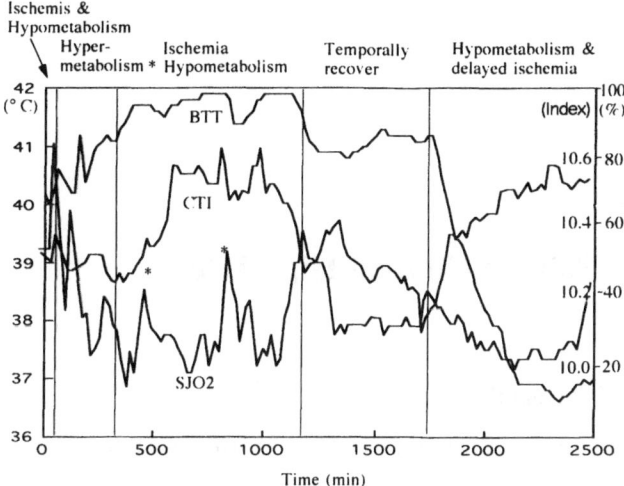

Fig. 2. The typical intracerebral pathophysiological changes in a critical diffuse brain injury patient. Cerebral ischemia and hypoxia at stage I, hypermetabolism at stage II, cerebral ischemia with or without hypometabolism at stage III, temporary recover at stage IV and progressive worsening of metabolism that follow delayed cerebral ischemia could be observed in the diffuse brain injury patient (GCS = 3)

vital signs were stabilized; and epileptic discharges, the Hoffman reflex, and C-reflex disappeared at a brain tissue temperature of 32–33°C. In this series of mild cerebral hypothermia treatment of critical diffuse brain injury patients with acute subdural hematoma who had bilaterally dilated pupils, 2 patients died and 8 patients survived. The prognosis was found to be excellent in 7 of the surviving patients. All 7 patients showed a good recovery at 3 months after treatment and the other surviving patient exhibited mild disability after 1 month (Table 1). No trouble was observed in coping with an ordinary life style. Previous papers have revealed high mortality and morbidity in these critical head injury patients[1,2,6]. The problems encountered with the mild cerebral hypothermia treatment included an unstable brain tissue temperature in a patient with a high body weight (> 80 kg) and vasoparalysis at the rewarming stage.

The changes in brain tissue temperature were influenced by four factors: 1. the cerebral metabolism, 2. the systemic excess energy metabolism, 3. the CPP that carries the systemic energy to the brain tissue, and 4. the cerebral blood flow leads to washout of the brain tissue temperature. The, compartment whole body cooling technique with stabilization of the brain tissue temperature at exactly 32–33°C and the systemic O_2ER is more useful than selective cooling of the head.

Recent studies have demonstrated that protection from glutamate release after cerebral ischemia can be achieved by cerebral hypothermia at 33°C[3]. The mechanism of successful treatment by application of mild cerebral hypothermia for injured brain tissue is based on protection against thermo-pooling, non-release of glutamate, lack of synaptic excitation, a low metabolic demand, stabilization of the vital signs and a low systemic energy transport into the brain.

References

1. Becker DP, Miller JD, Ward JD, Greenberg RP, Young HF, Sakalas R (1977) The outcome from severe head injury with early diagnosis and intensive management. J Neurosurgery 47: 491–502
2. Bricolo A, Turazzi S, Feriotti G(1980) Prolonged post-traumatic unconsciousness. J Neurosurg 52: 625–634
3. Busto R, Deitolich WD, Glogus MY-T, Martinez E, Valdes I, Ginsberg MD(1989) Effect of mild hypothermia on ischemia-induced release of neurotransmitters and free fatty acids in rat brain. Stroke 20: 904–910
4. Cruz J, Miner ME, Allen SJ, Alves WM, Gannarelli TA (1991) Continuous monitoring of cerebral oxygenation in acute brain injury: assessment of cerebral hemodynamic reserve. Neurosurgery 29: 743–749
5. Hayashi N, Hirayama T, Utagawa A, Ohata M (1993) Computed monitoring of systemic-cerebral thermal distribution, circulation and oxygen metabolism of the critical emergency patient in ICU. J Clin Monit, in press
6. Seeling JM, Greenberg RP, Becker DP, Miller JD, Choi SC (1981) Reversible brain stem dysfunction following acute subdural hematoma. J Neurosurg 55: 516–523

Correspondence: Nariyuki Hayashi, M.D., Nihon University Critical and Emergency Center, 30-1 Oyaguchi Mami-Machi, Itabashi-ku, Tokyo 173, Japan.

Acta Neurochir (1994) [Suppl] 60: 544–546
© Springer-Verlag 1994

Propofol Sedation in Severe Head Injury Fails to Control High ICP, but Reduces Brain Metabolism

L. Stewart, R. Bullock, C. Rafferty, W. Fitch, and **G. M. Teasdale**

Departments of Neurosurgery and Anaesthesia, Institute of Neurological Sciences, Southern General Hospital, Glasgow, U.K.

Summary

We have compared the effects of an intravenous infusion of propofol with those of morphine and midazolam on global brain metabolism ($AVDO_2$) and brain perfusion following severe head injury. Fifteen patients were sedated with either a continuous infusion of propofol (mean rate 232 mg/h, range 150–400 mg/h) or infusions of morphine (mean rate 2.3 mg/h, range 0–4 mg/h) and midazolam (mean rate 2.8 mg/h, range 0–5 mg/h). Both groups were well matched for sex, age and level of coma (Glasgow coma scale) prior to sedation.

Continuous data collection of $AVDO_2$, mean arterial blood pressure (MABP), intracranial pressure (ICP), and cerebral perfusion pressure (CPP) began at 12 hours post injury and continued for a mean period of 40 hours. Morphine and midazolam did not have a significant effect on any of the measured parameters. Propofol led to a fall in $AVDO_2$ from 6.0 ± 2.6 ml/dl to 3.0 ± 0.6 ml/dl at 4 hours. However, there was no effect on MABP, ICP or CPP. Outcome was similar in the 2 groups.

Keywords: Sedation; propofol; head injury; cerebral metabolism.

Introduction

Propofol depresses cerebral metabolism and oxygen consumption[8]. In combination with its rapid onset and short half-life, this should make it an ideal agent for sedation in head injury. However, when given as an intravenous bolus to patients with head injuries, propofol has been shown to reduce cerebral perfusion pressure and jugular venous oxygen saturation (SjO_2)[1].

We have compared the effects of an intravenous infusion of propofol with those of conventional opiate/benzodiazepine sedation on global cerebral metabolism, cerebral perfusion and ICP following severe head injury.

Materials and Methods

15 patients with severe head injury were studied. Each was intubated and ventilated as part of their clinical management. A radial arterial cannula, fibreoptic jugular bulb catheter (Oximetrix 3 system, Abbott Laboratories) and a fibreoptic intraparenchymal pressure transducer (Camino Laboratories) were inserted in all patients on admission to the intensive care unit.

9 were sedated with an infusion of propofol at a mean infusion rate of 232 mg/hour (range 150–400 mg/h). 6 patients sedated with infusions of morphine (mean rate 2.3 mg/h, range 0–4 mg/h) and midazolam (mean rate 2.8 mg/h, range 0–5 mg/h) acted as controls. Ventilation was controlled to maintain $PaCO_2$ between 3.5–4.5 KPa. Both groups were well matched for age, sex and level of coma (Glasgow Coma Scale) prior to admission (Table 1).

Physiological Measurements

The following parameters were continuously monitored from 12 hours post injury for a mean of 40 hours-arterial blood pressure via radial artery cannula, ICP, arterial oxygen saturation (SaO_2) and SjO_2. All parameters were time-locked and stored in a computerised data base. CPP was calculated from MABP-ICP and $AVDO_2$ was calculated 2 hourly from SjO_2, SaO_2 and haemoglobin concentration (Corning 2500 Co-oximeter.)

Data Analysis

Means and standard errors are given. Wilcoxon signed rank sum test has been applied to within group changes from baseline. $p < 0.05$ was considered to be significant.

Table 1. *Details of Patients Studied (Mean Range)*

		Propofol (n = 9)	Morphine and midazolam (n = 6)
Age (years)		30.5 (12–62)	30.5 (12–57)
Sex (M : F)		8 : 1	6 : 0
Admission GCS	3–8	7	5
	9–12	2	1
	13–15	0	0
Injury: focal		4	5
diffuse		5	1

Results

Morphine and midazolam had no effect on MABP, ICP, CPP or $AVDO_2$. Propofol led to a fall in $AVDO_2$ from 6.3 ± 2.6 ml/dl to 3.0 ± 0.6 ml/dl at 4 hours

Table 2. *Outcome at 6 Months in Each Group*

	Propofol (n = 9)	Morphine and midazolam (n = 6)
Good outcome	3 (33%)	2 (33%)
Moderate disability	2 (22%)	3 (50%)
Severe disability	2 (22%)	0 (0%)
Persistent vegetative	0 (0%)	0 (0%)
Dead	2 (22%)	1 (17%)

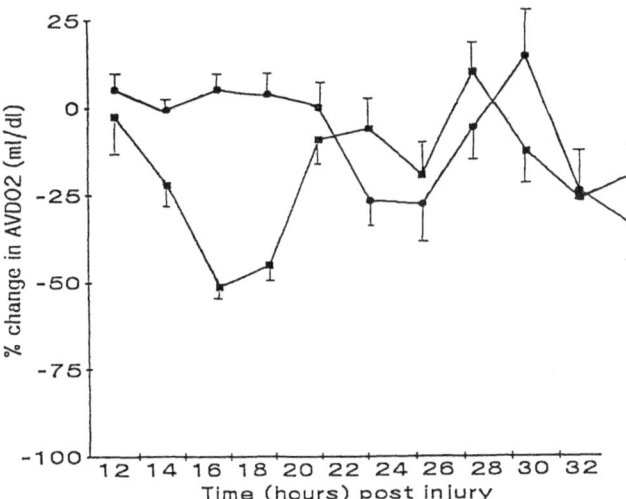

Fig. 1. Changes in AVDO2 following propofol ■ or morphine and midazolam ●. There was a statistically significant reduction from baseline in the propofol group at 16 hours post injury (4 hours on propofol ($p < 0.02$))

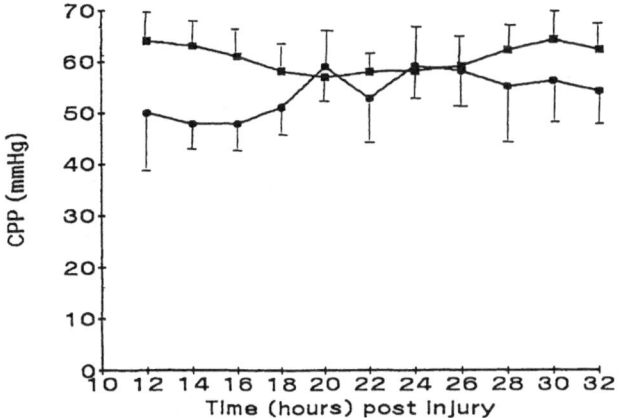

Fig. 2. Changes in CPP following propofol ■ or morphine and midazolam ●

($p < 0.02$), rising again towards pre-propofol levels at 8 hours (Fig. 1). Propofol had no effect on raised ICP (22 mmHg prior to propofol and 20 mmHg after 40 hours) and MABP and CPP (Fig. 2) were unchanged. Outcome did not differ between the 2 groups (Table 2).

Discussion

Sedation is essential in head injured patients requiring ventilation, but the "traditional" sedatives have disadvantages. Barbiturates and benzodiazepines have long half lives, with consequent delayed neurological recovery, and 'masking' of neurological status. Thiopentone also has a high incidence of side effects[7], including hypotension, hypokalaemia and hepatic and renal dysfunction. Propofol has become a popular neuro-sedative agent as it has a rapid onset and short duration of action. When used as an infusion there is lack of accumulation and rapid recovery. It may have neuroprotective effects by reducing cerebral metabolic rate for oxygen ($CMRO_2$)[8], and attenuating damage due to free oxygen radicals[5]. Hartung[3] suggested that it may also reduce ICP, but this has not been found in this study or by other workers[2,4]. Benefits offered by propofol have to be balanced against potential cardiac depressant effects[6] which may reduce MABP and CPP. This has been found to occur during intravenous bolus administration[1], but we found no significant effects during prolonged intravenous infusion.

In this study the reduction in $AVDO_2$ did not continue past the first 8 hours of a continuous infusion. We postulate that the initial fall in $AVDO_2$ is due predominantly to a fall in $CMRO_2$. The fact that $AVDO_2$ changes are not sustained may be due to either the development of "tolerance" to the $CMRO_2$ reducing effect of propofol or to a secondary, delayed reduction in cerebral blood flow. This merits further investigation. We conclude that propofol is a useful agent for sedation in head injury, however it has no lowering effect on raised ICP and the cerebroprotection offered by the reduction in $CMRO_2$ may only be transient.

References

1. Andrews PJD, Dearden NM, Miller JD (1991) Comparison of thiopentone and propofol at two rates of administration in patients with severe head injury. Br J Anaesth 67: 212
2. Farling PA, Johnston JR, Coppel DL (1989) Propofol infusion for sedation of patients with head injury in intensive care: a preliminary report. Anaesthesia 44: 222–226
3. Hartung HJ (1987) Effect of propofol on intracranial pressure: preliminary results. Anaesthesia 36: 66–68

4. Mergaert C, Herregods L, Rolly G, Collardyn F (1991) The effect of 24 hour propofol or fentanyl sedation on intracranial pressure. Eur J Anaesthesiol 8: 324–325

5. Murphy PG, Myers DS, Davies MJ, Webster NR, Jones JG (1992) Antioxidant potential of propofol (2,6,di-isopropyl phenol). Br J Anaesth 68: 613–618

6. Puttick RM, Diedericks J, Sear JW, Glen JB, Foex P, Pyder WA (1992) Effects of graded infusion rates of propofol on regional and global left ventricular function in the dog. Br J Anaesth 69: 375–381

7. Schalen W, Messeter K, Niordstrom CH (1992) Complications and side effects during thiopentone therapy in patients with severe head injury. Acta Anaesth Scand 3: 369–377

8. Stephan H, Sonntag H, Schenk HD, Kohlhausen S (1987) Effects of propofol on cerebral blood flow, cerebral oxygen consumption and cerebral vascular reactivity. Anaesthesia 36: 60–65

Correspondence: L. Stewart, M. D., Department of Neurosurgery and Anaesthesia, Institute of Neurological Sciences, Southern General Hospital, 1345 Goven Road, Glasgow, U.K.

Acta Neurochir (1994) [Suppl] 60: 547–549
© Springer-Verlag 1994

High Colloid Oncotic Therapy for Contusional Brain Edema

H. Tomita, U. Ito, O. Tone, H. Masaoka, and **B. Tominaga**

Department of Neurosurgery, Musashino Red-Cross Hospital, Tokyo, Japan

Summary

We investigated whether prolonged high colloid oncotic therapy for two weeks can suppress contusional brain edema. Eighteen patients with cerebral contusion were randomly divided into two groups of patients receiving high oncotic pressure (HOP; 26–30 mmHg) treatment and those receiving normal oncotic pressure (NOP; 22–26 mmHg) treatment. Oncotic pressure was maintained for two weeks with administration of a 25% albumin solution with additional use of furosemide. Edema volume was calculated by summation of all measured low-density areas in each CT slice multiplied by 1.0 cm of slice of thickness. We expressed contusional brain edema volume as a percent increase based on each patient's initial CT. The mean percent increase of contusional brain edema in the NOP group was significantly higher than that in the HOP group at 9–15 days (208.9% and 14.0%, respectively) and 16–25 days (188.8% and 10.0%, respectively). There were no complications such as heart failure or renal failure during treatment. All the patients in the HOP group recovered with minimal or no neurological deficit. On the other hand, 30% of patients in the NOP group remained in poor condition. With frequent measurement of oncotic pressure and adjustment of fluids and electrolytes, continuous oncotic therapy for two weeks effectively and safely reduced contusional brain edema.

Keywords: Brain edema; cerebral contusion; oncotic pressure; treatment.

Introduction

The effectiveness of colloid oncotic therapy for brain edema is still controversial. In past experimental studies, the duration of treatment has been less than 8 hours, and long-term treatment has not been studied[1,3,7]. In clinical studies, oncotic therapy has been employed to decrease intracranial pressure (ICP) in patients with increased ICP, but the maximum duration of treatment was 24 hours and results varied widely from effective to not effective[2,5,8].

Various osmotic diuretics, such as mannitol and urea have been used to lower ICP. However, the effect lasts only a few hours, and ICP rises again thereafter. Moreover, osmotic therapy could cause electrolyte imbalance and hypovolemia when repeatedly used.

The present study aims to investigate whether prolonged oncotic therapy can suppress contusional brain edema, by administering albumin to maintain an increased oncotic gradient between blood and parenchymal tissue for two weeks.

Material and Methods

Eighteen patients (13 males and 5 females) who had sustained closed brain injury and had cerebral contusion were studied. Severity of head injury was evaluated on admission according to the Glasgow Coma Scale (GCS) score. Contusional brain edema was diagnosed by CT scans taken within 2 days after injury. The patients were randomly divided into two groups. In the normal oncotic pressure (NOP) group, 10 patients were treated to keep the serum oncotic pressure between 22 and 26 mmHg. In the high oncotic pressure (HOP) group, 8 patients were treated to maintain the serum oncotic pressure between 26 and 30 mmHg. The treatment protocol included administration of 25% albumin (50–100 ml/day) with supplementary use of furosemide (20–40 mg/day). All the patients received glycerol (600 ml/day). Serum oncotic pressure was measured daily using a WESCOR 4400 colloid osmometer (Wescor, Logan, Utah). Fluids and electrolytes were carefully controlled and osmolality was maintained around 300 mOsm/liter in both groups. This treatment was continued for 2 weeks.

Axial scan using the GE 9200 CT scanner was repeated at 1–2, 3–5, 6–8, 9–15, 16–25 days after head injury. The cross-sectional area of low density adjacent to a cortical contusion with mixed density was planimetrically measured in each CT slice (1.0 cm in thickness) using the Color Image Analyzer (Image Command 5098, Olympus Optical Co. Ltd.). Edema volume was calculated by summation of all measured low-density areas in each CT slice multiplied by 1.0 cm of slice thickness. The brain edema volume was expressed by percent increase based on that of each patients initial CT. A temporal profile of percent increase of the edema volume was compared between two groups for three weeks. A temporal profile of the oncotic pressure and osmolality on the day when CT scan was taken was also compared between two groups during the same period. The outcome of the patients of both groups was evaluated using the Glasgow outcome scale 6 months after injury. All measurements are expressed as mean ± standard error of the mean. Oncotic pressure, osmolality and percent increase of contusional brain edema were compared between two groups using paired Student's t-test. Informed consent was obtained from all patients prior to the treatment.

Result

The age of the 18 patients ranged from 13 to 83 years (mean: 48.4 ± 5.3 years). The mean GCS score of the NOP group was 12.0 ± 3.7 and that of the HOP group was 11.9 ± 2.8. There was no significant difference in GCS score between the two groups (Table 1). Oncotic pressure of the NOP group was highest (27.1 ± 1.3 mmHg) 3–5 days after injury and stayed between 23 and 26 mmHg thereafter. Oncotic pressure of the HOP group was kept between 27 and 30 mmHg during the two weeks following injury (Fig. 1 top). The difference in oncotic pressure between the two groups was significant at 6–8 days and 9–15 days after injury. Osmolality was not significantly different between the two groups and was maintained around 300 mOsm/liter for two weeks after injury (Fig. 1 bottom).

Initial edema volume was 18.8 ± 3.8 cm³ and 32.7 ± 10.3 cm³, respectively, in the NOP and the HOP group, and there was no significant difference between them. The mean percent increase of the edema volume in the NOP group increased to 208.6 ± 38.3% 9–15 days after

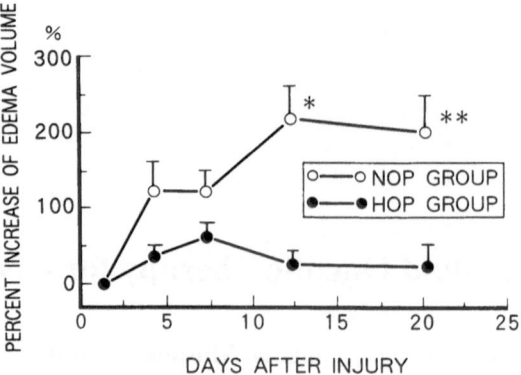

Fig. 2. Temporal profile of percentage increase of edema volume of high oncotic pressure (*HOP*) group and normal oncotic pressure (*NOP*) group. * p < 0.005, ** p < 0.05. Values are mean ± standard error of the mean

injury, while the maximum percent increase of contusional brain edema in the HOP group was 51.0 ± 18.5% 6–8 days after injury (Fig. 2); and there was significant difference between the two groups at 9–15 days and 16–25 days after injury. No systemic complications such as heart failure or renal failure were observed during this procedure.

All patients in the HOP group recovered with minimal or no neurologic deficit. In the NOP group, however, although seven patients (70%) recovered with minimal or no neurologic deficit, one (10%) died and two (20%) patients were severely disabled and in a vegetative state (Table 1).

Discussion

Oncotic therapy slowly extracts water from the extravascular brain parenchyma into the blood. Calculat-

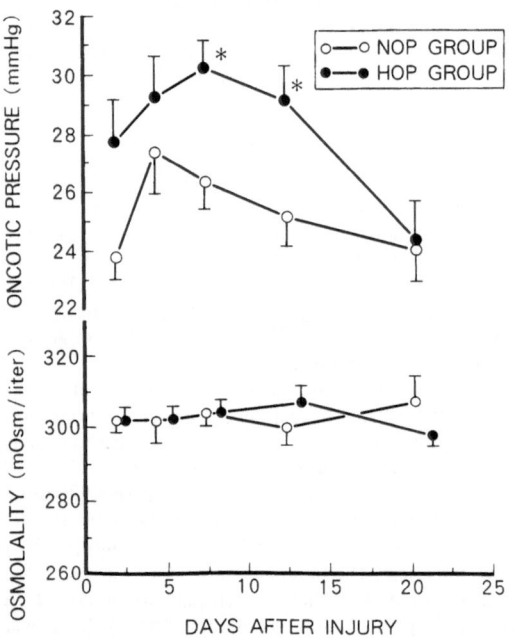

Fig. 1. Top: Temporal profile of oncotic pressure of the high oncotic pressure (*HOP*) group and the normal oncotic pressure (*NOP*) group. * p < 0.05. Values are mean ± standard error of the mean. Bottom: Temporal profile of osmolality of the high oncotic pressure (*HOP*) group and the normal oncotic pressure (*NOP*) group. Values are mean ± standard error of the mean

Table 1. *Clinical Summary*

	High oncotic pressure group	Normal oncotic pressure group
Number of patients	8	10
Glasgow coma score on admission	11.9 ± 1.0	12.0 ± 1.2
Glasgow outcome scale		
Good recovery	5	6
Moderate disability	3	1
Severe disability	0	1
Vegetative state	0	1
Dead	0	1

ing from Starling's law[9], approximately 20 ml/day of water can be extracted from the normal brain tissue to blood when the oncotic gradient is kept 5 mmHg higher than the normal level. As oncotic therapy is safe over a long period, when the same oncotic gradient situation is kept for 10 days, about 200 ml of water can be extracted from the brain into the blood during 10 days.

In the present study, oncotic pressure is kept significantly different between the two groups for two weeks after injury. On the other hand, osmolality is kept around 300 mOsm/l by adjusting fluid and electrolytes, and there is no significant difference in osmolality between the two groups. Thus the therapeutic effect of the contusional brain edema volume can be attributed to the oncotic pressure difference.

Albumin contributes 80% of the oncotic pressure of plasma protein[10]. The molecular weight of albumin is 66,300–69,000 and albumin stays in the blood for a long period and does not extravasate from cerebral vessels where the blood-brain-barrier (BBB) exists.

Although albumin extravasates the damaged BBB, it is unlikely that albumin extravasates the BBB so that the gradient of the oncotic pressure between blood and brain parenchyma is diminished greatly[4,6]. In addition, improvement of rCBF in the edematous tissue can be predicted, by albumin infusion, to enhance resolution of brain edema. Therefore, this therapy seems to prevent further edema development after start of this therapy (Fig. 2).

Administration of albumin increases plasma volume and the subsequent hypervolemia may induce heart failure. Furosemide is occasionally given to normalize plasma volume and to increase the oncotic pressure by hemoconcentration, but not to induce hypovolemia,

systemic dehydration and electrolyte imbalance[1]. In conclusion, with frequent measurement of oncotic pressure and adjustment of fluids and electrolyte, prolonged oncotic therapy effectively reduces the contusional brain edema. Oncotic therapy can be continued safely for two weeks.

References

1. Albright AL, Latchaw RE, Robinson AG (1984) Intracranial and systemic effects of osmotic and oncotic therapy in experimental cerebral edema. J Neurosurg 60: 481–489
2. Bakay L, Lee JC (1965) Cerebral edema. Thomas, Springfield, III, p 132
3. Clasen RA, Prouty RR, Bingham WG, Uartin FA, Hass GU (1982) Treatment of experimental cerebral edema with intravenous hypertonic glucose, albumin, and dextran. Surg Gynecol Obstet 104: 591–606
4. Clasen RA, Bezkorovainy A, Pandolfis S (1982) Protein and electrolyte changes in experimental cerebral edema. J Neuropathol Exp Neurol 41: 113–128
5. Fender FA, MacKenzie AS (1948) Effect of albumin solution on cerebrospinal fluid pressure. Arch Neuro Psychiatry 59: 529–531
6. Gazendam J, Go KG, van Zanten AK (1979) Composition of isolated edema fluid in cold-induced brain edema. J Neurosurg 51: 70–77
7. Kaieda R, Todd Uu, Cook LN, Warner DS (1989) Acute effects of changing plasma osmolality and colloid oncotic pressure on the formation of brain edema after cryogenic injury. Neurosurgery 24: 671–678
8. Uiyasaka Y, Nakayama K, Uatsumori K, Beppu T, Tanabe T (1983) Albumin (oncotic) therapy for patients with increased intracranial pressure. No-Shinkei-Geka 11: 947–954
9. Starling EH (1986) On the absorption of fluids from the connective tissue spaces. J Physiol (Lond) 9: 312–326
10. Tullis JL (1977) Albumin: background and use. JAMA 237: 355–360

Correspondence: Hiroki Tomita, M. D., Department of Neurosurgery, Musashino Red-Cross Hospital, 1-26-1 Kyonan-cho, Musashino-shi, Tokyo 180, Japan.

Acta Neurochir (1994) [Suppl] 60: 550–551
© Springer-Verlag 1994

Pharmacokinetics and Pharmacodynamics of Serum Glycerol in Patients with Brain Edema: Comparison of Oral and Intraveneous Administration

H. Tsuyusaki[1], **R. Kondo**[1], **S. Murase**[1], **Y. Yoshiyama**[3], **T. Kobayashi**[1], **H. Takagi**[4], **Y. Kitahara**[2], **S. Morii**[2], **F. Tomonaga**[1], and **T. Ohwada**[2]

[1] Department of Pharmacy, [2] Emergency and Critical Care Medicine, [3] School of Phamaceutical Science, Kitasato University Hospital, Kanagawa, and [4] Yamato Municipal Hospital, Yamato, Japan

Summary

To clarify the pharmacokinetics and effect of ICP reduction of glycerol administered orally, the serum concentration of glycerol was measured by the enzymatic method and ICP was measured by the subdural balloon method in severe head injured patients with brain edema and increased ICP. Sequential change of serum glycerol concentration and its relationship to the reduction of ICP were analyzed. The results showed that the pharmacokinetics of glycerol through oral administration were similar to that of intravenous glycerol administration and the changes of ICP were also similar to that of intravenous glycerol administration. We determined that glycerol can be administered either per oral or per venous to obtain the same results for treatment of brain edema with raised ICP.

Keywords: Glycerol; pharmacokinetics; brain edema; ICP.

Introduction

Intravenous administration of glycerol is widely used for the treatment of brain edema[1] with increased intracranial pressure (ICP). In our institute, oral administration of glycerol has been employed clinically to control increased ICP. We investigated the pharmacokinetics and the effect of oral glycerol administration on increased ICP. The results were compared with those obtained through intravenous glycerol administration to determine if there were any differences.

Materials and Methods

Seven severe head injured patients with brain edema whose ICP was monitored by the subdural balloon method were subjects for this study. The average ICP was 31 mmHg and 0.5 g/kg of 50% glycerol was administered in 1 minute through a naso-gastric tube (NGT) in 3 patients and through an entero-duodenal tube (EDT) in 4 patients.

Separately, 0.25 g/kg of 10% glycerol was administered intravenously within 15 minutes. After the glycerol administration, ICP was continously monitored and serum glycerol concentration was sequentially (1, 2, 3, 4, 5, 6, 7, 8, 9, 10, 15, 20, and 30 min) measured by the enzymatic method[3] using Glycerol-Lipid Test (Behringer-Manhaim). The pharmacokinetics of glycerol were calculated and analysed using the MULTI computer program.

Results

The time course of serum glycerol concentration after oral administration showed rapid elevation of serum glycerol level and reached a maximum (Cmax) within 10 minutes (Tmax).

Table 1. *Pharmacokinetics Parameters After oral Glycerol Administration*

Route of administration	Parameters			
	Tmax (min)	Cmax (µg/ml)	Ka (– h)	Δ ICP (mmHg)
Naso-gastric tube (n = 3)	8 (8–10)	823.3 (290–1500)	0.127 (0.252–0.024)	17.7 (14–20)
Entero-duodenal tube (n = 4)	6.7 (5–10)	1720 (960–3400)	0.218 (0.139–0.351)	25.0 (18–36)

Ka apparent first-order absorption rate constat, *Cmax* maximum concentration, *Tmax* time to reach the maximum glycerol concentration, *ICP* maximum reduction of ICP after oral glycerol administration.

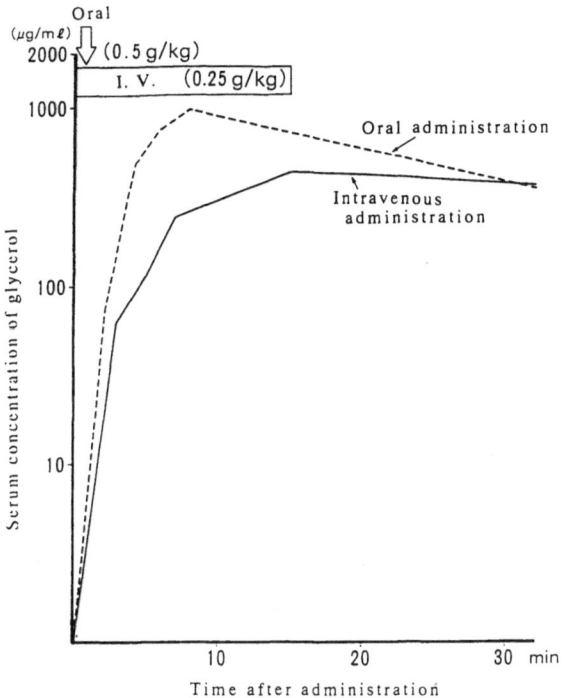

Fig. 1. Time course of serum glycerol concentration after glycerol administration (oral vs. intravenous)

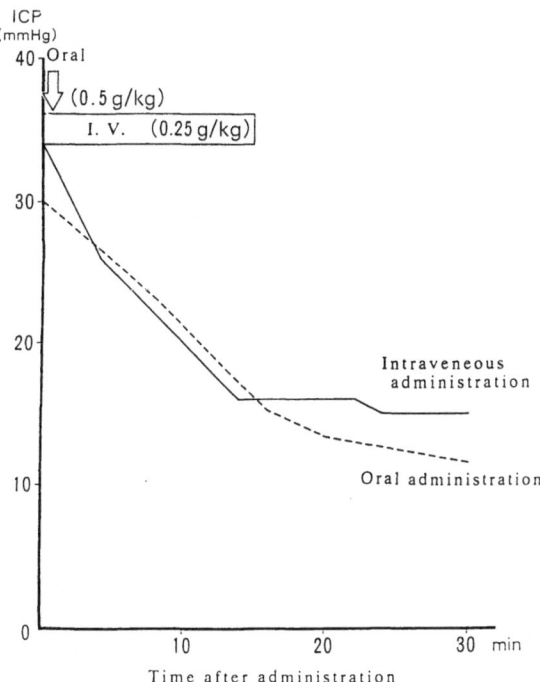

Fig. 2. Time course of ICP after glycerol administration (oral vs. intravenous)

This serum glycerol concentration change was more rapid and higher than that of intravenous glycerol administration (Fig. 1).

The ICP time course showed an immediate decrease of ICP and reached a maximum ICP reduction within 30 minutes.

This pattern of ICP change was quite similar to that of intravenous administration (Fig. 2). Pharamacokinetic parameters of serum glycerol seemed to be different between the different routes of administration (Table 1). Through EDT, the time to reach to the maximum concentration of glycerol was faster (Tmax = 6.7 min), maximum glycerol concentration was higher (Cmax = 1720 µg/ml) and maximum reduction of ICP was greater (* ICP = 25 mmHg) than those parameters obtained through NGT.

Discussion

The pharmacokinetic parameters of serum glycerol after oral administration through NGT and EDT showed changes similar to those[2] reported through intravenous administration. There was a strong positive correlation between serum glycerol kinetics and the decrease of ICP. The pattern of ICP reduction was quite similar to that of intravenous glycerol administration. These results indicated that glycerol was rapidly absorbed from gastrointestinal tract and produced effective ICP reduction with the same pharmacokinetics as intravenous glycerol administration. Thus, glycerol can be administered either per oral or per venous to obtain the same results for the treatment of increased ICP. The entero-duodenal route may be more effective and the benefits of oral administration can be appreciated in a clinical setting.

References

1. Node Y (1982) A study of glycerol on the reduction of raised intracranial pressure and its rebound phenomenon. J Nippon Med School 49: 752–758
2. Wald L, McLaurin S, Robert (1982) Oral glycerol for the treatment of traumatic intracranial hypertension. J Neurosurg 56: 323–331
3. Stinshoff K, Weisshaar D, Staehler F, Hesse D, Gruber W, Steier E (1977) Relation between concentrations of free glycerol and triglycerides in human sera. Clin Chem 23: 1029–1032

Correspondence: H. Tsuyusaki, M. D., Department of Pharmacy, Kitasato University Hospital, 1-15-1, Sagamihara-shi, Kanagawa Prefect 228, Japan.

Intracerebral Hematoma and Various Edema Pathophysiology

Acta Neurochir (1994) [Suppl] 60: 555–557
© Springer-Verlag 1994

The Effects of Blood or Plasma Clot on Brain Edema in the Rat with Intracerebral Hemorrhage

G. Y. Yang[1], **A. L. Betz**[1–3], and **J. T. Hoff**[1]

Department of [1] Surgery (Neurosurgery), [2] Pediatrics, and [3] Neurology, University of Michigan, Ann Arbor, MI, U.S.A.

Summary

The causes and characteristics of the brain edema which forms adjacent to an acute intracerebral hemorrhage (ICH) have not been explored thoroughly. This study was designed to examine the edema process in rat brain provoked by two different blood clot components. An intracerebral clot was produced by stereotactic injection of 100 μl of either blood (bICH) or cryoprecipitate/thrombin (pICH) into the right caudate nucleus. Water, Na^+, K^+, and Cl^- contents were measured at 0, 4, 24, 48, and 72 h after instillation of the clot. During the first 24 h, the water content of the ipsilateral caudate nucleus and cortex gradually increased in both groups. By 48 h brain edema was more severe in the bICH compared to that with pICH in the ipsilateral basal ganglia and cortex. The edema formation was accompanied by significant increase in sodium and chloride, as well as a decrease in potassium content by 48 h and sustained to 72 h. These results suggest that both blood and plasma clots can cause brain edema, but a plasma clot is less damaging than a blood clot in the immediate vicinity of the mass.

Keywords: Intracerebral hemorrhage; blood clot; plasma clot; brain edema.

Introduction

The mass effect of a hematoma is acknowledged to be disruptive. Brain edema secondary to spontaneous ICH is a major clinical problem and is usually refractory to treatment. While this edema is in part secondary to ischemia[5], blood constituents may also play a role in producing edema. Injection of autologous blood into the basal ganglia has been shown to produce greater ischemia and edema than injection of inert oil[4,11], These results suggest that edema is not simply caused by mechanical compression of the brain by mass effect. Whether the ICH produces some "toxic" effect on surrounding brain is less well documented[11]. It is still unclear which substances in the intracerebral blood clot could cause progressive brain edema formation after ICH.

This study compares edema formation during bICH or pICH as a first step toward determining whether edemagenic substances are released from a blood clot.

Materials and Methods

Sixty Sprague Dawley rats weighing 250–350 grams were anesthetized with Ketamine (50 mg/kg i.p.) and xylazine (10 mg/kg i.p.). A polyethylene catheter was inserted into a femoral artery for continuous monitoring of blood pressure, sampling for analysis of blood gases and blood pH, and drawing blood for the ICH. Body temperature was maintained at 37.5°C and blood pressure was maintained above 90 mmHg. Rats were given a 30% O_2/70% N_2O

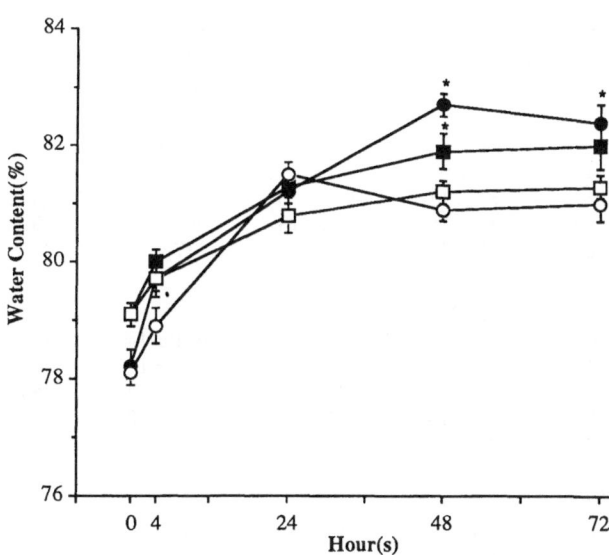

Fig. 1. Changes in brain water content following blood or plasma clot injection. Brain water content was measured in samples of (●) basal ganglia and (■) cortex after the injection of 100 μl of blood and (○) basal ganglia and (□) cortex after the injection of 100 μl of plasma clot. Values shown are means ± SE of 6 rats. *p < 0.05 comparing the bICH to the pICH at the same time point based upon a two-tailed Student's t-test

gas mixture via an inhalation mask to maintain a PaO_2 level of 90 mmHg or higher. Masuda's model of intracerebral hematoma was used with minor modification[7]. Two 30 gauge needles were implanted into the right caudate nucleus at coordinates A6.8, L3.0, H1.0. A syringe filled with fresh autologous blood was fixed on an infusion pump and 100 μl of fresh blood was injected at a rate of 20 μl/min. The fibrin clot was formed by simultaneously injecting the same volume of 2 different solutions into the caudate nucleus through separate needles. One solution contained cryoprecipitate from human plasma, the other thrombin. When mixed together in the brain, these components formed a rubbery mass within seconds.

The rats were divided into 10 groups of six each. Five groups underwent bICH for 0, 4, 24, 48, and 72 h; another five groups underwent pICH and were studied at the same time points. Tissue samples were obtained from both hemispheres in each rat. The hemisphere on the injected side (right) is referred to as ipsilateral and the other contralateral. Samples from each hemisphere were separated into cortex and basal ganglia areas.

Water and ion contents were measured in each sample. The water content, expressed as % wet weight, was calculated as $(W - D) / W \times 100$. The Na^+ and K^+ contents were measured by atomic absorption spectroscopy and Cl^- was measured using a digital chloridometer. Differences in water, Na^+, K^+, and Cl^- contents among the time points were evaluated using ANOVA with the Scheffe F test. Differences between the blood and plasma groups for each region were determined using a Student's t-test for unpaired samples.

Results

All physiological parameters were in the normal range. There were no significant differences between bICH and pICH groups in the mean arterial blood pressure, blood gases (PaO_2, $PaCO_2$ and HCO_3^-), blood pH, and blood glucose.

The average brain water and cation contents over the first 72 h after ICH are given in Table 1 and edema formation is shown in Fig. 1. The water content was significantly increased by 24 h in the ipsilateral basal ganglia and cortex of both groups. Brain edema formation was accompanied by significant increases in the sodium and chloride, as well as decreases in potassium content by 48 h which were sustained to 72 h in the

Table 1. *Water, Cation Concentrations After ICH Using Different Blood Components*

	Blood				Plasma			
	Basal ganglia		Cortex		Basal ganglia		Cortex	
	ipsi	contra	ipsi	contra	ipsi	contra	ipsi	contra
Water (%)								
0 h	78.2 ± 0.3	77.9 ± 0.2	79.1 ± 0.2	78.8 ± 0.2	78.1 ± 0.1	78.0 ± 0.2	79.1 ± 0.2	78.8 ± 0.2
4 h	79.7 ± 0.3	78.8 ± 0.1	80.0 ± 0.2	79.8 ± 0.1*	78.9 ± 0.3	78.1 ± 0.3	79.7 ± 0.2	79.3 ± 0.1
24 h	81.2 ± 0.2[c]	78.9 ± 0.4	81.3 ± 0.1[c]	79.4 ± 0.2	81.5 ± 0.2[c]	79.0 ± 0.2	80.8 ± 0.3	79.4 ± 0.2
48 h	82.7 ± 0.2[c]	79.3 ± 0.1[a]	81.9 ± 0.3[c]	79.8 ± 0.2	80.9 ± 0.2[c]	78.9 ± 0.1	81.2 ± 0.2[c]	79.6 ± 0.1
72 h	82.4 ± 0.3[c]	78.5 ± 0.3	82.0 ± 0.4[c]	79.7 ± 0.3	81.0 ± 0.3[c]	78.3 ± 0.1	81.3 ± 0.2[c]	79.2 ± 0.1
Sodium (μEq/gm dry wt.)								
0 h	210 ± 8	185 ± 3	211 ± 6	195 ± 5	189 ± 5	188 ± 5	209 ± 4	192 ± 6
4 h	227 ± 9	lg0 ± 7	215 ± 4	196 ± 4	214 ± 10	182 ± 11	217 ± 11	189 ± 13
24 h	285 ± 17	189 ± 10	272 ± 13	189 ± 3	297 ± 17[b]	208 ± 6	260 ± 12	198 ± 8
48 h	383 ± 18[c]	212 ± 13	295 ± 17[b]	209 ± 8	266 ± 13	211 ± 16	253 ± 13	196 ± 14
72 h	358 ± 21[c]	189 ± 8	317 ± 13[c]	205 ± 4	263 ± 9	188 ± 3	267 ± 6	196 ± 4
Potassium (μEq/gm dry wt.)								
0 h	419 ± 17	431 ± 16	461 ± 14	476 ± 12	428 ± 3	431 ± 5	462 ± 7	469 ± 7
4 h	398 ± 11	418 ± 8	445 ± 11	466 ± 7	381 ± 16	421 ± 20	446 ± 21	462 ± 22
24 h	378 ± 17	410 ± 9	398 ± 13	454 ± 8	371 ± 23	427 ± 12	438 ± 15	473 ± 12
48 h	315 ± 18*	406 ± 4	398 ± 12	438 ± 11[a]	397 ± 19	439 ± 15	422 ± 7	449 ± 12
72 h	316 ± 20*	400 ± 9	373 ± 10[a]	450 ± 9	390 ± 10	447 ± 24	418 ± 19	447 ± 11
Chloride (μEq/gm dry wt.)								
0 h	170 ± 10	153 ± 5	165 ± 8	154 ± 5	152 ± 3	144 ± 6	154 ± 1	145 ± 2
4 h	191 ± 9	153 ± 4	171 ± 4	151 ± 4	170 ± 15	139 ± 7	168 ± 12	142 ± 11
24 h	235 ± 11	154 ± 8	223 ± 12	145 ± 6[a]	239 ± 7	170 ± 6	192 ± 9	150 ± 11[c]
48 h	289 ± 18*	159 ± 5	235 ± 11[a]	160 ± 6[c]	223 ± 7	165 ± 16	197 ± 16	158 ± 14[a]
72 h	287 ± 9*	160 ± 8	259 ± 16[c]	158 ± 6[c]	233 ± 7	153 ± 6	211 ± l0	160 ± 5[b]

Values are mean ± SE, ml/100 mg wet tissue weight for water and μmol/g dry tissue weight for Na, K, Cl, from six rats in each.
[a] $p < 0.05$, [b] $p < 0.01$, [c] $p < 0.002$ compared to zero-time group using ANOVA with the Scheffe F-test. * $p < 0.05$ indicate the level of significance between bICH and pICH groups. *ipsi* lesion hemisphere; *contra* contralateral hemisphere.

ipsilateral basal ganglia. Brain edema was most severe in the ipsilateral basal ganglia and was significantly greater 48 h after bICH as compared to pICH.

Discussion

This study demonstrates that edema formation is more severe in the bICH than in the pICH. At an early stage after ICH, edema develops at a similar rate in both groups. This edema may be caused by mechanical compression and local ischemia because the same volume of fresh blood or plasma was injected. CBF is significantly decreased in areas surrounding the clot and more distant regions immediately after ICH[9,10,14]. After 48 h, brain edema is more severe in bICH than pICH, and the blood clot is bigger than the plasma clot, suggesting that more blood components remained in the bICH animals, possibly accounting for later secondary edema.

Ferric iron plays an important role in free radical-mediated injury since it catalyzes the conversion of superoxide and hydrogen peroxide to the more toxic hydroxyl radical which has been identified as an important mediator of ischemic brain edema[1,3,6]. Since red blood cells are an extremely rich source of iron, this ion may play an important role in brain injury following ICH. Indeed, extravascular blood in brain has been shown to generate superoxide anion[2,13]. Platelets that participate in clot formation following ICH may also be the source of important toxins. Platelet cyclooxygenase has been implicated in superoxide production when blood is placed in contact with the cerebral cortex[8]. In addition, several types of biochemical mediators, such as bradykinin, serotonin, histamine, arachidonic acid, and leukotrienes have been associated with breakdown of the BBB and vasogenic edema during brain injury[12]. More severe edema in the later stage in bICH animals suggests that one or more of these components of whole blood may play an important role as edema mediators. Identification of the mediators which are most important in edema formation after ICH requires additional investigation.

Acknowledgements

This work was supported by grants from the National Institutes of Health (NS-17760 and NS-23870).

References

1. Betz AL, Iannotti F, Hoff JT (1989) Brain edema: a classification based on blood-brain barrier integrity. Cerebrovasc Brain Metab Rev 1: 133–154
2. Ikeda Y, Ikeda K, Long DM (1989) Protective effect of the iron chelator deferoxamine on cold-induced brain edema. J Neurosurg 71: 233–238
3. Ikeda Y, Long DM (1990) The molecular basis of brain injury and brain edema: the role of oxygen free radicals. Neurosurgery 27: 1–11
4. Jenkins A, Mendelow AD, Graham DI, Nath FP, Teasdale GM (1990) Experimental intracerebral haematoma: the role of blood constituents in early ischemia. Br J Neurosurg 4: 45–51
5. Kingman TA, Mendelow AD, Graham DI, Teasdale GM (1988) Experimental intracerebral mass: description of model, intracranial pressure changes and neuropathology. J Neuropathol Exp Neurol 47: 128–137
6. Martz D, Beer M, Betz AL (1990) Dimethylthiourea reduces ischemic brain edema without affecting cerebral blood flow. J Cereb Blood Flow Metab 10: 352–357
7. Masuda M, Dohrman GJ, Kwaan HC, Erickson RK, Wollman RL (1988) Fibrinolytic activity in experimental intracerebral hematoma. J Neurosurg 68: 274–278
8. Mirro R, Armstead WM, Mirro J Jr., Busija DW, Leffler CW (1989) Blood-induced superoxide anion generation on the cerebral cortex of newborn pigs. Am J Physiol 257: H1560–H1564
9. Nath FP, Jenkins A, Mendelow AD, Graham DI, Teasdale GM (1986) Early hemodynamic changes in experimental intracerebral hemorrhage. J Neurosurg 65: 697–703
10. Nath FP, Kelly PT, Jenkins A, Mendelow AD, Graham DI, Teasdale GM (1987) Effects of experimental intracerebral hemorrhage on blood flow, capillary permeability, and histochemistry. J Neurosurg 66: 555–562
11. Suzuki J, Ebina T (1980) Sequential changes in tissue surrounding ICH. In: Pia HW, Langmaid C, Zierski J (eds) Spontaneous intracerebral haematomas: advances in diagnosis and therapy. Springer, Berlin Heidelberg New York, p 121–128
12. Wahl M, Unterberg A, Baethmann A, Schilling L (1988) Mediators of blood-brain barrier dysfunction and formation of vasogenic brain edema. J Cereb Blood Flow Metab 8: 621–634
13. Willmore JJ, Rubin JJ (1982) Formation of malonaldehyde and focal brain edema induced by subpial injection of $FeCl_2$ into rat isocortex. Brain Res 246: 113–119
14. Yang G-Y, Betz AL, Hoff JT (1994) Experimental intracerebral hemorrhage. Part 2: relationship between brain edema, blood flow, and blood-brain barrier permeability. J Neurosurg

Correspondence: Guo-Yuan Yang, M.D., Ph.D., University of Michigan, 5605 Kresge I, Ann Arbor, MI 48109-0532, U.S.A.

Acta Neurochir (1994) [Suppl] 60: 558–560
© Springer-Verlag 1994

Chronological Changes in Brain Edema Induced by Experimental Intracerebral Hematoma in Cats

H. Tomita[1], **U. Ito**[1], **K. Ohno**[2], and **K. Hirakawa**[2]

Department of Neurosurgery, [1] Musashino Red Cross Hospital, and [2] Tokyo Medical and Dental University, Tokyo, Japan

Summary

To investigate the temporary profile of changing perifocal brain edema around an intracerebral hematoma (ICH), we developed an experimental ICH model using cats. The developing perifocal edema in the white matter around the hematoma was measured by means of a gravimetric technique. Edema was more severe near the ICH, and declined with increasing distance. Edema was mild 2 hours after the onset of the ICH, and was most severe in all the regions examined 3 days later. Edema decreased but still existed in all regions 7 days after the onset of the ICH. The results suggest that the mechanism of the development of edema associated with ICH seems to differ from that associated with a cold injury. This experimental ICH model proved to be useful for the study of formation, expansion, and resolution of edema associated with ICH.

Keywords: Brain edema; cat; experimental intracerebral hematoma; specific gravity.

Introduction

Spontaneous intracerebral hematoma (ICH) induces acute neurological deficits due to the hematoma itself. Neurological deterioration often develops clinically after the onset of stroke especially in patients with a large hematoma. The pathophysiology of this deterioration remains unclear, although it may be attributed to the secondary changes such as expanding perifocal cerebral edema[9,10], decreased cerebral blood flow[10], increased intracranial pressure[8], or chemical stimulation. Among these factors, perifocal edema aggravates the mass effect of the hematoma itself and seems closely related to the other changes. However, the pathomechanism of perifocal edema around hematoma is still obscure. Only a few experimental studies on cerebral edema associated with ICH have been reported[3,9,11], probably because no good model of ICH has been available to date. Rodents are unsuitable for the investigation on brain edema due to ICH because of their scarceness of the white matter. Therefore, in the present study, we developed an experimental ICH model in cats, and investigated the development of the perifocal edema around the hematoma.

Materials and Methods

Production of Hematoma

Nine adult mongrel cats weighing from 2.0 kg to 5.2 kg were anesthetized with ketamine (25 mg/kg) and given a subcutaneous injection of atropine sulfate (0.025 mg/kg). The animals were mounted in a prone position on a stereotactic instrument. After a small craniectomy, a No. 22 needle was inserted through the dura mater, 20 mm anterior to the external acoustic meatus, 8 mm lateral from the midline to the right side and 8 mm deep from the dura level, with the bevel of the needle pointing in the anterolateral direction. Thus, the needle tip was located at the center of the right frontal subcortical space. About 1.0 ml of fresh blood was taken from the femoral vein and 0.6 ml of the blood was manually injected into the brain over more than 5 minutes. Ten minutes later, the needle was slowly withdrawn over a span of 3 minutes. Animals were sacrificed either 2 hours later or on the 3rd or 7th postoperative day. Evans blue (2 ml/kg) was injected intravenously 1 hour before sacrifice. The brain was removed and cut horizontally at the level of the hematoma center. One cut surface was used for histological studies and the other for the measurement of specific gravity.

Measurement of Specific Gravity

Approximately one-cubic-mm-sized pieces of white matter were taken sequentially from adjacent regions up to 10 mm from the hematoma toward the internal capsule (Fig. 1), and the specific gravity of 8 pieces was measured according to the gravimetric technique described by Fujiwara *et al.*[2].

Results

The reproducibility of the ICH in size and location by our technique was excellent. The hematomas were always located in the frontal subcortical space with no penetration into the lateral ventricle (Fig. 1). All cats

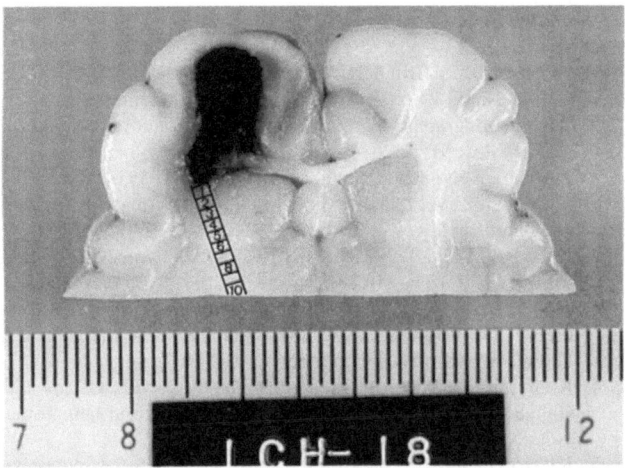

Fig. 1. A horizontal section at the level of the hematoma. The hematoma is localized in the frontal subcortical region. Approximately 1-cubic-mm-sized pieces of white matter were taken sequentially from the hematoma up to a distance of 10 mm towards the internal capsule to measure the specific gravity

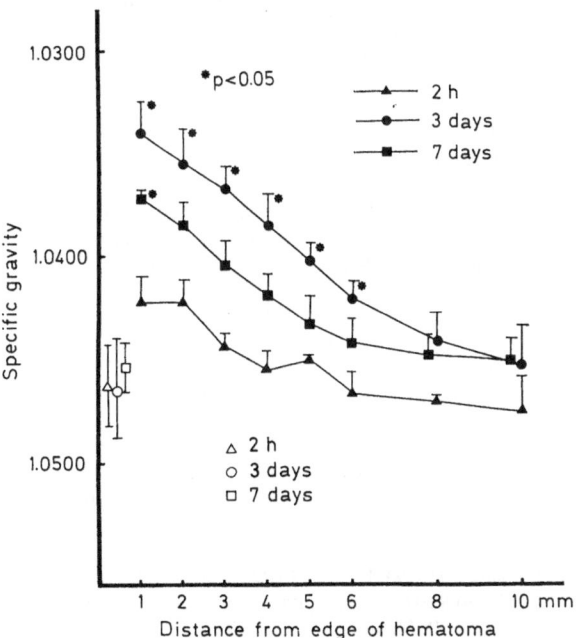

Fig. 2. Chronological changes in the specific gravity. Closed symbols indicate the specific gravity for the eight regions from the hematoma to a distance of 10 mm at 2 hours, 3 days and 7 days after induction of ICH. Open symbols indicate the specific gravity for the white matter in the contralateral hemisphere. Asterisks indicate significant differences of the specific gravity from the group examined 2 hours after induction of ICH ($p < 0.05$)

developed a transient contralateral hemiparesis, but there was no mortality.

Figure 2 shows the specific gravity in all regions in the three groups. The degree of edema was more severe near the ICH in all groups, and declined with increas-

ing distance. Edema was mild 2 hours after the onset of the ICH, and was most severe in all 8 regions in the group examined 3 days after the onset of the ICH. The edema decreased but still persisted in all regions even on the 7th day after the onset of the ICH.

Discussion

In recent reports, dogs, rabbits, rats, or monkeys have been used as experimental models for studying the pathophysiology of spontaneous ICH[4]. Among them, rodents are inappropriate for the study on perifocal edema which develops after the onset of the ICH, because of scarceness of the white matter. Kaufman *et al.*[4] reviewed the models of ICH previously reported, and suggested that animals with a larger brain should be employed for the study of the pathophysiology of ICH including perifocal edema, because such animals produce a relatively regular hematoma of predictable shape and size. In this study, we used cats and found them to be a suitable model in terms of the reproducibility and development of the perifocal edema. We tested both 1.0 ml and 0.6 ml autologous blood injections in a preliminary experiment. The latter volume was found to be more suitable, because when 1.0 ml of blood was injected, the animals became severely ill with hemiplegia or died soon after the procedure, and ventricular penetration of the blood often occurred.

Only a few studies on the temporary profile of perifocal edema developing around the hematoma have been reported[9–11]. Histopathological studies on perifocal edema in experimental ICH have shown that the most severe perifocal edema developed 3 to 4 days after the onset of the ICH[1,11]. From clinical studies, Suzuki *et al.*[10] reported that the mass effect due to perifocal edema after a putaminal hemorrhage became maximum on CT between 10 and 20 days after the hemorrhage. In the present study, using the gravimetric method, we found that the time in which edema was most severe was 3 days after the onset of the ICH, which is in agreement with previous experimental studies[1,11]. However, it should be noted that the perifocal edema still existed even 7 days after the onset of the ICH.

According to Klatzo *et al.*[5], edema propagating from a lesion due to cold injury reaches a maximum in the region closest to the lesion around 2 hours after the injury, and in the regions more than 2–3 mm away from the lesion 24 hours after the injury. Thereafter, the edema decreases gradually over time in all regions. On the other hand, Nagashima *et al.*[6] reported that the edema after a cold injury was more severe in all regions at 48 hours than at 24 hours after the injury. As for ICH, the edema reached a maximum in all regions around 3 days after the onset of the ICH and decreased slowly during the period of more than 7 days after the onset. This study suggested that the evolution and resolution of edema occurred more slowly after induction of the ICH than the same processes after a cold injury. The pathomechanism for the development of brain edema associated with ICH seemed to be quite different from that of cold lesion edema.

Various mechanisms for the development of perifocal edema due to an ICH have been reported. Nehls *et al.*[7] showed that microballoon inflation in the caudate nucleus as an intracerebral mass significantly reduced cerebral blood flow in the caudate nucleus with significant reductions in the specific gravity of the brain tissue around the balloon. Suzuki and Ebina[9] histologically compared the effects between the injection of oil-wax and autologous blood into the basal ganglia of dogs on the surrounding brain tissue, and found that brain edema was much more extensive in the hematoma group than in the oil-wax group. They concluded that the edema associated with the hematoma resulted from not merely compression of the surrounding brain tissue, but the infiltration of plasma components from the hematoma into the surrounding brain tissue. Kane *et al.*[3] demonstrated, using an experimental rat model of ICH, that global depletion of circulating leucocytes and platelets by whole-body irradiation protected against both ischemia and edema formation associated with ICH, and speculated the contribution of an immune response to edema formation. Various mediators of edema such as cytokines, free radicals, vascular permeability factors, and other chemical substances seem to be involved in edema formation. We believe that the present model is adequate for various studies of perifocal edema around the ICH.

References

1. Enzmann DR, Britt RH, Lyons BE, Buxton JL, Wilson DA (1981) Natural history of experimental intracerebral hemorrhage: sonography, computed tomography and neuropathology. AJNR 2: 517–526
2. Fujiwara K, Nitsch C, Suzuki R, Klatzo I (1981) Factors in the reproducibility of the gravimetric method for evaluation of edematous changes in the brain. Neurol Res 3: 345–361
3. Kane PJ, Modha P, Strachan RD, Cook S, Chambers IR, Clayton CB (1992) The effect of immunosuppression on the development of cerebral oedema in an experimental model of intracerebral haemorrhage: wole body and regional irradiation. J Neurol Neurosurg Psychiatry 55: 781–786
4. Kaufman HH, Pruessner JL, Bernstein DP, Borit A, Ostrow PT, Cahall DL (1985) A rabbit model of intracerebral hematoma. Acta Neuropathol (Berl) 65: 318–321
5. Klatzo I, Chui E, Fujiwara K, Spatz M (1980) Resolution of vasogenic brain edema. In: Cervós-Navaro J, Ferszt R (eds) Advances in neurology, Vol 28. Brain edema. Raven, New York, pp 359–373
6. Nagashima T, Tamaki N, Shirakuni T, Matsumoto S, Seguchi Y, Tamura T (1985) Biomechanics of vasogenic brain edema: application of Biot's consolidation theory and the finite element method. In: Inaba Y, Klatzo I, Spatz M (eds) Brain edema. Springer, Berlin Heidelberg New York Tokyo, pp 92–98
7. Nehls DG, Mendelow AD, Graham DI, Teasdale GM (1990) Experimental intracerebral hemorrhage: early removal of a spontaneous mass lesion improves late outcome. Neurosurgery 27: 674–682
8. Papo I, Janny P, Caruselli, Colnet G, Luongo A (1979) Intracranial pressure time course in primary intracerebral hemorrhage. Neurosurgery 4: 504–511
9. Suzuki J, Ebina T (1980) Sequential changes in tissue surrounding ICH. In: Pia HW, Langmaid C, Zierski J (eds) Spontaneous intracerebral hematomas. Springer, Berlin Heidelberg New York, pp 121–128
10. Suzuki R, Ohno K, Hiratsuka H, Inaba Y (1985) Chronological changes in brain edema in hypertensive intracerebral hemorrhage observed by CT and xenon-enhanced CT. In: Inaba Y, Klatzo I, Spatz M (eds) Brain edema. Springer, Berlin Heidelberg New York Tokyo, pp 613–620
11. Tochio H (1988) Histological study on acute and chronic experimental intracerebral hemorrhage: is there any difference between surgically treated group and non-surgical group? Mie Igaku 32: 1–8 (Jpn)

Correspondence: Hiroki Tomita, M. D., Department of Neurosurgery, Musashino Red Cross Hospital, 1-26-1 Kyonancho, Musashino City, Tokyo 180, Japan.

Acta Neurochir (1994) [Suppl] 60: 561–563
© Springer-Verlag 1994

Cerebral Oedema Following Intracerebral Haemorrhage: the Effect of the NMDA Receptor Antagonists MK-801 and D-CPPene

P. J. Kane, S. Cook, I. R. Chambers[1], and **A. D. Mendelow**

Department of Surgery (Neurosurgery), University of Newcastle upon Tyne, and [1] Regional Medical Physics Department, Newcastle General Hospital, Newcastle upon Tyne, England, U.K.

Summary

The effects of pre-treatment with the NMDA receptor antagonists MK-801 and D-CPPene on the development of brain oedema were investigated in a rodent model of intracerebral haemorrhage. In acute experiments (4 hour survival) both drugs caused significant hypotension and had significant anaesthetic effects whilst conferring no protection against oedema formation. In chronic experiments (24 hour survival) MK-801 conferred no protection against brain oedema. With D-CPPene marginal protection against cortical oedema may have been conferred but this result should be interpreted with caution.

Keywords: Intracerebral haemorrhage; NMDA antagonist; brain oedema; rodent model.

Introduction

The role of excitotoxicity as a cause of neuronal death following an ischaemic insult is well established. Excitatory amino acids such as glutamate, released in excessive amounts during cerebral ischaemia, overstimulate their postsynaptic receptors leading to abnormal transmembrane calcium fluxes and subsequent cell death[1]. Blockade of glutamate receptors, especially the N-methyl-D-aspartate (NMDA) subtype, has been shown to provide significant protection against ischaemic brain damage in animal models of ischaemic stroke[4]. Most experimental work has involved the use of the non competitive NMDA receptor antagonist MK-801. Its clinical use may be limited by its side effects[6]. Competitive NMDA receptor antagonists have been formulated which are believed to have an equivalent anti-ischaemic effect to the non-competitive antagonists but with fewer side effects.

Data relating to the use of the NMDA receptor antagonists in models of haemorrhagic stroke are limited. Following a spontaneous intracerebral haemorrhage the clot is surrounded by a penumbra of ischaemic brain and associated brain oedema. In clinical practice this ischaemic oedema is a significant cause of morbidity and mortality. The aim of this study was to determine whether the use of MK-801 or the competitive NMDA antagonist D-CPPene would prevent the development of cerebral oedema in a rodent model of intracerebral haemorrhage and to compare their effects.

Methods

Adult male Wistar rats were studied. All experiments were performed in accordance with the Animals (Scientific Procedures) Act 1986. Experiments were performed in two groups:

Group 1: Four Hour Brain Oedema Measurement

The rats were anaesthetised with halothane in nitrous oxide/oxygen (70% : 30%) via a tracheostomy with mechanical ventilation. The right femoral artery and vein were cannulated to enable: serial measurement of arterial blood gases, glucose and haematocrit; continous blood pressure monitoring; the intravenous administration of fluids/drugs. Body temperature was maintained at 37°C using a heating unit. The rat was placed in a stereotaxic frame and a burr hole made in the skull (right side, 2.8 mm lateral and 1.1 mm anterior to the bregma) to allow insertion of a microballoon (Balt Laboratories) into the right caudate nucleus. The dura was unopened until after the administration of the treatment solutions. The blood pressure and arterial blood gases were allowed to stabilise before the rats were treated with MK-801 (3 mg/kg in saline) or D-CPPene (15 mg/kg in saline) by infusion over ten minutes. Control rats received an equivalent volume of saline. A microballoon mounted on the tip of a blunted 25 gauge butterfly needle was inflated to 50 microlitres with saline to test its integrity. It was then deflated and inserted through the burr hole into the brain to a depth of 5.5 mm and then reinflated to 50 microlitres with saline to simulate an intracerebral haematoma[3]. It remained inflated for five minutes and was then deflated and removed. The burr hole was sealed with bone wax. The arterial blood gases, blood pressure and body temperature were maintained in the normal range for four hours and then cardiac arrest was induced with intravenous pentobarbitone. The brain was then rapidly removed for oedema measurements. Paired samples of tissue (1–2 mm³) were taken from the cortex overlying the simulated haematoma, adjacent white matter and the caudate nucleus in the penumbra of the lesion. Equivalent samples were also taken from the left hemisphere and also samples from both cerebellar hemispheres. Brain oedema measurements were made using a gravimetric technique[5]. The samples were placed in the gravimetric column and their level of descent measured after 1 minute. Specific gravity of the sample was calculated using a linear regression analysis derived from calibration of the column with droplets of potassium sulphate of known specific gravity. All columns had a correlation co-efficient > 0.995.

Group 2: Twenty-Four Hour Brain Oedema Measurement

The rats were anaesthetized with halothane in nitrous oxide/oxygen (70% : 30%) administered by face mask. Pre-treatment was as follows:

a) MK-801: Prior to anaesthesia rats received MK-801 (dose 3 mg/kg) intraperitoneally or an equivalent volume of saline, or

b) D-CPPene: After anaesthesia the right femoral vein was cannulated. Rats received D-CPPene (15 mg/kg) or an equivalent volume of saline by infusion over 10 minutes.

Body temperature was maintained at 37°C using a heating unit. Oxygen saturation was monitored continuously using a pulse oximeter probe on the shaved left hind limb and maintained > 95%. The rat was placed in the stereotaxic frame and a burr hole was made as in Group 1. The microballoon was then inserted into the right caudate nucleus as for Group 1 animals and inflated for five minutes. It was then deflated and removed. The burr hole was sealed with bone wax and the scalp wound sutured. In the D-CPPene study the venous cannula was removed, the vessel coagulated and the groin wound sutured. All rats were recovered in a warmed incubator in an oxygen enriched atmosphere. Postoperatively all rats received buprenorphine analgesia. In the D-CPPene limb of the study treated animals received D-CPPene (10 mg/kg), intraperitoneally five hours after the initial infusion and every three hours thereafter. Control animals received an equivalent volume of intraperitoneal saline. At 24 hours all rats were re-anaesthetized with halothane and then killed with an intracardiac injection of pentobarbitone. Brain oedema measurements were then made as in Group 1.

Statistical analysis in all groups of experiments was carried out using the unpaired Student's t-test (Minitab statistical package).

Results

Group 1: Four Hour Brain Oedema Measurement

1) Physiological variables: No significant differences were found between the MK-801 and saline treated rats or the D-CPPene and saline treated rats.

2) Haemodynamic effects: On administration of the MK-801 and D-CPPene infusion there was a progressive and significant fall (p > 0.001 in both groups) in mean arterial blood pressure (MAP) in all animals (Fig 1: MK-801). This drop in MAP did not occur in the saline treated control rats. The hypotension did not improve with intravenous fluid replacement in either

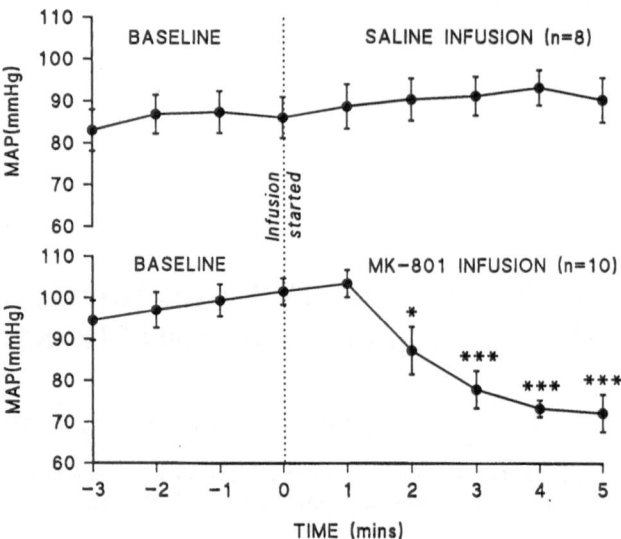

Fig. 1. Change in mean arterial blood pressure (MAP) with time during infusion of MK-801 compared to saline treated controls. Similar effects were observed in D-CPPene treated rats. All values are means ± sem. * = p < 0.05; *** = p < 0.001; Student's t-test (Comparing value with T_0 baseline value)

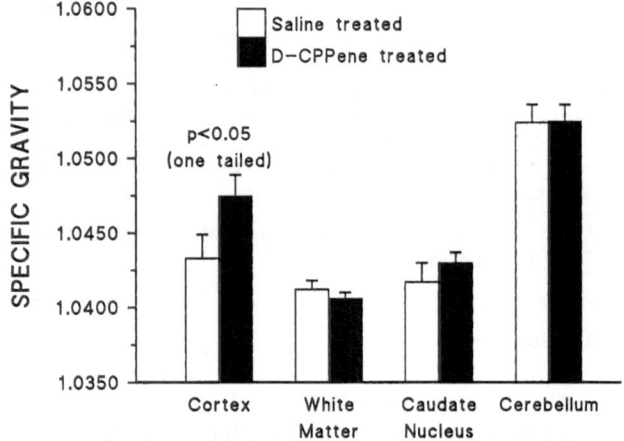

Fig. 2. 24 hour brain oedema (specific gravity) measurements in right hemisphere adjacent to the simulated haematoma in D-CPPene treated rats and saline treated controls. All values are means ± SEM

group. To prevent the MAP falling below 60 mmHg it was necessary to decrease the inspired halothane concentration by at least 50%. This resulted in the MAP returning to pretreatment levels within 5 minutes. The rats remained fully anaesthetized, as evidenced by the absence of whisker twitching, blink reflex and limb withdrawal reflexes.

3) Brain oedema measurements: All MK-801, D-CPPene and saline treated rats developed significant oedema in the right caudate nucleus at four hours. The amount of oedema generated was not significantly different between the MK-801 and saline treated rats nor D-CPPene and saline treated rats.

Group 2: Twenty-Four Hour Brain Oedema Measurements

1) Behavioural effects: All saline treated control rats were fully recovered and mobile in their cages within 30 minutes of discontinuing anaesthesia. MK-801 and D-CPPene animals recovered more slowly. At 30 minutes after anaesthesia most showed evidence of whisker twitching and limb withdrawal reflexes but absence of the postural reflexes. These reflexes returned at 60–90 minutes. Occasionally the animals would move slowly about their cages but the majority remained lying in one position. Most lay with their eyes open and showed no spontaneous blinking. Occasionally "head weaving" was observed.

2) Brain oedema measurements: All animals developed significant cortical and caudate oedema in the right hemisphere adjacent to the haematoma at 24 hours. Treatment with MK-801 did not confer any protection against oedema formation when compared to saline treated rats. In rats treated with D-CPPene there was marginally less cortical oedema when compared to saline treated rats ($p < 0.05$, one tailed Student's t-test, Fig. 2). However three D-CPPene treated rats died in the postoperative period between the administration of the first and second intraperitoneal doses whilst there were no deaths in the saline treated group. If the specific gravity values for D-CPPene treated rats are "corrected" to include dead animals (assuming that the three dead rats had brain oedema at least as severe as that found in the "worst case" survivor) then no protection is conferred.

Discussion

The NMDA antagonists MK-801 and D-CPPene have been shown to have potent anti-ischaemic effects in experimental models of ischaemic stroke[2,4]. Little data relates to their ability to prevent brain oedema. In this study pretreatment with D-CPPene may have conferred marginal protection against the development of cortical oedema but this result must be interpreted with caution in view of the increased death rate amongst spontaneously breathing D-CPPene treated rats. Both drugs had a significant hypotensive and anaesthetic effect and may be related to the ketamine like properties of the NMDA receptor antagonists[6]. These physiological considerations should be considered if these drugs or their derivatives are to be used in clinical practice.

Acknowledgements

PJ Kane was supported by the Harker Foundation Fellowship, University of Newcastle upon Tyne. We thank: Merck Sharpe and Dohme UK for supplying the MK-801; Sandoz, Bern for supplying the D-CPPene and dosage regimen.

References

1. Benveniste H (1991) The excitotoxin hypothesis in relation to cerebral ischaemia Cerebrovasc Brain Metab Rev 3: 213–245
2. Bullock R, Graham DI, Chen MH, Lowe D, McCulloch J (1990) Focal cerebral ischaemia in the cat: pre-treatment with a competitive NMDA receptor antagonist SDZ-EAA-494. J Cereb Blood Flow Metab 10: 668–675
3. Kingman TA, Mendelow AD, Graham DI, Teasdale GM (1987) Experimental intracerebral mass: time related effects on local cerebral blood flow. J Neurosurg 67: 732–738
4. McCulloch J, Bullock R, Teasdale GM (1991) Excitatory amino acid antagonists: opportunities for the treatment of ischaemic brain damage in man. In: Meldrum B (ed) Excitatory amino acid antagonists. Frontiers in pharmacology and therapeutics. Blackwell, Oxford, pp 287–326
5. Shigeno T, Brock M, Shigeno S, Fritschka E, Cervós-Navarro (1982) The determination of brain water content: microgravimetry versus dry-weighing method. J Neurosurg 57: 99–107
6. Willetts J, Balster RL, Leander JD (1990) The behavioural pharmacology of NMDA receptor antagonists. In: Lodge D, Collingridge G (eds) The pharmacology of excitatory amino acids. Elsevier, Cambridge, pp 62–67

Correspondence: Mr. P. J. Kane, Department of Neurosurgery, Regional Neurosciences Centre, Newcastle General Hospital, Westgate Road, Newcastle upon Tyne, England, NE4 6BE, U.K.

Acta Neurochir (1994) [Suppl] 60: 564–567
© Springer-Verlag 1994

Sequential Change of Brain Edema by Semiquantitative Measurement on MRI in Patients with Hypertensive Intracerebral Hemorrhage

S. Suga, S. Sato, K. Yunoki, and **B. Mihara**

Institute of Brain and Blood Vessels, Mihara Memorial Hospital, Gunma, Japan

Summary

The progression of brain edema in seven patients with hypertensive intracerebral hemorrhage (ICH) was evaluated. Five were of putaminal and two were of thalamic hemorrhage. The hematoma volume in the patients was 4 ~ 40 ml (18.9 ± 8.0 ml). Sequential MRI (SE: 2000/40) was performed at one, two and four weeks after onset. The edema volume (EV) was calculated as 1/2 · (long diameter) · (short diameter) · (thickness) of the high intensity area (HIA) on MRI. In comparison with the EV at one week after onset, the EV at two weeks was increased and the EV at four weeks returned to the same level of that at one week (132.3 ± 26.1%, 100 ± 10.6%, respectively). In contrast, the consciousness level and motor weakness of the patients had already improved at two weeks after onset. Our results demonstrate that progression of brain edema after small or medium size ICH may not bring about a deterioration of the clinical course. Moreover, progression of brain edema to the cerebral cortex and ventricle as indicated by MRI suggested an absorption pathway for the edema fluid, and implying that brain edema following ICH could play a part in the healing process after ICH.

Keywords: Brain edema; intracerebral hemorrhage; MRI.

Introduction

Progression of brain edema following hypertensive intracerebral hemorrhage (ICH) has been reported to be one of the crucial factors which can lead to deterioration of the clinical course in patients with ICH. In our clinical experiences, patients with small or medium size ICH recovered within a week after onset, although CT scans showed a marked midline shift due to brain edema for a few weeks. However, it is difficult to evaluate the precise extension of the brain edema revealed by low density areas on CT scans. In this study, we examined the serial semiquantitative changes of brain edema by MRI, which can demonstrate the area of brain edema clearly, and investigated the relationship between the clinical course and edema progression.

Materials and Methods

Seven patients with ICH were evaluated (sex: three males and four females, age: 52 ~ 76 y/o). Five were of putaminal and two were of thalamic hemorrhage. All patients were admitted to our hospital within a few days after onset, and supratentorial ICH was determined by CT scan. They were treated medically by intravenous injection of 400 ~ 800 ml of glycerol in physiological saline per day for seven to ten days.

MRI was performed with a TOSHIBA MRT-22A using a routine proton density (2000/40) spin echo sequence. All patients underwent an initial CT scan on admission and MRI at one, two and four weeks after onset. The hematoma volume (HV) was measured as 1/2 · long diameter · short diameter · thickness of the high density area on CT. The edema volume (EV) was estimated with the same formula from the high intensity area (HIA) on MRI[2]. We examined the relationship between serial changes of EV and clinical improvement such as of the consciousness level and motor weakness. We also evaluated the direction of brain edema progression.

Results

A typical case will be presented first (Fig. 1) . This 54-year-old male suffered rightsided, numbness suddenly and was admitted to our hospital on the day of onset. A CT scan demonstrated small left putaminal hemorrhage (HV: 7 ml). Serial MRI at one, two and four weeks after onset showed that the brain edema presented by the HIA had increased at two weeks but returned to the one-week level at four weeks after onset (EV: 42.9, 65.8 and 48.8 ml, respectively).

Compared to the EV at one week after onset, the EV at two weeks in all seven patients had increased but returned to the one-week level at four weeks after onset (132.3 ± 26.1, 100.0 ± 10.6%, respectively) (Table 1). In particular, the brain edema increased after small ICH increments. Concerning the consciousness level, one patient was in a state of stupor, two were somnolent and four were alert at one week after onset, but all patients were alert at two weeks. With regards the motor weakness of the upper limbs, one patient displayed no weakness, one had slight weakness, one had moderate weakness and four had severe weakness at one week. After two weeks from onset, three patients exhibited improvement of their motor weakness, but four showed no change (Table 1). These results indicated a discrepancy between edema progression and clinical improvement.

The direction of edema progression depended on the HV and site of the hematoma (Fig. 2, Table 2). The brain edema following small putaminal hemorrhage extended to the lateral ventricle and insular cortex; and that following medium size putaminal hemorrhage extended more to the temporoparietal cortex. In the case of thalamic hemorrhage, the brain edema progressed to the ventricle.

Discussion

Progression of brain edema after ICH has been considered to lead to the deterioration of the clinical course in patients with ICH. Wallenfang et al.[6] described the pathomechanism of brain edema following ICH experimentally. We sometimes experience a deterioration in patients after a large ICH within a few days of onset. In such cases, secondary disturbance of the microcirculation by the compression of the edema caused by the large ICH or an increased intracranial pressure by hydrocephalus due to ventricular rupture of the ICH could be a worsening factor. On the other hand, we experienced patients with small ICH who recovered quickly from their neurological impairment, even though CT scans demonstrated a marked midline shift for a long period. Some other authors have also de-

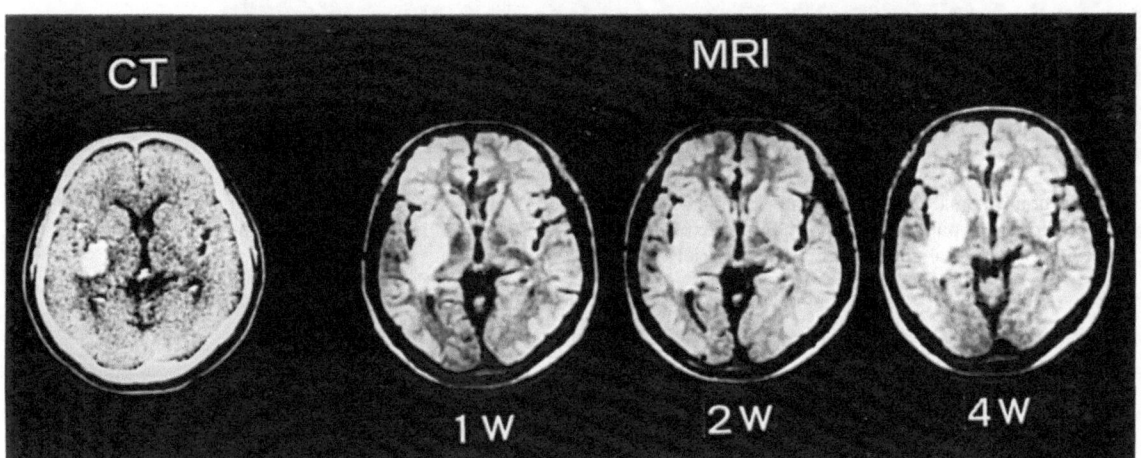

Fig. 1. CT scan on the day of onset and serial MRI at one, two and four weeks after onset in a patient with putaminal hemorrhage. The brain edema shown by HIA on MRI was increased at two weeks after onset

Table 1. *Summary of Serial Changes in Edema Volume, Consciousness Level and Motor Weakness in Patients with ICH*

Site of ICH	Hematoma volume (ml)	Edema volume (ml/%)			Consc. level[a] (1 W / 2 W)	Motor weakness[b] (1 W / 2 W)
		1 week	2 weeks	4 weeks		
(1) Thalamus	4	41/100	71/174	35/85	1/1	4/2
(2) Putamen	7	43/100	66/153	49/114	1/1	1/1
(3) Putamen	12	96/100	150/157	n.a.*	1/1	2/1
(4) Thalamus	15	53/100	55/103	n.a.*	3/1	5/5
(5) Putamen	19	94/100	102/109	98/104	1/1	2/1
(6) Putamen	35	136/100	158/116	132/97	2/1	5/5
(7) Putamen	40	208/100	237/114	n.a.*	2/1	5/5

* *n. a.* not available; [a] *Consc. level* consciousness level at one and two weeks. *1* alert, *2* somnolence, *3* stupor, *4* semicoma, *5* coma. [b] Motor weakness. *1* no weakness, *2* slight weakness, *3* moderate weakness, *4* severe weakness, *5* no movement.

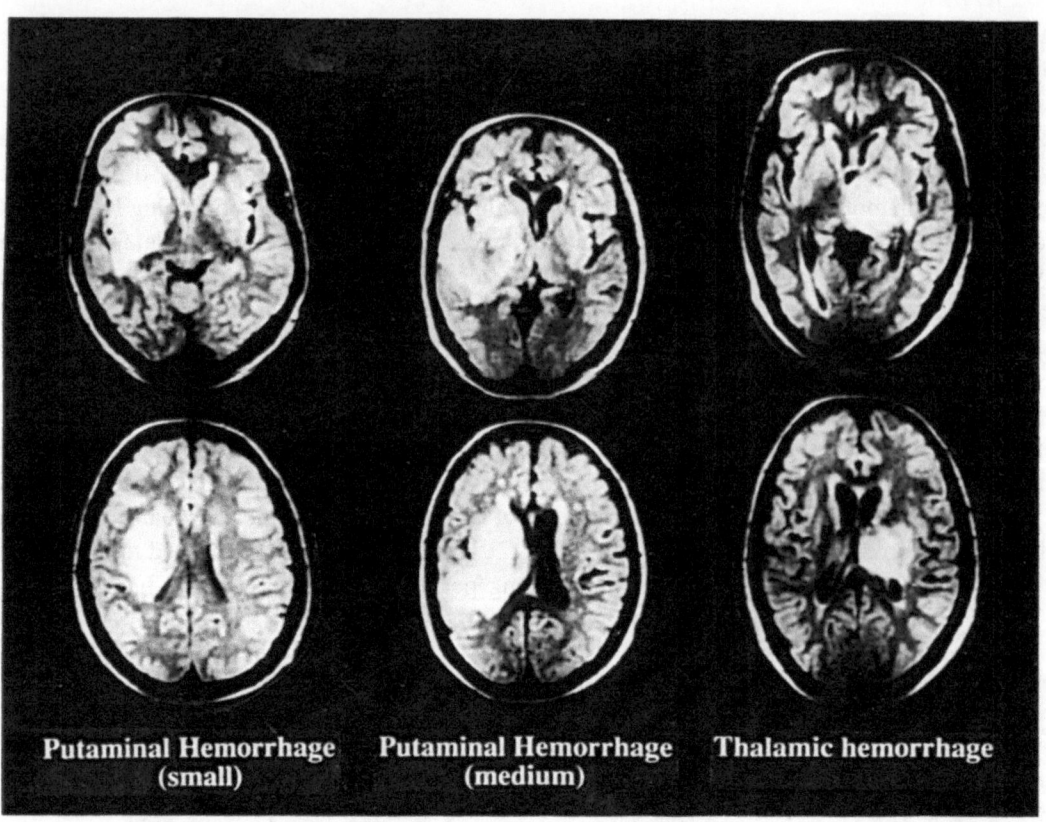

Fig. 2. Progression of brain edema. Following small putaminal hemorrhage, the brain edema extended to the lateral ventricle and insular cortex. Following medium size putaminal hemorrhage, it extended to the lateral ventricle, insular cortex and temporoparietal cortex. Following thalamic hemorrhage, it extended only to the ventricle

Table 2. *Hematoma Volume and Progression of Brain Edema in Patients with ICH*

Site of ICH	Hematoma volume (ml)	Progression of edema		
		Ventricle	Insular cortex	Temporoparietal cortex
(1) Thalamus	4	+	−	−
(2) Putamen	7	+	+	−
(3) Putamen	12	+	+	−
(4) Thalamus	15	+	−	−
(5) Putamen	19	+	+	−
(6) Putamen	35	+	+	+
(7) Putamen	40	+	+	+

scribed a prolonged midline shift based on CT scans or angiography[5,7]. In our study, the peak brain edema volume following ICH was detected at two weeks after onset, although the consciousness level and motor weakness of the patients had already undergone improvement by this time. These findings indicated that brain edema caused by small or medium size ICH may not bring about a deterioration of the clinical course itself. Moreover, progression of the brain edema to the cerebral cortex and ventricle suggested an absorption pathway for the edema fluid. In the case of medium size putaminal hemorrhage, the brain edema extended to the temporoparietal cortex, but no clinical symptoms of the temporoparietal cortex were observed. This also implied that edema progression did not impair neurological function.

After two weeks from onset of ICH, angiogenesis occurs around the hematoma to absorb the damaged tissue and hematoma, and such neovessels do not have a blood-brain barrier (BBB)[1]. Ring enhancement on CT scans can be observed at this time[4]. Through the capillary vessels without a BBB, vasogenic type brain edema is known to occur. This could explain why the peak brain edema following ICH appears so late. The significance of this vasogenic edema has not yet been determined, although Ikuta et al.[3] has suggested that vasogenic edema might not act as a worsening factor but could play an important role in the repair process of damaged tissue.

In conclusion, it should be emphasized that the peak brain edema following small or medium size ICH was observed at two weeks after onset, although the neurological impairment had already undergone improvement at that time.

References

1. Fukazawa H, Uemura K (1980) Computed tomography and pathology of cerebrovascular diseases. 1. Histopathology of cerebral hemorrhage and infarction. Progr Comput Tomogr 2: 5–13
2. Gomori JM, Grossman RI, Goldberg HI, et al (1985) Intracranial hematomas: imaging by high-field MR. Radiology 157: 87–93
3. Ikuta F, Yoshida Y, Ohama E, et al (1984) Brain and peripheral nerve edema as an initial stage of the lesion repair: revival of the mechanisms of the normal development in the fetal brain. Shinkei Kenkyu No Shinpo 28: 599–628
4. Laster DW, Moody DM, Ball MR (1978) Resolving intracerebral hematoma: alternation of the "ring sign" with steroid. Am J Roentgenol 130: 935–939
5. Uemura K, Gotoh K, Ishii K, et al (1978) Computed tomography in evaluation of intracranial hemorrhage. Shinkei Kenkyu No Shinpo 22: 200–218
6. Wallenfang T, Fries G, Jantzen JP, et al (1988) Pathomechanism of brain oedema in experimental intracerebral mass haemorrhage. Acta Neurochir (Wien) [Suppl] 43: 182–185
7. Yamaguchi K, Uemura K, Takahashi H (1973) An angiographic study of sequential change in hypertensive intracerebral haemorrhage. Br J Radiol 46: 125–130

Correspondence: Sadao Suga, M. D., Institute of Brain and Blood Vessels, Mihara Memorial Hospital, 366 Ohtamachi, Isesaki, Gunma 373, Japan.

Acta Neurochir (1994) [Suppl] 60: 568–570
© Springer-Verlag 1994

High Colloid Oncotic Therapy for Brain Edema with Cerebral Hemorrhage

O. Tone, U. Ito, H. Tomita, H. Masaoka, and B. Tominaga

Department of Neurosurgery, Musashino Red Cross Hospital, Tokyo, Japan

Summary

We examined the effectiveness of high colloid oncotic pressure (COP) therapy to suppress and/or reduce brain edema associated with putaminal hemorrhage of patients whose clinical grades were grade 3 or 4a classified according to the Japanese neurological grading for putaminal hemorrhage. In the treated group of 11 patients, 25% albumin solution was intravenously administered (50–100 ml/day) with additional use of furosemide (20–40 mg/day) following hematoma removal. The serum COP was maintained at 25–30 mmHg for 2 weeks. In the untreated group of 11 patients, the COP therapy was not applied following hematoma removal. The serum COP was 20–25 mmHg for 2 weeks thereafter. During the 2-week observation period, serum osmolality, electrolyte, and hematocrit levels did not significantly differ between the two groups. The midline structure shift on CT of the treated group was 4.5 mm, which was significantly smaller than that of the untreated group (p < 0.05). The numbers of patients either in the vegetative state or death were 0 and 3, respectively, in the treated and the untreated groups. We concluded that high COP therapy for 2 weeks following hematoma removal was effective to suppress and/or reduce brain edema associated with putaminal hemorrhage, and that this therapy could be continued for 2 weeks without systemic complications.

Keywords: Brain edema; colloid oncotic pressure; cerebral hemorrhage.

Introduction

Osmotic diuretics such as mannitol or urea, have been used as treatment for brain edema. However, the effect of these osmotic diuretics is rather short because of their small molecular weight and low reflection coefficient. Moreover, therapy utilizing these diuretics is often followed by a rebound phenomenon.

On the other hand, large molecules such as albumin have a long half-life, and maintain prolonged high serum colloid oncotic pressure (COP), even though they have less of an effect on serum osmolality. It is conceivable that the increase of COP in the plasma can lead to a continuous withdrawal of water from the edematous brain.

Albright et al.[1] have reported that the increased COP induced by administration of albumin and furosemide decreased water content in brain and decreased intracranial pressure on the rat brain with cold lesion. Korosue et al.[5] have also showed that the increased serum COP was effective to reduce the size of cerebral infarction on an ischemic model of dog.

In this study the object of high COP therapy was to maintain serum COP level 3–5 mmHg higher than the normal value for 2 weeks, by administering albumin with supplementary use of furosemide. We applied this therapy to the patients with putaminal hemorrhage.

Materials and Methods

We investigated the effects of high COP therapy on the patients of grades 3 and 4a classified according to the Japanese neurological grading for putaminal hemorrhage[4].

The numbers of patients in this study are shown in Table 1. Group A included eleven patients suffering from putaminal hemorrhage who received high COP therapy. Informed consent prior to the treatment was obtained from all families of the patients in this group. Group B included eleven patients suffering from putaminal hemorrhage who did not receive oncotic therapy. The mean age of group A was 57.7 ± 5.9 (SD) years, and that of group B was 53.4 ± 7.9 (SD) years. There was no significant difference between the mean ages of the two groups.

Following hematoma removal, the patients in group A received 25% albumin solution (50–100 ml/day) intravenously with supplementary use of furosemide (20–40 mg/day), for two weeks. Electrolyte fluid transfusion (1500–1800 ml/day) and glycerol (600 ml/day) were also given. The patients in group B received electrolyte fluid and glycerol in the same amounts as in group A. As the parameters, serum COP, water balance, body weight, central venous pressure (CVP), hematocrit, and serum osmolality were measured every day. In group A, COP was maintained between 25–30 mmHg. Insensible water loss was not included in water balance.

In order to assess the brain edema, axial CT scans (GE 9200) were repeatedly taken on the 1st, 3rd, 7th, and 14th day after the disease onset, and the midline structure shift was measured from the same two slice levels including the basal ganglia. The Glasgow outcome scale score was used to evaluate the patients' condition on the 3rd month after hemorrhage. Statistical analysis was done by Student's t-test at the significance level of p < 0.05.

Results

A negative water balance giving a total value of about 1 liter was obtained in a 2-week period for group A, whereas a positive water balance of about 3 liters was obtained for the same given period for group B. Values were significantly different between the two groups.

The CVP of group A was maintained at around 4–6 cm H_2O, and that of group B was around 6–9 cm H_2O. There was no significant difference between the two groups in terms of hematocrit or serum electrolyte levels. The serum osmolality level was 300–320 mOsm/l for the 2-week period in both groups, thus, no significant difference was obtained.

The time courses of serum oncotic pressure are shown in Fig. 1. The COP of group A was maintained

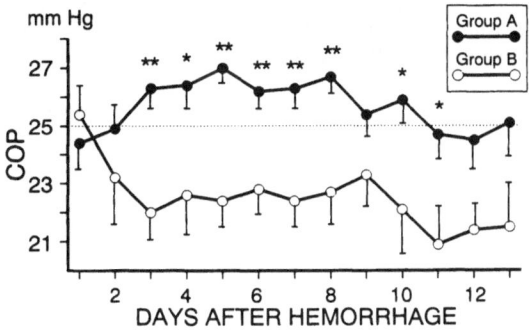

Fig. 1. Time courses of serum colloid oncotic pressure of groups A and B. Values are means ± SEM. * p < 0.05; ** p < 0.01; Groups A and B: see Table 1

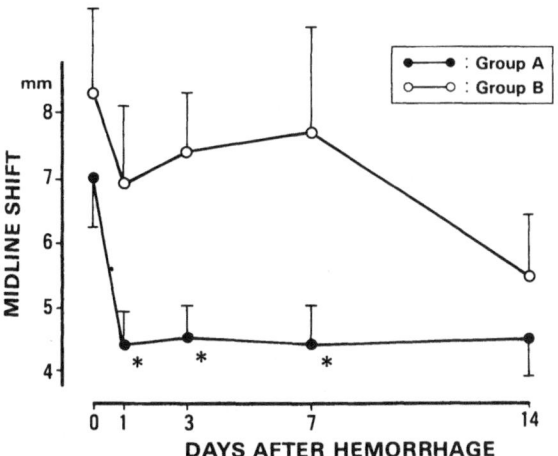

Fig. 2. Chronological changes in the midline structure shift on CT scan. Values are means ± SEM. * p < 0.05

Table 1. *Neurological Grading and the Number of Patients in Group A and B*

Neurological grade	Number of patients	
	Group A (n = 11)	Group B (n = 11)
3	6	5
4a	5	6

Group A: with high colloid oncotic therapy.
Group B: without high colloid oncotic therapy.

above 25 mmHg from 3 to 10 days after surgery, which was significantly higher than that of group B.

Figure 2 illustrates the temporal profile of the midline structure shift on CT scan for 2 weeks. The midline structure shift of group A was as low as 4.5 mm, and did not increase during the 2-week period. In group B, however, it increased within 1 week after hemorrhage, and was significantly higher than the corresponding values in group A from day 1 to 7.

According to the Glasgow outcome scale score evaluated 3 months after hemorrhage, the number of patients either in the vegetative state or death was zero in group A, but 3 in group B.

Discussion

Our results indicate that persistent high COP therapy for 2 weeks reduces brain edema associated with putaminal hemorrhage, and even prevents development of brain edema. Bulk flow through the capillary wall between plasma and extracellular space is effected, according to Starling's law[8], by hydrostatic pressure gradient and colloid oncotic and osmotic gradient between the both compartments. As long as blood brain barrier is intact, the COP of the brain tissue is almost 0 mmHg. The serum COP is counterbalanced by the hydrostatic gradient between the two compartments[7]. Theoretically, the continuous increase of serum COP to values 5 mmHg higher than the normal value leads to withdrawal of about 20 ml/day of water from the entire human brain, if the hydrostatic pressure is constant. Although the role of colloid oncotic pressure gradient is observed to be small for water movement during the short duration of previously reported experiments[3,10], it is thought that the continuous increase of serum COP is effective in reducing brain edema associated with putaminal hemorrhage.

In the edematous brain tissue, when blood brain barrier is destroyed, leakage of plasma protein into brain parenchyma[2,6] could reduce the effectivity of high COP therapy. However, no evidence of the serum protein leakage has been reported in the edematous brain tissue associated with putaminal hemorrhage to date[9].

Furthermore, utilizing this therapy, the improvement of regional cerebral blood flow in the edematous tissue could have led to the prevention of edema development from the very next day after the start of high COP therapy (Fig. 2). These observations still require further investigation.

Albright *et al.*[1] have reported that furosemide alone increased serum COP, but could increase serum viscosity due to hemoconcentration. However, administration of albumin and furosemide created a normovolemic dehydration and maintained hematocrit and serum electrolytes levels within normal range. Therefore, high COP therapy could be safely continued systemically for a prolonged period as 2 weeks.

References

1. Albright AL, Latchaw RE, Robinson AG (1984) Intracranial and systemic effects of osmotic and oncotic therapy in experimental cerebral edema. J Neurosurg 60: 481–489
2. Bakay L (1965) The movement of electrolytes and albumin in different types of cerebral edema. Prog Brain Res 15: 155–183
3. Kaieda R, Todd ML, Cook LN (1989) Acute effects of changing plasma osmolality and colloid oncotic pressure on the formation of brain edema after cryogenic injury. Neurosurgery 24: 671–678
4. Kanaya H, Yukawa H, Itoh Z, *et al* (1978) A neurological grading for patients with hypertensive intracerebral hemorrhage and a classification for hematoma location on computed tomography. Proceedings of The 7th Conference of Surgical Treatment of Stroke, pp 265–270
5. Korosue K, Heros RC, Ogilvy CS (1990) Comparison of crystalloids and colloids for hemodilution in a model of focal cerebral ischemia. J Neurosurg 73: 576–584
6. Pollay M, Robert PA (1980) Blood-brain barrier. A definition of normal and altered function. Neurosurgery 6: 675–685
7. Rapoport SI (1976) Blood-brain barrier in physiology and medicine. Raven, New York, pp 122–125.
8. Starling EH (1896) On the absorption of fluids from the connective tissue spaces. J Physiol (Lond) 19: 312–326
9. Wakai S, Andoh Y, Ochiai C, *et al* (1990) Postoperative contrast enhancement in brain tumors and intracerebral hematomas: CT study. J Comput Assist Tomogr 14: 267–271
10. Zornow MH, Todd MM, Moore SS (1987) The acute cerebral effects of changes in plasma osmolality and oncotic pressure. Anesthesiology 7: 936–941

Correspondence: Osamu Tone, M. D., Department of Neurosurgery, Musashino Red Cross Hospital, 1-26-1, Kyonan-cho, Musashino City, Tokyo 180, Japan.

Acta Neurochir (1994) [Suppl] 60: 571–573
© Springer-Verlag 1994

Brain Osmolyte Content and Blood-Brain Barrier Water Permeability Surface Area Product in Osmotic Edema

J. E. Olson[1,2], **J. A. Evers**[1], and **M. Banks**[1]

[1] Department of Emergency Medicine, and [2] Department of Physiology and Biophysics, Wright State University School of Medicine, Dayton, OH, U.S.A.

Summary

Brain edema was induced in adult rats by intraperitoneal injection of distilled water equivalent to 15% of the animal's body weight. Mean ± SEM serum osmolality fell from 291 ± 3 mOsm to 253 ± 4 mOsm during the next hour while cerebral gray matter water content increased from 79.5 ± 0.2% to 80.9 ± 0.2%. Gray matter content of sodium, potassium, taurine, glycine, glutamine, and glutamate were unchanged. However, the blood-brain barrier permeability/surface area product for water decreased by 40%. This alteration in water permeability may represent a response to limit water influx during the first hour of hypoosmotic brain edema.

Keywords: Sodium; potassium; amino acids; specific gravity.

Introduction

Animals with serum hypoosmolality develop brain edema; however, the resulting increase in brain water content is lower than that predicted from osmotic equilibrium across the blood-brain barrier[6,18]. Following several hours of hypoosmotic exposure, brain tissue looses sufficient sodium, potassium and chloride to account for osmotic protection from swelling[6]. After 12 to 24 h, loss of organic osmolytes contribute to the recovery of the brain to normal water content despite persistent hypoosmotic conditions[14–16]. Mechanisms which prevent excess swelling during the first hour of hypoosmotic exposure remain unclear.

We explored the role that amino acid osmolytes play in cerebral volume regulation during the first hour of serum hypoosmolality. In a separate series of experiments, we examined alterations in cerebral blood flow and water permeability of the blood-brain barrier induced by hypoosmotic treatment.

Materials and Methods

All studies using live animals were approved by the Laboratory Animal Utilization Committee of Wright State University. Adult Sprague-Dawley rats (250–350 g) were anesthestized with 55 mg/kg pentobarbital. Femoral artery and venous catheters were placed for blood sampling and infusions. Arterial blood pressure and blood gas values were determined periodically. Experimental animals were injected intraperitoneally (IP) with distilled water equivalent to 15% of their body weight and sacrificed 60 min later. Control rats did not receive IP water and were sacrificed at approximately the same time after induction of anesthesia that water was injected into experimental rats. For analysis of gray matter cation and amino acid contents, the brain was frozen in situ by placing liquid nitrogen into a funnel attached to the skull. After 5 min, the head was removed and frozen at −70°C. To determine cerebral blood flow and blood-brain barrier water permeability/surface area (PS) product, 25 μCi of [14C]-iodoantipyrine plus 50 μCi of [3H]-water in saline was infused via the femoral vein over 1 min while blood was withdrawn from the arterial catheter. The infusion rate was adjusted to give a linearly increasing arterial content of radiolabeled compounds during the infusion period. The animal then was decapitated and the head frozen by immersion in liquid nitrogen.

Samples (1–4 mg) of parietal gray matter were removed from a coronal section just posterior to Bregma[9]. Tissue specific gravity was measured using a hydrophobic density gradient column[7]. Other samples of gray matter were placed in 1 ml of 0.6 M $HClO_4$ plus 4 mM $CsCl_2$ for determination of sodium, potassium, amino acid, and protein contents. In animals infused with radioisotopes, brain samples were weighed and then dissolved for liquid scintillation counting. Cation content was determined by atomic absorption spectroscopy of perchloric acid extracts. Amino acids were measured by HPLC in the same extract after neutralizing the acid with KOH[8]. Protein was measured in the perchloric acid pellet using the method of Lowry *et al.*[5]. Cerebral blood flow and blood-brain barrier PS product for water was determined from the quench-corrected activities of 14C and 3H in blood and brain samples using the method of Reid *et al.*[11].

Results

During the hour after IP water injection, mean ± SEM serum osmolality and sodium concentration dropped from 291 ± 3 mOsm and 131 ± 13 mequiv/l to 253 ± 4 mOsm and 75 ± 6 mequiv/l, respectively. A significant decrease in tissue specific gravity was observed (Table 1). In contrast, brain content of sodium, potassium, taurine, glutamate, glutamine, and glycine were unchanged. The sum of these osmolytes also was not significantly affected by the hypoosmotic treatment (1538 ± 79 nmol/mg protein for control rats and 1380 ± 64 nmol/mg protein for water injected rats, mean $\pm$ SEM).

A second series of animals had a similar decrease in tissue specific gravity after water injection (Table 2). Tissue blood flow and water PS product was significantly lower in animals injected with distilled water. The systolic and diastolic blood pressures for control animals were 149 ± 9 mmHg and 99 ± 10 mmHg at the start of radioisotope infusion (mean $\pm$ SEM). At this time, the mean $\pm$ SEM arterial pO_2, pCO_2 and pH values were 84 ± 3 mmHg, 35 ± 2 mmHg, and 7.38 ± 0.01, respectively. These values were not different for animals sacrificed 1 h after IP water injection. During infusion of radio-isotopes and withdrawal of blood, the mean arterial pressure fell approximately 20 mmHg in both groups.

Discussion

From the specific gravity data in Table 1, we calculate that brain water content was $79.5 \pm 0.2\%$ in control rats and $80.9 \pm 0.2\%$ in water-injected rats[7]. However, using the measured values of serum osmolality, the final brain water content should have increased to 81.7% assuming osmotic equilibration across the blood-brain barrier and no net change in brain osmolyte content. This disparity of predicted and measured results suggests that one or both of these assumptions are not valid.

Our results indicate that the major osmolytes implicated in volume regulation of isolated brain cells[2,10] and in control of brain water content in chronic hypoosmotic conditions[14] do not contribute to brain volume regulation during the first hour of hypoosmotic exposure. Melton *et al.*[6] reported a decrease in the cerebral content of sodium, 1 h after IP water injection. We found that the mean sodium content in water-injected rats was less than that of controls; however, the difference was not statistically significant. This may be due to the greater variance we obtained using small tissue samples.

Table 1. *Cerebral Specific Gravity and Cerebral Cation and Amino Acid Contents*

Animal group	Tissue specific gravity	Tissue contents (nmol/mg protein)					
		Na	K	Tau	Glu	Gln	Gly
Control	1.0445 ± 0.0004	383 ± 40	1084 ± 72	48 ± 8	82 ± 8	34 ± 2	9 ± 1
Water-injected	1.0412 * ± 0.0004	322 ± 16	979 ± 57	43 ± 4	70 ± 9	38 ± 3	10 ± 2

Values are the mean ± SEM of 21–37 animals.
* Indicates mean values significantly different from that obtained from control animals (Student's test, p < 0.5).

Table 2. *Cerebral Blood Flow and Blood-Brain Barrier Water Permeability × Surface Area Product*

Animal group	Specific gravity	Blood flow (ml/gm min)	Water permeability × surface area product (ml/gm min)
Control	1.0438 ± 0.0006	1.00 ± 0.15	0.78 ± 0.09
Water-injected	1.0393* ± 0.0007	0.48* ± 0.02	0.46* ± 0.09

Values are the mean ± SEM of 5 animals.
* Indicates mean values significantly different from that obtained from control animals (Student's test, p < 0.5).

Cerebral amino acid contents were not affected by hypoosmotic exposure. Numerous studies have demonstrated increased concentrations of taurine and glutamate in brain microdialysis perfusates of water-injected rats[3,13,17]. Lehmann demonstrated large increases in plasma taurine in a similar model of serum hypoosmolality and suggested that the gradient for taurine across the blood-brain barrier may favor transport into the brain[4]. Our data indicate that serum hypoosmolality results in no net efflux of taurine from cerebral gray matter. Taurine may be mobilized during hypoosmotic conditions to transport osmolytes (and accompanying water) between brain compartments. The release of taurine by hypoosmotically swollen astrocytes[10] and the uptake of taurine by diffusion from media containing high taurine concentrations[12] support this hypothesized role for taurine in the brain.

The half-time for osmotic equilibration of water across the blood-brain barrier is 12–15 min in rabbits[1]. Assuming a similar value for rats, the blood and brain osmolality should have come to near equilibrium during the 1 h time course of this experiment. However, the data of Table 2 indicate that water permeability of the blood-brain barrier decreases during hypoosmotic exposure. If this lowered permeability correlates with decreased osmotic water movement into the brain, it would indicate a lengthening of the osmotic equilibration half-time. As a result, the rate of brain swelling during systemic hypoosmolality would be slowed, providing a temporizing measure to limit brain volume changes until osmolytes can be mobilized and transported out of the brain parenchyma. Mechanisms which regulate this osmotically induced permeability change and its functional significance for brain volume control must await further experimentation.

Acknowledgements

Supported by the Emergency Medicine Foundation and NIH (NS 23218).

References

1. Fenstermacher JD (1984) Volume regulation of the central nervous system. In: Staub NC, Taylor AE (eds) Edema. Raven, New York
2. Kimelberg HK, Frangakis MV (1985) Furosemide- and bumetanide-sensitive ion transport and volume control in primary astrocyte cultures from rat brain. Brain Res 361: 125–134
3. Lehmann A (1989) Effects of microdialysis-perfusion with anisoosmotic media on extracellular amino acids in the rat hippocampus and skeletal muscle. J Neurochem 53: 525–535
4. Lehmann A, Carlström C, Nagelhus EA, Ottersen OP (1991) Elevation of taurine in hippocampal extracellular fluid and cerebrospinal fluid of acutely hypoosmotic rats: contribution by influx from blood? J Neurochem 56: 690–697
5. Lowry OH, Rosebrough NJ, Farr AL, Randall RJ (1951) Protein measurement with the Folin phenol reagent. J Biol Chem 193: 265–275
6. Melton JE, Patlak CS, Pettigrew KD, Cserr HF (1987) Volume regulatory loss of Na, Cl, and K from rat brain during acute hyponatremia. Am J Physiol 252: F661–F669
7. Nelson SR, Mantz ML, Maxwell JA (1971) Use of specific gravity in the measurement of cerebral edema. J Appl Physiol 30: 268–271
8. Olson J, Goldfinger MD (1990) Amino acid content of rat cerebral astrocytes adapted to hyperosmotic medium in vitro. J Neurosci Res 27: 241–246
9. Olson J, Mishler L, Dimlich R (1990) Brain water content, brain blood volume, blood chemistry and histopathology in a model of cerebral edema. Ann Emerg Med 19: 1113–1121
10. Pasantes-Morales H, Schousboe A (1988) Volume regulation in astrocytes: a role for taurine as an osmoeffector. J Neurosci 20: 505–509
11. Reid AC, McCulloch J (1983) The effects of dexamethasone administration and withdrawal on water permeability across the blood-brain barrier. Ann Neurol 13: 28–31
12. Sanchez-Olea R, Moran J, Schousboe A, Pasantes-Morales H (1991) Hypoosmolarity-activated fluxes of taurine in astrocytes are mediated by diffusion. Neurosci Lett 130: 233–236
13. Solis JM, Herranz AS, Herreras O, Lerma J, Martin del Rio R (1988) Does taurine act as an osmoregulatory substance in the rat brain? Neurosci Lett 91: 53–58
14. Thurston JH, Hauhart RE, Nelson JS (1987) Adaptive decreases in amino acids (taurine in particular), creatine, and electrolytes prevent cerebral edema in chronically hyponatremic mice: rapid correction (experimental model of central pontine myelinolysis) causes dehydration and shrinkage of brain. Metab Brain Dis 2: 223–241
15. Trachtman H, Futterweit S, Hammer E, Siegel TW, Oates P (1991) The role of polyols in cerebral cell volume regulation in hypernatremic and hyponatremic states. Life Sci 49: 677–688
16. Verbalis JG, Gullans SR (1991) Hyponatremia causes large sustained reduction in brain content of multiple organic osmolytes in rats. Brain Res 587: 274–282.
17. Wade JV, Olson JP, Samson FE, Nelson SR, Pazdernik TL (1988) A possible role for taurine in osmoregulation within the brain. J Neurochem 51: 740–745
18. Wasterlain CG, Torack RM (1968) Cerebral edema in water intoxication. Arch Neurol 19: 79–87

Correspondence: James E. Olson, Ph.D., Department of Emergency Medicine, Cox Institute, 3525 Southern Boulevard, Kettering, Ohio 4529, U.S.A.

Acta Neurochir (1994) [Suppl] 60: 574–576
© Springer-Verlag 1994

A Square Signal Wave Method for Measurement of Brain Extra- and Intracellular Water Content

M. L. Itkis, J. K. Roberts, J. B. G. Ghajar, and **R. J. Hariri**

The Aitken Neurosurgery Laboratory, Division of Neurosurgery, Department of Surgery, Cornell University Medical College, New York, NY, U.S.A.

Summary

Brain tissue electrical impedance is a commonly used method to evaluate the dynamics of brain edema. We have found the square wave impedance method simpler and more cost-effective than the currently used sine wave impedance method. This square wave method avoids the necessity for expensive frequency control and amplitude-phase measuring devices as well as simplifying on-line data processing. In our experiments the electrical impulse was generated by a pulse generator of Macintosh data acquisition system. The signal ($I = 11\,\mu A$, $t = 2$–$20\,ms$) was delivered every 2–3 s external electrodes of a tetrapolar system through a specially designed isolation-calibration device. This electrode system was inserted into the cerebral cortex of experimental animals (rat). The cerebral cortex was found to have linear electrical properties in the 5–30 μA range. Our impedance measurement system was tested in calibration trials, and showed system reliability and accuracy. The system was also tested in pilot experiments, in vivo, in a rat brain osmotic edema model.

Keywords: Brain edema; impedance measurements.

Introduction

The continuous measurement of brain water content (WC) is one of the most important tasks in the study of Brain Edema. The impedance method still remains the best approach for WC dynamics and extracellular space volume evaluation[5,7,9].

The experimental study of the electrical properties of soft tissue generally showed decreasing resistance, with increasing frequency[1,2,8]. Low frequency signals propagate mostly through the ECS, while higher frequencies propagate through the cell membrane as well, due to its higher capacitance. These tissue properties were simulated using a simple, electrical equivalent circuit[1,2,4,8]. Such circuits have to include the resistance of extracellular (Re) space, which is parallel to serially connected intracellular (Ri) resistance and membrane capacitance Cm (CC on Fig. 1A).

Impedance of living tissues was studied through the in vivo experiments using sine wave signals[4,12], as well as square wave signals[6,7]. Both approaches have their own advantages and limitations. Thus, square wave methods are relatively simpler and cost effective. More information could be obtained through in vivo experiments using the sine wave signal of changing frequency and the so called "admittance locus" method[1,4,8]. By plotting values of reactive, versus real components of admittance, the above authors obtained absolute values of extra (Ri), and intracellular (Re) resistances and membrane capacitance (Cm), employing the semi-circular "admittance locus". The main limitation of this method is such that it requires a special automatic device for frequency control. Precise measurements of signal amplitude and phase angle difference also require additional more costly equipment. As well, the algorithm for data analysis and "admittance locus" plotting is quite complex, especially for on-line data processing[9,10].

Literature survey shows only one trial, using a square wave signal to plot "admittance locus" for (Ri) and (Re) determination[12]. Unfortunately, the wrong electrical model of tissue, where (Ri) and (Re) were connected serially, was applied.

We are proposing a method which has advantages of simplicity, cost effectiveness and informativeness over both approaches. This method also allows for on-line data processing.

Theory

Extra (Re) and intracellular resistance (Ri) values could be obtained by applying the "Hevisaid func-

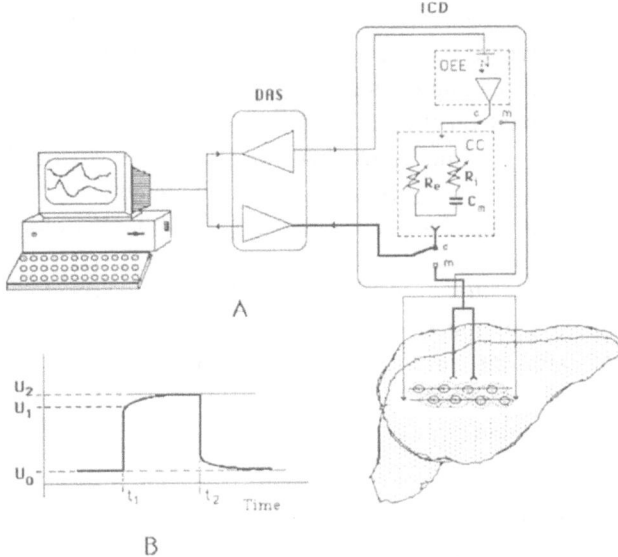

Fig. 1. Tetrapolar method for an impedance measurement using square wave signal (A) and typical experimental curve (B). *DAS* computer Data Acquisition System, *ICD* isolation-calibration device; *OEE* optoelectrollic element for isolation animal brain from electrical circuit, *CC* calibration component – circuit simulating the brain tissue electrical properties, *Re, Ri* extra- and intracellular space resistances, *Cm* cell membrane capacitance. Arrows show the current directions. Thin lines show stimulation (external) electrodes on the animal brain. Thick lines show measurement (internal) electrodes. Calibration/measurement (*c/m*) switch is shown in calibration mode. In the animal brain straight lines show pathways of high frequency signals and dashed curve lines show pathways of low frequency signals through the ECS. (B) *Uo* zero signal before the impulse beginning, on internal electrodes, *U₁* instant level of voltage loss after beginning of impulse, *U₂* the maximum level of voltage loss on measurement electrodes in the impulse end (see explanation in the text)

tion"[1,2,8] to the tissue equivalent circuit (see CC on Fig. 1A). The value of impedance for such a circuit is shown in the equation below:

$$Z = R\infty + (Ro - R\infty)\,[1 - \exp(-t/To)] \quad (1)$$

Here: Z – impedance, $R\infty$ is the resistance at frequency F = ∞ when t is close to 0 (t ≪ To = Ri · Cm), $R\infty = Re \cdot Ri / (Ri + Re)$, and Ro = Re, when Ro is the resistance at frequency F = 0, (t ≫ To).

This equation is correct only for low current intensities where Ohm's law is fulfilled[8]. From Fig. 1A and 1B it is clear that:

$$\Delta U_1 = U_1 - Uo = iR\infty = i \cdot (Re \cdot Ri) / (Ri + Re)$$

and

$$\Delta U_2 = U_2 - Uo = i \cdot Ro = i \cdot Re$$

Thus, the algorithm for the Re and Ri evaluation has to include:

1. The measurement of a maximal value of voltage loss close to the end of the impulse – ΔU_2, which gives us a value proportional to Re.
2. The measurement of the instant value of a voltage loss after the beginning of the impulse – ΔU_1, as a function of Ri and Re.

Equipment

In these experiments (Fig. 1A), generation and registration of square wave signals were produced by means of a data acquisition system – Analog Connections WorkBench (Macintosh), with a special terminal panel T-51 for all inputs and outputs.

Signals of 1–1.5 v intensity (2 < t < 20 ms duration every 2–3 s) from computer analog output, through the Isolation-Calibration Device (ICD), (specially developed for the present studies), were delivered to the stainless-steel external electrodes of the tetrapolar impedance measurement device[6].

This ICD has such options, as the ability to:

1. Electrically isolate the circuit from the animal brain.
2. Deliver, via external stimulation electrodes, an impulse signal of stable current which could be adjusted in definite limits (from 5 to 50 microA).
3. Simulate the work of an equivalent electrical circuit of the brain (CC on Fig. 1A) and change, for calibration aims, the Re and Ri resistances, without changing the current through these calibration resistors.

A special program for Re and Ri evaluation was developed using the Omega Engineering Omegalog software (Analog Connection WorkBench™ Release 3.0, Strawberry Tree Inc.). This program also allowed us to record our physiological data (8 analog channels).

In the present experiments, stainless-steel electrodes of 0.49 mm diameter, with a 2–4 mm distance between tips, were used.

Calibration and Testing

Because of the relatively low frequency of AD converters, the paper shows only the results obtained using the low-frequency portion of the signal.

Preliminary results were obtained through in vivo

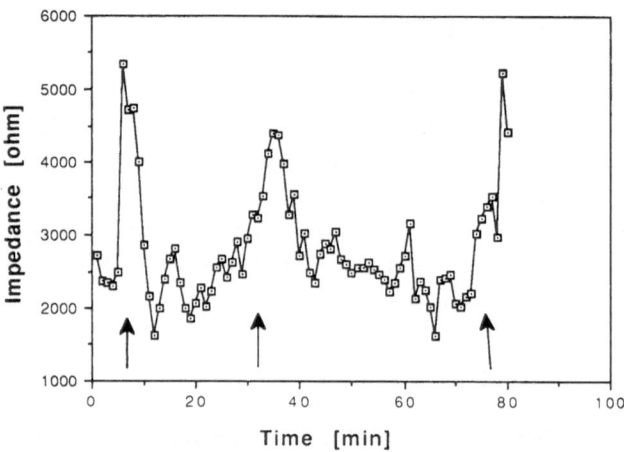

Fig. 2. The results of the impedance study on rat brain cortex in vivo. Arrows show the increases of impedance, corresponds to water injections in animal blood stream (see explanation in the text)

experiments on rat brain. Testing of the above mentioned devices and program accuracy was done in a series of calibration model experiments on resistors of ICD (Fig. 1A)

In the first in vivo experiments (n = 5) the relationship between the input current (U_{inp}) and loss of voltage (U_{out}) was studied. For the current range from I = 5 μA to I = 30 μA experiments showed that rat brain has practically linear (r = 0.928 ± 0.04) volt-amper characteristic, thus following Ohm's law.

Discussion

The program reliability and accuracy of impedance measurements using the calibration resistors of ICD were tested in the present work, simulating the electrical properties of brain tissue. In these model experiments (n = 9), linearity of input-output characteristics were also studied. The input characteristics in this case were also Uinp values, while the output characteristics were served Uout-loss of voltage on calibration resistors of ICD. The calibration resistances were changed from 0 ohm to 11 Kohm at nine current levels from 5 to 25 μA. The results, which were obtained in this run of experiments, allow us to choose the best range for working current – from 10.9 to 12.8 microA. In this

range the measured and theoretical values of impedance have no significant differences (p > 0.05).

As an example, the average data from one pilot experiment on the open rat brain is shown (Fig. 2). Changes of impedance here are related to artificial distilled water injection in the blood stream. This increase in impedance corresponds to the dilution of ions in ECS[3,5].

The results of this study allow us to offer this method, of impedance measurement and the evaluation of extra- and intracellular water content changes, for use in studies of various physiological events in the animal brain.

References

1. Ackman J, Seitz MA (1984) Methods of complex impedance measurements in biological tissue. Crit Rev Biomed Eng 281–311
2. Geddes LA, Baker LE (1989) Principles of applied biomedical instrumentation, 3rd ed. Wiley, New York, p 961
3. Herschkowitz N, MacGillvary BB, Cumings JN (1965) Biochemical and electrophysiological studies in experimental cerebral edema. Brain 88: 557–584
4. Kanai H, Sakamoto K, Haeno M (1983) Electrical measurement of fluid distribution in human legs: estimation of extra- and intracellular fluid volume. J Michrow Power Electromagn Energy 18: 233–243
5. Kao HP, Shwedyk E, Cardoso ER (1991) Measurement of canine cerebral oedema using vector impedance method. Neurol Res 13: 233–236
6. Li CH, Bak A, Parker L (1968) Specific resistivity of the cerebral cortex and white matter. Exp Neurol 20: 544–557
7. Matsuoka Y, Hossmann KA (1982) Cortical impedance and extracellular volume changes following middle cerebral artery occlusion in cats. Cereb Blood Flow Metab 2: 466–4744
8. Schwan HP (1957) Electrical properties of tissue and cell suspensions. Adv Biol Med Phys 5: 147–207
9. Suga S, Mitani S, Shimamoto Y, Kawase T, Toya S, Sakamoto K, Kanai H, Fukui M, Takeneka N (1990) In vivo measurement of intra- and extracellular space of brain tissue by electrical impedance method. Acta Neurochir (Wien) [Suppl] 51: 22–24
10. Tender BT (1978) Automatic recording of biological impedances. J Med Eng Technol 2: 70–75
11. Teorell T (1947) Application of "square wave analysis" to bioelectric studies. Acta Phys Scand 12: 235–254
12. Van Harreveld A, Ochs S (1957) Cerebral impedance changes after circulatory arrest. Am J Phys 187: 180–192

Correspondence: Robert J. Hariri, M. D., Ph. D., Division of Neurosurgery, Cornell University Medical College, 1300 York Avenue, Room LC-807, New York, NY 10021, U.S.A.

Acta Neurochir (1994) [Suppl] 60: 577–581
© Springer-Verlag 1994

The Efficacy of Shunting the Hydrocephalic Edema

F. Takei[1], K. Shapiro[2], S. Oi[1], and O. Sato[1]

[1] Department of Neurosurgery, Tokai University School of Medicine, Bohseidai, Isehara, Kanagawa, Japan, and
[2] Department of Neurosurgery, Humana Hospital, Medical City, Dallas, TX, U.S.A.

Summary

The efficacy of shunting the hydrocephalic edema was evaluated by means of transmission electromicroscopical observation (TEM) comparing ultrastructual alterations seen in either valid or invalid shunted feline hydrocephalus. Owning to shunt placement, deteriorated clinical symptoms recovered and one of the observed morphological alterations such as gliosis developed chronologically after the initiation of hydrocephalus. On the other hand, so called hydrocephalic edema observed in the region of periependymal tissue after shunt placement was improved not only in both valid and invalid shunted but also even in sham group however the extent of these alterations were different from each other. Among such observation, a distinct chronological linkage between the morphological alterations and clinical outcome was not noted. Our conclusions in present study were 1) shunt implantation results in the promotion of gliosis in the region of periventricular tissue in spite of the expected efficacy of shunt, 2) recovery of hydrocephalic edema was influenced not only by effective shunt implantation but also by the processes in chronological tissue reconstructions occurred in the natural course of hydrocephalus, and 3) early treatment of periventricular edema in hydrocephalus through shunting corrects the ill-fated neuro axis environments and reforming the intracranial conditions such as normalizing ICP and relieving hemodynamic distress seems to be more important than morphological recovery in treating hydrocephalus.

Keywords: Feline hydrocephalus; shunt; ultrastructure; gliosis.

Introduction

In hydrocephalic research, extrathecal shunting is generally accepted as a reliable technique for treating hydrocephalus. However such a shunting technique is emphasized, only a limited number of the morphological investigations focused on the efficacy of shunt are available[1,5,6]. Therefore the purpose of present study is to evaluate the efficacy of shunting in the hydrocephalic edema by means of the ultrastructual observation comparing the morphological differences seen on both valid and invalid shunted feline hydrocephalus.

Material and Method

Animal Preparations

Detailed descriptions for animal preparation was documented elsewhere[10]. Thirty five adult mongrel cats were used in this study. After bilateral craniectomies were performed, the animals were categorized as the group of dura removed (DR: n = 16) and the dura left intact (DI: n = 14) for purposes of obtaining the different rates of the ventricular dilatations on a process of hydrocephalus.

Following the surgery described above, one ml of kaolin suspension was injected into the cisterna magna. The valveless ventriculo-pleural shunt placement[2] was carried out three weeks after the initial preparation when the cat developed hydrocephalic signs such as lethargy, ataxic gait, anorexia and bulging of the brain through the open calvaria. After 3 to 5 weeks following the shunt operation, in vivo perfusion was performed in all animals for transmission electromicroscopical (TEM) study. At the time of the perfusion, shunted hydrocephalics were categorized into the valid shunt (DI: n = 4, DR: n = 4) and invalid shunt groups (DI: n = 3, DR: n = 4). The valid shunt group fits the following conditions such as 1) a flat craniectomy site in visual inspection, 2) improved clinical behavior, 3) non-stagnant CSF flow through the shunt tube when evaluated immediately prior to the in vivo perfusion and 4) an A/B ratio of less than 0.3^{11}. The invalid shunt group failed to clear these conditions with the A/B ratio of more than 0.4. Preparation of the samples for the TEM was made by the technique reported elsewhere[10]. Ultrastructual observation was made at three different locations in the periventricular tissue: W1 (0–1 mm), W2 (1–2 mm), and W3 (2–3 mm). All animals observed here were compared to the normal (n = 5), pre-shunt (DI: n = 4, DR: n = 4: three weeks after the initiation of hydrocephalus), and time-matched sham group (DI: n = 3, DR: n = 4: 6–8 weeks after initiation of hydrocephalus). None of the animals were used when the ventriculomegaly, assessed by individual A/B ratio, failed to fit in the standard curve which we have reported[11].

Result

Eight shunted animals were classified as valid while 19 cats received shunt.

At the time of shunt placement, once deteriorated clinical symptoms of cats after kaolin injection were somewhat improved showing no distinct difference between the DI and DR groups however the mortality rate and clinical distress within a first week after the initial preparation of the animals were greater in DR group than DI. After effective shunt was established, cats recovered from lethargic condition and activities improved. In some effective shunt cats, however, ataxic gait remained unchanged due to the syringomyelia which usually coincided in a process of kaolin induced hydrocephalus. In chronic stage, clinical symptoms in invalid shunted groups were grossly as same as the level of sham group but were much worse than in the valid shunted group.

Ultrastructual Alteration in Observed Animals

Pre-Shunted Group (3 Weeks)

Detailed ultrastructual observations of these animals have been reported previously elsewhere[10].

Group with the dura left intact (DI): The extracellular space (ECS) in W1 was widely expanded and considerable glial reaction was found in W1. In W2, the expansion of ECS was less prominent compared to W1. Profound gliosis was also observed. Neither the expansion of the ECS nor the glial reaction was found in W3.

Group with the dura removed (DR): The expansion of ECS was severe and glial fibers were scarcely observed in W1. In W2, ECS was also expanded, however, glial fibers was not obvious. Normal configurations was preserved in W3.

The Valid Shunted Group (6–8 Weeks): Fig. 1

Gross aspect of morphological alterations seen in DI group was similar to that of DR group.

Group with the dura left intact(DI): The ventricles were similar in size however not as slit-like as usually compared to the normal cats. The continuity of the ependymal cells was preserved and the intercellular connection was kept tight and compact. The widely expanded ECS was no longer observed, however, the gliofibrils were dense and abundant in W1. The ECS had not expanded and development of gliofibrils was observed in W2 but its extent was less prominent than that of W1. In W3, histological alterations were not observed.

Group with the dura removed (DR): The diminished ECS and development of gliosis were observed in W1 but this finding was somewhat less evident than DI group of valid shunt. In W2, obvious expansion of ECS was not noted and observed gliofibrils were less prom-

inent than the group of DI of valid shunt animals. The histological findings in W3 remained in normal configurations.

Invalid Shunted Group (6–8 Weeks): Fig. 2

The morphological appearance in this group was basically similar to sham group.

Group with the dura left intact (DI): The expansion of ECS was observed in W1 and this was less prominent comparing to the pre-shunted condition. The gliosis was also observed and extended into the deeper region of the W2, however, ECS expansion in this foci was not obvious. In W3, definitive ultrastructual alterations were not found.

Group with the dura removed (DR): The reduction of expanded ECS and development of gliosis either in W1 or W2 were observed. This was somewhat less than that of DI group in invalid shunted however the other morphological alterations remained in same extent. Normal configuration was preserved in W3.

Sham Group (6–8 Weeks)

Morphological details in both group was previously reported elsewhere[10].

Group with the dura left intact (DI): The expanded ECS reduced in volume in this group. A large amount of gliofibrils and increased number of reactive astrocytes were observed (W1).These findings were more advanced than pre-shunted group but not more than that of both shunted group. In W2, glial fibers are compactly arranged and ECS was somewhat visible. Normal configuration was preserved in W3.

Group with the dura removed (DR): No distinct morphological differences were found compared to the sham control in this region. Glial fibers and reactive astrocytes were abundant in W1 however ultrastructural alterations both W2 and W3 were not obvious.

Discussion

Since the era of Hippocrates, remarkable number of trials and errors have been made to treat hydrocephalus by means of either the medical or surgical approaches. Among such historical background, the "modern era of the treatment of hydrocephalus" was conducted by Ingraham *et al.*[3], Nulsen and Spitz *et al.*[4] and after this point, extrathecal shunting of hydrocephalus has been generally accepted as one of the established treatments for hydrocephalus.

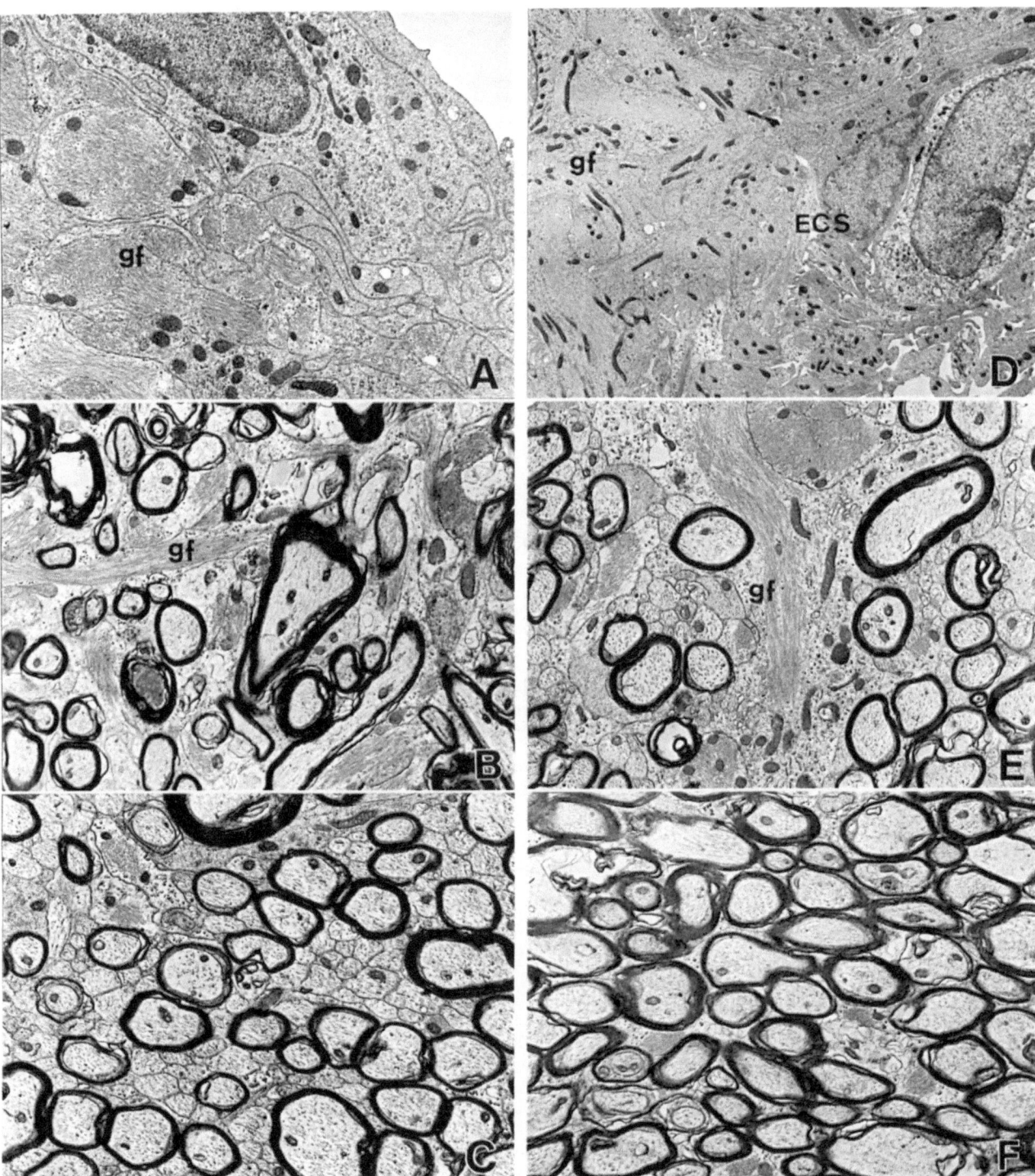

Fig. 1. Valid shunted group. Group with the dura left intact (DI): left: (A) W1: The continuity of the ependymal cell is preserved. The expanded extracellular space is not observed and dense layer of the glial fibers (*gf*) are developed. (B) W2: The expanded ECS is not observed. Glial fibers are rich in volume and observed among the myelinated axons. (C) W3: Obvious alterations are not found. Group with dura removed (DR): right: (D) W1: The continuity of the ependymal cell is preserved. The slightly expanded ECS and dense layer of the glial fibers (*gf*) are observed. (E) W2: The ECS is not obvious however developed glial fibers (*gf*) are observed among the myelinated axons. (F) W3: The tissue remains normal configuration

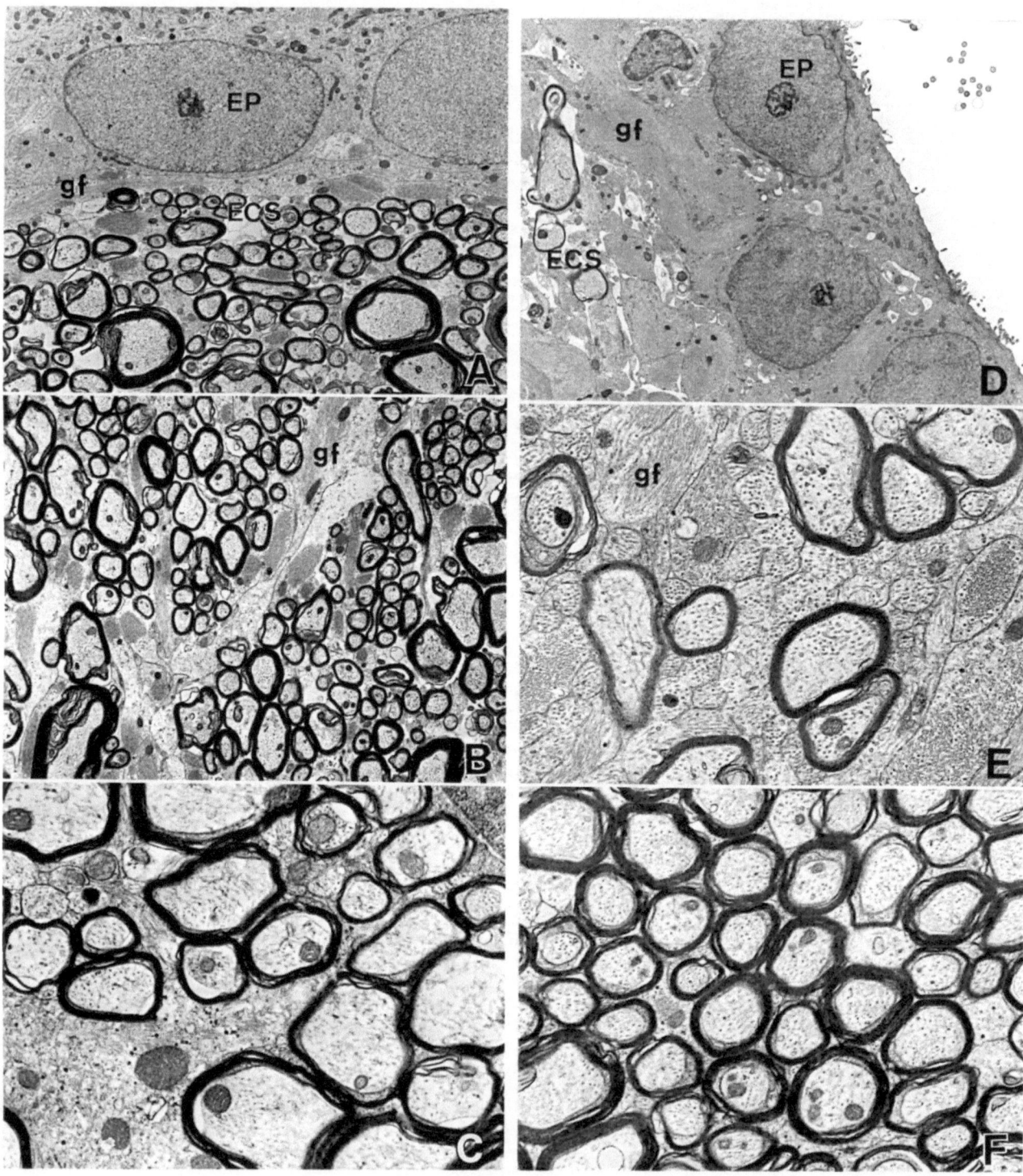

Fig. 2. Invalid shunted group. Group with the dura left intact (Dl): left: (A) W1: Ependymal lining (*EP*) is preserved. Slightly expanded ECS and developed glia fibers (*gf*) underneath the ependymal cell are observed. (B) W2: Slightly expanded ECS and developed gliofibrils (*gf*) are observed among the myelinated axons. (C) W3: Tissue configuration is similar to normal. Group with dura removed (DR): right: (D) W1: Expansion of the ECS is observed underneath the ependymal lining. Glial fibers (*gf*) are abundant. (E) W2: Glial fibers (*gf*) are observed and no obvious expanded ECS is seen. (F) W3: Tissue configuration is similar to normal

The most striking outcome in present study is the chronological development of gliosis in the periventricular region. In shunted animals, the observed gliosis was more developed than the other group such as pre-shunt and sham group. The gliosis in shunted DI group was more advanced than shunted DR group and this gliosis was transmigrated into the deeper region of the white matter. Furthermore, the gliosis observed in valid shunted was more developed than invalid shunted group. These findings indicated that the shunt placement promotes the progression of gliosis in the region of periventricular tissue. Considering the valid ventriculopleural shunt, the negative pressure created by intrathoracic placement of the shunt tube might cause an overdrainage of CSF from the ventricle resulting in an excessive reduction of ventricular size. This kind of barometric alterations, resulting in the rapid reduction of ventricles and evoking the gliosis, seem to insult the periventricular tissue even in the condition of decreased ICP as well as in the condition of increased ICP and, moreover, seem to be similar to the situation seen in the slit ventricle syndrome. This resemblance might be appreciated in interpreting the biomechanism of glial reactions in shunted animals. Del et al. observed the developed gliosis around the non-functioning shunt tube and concluded that the shunt placement might be a inducing factor for gliosis in the region of the periependymal tissue[1]. Most of the animals observed in acute stage of hydrocephalus, clinically deteriorated initially and recovered afterward while the group with dura removed showed high mortality rate and advanced clinical deterioration. The animals at the time of shunt operation, eventually no significant difference of symptoms were observed in spite of the ultrastructual difference between DI and DR groups were distinct. On the contrary to such a relationship, in chronic phase, either the clinical symptoms or histological alterations became similar in both groups. Therefore we were not able to find any affirmative linkage between clinical symptoms and ultrastructual alterations in both DI and DR group while the profile of chronological changes in clinical symptom and ventriculomegaly showed good accordance throughout the observation period.

As far as the factor contributing the histological differences observed in both DI and DR group, the initial tissue insult, which might be affected by the rate of enlarging ventricles in early stage of hydrocephalus, should play an important role in alteration of the tissue integrity. The biomechanical changes created after altering the container property of the brain[8] might have also affected in addition to other factors such as intracranial pressure[2,8,9] and hemodynamic distress[1,7]. These results indicate that the histological alterations might not being solely related to clinical outcome and careful chronological evaluation should be more important in treating hydrocephalus. Therefore, as long as the discrepancy between the clinical and histological alterations exists, the critical point in treating hydrocephalus is in the initial stage of hydrocephalus and treatment has to be started as early as possible. The correction of environmental disturbance of the neural axis should be emphasized more than that of the morphological recovery for the sake of prevention of wasting tissue integrity.

References

1. Del Bigio MR, Bruni JE (1988) Changes in periventricular vasculature of rabbit brain following induction of hydrocephalus and after shunting. J Neurosurg 69: 115–120
2. Fried A, Shapiro K, Takei F, Kohn I (1987) A laboratory model of shunt dependent hydrocephalus. Development and biomechanical characterization. J Neurosurg 66: 734–740
3. Ingraham FD, Matson DD, Alexander E Jr (1948) Studies in the treatment of experimental hydrocephalus. J Neuropathol Exp Neurol 7: 123
4. Nulsen FF, Spitz EB (1952) Treatment of hydrocephalus by direct shunt from ventricle to jugular vein. Surg Forum 2: 399
5. Oi S, Matsumoto S (1986) Morphological findings of post-shunt slit ventricle syndrome in experimental canine hydrocephalus. A aspect of causative factors of isolated ventricles and slit ventricle syndrome. Nerv Syst Child 2: 179–184
6. Rubin RC, Hochwald GM, Tiell M, et al (1976) Hydrocephalus: III Reconstitution of the cerebral cortical mantle following ventricular shunting. Surg Neurol 5: 179–183
7. Sato O, Ohya M, Nojiri K, Tsugane R (1984) Microcirculatory changes in experimental hydrocephalus: morphological and physiological studies. In: Shapiro K, Marmarou A, Portnoy H (eds) Hydrocephalus. Raven, New York, pp 215–230
8. Shapiro K, Takei F, Fried A, Kohn I (1985) Experimental feline hydro-cephalus. The role of biomechanical changes in ventricular enlargement in cats. J Neurosurgery 63: 82–87
9. Shapiro K, Fried A (1986) Pressure-volume relationship in shunt-dependent childhood hydrocephalus. The zero of pressure instability in children with acute deterioration. J Neurosurg 64: 390–396
10. Takei F, Shapiro K, Hirano A, Kohn I (1987) Influence of the rate of ventri enlargement on the ultrastructual morphology of the white matter in experimental hydrocephalus. Neurosurgery 21: 645–650
11. Takei F, Shapiro K, Kohn I (1987) Influence of the ventricular enlargement on the white matter water content in progressive feline hydrocephalus. J Neurosurgery 66: 577–583

Correspondence; Futoshi Takei, M. D., Department of Neurosurgery, Tokai University School of Medicine, Boseidai, Isehara, Kanagawa 259-11, Japan.

Author Index

Subject Index*

* Numbers indicate the first page of each article in which subjects are included.

A.W. Unterberg, G.-H. Schneider, W. R. Lanksch (eds.)

Monitoring of Cerebral Blood Flow and Metabolism in Intensive Care

1993. 76 figures. VIII, 125 pages.
Cloth DM 140,–, öS 980,–
Reduced price for subscribers to "Acta Neurochirurgica":
Cloth DM 126,–, öS 882,–
ISBN 3-211-82484-7
(Acta Neurochirurgica / Supplementum 59)

A. Baethmann, O. Kempski, L. Schürer (eds.)

Mechanisms of Secondary Brain Damage
Current State

1993. 76 figures. VIII, 165 pages.
Cloth DM 160,–, öS 1120,–
Reduced price for subscribers to "Acta Neurochirurgica":
Cloth DM 144,–, öS 1008,–
ISBN 3-211-82421-9
(Acta Neurochirurgica / Supplementum 57)

H.-J. Reulen, A. Baethmann, J. Fenstermacher, A. Marmarou, Maria Spatz (eds.)

Brain Edema VIII
Proceedings of the Eighth International Symposium Bern, June 17-20, 1990

1990. 203 figures. XIII, 416 pages.
Cloth DM 250,–, öS 1750,–
Reduced price for subscribers to "Acta Neurochirurgica":
Cloth DM 225,–, öS 1575,–
ISBN 3-211-82240-2
(Acta Neurochirurgica / Supplementum 51)

Prices are subject to change without notice

Sachsenplatz 4–6, P.O.Box 89, A-1201 Wien · 175 Fifth Avenue, New York, NY 10010, USA
Heidelberger Platz 3, D-14197 Berlin · 37-3, Hongo 3-chome, Bunkyo-ku, Tokyo 113, Japan